T0185820

Putnam and Beyond

Răzvan Gelca · Titu Andreescu

Putnam and Beyond

Second Edition

 Springer

Răzvan Gelca
Department Mathematics and Statistics
Texas Tech University
Lubbock, TX
USA

Titu Andreescu
Mathematics
University of Texas at Dallas
Richardson, TX
USA

ISBN 978-3-319-58986-2   ISBN 978-3-319-58988-6   (eBook)
DOI 10.1007/978-3-319-58988-6

Library of Congress Control Number: 2017943241

1st edition: © Springer Science+Business Media, LLC 2007
2nd edition: © Springer International Publishing AG 2017

Printed on acid-free paper

This Springer imprint is published by Springer Nature
The registered company is Springer International Publishing AG
The registered company address is: Gewerbestrasse 11, 6330 Cham, Switzerland

*Life is good for only two things, discovering
mathematics and teaching mathematics.*
Siméon Poisson

Preface to the Second Edition

The first edition of the book has generated great interest and a large amount of input from the readers, who have expressed their views and have suggested corrections and improvements. We are deeply grateful to all, as their contributions have significantly impacted the book.

At the same time, the authors have remained involved in mathematics competitions, being exposed to the constant flow of problems and to the evolution of ideas. The first author has continued training the United States International Mathematical Olympiad team. The second author has started and perfected the *Awesome Math* Summer program, and as editor-in-chief of *Mathematical Reflections*, has established it as one of the important journals promoting problem solving at high-school and college levels. Both authors have continued writing problems for mathematics contests at high school and collegiate levels. Some new perspectives in problem solving that appeared since the publication of the first edition have thus found their way into the book.

Also, by using *Putnam and Beyond* in teaching and coaching, the authors have realized that some material had to be improved and expanded to make it more accessible and complete.

The new edition adds 180 new problems and examples, eight new sections, some new solutions to previously existing problems, and corrects all the errors and typos that have been found in the first edition. It gives more substance to some topics that had a rather shallow treatment before. The experience of ten years of use gave rise to a more polished product.

April 2017

Răzvan Gelca
Texas Tech University

Titu Andreescu
University of Texas at Dallas

Preface to the First Edition

A problem book at the college level. A study guide for the Putnam competition. A bridge between high school problem solving and mathematical research. A friendly introduction to fundamental concepts and results. All these desires gave life to the pages that follow.

The William Lowell Putnam Mathematical Competition is the most prestigious mathematics competition at the undergraduate level in the world. Historically, this annual event began in 1938, following a suggestion of William Lowell Putnam, who realized the merits of an intellectual intercollegiate competition. Nowadays, over 2500 students from more than 300 colleges and universities in the USA and Canada take part in it. The name Putnam has become synonymous with excellence in undergraduate mathematics.

Using the Putnam competition as a symbol, we lay the foundations of higher mathematics from a unitary, problem-based perspective. As such, *Putnam and Beyond* is a journey through the world of college mathematics, providing a link between the stimulating problems of the high school years and the demanding problems of scientific investigation. It gives motivated students a chance to learn concepts and acquire strategies, hone their skills and test their knowledge, seek connections, and discover real world applications. Its ultimate goal is to build the appropriate background for graduate studies, whether in mathematics or applied sciences.

Our point of view is that in mathematics it is more important to understand *why* than to know *how*. Because of this we insist on proofs and reasoning. After all, mathematics means, as the Romanian mathematician Grigore Moisil once said, "correct reasoning". The ways of mathematical thinking are universal in today's science.

Putnam and Beyond targets primarily Putnam training sessions, problem-solving seminars, and math clubs at the college level, filling a gap in the undergraduate curriculum. But it does more than that. Written in the structured manner of a textbook, but with strong emphasis on problems and individual work, it covers what we think are the most important topics and techniques in undergraduate mathematics, brought together within the confines of a single book in order to strengthen one's belief in the unitary nature of mathematics. It is assumed that the reader possesses a moderate background, familiarity with the subject, and a certain level of sophistication, for what we cover reaches beyond the usual textbook, both in difficulty and in depth. When organizing the material, we were inspired by Georgia O'Keeffe's words: "Details are confusing. It is only by selection, by elimination, by emphasis that we get at the real meaning of things."

The book can be used to enhance the teaching of any undergraduate mathematics course, since it broadens the database of problems for courses in real analysis, linear algebra, trigonometry, analytical geometry, differential equations, number theory, combinatorics, and probability. Moreover, it can be used by graduate students and educators alike to expand their mathematical horizons, for many concepts of more advanced mathematics can be found here disguised in elementary language, such as the Gauss-Bonnet theorem, the linear propagation of errors in quantum mechanics, knot invariants, or the Heisenberg group. The way of thinking nurtured in this book opens the door for true scientific investigation.

As for the problems, they are in the spirit of mathematics competitions. Recall that the Putnam competition has two parts, each consisting of six problems, numbered A1 through A6, and B1 through B6. It is customary to list the problems in increasing order of difficulty, with A1 and B1 the easiest, and A6 and B6 the hardest. We keep the same ascending pattern but span a range from A0 to B7. This means that we start with some inviting problems below the difficulty of the test, then move forward into the depths of mathematics.

As sources of problems and ideas we used the Putnam exam itself, the International Competition in Mathematics for University Students, the International Mathematical Olympiad, national contests from the USA, Romania, Russia, China, India, Bulgaria, mathematics journals such as the *American Mathematical Monthly*, *Mathematics Magazine*, *Revista Matematică din Timişoara* (*Timişsoara Mathematics Gazette*), *Gazeta Matematică* (*Mathematics Gazette, Bucharest*), *Kvant* (*Quantum*), *Középiskolai Matematikai Lapok* (*Mathematical Magazine for High Schools* (*Budapest*)), and a very rich collection of Romanian publications. Many problems are original contributions of the authors. Whenever possible, we give the historical background and indicate the source and author of the problem. Some of our sources are hard to find; this is why we offer you their most beautiful problems. Other sources are widely circulated, and by selecting some of their most representative problems we bring them to your attention.

Here is a brief description of the contents of the book. The first chapter is introductory, giving an overview of methods widely used in proofs. The other five chapters reflect areas of mathematics: algebra, real analysis, geometry and trigonometry, number theory, combinatorics and probability. The emphasis is placed on the first two of these chapters, since they occupy the largest part of the undergraduate curriculum.

Within each chapter, problems are clustered by topic. We always offer a brief theoretical background illustrated by one or more detailed examples. Several problems are left for the reader to solve. And since our problems are true brainteasers, complete solutions are given in the second part of the book. Considerable care has been taken in selecting the most elegant solutions and writing them so as to stir imagination and stimulate research. We always "judged mathematical proofs", as Andrew Wiles once said, "by their beauty".

Putnam and Beyond is the fruit of work of the first author as coach of the University of Michigan and Texas Tech University Putnam teams and of the International Mathematical Olympiad teams of the USA and India, as well as the product of the vast experience of the second author as head coach of the United States International Mathematical Olympiad team, coach of the Romanian International Mathematical Olympiad team, director of the American Mathematics Competitions, and member of the Question Writing Committee of the William Lowell Putnam Mathematical Competition.

In conclusion, we would like to thank Elgin Johnston, Dorin Andrica, Chris Jeuell, Ioan Cucurezeanu, Marian Deaconescu, Gabriel Dospinescu, Ravi Vakil, Vinod Grover, V.V. Acharya, B.J. Venkatachala, C.R. Pranesachar, Bryant Heath, and the students of the International Mathematical Olympiad training programs of the USA and India for their suggestions and contributions. Most of all, we are deeply grateful to Richard Stong, David Kramer, and Paul Stanford for carefully reading the manuscript and considerably improving its quality. We would be delighted to receive further suggestions and corrections; these can be sent to rgelca@gmail.com.

May 2007

Răzvan Gelca
Texas Tech University

Titu Andreescu
University of Texas at Dallas

Contents

A Study Guide

The book has six chapters: Methods of Proof, Algebra, Real Analysis, Geometry and Trigonometry, Number Theory, Combinatorics and Probability, divided into subchapters such as Linear Algebra, Sequences and Series, Geometry, and Arithmetic. All subchapters are self-contained and independent of each other and can be studied in any order. In most cases they reflect standard undergraduate courses or fields of mathematics. The sections within each subchapter are best followed in the prescribed order.

If you are an *undergraduate student* trying to acquire skills or test your knowledge in a certain field, study first a regular textbook and make sure that you understand it very well. Then choose the appropriate chapter or subchapter of this book and proceed section by section. Read first the theoretical background and the examples from the introductory part; then do the problems. These are listed in increasing order of difficulty, but even the very first can be tricky. Don't get discouraged; put effort and imagination into each problem; and only if all else fails, look at the solution from the back of the book. But even if you are successful, you should read the solution, since many times it gives a new insight and, more important, opens the door toward more advanced mathematics.

Beware! The last few problems of each section can be very hard. It might be a good idea to skip them at the first encounter and return to them as you become more experienced.

If you are a *Putnam competitor*, then as you go on with the study of the book try your hand at the true Putnam problems (which have been published in three excellent volumes). Identify your weaknesses and insist on the related chapters of *Putnam and Beyond*. Every once in a while, for a problem that you have solved, write down the solution in detail, then compare it to the one given at the end of the book. It is very important that your solutions be correct, structured, convincing, and easy to follow.

Mathematical Olympiad competitors can also use this book. Appropriate chapters are Methods of Proof, Number Theory, and Combinatorics, as well as the subchapters 2.1 and 4.2.

An *instructor* can add some of the problems from the book to a regular course in order to stimulate and challenge the better students. Some of the theoretical subjects can also be incorporated in the course to give better insight and a new perspective. *Putnam and*

Beyond can be used as a textbook for problem-solving courses, in which case we recommend beginning with the first chapter. Students should be encouraged to come up with their own original solutions.

If you are a *graduate student* in mathematics, it is important that you know and understand the contents of this book. First, mastering problems and learning how to write down arguments are essential matters for good performance in doctoral examinations. Second, most of the presented facts are building blocks of graduate courses; knowing them will make these courses natural and easy.

It is important to keep in mind that detailed solutions to all problems are given in the second part of the book. After the solution we list the author of the problem and/or the place where it was published. In some cases we also describe how the problem fits in the big picture of mathematics.

"Don't bother to just be better than your contemporaries or predecessors. Try to be better than yourself" (W. Faulkner).

1

Methods of Proof

In this introductory chapter we explain some methods of mathematical proof. They are argument by contradiction, the principle of mathematical induction, the pigeonhole principle, the use of an ordering on a set, and the principle of invariance.

The basic nature of these methods and their universal use throughout mathematics makes this separate treatment necessary. In each case we have selected what we think are the most appropriate examples, solving some of them in detail and asking you to train your skills on the others. And since these are fundamental methods in mathematics, you should try to understand them in depth, for "it is better to understand many things than to know many things" (Gustave Le Bon).

1.1 Argument by Contradiction

The method of *argument by contradiction* proves a statement in the following way:

First, the statement is assumed to be false. Then, a sequence of logical deductions yields a conclusion that contradicts either the hypothesis (indirect method), or a fact known to be true (reductio ad absurdum). This contradiction implies that the original statement must be true.

This is a method that Euclid loved, and you can find it applied in some of the most beautiful proofs from his *Elements*. Euclid's most famous proof is that of the infinitude of prime numbers.

Euclid's theorem. *There are infinitely many prime numbers.*

Proof. Assume, to the contrary, that only finitely many prime numbers exist. List them as $p_1 = 2$, $p_2 = 3$, $p_3 = 5$, ..., p_n. The number $N = p_1 p_2 \ldots p_n + 1$ is divisible by a prime p, yet is coprime to $p_1, p_2, \ldots, p_n$. Therefore, p does not belong to our list of all prime numbers, a contradiction. Hence the initial assumption was false, proving that there are infinitely many primes.

© Springer International Publishing AG 2017
R. Gelca and T. Andreescu, *Putnam and Beyond*, DOI 10.1007/978-3-319-58988-6_1

Here is a variation of this proof using repunits. If there are only finitely many primes, then the terms of the sequence

$$x_1 = 1, x_2 = 11, x_3 = 111, x_4 = 1111, \dots$$

have only finitely many prime divisors, so there are finitely many terms of the sequence that exhaust them. Assume that the first n terms exhaust all prime divisors. Then $x_{n!}$ is divisible by $x_1, x_2, \dots, x_n$, since for each $k \leq n$, we can group the digits of $x_{n!}$ in strings of k. Then $x_{n!+1} = 10x_{n!} + 1$ is coprime with all of $x_1, x_2, \dots, x_n$. This is a contradiction because all prime divisors of terms of the sequence were exhausted by $x_1, x_2, \dots, x_n$. So there are infinitely many primes. $\square$

We continue our illustration of the method of argument by contradiction with an example of Euler.

Example. Prove that there is no polynomial

$$P(x) = a_n x^n + a_{n-1} x^{n-1} + \dots + a_0$$

with integer coefficients and of degree at least 1 with the property that $P(0), P(1), P(2), \dots$ are all prime numbers.

Solution. Assume the contrary and let $P(0) = p$, p prime. Then $a_0 = p$ and $P(kp)$ is divisible by p for all $k \geq 1$. Because we assumed that all these numbers are prime, it follows that $P(kp) = p$ for $k \geq 1$. Therefore, $P(x)$ takes the same value infinitely many times, a contradiction. Hence the conclusion. $\square$

The last example comes from I. Tomescu's book Problems in Combinatorics (Wiley, 1985).

Example. Let $F = \{E_1, E_2, \dots, E_s\}$ be a family of subsets with r elements of some set X. Show that if the intersection of any $r + 1$ (not necessarily distinct) sets in F is nonempty, then the intersection of all sets in F in nonempty.

Solution. Again we assume the contrary, namely that the intersection of all sets in F is empty. Consider the set $E_1 = \{x_1, x_2, \dots, x_r\}$ Because none of the x_i, $i = 1, 2, \dots, r$, lies in the intersection of all the e_j's (this intersection being empty), it follows that for each i we can find some E_{j_i} such that $x_i \notin E_{j_i}$. Then

$$E_1 \cap E_{j_1} \cap E_{j_2} \cap \dots \cap E_{j_r} = \emptyset,$$

since, at the same time, this intersection is included in E_1 and does not contain any element of E_1. But this contradicts the hypothesis. It follows that our initial assumption was false, and hence the sets from the family F have a nonempty intersection. $\square$

The following problems help you practice this method, which will be used often in the book.

1. Prove that $\sqrt{2} + \sqrt{3} + \sqrt{5}$ is an irrational number.

2. Show that no set of nine consecutive integers can be partitioned into two sets with the product of the elements of the first set equal to the product of the elements of the second set.

3. Find the least positive integer n such that any set of n pairwise relatively prime integers greater than 1 and less than 2005 contains at least one prime number.

4. Let $\mathcal{F} = \{E_1, E_2, \ldots, E_m\}$ be a family of subsets with $n - 2$ elements of a set S with n elements, $n \geq 3$. Show that if the union of any three subsets from $\mathcal{F}$ is not equal to S, then the union of all subsets from $\mathcal{F}$ is different from S.

5. Every point of three-dimensional space is colored red, green, or blue. Prove that one of the colors attains all distances, meaning that any positive real number represents the distance between two points of this color.

6. The union of nine planar surfaces, each of area equal to 1, has a total area equal to 5. Prove that the overlap of some two of these surfaces has an area greater than or equal to $\frac{1}{9}$.

7. Show that there does not exist a function $f : \mathbb{Z} \rightarrow \{1, 2, 3\}$ satisfying $f(x) \neq f(y)$ for all $x, y \in \mathbb{Z}$ such that $|x - y| \in \{2, 3, 5\}$.

8. Show that there does not exist a strictly increasing function $f : \mathbb{N} \rightarrow \mathbb{N}$ satisfying $f(2) = 3$ and $f(mn) = f(m)f(n)$ for all $m, n \in \mathbb{N}$.

9. Determine all functions $f : \mathbb{N} \rightarrow \mathbb{N}$ satisfying

$$xf(y) + yf(x) = (x + y)f(x^2 + y^2)$$

for all positive integers x and y.

10. Show that the interval $[0, 1]$ cannot be partitioned into two disjoint sets A and B such that $B = A + a$ for some real number a.

11. Let $n > 1$ be an arbitrary real number and let k be the number of positive prime numbers less than or equal to n. Select $k + 1$ positive integers such that none of them divides the product of all the others. Prove that there exists a number among the chosen $k + 1$ that is bigger than n.

1.2 Mathematical Induction

The principle of *mathematical induction*, which lies at the very heart of Peano's axiomatic construction of the set of positive integers, is stated as follows.

Induction principle. *Given $P(n)$, a property depending on a positive integer n,*

(i) if $P(n_0)$ is true for some positive integer n_0, and
(ii) if for every $k \geq n_0$, $P(k)$ true implies $P(k + 1)$ true,
*   then $P(n)$ is true for all $n \geq n_0$.*

This means that when proving a statement by mathematical induction you should (i) check the base case and (ii) verify the inductive step by showing how to pass from an arbitrary integer to the next. Here is a simple example from combinatorial geometry.

Example. Finitely many lines divide the plane into regions. Show that these regions can be colored by two colors in such a way that neighboring regions have different colors.

Solution. We prove this by induction on the number n of lines. The base case $n = 1$ is straightforward, color one half-plane black, the other white.

For the inductive step, assume that we know how to color any map defined by k lines. Add the $(k + 1)$st line to the picture; then keep the color of the regions on one side of this line the same while changing the color of the regions on the other side. The inductive step is illustrated in Figure 1.

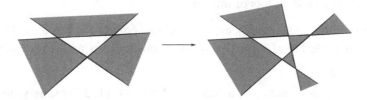

Figure 1

Regions that were adjacent previously still have different colors. Regions that share a segment of the $(k + 1)$st line, which were part of the same region previously, now lie on opposite sides of the line. So they have different colors, too. This shows that the new map satisfies the required property and the induction is complete.   □

A classical proof by induction is that of Fermat's so-called little theorem.

Fermat's little theorem. *Let p be a prime number, and n a positive integer. Then $n^p - n$ is divisible by p.*

Proof. We prove the theorem by induction on n. The base case $n = 1$ is obvious. Let us assume that the property is true for $n = k$ and prove it for $n = k + 1$. Using the induction hypothesis, we obtain

$$(k + 1)^p - (k + 1) \equiv k^p + \sum_{j=1}^{p-1} \binom{p}{j} k^j + 1 - k - 1 \equiv \sum_{j=1}^{p-1} \binom{p}{j} k^j \pmod{p}.$$

The key observation is that for $1 \le j \le p - 1$, $\binom{p}{j}$ is divisible by p. Indeed, examining

$$\binom{p}{j} = \frac{p(p - 1) \cdots (p - j + 1)}{1 \cdot 2 \cdots j},$$

it is easy to see that when $1 \le j \le p-1$, the numerator is divisible by p while the denominator is not. Therefore, $(k + 1)^p - (k + 1) \equiv 0 \pmod{p}$, which completes the induction.   □

The third example is a problem from the 5th W.L. Putnam Mathematical Competition, and it was selected because its solution combines several proofs by induction. If you find it too demanding, think of Vincent van Gogh's words: "The way to succeed is to keep your courage and patience, and to work energetically".

Example. For m a positive integer and n an integer greater than 2, define $f_1(n) = n$, $f_2(n) = n^{f_1(n)}, \ldots, f_{i+1}(n) = n^{f_i(n)}, \ldots$. Prove that

$$f_m(n) < n!! \cdots ! < f_{m+1}(n),$$

where the term in the middle has m factorials.

Solution. For convenience, let us introduce $g_0(n) = n$, and recursively $g_{i+1}(n) = (g_i(n))!$. The double inequality now reads

$$f_m(n) < g_m(n) < f_{m+1}(n).$$

For $m = 1$ this is obviously true, and it is only natural to think of this as the base case. We start by proving the inequality on the left by induction on m. First, note that if $t > 2n^2$ is a positive integer, then

$$t! > (n^2)^{t-n^2} = n^t n^{t-2n^2} > n^t.$$

Now, it is not hard to check that $g_m(n) > 2n^2$ for $m \geq 2$ and $n \geq 3$. With this in mind, let us assume the inequality to be true for $m = k$. Then

$$g_{k+1}(n) = (g_k(n))! > n^{g_k(n)} > n^{f_k(n)} = f_{k+1}(n),$$

which proves the inequality for $m = k + 1$. This verifies the inductive step and solves half of the problem.

Here we pause for a short observation. Sometimes the proof of a mathematical statement becomes simpler if the statement is strengthened. This is the case with the second inequality, which we replace by the much stronger

$$g_0(n)g_1(n) \cdots g_m(n) < f_{m+1}(n),$$

holding true for m and n as above.

As an intermediate step, we establish, by induction on m, that

$$g_0(n)g_1(n) \ldots g_m(n) < n^{g_0(n)g_1(n)\cdots g_{m-1}(n)},$$

for all m and all $n \geq 3$. The base case $m = 1$ is the obvious $n \cdot n! < n^n$. Now assume that the inequality is true for $m = k$, and prove it for $m = k + 1$. We have

$$g_0(n)g_1(n) \cdots g_{k+1}(n) = g_0(n)g_0(n!) \cdots g_k(n!) < g_0(n)(n!)^{g_0(n!)g_1(n!)\cdots g_{k-1}(n!)}$$
$$< n(n!)^{g_1(n)\cdots g_k(n)} < (n \cdot n!)^{g_1(n)\cdots g_k(n)}$$
$$< (n^n)^{g_1(n)\cdots g_k(n)} = n^{g_0(n)g_1(n)\cdots g_k(n)},$$

completing this induction, and proving the claim.

Next, we show, also by induction on m, that $g_0(n)g_1(n)\cdots g_m(n) < f_{m+1}(n)$ for all n. The base case $m = 1$ is $n \cdot n! < n^n$; it follows by multiplying $1 \cdot 2 < n$ and $3 \cdot 4 \cdots n < n^{n-2}$. Let's see the inductive step. Using the inequality for the g_m's proved above and the assumption that the inequality holds for $m = k$, we obtain

$$g_0(n) \cdots g_m(n)g_{m+1}(n) < n^{g_0(n)\cdots g_m(n)} < n^{f_{m+1}(n)} = f_{m+2}(n),$$

which is the inequality for $m = k + 1$. This completes the last induction, and with it the solution to the problem. No fewer than three inductions were combined to solve the problem!   □

Listen and you will forget, learn and you will remember, do it yourself and you will understand. Practice induction with the following examples.

12. Prove for all positive integers n the identity

$$\frac{1}{n+1} + \frac{1}{n+2} + \cdots + \frac{1}{2n} = 1 - \frac{1}{2} + \frac{1}{3} - \cdots + \frac{1}{2n-1} - \frac{1}{2n}.$$

13. Prove that $|\sin nx| \leq n|\sin x|$ for any real number x and positive integer n.

14. Prove that for any real numbers $x_1, x_2, \ldots, x_n, n \geq 1$,

$$|\sin x_1| + |\sin x_2| + \cdots + |\sin x_n| + |\cos(x_1 + x_2 + \cdots + x_n)| \geq 1.$$

15. Prove that $3^n \geq n^3$ for all positive integers n.

16. Let $n \geq 6$ be an integer. Show that

$$\left(\frac{n}{3}\right)^n < n! < \left(\frac{n}{2}\right)^n.$$

17. Let n be a positive integer. Prove that

$$1 + \frac{1}{2^3} + \frac{1}{3^3} + \cdots + \frac{1}{n^3} < \frac{3}{2}.$$

18. Prove that for any positive integer n there exists an n-digit number
   (a) divisible by 2^n and containing only the digits 2 and 3;
   (b) divisible by 5^n and containing only the digits 5, 6, 7, 8, 9.

19. Prove that for any $n \geq 1$, a $2^n \times 2^n$ checkerboard with a 1×1 corner square removed can be tiled by pieces of the form described in Figure 2.

Figure 2

20. Given a sequence of integers $x_1, x_2, \ldots, x_n$ whose sum is 1, prove that exactly one of the cyclic shifts

$$x_1, x_2, \ldots, x_n; \quad x_2, \ldots, x_n, x_1; \quad \ldots; \quad x_n, x_1, \ldots, x_{n-1}$$

has all of its partial sums positive. (By a partial sum we mean the sum of the first k terms, $k \leq n$.)

21. Let $x_1, x_2, \ldots, x_n, y_1, y_2, \ldots, y_m$ be positive integers, $n, m > 1$. Assume that

$$x_1 + x_2 + \cdots + x_n = y_1 + y_2 + \cdots + y_m < mn.$$

Prove that in the equality

$$x_1 + x_2 + \cdots + x_n = y_1 + y_2 + \cdots + y_m$$

one can suppress some (but not all) terms in such a way that the equality is still satisfied.

22. Prove that any function defined on the entire real axis can be written as the sum of two functions whose graphs admit centers of symmetry.

23. Prove that for any positive integer $n \geq 2$ there is a positive integer m that can be written simultaneously as a sum of $2, 3, \ldots, n$ squares of nonzero integers.

24. Let n be a positive integer, $n \geq 2$, and let $a_1, a_2, \ldots, a_{2n+1}$ be positive real numbers such that $a_1 < a_2 < \cdots < a_{2n+1}$. Prove that

$$\sqrt[n]{a_1} - \sqrt[n]{a_2} + \sqrt[n]{a_3} - \cdots - \sqrt[n]{a_{2n}} + \sqrt[n]{a_{2n+1}} < \sqrt[n]{a_1 - a_2 + a_3 - \cdots - a_{2n} + a_{2n+1}}.$$

25. It is given a finite set A of lines in a plane. It is known that, for some positive integer $k \geq 3$, for every subset B of A consisting of $k^2 + 1$ lines there are k points in the plane such that each line in B passes through at least one of them. Prove that there are k points in the plane such that every line in A passes through at least one of them.

Even more powerful is strong induction.

Induction principle (strong form). *Given $P(n)$ a property that depends on an integer n,*

(i) *if $P(n_0), P(n_0 + 1), \ldots, P(n_0 + m)$ are true for some positive integer n_0 and nonnegative integer m, and*

(ii) *if for every $k > n_0 + m$, $P(j)$ true for all $n_0 \leq j < k$ implies $P(k)$ true, then $P(n)$ is true for all $n \geq n_0$.*

We use strong induction to solve a problem from the 24th W.L. Putnam Mathematical Competition.

Example. Let $f : \mathbb{N} \to \mathbb{N}$ be a strictly increasing function such that $f(2) = 2$ and $f(mn) = f(m)f(n)$ for every relatively prime pair of positive integers m and n. Prove that $f(n) = n$ for every positive integer n.

Solution. The proof is of course by induction on n. Monotonicity implies right away that $f(1) = 1$. However, the base case is not the given $f(2) = 2$, but $f(2) = 2$ and $f(3) = 3$.

So let us find $f(3)$. Because f is strictly increasing, $f(3)f(5) = f(15) < f(18) = f(2)f(9)$. Hence $f(3)f(5) < 2f(9)$ and $f(9) < f(10) = f(2)f(5) = 2f(5)$. Combining these inequalities, we obtain $f(3)f(5) < 4f(5)$, so $f(3) < 4$. But we know that $f(3) > f(2) = 2$, which means that $f(3)$ can only be equal to 3.

The base case was the difficult part of the problem; the induction step is rather straightforward. Let $k > 3$ and assume that $f(j) = j$ for $j < k$. Consider $2^r(2m + 1)$ to be the smallest even integer greater than or equal to k that is not a power of 2. This number is equal to either k, $k + 1$, $k + 2$, or $k + 3$, and since $k > 3$, both 2^r and $2m + 1$ are strictly less than k. From the induction hypothesis, we obtain $f(2^r(2m + 1)) = f(2^r)f(2m + 1) = 2^r(2m + 1)$. Monotonicity, combined with the fact that there are at most $2^r(2m+1)$ values that the function can take in the interval $[1, 2^r(2m+1)]$, implies that $f(l) = l$ for $l \leq 2^r(2m+1)$. In particular, $f(k) = k$. We conclude that $f(n) = n$ for all positive integers n. $\qquad\square$

A function $f : \mathbb{N} \to \mathbb{C}$ with the property that $f(1) = 1$ and $f(mn) = f(m)f(n)$ whenever m and n are coprime is called a multiplicative function. Examples include the Euler totient function and the Möbius function. In the case of our problem, the multiplicative function is also strictly increasing. A more general result of P. Erdös shows that any increasing multiplicative function that is not constant is of the form $f(n) = n^\alpha$ for some $\alpha > 0$.

The second example is from the 1999 Balkan Mathematical Olympiad, being proposed by B. Enescu.

Example. Let $0 \leq x_0 \leq x_1 \leq x_2 \leq \cdots \leq x_n \leq \cdots$ be a sequence of non-negative integers such that for every index k, the number of the terms of the sequence that are less than or equal to k is finite. We denote this number by y_k. Prove that for any two positive integer numbers m and n, the following inequality holds

$$\sum_{i=0}^{n} x_i + \sum_{j=0}^{m} y_j \geq (n + 1)(m + 1).$$

Solution. We will prove this by strong induction on $s = m + n$.

The base case $s = 0$ is obvious, since either $x_0 > 0$, in which case the first sum is at least 1, or $x_0 = 0$, in which case $y_0 \geq 1$ and the second sum is at least 1. Let us now assume that the inequality holds for all $s \leq N - 1$ and let us prove it for $s = N$.

If $x_n \geq m + 1$, then

$$\sum_{i=0}^{n} x_i + \sum_{j=0}^{m} y_j = \sum_{i=0}^{n-1} x_i + \sum_{j=0}^{m} y_j + x_n,$$

where $\sum_{i=0}^{n-1} x_i$ is taken to be zero if $n = 0$. The induction hypothesis implies that this is greater than or equal to $n(m + 1) + (m + 1) = (n + 1)(m + 1)$, and we are done.

If $x_n < m + 1$, then $y_m \geq n + 1$ and so

$$\sum_{i=0}^{n} x_i + \sum_{j=0}^{m} y_j = \sum_{i=0}^{n} x_i + \sum_{j=0}^{m-1} y_j + y_m \geq (n + 1)m + (n + 1) = (n + 1)(m + 1),$$

where for the inequality we used again the induction hypothesis. This completes the induction and we are done. □

26. Show that every positive integer can be written as a sum of distinct terms of the Fibonacci sequence. (The Fibonacci sequence $(F_n)_n$ is defined by $F_0 = 0$, $F_1 = 1$, and $F_{n+1} = F_n + F_{n-1}$, $n \geq 1$.)

27. Prove that the Fibonacci sequence satisfies the identity

$$F_{2n+1} = F_{n+1}^2 + F_n^2, \text{ for } n \geq 0.$$

28. Prove that the Fibonacci sequence satisfies the identity

$$F_{3n} = F_{n+1}^3 + F_n^3 - F_{n-1}^3, \text{ for } n \geq 0.$$

29. Show that an isosceles triangle with one angle of $120°$ can be dissected into $n \geq 4$ triangles similar to it.

30. Show that for all $n > 3$ there exists an n-gon whose sides are not all equal and such that the sum of the distances from any interior point to each of the sides is constant. (An n-gon is a polygon with n sides.)

31. The vertices of a convex polygon are colored by at least three colors such that no two consecutive vertices have the same color. Prove that one can dissect the polygon into triangles by diagonals that do not cross and whose endpoints have different colors.

32. Prove that any polygon (convex or not) can be dissected into triangles by interior diagonals.

33. Prove that any positive integer can be represented as $\pm 1^2 \pm 2^2 \pm \ldots \pm n^2$ for some positive integer n and some choice of the signs.

Now we demonstrate a less frequently encountered form of induction that can be traced back to Cauchy's work, where it was used to prove the arithmetic mean–geometric mean inequality. We apply this method to solve a problem from D. Buşneag, I. Maftei, *Themes for Mathematics Circles and Contests* (Scrisul Românesc, Craiova, 1983).

Example. Let $a_1, a_2, \ldots, a_n$ be real numbers greater than 1. Prove the inequality

$$\sum_{i=1}^n \frac{1}{1+a_i} \geq \frac{n}{1 + \sqrt[n]{a_1 a_2 \cdots a_n}}.$$

Solution. As always, we start with the base case:

$$\frac{1}{1+a_1} + \frac{1}{1+a_2} \geq \frac{2}{1 + \sqrt{a_1 a_2}}.$$

Multiplying out the denominators yields the equivalent inequality

$$(2 + a_1 + a_2)(1 + \sqrt{a_1 a_2}) \geq 2(1 + a_1 + a_2 + a_1 a_2).$$

After multiplications and cancellations, we obtain

$$2\sqrt{a_1 a_2} + (a_1 + a_2)\sqrt{a_1 a_2} \geq a_1 + a_2 + 2a_1 a_2.$$

This can be rewritten as

$$2\sqrt{a_1 a_2}(1 - \sqrt{a_1 a_2}) + (a_1 + a_2)(\sqrt{a_1 a_2} - 1) \geq 0,$$

or

$$(\sqrt{a_1 a_2} - 1)(a_1 + a_2 - 2\sqrt{a_1 a_2}) \geq 0.$$

The inequality is now obvious since $a_1 a_2 \geq 1$ and $a_1 + a_2 \geq 2\sqrt{a_1 a_2}$.

Now instead of exhausting all positive integers n, we downgrade our goal and check just the powers of 2. So we prove that the inequality holds for $n = 2^k$ by induction on k. Assuming it true for k, we can write

$$\sum_{i=1}^{2^{k+1}} \frac{1}{1 + a_i} = \sum_{i=1}^{2^k} \frac{1}{1 + a_i} + \sum_{i=2^k+1}^{2^{k+1}} \frac{1}{1 + a_i}$$

$$\geq 2^k \left(\frac{1}{1 + \sqrt[2^k]{a_1 a_2 \dots a_{2^k}}} + \frac{1}{1 + \sqrt[2^k]{a_{2^k+1} a_{2^k+2} \cdots a_{2^{k+1}}}} \right)$$

$$\geq 2^k \frac{2}{1 + \sqrt[2^{k+1}]{a_1 a_2 \cdots a_{2^{k+1}}}},$$

where the first inequality follows from the induction hypothesis, and the second is just the base case. This completes the induction.

Now we have to cover the cases in which n is not a power of 2. We do the induction backward, namely, we assume that the inequality holds for $n + 1$ numbers and prove it for n. Let $a_1, a_2, \dots, a_n$ be some real numbers greater than 1. Attach to them the number $\sqrt[n]{a_1 a_2 \cdots a_n}$. When writing the inequality for these $n + 1$ numbers, we obtain

$$\frac{1}{1 + a_1} + \cdots + \frac{1}{1 + \sqrt[n]{a_1 a_2 \cdots a_n}} \geq \frac{n + 1}{1 + \sqrt[n+1]{a_1 \cdots a_n \sqrt[n]{a_1 a_2 \cdots a_n}}}.$$

Recognize the complicated radical on the right to be $\sqrt[n]{a_1 a_2 \dots a_n}$. After cancelling the last term on the left, we obtain

$$\frac{1}{1 + a_1} + \frac{1}{1 + a_2} + \cdots + \frac{1}{1 + a_n} \geq \frac{n}{1 + \sqrt[n]{a_1 a_2 \cdots a_n}},$$

as desired. The inequality is now proved, since we can reach any positive integer n by starting with a sufficiently large power of 2 and working backward. $\square$

Try to apply the same technique to the following problems.

34. Let $f : \mathbb{R} \to \mathbb{R}$ be a function satisfying

$$f\left(\frac{x_1 + x_2}{2} \right) = \frac{f(x_1) + f(x_2)}{2} \quad \text{for any } x_1, x_2.$$

Prove that

$$f\left(\frac{x_1 + x_2 + \cdots + x_n}{n}\right) = \frac{f(x_1) + f(x_2) + \cdots + f(x_n)}{n}$$

for any $x_1, x_2, \ldots, x_n$.

35. Show that if $a_1, a_2, \ldots, a_n$ are nonnegative numbers, then

$$(1 + a_1)(1 + a_2) \cdots (1 + a_n) \geq (1 + \sqrt[n]{a_1 a_2 \cdots a_n})^n.$$

1.3 The Pigeonhole Principle

The *pigeonhole principle* (or *Dirichlet's box principle*) is usually applied to problems in combinatorial set theory, combinatorial geometry, and number theory. In its intuitive form, it can be stated as follows.

Pigeonhole principle. *If $kn + 1$ objects ($k \geq 1$ not necessarily finite) are distributed among n boxes, one of the boxes will contain at least $k + 1$ objects.*

This is merely an observation, and it was Dirichlet who first used it to prove nontrivial mathematical results. The name comes from the intuitive image of several pigeons entering randomly in some holes. If there are more pigeons than holes, then we know for sure that one hole has more than one pigeon. We begin with an easy problem, which was given at the International Mathematical Olympiad in 1972, proposed by Russia.

Example. Prove that every set of 10 two-digit integer numbers has two disjoint subsets with the same sum of elements.

Solution. Let S be the set of 10 numbers. It has $2^{10} - 2 = 1022$ subsets that differ from both S and the empty set. They are the "pigeons". If $A \subset S$, the sum of elements of A cannot exceed $91 + 92 + \cdots + 99 = 855$. The numbers between 1 and 855, which are all possible sums, are the "holes". Because the number of "pigeons" exceeds the number of "holes", there will be two "pigeons" in the same "hole". Specifically, there will be two subsets with the same sum of elements. Deleting the common elements, we obtain two disjoint sets with the same sum of elements. $\square$

Here is a more difficult problem from the 26th International Mathematical Olympiad, proposed by Mongolia.

Example. Given a set M of 1985 distinct positive integers, none of which has a prime divisor greater than 26, prove that M contains at least one subset of four distinct elements whose product is the fourth power of an integer.

Solution. We show more generally that if the prime divisors of elements in M are among the prime numbers $p_1, p_2, \ldots, p_n$ and M has at least $3 \cdot 2^n + 1$ elements, then it contains a subset of four distinct elements whose product is a fourth power.

To each element m in M we associate an n-tuple $(x_1, x_2, \ldots, x_n)$, where x_i is 0 if the exponent of p_i in the prime factorization of m is even, and 1 otherwise. These n-tuples are the "objects". The "boxes" are the 2^n possible choices of 0's and 1's. Hence, by the pigeonhole principle, every subset of $2^n + 1$ elements of M contains two distinct elements with the same associated n-tuple, and the product of these two elements is then a square.

We can repeatedly take aside such pairs and replace them with two of the remaining numbers. From the set M, which has at least $3 \cdot 2^n + 1$ elements, we can select $2^n + 1$ such pairs or more. Consider the $2^n + 1$ numbers that are products of the two elements of each pair. The argument can be repeated for their square roots, giving four elements a, b, c, d in M such that $\sqrt{ab}\sqrt{cd}$ is a perfect square. Then $abcd$ is a fourth power and we are done. For our problem $n = 9$, while $1985 > 3 \cdot 2^9 + 1 = 1537$.   □

The third example comes from the 67th W.L. Putnam Mathematical Competition, 2006.

Example. Prove that for every set $X = \{x_1, x_2, \ldots, x_n\}$ of n real numbers, there exists a nonempty subset S of X and an integer m such that

$$\left| m + \sum_{s \in S} s \right| \leq \frac{1}{n+1}.$$

Solution. Recall that the fractional part of a real number x is $x - \lfloor x \rfloor$. Let us look at the fractional parts of the numbers $x_1, x_1 + x_2, \ldots, x_1 + x_2 + \ldots + x_n$. If any of them is either in the interval $\left[0, \frac{1}{n+1}\right]$ or $\left[\frac{n}{n+1}, 1\right]$, then we are done. If not, we consider these n numbers as the "pigeons" and the $n - 1$ intervals $\left[\frac{1}{n+1}, \frac{2}{n+1}\right], \left[\frac{2}{n+1}, \frac{3}{n+1}\right], \ldots, \left[\frac{n-1}{n+1}, \frac{n}{n+1}\right]$ as the "holes". By the pigeonhole principle, two of these sums, say $x_1 + x_2 + \cdots + x_k$ and $x_1 + x_2 + \cdots + x_{k+m}$, belong to the same interval. But then their difference $x_{k+1} + \cdots + x_{k+m}$ lies within a distance of $\frac{1}{n+1}$ of an integer, and we are done.   □

More problems are listed below.

36. Given 50 distinct positive integers strictly less than 99, prove that some two of them sum to 99.

37. A sequence of m positive integers contains exactly n distinct terms. Prove that if $2^n \leq m$ then there exists a block of consecutive terms whose product is a perfect square.

38. Let $x_1, x_2, \ldots, x_3, \ldots$ be a sequence of integers such that

$$1 = x_1 < x_2 < x_3 < \cdots \text{ and } x_{n+1} \leq 2n \text{ for } n = 1, 2, 3, \ldots.$$

Show that every positive integer k is equal to $x_i - x_j$ for some i and j.

39. Let p be a prime number and a, b, c integers such that a and b are not divisible by p. Prove that the equation $ax^2 + by^2 \equiv c \pmod{p}$ has integer solutions.

40. In each of the unit squares of a 10×10 checkerboard, a positive integer not exceeding 10 is written. Any two numbers that appear in adjacent or diagonally adjacent squares of the board are relatively prime. Prove that some number appears at least 17 times.

41. Show that there is a positive term of the Fibonacci sequence that is divisible by 1000.

42. Let $x_1 = x_2 = x_3 = 1$ and $x_{n+3} = x_n + x_{n+1}x_{n+2}$ for all positive integers n. Prove that for any positive integer m there is an index k such that m divides x_k.

43. A chess player trains by playing at least one game per day, but, to avoid exhaustion, no more than 12 games a week. Prove that there is a group of consecutive days in which he plays exactly 20 games.

44. Let m be a positive integer. Prove that among any $2m + 1$ distinct integers of absolute value less than or equal to $2m - 1$ there exist three whose sum is equal to zero.

45. There are n people at a party. Prove that there are two of them such that among the remaining $n - 2$ people there are at least $\lfloor \frac{n}{2} \rfloor - 1$, each of whom knows both or else knows neither of the two.

46. Let $x_1, x_2, \ldots, x_k$ be real numbers such that the set

$$A = \{\cos(n\pi x_1) + \cos(n\pi x_2) + \cdots + \cos(n\pi x_k) \mid n \geq 1\}$$

is finite. Prove that all the x_i are rational numbers.

Particularly attractive are the problems in which the pigeons and holes are geometric objects. Here is a problem from a Chinese mathematical competition.

Example. Given nine points inside the unit square, prove that some three of them form a triangle whose area does not exceed $\frac{1}{8}$.

Solution. Divide the square into four equal squares, which are the "boxes". From the $9 = 2 \times 4 + 1$ points, at least $3 = 2 + 1$ will lie in the same box. We are left to show that the area of a triangle placed inside a square does not exceed half the area of the square.

Cut the square by the line passing through a vertex of the triangle, as in Figure 3. Since the area of a triangle is $\frac{\text{base} \times \text{height}}{2}$ and the area of a rectangle is base $\times$ height, the inequality holds for the two smaller triangles and their corresponding rectangles. Adding up the two incqualitics, we obtain the inequality for the square. This completes the solution.   □

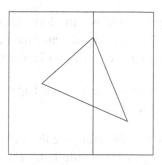

Figure 3

47. Inside a circle of radius 4 are chosen 61 points. Show that among them there are two at distance at most $\sqrt{2}$ from each other.

48. Each of nine straight lines divides a square into two quadrilaterals with the ratio of their areas equal to $r > 0$. Prove that at least three of these lines are concurrent.

49. Show that any convex polyhedron has two faces with the same number of edges.

50. Draw the diagonals of a 21-gon. Prove that at least one angle of less than $1°$ is formed. (Angles of $0°$ are allowed in the case that two diagonals are parallel.)

51. Let $P_1, P_2, \ldots, P_{2n}$ be a permutation of the vertices of a regular polygon. Prove that the closed polygonal line $P_1 P_2 \ldots P_{2n}$ contains a pair of parallel segments.

52. Let S be a convex set in the plane that contains three noncollinear points. Each point of S is colored by one of p colors, $p > 1$. Prove that for any $n \geq 3$ there exist infinitely many congruent n-gons whose vertices are all of the same color.

53. The points of the plane are colored by finitely many colors. Prove that one can find a rectangle with vertices of the same color.

54. Inside the unit square lie several circles the sum of whose circumferences is equal to 10. Prove that there exist infinitely many lines each of which intersects at least four of the circles.

1.4 Ordered Sets and Extremal Elements

An *order* on a set is a relation $\leq$ with three properties: (i) $a \leq a$; (ii) if $a \leq b$ and $b \leq a$, then $a = b$; (iii) $a \leq b$ and $b \leq c$ implies $a \leq c$. The order is called total if any two elements are comparable, that is, if for every a and b, either $a \leq b$ or $b \leq a$. The simplest example of a total order is $\leq$ on the set of real numbers. The existing order on a set can be useful when solving problems. This is the case with the following two examples, the second of which is a problem of G. Galperin published in the Russian journal *Quantum*.

Example. Prove that among any 50 distinct positive integers strictly less than 100 there are two that are coprime.

Solution. Order the numbers: $x_1 < x_2 < \ldots < x_{50}$. If in this sequence there are two consecutive integers, they are coprime and we are done. Otherwise, $x_{50} \geq x_1 + 2 \cdot 49 = 99$. Equality must hold, since $x_{50} < 100$, and in this case the numbers are precisely the 50 odd integers less than 100. Among them 3 is coprime to 7. The problem is solved.

Example. Given finitely many squares whose areas add up to 1, show that they can be arranged without overlaps inside a square of area 2.

Solution. The guess is that a tight way of arranging the small squares inside the big square is by placing the squares in decreasing order of side-lengths.

To prove that this works, denote by x the side length of the first (that is, the largest) square. Arrange the squares inside a square of side $\sqrt{2}$ in the following way. Place the first in the lower-left corner, the next to its right, and so on, until obstructed by the right side of the big square. Then jump to height x, and start building the second horizontal layer of squares by the same rule. Keep going until the squares have been exhausted (see Figure 4).

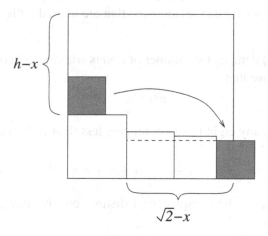

Figure 4

Let h be the total height of the layers. We are to show that $h \le \sqrt{2}$, which in turn will imply that all the squares lie inside the square of side $\sqrt{2}$. To this end, we will find a lower bound for the total area of the squares in terms of x and h. Let us mentally transfer the first square of each layer to the right side of the previous layer. Now each layer exits the square, as shown in Figure 4.

It follows that the sum of the areas of all squares but the first is greater than or equal to $(\sqrt{2} - x)(h - x)$. This is because each newly obtained layer includes rectangles of base $\sqrt{2} - x$ and with the sum of heights equal to $h - x$. From the fact that the total area of the squares is 1, it follows that

$$x^2 + (\sqrt{2} - x)(h - x) \le 1.$$

This implies that

$$h \le \frac{2x^2 - \sqrt{2}x - 1}{x - \sqrt{2}}.$$

That $h \le \sqrt{2}$ will follow from

$$\frac{2x^2 - \sqrt{2}x - 1}{x - \sqrt{2}} \le \sqrt{2}.$$

This is equivalent to

$$2x^2 - 2\sqrt{2}x + 1 \ge 0,$$

or $(x\sqrt{2} - 1)^2 \ge 0$, which is obvious and we are done. $\qquad\square$

What we particularly like about the shaded square from Figure 4 is that it plays the role of the "largest square" when placed on the left, and of the "smallest square" when placed on the right. Here are more problems.

55. Given $n \geq 3$ points in the plane, prove that some three of them form an angle less than or equal to $\frac{\pi}{n}$.

56. Consider a planar region of area 1, obtained as the union of finitely many disks. Prove that from these disks we can select some that are mutually disjoint and have total area at least $\frac{1}{9}$.

57. Suppose that $n(r)$ denotes the number of points with integer coordinates on a circle of radius $r > 1$. Prove that

$$n(r) < 2\pi \sqrt[3]{r^2}.$$

58. Prove that among any eight positive integers less than 2004 there are four, say a, b, c, and d, such that

$$4 + d \leq a + b + c \leq 4d.$$

59. Let $a_1, a_2, \ldots, a_n, \ldots$ be a sequence of distinct positive integers. Prove that for any positive integer n,

$$a_1^2 + a_2^2 + \cdots + a_n^2 \geq \frac{2n+1}{3}(a_1 + a_2 + \cdots + a_n).$$

60. Let X be a subset of the positive integers with the property that the sum of any two not necessarily distinct elements in X is again in X. Suppose that $\{a_1, a_2, \ldots, a_n\}$ is the set of all positive integers not in X. Prove that $a_1 + a_2 + \cdots + a_n \leq n^2$.

61. Let $P(x)$ be a polynomial with integer coefficients, of degree $n \geq 2$. Prove that the set $A = \{x \in \mathbb{Z} \mid P(P(x)) = x\}$ has at most n elements.

An order on a finite set has *maximal* and *minimal* elements. If the order is total, the maximal (respectively, minimal) element is unique. Quite often it is useful to look at such extremal elements, like in the solution to the following problem.

Example. Prove that it is impossible to dissect a cube into finitely many cubes, no two of which are the same size.

Solution. For the solution, assume that such a dissection exists, and look at the bottom face. It is cut into squares. Take the smallest of these squares. It is not hard to see that this square lies in the interior of the face, meaning that it does not touch any side of the bottom face. Look at the cube that lies right above this square! This cube is surrounded by bigger cubes, so its upper face must again be dissected into squares by the cubes that lie on top of it. Take the smallest of the cubes and repeat the argument. This process never stops, since the cubes that lie on top of one of these little cubes cannot end up all touching the upper face of the original cube. This contradicts the finiteness of the decomposition. Hence the conclusion.   □

By contrast, a square can be dissected into finitely many squares of distinct size. Why does the above argument not apply in this case?

And now an example of a more exotic kind.

Example. Given is a finite set of spherical planets, all of the same radius and no two inter-secting. On the surface of each planet consider the set of points not visible from any other planet. Prove that the total area of these sets is equal to the surface area of one planet.

Solution. The problem was on the short list of the 22nd International Mathematical Olympiad, proposed by the Soviet Union. The solution below we found in I. Cuculescu's book on the *International Mathematical Olympiads* (Editura Tehnică, Bucharest, 1984).

Choose a preferential direction in space, which defines the north pole of each planet. Next, define an order on the set of planets by saying that planet A is greater than planet B if on removing all other planets from space, the north pole of B is visible from A. Figure 5 shows that for two planets A and B, either $A < B$ or $B < A$, and also that for three planets A, B, C, if $A < B$ and $B < C$ then $A < C$. The only case in which something can go wrong is that in which the preferential direction is perpendicular to the segment joining the centers of two planets. If this is not the case, then $<$ defines a total order on the planets. This order has a unique maximal element M. The north pole of M is the only north pole not visible from another planet.

Now consider a sphere of the same radius as the planets. Remove from it all north poles defined by directions that are perpendicular to the axes of two of the planets. This is a set of area zero. For every other point on this sphere, there exists a direction in space that makes it the north pole, and for that direction, there exists a unique north pole on one of the planets that is not visible from the others. As such, the surface of the newly introduced sphere is covered by patches translated from the other planets. Hence the total area of invisible points is equal to the area of this sphere, which in turn is the area of one of the planets.   □

62. Complete the square in Figure 6 with integers between 1 and 9 such that the sum of the numbers in each row, column, and diagonal is as indicated.

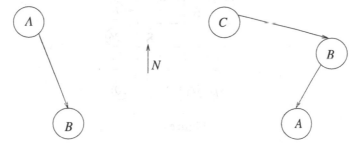

Figure 5

63. Given n points in the plane, no three of which are collinear, show that there exists a closed polygonal line with no self-intersections having these points as vertices.

64. Show that any polygon in the plane has a vertex, and a side not containing that vertex, such that the projection of the vertex onto the side lies in the interior of the side or at one of its endpoints.

65. In some country all roads between cities are one-way and such that once you leave a city you cannot return to it again. Prove that there exists a city into which all roads enter and a city from which all roads exit.

66. At a party assume that no boy dances with all the girls, but each girl dances with at least one boy. Prove that there are two girl-boy couples gb and $g'b'$ who dance, whereas b does not dance with g', and g does not dance with b'.

67. In the plane we have marked a set S of points with integer coordinates. We are also given a finite set V of vectors with integer coordinates. Assume that S has the property that for every marked point P, if we place all vectors from V with origin are P, then more of their ends are marked than unmarked. Show that the set of marked points is infinite.

68. The entries of a matrix are real numbers of absolute value less than or equal to 1, and the sum of the elements in each column is 0. Prove that we can permute the elements of each column in such a way that the sum of the elements in each row will have absolute value less than or equal to 2.

69. Find all odd positive integers n greater than 1 such that for any coprime divisors a and b of n, the number $a + b - 1$ is also a divisor of n.

70. The positive integers are colored by two colors. Prove that there exists an infinite sequence of positive integers $k_1 < k_2 < \cdots < k_n < \cdots$ with the property that the terms of the sequence $2k_1 < k_1 + k_2 < 2k_2 < k_2 + k_3 < 2k_3 < \cdots$ are all of the same color.

71. Let $P_1 P_2 \ldots P_n$ be a convex polygon in the plane. Assume that for any pair of vertices P_i and P_j, there exists a vertex P_k of the polygon such that $\angle P_i P_k P_j = \pi/3$. Show that $n = 3$.

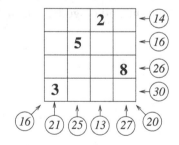

Figure 6

1.5 Invariants and Semi-Invariants

In general, a mathematical object can be studied from many points of view, and it is always desirable to decide whether various constructions produce the same object. One usually distinguishes mathematical objects by some of their properties. An elegant method is to associate to a family of mathematical objects an *invariant*, which can be a number, an algebraic structure, or some property, and then distinguish objects by the different values of the invariant.

The general framework is that of a set of objects or configurations acted on by transformations that identify them (usually called isomorphisms). Invariants then give obstructions to transforming one object into another. Sometimes, although not very often, an invariant is able to tell precisely which objects can be transformed into one another, in which case the invariant is called complete.

An example of an invariant (which arises from more advanced mathematics yet is easy to explain) is the property of a knot to be 3-colorable. Formally, a knot is a simple closed curve in $\mathbb{R}^3$. Intuitively it is a knot on a rope with connected endpoints, such as the right-handed trefoil knot from Figure 7.

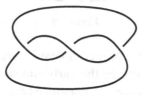

Figure 7

How can one prove *mathematically* that this knot is indeed "knotted"? The answer is, using an invariant. To define this invariant, we need the notion of a knot diagram. Such a diagram is the image of a regular projection (all self-intersections are nontangential and are double points) of the knot onto a plane with crossing information recorded at each double point, just like the one in Figure 7. But a knot can have many diagrams (pull the strands around, letting them pass over each other).

A deep theorem of K. Reidemeister states that two diagrams represent the same knot if they can be transformed into one another by the three types of moves described in Figure 8.

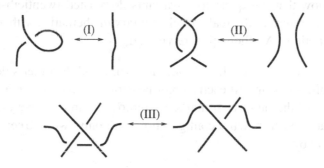

Figure 8

The simplest knot invariant was introduced by the same Reidemeister, and is the property of a knot diagram to be 3-colorable. This means that you can color each strand in the knot diagram by a residue class modulo 3 such that

(i) at least two distinct residue classes modulo 3 are used, and

(ii) at each crossing, $a + c \equiv 2b$ (mod 3), where b is the color of the arc that crosses over, and a and c are the colors of the other two arcs (corresponding to the strand that crosses under).

It is rather easy to prove, by examining the local picture, that this property is invariant under Reidemeister moves. Hence this is an invariant of knots, not just of knot diagrams.

The trefoil knot is 3-colorable, as demonstrated in Figure 9. On the other hand, the unknotted circle is not 3-colorable, because its simplest diagram, the one with no crossings, cannot be 3-colored. Hence the trefoil knot is knotted.

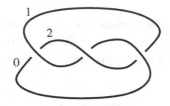

Figure 9

This 3-colorability is, however, not a complete invariant. We now give an example of a complete invariant from geometry. In the early nineteenth century, F. Bolyai and a less well-known mathematician Gerwin proved that given two polygons of equal area, the first can be dissected by finitely many straight cuts and then assembled to produce the second polygon. In his list of 23 problems presented to the International Congress of Mathematicians, D. Hilbert listed as number 3 the question whether the same property remains true for solid polyhedra of the same volume, and if not, what would the obstruction be.

The problem was solved by M. Dehn, a student of Hilbert. Dehn defined an invariant that associates to a finite disjoint union of polyhedra P the sum $I(P)$ of all their dihedral angles reduced modulo rational multiples of π (viewed as an element in $\mathbb{R}/\pi\mathbb{Q}$). He showed that two polyhedra P_1 and P_2 having the same volume can be transformed into one another if and only if $I(P_1) = I(P_2)$, i.e., if and only if the sums of their dihedral angles differ by a rational multiple of π.

It is good to know that the quest for invariants dominated twentieth-century geometry. That being said, let us return to the realm of elementary mathematics with a short list problem from the 46th International Mathematical Olympiad.

Example. There are n markers, each with one side white and the other side black, aligned in a row with their white sides up. At each step, if possible, we choose a marker with the white side up (but not one of the outermost markers), remove it, and reverse the two neighboring markers. Prove that one can reach a configuration with only two markers left if and only if $n - 1$ is not divisible by 3.

Solution. We refer to a marker by the color of its visible face. Note that the parity of the number of black markers remains unchanged during the game. Hence if only two markers are left, they must have the same color.

We define an invariant as follows. To a white marker with t black markers to its left we assign the number $(-1)^t$. Only white markers have numbers assigned to them. The invariant S is the residue class modulo 3 of the sum of all numbers assigned to the white markers.

It is easy to check that S is invariant under the operation defined in the statement. For instance, if a white marker with t black markers on the left and whose neighbors are both black is removed, then S increases by $-(-1)^t + (-1)^{t-1} + (-1)^{t-1} = 3(-1)^{t-1}$, which is zero modulo 3. The other three cases are analogous.

If the game ends with two black markers then S is zero; if it ends with two white markers, then S is 2. This proves that $n - 1$ is not divisible by 3.

Conversely, if we start with $n \geq 5$ white markers, $n \equiv 0$ or 2 modulo 3, then by removing in three consecutive moves the leftmost allowed white markers, we obtain a row of $n - 3$ white markers. Working backward, we can reach either 2 white markers or 3 white markers. In the latter case, with one more move we reach 2 black markers as desired. □

Now try to find the invariants that lead to the solutions of the following problems.

72. An ordered triple of numbers is given. It is permitted to perform the following operation on the triple: to change two of them, say a and b, to $(a+b)/\sqrt{2}$ and $(a-b)/\sqrt{2}$. Is it possible to obtain the triple $(1, \sqrt{2}, 1 + \sqrt{2})$ from the triple $(2, \sqrt{2}, 1/\sqrt{2})$ using this operation?

73. There are 2000 white balls in a box. There are also unlimited supplies of white, green, and red balls, initially outside the box. During each turn, we can replace two balls in the box with one or two balls as follows: two whites with a green, two reds with a green, two greens with a white and red, a white and a green with a red, or a green and red with a white.

   (a) After finitely many of the above operations there are three balls left in the box. Prove that at least one of them is green.
   (b) Is it possible that after finitely many operations only one ball is left in the box?

74. There is a heap of 1001 stones on a table. You are allowed to perform the following operation: you choose one of the heaps containing more than one stone, throw away a stone from the heap, then divide it into two smaller (not necessarily equal) heaps. Is it possible to reach a situation in which all the heaps on the table contain exactly 3 stones by performing the operation finitely many times?

75. Starting with an ordered quadruple of positive integers, a generalized Euclidean algorithm is applied successively as follows: if the numbers are x, y, u, v and $x > y$, then the quadruple is replaced by $x - y, y, u + v, v$. Otherwise, it is replaced by $x, y - x, u, v + u$. The algorithm stops when the numbers in the first pair become equal (in which case they are equal to the greatest common divisor of x and y). Assume that we start with m, n, m, n. Prove that when the algorithm ends, the arithmetic mean of the numbers in the second pair equals the least common multiple of m and n.

76. On an arbitrarily large chessboard consider a generalized knight that can jump p squares in one direction and q in the other, $p, q > 0$. Show that such a knight can return to its initial position only after an *even* number of jumps.

77. Prove that the figure eight knot described in Figure 10 is knotted.

78. In the squares of a 3×3 chessboard are written the signs + and − as described in Figure 11(a). Consider the operations in which one is allowed to simultaneously change all signs in some row or column. Can one change the given configuration to the one in Figure 11(b) by applying such operations finitely many times?

79. The number $99 \ldots 99$ (having 1997 nines) is written on a blackboard. Each minute, one number written on the blackboard is factored into two factors and erased, each factor is (independently) increased or decreased by 2, and the resulting two numbers are written. Is it possible that at some point all of the numbers on the blackboard are equal to 9?

80. Four congruent right triangles are given. One can cut one of them along the altitude and repeat the operation several times with the newly obtained triangles. Prove that no matter how we perform the cuts, we can always find among the triangles two that are congruent.

81. For an integer $n \geq 4$, consider an n-gon inscribed in a circle. Dissect the n-gon into $n - 2$ triangles by nonintersecting diagonals. Prove that the sum of the radii of the incircles of these $n - 2$ triangles does not depend on the dissection.

Figure 10

 (a)　　　　　　　**(b)**

Figure 11

In some cases a semi-invariant will do. A *semi-invariant* (also known as monovariant) is a quantity that, although not constant under a specific transformation, keeps increasing (or decreasing). As such it provides a unidirectional obstruction.

For his solution to the following problem from the 27th International Mathematical Olympiad, J. Keane, then a member of the US team, was awarded a special prize.

Example. To each vertex of a regular pentagon an integer is assigned in such a way that the sum of all of the five numbers is positive. If three consecutive vertices are assigned the numbers x, y, z, respectively, and $y < 0$, then the following operation is allowed: the numbers x, y, z are replaced by $x + y, -y, z + y$, respectively. Such an operation is performed repeatedly as long as at least one of the five numbers is negative. Determine whether this procedure necessarily comes to an end after a finite number of steps.

Solution. The answer is yes. The key idea of the proof is to construct an integer-valued semi-invariant whose value decreases when the operation is performed. The existence of such a semi-invariant will guarantee that the operation can be performed only finitely many times.

Notice that the sum of the five numbers on the pentagon is preserved by the operation, so it is natural to look at the sum of the absolute values of the five numbers. When the operation is performed this quantity decreases by $|x| + |z| - |x + y| - |y + z|$. Although this expression is not always positive, it suggests a new choice. The desired semi-invariant should include the absolute values of pairwise sums as well. Upon testing the new expression and continuing this idea, we discover in turn that the desired semi-invariant should also include absolute values of sums of triples and foursomes. At last, with a pentagon numbered v, w, x, y, z and the semi-invariant defined by

$$\begin{aligned} S(v, w, x, y, z) = {}& |v| + |w| + |x| + |y| + |z| + |v + w| + |w + x| + |x + y| \\ & + |y + z| + |z + v| + |v + w + x| + |w + x + y| + |x + y + z| \\ & + |y + z + v| + |z + v + w| + |v + w + x + y| + |w + x + y + z| \\ & + |x + y + z + v| + |y + z + v + w| + |z + v + w + x|, \end{aligned}$$

we find that the operation reduces the value of S by the simple expression

$$|z + v + w + x| - |z + v + w + x + 2y| - |s - y| \quad |s + y|,$$

where $s = v + w + x + y + z$. Since $s > 0$ and $y < 0$, we see that $|s - y| - |s + y| > 0$, so S has the required property. It follows that the operation can be performed only finitely many times. $\square$

Using the semi-invariant we produced a proof based on Fermat's infinite descent method. This method will be explained in the Number Theory chapter of this book. Here the emphasis was on the guess of the semi-invariant. And now some problems.

82. A real number is written in each square of an $n \times n$ chessboard. We can perform the operation of changing all signs of the numbers in a row or a column. Prove that by performing this operation a finite number of times we can produce a new table for which the sum of each row or column is positive.

83. Starting with an ordered quadruple of integers, perform repeatedly the operation

$$(a, b, c, d) \xrightarrow{T} (|a - b|, |b - c|, |c - d|, |d - a|).$$

Prove that after finitely many steps, the quadruple becomes $(0, 0, 0, 0)$.

84. Several positive integers are written on a blackboard. One can erase any two distinct integers and write their greatest common divisor and least common multiple instead. Prove that eventually the numbers will stop changing.

85. Consider the integer lattice in the plane, with one pebble placed at the origin. We play a game in which at each step one pebble is removed from a node of the lattice and two new pebbles are placed at two neighboring nodes, provided that those nodes are unoccupied. Prove that at any time there will be a pebble at distance at most 5 from the origin.

2

Algebra

It is now time to split mathematics into branches. First, algebra. A section on algebraic identities hones computational skills. It is followed naturally by inequalities. In general, any inequality can be reduced to the question of finding the minimum of a function. But this is a highly nontrivial matter, and that makes the subject exciting. We discuss the fact that squares are nonnegative, the Cauchy-Schwarz inequality, the triangle inequality, the arithmetic mean-geometric mean inequality, and also Sturm's method for proving inequalities.

Our treatment of algebra continues with polynomials. We focus on quadratic polynomials, the relations between zeros and coefficients, the properties of the derivative of a polynomial, problems about the location of the zeros in the complex plane or on the real axis, and methods for proving irreducibility of polynomials (such as the Eisenstein criterion). From all special polynomials we present the most important, the Chebyshev polynomials.

Linear algebra comes next. The first three sections, about operations with matrices, determinants, and the inverse of a matrix, insist on both the array structure of a matrix and the ring structure of the set of matrices. They are more elementary, as is the section on linear systems. The last three sections, about vector spaces and linear transformations, are more advanced, covering among other things the Cayley-Hamilton Theorem and the Perron-Frobenius Theorem.

The chapter concludes with a brief incursion into abstract algebra: binary operations, groups, and rings, really no further than the definition of a group or a ring.

2.1 Identities and Inequalities

2.1.1 Algebraic Identities

The scope of this section is to train algebraic skills. Our idea is to hide behind each problem an important algebraic identity. We commence with three examples, the first and the last written by the second author of the book, and the second given at a Soviet Union college entrance exam and suggested to us by A. Soifer.

© Springer International Publishing AG 2017
R. Gelca and T. Andreescu, *Putnam and Beyond*, DOI 10.1007/978-3-319-58988-6_2

Example. Solve in real numbers the system of equations

$$(3x + y)(x + 3y)\sqrt{xy} = 14,$$
$$(x + y)(x^2 + 14xy + y^2) = 36.$$

Solution. By substituting $\sqrt{x} = u$, $\sqrt{y} = v$, we obtain the equivalent form

$$uv(3u^4 + 10u^2v^2 + 3v^4) = 14,$$
$$u^6 + 15u^4v^2 + 14u^2v^4 + v^6 = 36.$$

Here we should recognize elements of the binomial expansion with exponent equal to 6. Based on this observation we find that

$$36 + 2 \cdot 14 = u^6 + 6u^5v + 15y^4v^2 + 20u^3v^3 + 15u^2v^4 + 6uv^5 + v^6$$

and

$$36 - 2 \cdot 14 = u^6 - 6u^5v + 15y^4v^2 - 20u^3v^3 + 15u^2v^4 - 6uv^5 + v^6.$$

Therefore, $(u + v)^6 = 64$ and $(u - v)^6 = 8$, which implies $u + v = 2$ and $u - v = \pm\sqrt{2}$ (recall that u and v have to be positive). So $u = 1 + \frac{\sqrt{2}}{2}$ and $v = 1 - \frac{\sqrt{2}}{2}$ or $u = 1 - \frac{\sqrt{2}}{2}$ and $v = 1 + \frac{\sqrt{2}}{2}$. The solutions to the system are

$$(x, y) = \left(\frac{3}{2} + \sqrt{2}, \frac{3}{2} - \sqrt{2}\right) \quad \text{and} \quad (x, y) = \left(\frac{3}{2} - \sqrt{2}, \frac{3}{2} + \sqrt{2}\right). \qquad \square$$

Example. Given two segments of lengths a and b, construct with a straightedge and a compass a segment of length $\sqrt[4]{a^4 + b^4}$.

Solution. The solution is based on the following version of the Sophie Germain identity:

$$a^4 + b^4 = (a^2 + \sqrt{2}ab + b^2)(a^2 - \sqrt{2}ab + b^2).$$

Write

$$\sqrt[4]{a^4 + b^4} = \sqrt{\sqrt{a^2 + \sqrt{2}ab + b^2} \cdot \sqrt{a^2 - \sqrt{2}ab + b^2}}.$$

Applying the law of cosines, we can construct segments of lengths $\sqrt{a^2 \pm \sqrt{2}ab + b^2}$ using triangles of sides a and b with the angle between them 135°, respectively, 45°.

On the other hand, given two segments of lengths x, respectively y, we can construct a segment of length $\sqrt{xy}$ (their geometric mean) as the altitude AD in a right triangle ABC ($\angle A = 90°$) with $BD = x$ and $CD = y$. These two steps combined give the method for constructing $\sqrt[4]{a^4 + b^4}$. $\qquad \square$

Example. Let x, y, z be distinct real numbers. Prove that

$$\sqrt[3]{x-y} + \sqrt[3]{y-z} + \sqrt[3]{z-x} \neq 0.$$

Solution. The solution is based on the identity

$$a^3 + b^3 + c^3 - 3abc = (a+b+c)(a^2 + b^2 + c^2 - ab - bc - ca).$$

This identity arises from computing the circulant determinant

$$D = \begin{vmatrix} a & b & c \\ c & a & b \\ b & c & a \end{vmatrix}$$

in two ways: first by expanding with Sarrus' rule, and second by adding up all columns to the first, factoring $(a+b+c)$, and then expanding the remaining determinant. Note that this identity can also be written as

$$a^3 + b^3 + c^3 - 3abc = \frac{1}{2}(a+b+c)[(a-b)^2 + (b-c)^2 + (c-a)^2].$$

Returning to the problem, let us assume the contrary, and set $\sqrt[3]{x-y} = a$, $\sqrt[3]{y-z} = b$, $\sqrt[3]{z-x} = c$. By assumption, $a + b + c = 0$, and so $a^3 + b^3 + c^3 = 3abc$. But this implies

$$0 = (x-y) + (y-z) + (z-x) = 3\sqrt[3]{x-y}\sqrt[3]{y-z}\sqrt[3]{z-x} \neq 0,$$

since the numbers are distinct. The contradiction we have reached proves that our assumption is false, and so the sum is nonzero. $\square$

And now the problems.

86. Show that for no positive integer n can both $n+3$ and $n^2 + 3n + 3$ be perfect cubes.

87. Let A and B be two $n \times n$ matrices that commute and such that for some positive integers p and q, $A^p = I_n$ and $B^q = O_n$. Prove that $A + B$ is invertible, and find its inverse.

88. Prove that any polynomial with real coefficients that takes only nonnegative values can be written as the sum of the squares of two polynomials.

89. Prove that for any nonnegative integer n, the number $5^{5^{n+1}} + 5^{5^n} + 1$ is not prime.

90. Show that for an odd integer $n \geq 5$,

$$\binom{n}{0}5^{n-1} - \binom{n}{1}5^{n-2} + \binom{n}{2}5^{n-3} - \cdots + \binom{n}{n-1}$$

is not a prime number.

91. Factor $5^{1985} - 1$ into a product of three integers, each of which is greater than 5^{100}.

92. Prove that the number $\dfrac{5^{125} - 1}{5^{25} - 1}$ is not prime.

93. Let a and b be coprime integers greater than 1. Prove that for $n \geq 0$ is $a^{2n} + b^{2n}$ divisible by $a + b$.

94. Prove that any integer can be written as the sum of five perfect cubes.

95. Prove that

$$\sum_{k=1}^{31} \frac{1}{(k-1)^{4/5} - k^{4/5} + (k-1)^{4/5}} < \frac{3}{2} + \sum_{k=1}^{31} (k-1)^{1/5}.$$

96. Solve in real numbers the equation

$$\sqrt[3]{x-1} + \sqrt[3]{x} + \sqrt[3]{x+1} = 0.$$

97. Find all triples (x, y, z) of positive integers such that

$$x^3 + y^3 + z^3 - 3xyz = p,$$

where p is a prime number greater than 3.

98. Let a, b, c be distinct positive integers such that $ab + bc + ca \geq 3k^2 - 1$, where k is a positive integer. Prove that

$$a^3 + b^3 + c^3 \geq 3(abc + 3k).$$

99. Show that the expression

$$(x^2 - yz)^3 + (y^2 - zx)^3 + (x^2 - yz)^3 - 3(x^2 - yz)(y^2 - zx)(z^2 - xy)$$

is a perfect square.

100. Find all triples (m, n, p) of positive integers such that $m + n + p = 2002$ and the system of equations

$$\frac{x}{y} + \frac{y}{x} = m, \quad \frac{y}{z} + \frac{z}{y} = n, \quad \frac{z}{x} + \frac{x}{z} = p$$

has at least one solution in nonzero real numbers.

2.1.2 $x^2 \geq 0$

We now turn to inequalities. The simplest inequality in algebra says that the square of any real number is nonnegative, and it is equal to zero if and only if the number is zero. We illustrate how this inequality can be used with an example by the second author of the book.

Example. Find the minimum of the function $f : (0, \infty)^3 \to \mathbb{R}$,

$$f(x, y, z) = x^z + y^z - (xy)^{z/4}.$$

Solution. Rewrite the function as

$$f(x, y, z) = (x^{z/2} - y^{z/2})^2 + 2\left[(xy)^{z/4} - \frac{1}{4}\right]^2 - \frac{1}{8}.$$

We now see that the minimum is $-\frac{1}{8}$, achieved if and only if

$$(x, y, z) = \left(a, a, \log_a \frac{1}{16}\right),$$

where $a \in (0, 1) \cup (1, \infty)$.   □

We continue with a problem from the 2001 USA team selection test proposed also by the second author of the book.

Example. Let $(a_n)_{n \geq 0}$ be a sequence of real numbers such that

$$a_{n+1} \geq a_n^2 + \frac{1}{5}, \text{ for all } n \geq 0.$$

Prove that $\sqrt{a_{n+5}} \geq a_{n-5}^2$, for all $n \geq 5$.

Solution. It suffices to prove that $a_{n+5} \geq a_n^2$, for all $n \geq 0$. Let us write the inequality for five consecutive indices:

$$a_{n+1} \geq a_n^2 + \frac{1}{5},$$

$$a_{n+2} \geq a_{n+1}^2 + \frac{1}{5},$$

$$a_{n+3} \geq a_{n+2}^2 + \frac{1}{5},$$

$$a_{n+4} \geq a_{n+3}^2 + \frac{1}{5},$$

$$a_{n+5} \geq a_{n+4}^2 + \frac{1}{5}.$$

If we add these up, we obtain

$$a_{n+5} - a_n^2 \geq (a_{n+1}^2 + a_{n+2}^2 + a_{n+3}^2 + a_{n+4}^2) - (a_{n+1} + a_{n+2} + a_{n+3} + a_{n+4}) + 5 \cdot \frac{1}{5}$$

$$= \left(a_{n+1} - \frac{1}{2}\right)^2 + \left(a_{n+2} - \frac{1}{2}\right)^2 + \left(a_{n+3} - \frac{1}{2}\right)^2 + \left(a_{n+4} - \frac{1}{2}\right)^2 \geq 0.$$

The conclusion follows.   □

And finally a more challenging problem from the 64th W.L. Putnam Mathematics Competition.

Example. Let f be a continuous function on the unit square. Prove that

$$\int_0^1 \left(\int_0^1 f(x, y)dx \right)^2 dx + \int_0^1 \left(\int_0^1 f(x, y)dy \right)^2 dx$$

$$\leq \left(\int_0^1 \int_0^1 f(x, y)dxdy \right)^2 + \int_0^1 \int_0^1 f(x, y)^2 dxdy.$$

Solution. To make this problem as simple as possible, we prove the inequality for a Riemann sum, and then pass to the limit. Divide the unit square into n^2 equal squares, then pick a point (x_i, y_j) in each such square and define $a_{ij} = f(x_i, y_j)$, $i, j = 1, 2, \ldots, n$. Written for the Riemann sum, the inequality becomes

$$\frac{1}{n^3} \sum_i \left(\left(\sum_j a_{ij} \right)^2 + \left(\sum_j a_{ji} \right)^2 \right) \leq \frac{1}{n^4} \left(\sum_{ij} a_{ij} \right)^2 + \frac{1}{n^2} \left(\sum_{ij} a_{ij}^2 \right).$$

Multiply this by n^4, then move everything to one side. After cancellations, the inequality becomes

$$(n-1)^2 \sum_{ij} a_{ij}^2 + \sum_{i \neq k, j \neq l} a_{ij} a_{kl} - (n-1) \sum_{ijk, j \neq k} (a_{ij} a_{ik} + a_{ji} a_{ki}) \geq 0.$$

Here we have a quadratic function in the a_{ij}'s that should always be nonnegative. In general, such a quadratic function can be expressed as an algebraic sum of squares, and it is nonnegative precisely when all squares appear with a positive sign. We are left with the problem of representing our expression as a sum of squares. To boost your intuition, look at the following tableau:

$$
\begin{matrix}
a_{11} & \cdots & \cdots & \cdots & \cdots & \cdots & a_{1n} \\
\vdots & \ddots & \vdots & \ddots & \vdots & \ddots & \vdots \\
\cdots & \cdots & a_{ij} & \cdots & a_{il} & \cdots & \cdots \\
\vdots & \ddots & \vdots & \ddots & \vdots & \ddots & \vdots \\
\cdots & \cdots & a_{kj} & \cdots & a_{kl} & \cdots & \cdots \\
\vdots & \ddots & \vdots & \ddots & \vdots & \ddots & \vdots \\
a_{n1} & \cdots & \cdots & \cdots & \cdots & \cdots & a_{nn}
\end{matrix}
$$

The expression

$$(a_{ij} + a_{kl} - a_{il} - a_{kj})^2$$

when expanded gives rise to the following terms:

$$a_{ij}^2 + a_{kl}^2 + a_{il}^2 + a_{kj}^2 + 2a_{ij}a_{kl} + 2a_{il}a_{kj} - 2a_{il}a_{ij} - 2a_{ij}a_{kj} - 2a_{kl}a_{il} - 2a_{kl}a_{kj}.$$

For a fixed pair (i, j), the term a_{ij} appears in $(n-1)^2$ such expressions. The products $2a_{ij}a_{kl}$ and $2a_{il}a_{kj}$ appear just once, while the products $2a_{il}a_{ij}$, $2a_{ij}a_{kj}$, $2a_{kl}a_{il}$, $2a_{kl}a_{kj}$ appear $(n-1)$

times (once for each square of the form (i, j), (i, l), (k, j), (k, l)). It follows that the expression that we are trying to prove is nonnegative is nothing but

$$\sum_{ijkl}(a_{ij} + a_{kl} - a_{il} - a_{kj})^2,$$

which is of course nonnegative. This proves the inequality for all Riemann sums of the function f, and hence for f itself.   □

101. Find $\min\limits_{a,b\in\mathbb{R}} \max(a^2 + b, b^2 + a)$.

102. Prove that for all real numbers x,

$$2^x + 3^x - 4^x + 6^x - 9^x \le 1.$$

103. Find all positive integers n for which the equation

$$nx^4 + 4x + 3 = 0$$

has a real root.

104. Find all triples (x, y, z) of real numbers that are solutions to the system of equations

$$\frac{4x^2}{4x^2 + 1} = y,$$

$$\frac{4y^2}{4y^2 + 1} = z,$$

$$\frac{4z^2}{4z^2 + 1} = x.$$

105. Find the minimum of

$$\log_{x_1}\left(x_2 - \frac{1}{4}\right) + \log_{x_2}\left(x_3 - \frac{1}{4}\right) + \cdots + \log_{x_n}\left(x_1 - \frac{1}{4}\right),$$

over all $x_1, x_2, \ldots, x_n \in \left(\frac{1}{4}, 1\right)$.

106. Let a and b be real numbers such that

$$9a^2 + 8ab + 7b^2 \le 6.$$

Prove that $7a + 5b + 12ab \le 9$.

107. Let $a_1, a_2, \ldots, a_n$ an be real numbers such that $a_1 + a_2 + \cdots + a_n \ge n^2$ and $a_1^2 + a_2^2 + \cdots + a_n^2 \le n^3 + 1$. Prove that $n - 1 \le a_k \le n + 1$ for all k.

108. Find all pairs (x, y) of real numbers that are solutions to the system

$$x^4 + 2x^3 - y = -\frac{1}{4} + \sqrt{3},$$

$$y^4 + 2y^3 - x = -\frac{1}{4} - \sqrt{3}.$$

109. Let n be an even positive integer. Prove that for any real number x there are at least $2^{n/2}$ choices of the signs $+$ and $-$ such that

$$\pm x^n \pm x^{n-1} \pm \cdots \pm x < \frac{1}{2}.$$

2.1.3 The Cauchy-Schwarz Inequality

A direct application of the discussion in the previous section is the proof of the Cauchy-Schwarz (or Cauchy-Bunyakovski-Schwarz) inequality

$$\sum_{k=1}^{n} a_k^2 \sum_{k=1}^{n} b_k^2 \geq \left(\sum_{k=1}^{n} a_k b_k \right)^2,$$

where the equality holds if and only if the a_i's and the b_i's are proportional. The expression

$$\sum_{k=1}^{n} a_k^2 \sum_{k=1}^{n} b_k^2 - \left(\sum_{k=1}^{n} a_k b_k \right)^2$$

is a quadratic function in the a_i's and b_i's. For it to have only nonnegative values, it should be a sum of squares. And this is true by the Lagrange identity

$$\sum_{k=1}^{n} a_k^2 \sum_{k=1}^{n} b_k^2 - \left(\sum_{k=1}^{n} a_k b_k \right)^2 = \sum_{i<k} (a_i b_k - a_k b_i)^2.$$

Sadly, this proof works only in the finite-dimensional case, while the Cauchy-Schwarz inequality is true in far more generality, such as for square integrable functions. Its correct framework is that of a real or complex vector space, which could be finite or infinite dimensional, endowed with an inner product $\langle \cdot, \cdot \rangle$.

By definition, an inner product is subject to the following conditions:

(i) $\langle x, x \rangle \geq 0$, with equality if and only if $x = 0$,

(ii) $\langle x, y \rangle = \overline{\langle y, x \rangle}$, for any vectors x, y (here the bar stands for complex conjugation if the vector space is complex),

(iii) $\langle \lambda_1 x_1 + \lambda_2 x_2, y \rangle = \lambda_1 \langle x_1, y \rangle + \lambda_2 \langle x_2, y \rangle$, for any vectors x_1, x_2, y and scalars λ_1 and λ_2.

The quantity $\|x\| = \sqrt{\langle x, x \rangle}$ is called the norm of x. Examples of inner product spaces are $\mathbb{R}^n$ with the usual dot product, $\mathbb{C}^n$ with the inner product

$$\langle (z_1, z_2, \ldots, z_n), (w_1, w_2, \ldots, w_n) \rangle = z_1 \overline{w_1} + z_2 \overline{w_2} + \ldots + z_n \overline{w_n},$$

but also the space of square integrable functions on an interval $[a, b]$ with the inner product

$$\langle f, g \rangle = \int_a^b f(t) \overline{g(t)} dt.$$

The Cauchy-Schwarz inequality. *Let x, y be two vectors. Then*

$$\|x\| \cdot \|y\| \geq |\langle x, y \rangle|,$$

with equality if and only if the vectors x and y are parallel and point in the same direction.

Proof. We have

$$0 \le \langle \|y\|x - \|x\|y, \|y\|x - \|x\|y \rangle = 2\|x\|^2\|y\|^2 - \|x\| \cdot \|y\|(\langle x, y \rangle + \langle y, x \rangle),$$

hence $2\|x\| \cdot \|y\| \ge (\langle x, y \rangle + \langle y, x \rangle)$. Yet another trick: rotate y by $\langle x, y \rangle/|\langle x, y \rangle|$. The left-hand side does not change, but because of property (ii) the right-hand side becomes $\frac{1}{|\langle x,y \rangle|}(\langle x, y \rangle \overline{\langle x, y \rangle} + \overline{\langle x, y \rangle}\langle x, y \rangle)$, which is the same as $2|\langle x, y \rangle|$. It follows that

$$\|x\| \cdot \|y\| \ge |\langle x, y \rangle|,$$

which is the Cauchy-Schwarz inequality in its full generality. In our sequence of deductions, the only inequality that showed up holds with equality precisely when the vectors are parallel and point in the same direction.   □

As an example, if f and g are two complex-valued continuous functions on the interval $[a, b]$, or more generally two square integrable functions, then

$$\int_a^b |f(t)|^2 dt \int_a^b |g(t)|^2 dt \ge \left| \int_a^b f(t)\overline{g(t)}dt \right|^2.$$

Let us turn to more elementary problems.

Example. Find the maximum of the function $f(x, y, z) = 5x - 6y + 7z$ on the ellipsoid

$$2x^2 + 3y^2 + 4z^2 = 1.$$

Solution. For a point (x, y, z) on the ellipsoid,

$$(f(x, y, z))^2 = (5x - 6y + 7z)^2 = \left(\frac{5}{\sqrt{2}} \cdot \sqrt{2}x - \frac{6}{\sqrt{3}} \cdot \sqrt{3}y + \frac{7}{2} \cdot 2z \right)^2$$

$$\le \left(\left(\frac{5}{\sqrt{2}}\right)^2 + \left(-\frac{6}{\sqrt{3}}\right)^2 + \left(\frac{7}{2}\right)^2 \right) \left((\sqrt{2}x)^2 + (\sqrt{3}y)^2 + (2z)^2 \right)$$

$$= \frac{147}{4}(2x^2 + 3y^2 + 4z^2) = \frac{147}{4}.$$

Hence the maximum of f is $\sqrt{147}/2$, reached at the point (x, y, z) on the ellipsoid for which $x, z > 0$, $y < 0$, and $x : y : z = \frac{5}{\sqrt{2}} : -\frac{6}{\sqrt{3}} : \frac{7}{2}$.   □

The next problem was on the short list of the 1993 International Mathematical Olympiad, being proposed by the second author of the book.

Example. Prove that

$$\frac{a}{b + 2c + 3d} + \frac{b}{c + 2d + 3a} + \frac{c}{b + 2a + 3b} + \frac{d}{a + 2b + 3c} \ge \frac{2}{3},$$

for all $a, b, c, d > 0$.

Solution. Denote by E the expression on the left. Then

$$4(ab + ac + ad + bc + bd + cd)E$$
$$= (a(b + 2c + 3d) + b(c + 2d + 3a) + c(d + 2a + 3b) + d(a + 2b + 3c))$$
$$\times \left(\frac{a}{b + 2c + 3s} + \frac{b}{c + 2d + 3a} + \frac{c}{b + 2a + 3b} + \frac{d}{a + 2b + 3c} \right)$$
$$\geq (a + b + c + d)^2,$$

where the last inequality is a well-disguised Cauchy-Schwarz. Finally,

$$3(a + b + c + d)^2 \geq 8(ab + ac + ad + bc + bd + cd),$$

because it reduces to

$$(a - b)^2 + (a - c)^2 + (a - d)^2 + (b - c)^2 + (b - d)^2 + (c - d)^2 \geq 0.$$

Combining these two and cancelling the factor $ab + ac + ad + bc + bd + cd$, we obtain the inequality from the statement. $\qquad\square$

And now a list of problems, all of which are to be solved using the Cauchy-Schwarz inequality.

110. If a, b, c are positive numbers, prove that

$$9a^2b^2c^2 \leq (a^2b + b^2c + c^2a)(ab^2 + bc^2 + ca^2).$$

111. If $a_1 + a_2 + \cdots + a_n = n$ prove that $a_1^4 + a_2^4 + \cdots + a_n^4 \geq n$.

112. Let $a_1, a_2, \ldots, a_n$ be distinct real numbers. Find the maximum of

$$a_1 a_{\sigma(a)} + a_2 a_{\sigma(2)} + \cdots + a_n a_{\sigma(n)}$$

over all permutations of the set $\{1, 2, \ldots, n\}$.

113. Let $f_1, f_2, \ldots, f_n$ be positive real numbers. Prove that for any real numbers $x_1, x_2, \ldots, x_n$, the quantity

$$f_1 x_1^2 + f_2 x_2^2 + \cdots + f_n x_n^2 - \frac{(f_1 x_1 + f_2 x_2 + \cdots + f_n x_n)^2}{f_1 + f_2 + \cdots + f_n}$$

is nonnegative.

114. Find all positive integers $n, k_1, \ldots, k_n$ such that $k_1 + \cdots + k_n = 5n - 4$ and

$$\frac{1}{k_1} + \cdots + \frac{1}{k_n} = 1.$$

115. Prove that the finite sequence $a_0, a_1, \ldots, a_n$ of positive real numbers is a geometric progression if and only if

$$(a_0 a_1 + a_1 a_2 + \cdots + a_{n-1} a_n)^2 = (a_0^2 + a_1^2 + \cdots + a_{n-1}^2)(a_1^2 + a_2^2 + \cdots + a_n^2).$$

116. Let $P(x)$ be a polynomial with positive real coefficients. Prove that

$$\sqrt{P(a)P(b)} \geq P(\sqrt{ab}),$$

for all positive real numbers a and b.

117. Consider the real numbers $x_0 > x_1 > x_2 > \cdots > x_n$. Prove that

$$x_0 + \frac{1}{x_0 - x_1} = \frac{1}{x_1 - x_2} + \cdots + \frac{1}{x_{n-1} - x_n} \geq x_n + 2n.$$

When does equality hold?

118. Prove that

$$\frac{\sin^3 a}{\sin b} + \frac{\cos^3 a}{\cos b} \geq \sec(a - b),$$

for all $a, b \in \left(0, \frac{\pi}{2}\right)$.

119. Prove that

$$\frac{1}{a + b} + \frac{1}{b + c} + \frac{1}{c + a} + \frac{1}{2\sqrt[3]{abc}} \geq \frac{(a + b + c + \sqrt[3]{abc})^2}{(a + b)(b + c)(c + a)},$$

for all $a, b, c > 0$.

2.1.4 The Triangle Inequality

In its most general form, the triangle inequality states that in a metric space X the distance function δ satisfies

$$\delta(x, y) \leq \delta(x, z) + \delta(z, y), \text{ for any } x, y, z \in X.$$

An equivalent form is

$$|\delta(x, y) - \delta(y, z)| \leq \delta(x, z).$$

Here are some familiar examples of distance functions: the distance between two real or complex numbers as the absolute value of their difference, the distance between two vectors in n-dimensional Euclidean space as the length of their difference $\|v - w\|$, the distance between two matrices as the norm of their difference, the distance between two continuous functions on the same interval as the supremum of the absolute value of their difference. In all these cases the triangle inequality holds.

Let us see how the triangle inequality can be used to solve a problem from T.B. Soulami's book *Les olympiades de mathématiques: Réflexes et stratégies* (Ellipses, 1999).

Example. For positive numbers a, b, c prove the inequality

$$\sqrt{a^2 - ab + b^2} + \sqrt{b^2 - bc + c^2} \geq \sqrt{a^2 + ac + c^2}.$$

Solution. The inequality suggests the following geometric construction. With the same origin O, draw segments OA, OB, and OC of lengths a, b, respectively c, such that OB makes 60° angles with OA and OC (see Figure 12).

The law of cosines in the triangles OAB, OBC, and OAC gives $AB^2 = a^2 - ab + b^2$, $BC^2 = b^2 - bc + c^2$, and $AC^2 = a^2 + ac + c^2$. Plugging these formulas into the triangle inequality $AB + BC \geq AC$ produces the inequality from the statement. $\square$

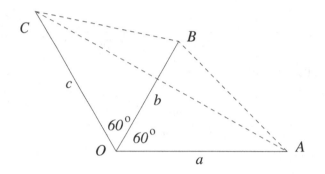

Figure 12

Example. Let $P(x)$ be a polynomial whose coefficients lie in the interval $[1, 2]$, and let $Q(x)$ and $R(x)$ be two nonconstant polynomials such that $P(x) = Q(x)R(x)$, with $Q(x)$ having the dominant coefficient equal to 1. Prove that $|Q(3)| > 1$.

Solution. Let $P(x) = a_n x^n + a_{n-1} x^{n-1} + \cdots + a_0$. We claim that the zeros of $P(x)$ lie in the union of the half-plane $\operatorname{Re} z = 0$ and the disk $|z| < 2$.

Indeed, suppose that $P(x)$ has a zero z such that $\operatorname{Re}, z > 0$ and $|z| = 2$. From $P(z) = 0$, we deduce that

$$a_n z^n + a_{n-1} z^{n-1} = -a_{n-2} z^{n-2} - a_{n-3} z^{n-3} - \cdots - a_0.$$

Dividing through by z^n, which is not equal to 0, we obtain

$$a_n + \frac{a_{n-1}}{z} = -\frac{a_{n-2}}{z^2} - \frac{a_{n-3}}{z^3} - \cdots - \frac{a_0}{z^n}.$$

Note that $\operatorname{Re} z > 0$ implies that $\operatorname{Re} \frac{1}{z} > 0$. Hence

$$1 \leq a_n \leq \operatorname{Re}\left(a_n + \frac{a_{n-1}}{z}\right) = \operatorname{Re}\left(-\frac{a_{n-2}}{z^2} - \frac{a_{n-3}}{z^3} - \cdots - \frac{a_0}{z^n}\right)$$

$$\leq \left|-\frac{a_{n-2}}{z^2} - \frac{a_{n-3}}{z^3} - \cdots - \frac{a_0}{z^n}\right| \leq \frac{a_{n-2}}{|z|^2} + \frac{a_{n-3}}{|z|^3} + \cdots + \frac{a_0}{|z|^n},$$

where for the last inequality we used the triangle inequality. Because the a_i's are in the interval $[1, 2]$, this is strictly less than

$$2|z|^{-2}(1 + |z|^{-1} + |z|^{-2} + \cdots) = \frac{2|z|^{-2}}{1 - |z|^{-1}}.$$

The last quantity must therefore be greater than 1. But this cannot happen if $|z| \geq 2$, because the inequality reduces to $\left(\frac{2}{|z|} - 1\right)\left(\frac{1}{|z|} + 1\right) > 0$, impossible. This proves the claim.

Returning to the problem, $Q(x) = (x - z_1)(x - z_2) \cdots (x - z_k)$, where $z_1, z_2, \ldots, z_k$ are some of the zeros of $P(x)$. Then

$$|Q(3)| = |3 - z_1| \cdot |3 - z_2| \cdots |3 - z_k|.$$

If $\operatorname{Re} z_i \leq 0$, then $|3 - z_i| \geq 0$. On the other hand, if $|z_i| < 2$, then by the triangle inequality $|3 - z_i| \geq 3 - |z_i| > 1$. Hence $|Q(3)|$ is a product of terms greater than 1, and the conclusion follows. $\square$

More applications follow.

120. Let a, b, c be the side lengths of a triangle with the property that for any positive integer n, the numbers a^n, b^n, c^n can also be the side lengths of a triangle. Prove that the triangle is necessarily isosceles.

121. Given the vectors $\vec{a}, \vec{b}, \vec{c}$ in the plane, show that

$$\|\vec{a}\| + \|\vec{b}\| + \|\vec{c}\| + \|\vec{a} + \vec{b} + \vec{c}\| \geq \|\vec{a} + \vec{b}\| + \|\vec{a} + \vec{c}\| + \|\vec{b} + \vec{c}\|.$$

122. Let $P(z)$ be a polynomial with real coefficients whose roots can be covered by a disk of radius R. Prove that for any real number k, the roots of the polynomial $nP(z) - kP'(z)$ can be covered by a disk of radius $R + |k|$, where n is the degree of $P(z)$, and $P'(z)$ is the derivative.

123. Prove that the positive real numbers a, b, c are the side lengths of a triangle if and only if

$$a^2 + b^2 + c^2 < 2\sqrt{a^2b^2 + b^2c^2 + c^2a^2}.$$

124. Let $ABCD$ be a convex cyclic quadrilateral. Prove that

$$|AB - CD| + |AD - BC| \geq 2|AC - BD|.$$

125. Let $V_1, V_2, \ldots, V_m$ and $W_1, W_2, \ldots, W_m$ be isometries of $\mathbb{R}^n$ (m, n positive integers). Assume that for all x with $\|x\| \leq 1$, $\|V_i x - W_i x\| \leq 1$, $i = 1, 2, \ldots, m$. Prove that

$$\left\|\left(\prod_{i=1}^m V_i\right)x - \left(\prod_{i=1}^m W_i\right)x\right\| \leq m,$$

for all x with $\|x\| \leq 1$.

126. Given an equilateral triangle ABC and a point P that does not lie on the circumcircle of ABC, show that one can construct a triangle with sides the segments PA, PB, and PC. If P lies on the circumcircle, show that one of these segments is equal to the sum of the other two.

127. Let M be a point in the plane of the triangle ABC whose centroid is G. Prove that

$$MA^3 \cdot BC + MB^3 \cdot AC + MC^3 \cdot AB \geq 3MG \cdot AB \cdot BC \cdot CA.$$

2.1.5 The Arithmetic Mean-Geometric Mean Inequality

Jensen's inequality, which will be discussed in the section about convex functions, states that if f is a real-valued concave function, then

$$f(\lambda_1 x_1 + \lambda_2 x_2 + \cdots + \lambda_n x_n) \geq \lambda_1 f(x_1) + \lambda_2 f(x_2) + \cdots + \lambda_n f(x_n),$$

for any $x_1, x_2, \ldots, x_n$ in the domain of f and for any positive weights $\lambda_1, \lambda_2, \ldots, \lambda_n$ with $\lambda_1 + \lambda_2 + \cdots + \lambda_n = 1$. Moreover, if the function is nowhere linear (that is, if it is strictly concave) and the numbers $\lambda_1, \lambda_2, \ldots, \lambda_n$ are nonzero, then equality holds if and only if $x_1 = x_2 = \cdots = x_n$.

Applying this to the concave function $f(x) = \ln x$, the positive numbers $x_1, x_2, \ldots, x_n$, and the weights $\lambda_1 = \lambda_2 = \cdots = \lambda_n = \frac{1}{n}$, we obtain

$$\ln \frac{x_1 + x_2 + \cdots + x_n}{n} \geq \frac{\ln x_1 + \ln x_2 + \cdots + \ln x_n}{n}.$$

Exponentiation yields the following inequality.

The arithmetic mean-geometric mean inequality. *Let $x_1, x_2, \ldots, x_n$ be nonnegative real numbers. Then*

$$\frac{x_1 + x_2 + \cdots + x_n}{n} \geq \sqrt[n]{x_1 x_2 \cdots x_n},$$

with equality if and only if all numbers are equal.

Proof. We will call this inequality AM-GM for short. We give it an alternative proof using derivatives, a proof by induction on n. For $n = 2$ the inequality is equivalent to the obvious $(\sqrt{a_1} - \sqrt{a_2})^2 \geq 0$. Next, assume that the inequality holds for any $n - 1$ positive numbers, meaning that

$$\frac{x_1 + x_2 + \cdots + x_{n-1}}{n - 1} \geq \sqrt[n-1]{x_1 x_2 \cdots x_{n-1}},$$

with equality only when $x_1 = x_2 = \cdots = x_{n-1}$. To show that the same is true for n numbers, consider the function $f : (0, \infty) \to \mathbb{R}$,

$$f(x) = \frac{x_1 + x_2 + \cdots + x_{n-1} + x}{n} - \sqrt[n]{x_1 x_2 \cdots x_{n-1} x}.$$

To find the minimum of this function we need the critical points. The derivative of f is

$$f'(x) = \frac{1}{n} - \frac{\sqrt[n]{x_1 x_2 \cdots x_{n-1}}}{n} x^{\frac{1}{n}-1} = \frac{x^{\frac{1}{n}-1}}{n}\left(x^{1-\frac{1}{n}} - \sqrt[n]{x_1 x_2 \cdots x_{n-1}}\right).$$

Setting this equal to zero, we find the unique critical point $x = {}^{n-1}\!\sqrt{x_1 x_2 \cdots x_n}$, since in this case $x^{1-\frac{1}{n}} = \sqrt[n]{x_1 x_2 \cdots x_{n-1}}$. Moreover, the function $x^{1-\frac{1}{n}}$ is increasing on $(0, \infty)$; hence $f'(x) < 0$ for $x < {}^{n-1}\!\sqrt{x_1 x_2 \cdots x_{n-1}}$, and $f'(x) > 0$ for $x > {}^{n-1}\!\sqrt{x_1 x_2 \cdots x_{n-1}}$. We find that f has a global minimum at $x = {}^{n-1}\!\sqrt{x_1 x_2 \cdots x_{n-1}}$, where it takes the value

$$
\begin{aligned}
f\left({}^{n-1}\!\sqrt{x_1 x_2 \cdots x_{n-1}}\right) &= \frac{x_1 + x_2 + \cdots + x_{n-1} + {}^{n-1}\!\sqrt{x_1 x_2 \cdots x_{n-1}}}{n} \\
&\quad - \sqrt[n]{x_1 x_2 \cdots x_{n-1} \cdot {}^{n(n-1)}\!\sqrt{x_1 x_2 \cdots x_{n-1}}} \\
&= \frac{x_1 + x_2 + \cdots + x_{n-1} + {}^{n-1}\!\sqrt{x_1 x_2 \cdots x_{n-1}}}{n} \\
&\quad - {}^{n-1}\!\sqrt{x_1 x_2 \cdots x_{n-1}} \\
&= \frac{x_1 + x_2 + \cdots + x_{n-1} - (n-1)\,{}^{n-1}\!\sqrt{x_1 x_2 \cdots x_{n-1}}}{n}.
\end{aligned}
$$

By the induction hypothesis, this minimum is nonnegative, and is equal to 0 if and only if $x_1 = x_2 = \cdots = x_{n-1}$. We conclude that $f(x_n) > 0$ with equality if and only if $x_1 = x_2 = \cdots = x_{n-1}$ and $x_n = {}^{n-1}\!\sqrt{x_1 x_2 \cdots x_{n-1}} = x_1$. This completes the induction. $\square$

We apply the AM-GM inequality to solve two problems composed by the second author of the book.

Example. Find the global minimum of the function $f : \mathbb{R}^2 \to \mathbb{R}$,

$$
f(x, y) = 3^{x+y}(3^{x-1} + 3^{y-1} - 1).
$$

Solution. The expression

$$
3f(x, y) + 1 = 3^{2x+y} + 3^{x+2y} + 1 - 3 \cdot 3^{x+y}
$$

is of the form $a^3 + b^3 + c^3 - 3abc$, where $a = \sqrt[3]{3^{2x+y}}$, $b = \sqrt[3]{3^{x+2y}}$, and $c = 1$, all of which are positive. By the AM-GM inequality, this expression is nonnegative. It is equal to zero only when $a = b = c$, that is, when $2x + y = x + 2y = 0$. We conclude that the minimum of f is $f(0, 0) = -\frac{1}{3}$. $\square$

Example. Let a, b, c, d be positive real numbers with $abcd = 1$. Prove that

$$
\frac{a}{b+c+d+1} + \frac{b}{c+d+a+1} + \frac{c}{d+a+b+1} + \frac{d}{a+b+c+1} \geq 1.
$$

Solution. A first idea is to homogenize this inequality, and for that we replace the 1 in each denominator by $\sqrt[4]{abcd}$, transforming the inequality into

$$
\frac{a}{b+c+d+\sqrt[4]{abcd}} + \frac{b}{c+d+a+\sqrt[4]{abcd}} + \frac{c}{d+a+b+\sqrt[4]{abcd}}
$$

$$
+ \frac{d}{a+b+c+\sqrt[4]{abcd}} \geq 1.
$$

Then we apply the AM-GM inequality to the last term in each denominator to obtain the stronger inequality

$$\frac{4a}{a+5(b+c+d)} + \frac{4b}{b+5(c+d+a)} + \frac{4c}{c+5(d+a+b)} + \frac{4d}{d+5(a+b+c)} \geq 1,$$

which we proceed to prove.

In order to simplify computations, it is better to denote the four denominators by $16x$, $16y$, $16z$, $16w$, respectively. Then $a+b+c+d = x+y+z+w$, and so $4a + 16x = 4b + 16y = 4c + 16z = 4d + 16w = 5(x+y+z+w)$. The inequality becomes

$$\frac{-11x + 5(y+z+w)}{16x} + \frac{-11y + 5(z+w+x)}{16y} + \frac{-11z + 5(w+x+y)}{16z}$$
$$+ \frac{-11w + 5(x+y+z)}{16w} \geq 1,$$

or

$$-4 \cdot 11 + 5\left(\frac{y}{x} + \frac{z}{x} + \frac{w}{x} + \frac{z}{y} + \frac{w}{y} + \frac{x}{y} + \frac{w}{z} + \frac{x}{z} + \frac{y}{z} + \frac{x}{w} + \frac{y}{w} + \frac{z}{w}\right) \geq 16.$$

And this follows by applying the AM-GM inequality to the twelve summands in the parentheses. □

We continue with a third example, which is an problem of A. Basyoni that was given in 2015 at a preliminary selection test for the team that represented the United States at the International Mathematical Olympiad in 2016.

Example. Let x, y, z be real numbers satisfying $x^4 + y^4 + z^4 + xyz = 4$. Show that

$$\sqrt{2-x} \geq \frac{y+z}{2}.$$

Solution. We have selected the problem for the book because of this elegant solution based on the AM-GM inequality found by the member of the Canadian team Zh.Q. (Alex) Song. It suffices to show that

$$\sqrt{2-x} \geq \left|\frac{y+z}{2}\right|.$$

This inequality and the fact that the square root is well defined follow simultaneously if we prove that

$$x + \left(\frac{y+z}{2}\right)^2 \leq 2.$$

Apply the AM-GM inequality three times:

$$\frac{x^4}{8} + \frac{y^4}{8} + \frac{y^4}{8} + \frac{1}{8} \geq \frac{xy^2}{2}$$
$$\frac{x^4}{8} + \frac{z^4}{8} + \frac{z^4}{8} + \frac{1}{8} \geq \frac{xz^2}{2}$$
$$\frac{3x^4}{4} + \frac{3}{4} \geq \frac{3x^2}{2}.$$

Then apply the power-mean inequality:

$$\frac{y^4 + z^4}{2} \geq \left(\frac{y+z}{2}\right)^4,$$

to write

$$\frac{3y^4}{4} + \frac{3z^4}{4} \geq \frac{3}{2}\left(\frac{y+z}{2}\right)^4.$$

Now add the four inequalities and use the relation from the statement to obtain

$$5 \geq \frac{3}{2}x^2 + \frac{3}{2}\left[\left(\frac{y+z}{2}\right)^2\right]^2 + 2\left[x\left(\frac{y+z}{2}\right)\right]^2.$$

Finally, noticing that the AM-GM inequality implies

$$\frac{1}{4}\left(x^2 + \left[\left(\frac{y+z}{2}\right)^2\right]^2\right) \geq \frac{1}{2}\left[x\left(\frac{y+z}{2}\right)^2\right],$$

we obtain

$$5 \geq \frac{5}{4}\left[x + \left(\frac{y+z}{2}\right)^2\right]^2,$$

and the conclusion follows. □

For completeness let us prove this particular case of the power mean inequality:

$$\frac{y^4 + z^4}{2} \geq \left(\frac{y^2 + z^2}{2}\right)^2 \geq \left[\left(\frac{y+z}{2}\right)^2\right]^2 = \left(\frac{y+z}{2}\right)^4.$$

It becomes clear after expanding the square that the first inequality is a consequence of the AM-GM inequality. Taking the square root of the second inequality, we recognize that it is of the same type. So we are done.

Try your hand at the following problems.

128. Show that all real roots of the polynomial $P(x) = x^5 - 10x + 35$ are negative.

129. Find all real numbers that satisfy

$$x \cdot 2^{\frac{1}{x}} + \frac{1}{x} \cdot 2^x = 4.$$

130. Let $a_1, a_2, \ldots, a_n$ and $b_1, b_2, \ldots, b_n$ be nonnegative numbers. Show that

$$(a_1 a_2 \cdots a_n)^{1/n} + (b_1 b_2 \cdots b_n)^{1/n} \leq ((a_1 + b_1)(a_2 + b_2) \cdots (a_n + b_n))^{1/n}.$$

131. Let a, b, c be the side lengths of a triangle with semiperimeter 1. Prove that

$$1 < ab + bc + ca - abc \leq \frac{28}{27}.$$

132. Which number is larger,

$$\prod_{n=1}^{25} \left(1 - \frac{n}{365}\right) \quad \text{or} \quad \frac{1}{2}?$$

133. On a sphere of radius 1 are given four points A, B, C, D such that

$$AB \cdot AC \cdot AD \cdot BC \cdot BD \cdot CD = \frac{2^9}{3^3}.$$

Prove that the tetrahedron $ABCD$ is regular.

134. Prove that

$$\frac{y^2 - x^2}{2x^2 + 1} + \frac{z^2 - y^2}{2y^2 + 1} + \frac{x^2 - z^2}{2z^2 + 1} \geq 0,$$

for all real numbers x, y, z.

135. Let $a_1, a_2, \ldots, a_n$ be positive real numbers such that $a_1 + a_2 + \cdots + a_n < 1$. Prove that

$$\frac{a_1 a_2 \cdots a_n (1 - (a_1 + a_2 + \cdots + a_n))}{(a_1 + a_2 + \cdots + a_n)(1 - a_1)(1 - a_2) \cdots (1 - a_n)} \leq \frac{1}{n^{n+1}}.$$

136. Consider the positive real numbers $x_1, x_2, \ldots, x_n$ with $x_1 x_2 \cdots x_n = 1$. Prove that

$$\frac{1}{n - 1 + x_1} + \frac{1}{n - 1 + x_2} + \cdots + \frac{1}{n - 1 + x_n} \leq 1.$$

2.1.6 Sturm's Principle

In this section we present a method for proving inequalities that is based on real analysis. It is based on a principle attributed to R. Sturm, phrased as follows.

Sturm's principle. *Given a function f defined on a set M and a point $x_0 \in M$, if*

(i) f has a maximum (minimum) on M, and

(ii) if no other point x in M is a maximum (minimum) of f,

then x_0 is the maximum (minimum) of f.

But how to decide whether the function f has a maximum or a minimum? Two results from real analysis come in handy.

Theorem. *A continuous function on a compact set always attains its extrema.*

Theorem. *A closed and bounded subset of $\mathbb{R}^n$ is compact.*

Let us see how Sturm's principle can be applied to a problem from the first Balkan Mathematical Olympiad in 1984.

Example. Let $\alpha_1, \alpha_2, \ldots, \alpha_n$ be positive real numbers, $n \geq 2$, such that $\alpha_1 + \alpha_2 + \cdots + \alpha_n = 1$. Prove that

$$\frac{\alpha_1}{1 + \alpha_2 + \cdots + \alpha_n} + \frac{\alpha_2}{1 + \alpha_1 + \cdots + \alpha_n} + \cdots + \frac{\alpha_n}{1 + \alpha_1 + \cdots + \alpha_{n-1}} \geq \frac{n}{2n - 1}.$$

Solution. Rewrite the inequality as

$$\frac{\alpha_1}{2 - \alpha_1} + \frac{\alpha_2}{2 - \alpha_2} + \cdots + \frac{\alpha_n}{2 - \alpha_n} \geq \frac{n}{2n - 1},$$

and then define the function

$$f(\alpha_1, \alpha_2, \ldots, \alpha_n) = \frac{\alpha_1}{2 - \alpha_1} + \frac{\alpha_2}{2 - \alpha_2} + \cdots + \frac{\alpha_n}{2 - \alpha_n}.$$

As said in the statement, this function is defined on the subset of $\mathbb{R}^n$ consisting of points whose coordinates are positive and add up to 1. We would like to show that on this set f is greater than or equal to $\frac{n}{2n-1}$.

Does f have a minimum? The domain of f is bounded but is not closed, being the interior of a tetrahedron. We can enlarge it, though, by adding the boundary, to the set

$$M = \{(\alpha_1, \alpha_2, \ldots, \alpha_n) \mid \alpha_1 + \alpha_2 + \cdots + \alpha_n = 1, \ \alpha_i \geq 0, \ i = 1, 2, \ldots, n\}.$$

We now know that f has a minimum on M.

A look at the original inequality suggests that the minimum is attained when all the α_i's are equal. So let us choose a point $(\alpha_1, \alpha_2, \ldots, \alpha_n)$ for which $\alpha_i \neq \alpha_j$ for some indices i, j. Assume that $\alpha_i < \alpha_j$ and let us see what happens if we substitute $\alpha_i + x$ for α_i and $\alpha_j - x$

for α_j, with $0 < x < \alpha_j - \alpha_i$. In the defining expression of f, only the ith and jth terms change. Moreover,

$$\frac{\alpha_i}{2 - \alpha_i} + \frac{\alpha_j}{2 - \alpha_j} - \frac{\alpha_i + x}{2 - \alpha_i - x} - \frac{\alpha_j - x}{2 - \alpha_j + x}$$

$$= \frac{2x(\alpha_j - \alpha_i - x)(4 - \alpha_i - \alpha_j)}{(2 - \alpha_i)(2 - \alpha_j)(2 - \alpha_i - x)(2 - \alpha_j - x)} > 0,$$

so on moving the numbers closer, the value of f decreases. It follows that the point that we picked was not a minimum. Hence the only possible minimum is $\left(\frac{1}{n}, \frac{1}{n}, \ldots, \frac{1}{n}\right)$, in which case the value of f is $\frac{n}{2n-1}$. This proves the inequality. $\qquad\square$

However, in most situations, as is the case with this problem, we can bypass the use of real analysis and argue as follows. If the a_i's were not all equal, then one of them must be less than $\frac{1}{n}$ and one of them must be greater. Take these two numbers and move them closer until one of them reaches $\frac{1}{n}$. Then stop and choose another pair. Continue the algorithm until all numbers become $\frac{1}{n}$. At this very moment, the value of the expression is

$$\frac{1}{n}\left(2 - \frac{1}{n}\right)^{-1} \cdot n = \frac{n}{2n-1}.$$

Since during the process the value of the expression kept decreasing, initially it must have been greater than or equal to $\frac{n}{2n-1}$. This proves the inequality.

Let us summarize the last idea. We want to maximize (or minimize) an n-variable function, and we have a candidate for the extremum. If we can move the variables one by one toward the maximum without decreasing (respectively, increasing) the value of the function, than the candidate is indeed the desired extremum. This approach is more elementary but can be more time consuming than the application of the principle itself.

Let us revisit the AM-GM inequality with a proof using Sturm's principle.

The arithmetic mean-geometric mean inequality. Let $x_1, x_2, \ldots, x_n$ be nonnegative real numbers. Then

$$\frac{x_1 + x_2 + \cdots + x_n}{n} \geq \sqrt[n]{x_1 x_2 \cdots x_n}.$$

with equality if and only if $x_1 = x_2 = \cdots = x_n$.

Proof. The inequality is homogeneous in the variables, so the general case follows if we check the inequality for a fixed value of the sum of the numbers, say $x_1 + x_2 + \cdots + x_n = 1$. This amounts to checking that $\sqrt[n]{x_1 x_2 \cdots x_n} \leq \frac{1}{n}$ if $x_1 + x_2 + \cdots + x_n = 1$ with equality only when $x_1 = x_2 = \cdots = x_n$, and this is equivalent to checking $x_1 x_2 \cdots x_n \leq \frac{1}{n^n}$ with equality as specified.

The set

$$K = \{(x_1, x_2, \ldots, x_n) \subset \mathbb{R}^n \mid x_j \geq 0, x_1 + x_2 + \cdots + x_n = 1\}$$

contains all its limit points, so it is closed. It also lies in the hypercube $[0, 1]^n$ so it is bounded, thus it is compact. The function

$$f : K \to \mathbb{R}, \quad f(x_1, x_2, \ldots, x_n) = x_1 x_2 \cdots x_n$$

is continuous, being a polynomial, so it attains its maximum. This maximum is not attained at a point where not all x_j are equal, because if $x_j < x_k$ and we replace x_j by $x_j + \varepsilon$ and x_k by $x_k - \varepsilon$, where $\varepsilon = \frac{x_k - x_j}{2}$, then the value of f increases to

$$\prod_{i \neq j,k} x_i (x_j + \varepsilon)(x_k - \varepsilon) = \prod_{i \neq j,k} x_i (x_j x_k + \varepsilon(x_k - x_j) + \varepsilon^2)$$

$$= \prod_{i \neq j,k} x_i (x_j x_k + \varepsilon^2) = f(x_1, x_2, \cdots, x_n) + \varepsilon^2 \prod_{i \neq j,k} x_i.$$

Thus the only candidate for the maximum is $(\frac{1}{n}, \frac{1}{n}, \cdots, \frac{1}{n})$ and in this case the inequality holds with equality. $\qquad \square$

You can find more applications of Sturm's principle below.

137. Let a, b, c be nonnegative real numbers such that $a + b + c = 1$. Prove that

$$4(ab + bc + ac) - 9abc \leq 1.$$

138. Let $x_1, x_2, \ldots, x_n$, $n \geq 2$, be positive numbers such that

$$x_1 + x_2 + \cdots + x_n = 1.$$

Prove that

$$\left(1 + \frac{1}{x_1}\right)\left(1 + \frac{1}{x_2}\right) \cdots \left(1 + \frac{1}{x_n}\right) \geq (n+1)^n.$$

139. Prove that a necessary and sufficient condition that a triangle inscribed in an ellipse have maximum area is that its centroid coincide with the center of the ellipse.

140. Let $n > 2$ be an integer. A convex n-gon of area 1 is inscribed in a circle. What is the minimum that the radius of the circle can be?

141. Let $a, b, c > 0$, $a + b + c = 1$. Prove that

$$0 \leq ab + bc + ac - 2abc \leq \frac{7}{27}.$$

142. Let $x_1, x_2, \ldots, x_n$ be n real numbers such that $0 < x_j \leq \frac{1}{2}$, for $1 \leq j \leq n$. Prove the inequality

$$\frac{\prod_{j=1}^{n} x_j}{\left(\sum_{j=1}^{n} x_j\right)^n} \leq \frac{\prod_{j=1}^{n}(1 - x_j)}{\left(\sum_{j=1}^{n}(1 - x_j)\right)^n}.$$

143. Let a, b, c, and d be nonnegative numbers such that $a \le 1, a+b \le 5, a+b+c \le 14$, $a+b+c+d \le 30$. Prove that

$$\sqrt{a} + \sqrt{b} + \sqrt{c} + \sqrt{d} \le 10.$$

144. What is the maximal value of the expression $\sum_{i<j} x_i x_j$ if $x_1, x_2, \ldots, x_n$ are nonnegative integers whose sum is equal to m?

145. Given the $n \times n$ array $(a_{ij})_{ij}$ with $a_{ij} = i + j - 1$, what is the smallest product of n elements of the array provided that no two lie on the same row or column?

146. Given a positive integer n, find the minimum value of

$$\frac{x_1^3 + x_2^3 + \cdots + x_n^3}{x_1 + x_2 + \cdots + x_n}$$

subject to the condition that $x_1, x_2, \ldots, x_n$ be distinct positive integers.

2.1.7 Other Inequalities

We conclude with a section for the inequalities aficionado. Behind each problem hides a famous inequality.

147. If x and y are positive real numbers, show that $x^y + y^x > 1$.

148. Prove that for all $a, b, c \ge 0$,

$$(a^5 - a^2 + 3)(b^5 - b^2 + 3)(c^5 - c^2 + 3) \ge (a+b+c)^3.$$

149. Assume that all the zeros of the polynomial $P(x) = x^n + a_1 x^{n-1} + \cdots + a_n$ are real and positive. Show that if there exist $1 \le m < p \le n$ such that $a_m = (-1)^m \binom{n}{m}$ and $a_p = (-1)^p \binom{n}{p}$, then $P(x) = (x-1)^n$.

150. Let $n > 2$ be an integer, and let $x_1, x_2, \ldots, x_n$ be positive numbers with the sum equal to 1. Prove that

$$\prod_{i=1}^{n} \left(1 + \frac{1}{x_i}\right) \ge \prod_{i=1}^{n} \left(\frac{n - x_i}{1 - x_i}\right).$$

151. Let $a_1, a_2, \ldots, a_n, b_1, b_2, \ldots, b_n$ be real numbers such that

$$(a_1^2 + a_2^2 + \cdots + a_n^2 - 1)(b_1^2 + b_2^2 + \cdots + b_n^2 - 1) > (a_1 b_1 + a_2 b_2 + \cdots + a_n b_n - 1)^2.$$

Prove that $a_1^2 + a_2^2 + \cdots + a_n^2 > 1$ and $b_1^2 + b_2^2 + \cdots + b_n^2 > 1$.

152. Let a, b, c, d be positive numbers such that $abc = 1$. Prove that

$$\frac{1}{a^3(b+c)} + \frac{1}{b^3(c+a)} + \frac{1}{c^3(a+b)} \ge \frac{3}{2}.$$

2.2 Polynomials

2.2.1 A Warmup in One-Variable Polynomials

A polynomial is a sum of the form

$$P(x) = a_n x^n + a_{n-1} x^{n-1} + \cdots + a_0,$$

where x is the variable, and $a_n, a_{n-1}, \ldots, a_0$ are constant coefficients. If $a_n \neq 0$, the number n is called the degree, denoted by $\deg(P(x))$. If $a_n = 1$, the polynomial is called monic. The sets, which, in fact, are rings, of polynomials with integer, rational, real, or complex coefficients are denoted, respectively, by $\mathbb{Z}[x]$, $\mathbb{Q}[x]$, $\mathbb{R}[x]$, and $\mathbb{C}[x]$. A number r such that $P(r) = 0$ is called a zero of $P(x)$, or a root of the equation $P(x) = 0$. By the Gauss-d'Alembert theorem, also called the fundamental theorem of algebra, every nonconstant polynomial with complex coefficients has at least one complex zero. Consequently, the number of zeros of a polynomial equals the degree, multiplicities counted. For a number α, $P(\alpha) = a_n \alpha^n + a_{n-1} \alpha^{n-1} + \cdots + a_0$ is called the value of the polynomial at α.

We begin the section on polynomials with an old problem from the 1943 competition of the *Mathematics Gazette, Bucharest*, proposed by Gh. Buicliu.

Example. Verify the equality

$$\sqrt[3]{20 + 14\sqrt{2}} + \sqrt[3]{20 - 14\sqrt{2}} = 4.$$

Solution. Apparently, this problem has nothing to do with polynomials. But let us denote the complicated irrational expression by x and analyze its properties. Because of the cube roots, it becomes natural to raise x to the third power:

$$x^3 = 20 + 14\sqrt{2} + 20 - 14\sqrt{2}$$
$$+ 3\sqrt[3]{(20 + 14\sqrt{2})(20 - 14\sqrt{2})} \left(\sqrt[3]{20 + 14\sqrt{2}} + \sqrt[3]{20 - 14\sqrt{2}} \right)$$
$$= 40 + 3x\sqrt[3]{400 - 392} = 40 + 6x.$$

And now we see that x satisfies the polynomial equation

$$x^3 - 6x - 40 = 0.$$

We have already been told that 4 is a root of this equation. The other two roots are complex, and hence x can only equal 4, the desired answer.   □

Of course, one can also recognize the quantities under the cube roots to be the cubes of $2 + \sqrt{2}$ and $2 - \sqrt{2}$, but that is just a lucky strike.

The second example is a problem from the Russian Journal *Kvant (Quantum)*, proposed by A. Alexeev.

Example. Prove that for every odd positive integer n, there is a constant c_n such that

$$\frac{\tan x + \tan \left(x + \frac{\pi}{n}\right) + \cdots + \tan \left(x + \frac{(n-1)\pi}{n}\right)}{\tan x \tan \left(x + \frac{\pi}{n}\right) \cdots \tan \left(x + \frac{(n-1)\pi}{n}\right)} = c_n,$$

for all x for which the denominator is nonzero. Find the value of c_n.

Solution. Since the tangent function is periodic with period π, it suffices to look at $x \in [0, \pi]$. Consider the function $f : [0, \pi] \rightarrow \mathbb{R}$,

$$f(x) = \frac{\tan x + \tan \left(x + \frac{\pi}{n}\right) + \cdots + \tan \left(x + \frac{(n-1)\pi}{n}\right)}{\tan x \tan \left(x + \frac{\pi}{n}\right) \cdots \tan \left(x + \frac{(n-1)\pi}{n}\right)}.$$

Denote $\tan x = \xi$ and $\tan \frac{k\pi}{n} = t_k$, $k = 0, 1, \ldots n - 1$. Then the numerator and the denominator are of the form

$$\frac{P_1(\xi)}{Q(\xi)} = \xi + \frac{\xi + t_1}{1 - \xi t_1} + \cdots + \frac{\xi + t_{n-1}}{1 - \xi t_{n-1}}$$

$$\frac{P_2(\xi)}{Q(\xi)} = \xi \cdot \frac{\xi + t_1}{1 - \xi t_1} \cdots \frac{\xi + t_{n-1}}{1 - \xi t_{n-1}}$$

where $P_1(\xi)$, $P_2(\xi)$, $Q(\xi)$ are polynomials, and $Q(\xi) = (1 - \xi t_1)(1 - \xi t_2) \cdots (1 - \xi t_{n-1})$.

The polynomials $P_1(\xi)$, $P_2(\xi)$ have nth degree. Because of the fact that n is odd, and of the trigonometric identity $\tan(\pi - x) = -\tan x$, the roots of $P_1(\xi)$ must be $0, t_1, t_2, \ldots, t_{n-1}$. Of course these are also the roots of $P_2(\xi)$. It follows that one of the polynomials is a constant multiple of the other. This proves the existence of the constant c_n.

To find c_n, note that it is equal to the ratio of the dominant coefficient of the polynomials $P_1(\xi)$ and $P_2(\xi)$. In the case of the first polynomial, this coefficient is

$$\tan \frac{\pi}{n} \tan \frac{2\pi}{n} \cdots \tan \frac{(n-1)\pi}{n} = (-1)^{\frac{n-1}{2}} n \quad \text{(See Problem 207)}.$$

For the second polynomial this number is equal to 1. Hence $c_n = (-1)^{\frac{n-1}{2}} n$. □

And now the problems.

153. Find all solutions to the equation

$$(x + 1)(x + 2)(x + 3)^2(x + 4)(x + 5) = 360.$$

154. Solve the polynomial equation

$$x^3 - (7 + 2\sqrt{5})x + \sqrt{5} + 1 = 0.$$

155. Let a, b, c be real numbers. Prove that three roots of the equation

$$\frac{b+c}{x-a} + \frac{c+a}{x-b} + \frac{a+b}{x-c} = 3$$

are real.

156. Find all polynomials satisfying the functional equation

$$(x+1)P(x) = (x-10)P(x+1).$$

157. Let $n > 1$ be an integer and $x, a_1, a_2, \ldots, a_n$ be distinct real numbers. Show that

$$\frac{(x-a_2)(x-a_3)\cdots(x-a_n)}{(a_1-a_2)(a_1-a_3)\cdots(a_1-a_n)} + \frac{(x-a_1)(x-a_3)\cdots(x-a_n)}{(a_2-a_1)(a_2-a_3)\cdots(a_2-a_n)}$$
$$+ \cdots + \frac{(x-a_1)(x-a_2)\cdots(x-a_{n-1})}{(a_n-a_1)(a_n-a_2)\cdots(a_n-a_{n-1})} = 1.$$

158. Let $P(x)$ be a polynomial of odd degree with real coefficients. Show that the equation $P(P(x)) = 0$ has at least as many real roots as the equation $P(x) = 0$, counted without multiplicities.

159. Let $P(x) = x^2 + 2007x + 1$. Prove that for every positive integer n, $P^{(n)}(x) = 0$ has at least one real root, where $P^{(n)}$ denotes P composed with itself n times.

160. Determine all polynomials $P(x)$ with real coefficients for which there exists a positive integer n such that for all x,

$$P\left(x + \frac{1}{n}\right) + P\left(x - \frac{1}{n}\right) = 2P(x).$$

161. Find a polynomial with integer coefficients that has the zero $\sqrt{2} + \sqrt[3]{3}$.

162. Let $P(x)$ be a polynomial with real coefficients that satisfies the functional equation

$$(x-1)P(x+2) = (x+1)P(x-1) + 2, \text{ for all } x \in \mathbb{R}.$$

Compute $P(-1989)$.

163. Consider the polynomial with real coefficients $P(x) = x^6 + ax^5 + bx^4 + cx^3 + bx^2 + ax + 1$, and let $x_1, x_2, \ldots, x_6$ be its zeros. Prove that

$$\prod_{k=1}^{6}(x_k^2 + 1) = (2a - c)^2.$$

164. Let $P(z) = (z - z_1)(z - z_2)\cdots(z - z_n)$ with $|z_i| \geq 1, i = 1, 2, \ldots, n$. Prove that if $0 < r < 1$, then for any z, with $|z| = 1$,

$$\left|\frac{P(z)}{P(rz)}\right| \leq \left(\frac{2}{1+r}\right)^n.$$

165. Let $P(x) = x^4 + ax^3 + bx^2 + cx + d$ and $Q(x) = x^2 + px + q$ be two polynomials with real coefficients. Suppose that there exists an interval (r, s) of length greater than 2 such that both $P(x)$ and $Q(x)$ are negative for $x \in (r, s)$ and both are positive for $x < r$ or $x > s$. Show that there is a real number x_0 such that $P(x_0) < Q(x_0)$.

166. Let $P(x)$ be a polynomial of degree n. Knowing that

$$P(k) = \frac{k}{k+1}, \quad k = 0, 1, \ldots, n,$$

find $P(m)$ for $m > n$.

167. Consider the polynomials with complex coefficients

$$P(x) = x^n + a_1 x^{n-1} + \cdots + a_n$$

with zeros $x_1, x_2, \ldots, x_n$ and

$$Q(x) = x^n + b_1 x^{n-1} + \cdots + b_n$$

with zeros $x_1^2, x_2^2, \ldots, x_n^2$. Prove that if $a_1 + a_3 + a_5 + \cdots$ and $a_2 + a_4 + a_6 + \cdots$ are both real numbers, then so is $b_1 + b_2 + \cdots + b_n$.

168. Let $P(x)$ be a polynomial with complex coefficients. Prove that $P(x)$ is an even function if and only if there exists a polynomial $Q(x)$ with complex coefficients satisfying

$$P(x) = Q(x)Q(-x).$$

2.2.2 Polynomials in Several Variables

Let us switched to polynomials in several variables. The first example was published by the first author in *Mathematical Reflections*.

Example. Given that the real numbers x, y, z satisfy $x + y + z = 0$ and

$$\frac{x^4}{2x^2 + yz} + \frac{y^4}{2y^2 + xz} + \frac{z^4}{2z^2 + xy} = 1,$$

determine, with proof, all possible values of $x^4 + y^4 + z^4$.

Solution. First note that x, y, z have to be distinct, or else one of the denominators will be zero. We have

$$2x^2 + yz = x^2 + x^2 + yz = x^2 - (y + z)x + yz = (x - y)(x - z).$$

Similarly $2y^2 + xz = (y - z)(y - x)$ and $2z^2 + xy = (z - x)(z - y)$. Hence the second equation from the statement can be written as

$$\frac{x^4}{(x - y)(z - x)} + \frac{y^4}{(x - y)(y - z)} + \frac{z^4}{(z - x)(y - z)} = -1,$$

which gives the following equality

$$x^4(y-z) + y^4(z-x) + z^4(x-y) = -(x-y)(y-z)(z-x).$$

Viewing the left-hand side as a polynomial in x, the zeros of the polynomial are y and z, and its coefficients are divisible by $y - z$. Hence there is a quadratic homogeneous symmetric polynomial $Q(x, y, z)$ such that

$$x^4(y-z) + y^4(z-x) + z^4(x-y) = (x-y)(y-z)(z-x)Q(x, y, z).$$

Write $Q(x, y, z) = \alpha(x^2 + y^2 + z^2) + \beta(xy + xz + yz)$. Equating the coefficients of x^4 on both sides gives $\alpha = -1$. Equating the coefficients of $x^3 y^2$ on both sides gives $0 = -1 - \beta$, hence $\beta = -1$. We conclude that $Q(x, y, z) = -(x^2 + y^2 + z^2 + xy + xz + yz)$. Hence

$$(x-y)(y-z)(z-x)(x^2 + y^2 + x^2 + xy + xz + yz) = -(x-y)(y-z)(z-x).$$

Given that $(x + y + z)^2 = 0$, we have $x^2 + y^2 + z^2 = -2xy - 2xz - 2yz$, and we obtain

$$xy + xz + yz = -1,$$

or

$$x^2 + y^2 + z^2 = 2.$$

Then

$$1 = (xy + xz + yz)^2 = x^2 y^2 + x^2 z^2 + y^2 z^2 + 2xyz(x + y + z) = x^2 y^2 + x^2 z^2 + y^2 z^2,$$

and hence

$$x^4 + y^4 + z^4 = (x^2 + y^2 + z^2)^2 - 2(x^2 y^2 + x^2 z^2 + y^2 z^2) = 4 - 2 = 2.$$

We conclude that the answer to the question is 2. □

We continue with problems left to the reader.

169. Given the polynomial $P(x, y, z)$ prove that the polynomial

$$\begin{aligned} Q(x, y, z) = {} & P(x, y, z) + P(y, z, x) + P(z, x, y) \\ & - P(x, z, y) - P(y, x, z) - P(z, y, x) \end{aligned}$$

is divisible by $(x - y)(y - z)(z - x)$.

170. Let x, y, z be positive integers greater than 1. Prove that the expression

$$(x + y + z)^3 - (-x + y + z)^3 - (x - y + z)^3 - (x + y - z)^3$$

is the product of seven (not necessarily distinct) integers each of which is greater than one.

171. Factor completely the expression

$$(x + y + z)^5 - (-x + y + z)^5 - (x - y + z)^5 - (x + y - z)^5.$$

172. Factor the expression

$$E = a^3(b - c) + b^3(c - a) + c^3(a - b).$$

173. What conditions should the real numbers a, b, c, d satisfy in order for the equation

$$\frac{(x - b)(x - c)(x - d)}{(a - b)(a - c)(a - d)} + \frac{(x - a)(x - c)(x - d)}{(b - a)(b - c)(b - d)} + \frac{(x - a)(x - b)(x - d)}{(c - a)(c - b)(c - d)}$$
$$+ \frac{(x - a)(x - b)(x - c)}{(d - a)(d - b)(d - c)} = abcd$$

to admit real solutions.

174. Is there a polynomial $P(x, y, z)$ with integer coefficients such that $P(x, y, z)$ and $x + \sqrt[3]{2}y + \sqrt[3]{3}z$ have the same sign for all integers x, y, z?

175. Let $f(x, y, z) = x^2 + y^2 + z^2 + xyz$. Let $p(x, y, z), q(x, y, z), r(x, y, z)$ be polynomials with real coefficients satisfying

$$f(p(x, y, z), q(x, y, z), r(x, y, z)) = f(x, y, z).$$

Prove or disprove the assertion that the sequence p, q, r consists of some permutation of $\pm x, \pm y, \pm z$ where the number of minus signs is 0 or 2.

176. Find all positive integers p, q, with $p > 2q$, and real numbers a such that the two-variable polynomial

$$x^p + ax^{p-q}y^q + ax^{p-2q}y^{2q} + y^p$$

is divisible by $(x + y)^2$.

177. Find all polynomials of two variables satisfying

$$P(a, b)P(c, d) = P(ac + bd, ad + bc)$$

for all real numbers a, b, c, d.

2.2.3 Quadratic Polynomials

We continue our discussion of polynomials with the case of polynomials of second degree. We start with the following problem due to I. Cucurezeanu, whose solution is based just on the formula for the roots of a quadratic equation.

Example. Let a, b, c be integer numbers that are the sides of a triangle. Show that if the equation

$$x^2 + (a + 1)x + b - c = 0$$

has integer roots, then the triangle is isosceles.

Solution. The quadratic equation has solutions

$$\frac{-(a+1) \pm \sqrt{(a+1)^2 - 4(b-c)}}{2}.$$

For it to admit integer roots, it is necessary that the discriminant is a rational number. But then the discriminant has to be an integer number. If $b > c$ then $(a+1)^2 - 4(b-c)$ is a perfect square $< (a+1)^2$ and of the same parity with this number. Hence

$$(a+1)^2 - 4(b-c) \le (a-1)^2$$

We conclude that $a + c \le b$, which contradicts the triangle inequality. The case $b < c$ is similar. So the only possibility is $b = c$; the triangle is isosceles. $\square$

Here is a problem that uses the sign of a quadratic function. Recall that a quadratic function changes sign only if it has two distinct real zeros, and in that case it has the sign of the dominant coefficient outside of the interval formed by the zeros and opposite sign between the zero. If it has a double zero, or complex zeros, than it always has the sign of the dominant coefficient.

Example. Let a, b, c be distinct real numbers. Show that there is a real number x such that

$$x^2 + 2(a+b+c)x + 3(ab + bc + ac)$$

is negative.

Solution. We compute the discriminant

$$\Delta = 4(a^2 + b^2 + c^2 - ab - bc - ac) = 2[(a-b)^2 + (b-c)^2 + (c-a)^2] > 0.$$

Hence the quadratic function has two distinct real zeros. Between the zeros this function is negative. $\square$

From the equality

$$(x - x_1)(x - x_2) = x^2 + ax + b,$$

we see that the for the quadratic equation $x^2 + ax + b$ the sum of the roots is $-a$ and the product of the roots is b. This is a particular case of Viète's relations, which will be studied in general in the next section. Here is a problem.

Example. Find all positive integers a, b, c such that the equations

$$x^2 - ax + b = 0, \quad x^2 - bx + c = 0, \quad x^2 - cx + a = 0$$

have integer roots.

Solution. The roots must also be positive. Write $x_1 + x_2 = a = x_5 x_6$, $x_3 + x_4 = b = x_1 x_2$, $x_5 + x_6 = c = x_3 x_4$. Adding we obtain

$$x_1 + x_2 + x_3 + x_4 + x_5 + x_6 = x_1 x_2 + x_3 x_4 + x_5 x_6.$$

This is equivalent to

$$(x_1 - 1)(x_2 - 1) + (x_3 - 1)(x_4 - 1) + (x_5 - 1)(x_6 - 1) = 3.$$

On the left there are only non-negative integers, so they can only be $(0, 0, 3)$, $(0, 1, 2)$, or $(1, 1, 1)$. In the first case, if say the third term is 3 then $\{x_5, x_6\} = \{4, 2\}$, so $a = 8$, $c = 6$. Also one of x_1, x_2 is 1, so the other is $a - 1 = 7$, and thus $b = 7$. We obtain $(a, b, c) = (8, 7, 6)$ and its circular permutations.

If, say, the second term is 1 and the third term is 2, then on the one hand $x_3 = x_4 = 2$, so $b = c = 4$, and on the other hand $\{x_5, x_6\} = \{2, 3\}$ and so $c = 5$, impossible. A similar argument rules out the case where the second term is 1 and the first term is 2.

Finally, if each term is 1, then $x_i = 2$, $i = 1, 2, 3$, and so we obtain the triple $(a, b, c) = (4, 4, 4)$.   □

178. Let $a > 2$ be a real number. Solve the equation

$$x^3 - 2ax^2 + (a^2 + 1)x + 2 - 2a = 0.$$

179. Does there exist a positive integer n such that the quadratic equation

$$(n^3 - n + 1)x^2 - (n^5 - n + 1)x - (n^7 - n + 1) = 0$$

has rational solutions?

180. Assume that the quadratic function $f(x) = x^2 + ax + b$ has integer zeros, and has the property that there is an integer number n such that $f(n) = 13$. Prove that either $f(n + 1)$ or $f(n - 1)$ is equal to 28.

181. Let a, b, c be integer numbers that are the sides of a triangle.
(a) Show that if the equation

$$x^2 + (2ab + 1)x + a^2 + b^2 = c^2$$

has integer roots, the the triangle is right.
(b) Show that if the equation

$$x^2 + (a^2 + b^2 + c^2 + 1)x + ab + bc + ac = 0$$

has integer roots, then the triangle is equilateral.

182. Let $a < b < c < d$ be nonzero real numbers. Show that the equations

$$ax^2 + (b + d)x + c = 0$$
$$bx^2 + (c + d)x + a = 0$$
$$cx^2 + (a + d)x + b = 0$$

have a common root if and only if $a + b + c + d = 0$.

183. Find all real numbers a such that for all $x, y \in \mathbb{R}$ one has

$$2a(x^2 + y^2) + 4axy - y^2 - 2xy - 2x + 1 \geq 0.$$

184. Show that if the equation $x^2 + ax + b = 0$ has real roots, then so does the equation
$x^2 - (a^2 - 2b + 2)x + a^2 + b^2 + 1 = 0$.

185. Prove that

$$\log_2 3 + \log_3 4 + \log_4 5 + \log_5 6 > 5.$$

186. Let a, b be integer numbers. Decide when the equation

$$(ax - b)^2 + (bx - a)^2 = x$$

has an integer solution.

187. Prove that if the real numbers p_1, p_2, q_1, q_2 satisfy

$$(q_1 - q_2)^2 + (p_1 - p_2)(p_1 q_2 - p_2 q_1) < 0,$$

then the quadratic equations

$$x^2 + p_1 x + q_1 = 0 \text{ and } x^2 + p_2 x + q_2 = 0$$

have real roots and between the roots of one there is a root of the other.

188. Prove that if the inequality $a^2 + ab + ac < 0$ holds, then so does $b^2 - 4ac > 0$.

189. Let a and b be positive integers such that $a^2 + b^2$ is a prime number. Prove that the
equation $x^2 + ax + b + 1 = 0$ does not have integer roots.

190. Find all positive integers a, b, c such that the equations

$$x^2 - ax + b = 0, \quad x^2 - bx + c = 0, \quad x^2 - cx + a = 0$$

have integer roots.

191. Let ABC be a triangle. Show that there exists a point D inside the segment BC such
that $AD^2 = BD \cdot DC$ if and only if $b + c \leq \sqrt{2}a$.

192. Let $a_1 \leq a_2 \leq \cdots \leq a_n$ and $b_1 \geq b_2 \geq \cdots \geq b_n$ be real numbers such that

$$\sum_{i=1}^{n}(n - i)a_i b_i \text{ and } \sum_{j=1}^{n}(j - 1)a_j b_j$$

are both positive. Prove the inequality

$$\left[\left(\sum_{i=1}^{n} a_i \right) \left(\sum_{i=1}^{n} b_i \right) - \left(\sum_{i=1}^{n} a_i b_i \right) \right]^2 \geq 4 \left(\sum_{i=1}^{n}(n - i)a_i b_i \right) \left(\sum_{j=1}^{n}(j - 1)a_j b_j \right).$$

193. Let $a, a_1, a_2, \ldots, a_{2n}, b, b_1, \ldots, b_{2n}$ be real numbers such that

$$a^2 > 2 \max \left(a_1^2 + a_3^2 + \cdots + a_{2n-1}^2, a_2^2 + a_4^2 + \cdots + a_{2n}^2 \right).$$

Show that $(ab - a_1 b_1 - a_2 b_2 - \cdots a_{2n} b_{2n})^2$ is greater than or equal to the smaller of the quantities $(a^2 - 2a_1^2 - 2a_3^2 - \cdots - 2a_{2n-1}^2)(b^2 - 2b_1^2 - 2b_3^2 - \cdots - 2b_{2n-1}^2)$ and $(a^2 - 2a_2^2 - 2a_4^2 - \cdots - 2a_{2n}^2)(b^2 - 2b_2^2 - 2b_4^2 - \cdots - 2b_{2n}^2)$.

194. A sphere is inscribed in a regular cone. Around the sphere a cylinder is circumscribed so that its base is in the same plane as the base of the cone. Let V_1 be the volume of the cone, and V_2 the volume of the cylinder.

(a) Prove that V_1 cannot equal V_2.

(b) Find the smallest positive number k such that $V_1 = k V_2$.

2.2.4 Viète's Relations

From the Gauss-d'Alembert fundamental theorem of algebra it follows that a polynomial

$$P(x) = a_n x^n + a_{n-1} x^{n-1} + \cdots + a_0$$

can be factored over the complex numbers as

$$P(x) = a_n (x - x_1)(x - x_2) \cdots (x - x_n).$$

Equating the coefficients of x in the two expressions, we obtain

$$x_1 + x_2 + \cdots + x_n = -\frac{a_{n-1}}{a_n},$$

$$x_1 x_2 + x_1 x_3 + \cdots + x_{n-1} x_n = \frac{a_{n-2}}{a_n},$$

$$\cdots$$

$$x_1 x_2 \cdots x_n = (-1)^n \frac{a_0}{a_n}.$$

These relations carry the name of the French mathematician F. Viète. They combine two ways of looking at a polynomial: as a sum of monomials and as a product of linear factors. As a first application of these relations, we have selected a problem from a 1957 Chinese mathematical competition.

Example. If $x + y + z = 0$, prove that

$$\frac{x^2 + y^2 + z^2}{2} \cdot \frac{x^5 + y^5 + z^5}{5} = \frac{x^7 + y^7 + z^7}{7}.$$

Solution. Consider the polynomial $P(X) = X^3 + pX + q$, whose zeros are x, y, z. Then

$$x^2 + y^2 + z^2 = (x + y + z)^2 - 2(xy + xz + yz) = -2p.$$

Adding the relations $x^3 = -px - q$, $y^3 = -py - q$, and $z^3 = -pz - q$, which hold because x, y, z are zeros of $P(X)$, we obtain

$$x^3 + y^3 + z^3 = -3q.$$

Similarly,

$$x^4 + y^4 + z^4 = -p(x^2 + y^2 + z^2) - q(x + y + z) = 2p^2,$$

and therefore

$$x^5 + y^5 + z^5 = -p(x^3 + y^3 + z^3) - q(x^2 + y^2 + z^2) = 5pq,$$
$$x^7 + y^7 + z^7 = -p(x^5 + y^5 + z^5) - q(x^4 + y^4 + z^4) = -5p^2q - 2p^2q = -7p^2q.$$

The relation from the statement reduces to the obvious

$$\frac{-2p}{2} \cdot \frac{5pq}{5} = \frac{-7p^2q}{7}. \qquad \square$$

Viète's relations can be used to solve, or analyze, the roots of a polynomial equation when additional information about the roots is given, as the following problem of B. Enescu shows.

Example. Let $P(x) = x^3 + ax^2 + bx + c$ be a polynomial with rational coefficients, having the roots x_1, x_2, x_3. Show that if $\frac{x_1}{x_2}$ is a rational number different from 0 and -1, then x_1, x_2, x_3 are all rational.

Solution. Set $\frac{x_1}{x_2} = t$. Let us observe that if either x_1 or x_2 is rational, so is the other, and by Viète's relations x_3 is rational as well. Also, if x_3 is rational, then $x_1 + x_2 = x_2(1 + \frac{x_1}{2})$ is rational, so x_2 is rational, and x_1 is rational as well. Hence it suffices to show that $P(x)$ has a rational root.

Substituting $x_1 = tx_2$ in Viète's relations we obtain

$$(t + 1)x_2 + x_3 = -a$$
$$x_2[tx_2 + (t + 1)x_3] = b.$$

Substituting x_3 from the first equation we obtain the quadratic equation in x_2,

$$(t^2 + t + 1)x_2^2 + (t + 1)ax_2 + b = 0.$$

Thus x_2 is a zero of the quadratic polynomial with rational coefficients $Q(x) = (t^2 + t + 1)x^2 + (t + 1)ax + b$. We deduce that the greatest common divisor of $P(x)$ and $Q(x)$ is a non-constant polynomial. Moreover, because both $P(x)$ and $Q(x)$ have rational coefficients their greatest common divisor must have rational coefficients as well. So $P(x)$ can be written as a product of two polynomials with rational coefficients. One of the factors must be a linear polynomial, showing that $P(x)$ has a rational zero. Hence the conclusion. $\qquad \square$

Next, a problem from the short list of the 2005 Ibero-American Mathematical Olympiad.

Example. Find the largest real number k with the property that for all fourth-degree polynomials $P(x) = x^4 + ax^3 + bx^2 + cx + d$ whose zeros are all real and positive, one has

$$(b - a - c)^2 \geq kd,$$

and determine when equality holds.

Solution. Let r_1, r_2, r_3, r_4 be the zeros of $P(x)$. Viète's relations read

$$
\begin{aligned}
a &= -(r_1 + r_2 + r_3 + r_4), \\
b &= r_1r_2 + r_1r_3 + r_1r_4 + r_2r_3 + r_2r_4 + r_3r_4, \\
c &= -(r_1r_2r_3 + r_1r_2r_4 + r_1r_3r_4 + r_2r_3r_4), \\
d &= r_1r_2r_3r_4.
\end{aligned}
$$

From here we obtain

$$
\begin{aligned}
b - a - c &= (r_1r_2 + r_1r_3 + r_1r_4 + r_2r_3 + r_2r_4 + r_3r_4) + (r_1 + r_2 + r_3 + r_4) \\
&\quad + (r_1r_2r_3 + r_1r_2r_4 + r_1r_3r_4 + r_2r_3r_4).
\end{aligned}
$$

By the AM-GM inequality this is greater than or equal to

$$14 \sqrt[14]{(r_1r_2r_3r_4)^7} = 14\sqrt{d}.$$

Since equality can hold in the AM-GM inequality, we conclude that $k = 196$ is the answer to the problem. Moreover, equality holds exactly when $r_1 = r_2 = r_3 = r_4 = 1$, that is, when $P(x) = x^4 - 4x^3 + 6x^2 - 4x + 1$. $\square$

And now a challenging problem from A. Krechmar's *Problem Book in Algebra* (Mir Publishers, 1974).

Example. Prove that

$$\sqrt[3]{\cos \frac{2\pi}{7}} + \sqrt[3]{\cos \frac{4\pi}{7}} + \sqrt[3]{\cos \frac{8\pi}{7}} = \sqrt[3]{\frac{1}{2}(5 - 3\sqrt[3]{7})}.$$

Solution. We would like to find a polynomial whose zeros are the three terms on the left. Let us simplify the problem and forget the cube roots for a moment. In this case we have to find a polynomial whose zeros are $\cos \frac{2\pi}{7}$, $\cos \frac{4\pi}{7}$, $\cos \frac{8\pi}{7}$. The seventh roots of unity come in handy. Except for $x = 1$, which we ignore, these are also roots of the equation $x^6 + x^5 + x^4 + x^3 + x^2 + x + 1 = 0$, and are $\cos \frac{2k\pi}{7} + i \sin \frac{2k\pi}{7}$, $k = 1, 2, \ldots, 6$. We see that the numbers $2\cos \frac{2\pi}{7}$, $2\cos \frac{4\pi}{7}$, and $2\cos \frac{8\pi}{7}$ are of the form $x + \frac{1}{x}$, with x one of these roots.

If we define $y = x + \frac{1}{x}$, then $x^2 + \frac{1}{x^2} = y^2 - 2$ and $x^3 + \frac{1}{x^3} = y^3 - 3y$. Dividing the equation $x^6 + x^5 + x^4 + x^3 + x^2 + x + 1 = 0$ through by x^3 and substituting y in it, we obtain the cubic equation

$$y^3 + y^2 - 2y - 1 = 0.$$

The numbers $2\cos\frac{2\pi}{7}$, $2\cos\frac{4\pi}{7}$, and $2\cos\frac{8\pi}{7}$ are the three roots of this equation. The simpler task is fulfilled.

But the problem asks us to find the sum of the cube roots of these numbers. Looking at symmetric polynomials, we have

$$X^3 + Y^3 + Z^3 - 3XYZ = (X + Y + Z)^3 - 3(X + Y + Z)(XY + YZ + ZX)$$

and

$$X^3Y^3 + Y^3Z^3 + Z^3X^3 - 3(XYZ)^2 = (XY + YZ + XZ)^3$$
$$- 3XYZ(X + Y + Z)(XY + YZ + ZX).$$

Because X^3, Y^3, Z^3 are the roots of the equation $y^3 + y^2 - 2y - 1 = 0$, by Viète's relations, $X^3Y^3Z^3 = 1$, so $XYZ = \sqrt[3]{1} = 1$, and also $X^3 + Y^3 + Z^3 = -1$ and $X^3Y^3 + X^3Z^3 + Y^3Z^3 = -2$. In the above two equalities we now know the left-hand sides. The equalities become a system of two equations in the unknowns $u = X + Y + Z$ and $v = XY + YZ + ZX$, namely

$$u^3 - 3uv = -4,$$
$$v^3 - 3uv = -5.$$

Writing the two equations as $u^3 = 3uv - 4$ and $v^3 = 3uv - 5$ and multiplying them, we obtain $(uv)^3 = 9(uv)^2 - 27uv + 20$. With the substitution $m = uv$ this becomes $m^3 = 9m^2 + 27m - 20$ or $(m - 3)^3 + 7 = 0$. This equation has the unique solution $m = 3 - \sqrt[3]{7}$. Hence $u = \sqrt[3]{3m - 4} = \sqrt[3]{5 - 3\sqrt[3]{7}}$. We conclude that

$$\sqrt[3]{\cos\frac{2\pi}{7}} + \sqrt[3]{\cos\frac{4\pi}{7}} + \sqrt[3]{\cos\frac{8\pi}{7}} = X + Y + Z = \frac{1}{\sqrt[3]{2}}u = \sqrt[3]{\frac{1}{2}(5 - 3\sqrt[3]{7})},$$

as desired.   □

All problems below can be solved using Viète's relations.

195. Find the zeros of the polynomial

$$P(x) = x^4 - 6x^3 + 18x^2 - 30x + 25$$

knowing that the sum of two of them is 4.

196. Let a, b, c be real numbers. Show that $a \geq 0$, $b \geq 0$, and $c \geq 0$ if and only if $a + b + c \geq 0$, $ab + bc + ca \geq 0$, and $abc \geq 0$.

197. Solve the system

$$x + y + z = 1,$$
$$xyz = 1,$$

knowing that x, y, z are complex numbers of absolute value equal to 1.

198. Let x_1, x_2, x_3 be the roots of the equation

$$x^3 - x^2 - 2x + 4 = 0,$$

with $|x_1| \geq |x_2| \geq |x_3|$. Find a polynomial with integer coefficients of minimal degree that has the root $x_1^5 + x_2^3 + x_3^2$.

199. Find all real numbers r for which there is at least one triple (x, y, z) of nonzero real numbers such that

$$x^2y + y^2z + z^2x = xy^2 + yz^2 + zx^2 = rxyz.$$

200. Let a, b, c, d be real numbers with $a + b + c + d = 0$. Prove that

$$a^3 + b^3 + c^3 + d^3 = 3(abc + bcd + cda + dab)$$

201. Given the real numbers x, y, z, t such that

$$x + y + z + t = x^7 + y^7 + z^7 + t^7 = 0,$$

prove that

$$x(x + y)(x + z)(x + t) = 0.$$

202. For five integers a, b, c, d, e we know that the sums $a + b + c + d + e$ and $a^2 + b^2 + c^2 + d^2 + e^2$ are divisible by an odd number n. Prove that the number $a^5 + b^5 + c^5 + d^5 + e^5 - 5abcde$ is also divisible by n.

203. Find all polynomials whose coefficients are equal either to 1 or -1 and whose zeros are all real.

204. Let $P(z) = az^4 + bz^3 + cz^2 + dz + e = a(z - r_1)(z - r_2)(z - r_3)(z - r_4)$, where a, b, c, d, e are integers, $a \neq 0$. Show that if $r_1 + r_2$ is a rational number, and if $r_1 + r_2 \neq r_3 + r_4$, then $r_1 r_2$ is a rational number.

205. Let $P(x) = x^3 + ax^2 + bx + c$ be a polynomial with rational coefficients, having the roots x_1, x_2, x_3. Show that if $\frac{x_1}{x_2}$ is a rational number different from 0 and -1, then x_1, x_2, x_3 are all rational.

206. The zeros of the polynomial $P(x) = x^3 - 10x + 11$ are u, v, and w. Determine the value of $\arctan u + \arctan v + \arctan w$.

207. Prove that for every positive integer n,

$$\tan \frac{\pi}{2n+1} \tan \frac{2\pi}{2n+1} \cdots \tan \frac{n\pi}{2n+1} = \sqrt{2n+1}.$$

208. Let $P(x) = x^n + a_{n-1}x^{n-1} + \cdots + a_0$ be a polynomial of degree $n \geq 3$. Knowing that $a_{n-1} = -\binom{n}{1}$, $a_{n-2} = \binom{n}{2}$, and that all roots are real, find the remaining coefficients.

209. Determine the maximum value of λ such that whenever $P(x) = x^3 + ax^2 + bx + c$ is a cubic polynomial with all zeros real and nonnegative, then

$$P(x) \geq \lambda(x - a)^3$$

for all $x \geq 0$. Find the equality condition.

210. Prove that there are unique positive integers a, n such that

$$a^{n+1} - (a + 1)^n = 2001.$$

2.2.5 The Derivative of a Polynomial

This section adds some elements of real analysis. We remind the reader that the derivative of a polynomial

$$P(x) = a_n x^n + a_{n-1} x^{n-1} + \cdots + a_1 x + a_0$$

is the polynomial

$$P'(x) = n a_n x^{n-1} + (n - 1) a_{n-1} x^{n-2} + \cdots + a_1.$$

We also recall the product rule: $(P(x)Q(x))' = P'(x)Q(x) + P(x)Q'(x)$. If $x_1, x_2, \ldots, x_n$ are the zeros of $P(x)$, then by using the product rule we obtain

$$\frac{P'(x)}{P(x)} = \frac{1}{x - x_1} + \frac{1}{x - x_2} + \cdots + \frac{1}{x - x_n}.$$

If a zero of $P(x)$ has multiplicity greater than 1, then it is also a zero of $P'(x)$, and the converse is also true. By Rolle's theorem, if all zeros of $P(x)$ are real, then so are those of $P'(x)$. Let us discuss in detail two problems, the first of which belonging to the second author of the book, and the second to R. Gologan.

Example. Let $P(x)$ be a polynomial with real zeros and let $a < b$ be two real numbers that are smaller than any of the zeros of $P(x)$. Prove that

$$\exp\left(\int_a^b \frac{P'''(x)P(x)}{P'(x)^2} dx\right) < \left|\frac{P(a)^2 P'(b)^3}{P'(a)^3 P(b)^2}\right|.$$

Solution. Differentiate the identity

$$\frac{P'(x)}{P(x)} = \sum_{k=1}^n \frac{1}{x - x_k}$$

to obtain

$$\frac{P''(x)P(x) - P'(x)^2}{P(x)^2} = -\sum_{k=1}^n \frac{1}{(x - x_k)^2}.$$

Differentiate one more time and obtain

$$\frac{(P'''(x)P(x) - P''(x)P'(x))P(x)^2 - (P''(x)P(x) - P'(x)^2)2P(x)P'(x)}{(P(x))^4}$$

$$= \sum_{k=1}^{n} \frac{2}{(x - x_k)^3}.$$

Notice that the right-hand side is negative for $a \le x \le b < \min(x_1, \ldots, x_n)$. Hence

$$P'''(x)P(x)^3 - P''(x)P'(x)P(x)^2 - 2P''(x)P'(x)P(x)^2 + 2P'(x)^3P(x) < 0,$$

that is

$$P'''(x)P(x)^3 - 3P''(x)P'(x)P(x)^2 + 2P'(x)^3P(x) < 0.$$

Dividing by $P(x)^2 P'(x)^2$, we obtain

$$\frac{P'''(x)P(x)}{P'(x)^2} < \frac{3P''(x)}{P'(x)} - \frac{2P'(x)}{P(x)}.$$

Integrating we obtain

$$\int_a^b \frac{P'''(x)P(x)}{P'(x)^2} dx < 3\ln|P'(b)| - \ln|P'(a)| - 2\ln|P(b)| - \ln|P(a)|$$

$$= \ln\left|\frac{P(a)^2 P'(b)^3}{P'(a)^3 P(b)^2}\right|.$$

After exponentiation we obtain the inequality from the statement. □

Example. Let $P(x) \in \mathbb{Z}[x]$ be a polynomial with n distinct integer zeros. Prove that the polynomial $(P(x))^2 + 1$ has a factor of degree at least $2\left\lfloor\frac{n+1}{2}\right\rfloor$ that is irreducible over $\mathbb{Z}[x]$.

Solution. The statement apparently offers no clue about derivatives. The standard approach is to assume that

$$(P(x))^2 + 1 = P_1(x)P_2(x)\cdots P_k(x)$$

is a decomposition into factors that are irreducible over $\mathbb{Z}[x]$. Letting $x_1, x_2, \ldots, x_n$ be the integer zeros of $P(x)$, we find that

$$P_1(x_j)P_2(x_j)\cdots P_k(x_j) = 1, \text{ for } j = 1, 2, \ldots, n.$$

Hence $P_i(x_j) = \pm 1$, which then implies $\frac{1}{P_i(x_j)} = P_i(x_j)$, $i = 1, 2, \ldots, k$, $j = 1, 2, \ldots, n$.

Now let us see how derivatives come into play. The key observation is that the zeros x_j of $(P(x))^2$ appear with multiplicity greater than 1, and so they are zeros of the derivative. Differentiating with the product rule, we obtain

$$\sum_{i=1}^{k} P_1(x_j)\cdots P_i'(x_j)\cdots P_k(x_j) = 0, \text{ for } j = 1, 2, \ldots, n.$$

This sum can be simplified by taking into account that $P_1(x_j)P_2(x_j)\cdots P_k(x_j) = 1$ and $\frac{1}{P_i(x_j)} = P_i(x_j)$ as

$$\sum_{i=1}^{k} P_i'(x_j)P_i(x_j) = 0, \text{ for } j = 1, 2, \ldots, n.$$

It follows that x_j is a zero of the polynomial

$$\sum_{i=1}^{k} 2P_i'(x)P_i(x) = \left(\sum_{i=1}^{k} P_i^2(x) \right)'.$$

Let us remember that $P_i(x_j) = \pm 1$, which then implies $\sum_{i=1}^{k} P_i^2(x_j) - n = 0$ for $j = 1, 2, \ldots, n$. The numbers x_j, $j = 1, 2, \ldots, n$, are zeros of both $\sum_{i=1}^{k} P_i^2(x) - n$ and its derivative, so they are zeros of order at least 2 of this polynomial. Therefore,

$$\sum_{i=1}^{k} P_i^2(x) = (x - x_1)^2(x - x_2)^2 \cdots (x - x_n)^2 Q(x) + n,$$

for some polynomial $Q(x)$ with integer coefficients. We deduce that there exists an index i_0 such that the degree of $P_{i_0}(x)$ is greater than or equal to n. For n even, $n = 2\lfloor \frac{n+1}{2} \rfloor$, and we are done. For n odd, since $(P(x))^2 + 1$ does not have real zeros, neither does $P_{i_0}(x)$, so this polynomial has even degree. Thus the degree of $P_{i_0}(x)$ is at least $n + 1 = 2\lfloor \frac{n+1}{2} \rfloor$. This completes the solution. $\qquad\square$

211. Find all polynomials $P(x)$ with integer coefficients satisfying $P(P'(x)) = P'(P(x))$ for all $x \in \mathbb{R}$.

212. Determine all polynomials $P(x)$ with real coefficients satisfying $(P(x))^n = P(x^n)$ for all $x \in \mathbb{R}$, where $n > 1$ is a fixed integer.

213. Let $P(x)$ and $Q(x)$ be polynomials with complex coefficients and let a be a nonzero complex number. Prove that if

$$P(x)^3 = Q(x)^2 + a,$$

for all $x \in \mathbb{C}$, then $P(x)$ and $Q(x)$ are constant polynomials.

214. Let $P(z)$ and $Q(z)$ be polynomials with complex coefficients of degree greater than or equal to 1 with the property that $P(z) = 0$ if and only if $Q(z) = 0$ and $P(z) = 1$ if and only if $Q(z) = 1$. Prove that the polynomials are equal.

215. Let $P(x)$ be a polynomial with all zeros real and distinct and such that none of its zeros is equal to 0. Prove that the polynomial $x^2 P''(x) + 3x P'(x) + P(x)$ also has all roots real and distinct.

216. Let $P(x)$ be a polynomial of degree 5, with real coefficients, all of whose zeros are real. Prove that for each real number a that is not a zero of $P(x)$ or $P'(x)$, there is a real number b such that

$$b^2 P(a) + 4b P'(a) + 5 P''(a) = 0.$$

217. Let $P_n(x) = (x^n - 1)(x^{n-1} - 1) \cdots (x - 1)$, $n \geq 1$. Prove that for $n \geq 2$, $P_n'(x)$ is divisible by $P_{\lfloor n/2 \rfloor}$ in the ring of polynomials with integer coefficients.

218. The zeros of the nth-degree polynomial $P(x)$ are all real and distinct. Prove that the zeros of the polynomial $G(x) = n P(x) P''(x) - (n-1)(P'(x))^2$ are all complex.

219. Let $P(x)$ be a polynomial of degree $n > 3$ whose zeros $x_1 < x_2 < x_3 < \cdots < x_{n-1} < x_n$ are real. Prove that

$$P'\left(\frac{x_1 + x_2}{2}\right) \cdot P'\left(\frac{x_{n-1} + x_n}{2}\right) \neq 0.$$

220. A polynomial $P(x)$ with real coefficients is called a mirror polynomial if $|P(a)| = |P(-a)|$ for all real numbers a. Let $F(x)$ be a polynomial with real coefficients, and consider polynomials with real coefficients $P(x)$ and $Q(x)$ such that $P(x) - P'(x) = F(x)$ and $Q(x) + Q'(x) = F(x)$. Prove that $P(x) + Q(x)$ is a mirror polynomial if and only if $F(x)$ is a mirror polynomial.

2.2.6 The Location of the Zeros of a Polynomial

Since not all polynomial equations can be solved by radicals, methods of approximation are necessary. Results that allow you to localize the roots in certain regions of the real axis or complex plane are therefore useful.

The qualitative study of the position of the zeros of a polynomial has far-reaching applications. For example, the solutions of a homogeneous ordinary linear differential equation with constant coefficients are stable (under errors of measuring the coefficients) if and only if the roots of the characteristic equation lie in the open left half-plane (i.e., have negative real part). Stability is, in fact, an essential question in control theory, where one is usually interested in whether the zeros of a particular polynomial lie in the open left half-plane (Hurwitz stability) or in the open unit disk (Schur stability). Here is a famous result.

Lucas' theorem. *The zeros of the derivative $P'(z)$ of a polynomial $P(z)$ lie in the convex hull of the zeros of $P(z)$.*

Proof. Because any convex domain can be obtained as the intersection of half-planes, it suffices to show that if the zeros of $P(z)$ lie in an open half-plane, then the zeros of $P'(z)$ lie in that half-plane as well. Moreover, by rotating and translating the variable z we can further reduce the problem to the case in which the zeros of $P(z)$ lie in the upper half-plane $\operatorname{Im} z > 0$. Here $\operatorname{Im} z$ denotes the imaginary part.

So let $z_1, z_2, \ldots, z_n$ be the (not necessarily distinct) zeros of $P(z)$, which by hypothesis have positive imaginary part. If $\operatorname{Im} w \leq 0$, then $\operatorname{Im} \frac{1}{w-z_k} > 0$, for $k = 1, \ldots, n$, and therefore

$$\operatorname{Im} \frac{P'(w)}{P(w)} = \sum_{k=1}^{n} \operatorname{Im} \frac{1}{w - z_k} > 0.$$

This shows that w is not a zero of $P'(z)$ and so all zeros of $P'(z)$ lie in the upper half-plane. The theorem is proved. $\qquad\square$

221. Let $a_1, a_2, \ldots, a_n$ be positive real numbers. Prove that the polynomial

$$P(x) = x^n - a_1 x^{n-1} - a_2 x^{n-2} - \cdots - a_n$$

has a unique positive zero.

222. Prove that the zeros of the polynomial

$$P(z) = z^7 + 7z^4 + 4z + 1$$

lie inside the disk of radius 2 centered at the origin.

223. Prove that if the complex coefficients p, q of the quadratic equation $x^2 + px + q = 0$ satisfy $|p| + |q| < 1$, then the roots of this equation lie in the interior of the unit disk.

224. Let $P(x)$ be a polynomial with integer coefficients all of whose roots are real and lie in the interval $(0, 3)$. Prove that the roots of this polynomial lie in the set

$$\left\{ 1, 2, \frac{3 - \sqrt{5}}{2}, \frac{3 + \sqrt{5}}{2} \right\}.$$

225. For $a \neq 0$ a real number and $n > 2$ an integer, prove that every nonreal root z of the polynomial equation $x^n + ax + 1 = 0$ satisfies the inequality $|z| \geq \sqrt[n]{\frac{1}{n-1}}$.

226. Let $a \in \mathbb{C}$ and $n \geq 2$. Prove that the polynomial equation $ax^n + x + 1 = 0$ has a root of absolute value less than or equal to 2.

227. Let $P(z)$ be a polynomial of degree n, all of whose zeros have absolute value 1 in the complex plane. Set $g(z) = \frac{P(z)}{z^{n/2}}$. Show that all roots of the equation $g'(z) = 0$ have absolute value 1.

228. The polynomial $x^4 - 2x^2 + ax + b$ has four distinct real zeros. Show that the absolute value of each zero is smaller than $\sqrt{3}$.

229. Let $P_n(z)$, $n \geq 1$, be a sequence of monic kth-degree polynomials whose coefficients converge to the coefficients of a monic kth-degree polynomial $P(z)$. Prove that for any $\varepsilon > 0$ there is n_0 such that if $n \geq n_0$ then $|z_i(n) - z_i| < \varepsilon$, $i = 1, 2, \ldots, k$, where $z_i(n)$ are the zeros of $P_n(z)$ and z_i are the zeros of $P(z)$, taken in the appropriate order.

230. Let $P(x) = a_n x^n + a_{n-1} x^{n-1} + \cdots + a_0$ be a polynomial with complex coefficients, with $a_0 \neq 0$, and with the property that there exists an m such that

$$\left| \frac{a_m}{a_0} \right| > \binom{n}{m}.$$

Prove that $P(x)$ has a zero of absolute value less than 1.

231. For a polynomial $P(x) = (x - x_1)(x - x_2) \cdots (x - x_n)$, with distinct real zeros $x_1 < x_2 < \cdots < x_n$, we set $\delta(P(x)) = \min_i (x_{i+1} - x_i)$. Prove that for any real number k,

$$\delta(P'(x) - kP(x)) > \delta(P(x)),$$

where $P'(x)$ is the derivative of $P(x)$. In particular, $\delta(P'(x)) > \delta(P(x))$.

2.2.7 Irreducible Polynomials

A polynomial is irreducible if it cannot be written as a product of two polynomials in a nontrivial manner. The question of irreducibility depends on the ring of coefficients. When the coefficients are complex numbers, only linear polynomials are irreducible. For real numbers some quadratic polynomials are irreducible as well. Both these cases are rather dull. The interesting situations occur when the coefficients are rational or integer, in which case there is an interplay between polynomials and arithmetic. The cases of rational and integer coefficients are more or less equivalent, with minor differences such as the fact that $2x + 2$ is irreducible over $\mathbb{Q}[x]$ but reducible over $\mathbb{Z}[x]$. For matters of elegance we focus on polynomials with integer coefficients. We will assume implicitly from now on that for any polynomial with integer coefficients, the greatest common divisor of its coefficients is 1.

Definition. A polynomial $P(x) \in \mathbb{Z}[x]$ is called irreducible over $\mathbb{Z}[x]$ if there do not exist polynomials $Q(x), R(x) \in \mathbb{Z}[x]$ different from ± 1 such that $P(x) = Q(x)R(x)$. Otherwise, $P(x)$ is called reducible.

We commence with an easy problem.

Example. Let $P(x)$ be an nth-degree polynomial with integer coefficients with the property that $|P(x)|$ is a prime number for $2n + 1$ distinct integer values of the variable x. Prove that $P(x)$ is irreducible over $\mathbb{Z}[x]$.

Solution. Assume the contrary and let $P(x) = Q(x)R(x)$ with $Q(x), R(x) \in \mathbb{Z}[x]$, $Q(x)$, $R(x) \neq \pm 1$. Let $k = \deg(Q(x))$. Then $Q(x) = 1$ at most k times and $Q(x) = -1$ at most $n-k$ times. Also, $R(x) = 1$ at most $n-k$ times and $R(x) = -1$ at most k times. Consequently, the product $|Q(x)R(x)|$ is composite except for at most $k + (n - k) + (n - k) + k = 2n$ values of x. This contradicts the hypothesis. Hence $P(x)$ is irreducible. $\qquad\square$

The bound is sharp. For example, $P(x) = (x + 1)(x + 5)$ has $|P(-2)| = |P(-4)| = 3$, $P(0) = 5$, and $|P(-6)| = 7$.

Probably the most beautiful criterion of irreducibility of polynomials is that discovered independently by F.G.M. Eisenstein in 1850 and T. Schönemann in 1846. We present it below.

Eisenstein-Schönemann theorem. *Given a polynomial* $P(x) = a_n x^n + a_{n-1} x^{n-1} + \cdots + a_0$ *with integer coefficients, suppose that there exists a prime number p such that a_n is not divisible by p, a_k is divisible by p for $k = 0, 1, \ldots, n - 1$, and a_0 is not divisible by p^2. Then $P(x)$ is irreducible over $\mathbb{Z}[x]$.*

Proof. We argue by contradiction. Suppose that $P(x) = Q(x)R(x)$, with $Q(x)$ and $R(x)$ not identically equal to ± 1. Let

$$Q(x) = b_k x^k + b_{k-1} x^{k-1} + \cdots + b_0,$$
$$R(x) = c_{n-k} x^{n-k} + c_{n-k-1} x^{n-k-1} + \cdots + c_0.$$

Let us look closely at the equalities

$$\sum_{j=0}^{i} b_j c_{i-j} = a_i, \quad i = 0, 1, \ldots, n,$$

obtained by identifying the coefficients in the equality $P(x) = Q(x)R(x)$. From the first of them, $b_0 c_0 = a_0$, and because a_0 is divisible by p but not by p^2 it follows that exactly one of b_0 and c_0 is divisible by p. Assume that b_0 is divisible by p and take the next equality $b_0 c_1 + b_1 c_0 = a_1$. The right-hand side is divisible by p, and the first term on the left is also divisible by p. Hence $b_1 c_0$ is divisible by p, and since c_0 is not, b_1 must be divisible by p.

This reasoning can be repeated to prove that all the b_i's are divisible by p. It is important that both $Q(x)$ and $R(x)$ have degrees greater than or equal to 1, for the fact that b_k is divisible by p follows from

$$b_k c_0 + b_{k-1} c_1 + \cdots = a_k,$$

where a_k is divisible by p for $k < n$. The contradiction arises in the equality $a_n = b_k c_{n-k}$, since the right-hand side is divisible by p, while the left-hand side is not. This proves the theorem.

The first three problems listed below use this result, while the others apply similar ideas.

232. Prove that the polynomial

$$P(x) = x^{101} + 101 x^{100} + 102$$

is irreducible over $\mathbb{Z}[x]$.

233. Prove that for every prime number p, the polynomial

$$P(x) = x^{p-1} + x^{p-2} + \cdots + x + 1$$

is irreducible over $\mathbb{Z}[x]$.

234. Prove that for every positive integer n, the polynomial $P(x) = x^{2^n} + 1$ is irreducible over $\mathbb{Z}[x]$.

235. Prove that for any distinct integers $a_1, a_2, \ldots, a_n$ the polynomial

$$P(x) = (x - a_1)(x - a_2) \cdots (x - a_n) - 1$$

cannot be written as a product of two nonconstant polynomials with integer coefficients.

236. Prove that for any distinct integers $a_1, a_2, \ldots, a_n$ the polynomial

$$P(x) = (x - a_1)^2(x - a_2)^2 \cdots (x - a_n)^2 + 1$$

cannot be written as a product of two nonconstant polynomials with integer coefficients.

237. Associate to a prime the polynomial whose coefficients are the decimal digits of the prime (for example, for the prime 7043 the polynomial is $P(z) = 7x^3 + 4x + 3$). Prove that this polynomial is always irreducible over $\mathbb{Z}[x]$.

238. Let p be a prime number of the form $4k + 3$, k an integer. Prove that for any positive integer n, the polynomial $(x^2 + 1)^n + p$ is irreducible in the ring $\mathbb{Z}[x]$.

239. Let p be a prime number. Prove that the polynomial

$$P(x) = x^{p-1} + 2x^{p-2} + 3x^{p-3} + \cdots + (p - 1)x + p$$

is irreducible in $\mathbb{Z}[x]$.

240. Let $P(x)$ be a monic polynomial in $\mathbb{Z}[x]$, irreducible over this ring, and such that $|P(0)|$ is not the square of an integer. Prove that the polynomial $Q(x)$ defined by $Q(x) = P(x^2)$ is also irreducible over $\mathbb{Z}[x]$.

2.2.8 Chebyshev Polynomials

The nth Chebyshev polynomial $T_n(x)$ expresses $\cos n\theta$ as a polynomial in $\cos \theta$. This means that $T_n(x) = \cos(n \arccos x)$, for $n \geq 0$. These polynomials satisfy the recurrence

$$T_0(x) = 1, \ T_1(x) = x, \ T_{n+1}(x) = 2xT_n(x) - T_{n-1}(x), \ \text{for } n \geq 1.$$

For example, $T_2(x) = 2x^2 - 1$, $T_3(x) = 4x_36 - 3x$, $T_4(x) = 8x^4 - 8x^2 + 1$.

One usually calls these the Chebyshev polynomials of the first kind, to distinguish them from the Chebyshev polynomials of the second kind $U_n(x)$ defined by

$$U_0(x) = 1, \ U_1(x) = 2x, \ U_{n+1}(x) = 2xU_n(x) - U_{n-1}(x), \ \text{for } n \geq 1$$

(same recurrence relation but different initial condition). Alternatively, $U_n(x)$ can be defined by the equality

$$U_n(\cos \theta) = \frac{\sin(n + 1)\theta}{\sin \theta}.$$

Chebyshev's theorem. *Forfixed $n \geq 1$, the polynomial $2^{-n+1}T_n(x)$ s the unique monic nth-degree polynomial satisfying*

$$\max_{-1 \leq x \leq 1} |2^{-n+1}T(x)| \leq \max_{-1 \leq x \leq 1} |P(x)|,$$

for any other monic nth-degree polynomial $P(x)$.

One says that among all monic nth-degree polynomials, $2^{-n+1}T_n(x)$ has the smallest variation away from zero on $[-1, 1]$. This variation is $\frac{1}{2^{n-1}}$. Let us see how Chebyshev's theorem applies to a problem from *Challenging Mathematical Problems with Elementary Solutions* by A.M. Yaglom and I.M. Yaglom.

Example. Let $A_1, A_2, \ldots, A_n$ be points in the plane. Prove that on any segment of length l there is a point M such that

$$MA_1 \cdot MA_2 \cdots MA_n \geq 2 \left(\frac{l}{4}\right)^n.$$

Solution. Rescaling, we can assume that $l = 2$. Associate complex coordinates to points in such a way that the segment coincides with the interval $[-1, 1]$. Then

$$MA_1 \cdot MA_2 \cdots MA_n = |z - z_1| \cdot |z - z_2| \cdots |z - z_n| = |P(z)|,$$

where $P(z)$ is a monic polynomial with complex coefficients, and $z \in [-1, 1]$. Write $P(z) = R(z) + iQ(z)$, where $R(z)$ is the real part and $Q(z)$ is the imaginary part of the polynomial. Since z is real, we have $|P(z)| \geq |R(z)|$. The polynomial $R(z)$ is monic, so on the interval $[-1, 1]$ it varies away from zero at least as much as the Chebyshev polynomial. Thus we can find z in this interval such that $|R(z)| \geq \frac{1}{2^{n-1}}$. This implies $|P(z)| \geq 2 \cdot \frac{1}{2^n}$, and rescaling back we deduce the existence in the general case of a point M satisfying the inequality from the statement. $\square$

Stepping aside from the classical picture, let us also consider the families of polynomials $\mathcal{T}_n(x)$ and $\mathcal{U}_n(x)$ defined by $\mathcal{T}_0(x) = 2$, $\mathcal{T}_1(x) = x$, $\mathcal{T}_{n+1}(x) = x\mathcal{T}_n(x) - \mathcal{T}_{n-1}(x)$, and $\mathcal{U}_0(x) = 1, \mathcal{U}_1(x) = x, \mathcal{U}_{n+1}(x) = x\mathcal{U}_n(x) - \mathcal{U}_{n-1}(x)$. These polynomials are determined by the equalities

$$\mathcal{T}_n\left(z + \frac{1}{z}\right) = z^n + \frac{1}{z^n} \quad \text{and} \quad \mathcal{U}_n\left(z + \frac{1}{z}\right) = \left(z^{n+1} - \frac{1}{z^{n+1}}\right) \Big/ \left(z - \frac{1}{z}\right).$$

Also, $T_n(x) = \frac{1}{2}\mathcal{T}_n(2x)$ and $U_n(x) = \mathcal{U}_n(2x)$. Here is a quickie that uses $\mathcal{T}_n(x)$.

Example. Let a be a real number such that $a + a^{-1}$ is an integer. Prove that for any $n \geq 1$, the number $a^n + a^{-n}$ is an integer.

Solution. An inductive argument based on the recurrence relation shows that $\mathcal{T}_n(x)$ is a polynomial with integer coefficients. And since $a^n + a^{-n} = \mathcal{T}_n(a + a^{-1})$, it follows that this number is an integer. $\square$

241. Prove that for $n \geq 1$,

$$T_{n+1}(x) = xT_n(x) - (1 - x^2)U_{n-1}(x),$$
$$U_n(x) = xU_{n-1}(x) + T_n(x).$$

242. Compute the $n \times n$ determinants

$$\begin{vmatrix} x & 1 & 0 & 0 & \ldots & 0 \\ 1 & 2x & 1 & 0 & \ldots & 0 \\ 0 & 1 & 2x & 1 & \ldots & 0 \\ \vdots & \vdots & \vdots & \vdots & \ddots & \vdots \\ 0 & 0 & 0 & 0 & \ldots & 1 \\ 0 & 0 & 0 & 0 & \ldots & 2x \end{vmatrix} \quad \text{and} \quad \begin{vmatrix} 2x & 1 & 0 & 0 & \ldots & 0 \\ 1 & 2x & 1 & 0 & \ldots & 0 \\ 0 & 1 & 2x & 1 & \ldots & 0 \\ \vdots & \vdots & \vdots & \vdots & \ddots & \vdots \\ 0 & 0 & 0 & 0 & \ldots & 1 \\ 0 & 0 & 0 & 0 & \ldots & 2x \end{vmatrix}.$$

243. Prove Chebyshev's theorem for $n = 4$: namely, show that for any monic fourth-degree polynomial $P(x)$,

$$\max_{-1 \leq x \leq 1} |P(x)| \geq \max_{-1 \leq x \leq 1} |2^{-3} T_4(x)|,$$

with equality if and only if $P(x) = 2^{-3} T_4(x)$.

244. Let r be a positive real number such that $\sqrt[6]{r} + \frac{1}{\sqrt[6]{r}} = 6$. Find the maximum value of $\sqrt[4]{r} - \frac{1}{\sqrt[4]{r}}$.

245. Let $\alpha = \frac{2\pi}{n}$. Prove that the matrix

$$\begin{pmatrix} 1 & 1 & \ldots & 1 \\ \cos \alpha & \cos 2\alpha & \ldots & \cos n\alpha \\ \cos 2\alpha & \cos 4\alpha & \ldots & \cos 2n\alpha \\ \vdots & \vdots & \ddots & \vdots \\ \cos(n-1)\alpha & \cos 2(n-1)\alpha & \ldots & \cos(n-1)n\alpha \end{pmatrix}$$

is invertible.

246. Find all quintuples (x, y, z, v, w) with $x, y, z, v, w \in [-2, 2]$ satisfying the system of equations

$$x + y + z + v + w = 0,$$
$$x^3 + y^3 + z^3 + v^3 + w^3 = 0,$$
$$x^5 + y^5 + z^5 + v^5 + w^5 = -10.$$

247. Let $x_1, x_2, \ldots, x_n$, $n \geq 2$, be distinct real numbers in the interval $[-1, 1]$. Prove that

$$\frac{1}{t_1} + \frac{1}{t_2} + \cdots + \frac{1}{t_n} \geq 2^{n-2},$$

where $t_k = \prod_{j \neq k} |x_j - x_k|$, $k = 1, 2, \ldots, n$.

248. Let $n \geq 3$ be an odd integer. Evaluate

$$\sum_{k=1}^{\frac{n-1}{2}} \sec \frac{2k\pi}{n}.$$

249. For $n \geq 1$, prove the following identities:

$$\frac{T_n(x)}{\sqrt{1 - x^2}} = \frac{(-1)^n}{1 \cdot 3 \cdot 5 \cdots (2n - 1)} \frac{d^n}{dx^n} (1 - x^2)^{n - \frac{1}{2}},$$

$$U_n(x)\sqrt{1 - x^2} = \frac{(-1)^n (n + 1)}{1 \cdot 3 \cdot 5 \cdots (2n + 1)} \frac{d^n}{dx^n} (1 - x^2)^{n + \frac{1}{2}}.$$

2.3 Linear Algebra

2.3.1 Operations with Matrices

An $m \times n$ matrix is an array with m rows and n columns. The standard notation is $A = (a_{ij})_{i,j}$, where a_{ij} is the entry (element) in the ith row and jth column. We denote by $\mathcal{I}_n$ the $n \times n$ identity matrix (for which $a_{ij} = 1$ if $i = j$, and 0 otherwise) and by $\mathcal{O}_n$ the $n \times n$ zero matrix (for which $a_{ij} = 0$ for all i, j).

Given the matrix $A = (a_{ij})_{i,j}$, A^t denotes the transpose of A, in which the i, j entry is a_{ji}, and $\overline{A}$ denotes the complex conjugate, whose entries are the complex conjugates of the entries of A. Also, tr A is the trace of A, namely the sum of the elements on the main diagonal: $a_{11} + a_{22} + \cdots + a_{nn}$.

We illustrate how matrix multiplication can be used to prove an identity satisfied by the Fibonacci sequence ($F_0 = 0$, $F_1 = 1$, $F_{n+1} = F_n + F_{n-1}$, $n \geq 1$). The identity we have in mind has already been discussed in the introductory chapter in the solution to Problem 27; we put it here in a new perspective.

Example. Prove that

$$F_{m+n+1} = F_{m+1} F_{n+1} + F_m F_n, \text{ for } m, n \geq 0.$$

Solution. Consider the matrix

$$M = \begin{pmatrix} 1 & 1 \\ 1 & 0 \end{pmatrix}.$$

An easy induction shows that for $n \geq 1$,

$$M^n = \begin{pmatrix} F_{n+1} & F_n \\ F_n & F_{n-1} \end{pmatrix}.$$

The equality $M^{m+n} = M^m M^n$ written in explicit form is

$$\begin{pmatrix} F_{m+n+1} & F_{m+n} \\ F_{m+n} & F_{m+n-1} \end{pmatrix} = \begin{pmatrix} F_{m+1} & F_m \\ F_m & F_{m-1} \end{pmatrix} \begin{pmatrix} F_{n+1} & F_n \\ F_n & F_{n-1} \end{pmatrix}.$$

We obtain the identity by setting the upper left corners of both sides equal. $\quad \square$
Here are some problems for the reader.

250. Let M be an $n \times n$ complex matrix. Prove that there exist Hermitian matrices A and B such that $M = A + iB$. (A matrix X is called Hermitian if $\overline{X^t} = X$).

251. Do there exist $n \times n$ matrices A and B such that $AB - BA = \mathcal{I}_n$?

252. Let A and B be 2×2 matrices with real entries satisfying $(AB - BA)^n = \mathcal{I}_2$ for some positive integer n. Prove that n is even and $(AB - BA)^4 = \mathcal{I}_2$.

253. Let A and B be two $n \times n$ matrices that do not commute and for which there exist nonzero real numbers p, q, r such that $pAB + qBA = \mathcal{I}_n$ and $A^2 = rB^2$. Prove that $p = q$.

254. Let a, b, c, d be real numbers such that $c \neq 0$ and $ad - bc = 1$. Prove that there exist u and v such that

$$\begin{pmatrix} a & b \\ c & d \end{pmatrix} = \begin{pmatrix} 1 & -u \\ 0 & 1 \end{pmatrix} \begin{pmatrix} 1 & 0 \\ c & 1 \end{pmatrix} \begin{pmatrix} 1 & -v \\ 0 & 1 \end{pmatrix}.$$

255. Compute the nth power of the $m \times m$ matrix

$$J_m(\lambda) = \begin{pmatrix} \lambda & 1 & 0 & \cdots & 0 \\ 0 & \lambda & 1 & \cdots & 0 \\ 0 & 0 & \lambda & \cdots & 0 \\ \vdots & \vdots & \vdots & \ddots & \vdots \\ 0 & 0 & 0 & \cdots & 1 \\ 0 & 0 & 0 & \cdots & \lambda \end{pmatrix}, \quad \lambda \in \mathbb{C}.$$

256. Let A and B be $n \times n$ matrices with real entries satisfying

$$\operatorname{tr}(AA^t + BB^t) = \operatorname{tr}(AB + A^t B^t).$$

Prove that $A = B^t$.

2.3.2 Determinants

The determinant of an $n \times n$ matrix $A = (a_{ij})_{i,j}$, denoted by $\det A$ or $|a_{ij}|$, is the volume taken with sign of the n-dimensional parallelepiped determined by the row (or column) vectors of A. Formally, the determinant can be introduced as follows. Let $e_1 = (1, 0, \ldots, 0)$, $e_2 = (0, 1, \ldots, 0), \ldots, e_n = (0, 0, \ldots, 1)$ be the canonical basis of $\mathbb{R}^n$. The exterior algebra of $\mathbb{R}^n$ is the vector space spanned by products of the form $e_{i_1} \wedge e_{i_2} \wedge \ldots \wedge e_{i_k}$, where the multiplication $\wedge$ is distributive with respect to sums and is subject to the noncommutativity rule $e_i \wedge e_j = -e_j \wedge e_i$ for all i, j (which then implies $e_i \wedge e_i = 0$, for all i). If the row vectors of the matrix A are $r_1, r_2, \ldots, r_n$, then the determinant is defined by the equality

$$r_1 \wedge r_2 \wedge \cdots \wedge r_n = (\det A) e_1 \wedge e_2 \wedge \cdots \wedge e_n.$$

The explicit formula is

$$\det A = \sum_{\sigma} \operatorname{sign}(\sigma) a_{1\sigma(1)} a_{2\sigma(2)} \cdots a_{n\sigma(n)},$$

with the sum taken over all permutations σ of $\{1, 2, \ldots, n\}$.

To compute the determinant of a matrix, one applies repeatedly the row operation that adds to one row a multiple of another until the matrix either becomes diagonal or has a row of zeros. In the first case this transforms the parallelepiped determined by the row vectors into a right parallelepiped in standard position without changing its volume, as suggested in Figure 13.

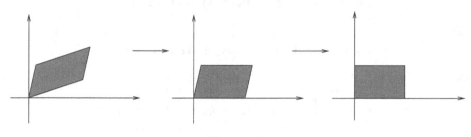

Figure 13

But it is not our purpose to teach the basics. We insist only on nonstandard tricks and methods. A famous example is the computation of the Vandermonde determinant.

Example. Let $x_1, x_2, \ldots, x_n$ be arbitrary numbers ($n \geq 1$). Compute the determinant

$$\begin{vmatrix} 1 & 1 & \ldots & 1 \\ x_1 & x_2 & \ldots & x_n \\ \vdots & \vdots & \ddots & \vdots \\ x_1^{n-1} & x_2^{n-1} & \ldots & x_n^{n-1} \end{vmatrix}.$$

Solution. The key idea is to view x_n as a variable and think of the determinant as an $(n-1)$st-degree polynomial in x_n. The leading coefficient is itself a Vandermonde determinant of order $n-1$, while the $n-1$ roots are obviously $x_2, x_3, \ldots, x_{n-1}$. The determinant is therefore equal to

$$\begin{vmatrix} 1 & 1 & \ldots & 1 \\ x_1 & x_2 & \ldots & x_n \\ \vdots & \vdots & \ddots & \vdots \\ x_1^{n-2} & x_2^{n-2} & \ldots & x_n^{n-2} \end{vmatrix} (x_n - x_1)(x_n - x_2) \cdots (x_n - x_{n-1}).$$

Now we can induct on n to prove that the Vandermonde determinant is equal to

$$\prod_{i>j}(x_i - x_j).$$

This determinant is equal to zero if and only if two of the x_i's are equal. □

We continue with a problem of D. Andrica.

Example. (a) Consider the real numbers a_{ij}, $i = 1, 2, \ldots, n-2$, $j = 1, 2, \ldots, n$, $n \geq 3$, and the determinants

$$A_k = \begin{vmatrix} 1 & \cdots & 1 & 1 & \cdots & 1 \\ a_{11} & \cdots & a_{1,k-1} & a_{1,k+1} & \cdots & a_{1n} \\ \vdots & \ddots & \vdots & \vdots & \ddots & \vdots \\ a_{n-2,1} & \cdots & a_{n-2,k-1} & a_{n-2,k+1} & \cdots & a_{n-2,n} \end{vmatrix}, \quad k \geq 1.$$

Prove that

$$A_1 + A_3 + A_5 + \cdots = A_2 + A_4 + A_6 + \cdots$$

(b) Define

$$p_k = \prod_{i=0}^{n-(k+1)} (x_{n-i} - x_k), \quad q_k = \prod_{i=1}^{k-1} (x_k - x_i),$$

where x_i, $i = 1, 2, \ldots, n$, are some distinct real numbers. Prove that

$$\sum_{k=1}^{n} \frac{(-1)^k}{p_k q_k} = 0.$$

(c) Prove that for any positive integer $n \geq 3$ the following identity holds:

$$\sum_{k=1}^{n} \frac{(-1)^k k^2}{(n-k)!(n+k)!} = 0.$$

Solution. We have

$$\begin{vmatrix} 1 & 1 & \cdots & 1 & 1 \\ 1 & 1 & \cdots & 1 & 1 \\ a_{11} & a_{12} & \cdots & a_{1,n-1} & a_{1n} \\ a_{21} & a_{22} & \cdots & a_{2,n-1} & a_{2n} \\ \vdots & \vdots & \ddots & \vdots & \vdots \\ a_{n-2,1} & a_{n-2,2} & \cdots & a_{n-2,n-1} & a_{n-2,n} \end{vmatrix} = 0.$$

Expanding by the first row, we obtain

$$A_1 - A_2 + A_3 - A_4 + \cdots = 0.$$

This implies

$$A_1 + A_3 + A_5 + \cdots = A_2 + A_4 + A_6 + \cdots,$$

and (a) is proved.

For (b), we substitute $a_{ij} = x_i^j$, $i = 1, 2, \ldots, n-2$, $j = 1, 2, \ldots, n$. Then

$$A_k = \begin{vmatrix} 1 & \cdots & 1 & 1 & \cdots & 1 \\ x_1 & \cdots & x_{k-1} & x_{k+1} & \cdots & x_n \\ \vdots & \ddots & \vdots & \vdots & \ddots & \vdots \\ x_1^{n-2} & \cdots & x_{k-1}^{n-2} & x_{k+1}^{n-2} & \cdots & x_n^{n-2} \end{vmatrix},$$

which is a Vandermonde determinant. Its value is equal to

$$\prod_{\substack{i>j \\ i,j \neq k}} (x_j - x_i) = \frac{\prod_{i>j}(x_j - x_i)}{p_k q_k}.$$

The equality proved in (a) becomes, in this particular case,

$$\sum_{k=1}^{n} \frac{(-1)^k}{p_k q_k} = 0,$$

as desired.

Finally, if in this we let $x_k = k^2$, then we obtain the identity from part (c), and the problem is solved. $\qquad\square$

And here comes a set of problems for the reader.

257. Prove that

$$\begin{vmatrix} (x^2 + 1)^2 & (xy + 1)^2 & (xz + 1)^2 \\ (xy + 1)^2 & (y^2 + 1)^2 & (yz + 1)^2 \\ (xz + 1)^2 & (yz + 1)^2 & (z^2 + 1)^2 \end{vmatrix} = 2(y - z)^2(z - x)^2(x - y)^2.$$

258. Let $(F_n)_n$ be the Fibonacci sequence. Using determinants, prove the identity

$$F_{n+1}F_{n-1} - F_n^2 = (-1)^n, \quad \text{for all } n \geq 1.$$

259. Let $p < m$ be two positive integers. Prove that

$$\begin{vmatrix} \binom{m}{0} & \binom{m}{1} & \cdots & \binom{m}{p} \\ \binom{m+1}{0} & \binom{m+1}{1} & \cdots & \binom{m+1}{p} \\ \vdots & \vdots & \ddots & \vdots \\ \binom{m+p}{0} & \binom{m+p}{1} & \cdots & \binom{m+p}{p} \end{vmatrix} = 1.$$

260. Given distinct integers $x_1, x_2, \ldots, x_n$, prove that $\prod_{i>j}(x_i - x_j)$ is divisible by $1!2!\cdots(n-1)!$.

261. Find all numbers in the interval $[-2015, 2015]$ that can be equal to the determinant of an 11×11 matrix with entries equal to 1 or -1.

262. Prove the formula for the determinant of a circulant matrix

$$\begin{vmatrix} x_1 & x_2 & x_3 & \cdots & x_n \\ x_n & x_1 & x_2 & \cdots & x_{n-1} \\ \vdots & \vdots & \vdots & \ddots & \vdots \\ x_3 & x_4 & x_5 & \cdots & x_2 \\ x_2 & x_3 & x_4 & \cdots & x_1 \end{vmatrix} = (-1)^{n-1} \prod_{j=0}^{n-1} \left(\sum_{k=1}^{n} \zeta^{jk} x_k \right),$$

where $\zeta = e^{2\pi i/n}$.

263. Let a and b be integers such that $a + b = 2014$. Prove that the determinant

$$\begin{vmatrix} a^3 & b^3 & 3ab & -1 \\ -1 & a^2 & b^2 & 2ab \\ 2b & -1 & a^2 & -b^2 \\ 0 & b & -1 & a \end{vmatrix}$$

is a multiple of 61.

264. Compute the determinant of the $n \times n$ matrix $A = (a_{ij})_{ij}$, where

$$a_{ij} = \begin{cases} (-1)^{|i-j|} & \text{if } i \neq j, \\ 2 & \text{if } i = j. \end{cases}$$

265. Prove that for any integers $x_1, x_2, \ldots, x_n$ and positive integers $k_1, k_2, \ldots, k_n$, the determinant

$$\begin{vmatrix} x_1^{k_1} & x_2^{k_1} & \cdots & x_n^{k_1} \\ x_1^{k_2} & x_2^{k_2} & \cdots & x_n^{k_2} \\ \vdots & \vdots & \ddots & \vdots \\ x_1^{k_n} & x_2^{k_n} & \cdots & x_n^{k_n} \end{vmatrix}$$

is divisible by $n!$.

266. Let A and B be 3×3 matrices with real elements such that

$$\det A = \det B = \det(A + B) = \det(A - B) = 0.$$

Prove that $\det(xA + yB) = 0$ for any real numbers x and y.

Sometimes it is more convenient to work with blocks instead of entries. For that we recall the rule of Laplace, which is the direct generalization of the row or column expansion. The determinant is computed by expanding over all $k \times k$ minors of some k rows or columns. Explicitly, given $A = (a_{ij})_{i,j=1}^{n}$, when expanding by the rows $i_1, i_2, \ldots, i_k$, the determinant is given by

$$\det A = \sum_{j_1 < j_2 < \cdots < j_k} (-1)^{i_1 + \cdots + i_k + j_1 + \cdots + j_k} M_k N_k,$$

where M_k is the determinant of the $k \times k$ matrix whose entries are a_{ij}, with $i \in \{i_1, i_2, \ldots, i_k\}$ and $j \in \{j_1, j_2, \ldots, j_k\}$, while N_k is the determinant of the $(n - k) \times (n - k)$ matrix whose entries are a_{ij} with $i \notin \{i_1, i_2, \ldots, i_k\}$ and $j \notin \{j_1, j_2, \ldots, j_k\}$. We exemplify this rule with a problem from the 4th International Competition in Mathematics for University Students (1997).

Example. Let M be an invertible $2n \times 2n$ matrix, represented in block form as

$$M = \begin{pmatrix} A & B \\ C & D \end{pmatrix} \quad \text{and} \quad M^{-1} = \begin{pmatrix} E & F \\ G & H \end{pmatrix}.$$

Show that $\det M \cdot \det H = \det A$.

Solution. The idea of the solution is that the relation between determinants should come from a relation between matrices. To this end, we would like to find three matrices X, Y, Z such that $XY = Z$, while $\det X = \det M$, $\det Y = \det H$, and $\det Z = \det A$. Since among M, H, and A, the matrix M has the largest dimension, we might try to set $X = M$ and find $2n \times 2n$ matrices Y and Z. The equality $M \cdot M^{-1} = \mathcal{I}_{2n}$ yields two relations involving H, namely $AF + BH = 0$ and $CF + DH = \mathcal{I}_n$. This suggests that we should use both F and H in the definition of Y. So we need an equality of the form

$$\begin{pmatrix} A & B \\ C & D \end{pmatrix} \begin{pmatrix} * & F \\ * & H \end{pmatrix} = \begin{pmatrix} * & 0 \\ * & \mathcal{I}_n \end{pmatrix}.$$

We can try

$$Y = \begin{pmatrix} \mathcal{I}_n & F \\ 0 & H \end{pmatrix}.$$

The latter has determinant equal to $\det H$, as desired. Also,

$$Z = \begin{pmatrix} A & 0 \\ C & \mathcal{I}_n \end{pmatrix}.$$

According to the rule of Laplace, the determinant of Z can be computed by expanding along the $n \times n$ minors from the top n rows, and all of them are zero except for the first. Hence $\det Z = \det A \cdot \det \mathcal{I}_n = \det A$, and so the matrices X, Y, Z solve the problem. $\qquad \square$

267. Show that if

$$x = \begin{vmatrix} a & b \\ c & d \end{vmatrix} \quad \text{and} \quad x' = \begin{vmatrix} a' & b' \\ c' & d' \end{vmatrix},$$

then

$$(xx')^2 = \begin{vmatrix} ab' & cb' & ba' & da' \\ ad' & cd' & bc' & dc' \\ bb' & db' & aa' & ca' \\ bd' & dd' & ac' & cc' \end{vmatrix}.$$

268. Let A, B, C, D be $n \times n$ matrices such that $AC = CA$. Prove that

$$\det \begin{pmatrix} A & B \\ C & D \end{pmatrix} = \det(AD - CB).$$

269. Let X and Y be $n \times n$ matrices. Prove that

$$\det(\mathcal{I}_n - XY) = \det(\mathcal{I}_n - YX).$$

A property exploited often in Romanian mathematics competitions states that for any $n \times n$ matrix A with real entries,

$$\det(\mathcal{I}_n + A^2) \geq 0.$$

The proof is straightforward:

$$\det(\mathcal{I}_n + A^2) = \det((\mathcal{I}_n + iA)(\mathcal{I}_n - iA)) = \det(\mathcal{I}_n + iA)\det(\mathcal{I}_n - iA)$$
$$= \det(\mathcal{I}_n + iA)\det(\overline{\mathcal{I}_n + iA}) = \det(\mathcal{I}_n + iA)\overline{\det(\mathcal{I}_n + iA)}.$$

In this computation the bar denotes the complex conjugate, and the last equality follows from the fact that the determinant is a polynomial in the entries. The final expression is nonnegative, being equal to $|\det(\mathcal{I}_n + iA)|^2$.

Use this property to solve the following problems, while assuming that all matrices have real entries.

270. Let A and B be $n \times n$ matrices that commute. Prove that if $\det(A + B) = 0$, then $\det(A^k + B^k) \geq 0$ for all $k \geq 1$.

271. Let A be an $n \times n$ matrix such that $A + A^t = \mathcal{O}_n$. Prove that

$$\det(\mathcal{I}_n + \lambda A^2) \geq 0,$$

for all $\lambda \in \mathbb{R}$.

272. Let $P(t)$ be a polynomial of even degree with real coefficients. Prove that the function $f(X) = P(X)$ defined on the set of $n \times n$ matrices is not onto.

273. Let n be an odd positive integer and A an $n \times n$ matrix with the property that $A^2 = \mathcal{O}_n$ or $A^2 = \mathcal{I}_n$. Prove that $\det(A + \mathcal{I}_n) \geq \det(A - \mathcal{I}_n)$.

2.3.3 The Inverse of a Matrix

An $n \times n$ matrix A is called invertible if there exists an $n \times n$ matrix A^{-1} such that $AA^{-1} = A^{-1}A = \mathcal{I}_n$. The inverse of a matrix can be found either by using the adjoint matrix, which amounts to computing several determinants, or by performing row and column operations. We illustrate how the latter method can be applied to a problem from the first International Competition in Mathematics for University Students (1994).

Example. (a) Let A be an $n \times n$ symmetric invertible matrix with positive real entries, $n \geq 2$. Show that A^{-1} has at most $n^2 - 2n$ entries equal to zero.
(b) How many entries are equal to zero in the inverse of the $n \times n$ matrix

$$A = \begin{pmatrix} 1 & 1 & 1 & 1 & \ldots & 1 \\ 1 & 2 & 2 & 2 & \ldots & 2 \\ 1 & 2 & 1 & 1 & \ldots & 1 \\ 1 & 2 & 1 & 2 & \ldots & 2 \\ \vdots & \vdots & \vdots & \vdots & \ddots & \vdots \\ 1 & 2 & 1 & 2 & \ldots & 1 \end{pmatrix} ?$$

Solution. Denote by a_{ij} and b_{ij} the entries of A, respectively, A^{-1}. Then we have $\sum_{i=0}^{n} a_{mi} b_{im} = 1$, so for fixed m not all the b_{im}'s are equal to zero. For $k \neq m$ we have $\sum_{i=0}^{n} a_{ki} b_{im} = 0$, and from the positivity of the a_{ki}'s we conclude that at least one b_{im} is negative, and at least one is positive. Hence every column of A^{-1} contains at least two nonzero elements. This proves part (a).

To compute the inverse of the matrix in part (b), we consider the extended matrix $(A\mathcal{I}_n)$, and using row operations we transform it into the matrix $(\mathcal{I}_n A^{-1})$. We start with

$$\begin{pmatrix} 1\ 1\ 1\ 1 \ldots\ 1 & 1\ 0\ 0\ 0 \ldots\ 0 \\ 1\ 2\ 2\ 2 \ldots\ 2 & 0\ 1\ 0\ 0 \ldots\ 0 \\ 1\ 2\ 1\ 1 \ldots\ 1 & 0\ 0\ 1\ 0 \ldots\ 0 \\ 1\ 2\ 1\ 2 \ldots\ 2 & 0\ 0\ 0\ 1 \ldots\ 0 \\ \vdots\ \vdots\ \vdots\ \vdots\ \ddots\ \vdots & \vdots\ \vdots\ \vdots\ \vdots\ \ddots\ \vdots \\ 1\ 2\ 1\ 2 \ldots\ \ldots\ 0 & 0\ 0\ 0 \ldots\ 1 \end{pmatrix}.$$

Subtracting the first row from each of the others, then the second row from the first, we obtain

$$\begin{pmatrix} 1\ 0\ 0\ 0 \ldots\ 0 & 2\ -1\ 0\ 0 \ldots\ 0 \\ 0\ 1\ 1\ 1 \ldots\ 1 & -1\ 1\ 0\ 0 \ldots\ 0 \\ 0\ 1\ 0\ 0 \ldots\ 0 & -1\ 0\ 1\ 0 \ldots\ 0 \\ 0\ 1\ 0\ 1 \ldots\ 1 & -1\ 0\ 0\ 1 \ldots\ 0 \\ \vdots\ \vdots\ \vdots\ \vdots\ \ddots\ \vdots & \vdots\ \vdots\ \vdots\ \vdots\ \ddots\ \vdots \\ 0\ 1\ 0\ 1 \ldots\ \ldots\ & -1\ 0\ 0 \ldots\ 1 \end{pmatrix}.$$

We continue as follows. First, we subtract the second row from the third, fourth, and so on. Then we add the third row to the second. Finally, we multiply all rows, beginning with the third, by -1. This way we obtain

$$\begin{pmatrix} 1\ 0\ 0\ 0 \ldots\ 0 & 2\ -1\ 0\ 0 \ldots\ 0 \\ 0\ 1\ 0\ 0 \ldots\ 0 & -1\ 0\ 1\ 0 \ldots\ 0 \\ 0\ 0\ 1\ 1 \ldots\ 1 & 0\ 1\ -1\ 0 \ldots\ 0 \\ 0\ 0\ 1\ 0 \ldots\ 0 & 0\ 1\ 0\ -1 \ldots\ 0 \\ \vdots\ \vdots\ \vdots\ \vdots\ \ddots\ \vdots & \vdots\ \vdots\ \vdots\ \vdots\ \ddots\ \vdots \\ 0\ 0\ 1\ 0 \ldots\ \ldots\ & 1\ 0\ 0\ 0 \ldots\ -1 \end{pmatrix}.$$

Now the inductive pattern is clear. At each step we subtract the kth row from the rows below, then subtract the $(k+1)$st from the kth, and finally multiply all rows starting with the $(k+1)$st by -1. In the end we find that the entries of A^{-1} are $b_{1,1} = 2$, $b_{n,n} = (-1)^n$, $b_{i,i+1} = b_{i+1,i} = (-1)^i$, and $b_{ij} = 0$, for $|i - j| \geq 2$. This example shows that equality can hold in part (a). □

274. For distinct numbers $x_1, x_2, \ldots, x_n$, consider the matrix

$$A = \begin{pmatrix} 1 & 1 & \ldots & 1 \\ x_1 & x_2 & \ldots & x_n \\ \vdots & \vdots & \ddots & \vdots \\ x_1^{n-1} & x_2^{n-1} & \ldots & x_n^{n-1} \end{pmatrix}.$$

It is known that det A is the Vandermonde determinant

$$\Delta(x_1, x_2, \ldots, x_n) = \prod_{i>j}(x_i - x_j).$$

Prove that the inverse of A is $B = (b_{km})_{1 \leq k, m \leq n}$, where

$$b_{km} = (-1)^{k+m} \Delta(x_1, x_2, \ldots, x_n)^{-1} \Delta(x_1, \ldots, x_{k-1}, x_{k+1}, \ldots, x_n)$$
$$\times S_{n-1}(x_1, \ldots, x_{k-1}, x_{k+1}, \ldots, x_n).$$

Here S_{n-1} denotes the $(n-1)$st symmetric polynomial in $n-1$ variables.

275. Let A and B be 2×2 matrices with integer entries such that A, $A+B$, $A+2B$, $A+3B$, and $A + 4B$ are all invertible matrices whose inverses have integer entries. Prove that $A + 5B$ is invertible and that its inverse has integer entries.

276. Determine the matrix A knowing that its adjoint matrix (the one used in the computation of the inverse) is

$$A^* = \begin{pmatrix} m^2 - 1 & 1 - m & 1 - m \\ 1 - m & m^2 - 1 & 1 - m \\ 1 - m & 1 - m & m^2 - 1 \end{pmatrix}, \quad m \neq 1, -2.$$

277. Let $A = (a_{ij})_{ij}$ be an $n \times n$ matrix such that $\displaystyle\sum_{j=1}^{n} |a_{ij}| < 1$ for each i. Prove that $\mathcal{I}_n - A$ is invertible.

278. Let $\alpha = \frac{\pi}{n+1}$, $n > 2$. Prove that the $n \times n$ matrix

$$\begin{pmatrix} \sin \alpha & \sin 2\alpha & \ldots & \sin n\alpha \\ \sin 2\alpha & \sin 4\alpha & \ldots & \sin 2n\alpha \\ \vdots & \vdots & \ddots & \vdots \\ \sin n\alpha & \sin 2n\alpha & \ldots & \sin n^2\alpha \end{pmatrix}$$

is invertible.

279. Assume that A and B are invertible complex $n \times n$ matrices such that $i(A^\dagger B - B^\dagger A)$ is positive semidefinite, where $X^\dagger = \overline{X}^t$, the transpose conjugate of X. Prove that $A + iB$ is invertible. (A matrix T is positive semidefinite if $\langle Tv, v \rangle \geq 0$ for all vectors v, where $\langle v, w \rangle = v^t \overline{w}$ the complex inner product.)

We continue with problems that exploit the ring structure of the set of $n \times n$ matrices. There are some special properties of matrices that do not hold in arbitrary rings. For example, an $n \times n$ matrix A is either a zero divisor (there exist nonzero matrices B and C such that $AB = CA = \mathcal{O}_n$), or it is invertible. Also, if a matrix has a left (or right) inverse, then the matrix is invertible, which means that if $AB = \mathcal{I}_n$ then also $BA = I_n$.

A good example is a problem of I.V. Maftei that appeared in the 1982 Romanian Mathematical Olympiad.

Example. Let A, B, C be $n \times n$ matrices, $n \geq 1$, satisfying

$$ABC + AB + BC + AC + A + B + C = \mathcal{O}_n.$$

Prove that A and $B + C$ commute if and only if A and BC commute.

Solution. If we add $\mathcal{I}_n$ to the left-hand side of the identity from the statement, we recognize this expression to be the polynomial $P(X) = (X + A)(X + B)(X + C)$ evaluated at the identity matrix. This means that

$$(\mathcal{I}_n + A)(\mathcal{I}_n + B)(\mathcal{I}_n + C) = \mathcal{I}_n.$$

This shows that $\mathcal{I}_n + A$ is invertible, and its inverse is $(\mathcal{I}_n + B)(\mathcal{I}_n + C)$. It follows that

$$(\mathcal{I}_n + B)(\mathcal{I}_n + C)(\mathcal{I}_n + A) = \mathcal{I}_n,$$

or

$$BCA + BC + BA + CA + A + B + C = \mathcal{O}_n.$$

Subtracting this relation from the one in the statement and grouping the terms appropriately, we obtain

$$ABC - BCA = (B + C)A - A(B + C).$$

The conclusion follows.   □

   Here are other examples.

280. Let A be an $n \times n$ matrix such that there exists a positive integer k for which

$$kA^{k+1} = (k + 1)A^k.$$

Prove that the matrix $A - \mathcal{I}_n$ is invertible and find its inverse.

281. Let A be an invertible $n \times n$ matrix, and let $B = XY$, where X and Y are $1 \times n$, respectively, $n \times 1$ matrices. Prove that the matrix $A + B$ is invertible if and only if $\alpha = YA^{-1}X \neq -1$, and in this case its inverse is given by

$$(A + B)^{-1} = A^{-1} - \frac{1}{\alpha + 1}A^{-1}BA^{-1}.$$

282. Given two $n \times n$ matrices A and B for which there exist nonzero real numbers a and b such that $AB = aA + bB$, prove that A and B commute.

283. Let A and B be $n \times n$ matrices, $n \geq 1$, satisfying $AB - B^2A^2 = \mathcal{I}_n$ and $A^3 + B^3 = \mathcal{O}_n$. Prove that $BA - A^2B^2 = \mathcal{I}_n$.

2.3.4 Systems of Linear Equations

A system of m linear equations with n unknowns can be written as

$$Ax = b,$$

where A is an $m \times n$ matrix called the coefficient matrix, and b is an m-dimensional vector. If $m = n$, the system has a unique solution if and only if the coefficient matrix A is invertible. If A is not invertible, the system can have either infinitely many solutions or none at all. If additionally $b = 0$, then the system does have infinitely many solutions and the codimension of the space of solutions is equal to the rank of A.

We illustrate this section with two problems that apparently have nothing to do with the topic. The first was published in *Mathematics Gazette, Bucharest*, by L. Pîrşan.

Example. Consider the matrices

$$A = \begin{pmatrix} a & b \\ c & d \end{pmatrix}, \quad B = \begin{pmatrix} \alpha & \beta \\ \gamma & \delta \end{pmatrix}, \quad C = \begin{pmatrix} a\alpha & b\alpha & a\gamma & b\gamma \\ a\beta & b\beta & a\delta & b\delta \\ c\alpha & d\alpha & c\gamma & d\gamma \\ c\beta & d\beta & c\delta & d\delta \end{pmatrix},$$

where $a, b, c, d, \alpha, \beta, \gamma, \delta$ are real numbers. Prove that if A and B are invertible, then C is invertible as well.

Solution. Let us consider the matrix equation $AXB = D$, where

$$X = \begin{pmatrix} x & z \\ y & t \end{pmatrix} \quad \text{and} \quad D = \begin{pmatrix} m & n \\ p & q \end{pmatrix}.$$

Solving it for X gives $X = A^{-1}DB^{-1}$, and so X is uniquely determined by A, B, and D. Multiplying out the matrices in this equation,

$$\begin{pmatrix} a & b \\ c & d \end{pmatrix} \begin{pmatrix} x & z \\ y & t \end{pmatrix} \begin{pmatrix} \alpha & \beta \\ \gamma & \delta \end{pmatrix} = \begin{pmatrix} m & n \\ p & q \end{pmatrix},$$

we obtain

$$\begin{pmatrix} a\alpha x + b\alpha y + a\gamma z + b\gamma t & a\beta x + b\beta y + a\delta z + b\delta t \\ c\alpha x + d\alpha y + c\gamma z + d\gamma t & c\beta x + d\beta y + c\delta z + d\delta t \end{pmatrix} = \begin{pmatrix} m & n \\ p & q \end{pmatrix}.$$

This is a system in the unknowns x, y, z, t:

$$a\alpha x + b\alpha y + a\gamma z + b\gamma t = m,$$
$$a\beta x + b\beta y + a\delta z + b\delta t = n,$$
$$c\alpha x + d\alpha y + c\gamma z + d\gamma t = p,$$
$$c\beta x + d\beta y + c\delta z + d\delta t = q.$$

We saw above that this system has a unique solution, which implies that its coefficient matrix is invertible. This coefficient matrix is C. $\qquad\square$

The second problem we found in an old textbook on differential and integral calculus.

Example. Given the distinct real numbers a_1, a_2, a_3, let x_1, x_2, x_3 be the three roots of the equation

$$\frac{u_1}{a_1 + t} + \frac{u_2}{a_2 + t} + \frac{u_3}{a_3 + t} = 1,$$

where u_1, u_2, u_3 are real parameters. Prove that u_1, u_2, u_3 are smooth functions of x_1, x_2, x_3 and that

$$\det\left(\frac{\partial u_i}{\partial x_j}\right) = -\frac{(x_1 - x_2)(x_2 - x_3)(x_3 - x_1)}{(a_1 - a_2)(a_2 - a_3)(a_3 - a_1)}.$$

Solution. After eliminating the denominators, the equation from the statement becomes a cubic equation in t, so x_1, x_2, x_3 are well defined. The parameters u_1, u_2, u_3 satisfy the system of equations

$$\frac{1}{a_1 + x_1}u_1 + \frac{1}{a_2 + x_1}u_2 + \frac{1}{a_3 + x_1}u_3 = 1,$$

$$\frac{1}{a_1 + x_2}u_1 + \frac{1}{a_2 + x_2}u_2 + \frac{1}{a_3 + x_2}u_3 = 1,$$

$$\frac{1}{a_1 + x_3}u_1 + \frac{1}{a_2 + x_3}u_2 + \frac{1}{a_3 + x_3}u_3 = 1.$$

When solving this system, we might end up entangled in algebraic computations. Thus it is better instead to take a look at the two-variable situation. Solving the system

$$\frac{1}{a_1 + x_1}u_1 + \frac{1}{a_2 + x_1}u_2 = 1,$$

$$\frac{1}{a_1 + x_2}u_1 + \frac{1}{a_2 + x_2}u_2 = 1,$$

with Cramer's rule we obtain

$$u_1 = \frac{(a_1 + x_1)(a_1 + x_2)}{(a_1 - a_2)} \quad \text{and} \quad u_2 = \frac{(a_2 + x_1)(a_2 + x_2)}{(a_2 - a_1)}.$$

Now we can extrapolate to the three-dimensional situation and guess that

$$u_i = \frac{\displaystyle\prod_{k=1}^{3}(a_i + x_k)}{\displaystyle\prod_{k \neq i}(a_i - a_k)}, \quad i = 1, 2, 3.$$

It is not hard to check that these satisfy the system of equations. Observe that

$$\frac{\partial u_i}{\partial x_j} = \frac{\displaystyle\prod_{k \neq j}(a_i + x_k)}{\displaystyle\prod_{j \neq i}(a_i - a_j)}, \quad \text{and so} \quad \frac{\partial u_i}{\partial x_j} = \frac{1}{a_i + x_j}u_i, \; i, j = 1, 2, 3.$$

The determinant in question looks again difficult to compute. Some tricks simplify the task. An observation is that the sum of the columns is 1. Indeed, these sums are

$$\frac{\partial u_1}{\partial x_i} + \frac{\partial u_2}{\partial x_i} + \frac{\partial u_3}{\partial x_i}, \ i = 1, 2, 3,$$

which we should recognize as the left-hand sides of the linear system. So the determinant becomes much simpler if we add the first and second rows to the last. Another observation is that the determinant is a 3-variable polynomial in x_1, x_2, x_3. Its total degree is 3, and it becomes zero if $x_i = x_j$ for some $i \neq j$. Consequently, the determinant is a number not depending on x_1, x_2, x_3 times $(x_1 - x_2)(x_2 - x_3)(x_3 - x_1)$. This number can be determined by looking just at the coefficient of $x_2^2 x_3$. And an easy computation shows that this coefficient is equal to $\frac{1}{(a_1 - a_2)(a_2 - a_3)(a_3 - a_1)}$. $\qquad\square$

From the very many practical applications of the theory of systems of linear equations, let us mention the Global Positioning System (GPS). The principle behind the GPS is the measurement of the distances between the receiver and 24 satellites (in practice some of these satellites might have to be ignored in order to avoid errors due to atmospheric phenomena). This yields 24 quadratic equations $d(P, S_i)^2 = r_i^2, \ i = 1, 2, \ldots, 24$, in the three spatial coordinates of the receiver. Subtracting the first of the equations from the others cancels the quadratic terms and gives rise to an overdetermined system of 23 linear equations in three unknowns. Determining the location of the receiver is therefore a linear algebra problem.

284. Solve the system of linear equations

$$x_1 + x_2 + x_3 = 0,$$
$$x_2 + x_3 + x_4 = 0,$$
$$\cdots$$
$$x_{99} + x_{100} + x_1 = 0,$$
$$x_{100} + x_1 + x_2 = 0.$$

285. Find the solutions x_1, x_2, x_3, x_4, x_5 to the system of equations

$$x_5 + x_2 = yx_1, \quad x_1 + x_3 = yx_2, \quad x_2 + x_4 = yx_3,$$

$$x_3 + x_5 = yx_1, \quad x_4 + x_1 = yx_5,$$

where y is a parameter.

286. Let a, b, c, d be positive numbers different from 1, and x, y, z, t real numbers satisfying $a^x = bcd, b^y = cda, c^z = dab, d^t = abc$. Prove that

$$\begin{vmatrix} -x & 1 & 1 & 1 \\ 1 & -y & 1 & 1 \\ 1 & 1 & -z & 1 \\ 1 & 1 & 1 & -t \end{vmatrix} = 0.$$

287. Given the system of linear equations

$$a_{11}x_1 + a_{12}x_2 + a_{13}x_3 = 0,$$
$$a_{21}x_1 + a_{22}x_2 + a_{23}x_3 = 0,$$
$$a_{31}x_1 + a_{32}x_2 + a_{33}x_3 = 0,$$

whose coefficients satisfy the conditions

(i) a_{11}, a_{22}, a_{33} are positive,
(ii) all other coefficients are negative,
(iii) in each equation, the sum of the coefficients is positive,

prove that the system has the unique solution $x_1 = x_2 = x_3 = 0$.

288. Let $P(x) = x^n + x^{n-1} + \cdots + x + 1$. Find the remainder obtained when $P(x^{n+1})$ is divided by $P(x)$.

289. Find all functions $f : \mathbb{R} \setminus \{-1, 1\} \to \mathbb{R}$ satisfying

$$f\left(\frac{x-3}{x+1}\right) + f\left(\frac{3+x}{1-x}\right) = x \text{ for all } x \neq \pm 1.$$

290. Find all positive integer solutions (x, y, z, t) to the Diophantine equation

$$(x + y)(y + z)(z + x) = txyz$$

such that $\gcd(x, y) = \gcd(y, z) = \gcd(z, x) = 1$.

291. We have n coins of unknown masses and a balance. We are allowed to place some of the coins on one side of the balance and an equal number of coins on the other side. After thus distributing the coins, the balance gives a comparison of the total mass of each side, either by indicating that the two masses are equal or by indicating that a particular side is the more massive of the two. Show that at least $n - 1$ such comparisons are required to determine whether all of the coins are of equal mass.

292. Let $a_0 = 0, a_1, \ldots, a_n, a_{n+1} = 0$ be a sequence of real numbers that satisfy

$$|a_{k-1} - 2a_k + a_{k+1}| \leq 1 \text{ for } k = 1, 2, \ldots, n - 1.$$

Prove that

$$|a_k| \leq \frac{k(n - k + 1)}{2} \text{ for } k = 1, 2, \ldots, n - 1.$$

293. Prove that the Hilbert matrix

$$\begin{pmatrix} 1 & \frac{1}{2} & \frac{1}{3} & \cdots & \frac{1}{n} \\ \frac{1}{2} & \frac{1}{3} & \frac{1}{4} & \cdots & \frac{1}{n+1} \\ \vdots & \vdots & \vdots & \ddots & \vdots \\ \frac{1}{n} & \frac{1}{n+1} & \frac{1}{n+2} & \cdots & \frac{1}{2n-1} \end{pmatrix}$$

is invertible. Prove also that the sum of the entries of the inverse matrix is n^2.

2.3.5 Vector Spaces, Linear Combinations of Vectors, Bases

In general, a vector space V over a field of scalars (which in our book will be only $\mathbb{C}$, $\mathbb{R}$, or $\mathbb{Q}$) is a set endowed with a commutative addition and a scalar multiplication that have the same properties as those for vectors in Euclidean space.

A linear combination of the vectors $v_1, v_2, \ldots, v_m$ is a sum $c_1 v_1 + c_2 v_2 + \cdots + c_m v_m$ with scalar coefficients. The vectors are called linearly independent if a combination of these vectors is equal to zero only when all coefficients are zero. Otherwise, the vectors are called linearly dependent. If $v_1, v_2, \ldots, v_n$ are linearly independent and if every vector in V is a linear combination of these vectors, then $v_1, v_2, \ldots, v_n$ is called a basis of V. The number of elements of a basis of a vector space depends only on the vector space, and is called the dimension of the vector space. We will be concerned only with finite-dimensional vector spaces. We also point out that if in a vector space there are given more vectors than the dimension, then these vectors must be linearly dependent.

The rank of a matrix is the dimension of its row vectors, which is the same as the dimension of the column vectors. A square matrix is invertible if and only if its rank equals its size.

Let us see some examples. The first appeared in the Soviet University Student Mathematical Competition in 1977.

Example. Let X and B_0 be $n \times n$ matrices, $n \geq 1$. Define $B_i = B_{i-1} X - X B_{i-1}$, for $i \geq 1$. Prove that if $X = B_{n^2}$, then $X = \mathcal{O}_n$.

Solution. Because the space of $n \times n$ matrices is n^2-dimensional, $B_0, B_1, \ldots, B_{n^2}$ must be linearly dependent, so there exist scalars $c_0, c_1, \ldots, c_{n^2}$ such that

$$c_0 B_0 + c_1 B_1 + \cdots + c_{n^2} B^{n^2} = \mathcal{O}_n.$$

Let k be the smallest index for which $c_k \neq 0$. Then

$$B_k = a_1 B_{k+1} + a_2 B_{k+2} + \cdots + a_{n^2 - k} B_{n^2},$$

where $a_j = -\frac{c_{k+j}}{c_k}$. Computing $B_{k+1} = B_k X - X B_k$, we obtain

$$B_{k+1} = a_1 B_{k+2} + a_2 B_{k+3} + \cdots + a_{n^2 - k} B_{n^2 + 1},$$

and inductively

$$B_{k+j} = a_1 B_{k+j+1} + a_2 B_{k+j+2} + \cdots + a_{n^2 - k} B_{n^2 + j}, \text{ for } j \geq 1.$$

In particular,

$$B_{n^2} = a_1 B_{n^2 + 1} + a_2 B_{n^2 + 2} + \cdots + a_{n^2 - k} B_{n^2 + k}.$$

But $B_{n^2 + 1} = B_{n^2} X - X B_{n^2} = X^2 - X^2 = \mathcal{O}_n$, and hence $B_{n^2 + j} = \mathcal{O}_n$, for $j \geq 1$. It follows that X, which is a linear combination of $B_{n^2 + 1}, B_{n^2 + 2}, \ldots, B_{n^2 + k}$ is the zero matrix. And we are done. $\qquad\square$

The second example was given at the 67th W.L. Putnam Mathematical Competition in 2006, and the solution that we present was posted by C. Zara on the Internet.

Example. Let Z denote the set of points in $\mathbb{R}^n$ whose coordinates are 0 or 1. (Thus Z has 2^n elements, which are the vertices of a unit hypercube in $\mathbb{R}^n$.) Let k be given, $0 \le k \le n$. Find the maximum, over all vector subspaces $V \subseteq \mathbb{R}^n$ of dimension k, of the number of points in $Z \cap V$.

Solution. Let us consider the matrix whose rows are the elements of $V \cap Z$. By construction it has row rank at most k. It thus also has column rank at most k; in particular, there are k columns such that any other column is a linear combination of these k. It means that the coordinates of each point of $V \cap Z$ are determined by the k coordinates that lie in these k columns. Since each such coordinate can have only two values, $V \cap Z$ can have at most 2^k elements.

   This upper bound is reached for the vectors that have all possible choices of 0 and 1 for the first k entries, and 0 for the remaining entries.   □

294. Prove that every odd polynomial function of degree equal to $2m - 1$ can be written as

$$P(x) = c_1 \binom{x}{1} + c_2 \binom{x+1}{3} + c_3 \binom{x+2}{5} + \ldots + c_m \binom{x+m-1}{2m-1},$$

where $\binom{x}{m} = \dfrac{x(x-1)\cdots(x-m+1)}{n!}$.

295. Let n be a positive integer and $P(x)$ an nth-degree polynomial with complex coefficients such that $P(0), P(1), \ldots, P(n)$ are all integers. Prove that the polynomial $n!P(x)$ has integer coefficients.

296. Let A be the $n \times n$ matrix whose i, j entry is $i + j$ for all $i, j = 1, 2, \ldots, n$. What is the rank of A?

297. For integers $n \ge 2$ and $0 \le k \le n - 2$, compute the determinant

$$\begin{vmatrix} 1^k & 2^k & 3^k & \ldots & n^k \\ 2^k & 3^k & 4^k & \ldots & (n+1)^k \\ 3^k & 4^k & 5^k & \ldots & (n+2)^k \\ \vdots & \vdots & \vdots & \ddots & \vdots \\ n^k & (n+1)^k & (n+2)^k & \ldots & (2n-1)^k \end{vmatrix}.$$

298. Let V be a vector space and let $f, f_1, f_2, \ldots, f_n$ be linear maps from V to $\mathbb{R}$. Suppose that $f(x) = 0$ whenever $f_1(x) = f_2(x) = \cdots = f_n(x) = 0$. Prove that f is a linear combination of $f_1, f_2, \ldots, f_n$.

299. Given a set S of $2n - 1$ different irrational numbers, $n \ge 1$, prove that there exist n distinct elements $x_1, x_2, \ldots, x_n \in S$ such that for all nonnegative rational numbers $a_1, a_2, \ldots, a_n$ with $a_1 + a_2 + \cdots + a_n > 0$, the number $a_1 x_1 + a_2 x_2 + \cdots + a_n x_n$ is irrational.

300. There are given $2n + 1$ real numbers, $n \ge 1$, with the property that whenever one of them is removed, the remaining $2n$ can be split into two sets of n elements that have the same sum of elements. Prove that all the numbers are equal.

301. Let V be an infinite set of vectors in $\mathbb{R}^n$ containing n linearly independent vectors. A finite subset $S \subset V$ is called crucial if the set $V \backslash S$ contains no n linearly independent vectors, but every set $V \backslash T$, with T a subset of S does. Prove there are only finitely many crucial subsets of V.

2.3.6 Linear Transformations, Eigenvalues, Eigenvectors

A linear transformation between vector spaces is a map $T : V \to W$ that satisfies $T(\alpha_1 v_1 + \alpha_2 v_2) = \alpha_1 T(v_1) + \alpha_2 T(v_2)$ for any scalars α_1, α_2 and vectors v_1, v_2. A matrix A defines a linear transformation by $v \to Av$, and any linear transformation between finite-dimensional vector spaces with specified bases is of this form. An eigenvalue of a matrix A is a zero of the characteristic polynomial $P_A(\lambda) = \det(\lambda \mathcal{I}_n - A)$. Alternatively, it is a scalar λ for which the equation $Av = \lambda v$ has a nontrivial solution v. In this case v is called an eigenvector of the eigenvalue λ. If $\lambda_1, \lambda_2, \ldots, \lambda_m$ are distinct eigenvalues and $v_1, v_2, \ldots, v_m$ are corresponding eigenvectors, then $v_1, v_2, \ldots, v_m$ are linearly independent. Moreover, if the matrix A is Hermitian, meaning that A is equal to its transpose conjugate, then $v_1, v_2, \ldots, v_m$ may be chosen to be pairwise orthogonal.

The set of eigenvalues of a matrix is called its spectrum. The reason for this name is that in quantum mechanics observable quantities are modelled by matrices. Physical spectra, such as the emission spectrum of the hydrogen atom, become spectra of matrices. Among all results in spectral theory we stopped at the spectral mapping theorem, mainly because we want to bring to your attention the method used in the proof.

The spectral mapping theorem. *Let A be an $n \times n$ matrix with not necessarily distinct eigenvalues $\lambda_1, \lambda_2, \ldots, \lambda_n$, and let $P(x)$ be a polynomial. Then the eigenvalues of the matrix $P(A)$ are $P(\lambda_1), P(\lambda_2), \ldots, P(\lambda_n)$.*

Proof. To prove this result we will apply a widely used idea (see for example the splitting principle in algebraic topology). We will first assume that the eigenvalues of A are all distinct. Then A can be diagonalized by eigenvectors as

$$\begin{pmatrix} \lambda_1 & 0 & \ldots & 0 \\ 0 & \lambda_2 & \ldots & 0 \\ \vdots & \vdots & \ddots & \vdots \\ 0 & 0 & \ldots & \lambda_n \end{pmatrix},$$

and in the basis formed by the eigenvectors of A, the matrix $P(A)$ assumes the form

$$\begin{pmatrix} P(\lambda_1) & 0 & \ldots & 0 \\ 0 & P(\lambda_2) & \ldots & 0 \\ \vdots & \vdots & \ddots & \vdots \\ 0 & 0 & \ldots & P(\lambda_n) \end{pmatrix}.$$

The conclusion is now straightforward. In general, the characteristic polynomial of a matrix depends continuously on the entries. Problem 229 in Section 2.2.6 proved that the roots of a polynomial depend continuously on the coefficients. Hence the eigenvalues of a matrix depend continuously on the entries.

The set of matrices with distinct eigenvalues is dense in the set of all matrices. To prove this claim we need the notion of the discriminant of a polynomial. By definition, if the zeros of a polynomial are $x_1, x_2, \ldots, x_n$, the discriminant is $\prod_{i<j}(x_i - x_j)^2$. It is equal to zero if and only if the polynomial has multiple zeros. Being a symmetric polynomial in the x_i's, the discriminant is a polynomial in the coefficients. Therefore, the condition that the eigenvalues of a matrix be not all distinct can be expressed as a polynomial equation in the entries. By slightly varying the entries, we can violate this condition. Therefore, arbitrarily close to any matrix there are matrices with distinct eigenvalues.

The conclusion of the spectral mapping theorem for an arbitrary matrix now follows by a limiting argument.   □

We continue with two more elementary examples.

Example. Let $A : V \to W$ and $B : W \to V$ be linear maps between finite-dimensional vector spaces. Prove that the linear maps AB and BA have the same set of nonzero eigenvalues, counted with multiplicities.

Solution. Choose a basis that identifies V with $\mathbb{R}^m$ and W with $\mathbb{R}^n$. Associate to A and B their matrices, denoted by the same letters. The problem is solved if we prove the equality

$$\det(\lambda \mathcal{I}_n - AB) = \lambda^k \det(\lambda \mathcal{I}_m - BA),$$

where k is of course $n - m$. The relation being symmetric, we may assume that $n \geq m$. In this case, complete the two matrices with zeros to obtain two $n \times n$ matrices A' and B'. Because $\det(\lambda \mathcal{I}_n - A'B') = \det(\lambda \mathcal{I}_n - AB)$ and $\det(\lambda \mathcal{I}_n - B'A') = \lambda^{n-m} \det(\lambda \mathcal{I}_n - BA)$, the problem reduces to proving that $\det(\lambda \mathcal{I}_n - A'B') - \det(\lambda \mathcal{I}_n - B'A')$. And this is true for arbitrary $n \times n$ matrices A' and B'. For a proof of this fact we refer the reader to problem 269 in Section 2.3.2.   □

If $B = A^\dagger$, the transpose conjugate of A, then this example shows that $AA^\dagger$ and $A^\dagger A$ have the same nonzero eigenvalues. The square roots of these eigenvalues are called the singular values of A. The second example comes from the first International Mathematics Competition (for university students), 1994.

Example. Let α be a nonzero real number and n a positive integer. Suppose that F and G are linear maps from $\mathbb{R}^n$ into $\mathbb{R}^n$ satisfying $F \circ G - G \circ F = \alpha F$.

(a) Show that for all $k \geq 1$ one has $F^k \circ G - G \circ F^k = \alpha k F^k$.
(b) Show that there exists $k \geq 1$ such that $F^k = \mathcal{O}_n$.

Here $F \circ G$ denotes F composed with G, and F^k denotes F composed with itself k times.

Solution. Expand $F^k \circ G - G \circ F^k$ using a telescopic sum as follows:

$$F^k \circ G - G \circ F^k = \sum_{i=1}^{k} (F^{k-i+1} \circ G \circ F^{i-1} - F^{k-i} \circ G \circ F^i)$$

$$= \sum_{i=1}^{k} F^{k-i} \circ (F \circ G - G \circ F) \circ F^{i-1}$$

$$= \sum_{i=1}^{k} F^{k-i} \circ \alpha F \circ F^{i-1} = \alpha k F^k.$$

This proves (a). For (b), consider the linear map $L(F) = F \circ G - G \circ F$ acting on all $n \times n$ matrices F. Assuming $F^k \neq \mathcal{O}_n$ for all k, we deduce from (a) that αk is an eigenvalue of L for all k. This is impossible since the linear map L acts on an n^2-dimensional space, so it can have at most n^2 eigenvalues. This contradiction proves (b).   □

302. Let A be a 2×2 matrix with complex entries and let $C(A)$ denote the set of 2×2 matrices that commute with A. Prove that $|\det(A + B)| \geq |\det B|$ for all $B \in C(A)$ if and only if $A^2 = \mathcal{O}_2$.

303. Let A, B be 2×2 matrices with integer entries, such that $AB = BA$ and $\det B = 1$. Prove that if $\det(A^3 + B^3) = 1$, then $A^2 = \mathcal{O}_2$.

304. Consider the $n \times n$ matrix $A = (a_{ij})$ with $a_{ij} = 1$ if $j - i \equiv 1 \pmod{n}$ and $a_{ij} = 0$ otherwise. For real numbers a and b find the eigenvalues of $aA + bA^t$.

305. Let A be an $n \times n$ matrix such that $\det A = 1$ and $A^t A = \mathcal{I}_n$. Show that 1 is an eigenvalue of A.

306. Let A be an $n \times n$ matrix that has zeros on the main diagonal and all other entries from the set $\{-1, 1\}$. Is it possible that $\det A = 0$ for $n = 2007$? What about for $n = 2008$?

307. Let A be an $n \times n$ skew-symmetric matrix (meaning that for all $i, j, a_{ij} = -a_{ji}$) with real entries. Prove that

$$\det(A + x\mathcal{I}_n) \cdot \det(A + y\mathcal{I}_n) \geq \det(A + \sqrt{xy}\mathcal{I}_n)^2,$$

for all $x, y \in [0, \infty)$.

308. Let A be an $n \times n$ matrix. Prove that there exists an $n \times n$ matrix B such that $ABA = A$.

309. Consider the angle formed by two half-lines in three-dimensional space. Prove that the average of the measure of the projection of the angle onto all possible planes in the space is equal to the angle.

310. A linear map A on the n-dimensional vector space V is called an involution if $A^2 = \mathcal{I}$.

   (a) Prove that for every involution A on V there exists a basis of V consisting of eigenvectors of A.
   (b) Find the maximal number of distinct pairwise commuting involutions.

311. Let A be a 3×3 real matrix such that the vectors Au and u are orthogonal for each column vector $u \in \mathbb{R}^3$. Prove that

(a) $A^t = -A$, where A^t denotes the transpose of the matrix A;

(b) there exists a vector $v \in \mathbb{R}^3$ such that $Au = v \times u$ for every $u \in \mathbb{R}^3$.

312. Denote by $M_n(\mathbb{R})$ the set of $n \times n$ matrices with real entries and let $f : M_n(\mathbb{R}) \to \mathbb{R}$ be a linear function. Prove that there exists a unique matrix $C \in M_n(\mathbb{R})$ such that $f(A) = \mathrm{tr}(AC)$ for all $A \in M_n(\mathbb{R})$. In addition, if $f(AB) = f(BA)$ for all matrices A and B, prove that there exists $\lambda \in \mathbb{R}$ such that $f(A) = \lambda \mathrm{tr} A$ for any matrix A.

313. Let U and V be isometric linear transformations of $\mathbb{R}^n$, $n \geq 1$, with the property that $\|Ux - x\| \leq \frac{1}{2}$ and $\|Vx - x\| \leq \frac{1}{2}$ for all $x \in \mathbb{R}^n$ with $\|x\| = 1$. Prove that

$$\|UVU^{-1}V^{-1}x - x\| \leq \frac{1}{2},$$

for all $x \in \mathbb{R}^n$ with $\|x\| = 1$.

314. For an $n \times n$ matrix A denote by $\phi_k(A)$ the symmetric polynomial in the eigenvalues $\lambda_1, \lambda_2, \ldots, \lambda_n$ of A,

$$\phi_k(A) = \sum_{i_1 i_2 \ldots i_k} \lambda_{i_1} \lambda_{i_2} \cdots \lambda_{i_k}, \ k = 1, 2, \ldots, n.$$

For example, $\phi_1(A)$ is the trace and $\phi_n(A)$ is the determinant. Prove that for two $n \times n$ matrices A and B, $\phi_k(AB) = \phi_k(BA)$ for all $k = 1, 2, \ldots, n$.

2.3.7 The Cayley-Hamilton and Perron-Frobenius Theorems

We devote this section to two more advanced results, which seem to be relevant to mathematics competitions. All matrices below are assumed to have complex entries.

The Cayley-Hamilton Theorem. *Any $n \times n$ matrix A satisfies its characteristic equation, which means that if $P_A(\lambda) = \det(\lambda \mathcal{I}_n - A)$, then $P_A(A) = \mathcal{O}_n$.*

Proof. Let $P_A(\lambda) = \lambda^n + a_{n-1}\lambda^{n-1} + \cdots + a_0$. Denote by $(\lambda \mathcal{I}_n - A)^*$ the adjoint of $(\lambda \mathcal{I}_n - A)$ (the one used in the computation of the inverse). Then

$$(\lambda \mathcal{I}_n - A)(\lambda \mathcal{I}_n - A)^* = \det(\lambda \mathcal{I}_n - A)\mathcal{I}_n.$$

The entries of the adjoint matrix $(\lambda \mathcal{I}_n - A)^*$ are polynomials in λ of degree at most $n - 1$. Splitting the matrix by the powers of λ, we can write

$$(\lambda \mathcal{I}_n - A)^* = B_{n-1}\lambda^{n-1} + B_{n-2}\lambda^{n-2} + \cdots + B_0.$$

Equating the coefficients of λ on both sides of

$$(\lambda \mathcal{I}_n - A)(B_{n-1}\lambda^{n-1} + B_{n-2}\lambda^{n-2} + \cdots + B_0) = \det(\lambda \mathcal{I}_n - A)\mathcal{I}_n,$$

we obtain the equations

$$B_{n-1} = \mathcal{I}_n,$$
$$-AB_{n-1} + B_{n-2} = a_{n-1}\mathcal{I}_n,$$
$$-AB_{n-2} + B_{n-3} = a_{n-2}\mathcal{I}_n,$$
$$\cdots$$
$$-AB_0 = a_0\mathcal{I}_n.$$

Multiply the first equation by A^n, the second by A^{n-1}, the third by A^{n-2}, and so on, then add the $n + 1$ equations to obtain

$$\mathcal{O}_n = A^n + a_{n-1}A^{n-1} + a_{n-2}A^{n-2} + \cdots + a_0\mathcal{I}_n.$$

This equality is just the desired $P_A(A) = \mathcal{O}_n$. $\qquad\square$

As a corollary we prove the trace identity for $SL(2, \mathbb{C})$ matrices. This identity is important in the study of characters of group representations.

Example. Let A and B be 2×2 matrices with determinant equal to 1. Prove that

$$\mathrm{tr}(AB) - (\mathrm{tr}A)(\mathrm{tr}B) + \mathrm{tr}(AB^{-1}) = 0.$$

Solution. By the Cayley-Hamilton Theorem,

$$B^2 - (\mathrm{tr}B)B + \mathcal{I}_2 = \mathcal{O}_2.$$

Multiply on the left by AB^{-1} to obtain

$$AB - (\mathrm{tr}B)A + AB^{-1} = \mathcal{O}_2,$$

and then take the trace to obtain the identity from the statement. $\qquad\square$

Five more examples are left to the reader.

315. Let A be a 2×2 matrix. Show that if for some complex numbers u and v the matrix $u\mathcal{I}_2 + vA$ is invertible, then its inverse is of the form $u'\mathcal{I}_2 + v'A$ for some complex numbers u' and v'.

316. Find the 2×2 matrices X with real entries that satisfy the equation

$$X^3 - 3X^2 = \begin{pmatrix} -2 & -2 \\ -2 & -2 \end{pmatrix}.$$

317. Let A, B, C, D be 2×2 matrices. Prove that the matrix $[A, B] \cdot [C, D] + [C, D] \cdot [A, B]$ is a multiple of the identity matrix (here $[A, B] = AB - BA$, the commutator of A and B).

318. Let A and B be two 2×2 matrices that do not commute. Assume that there is a nonconstant polynomial $P(x)$ with real coefficients such that $P(AB) = P(BA)$. Prove that there exists a real number a such that $P(AB) = a\mathcal{I}_2$.

319. Let A and B be 3×3 matrices. Prove that

$$\det(AB - BA) = \frac{\operatorname{tr}((AB - BA)^3)}{3}.$$

320. Show that there do not exist real 2×2 matrices A and B such that their commutator is nonzero and commutes with both A and B.

Here is the simplest version of the other result that we had in mind.

The Perron-Frobenius theorem. *Any square matrix with positive entries has a unique eigenvector with positive entries (up to a multiplication by a positive scalar), and the corresponding eigenvalue has multiplicity one and is strictly greater than the absolute value of any other eigenvalue.*

Proof. The proof uses real analysis. Let $A = (a_{ij})_{i,j=1}^n$, $n \geq 1$. We want to show that there is a unique $v \in [0, \infty)^n$, $v \neq 0$, such that $Av = \lambda v$ for some λ. Of course, since A has positive entries and v has positive coordinates, λ has to be a positive number. Denote by K the intersection of $[0, \infty)^n$ with the $n - 1$-dimensional unit sphere. Reformulating the problem, we want to show that the function $f : K \to K$, $f(v) = \frac{Av}{\|Av\|}$ has a fixed point.

Now, there is a rather general result that states that a contractive function on a compact metric space has a unique fixed point (see Section 3.2.3). Recall that a metric space is a set X endowed with a function $\delta : X \times X \to [0, \infty)$ satisfying

(i) $\delta(x, y) = 0$ if and only if $x = y$,

(ii) $\delta(x, y) = \delta(y, x)$ for all $x, y \in X$,

(iii) $\delta(x, y) + \delta(y, z) \geq \delta(x, z)$ for all $x, y, z \in X$.

We use the property in the case of a compact set in $\mathbb{R}^n$, where compact sets are characterized by being closed and bounded. A function $f : X \to X$ is contractive if

$$\delta(f(x), f(y)) < \delta(x, y), \quad \text{for every } x \neq y.$$

With this in mind, we want to find a distance on the set K that makes the function f defined above contractive. This is the Hilbert metric defined by the formula

$$\delta(v, w) = \ln\left(\max_i\left\{\frac{v_i}{w_i}\right\} \Big/ \min_i\left\{\frac{v_i}{w_i}\right\}\right),$$

for $v = (v_1, v_2, \dots, v_n)$ and $w = (w_1, w_2, \dots, w_n) \in K$. That this satisfies the triangle inequality $\delta(u, w) + \delta(w, u) \geq \delta(v, w)$ is a consequence of the inequalities

$$\max_i\left\{\frac{v_i}{w_i}\right\} \cdot \max_i\left\{\frac{w_i}{u_i}\right\} \geq \max_i\left\{\frac{v_i}{w_i}\right\},$$

$$\min_i \left\{ \frac{v_i}{w_i} \right\} \cdot \min_i \left\{ \frac{w_i}{u_i} \right\} \geq \min_i \left\{ \frac{v_i}{w_i} \right\}.$$

Let us show that f is contractive. If $v = (v_1, v_2, \ldots, v_n)$ and $w = (w_1, w_2, \ldots, w_n)$ are in K, $v \neq w$, and if $\alpha_i > 0$, $i = 1, 2, \ldots, n$, then

$$\min_i \left\{ \frac{v_i}{w_i} \right\} < \frac{\alpha_1 v_1 + \alpha_2 v_2 + \cdots + \alpha_n v_n}{\alpha_1 w_1 + \alpha_2 w_2 + \cdots + \alpha_n w_n} < \max_i \left\{ \frac{v_i}{w_i} \right\}.$$

Indeed, to prove the first inequality, add the obvious inequalities

$$\alpha_j w_j \min_i \left\{ \frac{v_i}{w_i} \right\} \leq \alpha_j v_j, \ j = 1, 2, \ldots, n.$$

Because $v \neq w$ and both vectors are on the unit sphere, at least one inequality is strict. The second inequality follows from

$$\alpha_j w_j \max_i \left\{ \frac{v_i}{w_i} \right\} \geq \alpha_j v_j, \ j = 1, 2, \ldots, n,$$

where again at least one inequality is strict.

Using this fact, we obtain for all j, $1 \leq j \leq n$,

$$\frac{a_{j1} v_1 + \cdots + a_{jn} v_n}{a_{j1} w_1 + \cdots + a_{jn} w_n} \Big/ \max_i \left\{ \frac{v_i}{w_i} \right\} < 1 < \frac{a_{j1} v_1 + \cdots + a_{jn} v_n}{a_{j1} w_1 + \cdots + a_{jn} w_n} \Big/ \min_i \left\{ \frac{v_i}{w_i} \right\}.$$

Therefore,

$$\frac{\max_j \left\{ \dfrac{a_{j1} v_1 + \cdots + a_{jn} v_n}{a_{j1} w_1 + \cdots + a_{jn} w_n} \right\}}{\max_i \left\{ \dfrac{v_i}{w_i} \right\}} < \frac{\min_j \left\{ \dfrac{a_{j1} v_1 + \cdots + a_{jn} v_n}{a_{j1} w_1 + \cdots + a_{jn} w_n} \right\}}{\min_i \left\{ \dfrac{v_i}{w_i} \right\}}.$$

It follows that for $v, w \in K$, $v \neq w$, $\delta(f(v), f(w)) < \delta(v, w)$.

Now, K is closed and but is not bounded in the Hilbert metric; some points are infinitely far apart. But even if K is not bounded in the Hilbert metric, $f(K)$ is (prove it!). If we denote by K_0 the closure of $f(K)$ in the Hilbert metric, then this space is closed and bounded. On K_0, f is contractive, and so it has a unique fixed point. Note that all fixed points of f are necessarily in K_0 (because if $f(v) = v$, then $v = f(v) \in f(K)$).

We are done with the first half of the proof. Now let us show that the eigenvalue of this positive vector is larger than the absolute value of any other eigenvalue. Let $r(A)$ be the largest of the absolute values of the eigenvalues of A and let λ be an eigenvalue with $|\lambda| = r(A)$. In general, for a vector v we denote by $|v|$ the vector whose coordinates are the absolute values of the coordinates of v. Also, for two vectors v, w we write $v \geq w$ if each coordinate of v is greater than the corresponding coordinate of w. If v is an eigenvector of A corresponding to the eigenvalue λ, then $|Av| = |\lambda| \cdot |v|$. The triangle inequality implies $A|v| \geq |Av| = r(A)|v|$. It follows that the set

$$K_1 = \{v \mid \|v\| = 1, \ v \geq 0, \ Av \geq r(A)v\},$$

is nonempty. Because A has positive entries, $A(Av - r(A)v) \geq 0$ for $v \in K_1$. So $A(Av) \geq r(A)(Av)$, for $v \in K_1$, proving that $f(K_1) \subset K_1$. Again K_1 is closed and $f(K_1)$ is bounded, so we can reason as above to prove that f restricted to K_1 has a fixed point, and because $K_1 \subset K$, this is the fixed point that we detected before. Thus $r(A)$ is the unique positive eigenvalue.

There cannot exist another eigenvalue λ with $|\lambda| = r(A)$, for otherwise, for a small $\varepsilon > 0$ the matrix $A - \varepsilon \mathcal{I}_n$ would still have positive entries, but its positive eigenvalue $r(A) - \varepsilon$ would be smaller than the absolute value of the other eigenvalue contradicting what we just proved. This concludes the proof of the theorem. $\square$

Nowhere in the book are more appropriate the words of Sir Arthur Eddington: "Proof is an idol before which the mathematician tortures himself."

The conclusion of the theorem still holds in the more general setting of irreducible matrices with nonnegative entries (*irreducible* means that there is no reordering of the rows and columns that makes it block upper triangular). This more general form of the Perron-Frobenius Theorem is currently used by the Internet browser Google to sort the entries of a search. The idea is the following: Write the adjacency matrix of the Internet with a link highlighted if it is related to the subject. Then multiply each nonzero entry by a larger or smaller number that takes into account how important the subject is in that page. The Perron-Frobenius vector of this new matrix assigns a positive weight to each site on the Internet. The Internet browser then lists the sites in decreasing order of their weights.

We now challenge you with some problems.

321. Let A be a square matrix whose off-diagonal entries are positive. Prove that the rightmost eigenvalue of A in the complex plane is real and all other eigenvalues are strictly to its left in the complex plane.

322. Let a_{ij}, $i, j = 1, 2, 3$, be real numbers such that a_{ij} is positive for $i = j$ and negative for $i \neq j$. Prove that there exist positive real numbers c_1, c_2, c_3 such that the numbers

$$a_{11}c_1 + a_{12}c_2 + a_{13}c_3, \quad a_{21}c_1 + a_{22}c_2 + a_{23}c_3, \quad a_{31}c_1 + a_{32}c_2 + a_{33}c_3$$

are all negative, all positive, or all zero.

323. Let $x_1, x_2, \ldots, x_n$ be differentiable (real-valued) functions of a single variable t that satisfy

$$\frac{dx_1}{dt} = a_{11}x_1 + a_{12}x_2 + \cdots + a_{1n}x_n,$$
$$\frac{dx_2}{dt} = a_{21}x_1 + a_{22}x_2 + \cdots + a_{2n}x_n,$$
$$\cdots$$
$$\frac{dx_n}{dt} = a_{n1}x_1 + a_{n2}x_2 + \cdots + a_{nn}x_n,$$

for some constants $a_{ij} > 0$. Suppose that for all i, $x_i(t) \to 0$ as $t \to \infty$. Are the functions $x_1, x_2, \ldots, x_n$ necessarily linearly independent?

324. For a positive integer n and any real number c, define $(x_k)_{k \geq 0}$ recursively by $x_0 = 0$, $x_1 = 1$, and for $k \geq 0$,

$$x_{k+2} = \frac{cx_{k+1} - (n-k)x_k}{k+1}.$$

Fix n and then take c to be the largest value for which $x_{n+1} = 0$. Find x_k in terms of n and k, $1 \leq k \leq n$.

2.4 Abstract Algebra

2.4.1 Binary Operations

A binary operation $*$ on a set S associates to each pair $(a, b) \in S \times S$ an element $a * b \in S$. The operation is called associative if $a * (b * c) = (a * b) * c$ for all $a, b, c \in S$, and commutative if $a * b = b * a$ for all $a, b \in S$. If there exists an element e such that $a * e = e * a = a$ for all $a \in S$, then e is called an identity element. If an identity exists, it is unique. In this case, if for an element $a \in S$ there exists $b \in S$ such that $a * b = b * a = e$, then b is called the inverse of a and is denoted by a^{-1}. If an element has an inverse, the inverse is unique.

Just as a warmup, we present a problem from the 62nd W.L. Putnam Competition, 2001.

Example. Consider a set S and a binary operation $*$ on S. Assume that $(a * b) * a = b$ for all $a, b \in S$. Prove that $a * (b * a) = b$ for all $a, b \in S$.

Solution. Substituting $b * a$ for a, we obtain

$$((b * a) * b) * (b * a) = b.$$

The expression in the first set of parentheses is a. Therefore,

$$a * (b * a) = b,$$

as desired. $\square$

Often, problems about binary operations look like innocent puzzles, yet they can have profound implications. This is the case with the following example.

Example. For three-dimensional vectors $X = (p, q, t)$ and $Y = (p', q', t')$ define the operations $(p, q, t) * (p', q', t') = (0, 0, pq' - qp')$, and $X \circ Y = X + Y + \frac{1}{2}X * Y$, where $+$ denotes the addition in $\mathbb{R}^3$.

(a) Prove that $(\mathbb{R}^3, \circ)$ is a group.

(b) Let $\alpha : (\mathbb{R}^3, \circ) \to (\mathbb{R}^3, \circ)$ be a continuous map satisfying $\alpha(X \circ Y) = \alpha(X) \circ \alpha(Y)$ for all X, Y (which means that α is a homomorphism). Prove that

$$\alpha(X + Y) = \alpha(X) + \alpha(Y) \quad \text{and} \quad \alpha(X * Y) = \alpha(X) * \alpha(Y).$$

Solution. (a) Associativity can be verified easily, the identity element is $(0, 0, 0)$, and the inverse of (p, q, t) is $(-p, -q, -t)$.

(b) First, note that $X * Y = -Y * X$. Therefore, if X is a scalar multiple of Y, then $X * Y = Y * X = 0$. In general, if $X * Y = 0$, then $X \circ Y = X + Y = Y \circ X$. Hence in this case,

$$\alpha(X + Y) = \alpha(X \circ Y) = \alpha(X) \circ \alpha(Y) = \alpha(X) + \alpha(Y) + \frac{1}{2}\alpha(X) * \alpha(Y)$$

on the one hand, and

$$\alpha(X + Y) = \alpha(Y \circ X) = \alpha(Y) \circ \alpha(X) = \alpha(Y) + \alpha(X) + \frac{1}{2}\alpha(Y) * \alpha(X).$$

Because $\alpha(X) * \alpha(Y) = -\alpha(Y) * \alpha(X)$, this implies that $\alpha(X) * \alpha(Y) = 0$. and consequently $\alpha(X+Y) = \alpha(X) + \alpha(Y)$. In particular, α is additive on every one-dimensional space, whence $\alpha(rX) = r\alpha(X)$, for every rational number r. But α is continuous, so $\alpha(sX) = s\alpha(X)$ for every real number s. Applying this property we find that for any $X, Y \in \mathbb{R}^3$ and $s \in \mathbb{R}$,

$$s\alpha\left(X + Y + \frac{1}{2}sX * Y\right) = \alpha\left(sX + sY + \frac{1}{2}s^2X * Y\right) = \alpha((sX) \circ (sY))$$

$$= \alpha(sX) \circ \alpha(sY) = (s\alpha(X)) \circ (s\alpha(Y))$$

$$= s\alpha(X) + s\alpha(Y) + \frac{1}{2}s^2\alpha(X) * \alpha(Y).$$

Dividing both sides by s, we obtain

$$\alpha\left(X + Y + \frac{1}{2}sX * Y\right) = \alpha(X) + \alpha(Y) + \frac{1}{2}s\alpha(X) * \alpha(Y).$$

In this equality if we let $s \to 0$, we obtain $\alpha(X + Y) = \alpha(X) + \alpha(Y)$. Also, if we let $s = 1$ and use the additivity we just proved, we obtain $\alpha(X * Y) = \alpha(X) * \alpha(Y)$. The problem is solved. $\qquad\square$

Traditionally, $X * Y$ is denoted by $[X, Y]$ and $\mathbb{R}^3$ endowed with this operation is called the Heisenberg Lie algebra. Also, $\mathbb{R}^3$ endowed with $\circ$ is called the Heisenberg group. And we just proved a famous theorem showing that a continuous automorphism of the Heisenberg group is also an automorphism of the Heisenberg Lie algebra. The Heisenberg group and algebra are fundamental concepts of quantum mechanics.

325. With the aid of a calculator that can add, subtract, and determine the inverse of a nonzero number, find the product of two nonzero numbers using at most 20 operations.

326. Invent a binary operation from which $+$, $-$, $\times$, and $/$ can be derived.

327. A finite set S with at least four elements is endowed with an associative binary operation $*$ that satisfies

$$(a * a) * b = b * (a * a) = b \text{ for all } a, b \in S.$$

Prove that the set of all elements of the form $a * (b * c)$ with a, b, c distinct elements of S coincides with S.

328. Let S be the smallest set of rational functions containing $f(x, y) = x$ and $g(x, y) = y$ and closed under subtraction and taking reciprocals. Show that S does not contain the nonzero constant functions.

329. Let $*$ and $\circ$ be two binary operations on the set M, with identity elements e, respectively, e', and with the property that for every $x, y, u, v \in M$,

$$(x * y) \circ (u * v) = (x \circ u) * (y \circ v).$$

Prove that

(a) $e = e'$;
(b) $x * y = x \circ y$, for every $x, y \in M$;
(c) $x * y = y * x$, for every $x, y \in M$.

330. Consider a set S and a binary operation $*$ on S such that $x * (y * x) = y$ for all x, y in S. Prove that each of the equations $a * x = b$ and $x * a = b$ has a unique solution in S.

331. On a set M an operation $*$ is given satisfying the properties

(i) there exists an element $e \in M$ such that $x * e = x$ for all $x \in M$;
(ii) $(x * y) * z = (z * x) * y$ for all $x, y, z \in M$.

Prove that the operation $*$ is both associative and commutative.

332. Prove or disprove the following statement: If F is a finite set with two or more elements, then there exists a binary operation $*$ on F such that for all $x, y, z \in F$,

(i) $x * z = y * z$ implies $x = y$ (right cancellation holds), and
(ii) $x * (y * z) \neq (x * y) * z$ (no case of associativity holds).

333. Let $*$ be an associative binary operation on a set S satisfying $a * b = b * a$ only if $a = b$. Prove that $a * (b * c) = a * c$ for all $a, b, c \in S$. Give an example of such an operation.

334. Let S be a set and $*$ a binary operation on S satisfying the laws

(i) $x * (x * y) = y$ for all $x, y \in S$,
(ii) $(y * x) * x = y$ for all $x, y \in S$.

Show that $*$ is commutative but not necessarily associative.

335. Let $*$ be a binary operation on the set $\mathbb{Q}$ of rational numbers that is associative and commutative and satisfies $0 * 0 = 0$ and $(a + c) * (b + c) = a * b + c$ for all $a, b, c \in \mathbb{Q}$. Prove that either $a * b = \max(a, b)$ for all $a, b \in \mathbb{Q}$, or $a * b = \min(a, b)$ for all $a, b \in \mathbb{Q}$.

2.4.2 Groups

Definition. A group is a set of transformations (of some space) that contains the identity transformation and is closed under composition and under the operation of taking the inverse.

The isometries of the plane, the permutations of a set, the continuous bijections on a closed bounded interval all form groups.

There is a more abstract, and apparently more general definition, which calls a group a set G endowed with a binary operation $\cdot$ that satisfies

(i) (associativity) $x(yz) = (xy)z$ for all $x, y, z \in S$;

(ii) (identity element) there is $e \in G$ such that for any $x \in G$, $ex = xe = x$;

(iii) (existence of the inverse) for every $x \in G$ there is $x^{-1} \in G$ such that

$$xx^{-1} = x^{-1}x = e.$$

But Cayley observed the following fact.

Theorem. *Any group is a group of transformations.*

Proof. Indeed, any group G acts on itself on the left. Specifically, $x \in G$ acts as a transformation of G by $y \to xy$, $y \in G$. $\qquad\square$

A group G is called Abelian (after N. Abel) if the operation is commutative, that is, if $xy = yx$ for all $x, y \in G$. An example of an Abelian group is the Klein four-group, introduced abstractly as $K = \{a, b, c, e \mid a^2 = b^2 = c^2 = e,\ ab = ac,\ ac = b,\ bc = a\}$, or concretely as the group of the symmetries of a rectangle (depicted in Figure 14).

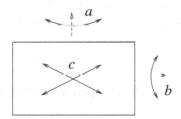

Figure 14

A group is called cyclic if it is generated by a single element, that is, if it consists of the identity element and the powers of some element.

Let us turn to problems and start with one published by L. Daia in the *Mathematics Gazette, Bucharest.*

Example. A certain multiplicative operation on a nonempty set G is associative and allows cancellations on the left, and there exists $a \in G$ such that $x^3 = axa$ for all $x \in G$. Prove that G endowed with this operation is an Abelian group.

Solution. Replacing x by ax in the given relation, we obtain $axaxax = a^2xa$. Cancelling a on the left, we obtain $x(axa)x = axa$. Because $axa = x^3$, it follows that $x^5 = x^3$, and cancelling an x^2, we obtain

$$x^3 = x \text{ for all } x \in G.$$

In particular, $a^3 = a$, and hence $a^3x = ax$ for all $x \in G$. Cancel a on the left to find that

$$a^2x = x \text{ for all } x \in G.$$

Substituting x by xa, we obtain $a^2xa = xa$, or $ax^3 = xa$, and since $x^3 = x$, it follows that a commutes with all elements in G. We can therefore write

$$a^2x = a(ax) = a(xa) = (xa)a = xa^2,$$

whence $xa^2 = a^2x = x$. This shows that a^2 is the identity element of the multiplicative operation; we denote it by e. The relation from the statement implies $x^3 = axa = xa^2 = xe$; cancelling x, we obtain $x^2 = e$; hence for all $x \in G$, $x^{-1} = x$. It follows that G is a group. It is Abelian by the well-known computation

$$xy = (xy)^{-1} = y^{-1}x^{-1} = yx. \qquad \square$$

Here are more examples of the kind.

336. Prove that in order for a set G endowed with an associative operation to be a group, it suffices for it to have a left identity, and for each element to have a left inverse. This means that there should exist $e \in G$ such that $ex = x$ for all $x \in G$, and for each $x \in G$, there should exist $x' \in G$ such that $x'x = e$. The same conclusion holds if "left" is replaced by "right".

337. Let $(G, \perp)$ and $(G, *)$ be two group structures defined on the same set G. Assume that the two groups have the same identity element and that their binary operations satisfy

$$a * b = (a \perp a) \perp (a \perp b),$$

for all $a, b \in G$. Prove that the binary operations coincide and the group they define is Abelian.

338. Let r, s, t be positive integers that are pairwise relatively prime. If the elements a and b of an Abelian group with identity element e satisfy $a^r = b^s = (ab)^t = e$, prove that $a = b = e$. Does the same conclusion hold if a and b are elements of an arbitrary nonAbelian group?

339. A is a subset of a finite group G which contains more than one half of the elements of G. Prove that every element of G is the product of two elements of A.

340. On the set $M = \mathbb{R}\backslash\{3\}$ the following binary operation is defined:

$$x * y = 3(xy - 3x - 3y) + m,$$

where $m \in \mathbb{R}$. Find all possible values of m for which $(M, *)$ is a group.

341. Assume that a and b are elements of a group with identity element e satisfying $(aba^{-1})^n = e$ for some positive integer n. Prove that $b^n = e$.

342. Let G be a group with the following properties:

(i) G has no element of order 2,

(ii) $(xy)^2 = (yx)^2$, for all $x, y \in G$.

Prove that G is Abelian.

343. A multiplicative operation on a set M satisfies

(i) $a^2 = b^2$, (ii) $ab^2 = a$, (iii) $a^2(bc) = cb$, (iv) $(ac)(bc) = ab$, for all $a, b, c \in M$.

Define on M the operation

$$a * b = a(b^2 b).$$

Prove that $(M, *)$ is a group.

We would like to point out the following property of the set of real numbers.

Kronecker's theorem. *A nontrivial subgroup of the additive group of real numbers is either cyclic or it is dense in the set of real numbers.*

Proof. Denote the group by G. It is either discrete, or it has an accumulation point on the real axis. If it is discrete, let a be its smallest positive element. Then any other element is of the form $b = ka + \alpha$ with $0 \le \alpha < a$. But b and ka are both in G; hence α is in G as well. By the minimality of a, α can only be equal to 0, and hence the group is cyclic.

If there is a nonconstant sequence $(x_n)_n$ in G converging to some real number, then $\pm(x_n - x_m)$ approaches zero as $n, m \to \infty$. Choosing the indices m and n appropriately, we can find a sequence of positive elements in G that converges to 0. Thus for any $\varepsilon > 0$ there is an element $c \in G$ with $0 < c < \varepsilon$. For some integer k, the distance between kc and $(k+1)c$ is less than ε; hence any interval of length ε contains some multiple of c. Varying ε, we conclude that G is dense in the real axis. $\qquad\square$

Try to use this result to solve the following problems.

344. Let $f : \mathbb{R} \to \mathbb{R}$ be a continuous function satisfying

$$f(x) + f(x + \sqrt{2}) = f(x + \sqrt{3}) \text{ for all } x.$$

Prove that f is constant.

345. Prove that the sequence $(\sin n)_{n \ge 1}$ is dense in the interval $[-1, 1]$.

346. Show that infinitely many powers of 2 start with the digit 7.

347. Given a rectangle, we are allowed to fold it in two or in three, parallel to one side or the other, in order to form a smaller rectangle. Prove that for any $\varepsilon > 0$ there are finitely many such operations that produce a rectangle with the ratio of the sides lying in the interval $(1 - \varepsilon, 1 + \varepsilon)$ (which means that we can get arbitrarily close to a square).

348. A set of points in the plane is invariant under the reflections across the sides of some given regular pentagon. Prove that the set is dense in the plane.

We continue with problems about groups of matrices.

349. Prove that the group of invertible 4×4 matrices with rational entries has no elements of order 7.

350. Given Γ a finite multiplicative group of invertible matrices with complex entries, denote by M the sum of the matrices in Γ. Prove that $\det M$ and $\operatorname{tr} M$ are integers.

351. Let n be a positive integer. What is the size of the largest multiplicative group of invertible $n \times n$ matrices with integer entries such that for every matrix A in the group all the entries of $A - I_n$ are even?

352. For an $n \times n$ matrix with complex entries, A, we define its norm to be

$$\|A\| = \sup_{\|x\| \leq 1} \|Ax\|,$$

where $\|x\|$ denotes the usual norm on $\mathbb{C}^n$ (the square root of the sum of the squares of the absolute values of the coordinates). Let $a < 2$, and let G be a multiplicative group of invertible $n \times n$ matrices such that

$$\|A - \mathcal{I}_n\| \leq a \text{ for all } A \in G.$$

Prove that G is finite.

"There is no certainty in sciences where one of the mathematical sciences cannot be applied, or which are not in relation with this mathematics." This thought of Leonardo da Vinci motivated us to include an example of how groups show up in natural sciences.

The groups of symmetries of three-dimensional space play an important role in chemistry and crystallography. In chemistry, the symmetries of molecules give rise to physical properties such as optical activity. The point groups of symmetries of molecules were classified by A. Schönflies as follows:

- C_s: a reflection with respect to a plane, isomorphic to $\mathbb{Z}_2$,

- C_i: a reflection with respect to a point, isomorphic to $\mathbb{Z}_2$,

- C_n: the rotations by multiples of $\frac{2\pi}{n}$ about an axis, isomorphic to $\mathbb{Z}_n$,

- C_{nv}: generated by a C_n and a C_s with the reflection plane containing the axis of rotation; in mathematics this is called the dihedral group,

- C_{nh}: generated by a C_n and a C_s with the reflection plane perpendicular to the axis of rotation, isomorphic to $C_n \times C_2$,

- D_n: generated by a C_n and a C_2, with the rotation axes perpendicular to each other, isomorphic to the dihedral group,

- D_{nd}: generated by a C_n and a C_2, together with a reflection across a plane that divides the angle between the two rotation axes,

- D_{nh}: generated by a C_n and a C_2 with perpendicular rotation axes, together with a reflection with respect to a plane perpendicular to the first rotation axis,

- S_n: improper rotations by multiples of $\frac{2\pi}{n}$, i.e., the group generated by the element that is the composition of the rotation by $\frac{2\pi}{n}$ and the reflection with respect to a plane perpendicular to the rotation axis,

- Special point groups: $C_{\infty v}$'s and $D_{\infty h}$'s (same as C_{nv} and D_{nh} but with all rotations about the axis allowed), together with the symmetry groups of the five Platonic solids.

When drawing a molecule, we use the convention that all segments represent bonds in the plane of the paper, all bold arrows represent bonds with the tip of the arrow below the tail of the arrow. The molecules from Figure 15 have respective symmetry point groups the octahedral group and C_{3h}.

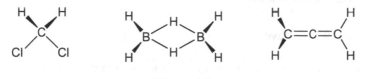

Figure 15

353. Find the symmetry groups of the molecules depicted in Figure 16.

Figure 16

2.4.3 Rings

Rings mimic in the abstract setting the properties of the sets of integers, polynomials, or matrices.

Definition. A ring is a set R endowed with two operations $+$ and $\cdot$ (addition and multiplication) such that $(R, +)$ is an Abelian group with identity element 0 and the multiplication satisfies

(i) (associativity) $x(yz) = (xy)z$ for all $x, y, z \in R$, and

(ii) (distributivity) $x(y + z) = xy + xz$ and $(x + y)z = xz + yz$ for all $x, y, z \in R$.

A ring is called commutative if the multiplication is commutative. It is said to have identity if there exists $1 \in R$ such that $1 \cdot x = x \cdot 1 = x$ for all $x \in R$. An element $x \in R$ is called invertible if there exists $x^{-1} \in R$ such that $xx^{-1} = x^{-1}x = 1$.

We consider two examples, the second of which appeared many years ago in the Balkan Mathematics Competition for university students.

Example. Let x and y be elements in a ring with identity. Prove that if $1 - xy$ is invertible, then so is $1 - yx$.

Solution. If we naively use the expansion $(1 - x)^{-1} = 1 + x + x^2 + x^3 + \cdots$ to write

$$(1 - xy)^{-1} = 1 + xy + xyxy + xyxyxy + \cdots$$
$$(1 - yx)^{-1} = 1 + yx + yxyx + yxyxyx + \cdots,$$

we can rearrange the second as

$$(1 - yx)^{-1} = 1 + y(1 + xy + xyxy + xyxyxy + \cdots)x$$

So we can gess that if v be the inverse of $1 - xy$ then $1 + yvx$ is the inverse of $1 - yx$. We have $v(1 - xy) = (1 - xy)v = 1$; hence $vxy = xyv = v - 1$. We compute

$$(1 + yvx)(1 - yx) = 1 - yx + yvx - yvxyx = 1 - yx + yvx - y(v - 1)x = 1.$$

A similar verification shows that $(1 - yx)(1 + yvx) = 1$. It follows that $1 - yx$ is invertible and its inverse is $1 + yvx$.   □

Example. Prove that if in a ring R (not necessarily with identity element) $x^3 = x$ for all $x \in R$, then the ring is commutative.

Solution. For $x, y \in R$, we have

$$xy^2 - y^2xy^2 = (xy^2 - y^2xy^2)^3 = xy^2xy^2xy^2 - xy^2xy^2y^2xy^2 - xy^2y^2xy^2xy^2$$
$$- y^2xy^2xy^2xy^2 + y^2xy^2xy^2y^2xy^2 + y^2xy^2y^2xy^2xy^2$$
$$- y^2xy^2y^2xy^2y^2xy^2 + xy^2y^2xy^2y^2xy^2.$$

Using the fact that $y^4 = y^2$, we see that this is equal to zero, and hence $xy^2 - y^2xy^2 = 0$, that is, $xy^2 = y^2xy^2$. A similar argument shows that $y^2x = y^2xy^2$, and so $xy^2 = y^2x$ for all $x, y \in R$.

Using this we obtain

$$xy = xyxyxy = xy(xy)^2 = x(xy)^2y = x^2yxy^2 = y^3x^3 = yx.$$

This proves that the ring is commutative, as desired.   □

We remark that both this and the third problem below are particular cases of the following result by N. Jacobson:

Jacobson theorem. If a ring (with or without identity) has the property that for every element x there exists an integer $n(x) > 1$ such that $x^{n(x)} = x$, then the ring is commutative.

Try your hand at the following problems.

354. Let a, b, c be elements of a ring with identity.

(a) Show that if $I_n - abc$ is invertible, then $I_n - cab$ is invertible.

(b) Can it happen that $I_n - abc$ is invertible but $I_n - cba$ is not?

355. Let R be a nontrivial ring with identity, and $M = \{x \in R \mid x = x^2\}$ the set of its idempotents. Prove that if M is finite, then it has an even number of elements.

356. Let R be a ring with identity such that $x^6 = x$ for all $x \in R$. Prove that $x^2 = x$ for all $x \in R$. Prove that any such ring is commutative.

357. Let R be a ring with identity with the property that $(xy)^2 = x^2 y^2$ for all $x, y \in R$. Show that R is commutative.

358. Let R be a finite ring with unit, having n elements and such that the equation $x^n = 1$ has the unique solution $x = 1$ in R. Prove that

(a) 0 is the unique nilpotent element of R;

(b) there is a positive integer $k \geq 2$ such that the equation x^k has n solutions in R.

($x \in R$ is called nilpotent if there is a positive integer m such that $x^m = 0$.)

359. Let R be a finite ring such that $1 + 1 = 0$. Prove that the number of solutions to the equation $x^2 = 0$ is equal to the number of solutions to the equation $x^2 = 1$.

360. Let x and y be elements in a ring with identity and n a positive integer. Prove that if $1 - (xy)^n$ is invertible, then so is $1 - (yx)^n$.

361. Let R be a ring with the property that if $x \in R$ and $x^2 = 0$, then $x = 0$.

(a) Prove that if $x, z \in R$ and $z^2 = z$, then $zxz - xz = 0$.

(b) Prove that any idempotent of R belongs to the center of R (the center of a ring consists of those elements that commute with all elements of the ring).

362. Show that if a ring R with identity has three elements a, b, c such that

(i) $ab = ba, bc = cb$;

(ii) for any $x, y \in R$, $bx = by$ implies $x = y$;

(iii) $ca = b$ but $ac \neq b$,

then the ring cannot be finite.

3

Real Analysis

The chapter on real analysis groups material covering differential and integral calculus, ordinary differential equations, and also a rigorous introduction to real analysis with $\varepsilon - \delta$ proofs.

We found it natural, and also friendly, to begin with sequences. As you will discover, the theory of linear recurrences parallels that of linear ordinary differential equations. The theory of limits is well expanded, covering for example Cauchy's criterion for convergence, the convergence of bounded monotone sequences, the Cesàro-Stolz theorem, and Cantor's nested intervals theorem. It is followed by some problems about series, with particular attention given to the telescopic method for computing sums and products.

A long discussion is devoted to one-variable functions. You might find the sections on limits, continuity, and the intermediate value property rather theoretical. Next, you will be required to apply derivatives and their properties to a wide range of examples. Then come integrals, with emphasis placed on computations and inequalities. One-variable real analysis ends with Taylor and Fourier series.

From multivariable differential and integral calculus we cover partial derivatives and their applications, computations of integrals, focusing on change of variables and on Fubini's theorem, all followed by a section of geometric flavor devoted to Green's theorem, the Kelvin-Stokes theorem, and the Gauss-Ostrogradsky (divergence) theorem.

The chapter concludes with functional equations, among which will be found Cauchy's equation, and with ordinary differential and integral equations.

This is a long chapter, with many challenging problems. Now, as you start it, think of T. Edison's words: "Opportunity is missed by many people because it is dressed in overalls and looks like work."

© Springer International Publishing AG 2017
R. Gelca and T. Andreescu, *Putnam and Beyond*, DOI 10.1007/978-3-319-58988-6_3

3.1 Sequences and Series

3.1.1 Search for a Pattern

In this section we train guessing. In each problem you should try particular cases until you guess either the general term of a sequence, a relation that the terms satisfy, or an appropriate construction. The idea to write such a section came to us when we saw the following Putnam problem.

Example. Consider the sequence $(u_n)_n$ defined by $u_0 = u_1 = u_2 = 1$, and

$$\det \begin{pmatrix} u_{n+3} & u_{n+2} \\ u_{n+1} & u_n \end{pmatrix} = n!, \ n \geq 0.$$

Prove that u_n is an integer for all n.

Solution. The recurrence relation of the sequence is

$$u_{n+3} = \frac{u_{n+2}u_{n+1}}{u_n} + \frac{n!}{u_n}.$$

Examining some terms:

$$u_3 = \frac{1 \cdot 1}{1} + \frac{1}{1} = 2,$$

$$u_4 = \frac{2 \cdot 1}{1} + \frac{1}{1} = 3,$$

$$u_5 = \frac{3 \cdot 2}{1} + \frac{2}{1} = 4 \cdot 2,$$

$$u_6 = \frac{4 \cdot 2 \cdot 3}{2} + \frac{3 \cdot 2}{2} = 4 \cdot 3 + 1 \cdot 3 = 5 \cdot 3,$$

$$u_7 = \frac{5 \cdot 3 \cdot 4 \cdot 2}{3} + \frac{4 \cdot 3 \cdot 2}{3} = 5 \cdot 4 \cdot 2 + 4 \cdot 2 = 6 \cdot 4 \cdot 2,$$

$$u_8 = \frac{6 \cdot 4 \cdot 2 \cdot 5 \cdot 3}{4 \cdot 2} + \frac{5 \cdot 4 \cdot 3 \cdot 2}{4 \cdot 2} = 6 \cdot 5 \cdot 3 + 5 \cdot 3 = 7 \cdot 5 \cdot 3,$$

we conjecture that

$$u_n = (n-1)(n-3)(n-5) \cdots .$$

This formula can be proved by induction. Assuming the formula true for u_n, u_{n+1}, and u_{n+2}, we obtain

$$u_{n+3} = \frac{u_{n+2}u_{n+1} + n!}{u_n} = \frac{(n+1)(n-1)(n-3) \cdots n(n-2)(n-4) \cdots + n!}{(n-1)(n-3)(n-5) \cdots}$$

$$= \frac{(n+1) \cdot n! + n!}{(n-1)(n-3)(n-5) \cdots} = \frac{(n+2)n!}{(n-1)(n-3)(n-5) \cdots}$$

$$= (n+2)n(n-2)(n-4) \cdots$$

This completes the induction, and the problem is solved.   □

363. Find a formula for the general term of the sequence

$$1, 2, 2, 3, 3, 3, 4, 4, 4, 4, 5, 5, 5, 5, 5, \ldots$$

364. Find a formula in compact form for the general term of the sequence defined recursively by $x_1 = 1$, $x_n = x_{n-1} + n$ if n is odd, and $x_n = x_{n-1} + n - 1$ if n is even.

365. Define the sequence $(a_n)_{n \geq 0}$ by $a_0 = 0$, $a_1 = 1$, $a_2 = 2$, $a_3 = 6$, and

$$a_{n+4} = 2a_{n+3} + a_{n+2} - 2a_{n+1} - a_n, \text{ for } n \geq 0.$$

Prove that n divides a_n for all $n \geq 1$.

366. Let $n > 1$ be an integer. Find, with proof, all sequences $x_1 < x_2 < \cdots < x_{n-1}$ of positive integers with the following two properties:

(i) $x_i + x_{n-i} = 2n$ for all $i = 1, 2, \ldots, n - 1$;
(ii) for every not necessarily distinct indices i and j for which $x_i + x_j < 2n$, there is an index k such that $x_i + x_j = x_k$.

367. The sequence $a_0, a_1, \ldots, a_2, \ldots$ satisfies

$$a_{m+n} + a_{m-n} = \frac{1}{2}(a_{2m} + a_{2n}),$$

for all nonnegative integers m and n with $m \geq n$. If $a_1 = 1$, determine a_n.

368. Consider the sequences $(a_n)_n$, $(b_n)_n$, defined by

$$a_0 = 0, \ a_1 = 2, \ a_{n+1} = 4a_n + a_{n-1}, \qquad n \geq 0,$$
$$b_0 = 0, \ b_1 = 1, \ b_{n+1} = a_n - b_n + b_{n-1}, \ n \geq 0.$$

Prove that $(a_n)^3 = b_{3n}$ for all n.

369. A sequence u_n is defined by

$$u_0 = 2, \ u_1 = \frac{5}{2}, \ u_{n+1} = u_n(u_{n-1}^2 - 2) - u_1, \text{ for } n \geq 1.$$

Prove that for all positive integers n,

$$\lfloor u_n \rfloor = 2^{(2^n - (-1)^n)/3},$$

where $\lfloor \cdot \rfloor$ denotes the greatest integer function.

370. Consider the sequences $(a_n)_n$ and $(b_n)_n$ defined by $a_1 = 3$, $b_1 = 100$, $a_{n+1} = 3^{a_n}$, $b_{n+1} = 100^{b_n}$. Find the smallest number m for which $b_m > a_{100}$.

3.1.2 Linear Recursive Sequences

In this section we give an overview of the theory of linear recurrence relations with constant coefficients. You should notice the analogy with the theory of ordinary differential equations. This is not an accident, since linear recurrence relations are discrete approximations of differential equations.

A kth-order linear recurrence relation with constant coefficients is a relation of the form

$$x_n = a_1 x_{n-1} + a_2 x_{n-2} + \cdots + a_k x_{n-k}, \ n \geq k,$$

satisfied by a sequence $(x_n)_{n \geq 0}$.

The sequence $(x_n)_n$ is completely determined by $x_0, x_1, \ldots, x_{k-1}$ (the initial condition). To find the formula for the general term, we introduce the vector-valued first-order linear recursive sequence $\mathbf{v}_n = (v_n^1, v_n^2, \ldots, v_n^k)$ defined by $v_n^1 = x_{n+k-1}, v_n^2 = x_{n+k-2}, \ldots, v_n^k = x_n$. This new sequence satisfies the recurrence relation $\mathbf{v}_{n+1} = A\mathbf{v}_n, n \geq 0$, where

$$A = \begin{pmatrix} a_1 & a_2 & a_3 & \ldots & a_{k-1} & a_k \\ 1 & 0 & 0 & \ldots & 0 & 0 \\ 0 & 1 & 0 & \ldots & 0 & 0 \\ 0 & 0 & 1 & \ldots & 0 & 0 \\ \vdots & \vdots & \vdots & \ddots & \vdots & \vdots \\ 0 & 0 & 0 & \ldots & 1 & 0 \end{pmatrix}.$$

It follows that $\mathbf{v}_n = A^n \mathbf{v}_0$, and the problem reduces to the computation of the nth power of A. A standard method employs the Jordan canonical form.

First, we determine the eigenvalues of A. The characteristic polynomial is

$$P_A(\lambda) = \begin{vmatrix} \lambda - a_1 & -a_2 & -a_3 & \ldots & -a_{k-1} & -a_k \\ -1 & \lambda & 0 & \ldots & 0 & 0 \\ 0 & -1 & \lambda & \ldots & 0 & 0 \\ 0 & 0 & -1 & \ldots & 0 & 0 \\ \vdots & \vdots & \vdots & \ddots & \vdots & \vdots \\ 0 & 0 & 0 & \ldots & -1 & \lambda \end{vmatrix}.$$

When expanding by the first row it is easy to remark that all minors are triangular, so the determinant is equal to $\lambda^k - a_1 \lambda^{k-1} - a_2 \lambda^{k-2} - \cdots - a_k$. The equation

$$P_A(\lambda) = \lambda^k - a_1 \lambda^{k-1} - a_2 \lambda^{k-2} - \cdots - a_k = 0$$

is called the characteristic equation of the recursive sequence.

Let $\lambda_1, \lambda_2, \ldots, \lambda_k$ be the roots of the characteristic equation, which are, in fact, the eigenvalues of A. If these roots are all distinct, the situation encountered most often, then A is diagonalizable. There exists an invertible matrix S such that $A = SDS^{-1}$, where D is diagonal with diagonal entries equal to the eigenvalues of A. From the equality

$$\mathbf{v}_n = SD^n S^{-1} \mathbf{v}_0,$$

we conclude that the entries of $\mathbf{v}_n$ are linear combinations of $\lambda_1^n, \lambda_2^n, \ldots, \lambda_k^n$. In particular, for x_n, which is the first coordinate of $\mathbf{v}_n$, there exist constants $\alpha_1, \alpha_2, \ldots, \alpha_k$ such that

$$x_n = \alpha_1 \lambda_1^n + \alpha_2 \lambda_2^n + \cdots + \alpha_k \lambda_k^n, \text{ for } n \geq 0.$$

The numbers $\alpha_1, \alpha_2, \ldots, \alpha_k$ are found from the initial condition, by solving the linear system

$$\alpha_1 + \alpha_2 + \cdots + \alpha_k = x_0,$$
$$\lambda_1 \alpha_1 + \lambda_2 \alpha_2 + \cdots + \lambda_k \alpha_k = x_1,$$
$$\lambda_1^2 \alpha_1 + \lambda_2^2 \alpha_2 + \cdots + \lambda_k^2 \alpha_k = x_2,$$
$$\cdots$$
$$\lambda_1^{k-1} \alpha_1 + \lambda_2^{k-1} \alpha_2 + \cdots + \lambda_k^{k-1} \alpha_k = x_{k-1}.$$

Note that the determinant of the coefficient matrix is Vandermonde, so the system has a unique solution!

If the roots of the characteristic equation have multiplicities greater than 1, it might happen that A is not diagonalizable. The Jordan canonical form of A has blocks of the form

$$J_m(\lambda_i) = \begin{pmatrix} \lambda_i & 1 & 0 & \ldots & 0 \\ 0 & \lambda_i & 1 & \ldots & 0 \\ 0 & 0 & \lambda_i & \ldots & 0 \\ \vdots & \vdots & \vdots & \ddots & \vdots \\ 0 & 0 & 0 & \ldots & \lambda_i \end{pmatrix}.$$

An exercise in Section 2.3.1 shows that for $j \geq i$, the entry of $J_m(\lambda_i)^n$ is $\binom{n}{j-i} \lambda_i^{n+i-j}$. We conclude that if the roots of the characteristic equations are $\lambda_1, \lambda_2, \ldots, \lambda_t$ and $m_1, m_2, \ldots, m_t$ their respective multiplicities, then there exist constants $\alpha_{ij}, i = 1, 2, \ldots, t, j = 0, 1, \ldots, m_i - 1$, such that

$$x_n = \sum_{i=1}^{t} \sum_{j=0}^{m_i-1} \alpha_{ij} \binom{n}{j} \lambda_i^{n-j}, \text{ for } n \geq 0.$$

It might be more useful to write this as

$$x_n = \sum_{i=1}^{t} \sum_{j=0}^{m_i} \beta_{ij} n^h \lambda_i^{n-j}, \text{ for } n \geq 0.$$

As is the case with differential equations, to find the general term of an inhomogeneous linear recurrence relation

$$x_n = a_1 x_{n-1} + a_2 x_{n-2} + \cdots + a_k x_{n-k} + f(n), \ n \geq 1,$$

one has to find a particular solution to the recurrence, then add to it the general term of the associated homogeneous recurrence relation.

Putting these ideas together, let us compute the general-term formula of the Fibonacci sequence. The recurrence relation $F_{n+1} = F_n + F_{n-1}$ has characteristic equation $\lambda^2 - \lambda - 1 = 0$, with roots $\lambda_1 = \frac{1-\sqrt{5}}{2}$ and $\lambda_2 = \frac{1+\sqrt{5}}{2}$. Writing $F_n = \alpha_1 \lambda_1^n + \alpha_2 \lambda_2^n$ and solving the system

$$\alpha_1 + \alpha_2 = F_0 = 0,$$
$$\alpha_1 \lambda_1 + \alpha_2 \lambda_2 = F_1 = 1,$$

we obtain $\alpha_1 = -\alpha_2 = -\dfrac{1}{\sqrt{5}}$. We rediscover the well-known Binet formula

$$F_n = \frac{1}{\sqrt{5}} \left(\left(\frac{1+\sqrt{5}}{2} \right)^n - \left(\frac{1-\sqrt{5}}{2} \right)^n \right).$$

In the same vein, let us solve a problem published in the *American Mathematical Monthly* by I. Tomescu.

Example. In how many ways can one tile a $2n \times 3$ rectangle with 2×1 tiles?

Solution. Denote by u_n the number of such tilings. Start tiling the rectangle from the short side of length 3, as shown in Figure 17.

Figure 17

In the last two cases from the figure, an uncovered 1×1 square can be covered in a single way: by the shaded rectangle. We thus obtain

$$u_{n+1} = 3u_n + 2v_n,$$

where v_n is the number of tilings of a $(2n - 1) \times 3$ rectangle with a 1×1 square missing in one corner, like the one in Figure 18. That figure shows how to continue tiling this kind of rectangle, giving rise to the recurrence

$$v_{n+1} = u_n + v_n.$$

Combining the two, we obtain the (vector-valued) recurrence relation

$$\begin{pmatrix} u_{n+1} \\ v_{n+1} \end{pmatrix} = \begin{pmatrix} 3 & 2 \\ 1 & 1 \end{pmatrix} \begin{pmatrix} u_n \\ v_n \end{pmatrix}.$$

The characteristic equation, of the coefficient matrix but also of the sequences u_n and v_n, is

$$\begin{vmatrix} \lambda - 3 & -2 \\ -1 & \lambda - 1 \end{vmatrix} = \lambda^2 - 4\lambda + 1 = 0.$$

Its roots are $\lambda_{1,2} = 2 \pm \sqrt{3}$. We compute easily $u_1 = 3$ and $v_1 = 1$, so $u_2 = 3 \cdot 3 + 2 \cdot 1 = 11$. The desired general-term formula is then

$$u_n = \frac{1}{2\sqrt{3}} \left(\left(\sqrt{3} + 1 \right) \left(2 + \sqrt{3} \right)^n + \left(\sqrt{3} - 1 \right) \left(2 - \sqrt{3} \right)^n \right). \qquad \square$$

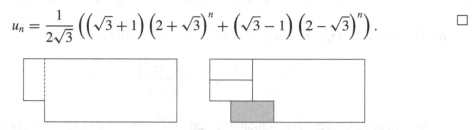

Figure 18

Below are listed more problems of this kind.

371. Let $p(x) = x^2 - 3x + 2$. Show that for any positive integer n there exist unique numbers a_n and b_n such that the polynomial $q_n(x) = x^n - a_n x - b_n$ is divisible by $p(x)$.

372. Find the general term of the sequence given by $x_0 = 3$, $x_1 = 4$, and

$$(n+1)(n+2)x_n = 4(n+1)(n+3)x_{n-1} - 4(n+2)(n+3)x_{n-2}, \ n \geq 2.$$

373. Let $(x_n)_{n\geq 0}$ be defined by the recurrence relation $x_{n+1} = ax_n + bx_{n-1}$, with $x_0 = 0$. Show that the expression $x_n^2 - x_{n-1}x_{n+1}$ depends only on b and x_1, but not on a.

374. Define the sequence $(a_n)_n$ recursively by $a_1 = 1$ and

$$a_{n+1} = \frac{1 + 4a_n + \sqrt{1 + 24a_n}}{16}, \ \text{for } n \geq 1.$$

Find an explicit formula for a_n in terms of n.

375. Let $a = 4k - 1$, where k is an integer. Prove that for any positive integer n the number

$$1 - \binom{n}{2}a + \binom{n}{4}a^2 - \binom{n}{6}a^3 + \cdots$$

is divisible by 2^{n-1}.

376. Let A and E be opposite vertices of a regular octagon. A frog starts jumping at vertex A. From any vertex of the octagon except E, it may jump to either of the two adjacent vertices. When it reaches vertex E, the frog stops and stays there. Let a_n be the number of distinct paths of exactly n jumps ending at E. Prove that $a_{2n-1} = 0$ and

$$a_{2n} = \frac{1}{\sqrt{2}}(x^{n-1} - y^{n-1}), \ n = 1, 2, 3, \ldots$$

where $x = 2 + \sqrt{2}$ and $y = 2 - \sqrt{2}$.

377. Find all functions $f : \mathbb{N} \to \mathbb{N}$ satisfying

$$f(f(f(n))) + 6f(n) = 3f(f(n)) + 4n + 2001, \text{ for all } n \in \mathbb{N}.$$

378. The sequence $(x_n)_n$ is defined by $x_1 = 4$, $x_2 = 19$, and for $n \geq 2$,

$$x_{n+1} = \left\lceil \frac{x_n^2}{x_{n-1}} \right\rceil,$$

the smallest integer greater than or equal to $\frac{x_n^2}{x_{n-1}}$. Prove that $x_n - 1$ is always a multiple of 3.

379. Consider the sequences given by

$$a_0 = 1, \; a_{n+1} = \frac{3a_n + \sqrt{5a_n^2 - 4}}{2}, \quad n \geq 1,$$
$$b_0 = 0, \; b_{n+1} = a_n - b_n, \qquad\qquad n \geq 1.$$

Prove that $(a_n)^2 = b_{2n+1}$ for all n.

3.1.3 Limits of Sequences

There are three methods for determining the limit of a sequence. The first of them is based on the following definition.

Cauchy's definition. (a) *A sequence $(x_n)_n$ converges to a finite limit L if and only if for every $\varepsilon > 0$ there exists $n(\varepsilon)$ such that for every $n > n(\varepsilon)$, $|x_n - L| < \varepsilon$.*

(b) *A sequence $(x_n)_n$ tends to infinity if for every $\varepsilon > 0$ there exists $n(\varepsilon)$ such that for $n > n(\varepsilon)$, $x_n > \varepsilon$.*

The definition of convergence is extended to $\mathbb{R}^n$, and in general to any metric space, by replacing the absolute value with the distance. The second method for finding the limit is called the squeezing principle.

The squeezing principle. (a) *If $a_n \leq b_n \leq c_n$ for all n, and if $(a_n)_n$ and $(c_n)_n$ converge to the finite limit L, then $(b_n)_n$ also converges to L.*

(b) *If $a_n \leq b_n$ for all n and if $(a_n)_n$ tends to infinity, then $(b_n)_n$ also tends to infinity.*

Finally, the third method reduces the problem via algebraic operations to sequences whose limits are known. We illustrate each method with an example. The first is from P.N. de Souza, J.N. Silva, *Berkeley Problems in Mathematics* (Springer, 2004).

Example. Let $(x_n)_n$ be a sequence of real numbers such that

$$\lim_{n \to \infty} (2x_{n+1} - x_n) = L.$$

Prove that the sequence $(x_n)_n$ converges and its limit is L.

Solution. By hypothesis, for every ε there is $n(\varepsilon)$ such that if $n \geq n(\varepsilon)$, then

$$L - \varepsilon < 2x_{n+1} - x_n < L + \varepsilon.$$

For such n and some $k > 0$ let us add the inequalities

$$L - \varepsilon < 2x_{n+1} - x_n < L + \varepsilon,$$
$$2(L - \varepsilon) < 4x_{n+2} - 2x_{n+1} < 2(L + \varepsilon),$$
$$\cdots$$
$$2^{k-1}(L - \varepsilon) < 2^k x_{n+k} - 2^{k-1}x_{n+k-1} < 2^{k-1}(L + \varepsilon).$$

We obtain

$$(1 + 2 + \cdots + 2^{k-1})(L - \varepsilon) < 2^k x_{n+k} - x_n < (1 + 2 + \cdots + 2^{k-1})(L + \varepsilon),$$

which after division by 2^k becomes

$$\left(1 - \frac{1}{2^k}\right)(L - \varepsilon) < x_{n+k} - \frac{1}{2^k}x_n < \left(1 - \frac{1}{2^k}\right)(L + \varepsilon).$$

Now choose k such that $\left|\frac{1}{2^k}x_n\right| < \varepsilon$ and $\left|\frac{1}{2^k}(L \pm \varepsilon)\right| < \varepsilon$. Then for $m \geq n + k$,

$$L - 3\varepsilon < x_m < L + 3\varepsilon,$$

and since ε was arbitrary, this implies that $(x_n)_n$ converges to L. $\qquad\square$

Example. Prove that $\lim\limits_{n\to\infty} \sqrt[n]{n} = 1$.

Solution. The sequence $x_n = \sqrt[n]{n} - 1$ is clearly positive, so we only need to bound it from above by a sequence converging to 0. For that we employ the binomial expansion

$$n = (1 + x_n)^n = 1 + \binom{n}{1}x_n + \binom{n}{2}x_n^2 + \cdots + \binom{n}{n-1}x_n^{n-1} + x_n^n.$$

Forgetting all terms but the third in this expansion, we can write

$$n > \binom{n}{2}x_n^2,$$

which translates to $x_n < \sqrt{\frac{2}{n-1}}$, for $n \geq 2$. The sequence $\sqrt{\frac{2}{n-1}}$, $n \geq 2$, converges to 0, and hence by the squeezing principle, $(x_n)_n$ itself converges to 0, as desired. $\qquad\square$

The third example was published by the Romanian mathematician T. Lalescu in 1901 in the *Mathematics Gazette, Bucharest.*

Example. Prove that the sequence $a_n = \sqrt[n+1]{(n+1)!} - \sqrt[n]{n!}$, $n \geq 1$, is convergent and find its limit.

Solution. The solution we present belongs to M. Ţena. It uses Stirling's formula

$$n! = \sqrt{2\pi n}\left(\frac{n}{e}\right)^n \cdot e^{\frac{\theta_n}{12n}}, \quad \text{with } 0 < \theta_n < 1,$$

which will be proved in Section 3.2.12. Taking the nth root and passing to the limit, we obtain

$$\lim_{n\to\infty} \frac{n}{\sqrt[n]{n!}} = e.$$

We also deduce that

$$\lim_{n\to\infty} \frac{n+1}{\sqrt[n]{n!}} = \lim_{n\to\infty} \frac{n+1}{n}\cdot\frac{n}{\sqrt[n]{n!}} = e.$$

Therefore,

$$\lim_{n\to\infty}\left(\frac{\sqrt[n+1]{(n+1)!}}{\sqrt[n]{n!}}\right)^n = \lim_{n\to\infty}\left(\sqrt[n(n+1)]{\frac{((n+1)!)^n}{(n!)^{n+1}}}\right)^n = \lim_{n\to\infty}\left(\sqrt[n(n+1)]{\frac{(n+1)^n}{n!}}\right)^n$$

$$= \lim_{n\to\infty}\left(\sqrt[n+1]{\frac{n+1}{\sqrt[n]{n!}}}\right)^n = \lim_{n\to\infty}\left(\frac{n+1}{\sqrt[n]{n!}}\right)^{\frac{n}{n+1}}$$

$$= \left(\lim_{n\to\infty}\frac{n+1}{\sqrt[n]{n!}}\right)^{\lim_{n\to\infty}\frac{n}{n+1}} = e.$$

Taking the nth root and passing to the limit, we obtain

$$\lim_{n\to\infty}\frac{\sqrt[n+1]{(n+1)!}}{\sqrt[n]{n!}} = 1,$$

and hence

$$\lim_{n\to\infty}\frac{a_n}{\sqrt[n]{n!}} = \lim_{n\to\infty}\frac{\sqrt[n+1]{(n+1)!}}{\sqrt[n]{n!}} - 1 = 0.$$

Thus, if we set

$$b_n = \left(1 + \frac{a_n}{\sqrt[n]{n!}}\right)^{\frac{\sqrt[n]{n!}}{a_n}},$$

then $\lim_{n\to\infty} b_n = e$. From the equality

$$\left(\frac{\sqrt[n+1]{(n+1)!}}{\sqrt[n]{n!}}\right) = b_n^{a_n\frac{n}{\sqrt[n]{n!}}},$$

we obtain

$$a_n = \ln\left(\frac{\sqrt[n+1]{(n+1)!}}{\sqrt[n]{n!}}\right)^n (\ln b_n)^{-1}\left(\frac{n}{\sqrt[n]{n!}}\right)^{-1}.$$

The right-hand side is a product of three sequences that converge, respectively, to $1 = \ln e$, $1 = \ln e$, and $\frac{1}{e}$. Therefore, the sequence $(a_n)_n$ converges to the limit $\frac{1}{e}$. $\square$

Apply these methods to the problems below.

380. Find the limit of the sequence $x_0 = 0$, $x_1 = 1$, $x_{n+1} = \frac{1}{2}(x_n + x_{n-1})$.

381. Compute

$$\lim_{n \to \infty} \left| \sin \left(\pi \sqrt{n^2 + n + 1} \right) \right|.$$

382. Compute $\lim_{n \to \infty} \{(\sqrt{2} + 1)^{2n}\}$ where $\{a\}$ denotes the fractional part of a, i.e. $\{a\} = a - \lfloor a \rfloor$ (for example the fractional part of 1.32 is 0.32).

383. Let k be a positive integer and μ a positive real number. Prove that

$$\lim_{n \to \infty} \binom{n}{k} \left(\frac{\mu}{n} \right)^k \left(1 - \frac{\mu}{n} \right)^{n-k} = \frac{\mu^k}{e^\mu \cdot k!}.$$

384. Let $(x_n)_n$ be a sequence of positive integers such that $x_{x_n} = n^4$ for all $n \geq 1$. It is true that $\lim_{n \to \infty} x_n = \infty$?

385. Let a and b be integers such that $a \cdot 2^n + b$ is a perfect square for all positive integers n. Prove that $a = 0$.

386. Let a, b, c be integers with $a \neq 0$ such that

$$an^2 + bn + c$$

is a perfect square for any positive integer n. Prove that there exist integers x and y such that $a = x^2$, $b = 2xy$, $c = y^2$.

387. Let $(a_n)_n$ be a sequence of real numbers with the property that for any $n \geq 2$ there exists an integer k, $\frac{n}{2} \leq k < n$, such that $a_n = \frac{a_k}{2}$. Prove that $\lim_{n \to \infty} a_n = 0$.

388. Given two natural numbers k and m let $a_1, a_2, \ldots, a_k, b_1, b_2, \ldots, b_m$ be positive numbers such that

$$\sqrt[n]{a_1} + \sqrt[n]{a_2} + \cdots + \sqrt[n]{a_k} = \sqrt[n]{b_1} + \sqrt[n]{b_2} + \cdots + \sqrt[n]{b_m},$$

for all positive integers n. Prove that $k = m$ and $a_1 a_2 \cdots a_k = b_1 b_2 \cdots b_m$.

389. Prove that

$$\lim_{n \to \infty} n^2 \int_0^{\frac{1}{n}} x^{x+1} dx = \frac{1}{2}.$$

390. Let a be a positive real number and $(x_n)_{n \geq 1}$ a sequence of real numbers such that $x_1 = a$ and

$$x_{n+1} \geq (n+2)x_n - \sum_{k=1}^{n-1} kx_k, \quad \text{for all } n \geq 1.$$

Find the limit of the sequence.

391. Let $(x_n)_{n \geq 1}$ be a sequence of real numbers satisfying

$$x_{n+m} \leq x_n + x_m, \quad n, m \geq 1.$$

Show that $\lim_{n \to \infty} \frac{x_n}{n}$ exists and is equal to $\inf_{n \geq 1} \frac{x_n}{n}$.

392. Compute

$$\lim_{n \to \infty} \sum_{k=1}^{n} \left(\frac{k}{n^2} \right)^{\frac{k}{n^2}+1}.$$

393. Let b be an integer greater than 5. For each positive integer n, consider the number

$$x_n = \underbrace{11 \ldots 1}_{n-1} \underbrace{22 \ldots 2}_{n} 5,$$

written in base b. Prove that the following condition holds if and only if $b = 10$:
There exists a positive integer M such that for any integer n greater than M, the number x_n is a perfect square.

We exhibit two criteria for proving that a sequence is convergent without actually computing the limit. The first is due to K. Weierstrass.

Weierstrass' theorem. *A monotonic bounded sequence of real numbers is convergent.*

Below are some instances in which this theorem is used.

394. Prove that the sequence $(a_n)_{n \geq 1}$ defined by

$$a_n = 1 + \frac{1}{2} + \frac{1}{3} + \cdots + \frac{1}{n} - \ln(n+1), \quad n \geq 1,$$

is convergent.

395. Prove that the sequence

$$a_n = \sqrt{1 + \sqrt{2 + \sqrt{3 + \cdots + \sqrt{n}}}}, \quad n \geq 1,$$

is convergent.

396. Let $(a_n)_n$ be a sequence of real numbers that satisfies the recurrence relation

$$a_{n+1} = \sqrt{a_n^2 + a_n - 1}, \quad \text{for } n \geq 1.$$

Prove that $a_1 \notin (-2, 1)$.

397. Using the Weierstrass theorem, prove that any bounded sequence of real numbers has a convergent subsequence.

Widely used in higher mathematics is the following convergence test.

Cauchy's criterion for convergence. *A sequence $(x_n)_n$ of points in $\mathbb{R}^n$ (or, in general, in a complete metric space) is convergent if and only if for any $\varepsilon > 0$ there is a positive integer n_ε such that whenever $n, m \geq n_\varepsilon$, $\|x_n - x_m\| < \varepsilon$.*

A sequence satisfying this property is called Cauchy, and it is the completeness of the space (the fact that it has no gaps) that forces a Cauchy sequence to be convergent. This property is what essentially distinguishes the set of real numbers from the rationals. In fact, the set of real numbers can be defined as the set of Cauchy sequences of rational numbers, with two such sequences identified if the sequence formed from alternating terms of the two sequences is also Cauchy.

398. Let $(a_n)_{n \geq 1}$ be a decreasing sequence of positive numbers converging to 0. Prove that the series $S = a_1 - a_2 + a_3 - a_4 + \cdots$ is convergent.

399. Let a_0, b_0, c_0 be real numbers. Define the sequences $(a_n)_n$, $(b_n)_n$, $(c_n)_n$ recursively by

$$a_{n+1} = \frac{a_n + b_n}{2}, \quad b_{n+1} = \frac{b_n + c_n}{2}, \quad c_{n+1} = \frac{c_n + a_n}{2}, \quad n \geq 0.$$

Prove that the sequences are convergent and find their limits.

400. Show that if the series $\sum a_n$ converges, where $(a_n)_n$ is a decreasing sequence, then $\lim_{n \to \infty} n a_n = 0$.

The following fixed point theorem is a direct application of Cauchy's criterion for convergence.

Theorem. *Let X be a closed subset of $\mathbb{R}^n$ (or in general of a complete metric space) and $f : X \to X$ a function with the property that $\|f(x) - f(y)\| \leq c\|x - y\|$ for any $x, y \in X$, where $0 < c < 1$ is a constant. Then f has a unique fixed point in X.*

Such a function is called contractive. Recall that a set is closed if it contains all its limit points.

Proof. Let $x_0 \in X$. Recursively define the sequence $x_n = f(x_{n-1})$, $n \geq 1$. Then

$$\|x_{n+1} - x_n\| \leq c\|x_n - x_{n-1}\| \leq \cdots \leq c^n\|x_1 - x_0\|.$$

Applying the triangle inequality, we obtain

$$\|x_{n+p} - x_n\| \leq \|x_{n+p} - x_{n+p-1}\| + \|x_{n+p-1} - x_{n+p-2}\| + \cdots + \|x_{n+1} - x_n\|$$
$$\leq (c^{n+p-1} + c^{n+p-2} + \cdots + c^n)\|x_1 - x_0\|$$
$$= c^n(1 + c + \cdots + c^{p-1})\|x_1 - x_0\| \leq \frac{c^n}{1-c}\|x_1 - x_0\|.$$

This shows that the sequence $(x_n)_n$ is Cauchy. Its limit x^* satisfies $f(x^*) = \lim_{n \to \infty} f(x_n) = \lim_{n \to \infty} x_n = x^*$; it is a fixed point of f. A second fixed point y^* would give rise to the contradiction $\|x^* - y^*\| = \|f(x^*) - f(y^*)\| \leq c\|x^* - y^*\|$. Therefore, the fixed point is unique. $\qquad\square$

Use this theorem to solve the next three problems.

401. Two maps of the same region drawn to different scales are superimposed so that the smaller map lies entirely inside the larger. Prove that there is precisely one point on the small map that lies directly over a point on the large map that represents the same place of the region.

402. Let t and ε be real numbers with $|\varepsilon| < 1$. Prove that the equation $x - \varepsilon \sin x = t$ has a unique real solution.

403. Let c and x_0 be fixed positive numbers. Define the sequence

$$x_n = \frac{1}{2}\left(x_{n-1} + \frac{c}{x_{n-1}}\right), \quad \text{for } n \geq 1.$$

Prove that the sequence converges and that its limit is $\sqrt{c}$.

3.1.4 More About Limits of Sequences

We continue our discussion about limits of sequences with three more topics: the method of passing to the limit in a recurrence relation, the Cesàro-Stolz theorem, and Cantor's nested intervals theorem. We illustrate the first with the continued fraction expansion of the golden ratio.

Example. Prove that

$$\frac{1 + \sqrt{5}}{2} = 1 + \cfrac{1}{1 + \cfrac{1}{1 + \cfrac{1}{1 + \cfrac{1}{1 + \cdots}}}}.$$

Solution. A close look at the right-hand side shows that it is the limit of a sequence $(x_n)_n$ subject to the recurrence relation $x_1 = 1, x_{n+1} = 1 + \frac{1}{x_n}$. If this sequence has a finite limit L, then passing to the limit on both sides of the recurrence relation yields $L = 1 + \frac{1}{L}$. Because L can only be positive, it must be equal to the golden ratio.

But does the limit exist? Investigating the first terms of the sequence we see that

$$x_1 < x_3 < \frac{1 + \sqrt{5}}{2} < x_4 < x_2,$$

and we expect the general situation to be

$$x_1 < x_3 < \cdots < x_{2n+1} < \cdots < \frac{1 + \sqrt{5}}{2} < \cdots < x_{2n} < x_{2n-2} < \cdots < x_2.$$

This can be proved by induction. Firstly, if $x_{2n+1} < \frac{1+\sqrt{5}}{2}$, then

$$x_{2n+2} = 1 + \frac{1}{x_{2n+1}} > 1 + \frac{2}{1 + \sqrt{5}} = 1 + \frac{\sqrt{5} - 1}{2} = \frac{1 + \sqrt{5}}{2},$$

and by a similar computation, if $x_{2n+2} > \frac{1+\sqrt{5}}{2}$, then $_{2n+3} < \frac{1+\sqrt{5}}{2}$. Secondly,

$$x_{n+2} = 2 - \frac{1}{x_n + 1},$$

and the inequality $x_{n+2} > x_m$ is equivalent to $x_n^2 - x_n - 1 < 0$, which holds if and only if $x_n < \frac{1+\sqrt{5}}{2}$. Now an inductive argument shows that $(x_{2n+1})_n$ is increasing and $(x_{2n+2})_n$ is decreasing. Being bounded, both sequences are convergent by the Weierstrass theorem. Their limits are positive, and both should satisfy the equation $L = 2 - \frac{1}{L+1}$. The unique positive solution to this equation is the golden ratio, which is therefore the limit of both sequences, and consequently the limit of the sequence $(x_n)_n$. $\square$

Next, we present a famous identity of S.A. Ramanujan.

Example. Prove that

$$\sqrt{1 + 2\sqrt{1 + 3\sqrt{1 + 4\sqrt{1 + \cdots}}}} = 3.$$

Solution. We approach the problem in more generality by introducing the function $f :$ $[1, \infty) \to \mathbb{R}$,

$$f(x) = \sqrt{1 + x\sqrt{1 + (x+1)\sqrt{1 + (x+2)\sqrt{1 + \cdots}}}}.$$

Is this function well defined? Truncating to n square roots, we obtain an increasing sequence. All we need to show is that this sequence is bounded from above. And it is, because

$$f(x) \leq \sqrt{(x+1)\sqrt{(x+2)\sqrt{(x+3)\cdots}}}$$

$$\leq \sqrt{2x\sqrt{3x\sqrt{4x\cdots}}} \leq \sqrt{2x\sqrt{4x\sqrt{8x\cdots}}}$$

$$= 2^{\sum \frac{k}{2^k}} x^{\sum \frac{1}{2^k}} \leq 2^{\frac{1}{2}+\frac{1}{2}+\frac{1}{4}+\frac{1}{4}+\frac{1}{8}+\frac{1}{8}+\cdots} x = 2x.$$

This shows, moreover, that $f(x) \leq 2x$, for $x \geq 1$. Note also that

$$f(x) \geq \sqrt{x\sqrt{x\sqrt{x\cdots}}} = x.$$

For reasons that will become apparent, we weaken this inequality to $f(x) \geq \frac{1}{2}(x + 1)$. We then square the defining relation and obtain the functional equation

$$(f(x))^2 = xf(x+1) + 1.$$

Combining this with

$$\frac{1}{2}(x + 2) \leq f(x + 1) \leq 2(x + 2),$$

we obtain

$$x \cdot \frac{x+2}{2} + 1 \leq (f(x))^2 \leq 2x(x+2) + 1,$$

which yields the sharper double inequality

$$\frac{1}{\sqrt{2}}(x+1) \le f(x) \le \sqrt{2}(x+1).$$

Repeating successively the argument, we find that

$$2^{-\frac{1}{2^n}}(x+1) \le f(x) \le 2^{\frac{1}{2^n}}(x+1), \ \text{for } n \ge 1.$$

If in this double inequality we let $n \to \infty$, we obtain $x + 1 \le f(x) \le x + 1$, and hence $f(x) = x + 1$. The particular case $x = 2$ yields Ramanujan's formula

$$\sqrt{1 + 2\sqrt{1 + 3\sqrt{1 + 4\sqrt{1 + \cdots}}}} = 3,$$

and we are done. $\square$

Here are some problems of this kind.

404. Compute

$$\sqrt{1 + \sqrt{1 + \sqrt{1 + \sqrt{1 + \cdots}}}}$$

405. Let a and b be real numbers. Prove that the recurrence sequence $(x_n)_n$ defined by $x_1 > 0$ and $x_{n+1} = \sqrt{a + bx_n}$, $n \ge 1$, is convergent, and find its limit.

406. Let $0 < a < b$ be two real numbers. Define the sequences $(a_n)_n$ and $(b_n)_n$ by $a_0 = a$, $b_0 = b$, and

$$a_{n+1} = \sqrt{a_n b_n}, \ b_{n+1} = \frac{a_n + b_n}{2}, \ n \ge 0.$$

Prove that the two sequences are convergent and have the same limit.

407. Prove that for $n \ge 2$, the equation $x^n + x - 1 = 0$ has a unique root in the interval $[0, 1]$. If x_n denotes this root, prove that the sequence $(x_n)_n$ is convergent and find its limit.

408. Compute up to two decimal places the number

$$\sqrt{1 + 2\sqrt{1 + 2\sqrt{1 + \cdots + 2\sqrt{1 + 2\sqrt{1969}}}}},$$

where the expression contains 1969 square roots.

409. Find the positive real solutions to the equation

$$\sqrt{x + 2\sqrt{x + \cdots + 2\sqrt{x + 2\sqrt{3x}}}} = x.$$

410. Show that the sequence

$$\sqrt{7}, \; \sqrt{7-\sqrt{7}}, \; \sqrt{7-\sqrt{7+\sqrt{7}}}, \; \sqrt{7-\sqrt{7+\sqrt{7-\sqrt{7}}}}, \ldots$$

converges, and evaluate its limit.

411. (a) What is

$$\sqrt{4}^{\sqrt{4}^{\sqrt{4}^{\sqrt{4}^{\cdots}}}} \quad ?$$

(b) What is

$$\sqrt{2}^{\sqrt{2}^{\sqrt{2}^{\sqrt{2}^{\cdots}}}} \quad ?$$

(c) For what numbers $a > 1$ is

$$a^{a^{a^{a^{\cdots}}}}$$

a finite number? (In this problem we are evaluating the limit of $(x_n)_n$ defined recursively by $x_1 = a$, $x_{n+1} = a^{x_n}$, $n \geq 1$.)

There is a vocabulary for translating the language of derivatives to the discrete framework of sequences. The first derivative of a sequence $(x_n)_n$, usually called the first difference, is the sequence $(\Delta x_n)_n$ defined by $\Delta x_n = x_{n+1} - x_n$. The second derivative, or second difference, is $\Delta^2 x_n = \Delta(\Delta x_n) = x_{n+2} - 2x_{n+1} + x_n$. A sequence is increasing if the first derivative is positive; it is convex if the second derivative is positive. The Cesàro-Stolz theorem, which we discuss below, is the discrete version of L'Hôpital's theorem.

The Cesàro-Stolz theorem. *Let $(x_n)_n$ and $(y_n)_n$ be two sequences of real numbers with $(y_n)_n$ strictly positive, increasing, and unbounded. If*

$$\lim_{n\to\infty} \frac{x_{n+1} - x_n}{y_{n+1} - y_n} = L,$$

then the limit

$$\lim_{n\to\infty} \frac{x_n}{y_n}$$

exists and is equal to L.

Proof. We apply the same $\varepsilon - \delta$ argument as for L'Hôpital's theorem. We do the proof only for L finite, the cases $L = \pm\infty$ being left to the reader.

Fix $\varepsilon > 0$. There exists n_0 such that for $n \geq n_0$,

$$L - \frac{\varepsilon}{2} < \frac{x_{n+1} - x_n}{y_{n+1} - y_n} < L + \frac{\varepsilon}{2}.$$

Because $y_{n+1} - y_n \geq 0$, this is equivalent to

$$\left(L - \frac{\varepsilon}{2}\right)(y_{n+1} - y_n) < x_{n+1} - x_n < \left(L + \frac{\varepsilon}{2}\right)(y_{n+1} - y_n).$$

We sum all these inequalities for n ranging between n_0 and $m-1$, for some m. After cancelling terms in the telescopic sums that arise, we obtain

$$\left(L - \frac{\varepsilon}{2}\right)(y_m - y_{n_0}) < x_m - x_{n_0} < \left(L + \frac{\varepsilon}{2}\right)(y_m - y_{n_0}).$$

We divide by y_m and write the answer as

$$L - \frac{\varepsilon}{2} + \left(-L\frac{y_{n_0}}{y_m} + \frac{\varepsilon}{2} \cdot \frac{y_{n_0}}{y_m} + \frac{x_{n_0}}{y_m}\right) < \frac{x_m}{y_m} < L + \frac{\varepsilon}{2} + \left(-L\frac{y_{n_0}}{y_m} - \frac{\varepsilon}{2} \cdot \frac{y_{n_0}}{y_m} + \frac{x_{n_0}}{y_m}\right).$$

Because $y_n \to \infty$. there exists $n_1 > n_0$ such that for $m \geq n_1$, the absolute values of the terms in the parentheses are less than $\frac{\varepsilon}{2}$. Hence for $m \geq n_1$,

$$L - \varepsilon < \frac{x_m}{y_m} < L + \varepsilon.$$

Since ε was arbitrary, this proves that the sequence $\left(\dfrac{x_n}{y_n}\right)_n$ converges to L. $\square$

We continue this discussion with an application to Cesàro means. By definition, the Cesàro means of a sequence $(a_n)_{n \geq 1}$ are

$$s_n = \frac{a_1 + a_2 + \cdots + a_n}{n}, \quad n \geq 1.$$

Theorem. *If $(a_n)_{n \geq 1}$ converges to L, then $(s_n)_{n \geq 1}$ also converges to L.*

Proof. Apply the Cesàro-Stolz theorem to the sequences $x_n = a_1 + a_2 + \cdots + a_n$ and $y_n = n$, $n \geq 1$. $\square$

The Cesàro-Stolz theorem can be used to solve the following problems.

412. If $(u_n)_n$ is a sequence of positive real numbers and if $\lim\limits_{n \to \infty} \dfrac{u_{n+1}}{u_n} = u > 0$, then $\lim\limits_{n \to \infty} \sqrt[n]{u_n} = u$.

413. Let p be a real number, $p \neq -1$. Compute

$$\lim_{n \to \infty} \frac{1^p + 2^p + \cdots + n^p}{n^{p+1}}.$$

414. Let $0 < x_0 < 1$ and $x_{n+1} = x_n - x_n^2$ for $n \geq 0$. Compute $\lim\limits_{n \to \infty} n x_n$.

415. Let $x_0 \in [-1, 1]$ and $x_{n+1} = x_n - \arcsin(\sin^2 x_n)$ for $n \geq 0$. Compute $\lim\limits_{n \to \infty} \sqrt{n} x_n$.

416. For an arbitrary number $x_0 \in (0, \pi)$ define recursively the sequence $(x_n)_n$ by

$$x_{n+1} = \sin x_n, \ n \geq 0.$$

Compute $\lim_{n \to \infty} \sqrt{n} x_n$.

417. Let $f : \mathbb{R} \to \mathbb{R}$ be a continuous function such that the sequence $(a_n)_{n \geq 0}$ defined by

$$a_n = \int_0^1 f(n+x)dx$$

is convergent. Prove that the sequence $(b_n)_{n \geq 0}$, with

$$b_n = \int_0^1 f(nx)dx$$

is also convergent.

418. Consider the polynomial

$$P(x) = a_m x^m + a_{m-1} x^{m-1} + \cdots + a_0, \ a_i > 0, \ i = 0, 1, \ldots, m.$$

Denote by A_n and G_n the arithmetic and, respectively, geometric means of the numbers $P(1), P(2), \ldots, P(n)$. Prove that

$$\lim_{n \to \infty} \frac{A_n}{G_n} = \frac{e^m}{m+1}.$$

419. Let k be an integer greater than 1. Suppose $a_0 > 0$, and define

$$a_{n+1} = a_n + \frac{1}{\sqrt[k]{a_n}} \text{ for } n > 0.$$

Evaluate

$$\lim_{n \to \infty} \frac{a_n^{k+1}}{n^k}.$$

We conclude the discussion about limits of sequences with a theorem of G. Cantor.

Cantor's nested intervals theorem. *Given a decreasing sequence of closed intervals* $I_1 \supset I_2 \supset \cdots \supset I_n \supset \cdots$ *with lengths converging to zero, the intersection* $\bigcap_{n=1}^{\infty} I_n$ *consists of exactly one point.*

This theorem is true in general if the intervals are replaced by closed and bounded subsets of $\mathbb{R}^n$ with diameters converging to zero. As an application of this theorem we prove the compactness of a closed bounded interval. A set of real numbers is called compact if from every family of open intervals that cover the set one can choose finitely many that still cover it.

The Heine-Borel theorem. *A closed and bounded interval of real numbers is compact.*

Proof. Let the interval be $[a, b]$ and assume that for some family of open intervals $(I_\alpha)_\alpha$ that covers $[a, b]$ one cannot choose finitely many that still cover it. We apply the dichotomic (division into two parts) method. Cut the interval $[a, b]$ in half. One of the two intervals thus obtained cannot be covered by finitely many I_α's. Call this interval J_1. Cut J_1 in half. One of the newly obtained intervals will again not be covered by finitely many I_α'a. Call it J_2. Repeat the construction to obtain a decreasing sequence of intervals $J_1 \supset J_2 \supset J_3 \supset \cdots$, with the length of J_k equal to $\frac{b-a}{2^k}$ and such that none of these intervals can be covered by finitely many I_α's. By Cantor's nested intervals theorem, the intersection of the intervals J_k, $k \geq 1$, is some point x. This point belongs to an open interval I_{α_0}, and so an entire ε-neighborhood of x is in I_{α_0}. But then $J_k \subset I_{\alpha_0}$ for k large enough, a contradiction. Hence our assumption was false, and a finite subcover always exists. $\square$

Recall that the same dichotomic method can be applied to show that any sequence in a closed and bounded interval (and more generally in a compact metric space) has a convergent subsequence. And if you find the following problems demanding, remember Charlie Chaplin's words: "Failure is unimportant. It takes courage to make a fool of yourself."

420. Let $f : [a, b] \to [a, b]$ be an increasing function. Show that there exists $\xi \in [a, b]$ such that $f(\xi) = \xi$.

421. For every real number x_1 construct the sequence $x_1, x_2, x_3, \ldots$ by setting

$$x_{n+1} = x_n \left(x_n + \frac{1}{n} \right) \text{ for each } n \geq 1.$$

Prove that there exists exactly one value of x_1 for which $0 < x_n < x_{n+1} < 1$ for every n.

422. Given a sequence $(a_n)_n$ such that for any $\gamma > 1$ the subsequence $(a_{\lfloor \gamma^n \rfloor})_n$ converges to zero, does it follow that the sequence $(a_n)_n$ itself converges to zero?

423. Let $f : (0, \infty) \to \mathbb{R}$ be a continuous function with the property that for any $x > 0$,

$$\lim_{n \to \infty} f(nx) = 0.$$

Prove that $\lim_{x \to \infty} f(x) = 0$.

3.1.5 Series

A series is a sum

$$\sum_{n=1}^{\infty} a_n = a_1 + a_2 + \cdots + a_n + \cdots$$

The first question asked about a series is whether it converges. Convergence can be decided using Cauchy's $\varepsilon - \delta$ criterion, or by comparing it with another series. For comparison, two families of series are most useful:

(i) geometric series

$$1 + x + x^2 + \cdots + x^n + \cdots ,$$

which converge if $|x| < 1$ and diverge otherwise, and

(ii) *p*-series

$$1 + \frac{1}{2^p} + \frac{1}{3^p} + \cdots + \frac{1}{n^p} + \cdots ,$$

which converge if $p > 1$ and diverge otherwise.

The *p*-series corresponding to $p = 1$ is the harmonic series. Its truncation to the *n*th term approximates $\ln n$. Let us use the harmonic series to answer the following question.

Example. Does the series $\sum_{n=1}^{\infty} \frac{|\sin n|}{n}$ converge?

Solution. The inequality $|\sin x| > \frac{\sqrt{2 - \sqrt{2}}}{2}$ holds if and only if $\frac{1}{8} < \left\{\frac{x}{\pi}\right\} < \frac{7}{8}$, where $\{x\}$ denotes the fractional part of x (that is $x - \lfloor x \rfloor$). Because $\frac{1}{4} < \frac{1}{\pi}$, it follows that for any n, either $|\sin n|$ or $|\sin(n+1)|$ is greater than $\frac{\sqrt{2 - \sqrt{2}}}{2}$. Therefore

$$\frac{|\sin n|}{n} + \frac{|\sin(n+1)|}{n+1} \geq \frac{\sqrt{2 - \sqrt{2}}}{2} \cdot \frac{1}{n+1}.$$

Adding up these inequalities for all odd numbers n, we obtain

$$\sum_{n=1}^{\infty} \frac{|\sin n|}{n} \geq \frac{\sqrt{2 - \sqrt{2}}}{2} \sum_{n=1}^{\infty} \frac{1}{2n} = \frac{\sqrt{2 - \sqrt{2}}}{4} \sum_{n=1}^{\infty} \frac{1}{n} = \infty.$$

Hence the series diverges.   □

In fact, the so-called equidistribution criterion implies that if $f : \mathbb{R} \to \mathbb{R}$ is a continuous periodic function with irrational period and if $\sum_n \frac{|f(n)|}{n} < \infty$, then f is identically zero.

The comparison with a geometric series gives rise to d'Alembert's ratio test: $\sum_{n=0}^{\infty} a_n$ converges if $\limsup_n \left|\frac{a_{n+1}}{a_n}\right| < 1$ and diverges if $\liminf_n \left|\frac{a_{n+1}}{a_n}\right| > 1$. Here is a problem of P. Erdös from the *American Mathematical Monthly* that applies this test among other things.

Example. Let $(n_k)_{k \geq 1}$ be a strictly increasing sequence of positive integers with the property that

$$\lim_{k \to \infty} \frac{n_k}{n_1 n_2 \cdots n_{k-1}} = \infty.$$

Prove that the series $\sum_{k \geq 1} \frac{1}{n_k}$ is convergent and that its sum is an irrational number.

Solution. The relation from the statement implies in particular that $n_{k+1} \geq 3n_k$ for $k \geq 3$. By the ratio test the series $\sum_k \dfrac{1}{n_k}$ is convergent, since the ratio of two consecutive terms is less than or equal to $\frac{1}{3}$.

By way of contradiction, suppose that the sum of the series is a rational number $\dfrac{p}{q}$. Using the hypothesis we can find $k \geq 3$ such that

$$\frac{n_{j+1}}{n_1 n_2 \cdots n_j} \geq 3q, \ \text{if } j \geq k.$$

Let us start with the obvious equality

$$p(n_1 n_2 \cdots n_k) = q(n_1 n_2 \cdots n_k) \sum_{j=1}^{\infty} \frac{1}{n_j}.$$

From it we derive

$$p(n_1 n_2 \cdots n_k) - \sum_{j=1}^{k} \frac{q n_1 n_2 \cdots n_k}{n_j} = \sum_{j>k} \frac{q n_1 n_2 \cdots n_k}{n_j}.$$

Clearly, the left-hand side of this equality is an integer. For the right-hand side, we have

$$0 < \sum_{j>k} \frac{q n_1 n_2 \cdots n_k}{n_j} \leq \frac{q n_1 n_2 \cdots n_k}{n_{k+1}} + \frac{q n_1 n_2 \cdots n_k}{3 n_{k+1}} + \cdots \leq \frac{1}{3} + \frac{1}{9} + \frac{1}{27} + \cdots = \frac{1}{2}.$$

Here we used the fact that $\dfrac{n_1 n_2 \cdots n_k}{n_{k+1}} \leq \dfrac{1}{3q}$ and that $n_{j+1} \geq 3n_j$, for $j \geq k$ and k sufficiently large. This shows that the right-hand side cannot be an integer, a contradiction. It follows that the sum of the series is irrational. $\qquad\square$

We conclude our list of examples with a combinatorial proof of the fact that the harmonic series diverges. The lemma below, and the observation that it can be used to check the divergence of the harmonic series, have appeared in the Russian journal *Kvant (Quantum)*.

Example. If $a_1, a_2, a_3, \ldots$, is a sequence of positive numbers such that for every n, $a_n < a_{n+1} + a_{n^2}$, then $\sum a_n$ diverges. Consequently, the harmonic series

$$1 + \frac{1}{2} + \frac{1}{3} + \frac{1}{4} + \cdots$$

diverges.

Solution. We rely on the following result.

Lemma. *A triangular tableau is constructed as follows: the top row contains a natural number n. We pass from one row to the next by writing below a number k the numbers k^2 to the left and $k + 1$ to the right. Then the numbers on every row are distinct.*

Proof. Assume that the mth row is the first for which two numbers are equal. Let p and q be the two numbers that are equal. Because the previous row contains no equal numbers, p and q were obtained from the previous row by different procedures, say $p = r^2$ and $q = s + 1$. Then $s = r^2 - 1$, with r, s in the $m - 1$st row. Let us examine how s was obtained from n. Assume somewhere we performed a squaring of a number, and let k be the last number for which this happened. Because $s < r^2 - 1$, $k \leq r - 1$. But $s - k^2 \geq s - (r - 1)^2 \geq 2r - 2$. Hence after k we had to add $2r - 2$ units or more, so s was obtained from n in $2r - 1$ steps. Consequently, $m - 2 \geq 2r - 1$. But the numbers in the mth row are greater than or equal to $n + m - 1$, hence $r \geq n + 2r$, which is a contradiction. It follows that to get s no squarings were performed. The same is true for $q = s + 1$, so q is the left-most number of its row. But this is makes the equality $p = q$ impossible. $\quad\square$

With the lemma at hand, let $a_1, a_2, a_3, \ldots$, be a sequence of positive numbers such that for every n, $a_n < a_{n+1} + a_{n^2}$. In the sum $\sum_{n=1}^{N} a_n$ we can replace each a_n by a larger number of the form $\sum a_p$ over the elements p in the mth row of the above tableau. By spreading m's apart, we can make sure that there are no overlaps between the terms used for a_n and those for $a_{n'}$. We can also choose m's to exceed N. Consequently

$$\sum_{n=1}^{N} a_n \leq \sum_{n=N+1}^{\infty} a_n,$$

which implies that the series diverges. And because

$$\frac{1}{n} < \frac{1}{n+1} + \frac{1}{n^2},$$

the harmonic series diverges. $\quad\square$

424. Show that the series

$$\frac{1}{1+x} + \frac{2}{1+x^2} + \frac{4}{1+x^4} + \cdots + \frac{2^n}{1+x^{2^n}} + \cdots$$

converges when $|x| > 1$, and in this case find its sum.

425. For what positive x does the series

$$(x - 1) + (\sqrt{x} - 1) + (\sqrt[3]{x} - 1) + \cdots + (\sqrt[n]{x} - 1) + \cdots$$

converge?

426. Let $a_1, a_2, \ldots, a_n, \ldots$ be nonnegative numbers. Prove that $\sum_{n=1}^{\infty} a_n < \infty$ implies

$$\sum_{n=1}^{\infty} \sqrt{a_{n+1} a_n} < \infty.$$

427. Let $S = \{x_1, x_2, \ldots, x_n, \ldots\}$ be the set of all positive integers that do not contain the digit 9 in their decimal representation. Prove that

$$\sum_{n=1}^{\infty} \frac{1}{x_n} < 80.$$

428. Suppose that $(x_n)_n$ is a sequence of real numbers satisfying

$$x_{n+1} \leq x_n + \frac{1}{n^2}, \text{ for all } n \geq 1.$$

Prove that $\lim_{n\to\infty} x_n$ exists.

429. Does the series $\sum_{n=1}^{\infty} \sin \pi \sqrt{n^2 + 1}$ converge?

430. (a) Does there exist a pair of divergent series $\sum_{n=1}^{\infty} a_n$, $\sum_{n=1}^{\infty} b_n$, with $a_1 \geq a_2 \geq a_3 \geq \cdots \geq 0$ and $b_1 \geq b_2 \geq b_3 \geq \cdots \geq 0$, such that the series $\sum_n \min(a_n, b_n)$ is convergent?

(b) Does the answer to this question change if we assume additionally that

$$b_n = \frac{1}{n}, \ n = 1, 2, \ldots?$$

431. Given a sequence $(x_n)_n$ with $x_1 \in (0, 1)$ and $x_{n+1} = x_n - nx_n^2$ for $n \geq 1$, prove that the series $\sum_{n=1}^{\infty} x_n$ is convergent.

432. Is the number

$$\sum_{n=1}^{\infty} \frac{1}{2^{n^2}}$$

rational?

433. Let $(a_n)_{n\geq 0}$ be a strictly decreasing sequence of positive numbers, and let z be a complex number of absolute value less than 1. Prove that the sum

$$a_0 + a_1 z + a_2 z^2 + \cdots + a_n z^n + \cdots$$

is not equal to zero.

434. Let w be an irrational number with $0 < w < 1$. Prove that w has a unique convergent expansion of the form

$$w = \frac{1}{p_0} - \frac{1}{p_0 p_1} + \frac{1}{p_0 p_1 p_2} - \frac{1}{p_0 p_1 p_2 p_3} + \cdots,$$

where $p_0, p_1, p_2, \ldots$ are integers, $1 \leq p_0 < p_1 < p_2 < \cdots$.

435. The number q ranges over all possible powers with both the base and the exponent positive integers greater than 1, assuming each such value only once. Prove that

$$\sum_q \frac{1}{q-1} = 1.$$

436. Prove that for any $n \geq 2$,

$$\sum_{p \leq n, \, p \text{ prime}} \frac{1}{p} > \ln \ln n - 1.$$

Conclude that the sum of the reciprocals of all prime numbers is infinite.

3.1.6 Telescopic Series and Products

We mentioned earlier the idea of translating notions from differential and integral calculus to sequences. For example, the derivative of $(x_n)_n$ is the sequence whose terms are $x_{n+1} - x_n$, $n \geq 1$, while the definite integral is the sum $x_1 + x_2 + x_3 + \cdots$ The Leibniz-Newton fundamental theorem of calculus

$$\int_a^b f(t)dt = F(b) - F(a), \text{ where } F'(t) = f(t),$$

becomes the telescopic method for summing a series

$$\sum_{k=1}^n a_k = b_{n+1} - b_1, \text{ where } a_k = b_{k+1} - b_k, \ k \geq 1.$$

As in the case of integrals, when applying the telescopic method to a series, the struggle is to find the "antiderivative" of the general term. But compared to the case of integrals, here we lack an algorithmic way. This is what makes such problems attractive for mathematics competitions. A simple example that comes to mind is the following.

Example. Find the sum

$$\frac{1}{\sqrt{1} + \sqrt{2}} + \frac{1}{\sqrt{2} + \sqrt{3}} + \cdots + \frac{1}{\sqrt{n} + \sqrt{n+1}}.$$

Solution. The "antiderivative" of the general term of the sum is found by rationalizing the denominator:

$$\frac{1}{\sqrt{k} + \sqrt{k+1}} = \frac{\sqrt{k+1} - \sqrt{k}}{k+1-k} = \sqrt{k+1} - \sqrt{k}.$$

The sum is therefore equal to

$$(\sqrt{2} - \sqrt{1}) + (\sqrt{3} - \sqrt{2}) + \cdots + (\sqrt{n+1} - \sqrt{n}) = \sqrt{n+1} - 1. \qquad \square$$

Not all problems are so simple, as the next two examples show.

Example. Let $a_0 = 1$, $a_1 = 3$, $a_{n+1} = \dfrac{a_n^2 + 1}{2}$, $n \geq 1$. Prove that

$$\frac{1}{a_0 + 1} + \frac{1}{a_1 + 1} + \cdots + \frac{1}{a_n + 1} + \frac{1}{a_{n+1} - 1} = 1, \text{ for all } n \geq 1.$$

Solution. We have

$$a_{k+1} - 1 = \frac{a_k^2 - 1}{2},$$

so

$$\frac{1}{a_{k+1} - 1} = \frac{1}{a_k - 1} - \frac{1}{a_k + 1}, \text{ for } k \geq 1.$$

This allows us to express the terms of the sum from the statement as "derivatives":

$$\frac{1}{a_k + 1} = \frac{1}{a_k - 1} - \frac{1}{a_{k+1} - 1}, \text{ for } k \geq 1.$$

Summing up these equalities for $k = 1, 2, \ldots, n$ yields

$$\frac{1}{a_1 + 1} + \cdots + \frac{1}{a_n + 1} = \frac{1}{a_1 - 1} - \frac{1}{a_2 - 1} + \frac{1}{a_2 - 1} - \frac{1}{a_3 - 1} + \cdots$$

$$+ \frac{1}{a_n - 1} - \frac{1}{a_{n+1} - 1} = \frac{1}{2} - \frac{1}{a_{n+1} - 1}.$$

Finally, add $\dfrac{1}{a_0 + 1} + \dfrac{1}{a_{n+1} - 1}$ to both sides to obtain the identity from the statement.   □

Example. Express

$$\sum_{n=1}^{49} \frac{1}{\sqrt{n + \sqrt{n^2 - 1}}}$$

as $a + b\sqrt{2}$ for some integers a and b.

Solution. We have

$$\frac{1}{\sqrt{n + \sqrt{n^2 - 1}}} = \frac{1}{\sqrt{\left(\sqrt{\dfrac{n+1}{2}} + \sqrt{\dfrac{n-1}{2}}\right)^2}} = \frac{1}{\sqrt{\dfrac{n+1}{2}} + \sqrt{\dfrac{n-1}{2}}}$$

$$= \frac{\sqrt{\dfrac{n+1}{2}} - \sqrt{\dfrac{n-1}{2}}}{\dfrac{n+1}{2} - \dfrac{n-1}{2}} = \sqrt{\frac{n+1}{2}} - \sqrt{\frac{n-1}{2}}.$$

Hence the sum from the statement telescopes to

$$\sqrt{\frac{49+1}{2}} + \sqrt{\frac{48+1}{2}} - \sqrt{\frac{1}{2}} - 0 = 5 + \frac{7}{\sqrt{2}} - \frac{1}{\sqrt{2}} = 5 + 3\sqrt{2}.$$

   □

Apply the telescopic method to the following problems.

437. Prove the identity

$$\sum_{k=1}^{n}(k^2 + 1)k! = n(n + 1)!$$

438. Let ζ be a root of unity. Prove that

$$\zeta^{-1} = \sum_{n=0}^{\infty}\zeta^n(1 - \zeta)(1 - \zeta^2)\cdots(1 - \zeta^n),$$

with the convention that the 0th term of the series is 1.

439. For a nonnegative integer k, define $S_k(n) = 1^k + 2^k + \cdots + n^k$. Prove that

$$1 + \sum_{k=0}^{r-1}\binom{r}{k}S_k(n) = (n + 1)^r.$$

440. Let

$$a_n = \frac{4n + \sqrt{4n^2 - 1}}{\sqrt{2n + 1} + \sqrt{2n - 1}}, \quad \text{for } n \geq 1.$$

Prove that $a_1 + a_2 + \cdots + a_{40}$ is a positive integer.

441. Prove that

$$\sum_{k-1}^{n}\frac{(-1)^{k+1}}{1^2 - 2^2 + 3^2 - \cdots + (-1)^{k+1}k^2} = \frac{2n}{n + 1}.$$

442. Prove that

$$\sum_{n=1}^{9999}\frac{1}{(\sqrt{n} + \sqrt{n + 1})(\sqrt[4]{n} + \sqrt[4]{n + 1})} = 9.$$

443. Let $a_n = \sqrt{1 + \left(1 + \frac{1}{n}\right)^2} + \sqrt{1 + \left(1 - \frac{1}{n}\right)^2}$, $n \geq 1$. Prove that

$$\frac{1}{a_1} + \frac{1}{a_2} + \cdots + \frac{1}{a_{20}}$$

is a positive integer.

444. Evaluate in closed form

$$\sum_{m=0}^{\infty}\sum_{n=0}^{\infty}\frac{m!n!}{(m + n + 2)!}.$$

445. Let $a_n = 3n + \sqrt{n^2 - 1}$ and $b_n = 2(\sqrt{n^2 - n} + \sqrt{n^2 + n})$, $n \geq 1$. Show that

$$\sqrt{a_1 - b_1} + \sqrt{a_2 - b_2} + \cdots + \sqrt{a_{49} - b_{49}} = A + B\sqrt{2},$$

for some integers A and B.

446. Evaluate in closed form

$$\sum_{k=0}^{n} (-1)^k (n - k)!(n + k)!$$

447. Let $a_0 = 1994$ and $a_{n+1} = \dfrac{a_n^2}{a_n + 1}$ for each nonnegative integer n. Prove that for $0 \leq n \leq 998$, the number $1994 - n$ is the greatest integer less than or equal to a_n.

448. Fix k a positive integer and define the sequence

$$a_n = \left\lfloor (k + \sqrt{k^2 + 1})^n + \left(\frac{1}{2}\right)^n \right\rfloor, \quad n \geq 0.$$

Prove that

$$\sum_{n=1}^{\infty} \frac{1}{a_{n-1} a_{n+1}} = \frac{1}{8k^2}.$$

The telescopic method can be applied to products as well. Within the first, relatively easy, problem, the reader will recognize in disguise the Fermat numbers $2^{2^n} + 1$, $n \geq 1$.

Example. Define the sequence $(a_n)_n$ by $a_0 = 3$, and $a_{n+1} = a_0 a_1 \cdots a_n + 2$, $n \geq 0$. Prove that

$$a_{n+1} = 2(a_0 - 1)(a_1 - 1) \cdots (a_n - 1) + 1, \quad \text{for all } n \geq 0.$$

Solution. The recurrence relation gives $a_0 a_1 \ldots a_{k-1} = a_k - 2$, $k \geq 1$. Substitute this in the formula for a_{k+1} to obtain $a_{k+1} = (a_k - 2)a_k + 2$, which can be written as $a_{k+1} - 1 = (a_k - 1)^2$. And so

$$\frac{a_{k+1} - 1}{a_k - 1} = a_k - 1.$$

Multiplying these relations for $k = 0, 1, \ldots, n$, we obtain

$$\frac{a_{n+1} - 1}{a_n - 1} \cdot \frac{a_n - 1}{a_{n-1} - 1} \cdots \frac{a_1 - 1}{a_0 - 1} = (a_n - 1)(a_{n-1} - 1) \cdots (a_0 - 1).$$

Since the left-hand side telescopes, we obtain

$$\frac{a_{n+1} - 1}{a_0 - 1} = (a_0 - 1)(a_1 - 1) \cdots (a_n - 1),$$

and the identity follows.   □

A more difficult problem is the following.

Example. Compute the product

$$\prod_{n=1}^{\infty}\left(1 + \frac{(-1)^n}{F_n^2}\right),$$

where F_n is the nth Fibonacci number.

Solution. Recall that the Fibonacci numbers satisfy the Cassini identity

$$F_{n+1}F_{n-1} - F_n^2 = (-1)^n.$$

Hence

$$\prod_{n=1}^{\infty}\left(1 + \frac{(-1)^n}{F_n^2}\right) = \lim_{N\to\infty}\prod_{n=1}^{N}\frac{F_n^2 + (-1)^n}{F_n^2} = \lim_{N\to\infty}\prod_{n=1}^{N}\frac{F_{n-1}}{F_n}\cdot\frac{F_{n+1}}{F_n}$$

$$= \lim_{N\to\infty}\frac{F_0 F_{N+1}}{F_1 F_N} = \lim_{N\to\infty}\frac{F_{N+1}}{F_N}.$$

Because of the Binet formula

$$F_n = \frac{1}{\sqrt{5}}\left[\left(\frac{1+\sqrt{5}}{2}\right)^{n+1} - \left(\frac{1-\sqrt{5}}{2}\right)^{n+1}\right], \quad \text{for } n \geq 0,$$

the above limit is equal to $\dfrac{1+\sqrt{5}}{2}$. □

449. Compute the product

$$\left(1 - \frac{4}{1}\right)\left(1 - \frac{4}{9}\right)\left(1 - \frac{4}{25}\right)\cdots$$

450. Let x be a positive number less than 1. Compute the product

$$\prod_{n=0}^{\infty}(1 + x^{2^n}).$$

451. Let x be a real number. Define the sequence $(x_n)_{n\geq 1}$ recursively by $x_1 = 1$ and $x_{n+1} = x^n + na_n$ for $n \geq 1$. Prove that

$$\prod_{n=1}^{\infty}\left(1 - \frac{x^n}{x_{n+1}}\right) = e^{-x}.$$

3.2 Continuity, Derivatives, and Integrals

3.2.1 Functions

Before starting our discussion on differentiation and integration, let us warm ourselves up with some general problems about functions. We begin with an example given at a Romanian Team Selection Test for the International Mathematical Olympiad in 1982, proposed by S. Rădulescu and I. Tomescu.

Example. Let $f : \mathbb{R} \to \mathbb{R}$, a function with the property that

$$f(f(x)) = \frac{x^9}{(x^2 + 1)(x^6 + x^4 + 2x^2 + 1)},$$

for all $x \in \mathbb{R}$. Show that there is a unique point a such that $f(a) = a$.

Solution. If $a \in \mathbb{R}$ is such that $f(a) = a$, then $f(f(a)) = f(a) = a$, so

$$\frac{a^9}{(a^2 + 1)(a^6 + a^4 + 2a^2 + 1)} = a.$$

This can be rewritten as $a^9 = a^9 + 2a^7 + 3a^5 + 3a^3 + a$, which is equivalent to

$$a(2a^6 + 3a^4 + 3a^2 + 1) = 0.$$

The second factor is strictly positive, so this implies $a = 0$. Let us show that 0 is indeed a fixed point of f. Let $f(0) = b$. Then $f(b) = f(f(0)) = 0$. It follows that $f(f(b)) = f(0) = b$. But the above argument showed that $f(f(x))$ has a unique fixed point, namely $x = 0$. Hence $b = 0$, and we are done. □

Here is a second example.

Example. Solve in real numbers the system

$$3a = (b + c + d)^3$$
$$3b = (c + d + a)^3$$
$$3c = (d + a + b)^3$$
$$3d = (a + b + c)^3.$$

Solution. Taking the cube root of each equation, we deduce that

$$a + \sqrt[3]{3a} = b + \sqrt[3]{3b} = c + \sqrt[3]{3c} = d + \sqrt[3]{3d} = a + b + c + d.$$

Define the function $f(x) = x + \sqrt[3]{3x}$. This function is increasing, in particular injective. Hence $a = b = c = d$. From $3a = (3a)^3$ we obtain $a = 0$ or $a = \pm\frac{1}{3}$. □

452. Let a, b, c be positive real numbers. Solve the equation

$$\sqrt{a + bx} + \sqrt{b + cx} + \sqrt{c + ax} = \sqrt{b - ax} + \sqrt{c - bx} + \sqrt{a - cx}$$

453. Prove that for all positive integers n,

$$\sqrt[n]{3} + \sqrt[n]{7} > \sqrt[n]{4} + \sqrt[n]{5}.$$

454. Does there exist a function $f : \mathbb{R} \to \mathbb{R}$ such that the equation $f(f(x)) = x$ has exactly 5102 solutions and the equation $f(x) = x$ has exactly 2015 solutions?

455. Give an example of a function $f : \mathbb{R} \to \mathbb{R}$ whose graph is invariant under a 90° rotation about the origin.

456. Does there exist a function $f : \mathbb{R} \to \mathbb{R}$ such that

$$(f \circ f \circ f)(x) = x^3 \text{ and } (f \circ f \circ f \circ f \circ f)(x) = x^5,$$

for every $x \in \mathbb{R}$?

457. Find all real numbers x and y that are solutions to the system of equations

$$3^x - 3^y = 2^y$$
$$9^x - 6^y = 19^y.$$

458. Given a real number $a \in (0, 1)$ find all positive real solutions to the equation

$$x^{a^x} = a^{x^a}.$$

459. Find all positive real solutions to the system of equations

$$x^y = 8, \quad y^z = 81, \quad z^x = 16.$$

460. Let n be an odd integer greater than 1. Find the real solutions to the equation

$$\sqrt[n]{x^n + 1} + \sqrt[n]{1 - (x + 1)^n} = 1.$$

461. Let a, b, c be real numbers that satisfy $4ac \le (b-1)^2$, and let $f : \mathbb{R} \to \mathbb{R}$ be a function that satisfies

$$f(ax^2 + bx + c) = a(f(x))^2 + bf(x) + c, \text{ for all } x \in \mathbb{R}.$$

Prove that the equation $f(f(x)) = x$ has at least one solution.

462. Let ABC be a triangle with side-lengths a, b, c. Show that if $a^4 = b^4 + c^4$ then the measure of the angle $\angle A$ is greater than 72°.

3.2.2 Limits of Functions

Among the various ways to find the limit of a function, the most basic is the definition itself.

Definition. For x_0 an accumulation point of the domain of a function f, we say that $\lim\limits_{x \to x_0} f(x) = L$ if for every neighborhood V of L, there is a neighborhood U of x_0 such that $f(U \backslash \{x_0\}) \subset V$.

This definition is, however, seldom used in applications. Instead, it is more customary to use operations with limits, the squeezing principle (if $f(x) \le g(x) \le h(x)$ for all x and $\lim\limits_{x \to x_0} f(x) = \lim\limits_{x \to x_0} h(x) = L$, then $\lim\limits_{x \to x_0} g(x) = L$), continuity, or L'Hôpital's theorem, to be discussed later.

Example. Compute

$$\lim_{x \to \infty} \left(\sqrt{x + \sqrt{x + \sqrt{x}}} - \sqrt{x} \right).$$

Solution. The usual algorithm is to multiply and divide by the conjugate to obtain

$$\lim_{x \to \infty} \left(\sqrt{x + \sqrt{x + \sqrt{x}}} - \sqrt{x} \right) = \lim_{x \to \infty} \frac{x + \sqrt{x + \sqrt{x}} - x}{\sqrt{x + \sqrt{x + \sqrt{x}}} + \sqrt{x}}$$

$$= \lim_{x \to \infty} \frac{\sqrt{x + \sqrt{x}}}{\sqrt{x + \sqrt{x + \sqrt{x}}} + \sqrt{x}} = \lim_{x \to \infty} \frac{\sqrt{1 + \sqrt{\dfrac{1}{x}}}}{\sqrt{1 + \sqrt{\dfrac{1}{x} + \sqrt{\dfrac{1}{x^3}}}} + 1} = \frac{1}{2}. \qquad \square$$

And now an example of type 1^∞.

Example. Let $a_1, a_2, \ldots, a_n$ be positive real numbers. Prove that

$$\lim_{x \to 0} \left(\frac{a_1^x + a_2^x + \cdots + a_n^x}{n} \right)^{\frac{1}{x}} = \sqrt[n]{a_1 a_2 \cdots a_n}.$$

Solution. First, note that

$$\lim_{x \to 0} \frac{a^x - 1}{x} = \ln a.$$

Indeed, the left-hand side can be recognized as the derivative of the exponential at 0. Or to avoid a logical vicious circle, we can argue as follows: let $a^x = 1 + t$, with $t \to 0$. Then $x = \dfrac{\ln(1 + t)}{\ln a}$, and the limit becomes

$$\lim_{t \to 0} \frac{t \ln a}{\ln(1 + t)} = \lim_{t \to 0} \frac{\ln a}{\ln(1 + t)^{1/t}} = \frac{\ln a}{\ln e} = \ln a.$$

Let us return to the problem. Because the limit is of the form 1^∞, it is standard to write it as

$$\lim_{x\to 0} \left(1 + \frac{a_1^x + a_2^x + \cdots + a_n^x - n}{n}\right)^{\frac{n}{a_1^x + a_2^x + \cdots + a_n^x - n} \cdot \frac{a_1^x + a_2^x + \cdots + a_n^x - n}{nx}}.$$

Using the fact that $\lim_{t\to 0}(1+t)^{1/t} = e$, we find this to be equal to

$$\exp\left[\lim_{x\to 0}\left(\frac{a_1^x + a_2^x + \cdots + a_n^x - n}{nx}\right)\right] = \exp\left[\frac{1}{n}\lim_{x\to 0}\left(\frac{a_1^x - 1}{x} + \frac{a_2^x - 1}{x} + \cdots + \frac{a_n^x - 1}{x}\right)\right]$$

$$= \exp\left[\frac{1}{n}(\ln a_1 + \ln a_2 + \cdots + \ln a_n)\right] = \sqrt[n]{a_1 a_2 \ldots a_n},$$

the desired answer.   □

We continue with a problem of theoretical flavor that requires $\varepsilon - \delta$ argument. Written by M. Becheanu it was given at a Romanian competition in 2004.

Example. Let $a \in (0, 1)$ be a real number and $f : \mathbb{R} \to \mathbb{R}$ a function that satisfies the following conditions:

(i) $\lim_{x\to\infty} f(x) = 0$;

(ii) $\lim_{x\to\infty} \dfrac{f(x) - f(ax)}{x} = 0$.

Show that $\lim_{x\to\infty} \dfrac{f(x)}{x} = 0$.

Solution. The second condition reads: for any $\varepsilon > 0$, there exists $\delta > 0$ such that if $x \in (-\delta, \delta)$ then $|f(x) - f(ax)| \leq \varepsilon|x|$. Applying the triangle inequality, we find that for all positive integers n and all $x \in (-\delta, \delta)$,

$$|f(x) - f(a^n x)| \leq |f(x) - f(ax)| + |f(ax) - f(a^2 x)| + \cdots + |f(a^{n-1}x) - f(a^n x)|$$

$$< \varepsilon|x|(1 + a + a^2 + \cdots + a^{n-1}) = \varepsilon\frac{1 - a^n}{1 - a}|x| \leq \frac{\varepsilon}{1 - a}|x|.$$

Taking the limit as $n \to \infty$, we obtain

$$|f(x)| \leq \frac{\varepsilon}{1 - a}|x|.$$

Since $\varepsilon > 0$ was arbitrary, this proves that $\lim_{x\to\infty} \dfrac{f(x)}{x} = 0$.   □

463. Find the real parameters m and n such that the graph of the function $f : \mathbb{R} \to \mathbb{R}$,

$$f(x) = \sqrt[3]{8x^3 + mx^2} - nx$$

has the horizontal asymptote $y = 1$.

464. Does $\lim\limits_{x\to\pi/2} (\sin x)^{\frac{1}{\cos x}}$ exist?

465. For two positive integers m and n, compute

$$\lim_{x\to 0} \frac{\sqrt[m]{\cos x} - \sqrt[n]{\cos x}}{x^2}.$$

466. Does there exist a nonconstant function $f : (1, \infty) \to \mathbb{R}$ satisfying the relation

$$f(x) = f\left(\frac{x^2 + 1}{2}\right) \text{ for all } x > 1$$

and such that $\lim\limits_{x\to\infty} f(x)$ exists?

467. Let $f : (0, \infty) \to (0, \infty)$ be an increasing function with $\lim\limits_{t\to\infty} \frac{f(2t)}{f(t)} = 1$. Prove that

$$\lim_{t\to\infty} \frac{f(mt)}{f(t)} = 1 \text{ for any } m > 0.$$

468. Let $f(x) = \sum\limits_{k=1}^{n} a_k \sin kx$, with $a_1, a_2, \ldots, a_n \in \mathbb{R}$, $n \geq 1$. Prove that if $f(x) \leq |\sin x|$ for all $x \in \mathbb{R}$, then

$$\left|\sum_{k=1}^{n} ka_k\right| \leq 1.$$

3.2.3 Continuous Functions

A function f is continuous at x_0 if it has limit at x_0 and this limit is equal to $f(x_0)$. A function that is continuous at every point of its domain is simply called continuous.

Example. Find all continuous functions $f : \mathbb{R} \to \mathbb{R}$ satisfying $f(0) = 1$ and

$$f(2x) - f(x) = x, \text{ for all } x \in \mathbb{R}.$$

Solution. Write the functional equation as

$$f(x) - f\left(\frac{x}{2}\right) = \frac{x}{2},$$

then iterate

$$f\left(\frac{x}{2}\right) - f\left(\frac{x}{4}\right) = \frac{x}{4},$$
$$f\left(\frac{x}{4}\right) - f\left(\frac{x}{8}\right) = \frac{x}{8},$$
$$\cdots$$
$$f\left(\frac{x}{2^{n-1}}\right) - f\left(\frac{x}{2^n}\right) = \frac{x}{2^n}.$$

Summing up, we obtain

$$f(x) - f\left(\frac{x}{2^n}\right) = x\left(\frac{1}{2} + \frac{1}{4} + \cdots + \frac{1}{2^n}\right),$$

which, when n tends to infinity, becomes $f(x) - 1 = x$. Hence $f(x) = x + 1$ is the (unique) solution. □

We will now present the spectacular example of a continuous curve that covers a square completely. A planar curve $\phi(t) = (x(t), y(t))$ is called continuous if both coordinate functions $x(t)$ and $y(t)$ depend continuously on the parameter t.

Peano's theorem. *There exists a continuous surjection* $\phi : [0, 1] \to [0, 1] \times [0, 1]$.

Proof. G. Peano found an example of such a function in the early twentieth century. The curve presented below was constructed later by H. Lebesgue.

The construction of this "Peano curve" uses the Cantor set. This is the set C of all numbers in the interval $[0, 1]$ that can be written in base 3 with only the digits 0 and 2. For example, 0.1 is in C because it can also be written as $0.0222\ldots$, but 0.101 is not. The Cantor set is obtained by removing from $[0, 1]$ the interval $\left(\frac{1}{3}, \frac{2}{3}\right)$, then $\left(\frac{1}{9}, \frac{2}{9}\right)$ and $\left(\frac{7}{9}, \frac{8}{9}\right)$, then successively from each newly formed closed interval an open interval centered at its midpoint and $\frac{1}{3}$ of its size (Figure 19). The Cantor set is a *fractal*: each time we cut a piece of it and magnify it, the piece resembles the original set.

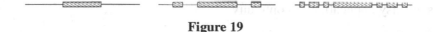

Figure 19

Next, we define a function $\phi : C \to [0, 1] \times [0, 1]$ in the following manner. For a number written in base 3 as $0.a_1a_2\ldots a_n\ldots$ with only the digits 0 and 2 (hence in the Cantor set), divide the digits by 2, then separate the ones in even positions from those in odd positions. Explicitly, if $b_n = \frac{a_n}{2}$, $n \geq 1$, construct the pair $(0.b_1b_3b_5\ldots, 0.b_2b_4b_6\ldots)$. This should be interpreted as a point in $[0, 1] \times [0, 1]$ with coordinates written in base 2. Then $\phi(0.a_1a_2a_3a_4\ldots) = (0.b_1b_3\ldots, 0.b_2b_4\ldots)$. The function ϕ is clearly onto. Is it continuous?

First, what does continuity mean in this case? It means that whenever a sequence $(x_n)_n$ in C converges to a point $x \in C$, the sequence $(\phi(x_n))_n$ should converge to $\phi(x)$. Note that since the complement of C is a union of open intervals, C contains all its limit points. Moreover, the Cantor set has the very important property that a sequence $(x_n)_n$ of points in it converges to $x \in C$ if and only if the base-3 digits of x_n successively become equal to the digits of x. It is essential that the base-3 digits of a number in C can equal only 0 or 2, so that the ambiguity of the ternary expansion is eliminated. This fundamental property of the Cantor set guarantees the continuity of ϕ.

The function ϕ is extended linearly over each open interval that was removed in the process of constructing C, to obtain a continuous surjection $\phi : [0, 1] \to [0, 1] \times [0, 1]$. This concludes the proof of the theorem. □

To visualize this Peano curve, consider the "truncations" of the Cantor set

$$C_1 = \left\{0, \frac{1}{3}, \frac{2}{3}, 1\right\}, \quad C_2 = \left\{0, \frac{1}{9}, \frac{2}{9}, \frac{1}{3}, \frac{2}{3}, \frac{7}{9}, \frac{8}{9}, 1\right\},$$

$$C_3 = \left\{0, \frac{1}{27}, \frac{2}{27}, \frac{1}{9}, \frac{2}{9}, \frac{7}{27}, \frac{8}{27}, \frac{1}{3}, \frac{2}{3}, \frac{19}{27}, \frac{7}{9}, \frac{8}{9}, \frac{25}{27}, \frac{26}{27}, 1\right\},$$

$$C_4 = \left\{0, \frac{1}{81}, \frac{2}{81}, \frac{1}{27}, \frac{2}{27}, \frac{7}{81}, \frac{8}{81}, \frac{1}{9}, \frac{2}{9}, \frac{19}{81}, \frac{20}{81}, \frac{7}{27}, \frac{25}{81}, \frac{26}{81}, \frac{1}{3}, \frac{2}{3}, \right.$$
$$\left. \frac{55}{81}, \frac{56}{81}, \frac{19}{27}, \frac{20}{27}, \frac{61}{81}, \frac{62}{81}, \frac{7}{9}, \frac{8}{9}, \frac{73}{81}, \frac{74}{81}, \frac{25}{27}, \frac{26}{27}, \frac{79}{81}, \frac{80}{81}, 1\right\}, \dots$$

and define $\phi_n : C_n \to [0, 1] \times [0, 1]$, $n \geq 1$, as above, and then extend linearly. This gives rise to the curves from Figure 20. The curve ϕ is their limit. It is a fractal: if we cut the unit square into four equal squares, the curve restricted to each of these squares resembles the original curve.

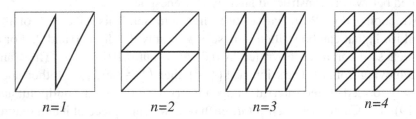

$n=1$ $n=2$ $n=3$ $n=4$

Figure 20

Here is an easy application of continuity.

Example. Let K be a closed, bounded set in $\mathbb{R}$ (or more generally, a compact set in some metric space). If $f : K \to K$ has the property that $|f(x) - f(y)| < |x - y|$ for all $x \neq y$, then f has a unique fixed point.

Solution. It is not hard to see that f is continuous. Let $g(x) = |f(x) - x|$. Then g is continuous, so g has a minimum a on K. Assume that $a \neq 0$. Then, if $x_0 \in K$ is such that $g(x_0) = a$, we have

$$g(f(x_0)) = |f(f(x_0)) - f(x_0)| < |f(x_0) - x_0| = g(x_0),$$

which contradicts minimality. Thus the minimum of g is 0, showing that f has a fixed point. The fixed point is unique since $f(x) = x$ and $f(y) = y$ yields $|x - y| = |f(x) - f(y)| < |x - y|$, impossible. $\square$

469. Let $f : \mathbb{R} \to \mathbb{R}$ be a continuous function satisfying $f(x) = f(x^2)$ for all $x \in \mathbb{R}$. Prove that f is constant.

470. Does there exist a continuous function $f : [0, 1] \to \mathbb{R}$ that assumes every element of its range an even (finite) number of times?

471. Let $f(x)$ be a continuous function defined on $[0, 1]$ such that

(i) $f(0) = f(1) = 0$;

(ii) $2f(x) + f(y) = 3f\left(\dfrac{2x+y}{3}\right)$ for all $x, y \in [0, 1]$.

Prove that $f(x) = 0$ for all $x \in [0, 1]$.

472. Let $f : \mathbb{R} \to \mathbb{R}$ be a continuous function with the property that

$$\lim_{h \to 0^+} \frac{f(x+2h) - f(x+h)}{h} = 0, \text{ for all } x \in \mathbb{R}.$$

Prove that f is constant.

473. Let a and b be real numbers in the interval $\left(0, \frac{1}{2}\right)$ and let f be a continuous real-valued function such that

$$f(f(x)) = af(x) + bx, \text{ for all } x \in \mathbb{R}.$$

Prove that $f(0) = 0$.

474. Let $f : [0, 1] \to \mathbb{R}$ be a continuous function. Prove that the series $\sum\limits_{n=1}^{\infty} \dfrac{f(x^n)}{2^n}$ is convergent for every $x \in [0, 1]$. Find a function f satisfying

$$f(x) = \sum_{n=1}^{\infty} \frac{f(x^n)}{2^n}, \text{ for all } x \in [0, 1].$$

475. Prove that there exists a continuous surjective function $\psi : [0, 1] \to [0, 1] \times [0, 1]$ that takes each value infinitely many times.

476. Give an example of a continuous function on an interval that is nowhere differentiable.

3.2.4 The Intermediate Value Property

A real-valued function f defined on an interval is said to have the intermediate value property (also known as the Darboux property) if for every $a < b$ in the interval and for every λ between $f(a)$ and $f(b)$, there exists c between a and b such that $f(c) = \lambda$. Equivalently, a real-valued function has the intermediate property if it maps intervals to intervals. The higher-dimensional analogue requires the function to map connected sets to connected sets. Continuous functions and derivatives of functions are known to have this property, although the class of functions with the intermediate value property is considerably larger.

Here is a problem from the 1982 Romanian Mathematical Olympiad, proposed by M. Chiriță.

Example. Let $f : [0, 1] \to \mathbb{R}$ be a continuous function with the property that

$$\int_0^1 f(x)dx = \frac{\pi}{4}.$$

Prove that there exists $x_0 \in (0, 1)$ such that

$$\frac{1}{1+x_0} < f(x_0) < \frac{1}{2x_0}.$$

Solution. Note that

$$\int_0^1 \frac{1}{1+x^2} dx = \frac{\pi}{4}.$$

Consequently, the integral of the function $g(x) = f(x) - \dfrac{1}{1+x^2}$ on the interval $[0, 1]$ is equal to 0. If $g(x)$ is identically 0, choose x_0 to be any number between 0 and 1. Otherwise, $g(x)$ assumes both positive and negative values on this interval. Being continuous, g has the intermediate value property, so there is some $x_0 \in (0, 1)$ for which $g(x_0) = 0$. We have thus found $x_0 \in (0, 1)$ such that $f(x_0) = \frac{1}{1+x_0^2}$. The double inequality from the statement follows from $2x_0 < 1 + x_0^2 < 1 + x_0$, which clearly holds since on the one hand, $x_0^2 - 2x_0 + 1 = (x_0 - 1) > 0$, and on the other, $x_0^2 < x_0$. $\qquad\square$

Example. Prove that every continuous mapping of a circle into a line carries some pair of diametrically opposite points to the same point.

Solution. Yes, this problem uses the intermediate value property, or rather the more general property that the image through a continuous map of a connected set is connected. The circle is connected, so its image must be an interval. This follows from a more elementary argument, if we think of the circle as the gluing of two intervals along their endpoints. The image of each interval is another interval, and the two images overlap, forming an interval.

Identify the circle with the set $S^1 = \{z \in \mathbb{C} \mid |z| = 1\}$. If $f : S^1 \to \mathbb{R}$ is the continuous mapping from the statement, then $\psi : S^1 \to \mathbb{R}$, $\psi(z) = f(z) - f(-z)$ is also continuous ($-z$ is diametrically opposite to z).

Pick $z_0 \in S^1$. If $\psi(x_0) = 0$, then z_0 and $-z_0$ map to the same point on the line. Otherwise,

$$\psi(-z_0) = f(-z_0) - f(z) = -\psi(z_0).$$

Hence ψ takes a positive and a negative value, and by the intermediate value property it must have a zero. The property is proved. A more difficult theorem of Borsuk and Ulam states that any continuous map of the sphere into the plane sends two antipodal points on the sphere to the same point in the plane. A nice interpretation of this fact is that at any time there are two antipodal points on earth with the same temperature and barometric pressure.

We conclude our list of examples with a surprising fact discovered by Lebesgue.

Theorem. *There exists a function $f : [0, 1] \to [0, 1]$ that has the intermediate value property and is discontinuous at every point.*

Proof. Lebesgue's function acts like an automaton. The value at a certain point is produced from information read from the digital expansion of the variable.

The automaton starts acting once it detects that all even-order digits have become 0. More precisely, if $x = 0.a_0 a_1 a_2 \ldots$, the automaton starts acting once $a_{2k} = 0$ for all $k \geq n$. It then reads the odd-order digits and produces the value $f(x) = 0.a_{2n+1} a_{2n+3} a_{2n+5} \ldots$ If the even-order digits do not eventually become zero, the automaton remains inactive, producing the value 0. Because only the rightmost digits of the numbers count, for any value of y and any interval $I \subset [0, 1]$, one can find a number $x \in I$ such that $f(x) = y$. Hence the function f maps any subinterval of $[0, 1]$ onto $[0, 1]$. It satisfies the intermediate value property trivially. And because any neighborhood of a point is mapped to the entire interval $[0, 1]$, the function is discontinuous everywhere. $\square$

As the poet Paul Valéry said: "a dangerous state is to think that you understand." To make sure that you *do* understand the intermediate value property, solve the following problems.

477. Let $f : [a, b] \to [a, b]$ be a continuous function. Prove that f has a fixed point.

478. One day, a Buddhist monk climbed from the valley to the temple up on the mountain. The next day, the monk came down, on the same trail and during the same time interval. Prove that there is a point on the trail that the monk reached at precisely the same moment of time on the two days.

479. Let $f : \mathbb{R} \to \mathbb{R}$ be a continuous decreasing function. Prove that the system

$$x = f(y),$$
$$y = f(z),$$
$$z = f(x)$$

has a unique solution.

480. Let $f : \mathbb{R} \to \mathbb{R}$ be a continuous function such that $|f(x) - f(y)| \geq |x - y|$ for all $x, y \in \mathbb{R}$. Prove that the range of f is all of $\mathbb{R}$.

481. A cross-country runner runs a six-mile course in 30 minutes. Prove that somewhere along the course the runner ran a mile in exactly 5 minutes.

482. Let A and B be two cities connected by two different roads. Suppose that two cars can travel from A to B on different roads keeping a distance that does not exceed one mile between them. Is it possible for the cars to travel the first one from A to B and the second one from B to A in such a way that the distance between them is always greater than one mile?

483. Let

$$P(x) = \sum_{k=1}^{n} a_k x^k \quad \text{and} \quad Q(x) = \sum_{k=1}^{n} \frac{a_k}{2^k - 1} x^k,$$

where $a_1, a_2, \ldots, a_n$ are real numbers, $n \geq 1$. Show that if 1 and 2^{n+1} are zeros of the polynomial $Q(x)$, then $P(x)$ has a positive zero less than 2^n.

484. Prove that any convex polygonal surface can be divided by two perpendicular lines into four regions of equal area.

485. Let $f : I \to \mathbb{R}$ be a function defined on an interval. Show that if f has the intermediate value property and for any $y \in \mathbb{R}$ the set $f^{-1}(y)$ is closed, then f is continuous.

486. Show that the function

$$f_a(x) = \begin{cases} \cos \dfrac{1}{x} & \text{for } x \neq 0 \\ a & \text{for } x = 0, \end{cases}$$

has the intermediate value property if $a \in [-1, 1]$ but is the derivative of a function only if $a = 0$.

3.2.5 Derivatives and Their Applications

A function f defined in an open interval containing the point x_0 is called differentiable at x_0 if

$$\lim_{h \to 0} \frac{f(x_0 + h) - f(x_0)}{h}$$

exists. In this case, the limit is called the derivative of f at x_0 and is denoted by $f'(x_0)$ or $\frac{df}{dx}(x_0)$. If the derivative is defined at every point of the domain of f, then f is simply called differentiable.

The derivative is the instantaneous rate of change. Geometrically, it is the slope of the tangent to the graph of the function. Because of this, where the derivative is positive the function is increasing, where the derivative is negative the function is decreasing, and on intervals where the derivative is zero the function is constant. Moreover, the maxima and minima of a differentiable function show up at points where the derivative is zero, the so-called critical points.

Let us present some applications of derivatives. We begin with an observation made by F. Pop during the grading of USA Mathematical Olympiad 1997 about a student's solution. The student reduced one of the problems to a certain inequality, and the question was whether this inequality is easy or difficult to prove. Here is the inequality and Pop's argument.

Example. Let a, b, c be positive real numbers such that $abc = 1$. Prove that

$$a^2 + b^2 + c^2 \leq a^3 + b^3 + c^3,$$

where equality holds if and only if $a = b = c = 1$.

Solution. If $a = b = c = 1$ there is nothing to check. So let us assume that at least one of a, b, c is not 1. We prove that the function

$$f(t) = a^t + b^t + c^t$$

is strictly increasing for $t \geq 0$. Its first derivative is

$$f'(t) = a^t \ln a + b^t \ln b + c^t \ln c,$$

for which we can tell only that $f'(0) = \ln abc = \ln 1 = 0$. However, the second derivative is $f''(t) = a^t (\ln a)^2 + b^t (\ln b)^2 + c^t (\ln c)^2$, which is clearly positive. We thus deduce that f' is strictly increasing, and so $f'(t) > f'(0) = 0$ for $t > 0$. Therefore f itself is strictly increasing for $t \geq 0$, and the conclusion follows. $\square$

And now an exciting example found in D. Buşneag, I. Maftei, *Themes for Mathematics Circles and Contests* (Scrisul Românesc, Craiova).

Example. Prove that

$$
\begin{vmatrix}
1+a_1 & 1 & \cdots & 1 \\
1 & 1+a_2 & \cdots & 1 \\
\vdots & \vdots & \ddots & \vdots \\
1 & 1 & \cdots & 1+a_n
\end{vmatrix}
= a_1 a_2 \cdots a_n \left(1 + \frac{1}{a_1} + \frac{1}{a_2} + \cdots + \frac{1}{a_n} \right).
$$

Solution. In general, if the entries of a matrix depend in a differentiable manner on a parameter x,

$$
\begin{pmatrix}
a_{11}(a) & a_{12}(x) & \cdots & a_{1n}(x) \\
a_{21}(a) & a_{22}(x) & \cdots & a_{2n}(x) \\
\vdots & \vdots & \ddots & \vdots \\
a_{n1}(a) & a_{n2}(x) & \cdots & a_{nn}(x)
\end{pmatrix},
$$

then the determinant is a differentiable function of x, and its derivative is equal to

$$
\begin{vmatrix}
a'_{11}(a) & a'_{12}(x) & \cdots & a'_{1n}(x) \\
a_{21}(a) & a_{22}(x) & \cdots & a_{2n}(x) \\
\vdots & \vdots & \ddots & \vdots \\
a_{n1}(a) & a_{n2}(x) & \cdots & a_{nn}(x)
\end{vmatrix}
+
\begin{vmatrix}
a_{11}(a) & a_{12}(x) & \cdots & a_{1n}(x) \\
a'_{21}(a) & a'_{22}(x) & \cdots & a'_{2n}(x) \\
\vdots & \vdots & \ddots & \vdots \\
a_{n1}(a) & a_{n2}(x) & \cdots & a_{nn}(x)
\end{vmatrix}
+ \cdots
$$

$$
+
\begin{vmatrix}
a_{11}(a) & a_{12}(x) & \cdots & a_{1n}(x) \\
a_{21}(a) & a_{22}(x) & \cdots & a_{2n}(x) \\
\vdots & \vdots & \ddots & \vdots \\
a'_{n1}(a) & a'_{n2}(x) & \cdots & a'_{nn}(x)
\end{vmatrix}.
$$

This follows by applying the product rule to the formula of the determinant. For our problem, consider the function

$$
f(x) =
\begin{vmatrix}
x+a_1 & x & \cdots & x \\
x & x+a_2 & \cdots & x \\
\vdots & \vdots & \ddots & \vdots \\
x & x & \cdots & x+a_n
\end{vmatrix}.
$$

Its first derivative is

$$
f'(x) =
\begin{vmatrix}
1 & 1 & \cdots & 1 \\
x & x+a_2 & \cdots & x \\
\vdots & \vdots & \ddots & \vdots \\
x & x & \cdots & x+a_n
\end{vmatrix}
+
\begin{vmatrix}
x+a_1 & x & \cdots & x \\
1 & 1 & \cdots & 1 \\
\vdots & \vdots & \ddots & \vdots \\
x & x & \cdots & x+a_n
\end{vmatrix}
+ \cdots
$$

$$
+
\begin{vmatrix}
x+a_1 & x & \cdots & x \\
x & x+a_2 & \cdots & x \\
\vdots & \vdots & \ddots & \vdots \\
1 & 1 & \cdots & 1
\end{vmatrix}.
$$

Proceeding one step further, we see that the second derivative of f consists of two types of determinants: some that have a row of 0's, and others that have two rows of 1's. In both cases the determinants are equal to zero, showing that $f''(x) = 0$. It follows that f itself must be a linear function,

$$f(x) = f(0) + f'(0)x.$$

One finds immediately that $f(0) = a_1 a_2 \cdots a_n$. To compute

$$f'(0) = \begin{vmatrix} 1 & 1 & \ldots & 1 \\ 0 & a_2 & \ldots & 0 \\ \vdots & \vdots & \ddots & \vdots \\ 0 & 0 & \ldots & a_n \end{vmatrix} + \begin{vmatrix} a_1 & 0 & \ldots & 0 \\ 1 & 1 & \ldots & 1 \\ \vdots & \vdots & \ddots & \vdots \\ 0 & 0 & \ldots & a_n \end{vmatrix} + \cdots + \begin{vmatrix} a_1 & 0 & \ldots & 0 \\ 0 & a_2 & \ldots & 0 \\ \vdots & \vdots & \ddots & \vdots \\ 1 & 1 & \ldots & 1 \end{vmatrix}$$

expand each determinant along the row of 1's. The answer is

$$f'(0) = a_2 a_3 \cdots a_n + a_1 a_3 \cdots a_n + \cdots + a_1 a_2 \cdots a_{n-1},$$

whence

$$f(x) = a_1 a_2 \cdots a_n \left[1 + \left(\frac{1}{a_1} + \frac{1}{a_2} + \cdots + \frac{1}{a_n} \right) x \right].$$

Substituting $x = 1$, we obtain the formula from the statement.   □

487. Find all positive real solutions to the equation $2^x = x^2$.

488. Let $f : \mathbb{R} \to \mathbb{R}$ be given by

$$f(x) = (x - a_1)(x - a_2) + (x - a_2)(x - a_3) + (x - a_3)(x - a_1)$$

with a_1, a_2, a_3 real numbers. Prove that $f(x) \geq 0$ for all real numbers x if and only if $a_1 = a_2 = a_3$.

489. Let a and b be positive real numbers. Show that for all positive integers n,

$$(n - 1)a^n + b^n \geq na^{n-1}b,$$

with equality if and only if $a = b$.

490. Determine $\max\limits_{z \in \mathbb{C}, |z|=1} |z^3 - z + 2|$.

491. Find the minimum of the function $f : \mathbb{R} \to \mathbb{R}$,

$$f(x) = \frac{(x^2 - x + 1)^3}{x^6 - x^3 + 1}.$$

492. How many real solutions does the equation

$$\sin(\sin(\sin(\sin(\sin x)))) = \frac{x}{3}$$

have?

493. Let $(a_n)_n$ be a sequence of real numbers satisfying $e^{a_n} + na_n = 2$, for all $n \geq 1$. Evaluate

$$\lim_{n \to \infty} n(1 - na_n).$$

494. Let n be an even integer greater than 2 and x, y real numbers such that

$$x^n + y^{n+1} > n^n \text{ and } y^n + x^{n+1} > n^n.$$

Show that $x + y > 1$.

495. Find all functions $f : \mathbb{R} \to \mathbb{R}$, satisfying

$$|f(x) - f(y)| \leq |x - y|^2$$

for all $x, y \in \mathbb{R}$.

496. Let $f : \mathbb{R} \to \mathbb{R}$ be a differentiable function. Show that if $\lim_{x \to \infty}(f(x) + f'(x)) = 0$, then $\lim_{x \to \infty} f(x) = \lim_{x \to \infty} f'(x) = 0$.

497. Let $f : \mathbb{R} \to \mathbb{R}$ be a continuous function. For $x \in \mathbb{R}$ we define

$$g(x) = f(x) \int_0^x f(t)dt.$$

Show that if g is a nonincreasing function, then f is identically equal to zero.

498. Let f be a function having a continuous derivative on $[0, 1]$ and with the property that $0 < f'(x) \leq 1$. Also, suppose that $f(0) = 0$. Prove that

$$\left[\int_0^1 f(x)dx \right]^2 \geq \int_0^1 [f(x)]^3 dx.$$

Give an example in which equality occurs.

499. Find all functions $f : [0, \infty) \to [0, \infty)$ differentiable at $x = 1$ and satisfying

$$f(x^3) + f(x^2) + f(x) = x^3 + x^2 + x \text{ for all } x \geq 0.$$

500. Let x, y, z be nonnegative real numbers. Prove that
(a) $(x + y + z)^{x+y+z} x^x y^y z^z \leq (x + y)^{x+y} (y + z)^{y+z} (z + x)^{z+x}$.
(b) $(x + y + z)^{(x+y+z)^2} x^{x^2} y^{y^2} z^{z^2} \geq (x + y)^{(x+y)^2} (y + z)^{(y+z)^2} (z + x)^{(z+x)^2}$.

Derivatives have an important application to the computation of limits.

L'Hôpital's rule. *For an open interval I, if the functions f and g are differentiable on $I \setminus \{x_0\}$, $g'(x) \neq 0$ for $x \in I$, $x \neq x_0$, and either $\lim_{x \to x_0} f(x) = \lim_{x \to x_0} g(x) = 0$ or $\lim_{x \to x_0} |f(x)| = \lim_{x \to x_0} |g(x)| = \infty$, and if additionally $\lim_{x \to x_0} \frac{f'(x)}{g'(x)}$ exists, then $\lim_{x \to x_0} \frac{f(x)}{g(x)}$ exists and*

$$\lim_{x \to x_0} \frac{f(x)}{g(x)} = \lim_{x \to x_0} \frac{f'(x)}{g'(x)}.$$

Let us see how L'Hôpital's rule is applied.

Example. Prove that if $f : \mathbb{R} \to \mathbb{R}$ is a differentiable function with the property that $\lim\limits_{x \to x_0} f(x)$ exists and is finite, and if $\lim\limits_{x \to x_0} x f'(x)$ exists, then this limit is equal to zero.

Solution. If the limit $\lim\limits_{x \to x_0} x f'(x)$ exists, then so does $\lim\limits_{x \to x_0} (x f(x))'$, and the latter is equal to $\lim\limits_{x \to x_0} f(x) + \lim\limits_{x \to x_0} x f'(x)$. Applying L'Hôpital's rule yields

$$\lim_{x \to x_0} (x f(x))' == \lim_{x \to x_0} \frac{(x f(x))'}{x'} = \lim_{x \to x_0} \frac{x f(x)}{x} = \lim_{x \to x_0} f(x).$$

Therefore,

$$\lim_{x \to x_0} f(x) = \lim_{x \to x_0} f(x) + \lim_{x \to x_0} x f'(x),$$

and it follows that $\lim\limits_{x \to x_0} x f'(x) = 0$, as desired.   □

More problems follow.

501. Let f and g be n-times continuously differentiable functions in a neighborhood of a point a, such that $f(a) = g(a) = \alpha$, $f'(a) = g'(a), \ldots, f^{(n-1)}(a) = g^{(n-1)}(a)$, and $f^{(n)}(a) \neq g^{(n)}(a)$. Find, in terms of α,

$$\lim_{x \to a} \frac{e^{f(x)} - e^{g(x)}}{f(x) - g(x)}.$$

502. For any real number $\lambda \geq 1$, denote by $f(\lambda)$ the real solution to the equation

$$x(1 + \ln x) = \lambda.$$

Prove that

$$\lim_{\lambda \to \infty} \frac{f(\lambda)}{\dfrac{\lambda}{\ln \lambda}} = 1.$$

503. Let $f_0(x) = x$, $g_0(x) = x$ and for $n \geq 0$,

$$f_{n+1}(x) = \ln[1 + 3(f_n(x))^2], \quad g_{n+1}(x) = \ln[1 + 5(g_n(x))^2].$$

Prove that the limit

$$\lim_{x \to 0} \frac{f_{2014}(x)}{g_{2014}(x)}$$

exists and compute it.

3.2.6 The Mean Value Theorem

In the old days, when mathematicians were searching for methods to solve polynomial equations, an essential tool was Rolle's theorem.

Rolle's theorem. *If $f : [a, b] \to \mathbb{R}$ is continuous on $[a, b]$, differentiable on (a, b), and satisfies $f(a) = f(b)$, then there exists $c \in (a, b)$ such that $f'(c) = 0$.*

Its standard use was on problems like the following.

Example. Prove that the Legendre polynomial

$$P_n(x) = \frac{d^n}{dx^n}(x^2 - 1)^n$$

has n distinct zeros in the interval $(-1, 1)$.

Solution. Consider the polynomial function $Q_n(x) = (x^2 - 1)^n$. Its zeros $x = 1$ and $x = -1$ have multiplicity n. Therefore, for every $k < n$, the kth derivative $Q_n^{(k)}(x)$ has 1 and -1 as zeros. We prove by induction on k that for $1 < k \le n$, $Q_n^{(k)}(x)$ has k distinct zeros in $(-1, 1)$.

By Rolle's theorem this is true for $k = 1$. Assume that the property is true for $k < n$, and let us prove it for $k + 1$. The polynomial $Q_n^{(k)}(x)$ has $k + 2$ zeros $x_0 = -1 < x_1 < \cdots < x_k < x_{k+1} = 1$. By Rolle's theorem, between any two consecutive zeros of the function there is a zero of the derivative $Q_n^{(k+1)}(x)$. Hence $Q_n^{(k+1)}(x)$ has $k + 1$ distinct zeros between -1 and 1. This completes the induction.

In particular, $Q_n^{(n)}(x) = P_n(x)$ has n distinct zeros between -1 and 1, as desired. $\square$

Rolle's theorem applied to the function $\phi : [a, b] \to \mathbb{R}$,

$$\phi(x) = \begin{vmatrix} f(x) & g(x) & 1 \\ f(a) & g(a) & 1 \\ f(b) & g(b) & 1 \end{vmatrix},$$

yields the following theorem.

Cauchy's theorem. *If $f, g : [a, b] \to \mathbb{R}$ are two functions, continuous on $[a, b]$ and differentiable on (a, b), then there exists a point $c \in (a, b)$ such that*

$$(f(b) - f(a))g'(c) = (g(b) - g(a))f'(c).$$

In the particular case $g(x) = x$, we have the following.

The mean value theorem (Lagrange). *If $f : [a, b] \to \mathbb{R}$ is a function that is continuous on $[a, b]$ and differentiable on (a, b), then there exists $c \in (a, b)$ such that*

$$f'(c) = \frac{f(b) - f(a)}{b - a}.$$

It should be noted that the mean value theorem was already used, implicitly, in the previous section; the monotonicity test and l'Hospital's theorem both rely on it, but in this section we make explicit use of it, such as in solving the following problem of D. Andrica.

Example. Let $f : \mathbb{R} \to \mathbb{R}$ be a twice-differentiable function, with positive second derivative. Prove that

$$f(x + f'(x)) \geq f(x),$$

for any real number x.

Solution. If x is such that $f'(x) = 0$, then the relation holds with equality. If for a certain x, $f'(x) < 0$, then the mean value theorem applied on the interval $[x + f'(x), x]$ yields

$$f(x) - f(x + f'(x)) = f'(c)(-f'(x)),$$

for some c with $x + f'(x) < c < x$. Because the second derivative is positive, f' is increasing; hence $f'(c) < f'(x) < 0$. Therefore, $f(x) - f(x + f'(x)) < 0$, which yields the required inequality.

In the case $f'(x) > 0$, by the same argument $f(x + f'(x)) - f(x) = f'(x)f'(c)$ for c between x and $x + f'(x)$, and $f'(c) > f'(x) > 0$. We obtain again $f(x) - f(x + f'(x)) < 0$, as desired. $\square$

Example. Find all real solutions to the equation

$$4^x + 6^{x^2} = 5^x + 5^{x^2}.$$

Solution. This problem was given at the 1984 Romanian Mathematical Olympiad, being proposed by M. Chiriţă. The solution runs as follows.

Note that $x = 0$ and $x = 1$ satisfy the equation from the statement. Are there other solutions? The answer is no, but to prove it we use the amazing idea of treating the numbers 4, 5, 6 as variables and the presumably new solution x as a constant.

Thus let us consider the function $f(t) = t^{x^2} + (10 - t)^x$. The fact that x satisfies the equation from the statement translates to $f(5) = f(6)$. By Rolle's theorem there exists $c \in (5, 6)$, such that $f'(c) = 0$. This means that

$$x^2 c^{x^2 - 1} - x(10 - c)^{x-1} = 0, \quad \text{or} \quad xc^{x^2-1} = (10 - c)^{x-1}.$$

Because exponentials are positive, this implies that x is positive.

If $x > 1$, then $xc^{x^2-1} > c^{x^2-1} > c^{x-1} > (10 - c)^{x-1}$, which is impossible since the first and the last terms in this chain of inequalities are equal. Here we used the fact that $c > 5$.

If $0 < x < 1$, then $xc^{x^2-1} < xc^{x-1}$. Let us prove that $xc^{x-1} < (10 - c)^{x-1}$. With the substitution $y = x - 1$, $y \in (-1, 0)$, the inequality can be rewritten as $y + 1 < \left(\frac{10-c}{c}\right)^y$. The exponential has base less than 1, so it is decreasing, while the linear function on the left is increasing. The two meet at $y = 0$. The inequality follows. Using it we conclude again that xc^{x^2-1} cannot be equal to $(10 - c)^{x-1}$. This shows that a third solution to the equation from the statement does not exist. So the only solutions to the given equation are $x = 0$ and $x = 1$. $\square$

Below you will find a variety of problems based on the above-mentioned theorems (Rolle, Lagrange, Cauchy). Try to solve them, remembering that "good judgment comes from experience, and experience comes from bad judgment" (Barry LePatner).

504. Prove that not all zeros of the polynomial $P(x) = x^4 - \sqrt{7}x^3 + 4x^2 - \sqrt{22}x + 15$ are real.

505. Let $f : [a, b] \to \mathbb{R}$ be a function, continuous on $[a, b]$ and differentiable on (a, b). Prove that if there exists $c \in (a, b)$ such that

$$\frac{f(b) - f(c)}{f(c) - f(a)} < 0,$$

then there exists $\xi \in (a, b)$ such that $f'(\xi) = 0$.

506. For $x \geq 2$ prove the inequality

$$(x + 1) \cos \frac{\pi}{x + 1} - x \cos \frac{\pi}{x} > 1.$$

507. Let $n > 1$ be an integer, and let $f : [a, b] \to \mathbb{R}$ be a continuous function, n-times differentiable on (a, b), with the property that the graph of f has $n + 1$ collinear points. Prove that there exists a point $c \in (a, b)$ with the property that $f^{(n)}(c) = 0$.

508. Let $f : [a, b] \to \mathbb{R}$ be a function, continuous on $[a, b]$ and differentiable on (a, b). Let $M(\alpha, \beta)$ be a point on the line passing through the points $(a, f(a))$ and $(b, f(b))$ with $\alpha \notin [a, b]$. Prove that there exists a line passing through M that is tangent to the graph of f.

509. Let $f : [a, b] \to \mathbb{R}$ be a function, continuous on $[a, b]$ and twice differentiable on (a, b). If $f(a) = f(b)$ and $f'(a) = f'(b)$, prove that for every real number λ the equation

$$f''(x) - \lambda(f'(x))^2 = 0$$

has at least one solution in the interval (a, b).

510. Prove that there are no positive numbers x and y such that

$$x2^y + y2^{-x} = x + y.$$

511. Let α be a real number such that n^α is an integer for every positive integer n. Prove that α is a nonnegative integer.

512. Find all real solutions to the equation

$$6^x + 1 = 8^x - 27^{x-1}.$$

513. Let $P(x)$ be a polynomial with real coefficients such that for every positive integer n, the equation $P(x) = n$ has at least one rational root. Prove that $P(x) = ax + b$ with a and b rational numbers.

3.2.7 Convex Functions

A function is called convex if any segment with endpoints on its graph lies above the graph itself. The picture you should have in mind is Figure 21. Formally, if D is an interval of the real axis, or more generally a convex subset of a vector space, then a function $f : D \to \mathbb{R}$ is called convex if

$$f(\lambda x + (1 - \lambda)y) \leq \lambda f(x) + (1 - \lambda)f(y), \text{ for all } x, y \in D, \ \lambda \in (0, 1).$$

Here we should remember that a set D is called convex if for any $x, y \in D$ and $\lambda \in (0, 1)$ the point $\lambda x + (1 - \lambda)y$ is also in D, which geometrically means that D is an intersection of half-spaces.

A function f is called concave if $-f$ is convex. If f is both convex and concave, then f is linear, i.e., $f(x) = ax + b$ for some constants a and b.

Proposition. *A twice-differentiable function on an interval is convex if and only if its second derivative is nonnegative.*

In general, a twice-differentiable function defined on a convex domain in $\mathbb{R}^n$ is convex if at any point its Hessian matrix is semipositive definite. This is a way of saying that modulo a local change of coordinates, around each point the function f is of the form

$$f(x_1, x_2, \ldots, x_n) = \phi(x_1, x_2, \ldots, x_n) + x_1^2 + x_2^2 + \cdots + x_k^2,$$

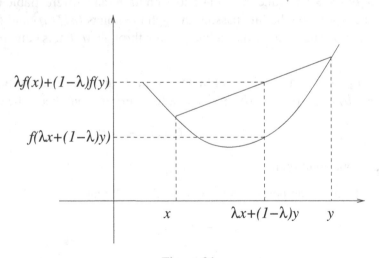

Figure 21

where $k \leq n$ and $\phi(x_1, x_2, \ldots, x_n)$ is linear.

As an application, we use convexity to prove Hölder's inequality.

Hölder's inequality. *If $x_1, x_2, \ldots, x_n, y_1, y_2, \ldots, y_n, p$ and q are positive numbers with $\frac{1}{p} + \frac{1}{q} = 1$, then*

$$\sum_{i=1}^{n} x_i y_i \leq \left(\sum_{i=1}^{n} x_i^p \right)^{1/p} \left(\sum_{i=1}^{n} y_i^q \right)^{1/q},$$

with equality if and only if the two vectors $(x_1, x_2, \ldots, x_n)$ and $(y_1, y_2, \ldots, y_n)$ are parallel.

Proof. The second derivative of $f : (0, \infty) \to \mathbb{R}$, $f(x) = \ln x$, is $f''(x) = -\frac{1}{x^2}$, which is negative. So this function is concave. Setting $\lambda = \frac{1}{p}$, we obtain

$$\ln X^{1/p} Y^{1/q} = \frac{1}{p} \ln X + \frac{1}{q} \ln Y \leq \ln \left(\frac{1}{p} X + \frac{1}{q} Y \right), \quad \text{for all } X, Y > 0;$$

hence

$$X^{1/p} Y^{1/q} \leq \frac{1}{p} X + \frac{1}{q} Y.$$

Using this fact, if we let $X = \sum_i x_i^p$ and $Y = \sum_i y_i^q$, then

$$\frac{1}{X^{1/p} Y^{1/q}} \sum_{i=1}^{n} x_i y_i = \sum_{i=1}^{n} \left(\frac{x_i^p}{X} \right)^{1/p} \left(\frac{y_i^q}{Y} \right)^{1/q} \leq \sum_{i=1}^{n} \left(\frac{1}{p} \cdot \frac{x_i^p}{X} + \frac{1}{q} \cdot \frac{y_i^q}{Y} \right)$$

$$= \left(\frac{1}{p} + \frac{1}{q} \right) = 1.$$

Hence

$$\sum_{i=1}^{n} x_i y_i \leq X^{1/p} Y^{1/q} = \left(\sum_{i=1}^{n} x_i^p \right)^{1/p} \left(\sum_{i=1}^{n} y_i^q \right)^{1/q},$$

and the inequality is proved. $\square$

By analogy, a sequence $(a_n)_{n \geq 0}$ is called convex if

$$a_n \leq \frac{a_{n+1} + a_{n-1}}{2}, \quad \text{for all } n \geq 1,$$

and concave if $(-a_n)_n$ is convex. Equivalently, a sequence is convex if its second difference (derivative) is nonnegative, and concave if its second difference is nonpositive. The following example motivates why convex sequences and functions should be studied together.

Example. Let $(a_n)_n$ be a bounded convex sequence. Prove that

$$\lim_{n \to \infty} (a_{n+1} - a_n) = 0.$$

Solution. A bounded convex function on $(0, \infty)$ has a horizontal asymptote, so its derivative tends to zero at infinity. Our problem is the discrete version of this result. The first derivative of the sequence is $b_n = a_{n+1} - a_n$, $n \geq 1$. The convexity condition can be written as $a_{n+1} - a_n \geq a_n - a_{n-1}$, which shows that $(b_n)_n$ is increasing. Since $(a_n)_n$ is bounded, $(b_n)_n$ is bounded too, and being monotonic, by the Weierstrass theorem it converges at a finite limit L. If $L > 0$, then b_n eventually becomes positive, so a_n becomes increasing because it has a positive derivative. Again by the Weierstrass theorem, a_n converges to some limit l, and then $L = l - l = 0$, a contradiction. A similar argument rules out the case $L < 0$. We are left with the only possibility $L = 0$. $\square$

And now some problems.

514. Prove that

$$\sqrt[3]{3 + \sqrt[3]{3}} + \sqrt[3]{3 - \sqrt[3]{3}} < 2\sqrt[3]{3}.$$

515. Let $x_1, x_2, \ldots, x_n$ be real numbers. Find the real numbers a that minimize the expression

$$|a - x_1| + |a - x_2| + \cdots + |a - x_n|.$$

516. Let $a, b > 0$ and $x, c > 1$. Prove that

$$x^{a^c} + x^{b^c} \geq 2x^{(ab)^{c/2}}.$$

517. Prove that

$$(\sin x)^{\sin x} < (\cos x)^{\cos x}$$

for all $x \in (0, \frac{\pi}{4})$.

518. A triangle has side lengths $a \geq b \geq c$ and vertices of measures $A, B,$ and C, respectively. Prove that

$$Ab + Bc + Ca \geq Ac + Ba + Cb.$$

519. Prove that for $a, b \geq \frac{1}{2}$,

$$\left(\frac{a^2 - b^2}{2}\right)^2 \geq \sqrt{\frac{a^2 + b^2}{2}} - \frac{a + b}{2}.$$

520. Show that if a function $f : [a, b] \to \mathbb{R}$ is convex, then it is continuous on (a, b).

521. Prove that a continuous function defined on a convex domain (for example, on an interval of the real axis) is convex if and only if

$$f\left(\frac{x + y}{2}\right) \leq \frac{f(x) + f(y)}{2}, \quad \text{for all } x, y \in D.$$

522. Call a real-valued function *very convex* if

$$\frac{f(x) + f(y)}{2} \geq f\left(\frac{x + y}{2}\right) + |x - y|$$

holds for all real numbers x and y. Prove that no very convex function exists.

523. Let $f : [a, b] \to \mathbb{R}$ be a convex function. Prove that

$$f(x) + f(y) + f(z) + 3f\left(\frac{x + y + z}{3}\right) \geq 2\left[f\left(\frac{x + y}{2}\right) + f\left(\frac{y + z}{2}\right) + f\left(\frac{z + x}{2}\right)\right],$$

for all $x, y, z \in [a, b]$.

524. Prove that if a sequence of positive real numbers $(b_n)_n$ has the property that $(a^n b_n)_n$ is a convex sequence for all real numbers a, then the sequence $(\ln b_n)_n$ is also convex.

525. Find the largest constant C such that for every $n \geq 3$ and every positive concave sequence $(a_k)_{k=1}^n$,

$$\left(\sum_{k=1}^n a_k \right)^2 \geq C(n-1) \sum_{k=1}^n a_k^2.$$

A convex function on a closed interval attains its maximum at an endpoint of the interval. We illustrate how this fact can be useful with a problem from *Timişoara Mathematics Gazette*, proposed by V. Cârtoaje and M. Lascu.

Example. Let $a, b, c, d \in [1, 3]$. Prove that

$$(a + b + c + d)^2 \geq 3(a^2 + b^2 + c^2 + d^2).$$

Solution. Divide by 2 and move everything to one side to obtain the equivalent inequality

$$a^2 + b^2 + c^2 + d^2 - 2ab - 2ac - 2ad - 2bc - 2bd - 2cd \leq 0.$$

Now we recognize the expression on the left to be a convex function in each variable. So the maximum is attained for some choice of $a, b, c, d = 1$ or 3. If k of these numbers are equal to 3, and $4 - k$ are equal to 1, where k could be 1, 2, 3, or 4, then the original inequality becomes

$$(3k + 4 - k)^2 = 3(9k + 4 - k).$$

Dividing by 3, we obtain $k^2 + 4k + 3 \geq 6k + 3$, or $(k - 1)^2 \geq 0$, which is clearly true. The inequality is proved. Equality occurs when one of the numbers a, b, c, d is equal to 3 and the other three are equal to 1. $\square$

Here are additional problems of this kind.

526. Let α, β and γ be three fixed positive numbers and $[a, b]$ a given interval. Find x, y, z in $[a, b]$ for which the expression

$$E(x, y, z) = \alpha(x - y)^2 + \beta(y - z)^2 + \gamma(z - x)^2$$

has maximal value.

527. Let $0 < a < b$ and $t_i \geq 0$, $i = 1, 2, \ldots, n$. Prove that for any $x_1, x_2, \ldots, x_n \in [a, b]$,

$$\left(\sum_{i=1}^n t_i x_i \right) \left(\sum_{i=1}^n \frac{t_i}{x_i} \right) \leq \frac{(a+b)^2}{4ab} \left(\sum_{i=1}^n t_i \right)^2.$$

528. Prove that for any natural number $n \geq 2$ and any $|x| \leq 1$,

$$(1 + x)^n + (1 - x)^n \leq 2^n.$$

529. Prove that for any positive real numbers a, b, c the following inequality holds

$$\frac{a+b+c}{3} - \sqrt[3]{abc} \le \max\{(\sqrt{a} - \sqrt{b})^2, (\sqrt{b} - \sqrt{c})^2, (\sqrt{c} - \sqrt{a})^2\}.$$

530. Let f be a real-valued continuous function on $\mathbb{R}$ satisfying

$$f(x) \le \frac{1}{2h} \int_{x-h}^{x+h} f(y)dy, \text{ for all } x \in \mathbb{R} \text{ and } h > 0.$$

Prove that (a) the maximum of f on any closed interval is assumed at one of the endpoints, and (b) the function f is convex.

An important property of convex (respectively, concave) functions is known as Jensen's inequality.

Jensen's inequality. *For a convex function f let $x_1, x_2, \ldots, x_n$ be points in its domain and let $\lambda_1, \lambda_2, \ldots, \lambda_n$ be positive numbers with $\lambda_1 + \lambda_2 + \cdots + \lambda_n = 1$. Then*

$$f(\lambda_1 x_1 + \lambda_2 x_2 + \cdots + \lambda_n x_n) \le \lambda_1 f(x_1) + \lambda_2 f(x_2) + \cdots + \lambda_n f(x_n).$$

If f is nowhere linear and the x_i's are not all equal, then the inequality is strict. The inequality is reversed for a concave function.

Proof. The proof is by induction on n. The base case is the definition of convexity. Let us assume that the inequality is true for any $n - 1$ points x_i and any $n - 1$ weights λ_i. Consider n points and weights, and let $\lambda = \lambda_1 + \ldots + \lambda_{n-1}$. Note that $\lambda + \lambda_n = 1$ and

$$\frac{\lambda_1}{\lambda} + \frac{\lambda_2}{\lambda} + \cdots + \frac{\lambda_{n-1}}{\lambda} = 1.$$

Using the base case and the inductive hypothesis we can write

$$
\begin{aligned}
f(\lambda_1 x_1 + \cdots + \lambda_{n-1} x_{n-1} + \lambda_n x_n) &= f\left(\lambda\left(\frac{\lambda_1}{\lambda} x_1 + \cdots + \frac{\lambda_{n-1}}{\lambda} x_{n-1}\right) + \lambda_n x_n\right) \\
&\le \lambda f\left(\frac{\lambda_1}{\lambda} x_1 + \cdots + \frac{\lambda_{n-1}}{\lambda} x_{n-1}\right) + \lambda_n f(x_n) \\
&\le \lambda\left(\frac{\lambda_1}{\lambda} f(x_1) + \cdots + \frac{\lambda_{n-1}}{\lambda} f(x_{n-1})\right) + \lambda_n f(x_n) \\
&= \lambda_1 f(x_1) + \cdots + \lambda_{n-1} f(x_{n-1}) + \lambda_n f(x_n),
\end{aligned}
$$

as desired. For the case of concave functions, reverse the inequalities. $\square$

As an application, we prove the following.

The generalized mean inequality. *Given the positive numbers $x_1, x_2, \ldots, x_n$ and the positive weights $\lambda_1, \lambda_2, \ldots, \lambda_n$ with $\lambda_1 + \lambda_2 + \cdots + \lambda_n = 1$, the following inequality holds:*

$$\lambda_1 x_1 + \lambda_2 x_2 + \cdots + \lambda_n x_n \ge x_1^{\lambda_1} x_2^{\lambda_2} \cdots x_n^{\lambda_n}.$$

Solution. Simply write Jensen's inequality for the concave function $f(x) = \ln x$, then exponentiate.   □

For $\lambda_1 = \lambda_2 = \cdots = \lambda_n = \dfrac{1}{n}$ one obtains the AM-GM inequality. The Cauchy-Schwarz inequality is also a direct consequence of Jensen's inequality.

Cauchy-Schwarz inequality. *If $a_1, a_2, \ldots, a_n$ and $b_1, b_2, \ldots, b_n$ are real numbers, then*

$$(a_1^2 + a_2^2 + \cdots + a_n^2)(b_1^2 + b_2^2 + \cdots + b_n^2) \geq (a_1b_1 + a_2b_2 + \cdots + a_nb_n)^2.$$

Proof. We will apply Jensen's inequality to the convex function $f(x) = x^2$. In this case, Jensen's inequality reads

$$\lambda_1 x_1^2 + \lambda_2 x_2^2 + \cdots + \lambda_n x_n^2 \geq (\lambda_1 x_1 + \lambda_2 x_2 + \cdots + \lambda_n x_n)^2,$$

for all nonnegative λ_i with the property that $\lambda_1 + \lambda_2 + \cdots + \lambda_n = 1$.
Rewrite the Cauchy-Schwarz inequality as

$$a_1^2(b_1^2 + b_2^2 + \cdots + b_n^2) + a_2^2(b_1^2 + b_2^2 + \cdots + b_n^2)^2 + \cdots + a_n^2(b_1^2 + b_2^2 + \cdots + b_n^2)^2$$
$$\geq (a_1b_1 + a_2b_2 + \cdots + a_nb_n)^2,$$

or

$$\sum_{k=1}^{n} \frac{b_k^2}{b_1^2 + b_2^2 + \cdots + b_n^2} \cdot \frac{a_k^2}{b_k^2}(b_1^2 + b_2^2 + \cdots + b_n^2)$$

$$\geq \left(\sum_{k=1}^{n} \frac{b_k^2}{b_1^2 + b_2^2 + \cdots + b_n^2} \cdot \frac{a_k}{b_k}(b_1^2 + b_2^2 \cdots + b_n^2) \right)^2.$$

This is Jensen's inequality with

$$x_i = \frac{a_i}{b_i}(b_1^2 + b_2^2 + \cdots + b_n^2)$$

$$\lambda_i = \frac{b_i^2}{b_1^2 + b_2^2 + \cdots + b_n^2},$$

for $i = 1, 2, \ldots, n$.   □

All inequalities below are supposed to be proved using Jensen's inequality. One of these problems has appeared also in Section 2.1.6 where you were supposed to solve it using a different method.

531. Show that if A, B, C are the angles of a triangle, then

$$\sin A + \sin B + \sin C \geq \frac{3\sqrt{3}}{2}.$$

532. Let a_i, $i = 1, 2, \ldots, n$ be nonnegative numbers with $\sum_{i=1}^{n} a_i = 1$, and let $0 < x_i \leq 1$, $i = 1, 2, \ldots, n$. Prove that

$$\sum_{i=1}^{n} \frac{a_i}{1 + x_i} \leq \frac{1}{1 + x_1^{a_1} x_2^{a_2} \cdots x_n^{a_n}}.$$

533. Prove that for any three positive real numbers a_1, a_2, a_3,

$$\frac{a_1^2 + a_2^2 + a_3^2}{a_1^3 + a_2^3 + a_3^3} \geq \frac{a_1^3 + a_2^3 + a_3^3}{a_1^4 + a_2^4 + a_3^4}.$$

534. Let $0 < x_i < \pi$, $i = 1, 2, \ldots, n$, and set $x = \dfrac{x_1 + x_2 + \cdots + x_n}{n}$. Prove that

$$\prod_{i=1}^{n} \left(\frac{\sin x_i}{x_i} \right) \leq \left(\frac{\sin x}{x} \right)^n.$$

535. Let $n > 1$ and $x_1, x_2, \ldots, x_n > 0$ be such that $x_1 + x_2 + \cdots + x_n = 1$. Prove that

$$\frac{x_1}{\sqrt{1 - x_1}} + \frac{x_2}{\sqrt{1 - x_2}} + \cdots + \frac{x_n}{\sqrt{1 - x_n}} \geq \frac{\sqrt{x_1} + \sqrt{x_2} + \cdots + \sqrt{x_n}}{\sqrt{n - 1}}.$$

536. Prove that if $a, b, c, d > 0$ and $a \leq 1, a + b \leq 5, a + b + c \leq 14, a + b + c + d \leq 30$, then

$$\sqrt{a} + \sqrt{b} + \sqrt{c} + \sqrt{d} \leq 10.$$

3.2.8 Indefinite Integrals

"Anyone who stops learning is old, whether at twenty or eighty. Anyone who keeps learning stays young. The greatest thing in life is to keep your mind young." Following this advice of Henry Ford, let us teach you some clever tricks for computing indefinite integrals.

We begin by recalling the basic facts about indefinite integrals. Integration is the inverse operation to differentiation. The fundamental methods for computing integrals are the backward application of the chain rule, which takes the form

$$\int f(u(x))u'(x)dx = \int f(u)du$$

and shows up in the guise of the first and second substitutions, and integration by parts

$$\int u\,dv = uv - \int v\,du,$$

which comes from the product rule for derivatives. Otherwise, there is Jacobi's partial fraction decomposition method for computing integrals of rational functions, as well as standard substitutions such as the trigonometric and Euler's substitutions.

Now let us turn to our nonstandard examples.

Example. Compute

$$I_1 = \int \frac{\sin x}{\sin x + \cos x} dx \quad \text{and} \quad I_2 = \int \frac{\cos x}{\sin x + \cos x} dx.$$

Solution. The well-known approach is to use the substitution $\tan \frac{x}{2} = t$. But it is much simpler to write the system

$$I_1 + I_2 = \int \frac{\sin x + \cos x}{\sin x + \cos x} dx = \int 1 dx = x + C_1,$$

$$-I_1 + I_2 = \int \frac{\cos x - \sin x}{\sin x + \cos x} dx = \ln|\sin x + \cos x| + C_2,$$

and then solve to obtain

$$I_1 = \frac{1}{2}x - \frac{1}{2}\ln|\sin x + \cos x| + C_1' \quad \text{and} \quad I_2 = \frac{1}{2}x + \frac{1}{2}\ln|\sin x + \cos x| + C_2'. \qquad \square$$

We continue with a more difficult computation based on a substitution.

Example. For $a > 0$ compute the integral

$$\int \frac{1}{x\sqrt{x^{2a} + x^a + 1}} dx, \quad x > 0.$$

Solution. Factor an x^{2a} under the square root to transform the integral into

$$\int \frac{1}{x^{a+1}\sqrt{1 + \dfrac{1}{x^a} + \dfrac{1}{x^{2a}}}} dx = \int \frac{1}{\sqrt{\left(\dfrac{1}{x^a} + \dfrac{1}{2}\right)^2 + \dfrac{3}{4}}} \cdot \frac{1}{x^{a+1}} dx.$$

With the substitution $u = \dfrac{1}{x^a} + \dfrac{1}{2}$ the integral becomes

$$-\frac{1}{a}\int \frac{1}{\sqrt{u^2 + \dfrac{3}{4}}} du = -\frac{1}{a}\ln\left(u + \sqrt{u^2 + \dfrac{3}{4}}\right) + C$$

$$= -\frac{1}{a}\ln\left(\frac{1}{x^a} + \frac{1}{2} + \sqrt{\frac{1}{x^{2a}} + \frac{1}{x^a} + 1}\right) + C. \qquad \square$$

537. Compute the integral

$$\int (1 + 2x^2)e^{x^2} dx.$$

538. Compute

$$\int \frac{x + \sin x - \cos x - 1}{x + e^x + \sin x}\,dx.$$

539. Find

$$\int (x^6 + x^3)\sqrt[3]{x^3 + 2}\,dx.$$

540. Compute the integral

$$\int \frac{x^2 + 1}{x^4 - x^2 + 1}\,dx.$$

541. Compute

$$\int \sqrt{\frac{e^x - 1}{e^x + 1}}\,dx, \ x > 0.$$

542. Evaluate

$$\int \frac{1 + x^2 \ln x}{x + x^2 \ln x}\,dx$$

543. Find the antiderivatives of the function $f : [0, 2] \to \mathbb{R}$,

$$f(x) = \sqrt{x^3 + 2 - 2\sqrt{x^3 + 1}} + \sqrt{x^3 + 10 - 6\sqrt{x^3 + 1}}.$$

544. For a positive integer n, compute the integral

$$\int \frac{x^n}{1 + x + \dfrac{x^2}{2!} + \cdots + \dfrac{x^n}{n!}}\,dx.$$

545. Compute the integral

$$\int \frac{dx}{(1 - x^2)\sqrt[4]{2x^2 - 1}}.$$

546. Compute

$$\int \frac{x^4 + 1}{x^6 + 1}\,dx.$$

Give the answer in the form $\alpha \arctan \dfrac{P(x)}{Q(x)} + C$, $\alpha \in \mathbb{Q}$, and $P(x), Q(x) \in \mathbb{Z}[x]$.

3.2.9 Definite Integrals

Next, definite integrals. Here the limits of integration also play a role.

Example. Let $f : [0, 1] \to \mathbb{R}$ be a continuous function. Prove that

$$\int_0^\pi xf(\sin x)dx = \pi \int_0^{\frac{\pi}{2}} f(\sin x)dx.$$

Solution. We have

$$\int_0^\pi xf(\sin x)dx = \int_0^{\frac{\pi}{2}} xf(\sin x)dx + \int_{\frac{\pi}{2}}^\pi xf(\sin x)dx.$$

We would like to transform both integrals on the right into the same integral, and for that we need a substitution in the second integral that changes the limits of integration. This substitution should leave $f(\sin x)$ invariant, so it is natural to try $t = \pi - x$. The integral becomes

$$\int_0^{\frac{\pi}{2}} (\pi - t)f(\sin t)dt.$$

Adding the two, we obtain $\pi \int_0^{\frac{\pi}{2}} f(\sin x)dx$, as desired. $\square$

547. Compute the integral

$$\int_{-1}^1 \frac{\sqrt[3]{x}}{\sqrt[3]{1-x} + \sqrt[3]{1+x}} dx.$$

548. Compute

$$\int_0^\pi \frac{x \sin x}{1 + \sin^2 x} dx.$$

549. Compute

$$\int_0^{\sqrt{\frac{\pi}{3}}} \sin x^2 dx + \int_{-\sqrt{\frac{\pi}{3}}}^{\sqrt{\frac{\pi}{3}}} x^2 \cos x^2 dx.$$

550. Let a and b be positive real numbers. Compute

$$\int_a^b \frac{e^{\frac{x}{a}} - e^{\frac{b}{x}}}{x} dx.$$

551. Compute the integral

$$I = \int_0^1 \sqrt[3]{2x^3 - 3x^2 - x + 1} dx.$$

552. Compute the integral

$$\int_0^a \frac{dx}{x + \sqrt{a^2 - x^2}} \quad (a > 0).$$

553. Compute the integral

$$\int_0^{\frac{\pi}{4}} \ln(1 + \tan x)dx.$$

554. Find

$$\int_0^1 \frac{\ln(1 + x)}{1 + x^2} dx.$$

555. Compute

$$\int_0^\infty \frac{\ln x}{x^2 + a^2} dx,$$

where a is a positive constant.

556. Compute the integral

$$\int_0^{\frac{\pi}{2}} \frac{x \cos x - \sin x}{x^2 + \sin^2 x} dx.$$

557. Let α be a real number. Compute the integral

$$I(\alpha) = \int_{-1}^1 \frac{\sin \alpha \, dx}{1 - 2x \cos \alpha + x^2}.$$

558. Give an example of a function $f : (2, \infty) \to (0, \infty)$ with the property that

$$\int_2^\infty f^p(x)dx$$

is finite if and only if $p \in [2, \infty)$.

559. Let $f : \left[-\frac{\pi}{2}, \frac{\pi}{2}\right] \to (-1, 1)$ be a differentiable function whose derivative is continuous and nonnegative. Prove that there is $x_0 \in \left[-\frac{\pi}{2}, \frac{\pi}{2}\right]$ such that

$$(f(x_0))^2 + (f'(x_0))^2 \le 1.$$

There are special types of integrals that are computed recursively. We illustrate this with a proof of the Leibniz formula.

The Leibniz formula.

$$\frac{\pi}{4} = 1 - \frac{1}{3} + \frac{1}{5} - \frac{1}{7} + \cdots$$

Proof. To prove the formula we start by computing recursively the integral

$$I_n = \int_0^{\frac{\pi}{4}} \tan^{2n} x\, dx, \ n \geq 1.$$

We have

$$I_n = \int_0^{\frac{\pi}{4}} \tan^{2n} x\, dx = \int_0^{\frac{\pi}{4}} \tan^{2n-2} x \tan^2 x\, dx$$

$$= \int_0^{\frac{\pi}{4}} \tan^{2n-2} x(1 + \tan^2 x)dx - \int_0^{\frac{\pi}{4}} \tan^{2n-2} x\, dx$$

$$= \int_0^{\frac{\pi}{4}} \tan^{2n-2} x \sec^2 x\, dx - I_{n-1}.$$

The remaining integral can be computed using the substitution $\tan x = t$. In the end, we obtain the recurrence

$$I_n = \frac{1}{2n-1} - I_{n-1}, \ n \geq 1.$$

So for $n \geq 1$,

$$I_n = \frac{1}{2n-1} - \frac{1}{2n-3} + \cdots + \frac{(-1)^{n-2}}{3} + (-1)^{n-1} I_1,$$

with

$$I_1 = \int_0^{\frac{\pi}{4}} \tan^2 x\, dx = \int_0^{\frac{\pi}{4}} \sec^2 x\, dx - \int_0^{\frac{\pi}{4}} 1 dx = \tan x \Big|_0^{\frac{\pi}{4}} - \frac{\pi}{4} = 1 - \frac{\pi}{4}.$$

We find that

$$I_n = \frac{1}{2n-1} - \frac{1}{2n-3} + \cdots + \frac{(-1)^{n-2}}{3} + (-1)^{n-1} + (-1)^n \frac{\pi}{4}.$$

Because $\tan^{2n} x \to 0$ as $n \to \infty$ uniformly on any interval of the form $[0, a)$, $a < \frac{\pi}{4}$, it follows that $\lim_{n \to \infty} I_n = 0$. The Leibniz formula follows. $\qquad \square$

Below are more examples of this kind.

560. Let $P(x)$ be a polynomial with real coefficients. Prove that

$$\int_0^{\infty} e^{-x} P(x)dx = P(0) + P'(0) + P''(0) + \cdots$$

561. Let $n \geq 0$ be an integer. Compute the integral

$$\int_0^{\pi} \frac{1 - \cos nx}{1 - \cos x} dx.$$

562. Compute the integral

$$I_n = \int_0^{\frac{\pi}{2}} \sin^n x \, dx.$$

Use the answer to prove the Wallis formula

$$\lim_{n \to \infty} \left[\frac{2 \cdot 4 \cdot 6 \cdots 2n}{1 \cdot 3 \cdot 5 \cdots (2n-1)} \right]^2 \cdot \frac{1}{n} = \pi.$$

563. Compute

$$\int_{-\pi}^{\pi} \frac{\sin nx}{(1 + 2^x) \sin x} dx, \ n \geq 0.$$

3.2.10 Riemann Sums

The definite integral of a function is the area under the graph of the function. In approximating the area under the graph by a family of rectangles, the sum of the areas of the rectangles, called a Riemann sum, approximates the integral. When these rectangles have equal width, the approximation of the integral by Riemann sums reads

$$\lim_{n \to \infty} \frac{1}{n} \sum_{i=1}^{n} f(\xi_i) = \int_a^b f(x) dx,$$

where each ξ_i is a number in the interval $\left[a + \frac{i-1}{n}(b-a), a + \frac{i}{n}(b-a) \right]$.

Since the Riemann sum depends on the positive integer n, it can be thought of as the term of a sequence. Sometimes the terms of a sequence can be recognized as the Riemann sums of a function, and this can prove helpful for finding the limit of the sequence. Let us show how this works, following Hilbert's advice: "always start with an easy example."

Example. Compute the limit

$$\lim_{n \to \infty} \left(\frac{1}{n+1} + \frac{1}{n+2} + \cdots + \frac{1}{2n} \right).$$

Solution. If we rewrite as

$$\frac{1}{n} \left[\frac{1}{1 + \frac{1}{n}} + \frac{1}{1 + \frac{2}{n}} + \cdots + \frac{1}{1 + \frac{n}{n}} \right],$$

we recognize the Riemann sum of the function $f : [0, 1] \to \mathbb{R}, f(x) = \frac{1}{1+x}$ associated to the subdivision $x_0 = 0 < x_1 = \frac{1}{n} < x_2 = \frac{2}{n} < \cdots < x_n = \frac{n}{n} = 1$, with the intermediate points $\xi_i = \frac{i}{n} \in [x_i, x_{i+1}]$. It follows that

$$\lim_{n \to \infty} \left(\frac{1}{n+1} + \frac{1}{n+2} + \cdots + \frac{1}{2n} \right) = \int_0^1 \frac{1}{1+x} = \ln(1+x) \Big|_0^1 = \ln 2,$$

and the problem is solved.   □

We continue with a beautiful example from the book of G. Pólya, G. Szegö, *Aufgaben und Lehrsatze aus der Analysis* (Springer-Verlag, 1964).

Example. Denote by G_n the geometric mean of the binomial coefficients

$$\binom{n}{0}, \binom{n}{1}, \ldots, \binom{n}{n}.$$

Prove that

$$\lim_{n \to \infty} \sqrt[n]{G_n} = \sqrt{e}.$$

Solution. We have

$$\binom{n}{0}\binom{n}{1} \cdots \binom{n}{n} = \prod_{k=0}^{n} \frac{n!}{k!(n-k)!} = \frac{(n!)^{n+1}}{(1!2! \cdots n!)^2}$$

$$= \prod_{k=1}^{n} (n+1-k)^{n+1-2k} = \prod_{k=1}^{n} \left(\frac{n+1-k}{n+1}\right)^{n+1-2k}.$$

The last equality is explained by $\sum_{k=1}^{n}(n+1-2k) = 0$, which shows that the denominator is just $(n+1)^0 = 1$. Therefore,

$$G_n = \sqrt[n+1]{\binom{n}{0}\binom{n}{1} \cdots \binom{n}{n}} = \prod_{k=1}^{n} \left(1 - \frac{k}{n+1}\right)^{1 - \frac{2k}{n+1}}.$$

Taking the natural logarithm, we obtain

$$\frac{1}{n} \ln G_n = \frac{1}{n} \sum_{k=1}^{n} \left(1 - \frac{2k}{n+1}\right) \ln \left(1 - \frac{k}{n+1}\right).$$

This is just a Riemann sum of the function $(1 - 2x) \ln(1 - x)$ over the interval $[0, 1]$. Passing to the limit, we obtain

$$\lim_{n \to \infty} \frac{1}{n} \ln G_n = \int_0^1 (1 - 2x) \ln(1 - x)dx.$$

The integral is computed by parts as follows:

$$\int_0^1 (1 - 2x) \ln(1 - x)dx = 2 \int_0^1 (1 - x) \ln(1 - x)dx - \int_0^1 \ln(1 - x)dx$$

$$= -(1 - x)^2 \ln(1 - x) \Big|_0^1 - 2 \int_0^1 \frac{(1 - x)^2}{2} \cdot \frac{1}{1 - x}dx + (1 - x) \ln(1 - x) \Big|_0^1 + x \Big|_0^1$$

$$= -\int_0^1 (1 - x)dx + 1 = \frac{1}{2}.$$

Exponentiating back, we obtain $\lim_{n \to \infty} \sqrt[n]{G_n} = \sqrt{e}$. $\qquad \square$

564. Compute

$$\lim_{n\to\infty}\left[\frac{1}{\sqrt{4n^2-1^2}}+\frac{1}{\sqrt{4n^2-2^2}}+\cdots+\frac{1}{\sqrt{4n^2-n^2}}\right].$$

565. Prove that for every positive integer n,

$$0.785n^2-n<\sqrt{n^2-1^2}+\sqrt{n^2-2^2}+\cdots+\sqrt{n^2-(n-1)^2}<0.79n^2.$$

566. Define the sequence

$$x_n=\sum_{k=1}^n\frac{k}{n^2+2k^2},\quad n\ge1.$$

Prove that the sequence x_n converges and find its limit.

567. Prove that for $n\ge1$,

$$\frac{1}{\sqrt{2+5n}}+\frac{1}{\sqrt{4+5n}}+\frac{1}{\sqrt{6+5n}}+\cdots+\frac{1}{\sqrt{2n+5n}}<\sqrt{7n}-\sqrt{5n}.$$

568. Compute

$$\lim_{n\to\infty}\left(\frac{2^{1/n}}{n+1}+\frac{2^{2/n}}{n+\frac{1}{2}}+\cdots+\frac{2^{n/n}}{n+\frac{1}{n}}\right).$$

569. Compute the integral

$$\int_0^\pi\ln(1-2a\cos x+a^2)dx.$$

570. Find all continuous functions $f:\mathbb{R}\to[1,\infty)$ for which there exist $a\in\mathbb{R}$ and k a positive integer such that

$$f(x)f(2x)\cdots f(nx)\le an^k,$$

for every real number x and positive integer n.

3.2.11 Inequalities for Integrals

A very simple inequality states that if $f:[a,b]\to\mathbb{R}$ is a nonnegative continuous function, then

$$\int_a^b f(x)dx\ge0,$$

with equality if and only if f is identically equal to zero. Easy as this inequality looks, its applications are often tricky. This is the case with a problem from the 1982 Romanian Mathematical Olympiad, proposed by the second author of the book.

Example. Find all continuous functions $f:[0,1]\to\mathbb{R}$ satisfying

$$\int_0^1 f(x)dx=\frac{1}{3}+\int_0^1 f(x^2)^2dx.$$

Solution. First, we would like the functions in both integrals to have the same variable. A substitution in the first integral changes it to $\int_0^1 f(x^2)2x\,dx$. Next, we would like to express the number $\frac{1}{3}$ as an integral, and it is natural to choose $\int_0^1 x^2\,dx$. The condition from the statement becomes

$$\int_0^1 2xf(x^2)\,dx = \int_0^1 x^2 + \int_0^1 f(x^2)^2\,dx.$$

This is the same as

$$\int_0^1 [f(x^2)^2 - 2xf(x^2) + x^2]\,dx = 0.$$

Note that the function under the integral, $f(x^2)^2 - 2xf(x^2) + x^2 = (f(x^2) - x)^2$, is a perfect square, so it is nonnegative. Therefore, its integral on $[0, 1]$ is nonnegative, and it can equal zero only if the function itself is identically zero. We find that $f(x^2) = x$. So $f(x) = \sqrt{x}$ is the unique function satisfying the condition from the statement. $\qquad \square$

571. Determine the continuous functions $f : [0, 1] \to \mathbb{R}$ that satisfy

$$\int_0^1 f(x)(x - f(x))\,dx = \frac{1}{12}.$$

572. Let n be an odd integer greater than 1. Determine all continuous functions $f : [0, 1] \to \mathbb{R}$ such that

$$\int_0^1 (f(x^{\frac{1}{k}}))^{n-k}\,dx = \frac{k}{n}, \quad k = 1, 2, \ldots, n - 1.$$

573. Let $f : [0, 1] \to \mathbb{R}$ be a continuous function such that

$$\int_0^1 f(x)\,dx = \int_0^1 xf(x)\,dx = 1.$$

Prove that

$$\int_0^1 f(x)^2\,dx \geq 4.$$

574. For each continuous function $f : [0, 1] \to \mathbb{R}$, we define

$$I(f) = \int_0^1 x^2 f(x)\,dx \quad \text{and} \quad J(f) = \int_0^1 x(f(x))^2\,dx.$$

Find the maximum value of $I(f) - J(f)$ over all such functions f.

575. Let $a_1, a_2, \ldots, a_n$ be positive real numbers and let $x_1, x_2, \ldots, x_n$ be real numbers such that $a_1x_1 + a_2x_2 + \cdots + a_nx_n = 0$. Prove that

$$\sum_{i,j} x_i x_j |a_i - a_j| \leq 0.$$

Moreover, prove that equality holds if and only if there exists a partition of the set

$\{1, 2, \ldots, n\}$ into the disjoint sets $A_1, A_2, \ldots, A_k$ such that if i and j are in the same set, then $a_i = a_j$ and also $\sum_{j \in A_i} x_j = 0$ for $i = 1, 2, \ldots, k$.

We now list some fundamental inequalities. We will be imprecise as to the classes of functions to which they apply, because we want to avoid the subtleties of Lebesgue's theory of integration. The novice mathematician should think of piecewise continuous, real-valued functions on some domain D that is an interval of the real axis or some region in $\mathbb{R}^n$.

The Cauchy-Schwarz inequality. *Let f and g be square integrable functions. Then*

$$\left(\int_D f(x)g(x)dx \right)^2 \leq \left(\int_D f(x)^2 dx \right) \left(\int_D g(x)^2 dx \right).$$

Minkowski's inequality. *If $p > 1$, then*

$$\left(\int_D |f(x) + g(x)|^p dx \right)^{\frac{1}{p}} \leq \left(\int_D |f(x)|^p dx \right)^{\frac{1}{p}} + \left(\int_D |g(x)|^p dx \right)^{\frac{1}{p}}.$$

Hölder's inequality. *If $p, q > 1$ such that $\dfrac{1}{p} + \dfrac{1}{q} = 1$, then*

$$\int_D |f(x)g(x)| dx \leq \left(\int_D |f(x)|^p dx \right)^{\frac{1}{p}} \left(\int_D |g(x)|^q dx \right)^{\frac{1}{q}}.$$

As an instructive example we present in detail the proof of another famous inequality.

Chebyshev's inequality. *Let f and g be two increasing functions on $\mathbb{R}$. Then for any real numbers $a < b$,*

$$(b - a) \int_a^b f(x)g(x)dx \geq \left(\int_a^b f(x)dx \right) \left(\int_a^b g(x)dx \right).$$

Proof. Because f and g are both increasing,

$$(f(x) - f(y))(g(x) - g(y)) \geq 0.$$

Integrating this inequality over $[a, b] \times [a, b]$, we obtain

$$\int_a^b \int_a^b (f(x) - f(y))(g(x) - g(y))dxdy \geq 0.$$

Expanding, we obtain

$$\int_a^b \int_a^b f(x)g(x)dxdy + \int_a^b \int_a^b f(y)g(y)dxdy - \int_a^b \int_a^b f(x)g(y)dxdy$$
$$- \int_a^b \int_a^b f(y)g(x)dxdy \geq 0.$$

By eventually renaming the integration variables, we see that this is equivalent to

$$(b-a)\int_a^b f(x)g(x)dx - \left(\int_a^b f(x)dx\right)\left(\int_a^b g(x)dx\right) \geq 0,$$

and the inequality is proved.   □

576. Let $f : [0, 1] \to \mathbb{R}$ be a continuous function. Prove that

$$\left(\int_0^1 f(t)dt\right)^2 \leq \int_0^1 f(t)^2 dt.$$

577. Find the maximal value of the ratio

$$\left(\int_0^3 f(x)dx\right)^3 / \int_0^3 f(x)^3 dx,$$

as f ranges over all positive continuous functions on $[0, 1]$.

578. Let $f : [0, \infty) \to [0, \infty)$ be a continuous, strictly increasing function with $f(0) = 0$. Prove that

$$\int_0^a f(x)dx + \int_0^b f^{-1}(x)dx \geq ab$$

for all positive numbers a and b, with equality if and only if $b = f(a)$. Here f^{-1} denotes the inverse of the function f.

579. Prove that for any positive real numbers x, y and any positive integers m, n,

$$(n-1)(m-1)(x^{m+n} + y^{m+n}) + (m+n-1)(x^m y^n + x^n y^m)$$
$$\geq mn(x^{m+n-1}y + y^{m+n-1}x).$$

580. Let f be a nonincreasing function on the interval $[0, 1]$. Prove that for any $\alpha \in (0, 1)$,

$$\alpha \int_0^1 f(x)dx \leq \int_0^\alpha f(x)dx.$$

581. Let $f : [0, 1] \to [0, \infty)$ be a differentiable function with decreasing first derivative, and such that $f(0) = 0$ and $f'(1) > 0$. Prove that

$$\int_0^1 \frac{dx}{f(x)^2 + 1} \leq \frac{f(1)}{f'(1)}.$$

Can equality hold?

582. Prove that any continuously differentiable function $f : [a, b] \to \mathbb{R}$ for which $f(a) = 0$ satisfies the inequality

$$\int_a^b f(x)^2 dx \le (b - a)^2 \int_a^b f'(x)^2 dx.$$

583. Let $f(x)$ be a continuous real-valued function defined on the interval $[0, 1]$. Show that

$$\int_0^1 \int_0^1 |f(x) + f(y)| dx dy \ge \int_0^1 |f(x)| dx.$$

584. Let $f : [a, b] \to \mathbb{R}$ be a continuous convex function. Prove that

$$\int_a^b f(x) dx \ge 2 \int_{\frac{3a+b}{4}}^{\frac{3b+a}{4}} f(x) dx \ge (b - a) f\left(\frac{a + b}{2}\right).$$

3.2.12 Taylor and Fourier Series

Some functions, called analytic, can be expanded around each point of their domain in a Taylor series

$$f(x) = f(a) + \frac{f'(a)}{1!}(x - a) + \frac{f''(a)}{2!}(x - a)^2 + \cdots + \frac{f^{(n)}(a)}{n!}(x - a)^n + \cdots$$

If $a = 0$, the expansion is also known as the Maclaurin series. Rational functions, trigonometric functions, the exponential and the natural logarithm are examples of analytic functions. A particular example of a Taylor series expansion is

Newton's binomial formula. For all real numbers a and $|x| < 1$, one has

$$(x + 1)^a = \sum_{n=0}^{\infty} \binom{a}{n} x^n = \sum_{n=0}^{\infty} \frac{a(a - 1) \cdots (a - n + 1)}{n!} x^n,$$

Here we make the usual convention that $\binom{a}{0} = 1$.

We begin our series of examples with a widely circulated problem.

Example. Compute the integral

$$\int_0^1 \ln x \ln(1 - x) dx.$$

Solution. Because

$$\lim_{x \to 0} \ln x \ln(1 - x) = \lim_{x \to 1} \ln x \ln(1 - x) = 0,$$

this is, in fact, a definite integral.

We will expand one of the logarithms in Taylor series. Recall the Taylor series expansion

$$\ln(1 - x) = -\sum_{n=1}^{\infty} \frac{x^n}{n}, \text{ for } x \in (-1, 1).$$

It follows that on the interval $(0, 1)$, the antiderivative of the function $f(x) = \ln x \ln(1 - x)$ is

$$\int \ln(1 - x) \ln x \, dx = -\int \sum_{n=1}^{\infty} \frac{x^n}{n} \ln x \, dx = -\sum_{n=1}^{\infty} \frac{1}{n} \int x^n \ln x \, dx.$$

Integrating by parts, we find this is to be equal to

$$-\sum_{n=1}^{\infty} \frac{1}{n} \left(\frac{x^{n+1}}{n+1} \ln x - \frac{x^{n+1}}{(n+1)^2} \right) + C.$$

Taking the definite integral over an interval $[\varepsilon, 1 - \varepsilon]$, then letting $\varepsilon \to 0$, we obtain

$$\int_0^1 \ln x \ln(1 - x) dx = \sum_{n=1}^{\infty} \frac{1}{n(n+1)^2}.$$

Using a telescopic sum and the well-known formula for the sum of the inverses of squares of positive integers, we compute this as follows:

$$\sum_{n=1}^{\infty} \frac{1}{n(n+1)^2} = \sum_{n=1}^{\infty} \left(\frac{1}{n(n+1)} - \frac{1}{(n+1)^2} \right) = \sum_{n=1}^{\infty} \left(\frac{1}{n} - \frac{1}{n+1} \right) - \sum_{n=2}^{\infty} \frac{1}{n^2}$$

$$= 1 - \left(\frac{\pi^2}{6} - 1 \right) = 2 - \frac{\pi^2}{6},$$

which is the answer to the problem. Note that in the above computation all series are absolutely convergent, so they can be reordered. □

Next, a problem that we found in S. Rădulescu, M. Rădulescu, *Theorems and Problems in Mathematical Analysis* (Editura Didactică şi Pedagogică, Bucharest, 1982).

Example. Prove that for $|x| < 1$,

$$(\arcsin x)^2 = \sum_{k=1}^{\infty} \frac{1}{k^2 \binom{2k}{k}} 2^{2k-1} x^{2k}.$$

Solution. The function $g : (-1, 1) \to \mathbb{R}, g(x) = (\arcsin x)^2$ satisfies the initial value problem

$$(1 - x^2)y'' - xy' - 2 = 0, \ y(0) = y'(0) = 0.$$

Looking for a solution of the form $y(x) = \sum_{k=0}^{\infty} a_k x^k$, we obtain the recurrence relation

$$(k+1)(k+2)a_{k+2} - k^2 a_k = 0, \ k \geq 1.$$

It is not hard to see that $a_1 = 0$; hence $a_{2k+1} = 0$ for all k. Also, $a_0 = 0$, $a_2 = 1$, and inductively we obtain

$$a_{2k} = \frac{1}{k^2 \binom{2k}{k}} 2^{2k-1}, \ k \geq 1.$$

The series

$$\sum_{k=1}^{\infty} \frac{1}{k^2 \binom{2k}{k}} 2^{2k-1} x^{2k}$$

is dominated by the geometric series $\displaystyle\sum_{k=1}^{\infty} x^{2k}$, so it converges absolutely for $|x| < 1$. Its term-by-term derivatives of first and second order also converge absolutely. We deduce that the series defines a solution to the differential equation. The uniqueness of the solution for the initial value problem implies that this function must equal g. $\qquad\square$

We conclude the list of examples with the proof of Stirling's formula.

Stirling's formula.

$$n! = \sqrt{2\pi n} \left(\frac{n}{e}\right)^n \cdot e^{\frac{\theta_n}{12n}}, \ for \ some \ 0 < x_n < 1.$$

Proof. We begin with the Taylor series expansions

$$\ln(1 \pm x) = \pm x - \frac{x^2}{2} \pm \frac{x^3}{3} - \frac{x^4}{4} \pm \frac{x^5}{5} + \cdots, \ \text{for } x \in (-1, 1).$$

Combining these two, we obtain the Taylor series expansion

$$\ln \frac{1+x}{1-x} = 2x + \frac{2}{3}x^3 + \frac{2}{5}x^5 + \cdots + \frac{2}{2m+1}x^{2m+1} + \cdots,$$

again for $x \in (-1, 1)$. In particular, for $x = \dfrac{1}{2n+1}$, where n is a positive integer, we have

$$\ln \frac{n+1}{n} = \frac{2}{2n+1} + \frac{2}{3(2n+1)^3} + \frac{2}{5(2n+1)^5} + \cdots$$

which can be written as

$$\left(n + \frac{1}{2}\right) \ln \frac{n+1}{n} = 1 + \frac{1}{3(2n+1)^2} + \frac{1}{5(2n+1)^4} + \cdots$$

The right-hand side is greater than 1. It can be bounded from above by a geometric series as follows:

$$1 + \frac{1}{3(2n+1)^2} + \frac{1}{5(2n+1)^4} + \cdots < 1 + \frac{1}{3} \sum_{k=1}^{\infty} \frac{1}{(2n+1)^{2k}}$$

$$= 1 + \frac{1}{3(2n+1)^2} \cdot \frac{1}{1 - \frac{1}{(2n+1)^2}}$$

$$= 1 + \frac{1}{12n(n+1)}.$$

So using Taylor series we have obtained the double inequality

$$1 \le \left(n + \frac{1}{2}\right) \ln \frac{n+1}{n} < 1 + \frac{1}{12n(n+1)}.$$

This transforms by exponentiating and dividing through by e into

$$1 < \frac{1}{e}\left(\frac{n+1}{n}\right)^{n+\frac{1}{2}} < e^{\frac{1}{12n(n+1)}}.$$

To bring this closer to Stirling's formula, note that the term in the middle is equal to

$$\frac{e^{-n-1}(n+1)^{n+1}((n+1)!)^{-1}\sqrt{n+1}}{e^{-n}n^n(n!)^{-1}\sqrt{n}} = \frac{x_{n+1}}{x_n},$$

where $x_n = e^{-n}n^n(n!)^{-1}\sqrt{n}$, a number that we want to prove is equal to $\sqrt{2\pi}\, e^{-\frac{\theta_n}{12n}}$ with $0 < \theta_n < 1$. In order to prove this, we write the above double inequality as

$$1 \le \frac{x_n}{x_{n+1}} \le \frac{e^{\frac{1}{12n}}}{e^{\frac{1}{12(n+1)}}}.$$

We deduce that the sequence x_n is positive and decreasing, while the sequence $e^{-\frac{1}{12n}}x_n$ is increasing. Because $e^{-\frac{1}{12n}}$ converges to 1, and because $(x_n)_n$ converges by the Weierstrass criterion, both x_n and $e^{-\frac{1}{12n}}x_n$ must converge to the same limit L. We claim that $L = \sqrt{2\pi}$. Before proving this, note that

$$e^{-\frac{1}{12n}}x_n < L < x_n,$$

so by the intermediate value property there exists $\theta_n \in (0, 1)$ such that $L = e^{-\frac{\theta_n}{12n}}x_n$, i.e. $x_n = e^{\frac{\theta_n}{12n}}L$.

The only thing left is the computation of the limit L. For this we employ the Wallis formula

$$\lim_{n\to\infty}\left[\frac{2\cdot 4\cdot 6\ldots 2n}{1\cdot 3\cdot 5\ldots(2n-1)}\right]^2\frac{1}{n} = \pi,$$

proved in problem 562 from Section 3.2.9 (the one on definite integrals). We rewrite this limit as

$$\lim_{n\to\infty}\frac{2^{2n}(n!)^2}{(2n)!}\cdot\frac{1}{\sqrt{n}} = \sqrt{\pi}.$$

Substituting $n!$ and $(2n)!$ by the formula found above gives

$$\lim_{n\to\infty}\frac{nL^2\left(\frac{n}{e}\right)^{2n}e^{\frac{2\theta_n}{12n}}2^{2n}}{\sqrt{2n}L\left(\frac{2n}{e}\right)^{2n}e^{\frac{\theta_{2n}}{24n}}}\cdot\frac{1}{\sqrt{n}} = \lim_{n\to\infty}\frac{1}{\sqrt{2}}Le^{\frac{4\theta_n-\theta_{2n}}{24n}} = \sqrt{\pi}.$$

Hence $L = \sqrt{2\pi}$, and Stirling's formula is proved. $\qquad\square$

Try your hand at the following problems.

585. Prove that for any real number x, the series

$$1 + \frac{x^4}{4!} + \frac{x^8}{8!} + \frac{x^{12}}{12!} + \cdots$$

is convergent and find its limit.

586. Compute the ratio

$$\frac{1 + \dfrac{\pi^4}{5!} + \dfrac{\pi^8}{9!} + \dfrac{\pi^{12}}{13!} + \cdots}{\dfrac{1}{3!} + \dfrac{\pi^4}{7!} + \dfrac{\pi^8}{11!} + \dfrac{\pi^{12}}{15!} + \cdots}$$

587. Compute

$$\frac{1}{\sqrt{3}} - \frac{1}{3}\frac{1}{\sqrt{3}^3} + \frac{1}{5}\frac{1}{\sqrt{3}^5} - \frac{1}{7}\frac{1}{\sqrt{3}^7} + \cdots$$

588. For $a > 0$, prove that

$$\int_{-\infty}^{\infty} e^{-x^2} \cos axdx = \sqrt{\pi}e^{-a^2/4}.$$

589. Find a quadratic polynomial $P(x)$ with real coefficients such that

$$\left| P(x) + \frac{1}{x - 4} \right| \leq 0.01, \quad \text{for all } x \in [-1, 1].$$

590. Without using a calculator, find the solution to the equation

$$x^2 \sin \frac{1}{x} = 2x - 1997$$

with an error less than 0.01.

591. Compute to three decimal places

$$\int_0^1 \cos \sqrt{x}dx.$$

592. Prove that for $|x| < 1$,

$$\arcsin x = \sum_{k=0}^{\infty} \frac{1}{2^{2k}(2k + 1)} \binom{2k}{k} x^{2k+1}.$$

593. (a) Prove that for $|x| < 2$,

$$\sum_{k=1}^{\infty} \frac{1}{\binom{2k}{k}} x^{2k} = \frac{x\left(4\arcsin\left(\frac{x}{2}\right) + x\sqrt{4 - x^2}\right)}{(4 - x^2)\sqrt{4 - x^2}}.$$

(b) Prove the identity

$$\sum_{k=1}^{\infty} \frac{1}{\binom{2k}{k}} = \frac{2\pi\sqrt{3} + 36}{27}.$$

In a different perspective, we have the Fourier series expansions. The Fourier series allows us to write an arbitrary oscillation as a superposition of sinusoidal oscillations. Mathematically, a function $f : \mathbb{R} \to \mathbb{R}$ that is continuous and periodic of period T admits a Fourier series expansion

$$f(x) = a_0 + \sum_{n=1}^{\infty} a_n \cos \frac{2n\pi}{T}x + \sum_{n=1}^{\infty} b_n \sin \frac{2n\pi}{T}x.$$

This expansion is unique, and

$$a_0 = \frac{1}{2\pi} \int_0^T f(x)dx,$$

$$a_n = \frac{1}{\pi} \int_0^T f(x) \cos \frac{2n\pi}{T}xdx,$$

$$b_n = \frac{1}{\pi} \int_0^T f(x) \sin \frac{2n\pi}{T}xdx.$$

Of course, we can require f to be defined only on an interval of length T, and then extend it periodically, but if the values of f at the endpoints of the interval differ, then the convergence of the series is guaranteed only in the interior of the interval.

Let us discuss a problem from the Soviet Union University Student Contest.

Example. Compute the sum

$$\sum_{n=1}^{\infty} \frac{\cos n}{1 + n^2}.$$

Solution. The sum looks like a Fourier series evaluated at 1. For this reason we concentrate on the general series

$$\sum_{n=0}^{\infty} \frac{1}{n^2 + 1} \cos nx.$$

The coefficients $\frac{1}{n^2+1}$ should remind us of the integration formulas

$$\int e^x \cos nxdx = \frac{1}{n^2 + 1} e^x(\cos nx + n \sin nx),$$

$$\int e^x \sin nxdx = \frac{n}{n^2 + 1} e^x(\sin nx + n \cos nx).$$

These give rise to the Fourier series expansion

$$e^x = \frac{1}{2\pi}(e^{2\pi} - 1) + \frac{1}{\pi}(e^{2\pi} - 1) \sum_{n=1}^{\infty} \frac{1}{n^2 + 1} \cos nx + \frac{1}{\pi}(e^{2\pi} - 1) \sum_{n=1}^{\infty} \frac{n}{n^2 + 1} \sin nx,$$

which holds true for $x \in (0, 2\pi)$. Similarly, for e^{-x} and $x \in (0, 2\pi)$, we have

$$e^{-x} = \frac{1}{2\pi}(1 - e^{2-\pi}) + \frac{1}{\pi}(1 - e^{-2\pi}) \sum_{n=1}^{\infty} \frac{1}{n^2 + 1} \cos nx - \frac{1}{\pi}(1 - e^{-2\pi}) \sum_{n=1}^{\infty} \frac{n}{n^2 + 1} \sin nx.$$

Let

$$C_n(x) = \sum_{n=-1}^{\infty} \frac{1}{n^2 + 1} \cos nx \quad \text{and} \quad S_n(x) = \sum_{n=1}^{\infty} \frac{n}{n^2 + 1} \sin nx.$$

They satisfy

$$\frac{1}{2} + C_n(x) + S_n(x) = \frac{\pi e^x}{e^{2\pi} - 1},$$

$$\frac{1}{2} + C_n(x) - S_n(x) = \frac{\pi e^{-x}}{1 - e^{-2\pi}}.$$

Solving this linear system, we obtain

$$C_n(x) = \frac{1}{2}\left[\frac{\pi e^x}{e^{2\pi} - 1} + \frac{\pi e^{-x}}{1 - e^{-2\pi}} - 1\right].$$

The sum from the statement is $C(1)$. The answer to the problem is therefore

$$C(1) = \frac{1}{2}\left[\frac{\pi e}{e^{2\pi} - 1} + \frac{\pi e^{-1}}{1 - e^{-2\pi}} - 1\right].$$

$\square$

We find even more exciting a fundamental result of ergodic theory that proves that for an irrational number α, the fractional parts of $n\alpha$, $n \geq 1$, are uniformly distributed in $[0, 1]$. For example, when $\alpha = \log_{10} 2$, we obtain as a corollary that on average, the first digit of a power of 2 happens to be 7 as often as it happens to be 1. Do you know a power of 2 whose first digit is 7?

Theorem. *Let $f : \mathbb{R} \to \mathbb{R}$ be a continuous function of period 1 and let α be an irrational number. Then*

$$\lim_{n \to \infty} \frac{1}{n}(f(\alpha) + f(2\alpha) + \cdots + f(n\alpha)) = \int_0^1 f(x)dx.$$

Proof. If we approximate f by a trigonometric polynomial with error less than ε, then both $\frac{1}{n}(f(\alpha) + f(2\alpha) + \ldots + f(n\alpha))$ and $\int_0^1 f(x)dx$ are evaluated with error less than ε. Hence it suffices to check the equality term by term for the Fourier series of f. For the constant term the equality is obvious. To check that it holds for $f(x) = \cos 2\pi mx$ or $f(x) = \sin 2\pi mx$, with $m \geq 1$, combine these two using Euler's formula into

$$e^{2\pi imx} = \cos 2\pi mx + i \sin 2\pi mx.$$

We then have

$$\frac{1}{n}(e^{2\pi m\alpha} + e^{2\pi i2m\alpha} + \cdots + e^{2\pi inm\alpha}) = \frac{e^{2\pi i\alpha}(e^{2\pi inm\alpha} - 1)}{n(e^{2\pi im\alpha} - 1)},$$

which converges to 0 as $n \to \infty$. And for the right-hand side,

$$\int_0^1 e^{2\pi imx}dx = \frac{1}{2\pi im}e^{2\pi imx}\Big|_0^1 = 0.$$

Therefore, equality holds term by term for the Fourier series. The theorem is proved. $\square$

If after this example you don't love Fourier series, you never will. Below are listed more applications of the Fourier series expansion.

594. Prove that for every $0 < x < 2\pi$ the following formula is valid:

$$\frac{\pi - x}{2} = \frac{\sin x}{1} + \frac{\sin 2x}{2} + \frac{\sin 3x}{3} + \cdots$$

Derive the formula

$$\frac{\pi}{4} = \sum_{k=1}^{\infty} \frac{\sin(2k-1)x}{2k-1}, \quad x \in (0, \pi).$$

595. Use the Fourier series of the function of period 1 defined by $f(x) = \frac{1}{2} - x$ for $0 \leq x < 1$ to prove Euler's formula

$$\frac{\pi^2}{6} = 1 + \frac{1}{2^2} + \frac{1}{3^2} + \frac{1}{4^2} + \cdots$$

596. Prove that

$$\frac{\pi^2}{8} = 1 + \frac{1}{3^2} + \frac{1}{5^2} + \frac{1}{7^2} + \cdots$$

597. For a positive integer n find the Fourier series of the function

$$f(x) = \frac{\sin^2 nx}{\sin^2 x}.$$

598. Let $f : [0, \pi] \to \mathbb{R}$ be a C^∞ functions such that $(-1)^n f^{(2n)}(x) \geq 0$ for any $x \in [0, \pi]$ and $f^{(2n)}(0) = f^{(2n)}(\pi) = 0$ for any $n \geq 0$. Show that $f(x) = a \sin x$ for some $a > 0$.

3.3 Multivariable Differential and Integral Calculus

3.3.1 Partial Derivatives and Their Applications

This section and the two that follow it cover differential and integral calculus in two and three dimensions. Most of the ideas generalize easily to n-dimensions. All functions below are assumed to be differentiable. For a two-variable function this means that its graph (which is a surface in $\mathbb{R}^3$) admits a tangent plane at each point. For a three-variable function, the graph is a three-dimensional manifold in a four-dimensional space, and differentiability means that at each point the graph admits a three-dimensional tangent hyperplane.

The tilting of the tangent (hyper)plane is determined by the slopes in the directions of the coordinate axes, and these slopes are the partial derivatives of the function. We denote the partial derivatives of f by $\frac{\partial f}{\partial x}, \frac{\partial f}{\partial y}, \frac{\partial f}{\partial z}$. They are computed by differentiating with respect to the one variable while keeping the others fixed. This being said, let us start with an example.

Euler's theorem. *A function $z(x, y)$ is called n-homogeneous if $z(tx, ty) = t^n z(x, y)$ for all $x, y \in R$ and $t > 0$. Assume that $z(x, y)$ is n-homogeneous with n an integer. Then for all $k \leq n + 1$,*

$$\sum_{j=1}^{k} \binom{k}{j} x^j y^{k-j} \frac{\partial^k z}{\partial x^j \partial y^{k-j}} = n(n-1) \cdots (n-k+1)z.$$

Proof. We first prove the case $k = 1$. Differentiating the relation $z(tx, ty) = t^n z(x, y)$ with respect to y, we obtain

$$t \frac{\partial z}{\partial y}(tx, ty) = t^n \frac{\partial z}{\partial y}(x, y),$$

which shows that $\dfrac{\partial z}{\partial y}$ is $(n-1)$-homogeneous.

Replace x by 1, y by $\frac{y}{x}$, and t by x in the homogeneity condition, to obtain $z(x, y) = x^n z\left(1, \frac{y}{x}\right)$. Differentiating this with respect to x yields

$$\frac{\partial z}{\partial x}(x, y) = nx^{n-1} z\left(1, \frac{y}{x}\right) + x^n \frac{\partial z}{\partial y}\left(1, \frac{y}{x}\right)\left(-\frac{y}{x^2}\right).$$

Because $\frac{\partial z}{\partial y}$ is $(n-1)$-homogeneous, the last term is just $-\frac{y}{x} \frac{\partial z}{\partial y}(x, y)$. Moving it to the right and multiplying through by x gives the desired

$$x \frac{\partial z}{\partial x} + y \frac{\partial z}{\partial y} = nz.$$

Now we prove the general case by induction on k, with $k = 1$ the base case. To simplify the notation, set $\binom{k}{j} = 0$ if $j < 0$ or $j > k$. The induction hypothesis is

$$\sum_{j} \binom{k}{j} x^j y^{k-j} \frac{\partial^k z}{\partial x^j \partial y^{k-j}} = n(n-1) \cdots (n-k+1)z,$$

for some $k \leq n$. Apply the operator $x\frac{\partial}{\partial x} + y\frac{\partial}{\partial y}$ to both sides. The left-hand side becomes

$$\sum_j \binom{k}{j}\left(x\frac{\partial}{\partial x} + y\frac{\partial}{\partial y}\right) x^j y^{k-j} \frac{\partial^k z}{\partial x^j \partial y^{k-j}}$$

$$= \sum_j j\binom{k}{j} x^j y^{k-j} \frac{\partial^k z}{\partial x^j \partial y^{k-j}} + \sum_j \binom{k}{j} x^{j+1} y^{k-j} \frac{\partial^{k+1} z}{\partial x^{j+1} \partial y^{k-j}}$$

$$+ \sum_j (k-j)\binom{k}{j} x^j y^{k-j} \frac{\partial^k z}{\partial x^j \partial y^{k-j}} + \sum_j \binom{k}{j} x^j y^{k-j+1} \frac{\partial^{k+1} z}{\partial x^j \partial y^{k-j+1}}$$

$$= k\sum_j \binom{k}{j} x^j y^{k-j} \frac{\partial^k z}{\partial x^j \partial y^{k-j}} + \sum_j \left(\binom{k}{j-1} + \binom{k}{j}\right) x^j y^{k+1-j} \frac{\partial^{k+1} z}{\partial x^j \partial y^{k+1-j}}$$

$$= k \cdot n(n-1)\cdots(n-k+1)z + \sum_j \binom{k+1}{j} x^j y^{k+1-j} \frac{\partial^{k+1} z}{\partial x^j y^{k+1-j}},$$

where for the last step we used the induction hypothesis. The base case $k = 1$ implies that the right side equals $n \cdot n(n-1)\cdots(n-k+1)z$. Equating the two, we obtain

$$\sum_j \binom{k+1}{j} x^j y^{k+1-j} \frac{\partial^{k+1} z}{\partial x^j y^{k+1-j}} = n(n-1)\cdots(n-k+1)(n-k)z,$$

completing the induction. This proves the formula. $\square$

599. Prove that if the function $u(x, t)$ satisfies the equation

$$\frac{\partial u}{\partial t} = \frac{\partial^2 u}{\partial x^2}, \quad (x, t) \in \mathbb{R}^2,$$

then so does the function

$$v(x, t) = \frac{1}{\sqrt{t}} e^{-\frac{x^2}{4t}} u(xt^{-1}, -t^{-1}), \quad x \in \mathbb{R}, \ t > 0.$$

600. Assume that a nonidentically zero harmonic function $u(x, y)$ is n-homogeneous for some real number n. Prove that n is necessarily an integer. (The function u is called harmonic if $\frac{\partial^2 u}{\partial x^2} + \frac{\partial^2 u}{\partial y^2} = 0$).

601. Let $P(x, y)$ be a harmonic polynomial divisible by $x^2 + y^2$. Prove that $P(x, y)$ is identically equal to zero.

602. Let $f : \mathbb{R}^2 \to \mathbb{R}^2$ be a differentiable function with continuous partial derivatives and with $f(0, 0) = 0$. Prove that there exist continuous functions $g_1, g_2 : \mathbb{R}^2 \to \mathbb{R}$ such that

$$f(x, y) = xg_1(x, y) + yg_2(x, y).$$

If a differentiable multivariable function has a global extremum, then this extremum is found either among the critical points or on the boundary of the domain. We recall that a point is critical if the (hyper)plane tangent to the graph is horizontal, which is equivalent to the fact that all partial derivatives are equal to zero. Because any continuous function on a compact domain attains its extrema, the global maximum and minimum exist whenever the domain is closed and bounded. Let us apply these considerations to the following problems.

Example. Find the triangles inscribed in the unit circle that have maximal perimeter.

Solution. Without loss of generality, we may assume that the vertices of the triangle have the coordinates $(1, 0)$, $(\cos s, \sin s)$, $(\cos t, \sin t)$, $0 \leq s \leq t \leq 2\pi$. We are supposed to maximize the function

$$
\begin{aligned}
f(s, t) &= \sqrt{(\cos s - 1)^2 + (\sin s)^2} + \sqrt{(1 - \cos t)^2 + (\sin t)^2} \\
&\quad + \sqrt{(\cos t - \cos s)^2 + (\sin t - \sin s)^2} \\
&= \sqrt{2}(\sqrt{1 - \cos s} + \sqrt{1 - \cos t} + \sqrt{1 - \cos(t - s)}) \\
&= 2\left(\sin\frac{s}{2} + \sin\frac{t}{2} + \sin\frac{t - s}{2}\right).
\end{aligned}
$$

over the domain $0 \leq s \leq t \leq 2\pi$. To this end, we first find the critical points of f in the interior of the domain. The equation

$$
\frac{\partial f}{\partial s}(s, t) = \cos\frac{s}{2} - \cos\frac{t - s}{2} = 0
$$

gives $\cos\frac{s}{2} = \cos\frac{t-s}{2}$, and since both $\frac{s}{2}$ and $\frac{t-s}{2}$ are between 0 and π, it follows that $\frac{s}{2} = \frac{t-s}{2}$. The equation

$$
\frac{\partial f}{\partial t}(s, t) = \cos\frac{t}{2} + \cos\frac{t - s}{2} = 0
$$

implies additionally that $\cos s = -\cos\frac{s}{2}$, and hence $s = \frac{2\pi}{3}$. Consequently, $t = \frac{4\pi}{3}$, showing that the unique critical point is the equilateral triangle, with the corresponding value of the perimeter $3\sqrt{3}$.

On the boundary of the domain of f two of the three points coincide, and in that case the maximum is achieved when two sides of the triangle become diameters. The value of this maximum is 4, which is smaller than $3\sqrt{3}$. We conclude that equilateral triangles maximize the perimeter. $\qquad\square$

603. Find the global minimum of the function $f : \mathbb{R}^2 \to \mathbb{R}$,

$$
f(x, y) = x^4 + 6x^2y^2 + y^4 - \frac{9}{4}x - \frac{7}{4}y.
$$

604. Find the range of the function

$$f : [-1, 1] \times [-1, 1] \to \mathbb{R}, \quad f(x, y) = x^4 + 6x^2 y^2 + y^4 + 8xy.$$

605. Find the equation of the smallest sphere that is tangent to both of the lines

(i) $x = t + 1, y = 2t + 4, z = -3t + 5$, and
(ii) $x = 4t - 12, y = -t + 8, z = t + 17$.

606. Determine the maximum and the minimum of $\cos A + \cos B + \cos C$ when A, B and C are the angles of a triangle.

607. Prove that for $\alpha, \beta, \gamma \in \left[0, \frac{\pi}{2}\right)$,

$$\tan \alpha + \tan \beta + \tan \gamma \leq \frac{2}{\sqrt{3}} \sec \alpha \sec \beta \sec \gamma.$$

608. Given n points in the plane, suppose there is a unique line that minimizes the sum of the distances from the points to the line. Prove that the line passes through two of the points.

To find the maximum of a function subject to a constraint one can employ the following tool.

The Lagrange multipliers theorem. *If a function $f(x, y, z)$ subject to the constraint $g(x, y, z) = C$ has a maximum or a minimum, then this maximum or minimum occurs at a point (x, y, z) of the set $g(x, y, z) = C$ for which the gradients of f and g are parallel.*

So in order to find the maximum of f we have to solve the system of equations $\nabla f = \lambda \nabla g$ and $g(x, y, z) = C$. The number λ is called the Lagrange multiplier; to understand its significance, imagine that f is the profit and g is the constraint on resources. Then λ is the rate of change of the profit as the constraint is relaxed (economists call this the shadow price).

As an application of the method of Lagrange multipliers, we will prove the law of reflection.

Example. For a light ray reflected off a mirror, the angle of incidence equals the angle of reflection.

Solution. Our argument relies on the fundamental principle of optics, which states that light travels always on the fastest path. This is known in physics as Fermat's principle of least time. We consider a light ray that travels from point A to point B reflecting off a horizontal mirror represented schematically in Figure 22. Denote by C and D the projections of A and B onto the mirror, and by P the point where the ray hits the mirror. The angles of incidence and reflection are, respectively, the angles formed by AP and BP with the normal to the mirror. To prove that they are equal it suffices to show that $\angle APC = \angle BPD$. Let $x = CP$ and $y = DP$. We have to minimize $f(x, y) = AP + BP$ with the constraint $g(x, y) = x + y = CD$.

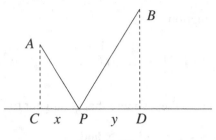

Figure 22

Using the Pythagorean theorem we find that

$$f(x, y) = \sqrt{x^2 + AC^2} + \sqrt{y^2 + BD^2}.$$

The method of Lagrange multipliers yields the system of equations

$$\frac{x}{\sqrt{x^2 + AC^2}} = \lambda,$$

$$\frac{y}{\sqrt{y^2 + BD^2}} = \lambda,$$

$$x + y = CD.$$

From the first two equations, we obtain

$$\frac{x}{\sqrt{x^2 + AC^2}} = \frac{y}{\sqrt{y^2 + BD^2}},$$

i.e., $\frac{CP}{AP} = \frac{DP}{BP}$. This shows that the right triangles CAP and DBP are similar, so $\angle APC = \angle BPD$ as desired.   □

The following example was proposed by C. Niculescu for *Mathematics Magazine*.

Example. Find the smallest constant $k > 0$ such that

$$\frac{ab}{a + b + 2c} = \frac{bc}{b + c + 2a} + \frac{ca}{c + a + 2b} \le k(a + b + c)$$

for every $a, b, c > 0$.

Solution. We will show that the best choice for k is $\frac{1}{4}$. To prove this fact, note that the inequality remains unchanged on replacing a, b, c by ta, tb, tc with $t > 0$. Consequently, the smallest value of k is the supremum of

$$f(a, b, c) = \frac{ab}{a + b + 2c} + \frac{bc}{b + c + 2a} + \frac{ca}{c + a + 2b}$$

over the domain $\Delta = \{(a, b, c) \mid a, b, c > 0, \ a + b + c = 1\}$. Note that on Δ,

$$f(a, b, c) = \frac{ab}{1 + c} + \frac{bc}{1 + a} + \frac{ca}{1 + b}.$$

To find the maximum of this function on Δ, we will apply the method of Lagrange multipliers with the constraint $g(a, b, c) = a + b + c = 1$. This yields the system of equations

$$\frac{b}{1+c} + \frac{c}{1+b} - \frac{bc}{(1+a)^2} = \lambda,$$

$$\frac{c}{1+a} + \frac{a}{1+c} - \frac{ca}{(1+b)^2} = \lambda,$$

$$\frac{a}{1+b} + \frac{b}{1+a} - \frac{ab}{(1+c)^2} = \lambda,$$

$$a + b + c = 1.$$

Subtracting the first two equations, we obtain

$$\frac{b-a}{1+c} + \frac{c}{1+b}\left[1 + \frac{a}{1+b}\right] - \frac{c}{1+a}\left[1 + \frac{b}{1+a}\right] = 0,$$

which after some algebraic manipulations transforms into

$$(b-a)\left[\frac{1}{1+c} + \frac{c(a+b+1)(a+b+2)}{(1+a)^2(1+b)^2}\right] = 0.$$

The second factor is positive, so this equality can hold only if $a = b$. Similarly, we prove that $b = c$. So the only extremum of f when restricted to the plane $a + b + c = 1$ is

$$f\left(\frac{1}{3}, \frac{1}{3}, \frac{1}{3}\right) = \frac{1}{4}.$$

But is this a *maximum*? Let us examine the behavior of f on the boundary of Δ (to which it can be extended). If say $c = 0$, then $f(a, b, 0) = ab$. When $a + b = 1$, the maximum of this expression is again $\frac{1}{4}$. We conclude that the maximum on Δ is indeed $\frac{1}{4}$, which is the desired constant. $\qquad\square$

609. Using the method of Lagrange multipliers prove Snell's law of optics: If a light ray passes between two media separated by a planar surface, then

$$\frac{\sin \theta_1}{\sin \theta_2} = \frac{v_1}{v_2},$$

where θ_1 and θ_2 are, respectively, the angle of incidence and the angle of refraction, and v_1 and v_2 are the speeds of light in the first and second media, respectively.

610. Let ABC be a triangle such that

$$\left(\cot \frac{A}{2}\right)^2 + \left(2\cot \frac{B}{2}\right)^2 + \left(3\cot \frac{C}{2}\right)^2 = \left(\frac{6s}{7r}\right)^2,$$

where s and r denote its semiperimeter and its inradius, respectively. Prove that triangle ABC is similar to a triangle T whose side lengths are all positive integers with no common divisors and determine these integers.

611. The angles of a certain triangle are measured in radians and the product of these measures is equal to $\pi^3/30$. Prove that the triangle is acute.

612. Prove that of all quadrilaterals that can be formed from four given sides, the one that is cyclic has the greatest area.

613. Of all triangles circumscribed about a given circle, find the one with the smallest area.

614. Prove that for non-negative x, y, z such that $x + y + z = 1$, the following inequality holds

$$0 \le xy + yz + xz - 2xyz \le \frac{7}{27}.$$

615. Let a, b, c, d be four nonnegative numbers satisfying $a + b + c + d = 1$. Prove the inequality

$$abc + bcd + cda + dab \le \frac{1}{27} + \frac{176}{27}abcd.$$

616. Given two triangles with angles α, β, γ, respectively, $\alpha_1, \beta_1, \gamma_1$, prove that

$$\frac{\cos \alpha_1}{\sin \alpha} + \frac{\cos \beta_1}{\sin \beta} + \frac{\cos \gamma_1}{\sin \gamma} \le \cot \alpha + \cot \beta + \cot \gamma,$$

with equality if and only if $\alpha = \alpha_1$, $\beta = \beta_1$, $\gamma = \gamma_1$.

3.3.2 Multivariable Integrals

For multivariable integrals, the true story starts with a change of coordinates.

Theorem. *Let $f : D \subset \mathbb{R}^n \to \mathbb{R}$ be an integrable function. Let also $x(u) = (x_i(u_j))_{i,j=1}^n$ be a change of coordinates, viewed as a map from some domain D^* to D, with Jacobian $\frac{\partial x}{\partial u} = \det\left(\frac{\partial x_i}{\partial u_j}\right)$. Then*

$$\int_D f(x)dx = \int_{D^*} f(x(u)) \left|\frac{\partial x}{\partial u}\right| du.$$

There are three special situations worth mentioning:

- The change in two dimensions from Cartesian to polar coordinates $x = r\cos\theta$, $y = r\sin\theta$, with the Jacobian $\frac{\partial(x,y)}{\partial(r,\theta)} = r$.

- The change in three dimensions from Cartesian to cylindrical coordinates $x = t\cos\theta$, $y = r\sin\theta$, $z = z$, with the Jacobian $\frac{\partial(x,y,z)}{\partial(r,\theta,z)} = r$.

- The change in three dimensions from Cartesian to spherical coordinates $x = \rho\sin\phi\cos\theta$, $y = \rho\sin\phi\sin\theta$, $z = \rho\cos\phi$, with the Jacobian $\frac{\partial(x,y,z)}{\partial(\rho,\theta,\phi)} = \rho^2\sin\phi$.

As an illustration, we show how multivariable integrals can be used for calculating the Fresnel integrals. These integrals arise in the theory of diffraction of light.

Example. Compute the Fresnel integrals

$$I = \int_0^\infty \cos x^2 dx \quad \text{and} \quad J = \int_0^\infty \sin x^2 dx.$$

Solution. For the computation of the first integral, we consider the surface $z = e^{-y^2} \cos x^2$ and determine the volume of the solid that lies below this surface in the octant $x, y, z \geq 0$. This will be done in both Cartesian and polar coordinates. We will also make use of the Gaussian integral

$$\int_0^\infty e^{-t^2} dt = \frac{\sqrt{\pi}}{2},$$

which is the subject of one of the exercises that follow.

In Cartesian coordinates,

$$V = \int_0^\infty \int_0^\infty e^{-y^2} \cos x^2 dy dx = \int_0^\infty \left(\int_0^\infty e^{-y^2} dy \right) \cos x^2 dx$$

$$= \int_0^\infty \frac{\sqrt{\pi}}{2} \cos x^2 dx = \frac{\sqrt{\pi}}{2} I.$$

In polar coordinates,

$$V = \int_0^{\frac{\pi}{2}} \int_0^\infty e^{-\rho^2 \sin^2 \theta} \cos(\rho^2 \cos^2 \theta) \rho d\rho d\theta$$

$$= \int_0^{\frac{\pi}{2}} \frac{1}{\cos^2 \theta} \int_0^\infty e^{-u \tan^2 \theta} \cos u du d\theta = \int_0^{\frac{\pi}{2}} \frac{1}{\cos^2 \theta} \cdot \frac{\tan^2 \theta}{1 + \tan^4 \theta} d\theta,$$

where we made the substitution $u = u(\rho) = \rho^2 \cos^2 \theta$. If in this last integral we substitute $\tan \theta = t$, we obtain

$$V = \frac{1}{2} \int_0^\infty \frac{t^2}{t^4 + 1} dt.$$

A routine but lengthy computation using Jacobi's method of partial fraction decomposition shows that the antiderivative of $\frac{t^2}{t^4+1}$ is

$$\frac{1}{2\sqrt{2}} \arctan \frac{x^2 - 1}{x\sqrt{2}} + \frac{1}{4\sqrt{2}} \ln \frac{x^2 - x\sqrt{2} + 1}{x^2 + x\sqrt{2} + 1} + C,$$

whence $V = \frac{\pi\sqrt{2}}{8}$. Equating the two values for V, we obtain $I = \frac{\sqrt{2\pi}}{4}$. A similar argument yields $J = \frac{\sqrt{2\pi}}{4}$. $\qquad \square$

The solutions to all but last problems below are based on appropriate changes of coordinates.

617. Compute the integral $\iint_D x dx dy$, where

$$D = \left\{ (x, y) \in \mathbb{R}^2 \mid x \geq 0, \ 1 \leq xy \leq 2, \ 1 \leq \frac{y}{x} \leq 2 \right\}.$$

618. Find the integral of the function

$$f(x, y, z) = \frac{x^4 + 2y^4}{x^4 + 4y^2 + z^4}$$

over the unit ball $B = \{(x, y, z) \mid x^2 + y^2 + z^2 \le 1\}$.

619. Compute the integral

$$\iint_D \frac{dxdy}{(x^2 + y^2)^2},$$

where D is the domain bounded by the circles

$$x^2 + y^2 - 2x = 0, \ x^2 + y^2 - 4x = 0,$$
$$x^2 + y^2 - 2y = 0, \ x^2 + y^2 - 6y = 0.$$

620. Compute the integral

$$I = \iint_D |xy|dxdy,$$

where

$$D = \left\{ (x, y) \in \mathbb{R}^2 \mid x \ge 0, \ \left(\frac{x^2}{a^2} + \frac{y^2}{b^2} \right)^2 \le \frac{x^2}{a^2} - \frac{y^2}{b^2} \right\}, \ a, b > 0.$$

621. Prove the Gaussian integral formula

$$\int_{-\infty}^{\infty} e^{-x^2} dx = \sqrt{\pi}.$$

622. Evaluate

$$\int_0^1 \int_0^1 \int_0^1 (1 + u^2 + v^2 + w^2)^{-2} du\, dv\, dw.$$

623. Let $D = \{(x, y) \in \mathbb{R}^2 \mid 0 \le x \le y \le \pi\}$. Prove that

$$\iint_D \ln|\sin(x - y)|dxdy = -\frac{\pi^2}{2} \ln 2.$$

Our next topic is the continuous analogue of the change of the order of summation in a double sum.

Fubini's theorem. *Let $f : \mathbb{R}^2 \to \mathbb{R}$ be a piecewise continuous function such that*

$$\int_c^d \int_a^b |f(x, y)|dxdy < \infty.$$

Then

$$\int_c^d \int_a^b f(x, y)dxdy = \int_a^b \int_c^d f(x, y)dydx.$$

The matter of convergence can be bypassed for positive functions, in which case we have the following result.

Tonelli's theorem. *Let $f : \mathbb{R}^2 \to \mathbb{R}$ be a positive piecewise continuous function. Then*

$$\int_a^b \int_c^d f(x, y) dx dy = \int_c^d \int_a^b f(x, y) dy dx.$$

The limits of integration can be finite or infinite. In the particular case that $f(x, y)$ is constant on the squares of an integer lattice, we recover the discrete version of Fubini's theorem, the change of order of summation in a double sum

$$\sum_{m=0}^{\infty} \sum_{n=0}^{\infty} f(m, n) = \sum_{n=0}^{\infty} \sum_{m=0}^{\infty} f(m, n).$$

A slightly more general situation occurs when f is a step function in one of the variables. In this case we recover the formula for commuting the sum with the integral:

$$\int_a^b \sum_{n=0}^{\infty} f(n, x) = \sum_{n=0}^{\infty} \int_a^b f(n, x).$$

Here we are allowed to commute the sum and the integral if either f is a positive function, or if $\int_a^b \sum_{n=0}^{\infty} |f(n, x)|$ (or equivalently $\sum_{n=0}^{\infty} \int_a^b |f(n, x)|$) is finite. It is now time for an application.

Example. Compute the integral

$$I = \int_0^{\infty} \frac{1}{\sqrt{x}} e^{-x} dx.$$

Solution. We will replace $\frac{1}{\sqrt{x}}$ by a Gaussian integral. Note that for $x > 0$,

$$\int_{-\infty}^{\infty} e^{-xt^2} dt = \int_{-\infty}^{\infty} e^{-(\sqrt{x}t)^2} dt = \frac{1}{\sqrt{x}} \int_{-\infty}^{\infty} e^{-u^2} du = \sqrt{\frac{\pi}{x}}.$$

Returning to the problem, we are integrating the positive function $\frac{1}{\sqrt{x}} e^{-x}$, which is integrable over the positive semiaxis because in a neighborhood of zero it is bounded from above by $\frac{1}{\sqrt{x}}$ and in a neighborhood of infinity it is bounded from above by $e^{-x/2}$.

Let us consider the two-variable function $f(x, y) = e^{-xt^2} e^{-x}$, which is positive and integrable over $\mathbb{R} \times (0, \infty)$. Using the above considerations and Tonelli's theorem, we can write

$$I = \int_0^{\infty} \frac{1}{\sqrt{x}} e^{-x} dx = \frac{1}{\sqrt{\pi}} \int_0^{\infty} \int_{-\infty}^{\infty} e^{-xt^2} e^{-x} dt dx = \frac{1}{\sqrt{\pi}} \int_{-\infty}^{\infty} \int_0^{\infty} e^{-(t^2+1)x} dx dt$$

$$= \frac{1}{\sqrt{\pi}} \int_{-\infty}^{\infty} \frac{1}{t^2 + 1} dt = \frac{\pi}{\sqrt{\pi}} = \sqrt{\pi}.$$

Hence the value of the integral in question is $I = \sqrt{\pi}$. $\qquad\square$

More applications are given below.

624. Let $a_1 \le a_2 \le \cdots \le a_n = m$ be positive integers. Denote by b_k the number of those a_i for which $a_i \ge k$. Prove that

$$a_1 + a_2 + \cdots + a_n = b_1 + b_2 + \cdots + b_m.$$

625. Show that for $s > 0$,

$$\int_0^\infty e^{-sx} x^{-1} \sin x \, dx = \arctan(s^{-1}).$$

626. Show that for $a, b > 0$,

$$\int_0^\infty \frac{e^{-ax} - e^{-bx}}{x} dx = \ln \frac{b}{a}.$$

627. Let $|x| < 1$. Prove that

$$\sum_{n=1}^\infty \frac{x^n}{n^2} = -\int_0^x \frac{1}{t} \ln(1-t) dt.$$

628. Let $F(x) = \sum_{n=1}^\infty \frac{1}{x^2 + n^4}$, $x \in \mathbb{R}$. Compute $\int_0^\infty F(t) dt$.

3.3.3 The Many Versions of Stokes' Theorem

We advise you that this is probably the most difficult section of the book. Yet Stokes' theorem plays such an important role in mathematics that it deserves an extensive treatment. As an encouragement, we offer you a quote by Marie Curie: "Nothing in life is to be feared. It is only to be understood."

In its general form, Stokes' theorem because is known as

$$\int_M d\omega = \int_{\partial M} \omega,$$

where ω is a "form", $d\omega$ its differential, and M a domain with boundary ∂M. The one-dimensional case is the most familiar; it is the Leibniz-Newton formula

$$\int_a^b f'(t) dt = f(b) - f(a).$$

Three versions of this result are of interest to us.

Green's theorem. *Let D be a domain in the plane with boundary C oriented such that D is to the left. If the vector field $\overrightarrow{F}(x, y) = P(x, y)\overrightarrow{i} + Q(x, y)\overrightarrow{j}$ is continuously differentiable on D, then*

$$\oint_C P dx + Q dy = \iint_D \left(\frac{\partial Q}{\partial x} - \frac{\partial P}{\partial y} \right) dx dy.$$

The Kelvin-Stokes (curl) theorem. *Let S be an oriented surface with normal vector* $\overrightarrow{n}$, *bounded by a closed, piecewise smooth curve C that is oriented such that if one travels on C with the upward direction* $\overrightarrow{n}$, *the surface is on the left. If* $\overrightarrow{F}$ *is a vector field that is continuously differentiable on S, then*

$$\oint_C \overrightarrow{F} \cdot d\overrightarrow{R} = \iint_S (\operatorname{curl} \overrightarrow{F} \cdot \overrightarrow{n})\, dS,$$

where dS is the area element on the surface.

The Gauss-Ostrogradsky (divergence) theorem. *Let S be a smooth, orientable surface that encloses a solid region V in space. If* $\overrightarrow{F}$ *is a continuously differentiable vector field on V, then*

$$\iint_S \overrightarrow{F} \cdot \overrightarrow{n}\, dS = \iiint_V \operatorname{div} \overrightarrow{F}\, dV,$$

where $\overrightarrow{n}$ *is the outward unit normal vector to the surface S, dS is the area element on the surface, and dV is the volume element inside of V.*

We recall that for a vector field $\overrightarrow{F} = (F_1, F_2, F_3)$, the divergence is

$$\operatorname{div} \overrightarrow{F} = \nabla \cdot \overrightarrow{F} = \frac{\partial F_1}{\partial x} + \frac{\partial F_2}{\partial y} + \frac{pF_3}{\partial z},$$

while the curl is

$$\operatorname{curl} \overrightarrow{F} = \nabla \times \overrightarrow{F} = \begin{vmatrix} \overrightarrow{i} & \overrightarrow{j} & \overrightarrow{k} \\ \dfrac{\partial}{\partial x} & \dfrac{\partial}{\partial y} & \dfrac{\partial}{\partial z} \\ F_1 & F_2 & F_3 \end{vmatrix}$$

$$= \left(\frac{\partial F_3}{\partial y} - \frac{\partial F_2}{\partial z} \right) \overrightarrow{i} + \left(\frac{\partial F_1}{\partial z} - \frac{\partial F_3}{\partial x} \right) \overrightarrow{j} + \left(\frac{\partial F_2}{\partial x} - \frac{\partial F_1}{\partial y} \right) \overrightarrow{k}.$$

The quantity $\iint_S \overrightarrow{F} \cdot \overrightarrow{n}\, dS$ is called the flux of $\overrightarrow{F}$ across the surface S.

Let us illustrate the use of these theorems with some examples. We start with an encouraging problem whose solution is based on the Kelvin-Stokes theorem.

Example. Compute

$$\oint_C y\, dx + z\, dy + x\, dz,$$

where C is the circle $x^2 + y^2 + z^2 = 1$, $x + y + z = 1$, oriented counterclockwise when seen from the positive side of the x-axis.

Solution. By the Kelvin-Stokes theorem,

$$\oint_C y\, dx + z\, dy + x\, dz = \iint_S \operatorname{curl} \overrightarrow{F} \cdot \overrightarrow{n}\, dS,$$

where S is the disk that the circle bounds. It is straightforward that curl $\overrightarrow{F} = (-1, -1, -1)$, while $\overrightarrow{n}$, the normal vector to the plane $x + y + z = 1$, is equal to $\left(\frac{1}{\sqrt{3}}, \frac{1}{\sqrt{3}}, \frac{1}{\sqrt{3}}\right)$. Therefore,

$$\oint_C y\,dx + z\,dy + x\,dz = -A\sqrt{3},$$

where A is the area of the disk bounded by C. Observe that C is the circumcircle of the triangle with vertices $(1, 0, 0)$, $(0, 1, 0)$, and $(0, 0, 1)$. The circumradius of this triangle is $\frac{\sqrt{6}}{3}$, so $A = \frac{2}{3}\pi$. The answer to the problem is therefore $-\frac{2\pi\sqrt{3}}{3}$. □

Example. Orthogonal to each face of a polyhedron construct an outward vector with length numerically equal to the area of the face. Prove that the sum of all these vectors is equal to zero.

Solution. We exhibit first an elementary solution based on vector operations. Consider the particular case of a tetrahedron $ABCD$. The four vectors are $\frac{1}{2}\overrightarrow{BC} \times \overrightarrow{BA}$, $\frac{1}{2}\overrightarrow{BA} \times \overrightarrow{BD}$, $\frac{1}{2}\overrightarrow{BD} \times \overrightarrow{BC}$, and $\frac{1}{2}\overrightarrow{DA} \times \overrightarrow{DC}$. Indeed, the lengths of these vectors are numerically equal to the areas of the corresponding faces, and the cross-product of two vectors is perpendicular to the plane determined by the vectors, and it points outward because of the right-hand rule. We have

$$\overrightarrow{BC} \times \overrightarrow{BA} + \overrightarrow{BA} \times \overrightarrow{BD} + \overrightarrow{BD} \times \overrightarrow{BC} + \overrightarrow{DA} \times \overrightarrow{DC}$$
$$= \overrightarrow{BC} \times \overrightarrow{BA} + \overrightarrow{BA} \times \overrightarrow{BD} + \overrightarrow{BD} \times \overrightarrow{BC} + (\overrightarrow{BA} - \overrightarrow{BD}) \times (\overrightarrow{BC} - \overrightarrow{BD})$$
$$= \overrightarrow{BC} \times \overrightarrow{BA} + \overrightarrow{BA} \times \overrightarrow{BD} + \overrightarrow{BD} \times \overrightarrow{BC} + \overrightarrow{BC} \times \overrightarrow{BA} - \overrightarrow{BA} \times \overrightarrow{BD} = \overrightarrow{BD} \times \overrightarrow{BC} + \overrightarrow{0} = \overrightarrow{0}.$$

This proves that the four vectors add up to zero.

In the general case, dissect the polyhedron into tetrahedra cutting the faces into triangles by diagonals and then joining the centroid of the polyhedron with the vertices. Sum up all vectors perpendicular to the faces of these tetrahedra, and note that the vectors corresponding to internal walls cancel out.

The elegant solution uses integrals. Let S be the polyhedron and assume that its interior V is filled with gas at a (not necessarily constant) pressure p. The force that the gas exerts on S is $\iint_S p\,\overrightarrow{n}\,A$, where $\overrightarrow{n}$ is the outward normal vector to the surface of the polyhedron and dA is the area element. The divergence theorem implies that

$$\iint_S p\,\overrightarrow{n}\,dA = \iiint_V \nabla p\,dV.$$

Here ∇p denotes the gradient of p. If the pressure p is constant, then the right-hand side is equal to zero. This is the case with our polyhedron, where $p = 1$. The double integral is exactly the sum of the vectors under discussion, these vectors being the forces exerted by pressure on the faces. □

As a corollary, we obtain the well-known fact that a container filled with gas under pressure is at equilibrium; a balloon will never move as a result of internal pressure.

We conclude our series of examples with an application of Green's theorem: the proof given by D. Pompeiu to Cauchy's formula for holomorphic functions. First, let us introduce some notation for functions of a complex variable $f(z) = f(x + iy) = u(x, y) + iv(x, y)$. If u and v are continuously differentiable, define

$$\frac{\partial f}{\partial \bar{z}} = \frac{1}{2}\left[\frac{\partial f}{\partial x} + i\frac{\partial f}{\partial y}\right] = \frac{1}{2}\left[\left(\frac{\partial u}{\partial x} - \frac{\partial v}{\partial y}\right) + i\left(\frac{\partial u}{\partial y} + \frac{\partial v}{\partial x}\right)\right].$$

The function f is called holomorphic if $\dfrac{\partial f}{\partial \bar{z}} = 0$. Examples are polynomials in z and any absolutely convergent power series in z. Also, let $dz = dx + idy$.

Cauchy's theorem. *Let Γ be an oriented curve that bounds a region Δ on its left, and let $a \in \Delta$. If $f(z) = f(x + iy) = u(x, y) + iv(x, y)$ is a holomorphic function on Δ such that u and v are continuous on $\Delta \cup \Gamma$ and continuously differentiable on Δ, then*

$$f(a) = \frac{1}{2\pi i}\oint_\Gamma \frac{f(z)}{z - a}dz.$$

Proof. Pompeiu's proof is based on Green's formula, applied on the domain Δ_ε obtained from Δ by removing a disk of radius ε around a as described in Figure 23 to $P = F$ and $Q = iF$, where F is a holomorphic function to be specified later. Note that the boundary of the domain consists of two curves, Γ and Γ_ε.

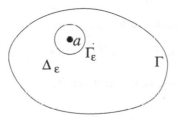

Figure 23

Green's formula reads

$$\oint_\Gamma Fdz - \oint_{\Gamma_\varepsilon} Fdz = \oint_\Gamma Fdx + iFdy - \oint_{\Gamma_\varepsilon} Fdx + iFdy$$

$$= \iint_{\Delta_\varepsilon} i\left(\frac{\partial F}{\partial x} + i\frac{\partial F}{\partial y}\right)dxdy = 2i\iint_{\Delta_\varepsilon}\frac{\partial F}{\partial \bar{z}}dxdy = 0.$$

Therefore,

$$\oint_\Gamma F(z)dz = \oint_{\Gamma_\varepsilon} F(z)dz.$$

We apply this to

$$F(z) = \frac{f(z)}{z - a} = \frac{(u(x, y) + iv(x, y))(x - iy + \alpha - i\beta)}{(x + \alpha)^2 + (y + \beta)^2},$$

where $a = \alpha + i\beta$. It is routine to check that F is holomorphic. We thus have

$$\oint_\Gamma \frac{f(z)}{z - a}dz = \oint_{\Gamma_\varepsilon} \frac{f(z)}{z - a}dz.$$

The change of variable $z = a + \varepsilon a^{it}$ on the right-hand side yields

$$\oint_{\Gamma_\varepsilon} \frac{f(z)}{z-a} dz = \int_{-\pi}^{\pi} \frac{f(a + \varepsilon a^{it})}{\varepsilon a^{it}} i \varepsilon e^{it} dt = i \int_{-\pi}^{\pi} f(a + \varepsilon e^{it}) dt.$$

When $\varepsilon \to 0$ this tends to $2\pi i f(a)$, and we obtain

$$\oint_{\Gamma_\varepsilon} \frac{f(z)}{z-a} dz = 2\pi i f(a).$$

Hence the desired formula. $\square$

629. Assume that a curve $(x(t), y(t))$ runs counterclockwise around a region D. Prove that the area of D is given by the formula

$$A = \frac{1}{2} \oint_{\partial D} (xy' - yx') dt.$$

630. There is given an n-gon in the plane, whose vertices have integer coordinates and whose sides, all of odd lengths, are parallel to the coordinate axes.

   (a) Show that n is a multiple of 4.
   (b) Show that if $n = 100$, then the area of this polygon is odd.

631. Compute the flux of the vector field

$$\overrightarrow{F}(x, y, z) = x(e^{xy} - e^{zx}) \overrightarrow{i} + y(e^{yz} - e^{xy}) \overrightarrow{j} + z(e^{zx} - e^{yz}) \overrightarrow{k}$$

across the upper hemisphere of the unit sphere.

632. Compute

$$\oint_C y^2 dx + z^2 dy + x^2 dz,$$

where C is the Viviani curve, defined as the intersection of the sphere $x^2 + y^2 + z^2 = a^2$ with the cylinder $x^2 + y^2 = ax$.

633. Let $\phi(x, y, z)$ and $\psi(x, y, z)$ be twice continuously differentiable functions in the region $\{(x, y, z) \mid \frac{1}{2} < \sqrt{x^2 + y^2 + z^2} < 2\}$. Prove that

$$\iint_S (\nabla\phi \times \nabla\psi) \cdot \overrightarrow{n} \, dS = 0,$$

where S is the unit sphere centered at the origin, $\overrightarrow{n}$ is the normal unit vector to this sphere, and $\nabla\phi$ denotes the gradient $\frac{\partial\phi}{\partial x} \overrightarrow{i} + \frac{\partial\phi}{\partial y} \overrightarrow{j} + \frac{\partial\phi}{\partial z} \overrightarrow{k}$.

634. Let $f, g : \mathbb{R}^3 \to \mathbb{R}$ be twice continuously differentiable functions that are constant along the lines that pass through the origin. Prove that on the unit ball $B = \{(x, y, z) \mid x^2 + y^2 + z^2 \le 1\}$,

$$\iiint_B f\nabla^2 g \, dV = \iiint_B g\nabla^2 f \, dV.$$

Here $\nabla^2 = \frac{\partial^2}{\partial x^2} + \frac{\partial^2}{\partial y^2} + \frac{\partial^2}{\partial z^2}$ is the Laplacian.

635. Prove Gauss' law, which states that the total flux of the gravitational field through a closed surface equals $-4\pi G$ times the mass enclosed by the surface, where G is the constant of gravitation. The mathematical formulation of the law is

$$\iint_S \overrightarrow{F} \cdot \overrightarrow{n}\, dS = -4\pi M G.$$

636. Let

$$\overrightarrow{G}(x, y) = \left(\frac{-y}{x^2 + 4y^2}, \frac{x}{x^2 + 4y^2}, 0 \right).$$

Prove or disprove that there is a vector field $\overrightarrow{F} : \mathbb{R}^3 \to \mathbb{R}^3$,

$$\overrightarrow{F}(x, y, z) = (M(x, y, z), N(x, y, z), P(x, y, z)),$$

with the following properties:

(i) M, N, P have continuous partial derivatives for all $(x, y, z) \neq (0, 0, 0)$;
(ii) $\text{curl}\,\overrightarrow{F} = \overrightarrow{0}$, for all $(x, y, z) \neq (0, 0, 0)$;
(iii) $\overrightarrow{F}(x, y, 0) = \overrightarrow{G}(x, y)$.

637. Let $\overrightarrow{F} : \mathbb{R}^2 \to \mathbb{R}^2$, $\overrightarrow{F}(x, y) = (F_1(x, y), F_2(x, y))$ be a vector field, and let $G : \mathbb{R}^3 \to \mathbb{R}$ be a smooth function whose first two variables are x and y, and the third is t, the time. Assume that for any rectangular surface D bounded by the curve C,

$$\frac{d}{dt} \iint_D G(x, y, t)\, dx\, dy = - \oint_C \overrightarrow{F} \cdot d\overrightarrow{R}.$$

Prove that

$$\frac{\partial G}{\partial t} + \frac{\partial F_2}{\partial x} + \frac{\partial F_1}{\partial y} = 0.$$

638. For two disjoint oriented curves C_1 and C_2 in three-dimensional space, parametrized by $\overrightarrow{v}_1(s)$ and $\overrightarrow{v}_2(t)$, define the linking number

$$\text{lk}(C_1, C_2) = \frac{1}{4\pi} \oint_{C_1} \oint_{C_2} \frac{\overrightarrow{v}_1 - \overrightarrow{v}_2}{\|\overrightarrow{v}_1 - \overrightarrow{v}_2\|^3} \cdot \left(\frac{d\overrightarrow{v}_1}{ds} \times \frac{d\overrightarrow{v}_2}{dt} \right) dt\, ds.$$

Prove that if the oriented curves C_1 and $-C_1'$ bound an oriented surface S such that S is to the left of each curve, and if the curve C_2 is disjoint from S, then $\text{lk}(C_1, C_2) = \text{lk}(C_1', C_2)$.

3.4 Equations with Functions as Unknowns

3.4.1 Functional Equations

We will now look at equations whose unknowns are functions. Here is a standard example that we found in B.J. Venkatachala, *Functional Equations: A Problem Solving Approach* (Prism Books PVT Ltd., 2002).

Example. Find all functions $f : \mathbb{R} \to \mathbb{R}$ satisfying the functional equation

$$f((x-y)^2) = f(x)^2 - 2xf(y) + y^2.$$

Solution. For $y = 0$, we obtain

$$f(x^2) = f(x)^2 - 2xf(0),$$

and for $x = 0$, we obtain

$$f(y^2) = f(0)^2 + y^2.$$

Setting $y = 0$ in the second equation, we find that $f(0) = 0$ or $f(0) = 1$. On the other hand, combining the two equalities, we obtain

$$f(x)^2 - 2xf(0) = f(0)^2 + x^2,$$

that is,

$$f(x)^2 = (x + f(0))^2.$$

Substituting this in the original equation yields

$$f(y) = \frac{f(x)^2 - f((x-y)^2) + y^2}{2x} = \frac{(x+f(0))^2 - (x-y)^2 - f(0)^2 + y^2}{2x} = y + f(0).$$

Thus the functional equation has two solutions: $f(x) = x$ and $f(x) = x + 1$.   □

But we like more the nonstandard functional equations. Here is one, which is a simplified version of a short-listed problem from the 42nd International Mathematical Olympiad. We liked about it the fact that the auxiliary function h from the solution mimics, in a discrete situation, harmonicity – a fundamental concept in mathematics. The solution applies the maximum modulus principle, which states that if h is a harmonic function then the maximum of $|h|$ is attained on the boundary of the domain of definition. Harmonic functions, characterized by the fact that the value at one point is the average of the values in a neighborhood of the point, play a fundamental role in geometry. For example, they encode geometric properties of their domain, a fact made explicit in Hodge theory.

Example. Find all functions $f : \{0, 1, 2, \ldots, \} \times \{0, 1, 2, \ldots\} \to \mathbb{R}$ satisfying

$$f(p, q) = \begin{cases} \frac{1}{2}(f(p+1, q-1) + f(p-1, q+1)) + 1 & \text{if } pq \neq 0, \\ 0 & \text{if } pq = 0. \end{cases}$$

Solution. We see that $f(1, 1) = 1$. The defining relation gives $f(1, 2) = 1 + f(2, 1)/2$ and $f(2, 1) = 1 + f(1, 2)/2$, and hence $f(2, 1) = f(1, 2) = 2$. Then $f(3, 1) = 1 + f(2, 2)/2$, $f(2, 2) = 1 + f(3, 1)/2 + f(1, 3)/2, f(1, 3) = 1 + f(2, 2)/2$. So $f(2, 2) = 4, f(3, 1) = 3$, $f(1, 3) = 3$. Repeating such computations, we eventually guess the explicit formula $f(p, q) = pq, p, q \geq 0$. And indeed, this function satisfies the condition from the statement. Are there other solutions to the problem? The answer is no, but we need to prove it.

Assume that f_1 and f_2 are both solutions to the functional equation. Let $h = f_1 - f_2$. Then h satisfies

$$h(p, q) = \begin{cases} \frac{1}{2}(h(p+1, q-1) + h(p-1, q+1)) & \text{if } pq \neq 0, \\ 0 & \text{if } pq = 0. \end{cases}$$

Fix a line $p + q = n$, and on this line pick (p_0, q_0) the point that maximizes the value of h. Because

$$h(p_0, q_0) = \frac{1}{2}(h(p_0 + 1, q_0 - 1) + h(p_0 - 1, q_0 + 1)),$$

it follows that $h(p_0 + 1, q_0 - 1) = h(p_0 - 1, q_0 + 1) = h(p_0, q_0)$. Shifting the point, we eventually conclude that h is constant on the line $p + q = n$, and its value is equal to $h(n, 0) = 0$. Since n was arbitrary, we see that h is identically equal to 0. Therefore, $f_1 = f_2$, the problem has a unique solution, and this solution is $f(p, q) = pq, p, q \geq 0$.   □

And now an example of a problem about a multivariable function, from the same short list, submitted by B. Enescu (Romania).

Example. Let $x_1, x_2, \ldots, x_n$ be arbitrary real numbers. Prove the inequality

$$\frac{x_1}{1 + x_1^2} + \frac{x_2}{1 + x_1^2 + x_2^2} + \cdots + \frac{x_n}{1 + x_1^2 + \cdots + x_n^2} < \sqrt{n}.$$

Solution. We introduce the function

$$f_n(x_1, x_2, \ldots, x_n) = \frac{x_1}{1 + x_1^2} + \frac{x_2}{1 + x_1^2 + x_2^2} + \cdots + \frac{x_n}{1 + x_1^2 + \cdots + x_n^2}.$$

If we set $r = \sqrt{1 + x_1^2}$, then

$$f_n(x_1, x_2, \ldots, x_n) = \frac{x_1}{r^2} + \frac{x_2}{r^2 + x_2^2} + \cdots + \frac{x_n}{r^2 + x_2^2 + \cdots + x_n^2}$$

$$= \frac{x_1}{r^2} + \frac{1}{r}\left(\frac{\frac{x_2}{r}}{1 + \left(\frac{x_2}{r}\right)^2} + \cdots + \frac{\frac{x_n}{r}}{1 + \left(\frac{x_2}{r}\right)^2 + \cdots + \left(\frac{x_n}{r}\right)^2} \right).$$

We obtain the functional equation

$$f_n(x_1, x_2, \ldots, x_n) = \frac{x_1}{1 + x_1^2} + \frac{1}{\sqrt{1 + x_1^2}} f_{n-1}\left(\frac{x_2}{r}, \frac{x_3}{r}, \ldots, \frac{x_n}{r} \right).$$

Writing $M_n = \sup f_n(x_1, x_2, \ldots, x_n)$, we observe that the functional equation gives rise to the recurrence relation

$$M_n = \sup_{x_1} \left(\frac{x_1}{1 + x_1^2} + \frac{M_{n-1}}{\sqrt{1 + x_1^2}} \right).$$

We will now prove by induction that $M_n < \sqrt{n}$. For $n = 1$, this follows from $\frac{x_1}{1+x_1^2} \leq \frac{1}{2} < 1$. Assume that the property is true for k and let us prove it for $k + 1$. From the induction hypothesis, we obtain

$$M_k < \sup_{x_1} \left(\frac{x_1}{1 + x_1^2} + \frac{\sqrt{k}}{\sqrt{1 + x_1^2}} \right).$$

We need to show that the right-hand side of the inequality is less than or equal to $\sqrt{k+1}$. Rewrite the desired inequality as

$$\frac{x}{\sqrt{1 + x^2}} + \sqrt{k} \le \sqrt{k + kx^2 + 1 + x^2}.$$

Increase the left-hand side to $x + \sqrt{k}$; then square both sides. We obtain

$$x^2 + k + 2x\sqrt{k} \le k + kx^2 + 1 + x^2,$$

which reduces to $0 \le (x\sqrt{k} - 1)^2$, and this is obvious. The induction is now complete.   □

639. Find all functions $f : \mathbb{R} \to \mathbb{R}$ satisfying

$$f(x^2 - y^2) = (x - y)(f(x) + f(y)).$$

640. Find all complex-valued functions of a complex variable satisfying

$$f(z) + zf(1 - z) = 1 + z, \text{ for all } z.$$

641. Find all functions $f : \mathbb{R} \setminus \{1\} \to \mathbb{R}$, continuous at 0, that satisfy

$$f(x) = f\left(\frac{x}{1 - x}\right), \text{ for } x \in \mathbb{R} \setminus \{1\}.$$

642. Find all increasing bijections $f : (0, \infty) \to (0, \infty)$ satisfying the functional equation

$$f(f(x)) - 3f(x) + 2x = 0$$

for which there exists $x_0 > 0$ such that $f(x_0) = 2x_0$.

643. Find all functions $f : \mathbb{R} \to \mathbb{R}$ that satisfy the inequality

$$f(x + y) + f(y + z) + f(z + x) \ge 3f(x + 2y + 3z)$$

for all $x, y, z \in \mathbb{R}$.

644. Does there exist a function $f : \mathbb{R} \to \mathbb{R}$ such that $f(f(x)) = x^2 - 2$ for all real numbers x?

645. Find all functions $f : \mathbb{R} \to \mathbb{R}$ satisfying

$$f(x + y) = f(x)f(y) - c \sin x \sin y,$$

for all real numbers x and y, where c is a constant greater than 1.

646. Let f and g be real-valued functions defined for all real numbers and satisfying the functional equation

$$f(x+y) + f(x-y) = 2f(x)g(y)$$

for all x and y. Prove that if $f(x)$ is not identically zero, and if $|f(x)| \le 1$ for all x, then $|g(y)| \le 1$ for all y.

647. Find all continuous functions $f : \mathbb{R} \to \mathbb{R}$ that satisfy the relation

$$3f(2x+1) = f(x) + 5x, \quad \text{for all } x.$$

648. Find all functions $f : (0, \infty) \to (0, \infty)$ subject to the conditions
(i) $f(f(f(x))) + 2x = f(3x)$, for all $x > 0$;
(ii) $\lim_{x \to \infty} (f(x) - x) = 0$.

649. Suppose that $f, g : \mathbb{R} \to \mathbb{R}$ satisfy the functional equation

$$g(x-y) = g(x)g(y) + f(x)f(y)$$

for x and y in $\mathbb{R}$, and that $f(t) = 1$ and $g(t) = 0$ for some $t = 0$. Prove that f and g satisfy

$$g(x+y) = g(x)g(y) - f(x)f(y)$$

and

$$f(x \pm y) = f(x)g(y) \pm g(x)f(y)$$

for all real x and y.

A famous functional equation, which carries the name of Cauchy, is

$$f(x+y) = f(x) + f(y).$$

We are looking for solutions $f : \mathbb{R} \to \mathbb{R}$.

It is straightforward that $f(2x) = 2f(x)$, and inductively $f(nx) = nf(x)$. Setting $y = nx$, we obtain $f\left(\frac{1}{n}y\right) = \frac{1}{n}f(y)$. In general, if m, n are positive integers, then

$$f\left(\frac{m}{n}\right) = mf\left(\frac{1}{n}\right) = \frac{m}{n}f(1).$$

On the other hand, $f(0) = f(0) + f(0)$ implies $f(0) = 0$, and $0 = f(0) = f(x) + f(-x)$ implies $f(-x) = -f(x)$. We conclude that for any rational number x, $f(x) = f(1)x$.

If f is continuous, then the linear functions of the form

$$f(x) = cx,$$

where $c \in \mathbb{R}$, are the only solutions. That is because a solution is linear when restricted to rational numbers and therefore must be linear on the whole real axis. Even if we assume the solution f to be continuous at just one point, it still is linear. Indeed, because $f(x+y)$ is the translate of $f(x)$ by $f(y)$, f must be continuous everywhere.

But if we do not assume continuity, the situation is more complicated. In set theory there is an independent statement called the *Axiom of choice*, which postulates that given a family of nonempty sets $(A_i)_{i \in I}$, there is a function $f : I \to \cup_i A_i$ with $f(i) \in A_i$. In other words, it is possible to select one element from each set.

Real numbers form an infinite-dimensional vector space over the rational numbers (vectors are real numbers, scalars are rational numbers). A corollary of the axiom of choice, Zorn's lemma, implies the existence of a basis for this vector space. If $(e_i)_{i \in I}$ this basis, then any real number x can be expressed uniquely as

$$x = r_1 e_{i_1} + r_2 e_{i_2} + \cdots + r_n e_{i_n},$$

where $r_1, r_2, \ldots, r_n$ are nonzero rational numbers. To obtain a solution to Cauchy's equation, make any choice for $f(e_i)$, $i \in I$, and then extend f to all reals in such a way that it is linear over the rationals. Most of these functions are discontinuous. As an example, for a basis that contains the real number 1, set $f(1) = 1$ and $f(e_i) = 0$ for all other basis elements. Then this function is not continuous.

The problems below are all about Cauchy's equation for continuous functions.

650. Let $f : \mathbb{R} \to \mathbb{R}$ be a continuous nonzero function, satisfying the equation

$$f(x + y) = f(x)f(y), \quad \text{for all } x, y \in \mathbb{R}.$$

Prove that there exists $c > 0$ such that $f(x) = c^x$ for all $x \in \mathbb{R}$.

651. Find all continuous functions $f : \mathbb{R} \to \mathbb{R}$ satisfying

$$f(x + y) = f(x) + f(y) + f(x)f(y), \quad \text{for all } x, y \in \mathbb{R}.$$

652. Determine all continuous functions $f : \mathbb{R} \to \mathbb{R}$ satisfying

$$f(x + y) = \frac{f(x) + f(y)}{1 + f(x)f(y)}, \quad \text{for all } x, y \in \mathbb{R}.$$

653. Find all continuous functions $f : \mathbb{R} \to \mathbb{R}$ satisfying the condition

$$f(xy) = xf(y) + yf(x), \quad \text{for all } x, y \in \mathbb{R}.$$

654. Find the continuous functions $\phi, f, g, h : \mathbb{R} \to \mathbb{R}$ satisfying

$$\phi(x + y + z) = f(x) + g(y) + h(z),$$

for all real numbers x, y, z.

655. Given a positive integer $n \geq 2$, find the continuous functions $f : \mathbb{R} \to \mathbb{R}$, property that for any real numbers $x_1, x_2, \ldots, x_n$,

$$\sum_i f(x_i) - \sum_{i<j} f(x_i + x_j) + \sum_{i<j<k} f(x_i + x_j + x_k) + \cdots$$

$$+ (-1)^{n-1} f(x_1 + x_2 + \cdots + x_n) = 0.$$

We conclude our discussion about functional equations with another instance in which continuity is important. The intermediate value property implies that a one-to-one continuous function is automatically monotonic. So if we can read from a functional equation that a function, which is assumed to be continuous, is also one-to-one, then we know that the function is monotonic, a much more powerful property to be used in the solution.

Example. Find all continuous functions $f : \mathbb{R} \to \mathbb{R}$ satisfying $(f \circ f \circ f)(x) = x$ for all $x \in \mathbb{R}$.

Solution. For any $x \in \mathbb{R}$, the image of $f(f(x))$ through f is x. This shows that f is onto. Also, if $f(x_1) = f(x_2)$ then $x_1 = f(f(f(x_1))) = f(f(f(x_2))) = x_2$, which shows that f is one-to-one. Therefore, f is a continuous bijection, so it must be strictly monotonic. If f is decreasing, then $f \circ f$ is increasing and $f \circ f \circ f$ is decreasing, contradicting the hypothesis. Therefore, f is strictly increasing.

Fix x and let us compare $f(x)$ and x. There are three possibilities. First, we could have $f(x) > x$. Monotonicity implies $f(f(x)) > f(x) > x$, and applying f again, we have $x = f(f(f(x))) > f(f(x)) > f(x) > x$, impossible. Or we could have $f(x) < x$, which then implies $f(f(x)) < f(x) < x$, and $x = f(f(f(x))) < f(f(x)) < f(x) < x$, which again is impossible. Therefore, $f(x) = x$. Since x was arbitrary, this shows that the unique solution to the functional equation is the identity function $f(x) = x$. $\square$

656. Do there exist continuous functions $f, g : \mathbb{R} \to \mathbb{R}$ such that $f(g(x)) = x^2$ and $g(f(x)) = x^3$ for all $x \in \mathbb{R}$?

657. Find all continuous functions $f : \mathbb{R} \to \mathbb{R}$ with the property that

$$f(f(x)) - 2f(x) + x = 0, \quad \text{for all } x, y \in \mathbb{R}.$$

3.4.2 Ordinary Differential Equations of the First Order

Of far greater importance than functional equations are the differential equations, because practically every evolutionary phenomenon of the real world can be modeled by a differential equation. This section is about first-order ordinary differential equations, namely equations expressed in terms of an unknown one-variable function, its derivative, and the variable. In their most general form, they are written as $F(x, y, y') = 0$, but we will be concerned with only two classes of such equations: separable and exact.

An equation is called separable if it is of the form $\frac{dy}{dx} + f(x)g(y)$. In this case we formally separate the variables and write

$$\int \frac{dy}{g(y)} = \int f(x)dx.$$

After integration, we obtain the solution in implicit form, as an algebraic relation between x and y. Here is a problem of I.V. Maftei from the 1971 Romanian Mathematical Olympiad that applies this method.

Example. Find all continuous functions $f : \mathbb{R} \to \mathbb{R}$ satisfying the equation

$$f(x) = \lambda(1 + x^2)\left[1 + \int_0^x \frac{f(t)}{1 + t^2}dt\right],$$

for all $x \in \mathbb{R}$. Here λ is a fixed real number.

Solution. Because f is continuous, the right-hand side of the functional equation is a differentiable function; hence f itself is differentiable. Rewrite the equation as

$$\frac{f(x)}{1+x^2} = \lambda \left[1 + \int_0^x \frac{f(t)}{1+t^2} dt \right],$$

and then differentiate with respect to x to obtain

$$\frac{f'(x)(1+x^2) - f(x)2x}{(1+x^2)^2} = \lambda \frac{f(x)}{1+x^2}.$$

We can separate the variables to obtain

$$\frac{f'(x)}{f(x)} = \lambda + \frac{2x}{1+x^2},$$

which, by integration, yields

$$\ln f(x) = \lambda x + \ln(1+x^2) + c.$$

Hence $f(x) = a(1 + x^2)e^{\lambda x}$ for some constant a. Substituting in the original relation, we obtain $a = \lambda$. Therefore, the equation from the statement has the unique solution

$$f(x) = \lambda(1+x^2)e^{\lambda x}. \qquad \square$$

A first-order differential equation can be written formally as

$$p(x, y)dx + q(x, y)dy = 0.$$

Physicists think of the expression on the left as the potential of a two-dimensional force field, with p and q the x and y components of the potential. Mathematicians call this expression a 1-form. The force field is called conservative if no energy is wasted in moving an object along any closed path. In this case the differential equation is called exact. For functions defined on the entire 2-dimensional plane, as a consequence of Green's theorem one can deduce that the field is conservative precisely when the exterior derivative

$$\left(\frac{\partial q}{\partial x} - \frac{\partial p}{\partial y} \right) dxdy$$

is equal to zero. This means that there exists a scalar function $u(x, y)$ whose differential is the field, i.e.,

$$\frac{\partial u}{\partial x} = p(x, y) \quad \text{and} \quad \frac{\partial u}{\partial y} = q(x, y).$$

If the domain has "holes", then there is an obstruction in de Rham cohomology for some equations to admit a potential. For a conservative field, the scalar potential solves the differential equation, giving the solution in implicit form as $u(x, y) = C$, with C a constant. Let us apply this method to a problem by the first author of the book.

Example. Does there exist a differentiable function y defined on the entire real axis that satisfies the differential equation

$$(2x + y - e^{-x^2})dx + (x + 2y - e^{-y^2})dy = 0?$$

Solution. Let us assume that such a y does exist. Because

$$\frac{\partial}{\partial x}(x + 2y - e^{-y^2}) = \frac{\partial}{\partial y}(2x + y - e^{-x^2}),$$

and because $2x + y - e^{-x^2}$ and $x + 2y - e^{-y^2}$ are defined everywhere, the equation can be integrated. The potential function is

$$u(x, y) = x^2 + xy + y^2 - \int_0^x e^{-s^2}ds - \int_0^y e^{-t^2}dt.$$

The differential equation translates into the algebraic equation

$$\left(x + \frac{1}{2}y\right)^2 + \frac{3}{4}y^2 = \int_0^x e^{-s^2}ds + \int_0^y e^{-t^2}dt + C$$

for some real constant C. The right-hand side is bounded from above by $\sqrt{8\pi} + C$ (note the Gaussian integrals). This means that both squares on the left must be bounded. In particular, y is bounded, but then $x + \frac{1}{2}y$ is unbounded, a contradiction. Hence the answer to the question is no; a solution can exist only on a bounded interval. □

Sometimes the field is not conservative but becomes conservative after the differential equation is multiplied by a function. This function is called an integrating factor. There is a standard method for finding integrating factors, which can be found in any textbook. In particular, any first-order linear equation

$$y' + p(x)y = q(x)$$

can be integrated after it is multiplied by the integrating factor $\exp\left(\int p(x)dx\right)$.

It is now time for problems. In the problems below, we denote by f^2 the product $f \cdot f$ (not the composition of f with itself).

658. A not uncommon mistake is to believe that the product rule for derivatives says that $(fg) = f'g'$. If $f(x) = e^{x^2}$, determine whether there exists an open interval (a, b) and a nonzero function g defined on (a, b) such that this wrong product rule is true for f and g on (a, b).

659. Find the functions $f, g : \mathbb{R} \to \mathbb{R}$ with continuous derivatives satisfying

$$f^2 + g^2 = f'^2 + g'^2, \quad f + g = g' - f',$$

and such that the equation $f = g$ has two real solutions, the smaller of them being zero.

660. Let f and g be differentiable functions on the real line satisfying the equation

$$(f^2 + g^2)f' + (fg)g' = 0.$$

Prove that f is bounded.

661. Let A, B, C, D, m, n be real numbers with $AD - BC \neq 0$. Solve the differential equation

$$y(B + Cx^m y^n)dx + x(A + Dx^m y^n)dy = 0.$$

662. Find all continuously differentiable functions $y : (0, \infty) \to (0, \infty)$ that are solutions to the initial value problem

$$y^{y'} = x, \quad y(1) = 1.$$

663. Find all differentiable functions $f : (0, \infty) \to (0, \infty)$ for which there is a positive real number a such that

$$f'\left(\frac{a}{x}\right) = \frac{x}{f(x)},$$

for all $x > 0$.

664. Prove that if the function $f(x, y)$ is continuously differentiable on the whole xy-plane and satisfies the equation

$$\frac{\partial f}{\partial x} + f\frac{\partial f}{\partial y} = 0,$$

then $f(x, y)$ is constant.

3.4.3 Ordinary Differential Equations of Higher Order

The field of higher-order ordinary differential equations is vast, and we assume that you are familiar at least with some of its techniques. In particular, we assume you are familiar with the theory of linear equations with fixed coefficients, from which we recall some basic facts. A linear equation with fixed coefficients has the general form

$$a_n\frac{d^n y}{dx^n} + \cdots + a_2\frac{d^2 y}{dx^2} + a_1\frac{dy}{dx} + a_0 = f(x).$$

If f is zero, the equation is called homogeneous. Otherwise, the equation is called inhomogeneous. In this case the general solution is found using the characteristic equation

$$a_l\lambda^n + a_{n-1}\lambda^{n-1} + \cdots + a_0 = 0.$$

If $\lambda_1, \lambda_2, \ldots, \lambda_r$ are the distinct roots, real or complex, of this equation, then the general solution to the homogeneous differential equation is of the form

$$y(x) = P_1(x)e^{\lambda_1 x} + P_2(x)e^{\lambda_2 x} + \cdots + P_r(x)e^{\lambda_r x},$$

where $P_i(x)$ is a polynomial of degree one less than the multiplicity of λ_i, $i = 1, 2, \ldots, r$. If the exponents are complex, the exponentials are changed into (damped) oscillations using Euler's formula ($e^{ix} = \cos x + i \sin x$).

The general solution depends on n parameters (the coefficients of the polynomials), so the space of solutions is an n-dimensional vector space V. For an inhomogeneous equation, the space of solutions is the affine space $y_0 + V$ obtained by adding a particular solution. This particular solution is found usually by the method of the variation of the coefficients.

We start with an example that exploits an idea that appeared once on a Putnam exam.

Example. Solve the system of differential equations

$$x'' - y' + x = 0,$$

$$y'' + x' + y = 0$$

in real-valued functions $x(t)$ and $y(t)$.

Solution. Multiply the second equation by i then add it to the first to obtain

$$(x + iy'') + i(x + iy)' + (x + iy) = 0.$$

With the substitution $z = x + iy$ this becomes the second-order homogeneous linear differential equation $z'' + iz' + z = 0$. The characteristic equation is $\lambda^2 + i\lambda + 1 = 0$, with solutions $\lambda_{1,2} = \dfrac{-1 \pm \sqrt{5}}{2} i$. We find the general solution to the equation

$$z(t) = (a + ib) \exp\left(\frac{-1 + \sqrt{5}}{2} it\right) + (c + id) \exp\left(\frac{-1 - \sqrt{5}}{2} it\right),$$

for arbitrary real numbers a, b, c, d. Since x and y are, respectively, the real and complex parts of the solution, they have the general form

$$x(t) = a \cos \frac{-1 + \sqrt{5}}{2} t - b \sin \frac{-1 + \sqrt{5}}{2} t + c \cos \frac{-1 - \sqrt{5}}{2} t - d \sin \frac{-1 - \sqrt{5}}{2} t,$$

$$y(t) = a \sin \frac{-1 + \sqrt{5}}{2} t + b \cos \frac{-1 + \sqrt{5}}{2} t + c \sin \frac{-1 - \sqrt{5}}{2} t + d \cos \frac{-1 - \sqrt{5}}{2} t.$$

The problem is solved. □

Our second example is an equation published by M. Ghermănescu in the *Mathematics Gazette, Bucharest*. Its solution combines several useful techniques.

Example. Solve the differential equation

$$2(y')^3 - yy'y'' - y^2 y''' = 0.$$

Solution. In a situation like this, where the variable x does not appear explicitly, one can reduce the order of the equation by taking y as the variable and $p = y'$ as the function. The higher-order derivatives of y'' are

$$y'' = \frac{d}{dx}y' = \frac{d}{dy}p\frac{dy}{dx} = p'p,$$

$$y''' = \frac{d}{dx}y'' = \left(\frac{d}{dy}pp'\right)\frac{dy}{dx} = ((p')^2 + pp'')p.$$

We end up with a second-order differential equation

$$2p^3 - yp^2p' - y^2pp'' - y^2p(p')^2 = 0.$$

A family of solutions is $p = 0$, that is, $y' = 0$. This family consists of the constant functions $y = C$. Dividing the equation by $-p$, we obtain

$$y^2p'' + y^2(p')^2 + ypp' - 2p^2 = 0.$$

The distribution of the powers of y reminds us of the Euler-Cauchy equation, while the last terms suggests the substitution $u = p^2$. And indeed, we obtain the Euler-Cauchy equation

$$y^2u'' + yu' - 4u = 0,$$

with general solution $u = C_1y^2 + C_2y^{-2}$. Remember that $u = p^2 = (y')^2$, from which we obtain the first-order differential equation

$$y' = \pm\sqrt{C_1y^2 + C_2y^{-2}} = \frac{\sqrt{C_1y^4 + C_2}}{y}.$$

This we solve by separation of variables

$$dx = \pm\frac{ydy}{\sqrt{C_1y^4 + C_2}},$$

which after integration gives

$$x = \pm\int\frac{ydy}{\sqrt{C_1y^4 + C_2}} = \pm\frac{1}{2}\int\frac{dz}{\sqrt{C_1z^2 + C_2}}.$$

This last integral is standard; it is equal to $\frac{1}{2\sqrt{C_1}}\ln\left|y + \sqrt{y^2 + C_2/C_1}\right|$ if $C_1 > 0$ and to $\frac{1}{2\sqrt{|C_1|}}\arcsin\left(\frac{|C_1|y}{C_2}\right)$ if $C_1 < 0$ and $C_2 > 0$. We obtain two other families of solutions given in implicit form by

$$x = \pm\frac{1}{2\sqrt{C_1}}\ln\left|y + \sqrt{y^2 + \frac{C_2}{C_1}}\right| + C_3 \quad\text{and}\quad x = \pm\frac{1}{2\sqrt{-C_1}}\arcsin\frac{|C_1|y}{C_2} + C_3,$$

that is,

$$x = A\ln\left|y + \sqrt{y^2 + B}\right| + C \quad\text{and}\quad x = E\arcsin Fy + G. \qquad \square$$

Here are more problems.

665. Solve the differential equation

$$xy'' + 2y' + xy = 0.$$

666. Find all twice-differentiable functions defined on the entire real axis that satisfy $f'(x)f''(x) = 0$ for all x.

667. Find all continuous functions $f : \mathbb{R} \to \mathbb{R}$ that satisfy

$$f(x) + \int_0^x (x - t)f(t)dt = 1, \quad \text{for all } x \in \mathbb{R}.$$

668. Solve the differential equation

$$(x - 1)y'' + (4x - 5)y' + (4x - 6)y = xe^{-2x}.$$

669. Let n be a positive integer. Show that the equation

$$(1 - x^2)y'' - xy' + n^2y = 0$$

admits as a particular solution an nth-degree polynomial.

670. Find the one-to-one, twice-differentiable solutions y to the equation

$$\frac{d^2y}{dx^2} + \frac{d^2x}{dy^2} = 0.$$

671. Show that all solutions to the differential equation $y'' + e^xy = 0$ remain bounded as $x \to \infty$.

3.4.4 Problems Solved with Techniques of Differential Equations

In this section we illustrate how tricks of differential equations can offer inspiration when one is tackling problems from outside this field.

Example. Let $f : [0, \infty) \to \mathbb{R}$ be a twice-differentiable function satisfying $f(0) \geq 0$ and $f'(x) > f(x)$ for all $x > 0$. Prove that $f(x) > 0$ for all $x > 0$.

Solution. To solve this problem we use an integrating factor. The inequality

$$f'(x) - f(x) > 0$$

can be "integrated" after multiplying it by e^{-x}. It simply says that the derivative of the function $e^{-x}f(x)$ is strictly positive on $(0, \infty)$. This function is therefore strictly increasing on $[0, \infty)$. So for $x > 0$ we have $e^{-x}f(x) > e^{-0}f(0) = f(0) \geq 0$, which then implies $f(x) > 0$, as desired. $\qquad\square$

Example. Compute the integral

$$y(x) = \int_0^\infty e^{-t^2/2} \cos \frac{x^2}{2t^2} dt.$$

Solution. We will show that the function $y(x)$ satisfies the ordinary differential equation $i^{iv} + y = 0$. To this end, we compute

$$y'(x) = \int_0^\infty e^{-t^2/2} \sin \frac{x^2}{2t^2} \cdot \frac{-x}{t^2} dt = -\int_0^\infty e^{-x^2/2u^2} \sin \frac{u^2}{2} du$$

and

$$y''(x) = -\int_0^\infty e^{-x^2/2u^2} \sin \frac{u^2}{2} \cdot \frac{-x}{u^2} du = \int_0^\infty e^{-t^2/2} \sin \frac{x^2}{2t^2} dt.$$

Iterating, we eventually obtain

$$y^{iv}(x) = -\int_0^\infty e^{-t^2/2} \cos \frac{x^2}{2t^2} dt = -y(x),$$

which proves that indeed y satisfies the differential equation $y^{iv} + y = 0$. The general solution to this differential equation is

$$y(x) = e^{\frac{x}{\sqrt{2}}} \left(C_1 \cos \frac{x}{\sqrt{2}} + C_2 \sin \frac{x}{\sqrt{2}} \right) + e^{-\frac{x}{\sqrt{2}}} \left(C_3 \cos \frac{x}{\sqrt{2}} + C_4 \sin \frac{x}{\sqrt{2}} \right).$$

To find which particular solution is the integral in question, we look at boundary values. To compute these boundary values we refer to Section 3.3.2, the one on multivariable integral calculus. We recognize that $y(0) = \int_0^\infty e^{-t^2/2} dt$ is a Gaussian integral equal to $\sqrt{\frac{\pi}{2}}$, $y'(0) = -\int_0^\infty \sin \frac{u^2}{2} du$ is a Fresnel integral equal to $-\frac{\sqrt{\pi}}{2}$, $y''(0) = 0$, while $y'''(0) = \int_0^\infty \cos \frac{u^2}{2} du$ is yet another Fresnel integral equal to $\frac{\sqrt{\pi}}{2}$. We find that $C_1 = C_2 = C_4 = 0$ and $C_3 = \sqrt{\frac{\pi}{2}}$. The value of the integral from the statement is therefore

$$y(x) = \sqrt{\frac{\pi}{2}} e^{-\frac{x}{\sqrt{2}}} \cos \frac{x}{\sqrt{2}}. \qquad \square$$

We leave the following examples to the reader.

672. Show that both functions

$$y_1(x) = \int_0^\infty \frac{e^{-tx}}{1+t^2} dt \quad \text{and} \quad y_2(x) = \int_0^\infty \frac{\sin t}{t+x} dt$$

satisfy the differential equation $y'' + y = \dfrac{1}{x}$. Prove that these two functions are equal.

673. Let f be a real-valued continuous nonnegative function on $[0, 1]$ such that

$$f(t)^2 \leq 1 + 2\int_0^t f(s)ds, \quad \text{for all } t \in [0, 1].$$

Show that $f(t) \leq 1 + t$ for every $t \in [0, 1]$.

674. Let $f : [0, 1] \to \mathbb{R}$ be a continuous function with $f(0) = f(1) = 0$. Assume that f'' exists on $(0, 1)$ and $f''(x) + 2f'(x) + f(x) \geq 0$ for all $x \in (0, 1)$. Prove that $f(x) \leq 0$ for all $x \in [0, 1]$.

675. Does there exist a continuously differentiable function $f : \mathbb{R} \to \mathbb{R}$ satisfying $f(x) > 0$ and $f'(x) = f(f(x))$ for every $x \in \mathbb{R}$?

676. Determine all nth-degree polynomials $P(x)$, with real zeros, for which the equality

$$\sum_{i=1}^{n} \frac{1}{P(x) - x_i} = \frac{n^2}{xP'(x)}$$

holds for all nonzero real numbers x for which $P'(x) \neq 0$, where x_i, $i = 1, 2, \ldots, n$, are the zeros of $P(x)$.

677. Let C be the class of all real-valued continuously differentiable functions f on the interval $[0, 1]$ with $f(0) = 0$ and $f(1) = 1$. Determine

$$u = \inf_{f \in C} \int_0^1 |f'(x) - f(x)|dx.$$

678. Let $f : \mathbb{R} \to \mathbb{R}$ be an infinitely differentiable function with the property that there are distinct positive real numbers a, b, c such that the function

$$g(x) = f(ax) + f(bx) + f(cx)$$

is a polynomial. Show that f is a polynomial function as well.

4

Geometry and Trigonometry

Geometry is the oldest of the mathematical sciences. Its age-old theorems and the sharp logic of its proofs make you think of the words of Andrew Wiles, "Mathematics seems to have a permanence that nothing else has".

This chapter is bound to take you away from the geometry of the ancients, with figures and pictorial intuition, and bring you to the science of numbers and equations that geometry has become today. In a dense exposition we have packed vectors and their applications, analytical geometry in the plane and in space, some applications of integral calculus to geometry, followed by a list of problems with Euclidean flavor but based on algebraic and combinatorial ideas. Special attention is given to conics, cubics, and quadrics, for their study already contains the germs of differential and algebraic geometry.

Four subsections are devoted to geometry's little sister, trigonometry. We insist on trigonometric identities, repeated in subsequent sections from different perspectives: Euler's formula, trigonometric substitutions, and telescopic summation and multiplication.

Since geometry lies at the foundation of mathematics, its presence could already be felt in the sections on linear algebra and multivariable calculus. It will resurface again in the chapter on combinatorics.

4.1 Geometry

4.1.1 Vectors

This section is about vectors in two and three dimensions. Vectors are oriented segments identified under translation.

There are four operations defined for vectors: scalar multiplication $\alpha \vec{v}$, addition $\vec{v} + \vec{w}$, dot product $\vec{v} \cdot \vec{w}$, and cross-product $\vec{v} \times \vec{w}$, the last being defined only in three dimensions. Scalar multiplication dilates or contracts a vector by a scalar. The sum of two vectors is computed with the parallelogram rule; it is the resultant of the vectors acting as forces on an object. The dot product of two vectors is a number equal to the product of the magnitudes of the vectors and the cosine of the angle between them. A dot product equal to zero tells us

© Springer International Publishing AG 2017
R. Gelca and T. Andreescu, *Putnam and Beyond*, DOI 10.1007/978-3-319-58988-6_4

that the vectors are orthogonal. The cross-product of two vectors is a vector orthogonal to the two vectors and of magnitude equal to the area of the parallelogram they generate. The orientation of the cross-product is determined by the right-hand rule: place your hand so that you can bend your palm from the first vector to the second, and your thumb will point in the direction of the cross-product. A cross-product equal to zero tells us that the vectors are parallel (although they might point in opposite directions).

The dot and cross-products are distributive with respect to sum; the dot product is commutative, while the cross-product is not. For the three-dimensional vectors $\vec{u}$, $\vec{v}$, $\vec{w}$, the number $\vec{u} \cdot (\vec{v} \times \vec{w})$ is the volume taken with sign of the parallelepiped constructed with the vectors as edges. The sign is positive if the three vectors determine a frame that is oriented the same way as the orthogonal frame of the three coordinate axes, and negative otherwise. Equivalently, $\vec{u} \cdot (\vec{v} \times \vec{w})$ is the determinant with the coordinates of the three vectors as rows.

A useful computational tool is the formula for the triple cross product:

$$\vec{a} \times (\vec{b} \times \vec{c}) = (\vec{a} \cdot \vec{c})\vec{b} - (\vec{a} \cdot \vec{b})\vec{c},$$

also known as the BAC-CAB formula (because it is also written as $\vec{a} \times (\vec{b} \times \vec{c}) = \vec{b}(\vec{a} \cdot \vec{c}) - \vec{c}(\vec{a} \cdot \vec{b})$).

The quickest way to prove it is to check it for $\vec{a}$, $\vec{b}$, $\vec{c}$ chosen among the three unit vectors parallel to the coordinate axes $\vec{i}$, $\vec{j}$, and $\vec{k}$, and then use the distributivity of the cross-product with respect to addition. Here is an easy application of this identity.

Example. Prove that for any vectors $\vec{a}$, $\vec{b}$, $\vec{c}$, $\vec{d}$,

$$(\vec{a} \times \vec{b}) \times (\vec{c} \times \vec{d}) = (\vec{a} \cdot (\vec{b} \times \vec{d}))\vec{c} - (\vec{a} \cdot (\vec{b} \times \vec{c}))\vec{d}.$$

Solution. We have

$$(\vec{a} \times \vec{b}) \times (\vec{c} \times \vec{d}) = (\vec{d} \cdot (\vec{a} \times \vec{b}))\vec{c} - (\vec{c} \cdot (\vec{a} \times \vec{b}))\vec{d}$$
$$= (\vec{a} \cdot (\vec{b} \times \vec{d}))\vec{c} - (\vec{a} \cdot (\vec{b} \times \vec{c}))\vec{d}$$

In the computation we used the equality $\vec{u} \cdot (\vec{v} \times \vec{w}) = \vec{w} \cdot (\vec{u} \times \vec{v})$, which is straightforward if we write these as determinants. $\qquad \square$

Let us briefly point out a fundamental algebraic property of the cross-product. Denote by so(3) the set of 3×3 matrices A satisfying $A + A^t = O_3$ endowed with the operation $[A, B] = AB - BA$.

Theorem. *The map*

$$(a_1, a_2, a_3) \rightarrow \begin{pmatrix} 0 & -a_1 & -a_2 \\ a_1 & 0 & -a_3 \\ a_2 & a_3 & 0 \end{pmatrix}$$

establishes an isomorphism between $(\mathbb{R}^3, \times)$ *and* $(so(3), [\cdot, \cdot])$.

Proof. The proof is straightforward if we write the cross-product in coordinates. The result shows that the cross-product defines a Lie algebra structure on the set of three-dimensional vectors. Note that the isomorphism maps the sum of vectors to the sum of matrices, and the dot product of two vectors to the negative of half the trace of the product of the corresponding matrices. □

It is worth mentioning that $so(3)$ is the Lie algebra of the Lie group $SO(3)$ of rotations of $\mathbb{R}^3$ about the origin. And now the problems.

679. For any three-dimensional vectors $\vec{u}$, $\vec{v}$, $\vec{w}$, prove the identity

$$\vec{u} \times (\vec{v} \times \vec{w}) + \vec{v} \times (\vec{w} \times \vec{u}) + \vec{w} \times (\vec{u} \times \vec{v}) = \vec{0}.$$

680. Given three vectors $\vec{a}$, $\vec{b}$, $\vec{c}$, define

$$\vec{u} = (\vec{b} \cdot \vec{c})\vec{a} - (\vec{c} \cdot \vec{a})\vec{b},$$

$$\vec{v} = (\vec{a} \cdot \vec{c})\vec{b} - (\vec{a} \cdot \vec{b})\vec{c},$$

$$\vec{w} = (\vec{b} \cdot \vec{a})\vec{c} - (\vec{b} \cdot \vec{c})\vec{a}.$$

Prove that if $\vec{a}$, $\vec{b}$, $\vec{c}$ form a triangle, then $\vec{u}$, $\vec{v}$, $\vec{w}$ also form a triangle, and this triangle is similar to the first.

681. Let $\vec{a}$, $\vec{b}$, $\vec{c}$ be vectors such that $\vec{b}$ and $\vec{c}$ are perpendicular, but $\vec{a}$ and $\vec{b}$ are not. Let m be a real number. Solve the system

$$\vec{x} \cdot \vec{a} = m,$$

$$\vec{x} \times \vec{b} = \vec{c}.$$

682. Consider three linearly independent vectors $\vec{a}$, $\vec{b}$, $\vec{c}$ in space, having the same origin. Prove that the plane determined by the endpoints of the vectors is perpendicular to the vector $\vec{a} \times \vec{b} + \vec{b} \times \vec{c} + \vec{c} \times \vec{a}$.

683. The vectors $\vec{a}$, $\vec{b}$, and $\vec{c}$ satisfy

$$\vec{a} \times \vec{b} = \vec{b} \times \vec{c} = \vec{c} \times \vec{a} \neq \vec{0}.$$

Prove that $\vec{a} + \vec{b} + \vec{c} = \vec{0}$.

684. Find the vector-valued functions $\vec{u}(t)$ satisfying the differential equation

$$\vec{u} \times \vec{u}' = \vec{v},$$

where $\vec{v} = \vec{v}(t)$ is a twice-differentiable vector-valued function such that both $\vec{v}$ and $\vec{v}'$ are never zero or parallel.

685. Does there exist a bijection f of (a) a plane with itself or (b) three-dimensional space with itself such that for any distinct points A, B the lines AB and $f(A)f(B)$ are perpendicular?

686. On $so(3)$ we define the operation $*$ such that if A and B are matrices corresponding to the vectors $\vec{a} = (a_1, a_2, a_3)$ and $\vec{b} = (b_1, b_2, b_3)$, then the ij entry of $A * B$ is equal to $(-1)^{i+j} a_{4-j} b_{4-i}$. Prove the identity

$$CBA - BCA = (A * C)B - (A * B)C.$$

687. Prove that there is a bijection f from $\mathbb{R}^3$ to the set $su(2)$ of 2×2 matrices with complex entries that are skew symmetric and have trace equal to zero such that

$$f(\vec{v} \times \vec{w}) = [f(\vec{v}), f(\vec{w})].$$

(Here $[A, B] = AB - BA$; the commutant.)

688. We are given 2015 unit vectors starting at the origin, with the property that every line passing through the origin has at least 515 vectors on each side. Show that the length of the sum of the vectors does not exceed 1015.

We present two applications of vector calculus to geometry, one with the dot product, one with the cross-product.

Example. Given two triangles ABC and $A'B'C'$ such that the perpendiculars from A, B, C onto $B'C', C'A', A'B'$ intersect, show that the perpendiculars from A', B', C' onto BC, CA, AB also intersect.

Solution. This is the *property of orthological triangles*. Denote by O the intersection of the first set of three perpendiculars, and by O' the intersection of perpendiculars from A' and B'. Note that if the vector $\overrightarrow{XY}$ is orthogonal to a vector $\overrightarrow{ZW}$, then for any point P in the plane,

$$(\overrightarrow{PX} - \overrightarrow{PY}) \cdot \overrightarrow{ZW} = \overrightarrow{XY} \cdot \overrightarrow{ZW} = 0;$$

hence $\overrightarrow{PX} \cdot \overrightarrow{ZW} = \overrightarrow{PY} \cdot \overrightarrow{ZW}$. Using this fact we can write

$$\overrightarrow{O'C'} \cdot \overrightarrow{OB} = \overrightarrow{O'A'} \cdot \overrightarrow{OB} = \overrightarrow{O'A'} \cdot \overrightarrow{OC} = \overrightarrow{O'B'} \cdot \overrightarrow{OC} = \overrightarrow{O'B'} \cdot \overrightarrow{OA} = \overrightarrow{O'C'} \cdot \overrightarrow{OA}.$$

Therefore, $\overrightarrow{O'C'} \cdot (\overrightarrow{OB} - \overrightarrow{OA}) = \overrightarrow{O'C'} \cdot \overrightarrow{AB} = 0$, which shows that $O'C'$ is perpendicular to AB. This proves that the second family of perpendiculars are concurrent. $\qquad\square$

Example. Let $ABCD$ be a convex quadrilateral, M, N on side AB and P, Q on side CD. Show that if $AM = NB$ and $CP = QD$, and if the quadrilaterals $AMQD$ and $BNPC$ have the same area, then AB is parallel to CD.

Solution. Throughout the solution we refer to Figure 24. We decompose the quadrilaterals into triangles, and then use the formula for the area in terms of the cross-product.

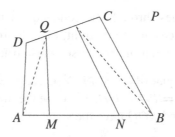

Figure 24

In general, the triangle determined by $\vec{v}_1$ and $\vec{v}_2$ has area equal to half the magnitude of $\vec{v}_1 \times \vec{v}_2$. Note also that $\vec{v}_1 \times \vec{v}_2$ is perpendicular to the plane of the triangle, so for a problem in plane geometry there is no danger in identifying the areas with the cross-products, provided that we keep track of the orientation. The hypothesis of the problem implies that

$$\frac{1}{2}(\overrightarrow{DA} \times \overrightarrow{DQ} + \overrightarrow{AM} \times \overrightarrow{AQ}) = \frac{1}{2}(\overrightarrow{CP} \times \overrightarrow{CB} + \overrightarrow{BP} \times \overrightarrow{BN}).$$

Hence

$$\overrightarrow{DA} \times \overrightarrow{DQ} + \overrightarrow{AM} \times (\overrightarrow{AD} + \overrightarrow{DQ}) = \overrightarrow{CP} \times \overrightarrow{CB} + (\overrightarrow{BC} + \overrightarrow{CP}) \times \overrightarrow{BN}.$$

Because $\overrightarrow{BN} = -\overrightarrow{AM}$ and $\overrightarrow{CP} = -\overrightarrow{DQ}$, this equality can be rewritten as

$$(\overrightarrow{AM} + \overrightarrow{DQ}) \times (\overrightarrow{AD} + \overrightarrow{CB}) = 2\overrightarrow{DQ} \times \overrightarrow{AM}.$$

Using the fact that $\overrightarrow{AD} + \overrightarrow{CB} = \overrightarrow{AB} + \overrightarrow{CD}$ (which follows from $\overrightarrow{AB} + \overrightarrow{BC} + \overrightarrow{CD} + \overrightarrow{DA} = \vec{0}$), we obtain

$$\overrightarrow{AM} \times \overrightarrow{CD} + \overrightarrow{DQ} \times \overrightarrow{AB} = 2\overrightarrow{DQ} \times \overrightarrow{AM}.$$

From here we deduce that $\overrightarrow{AM} \times \overrightarrow{QC} = \overrightarrow{DQ} \times \overrightarrow{MB}$. These two cross-products point in opposite directions, so equality can hold only if both are equal to zero, i.e., if AB is parallel to CD.

More applications of the dot and cross-products to geometry can be found below.

689. Given two triangles ABC and $A'B'C'$ with the same centroid, prove that one can construct a triangle with sides equal to the segments AA', BB', and CC'.

690. Given a quadrilateral $ABCD$, consider the points A', B', C', D' on the half-lines (i.e., rays) $|AB$, $|BC$, $|CD$, and $|DA$, respectively, such that $AB = BA'$, $BC = CB'$, $CD = DC'$, $DA = AD'$. Suppose now that we start with the quadrilateral $A'B'C'D'$. Using a straightedge and a compass only, reconstruct the quadrilateral $ABCD$.

691. On the sides of the triangle ABC construct in the exterior the rectangles ABB_1A_2, BCC_1B_2, CAA_1C_2. Prove that the perpendicular bisectors of A_1A_2, B_1B_2, and C_1C_2 intersect at one point.

692. Let $ABCD$ be a convex quadrilateral. The lines parallel to AD and CD through the orthocenter H of triangle ABC intersect AB and BC, respectively, at P and Q. Prove that the perpendicular through H to the line PQ passes through the orthocenter of triangle ACD.

693. Prove that if the four lines through the centroids of the four faces of a tetrahedron perpendicular to those faces are concurrent, then the four altitudes of the tetrahedron are also concurrent. Prove that the converse is also true.

694. Let $ABCD$ be a convex quadrilateral, $M, N \in AB$ such that $AM = MN = NB$, and $P, Q \in CD$ such that $CP = PQ = QD$. Let O be the intersection of AC and BD. Prove that the triangles MOP and NOQ have the same area.

695. Let ABC be a triangle, with D and E on the respective sides AC and AB. If M and N are the midpoints of BD and CE, prove that the area of the quadrilateral $BCDE$ is four times the area of the triangle AMN.

4.1.2 The Coordinate Geometry of Lines and Circles

Coordinate geometry was constructed by Descartes to translate Euclid's geometry into the language of algebra. In two dimensions one starts by fixing two intersecting coordinate axes and a unit on each of them. If the axes are perpendicular and the units are equal, the coordinates are called Cartesian (in the honor of Descartes); otherwise, they are called affine. A general affine change of coordinates has the form

$$\begin{pmatrix} x' \\ y' \end{pmatrix} = \begin{pmatrix} a & b \\ c & d \end{pmatrix} \begin{pmatrix} x \\ y \end{pmatrix} + \begin{pmatrix} e \\ f \end{pmatrix}, \quad \text{with} \quad \begin{pmatrix} a & b \\ c & d \end{pmatrix} \text{ invertible.}$$

If the change is between Cartesian systems of coordinates, a so-called Euclidean change of coordinates, it is required additionally that the matrix

$$\begin{pmatrix} a & b \\ c & d \end{pmatrix}$$

be orthogonal, meaning that its inverse is equal to the transpose.

Properties that can be formulated in the language of lines and ratios are invariant under affine changes of coordinates. Such are the properties of two lines being parallel or of a point to divide a segment in half. All geometric properties are invariant under Euclidean changes of coordinates. Therefore, problems about distances, circles, and angles should be modeled with Cartesian coordinates.

In this section we grouped problems that require only the knowledge of the theory of lines and circles. Recall that the general equation of a line (whether in a Cartesian or affine coordinate system) is $ax + by + c = 0$. That of a circle (in a Cartesian coordinate system) is $(x - h)^2 + (y - k)^2 = r^2$, where (h, k) is the center and r is the radius. Let us see two examples, one in affine and one in Cartesian coordinates. But before we do that let us recall that a complete quadrilateral is a quadrilateral in which the pairs of opposite sides have been extended until they meet. For that reason, a complete quadrilateral has six vertices and three diagonals.

Example. Prove that the midpoints of the three diagonals of a complete quadrilateral are collinear.

Solution. As said, we will work in affine coordinates. Choose the coordinate axes to be sides of the quadrilateral, as shown in Figure 25.

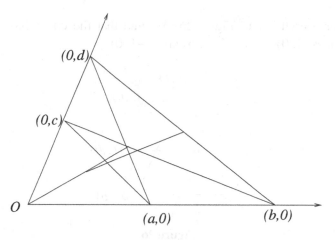

Figure 25

Five of the vertices have coordinates $(0, 0)$, $(a, 0)$, $(b, 0)$, $(0, c)$, and $(0, d)$, while the sixth is found as the intersection of the lines through $(a, 0)$ and $(0, d)$, respectively, $(0, c)$ and $(b, 0)$. For these two lines we know the $x-$ and $y-$ intercepts, so their equations are

$$\frac{1}{a}x + \frac{1}{d}y = 1 \quad \text{and} \quad \frac{1}{b}x + \frac{1}{c}y = 1.$$

The sixth vertex of the complete quadrilateral has therefore the coordinates

$$\left(\frac{ab(c - d)}{ac - bd}, \frac{cd(a - b)}{ac - bd}\right).$$

We find that the midpoints of the diagonals are

$$\left(\frac{a}{2}, \frac{c}{2}\right), \quad \left(\frac{b}{2}, \frac{d}{2}\right), \quad \left(\frac{ab(c - d)}{2(ac - bd)}, \frac{cd(a - b)}{2(ac - bd)}\right).$$

The condition that these three points be collinear translates to

$$\begin{vmatrix} \dfrac{a}{2} & \dfrac{c}{2} & 1 \\ \dfrac{b}{2} & \dfrac{d}{2} & 1 \\ \dfrac{ab(c - d)}{2(ac - bd)} & \dfrac{cd(a - b)}{2(ac - bd)} & 1 \end{vmatrix} = 0,$$

which is equivalent to

$$\begin{vmatrix} a & c & 1 \\ b & d & 1 \\ ab(c - d) & cd(a - b) & ac - bd \end{vmatrix} = 0.$$

This is verified by direct computation. $\qquad\qquad\qquad\qquad\qquad\qquad\qquad\qquad\qquad$ □

Example. In a circle are inscribed a trapezoid with one side as diameter and a triangle with sides parallel to sides of the trapezoid. Prove that the two have the same area.

Solution. We refer everything to Figure 26. Assume that the circle has radius 1, and the trapezoid has vertices $(1, 0)$, (a, b), $(-a, b)$ and $(-1, 0)$.

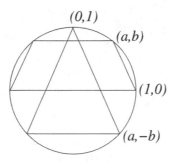

Figure 26

The triangle is isosceles and has one vertex at $(0, 1)$. We need to determine the coordinates of the other two vertices. One of them lies where the parallel through $(0, 1)$ to the line determined by $(1, 0)$ and (a, b) intersects the circle. The equation of the line is

$$y = \frac{b}{a-1}x + 1.$$

The relation $a^2 + b^2 = 1$ yields $b^2 = (1-a)(1+a)$, or $\frac{b}{1-a} = \frac{1+a}{b}$. So the equation of the line can be rewritten as

$$y = -\frac{1+a}{b}x + 1.$$

Now it is easy to guess that the intersection of this line with the circle is $(b, -a)$ (note that this point satisfies the equation of the circle). The other vertex of the triangle is $(-b, -a)$ so the area is $\frac{1}{2}(2b)(1+a) = b + ab$. And the area of the trapezoid is $\frac{1}{2}(2a + 2)b = b + ab$, the same number.   □

696. Prove that the midpoints of the sides of a quadrilateral form a parallelogram.

697. Let M be a point in the plane of triangle ABC. Prove that the centroids of the triangles MAB, MAC, and MCB form a triangle similar to triangle ABC.

698. Find the locus of points P in the interior of a triangle ABC such that the distances from P to the lines AB, BC, and CA are the side lengths of some triangle.

699. Let $A_1, A_2, \ldots, A_n$ be distinct points in the plane, and let m be the number of midpoints of all the segments they determine. What is the smallest value that m can have?

700. Given an acute-angled triangle ABC with altitude AD, choose any point M on AD, and then draw BM and extend until it intersects AC in E, and draw CM and extend until it intersects AB in F. Prove that $\angle ADE = \angle ADF$.

701. In a planar Cartesian system of coordinates consider a fixed point $P(a, b)$ and a variable line through P. Let A be the intersection of the line with the x-axis. Connect A with the midpoint B of the segment OP (O being the origin), and through C, which is the point of intersection of this line with the y-axis, take the parallel to OP. This parallel intersects PA at M. Find the locus of M as the line varies.

702. Let $ABCD$ be a parallelogram with unequal sides. Let E be the foot of the perpendicular from B to AC. The perpendicular through E to BD intersects BC in F and AB in G. Show that $EF = EG$ if and only if $ABCD$ is a rectangle.

703. Let ABC be a triangle with incircle Γ, and let D, E, F be the tangency points of Γ with sides BC, CA, AB, respectively. Furthermore, let K be the orthocenter of triangle DEF. Prove that $KB^2 - KC^2 = BE^2 - CF^2$.

704. Find all pairs of real numbers (p, q) such that the inequality

$$|\sqrt{1 - x^2} - px - q| \le \frac{\sqrt{2} - 1}{2}$$

holds for every $x \in [0, 1]$.

705. On the hyperbola $xy = 1$ consider four points whose x-coordinates are x_1, x_2, x_3 and x_4. Show that if these points lie on a circle, then $x_1 x_2 x_3 x_4 = 1$.

706. Let ABC and DAB be right isosceles triangles such that $\angle A = \angle D = 90°$, $AB = 1$, and C and D are separated by the line AB. Let M be a point on the segment AC, N the intersection of DM with BC and P the intersection of BM with AN. Show that when M varies on the side AC then P describes a smooth curve, and find the length of this curve.

The points of the plane can be represented as complex numbers. There are two instances in which complex coordinates come in handy: in problems involving "nice" angles (such as $\frac{\pi}{4}, \frac{\pi}{3}, \frac{\pi}{2}$), and in problems about regular polygons.

In complex coordinates the line passing through the points z_1 and z_2 has the parametric equation $z = tz_1 + (1 - t)z_2, t \in \mathbb{R}$. Also, the angle between the line passing through z_1 and z_2 and the line passing through z_3 and z_4 is the argument of the complex number $\frac{z_1 - z_2}{z_3 - z_4}$. The length of the segment determined by the points z_1 and z_2 is $|z_1 - z_2|$. The vertices of a regular n-gon can be chosen, up to a scaling factor, as $1, \varepsilon, \varepsilon^2, \ldots, \varepsilon^{n-1}$, where $\varepsilon = e^{2\pi i/n} = \cos \frac{2\pi}{n} + i \sin \frac{2\pi}{n}$.

Example. Let ABC and BCD be two equilateral triangles sharing one side. A line passing through D intersects AC at M and AB at N. Prove that the angle between the lines BM and CN is $\frac{\pi}{3}$.

Solution. In the complex plane, let B have the coordinate 0, and C the coordinate 1. Then A and D have the coordinates $e^{i\pi/3}$ and $e^{-i\pi/3}$, respectively, and N has the coordinate $te^{i\pi/3}$ for some real number t.

The parametric equations of ND and AC are, respectively,

$$z = \alpha t e^{i\pi/3} + (1 - \alpha)e^{-i\pi/3} \quad \text{and} \quad z = \beta e^{i\pi/3} + (1 - \beta), \ \alpha, \beta \in \mathbb{R}.$$

To find their intersection we need to determine the real numbers α and β such that

$$\alpha t e^{i\pi/3} + (1-\alpha)e^{-i\pi/3} = \beta e^{i\pi/3} + (1-\beta).$$

Explicitly, this equation is

$$\alpha t \frac{1+i\sqrt{3}}{2} + (1-\alpha)\frac{1-i\sqrt{3}}{2} = \beta \frac{1+i\sqrt{3}}{2} + (1-\beta).$$

Setting the real and imaginary parts equal, we obtain the system

$$\alpha t + (1-\alpha) = \beta + 2(1-\beta),$$
$$\alpha t - (1-\alpha) = \beta.$$

By adding the two equations, we obtain $\alpha = \frac{1}{t}$. So the complex coordinate of M is $e^{i\pi/3} + \left(1-\frac{1}{t}\right)e^{-i\pi/3}$.

The angle between the lines BM and CN is the argument of the complex number

$$\frac{e^{i\pi/3} + \left(1-\frac{1}{t}\right)e^{-i\pi/3}}{te^{i\pi/3}-1} = \frac{\left(e^{i\pi/3}+e^{-i\pi/3}\right) - \frac{1}{t}e^{-i\pi/3}}{te^{i\pi/3}-1} = \frac{1-\frac{1}{t}e^{-i\pi/3}}{te^{i\pi/3}-1} = \frac{1}{t}e^{-i\pi/3}.$$

The angle is therefore $\frac{\pi}{3}$, as claimed.

During the Mathematical Olympiad Summer Program of 2006, J. Bland discovered the following simpler solution:

Place the figure in the complex plane so that the coordinates of A, B, C, D are, respectively, $i\sqrt{3}, -1, 1$, and $-i\sqrt{3}$. Let MC have length $2t$, where t is a real parameter (positive if C is between A and M and negative otherwise). The triangles MCD and NBD have parallel sides, so they are similar. It follows that $BN = \frac{2}{t}$ (positive if B is between A and N and negative otherwise). The coordinates of M and N are

$$m = -\left(1+\frac{1}{t}\right) - \frac{1}{t}i\sqrt{3} \quad \text{and} \quad n = (t+1) - ti\sqrt{3}.$$

We compute

$$\frac{c-n}{b-m} = t\frac{2t+1+i\sqrt{3}}{-t-2+ti\sqrt{3}} = -te^{i\frac{\pi}{3}}.$$

It follows that the two lines form an angle of $\frac{\pi}{3}$, as desired.   □

The second example comes from the 15th W.L. Putnam Mathematical Competition, 1955.

Example. Let $A_1A_2A_3\ldots A_n$ be a regular polygon inscribed in the circle of center O and radius r. On the half-line $|OA_1$ choose the point P such that A_1 is between O and P. Prove that

$$\prod_{i=1}^{n} PA_i = PO^n - r^n.$$

Solution. Place the vertices in the complex plane such that $A_i = r\varepsilon^i$, $1 \leq i \leq n$, where ε is a primitive nth root of unity. The coordinate of P is a real number rx, with $x > 1$. We have

$$\prod_{i=1}^{n} PA_i = \prod_{i=1}^{n} |rx - r\varepsilon^i| = r^n \prod_{i=1}^{n} |x - \varepsilon^i| = r^n \left| \prod_{i=1}^{n} (x - \varepsilon^i) \right|$$
$$= r^n(x^n - 1) = (rx)^n - r^n = PO^n - r^n.$$

The identity is proved. □

707. Let $ABCDEF$ be a hexagon inscribed in a circle of radius r. Show that if $AB = CD = EF = r$, then the midpoints of BC, DE, and FA are the vertices of an equilateral triangle.

708. Prove that in a triangle the orthocenter H, centroid G, and circumcenter O are collinear. Moreover, G lies between H and O, and $\frac{OG}{GH} = \frac{1}{2}$.

709. On the sides of a convex quadrilateral $ABCD$ one draws outside the equilateral triangles ABM and CDP and inside the equilateral triangles BCN and ADQ. Describe the shape of the quadrilateral $MNPQ$.

710. Let ABC be a triangle. The triangles PAB and QAC are constructed outside of the triangle ABC such that $AP = AB$, $AQ = AC$, and $\angle BAP = \angle CAQ = \alpha$. The segments BQ and CP meet at R. Let O be the circumcenter of the triangle BCR. Prove that AO and PQ are orthogonal.

711. Let $A_1 A_2 \ldots A_n$ be a regular polygon with circumradius equal to 1. Find the maximum value of $\prod_{k=1}^{n} PA_k$ as P ranges over the circumcircle.

712. Let $A_0, A_1, \ldots, A_n$ be the vertices of a regular n-gon inscribed in the unit circle. Prove that
$$A_0 A_1 \cdot A_0 A_2 \cdots A_0 A_{n-1} = n.$$

713. Show that a positive integer p is prime if and only if every equiangular p-gon with rational side-lengths is regular.

4.1.3 Quadratic and Cubic Curves in the Plane

In what follows we introduce the reader to curves of degree two (other than the circle) and three, with some incursions into algebraic geometry.

The general equation of a quadratic curve is

$$ax^2 + by^2 + cxy + dx + ey + f = 0.$$

Such a curve is called a conic because (except for the degenerate case of two parallel lines) it can be obtained by sectioning a circular cone by a plane.

The degenerate conics are pairs of (not necessarily distinct) lines, single points, the entire plane, and the empty set. We ignore them. There are three types of nondegenerate conics, which up to a change of Cartesian coordinates are described in Figure 27.

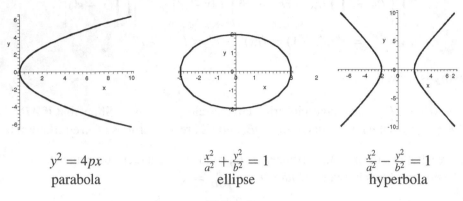

$$y^2 = 4px$$
parabola

$$\frac{x^2}{a^2} + \frac{y^2}{b^2} = 1$$
ellipse

$$\frac{x^2}{a^2} - \frac{y^2}{b^2} = 1$$
hyperbola

Figure 27

The parabola is the locus of the points at equal distance from the point $(p, 0)$ (focus) and the line $x = -p$ (directrix). The ellipse is the locus of the points with the sum of distances to the foci $(c, 0)$ and $(-c, 0)$ constant, where $c = \sqrt{|a^2 - b^2|}$. The hyperbola is the locus of the points with the difference of the distances to the foci $(c, 0)$ and $(-c, 0)$ constant, where $c = \sqrt{a^2 + b^2}$.

Up to an affine change of coordinates, the equations of the parabola, ellipse, and hyperbola are, respectively, $y^2 = x$, $x^2 + y^2 = 1$, and $x^2 - y^2 = 1$. Sometimes it is more convenient to bring the hyperbola into the form $xy = 1$ by choosing its asymptotes as the coordinate axes.

As conic sections, these curves are obtained by sectioning the circular cone $z^2 = x^2 + y^2$ by the planes $z - x = 1$ (parabola), $z = 1$ (ellipse), and $y = 1$ (hyperbola). The vertex of the cone can be thought of as the viewpoint of a person. The projections through this viewpoint of one plane to another are called projective transformations. Up to a projective transformation there is only one nondegenerate conic – the circle. Any projectively invariant property that can be proved for the circle is true for any conic (and by passing to the limit, even for degenerate conics). Such is the case with Pascal's theorem: The opposite sides of a hexagon inscribed in a conic meet at three collinear points. Note that when the conic degenerates into two parallel lines, this becomes Pappus' theorem.

To conclude our discussion, let us recall that the equation of the tangent line to a conic at a point (x_0, y_0) is obtained by replacing in the general equation of the conic x^2 and y^2 by xx_0, respectively yy_0, xy by $\frac{xy_0 + yx_0}{2}$, and x and y in the linear terms by $\frac{x + x_0}{2}$, respectively, $\frac{y + y_0}{2}$.

We now proceed with an example from A. Myller's *Analytical Geometry* (3rd ed., Editura Didactică şi Pedagogică, Bucharest, 1972).

Example. Find the locus of the centers of the equilateral triangles inscribed in the parabola $y^2 = 4px$.

Solution. Let us determine first some algebraic conditions that the coordinates (x_i, y_i), $i = 1, 2, 3$, of the vertices of a triangle should satisfy in order for the triangle to be equilateral.

The equation of the median from (x_3, y_3) is

$$\frac{x - y_3}{x - x_3} = \frac{y_1 + y_2 - 2y_3}{x_1 + x_2 - 2x_3}.$$

Requiring the median to be orthogonal to the side yields

$$\frac{y_1 + y_2 - 2y_3}{x_1 + x_2 - 2x_3} \cdot \frac{y_2 - y_1}{x_2 - x_1} = -1,$$

or

$$(x_1 - x_2)(x_1 + x_2 - 2x_3) + (y_1 - y_2)(y_1 + y_2 - 2y_3) = 0.$$

So this relation along with the two obtained by circular permutations of the indices are necessary and sufficient conditions for the triangle to be equilateral. Of course, the third condition is redundant. In the case of three points on the parabola, namely $\left(\frac{y_i^2}{4p}, y_i\right)$, $i = 1, 2, 3$, after dividing by $y_1 - y_2$, respectively, by $y_2 - y_3$ (which are both nonzero), we obtain

$$(y_1 + y_2)(y_1^2 + y_2^2 - 2y_3^2) + 16p^2(y_1 + y_2 - 2y_3) = 0,$$

$$(y_2 + y_3)(y_2^2 + y_3^2 - 2y_1^2) + 16p^2(y_2 + y_3 - 2y_1) = 0.$$

Subtracting the two gives

$$y_1^3 - y_3^3 + (y_1 - y_3)(y_2^2 - 2y_1y_3) + 48p^2(y_1 - y_3) = 0.$$

Divide this by $y_1 - y_3 \neq 0$ to transform it into

$$y_1^2 + y_2^2 + y_3^2 + 3(y_1y_2 + y_2y_3 + y_3y_1) + 48p^2 = 0.$$

This is the condition satisfied by the y-coordinates of the vertices of the triangle. Keeping in mind that the coordinates of the center of the triangle are

$$x = \frac{y_1^2 + y_2^2 + y_3^2}{12p}, \quad y = \frac{y_1 + y_2 + y_3}{3},$$

we rewrite the relation as

$$-\frac{1}{2}(y_1^2 + y_2^2 + y_3^2) + \frac{3}{2}(y_1 + y_2 + y_3)^2 + 48p^2 = 0,$$

then substitute $12px = y_1^2 + y_2^2 + y_3^2$ and $3y = y_1 + y_2 + y_3$ to obtain the equation of the locus

$$-6px + \frac{27}{2}y^2 + 48p^2 = 0,$$

or

$$y^2 = \frac{4p}{9}(x - 8p).$$

This is a parabola with vertex at $(8p, 0)$ and focus at $\left(\left(\frac{1}{9} + 8\right), p, 0\right)$.   □

The second problem was given at the 1977 Soviet Union University Student Mathematical Olympiad.

Example. Let P be a point on the hyperbola $xy = 4$, and Q a point on the ellipse $x^2 + 4y^2 = 4$. Prove that the distance from P to Q is greater than 1.

Solution. We will separate the conics by two parallel lines at a distance greater than 1. For symmetry reasons, it is natural to try the tangent to the hyperbola at the point $(2, 2)$. This line has the equation $y = 4 - x$.

Let us determine the point in the first quadrant where the tangent to the ellipse has slope -1. If (x_0, y_0) is a point on the ellipse, then the equation of the tangent at x is $xx_0 + 4yy_0 = 4$. Its slope is $-x_0/4y_0$. Setting $-x_0/4y_0 = -1$ and $x_0^2 + 4y_0^2 = 4$, we obtain $x_0 = 4/\sqrt{5}$ and $y_0 = 1/\sqrt{5}$. Consequently, the tangent to the ellipse is $y = \sqrt{5} - x$.

The distance between the lines $y = 4 - x$ and $y = \sqrt{5} - x$ is equal to $(4 - \sqrt{5})/\sqrt{2}$, which is greater than 1. Hence the distance between the arbitrary points P and Q is also greater than 1, and we are done. $\square$

714. Consider a circle of diameter AB and center O, and the tangent t at B. A variable tangent to the circle with contact point M intersects t at P. Find the locus of the point Q where the line OM intersects the parallel through P to the line AB.

715. On the axis of a parabola consider two fixed points at equal distance from the focus. Prove that the difference of the squares of the distances from these points to an arbitrary tangent to the parabola is constant.

716. With the chord PQ of a hyperbola as diagonal, construct a parallelogram whose sides are parallel to the asymptotes. Prove that the other diagonal of the parallelogram passes through the center of the hyperbola.

717. A straight line cuts the asymptotes of a hyperbola in points A and B and the hyperbola itself in P and Q. Prove that $AP = BQ$.

718. Consider the parabola $y^2 = 4px$. Find the locus of the points such that the tangents to the parabola from those points make a constant angle ϕ.

719. Let T_1, T_2, T_3 be points on a parabola, and t_1, t_2, t_3 the tangents to the parabola at these points. Compute the ratio of the area of triangle $T_1 T_2 T_3$ to the area of the triangle determined by the tangents.

720. Three points A, B, C are considered on a parabola. The tangents to the parabola at these points form a triangle MNP (NP being tangent at A, PM at B, and MN at C). The parallel through B to the symmetry axis of the parabola intersects AC at L.

   (a) Show that $LMNP$ is a parallelogram.
   (b) Show that the circumcircle of triangle MNP passes through the focus F of the parabola.
   (c) Assuming that L is also on this circle, prove that N is on the directrix of the parabola.

(d) Find the locus of the points L if AC varies in such a way that it passes through F and is perpendicular to BF.

721. Find all regular polygons that can be inscribed in an ellipse with unequal semiaxes.

722. We are given the parabola $y^2 = 2px$ with focus F. For an integer $n \geq 3$ consider a regular polygon $A_1 A_2 \ldots A_n$ whose center is F and such that none of its vertices is on the x-axis. The half-lines $|FA_1, |FA_2, \ldots, |FA_n$ intersect the parabola at $B_1, B_2, \ldots, B_n$. Prove that

$$FB_1 + FB_2 + \cdots + FB_n \geq np.$$

723. A *cevian* of a triangle is a line segment that joins a vertex to the line containing the opposite side. An *equicevian point* of a triangle ABC is a point P (not necessarily inside the triangle) such that the cevians on the lines AP, BP, and CP have equal lengths. Let SBC be an equilateral triangle, and let A be chosen in the interior of SBC, on the altitude dropped from S.

(a) Show that ABC has two equicevian points.
(b) Show that the common length of the cevians through either of the equicevian points is constant, independent of the choice of A.
(c) Show that the equicevian points divide the cevian through A in a constant ratio, which is independent of the choice of A.
(d) Find the locus of the equicevian points as A varies.
(e) Let S' be the reflection of S in the line BC. Show that (a), (b), and (c) hold if A varies on any ellipse with S and S' as its foci. Find the locus of the equicevian points as A varies on the ellipse.

A planar curve is called rational if it can be parametrized as $(x(t), y(t))$ with $x(t)$ and $y(t)$ rational functions of the real variable t. Here we have to pass to the closed real line, so t is allowed to be infinite, while the plane is understood as the projective plane, zero denominators giving rise to points on the line at infinity.

Theorem. *All conics are rational curves.*

Proof. The case of degenerate conics (i.e., pairs of lines) is trivial. The parabola $y^2 = 4px$ is parametrized by $\left(\dfrac{t^2}{4p}, t \right)$, the ellipse $\dfrac{x^2}{a^2} + \dfrac{y^2}{b^2} = 1$ by $\left(a \dfrac{1 - t^2}{1 + t^2}, b \dfrac{2t}{1 + t^2} \right)$, and the hyperbola $\dfrac{x^2}{a^2} - \dfrac{y^2}{b^2} = 1$ by $\left(a \dfrac{t + t^{-1}}{2}, b \dfrac{t - t^{-1}}{2} \right)$. The general case follows from the fact that coordinate changes are rational (in fact, linear) transformations. $\square$

Compare the standard parametrization of the circle $(\cos x, \sin x)$ to the rational parametrization $ds \left(\dfrac{1 - t^2}{1 + t^2}, \dfrac{2t}{1 + t^2} \right)$. This gives rise to the trigonometric substitution $\tan \dfrac{x}{2} = t$ and explains why integrals of the form

$$\int R(\cos x, \sin x) dx,$$

with R a two-variable rational function, can be reduced to integrals of rational functions.

Let us change slightly our point of view and take a look at the conic

$$y^2 = ax^2 + bx + c.$$

If we fix a point (x_0, y_0) on this conic, the line $y - y_0 = t(x - x_0)$ intersects the conic in exactly one more point (x, y). Writing the conditions that this point is both on the line and on the conic and eliminating y, we obtain the equation

$$[y_0 + t(x - x_0)]^2 = ax^2 + bx + c.$$

A few algebraic computations yield

$$2y_0 t + t^2(x - x_0) = a(x + x_0) + b.$$

This shows that x is a rational function of the slope t. The same is true for y. As t varies, (x, y) describes the whole conic. This is a rational parametrization of the conic, giving rise to Euler's substitutions. In their most general form, Euler's substitutions are

$$\sqrt{ax^2 + bx + c} - y_0 = t(x - x_0).$$

They are used for rationalizing integrals of the form

$$\int R(x, \sqrt{ax^2 + bx + c})dx,$$

where R is a two-variable rational function.

724. Compute the integral

$$\int \frac{dx}{a + b\cos x + c\sin x},$$

where a, b, c are real numbers, not all equal to zero.

725. Consider the system

$$x + y = z + u,$$

$$2xy = zu.$$

Find the greatest value of the real constant m such that $m \le \frac{x}{y}$ for any positive integer solution (x, y, z, u) of the system, with $x \ge y$.

We conclude this unit with problems about cubic curves, some of which, surprisingly, made the object of high school mathematical Olympiads despite their far reaching scope.

The first example is a problem from the 2014 USA Mathematical Olympiad, being proposed by S. Vandervelde.

Example. Prove that there is an infinite number of points

$$...P_{-3}, P_{-2}, P_{-1}, P_0, P_1, P_2, P_3, ...$$

in the plane with the following property: for every three distinct integer numbers a, b, c, the points P_a, P_b, P_c are collinear if and only if

$$a + b + c = 2014.$$

Solution. Translating the indices as

$$a \mapsto a - \frac{2014}{3}, \quad b \mapsto b - \frac{2014}{3}, \quad c \mapsto c - \frac{2014}{3},$$

the condition for the collinearity of the points P_a, P_b, P_c transforms into

$$a + b + c = 0.$$

This condition can be related to a fundamental property of one family of cubic curves, the *elliptic curves*.

An elliptic curve is defined by an equation of the form

$$f(x, y) = 0$$

where $f(x, y)$ is a polynomial of degree 3 in the variables x, y with the property that for no point (x_0, y_0) on the curve one has

$$\frac{\partial f}{\partial x}(x_0, y_0) = \frac{\partial f}{\partial y}(x_0, y_0) = 0.$$

Such a curve is nonsingular in the sense that its graph has no cusps and no intersections. By a change of coordinates, the curve can be changed into

$$y^2 = x^3 + ax + b, \quad a, b \in \mathbb{R}.$$

It is elliptic precisely when its discriminant $\Delta = -16(4a^2 + 27b^2)$ is nonzero. Depending on whether the equation $x^3 + ax + b = 0$ has one or three real roots, the elliptic curve has one or two components. The two cases are described in Figure 28. This curve admits the structure of an Abelian group, as we will now explain.

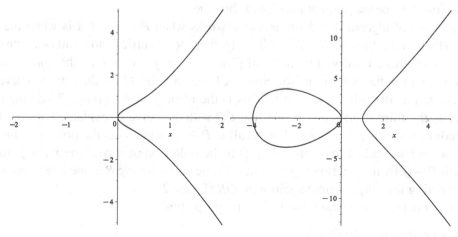

Figure 28

To define the sum of the points P and Q on this curve, consider the line that passes through P and Q, and let $\alpha x + \beta y + \gamma = 0$ be its equation. The intersections of this line with the curve can be obtained by solving the system

$$y^3 = x^3 + ax + b$$
$$\alpha x + \beta y + \gamma = 0.$$

Substituting $y = -(\alpha x + \gamma)/\beta$ in the first equation we obtain a cubic equation in x. This equation has two real solutions (the coordinates of P and Q), and hence has a third real solution, which gives us the third intersection point of the line with the curve. Call this point R.

Note that this construction works well for $\beta \neq 0$, but if $\beta = 0$ we obtain the equation of the vertical line $x = -\gamma/\alpha$. This only crosses the elliptic curve in two points, with coordinates

$$\left(-\frac{\gamma}{\alpha}, \pm\sqrt{\left(-\frac{\gamma}{\alpha}\right)^3 + a\left(-\frac{\gamma}{\alpha}\right) + b} \right),$$

and these are points P and Q.

This situation can be resolved by adding to the elliptic curve the point at infinity. We do this by passing to the projective plane, which is the extension (correctly called compactification) of the Euclidean plane in which any two lines intersect at one point. In the projective plane, every line has one point at infinity, and all the points at infinity are on the line at infinity. Two lines intersect at finite points when they are not parallel, when they are parallel they intersect at a point at infinity. The point at infinity of the elliptic curve (in the standard coordinates described above) is specified by the vertical direction. So in the case where PQ is vertical, we can define R to be the point at infinity of this line.

For reasons of algebra this definition also works when $P = Q$, that is when the line is tangent to the curve. Also if P or Q is at infinity, then PQ is vertical, so it intersect the elliptic curve one more time. Finally, if both P and Q are at infinity, we let R be the point at infinity.

Now we can define the group structure. Choose a point O on the elliptic curve (O is usually chosen as the point at infinity). This is the identity of the group. To define $P + Q$, choose R as the third intersection of the line PQ with the curve as explained above. Repeat the procedure with the pair (O, R). The result is $P + Q$ (when O is the point at infinity and the curve is in its standard form, then $P + Q$ is the reflection of R over the x-axis). In short, you obtain $P + Q$ by intersection PQ with the elliptic curve, taking R as the intersection point, and intersecting the elliptic curve again with OR (Figure 29).

One can check geometrically the following properties:

- $(P + Q) + S = P + (Q + S)$,

- $P + O = O + P = P$,

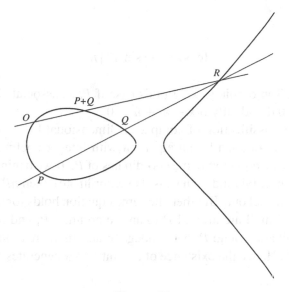

Figure 29

- $P + (-P) = O$, where $-P$ is the intersection of OP with the elliptic curve,
- $P + Q = Q + P$.

Note that three points P, Q, R are collinear if and only if

$$P + Q + R = 0.$$

To solve the Olympiad problem we have to show that the elliptic curve has a subgroup that is isomorphic to $\mathbb{Z}$. Here is one possible argument. After adding the point at infinity, the curve has either one or two closed components, which modulo a deformation are circles. Actually the point at infinity closes the unbounded component: one arrives at this point by following either the lower branch or the upper branch.

The operation of addition is a continuous two variable map with values in the curve, as it is not hard to verify geometrically that if $P' \to P$ and $Q' \to Q$, then $P' + Q' \to P + Q$. Also, the function that associates to a point its inverse, $P \to -P$ is continuous. We are in the presence of an Abelian *Lie group* of dimension 1, that is to say a curve with an Abelian group structure in which addition and taking the negative are continuous. Such groups are classified. If the group has one component, then it is the group of complex numbers of absolute value 1:

$$U(1) = \{z \in \mathbb{C} \mid |z| = 1\}$$

and if it has two components, then it is

$$U(1) \times \mathbb{Z}_2,$$

(where $\mathbb{Z}_2$ is the group whose elements are the two residue classes modulo 2). Here the addition is defined separately in each coordinate. In both cases the group contains a copy of $U(1)$, whose subgroup

$$G_\theta = \{\cos n\theta + i \sin n\theta \mid n \in \mathbb{Z}\}$$

with θ/π irrational is isomorphic to $\mathbb{Z}$. In this case if P_n is associated to $\cos n\theta + i \sin n\theta$, then $P_k + P_m + P_n = 0$ if and only if $k + m + n = 0$.

To avoid using the classification of compact 1-dimensional Lie groups, we can argue as follows. Choose a curve defined by an equation with integer coefficients. Then $nP = 0$ translates into an algebraic equation in the coordinates of P. If we begin with a point P whose x-coordinate is transcendental (and then so is its y-coordinate), then $nP = 0$ can only happen if the coordinates of P cancel out. But then the same equation holds for all points. Now notice that $P' \mapsto 2P'$ maps a small arc around P to an arc around $2P$, and repeating, we see that $P' \mapsto nP'$ maps a small arc around P to a nondegenerate small arc around nP. So we cannot have $nP' = 0$ for all P'. Hence the existence of a point P that generates an infinite group. The problem is solved. $\square$

We leave the following problems about cubic curves to the reader.

726. A cubic sequence is a sequence of integers given by $a_n = n^3 + bn^2 + cn + d$, where b, c, d are integer constants and n ranges over all integers, including negative integers.
(a) Show that there exists a cubic sequence such that the only terms of the sequence which are squares of integers are a_{2015} and a_{2016}.
(b) Determine the possible values of $a_{2015} \cdot a_{2016}$ for a cubic sequence satisfying the condition in part (a).

727. Solve in integers the equation

$$x^2 + xy + y^2 = \left(\frac{x+y}{3} + 1\right)^3.$$

728. Prove that the locus described by the equation $x^3 + 3xy + y^3 = 1$ contains precisely three noncollinear points A, B, C, equidistant to one another, and find the area of triangle ABC.

729. Prove that, for any integers a, b, c, there exists a positive integer n such that $\sqrt{n^3 + an^2 + bn + c}$ is not an integer.

4.1.4 Some Famous Curves in the Plane

We conclude our incursion into two-dimensional geometry with an overview of various planar curves that captured the imagination of mathematicians. The first answers a question of G.W. Leibniz.

Example. What is the path of an object dragged by a string of constant length when the end of the string not joined to the object moves along a straight line?

Solution. Assume that the object is dragged by a string of length 1, that its initial coordinates are $(0, 1)$, and that it is dragged by a vehicle moving along the x-axis in the positive direction. Observe that the slope of the tangent to the curve at a point (x, y) points toward the vehicle, while the distance to the vehicle is always equal to 1. These two facts can be combined in the differential equation

$$\frac{dy}{dx} = -\frac{y}{\sqrt{1 - y^2}}.$$

Separate the variables

$$dx = -\frac{\sqrt{1 - y^2}}{y} dy,$$

and then integrate to obtain

$$x = -\sqrt{1 - y^2} - \ln y - \ln(1 + \sqrt{1 - y^2}) + C.$$

The initial condition gives $C = 0$. The answer to the problem is therefore the curve

$$x = -\sqrt{1 - y^2} - \ln y - \ln(1 + \sqrt{1 - y^2}),$$

depicted in Figure 30. □

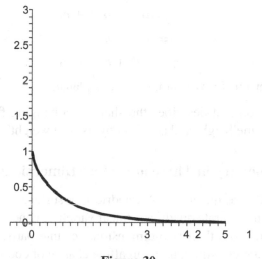

Figure 30

This curve is called a *tractrix*, a name given by Ch. Huygens. Clearly, it has the x-axis as an asymptote. E. Beltrami has shown that the surface of revolution of the tractrix around its asymptote provides a partial model for the hyperbolic plane of Lobachevskian geometry. This surface has been used in recent years for the shape of loudspeakers.

A variety of other curves show up in the problems below. In some of the solutions, polar coordinates might be useful. Recall the formulas for changing between Cartesian and polar coordinates: $x = r \cos \theta$, $y = r \sin \theta$.

730. Find the points where the tangent to the cardioid $r = 1 + \cos \theta$ is vertical.

731. Given a circle of diameter AB, a variable secant through A intersects the circle at C and the tangent through B at D. On the half-line AC a point M is chosen such that $AM = CD$. Find the locus of M.

732. Find the locus of the projection of a fixed point on a circle onto the tangents to the circle.

733. On a circle of center O consider a fixed point A and a variable point M. The circle of center A and radius AM intersects the line OM at L. Find the locus of L as M varies on the circle.

734. The endpoints of a variable segment AB lie on two perpendicular lines that intersect at O. Find the locus of the projection of O onto AB, provided that the segment AB maintains a constant length.

735. From the center of a rectangular hyperbola a perpendicular is dropped to a variable tangent. Find the locus in polar coordinates of the foot of the perpendicular. (A hyperbola is called rectangular if its asymptotes are perpendicular.)

736. Find a transformation of the plane that maps the unit circle $x^2 + y^2 = 1$ into a cardioid. (Recall that the general equation of a cardioid is $r = 2a(1 + \cos\theta)$.)

737. For n and p two positive integers consider the curve described by the parametric equations

$$x = a_1 t^n + b_1 t^p + c_1,$$
$$y = a_2 t^n + b_2 t^p + c_2,$$
$$z = a_3 t^n + b_3 t^p + c_3,$$

where t is a parameter. Prove that the curve is planar.

738. What is the equation that describes the shape of a hanging flexible chain with ends supported at the same height and acted on by its own weight?

4.1.5 Coordinate Geometry in Three and More Dimensions

In this section we emphasize quadrics. A quadric is a surface in space determined by a quadratic equation. The degenerate quadrics – linear varieties, cones, or cylinders over conics – add little to the picture from their two-dimensional counterparts, so we skip them. The nondegenerate quadrics are classified, up to an affine change of coordinates, as

- $x^2 + y^2 + z^2 = 1$, ellipsoid;
- $x^2 + y^2 - z^2 = 1$, hyperboloid of one sheet;
- $x^2 - y^2 - z^2 = 1$, hyperboloid of two sheets;
- $x^2 + y^2 = z$, elliptic paraboloid;
- $x^2 - y^2 = z$, hyperbolic paraboloid.

In Cartesian coordinates, in these formulas there is a scaling factor in front of each term. For example, the standard form of an ellipsoid in Cartesian coordinates is

$$\frac{x^2}{a^2} + \frac{y^2}{b^2} + \frac{z^2}{c^2} = 1.$$

As in the case of conics, the equation of the tangent plane to a quadric at a point (x_0, y_0, z_0) is obtained by replacing in the equation of the quadric x^2, y^2, and z^2, respectively, by xx_0, yy_0, and zz_0; xy, xz, and yz, respectively, by $\frac{xy_0 + yx_0}{2}$, $\frac{xz_0 + zx_0}{2}$, and $\frac{yz_0 + zy_0}{2}$; and x, y, and z in the linear terms, respectively, by $\frac{x + x_0}{2}$, $\frac{y + y_0}{2}$, and $\frac{z + z_0}{2}$.

Our first example comes from the 6th W.L. Putnam Mathematical Competition.

Example. Find the smallest volume bounded by the coordinate planes and by a tangent plane to the ellipsoid

$$\frac{x^2}{a^2} + \frac{y^2}{b^2} + \frac{z^2}{c^2} = 1.$$

Solution. The tangent plane to the ellipsoid at (x_0, y_0, z_0) has the equation

$$\frac{xx_0}{a^2} + \frac{yy_0}{b^2} + \frac{zz_0}{c^2} = 1.$$

Its x, y, and z intercepts are, respectively, $\frac{a^2}{x_0}$, $\frac{b^2}{y_0}$, and $\frac{c^2}{z_0}$. The volume of the solid cut off by the tangent plane and the coordinate planes is therefore

$$V = \frac{1}{6} \left| \frac{a^2 b^2 c^2}{x_0 y_0 z_0} \right|.$$

We want to minimize this with the constraint that (x_0, y_0, z_0) lie on the ellipsoid. This amounts to maximizing the function $f(x, y, z) = xyz$ with the constraint

$$g(x, y, z) = \frac{x^2}{a^2} + \frac{y^2}{b^2} + \frac{z^2}{c^2} = 1.$$

Because the ellipsoid is a closed bounded set, f has a maximum and a minimum on it. The maximum is positive, and the minimum is negative. The method of Lagrange multipliers yields the following system of equations in the unknowns x, y, z, and λ:

$$yz = 2\lambda \frac{x}{a^2},$$
$$xz = 2\lambda \frac{y}{b^2},$$
$$yz = 2\lambda \frac{z}{c^2},$$
$$\frac{x^2}{a^2} + \frac{y^2}{b^2} + \frac{z^2}{c^2} = 1.$$

Multiplying the first equation by x, the second by y, and the third by z, then summing up the three equations gives

$$3xyz = 2\lambda \left(\frac{x^2}{a^2} + \frac{y^2}{b^2} + \frac{z^2}{c^2} \right) = 2\lambda.$$

Hence $\lambda = \frac{3}{2} xyz$. Then multiplying the first three equations of the system together, we obtain

$$(xyz)^2 = 8\lambda^3 \frac{xyz}{a^2 b^2 c^2} = \frac{27(xyz)^4}{a^2 b^2 c^2}.$$

The solution $xyz = 0$ we exclude, since it does not yield a maximum or a minimum. Otherwise, $xyz = \pm\frac{abc}{\sqrt{27}}$. The equality with the plus sign is the maximum of f; the other is the minimum. Substituting in the formula for the volume, we find that the smallest volume is $\frac{\sqrt{3}}{2}abc$. □

Example. Find the nature of the surface defined as the locus of the lines parallel to a given plane and intersecting two given skew lines, neither of which is parallel to the plane.

Solution. We will work in affine coordinates. Call the plane π and the two skew lines l_1 and l_2. The x- and y-axes lie in π and the z-axis is l_1. The x-axis passes through $l_2 \cap \pi$. The y-axis is chosen to make l_2 parallel to the yz-plane. Finally, the orientation and the units are such that l_2 is given by $x = 1$, $y = z$ (see Figure 31).

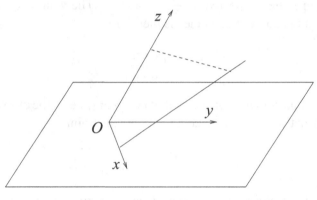

Figure 31

A line parallel to π and intersecting l_1 and l_2 passes through $(1, s, s)$ and $(0, 0, s)$, where s is some real parameter playing the role of the "height". Thus the locus consists of all points of the form $t(1, s, s) + (1 - t)(0, 0, s)$, where s and t are real parameters. The coordinates (X, Y, Z) of such a point satisfy $X = t$, $Y = ts$, $Z = s$. By elimination we obtain the equation $XZ = Y$, which is a hyperbolic paraboloid like the one from Figure 32. We stress once more that the type of a quadric is invariant under affine transformations. □

A surface generated by a moving line is called a ruled surface. Ruled surfaces are easy to build in real life. This together with its structural resistance makes the hyperbolic paraboloid popular as a roof in modern architecture (see for example Felix Candela's roof of the 1968 Olympic stadium in Mexico City). There is one more nondegenerate ruled quadric, which makes the object of one of the problems below. And if you find some of the problems below too difficult, remember Winston Churchill's words: "Success consists of going from failure to failure without loss of enthusiasm".

739. A cube is rotated about the main diagonal. What kind of surfaces do the edges describe?

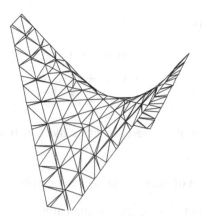

Figure 32

740. Prove that the plane

$$\frac{x}{a} + \frac{y}{b} - \frac{z}{c} = 1$$

is tangent to the hyperboloid of one sheet

$$\frac{x^2}{a^2} + \frac{y^2}{b^2} - \frac{z^2}{c^2} = 1.$$

741. Through a point M on the ellipsoid

$$\frac{x^2}{a^2} + \frac{y^2}{b^2} + \frac{z^2}{c^2} = 1$$

take planes perpendicular to the axes Ox, Oy, Oz. Let the areas of the planar sections thus obtained be S_x, S_y, respectively, S_z. Prove that the sum

$$aS_x + bS_y + cS_z$$

is independent of M.

742. Determine the radius of the largest circle that can lie on the ellipsoid

$$\frac{x^2}{a^2} + \frac{y^2}{b^2} + \frac{z^2}{c^2} = 1 \ (a > b > c).$$

743. Let a, b, c be distinct positive numbers. Prove that through each point of the three-dimensional space pass three surfaces described by equations of the form

$$\frac{x^2}{a^2 - \lambda} + \frac{y^2}{b^2 - \lambda} + \frac{z^2}{c^2 - \lambda} = 1.$$

Determine the nature of these surfaces and prove that they are pairwise orthogonal along their curves of intersection.

744. Show that the equations

$$x = u + v + w,$$
$$y = u^2 + v^2 + w^2,$$
$$z = u^3 + v^3 + w^3,$$

where the parameters u, v, w are subject to the constraint $uvw = 1$, define a cubic surface.

We conclude our discussion of coordinate geometry with some problems in n dimensions.

Example. Through a fixed point inside an n-dimensional sphere, n mutually perpendicular chords are drawn. Prove that the sum of the squares of the lengths of the chords does not depend on their directions.

Solution. We want to prove that the sum in question depends only on the radius of the sphere and the distance from the fixed point to the center of the sphere. Choose a coordinate system in which the chords are the n orthogonal axes and the radius of the sphere is $R > 0$. The fixed point, which we call P, becomes the origin. The endpoints of each chord have only one nonzero coordinate, and in the appropriate ordering, the kth coordinates of the endpoints X_k and Y_k of the kth chord are the nonzero numbers x_k and y_k, $k = 1, 2, \ldots, n$. The center of the sphere is then

$$O = \left(\frac{x_1 + y_1}{2}, \frac{x_2 + y_2}{2}, \ldots, \frac{x_n + y_n}{2} \right).$$

The conditions that the points X_k and Y_k lie on the sphere can be written as

$$\left(x_k - \frac{x_k + y_k}{2} \right)^2 + \sum_{j \neq k} \left(\frac{x_j + y_j}{2} \right)^2 = R^2,$$

$$\left(y_k - \frac{x_k + y_k}{2} \right)^2 + \sum_{j \neq k} \left(\frac{x_j + y_j}{2} \right)^2 = R^2,$$

with $k = 1, 2, \ldots, n$. This implies

$$\left(\frac{x_k - y_k}{2} \right)^2 = R^2 - \sum_{j \neq k} \left(\frac{x_j + y_j}{2} \right)^2, \quad k = 1, 2, \ldots, n.$$

The term on the left is one-fourth of the square of the length of $X_k Y_k$. Multiplying by 4 and summing up all these relations, we obtain

$$\sum_{k=1}^{n} \|X_k Y_k\|^2 = 4nR^2 - 4 \sum_{k=1}^{n} \sum_{j \neq k} \left(\frac{x_j + y_j}{2} \right)^2 = 4nR^2 - 4(n-1) \sum_{k=1}^{n} \left(\frac{x_k + y_k}{2} \right)^2$$

$$= 4nR^2 - 4(n-1)\|PO\|^2.$$

Hence the conclusion. $\square$

745. Let n be a positive integer. Prove that if the vertices of a $(2n + 1)$-dimensional cube have integer coordinates, then the length of the edge of the cube is an integer.

746. For a positive integer n denote by τ the permutation cycle $(n, \ldots, 2, 1)$. Consider the locus of points in $\mathbb{R}^n$ defined by the equation

$$\sum_{\sigma} \text{sign}(\sigma) x_{\sigma(1)} x_{\tau(\sigma(2))} \cdots x_{\tau^{n-1}(\sigma(n))} = 0,$$

where the sum is over all possible permutations of $\{1, 2, \ldots, n\}$. Prove that this locus contains a hperplane.

747. Prove that the intersection of an n-dimensional cube centered at the origin and with edges parallel to the coordinate axes with the plane determined by the vectors

$$\overrightarrow{a} = \left(\cos \frac{2\pi}{n}, \cos \frac{4\pi}{n}, \ldots, \cos \frac{2n\pi}{n}\right) \quad \text{and} \quad \overrightarrow{b} = \left(\sin \frac{2\pi}{n}, \sin \frac{4\pi}{n}, \ldots, \sin \frac{2n\pi}{n}\right)$$

is a regular $2n$-gon.

748. Find the maximal number of edges of an n-dimensional cube that are cut by a hyperplane. (By cut we mean intersected in exactly one point).

749. Find the maximum number of points on a sphere of radius 1 in $\mathbb{R}^n$ such that the distance between any two is strictly greater than $\sqrt{2}$.

4.1.6 Integrals in Geometry

We now present various applications of integral calculus to geometry problems. Here is a classic.

Example. A disk of radius R is covered by m rectangular strips of width 2. Prove that $m \geq R$.

Solution. Since the strips have different areas, depending on the distance to the center of the disk, a proof using areas will not work. However, if we move to three dimensions the problem becomes easy. The argument is based on the following property of the sphere.

Lemma. *The area of the surface cut from a sphere of radius R by two parallel planes at distance d from each other is equal to $2\pi Rd$.*

Proof. To prove this result, let us assume that the sphere is centered at the origin and the planes are perpendicular to the x-axis. The surface is obtained by rotating the graph of the function $f : [a, b] \to \mathbb{R}$, $f(x) = \sqrt{R^2 - x^2}$ about the x-axis, where $[a, b]$ is an interval of length d. The area of the surface is given by

$$2\pi \int_a^b f(x)\sqrt{(f'(x))^2 + 1}\, dx = 2\pi \int_a^b \sqrt{R^2 - x^2} \frac{R}{\sqrt{R^2 - x^2}}\, dx$$

$$= 2\pi \int_a^b R\, dx = 2\pi Rd,$$

and the lemma is proved. $\square$

Returning to the problem, the sphere has area $4\pi R^2$ and is covered by m surfaces, each having area $4\pi R$. The inequality $4\pi m R \geq 4\pi R^2$ implies that $m \geq R$, as desired.   □

The second example, suggested to us by Zh. Wang, is even more famous. We present the proof from H. Solomon, *Geometric Probability* (SIAM 1978).

Crofton's theorem. *Let D be a bounded convex domain in the plane. Through each point $P(x, y)$ outside D there pass two tangents to D. Let t_1 and t_2 be the lengths of the segments determined by P and the tangency points, and let α be the angle between the tangents, all viewed as functions of (x, y).*[1] *Then*

$$\iint_{P \notin D} \frac{\sin \alpha}{t_1 t_2} \, dx \, dy = 2\pi^2.$$

Proof. The proof becomes transparent once we examine the particular case in which D is the unit disk $x^2 + y^2 < 1$. Each point outside the unit disk can be parametrized by the pair of angles (ϕ_1, ϕ_2) where the tangents meet the unit circle S^1. Since there is an ambiguity in which tangent is considered first, the outside of the disk is in 1-to-2 correspondence with the set $S^1 \times S^1$. It so happens, and we will prove it in general, that on changing coordinates from (x, y) to (ϕ_1, ϕ_2) the integral from the statement becomes $\iint d\phi_1 d\phi_2$ (divided by 2 to take the ambiguity into account). The result follows.

In the general case we mimic the same argument, boosting your intuition with Figure 33. Fix a Cartesian coordinate system with the origin O inside D. For a point (x, y) denote by (ϕ_1, ϕ_2) the angles formed by the perpendiculars from O onto the tangents with the positive semiaxis. This is another parametrization of the exterior of D, again with the ambiguity of which tangent is considered first. Let $A_i(\varepsilon_i, \eta_i)$, $i = 1, 2$, tangency points.

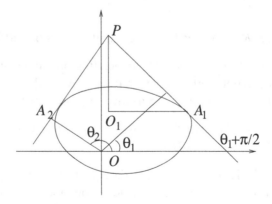

Figure 33

The main goal is to understand the change of coordinates $(x, y) \to (\phi_1, \phi_2)$ and in particular to write the Jacobian of this transformation. Writing the condition that the slope of the line A_1P is $\tan\left(\phi_1 + \frac{\pi}{2}\right)$, we obtain

$$(x - \varepsilon_1) \cos \phi_1 + (y - \eta_1) \sin \phi_1 = 0.$$

Taking the differential yields

$$\cos \phi_1 dx - \cos \phi_1 d\varepsilon_1 - (x - \varepsilon_1) \sin \phi_1 d\phi_1 + \sin \phi_1 dy - \sin \phi_1 d\eta_1$$
$$+ (y - \eta_1) \cos \phi_1 d\phi_1 = 0.$$

This expression can be simplified if we note that $\frac{d\eta_1}{d\varepsilon_1}$ is the slope of the tangent, namely $\tan\left(\phi_1 + \frac{\pi}{2}\right)$. Then $\cos \phi_1 d\varepsilon_1 + \sin \phi_1 d\eta_1 = 0$, so

$$\cos \phi_1 dx + \sin \phi_1 dy = [(x - \varepsilon_1) \sin \phi_1 - (y - \eta_1) \cos \phi_1] d\phi_1.$$

And now a little Euclidean geometry. Consider the right triangle O_1A_1P with legs parallel to the axes. The altitude from O_1 determines on A_1P two segments of lengths $(x - \varepsilon_1) \sin \phi_1$ and $-(y - \eta_1) \cos \phi_1$ (you can see by examining the picture that the signs are right). This allows us to further transform the identity obtained above into

$$\cos \phi_1 dx + \sin \phi_1 dy = t_1 d\phi_1.$$

The same argument shows that

$$\cos \phi_2 dx + \sin \phi_2 dy = t_2 d\phi_2.$$

The Jacobian of the transformation is therefore the absolute value of

$$\frac{1}{t_1 t_2}(\cos \phi_1 \sin \phi_2 - \sin \phi_1 \cos \phi_2) = \frac{1}{t_1 t_2} \sin(\phi_1 - \phi_2).$$

And $\phi_1 - \phi_2$ is, up to a sign, the supplement of α. We obtain

$$2\pi^2 = \frac{1}{2} \int_0^{2\pi} \int_0^{2\pi} d\phi_1 d\phi_2 = \iint_{P \notin D} \frac{\sin \alpha}{t_1 t_2} dx dy.$$

The theorem is proved.   □

750. A ring of height h is obtained by digging a cylindrical hole through the center of a sphere. Prove that the volume of the ring depends only on h and not on the radius of the sphere.

751. A polyhedron is circumscribed about a sphere. We call a face *big* if the projection of the sphere onto the plane of the face lies entirely within the face. Show that there are at most six big faces.

752. Let A and B be two finite sets of segments in three-dimensional space such that the sum of the lengths of the segments in A is larger than the sum of the lengths of the segments in B. Prove that there is a line in space with the property that the sum of the lengths of the projections of the segments in A onto that line is greater than the sum of the lengths of the projections of the segments in B.

753. Two convex polygons are placed one inside the other. Prove that the perimeter of the polygon that lies inside is smaller.

754. There are n line segments in the plane with the sum of the lengths equal to 1. Prove that there exists a straight line such that the sum of the lengths of the projections of the segments onto the line is equal to $\frac{2}{\pi}$.

755. In a triangle ABC for a variable point P on BC with $PB = x$ let $t(x)$ be the measure of $\angle PAB$. Compute

$$\int_0^a \cos t(x)dx$$

in terms of the sides and angles of triangle ABC.

756. Let $f : [0, a] \to \mathbb{R}$ be a continuous and increasing function such that $f(0) = 0$. Define by R the region bounded by $f(x)$ and the lines $x = a$ and $y = 0$. Now consider the solid of revolution obtained when R is rotated around the y-axis as a sort of dish. Determine f such that the volume of water the dish can hold is equal to the volume of the dish itself, this happening for all a.

757. Consider a unit vector starting at the origin and pointing in the direction of the tangent vector to a continuously differentiable curve in three-dimensional space. The endpoint of the vector describes the spherical image of the curve (on the unit sphere). Show that if the curve is closed, then its spherical image intersects every great circle of the unit sphere.

758. With the hypothesis of the previous problem, if the curve is twice differentiable, then the length of the spherical image of the curve is called the total curvature. Prove that the total curvature of a closed curve is at least 2π.

759. A rectangle R is tiled by finitely many rectangles each of which has at least one side of integral length. Prove that R has at least one side of integral length.

760. Show that if the distance between any two vertices of a polygon is less than or equal to 1, then the area of the polygon is less than $\pi/4$.

4.1.7 Other Geometry Problems

We conclude with problems from elementary geometry. They are less in the spirit of Euclid, being based on algebraic or combinatorial considerations. Here "imagination is more important than knowledge" (A. Einstein).

Example. Find the maximal number of triangles of area 1 with disjoint interiors that can be included in a disk of radius 1. Describe all such configurations.

Solution. Let us first solve the following easier problem:

Alternative problem. *Find all triangles of area 1 that can be placed inside a half-disk of radius* 1.

We will show that the only possible configuration is that in Figure 34. Consider a triangle that maximizes the area (such a triangle exists since the vertices vary on compact sets and the area depends continuously on the vertices). The vertices of this triangle must lie on the half-circle. If B lies between A and C, then A and C must be the endpoints of the diameter. Indeed, if say C is not an endpoint, then by moving it toward the closer endpoint of the diameter we increase both AC and the angle $\angle BAC$; hence we increase the area. Finally, among all triangles inscribed in a semicircle $\overset{\frown}{AC}$, the isosceles right triangle has maximal altitude, hence also maximal area. This triangle has area 1, and the claim is proved.

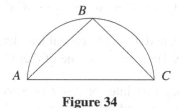

Figure 34

Returning to the problem, let us note that since the two triangles in question are convex sets, they can be separated by a line. That line cuts the disk into two regions, and one of them, containing one of the triangles, is included in a half-disk. By what we just proved, this region must itself be a half-disk. The only possible configuration consists of two isosceles triangles sharing the hypotenuse.   □

The next problem was published by the first author in the *Mathematics Magazine*.

Example. Let ABC be a right triangle ($\angle A = 90°$). On the hypotenuse BC construct in the exterior the equilateral triangle BCD. Prove that the lengths of the segments AB, AC, and AD cannot all be rational.

Solution. We will find a relation between AB, AC, and AD by placing them in a triangle and using the law of cosines.s For this, construct the equilateral triangle ACE in the exterior of ABC (Figure 35). We claim that $BE = AD$. This is a corollary of Napoleon's theorem, and can be proved in the following way. Let M be the intersection of the circumcircles of BCD and ACE. Then $\angle AMC = 120°$ and $\angle DMC = 60°$; hence $M \in AD$. Similarly, $M \in BE$. Ptolemy's theorem applied to quadrilaterals $AMCE$ and $BMCD$ shows that $ME = AM + CM$ and $MD = BM + CM$; hence $AD = AM + BM + CM = BE$.

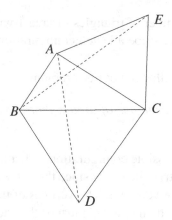

Figure 35

Applying the law of cosines in triangle ABE, we obtain $BE^2 = AB^2 + AE^2 + AB \cdot AE\sqrt{3}$, and since $BE = AD$ and $AE = AC$, it follows that

$$AD^2 = AB^2 + AC^2 + AB \cdot AC\sqrt{3}.$$

If all three segments AB, AC, and AD had rational lengths, this relation would imply that $\sqrt{3}$ is rational, which is not true. Hence at least one of these lengths is irrational.   □

761. Three lines passing through an interior point of a triangle and parallel to its sides determine three parallelograms and three triangles. If S is the area of the initial triangle and S_1, S_2 and S_3 are the areas of the newly formed triangles, prove that $S_1 + S_2 + S_3 \geq \frac{1}{3}S$.

762. Someone has drawn two squares of side 0.9 inside a disk of radius 1. Prove that the squares overlap.

763. A surface is generated by a segment whose midpoint rotates along the unit circle in the xy-plane such that for each $0 \leq \alpha < 2\pi$, at the point of coordinates $(\cos\alpha, \sin\alpha)$ on the circle the segment is in the same plane with the z-axis and makes with it an angle of $\frac{\alpha}{2}$. This surface, called a *Möbius band*, is depicted in Figure 36. What is the maximal length the segment can have so that the surface does not cross itself?

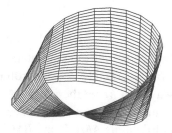

Figure 36

764. Let $ABCD$ be a convex quadrilateral and let O be the intersection of its diagonals. Given that the triangles OAB, OBC, OCD, and ODA have the same perimeter, prove that the quadrilateral is a rhombus. Does the property hold if O is some other point in the interior of the quadrilateral?

765. Prove that the plane cannot be covered by the interiors of finitely many parabolas.

766. Let ABC be a triangle with the largest angle at A. On line AB consider the point D such that A lies between B and D, and $AD = AB^3/AC^2$. Prove that $CD \leq \sqrt{3}BC^3/AC^2$.

767. Show that if all angles of an octagon are equal and all its sides have rational length, then the octagon has a center of symmetry.

768. Show that if each of the three main diagonals of a hexagon divides the hexagon into two parts with equal areas, then the three diagonals are concurrent.

769. Centered at every point with integer coordinates in the plane there is a disk with radius $\frac{1}{1000}$.

(a) Prove that there exists an equilateral triangle whose vertices lie in different disks.

(b) Prove that every equilateral triangle with vertices in different disks has side length greater than 96.

770. On a cylindrical surface of radius r, unbounded in both directions, consider n points and a surface S of area strictly less than 1. Prove that by rotating around the axis of the cylinder and then translating in the direction of the axis by at most $\frac{n}{4\pi r}$ units one can transform S into a surface that does not contain any of the n points.

4.2 Trigonometry

4.2.1 Trigonometric Identities

The beauty of trigonometry lies in its identities. There are two fundamental identities,

$$\sin^2 x + \cos^2 x = 1 \quad \text{and} \quad \cos(x - y) = \cos x \cos y - \sin x \sin y,$$

both with geometric origins, from which all the others can be derived. Our problems will make use of addition and subtraction formulas for two, three, even four angles, double- and triple-angle formulas, and product-to-sum formulas. While these identities are seen as very elementary today, we should remember that the quest to find their analogues led to the development of the theory of elliptic functions.

Example. Find all acute angles x satisfying the equation

$$2 \sin x \cos 40° = \sin(x + 20°).$$

Solution. Trying particular values we see that $x = 30°$ is a solution. Are there other solutions? Use the addition formula for sine to rewrite the equation as

$$\tan x = \frac{\sin 20°}{2 \cos 40° - \cos 20°}.$$

The tangent function is one-to-one on the interval $(0, 90°)$, which implies that the solution to the original equation is unique. □

Example. (a) Prove that if $\cos \pi a = \frac{1}{3}$ then a is an irrational number.

(b) Prove that a regular tetrahedron cannot be dissected into finitely many regular tetrahedra.

Solution. (a) Assume that a is rational, $a = \frac{m}{n}$. Then $\cos na\pi = \pm 1$. We will prove by induction that for all $k > 0$, $\cos ka\pi = \frac{m_k}{3^k}$, with m_k an integer that is not divisible by 3. This will then contradict the initial assumption.

The property is true for $k = 0$ and 1. The product-to-sum formula for cosines gives rise to the recurrence

$$\cos(k+1)a\pi = 2\cos a\pi \cos ka\pi - \cos(k-1)a\pi, \ k \geq 1.$$

Using the induction hypothesis, we obtain $\cos(k+1)a\pi = \frac{m_{k+1}}{3^{k+1}}$, with $m_{k+1} = 2m_k - 3m_{k-1}$. Since m_k is not divisible by 3, neither is m_{k+1}, and the claim is proved.

Part (b) is just a consequence of (a). To see this, let us compute the cosine of the dihedral angle of two faces of a regular tetrahedron $ABCD$. If AH is an altitude of the tetrahedron and AE is an altitude of the face ABC, then $\angle AEH$ is the dihedral angle of the faces ABC and BCD (see Figure 37). In the right triangle HAE, $\cos AEH = \frac{EH}{AD} = \frac{1}{3}$.

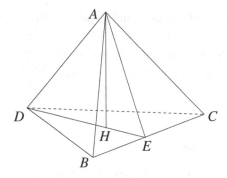

Figure 37

Now assume that there exists a dissection of a regular tetrahedron into regular tetrahedra. Several of these tetrahedra meet along a segment included in one of the faces of the initial tetrahedron. Their dihedral angles must add up to π, which implies that the dihedral angle of a regular tetrahedron is of the form $\frac{\pi}{n}$, for some integer n. This was shown above to be false. Hence no dissection of a regular tetrahedron into regular tetrahedra exists.   □

Remark. It is interesting to know that Leonardo da Vinci's manuscripts contain drawings of such decompositions. Later, however, Leonardo himself realized that the decompositions were impossible, and the drawings were mere optical illusions. Note also that Dehn's invariant mentioned in the first chapter provides an obstruction to the decomposition.

We conclude the introduction with a problem by the second author of the book.

Example. Let $a_0 = \sqrt{2} + \sqrt{3} + \sqrt{6}$ and let $a_{n+1} = \frac{a_n^2 - 5}{2(a_n+2)}$ for $n \geq 0$. Prove that

$$a_n = \cot\left(\frac{2^{n-3}\pi}{3}\right) - 2 \text{ for all } n.$$

Solution. We have

$$\cot\frac{\pi}{24} = \frac{\cos\dfrac{\pi}{24}}{\sin\dfrac{\pi}{24}} = \frac{2\cos^2\dfrac{\pi}{24}}{2\sin\dfrac{\pi}{24}\cos\dfrac{\pi}{24}} = \frac{1+\cos\dfrac{\pi}{12}}{\sin\dfrac{\pi}{12}} = \frac{1+\cos\left(\dfrac{\pi}{3}-\dfrac{\pi}{4}\right)}{\sin\left(\dfrac{\pi}{3}-\dfrac{\pi}{4}\right)}.$$

Using the subtraction formulas for sine and cosine we find that this is equal to

$$\frac{1+\dfrac{\sqrt{2}}{4}+\dfrac{\sqrt{6}}{4}}{\dfrac{\sqrt{6}}{4}-\dfrac{\sqrt{2}}{4}} = \frac{4+\sqrt{6}+\sqrt{2}}{\sqrt{6}-\sqrt{2}} = \frac{4(\sqrt{6}+\sqrt{2})+(\sqrt{6}+\sqrt{2})^2}{6-2}$$

$$= \frac{4(\sqrt{6}+\sqrt{2})+8+4\sqrt{3}}{4} = 2+\sqrt{2}+\sqrt{3}+\sqrt{6} = a_0 + 2.$$

Hence the equality $a_n = \cot\left(\frac{2^{n-3}\pi}{3}\right) - 2$ is true at least for $n = 0$.

To verify it in general, it suffices to prove that $b_n = \cot\left(\frac{2^{n-3}\pi}{3}\right)$, where $b_n = a_n + 2$, $n \geq 1$. The recurrence relation becomes

$$b_{n+1} - 2 = \frac{(b_n - 2)^2 - 5}{2b_n},$$

or $b_{n+1} = \frac{b_n^2 - 1}{2b_n}$. Assuming inductively that $b_k = \cot c_k$, where $c_k = \frac{2^{k-3}\pi}{3}$, and using the double-angle formula, we obtain

$$b_{k+1} = \frac{\cot^2 c_k - 1}{2\cot c_k} = \cot(2c_k) = \cot c_{k+1}.$$

This completes the proof.   □

771. Prove that

$$\sin 70° \cos 50° + \sin 260° \cos 280° = \frac{\sqrt{3}}{4}.$$

772. Show that the trigonometric equation

$$\sin(\cos x) = \cos(\sin x)$$

has no solutions.

773. Show that if the angles a and b satisfy

$$\tan^2 a \tan^2 b = 1 + \tan^2 a + \tan^2 b,$$

then

$$\sin a \sin b = \pm \sin 45°.$$

774. Find the range of the function $f : \mathbb{R} \to \mathbb{R}, f(x) = (\sin x + 1)(\cos x + 1)$.

775. Prove that

$$\sec^{2n} x + \csc^{2n} x \geq 2^{n+1},$$

for all integers $n \geq 0$, and for all $x \in \left(0, \frac{\pi}{2}\right)$.

776. Compute the integral

$$\int \sqrt{\frac{1-x}{1+x}}\,dx, \ x \in (-1, 1).$$

777. Find all integers k for which the two-variable function $f(x, y) = \cos(19x + 99y)$ can be written as a polynomial in $\cos x$, $\cos y$, $\cos(x + ky)$.

778. Let $a, b, c, d \in [0, \pi]$ be such that

$$2\cos a + 6\cos b + 7\cos c + 9\cos d = 0$$

and

$$2\sin a - 6\sin b + 7\sin c - 9\sin d = 0.$$

Prove that $3\cos(a + d) = 7\cos(b + c)$.

779. Let a be a real number. Prove that

$$5(\sin^3 a + \cos^3 a) + 3\sin a \cos a = 0.04$$

if and only if

$$5(\sin a + \cos a) + 2\sin a \cos a = 0.04.$$

780. Let $a_0, a_1, \ldots, a_n$ be numbers from the interval $\left(0, \frac{\pi}{2}\right)$ such that

$$\tan\left(a_0 - \frac{\pi}{4}\right) + \tan\left(a_1\frac{\pi}{4}\right) + \cdots + \tan\left(a_n - \frac{\pi}{4}\right) \geq n - 1.$$

Prove that

$$\tan a_0 \tan a_1 \cdots \tan a_n \geq n^{n+1}.$$

4.2.2 Euler's Formula

For a complex number z,

$$e^z = 1 + \frac{z}{1!} + \frac{z^2}{2!} + \cdots + \frac{z^n}{n!} + \cdots$$

In particular, for an angle x,

$$e^{ix} = 1 + i\frac{x}{1!} - \frac{x^2}{2!} - i\frac{x^3}{3!} + \frac{x^4}{4!} + i\frac{x^5}{5!} - \frac{x^6}{6!} - i\frac{x^7}{7!} + \cdots$$

The real part of e^{ix} is

$$1 - \frac{x^2}{2!} + \frac{x^4}{4!} - \frac{x^6}{6!} + \cdots ,$$

while the imaginary part is

$$\frac{x}{1!} - \frac{x^3}{3!} + \frac{x^5}{5!} - \frac{x^7}{7!} + \cdots$$

These are the Taylor series of $\cos x$ and $\sin x$. We obtain Euler's formula

$$e^{ix} = \cos x + i \sin x.$$

Euler's formula gives rise to one of the most beautiful identities in mathematics:

$$e^{i\pi} = -1,$$

which relates the number e from real analysis, the imaginary unit i from algebra, and π from geometry.

The equality $e^{nz} = (e^z)^n$ holds at least for z a real number. Two power series are equal for all real numbers if and only if they are equal coefficient by coefficient (since coefficients are computed using the derivatives at 0). So equality for real numbers means equality for complex numbers. In particular, $e^{inx} = (e^{ix})^n$, from which we deduce the de Moivre's formula

$$\cos nx + i \sin nx = (\cos x + i \sin x)^n.$$

We present an application of the de Moivre formula that we found in *Exercises and Problems in Algebra* by C. Năstăsescu, C. Niță, M. Brandiburu, and D. Joița (Editura Didactică și Pedagogică, Bucharest, 1983).

Example. Prove the identity

$$\binom{n}{0} + \binom{n}{k} + \binom{n}{2k} + \cdots = \frac{2^n}{k} \sum_{j=1}^{k} \cos^n \frac{j\pi}{k} \cos \frac{nj\pi}{k}.$$

Solution. Let $\varepsilon_1, \varepsilon_2, \ldots, \varepsilon_k$ be the kth roots of unity, that is, $\varepsilon_j = \cos \frac{2j\pi}{k} + i \sin \frac{2j\pi}{k}$, $j = 1, 2, \ldots, k$. The sum

$$\varepsilon_1^s + \varepsilon_2^s + \cdots + \varepsilon_k^s$$

is equal to k if k divides s, and to 0 if k does not divide s. We have

$$\sum_{j=1}^{k} (1 + \varepsilon_j)^n = \sum_{s=0}^{n} \binom{n}{s} \left(\sum_{j=1}^{k} \varepsilon_j^s \right) = k \sum_{j=0}^{\lfloor \frac{n}{k} \rfloor} \binom{n}{jk}.$$

Since

$$1 + \varepsilon_j = 2 \cos \frac{j\pi}{k} \left(\cos \frac{j\pi}{k} + i \sin \frac{j\pi}{k} \right),$$

it follows from the de Moivre formula that

$$\sum_{j=1}^{k}(1+\varepsilon_j)^n = \sum_{j=1}^{k} 2^n \cos^n \frac{j\pi}{k}\left(\cos \frac{nj\pi}{k} + i \sin \frac{nj\pi}{k}\right).$$

Therefore,

$$\binom{n}{0}+\binom{n}{k}+\binom{n}{2k}+\cdots = \frac{2^n}{k}\sum_{j=1}^{k} \cos^n \frac{j\pi}{k}\left(\cos \frac{nk\pi}{k} + i \sin \frac{nj\pi}{k}\right).$$

The left-hand side is real, so we can ignore the imaginary part and obtain the identity from the statement. □

And now a problem given at an Indian Team Selection Test for the International Mathematical Olympiad in 2005, proposed by the first author of the book.

Example. For real numbers a, b, c, d not all equal to zero, let $f : \mathbb{R} \to \mathbb{R}$,

$$f(x) = a + b\cos 2x + c \sin 5x + d \cos 8x.$$

Suppose that $f(t) = 4a$ for some real number t. Prove that there exists a real number s such that $f(s) < 0$.

Solution. Let $g(x) = be^{2ix} - ice^{5ix} + de^{8ix}$. Then $f(x) = a + \operatorname{Re} g(x)$. Note that

$$g(x) + g\left(x + \frac{2\pi}{3}\right) + g\left(x + \frac{4\pi}{3}\right) = g(x)(1 + e^{2\pi i/3} + e^{4\pi i/3}) = 0.$$

Therefore,

$$f(x) + f\left(x + \frac{2\pi}{3}\right) + f\left(x + \frac{4\pi}{3}\right) = 3a.$$

If $a < 0$, then $s = t$ would work. If $a = 0$, then for some x one of the terms of the above sum is negative. This is because $f(x)$ is not identically zero, since its Fourier series is not trivial. If $a > 0$, substituting $x = t$ in the identity deduced above and using the fact that $f(t) = 4a$, we obtain

$$f\left(t + \frac{2\pi}{3}\right) + f\left(t + \frac{4\pi}{3}\right) = -a < 0.$$

Hence either $f\left(t + \frac{2\pi}{3}\right)$ or $f\left(t + \frac{4\pi}{3}\right)$ is negative. The problem is solved. □

781. Prove the identity

$$\left(\frac{1 + i\tan t}{1 - i\tan t}\right)^n = \frac{1 + i\tan nt}{1 - i\tan nt}, \quad n \geq 1.$$

782. Prove the identity

$$1 - \binom{n}{2} + \binom{n}{4} - \binom{n}{6} + \cdots = 2^{n/2}\cos\frac{n\pi}{4}, \quad n \geq 1.$$

783. Compute the sum

$$\binom{n}{1} \cos x + \binom{n}{2} \cos 2x + \cdots + \binom{n}{n} \cos nx.$$

784. Find the Taylor series expansion at 0 of the function

$$f(x) = e^{x \cos \theta} \cos(x \sin \theta),$$

where θ is a parameter.

785. Let z_1, z_2, z_3 be complex numbers of the same absolute value, none of which is real and all distinct. Prove that if $z_1 + z_2 z_3$, $z_2 + z_3 z_1$ and $z_3 + z_1 z_2$ are all real, then $z_1 z_2 z_3 = 1$.

786. Let n be an odd positive integer and let θ be a real number such that $\frac{\theta}{\pi}$ is irrational. Set $a_k = \tan\left(\theta + \frac{k\pi}{n}\right)$, $k = 1, 2, \ldots, n$. Prove that

$$\frac{a_1 + a_2 + \cdots + a_n}{a_1 a_2 \cdots a_n}$$

is an integer and determine its value.

787. Find $(\cos \alpha)(\cos 2\alpha)(\cos 3\alpha) \cdots (\cos 999\alpha)$ with $\alpha = \frac{2\pi}{1999}$.

788. For positive integers n define $F(n) = x^n \sin(nA) + y^n \sin(nB) + z^n \sin(nC)$, where x, y, z, A, B, C are real numbers and $A + B + C = k\pi$ for some integer k. Prove that if $F(1) = F(2) = 0$, then $F(n) = 0$ for all positive integers n.

789. The continuous real-valued function $\phi(t)$ is defined for $t \geq 0$ and is absolutely integrable on every bounded interval. Define

$$P = \int_0^\infty e^{-(t+i\phi(t))} dt \quad \text{and} \quad Q = \int_0^\infty e^{-2(t+i\phi(t))} dt.$$

Prove that

$$|4P^2 - 2Q| \leq 3,$$

with equality if and only if $\phi(t)$ is constant.

4.2.3 Trigonometric Substitutions

The fact that the circle $x^2 + y^2 = 1$ can be parametrized by trigonometric functions as $x = \cos t$ and $y = \sin t$ gives rise to the standard substitution $x = a \cos t$ (or $x = a \sin t$) in expressions of the form $\sqrt{a^2 - x^2}$. Our purpose is to emphasize less standard substitutions, usually suggested by the similarity between an algebraic expression and a trigonometric formula. Such is the case with the following problem from the 61st W.L. Putnam Mathematical Competition, 2000.

Example. Let $f : [-1, 1] \to \mathbb{R}$ be a continuous function such that $f(2x^2 - 1) = 2xf(x)$ for all $x \in [-1, 1]$. Show that f is identically equal to zero.

Solution. Here the expression $2x^2 - 1$ should remind us of the trigonometric formula $2\cos^2 t - 1 = \cos 2t$, suggesting the substitution $x = \cos t$, $t \in [0, \pi]$. The functional equation from the statement becomes $f(\cos 2t) = 2 \cos t f(\cos t)$.

First, note that setting $x = 0$ and $x = 1$, we obtain $f(1) = f(-1) = 0$. Now let us define $g : \mathbb{R} \to \mathbb{R}$, $g(t) = \frac{f(\cos t)}{\sin t}$. Then for any t not a multiple of π,

$$g(2t) = \frac{f(2\cos^2 t - 1)}{\sin(2t)} = \frac{2 \cos t f(\cos t)}{2 \sin t \cos t} = \frac{f(\cos t)}{\sin t} = g(t).$$

Also, $g(t + 2\pi) = g(t)$. In particular, for any integers n and k,

$$g\left(1 + \frac{n\pi}{2^k}\right) = g(2^{k+1} + 2n\pi) = g(2^{k+1}) = g(1).$$

Because f is continuous, g is continuous everywhere except at multiples of π. The set $\left\{1 + \frac{n\pi}{2^k} \mid n, k \in \mathbb{Z}\right\}$ is dense on the real axis, and so g must be constant on its domain. Then $f(\cos t) = c \sin t$ for some constant c and t in $(0, \pi)$, i.e., $f(x) = c\sqrt{1 - x^2}$ for all $x \in (-1, 1)$. It follows that f is an even function. But then in the equation from the statement $f(2x^2 - 1) = 2xf(x)$ the left-hand side is an even function while the right-hand side is an odd function. This can happen only if both sides are identically zero. Therefore, $f(x) = 0$ for $x \in [-1, 1]$ is the only solution to the functional equation.   □

We continue with a problem that was proposed by Belgium for the 26th International Mathematical Olympiad in 1985.

Example. Let x, y, z be real numbers such that $x + y + z = xyz$. Prove that

$$x(1 - y^2)(1 - z^2) + y(1 - z^2)(1 - x^2) + z(1 - x^2)(1 - y^2) = 4xyz.$$

Solution. The conclusion is immediate if $xyz = 0$, so we may assume that $x, y, z \neq 0$. Dividing through by $4xyz$ we transform the desired equality into

$$\frac{1 - y^2}{2y} \cdot \frac{1 - z^2}{2z} + \frac{1 - z^2}{2z} \cdot \frac{1 - x^2}{2x} + \frac{1 - x^2}{2x} \cdot \frac{1 - y^2}{2y} = 1.$$

This, along with the condition from the statement, makes us think about the substitutions $x = \tan A$, $y = \tan B$, $z = \tan C$, where A, B, C are the angles of a triangle. Using the double-angle formula

$$\frac{1 - \tan^2 u}{2 \tan u} = \frac{1}{\tan 2u} = \cot 2u$$

we further transform the equality into

$$\cot 2B \cot 2C + \cot 2C \cot 2A + \cot 2A \cot 2B = 1.$$

But this is equivalent to

$$\tan 2A + \tan 2B + \tan 2C = \tan 2A \tan 2B \tan 2C,$$

which follows from $\tan(2A + 2B + 2C) = \tan 2\pi = 0$.   □

And now the problems.

790. Let $a, b, c \in [0, 1]$. Prove that

$$\sqrt{abc} + \sqrt{(1-a)(1-b)(1-c)} \le 1.$$

791. Solve the equation $x^3 - 3x = \sqrt{x+2}$ in real numbers.

792. Find the maximum value of

$$S = (1 - x_1)(1 - y_1) + (1 - x_2)(1 - y_2)$$

if $x_1^2 + x_2^2 = y_1^2 + y_2^2 = c^2$, where c is some positive number.

793. Prove for all real numbers a, b, c the inequality

$$\frac{|a-b|}{\sqrt{1+a^2}\sqrt{1+b^2}} \le \frac{|a-c|}{\sqrt{1+a^2}\sqrt{1+c^2}} + \frac{|b-c|}{\sqrt{1+b^2}\sqrt{1+c^2}}.$$

794. Let a, b, c be real numbers. Prove that

$$(ab + bc + ca - 1)^2 \le (a^2 + 1)(b^2 + 1)(c^2 + 1).$$

795. Prove that

$$\frac{x}{\sqrt{1+x^2}} + \frac{y}{\sqrt{1+y^2}} + \frac{z}{\sqrt{1+z^2}} \le \frac{3\sqrt{3}}{2}$$

if the positive real numbers x, y, z satisfy $x + y + z = xyz$.

796. Prove that

$$\frac{x}{1-x^2} + \frac{y}{1-y^2} = \frac{z}{1-z^2} \ge \frac{3\sqrt{3}}{2}$$

if $0 < x, y, z < 1$ and $xy + yz + xz = 1$.

797. Solve the system of equations

$$3\left(x + \frac{1}{x}\right) = 4\left(y + \frac{1}{y}\right) = 5\left(z + \frac{1}{z}\right)$$
$$xy + yz + zx = 1.$$

798. Solve the following system of equations in real numbers:

$$\frac{3x - y}{x - 3y} = x^2,$$

$$\frac{3y - z}{y - 3z} = y^2,$$

$$\frac{3z - x}{z - 3x} = z^2.$$

799. Let $a_0 = \sqrt{2}$, $b_0 = 2$, and

$$a_{n+1} = \sqrt{2 - \sqrt{4 - a_n^2}}, \quad b_{n+1} = \frac{2b_n}{2 + \sqrt{4 + b_n^2}}, \quad n \geq 0.$$

(a) Prove that the sequences $(a_n)_n$ and $(b_n)_n$ are decreasing and converge to zero.
(b) Prove that the sequence $(2^n a_n)_n$ is increasing, the sequence $(2^n b_n)_n$ is decreasing, and these two sequences converge to the same limit.
(c) Prove there is a positive constant C such that one has $0 < b_n - a_n < \frac{C}{8^n}$ for all n.

800. Let α be the greatest positive root of the equation

$$x^3 - 3x^2 + 1 = 0.$$

Show that both $\lfloor a^{1788} \rfloor$ and $\lfloor a^{1988} \rfloor$ are divisible by 17.

801. Two real sequences $x_1, x_2, \ldots$, and $y_1, y_2, \ldots$ are defined in the following way:

$$x_1 = y_1 = \sqrt{3}, \quad x_{n+1} = x_n + \sqrt{1 + x_n^2}, \quad y_{n+1} = \frac{y_n}{1 + \sqrt{1 + y_n^2}}, \quad \text{for } n \geq 1.$$

Prove that $2 < x_n y_n < 3$ for all $n > 1$.

802. Let a, b, c be real numbers different from $\pm\frac{1}{\sqrt{3}}$. Prove that the equality $abc = a+b+c$ holds only if

$$\frac{3a - a^3}{3a^2 - 1} \cdot \frac{3b - b^3}{3b^2 - 1} \cdot \frac{3c - c^3}{3c^2 - 1} = \frac{3a - a^3}{3a^2 - 1} + \frac{3b - b^3}{3b^2 - 1} + \frac{3c - c^3}{3c^2 - 1}.$$

803. Let a, b, $c > 0$. Find all triples (x, y, z) of positive real numbers such that

$$x + y + z = a + b + c$$
$$a^2 x + b^2 y + c^2 z + 4abc = 4xyz.$$

The parametrization of the hyperbola $x^2 - y^2 = 1$ by $x = \cosh t$, $y = \sinh t$ gives rise to the hyperbolic substitution $x = a \cosh t$ in expressions containing $\sqrt{a^2 - 1}$. We illustrate this with an example by the second author.

Example. Let $a_1 = a_2 = 97$ and

$$a_{n+1} = a_n a_{n-1} + \sqrt{(a_n^2 - 1)(a_{n-1}^2 - 1)}, \quad \text{for } n > 1.$$

Prove that

(a) $2 + 2a_n$ is a perfect square;

(b) $2 + \sqrt{2 + 2a_n}$ is a perfect square.

Solution. We are led to the substitution $a_n = \cosh t_n$ for some number t_n (which for the moment might be complex). The recurrence relation becomes

$$\cosh t_{n+1} = a_{n+1} = \cosh t_n \cosh t_{n-1} + \sinh t_n \sinh t_{n-1} = \cosh(t_n + t_{n-1}).$$

We deduce that the numbers t_n satisfy $t_0 = t_1$, and $t_{n+1} = t_n + t_{n-1}$ (in particular they are all real). And so $t_n = F_n t_0$, where $(F_n)_n$ is the Fibonacci sequence. Consequently, $a_n = \cosh(F_n t_0)$, $n \geq 1$.

Using the identity $2(\cosh t)^2 - 1 = \cosh 2t$, we obtain

$$2 + 2a_n = \left(2\cosh F_n \frac{t_0}{2}\right)^2.$$

The recurrence relation

$$2\cosh(k+1)t = (2\cosh t)(2\cosh kt) - 2\cosh(k-1)t$$

allows us to prove inductively that $2\cosh k\frac{t_0}{2}$ is an integer once we show that $2\cosh \frac{t_0}{2}$ is an integer. It would then follow that $2\cosh F_n \frac{t_0}{2}$ is an integer as well. And indeed $2\cosh \frac{t_0}{2} = \sqrt{2 + 2a_1} = 14$. This completes the proof of part (a).

To prove (b), we obtain in the same manner

$$2 + \sqrt{2 + 2a_n} = \left(2\cosh F_n \frac{t_0}{4}\right)^2,$$

and again we have to prove that $2\cosh \frac{t_0}{4}$ is an integer. We compute $2\cosh \frac{t_0}{4} = \sqrt{1 + \sqrt{2 + 2a_n}} = \sqrt{2 + 14} = 4$. The conclusion follows. $\square$

804. Compute the integral

$$\int \frac{dx}{x + \sqrt{x^2 - 1}}.$$

805. Let $n > 1$ be an integer. Prove that there is no irrational number a such that the number

$$\sqrt[n]{a + \sqrt{a^2 - 1}} + \sqrt[n]{a - \sqrt{a^2 - 1}}$$

is rational.

4.2.4 Telescopic Sums and Products in Trigonometry

The philosophy of telescopic sums and products in trigonometry is the same as in the general case, just that here we have more identities at hand. Let us take a look at a slightly modified version of an identity of C.A. Laisant.

Example. Prove that

$$\sum_{k=0}^{n} \left(-\frac{1}{3}\right)^k \cos^3(3^{k-n}\pi) = \frac{3}{4}\left[\left(-\frac{1}{3}\right)^{n+1} + \cos\frac{\pi}{3^n}\right].$$

Solution. From the identity $\cos 3x = 4\cos^3 x - 3\cos x$, we obtain

$$\cos^3 x = \frac{1}{4}(\cos 3x + 3\cos x).$$

Then

$$\sum_{k=0}^{n}\left(-\frac{1}{3}\right)^k \cos^3(3^k a) = \frac{1}{4}\sum_{k=0}^{n}\left[\left(-\frac{1}{3}\right)^k \cos(3^{k+1}a) - \left(-\frac{1}{3}\right)^{k-1}\cos(3^k a)\right].$$

This telescopes to

$$\frac{1}{4}\left[\left(-\frac{1}{3}\right)^n \cos(3^{n+1}a) - \left(-\frac{1}{3}\right)^{-1}\cos a\right].$$

For $a = 3^{-n}\pi$, we obtain the identity from the statement.   □

Test your skills against the following problems.

806. Prove that

$$27\sin^3 9° + 9\sin^3 27° + 3\sin^3 81° + \sin^3 243° = 20\sin 9°.$$

807. Prove that

$$\frac{1}{\cot 9° - 2\tan 9°} + \frac{3}{\cot 27° - 3\tan 27°} + \frac{9}{\cot 81° - 3\tan 81°}$$
$$+ \frac{27}{\cot 243° - 3\tan 243°} = 10\tan 9°.$$

808. Prove that

$$\frac{1}{\sin 45° \sin 46°} + \frac{1}{\sin 47° \sin 48°} + \cdots + \frac{1}{\sin 133° \sin 134°} = \frac{1}{\sin 1°}.$$

809. Obtain explicit values for the following series:

(a) $\displaystyle\sum_{n=1}^{\infty} \arctan\frac{2}{n^2}$,

(b) $\displaystyle\sum_{n=1}^{\infty} \arctan\frac{8n}{n^4 - 2n^2 + 5}$.

810. For $n \geq 0$ let

$$u_n = \arcsin\frac{\sqrt{n+1} - \sqrt{n}}{\sqrt{n+2}\sqrt{n+1}}.$$

Prove that the series

$$S = u_0 + u_1 + u_2 + \cdots + u_n + \cdots$$

is convergent and find its limit.

Now we turn to telescopic products.

Example. Prove that

$$\prod_{n=1}^{\infty} \frac{1}{1 - \tan^2 2^{-n}} = \tan 1.$$

Solution. The solution is based on the identity

$$\tan 2x = \frac{2 \tan x}{1 - \tan^2 x}.$$

Using it we can write

$$\prod_{n=1}^{N} \frac{1}{1 - \tan^2 2^{-n}} = \prod_{n=1}^{n} \frac{\tan 2^{-n+1}}{2 \tan 2^{-n}} = \frac{2^{-N}}{\tan 2^{-N}} \tan 1.$$

Since $\lim_{x \to 0} \frac{\tan x}{x} = 1$, when letting $N \to \infty$ this become $\tan 1$, as desired. $\qquad\square$

811. In a circle of radius 1 a square is inscribed. A circle is inscribed in the square and then a regular octagon in the circle. The procedure continues, doubling each time the number of sides of the polygon. Find the limit of the lengths of the radii of the circles.

812. Prove that

$$\left(1 - \frac{\cos 61°}{\cos 1°}\right)\left(1 - \frac{\cos 62°}{\cos 2°}\right) \cdots \left(1 - \frac{\cos 119°}{\cos 59°}\right) = 1.$$

813. Evaluate the product

$$(1 - \cot 1°)(1 - \cot 2°) \cdots (1 - \cot 44°).$$

814. Compute the product

$$(\sqrt{3} + \tan 1°)(\sqrt{3} + \tan 2°) \cdots (\sqrt{3} + \tan 29°).$$

815. Prove the identities

(a) $\left(\frac{1}{2} - \cos \frac{\pi}{7}\right)\left(\frac{1}{2} - \cos \frac{3\pi}{7}\right)\left(\frac{1}{2} - \cos \frac{9\pi}{7}\right) = -\frac{1}{8}$,

(b) $\left(\frac{1}{2} + \cos \frac{\pi}{20}\right)\left(\frac{1}{2} + \cos \frac{3\pi}{20}\right)\left(\frac{1}{2} + \cos \frac{9\pi}{20}\right)\left(\frac{1}{2} + \cos \frac{27\pi}{20}\right) = \frac{1}{16}$.

816. Prove the identities

(a) $\prod_{n=1}^{24} \sec(2^n)° = -2^{24} \tan 2°$,

(b) $\prod_{n=2}^{25} (2\cos(2^n)° - \sec(2^n)°) = -1$.

5

Number Theory

This chapter on number theory is truly elementary, although its problems are far from easy. (In fact, here, as elsewhere in the book, we tried to follow Felix Klein's advice: "Don't ever be absolutely boring".[1] We restricted ourselves to some basic facts about residue classes and divisibility: Fermat's little theorem and its generalization due to Euler, Wilson's theorem, the Chinese remainder theorem, and Polignac's formula, with just a short incursion into algebraic number theory. From all Diophantine equations we discuss linear equations in two variables and two types of quadratic equations: the Pythagorean equation and Pell's equation.

But first, three sections for which not much background is necessary.

5.1 Integer-Valued Sequences and Functions

5.1.1 Some General Problems

Here are some problems, not necessarily straightforward, that use only the basic properties of integers.

Example. Find all functions $f : \{0, 1, 2, \ldots\} \to \{0, 1, 2, \ldots\}$ with the property that for every $m, n \geq 0$,

$$2f(m^2 + n^2) = (f(m))^2 + (f(n))^2.$$

Solution. The substitution $m = n = 0$ yields

$$2f(0^2 + 0^2) = (f(0))^2 + (f(0))^2,$$

and this gives $f(0)^2 = f(0)$, hence $f(0) = 0$ or $f(0) = 1$.

We pursue the track of $f(0) = 0$ first. We have

$$2f(1^2 + 0^2) = (f(1))^2 + (f(0))^2,$$

so $2f(1) = f(1)^2$, and hence $f(1) = 0$ or $f(1) = 2$. Let us see what happens if $f(1) = 2$, since this is the most interesting situation. We find immediately

$$2f(2) = 2f(1^2 + 1^2) = (f(1))^2 + (f(1))^2 = 8,$$

---

[1] Seien Sie niemals absolut langweilig.

© Springer International Publishing AG 2017

R. Gelca and T. Andreescu, *Putnam and Beyond*, DOI 10.1007/978-3-319-58988-6_5

so $f(2) = 4$, and then

$$2f(4) = 2f(2^2 + 0^2) = (f(2))^2 + (f(0))^2 = 16,$$

$$2f(5) = 2f(2^2 + 1^2) = (f(2))^2 + (f(1))^2 = 20,$$

$$2f(8) = 2f(2^2 + 2^2) = (f(2))^2 + (f(2))^2 = 32.$$

So $f(4) = 8$, $f(5) = 10$, $f(8) = 16$. In fact, $f(n) = 2n$ for $n \leq 10$, but as we will see below, the proof is more involved. Indeed,

$$100 = (f(5))^2 + (f(0))^2 = 2f(5^2) = 2f(3^2 + 4^2) = (f(3))^2 + (f(4))^2$$
$$= (f(3))^2 + 64,$$

hence $f(3) = 6$. Then immediately

$$2f(9) = 2f(3^2 + 0^2) = (f(3))^2 + (f(0))^2 = 36,$$

$$2f(10) = 2f(3^2 + 1^2) = (f(3))^2 + (f(1))^2 = 40,$$

so $f(9) = 18$, $f(10) = 20$.

Applying an idea used before, we have

$$400 = (f(10))^2 + (f(0))^2 = 2f(10^2) = 2f(6^2 + 8^2) = (f(6))^2 + (f(8))^2$$
$$= (f(6))^2 + 256,$$

from which we obtain $f(6) = 12$. For $f(7)$ we use the fact that $7^2 + 1^2 = 5^2 + 5^2$ and the equality

$$(f(7))^2 + (f(1))^2 = (f(5))^2 + (f(6))^2$$

to obtain $f(7) = 14$.

We want to prove that $f(n) = 2n$ for $n > 10$ using strong induction. The argument is based on the identities

$$(5k + 1)^2 + 2^2 = (4k + 2)^2 + (3k - 1)^2,$$

$$(5k + 2)^2 + 1^2 = (4k + 1)^2 + (3k + 2)^2,$$

$$(5k + 3)^2 + 1^2 = (4k + 3)^2 + (3k + 1)^2,$$

$$(5k + 4)^2 + 2^2 = (4k + 2)^2 + (3k + 4)^2,$$

$$(5k + 5)^2 + 0^2 = (4k + 4)^2 + (3k + 3)^2.$$

Note that if $k \geq 2$, then the first term on the left is strictly greater then any of the two terms on the right, and this makes the induction possible. Assume that $f(m) = 2m$ for $m < n$ and let us prove $f(n) = 2n$. Let $n = 5k + j$, $1 \leq j \leq 5$, and use the corresponding identity to write $n^2 + m_1^2 = m_2^2 + m_3^2$, where m_1, m_2, m_3 are positive integers less than n. We then have

$$(f(n))^2 + (f(m_1))^2 = 2f(n^2 + m_1^2) = 2f(m_2^2 + m_3^2) = (f(m_2))^2 + (f(m_3))^2.$$

This then gives

$$(f(n))^2 = (2m_2)^2 + (2m_3)^2 - (2m_1)^2 = 4(m_2^2 + m_3^2 - m_1^2) = 4n^2.$$

Hence $f(n) = 2n$, completing the inductive argument. And indeed, this function satisfies the equation from the statement.

If we start with the assumption $f(1) = 0$, the exact same reasoning applied mutatis mutandis shows that $f(n) = 0$, $n \geq 0$. And the story repeats if $f(0) = 1$, giving $f(n) = 1$, $n \geq 0$. Thus the functional equation has three solutions: $f(n) = 2n$, $n \geq 0$, and the constant solutions $f(n) = 0$, $n \geq 0$, and $f(n) = 1$, $n \geq 0$. $\square$

With the additional hypothesis $f(m^2) \geq f(n^2)$ if $m \geq n$, this problem appeared at the 1998 Korean Mathematical Olympiad. The solution presented above was communicated to us by B.J. Venkatachala.

817. Let k be a positive integer. The sequence $(a_n)_n$ is defined by $a_1 = 1$, and for $n \geq 2$, a_n is the nth positive integer greater than a_{n-1} that is congruent to n modulo k. Find a_n in closed form.

818. Three infinite arithmetic progressions are given, whose terms are positive integers. Assuming that each of the numbers 1, 2, 3, 4, 5, 6, 7, 8 occurs in at least one of these progressions, show that 1980 necessarily occurs in one of them.

819. Find all functions $f : \mathbb{N} \to \mathbb{N}$ satisfying

$$f(n) + 2f(f(n)) = 3n + 5, \quad \text{for all } n \in \mathbb{N}.$$

820. Find all functions $f : \mathbb{Z} \to \mathbb{Z}$ with the property that

$$2f(f(x)) - 3f(x) + x = 0, \quad \text{for all } x \in \mathbb{Z}.$$

821. Prove that there exists no bijection $f : \mathbb{N} \to \mathbb{N}$ such that

$$f(mn) = f(m) + f(n) + 3f(m)f(n),$$

for all $m, n \geq 1$.

822. Show that there does not exist a sequence $(a_n)_{n \geq 1}$ of positive integers such that

$$a_{n-1} \leq (a_{n+1} - a_n)^2 \leq a_n, \quad \text{for all } n \geq 2.$$

823. Determine all functions $f : \mathbb{Z} \to \mathbb{Z}$ satisfying

$$f(x^3 + y^3 + z^3) = (f(x))^3 + (f(y))^3 + (f(z))^3, \quad \text{for all } x, y, z \in \mathbb{Z}.$$

5.1.2 Fermat's Infinite Descent Principle

Fermat's infinite descent principle states that there are no strictly decreasing infinite sequences of positive integers. Alternatively, any decreasing sequence of positive integers becomes stationary. This is a corollary of the fundamental property of the set of positive integers that every subset has a smallest element. To better understand this principle, let us apply it to an easy example.

Example. At each point of integer coordinates in the plane is written a positive integer number such that each of these numbers is the arithmetic mean of its four neighbors. Prove that all the numbers are equal.

Solution. The solution is an application of the maximum modulus principle. For $n \geq 1$, consider the square of side $2n$ centered at the origin. Among the numbers covered by it, the smallest must lie on its perimeter. Let this minimum be $m(n)$. If it is also attained in the interior of the square, then the four neighbors of that interior point must be equal, and step by step we show that all numbers inside that square are equal. Hence there are two possibilities. Either $m(1) > m(2) > m(3) > \cdots$ or $m(n) = m(n + 1)$ for infinitely many n. The former case is impossible, since the $m(n)$'s are positive integers; the latter case implies that all the numbers are equal.   □

We find even more spectacular this problem from the 2004 USA Mathematical Olympiad.

Example. Suppose that $a_1, \ldots, a_n$ are integers whose greatest common divisor is 1. Let S be a set of integers with the following properties:

   (i) For $i = 1, \ldots, n, a_i \in S$.

   (ii) For $i, j = 1, \ldots, n$ (not necessarily distinct), $a_i - a_j \in S$.

   (iii) For any integers $x, y \in S$, if $x + y \in S$, then $x - y \in S$.

Prove that S must equal the set of all integers.

Solution. This problem was submitted by K. Kedlaya and L. Ng. The solution below was discovered by M. Ince and earned him the Clay prize.

First thing, note that if $b_1, b_2, \ldots, b_m$ are some integers that generate S and satisfy the three conditions from the statement, then $b_i - 2b_j$ and $2b_i - b_j$ are also in S for any indices i and j. Indeed, since b_i, b_j, and $b_i - b_j$ are in S, by (iii) we have that $b_i - 2b_j \in S$. Moreover, for $i = j$ in (ii) we find that $0 = b_i - b_i \in S$. Hence applying (iii) to $x \in S$ and 0 we have that $-x \in S$ as well, and in particular $2b_i - b_j \in S$.

An n-tuple $(b_1, b_2, \ldots, b_n)$ as above can be substituted by $(b_1, b_2 - b_1, \ldots, b_n - b_1)$, which again generates S and, by what we just proved, satisfies (i), (ii), and (iii). Applying this step to $(|a_1|, |a_2|, \ldots, |a_n|)$ and assuming that $|a_1|$ is the smallest of these numbers, we obtain another n-tuple the sum of whose entries is smaller. Because we cannot have an infinite descent, we eventually reach an n-tuple with the first entry equal to 0. In the process we did not change the greatest common divisor of the entries. Ignoring the zero entries, we can repeat the procedure until there is only one nonzero number left. This number must be 1.

From the fact that $0, 1 \in S$ and then also $-1 \in S$, by applying (iii) to $x = 1, y = -1$ we find that $2 \in S$, and inductively we find that all positive, and also all negative, integers are in S. We conclude that $S = \mathbb{Z}$. As I. Kaplansky said, "An elegant proof hits you between your eyes with joy". $\qquad\square$

824. Show that no positive integers x, y, z can satisfy the equation

$$x^2 + 10y^2 = 3z^2.$$

825. Prove that the system of equations

$$x^2 + 5y^2 = z^2,$$

$$5x^2 + y^2 = t^2$$

does not admit nontrivial integer solutions.

826. Show that the equation

$$x^2 - y^2 = 2xyz$$

has no solutions x, y, z in the set of positive integers.

827. Prove that there is no infinite arithmetic progression whose terms are all perfect squares.

828. Let f be a bijection of the set of positive integers. Prove that there exist positive integers $a < a + d < a + 2d$ such that $f(a) < f(a+d) < f(a+2d)$.

829. Prove that for no integer $n > 1$ does n divide $2^n - 1$.

830. Find all pairs of positive integers (a, b) with the property that $ab + a + b$ divides $a^2 + b^2 + 1$.

831. Let x, y, z be positive integers such that $xy - z^2 = 1$. Prove that there exist nonnegative integers a, b, c, d such that

$$x = a^2 + b^2, \quad y = c^2 + d^2, \quad z = ac + bd.$$

5.1.3 The Greatest Integer Function

The greatest integer function associates to a number x the greatest integer less than or equal to x. The standard notation is $\lfloor x \rfloor$. Thus $\lfloor 2 \rfloor = 2$, $\lfloor 3.2 \rfloor = 3$, $\lfloor -2.1 \rfloor = -3$. This being said, let us start with the examples.

Beatty's theorem. *Let α and β be two positive irrational numbers satisfying $\frac{1}{\alpha} + \frac{1}{\beta} = 1$. Then the sequences $\lfloor \alpha n \rfloor$ and $\lfloor \beta n \rfloor$ are strictly increasing and determine a partition of the set of positive integers into two disjoint sets.*

Proof. In other words, each positive integer shows up in exactly one of the two sequences. Let us first prove the following result.

Lemma. *If x_n, $n \geq 1$, is an increasing sequence of positive integers with the property that for every n, the number of indices m such that $x_m < n$ is equal to $n - 1$, then $x_n = n$ for all n.*

Proof. We do the proof by induction. The base case is obvious: because the sequence is increasing, the only n for which $x_n < 2$ is $n = 1$. Now let us assume that $x_1 = 1, x_2 = 2, \ldots, x_{n-1} = n - 1$. From the hypothesis it also follows that there are no other indices m for which $x_m < n$. And because there is exactly one more term of the sequence that is less than $n + 1$, this term must be x_n and it is equal to n.   □

Returning to the problem, let us write all numbers of the form $\lfloor \alpha n \rfloor$ and $\lfloor \beta n \rfloor$ in an increasing sequence y_n. For every n there are exactly $\lfloor \frac{n}{\alpha} \rfloor$ numbers of the form $\lfloor k\alpha \rfloor$, and $\lfloor \frac{n}{\beta} \rfloor$ numbers of the form $\lfloor k\beta \rfloor$ that are strictly less than n (here we used the fact that α and β are irrational). We have

$$n - 1 = \left\lfloor \frac{n}{\alpha} + \frac{n}{\beta} \right\rfloor - 1 \le \left\lfloor \frac{n}{\alpha} \right\rfloor + \left\lfloor \frac{n}{\beta} \right\rfloor < \frac{n}{\alpha} + \frac{n}{\beta} = n.$$

Hence $\lfloor \frac{n}{\alpha} \rfloor + \lfloor \frac{n}{\beta} \rfloor = n - 1$, which shows that the sequence y_n satisfies the condition of the lemma. It follows that this sequence consists of all positive integers written in strictly increasing order. Hence the conclusion.   □

Our second example is a general identity discovered by the second author and D. Andrica. Note the similarity with Young's inequality for integrals (problem 578).

Theorem. *Let $a < b$ and $c < d$ be positive real numbers and let $f : [a, b] \to [c, d]$ be a continuous, bijective, and increasing function. Then*

$$\sum_{a \le k \le b} \lfloor f(k) \rfloor + \sum_{c \le k \le d} \lfloor f^{-1}(k) \rfloor - n(G_f) = \lfloor b \rfloor \lfloor d \rfloor - \alpha(a)\alpha(c),$$

where k is an integer, $n(G_f)$ is the number of points with integer coordinates on the graph of f, and $\alpha : \mathbb{R} \to \mathbb{R}$ is defined by

$$\alpha(x) = \begin{cases} \lfloor x \rfloor & \text{if } x \in \mathbb{R} \setminus \mathbb{Z}, \\ 0 & \text{if } x = 0, \\ x - 1 & \text{if } x \in \mathbb{Z} \setminus \{0\}. \end{cases}$$

Proof. The proof is by counting. For a region M of the plane, we denote by $n(M)$ the number of points with nonnegative integer coordinates in M. For our theorem, consider the sets

$$M_1 = \{(x, y) \in \mathbb{R}^2 \mid a \le x \le b, \ 0 \le y \le f(x)\},$$
$$M_2 = \{(x, y) \in \mathbb{R}^2 \mid c \le y \le d, \ 0 \le x \le f^{-1}(y)\},$$
$$M_3 = \{(x, y) \in \mathbb{R}^2 \mid 0 < x \le b, \ 0 < y \le d\},$$
$$M_4 = \{(x, y) \in \mathbb{R}^2 \mid 0 < x < a, \ 0 < y < c\}.$$

Then

$$n(M_1) = \sum_{a \le k \le b} \lfloor f(k) \rfloor, \ n(M_2) = \sum_{c \le k \le d} \lfloor f^{-1}(k) \rfloor,$$

$$n(M_3) = \lfloor b \rfloor \lfloor d \rfloor, \qquad n(M_4) = \alpha(a)\alpha(c).$$

By the inclusion-exclusion principle,

$$n(M_1 \cup M_2) = n(M_1) + n(M_2) - n(M_1 \cap M_2).$$

Note that $n(M_1 \cap M_2) = n(G_f)$ and $N(M_1 \cup M_2) = n(M_3) - n(M_4)$. The identity follows.

832. For a positive integer n and a real number x, prove the identity

$$\lfloor x \rfloor + \left\lfloor x + \frac{1}{n} \right\rfloor + \cdots + \left\lfloor x + \frac{1}{n-1} \right\rfloor = \lfloor nx \rfloor.$$

833. For a positive integer n and a real number x, compute the sum

$$\sum_{0 \leq i < j \leq n} \left\lfloor \frac{x+i}{j} \right\rfloor.$$

834. Find all pairs of real numbers x, y that satisfy

$$\lfloor x \rfloor (\lfloor x \rfloor + 1)(\lfloor x \rfloor x + 3)(\lfloor x \rfloor + 4) = \lfloor y \rfloor^2.$$

835. Prove that for every positive integer n,

$$\lfloor \sqrt{n} \rfloor = \left\lfloor \sqrt{n} + \frac{1}{\sqrt{n} + \sqrt{n+2}} \right\rfloor.$$

836. For what real numbers $x \geq 1$ is it true that

$$\left\lfloor \sqrt{\lfloor \sqrt{x} \rfloor} \right\rfloor = \left\lfloor \sqrt{\sqrt{x}} \right\rfloor ?$$

837. Express $\displaystyle\sum_{k=1}^{n} \lfloor \sqrt{k} \rfloor$ in terms of n and $a = \lfloor \sqrt{n} \rfloor$.

838. Prove the identity

$$\sum_{k=1}^{\frac{n(n+1)}{2}} \left\lfloor \frac{-1 + \sqrt{1+8k}}{2} \right\rfloor = \frac{n(n^2+2)}{3}, \quad n \geq 1.$$

839. Find all pairs of real numbers (a, b) such that $a \lfloor bn \rfloor = b \lfloor an \rfloor$ for all positive integers n.

840. Show that if $x \geq 1$ and $x \notin \mathbb{Z}$, then

$$\frac{1}{\lfloor x \rfloor} + \frac{1}{\{x\}} > \frac{7}{2x},$$

where $\{x\}$ is the fractional part of x ($\{x\} = x - \lfloor x \rfloor$).

841. For p and q coprime positive integers prove the reciprocity law

$$\left\lfloor \frac{p}{q} \right\rfloor + \left\lfloor \frac{2p}{q} \right\rfloor + \cdots + \left\lfloor \frac{(q-1)p}{q} \right\rfloor = \left\lfloor \frac{q}{p} \right\rfloor + \left\lfloor \frac{2q}{p} \right\rfloor + \cdots + \left\lfloor \frac{(p-1)q}{p} \right\rfloor.$$

842. Prove that for any real number x and for any positive integer n,

$$\lfloor nx \rfloor \geq \frac{\lfloor x \rfloor}{1} + \frac{\lfloor 2x \rfloor}{2} + \frac{\lfloor 3x \rfloor}{3} + \cdots + \frac{\lfloor nx \rfloor}{n}.$$

843. Does there exist a strictly increasing function $f : \mathbb{N} \to \mathbb{N}$ such that $f(1) = 2$ and $f(f(n)) = f(n) + n$ for all n?

844. Suppose that the strictly increasing functions $f, g : \mathbb{N} \to \infty$ partition $\mathbb{N}$ into two disjoint sets and satisfy

$$g(n) = f(f(kn)) + 1, \quad \text{for all } n \geq 1,$$

for some fixed positive integer k. Prove that f and g are unique with this property and find explicit formulas for them.

5.2 Arithmetic

5.2.1 Factorization and Divisibility

There isn't much to say here. An integer d divides another integer n if there is an integer d' such that $n = dd'$. In this case d is called a divisor of n. We denote by $\gcd(a, b)$ the greatest common divisor of a and b. For any positive integers a and b, Euclid's algorithm yields integers x and y such that $ax - by = \gcd(a, b)$. Two numbers are called coprime, or relatively prime, if their greatest common divisor is 1. For coprime numbers a and b there exist integers x and y such that $ax - by = 1$.

We begin with a problem from the Soviet Union Mathematical Olympiad for University Students in 1976.

Example. Prove that there is no polynomial with integer coefficients $P(x)$ with the property that $P(7) = 5$ and $P(15) = 9$.

Solution. Assume that such a polynomial $P(x) = a_n x^n + a_{n-1} x^{n-1} + \cdots + a_0$ does exist. Then $P(7) = a_n 7^n + a_{n-1} 7^{n-1} + \cdots + a_0$ and $P(15) = a_n 15^n + a_{n-1} 15^{n-1} + \cdots + a_0$. Subtracting, we obtain

$$4 = P(15) - P(7) = a_n(15^n - 7^n) + a_{n-1}(15^{n-1} - 7^{n-1}) + \cdots + a_1(15 - 7).$$

Since for any k, $15^k - 7^k$ is divisible by $15 - 7 = 8$, it follows that $P(15) - P(7) = 4$ itself is divisible by 8, a contradiction. Hence such a polynomial does not exist. □

The second problem was given at the Asia-Pacific Mathematical Olympiad in 1998.

Example. Show that for any positive integers a and b, the product $(36a + b)(a + 36b)$ cannot be a power of 2.

Solution. Assume that $(36a + b)(a + 36b)$ is a power of 2 for some integers a and b. Without loss of generality, we may assume that a and b are coprime and $a < b$. Let $36a + b = 2^m$ and $a + 36b = 2^n$. Adding and subtracting, we obtain $37(a + b) = 2^m(2^{n-m} + 1)$, respectively $35(a - b) = 2^m(2^{n-m} - 1)$. It follows that both $a + b$ and $a - b$ are divisible by $2m$. This can happen only if both a and b are divisible by 2^{m-1}. Our assumption that a and b are coprime implies that $m = 1$. But then $36a + b = 2$, which is impossible. Hence the conclusion. $\square$

845. Find the integers n for which $(n^3 - 3n^2 + 4)/(2n - 1)$ is an integer.

846. Prove that in the product $P = 1! \cdot 2! \cdot 3! \cdots 100!$ one of the factors can be erased so that the remaining product is a perfect square.

847. The sequence $a_1, a_2, a_3, \ldots$ of positive integers satisfies $\gcd(a_i, a_j) = \gcd(i, j)$ for $i \neq j$. Prove that $a_i - i$ for all i.

848. Let n, a, b be positive integers. Prove that

$$\gcd(n^a - 1, n^b - 1) = n^{\gcd(a,b)} - 1.$$

849. Let a and b be positive integers. Prove that the greatest common divisor of $2^a + 1$ and $2^b + 1$ divides $2^{\gcd(a,b)} + 1$.

850. Fix a positive integer k and define the sequence $(a_n)_n$ by $a_1 = k + 1$ and $a_{n+1} = a_n^2 - ka_n + k$ for $n \geq 1$. Prove that for any distinct positive integers m and n the numbers a_m and a_n are coprime.

851. Let a, b, c, d, e, and f be positive integers. Suppose that $S = a + b + c + d + e + f$ divides both $abc + def$ and $ab + bc + ca - de - ef - fd$. Prove that S is composite.

852. Let n be an integer greater than 2. Prove that $n(n - 1)^4 + 1$ is the product of two integers greater than 1.

853. Determine the functions $f : \{0, 1, 2, \ldots\} \to \{0, 1, 2, \ldots\}$ satisfying
(i) $(f(2n + 1))^2 - (f(2n))^2 = 6f(n) + 1$ and
(ii) $f(2n) \geq f(n)$ for all $n \geq 0$.

5.2.2 Prime Numbers

An integer greater than 1 is called prime if it has no other divisors than 1 and the number itself. Equivalently, a number is prime if whenever it divides a product it divides one of the factors. Any integer greater than 1 can be written as a product of primes in a unique way up to a permutation of the factors. This is known as the Fundamental theorem of arithmetic.

Euclid's theorem. *There are infinitely many prime numbers.*

Proof. From the more than one hundred proofs of this theorem we selected the fascinating topological proof given in 1955 by H. Furstenberg. It uses the concept of topology, which is an abstraction, in the spirit of Bourbaki, of the properties of open sets (i.e. unions of open intervals) on the real axis. By definition, a topology on a set X is a collection $\mathcal{T}$ of sets satisfying

(i) $\emptyset, X \in \mathcal{T}$;

(ii) for any family $(U_i)_{i \in I}$ of sets from $\mathcal{T}$, the union $\cup_{i \in I} U_i$ is also in $\mathcal{T}$;

(iii) for any $U_1, U_2, \ldots, U_n$ in $\mathcal{T}$, the intersection $U_1 \cap U_2 \cap \ldots \cap U_n$ is in $\mathcal{T}$.

The elements of $\mathcal{T}$ are called open sets; their complements are called closed sets.

Furstenberg's idea was to introduce a topology on $\mathbb{Z}$, namely the smallest topology in which any set consisting of all terms of a nonconstant arithmetic progression is open. As an example, in this topology both the set of odd integers and the set of even integers are open. Because the intersection of two arithmetic progressions is an arithmetic progression, the open sets of $\mathcal{T}$ are precisely the unions of arithmetic progressions. In particular, any open set is either infinite or empty.

If we define

$$A_{a,d} = \{\ldots, a - 2d, a - d, a, a + d, a + 2d, \ldots\}, \ a \in \mathbb{Z}, \ d > 0,$$

then $A_{a,d}$ is open by hypothesis, but it is also closed because it is the complement of the open set $A_{a+1,d} \cup A_{a+2,d} \cup \cdots \cup A_{a+d-1,d}$. Hence $\mathbb{Z} \setminus A_{a,d}$ is open.

Now let us assume that only finitely many primes exist, say $p_1, p_2, \ldots, p_n$. Then

$$A_{0,p_1} \cup A_{0,p_2} \cup \cdots \cup A_{0,p_n} = \mathbb{Z} \setminus \{-1, 1\}.$$

This union of open sets is the complement of the open set

$$(\mathbb{Z} \setminus A_{0,p_1}) \cap (\mathbb{Z} \setminus A_{0,p_2}) \cap \cdots \cap (\mathbb{Z} \setminus A_{0,p_n});$$

hence it is closed. The complement of this closed set, namely $\{-1, 1\}$, must therefore be open. We have reached a contradiction because this set is neither empty nor infinite. Hence our assumption was false, and so there are infinitely many primes. $\square$

Let us continue with the examples.

Example. Prove that for all positive integers n, the number

$$3^{3^n} + 1$$

is the product of at least $2n + 1$ not necessarily distinct primes.

Solution. We induct on n. The statement is clearly true if $n = 1$. Because

$$3^{3^{n+1}} + 1 = (3^{3^n} + 1)(3^{2 \cdot 3^n} - 3^{3^n} + 1),$$

it suffices to prove that $3^{2 \cdot 3^n} - 3^{3^n} + 1$ is composite for all $n \geq 1$. But this follows from the fact that

$$3^{2 \cdot 3^n} - 3^{3^n} + 1 = (3^{3^n} + 1)^2 - 3 \cdot 3^{3^n} = (3^{3^n} + 1)^2 - \left(3^{\frac{3^n+1}{2}}\right)^2$$

is the product of two integers greater than 1, namely,

$$3^{3^n} + 1 - 3^{\frac{3^n+1}{2}} \quad \text{and} \quad 3^{3^n} + 1 - 3^{\frac{3^n+1}{2}}.$$

This completes the induction. □

We proceed with a problem from the 35th International Mathematical Olympiad, 1994, followed by several others that are left to the reader.

Example. Prove that there exists a set A of positive integers with the property that for any infinite set S of primes, there exist two positive integers $m \in A$ and $n \notin A$ each of which is a product of k distinct elements of S for some $k \geq 2$.

Solution. The proof is constructive. Let $p_1 < p_2 < \cdots < p_n < \cdots$ be the increasing sequence of all prime numbers. Define A to be the set of numbers of the form $p_{i_1} p_{i_2} \cdots p_{i_k}$, where $i_1 < i_2 < \cdots < i_k$ and $k = p_{i_1}$. For example, $3 \cdot 5 \cdot 7 \in A$ and $5 \cdot 7 \cdot 11 \cdot 13 \cdot 17 \in A$, but $5 \cdot 7 \notin A$.

Let us show that A satisfies the desired condition. Consider an infinite set of prime numbers, say $q_1 < q_2 < \cdots < q_n < \cdots$ Take $m = q_2 q_3 \cdots q_{q_2}$ and $n = q_3 q_4 \cdots q_{q_2+1}$. Then $m \in A$, while $n \notin A$ because $q_2 \geq 3$ and so $q_2 + 1 \neq q_3$. □

854. Prove that there are infinitely many prime numbers of the form $4m + 3$, where $m \geq 0$ is an integer.

855. Let k be a positive integer such that the number $p = 3k + 1$ is prime and let

$$\frac{1}{1 \cdot 2} + \frac{1}{3 \cdot 4} + \cdots + \frac{1}{(2k-1)2k} = \frac{m}{n}$$

for some coprime positive integers m and n. Prove that p divides m.

856. Solve in positive integers the equation

$$x^{x+y} = y^{y-x}.$$

857. Show that each positive integer can be written as the difference of two positive integers having the same number of prime factors.

858. Find all composite positive integers n for which it is possible to arrange all divisors of n that are greater than 1 in a circle such that no two adjacent divisors are relatively prime.

859. Is it possible to place 1995 different positive integers around a circle so that for any two adjacent numbers, the ratio of the greater to the smaller is a prime?

860. Let p be a prime number. Prove that there are infinitely many multiples of p whose last ten digits are all distinct.

861. Let A be the set of positive integers representable in the form $a^2 + 2b^2$ for integers a, b with $b \neq 0$. Show that if $p^2 \in A$ for a prime p, then $p \in A$.

862. The positive divisors of an integer $n > 1$ are $1 = d_1 < d_2 < \cdots < d_k = n$. Let $s = d_1 d_2 + d_2 d_3 + \cdots + d_{k-1} d_k$. Prove that $s < n^2$ and find all n for which s divides n^2.

863. Prove that there exist functions $f, g : \{0, 1, 2, \ldots\} \times \{0, 1, 2, \ldots\} \to \{0, 1, 2, \ldots\}$ with the property that an odd number $n > 1$ is prime if and only if there do not exist nonnegative integers a and b such that $n = f(a, b) - g(a, b)$.

864. Let $n \geq 2$ be an integer. Prove that if $k^2 + k + n$ is a prime number for all $0 \leq k \leq \sqrt{\frac{n}{3}}$, then $k^2 + k + n$ is a prime number for all $0 \leq k \leq n - 2$.

The following formula is sometimes attributed to Legendre.

Polignac's formula. *If p is a prime number and n a positive integer, then the exponent of p in $n!$ is given by*

$$\left\lfloor \frac{n}{p} \right\rfloor + \left\lfloor \frac{n}{p^2} \right\rfloor + \left\lfloor \frac{n}{p^3} \right\rfloor + \cdots$$

Proof. Each multiple of p between 1 and n contributes a factor of p to $n!$. There are $\lfloor n/p \rfloor$ such factors. But the multiples of p^2 contribute yet another factor of p, so one should add $\lfloor n/p^2 \rfloor$. And then come the multiples of p^3 and so on. $\square$

Example. Let m be an integer greater than 1. Prove that the product of m consecutive terms in an arithmetic progression is divisible by $m!$ if the ratio of the progression is coprime to m.

Solution. Let p be a prime that divides $n!$. The exponent of p in $n!$ is given by Polignac's formula. On the other hand, in the product $a(a+r)(a+2r) \cdots (a+(m-1)r)$ of m consecutive terms in a progression of ratio r, with $\gcd(r, m) = 1$, at least terms are divisible by p^i. It follows that the power of p in this product is greater than or equal to the power of p in $m!$. Because this holds true for any prime factor in $m!$, the conclusion follows. $\square$

All problems below are based on Polignac's formula.

865. Find all positive integers n such that $n!$ ends in exactly 1000 zeros.

866. Prove that $n!$ is not divisible by 2^n for any positive integer n.

867. Show that for each positive integer n,

$$n! = \prod_{i=1}^{n} \text{lcm}(1, 2, \ldots, \lfloor n/i \rfloor),$$

where lcm denotes the least common multiple.

868. Prove that the expression

$$\frac{\gcd(m, n)}{n} \binom{n}{m}$$

is an integer for all pairs of integers $n \geq m \geq 1$.

869. Let k and n be integers with $0 \leq k \leq n^2/4$. Assume that k has no prime divisor greater than n. Prove that $n!$ is divisible by k.

5.2.3 Modular Arithmetic

A positive integer n partitions the set of integers $\mathbb{Z}$ into n equivalence classes by the remainders obtained on dividing by n. The remainders are called residues modulo n. We denote by $\mathbb{Z}_n = \{0, 1, \ldots, n-1\}$ the set of equivalence classes, indexed by their residues. Two numbers a and b are said to be congruent modulo n, which is written $a \equiv b \pmod{n}$, if they give the same remainder when divided by n, that is, if $a - b$ is divisible by n.

The ring structure of $\mathbb{Z}$ induces a ring structure on $\mathbb{Z}_n$. The latter ring is more interesting, since it has zero divisors whenever n is composite, and it has other invertible elements besides ± 1. To make this precise, for any divisor d of n the product of d and n/d is zero. On the other hand, the fundamental theorem of arithmetic, which states that whenever m and n are coprime there exist integers a and b such that $am - bn = 1$, implies that any number coprime to n has a multiplicative inverse modulo n. For a prime p, every nonzero element in $\mathbb{Z}_p$ has an inverse modulo p. This means that $\mathbb{Z}_p$ is a field. We also point out that the set of invertible elements in $\mathbb{Z}_n$ is closed under multiplication; it is an Abelian group.

A well-known property that will be used in some of the problems below is that modulo 9, a number is congruent to the sum of its digits. This is because the difference of the number and the sum of its digits is equal to 9 times the tens digit plus 99 times the hundreds digit plus 999 times the thousands digit, and so on. Here is an elementary application of this fact.

Example. The number 2^{29} has 9 distinct digits. Without using a calculator, tell which digit is missing.

Solution. As we have just observed, a number is congruent to the sum of its digits modulo 9. Note that $0 + 1 + 2 + \cdots + 9 = 45$, which is divisible by 9. On the other hand,

$$2^{29} \equiv 2^2(-1)^p \equiv -4 \pmod{9}.$$

So 2^{29} is off by 4 from a multiple of 9. The missing digit is 4. $\square$

We continue with a property of the harmonic series discovered by C. Pinzka.

Example. Let $p > 3$ be a prime number, and let

$$\frac{r}{ps} = 1 + \frac{1}{2} + \frac{1}{3} + \cdots + \frac{1}{p},$$

the sum of the first p terms of the harmonic series. Prove that p^3 divides $r - s$.

Solution. The sum of the first p terms of the harmonic series can be written as

$$\frac{\dfrac{p!}{1} + \dfrac{p!}{2} + \cdots + \dfrac{p!}{p}}{p!}.$$

Because the denominator is $p!$ and the numerator is not divisible by p, any common prime divisor of the numerator and the denominator is less than p. Thus it suffices to prove the property for $r = \frac{p!}{1} + \frac{p!}{2} + \cdots + \frac{p!}{p}$ and $s = (p-1)!$. Note that

$$r - s = p\left(\frac{(p-1)!}{1} + \frac{(p-1)!}{2} + \cdots + \frac{(p-1)!}{p-1}\right).$$

We are left with showing that

$$\frac{(p-1)!}{1} + \frac{(p-1)!}{2} + \cdots + \frac{(p-1)!}{p-1}$$

is divisible by p^2. This sum is equal to

$$\sum_{k-1}^{\frac{p-1}{2}}(k+p-k)\frac{(p-1)!}{k(p-k)} = p\sum_{k=1}^{\frac{p-1}{2}}\frac{(p-1)!}{k(p-k)}.$$

So let us show that

$$\sum_{k-1}^{\frac{p-1}{2}}\frac{(p-1)!}{k(p-k)}$$

is an integer divisible by p. Note that if k^{-1} denotes the inverse of k modulo p, then $p-k^{-1}$ is the inverse of $p-k$ modulo p. Hence the residue classes of $[k(p-k)]^{-1}$ represent just a permutation of the residue classes of $k(p-k)$, $k = 1, 2, \ldots, \frac{p-1}{2}$. Using this fact, we have

$$\sum_{k-1}^{\frac{p-1}{2}}\frac{(p-1)!}{k(p-k)} \equiv (p-1)!\sum_{k-1}^{\frac{p-1}{2}}[k(p-k)]^{-1} \equiv (p-1)!\sum_{k-1}^{\frac{p-1}{2}}k(p-k)$$

$$\equiv -(p-1)!\sum_{k-1}^{\frac{p-1}{2}}k^2 = -(p-1)!\frac{\frac{p-1}{2}\cdot\frac{p+1}{2}\cdot p}{6} \equiv 0 \pmod{p}.$$

This completes the proof.   □

We left the better problems as exercises.

870. Prove that among any three distinct integers we can find two, say a and b, such that the number $a^3b - ab^3$ is a multiple of 10.

871. Show that the number 2002^{2002} can be written as the sum of four perfect cubes, but not as the sum of three perfect cubes.

872. The last four digits of a perfect square are equal. Prove that they are all equal to zero.

873. Solve in positive integers the equation

$$2^x \cdot 3^y = 1 + 5^z.$$

874. Define the sequence $(a_n)_n$ recursively by $a_1 = 2$, $a_2 = 5$, and

$$a_{n+1} = (2 - n^2)a_n + (2 + n^2)a_{n-1} \text{ for } n \geq 2.$$

Do there exist indices p, q, r such that $a_p \cdot a_q = a_r$?

875. For some integer $k > 0$, assume that an arithmetic progression $an + b$, $n \geq 1$, with a and b positive integers, contains the kth power of an integer. Prove that for any integer $m > 0$ there exist an infinite number of values of n for which $an + b$ is the sum of m kth powers of nonzero integers.

876. Given a positive integer $n > 1000$, add the residues of 2^n modulo each of the numbers $1, 2, 3, \ldots, n$. Prove that this sum is greater than $2n$.

877. Prove that if $n \geq 3$ prime numbers form an arithmetic progression, then the common difference of the progression is divisible by any prime number $p < n$.

878. Let $P(x) = a_m x^m + a_{m-1} x^{m-1} + \cdots + a_0$ and $Q(x) = b_n x^n + b_{n-1} x^{n-1} + \cdots + b_0$ be two polynomials with each coefficient a_i and b_i equal to either 1 or 2002. Assuming that $P(x)$ divides $Q(x)$, show that $m + 1$ is a divisor of $n + 1$.

879. Prove that if n is a positive integer that is divisible by at least two primes, then there exists an n-gon with all angles equal and with side lengths the numbers $1, 2, \ldots, n$ in some order.

880. Find all prime numbers p having the property that when divided by every prime number $q < p$ yield a remainder that is a square-free integer.

5.2.4 Fermat's Little Theorem

A useful tool for solving problems about prime numbers is a theorem due to P. Fermat.

Fermat's little theorem. *Let p be a prime number, and n a positive integer. Then*

$$n^p - n \equiv 0 \pmod{p}.$$

Proof. We give a geometric proof discovered by J. Pedersen. Consider the set M of all possible colorings of the vertices of a regular p-gon by n colors (see Figure 38). This set has np elements. The group $\mathbb{Z}_p$ acts on this set by rotations of angles $\frac{2k\pi}{p}$, $k = 0, 1, \ldots, p - 1$.

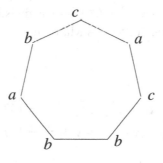

Figure 38

Consider the quotient space $M/\mathbb{Z}_p$ obtained by identifying colorings that become the same through a rotation. We want to count the number of elements of $M/\mathbb{Z}_p$. For that we need to understand the orbits of the action of the group, i.e., the equivalence classes of rotations under this identification.

The orbit of a monochromatic coloring has just one element: the coloring itself. There are n such orbits.

What if the coloring is not monochromatic? We claim that in this case its orbit has exactly p elements. Here is the place where the fact that p is prime comes into play. The additive group $\mathbb{Z}_p$ of residues modulo p is generated by any of its nonzero elements. Hence if the coloring coincided with itself under a rotation of angle $2k\pi/p$ for some $0 < k < p$, then it would coincide with itself under multiples of this rotation, hence under all rotations in $\mathbb{Z}_p$. But this is not possible, unless the coloring is monochromatic. This proves that rotations produce distinct colorings, so the orbit has p elements. We deduce that the remaining $n^p - n$ elements of M are grouped in (disjoint) equivalence classes each containing p elements. The counting of orbits gives

$$|M/\mathbb{Z}_p| = n + \frac{n^p - n}{p},$$

which shows that $(n^p - n)/p$ must be an integer. The theorem is proved.   □

In particular, if n and p are coprime, then $n^{p-1} - 1$ is divisible by p. However, this result alone cannot be used as a primality test for p. For example, L. Euler found that 341 divides $2^{340} - 1$, while $341 = 31 \times 11$. So the converse of Fermat's little theorem fails.

We illustrate the use of Fermat's little theorem with a problem from the 46th International Mathematical Olympiad, 2005.

Example. Show that for every prime p there is an integer n such that $2^n + 3^n + 6^n - 1$ is divisible by p.

Solution. The property is true for $p = 2$ and $p = 3$, since $2^2 + 3^2 + 6^2 - 1 = 48$. Let p be a prime greater than 3. By Fermat's little theorem, 2^{p-1}, 3^{p-1}, and 6^{p-1} are all congruent to 1 modulo p. Hence

$$3 \cdot 2^{p-1} + 2 \cdot 3^{p-1} + 6^{p-1} \equiv 3 + 2 + 1 = 6 \quad (\text{mod } p).$$

It follows that

$$6 \cdot 2^{p-2} + 6 \cdot 3^{p-2} + 6 \cdot 6^{p-2} \equiv 6 \quad (\text{mod } p).$$

Dividing by 6, we find that $2^{p-2} + 3^{p-2} + 6^{p-2} - 1$ is divisible by p, and we are done.   □

And here is a problem from the 2005 USA Mathematical Olympiad, proposed by the first author of the book.[2]

Example. Prove that the system

$$x^6 + x^3 + x^3 y + y = 147^{157},$$
$$x^3 + x^3 y + y^2 + y + z^9 = 157^{147}$$

has no solutions in integers x, y, and z.

Solution. Add the two equations, then add 1 to each side to obtain the Diophantine equation

$$(x^3 + y + 1)^2 + z^9 = 146^{157} + 157^{147} + 1.$$

---

[2]The statement was improved by R. Stong and E. Johnston to prevent a simpler solution.

The right-hand side is rather large, and it is natural to reduce modulo some number. And since the left-hand side is a sum of a square and a ninth power, it is natural to reduce modulo 19 because $2 \times 9 + 1 = 19$. By Fermat's little theorem, $a^{18} \equiv 1$ (mod 19) whenever a is not a multiple of 19, and so the order of a square is either 1, 3, or 9, while the order of a ninth-power is either 1 or 2.

Computed by hand, the quadratic residues mod 19 are $-8, -3, -2, 0, 1, 4, 5, 6, 7, 9$, while the residues of ninth powers are $-1, 0, 1$. Also, applying Fermat's little theorem we see that

$$147^{157} + 157^{147} + 1 \equiv 14^{13} + 5^3 + 1 \equiv 14 \quad (\text{mod } 19).$$

An easy verification shows that 14 cannot be obtained as a sum of a quadratic residue and a ninth-power residue. Thus the original system has no solution in integers x, y, and z.

A different solution is possible using reduction modulo 13. Fermat's little theorem implies $a^{12} \equiv 1$ (mod 13) when a is not a multiple of 13.

We start by producing the same Diophantine equation. Applying Fermat's little theorem, we can reduce the right-hand side modulo 13. We find that

$$147^{157} + 157^{147} + 1 \equiv 4^1 + 1^2 + 1 = 6 \quad (\text{mod } 13).$$

The cubes modulo 13 are $0, \pm 1$, and ± 5. Writing the first equation of the original system as

$$(x^3 + 1)(x^3 + y) = 4 \quad (\text{mod } 13),$$

it follows that $x^3 + y$ must be congruent to 4, 2, 5, or -1. Hence

$$(x^3 + y + 1)^2 \equiv 12, \ 9, \ 10 \quad \text{or} \quad 0 \quad (\text{mod } 13).$$

Note also that z^9 is a cube; hence z^9 must be 0, 1, 5, 8, or 12 modulo 13. It is easy to check that 6 (mod 13) cannot be obtained by adding one of 0, 9, 10, 12 to one of 0, 1, 5, 8, 12. As a remark, the second solution also works if z^9 is replaced by z^3. $\quad\square$

When solving the following problems, think that "work done with passion brings results" (Virgil).

881. Show that if n has $p - 1$ digits all equal to 1, where p is a prime not equal to 2, 3, or 5, then n is divisible by p.

882. Prove that for any prime $p > 17$, the number $p^{32} - 1$ is divisible by 16320.

883. Let p be an odd prime number. Show that if the equation $x^2 \equiv a$ (mod p) has a solution, then $a^{\frac{p-1}{2}} \equiv 1$ (mod p). Conclude that there are infinitely many primes of the form $4m + 1$.

884. Prove that the equation $x^2 = y^3 + 7$ has no integer solutions.

885. Let $n > 1$ be a positive integer. Prove that the equation $(x + 1)^n - x^n = ny$ has no positive integer solutions.

886. Prove that the sequence $2^n - 3$, $n \geq 1$, contains an infinite subsequence whose terms are pairwise relatively prime.

887. Let $(x_n)_n$ be a sequence of positive integers satisfying the recurrence relation $x_{n+1} = 5x_n - 6x_{n-1}$. Prove that infinitely many terms of the sequence are composite.

888. Let $f(x_1, x_2, \ldots, x_n)$ be a polynomial with integer coefficients of total degree less than n. Show that the number of ordered n-tuples $(x_1, x_2, \ldots, x_n)$ with $0 \le x_i \le 12$ such that $f(x_1, x_2, \ldots, x_n) \equiv 0 \pmod{13}$ is divisible by 13.

889. Determine all integers a such that $a^k + 1$ is divisible by 12321 for some appropriately chosen positive integer $k > 1$.

890. Find the greatest common divisor of the numbers
$$2^{561} - 2, \quad 3^{561} - 3, \quad \ldots, \quad 561^{561} - 561.$$

5.2.5 Wilson's Theorem

Another result about prime numbers is known as Wilson's theorem.

Wilson's theorem. *For every prime p, the number $(p-1)! + 1$ is divisible by p.*

Proof. We group the residue classes $1, 2, \ldots, p-1$ in pairs (a, b) such that $ab \equiv 1 \pmod{p}$. Let us see when $a = b$ in such a pair. The congruence $a^2 \equiv 1 \pmod{p}$ is equivalent to the fact that $a^2 - 1 = (a-1)(a+1)$ is divisible by p. This happens only when $a = 1$ or $a = p - 1$. For all other residue classes the pairs contain distinct elements. So in the product $2 \cdot 3 \cdots (p-2)$ the factors can be paired such that the product of the numbers in each pair is congruent to 1. Therefore,
$$1 \cdot 2 \cdots (p-2)(p-1) \equiv 1 \cdot (p-1) \equiv -1 \pmod{p}.$$
The theorem is proved. $\qquad\square$

The converse is also true, since n must divide $(n-1)!$ for composite n. And now an application.

Example. Let p be an odd prime. Prove that
$$1^2 \cdot 3^2 \cdots (p-2)^2 \equiv (-1)^{\frac{p+1}{2}} \pmod{p}$$
and
$$2^2 \cdot 4^2 \cdots (p-1)^2 \equiv (-1)^{\frac{p+1}{2}} \pmod{p}.$$

Solution. By Wilson's theorem,
$$(1 \cdot 3 \cdots (p-2))(2 \cdot 4 \cdots (p-1)) \equiv -1 \pmod{p}.$$
On the other hand,
$$1 \equiv -(p-1) \pmod{p}, \quad 3 \equiv -(p-3) \pmod{p}, \ldots, p-2 \equiv -(p-(p-2)) \pmod{p}.$$
Therefore,
$$1 \cdot 3 \cdots (p-2) \equiv (-1)^{\frac{p-1}{2}} (2 \cdot 4 \cdots (p-1)) \pmod{p}.$$
Multiplying the two congruences and canceling out the product $2 \cdot 4 \cdots (p-1)$, we obtain the first congruence from the statement. Switching the sides in the second and multiplying the congruences again, we obtain the second congruence from the statement. $\qquad\square$

Here are more examples.

891. For each positive integer n, find the greatest common divisor of $n! + 1$ and $(n + 1)!$.

892. Prove that there are no positive integers n such that the set $\{n, n + 1, n + 2, n + 3, n + 4, n + 5\}$ can be partitioned into two sets with the product of the elements of one set equal to the product of the elements of the other set.

893. Let p be an odd prime. Show that if the equation $x^2 \equiv a \pmod{p}$ has no solution then $a^{\frac{p-1}{2}} \equiv -1 \pmod{p}$.

894. Let p be an odd prime number. Show that the equation $x^2 \equiv -1 \pmod{p}$ has a solution if and only if $p \equiv 1 \pmod{4}$.

895. Let p be a prime number and n an integer with $1 \le n \le p$. Prove that

$$(p - n)!(n - 1)! \equiv (-1)^n \pmod{p}.$$

896. Let p be an odd prime and $a_1, a_2, \ldots, a_p$ an arithmetic progression whose common difference is not divisible by p. Prove that there exists an index i such that the number $a_1 a_2 \cdots a_p + a_i$ is divisible by p^2.

5.2.6 Euler's Totient Function

Euler's totient function associates to a positive integer n the number $\phi(n)$ of positive integers less than or equal to n that are coprime to n. It has a simple formula in terms of the prime factorization of n.

Proposition. *If the distinct prime factors of n are $p_1, p_2, \ldots, p_k$, then*

$$\phi(n) = n \left(1 - \frac{1}{p_1}\right) \left(1 - \frac{1}{p_2}\right) \cdots \left(1 - \frac{1}{p_k}\right).$$

Proof. This is just an easy application of the inclusion-exclusion principle (see Section 6.4.4). From the n numbers between 1 and n, we eliminate the n/p_i numbers that are divisible by p_i, for each $1 \le i \le n$. We are left with

$$n - n \left(\frac{1}{p_1} + \frac{1}{p_2} + \cdots + \frac{1}{p_k}\right)$$

numbers. But those divisible by both p_i and p_j have been subtracted twice, so we have to add them back, obtaining

$$n - n \left(\frac{1}{p_1} + \frac{1}{p_2} + \cdots + \frac{1}{p_k}\right) + n \left(\frac{1}{p_1 p_2} + \frac{1}{p_1 p_3} + \cdots + \frac{1}{p_{k-1} p_k}\right).$$

Again, we see that the numbers divisible by p_i, p_j, and p_l have been subtracted and then added back, so we need to subtract these once more. Repeating the argument, we obtain in the end

$$n - n \left(\frac{1}{p_1} + \frac{1}{p_2} + \cdots + \frac{1}{p_k}\right) + n \left(\frac{1}{p_1 p_2} + \frac{1}{p_1 p_3} + \cdots + \frac{1}{p_{k-1} p_k}\right) - \cdots \pm \frac{n}{p_1 p_2 \cdots p_k}.$$

Factoring this, we obtain the formula from the statement. $\square$

In particular, n is prime if and only if $\phi(n) = n - 1$, and if $n = p_1 p_2 \cdots p_k$, where p_i are distinct primes, $1 \le i \le k$, then $\phi(n) = (p_1 - 1)(p_2 - 1) \cdots (p_n - 1)$. Also, if m and n are coprime, then $\phi(mn) = \phi(m)\phi(n)$.

Fermat's little theorem admits the following generalization.

Euler's theorem. *Let $n > 1$ be an integer and a an integer coprime to n. Then*

$$a^{\phi(n)} \equiv 1 \pmod{n}.$$

Proof. The group of units $\mathbb{Z}_n^*$ in the ring $\mathbb{Z}_n$ consists of the residue classes coprime to n. Its order is $\phi(n)$. By the Lagrange theorem, the order of an element divides the order of the group. Hence the conclusion.

Here is a more elementary argument. Consider the set $S = \{a_1, a_2, \ldots, a_{\phi(n)}\}$ of all residue classes modulo n that are coprime to n. Because $\gcd(a, n) = 1$, it follows that, modulo n, $aa_1, aa_2, \ldots, aa_{\phi(n)}$ is a permutation of $a_1, a_2, \ldots, a_{\phi(n)}$. Then

$$(aa_1)(aa_2) \cdots (aa_{\phi(n)}) \equiv a_1 a_2 \cdots a_{\phi(n)} \pmod{n}.$$

Since $\gcd(a_k, n) = 1$, for $k = 1, 2, \ldots, \phi(n)$, we can divide both sides by $a_1 a_2 \ldots a_{\phi(n)}$ to obtain $a^{\phi(n)} \equiv 1 \pmod{n}$, as desired. $\square$

We apply Euler's theorem to a problem by I. Cucurezeanu.

Example. Let n be an even positive integer. Prove that $n^2 - 1$ divides $2^{n!} - 1$.

Solution. Let $n = m - 1$, so that m is odd. We must show that $m(m - 2)$ divides $2^{(m-1)!} - 1$. Because $\phi(m) < m$, $\phi(m)$ divides $(m - 1)!$, so $2^{\phi(m)} - 1$ divides $2^{(m-1)!} - 1$. On the other hand Euler's theorem implies that m divides $2^{\phi(m)} - 1$. Therefore, m divides $2^{(m-1)!} - 1$. Arguing similarly for $m - 2$, we see that $m - 2$ divides $2^{(m-1)!} - 1$ as well. The numbers m and $m - 2$ are relatively prime, so $m(m - 2)$ divides $2^{(m-1)!} - 1$, as desired. $\square$

A second example comes from the 1997 Romanian Mathematical Olympiad.

Example. Let $a > 1$ be an integer. Show that the set

$$S = \{a^2 + a - 1, a^3 + a^2 - 1, a^4 + a^3 - 1, \ldots\}$$

contains an infinite subset whose elements are pairwise coprime.

Solution. We show that any subset of S having n elements that are pairwise coprime can be extended to a set with $n + 1$ elements. Indeed, if N is the product of the elements of the subset, then since the elements of S are coprime to a, so must be N. By Euler's theorem,

$$a^{\phi(N)+1} + a^{\phi(N)} - 1 \equiv a + 1 - 1 \equiv a \pmod{N}.$$

It follows that $a^{\phi(N)+1} + a^{\phi(N)} - 1$ is coprime to N and can be added to S. We are done. $\square$

We now challenge you with the following problems.

897. Prove that for any positive integer n,

$$\sum_{k \mid n} \phi(k) = n.$$

Here $k \mid n$ means k divides n.

898. Prove that for any positive integer n other than 2 or 6,

$$\phi(n) \geq \sqrt{n}.$$

899. Prove that there are infinitely many positive integers n such that $(\phi(n))^2 + n^2$ is a perfect square.

900. Prove that there are infinitely many even positive integers m for which the equation $\phi(n) = m$ has no solutions.

901. Prove that for every positive integer s there exists a positive integer n divisible by s and with the sum of the digits equal to s.

902. Prove that the equation

$$2^x + 3 = z^3$$

does not admit positive integer solutions.

903. Prove for every positive integer n the identity

$$\phi(1)\left\lfloor\frac{n}{1}\right\rfloor + \phi(2)\left\lfloor\frac{n}{2}\right\rfloor + \phi(3)\left\lfloor\frac{n}{3}\right\rfloor + \cdots + \phi(n)\left\lfloor\frac{n}{n}\right\rfloor = \frac{n(n+1)}{2}.$$

904. Given the nonzero integers a and d, show that the sequence

$$a, a+d, a+2d, \ldots, a+nd, \ldots$$

contains infinitely many terms that have the same prime factors.

Euler's theorem is widely used in cryptography. The encryption scheme used nowadays, called the RSA algorithm, works as follows:

A merchant wants to obtain the credit card number of a customer over the Internet. The information traveling between the two can be viewed by anyone. The merchant is in possession of two large prime numbers p and q. It transmits to the customer the product $n = pq$ and a positive integer k coprime to $\phi(n) = (p-1)(q-1)$. The customer raises the credit card number α to the kth power, then reduces it modulo n and transmits the answer β to the merchant. Using the Euclidean algorithm for the greatest common divisor, the merchant determines positive integers m and a satisfying

$$mk - a(p-1)(q-1) = 1.$$

Then he computes the residue of β^m modulo n. By Euler's theorem,

$$\beta^m \equiv \alpha^{mk} = \alpha^{a(p-1)(q-1)+1} = (\alpha^{(p-1)(q-1)})^a \cdot \alpha = (\alpha^{\phi(n)})^a \cdot \alpha \equiv \alpha \pmod{n}.$$

For n sufficiently large, the residue class of α modulo n is α itself. The merchant was able to retrieve the credit card number.

As of this date there is no known algorithm for factoring numbers in polynomial time, while large primes can be found relatively quickly, and for this reason an eavesdropper cannot determine p and q from n in a reasonable amount of time, and hence cannot break the encryption.

905. Devise a scheme by which a bank can transmit to its customers secure information over the Internet. Only the bank (and not the customers) is in the possession of the secret prime numbers p and q.

906. A group of United Nations experts is investigating the nuclear program of a country. While they operate in that country, their findings should be handed over to the Ministry of Internal Affairs of the country, which verifies the document for leaks of classified information, then submits it to the United Nations. Devise a scheme by which the country can read the document but cannot modify its contents without destroying the information.

5.2.7 The Chinese Remainder Theorem

Mentioned for the first time in a fourth-century book of Sun Tsu Suan-Ching, this result can be stated as follows.

The Chinese remainder theorem. *Let $m_1, m_2, \ldots, m_k$ be pairwise coprime positive integers greater than 1. Then for any integers $a_1, a_2, \ldots, a_k$, the system of congruences*

$$x \equiv a_1 \pmod{m_1}, \ x \equiv a_2 \pmod{m_2}, \ldots, \ x \equiv a_k \pmod{m_k}$$

has solutions, and any two such solutions are congruent modulo $m = m_1 m_2 \cdots m_k$.

Proof. For any $1 \leq j \leq k$, the number m/m_j is coprime to m_j and hence invertible with respect to m_j. Let b_j be the inverse. Then

$$x_0 = \frac{m}{m_1} b_1 a_1 + \frac{m}{m_2} b_2 a_2 + \cdots + \frac{m}{m_k} b_k a_k$$

is a solution to the system. For any other solution x, the difference $x - x_0$ is divisible by m. It follows that the general solution is of the form $x_0 + mt$, with t an integer. □

We illustrate the use of the Chinese remainder theorem with an example from the classic book of W. Sierpiński, 250 *Problems in Elementary Number Theory* (Państwowe Wydawnictwo Naukowe, Warszawa, 1970).

Example. Prove that the system of Diophantine equations

$$x_1^2 + x_2^2 + x_3^2 + x_4^2 = y^5,$$
$$x_1^3 + x_2^3 + x_3^3 + x_4^3 = y^2,$$
$$x_1^5 + x_2^5 + x_3^5 + x_4^5 = y^3$$

has infinitely many solutions.

Solution. Let $a = 1^2 + 2^2 + 3^2 + 4^2$, $b = 1^3 + 2^3 + 3^3 + 4^3$, $c = 1^5 + 2^5 + 3^5 + 4^5$. We look for solutions of the form $x_1 = a^m b^n c^p$, $x_2 = 2a^m b^n c^p$, $x_3 = 3a^m b^n c^p$, $x_3 = 4a^m b^n c^p$. These satisfy

$$x_1^2 + x_2^2 + x_3^2 + x_4^2 = a^{2m+1} b^{2n} c^{2p},$$
$$x_1^3 + x_2^3 + x_3^3 + x_4^3 = a^{3m} b^{3n+1} c^{3p},$$
$$x_1^5 + x_2^5 + x_3^5 + x_4^5 = a^{5m} b^{5n} c^{5p+1}.$$

We would like the right-hand sides to be a fifth, second, and third power, respectively. Reformulating, we want to show that there exist infinitely many m, n, p such that

$$2m + 1 \equiv 2n \equiv 2p \equiv 0 \pmod{5},$$
$$3m \equiv 3n + 1 \equiv 3p \equiv 0 \pmod{2},$$
$$5m \equiv 5n \equiv 5p + 1 \equiv 0 \pmod{3}.$$

But this follows from the Chinese remainder theorem, and we are done.   □

907. An old woman went to the market and a horse stepped on her basket and smashed her eggs. The rider offered to pay for the eggs and asked her how many there were. She did not remember the exact number, but when she had taken them two at a time there was one egg left, and the same happened when she took three, four, five, and six at a time. But when she took them seven at a time, they came out even. What is the smallest number of eggs she could have had?

908. Prove that for every n, there exist n consecutive integers each of which is divisible by at least two different primes.

909. Let $P(x)$ be a polynomial with integer coefficients. For any positive integer m, let $N(m)$ denote the number of solutions to the equation $P(x) \equiv 0 \pmod{m}$. Show that if m_1 and m_2 are coprime integers, then $N(m_1 m_2) = N(m_1)N(m_2)$.

910. Alice and Bob play a game in which they take turns removing stones from a heap that initially has n stones. The number of stones removed at each turn must be one less than a prime number. The winner is the player who takes the last stone. Alice plays first. Prove that there are infinitely many n such that Bob has a winning strategy. (For example, if $n = 17$, then Alice might take 6 leaving 11; then Bob might take 1 leaving 10; then Alice can take the remaining stones to win.)

911. Show that there exists an increasing sequence $(a_n)_{n \geq 1}$ of positive integers such that for any $k \geq 0$, the sequence $k + a_n$, $n \geq 1$, contains only finitely many primes.

912. Is there a sequence of positive integers in which every positive integer occurs exactly once and for every $k = 1, 2, 3, \ldots$ the sum of the first k terms is divisible by k?

913. Prove that there exists a positive integer k such that $k \cdot 2^n + 1$ is composite for every positive integer n.

914. Let a and b be two positive integers such that for any positive integer n, $a^n + n$ divides $b^n + n$. Prove that $a = b$.

915. A lattice point $(x, y) \in \mathbb{Z}^2$ is visible from the origin if x and y are coprime. Prove that for any positive integer n there exists a lattice point (a, b) whose distance from every visible point is greater than n.

916. A set of positive integers is called fragrant if it contains at least two elements and each of its elements has a prime factor in common with at least one of the other elements.

Let $P(n) = n^2 + n + 1$. What is the least possible value of the positive integer b such that there exists a nonnegative integer a for which the set

$$\{P(a+1), P(a+2), \cdots, P(a+b)\}$$

is fragrant?

5.2.8 Quadratic Integer Rings

An algebraic integer is the root of a monic polynomial with integer coefficients. A quadratic integer ring is the ring of algebraic integers contained in $\mathbb{Q}[\sqrt{d}]$, for some square free integer d. Those algebraic integers are necessarily roots of quadratic equations of the form $x^2 + ax + b = 0$, $a, b \in \mathbb{Z}$. One has the following result:

Theorem. *If $d \equiv 2, 3 \pmod{4}$, then the ring of algebraic integers of $\mathbb{Q}[\sqrt{d}]$ is $\mathbb{Z}[\sqrt{d}]$. If $d \equiv 1 \pmod{4}$, then the ring of algebraic integers of $\mathbb{Q}[\sqrt{d}]$ is $\mathbb{Z}[(-1 + \sqrt{d})/2]$.*

Thus, every algebraic integer is of the form $a + b\sqrt{d}$, where a, b are both integers, or, only if $d \equiv 1 \pmod{4}$, both halves of odd integers. An algebraic integer $x = a + b\sqrt{d}$ has a conjugate $\bar{x} = a - b\sqrt{d}$. One defines the norm of an algebraic integer to be $N(x) = x\bar{x}$.

Proposition. *The norm is integer valued and multiplicative.*

Before proceeding with examples, we recall some terminology from ring theory. An integral domain is a commutative ring with an identity and without zero divisors.

An element u of an integral domain is called a unit if there is an element u' such that $uu' = 1$. An element u of a quadratic integer ring is a unit if and only if $N(u) = \pm 1$. An element p is called irreducible if $a|p$ implies a is a unit or p is the product of a and a unit. An nonzero element p that is not a unit is called prime if $p|ab$ implies $p|a$ or $p|b$.

A unique factorization domain is an integral domain in which every nonzero, nonunit element can be written as a product of primes, uniquely up to order and multiplication by units.

A fundamental question is the theory of quadratic integer rings is whether they are unique factorization domains. The answer is knows for $d < 0$:

Baker-Heegner-Stark theorem. *The ring of integers of $\mathbb{Q}[\sqrt{d}]$ with $d < 0$ is a unique factorization domain if and only if*

$$d \in \{-1, -2, -3, -7, -11, -19, -43, -67, -163\}.$$

Little is known for $d > 0$; finitely many integers for which the ring is a unique factorization domain are known, but it is not even known if the list is finite or infinite. A tool for proving unique factorization is by checking that the ring is Euclidean with the Euclidean function being the norm.

In what follows we will only be concerned with particular cases of the Baker-Heegner-Stark theorem, because when we have unique factorization, then quadratic integers behave exactly like the more familiar integers. On the list of rings from the Baker-Heegner-Stark theorem, the first is the ring of Gaussian integers

$$\mathbb{Z}[i] = \{a + bi \mid a, b \in \mathbb{Z}\}.$$

We illustrate how to use Gaussian integers to prove a well known result, stated by Fermat, and first proved by Euler.

Theorem. *Every prime number congruent to* 1 *modulo* 4 *can be written as the sum of two perfect squares.*

Proof. Let $p = 4k + 1$ be prime. Then by Wilson's theorem

$$-1 \equiv (p-1)! = 1 \cdot 2 \cdot 3 \cdots 4k \equiv (2k)!(2k+1) \cdots 4k$$
$$\equiv (2k)!(-1)^{2k}(p - (2k+1))(p - (2k+2)) \cdots (p - 4k) \equiv (2k)!(2k)(2k-1) \cdots 1$$
$$= ((2k)!)^2 \pmod{p}.$$

This means that p has a multiple of the form $m^2 + 1$, where m is an integer (here $m = \frac{p-1}{2}!$).

So p divides $(m + i)(m - i)$. But p does not divide $m + i$ because $p(a + bi) = m + i$ implies $pb = 1$, impossible. Similarly p does not divide $m - i$. Hence p is not prime in $\mathbb{Z}[i]$. Because the norm of p is $N(p) = p^2$, p is the product of exactly 2 factors. These factors must be one the complex conjugate of the other (otherwise the product is not real). Hence $p = (a + ib)(a - ib)$ with a and b integers. Multiplying we obtain

$$p = a^2 + b^2$$

as desired.   □

In fact as the first problem below shows, the fact that p divides $m^2 + 1$ immediately implies that it is the sum of two perfect squares regardless of whether it is prime or not. Each of the following problems uses one of the quadratic integer rings from the Baker-Heegner-Stark theorem.

917. Let m, n be integers such that m divides $n^2 + 1$. Show that m is the sum of two perfect squares.

918. Let $n > 1$ be an integer. Find all pairs of integers (x, y) such that

$$x^2 + 4 = y^3.$$

919. Let m be a positive integer such that $p = 4m - 1$ is prime. Let also x, y, z be relatively prime integers such that

$$x^2 + y^2 = z^{2m}.$$

Prove that p divides xy.

920. Find all integer solutions to the equation

$$x^3 - 2 = y^2.$$

921. Find all positive integer numbers x, y that satisfy the equation $x^2 + 11 = 3^y$.

5.3 Diophantine Equations

5.3.1 Linear Diophantine Equations

A linear Diophantine equation (named in the honor of Diophantus, who studied equations over the integers) is an equation of the form

$$a_1 x_1 + \cdots + a_n x_n = b,$$

where $a_1, \ldots, a_n$, and b are integers. We will discuss only the Diophantine equation

$$ax - by = c.$$

Theorem. *The equation*

$$ax - by = c$$

has solutions if and only if $\gcd(a, b)$ *divides* c. *If* (x_0, y_0) *is a solution, then all other solutions are of the form*

$$x = x_0 + \frac{b}{\gcd(a, b)}t, \text{ and } y = y_0 + \frac{a}{\gcd(a, b)}t, \quad t \in \mathbb{Z}.$$

Proof. For the equation to have solutions it is clearly necessary that c be divisible by $\gcd(a, b)$. Dividing through by $\gcd(a, b)$ we can assume that a and b are coprime.

To show that the equation has solutions, we first examine the case $c = 1$. The method of solving this equation is a consequence of Euclid's algorithm for finding the greatest common divisor. This algorithm consists of a successive series of divisions

$$a = q_1 b + r_1,$$
$$b = q_2 r_1 + r_2,$$
$$r_1 = q_3 r_2 + r_3,$$
$$\cdots$$
$$r_{n-2} = q_n r_{n-1} + r_n,$$

where r_n is the greatest common divisor of a and b, which in our case is 1. If we work backward, we obtain

$$1 = r_{n-1}(-q_n) - (-r_{n-2}) = r_{n-2}(1 - q_{n-1}) - r_{n-3}q_n = \cdots = ax_0 - by_0$$

for whatever numbers x_0 and y_0 arise at the last stage. This yields a particular solution (x_0, y_0).

For a general c, just multiply this solution by c. If (x_1, y_1) is another solution, then by subtracting $ax_0 - by_0 = c$ from $ax_1 - by_1 = c$, we obtain $a(x_1 - x_0) - b(y_1 - y_0) = 0$, hence $x_1 - x_0 = \dfrac{b}{\gcd(a, b)}t$, and $y_1 - y_0 = \dfrac{a}{\gcd(a, b)}t$ for some integer number t. This shows that the general solution is of the form $\left(x_0 + \frac{b}{\gcd(a,b)}t, y_0 + \frac{a}{\gcd(a,b)}t\right)$, t an integer. The theorem is proved. $\square$

The algorithm for finding a particular solution can be better visualized if we use the continued fraction expansion

$$\frac{a}{b} = -a_1 + \cfrac{1}{-a_2 + \cfrac{1}{-a_3 + \cdots + \cfrac{1}{-a_{n-1} + \cfrac{1}{-a_n}}}}.$$

In this, if we delete $\frac{1}{-a_n}$, we obtain a simpler fraction, and this fraction is nothing but $\frac{y_0}{x_0}$.

The equality $ax - by = 1$ shows that the matrix with integer entries

$$\begin{pmatrix} a & y \\ b & x \end{pmatrix}$$

has determinant 1. The matrices with this property form the special linear group $SL(2, \mathbb{Z})$. This group is generated by the matrices

$$S = \begin{pmatrix} 0 & -1 \\ 1 & 0 \end{pmatrix} \quad \text{and} \quad T = \begin{pmatrix} 1 & 1 \\ 0 & 1 \end{pmatrix}.$$

Explicitly,

$$\begin{pmatrix} a & y \\ b & x \end{pmatrix} = ST^{a_1} ST^{a_2} S \cdots ST^{a_n} S,$$

since matrix multiplication mimics the (backward) calculation of the continued fraction. We thus have a method of expressing the elements of $SL(2, \mathbb{Z})$ in terms of generators.

The special linear group $SL(2, \mathbb{Z})$ arises in non-Euclidean geometry. It acts on the upper half-plane, on which Poincaré modeled the "plane" of Lobachevskian geometry. The "lines" of this "plane" are the semicircles and half lines orthogonal to the real axis. A matrix

$$A \begin{pmatrix} a & b \\ c & d \end{pmatrix}$$

acts on the Lobachevski plane by

$$z \to \frac{az + b}{cz + d}, \quad ad - bc = 1.$$

All these transformations form a group of isometries of the Lobachevski plane. Note that A and $-A$ induce the same transformations; thus this group of isometries of the Lobachevski plane, also called the modular group, is isomorphic to $PSL(2, \mathbb{Z}) = SL(2, \mathbb{Z})/\{-\mathcal{I}_2, \mathcal{I}_2\}$. The matrices S and T become the inversion with respect to the unit circle $z \to -\frac{1}{z}$ and the translation $z \to z + 1$.

We stop here with the discussion and list some problems.

922. Write the matrix

$$\begin{pmatrix} 12 & 5 \\ 7 & 3 \end{pmatrix}$$

as the product of several copies of the matrices

$$\begin{pmatrix} 0 & 1 \\ 1 & 0 \end{pmatrix} \quad \text{and} \quad \begin{pmatrix} 1 & 1 \\ 0 & 1 \end{pmatrix}.$$

(No, there is no typo in the matrix on the left.)

923. Let a, b, c, d be integers with the property that for any two integers m and n there exist integers x and y satisfying the system

$$ax + by = m,$$

$$cx + dy = n.$$

Prove that $ad - bc = \pm 1$.

924. Let a, b, c, d be positive integers with $\gcd(a, b) = 1$. Prove that the system of equations

$$\begin{cases} ax - yz - c = 0, \\ bx - yt + d = 0 \end{cases}$$

has infinitely many solutions in positive integers (x, y, z, t).

We now ask for the nonnegative solutions to the equation $ax + by = c$, where a, b, c are positive numbers. This is a particular case, solved by Sylvester, of the Frobenius coin problem: what is the largest amount of money that cannot be paid using coins worth $a_1, a_2, \ldots, a_n$ cents? Here is the answer.

Sylvester's theorem. *Let a and b be coprime positive integers. Then $ab - a - b$ is the largest positive integer c for which the equation*

$$ax + by = c$$

is not solvable in nonnegative integers.

Proof. Let $N > ab - a - b$. The integer solutions to the equation $ax + by = N$ are of the form $(x, y) = (x_0 + bt, y_0 - at)$, with t an integer. Choose t such that $0 \le y_0 - at \le a - 1$. Then

$$(x_0 + bt)a = N - (y_0 - at)b > ab - a - b - (a - 1)b = -a,$$

which implies that $x_0 + bt > -1$, and so $x_0 + bt \ge 0$. Hence in this case the equation $ax + by = N$ admits nonnegative integer solutions.

On the other hand, if there existed $x, y \ge 0$ such that

$$ax + by = ab - a - b,$$

then we would have $ab = a(x + 1) + b(y + 1)$. Since a and b are coprime, this would imply that a divides $y + 1$ and b divides $x + 1$. But then $y + 1 \ge a$ and $x + 1 \ge b$, which would then lead to the contradiction

$$ab = a(x + 1) + b(y + 1) \ge 2ab.$$

This proves the theorem. $\square$

And now the problems.

925. Given a piece of paper, we can cut it into 8 or 12 pieces. Any of these pieces can be cut into 8 or 12, and so on. Show that we can obtain any number of pieces greater than 60. Can we obtain exactly 60 pieces?

926. Let a and b be positive integers. For a nonnegative integer n let $s(n)$ be the number of nonnegative integer solutions to the equation $ax + by = n$. Prove that the generating function of the sequence $(s(n))_n$ is

$$f(x) = \frac{1}{(1 - x^a)(1 - x^b)}.$$

927. Let $n > 6$ be a positive integer. Prove that the equation

$$x + y = n$$

admits a solution with x and y coprime positive integers both greater than 1.

928. Prove that the d-dimensional cube can be dissected into n d-dimensional cubes for all sufficiently large values of n.

5.3.2 The Equation of Pythagoras

The Diophantine equation

$$x^2 + y^2 = z^2,$$

has as solutions triples of positive integers that are the side lengths of a right triangle, whence the name. Let us solve it.

If x and z have a common factor, this factor divides y as well. Let us assume first that x and z are coprime. We can also assume that x and z have the same parity (both are odd); otherwise, exchange x and y.

In this situation, write the equation as

$$y^2 = (z + x)(z - x).$$

The factors $z + x$ and $z - x$ are both divisible by 2. Moreover, 2 is their greatest common divisor, since it is the greatest common divisor of their sum $2z$ and their difference $2x$. We deduce that y is even, and there exist coprime integers u and v such that $y = 2uv$, $z + x = 2u^2$ and $z - x = 2v^2$. We obtain $x = u^2 - v^2$ and $z = u^2 + v^2$. Incorporating the common factor of x, y, and z, we find that the solutions to the equation are parametrized by triples of integers (u, v, k) as $x = k(u^2 - v^2)$, $y = 2kuv$, and $z = k(u^2 + v^2)$. The positive solutions are called Pythagorean triples.

There is a more profound way to look at this equation. Dividing through by z^2, we obtain the equivalent form

$$\left(\frac{x}{z}\right)^2 + \left(\frac{y}{z}\right)^2 = 1.$$

This means that we are supposed to find the points of rational coordinates on the unit circle. Like any conic, the circle can be parametrized by rational functions. A parametrization is $\left(\frac{1-t^2}{1+t^2}, \frac{2t}{1+t^2}\right)$, $t \in \mathbb{R} \cup \{\infty\}$. The fractions $\frac{1-t^2}{1+t^2}$ and $\frac{2t}{1+t^2}$ are simultaneously rational if and only if t itself is rational. In that case $t = \frac{u}{v}$ for some coprime integers u and v. Thus we should have

$$\frac{x}{z} = \frac{1 - \left(\frac{u}{v}\right)^2}{1 + \left(\frac{u}{v}\right)^2} \quad \text{and} \quad \frac{y}{z} = \frac{2\frac{u}{v}}{1 + \left(\frac{u}{v}\right)^2},$$

where again we look at the case in which x, y, and z have no common factor, and x and z are both odd. Then y is necessarily even and

$$\frac{y}{z} = \frac{2uv}{u^2 + v^2}.$$

Because u and v are coprime, and because y is even, the fraction on the right-hand side is irreducible. Hence $y = 2uv$, $z = u^2 + v^2$, and consequently $x = u^2 - v^2$. Exchanging x and y, we obtain the other parametrization. In conclusion, we have the following theorem.

Theorem. *Any solution x, y, z to the equation $x^2 + y^2 = z^2$ in positive integers is of the form $x = k(u^2 - v^2)$, $y = 2kuv$, $z = k(u^2 + v^2)$, or $x = 2kuv$, $y = k(u^2 - v^2)$, $z = k(u^2 + v^2)$, where k is an integer and u, v are coprime integers with $u > v$ not both odd.*

We now describe an occurrence of Pythagorean triples within the Fibonacci sequence

$$1, 1, 2, \underbrace{3, 5}, 8, 13, 21, 34, 55, 89, 144, 233, \dots$$

Take the terms $F_4 = 3$ and $F_5 = 5$, multiply them, and double the product. Then take the product of $F_3 = 2$ and $F_6 = 8$. You obtain the numbers 30 and 16, and $30^2 + 16^2 = 1156$, which is the square of $F_9 = 34$.

Similarly, the double product of $F_5 = 5$ and $F_6 = 8$ is 80, and the product of $F_4 = 3$ and $F_7 = 13$ is 39. And $80^2 + 39^2 = 7921 = F_{11}^2$. One more check: the double product of $F_6 = 8$ and $F_7 = 13$ is 208, the product of $F_5 = 5$ and $F_8 = 21$ is 105, and $105^2 + 208^2 = 54289 = F_{13}^2$. In general, we may state the following.

Example. The numbers $2F_n F_{n+1}$, $F_{n-1} F_{n+2}$ and F_{2n+1} form a Pythagorean triple.

Solution. In our parametrization, it is natural to try $u = F_{n+1}$ and $v = F_n$. And indeed,

$$u^2 - v^2 = (u - v)(u + v) = (F_{n+1} - F_n)(F_{n+1} + F_n) = F_{n-1} F_{n+2},$$

while the identity

$$F_{2n+1} = u^2 + v^2 = F_{n+1}^2 + F_n^2$$

was established in Section 2.3.1. This proves our claim.   □

929. Given that the sides of a right triangle are coprime integers and the sum of the legs is a perfect square, show that the sum of the cubes of the legs can be written as the sum of two perfect squares.

930. Find all right triangles whose sides are positive integers and whose perimeter is numerically equal to their area.

931. Find all positive integers x, y, z satisfying the equation $3^x + y^2 = 5^z$.

932. Show that for no positive integers x and y can $2^x + 25^y$ be a perfect square.

933. Solve the following equation in positive integers:

$$x^2 + y^2 = 1997(x - y).$$

5.3.3 Pell's Equation

Euler, after reading Wallis' *Opera Mathematica*, mistakenly attributed the first serious study of nontrivial solutions to the equation

$$x^2 - Dy^2 = 1$$

to John Pell. However, there is no evidence that Pell, who taught at the University of Amsterdam, had ever considered solving such an equation. It should more aptly be called Fermat's equation, since it was Fermat who first investigated it. Nevertheless, equations of Pell type can be traced back to the Greeks. Theon of Smyrna used the ratio $\frac{x}{y}$ to approximate $\sqrt{2}$, where x and y are solutions to $x^2 - 2y^2 = 1$. A more famous equation is Archimedes' *problema bovinum* (cattle problem) posed as a challenge to Apollonius, which received a complete solution only in the twentieth century.

Indian mathematicians of the sixth century devised a method for finding solutions to Pell's equation. But the general solution was first explained by Lagrange in a series of papers presented to the Berlin Academy between 1768 and 1770.

Lagrange's theorem. *If D is a positive integer that is not a perfect square, then the equation*

$$x^2 - Dy^2 = 1$$

has infinitely many solutions in positive integers and the general solution $(x_n, y_n)_{n \geq 1}$ is computed from the relation

$$(x_n, y_n) = (x_1 + y_1\sqrt{D})^n,$$

where (x_1, y_1) is the fundamental solution (the minimal solution different from the trivial solution $(1, 0)$).

The fundamental solution can be obtained by trial and error. But there is an algorithm to find it. The continued fraction expansion or $\sqrt{D}$ is periodic, so that if n is the minimal period we write:

$$\sqrt{D} = a_0 + \cfrac{1}{a_1 + \cfrac{1}{a_2 + \cdots + \cfrac{1}{a_n + \cfrac{1}{a_1 + \cfrac{1}{a_2 + \cdots}}}}}.$$

When n is even, the fundamental solution is given by the numerator and the denominator of the fraction

$$a_0 + \cfrac{1}{a_1 + \cfrac{1}{a_2 + \cdots + \cfrac{1}{a_{n-1}}}}$$

while when n is odd, the fundamental solution is given by the numerator and the denominator of the fraction

$$a_0 + \cfrac{1}{a_1 + \cfrac{1}{a_2 + \cdots + \cfrac{1}{a_n + \cfrac{1}{a_1 + \cfrac{1}{a_2 + \cdots + \cfrac{1}{a_{n-1}}}}}}}$$

This algorithm is not as simple as it seems. The smallest solution (x_1, y_1) can depend exponentially on D. From the computational point of view, the challenge is to determine the number $R = \ln(x_1 + y_1 \sqrt{D})$, called the regulator, with a certain accuracy. At the time of the writing this book no algorithm has been found to solve the problem in polynomial time on a classical computer. If a computer governed by the laws of quantum physics could be built, then such an algorithm exists and was discovered by S. Hallgren.

We found the following application of Pell's equation published by M.N. Deshpande in the *American Mathematical Monthly*.

Example. Find infinitely many triples (a, b, c) of positive integers such that a, b, c are in arithmetic progression and such that $ab + 1$, $bc + 1$, and $ca + 1$ are perfect squares.

Solution. A slick solution is based on Pell's equation

$$x^2 - 3y^2 = 1.$$

Pell's equation, of course, has infinitely many solutions. If (r, s) is a solution, then the triple $(a, b, c) = (2s - r, 2s, 2s + r)$ is in arithmetic progression and satisfies $(2s - r)2s + 1 = (r - s)^2$, $(2s - r)(2s + r) + 1 = s^2$, and $2s(2s + r) + 1 = (r + s)^2$.   □

More examples follow.

934. Find a solution to the Diophantine equation

$$x^2 - (m^2 + 1)y^2 = 1,$$

where m is a positive integer.

935. Prove that there exist infinitely many squares of the form

$$1 + 2^{x^2} + 2^{y^2},$$

where x and y are positive integers.

936. Prove that there exist infinitely many integers n such that $n, n+1, n+2$ are each the sum of two perfect squares. (Example: $o = 0^2 + 0^2$, $1 = 0^2 + 1^2$, $2 = 1^2 + 1^2$.)

937. Prove that for no integer n can $n^2 - 2$ be a power of 7 with exponent greater than 1.

938. Find the positive solutions to the Diophantine equation

$$(x+1)^3 - x^3 = y^2.$$

939. Find the positive integer solutions to the equation

$$(x-y)^5 = x^3 - y^3.$$

940. Prove that the equation

$$x^3 + y^3 + z^3 + t^3 = 1999$$

has infinitely many integer solutions.

941. Prove that for every pair of positive integers m and n, there exists a positive integer p satisfying

$$(\sqrt{m} + \sqrt{m-1})^n = \sqrt{p} + \sqrt{p-1}.$$

5.3.4 Other Diophantine Equations

In conclusion, try your hand at the following Diophantine equations. Any method is allowed!

942. Find all integer solutions (x, y) to the equation

$$x^2 + 3xy + 4006(x+y) + 2003^2 = 0.$$

943. Prove that there do not exist positive integers x and y such that

$$x^2 + xy + y^2 = x^2 y^2.$$

944. Prove that there are infinitely many quadruples x, y, z, w of positive integers such that

$$x^4 + y^4 + z^4 = 2002^w.$$

945. Find all nonnegative integers x, y, z, w satisfying

$$4^x + 4^y + 4^z = w^2.$$

946. Prove that the equation

$$x^2 + y^2 + z^2 + 3(x+y+z) + 5 = 0$$

has no solutions in rational numbers.

947. Find all positive integers x satisfying

$$3^{2^{x!}} = 2^{3^{x!}} + 1.$$

948. Find all quadruples (u, v, x, y) of positive integers, where u and v are consecutive in some order, satisfying

$$u^x - v^y = 1.$$

6

Combinatorics and Probability

We conclude the book with combinatorics. First, we train combinatorial skills in set theory, number theory, and geometry, with a glimpse at permutations. Then we turn to some specific techniques: graphs, generating functions, counting arguments, the inclusion-exclusion principle. A strong accent is placed on binomial coefficients.

This is followed by probability, which, in fact, should be treated separately. But the level of this book restricts us to problems that use counting, classical schemes such as the Bernoulli and Poisson schemes and Bayes' theorem, recurrences, and some minor geometric considerations. It is only later in the development of mathematics that probability loses its combinatorial flavor and borrows the analytical tools of Lebesgue integration.

6.1 Combinatorial Arguments in Set Theory

6.1.1 Combinatorics of Sets

A first example comes from the 1971 German Mathematical Olympiad.

Example. Given 2^{n-1} subsets of a set with n elements with the property that any three have nonempty intersection, prove that the intersection of all the sets is nonempty.

Solution. Let $S = \{A_1, A_2, \ldots, A_{2^{n-1}}\}$ be the family of subsets of the set A with n elements. Because S has 2^{n-1} elements, for any subset B of A, either B or its complement B^c is in S. (They cannot both be in S by the other hypothesis.)

So if A_i and A_j are in S, then either $A_i \cap A_j$ is in S, or its complement is in S. If the complement is in S then $A_i \cap A_j \cap (A_i \cap A_j)^c$ is empty, contradicting the fact that the intersection of any three elements of S is nonempty. Hence $A_i \cap A_j \in S$.

We will now show by induction on k that the intersection of any k sets in S is nontrivial. We just proved the base case $k = 2$. Assume that the property is true for any $k - 1$ elements of S, and let us prove it for $A_{i_1}, A_{i_2}, \ldots, A_{i_k} \in S$. By the induction hypothesis, $A_{i_1} \cap \ldots \cap A_{i_{k-1}} \in S$, and also $A_{i_k} \in S$, so $(A_{i_1} \cap \ldots \cap A_{i_{k-1}}) \cap A_{i_k}$ is in S. This completes the induction. For $k = 2^{n-1}$, we obtain that the intersection of all sets in S is nontrivial. $\qquad\square$

© Springer International Publishing AG 2017
R. Gelca and T. Andreescu, *Putnam and Beyond*, DOI 10.1007/978-3-319-58988-6_6

We found the following problem in the *Mathematics Magazine for High Schools* (*Budapest*).

Example. Let A be a nonempty set and let $f : \mathcal{P}(A) \to \mathcal{P}(A)$ be an increasing function on the set of subsets of A, meaning that

$$f(X) \subset f(Y) \quad \text{if} \quad X \subset Y.$$

Prove that there exists T, a subset of A, such that $f(T) = T$.

Solution. Consider the family of sets

$$\mathcal{F} = \{K \in \mathcal{P}(A) \mid f(K) \subset K\}.$$

Because $A \in \mathcal{F}$, the family $\mathcal{F}$ is not empty. Let T be the intersection of all sets in $\mathcal{F}$. We will show that $f(T) = T$.

If $K \in \mathcal{F}$, then $f(T) \subset f(K) \subset K$, and by taking the intersection over all $K \in \mathcal{F}$, we obtain that $f(T) \subset T$. Hence $T \in \mathcal{F}$.

Because f is increasing it follows that $f(f(T)) \subset f(T)$, and hence $f(T) \in \mathcal{F}$. Since T is included in every element of $\mathcal{F}$, we have $T \subset f(T)$. The double inclusion proves that $f(T) = T$, as desired. $\qquad\square$

949. Let A and B be two sets. Find all sets X with the property that

$$A \cap X = B \cap X = A \cap B,$$

$$A \cup B \cup X = A \cup B.$$

950. Prove that a list can be made of all the subsets of a finite set such that

(i) the empty set is the first set;
(ii) each subset occurs once;
(iii) each subset is obtained from the preceding by adding or deleting an element.

951. Let S be a nonempty set and $\mathcal{F}$ a family of $m \geq 2$ subsets of S. Show that among the sets of the form $A \triangle B$ with $A, B \in \mathcal{F}$ there are at least m that are distinct. (Here $A \triangle B = (A \setminus B) \cup (B \setminus A)$.)

952. Consider the sequence of functions and sets

$$\cdots \to A_n \xrightarrow{f_{n-1}} A_{n-1} \xrightarrow{f_{n-2}} A_{n-2} \xrightarrow{f_{n-3}} \cdots \xrightarrow{f_3} A_3 \xrightarrow{f_2} A_2 \xrightarrow{f_1} A_1.$$

Prove that if the sets A_n are nonempty and finite for all n, then there exists a sequence of elements $x_n \in A_n$, $n = 1, 2, 3, \ldots$, with the property that $f_n(x_{n+1}) = x_n$ for all $n \geq 1$.

953. In a society of n people, any two persons who do not know each other have exactly two common acquaintances, and any two persons who know each other don't have other common acquaintances. Prove that in this society every person has the same number of acquaintances.

954. In the country Anchuria, led by president Miraflores, it is time for presidential elections. The country has 20 million voters, of which 1% support the president. The election is organized as follows: all voters are divided into equal groups, each group is divided itself in equal groups, and so on. At each level the groups have the same number of people. In the smallest group one elects one representative—the elector, these electors choose the representatives of the larger groups they are members of and so on. In the end, the representatives of the largest groups elect the president. If Miraflores is allowed to divide the voters in groups at his own discretion, can he win the elections? (Note: If in a group there is a tie, the opposition wins.)

6.1.2 Combinatorics of Numbers

We continue with problems about numbers, which are based on combinatorial thinking but also use algebraic operations and properties of numbers. The following example clarifies what we have in mind. It is a problem of B. Enescu and D. Ismailescu that was given at a Romanian Team Selection Test for the International Mathematical Olympiad in 1999.

Example. Find the number of sets of positive integers $A = \{a_1, a_2, \ldots, a_9\}$ with the property that for every positive integer n, $1 \leq n \leq 500$ there is $B \subset A$ such that the sum of the elements of B is n.

Solution. Note that $A = \{1, 2, 2^2, \ldots, 2^8\}$ has the required property, since $2^9 = 512$, and so every number less than 512 has a binary expansion involving only $1, 2, 2^2, \ldots, 2^8$.

On the other hand, a set with 9 elements has $2^9 - 1 = 511$ non-empty subsets. Because the sums of elements in these subsets must cover all numbers from 1 through 500, it means that the sums in the subsets of A must be "very different". For example if A contains three elements x, y, z with $x = y + z$, then for every $B \subset A \backslash \{x, y, z\}$, the subsets $B \cup \{x\}$ and $B \cup \{y, z\}$ have the same sum of elements, and there are $2^6 = 64$ subsets of this form. However, at most 12 subsets can have the same sum, or else the remaining 499 sums cannot be achieved. A similar argument shows that the sum of three or four elements of A cannot equal an element of A. However, there is no contradiction if the sum of five elements from A equals a number in A.

Let us try to construct A. Note that 1 and 2 are in A, or else these numbers cannot be obtained as sums of elements in A. Since $3 = 1 + 2$, $3 \notin A$, and therefore $4 \in A$. With 1, 2, and 4 we can produce the numbers 5, 6, 7, so neither of these is in A, and hence $8 \in A$. Same reasoning with 1, 2, 4, and 8 shows that $16 \in A$. We are left with finding four more elements of A. All numbers from 1 through 30 can be written as sums of four elements in A, but 31 is the sum of five elements. Hence 31 can be an element of A; if it is not, then 32 is. So the sixth element of A is either 31 or 32. For simplicity, let this element be $32 - a$, $a \in \{0, 1\}$.

The numbers $1, 2, 4, 8, 16, 32 - a$ generate the sums between 1 and $63 - a$. So the next element of A must be of the form $64 - a - b$, where b is a non-negative integer. The new numbers $1, 2, 4, 8, 16, 32 - a, 64 - a - b$ generate sums up to $127 - 2a - b$, so there is an element of the form $128 - 2a - b - c$, with c a nonnegative integer that is an element of A. At the next step we conclude that the ninth element of A is of the form $256 - 4a - 2b - c - d$, with d a non-negative integer. Therefore

$$A = \{1, 2, 4, 8, 16, 32 - a, 64 - a - b, 128 - 2a - b - c, 256 - 4a - 2b - c - d\}.$$

The sum of the nine elements must exceed 500, from where we deduce the inequality

$$511 - 8a - 4b - 2c - d \geq 500,$$

that is

$$8a + 4b + 2c + d \leq 11.$$

For any such a, b, c, d, the elements of A generate all sums from 1 through $511 - 8a - 4b - 2c - d$, so the set A has the required property. So we are left with counting the number of quadruples (a, b, c, d) satisfying $8a + 4b + 2c + d \leq 11$. Clearly a can only equal 0 or 1. If $a = 1$, then $b = 0$ and $2c + d \leq 3$, in which case we obtain 6 solutions. If $a = 0$, then $b \leq 2$. For $b = 2$ we should have $2c + d \leq 3$ and again we have 6 solutions. If $b = 1$, then $2c + d \leq 7$, i.e. $d \leq 7 - 2c$. For the values 0, 1, 2, 3 of c we obtain 8, 6, 4, 2 possible values for d, so the number of solutions in this case is $2 + 4 + 6 + 8 = 20$. Finally, if $b = 0$, then $c = 0, 1, 2, 3, 4, 5$, and we obtain 12, 10, 8, 6, 4, 2 possible values of d, so the number of solutions is 42. The total number of sets A with the required property is therefore $6 + 6 + 20 + 42 = 74$. $\square$

Below are listed more problems of this kind.

955. At the beginning of a game, a positive integer n is chosen. At each turn one of the players writes a positive integer that does not exceed n on the blackboard, the rule being that the player cannot write a divisor of a number already existing on the blackboard. The player who cannot continue loses.

(a) Find a winning strategy for the first player when $n = 10$.
(b) Does any of the players have a winning strategy if $n = 1000$?

956. Let M be a subset of $\{1, 2, 3, \ldots, 15\}$ such that the product of any three distinct elements of M is not a square. Determine the maximum number of elements in M.

957. Is it true that from any six positive integers one can either select three that are pairwise coprime, or three whose greatest common divisor is greater than 1?

958. Let $n \geq 2$ be an integer. Let S be a subset of $\{1, 2, \ldots, n\}$ such that S neither contains two elements one of which divides the other, nor contains two elements which are coprime. What is the maximum possible number of elements of such a set S?

959. A number with an even number of digits is called "acceptable" if on its odd positions there are as many even digits as there are on its even positions. Show that from any number with an odd number of digits one can erase one digit so that the new number is acceptable. For example, from 12345 we can eliminate the digit 3 obtaining 1245, which is acceptable.

960. An even number of people sit at a round table. After a break, the people return to the table and seat themselves randomly. Show that there are two people such that the number of people who sit between them remained unchanged. Does this property still hold if the number of people is odd?

961. For what n and k can the set $\{1, 2, 3, \ldots, nk\}$ be partitioned into n subsets of k elements each such that the sum of the elements in each set is the same?

962. For a positive integer n, $n \geq 3$, consider the points $A_1, A_2, \ldots, A_n$ in this order, and place the numbers $1, 2, \ldots, n$ randomly at these points.

  (a) Show that the sum of the absolute values of the differences of neighboring numbers is greater or equal to $2n - 2$.
  (b) For how many arrangements of these numbers is the sum exactly $2n - 2$?

963. For every positive integer m, denote by $f(m)$ the largest positive integer k with the property that there is a set $A = \{a_1, a_2, \ldots, a_k\} \subset \{1, 2, \ldots, m\}$, such that for every $1 \leq i < j \leq k$, $a_i + a_j$ is not a divisor of $a_i a_j$. Prove that $f(2^n) \geq 2^{n-1} + n$.

964. An equilateral triangle is divided into 16 equal equilateral triangles, and in each of these triangles one of the numbers from 1 to 16 is written. Show that there are two triangles sharing a side with the difference of the numbers written in them being at least 4.

965. The plane is partitioned into regions by a finite number of lines, no three of which are concurrent. Two regions are called neighbors if they share a common border. An integer is assigned to each region such that

  (i) the product of the integers assigned to any two neighbors is less than their sum;
  (ii) for each of the given lines, and each of the two half-planes determined by it, the sum of the integers assigned to all of the regions lying in this half-plane is zero.

Prove that this is possible if and only if not all of the lines are parallel.

6.1.3 Permutations

A permutation of a set S is a bijection $\sigma : S \to S$. Composition induces a group structure on the set of all permutations. We are concerned only with the finite case $S = \{1, 2, \ldots, n\}$. The standard notation for a permutation is

$$\sigma = \begin{pmatrix} 1 & 2 & 3 & \ldots & n \\ a_1 & a_2 & a_3 & \ldots & a_n \end{pmatrix},$$

with $a_i = \sigma(i)$, $i = 1, 2, \ldots, n$.

A permutation is a cycle $(i_1 i_2 \ldots i_n)$ if $\sigma(i_1) = i_2$, $\sigma(i_2) = i_3, \ldots, \sigma(i_n) = i_1$, and $\sigma(j) = j$ for $i \neq i_1, i_2, \ldots, i_n$. Any permutation is a product of disjoint cycles. A cycle of length two $(i_1 i_2)$ is called a transposition. Any permutation is a product of transpositions. For a given permutation σ, the parity of the number of transpositions in this product is always the same; the signature of σ, denoted by $\text{sign}(\sigma)$, is 1 if this number is even and -1 if this number is odd. An inversion is a pair (i, j) with $i < j$ and $\sigma(i) > \sigma(j)$.

Let us look at a problem from the 1979 Romanian Mathematical Olympiad, proposed by I. Raşa.

Example. Consider the permutations

$$\sigma_1 = \begin{pmatrix} 1 & 2 & 3 & 4 & \dots & 19 & 20 \\ a_1 & a_2 & a_3 & a_4 & \dots & a_{19} & a_{20} \end{pmatrix},$$

$$\sigma_2 = \begin{pmatrix} 1 & 2 & 3 & 4 & \dots & 19 & 20 \\ a_{19} & a_{20} & a_{17} & a_{18} & \dots & a_1 & a_2 \end{pmatrix}.$$

Prove that if σ_1 has 100 inversions, then σ_2 has at most 100 inversions.

Solution. Let us see what an inversion (a_i, a_j) of σ_1 becomes in σ_2. If i and j have the same parity, then $_i$ and a_j are switched in σ_2, and so (a_j, a_i) is no longer an inversion. If i is even and j is odd, then a_i and a_j are also switched in σ_2, so the inversion again disappears.

We investigate the case i odd and j even more closely. If $j > i + 1$, then in σ_2 the two elements appear in the order (a_j, a_i), which is again not an inversion. However, if i and j are consecutive, then the pair is not permuted in σ_2; the inversion is preserved. There are at most 10 such pairs, because i can take only the values $1, 3, 5, \dots, 19$. So at most 10 inversions are "transmitted" from σ_1 to σ_2. From the 100 inversions of σ_1, at most 10 become inversions of σ_2, while 90 are "lost": they are no longer inversions in σ_2.

It follows that from the $\binom{20}{2} = 190$ pairs (a_i, a_j) in σ_2 with $i < j$, at least 90 are not inversions, which means that at most $190 - 90 = 100$ are inversions. This completes the proof.

Here is a different way of saying this. Define

$$\sigma_3 = \begin{pmatrix} 1 & 2 & 3 & 4 & \dots & 19 & 20 \\ a_{20} & a_{19} & a_{18} & a_{17} & \dots & a_2 & a_1 \end{pmatrix}.$$

Then between them σ_1 and σ_3 have exactly $\binom{20}{2}$ inversions, since each pair is an inversion in exactly one. Hence σ_3 has at most 90 inversions. Because σ_2 differs from σ_3 by swapping 10 pairs of adjacent outputs, these are the only pairs in which it can differ from σ_3 in whether it has has an inversion. Hence σ_2 has at most 100 inversions. $\square$

And now an example with a geometric flavor.

Example. Let σ be a permutation of the set $\{1, 2, \dots, n\}$. Prove that there exist permutations σ_1 and σ_2 of the same set such that $\sigma = \sigma_1 \sigma_2$ and σ_1^2 and σ_2^2 are both equal to the identity permutation.

Solution. Decompose the permutation σ into a product of disjoint cycles. It suffices to prove the property for each of these cycles; therefore, we can assume from the beginning that σ itself is a cycle of length n. If $n = 1$ or 2, then we choose $\sigma_1 = \sigma$ and σ_2 the identity permutation. Otherwise, we think of σ as the rotation of a regular n-gon $A_1 A_2 \dots A_n$ by an angle of $\frac{2\pi}{n}$ around its center. Such a rotation can be written as the composition of two reflections that map the n-gon to itself, namely the reflection with respect to the perpendicular bisector of

$A_1 A_3$ and the reflection with respect to the perpendicular bisector of $A_2 A_3$ (see Figure 39). These reflections define the permutations σ_1 and σ_2. $\square$

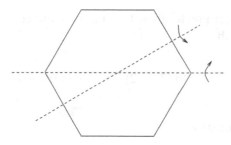

Figure 39

The following problems are left to the reader.

966. For each permutation $a_1, a_2, \ldots, a_{10}$ of the integers $1, 2, 3, \ldots 10$, form the sum

$$|a_1 - a_2| + |a_3 - a_4| + |a_5 - a_6| + |a_7 - a_8| + |a_9 - a_{10}|.$$

Find the average value of all such sums.

967. Find the number of permutations $a_1, a_2, a_3, a_4, a_5, a_6$ of the numbers $1, 2, 3, 4, 5, 6$ that can be transformed into $1, 2, 3, 4, 5, 6$ through exactly four transpositions (and not fewer).

968. Let $f(n)$ be the number of permutations $a_1, a_2, \ldots, a_n$ of the integers $1, 2, \ldots, n$ such that

(i) $a_1 = 1$ and
(ii) $|a_i - a_{i+1}| \leq 2, i = 1, 2, \ldots, n-1$.

Determine whether $f(1996)$ is divisible by 3.

969. Consider the sequences of real numbers $x_1 > x_2 > \cdots > x_n$ and $y_1 > y_2 > \cdots > y_n$, and let σ be a nontrivial permutation of the set $\{1, 2, \ldots, n\}$. Prove that

$$\sum_{i=1}^{n} (x_i - y_i)^2 < \sum_{i=1}^{n} (x_i - y_{\sigma(i)})^2.$$

970. Let $a_1, a_2, \ldots, a_n$ be a permutation of the numbers $1, 2, \ldots, n$. We call a_i a *large* integer if $a_i > a_j$ for all $i < j < n$. Find the average number of large integers over all permutations of the first n positive integers.

971. Given some positive real numbers $a_1 < a_2 < \cdots < a_n$ find all permutations σ with the property that

$$a_1 a_{\sigma(1)} < a_2 a_{\sigma(2)} < \cdots < a_n a_{\sigma(n)}.$$

972. Determine the number of permutations $a_1, a_2, \ldots, a_{2004}$ of the numbers $1, 2, \ldots, 2004$ for which

$$|a_1 - 1| = |a_2 - 2| = \cdots = |a_{2004} - 2004| > 0.$$

973. Let n be an odd integer greater than 1. Find the number of permutations σ of the set $\{1, 2, \ldots, n\}$ for which

$$|\sigma(1) - 1| + |\sigma(2) - 2| + \cdots + |\sigma(n) - n| = \frac{n^2 - 1}{2}.$$

6.2 Combinatorial Geometry

6.2.1 Tessellations

We begin our incursion in combinatorial geometry with problems about tilings of the plane (or of part the plane) by equal polygons. Tiling have aesthetic qualities that make them appealing to architects, and they also hide challenging mathematical problems. Our first example was published in the Russian journal *Kvant* (*Quantum*) by A.N. Kolmogorov.

Example. Consider the tessellation of the plane by unit squares. For what $n \geq 2$ is it possible to color the plane by n colors such that the centers of the squares colored by the same color form a square lattice of the plane? What if we require additionally that all these lattices have equal squares and parallel sides?

Solution. The answer to the first question is all n, since one can obtain inductively, from a coloring by n colors, a coloring by $n + 1$ colors if we "dilate" the coloring by n colors to occupy the black squares of a chess board, and use the $n + 1$st color for the white squares.

The second question is more interesting. Let us show that if such a coloring of the plane exists, then n can be represented as $m^2 + k^2$ where m and k are non-negative integers. Consider the lattice determined by the squares colored by one color. Take a square of this lattice and glue its opposite sides to obtain a torus (the torus is the surface of a "donut"). Then this torus contains precisely one unit square for each color, so its area is n. Thus the side-length of the square is $\sqrt{n}$. Examining the original lattice (of all squares), we see that the side of the square in question can be placed in a right triangle with integer sides, with, say, m units on the vertical and k units on the horizontal. By the Pythagorean theorem, $n = m^2 + k^2$.

Conversely, let us show that every positive integer n of the form $m^2 + k^2$ has an associated coloring. Color one unit square red, then move m units to the right and k units up. Color this square red as well. Then complete a square lattice containing the two red squares and color all its vertices red. Then choose an uncolored unit square, color it by a remaining color, and repeat the above construction. Repeating with each of the colors we obtain the desired coloring. $\qquad \square$

The second example is a problem given at the Romanian Master of Mathematics in 2016, which we selected for our exposition because of the elegant solution given by the US student J. Peng.

Example. Given positive integers m and $n \geq m$, determine the largest number of dominoes (1×2 or 2×1 rectangles) that can be placed on a rectangular board with m rows and $2n$ columns so that:

(i) each domino covers exactly two adjacent cells of the board;

(ii) no two dominoes overlap;

(iii) no two dominoes form a 2×2 square;

(iv) the bottom row of the board is completely covered by n dominoes.

Solution. The required maximum is $mn - \lfloor m/2 \rfloor$ and is achieved by the brick-like vertically symmetric arrangement of blocks of n and $n - 1$ horizontal dominoes placed on alternate rows, so that the bottom row of the board is completely covered by dominoes.

We are going to prove that one cannot exceed $mn - \lfloor m/2 \rfloor$. The possible locations of the centers of the dominoes are the midpoints of the edges that are not on the boundary of the board. These locations form a lattice (rotated by $45°$ compared to the original board), which, when viewed in the horizontal-vertical orientation, has $2n - 1$ points on the horizontal rows $1, 3, \ldots, 2m - 1$, and $2n$ points on the horizontal rows $2, 4, \ldots, 2m - 2$. Call this lattice Λ and let S be the set of its nodes that are centers of domino pieces. We will always use the vertical-horizontal reference to associate coordinates to the nodes. As such, the jth point from the left on the ith horizontal row from the top is (i, j).

A first observation is that if $P \in S$ than its 4 neighbors from the lattice Λ are not in S, because they lie on the sides of the domino centered at P. Since the bottom row is already covered, this observation shows that, since the following nodes are in S

$$(2m - 1, 1), (2m - 1, 3), \ldots, (2m - 1, 2n - 1),$$

the following nodes are not in S:

$$(2m - 1, 2), (2m - 1, 4), \ldots, (2m - 1, 2n - 2);$$
$$(2m - 2, 1), (2m - 2, 2), \ldots, (2m - 2, 2n);$$
$$(2m - 3, 1), (2m - 3, 3), \ldots, (2m - 3, 2n - 1).$$

We call these the forbidden nodes.

Lemma. *All nodes of Λ that are not forbidden can be partitioned into $mn - \lfloor m/n \rfloor$ groups of the types described in Figure 40.*

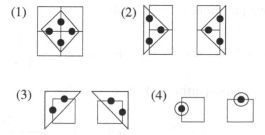

Figure 40

Proof. We first ignore the last row. We will argue on Figure 41, which depicts the case $m = n = 5$.

For every i, j with $i = 2, 3, \ldots, m - 1, j = 0, 2, 4, \ldots, 2n - 2i$ we let

$$(2m - 2i, i + j), (2m - 2i, i + j + 1), (2m - 2i + 1, i + j), (2m - 2i - 1, i + j)$$

be in a group. There are $\frac{1}{2}(2n - m + 1)(m - 2)$ such groups.

There are 3 families of nodes that are not yet split into groups:

I. $(2m - 2i, i - j), (2m - 2i - 1, i - j)$, with $2 \leq i \leq m - 1, 1 \leq j \leq i - 1$;

II. $(2m - 2i, 2n - i + j + 1), (2m - 2i - 1, 2n - i + j)$, where $2 \leq i \leq m - 1, 1 \leq j \leq i - 1$.

III. $(1, m + 2i)$, where $0 \leq i \leq n - m$.

We partition the family I into $\lfloor (m - 1)^2/4 \rfloor$ groups as shown in Figure 42. The nodes in this family form an isosceles triangle with the top vertex missing. Divide this triangle, starting from the base, into isosceles trapezoids by lines that are parallel to the base of the isosceles triangle, so that between the parallel sides of the trapezoid lie 2 rows of nodes (except maybe at the top where we might have just one row). Then divide each trapezoid but the one on top by lines perpendicular to the base into groups that consist of several rhombi and two right triangles so that in each rhombus exactly 4 nodes of Λ lie, and each of the two right triangles contains 2 nodes. If m is odd, then the trapezoid on top has only 2 nodes, they form a group. If m is even, cut the trapezoid on top into 2 groups which are right triangles.

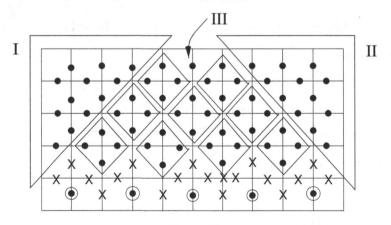

Figure 41

So if m is odd we have partitioned family I into

$$(m - 2) + (m - 4) + \cdots + 3 + 1 = \frac{(m - 1)^2}{4}$$

groups, and if m is even we have partitioned I into

$$(m - 2) + (m - 4) + \cdots + 4 + 2 = \frac{m(m - 2)}{4}$$

I

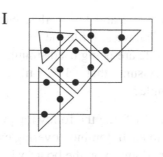

Figure 42

groups. Both these numbers are equal to $\lfloor (m-1)^2/4 \rfloor$. By symmetry we can partition the nodes in II into $\lfloor (m-1)^2/4 \rfloor$ groups as well. Trivially we can partition family III into $(n - m + 1)$ groups of type (4).

The total number of groups is

$$\frac{1}{2}(2n - m + 1)(m - 2) + 2 \times \left\lfloor \frac{(m-1)^2}{4} \right\rfloor + (n - m + 1) = mn - n - \left\lfloor \frac{m}{2} \right\rfloor.$$

Add the n nodes from the bottom row, each in a separate group to obtain a total of $mn - \lfloor m/2 \rfloor$ groups. The lemma is proved.   □

Returning to the problem, note that in each group there is at most one node in S, so the total number of nodes in S is at most $mn - \lfloor m/2 \rfloor$. We conclude that we can have at most $mn - \lfloor m/2 \rfloor$ dominoes, and the problem is solved.   □

The following problems are left to the reader.

974. An equilateral triangle is divided into n^2 equal equilateral triangles by lines parallel to the sides. We call *chain* a sequence of triangles in which no triangle appears twice and every two consecutive triangles share a side. What is the largest possible number of triangles in a chain.

975. An equilateral triangle of side length n is drawn with sides along a triangular grid of side length 1. What is the maximum number of grid segments on or inside the triangle that can be marked so that no three marked segments form a triangle?

976. An equilateral triangle is divided into n^2 equal equilateral triangles by lines parallel to the sides. From the vertices of the triangle obtained this way one chooses m such that for any two chosen vertices A and B, the segment AB is not parallel to any of the sides of the original triangle. What is the largest possible value that m can have?

977. Some of the squares of an infinite lattice are colored red, the others blue, such that inside every 2×3 rectangle there are exactly two red squares. How many red squares can a 9×11 rectangle contain?

978. Can one tile a 6×6 square by 1×2 tiles so that no segment that joins opposite sides of the square shows up in the tiling?

979. (a) A rectangle is tiled by 2×2 and 1×4 tiles. Now assume that we substitute one 2×2 tile by a 1×4 tile. Can we still tile the rectangle?

(b) In the tiling of a rectangle are being used simultaneously 1×3 tiles and L-shaped tiles with 3 unit squares. Assume that a 1×3 tile is substituted by an L-shaped tile. Can we still tile the rectangle?

980. Let n be a positive integer satisfying the following property: If n dominoes are placed on a 6×6 chessboard with each domino covering exactly two unit squares, then one can always place one more domino on the board without moving any other dominoes. Determine the maximum value of n.

6.2.2 Miscellaneous Combinatorial Geometry Problems

We grouped under this title various problems that are solved by analyzing configurations of geometric objects. We start with an easy problem that was proposed in 1999 for the Junior Balkan Mathematical Olympiad.

Example. In a regular $2n$-gon, n diagonals intersect at a point S, which is not a vertex. Prove that S is the center of the $2n$-gon.

Solution. Fix one of the n diagonals. The other $n - 1$ diagonals that run through S cross it, so there are $n - 1$ vertices on one side and $n - 1$ vertices on the other side of this diagonal. Hence this was a main diagonal. Repeating the argument we conclude that all n diagonals are main diagonals, so they meet at the center.   □

We continue with an example suggested to us by G. Galperin.

Example. Show that from any finitely many (closed) hemispheres that cover a sphere one can choose four that cover the sphere.

Solution. In what follows, by a half-line, half-plane, and half-space we will understand a closed half-line (ray), half-plane, respectively, half-space. The hemispheres are obtained by intersecting the sphere with half-spaces passing through the origin. This observation allows us to modify the statement so as to make an inductive argument on the dimension possible.

Alternative problem. *Show that from any finitely many half-spaces that cover the three-dimensional space one can choose four that cover the space.*

Let us analyze first the one- and two-dimensional cases. Among any finite set of half-lines (rays) covering a certain line one can choose two that cover it. Indeed, identifying the line with the real axis, the first of them can be chosen to be of the form $[a, \infty)$, with a smallest among the half-lines of this type in our set, and the other to be of the form $(-\infty, b]$, with b largest among the half-lines of this type in our set.

The two-dimensional analogue of this property states that from finitely many half-planes covering the two-dimensional plane one can choose three that cover the plane.

We prove this by induction on the number n of half-planes. For $n = 3$ there is nothing to prove. Assume that the property is true for n half-planes and let us prove it for $n+1$. Choose h_1 to be one of these half-planes.

If the boundary ∂h_1 of h_1 is contained in some other half-plane h_2, then either h_1 and h_2 cover the plane, or h_2 contains h_1. In the latter case we dispose of h_1 and use the induction hypothesis.

If the boundary ∂h_1 is not contained in any half-plane, then any other half-plane intersects it along a half-line. From the one-dimensional situation we know that two of these half-lines cover it completely. Let h_2 and h_3 be the half-planes corresponding to these two half-lines. There are two possibilities, described in Figure 43. In the first case h_1 is contained in the union of h_2 and h_3, so it can be removed, and then we can use the induction hypothesis. In the second case, h_1, h_2, and h_3 cover the plane. This completes the two-dimensional case.

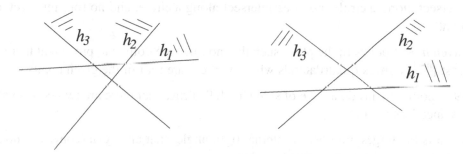

Figure 43

The proof can be extended to three dimensions. As before, we use induction on the number $n \geq 4$ of half-spaces. For the base case $n = 4$ there is nothing to prove. Now let us assume that the property is true for n half-spaces, and let us prove it for $n + 1$. Let H_1 be one of the half-spaces. If the boundary of H_1, ∂H_1, is included in another half-space H_2, then either H_1 and H_2 cover three-dimensional space, or H_1 is included in H_2 and then we can use the induction hypothesis.

In the other case we use the two-dimensional version of the result to find three half-spaces H_2, H_3, and H_4 that determine half-planes on ∂H_1 that cover ∂H_1. To simplify the discussion let us assume that the four boundary planes ∂H_i, $i = 1, 2, 3, 4$, are in general position. Then they determine a tetrahedron. If H_1 contains this tetrahedron, then H_1, H_2, H_3, H_4 cover three-dimensional space. If H_1 does not contain this tetrahedron, then it is contained in the union of H_2, H_3, and H_4, so it can be removed and we can apply the induction hypothesis to complete the argument.   □

Our third example was published by V.I. Arnol'd in the Russian journal *Quantum*.

Example. Prove that any n points in the plane can be covered by finitely many disks with the sum of the diameters less than n and the distance between any two disks greater than 1.

Solution. First, note that if two disks of diameters d_1 and d_2 intersect, then they can be included in a disk of diameter $d_1 + d_2$.

Let us place n disks centered at our points, of some radius $a > 1$ the size of which will be specified later. Whenever two disks intersect, we replace them with a disk that covers them, of diameter equal to the sum of their diameters. We continue this procedure until we have only disjoint disks.

We thus obtained a family of $k \leq n$ disks with the sum of diameters equal to na and such that they cover the disks of diameter a centered at the points. Now let us shrink the diameters of the disks by b, with $1 < b < a$. Then the new disks cover our points, the sum of their diameters is $na - kb \leq na - b$, and the distances between disks are at least b. Choosing a and b such that $1 < b < a$ and $na - b \leq n$ would then lead to a family of circles with the sum of diameters less than n and at distance greater than 1 from each other. For example, we can let $a = 1 + \frac{1}{n}$ and $b = 1 + \frac{1}{2n}$. □

981. In how many regions do n great circles, any three nonintersecting, divide the surface of a sphere?

982. In how many regions do n spheres divide the three-dimensional space if any two intersect along a circle, no three intersect along a circle, and no four intersect at one point?

983. Given $n > 4$ points in the plane such that no three are collinear, prove that there are at least $\binom{n-3}{2}$ convex quadrilaterals whose vertices are four of the given points.

984. 1981 points lie inside a cube of side length 9. Prove that there are two points within a distance less than 1.

985. What is the largest number of internal right angles that an n-gon (convex or not, with non-self-intersecting boundary) can have?

986. A circle of radius 1 rolls without slipping on the outside of a circle of radius $\sqrt{2}$. The contact point of the circles in the initial position is colored. Any time a point of one circle touches a colored point of the other, it becomes itself colored. How many colored points will the moving circle have after 100 revolutions?

987. Several chords are constructed in a circle of radius 1. Prove that if every diameter intersects at most k chords, then the sum of the lengths of the chords is less than $k\pi$.

988. We will call the deformation coefficient of a rectangle the ratio between the smallest and the largest side. Prove that for every decomposition of a square into rectangles, the sum of the deformation coefficients is at least 1. When does equality hold?

989. Inside a square of side 38 lie 100 convex polygons, each with an area at most π and the perimeter at most 2π. Prove that there exists a circle of radius 1 inside the square that does not intersect any of the polygons.

990. Several segments are drawn inside the unit square, parallel to the sides. The sum of the lengths of the segments is 18. These segments divide the square into several regions. Show that there is a region of area at least 0.01.

991. Given a set M of $n \geq 3$ points in the plane such that any three points in M can be covered by a disk of radius 1, prove that the entire set M can be covered by a disk of radius 1.

992. Several disks of the same radius are drawn in the plane so that no two overlap, although they might touch. Show that one can color the circles by four colors such that any two tangent circles have different colors. Do three colors suffice?

6.3 Graphs

6.3.1 Some Basic Graph Theory

A graph G consists of a set V of vertices and a set E of edges, where each edge is a pair of (not necessarily distinct) vertices. Vertices are usually drawn as points and edges as arcs joining those points. A graph is called complete if every two vertices are joined by an edge. A path is a sequence of edges which connect a sequence of vertices. A closed path is also called a circuit or a cycle. The length of a path is the number of edges of the path.

A Hamiltonian path is a path that visits each vertex exactly once. A graph that has a Hamiltonian path is called a traceable graph, while a graph that has a Hamiltonian circuit is called Hamiltonian. The problems of deciding if a graph is traceable or Hamiltonian are NP-complete.

A graph is called Eulerian if there is a circuit that visits each edge exactly once. We have the following result.

Euler's theorem. A graph is Eulerian if each vertex belongs to an even number of edges.

Proof. The property is clearly true if the graph consists of just one point with an even number of edges starting and ending at that point. We can then proceed by induction on the number of vertices, by deleting one vertex and joining in pairs the edges that enter that vertex.

Let us prove a classical fact: the graph the n-dimensional cube is Hamiltonian.

Example. Consider the graph whose vertices and edges the vertices and edges of the n-dimensional cube, $n \geq 2$. Then there is Hamiltonian circuit on this graph.

Solution. We consider the n-dimensional cube whose vertices are the points in $\mathbb{R}^n$ with coordinates 0 or 1. Two vertices are connected by an edge if all but one of the their coordinates coincide.

We prove the property by induction on n. For $n = 2$, the Hamiltonian circuit is

$$(0, 0) \to (0, 1) \to (1, 1) \to (1, 0) \to (0, 0).$$

Now assume that we have found a Hamiltonian path

$$a_1 \to a_2 \to a_3 \to \cdots \to a_{2^n} \to a_1$$

on the n-dimensional cube. Then

$$(0, a_1) \to (0, a_2) \to \cdots \to (0, a_{2^n}) \to (1, a_{2^n}) \to (1, a_{2^n-1}) \to \cdots \to (1, a_1) \to (0, a_1).$$

is a Hamiltonian circuit on the $n + 1$-dimensional cube. The induction is complete. The case $n = 3$ is shown in Figure 44.

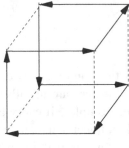

Figure 44

Do you see any similarity with a problem that appeared before in this chapter?   □

Next, a problem about Eulerian graphs, published in the journal *Mathematical Reflections* by O. Dobosevych.

Example. An equilateral triangle is divided into n^2 congruent equilateral triangles. Find for what n we can color all sides of these triangles black or white such that at every vertex an equal number of black and white edges meet.

Solution. Consider the graph whose vertices are the vertices of the triangles and whose edges are their sides. The total number of edges is $3n(n+1)/2$ (a fact that can be checked easily by induction on n). To have an equal number of black and white edges meet at each vertex, there should be an equal number of black and white altogether. Indeed, if we count edges by the vertices, we obtain twice the number of edges (every edge has two endpoints). And counting black and white edges by vertices gives equal numbers. We conclude that $3n(n+1)/2$ should be an even number. Thus a necessary condition is that either n or $n+1$ is a multiple of 4.

Let us show that this is also a sufficient condition. At every vertex meet an even number of edges, so the graph is Eulerian. Consider an Eulerian cycle. Because the number of edges is even, we can color the edges of the cycle alternatively black and white. This coloring satisfies the condition from the statement.   □

We continue with an application of graphs to a problem from the 29th International Mathematical Olympiad, communicated to us by I. Tomescu.

Example. Let n be a positive integer and $A_1, A_2, \ldots, A_{2n+1}$ be a family of subsets of a certain set, each containing $2n$ elements. Let $B = \cup_{i=1}^{2n+1} A_i$. Assume that

(i) For every $i \neq j$, the sets A_i and A_j have exactly one element in common.

(ii) Every element in B belongs to at least two of the sets A_i.

For what numbers n can one color every element in B by $+1$ and -1 such that each of the sets A_i contains exactly n elements colored by $+1$?

Solution. In this solution we denote, as it is customary, by $|X|$ the number of elements of the set X.

Let $B = \{x_1, x_2, \ldots, x_m\}$. Consider a graph G with vertices $x_1, x_2, \ldots, x_m, A_1, A_2, \ldots, A_{2n+1}$. The edges are defined by joining x_i to A_j whenever $x_i \in A_j$.

For a vertex v, we let $d(v)$ be the degree of v, namely the number of edges that contain v. By hypothesis, $d(x_i) \geq 2$ and $d(A_j) = 2n$ for all i, j. Condition (ii) also implies that $(2n + 1)2n \geq 2m$, hence $2n^2 + n \geq m$. On the other hand, the inclusion-exclusion principle (see Section 6.4.4) implies that

$$\left| \bigcup_{j=1}^{2n+1} A_j \right| \geq \sum_{j=1}^{2n+1} |A_j| - \sum_{1 \leq j < k \leq 2n+1} |A_j \cap A_k|,$$

which translates to

$$m \geq (2n + 1)2n - \binom{2n + 1}{2} = 2n^2 + n,$$

where we used the fact that $|A_j \cap A_k| = 1$ for all $j \neq k$. We conclude that $m = 2n^2 + n$, that is B has precisely $2n^2 + n$ elements.

Next we will show that a necessary condition for the coloring to exist is that n is even. Let $f : B \to \{-1, +1\}$ be the coloring. Indeed,

$$\sum_{i=1}^{2n^2+n} f(x_i) = \frac{1}{2} \sum_{j=1}^{2n+1} \sum_{x_i \in A_j} f(x_i) = \sum_{j=1}^{2n+1} 0 = 0.$$

Since $f(x_i) = \pm 1$, this can only happen if $2n^2 + n$ is even, that is if n is even.

Now we will show that the fact that n is even is also a sufficient condition. First note that the graph G is connected. Indeed, by (i), any two sets A_j and A_k are connected by a path (of length 2), and also every vertex x_i is connected to some vertex A_j. Every vertex has an even degree, since $d(x_i) = 2$ and $d(A_j) = 2n$. Therefore G contains an Eulerian cycle (i.e. a path that travels over each edge exactly once returning where it started). Let us start at x_1 and color the first edge, connecting x_1 to some A_j by $+1$, the second edge, from A_j to some x_i by -1, the third edge by -1 again, then by $+1$, the convention being that at the vertices A_j we change the sign, while at the vertices x_i we keep the sign. Then both edges entering one vertex x_i are colored by the same number; we associate this number to the vertex. Because half of the edges entering a vertex A_j are $+1$ and half are -1, it follows that for this particular coloring, half of the elements of A_j are colored by $+1$ and half by -1. The problem is solved. □

The following problems are left to the reader.

993. Prove that every graph has two vertices that are endpoints of the same number of edges.

994. Let A be a finite set and let $f : A \to A$ be a function. Prove that there exist the pairwise disjoint sets A_0, A_1, A_2, A_3 such that $A = A_0 \cup A_1 \cup A_2 \cup A_3$, $f(x) = x$ for any $x \in A_0$ and $f(A_i) \cap A_i = \emptyset$, $i = 1, 2, 3$. What if the set A is infinite?

995. One day, the students living in the dorms of a university got the flu. First some students got it, then they were taken care of by their friends, who got themselves sick as a result. Each student was sick exactly for one day, after which he was immune for a day and healthy for another, during which he could contact the flu if close to a sick friend and get sick the day after. While healthy, each student takes care of all sick friends. Once the epidemics started, the students forgot that vaccines are available.

(a) Show that if right before the epidemics started some students got vaccinated, so that they were immune on the first day, then it could happen that the epidemics lasts forever.

(b) If on the first day none of the students is immune, then the epidemics will eventually die out.

996. In a tournament $2n$ teams took part. On the first day, n pairs of teams competed. On the second day, other n pairs of teams competed. Show that at the end of the second day one can find n teams such that no two have competed with each other yet.

997. Let G be a connected graph with k edges. Show that it is possible to label the edges of this graph with the numbers $1, 2, \ldots, k$, so that for every vertex that belongs to at least two edges, the greatest common divisor of the integers that label the edges containing this vertex is equal to 1.

998. Let G be the complete graphs with 4 vertices from which we deleted one edge. Find the number of circuits of length n in G.

999. Let G be a graph with the property that for every connected subgraph H of G, the graph $G \backslash H$ is also connected. Prove that G is either a complete graph, or a cycle. (Here $G \backslash H$ is obtained from G by deleting all edges of H and all vertices of H that do not belong to edges that are not in H).

1000. A triangle is divided into smaller triangles that do not overlap such that any two triangles of the decomposition are either disjoint, have a vertex in common, or have an entire side in common. The three vertices of the original triangle are colored red, green, and blue, respectively. The vertices of the triangles from the decomposition are colored red, green, or blue, so that the vertices that lie on a side of the original triangle are only colored with the colors of the vertices of that side. Prove that among the triangles of the decomposition there is one whose vertices are colored by each of red, green, and blue.

1001. Prove that if a convex polyhedron has the property that every vertex belongs to an even number of edges, then any section determined by a plane that does not pass through a vertex is a polygon with an even number of sides.

1002. Consider a graph with the property that each vertex belongs to an odd number of edges. Initially, the vertices are colored red or blue, then at each step the following operation is performed repeatedly: if for a vertex more than half of its neighbors are colored by a different color then the vertex changes its color, otherwise it stays the same. Show that after awhile there will be some vertices that don't change color anymore and some that change color at every step.

6.3.2 Euler's Formula for Planar Graphs

This section is about a graph-theoretical result with geometric flavor, the famous Euler's formula. A planar graph is a graph embedded in the plane in such a way that edges do not cross. The connected components of the complement of a planar graph are called faces. For example, the graph in Figure 45 has four faces (this includes the infinite face). Unless otherwise specified, all our graphs are assumed to be connected.

Euler's theorem. *Given a connected planar graph, denote by V the number of vertices, by E the number of edges, and by F the number of faces (including the infinite face). Then*

$$V - E + F = 2.$$

Proof. The proof is an easy induction on F. If $F = 1$ the graph is a tree, and the number of vertices exceeds that of edges by 1. The formula is thus verified in this case.

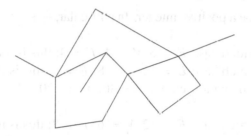

Figure 45

Let us now consider some $F > 1$ and assume that the formula holds for all graphs with at most $F - 1$ faces. Since there are at least two faces, the graph is not a tree. Therefore, it must contain cycles. Remove one edge from a cycle. The new graph is still connected. The number of edges has decreased by 1; that of faces has also decreased by 1. By the induction hypothesis,

$$V - (E - 1) + (F - 1) = 2;$$

hence Euler's formula holds for the original graph, too. This completes the proof.   □

This method of proof is called reduction of complexity, and is widely applied in a combinatorial branch of geometry called low-dimensional topology.

As a corollary, if V, E, and F are the numbers of vertices, edges, and faces of a convex polyhedron, then $V - E + F = 2$. As you can see, it was much easier to prove this formula for general planar graphs. The number 2 in Euler's formula is called the Euler (or Euler-Poincaré) characteristic of the sphere, since any convex polyhedron has the shape of a sphere. If a polyhedron has the shape of a sphere with g handles (a so-called surface of genus g), this number should be replaced by $2 - 2g$. The faces of such a graph should be planar polygons (no holes or handles). The Euler characteristic is an example of a "topological invariant"; it detects the number of handles of a polyhedral surface. The Euler characteristic has far reaching generalizations throughout algebraic topology.

As an application of Euler's formula, let us determine the Platonic solids. Recall that a Platonic solid (i.e., a regular polyhedron) is a polyhedron whose faces are congruent regular polygons and such that each vertex belongs to the same number of edges.

Example. Find all Platonic solids.

Solution. Let m be the number of edges that meet at a vertex and let n be the number of edges of a face. With the usual notation, when counting vertices by edges, we obtain $2E = mV$. When counting faces by edges, we obtain $2E = nF$. Euler's formula becomes

$$\frac{2}{m}E - E + \frac{2}{n}E = 2,$$

or

$$E = \left(\frac{1}{m} + \frac{1}{n} - \frac{1}{2}\right)^{-1}.$$

The right-hand side must be a positive integer. In particular, $\frac{1}{m} + \frac{1}{n} > \frac{1}{2}$. The only possibilities are the following:

1. $m = 3, n = 3$, in which case $E = 6$, $V = 4$, $F = 4$; this is the regular tetrahedron.
2. $m = 3, n = 4$, in which case $E = 12$, $V = 8$, $F = 6$; this is the cube.
3. $m = 3$, $n = 5$, in which case $E = 30$, $V = 20$, $F = 12$; this is the regular dodecahedron.
4. $m = 4, n = 3$, in which case $E = 12$, $V = 6$, $F = 8$; this is the regular octahedron.
5. $m = 5, n = 3$, in which case $E = 30$, $V = 12$, $F = 20$; this is the regular icosahedron.

We have proved the well-known fact that there are five Platonic solids. $\qquad\square$

1003. In the plane are given $n > 2$ points joined by segments, such that the interiors of any two segments are disjoint. Find the maximum possible number of such segments as a function of n.

1004. Three conflicting neighbors have three common wells. Can one draw nine paths connecting each of the neighbors to each of the wells such that no two paths intersect?

1005. Consider a polyhedron with at least five faces such that exactly three edges emerge from each vertex. Two players play the following game: the players sign their names alternately on precisely one face that has not been previously signed. The winner is the player who succeeds in signing the name on three faces that share a common vertex. Assuming optimal play, prove that the player who starts the game always wins.

1006. Denote by V the number of vertices of a convex polyhedron, and by Σ the sum of the (planar) angles of its faces. Prove that $2\pi V - \Sigma = 4\pi$.

1007. (a) Given a connected planar graph whose faces are polygons with at least three sides (no loops or bigons), prove that there is a vertex that belongs to at most five edges.
(b) Prove that any map in the plane can be colored by five colors such that adjacent regions have different colors (the regions are assumed to be polygons, two regions are adjacent if they share at least one side).

1008. Consider a convex polyhedron whose faces are triangles and whose edges are oriented. A *singularity* is a face whose edges form a cycle, a vertex that belongs only to incoming edges, or a vertex that belongs only to outgoing edges. Show that the polyhedron has at least two singularities.

6.3.3 Ramsey Theory

Ramsey theory is a difficult branch of combinatorics, which gathers results showing that when a sufficiently large set is partitioned into a fixed number of subsets, one of the subsets has a certain property. Finding sharp bounds on how large the set should be is a truly challenging question, unanswered in most cases. Because most of Ramsey theory is about finding structure on graphs, we placed it in the section about graphs.

The origins of this field lie in Ramsey's theorem, which states that for every pair of positive integers (p, q) there is a smallest integer $R(p, q)$, nowadays called the Ramsey number, such that whenever the edges of a complete graph with $R(p, q)$ vertices are colored red and blue, there is either a complete subgraph with p vertices whose edges are all red, or a complete subgraph with q vertices whose edges are all blue. (Recall that a complete graph is an unoriented graph in which any two vertices are connected by an edge.)

Here is a simple problem in Ramsey theory.

Example. Show that if the points of the plane are colored black or white, then there exists an equilateral triangle whose vertices are colored by the same color.

Solution. Suppose that there exists a configuration in which no monochromatic equilateral triangle is formed.

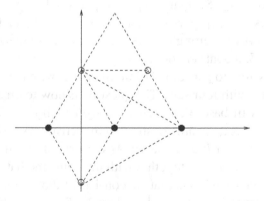

Figure 46

Start with two points of the same color, say black. Without loss of generality, we may assume that they are $(1, 0)$ and $(-1, 0)$. Then $(0, \sqrt{3})$ and $(0, -\sqrt{3})$ must both be white. Consequently, $(2, 0)$ is black, and so $(1, \sqrt{3})$ is white. Then on the one hand, $(1, 2\sqrt{3})$ cannot be black, and on the other hand it cannot be white, a contradiction. Hence the conclusion. This argument can be followed easily on Figure 46. $\qquad\square$

We now present a problem from the 2000 Belarus Mathematical Olympiad, which we particularly liked because the solution contains a nice interplay between combinatorics and number theory.

Example. Let $M = \{1, 2, \ldots, 40\}$. Find the smallest positive integer n for which it is possible to partition M into n disjoint subsets such that whenever a, b, and c (not necessarily distinct) are in the same subset, $a \neq b + c$.

Solution. We will show that $n = 4$. Assume first that it is possible to partition M into three such sets X, Y, and Z. First trick: order the sets in decreasing order of their cardinalities as $|X| \geq |Y| \geq |Z|$. Let $x_1, x_2, \ldots, x_{|X|}$ be the elements of X in increasing order. These numbers, together with the differences $x_i - x_1, i = 2, 3, \ldots, |X|$, must all be distinct elements of M. Altogether, there are $2|X| - 1$ such numbers, implying that $2|X| - 1 \leq 40$, or $|X| \leq 20$. Also, $3|X| \geq |X| + |Y| + |Z| = 40$, so $|X| \geq 14$.

There are $|X| \cdot |Y| \geq |X| \times \frac{1}{2}(40 - |X|)$ pairs in $X \times Y$. The sum of the numbers in each pair is at least 2 and at most 80, a total of 79 possible values. Because $14 \leq |X| \leq 20$ and the function $f(t) = \frac{1}{2}t(40 - t)$ is concave on the interval $[14, 20]$, we have that

$$\frac{|X|(20 - |X|)}{2} \geq \min\left\{\frac{14 \cdot 26}{2}, \frac{20 \cdot 20}{2}\right\} = 182 > 2 \cdot 79.$$

We can use the pigeonhole principle to find three distinct pairs $(x_1, y_1), (x_2, y_2), (x_3, y_3) \in X \times Y$ with $x_1 + y_1 = x_2 + y_2 = x_3 + y_3$.

If any of the x_i's were equal, then the corresponding y_i's would be equal, which is impossible because the pairs (x_i, y_i) are distinct. We may thus assume, without loss of generality, that $x_1 < x_2 < x_3$. For $1 \leq j < k \leq 3$, the value $x_k - x_j$ in M but cannot be in X because otherwise $x_j + (x_k - x_j) = x_k$. Similarly, $y_j - y_k \notin Y$ for $1 \leq j < k \leq 3$. Therefore, the three common differences $x_2 - x_1 = y_1 - y_2, x_3 - x_2 = y_2 - y_3$, and $x_3 - x_1 = y_1 - y_3$ are in $M \setminus (X \cup Y) = Z$. However, setting $a = x_2 - x_1, b = x_3 - x_2$, and $c = x_3 - x_1$, we have $a + b = c$ with $a, b, c \in \mathbb{Z}$, a contradiction.

Therefore, it is impossible to partition M into three sets with the desired property. Let us show that this can be done with four sets. The question is how to organize the 40 numbers.

We write the numbers in base 3 as $\ldots a_t \ldots a_3 a_2 a_1$ with only finitely many digits not equal to 0. The sets $A_1, A_2, A_3, \ldots$ are constructed inductively as follows. A_1 consists of all numbers for which $a_1 = 1$. For $k > 1$ the set A_k consists of all numbers with $a_k = 0$ that were not already placed in other sets, together with the numbers that have $a_k = 1$ and $a_i = 0$ for $i < k$. An alternative description is that A_k consists of those numbers that are congruent to some integer in the interval $\left(\frac{1}{2}3^{k-1}, 3^{k-1}\right]$ modulo 3^k. For our problem,

$A_1 = \{1, 11, 21, 101, 111, 121, 201, 211, 221, 1001, 1011, 1021, 1101, 1111\},$

$A_2 = \{2, 10, 102, 110, 202, 210, 1002, 1010, 1102, 1110\},$

$A_3 = \{12, 20, 22, 100, 1012, 1020, 1022, 1100\},$

$A_4 = \{112, 120, 122, 200, 212, 220, 222, 1000\}.$

Using the first description of these sets, we see that they exhaust all positive integers. Using the second description we see that $(A_k + A_k) \cap A_k = \emptyset, k \geq 1$. Hence A_1, A_2, A_3, A_4 provide the desired example, showing that the answer to the problem is $n = 4$. $\square$

Remark. In general, for positive integers n and k and a partition of $\{1, 2, \ldots, k\}$ into n sets, a triple (a, b, c) such that a, b, and c are in the same set and $a + b = c$ is called a Schur triple. Schur's theorem proves that for each n there exists a minimal number $S(n)$ such that for any partition of $\{1, 2, \ldots, S(n)\}$ into n sets one of the sets will contain a Schur triple. No general formula for $S(n)$ exists although upper and lower bounds have been found. Our problem proves that $S(4) > 40$. In fact, $S(4) = 45$.

We suggest that after solving Problems 1013 and 1014 below, the reader takes a look at the remark after the solution of the second of these problems given in the second part of the book. That remark will explain how the problem we just solved is actually a question in graph theory.

1009. What is the largest number of vertices that a complete graph can have so that its edges can be colored by two colors in such a way that no monochromatic triangle is formed?

1010. For the Ramsey numbers defined above, prove that $R(p, q) \leq R(p-1, q) + R(p, q-1)$. Conclude that for $p, q \geq 2$,

$$R(p, q) \leq \binom{p + q - 2}{p - 1}.$$

1011. A group of people is said to be n-balanced if in any subgroup of 3 people there exists (at least) a pair acquainted with each other and in any subgroup of n people there exists (at least) a pair not acquainted with each other. Prove that the number of people in an n-balanced group has an upper bound, and compute this upper bound for $n = 3, 4, 5$.

1012. The edges of a complete graph with $\lfloor k!e \rfloor + 1$ vertices are colored by k colors. Prove that there is a triangle whose edges are colored by the same color.

1013. An international society has members from six different countries. The list of members contains 1978 names, numbered $1, 2, \ldots, 1978$. Prove that there exists at least one member whose number is the sum of the numbers of two members from his/her own country, or twice as large as the number of one member from his/her country.

6.4 Binomial Coefficients and Counting Methods

6.4.1 Combinatorial Identities

The binomial coefficient $\binom{n}{k}$ counts the number of ways one can choose k objects from given n. Binomial coefficients show up in Newton's binomial expansion

$$(x + 1)^n = \binom{n}{0}x^n + \binom{n}{1}x^{n-1} + \cdots + \binom{n}{n-1}x + \binom{n}{n}.$$

Explicitly,

$$\binom{n}{k} = \frac{n!}{k!(n-k)!} = \frac{n(n-1)\cdots(n-k+1)}{k!} \quad \text{if} \ \ 0 \le k \le n.$$

The recurrence relation

$$\binom{n}{k} = \binom{n-1}{k} + \binom{n-1}{k-1}$$

allows the binomial coefficients to be arranged in *Pascal's triangle*:

$$
\begin{array}{ccccccccccc}
 & & & & & 1 & & & & & \\
 & & & & 1 & & 1 & & & & \\
 & & & 1 & & 2 & & 1 & & & \\
 & & 1 & & 3 & & 3 & & 1 & & \\
 & 1 & & 4 & & 6 & & 4 & & 1 & \\
1 & & 5 & & 10 & & 10 & & 5 & & 1 \\
\end{array}
$$

$$\cdots\ \cdots\ \cdots\ \cdots\ \cdots\ \cdots\ \cdots\ \cdots\ \cdots\ \cdots\ \cdots$$

Here every entry is obtained by summing the two entries just above it.

Let us familiarize ourselves with Pascal's triangle with the following problem published by A. Avramov in the Russian Journal *Kvant (Quantum)*.

Example. In the seventh row of Pascal's triangle, there are three consecutive binomial coefficients that form an arithmetic progression, namely $7, 21, 35$. Find all rows that contain 3-term arithmetic progressions.

Solution. Assume that for some n and k, $\binom{n}{k-1}$, $\binom{n}{k}$ and $\binom{n}{k+1}$ form an arithmetic progression. This means that

$$2\binom{n}{k} = \binom{n}{k-1} + \binom{n}{k+1}.$$

Writing

$$\frac{2n!}{k!(n-k)!} = \frac{n!}{(k-1)!(n-k+1)!} + \frac{n!}{(k+1)!(n-k-1)!},$$

we obtain, after some cancellations,

$$4k^2 - 4nk + n^2 - n - 2 = 0.$$

We solve this like a quadratic equation in k to obtain

$$k_{1,2} = \frac{n \pm \sqrt{n+2}}{2}.$$

We deduce that $n + 2$ must be a perfect square, say $n + 2 = m^2$. Then the solutions to this quadratic equation are

$$k_1 = \frac{(m+1)(m-2)}{2} \quad \text{and} \quad k_2 = \frac{(m-1)(m+2)}{2}.$$

The numerators are even numbers in both cases, so these are integers. Note also that since $n \geq 0$ and $k \geq 1$, we must have $m \geq 3$. In that case both k_1 and k_2 yield solutions, but they must be the same solution because of the fact that $\binom{n}{k} = \binom{n}{n-k}$.

We conclude that the ranks of the rows in which this happens are of the form $m^2 - 2$, with $m \geq 3$. In each row there are exactly two such progressions, one being the other written in reverse.   □

As a corollary we obtain that there are no rows of Pascal's triangle that contain four-term arithmetic progressions. And now a problem from the 2001 Hungarian Mathematical Olympiad.

Example. Let m and n be integers such that $1 \leq m \leq n$. Prove that m divides the number

$$n \sum_{k=0}^{m-1} (-1)^k \binom{n}{k}.$$

Solution. We would like to express the sum in closed form. To this end, we apply the recurrence formula for binomial coefficients and obtain

$$n \sum_{k=0}^{m-1} (-1)^k \binom{n}{k} = n \sum_{k=0}^{m-1} (-1)^k \left(\binom{n-1}{k} + \binom{n-1}{k-1} \right)$$

$$= n \sum_{k=0}^{m-1} (-1)^k \binom{n-1}{k} - n \sum_{k=0}^{m-2} (-1)^k \binom{n-1}{k}$$

$$= n(-1)^{m-1} \binom{n-1}{m-1} = m(-1)^{m-1} \binom{n}{m}.$$

The answer is clearly divisible by m.   □

The methods used in proving combinatorial identities can be applied to problems outside the field of combinatorics. As an example, let us take a fresh look at a property that we encountered elsewhere in the solution to a problem about polynomials.

Example. If k and m are positive integers, prove that the polynomial

$$(x^{k+m} - 1)(x^{k+m-1} - 1) \cdots (x^{k+1} - 1)$$

is divisible by

$$(x^m - 1)(x^{m-1} - 1) \cdots (x - 1)$$

in the ring of polynomials with integer coefficients.

Solution. Let us analyze the quotient

$$p_{k,m}(x) = \frac{(x^{k+m} - 1)(x^{k+m-1} - 1) \cdots (x^{k+1} - 1)}{(x^m - 1)(x^{m-1} - 1) \cdots (x - 1)},$$

which conjecturally is a polynomial with integer coefficients. The main observation is that

$$\lim_{x \to 1} p_{k,m}(x) = \lim_{x \to 1} \frac{(x^{k+m} - 1)(x^{k+m-1} - 1) \cdots (x^{k+1} - 1)}{(x^m - 1)(x^{m-1} - 1) \cdots (x - 1)}$$

$$= \lim_{x \to 1} \frac{x^{k+m} - 1}{x - 1} \cdots \frac{x^{k+1} - 1}{x - 1} \cdot \frac{x - 1}{x^m - 1} \cdots \frac{x - 1}{x - 1}$$

$$= \frac{(k + m)(k + m - 1) \cdots (k + 1)}{m(m - 1) \cdots 1} = \binom{k + m}{m}.$$

With this in mind, we treat $p_{k,m}(x)$ as some kind of binomial coefficient. Recall that one way of showing that $\binom{n}{m} = \dfrac{n!}{m!(n - m)!}$ is an integer number is by means of Pascal's triangle. We will construct a Pascal's triangle for the polynomials $p_{k,m}(x)$. The recurrence relation

$$\binom{k + m + 1}{m} = \binom{k + m}{m} + \binom{k + m}{m - 1}$$

has the polynomial analogue

$$\frac{(x^{k+m+1} - 1) \cdots (x^{k+2} - 1)}{(x^m - 1) \cdots (x - 1)} = \frac{(x^{k+m} - 1) \cdots (x^{k+1} - 1)}{(x^m - 1) \cdots (x - 1)}$$

$$+ x^{k+1} \frac{(x^{k+m} - 1) \cdots (x^{k+2} - 1)}{(x^{m-1} - 1) \cdots (x - 1)}.$$

Now the conclusion follows by induction on $m + k$, with the base case the obvious

$$\frac{x^{k+1} - 1}{x - 1} = x^k + x^{k-1} + \cdots + 1. \qquad \square$$

In quantum physics the variable x is replaced by $q = e^{i\hbar}$, where $\hbar$ is Planck's constant, and the polynomials $p_{n-m,m}(q)$ are denoted by $\binom{n}{m}_q$ and called quantum binomial coefficients (or Gauss polynomials). They arise in the context of the Heisenberg uncertainty principle. Specifically, if P and Q are the linear transformations that describe, respectively, the time evolution of the momentum and of the position of a particle, then $PQ = qQP$. The binomial formula for them reads

$$(Q + P)^n = \sum_{k=0}^{n} \binom{n}{k} Q^k P^{n-k}.$$

The recurrence relation we obtained a moment ago,

$$\binom{n}{m}_q = \binom{n - 1}{m}_q + q^{n-m} \binom{n - 1}{m - 1}_q,$$

gives rise to what is called the q-Pascal triangle.

1014. Prove that

$$\binom{2k}{k} = \frac{2}{\pi} \int_0^{\frac{\pi}{2}} (2 \sin \theta)^{2k} d\theta.$$

1015. Consider the triangular $n \times n$ matrix

$$A = \begin{pmatrix} 1 & 1 & 1 & \dots & 1 \\ 0 & 1 & 1 & \dots & 1 \\ 0 & 0 & 1 & \dots & 1 \\ \vdots & \vdots & \vdots & \ddots & \vdots \\ 0 & 0 & 0 & \dots & 1 \end{pmatrix}.$$

Compute the matrix A^k, $k \geq 1$.

1016. Let $(F_n)_n$ be the Fibonacci sequence, $F_1 = F_2 = 1$, $F_{n+1} = F_n + F_{n-1}$. Prove that for any positive integer n,

$$F_1 \binom{n}{1} + F_2 \binom{n}{2} + \cdots + F_n \binom{n}{n} = F_{2n}.$$

1017. For an arithmetic sequence $a_1, a_2, \ldots, a_n, \ldots$, let $S_n = a_1 + a_2 + \cdots + a_n$, $n \geq 1$. Prove that

$$\sum_{k=0}^{n} \binom{n}{k} a_{k+1} = \frac{2^n}{n+1} S_{n+1}.$$

1018. Show that for any positive integer n, the number

$$S_n = \binom{2n+1}{0} \cdot 2^{2n} + \binom{2n+1}{2} \cdot 2^{2n-2} \cdot 3 + \cdots + \binom{2n+1}{2n} \cdot 3^n$$

is the sum of two consecutive perfect squares.

1019. For a positive integer n define the integers a_n, b_n, and c_n by

$$a_n + b_n \sqrt[3]{2} + c_n \sqrt[3]{4} = (1 + \sqrt[3]{2} + \sqrt[3]{4})^n.$$

Prove that

$$2^{-\frac{n}{3}} \sum_{k=0}^{n} \binom{n}{k} a_k = \begin{cases} a_n & \text{if } n \equiv 0 \pmod 3, \\ b_n \sqrt[3]{2} & \text{if } n \equiv 2 \pmod 3, \\ c_n \sqrt[3]{4} & \text{if } n \equiv 1 \pmod 3. \end{cases}$$

1020. Prove the analogue of Newton's binomial formula

$$[x + y]_n = \sum_{k=0}^{n} \binom{n}{k} [x]_k [y]_{n-k},$$

where $[x]_n = x(x-1) \cdots (x-n+1)$.

1021. Prove that the quantum binomial coefficients $\binom{n}{k}_q$ previously defined satisfy

$$\sum_{k=0}^{n} (-1)^k q^{\frac{k(k-1)}{2}} \binom{n}{k}_q = 0.$$

6.4.2 Generating Functions

The terms of a sequence $(a_n)_{n \geq 0}$ can be combined into a function

$$(x) = a_0 + a_1 x + a_2 x^2 + \cdots + a_n x^n + \cdots,$$

called the generating function of the sequence. Sometimes this function can be written in closed form and carries useful information about the sequence. For example, if the sequence satisfies a second-order linear recurrence, say $a_{n+1} + u a_n + v a_{n-1} = 0$, then the generating function satisfies the functional equation

$$G(x) - a_0 - a_1 x + ux(G(x) - a_0) + vx^2 G(x) = 0.$$

This equation can be solved easily, giving

$$G(x) = \frac{a_0 + (ua_0 + a_1)x}{1 + ux + vx^2}.$$

If r_1 and r_2 are the roots of the characteristic equation $\lambda^2 + u\lambda + v = 0$, then by using the partial fraction decomposition, we obtain

$$G(x) = \frac{a_0 + (ua_0 + a_1)x}{(1 - r_1 x)(1 - r_2 x)} = \frac{\alpha}{1 - r_1 x} + \frac{\beta}{1 - r_2 x} = \sum_{n=0}^{\infty} (\alpha r_1^n + \beta r_2^n) x^n.$$

And we recover the general-term formula $a_n = \alpha r_1^n + \beta r_2^n$, $n \geq 0$, where α and β depend on the initial condition.

It is useful to notice the analogy with the method of the Laplace transform used for solving linear ordinary differential equations. Recall that the Laplace transform of a function $y(t)$ is defined as

$$\mathcal{L}y(s) = \int_0^{\infty} y(t) e^{ts} dt.$$

The Laplace transform applied to the differential equation

$$y'' + uy' + vy = 0$$

produces the algebraic equation

$$s^2 \mathcal{L}(y) - y'(0) - sy(0) + u(s\mathcal{L}(y) - y(0)) - v\mathcal{L}(y) = 0,$$

with the solution

$$\mathcal{L}(y) = \frac{sy(0) + uy(0) + y'(0)}{s^2 + us + v}.$$

Again the partial fraction decomposition comes in handy, since we know that the inverse Laplace transforms of $\frac{1}{s-r_1}$ and $\frac{1}{s-r_2}$ are $e^{r_1 x}$ and $e^{r_2 x}$. The similarity of these two methods is not accidental, for recursive sequences are discrete approximations of differential equations.

Let us return to problems and look at the classical example of the Catalan numbers.

Example. Prove that the number of ways one can insert parentheses into a product of $n + 1$ factors is the Catalan number

$$C_n = \frac{1}{n+1}\binom{2n}{n}.$$

Solution. Alternatively, the Catalan number C_n is the number of ways the terms of the product can be grouped for performing the multiplication. This is a better point of view, because the location of the final multiplication splits the product in two, giving rise to the recurrence relation

$$C_n = C_0 C_{n-1} + C_1 C_{n-2} + \cdots + C_{n-1} C_0, \quad n \geq 1.$$

Indeed, for every $k = 0, 1, \ldots, n - 1$, the first $k + 1$ terms can be grouped in C_k ways, while the last $n - k$ terms can be grouped in C_{n-k-1} ways. You can recognize that the expression on the right shows up when the generating function is squared. We deduce that the generating function satisfies the equation

$$G(x) = x G(x)^2 + 1.$$

This is a quadratic equation, with two solutions. And because $\lim_{x \to 0} G(x) = a_0$, we know precisely which solution to choose, namely

$$G(x) = \frac{1 - \sqrt{1 - 4x}}{2x}.$$

Expanding the square root with Newton's binomial formula, we have

$$\sqrt{1 - 4x} = (1 - 4x)^{1/2} = \sum_{n=0}^{\infty} \binom{1/2}{n}(-4x)^n$$

$$= \sum_{n=0}^{\infty} \frac{\left(\frac{1}{2}\right)\left(\frac{1}{2} - 1\right)\cdots\left(\frac{1}{2} - n + 1\right)}{n!}(-4x)^n$$

$$= 1 - \sum_{n=1}^{\infty} \frac{(2n - 3)(2n - 5)\cdots 1}{n!}(2x)^n$$

$$= 1 - 2\sum_{n=1}^{\infty} \frac{(2n - 2)!}{(n - 1)!(n - 1)!}\frac{x^n}{n}$$

$$= 1 - 2\sum_{n=1}^{\infty} \binom{2n - 2}{n - 1}\frac{x^n}{n}.$$

Substituting in the expression for the generating function and shifting the index, we obtain

$$G(x) = \sum_{n=0}^{\infty} \frac{1}{n+1}\binom{2n}{n}x^n,$$

which gives the formula for the Catalan number

$$C_n = \frac{1}{n+1}\binom{2n}{n}.$$

$\square$

The binomial coefficients $\binom{n}{k}$ are generated by a very simple function, $G(x) = (x + 1)^n$, and variations of this fact can be exploited to obtain combinatorial identities. This is the case with a problem published in the *American Mathematical Monthly* by N. Gonciulea.

Example. Prove that

$$\sum_{j=0}^{n} \binom{n}{j} 2^{n-j} \binom{j}{\lfloor j/2 \rfloor} = \binom{2n+1}{n}.$$

Solution. Observe that $\binom{j}{\lfloor j/2 \rfloor}$ is the constant term in $(1 + x)(x^{-1} + x)^j$. It follows that the sum is equal to the constant term in

$$\sum_{j=0}^{n} \binom{n}{j} 2^{n-j} (1 + x)(x^{-1} + x)^j = (1 + x) \sum_{j=0}^{n} \binom{n}{j} (x^{-1} + x)^j 2^{n-j}$$

$$= (1 + x)(2 + x^{-1} + x)^n$$

$$= \frac{1}{x^n}(1 + x)(2x + 1 + x^2)^n = \frac{1}{x^n}(1 + x)^{2n+1}.$$

And the constant term in this last expression is $\binom{2n+1}{n}$.   □

1022. Find the general-term formula for the sequence $(y_n)_{n \geq 0}$ with $y_0 = 1$ and $y_n = ay_{n-1} + b^n$ for $n \geq 1$, where a and b are two fixed distinct real numbers.

1023. Compute the sums

$$\sum_{k=1}^{n} k \binom{n}{k} \quad \text{and} \quad \sum_{k=1}^{n} \frac{1}{k+1} \binom{n}{k}.$$

1024. (a) Prove the identity

$$\binom{m + n}{k} = \sum_{j=0}^{k} \binom{m}{j} \binom{n}{k - j}.$$

(b) Prove that the quantum binomial coefficients defined in the previous section satisfy the identity

$$\binom{m + n}{k}_q = \sum_{j=0}^{k} q^{(m-j)(k-j)} \binom{m}{j}_q \binom{n}{k - j}_q.$$

1025. Compute the sum

$$\binom{n}{0} - \binom{n}{1} + \binom{n}{2} - \cdots + (-1)^m \binom{n}{m}.$$

1026. Write in short form the sum

$$\binom{n}{k} + \binom{n + 1}{k} + \binom{n + 2}{k} + \cdots + \binom{n + m}{k}.$$

1027. Prove that the Fibonacci numbers satisfy

$$F_{n+1} = \binom{n}{0} + \binom{n-1}{1} + \binom{n-2}{2} + \cdots$$

1028. Denote by $P(n)$ the number of partitions of the positive integer n, i.e., the number of ways of writing n as a sum of positive integers. Prove that the generating function of $P(n)$, $n \geq 1$, is given by

$$\sum_{n=0}^{\infty} P(n)x^n = \frac{1}{(1-x)(1-x^2)(1-x^3) \cdots}$$

with the convention $P(0) = 1$.

1029. Prove that the number of ways of writing n as a sum of distinct positive integers is equal to the number of ways of writing n as a sum of odd positive integers.

1030. Let p be an odd prime number. Find the number of subsets of $\{1, 2, \ldots, p\}$ with the sum of elements divisible by p.

1031. For a positive integer n, denote by $S(n)$ the number of choices of the signs "$+$" or "$-$" such that $\pm 1 \pm 2 \pm \cdots \pm n = 0$. Prove that

$$S(n) = \frac{2^{n-1}}{\pi} \int_0^{2\pi} \cos t \cos 2t \cdots \cos nt\, dt.$$

1032. The distinct positive integers $a_1, a_2, \ldots, a_n, b_1, b_2, \ldots, b_n$, with $n \geq 2$, have the property that the $\binom{n}{2}$ sums $a_i + a_j$ are the same as the $\binom{n}{2}$ sums $b_i + b_j$ (in some order). Prove that n is a power of 2.

1033. Let $A_1, A_2, \ldots, A_n, \ldots$ and $B_1, B_2, \ldots, B_n, \ldots$ be sequences of sets defined by $A_1 = \emptyset$, $B_1 = \{0\}$, $A_{n+1} = \{x + 1 \mid x \in B_n\}$, $B_{n+1} = (A_n \cup B_n) \setminus (A_n \cap B_n)$. Determine all positive integers n for which $B_n = \{0\}$.

6.4.3 Counting Strategies

We illustrate how some identities can be proved by counting the number of elements of a set in two different ways. For example, we give a counting argument to the well-known reciprocity law, which we have already encountered in Section 5.1.3, of the greatest integer function.

Example. Given p and q coprime positive integers, prove that

$$\left\lfloor \frac{p}{q} \right\rfloor + \left\lfloor \frac{2p}{q} \right\rfloor + \cdots + \left\lfloor \frac{(q-1)p}{q} \right\rfloor = \left\lfloor \frac{q}{p} \right\rfloor + \left\lfloor \frac{2q}{p} \right\rfloor + \cdots + \left\lfloor \frac{(p-1)q}{p} \right\rfloor.$$

Solution. Let us look at the points of integer coordinates that lie inside the rectangle with vertices $O(0, 0)$, $A(q, 0)$, $B(q, p)$, $C(0, p)$ (see Figure 47). There are $(p - 1)(q - 1)$ such points. None of them lies on the diagonal OB because p and q are coprime. Half of them lie above the diagonal and half below.

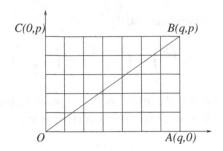

Figure 47

Now let us count by a different method the points underneath the line OB. The equation of this line is $y = \frac{p}{q}x$. For each $0 < k < q$ on the vertical segment $x = k$ there are $\lfloor kp/q \rfloor$ points below OB. Summing up, we obtain

$$\left\lfloor \frac{p}{q} \right\rfloor + \left\lfloor \frac{2p}{q} \right\rfloor + \cdots + \left\lfloor \frac{(q-1)p}{q} \right\rfloor = \frac{(p-1)(q-1)}{2}.$$

The expression on the right remains unchanged if we switch p and q, which proves the identity. $\qquad\square$

Next, a combinatorial identity.

Example. Let m and n be two integers, $m \le \dfrac{n-1}{2}$. Prove the identity

$$\sum_{k=m}^{\frac{n-1}{2}} \binom{n}{2k+1}\binom{k}{m} = 2^{n-2m-1}\binom{n-m-1}{m}.$$

Solution. The solution is a "Fubini-type" argument (counting the same same thing in two different ways). Consider the set $\mathcal{P}$ of pairs (A, B), where A is a subset of $\{1, 2, \ldots, n\}$ with an odd number of elements $a_1 < a_2 < \cdots < a_{2k+1}$ and B is a subset of $\{a_2, a_4, \ldots, a_{2k-2}, a_{2k}\}$ with m elements $b_1 < b_2 < \cdots < b_m$.

For a given k there are $\binom{n}{2k+1}$ such subsets A, and for each A there are $\binom{k}{m}$ subsets B, so the left-hand side of the identity is the number of elements of $\mathcal{P}$ counted by choosing A first.

Let us count the same number choosing B first. Note that if $(A, B) \in \mathcal{P}$, then B contains no pairs of consecutive numbers. More precisely, $B = \{b_1, b_2, \ldots, b_m\} \subset \{2, 3, \ldots, n-1\}$ with $b_{i+1} - b_i \ge 2$.

Fix B_0, a set with this property. We want to count the number of pairs (A, B_0) in X. Choose $c_0, c_1, \ldots, c_m$ such that

$$1 \le c_0 \le b_1 < c_1 < b_2 < \cdots < b_i < c_i < b_{i+1} < \cdots < b_m < c_m \le n.$$

Then for any subset E of $\{1, 2, \ldots, n\} \setminus \{c_0, b_1, c_1, b_2, \ldots, b_m, c_m\}$ there is a unique A such that $(A, B_0) \in \mathcal{P}$ and

$$E = A \cap (\{1, 2, \ldots, n\} \setminus \{c_0, b_1, c_1, b_2, \ldots, b_m, c_m\}).$$

Indeed, if $(A, B_0) \in P$ and $E \subset A$ we have to decide which c_i's are in A. Since the set $D_i = \{x \in A \mid b_i < x < b_{i+1}\}$ must contain an odd number of elements for each $0 \le i \le m+1$ (with $b_0 = 0$, $b_{m+1} = n+1$), and the set D_i is either $\{x \in E \mid b_i < x < b_{i+1}\}$ or $\{x \in E \mid b_i < x < b_{i+1}\} \cup \{c_i\}$, the parity condition on the cardinality of D_i decides whether c_i belongs to A. It is now clear that the number of pairs (A, B_0) in P is the same as the number of subsets of $\{1, 2, \ldots, n\} \setminus \{c_0, b_1, \ldots, b_m, c_m\}$ and the latter is 2^{n-2m-1}.

How many subsets B with m elements of $\{2, 3, \ldots, n-1\}$ do not contain consecutive numbers? If $B = \{b_1 < b_2 < \cdots < b_m\}$ is such a set, let $B' = \{b_1 - 1, b_2 - 2, \ldots, b_m - m\}$. It is easy to see that B' is an (arbitrary) subset of $\{1, 2, \ldots, n-m-1\}$ with m elements, and for each such subset $B' = \{b'_1 < b'_2 < \cdots < b'_m\}$, by letting $b_i = b'_i + i$, we obtain a set B as above. Hence the number of such B's is $\binom{n-m-1}{m}$, and by choosing B first we count the number of elements in P as $2^{n-2m-1}\binom{n-m-1}{m}$. The identity is proved. $\square$

Using similar ideas solve the following problems.

1034. Find in closed form

$$1 \cdot 2\binom{n}{2} + 2 \cdot 3\binom{n}{3} + \cdots + (n-1) \cdot n\binom{n}{n}.$$

1035. Prove the combinatorial identity

$$\sum_{k=1}^{n} k\binom{n}{k}^2 = n\binom{2n-1}{n-1}.$$

1036. Prove the identity

$$\sum_{k=0}^{m}\binom{m}{k}\binom{n+k}{m} = \sum_{k=0}^{m}\binom{m}{k}\binom{n}{k}2^k.$$

1037. For integers $0 \le k \le n$, $1 \le m \le n$, prove the identity

$$\sum_{i=0}^{m}\binom{m}{i}\binom{n-i}{k} = \sum_{i=0}^{m}\binom{m}{i}\binom{n-m}{k-i}2^{m-i}.$$

1038. Show that for any positive integers p and q,

$$\sum_{k=0}^{q}\frac{1}{2^{p+k}}\binom{p+k}{k} + \sum_{k=0}^{p}\frac{1}{2^{q+k}}\binom{q+k}{k} = 2.$$

1039. Let $c_n = \binom{n}{\lfloor n/2 \rfloor}$. Prove that

$$\sum_{k=0}^{n}\binom{n}{k}c_k c_{n-k} = c_n c_{n+1}.$$

1040. Let p and q be odd, coprime positive integers. Set $p' = \frac{p-1}{2}$ and $q' = \frac{q-1}{2}$. Prove the identity

$$\left(\left\lfloor \frac{q}{p} \right\rfloor + \left\lfloor \frac{2q}{p} \right\rfloor + \cdots + \left\lfloor \frac{p'q}{p} \right\rfloor\right) + \left(\left\lfloor \frac{p}{q} \right\rfloor + \left\lfloor \frac{2p}{q} \right\rfloor + \cdots + \left\lfloor \frac{q'p}{p} \right\rfloor\right) = p'q'.$$

Now we turn to more diverse counting arguments.

Example. What is the number of ways of writing the positive integer n as an ordered sum of m positive integers?

Solution. This is a way of saying that we have to count the number of m-tuples of positive integers $(x_1, x_2, \ldots, x_m)$ satisfying the equation $x_1 + x_2 + \cdots + x_m = n$. These m-tuples are in one-to-one correspondence with the strictly increasing sequences $0 < y_1 < y_2 < \cdots < y_m = n$ of positive integers, with the correspondence given by $y_1 = x_1$, $y_2 = x_1 + x_2, \ldots$, $y_m = x_1 + x_2 + \cdots + x_m$. The numbers $y_1, y_2, \ldots, y_{m-1}$ can be chosen in $\binom{n-1}{m-1}$ ways from $1, 2, \ldots, n-1$. Hence the answer to the question is $\binom{n-1}{m-1}$.

This formula can also be proved using induction on m for arbitrary n. The case $m = 1$ is obvious. Assume that the formula is valid for partitions of any positive integer into $k \leq m$ positive integers, and let us prove it for partitions into $m + 1$ positive integers. The equation $x_1 + x_2 + \cdots + x_m + x_{m+1} = n$ can be written as

$$x_1 + x_2 + \cdots + x_m = n - x_{m+1}.$$

As x_{m+1} ranges among $1, 2, \ldots, n-m$, we are supposed to count the total number of solutions of the equations $x_1 + x_2 + \cdots + x_m = r$, with $r = m, m+1, \ldots, n-1$. By the induction hypothesis, this number is

$$\sum_{r=m}^{n-1} \binom{r-1}{m-1}.$$

We have seen in Section 6.4.2 that this number is equal to $\binom{n-1}{m-1}$. This equality can also be proved using Pascal's triangle as follows:

$$\binom{m-1}{n-1} + \binom{m}{m-1} + \cdots + \binom{n-2}{m-1} = \binom{m}{m} + \binom{m}{m-1} + \cdots + \binom{n-2}{m-1}$$
$$= \binom{m+1}{m} + \binom{m+1}{m-1} + \cdots + \binom{n-2}{m-1}$$
$$= \binom{m+2}{m} + \cdots + \binom{n-2}{m-1} = \cdots$$
$$= \binom{n-2}{m} + \binom{n-2}{m-1} = \binom{n-1}{m}.$$

This proves that the formula is true for $m + 1$, and the induction is complete. $\square$

Example. There are n students at a university, n an odd number. Some students join together to form several clubs (a student may belong to different clubs). Some clubs join together to form several societies (a club may belong to different societies). There are k societies. Suppose that the following hold:

(i) each pair of students is in exactly one club,

(ii) for each student and each society, the student is in exactly one club of the society,

(iii) each club has an odd number of students; in addition, a club with $2m + 1$ students ($m > 0$) is in exactly m societies.

Find all possible values of k.

Solution. This is a short-listed problem from the 45th International Mathematical Olympiad, 2004, proposed by Puerto Rico, which was given a year later at an Indian team selection test. Here is an ingenious approach found by one of the Indian students, R. Shah.

Fix a student x and list the clubs to which the student belongs: $C_1, C_2, \ldots, C_r$. If C_i has $2m_i + 1$ students, then it belongs to m_i societies. Condition (ii) implies that for $i \neq j$ the societies to which C_i belongs are all different from the societies to which C_j belongs. Moreover, condition (ii) guarantees that any society will contain one of the clubs C_i. Therefore, $m_1 + m_2 + \cdots + m_r = k$.

From condition (i) we see that any two clubs C_i and C_j have in common exactly the student x. Therefore, in $C_1, C_2, \ldots, C_r$ there are altogether $2(m_1 + m_2 + \cdots + m_r) + 1$ students. But these are all the students, because by condition (i) any other student is in some club with x. We obtain

$$2(m_1 + m_2 + \cdots + m_r) + 1 = 2k + 1 = n.$$

Hence $k = \frac{n-1}{2}$ is the only possibility. And this situation can be achieved when all students belong to one club, which then belongs to $\frac{n-1}{2}$ societies.   □

Here is a third example.

Example. On an 8×8 chessboard whose squares are colored black and white in an arbitrary way we are allowed to simultaneously switch the colors of all squares in any 3×3 and 4×4 region. Can we transform any coloring of the board into one where all the squares are black?

Solution. We claim that the answer is no. It is a matter of counting into how many regions can an all-black board be transformed by applying the two moves several times. The total number of 3×3 regions is $(8 - 2) \times (8 - 2) = 36$, which is the same as the number of moves in which the colors in a 3×3 region are switched. As for the 4×4 regions, there are $(8 - 3) \times (8 - 3)25$ of them. Hence the total number of colorings that can be obtained from an all-black coloring by applying the specified operations does not exceed

$$2^{36} \times 2^{25} = 2^{61}.$$

This number is less than the total number of colorings, which is 2^{64}. Hence there are colorings that cannot be achieved. Since the operations are reversible, this actually proves our claim.   □

And now the problems.

1041. Two hundred students took part in a mathematics contest. They had 6 problems to solve. It is known that each problem was correctly solved by at least 120 participants. Prove that there exist two participants such that every problem was solved by at least one of them.

1042. Prove that the number of nonnegative integer solutions to the equation

$$x_1 + x_2 + \cdots + x_m = n$$

is equal to $\binom{m+n-1}{m-1}$.

1043. Find the number of subsets of the set $\{1, 2, \ldots, n\}$, including the empty set, that do not contain two consecutive integers.

1044. Consider a polyhedron whose faces are triangles. Color the vertices by $n \geq 3$ colors. Prove that the number of faces with vertices colored by three different colors is even.

1045. A number n of tennis players take part in a tournament in which each of them plays exactly one game with each of the others. If x_i and y_i denote the number of victories, respectively, losses, of the ith player, $i = 1, 2, \ldots, n$, show that

$$x_1^2 + x_2^2 + \cdots + x_n^2 = y_1^2 + y_2^2 + \cdots + y_n^2.$$

1046. Let A be a finite set and f and g two functions on A. Let m be the number of pairs $(x, y) \in A \times A$ for which $f(x) = g(y)$, n the number of pairs for which $f(x) = f(y)$, and k the number of pairs for which $g(x) = g(y)$. Prove that

$$2m \leq n + k.$$

1047. A set S containing four positive integers is called *connected* if for every $x \in S$ at least one of the numbers $x - 1$ and $x + 1$ belongs to S. Let C_n be the number of connected subsets of the set $\{1, 2, \ldots, n\}$.

(a) Evaluate C_7.
(b) Find a general formula for C_n.

1048. Prove that the set of numbers $\{1, 2, \ldots, 2005\}$ can be colored with two colors such that any of its 18-term arithmetic sequences contains both colors.

1049. For $A = \{1, 2, \ldots, 100\}$ let $A_1, A_2, \ldots, A_m$ be subsets of A with four elements with the property that any two have at most two elements in common. Prove that if $m \geq 40425$ then among these subsets there exist 49 whose union is equal to A but with the union of any 48 of them not equal to A.

1050. Let S be a finite set of points in the plane. A linear partition of S is an unordered pair $\{A, B\}$ of subsets of S such that $A \cup B = S$, $A \cap B = \emptyset$, and A and B lie on opposite sides of some straight line disjoint from S (A or B may be empty). Let L_S be the number of linear partitions of S. For each positive integer n, find the maximum of L_S over all sets S of n points.

1051. Let A be a 101-element subset of the set $S = \{1, 2, \ldots, 1000000\}$. Prove that there exist numbers $t_1, t_2, \ldots, t_{100}$ in S such that the sets

$$A_j = \{x + t_j \mid x \in A\}, \quad j = 1, 2, \ldots, 100,$$

are pairwise disjoint.

1052. Given a set A with n^2 elements, $n \geq 2$, and $\mathcal{F}$ a family of subsets of A each of which has n elements, suppose that any two sets of $\mathcal{F}$ have at most one element in common.

(a) Prove that there are at most $n^2 + n$ sets in $\mathcal{F}$.
(b) In the case $n = 3$, show with an example that this bound can be reached.

1053. A sheet of paper in the shape of a square is cut by a line into two pieces. One of the pieces is cut again by a line, and so on. What is the minimum number of cuts one should perform such that among the pieces one can find one hundred polygons with twenty sides?

1054. Twenty-one girls and twenty-one boys took part in a mathematics competition. It turned out that
(i) each contestant solved at most six problems, and
(ii) for each pair of a girl and a boy, there was at least one problem that was solved by both the girl and the boy.
Show that there is a problem that was solved by at least three girls and at least three boys.

1055. Is it possible to color the squares of a rectangular grid by black and white such that there are as many black squares as white squares and on each row and column more than 3/4 of the squares are of the same color?

6.4.4 The Inclusion-Exclusion Principle

A particular counting method that we emphasize is the inclusion-exclusion principle, also known as the Boole-Sylvester formula. It concerns the counting of the elements in a union of sets $A_1 \cup A_2 \cup \cdots \cup A_n$, and works as follows. If we simply wrote

$$|A_1 \cup A_2 \cup \cdots \cup A_n| = |A_1| + |A_2| + \cdots + |A_n|,$$

we would overcount the elements in the intersections $A_i \cap A_j$. Thus we have to subtract $|A_1 \cap A_2| + |A_1 \cap A_3| + \cdots + |A_{n-1} \cap A_n|$. But then the elements in the triple intersections $A_i \cap A_j \cap A_k$ were both added and subtracted. We have to put them back. Therefore, we must add $|A_1 \cap A_2 \cap A_3| + \cdots + |A_{n-2} \cap A_{n-1} \cap A_n|$. And so on. The final formula is

$$|A_1 \cup A_2 \cup \cdots \cup A_n| = \sum_i |A_i| - \sum_{i,j} |A_i \cap A_j| + \cdots + (-1)^{n-1} |A_1 \cap A_2 \cap \cdots \cap A_n|.$$

Example. How many integers less than 1000 are not divisible by 2, 3, or 5?

Solution. To answer the question, we will count instead how many integers between 1 and 1000 are divisible by 2, 3, or 5. Denote by A_2, A_3, and A_5 be the sets of integers divisible by 2, 3, respectively, 5. The Boole-Sylvester formula counts $|A_2 \cup A_3 \cup A_5|$ as

$$|A_2| + |A_3| + |A_5| - |A_2 \cap A_3| - |A_2 \cap A_5| - |A_3 \cap A_5| + |A_2 \cap A_3 \cap A_5|$$

$$= \left\lfloor \frac{1000}{2} \right\rfloor + \left\lfloor \frac{1000}{3} \right\rfloor + \left\lfloor \frac{1000}{5} \right\rfloor - \left\lfloor \frac{1000}{6} \right\rfloor - \left\lfloor \frac{1000}{10} \right\rfloor - \left\lfloor \frac{1000}{15} \right\rfloor + \left\lfloor \frac{1000}{30} \right\rfloor$$

$$= 500 + 333 + 200 - 166 - 100 - 66 + 33 = 734.$$

It follows that there are $1000 - 734 = 266$ integers less than 1000 that are not divisible by 2, 3, or 5.   □

Example. How many colorings of the faces of the cube by 6 different colors exist? (Two colorings are the same if they coincide after a rotation.)

Solution. Let three of the colors be red, blue, green. Look at the red face. It is adjacent to either the blue face or the green face, since the opposite face can only be of one color.

Assume that red is adjacent to blue. The other four faces can be colored in $4 \times 3 \times 2 \times 1 = 24$ ways, and any two ways are distinct, since once the location of adjacent red and blue faces is fixed, the cube is rigid. If the red is adjacent to green, there are also 24 ways to color the other four faces.

But blue and green can be simultaneously adjacent to red. The can be opposite to each other, and then there are $3 \times 2 \times 1 = 6$ ways to color the other three faces, and the colorings are distinct since fixing the locations of the red, blue, and green faces makes the cube rigid. Or the red-blue-green faces can be pairwise adjacent, in which case, modulo a rotation, we have two possibilities: they lie in the xy, yz, xz planes or they lie in the xy, xz, yz planes, respectively. For each of the two situations the other faces can be colored in $3 \times 2 \times 1 = 6$ ways, and the colorings are distinct. So when the blue and the green faces are adjacent to the red face, then we have $6 + 6 + 6 = 18$ colorings.

We conclude that the number of colorings of the faces of the cube is

$$24 + 24 - 18 = 30.$$

Here at the last step we applied the inclusion-exclusion principle.   □

The third example comes from I. Tomescu's book *Problems in Combinatorics* (Wiley, 1985).

Example. An alphabet consists of the letters $a_1, a_2, \ldots, a_n$. Prove that the number of all words that contain each of these letters twice, but with no consecutive identical letters, is equal to

$$\frac{1}{2^n} \left[(2n)! - \binom{n}{1} 2(2n-1)! + \binom{n}{2} 2^2 (2n-2)! - \cdots + (-1)^n 2^n n! \right].$$

Solution. The number of such words without imposing the restriction about consecutive letters is

$$\frac{(2n)!}{(2!)^n} = \frac{(2n)!}{2^n}.$$

This is so because the identical letters can be permuted.

Denote by A_i the number of words formed with the n letters, each occurring twice, for which the two letters a_i appear next to each other. The answer to the problem is then

$$\frac{(2n)!}{2^n} - |A_1 \cup A_2 \cup \cdots \cup A_n|.$$

We evaluate $|A_1 \cup A_2 \cup \cdots \cup A_n|$ using the inclusion-exclusion principle. To this end, let us compute $|A_{i_1} \cap A_{i_2} \cap \cdots \cap A_{i_k}|$ for some indices $i_1, i_2, \ldots, i_k, k \le n$. Collapse the consecutive letters a_{i_j}, $j = 1, 2, \ldots, k$. As such, we are, in fact, computing the number of words made of the letters $a_1, a_2, \ldots, a_n$ in which $a_{i_1}, a_{i_2}, \ldots, a_{i_k}$ appear once and all other letters appear twice. This number is clearly equal to

$$\frac{(2n - k)!}{2^{n-k}},$$

since such a word has $2n - k$ letters, and identical letters can be permuted. There are $\binom{n}{k}$ k-tuples $(i_1, i_2, \ldots, i_k)$. We thus have

$$|A_1 \cup A_2 \cup \cdots \cup A_n| = \sum_k \sum_{i_1 \ldots i_k} (-1)^{k-1} |A_{i_1} \cap A_{i_2} \cap \cdots \cap A_{i_k}|$$

$$= \sum_k (-1)^{k-1} \binom{n}{k} \frac{(2n - k)!}{2^{n-k}},$$

and the formula is proved. □

1056. Let m, n, p, q, r, s be positive integers such that $p < r < m$ and $q < s < n$. In how many ways can one travel on a rectangular grid from $(0, 0)$ to (m, n) such that at each step one of the coordinates increases by one unit and such that the path avoids the points (p, q) and (r, s)?

1057. Let E be a set with n elements and F a set with p elements, $p \le n$. How many surjective (i.e., onto) functions $f : E \to F$ are there?

1058. A permutation σ of a set S is called a derangement if it does not have fixed points, i.e., if $\sigma(x) \ne x$ for all $x \in S$. Find the number of derangements of the set $\{1, 2, \ldots, n\}$.

1059. Given a graph with n vertices, prove that either it contains a triangle, or there exists a vertex that is the endpoint of at most $\lfloor \frac{n}{2} \rfloor$ edges.

1060. In the plane are given two closed polygonal lines, each with an odd number of segments, so that the lines of support of the sides are distinct and no three such lines intersect. Show that one can choose one side of the first polygonal line and one side of the second, so that they are the opposite sides of a convex quadrilateral.

1061. Let $m \geq 5$ and n be given positive integers, and suppose that $\mathcal{P}$ is a regular $(2n + 1)$-gon. Find the number of convex m-gons having at least one acute angle and having vertices exclusive among the vertices of $\mathcal{P}$.

1062. Let $S^1 = \{z \in \mathbb{C} \mid |z| = 1\}$. For all functions $f : S^1 \to S^1$ set $f^1 = f$ and $f^{n+1} = f \circ f^n, n \geq 1$. Call $w \in S^1$ a periodic point of f of period n if $f^i(w) \neq w$ for $i = 1, \ldots, n - 1$ and $f^n(w) = w$. If $f(z) = z^m$, m a positive integer, find the number of periodic points of f of period 1989.

1063. For positive integers $x_1, x_2, \ldots, x_n$ denote by $[x_1, x_2, \ldots, x_n]$ their least common multiple and by $(x_1, x_2, \ldots, x_n)$ their greatest common divisor. Prove that for positive integers a, b, c,

$$\frac{[a, b, c]^2}{[a, b][b, c][c, a]} = \frac{(a, b, c)^2}{(a, b)(b, c)(c, a)}.$$

1064. A $150 \times 324 \times 375$ rectangular solid is made by gluing together $1 \times 1 \times 1$ cubes. An internal diagonal of this solid passes through the interiors of how many of the $1 \times 1 \times 1$ cubes?

6.5 Probability

6.5.1 Equally Likely Cases

In this section we consider experiments with finitely many outcomes each of which can occur with equal probability. In this case the probability of an event A is given by

$$P(A) = \frac{\text{number of favorable outcomes}}{\text{total number of possible outcomes}}.$$

The computation of the probability is purely combinatorial; it reduces to a counting problem.
We start with the example that gave birth to probability theory.

Example. Show that the probability of getting a six when a die is rolled four times is greater than the probability of getting a double six when two dice are rolled 24 times.

Here is a brief history of the problem. Chevalier de Méré, a gambler of the seventeenth century, observed while gambling that the odds of getting a six when rolling a die four times seem to be greater than $\frac{1}{2}$, while the odds of getting a double six when rolling two dice 24 times seem to be less than $\frac{1}{2}$. De Méré thought that this contradicted mathematics because $\frac{4}{6} = \frac{24}{36}$. He posed this question to B. Pascal and P. Fermat. They answered the question…and probability theory was born. Let us see the solution.

Solution. The probability that a six does not occur when rolling a die four times is $\left(\frac{5}{6}\right)^4$, and so the probability that a six occurs is $1 - \left(\frac{5}{6}\right)^4 \approx 0.5177$. The probability that a double six does not occur when rolling two dice 24 times is $\left(\frac{35}{36}\right)^{24}$, whence the probability that a double six occurs is $1 - \left(\frac{35}{36}\right)^{24} \approx 0.4914$. The second number is smaller. $\square$

Example. Consider n indistinguishable balls randomly distributed in m boxes. What is the probability that exactly k boxes remain empty?

Solution. Number the boxes $1, 2, \ldots, m$ and let x_i be the number of balls in the ith box. The number of ways one can distribute n balls in m boxes is equal to the number of nonnegative integer solutions to the equation

$$x_1 + x_2 + \cdots + x_m = n.$$

These solutions were counted in problem 1042 from Section 6.4.3 and were found to be $\binom{m+n-1}{m-1}$. This is the total number of cases.

If we fix k boxes and distribute the balls in the remaining $n - k$ boxes such that each box receives at least one ball, then the number of ways to do this is equal to the number of positive integer solutions to the equation

$$x_1 + x_2 + \cdots + x_{m-k} = n.$$

This was also computed in one of the examples from Section 6.4.3 and was shown to be $\binom{n-1}{m-k-1}$. The k boxes can be chosen in $\binom{m}{k}$ ways. We find the number of favorable cases to be $\binom{m}{k}\binom{n-1}{m-k-1}$. The required probability is therefore

$$\frac{\binom{m}{k}\binom{n-1}{m-k-1}}{\binom{m+n-1}{m-1}}. \qquad \square$$

If you grab n balls and place them one at a time randomly in boxes, you will find that they do not seem to fit the probabilities just calculated. This is because they are not really indistinguishable balls: the order of placement and the fact that they are macroscopic balls makes them distinguishable. However, this example does correspond to a real world situation, namely that about particles and states. The above considerations apply to bosons, particles that obey the Bose-Einstein statistics, which allows several particles to occupy the same state. Examples of bosons are photons, gluons, and the helium-4 atom. Electrons and protons, on the other hand, are fermions. They are subject to the Pauli exclusion principle: at most one can occupy a certain state. As such, fermions obey what is called the Fermi-Dirac statistics.

A third problem comes from C. Reischer, A. Sâmboan, *Collection of Problems in Probability Theory and Mathematical Statistics* (Editura Didactică şi Pedagogică, Bucharest, 1972). It shows how probabilities can be used to prove identities.

Example. Prove the identity

$$1 + \frac{n}{m+n-1} + \cdots + \frac{n(n-1)\cdots 1}{(m+n-1)(m+n-2)\cdots m} = \frac{m+n}{m}.$$

Solution. Consider a box containing n white balls and m black balls. Let A_i be the event of extracting the first white ball at the ith extraction. We compute

$$P(A_1) = \frac{m}{m+n},$$

$$P(A_2) = \frac{n}{m+n} \cdot \frac{m}{m+n-1},$$

$$P(A_3) = \frac{n}{m+n} \cdot \frac{n-1}{m+n-1} \cdot \frac{m}{m+n-2},$$

$$\cdots$$

$$P(A_m) = \frac{n}{m+n} \cdot \frac{n-1}{m+n-1} \cdots \frac{1}{m+1}.$$

The events $A_1, A_2, A_3, \ldots$ are disjoint, and therefore

$$1 = P(A_1) + P(A_2) + \cdots + P(A_m)$$

$$= \frac{m}{m+n} \left[1 + \frac{n}{m+n-1} + \cdots + \frac{n(n-1)\cdots 1}{(m+n-1)(m+n-2)\cdots m} \right].$$

The identity follows. □

Because it will be needed in several problems, let us recall the following definition.

Definition. The expected value of an experiment X with possible outcomes $a_1, a_2, \ldots, a_n$ is the weighted mean

$$E[X] = a_1 P(X = a_1) + a_2 P(X = a_2) + \cdots + a_n P(X = a_n).$$

If X is distributed in a "geometric" domain (like in the last section of this chapter), with probability density $p(x)$, then the expected value is

$$E[X] = \int x p(x) dx.$$

So let us see the problems.

1065. Let v and w be distinct, randomly chosen roots of the equation $z^{1997} - 1 = 0$. Find the probability that $\sqrt{2 + \sqrt{3}} \le |v + w|$.

1066. Find the probability that in a group of n people there are two with the same birthday. Ignore leap years.

1067. A solitaire game is played as follows. Six distinct pairs of matched tiles are placed in a bag. The player randomly draws tiles one at a time from the bag and retains them, except that matching tiles are put aside as soon as they appear in the player's hand. The game ends if the player ever holds three tiles, no two of which match; otherwise, the drawing continues until the bag is empty. Find the probability that the bag will be emptied.

1068. An urn contains n balls numbered $1, 2, \ldots, n$. A person is told to choose a ball and then extract m balls among which is the chosen one. Suppose he makes two independent extractions, where in each case he chooses the remaining $m - 1$ balls at random. What is the probability that the chosen ball can be determined?

1069. A bag contains 1993 red balls and 1993 black balls. We remove two balls at a time repeatedly and

 (i) discard them if they are of the same color,
 (ii) discard the black ball and return to the bag the red ball if they are of different colors.

What is the probability that this process will terminate with one red ball in the bag?

1070. The numbers $1, 2, 3, 4, 5, 6, 7$, and 8 are written on the faces of a regular octahedron so that each face contains a different number. Find the probability that no two consecutive numbers are written on faces that share an edge, where 8 and 1 are considered consecutive.

1071. What is the probability that a permutation of the first n positive integers has the numbers 1 and 2 within the same cycle.

1072. An unbiased coin is tossed n times. Find a formula, in closed form, for the expected value of $|H - T|$, where H is the number of heads, and T is the number of tails.

1073. Prove the identities

$$\sum_{k=1}^{n} \frac{1}{(k-1)!} \sum_{i=0}^{n-k} \frac{(-1)^i}{i!} = 1,$$

$$\sum_{k=1}^{n} \frac{1}{(k-1)!} \sum_{i=0}^{n-k} \frac{(-1)^i}{i!} = 2.$$

6.5.2 Establishing Relations Among Probabilities

We adopt the usual notation: $P(A)$ is the probability of the event A, $P(A \cap B)$ is the probability that A and B occur simultaneously, $P(A \cup B)$ is the probability that either A or B occurs, $P(A - B)$ is the probability that A occurs but not B, and $P(A/B)$ is the probability that A occurs given that B also occurs.

Recall the classical formulas:

- addition formula:

$$P(A \cup B) = P(A) + P(B) - P(A \cap B);$$

- multiplication formula:

$$P(A \cap B) = P(A)P(B/A);$$

- total probability formula: if $B_i \cap B_j = \emptyset$, $i, j = 1, 2, \ldots, n$ (meaning that they are independent), and $A \subset B_1 \cup B_2 \cup \cdots \cup B_n$, then

$$P(A) = P(A/B_1)P(B_1) + P(A/B_2)P(B_2) + \cdots + P(A/B_n)P(B_n);$$

- Bayes' formula: with the same hypothesis,

$$P(B_i/A) = \frac{P(A/B_i)P(B_i)}{P(A/B_1)P(B_1) + P(A/B_2)P(B_2) + \cdots + P(A/B_n)P(B_n)}.$$

In particular, if $B_1, B_2, \ldots, B_n$ cover the entire probability field, then

$$P(B_i/A) = \frac{P(B_i)}{P(A)} P(A/B_i).$$

The Bernoulli scheme. As a result of an experiment either the event A occurs with probability p or the contrary event $\overline{A}$ occurs with probability $q = 1 - p$. We repeat the experiment n times. The probability that A occurs exactly m times is $\binom{n}{m} p^m q^{n-m}$. This is also called the binomial scheme because the generating function of these probabilities is $(q + px)^n$.

The Poisson scheme. We perform n independent experiments. For each k, $1 \le k \le n$, in the kth experiment the event A can occur with probability p_k, or $\overline{A}$ can occur with probability $q_k = 1 - p_k$. The probability that A occurs exactly m times while the n experiments are performed is the coefficient of x^m in the expansion of

$$(p_1 x + q_1)(p_2 x + q_2) \cdots (p_n x + q_n).$$

Here is a problem from the 1970 Romanian Mathematical Olympiad that applies the Poisson scheme.

Example. In a selection test, each of three candidates receives a problem sheet with n problems from algebra and geometry. The three problem sheets contain, respectively, one, two, and three algebra problems. The candidates choose randomly a problem from the sheet and answer it at the blackboard. What is the probability that

(a) all candidates answer geometry problems;
(b) all candidates answer algebra problems;
(c) at least one candidate answers an algebra problem?

Solution. We apply the Poisson scheme. Define the polynomial

$$P(x) = \left(\frac{1}{m}x + \frac{n-1}{n}\right)\left(\frac{2}{n}x + \frac{n-2}{n}\right)\left(\frac{3}{n}x + \frac{n-3}{n}\right)$$

$$= \frac{1}{n^3}[6x^3 + (11n - 18)x^2 + (6n^2 - 22n + 18)x + (n-1)(n-2)(n-3)]$$

$$= P_3 x^3 + P_2 x^2 + P_1 x + P_0.$$

The answer to question (a) is the free term

$$P_0 = \frac{(n-1)(n-2)(n-3)}{n^3}.$$

The answer to (b) is the coefficient of x^3, namely,

$$P_3 = \frac{6}{n^3}.$$

The answer to (c) is

$$P = 1 - P_0 = \frac{6n^2 - 11n + 6}{n^3}. \qquad \square$$

And now another problem posed to Pascal and Fermat by the Chevalier de Méré.

Example. Two players repeatedly play a game in which the first wins with probability p and the second wins with probability $q = 1 - p$. They agree to stop when one of them wins a certain number of games. They are forced to interrupt their game when the first player has a more games to win and the second player has b more games to win. How should they divide the stakes correctly? Use the answer to prove the combinatorial identities

$$p^a \sum_{k=0}^{b-1} \binom{a-1+k}{a-1} q^k + q^b \sum_{k=0}^{a-1} \binom{b-1+k}{b-1} p^k = 1,$$

$$p^a \sum_{k=0}^{b-1} \binom{a-1+k}{a-1} q^k = \frac{(a+b-1)!}{a!(b-1)!} p^a q^{b-1} \left[1 + \sum_{k=1}^{b-1} \frac{(b-1)\cdots(b-k)}{(a+1)\cdots(a+k)} \left(\frac{p}{q}\right)^k \right].$$

Solution. Call P the probability that the first player wins the a remaining games before the second player wins the b games he needs, and $Q = 1 - P$, the probability that the second player wins b games before the first wins a. The players should divide the stakes in the ratio $\frac{P}{Q}$.

We proceed with the computation of P. The first player could have won the a games in several mutually exclusive ways: in exactly a games, in exactly $a + 1$ games,..., in exactly $a + b - 1$ games. In all cases the last game should be won by the first player.

Let us find the probability that the first player wins in exactly $a+k$ games, $k = 0, 1, \ldots, b - 1$. The probability that the first player wins $a - 1$ games out of $a + k - 1$ is computed using the Bernoulli scheme and is equal to $\binom{a+k-1}{a-1} p^{a-1} q^k$, and the probability of winning the $(a + k)$th is p. The probability of winning in exactly $a + k$ games is the product of the two, namely $\binom{a+k-1}{a-1} p^a q^k$.

We deduce that the probability of the first player winning the stakes is

$$P = \sum_{k=0}^{b-1} \binom{a+k-1}{a-1} p^a q^k,$$

while for the second player this is

$$Q = q^b \sum_{k=0}^{a-1} \binom{b-1+k}{b-1} p^k.$$

The stakes should be divided in the ratio

$$\frac{P}{Q} = \frac{p^a \sum_{k=0}^{b-1} \binom{a-1+k}{a-1} q^k}{q^b \sum_{k=0}^{a-1} \binom{b-1+k}{b-1} p^k}.$$

The first combinatorial identity is equivalent to $P + Q = 1$. For the second combinatorial identity, we look for a different way to compute P. Observe that after at most $a + b - 1$ games have been played, the winner is known. Let us assume that regardless of the results, the players kept playing all the $a + b - 1$ games. If the first player had won at least a of these games, he would have won the stakes as well. Hence P is the probability that the first player won $a, a + 1, \ldots, a + b - 1$ of the final $a + b - 1$ games. Each of these is computed using the Bernoulli scheme, and P is their sum, since the events are incompatible. We obtain

$$P = \sum_{k=0}^{b-1} \binom{a+b-1}{a+k} p^{a+k} q^{b-1-k}$$

$$= \frac{(a+b-1)!}{a!(b-1)!} p^a q^{b-1} \left[1 + \sum_{k=1}^{b-1} \frac{(b-1)\cdots(b-k)}{(a+1)\cdots(a+k)} \left(\frac{p}{q}\right)^k \right].$$

The second identity follows by equating the two formulas that we obtained for P.   □

This is yet another example of how probability theory can be used to prove identities. Since "wisdom is the daughter of experience" (Leonardo da Vinci), we let you train your probabilistic skills with the following problems.

1074. An exam consists of 3 problems selected randomly from a list of $2n$ problems, where n is an integer greater than 1. For a student to pass, he needs to solve correctly at least two of the three problems. Knowing that a certain student knows how to solve exactly half of the $2n$ problems, find the probability that the student will pass the exam.

1075. The probability that a woman has breast cancer is 1%. If a woman has breast cancer, the probability is 60% that she will have a positive mammogram. However, if a woman does not have breast cancer, the mammogram might still come out positive, with a probability of 7%. What is the probability for a woman with positive mammogram to actually have cancer?

1076. Find the probability that in the process of repeatedly flipping a coin, one will encounter a run of 5 heads before one encounters a run of 2 tails.

1077. The temperatures in Chicago and Detroit are $x°$ and $y°$, respectively. These temperatures are not assumed to be independent; namely, we are given the following:
(i) $P(x° = 70°) = a$, the probability that the temperature in Chicago is 70°,
(ii) $P(y° = 70°) = b$, and
(iii) $P(\max(x°, y°) = 70°) = c$.
Determine $P(\min(x°, y°) = 70°)$ in terms of a, b, and c.

1078. An urn contains both black and white marbles. Each time you pick a marble you return it to the urn. Let p be the probability of drawing a white marble and $q = 1 - p$ the probability of drawing a black marble. Marbles are drawn until n black marbles have been drawn. If $n + x$ is the total number of draws, find the probability that $x = m$.

1079. Three independent students took an exam. The random variable X, representing the students who passed, has the distribution

$$\begin{pmatrix} 0 & 1 & 2 & 3 \\ \frac{2}{5} & \frac{13}{30} & \frac{3}{20} & \frac{1}{60} \end{pmatrix}.$$

Find each student's probability of passing the exam.

1080. Given the independent events $A_1, A_2, \ldots, A_n$ with probabilities $p_1, p_2, \ldots, p_n$, find the probability that an odd number of these events occurs.

1081. Out of every batch of 100 products of a factory, 5 are quality checked. If one sample does not pass the quality check, then the whole batch of one hundred will be rejected. What is the probability that a batch is rejected if it contains 5% faulty products.

1082. There are two jet planes and a propeller plane at the small regional airport of Gauss City. A plane departs from Gauss City and arrives in Eulerville, where there were already five propeller planes and one jet plane. Later, a farmer sees a jet plane flying out of Eulerville. What is the probability that the plane that arrived from Gauss City was a propeller plane, provided that all events are equiprobable?

1083. A coin is tossed n times. What is the probability that two heads will turn up in succession somewhere in the sequence?

1084. Two people, A and B, play a game in which the probability that A wins is p, the probability that B wins is q, and the probability of a draw is r. At the beginning, A has m dollars and B has n dollars. At the end of each game, the winner takes a dollar from the loser. If A and B agree to play until one of them loses all his/her money, what is the probability of A winning all the money?

1085. We play the coin tossing game in which if tosses match, I get both coins; if they differ, you get both. You have m coins, I have n. What is the expected length of the game (i.e., the number of tosses until one of us is wiped out)?

6.5.3 Geometric Probabilities

In this section we look at experiments whose possible outcomes are parametrized by the points of a geometric region. Here we interpret "at random" to mean that the probability that a point lies in a certain region is proportional to the area or volume of the region. The probability of a certain event is then computed by taking the ratio of the area (volume) of the favorable region to the area (volume) of the total region. We start with the game of franc-carreau investigated by George-Louis Leclerc, Comte de Buffon, in his famous *Essai d'arithmetique morale*.

Example. A coin of diameter d is thrown randomly on a floor tiled with squares of side l. Two players bet that the coin will land on exactly one, respectively, more than one, square. What relation should l and d satisfy for the game to be fair?

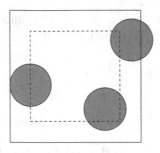

Figure 48

Solution. The center of the coin falls on some tile. For the coin to lie entirely on that tile, its center must fall inside the dotted square of side length $l - 2 \cdot \frac{d}{2} = l - d$ shown in Figure 48. This happens with probability

$$P = \frac{(l - d)^2}{l^2}.$$

For the game to be fair, P must be equal to $\frac{1}{2}$, whence the relation that d and l should satisfy is

$$d = \frac{1}{2}(2 - \sqrt{2})l.$$

□

Example. What is the probability that three randomly chosen points on a circle form an acute triangle?

Solution. The fact that the triangle is acute is equivalent to the fact that each of the arcs determined by the vertices is less than a semicircle.

Because of the rotational symmetry of the figure, we can assume that one of the points is fixed. Cut the circle at that point to create a segment. In this new framework, the problem asks us to find the probability that two randomly chosen points on a segment cut it in three parts, none of which is larger than half of the original segment.

Identify the segment with the interval $[0, 1]$, and let the coordinates of the two points be x and y. Then the possible choices can be identified with points (x, y) randomly distributed in the interior of the square $[0, 1] \times [0, 1]$. The area of the total region is therefore 1. The favorable region, namely, the set of points inside the square that yield an acute triangle, is

$$\left\{(x, y) \mid 0 < x < \frac{1}{2}, \frac{1}{2} < y < \frac{1}{2} + x\right\} \cup \left\{(x, y) \mid \frac{1}{2} < x < 1, x - \frac{1}{2} < y < \frac{1}{2}\right\}.$$

The area of this region is $\frac{1}{4}$. Hence the probability in question is $\frac{1}{4}$.

□

As an outcome of the solution we find that when cutting a segment into three random parts, the probability that the three segments can be the sides of an acute triangle is $\frac{1}{4}$.

In all problems below, points are chosen randomly with respect to the uniform distribution on segments, circles, spheres (the one that comes from the standard integration measure).

1086. What is the probability that the sum of two randomly chosen numbers in the interval $[0, 1]$ does not exceed 1 and their product does not exceed $\frac{2}{9}$?

1087. Let α and β be given positive real numbers, with $\alpha < \beta$. If two points are selected at random from a straight line segment of length β, what is the probability that the distance between them is at least α?

1088. A husband and wife agree to meet at a street corner between 4 and 5 o'clock to go shopping together. The one who arrives first will await the other for 15 minutes, and then leave. What is the probability that the two meet within the given time interval, assuming that they can arrive at any time with the same probability?

1089. Two airplanes are supposed to park at the same gate of a concourse. The arrival times of the airplanes are independent and randomly distributed throughout the 24 hours of the day. What is the probability that both can park at the gate, provided that the first to arrive will stay for a period of two hours, while the second can wait behind it for a period of one hour?

1090. Find the expected value of the square of the distance between two randomly chosen points on the unit sphere.

1091. What is the probability that three points selected at random on a circle lie on a semi-circle?

1092. Let $n \geq 4$ be given, and suppose that the points $P_1, P_2, \ldots, P_n$ are randomly chosen on a circle. Consider the convex n-gon whose vertices are these points. What is the probability that at least one of the vertex angles of this polygon is acute?

1093. Let C be the unit circle $x^2 + y^2 = 1$. A point p is chosen randomly on the circumference of C and another point q is chosen randomly from the interior of C (these points are chosen independently and uniformly over their domains). Let R be the rectangle with sides parallel to the x- and y-axes with diagonal pq. What is the probability that no point of R lies outside of C?

1094. If a needle of length 1 is dropped at random on a surface ruled with parallel lines at distance 2 apart, what is the probability that the needle will cross one of the lines?

1095. Four points are chosen uniformly and independently at random in the interior of a given circle. Find the probability that they are the vertices of a convex quadrilateral.

Methods of Proof

1. Assume the contrary, namely that $\sqrt{2} + \sqrt{3} + \sqrt{5} = r$, where r is a rational number. Square the equality $\sqrt{2} + \sqrt{3} = r - \sqrt{5}$ to obtain $5 + 2\sqrt{6} = r^2 + 5 - 2r\sqrt{5}$. It follows that $2\sqrt{6} + 2r\sqrt{5}$ is itself rational. Squaring again, we find that $24 + 20r^2 + 8r\sqrt{30}$ is rational, and hence $\sqrt{30}$ is rational, too. Pythagoras' method for proving that $\sqrt{2}$ is irrational can now be applied to show that this is not true. Write $\sqrt{30} = \frac{m}{n}$ in lowest terms; then transform this into $m^2 = 30n^2$. It follows that m is divisible by 2 and because $2\left(\frac{m}{2}\right)^2 = 15n^2$ it follows that n is divisible by 2 as well. So the fraction was not in lowest terms, a contradiction. We conclude that the initial assumption was false, and therefore $\sqrt{2} + \sqrt{3} + \sqrt{5}$ is irrational.

2. Assume that such numbers do exist, and let us look at their prime factorizations. For primes p greater than 7, at most one of the numbers can be divisible by p, and the partition cannot exist. Thus the prime factors of the given numbers can be only 2, 3, 5, and 7.

We now look at repeated prime factors. Because the difference between two numbers divisible by 4 is at least 4, at most three of the nine numbers are divisible by 4. Also, at most one is divisible by 9, at most one by 25, and at most one by 49. Eliminating these at most $3 + 1 + 1 + 1 = 6$ numbers, we are left with at least three numbers among the nine that do not contain repeated prime factors. They are among the divisors of $2 \cdot 3 \cdot 5 \cdot 7$, and so among the numbers

$$2, \ 3, \ 5, \ 6, \ 7, \ 10, \ 14, \ 15, \ 21, \ 30, \ 35, \ 42, \ 70, \ 105, \ 210.$$

Because the difference between the largest and the smallest of these three numbers is at most 9, none of them can be greater than 21. We have to look at the sequence $1, 2, 3, \ldots, 29$. Any subsequence of consecutive integers of length 9 that has a term greater than 10 contains a prime number greater than or equal to 11, which is impossible. And from $1, 2, \ldots, 10$ we cannot select nine consecutive numbers with the required property. This contradicts our assumption, and the problem is solved.

Remark. In the argument, the number 29 can be replaced by 27, namely by 21 plus the 6 numbers that can have repeated prime factor.

© Springer International Publishing AG 2017

R. Gelca and T. Andreescu, *Putnam and Beyond*, DOI 10.1007/978-3-319-58988-6

3. The example $2^2, 3^2, 5^2, \ldots, 43^2$, where we considered the squares of the first 14 prime numbers, shows that $n \geq 15$.

Assume that there exist $a_1, a_2, \ldots, a_{15}$, pairwise relatively prime integers greater than 1 and less than 2005, none of which is a prime. Let q_k be the least prime number in the factorization of a_k, $k = 1, 2, \ldots, 15$. Let q_i be the maximum of $q_1, q_2, \ldots, q_{15}$. Then $q_i \geq p_{15} = 47$. Because a_i is not a prime, $\frac{a_i}{q_i}$ is divisible by a prime number greater than or equal to q_i. Hence $a_i \geq q_i^2 = 47^2 > 2005$, a contradiction. We conclude that $n = 15$.

4. Let $X = \{x_1, x_2, \ldots, x_n\}$ and $E_1 = \{x_1, x_2, \ldots, x_{n-2}\}$. Arguing by contradiction, let us assume that $\cup_{k=1}^{m} E_k = S$. Choose E_j and E_k such that $x_{n-1} \in E_j$ and $x_n \in E_k$. Then $E_1 \cup E_j \cup E_k = S$, a contradiction.

(Romanian Mathematical Olympiad, 1986, proposed by I. Tomescu)

5. Arguing by contradiction, we assume that none of the colors has the desired property. Then there exist distances $r \geq g \geq b$ such that r is not attained by red points, g by green points, and b by blue points (for these inequalities to hold we might have to permute the colors).

Consider a sphere of radius r centered at a red point. Its surface has green and blue points only. Since $g, b \leq r$, the surface of the sphere must contain both green and blue points. Choose M a green point on the sphere. There exist two points P and Q on the sphere such that $MP = MQ = g$ and $PQ = b$. So on the one hand, either P or Q is green, or else P and Q are both blue. Then either there exist two green points at distance g, namely M and P, or Q, or there exist two blue points at distance b. This contradicts the initial assumption. The conclusion follows.

(German Mathematical Olympiad, 1985)

6. Arguing by contradiction, let us assume that the area of the overlap of any two surfaces is less than $\frac{1}{9}$. In this case, if $S_1, S_2, \ldots, S_9$ denote the nine surfaces, then the area of $S_1 \cup S_2$ is greater than $1 + \frac{8}{9}$, the area of $S_1 \cup S_2 \cup S_3$ is greater than $1 + \frac{8}{9} + \frac{7}{9}, \ldots$, and the area of $S_1 \cup S_2 \cup \cdots \cup S_9$ is greater than

$$1 + \frac{8}{9} + \frac{7}{9} + \cdots + \frac{1}{9} = \frac{45}{9} = 5$$

a contradiction. Hence the conclusion.

(L. Panaitopol, D. Şerbănescu, *Probleme de Teoria Numerelor şi Combinatorică pentru Juniori* (*Problems in Number Theory and Combinatorics for Juniors*), GIL, 2003)

7. Assume that such an f exists. We focus on some particular values of the variable. Let $f(0) = a$ and $f(5) = b$, $a, b \in \{1, 2, 3\}$, $a \neq b$. Because $|5 - 2| = 3$, $|2 - 0| = 2$, we have $f(2) \neq a, b$, so $f(2)$ is the remaining number, say c. Finally, because $|3 - 0| = 3$, $|3 - 5| = 2$, we must have $f(3) = c$. Therefore, $f(2) = f(3)$. Translating the argument to an arbitrary number x instead of 0, we obtain $f(x + 2) = f(x + 3)$, and so f is constant. But this violates the condition from the definition. It follows that such a function does not exist.

8. Arguing by contradiction, let us assume that such a function exists. Set $f(3) = k$. Using the inequality $2^3 < 3^2$, we obtain

$$3^3 = f(2)^3 = f(2^3) < f(3^2) = f(3)^2 = k^2,$$

hence $k > 5$. Similarly, using $3^3 < 2^5$, we obtain

$$k^3 = f(3)^3 = f(3^3) < f(2^5) = f(2)^5 = 3^5 = 243 < 343 = 7^3.$$

This implies that $k < 7$, and consequently k can be equal only to 6. Thus we should have $f(2) = 3$ and $f(3) = 6$. The monotonicity of f implies that $2^u < 3^v$ if and only if $3^u < 6^v$, u, v being positive integers. Taking logarithms this means that $\frac{v}{u} > \log_2 3$ if and only if $\frac{v}{u} > \log_3 6$. Since rationals are dense, it follows that $\log_2 3 = \log_3 6$. This can be written as $\log_2 3 = \frac{1}{\log_2 3} + 1$, and so $\log_2 3$ is the positive solution of the quadratic equation $x^2 - x - 1 = 0$, which is the golden ratio $\frac{1+\sqrt{5}}{2}$. The equality $2^{\frac{1+\sqrt{5}}{2}} = 3$ translates to $2^{1+\sqrt{5}} = 9$. But this would imply

$$65536 = 2^{5 \times 3.2} < 2^{5(1+\sqrt{5})} = 9^5 = 59049.$$

We have reached a contradiction, which proves that the function f cannot exist.

(B.J. Venkatachala, *Functional Equations: A Problem Solving Approach*, Prism Books PVT Ltd., Bangalore, 2002)

9. The constant function $f(x) = k$, where k is a positive integer, is the only possible solution. That any such function satisfies the given condition is easy to check.

Now suppose there exists a nonconstant solution f. There must exist two positive integers a and b such that $f(a) < f(b)$. This implies that $(a+b)f(a) < af(b) + bf(a) < (a+b)f(b)$, which by the given condition is equivalent to $(a+b)f(a) < (a+b)f(a^2+b^2) < (a+b)f(b)$. We can divide by $a + b > 0$ to find that $f(a) < f(a^2 + b^2) < f(b)$. Thus between any two different values of f we can insert another. But this cannot go on forever, since f takes only integer values. The contradiction shows that such a function cannot exist. Thus constant functions are the only solutions.

(Canadian Mathematical Olympiad, 2002)

10. Assume that A, B, and a satisfy $A \cup B = [0, 1]$, $A \cap B = \emptyset$, $B = A + a$. We can assume that a is positive; otherwise, we can exchange A and B. Then $(1 - a, 1] \subset B$; hence $(1 - 2a, 1 - a] \subset A$. An inductive argument shows that for any positive integer n, the interval $(1 - (2n + 1)a, 1 - 2na]$ is in B, the interval $(1 - (2n + 2)a, 1 - (2n + 1)a]$ is in A. However, at some point this sequence of intervals leaves $[0, 1]$. The interval of the form $(1 - na, 1 - (n - 1)a]$ that contains 0 must be contained entirely in either A or B, which is impossible since this interval exits $[0, 1]$. The contradiction shows that the assumption is wrong, and hence the partition does not exist.

(Austrian-Polish Mathematics Competition, 1982)

11. Assume the contrary. Our chosen numbers $a_1, a_2, \ldots, a_{k+1}$ must have a total of at most k distinct prime factors (the primes less than or equal to n). Let $o_p(q)$ denote the highest value of d such that $p^d | q$. Also, let $a = a_1 a_2 \cdots a_{k+1}$ be the product of the numbers. Then for each prime p,

$$o_p(a) = \sum_{i=1}^{k+1} o_p(a_i),$$

and it follows that there can be at most one *hostile* value of i for which $o_p(a_i) > \frac{o_p(a)}{2}$. Because there are at most k primes that divide a, there is some i that is not hostile for any such prime.

Then $2o_p(a_i) \leq o_p(a)$, so $o_p(a_i) \leq o_p\left(\frac{a}{a_i}\right)$ for each prime p dividing a. This implies that a_i divides $\frac{a}{a_i}$, which contradicts the fact that the a_i does not divide the product of the other a_j's. Hence our assumption was false, and the conclusion follows.

(Hungarian Mathematical Olympiad, 1999)

12. The base case $n = 1$ is $\frac{1}{2} = 1 - \frac{1}{2}$, true. Now the inductive step. The hypothesis is that

$$\frac{1}{k+1} + \frac{1}{k+2} + \cdots + \frac{1}{2k} = 1 - \frac{1}{2} + \cdots + \frac{1}{2k-1} - \frac{1}{2k}.$$

We are to prove that

$$\frac{1}{k+2} + \cdots + \frac{1}{2k} + \frac{1}{2k+1} + \frac{1}{2k+2} = 1 - \frac{1}{2} + \cdots - \frac{1}{2k} + \frac{1}{2k+1} - \frac{1}{2k+2}.$$

Using the induction hypothesis, we can rewrite this as

$$\frac{1}{k+2} + \cdots + \frac{1}{2k} + \frac{1}{2k+1} + \frac{1}{2k+2} = \frac{1}{k+1} + \frac{1}{k+2} + \cdots + \frac{1}{2k} + \frac{1}{2k+1} - \frac{1}{2k+2},$$

which reduces to

$$\frac{1}{2k+2} = \frac{1}{k+1} - \frac{1}{2k+2},$$

obvious. This completes the induction.

13. The base case is trivial. However, as I.M. Vinogradov once said, "it is the first nontrivial example that matters". And this is $n = 2$, in which case we have

$$|\sin 2x| = 2|\sin x||\cos x| \leq 2|\sin x|.$$

This suggests to us to introduce cosines as factors in the proof of the inductive step. Assuming the inequality for $n = k$, we can write

$$|\sin(k+1)x| = |\sin kx \cos x + \sin x \cos kx| \leq |\sin kx||\cos x| + |\sin x||\cos kx|$$
$$\leq |\sin kx| + |\sin x| \leq k|\sin x| + |\sin x| = (k+1)|\sin x|.$$

The induction is complete.

14. As in the solution to the previous problem we argue by induction on n using trigonometric identities. The base case holds because

$$|\sin x_1| + |\cos x_1| \geq \sin^2 x_1 + \cos^2 x_1 = 1.$$

Next, assume that the inequality holds for $n = k$ and let us prove it for $n = k + 1$. Using the inductive hypothesis, it suffices to show that

$$|\sin x_{n+1}| + |\cos(x_1 + x_2 + \cdots + x_{n+1})| \geq |\cos(x_1 + x_2 + \cdots + x_n)|.$$

To simplify notation let $x_{n+1} = x$ and $x_1 + x_2 + \cdots + x_n + x_{n+1} = y$, so that the inequality to be proved is $|\sin x| + |\cos y| \geq |\cos(y - x)|$. The subtraction formula gives

$$|\cos(y - x)| = |\cos y \cos x + \sin y \sin x| \leq |\cos y||\cos x| + |\sin y||\sin x|$$
$$\leq |\cos y| + |\sin x|.$$

This completes the inductive step, and concludes the solution.

(*Revista Mathematică din Timişoara* (*Timişoara Mathematics Gazette*), proposed by T. Andreescu)

15. We expect an inductive argument, with a possible inductive step given by

$$3^{n+1} = 3 \cdot 3^n \geq 3n^3 \geq (n+1)^3.$$

In order for this to work, the inequality $3n^3 \geq (n+1)^3$ needs to be true. This inequality is equivalent to $2n^3 \geq 3n^2 + 3n + 1$, which would, for example, follow from the separate inequalities $n^3 \geq 3n^2$ and $n^3 \geq 3n + 1$. These are both true for $n \geq 3$. Thus we can argue by induction starting with the base case $n = 3$, where equality holds. The cases $n = 0$, $n = 1$, and $n = 2$ can be checked by hand.

16. The base case $2^6 < 6! < 3^6$ reduces to $64 < 720 < 729$, which is true. Assuming the double inequality true for n we are to show that

$$\left(\frac{n+1}{3}\right)^{n+1} < (n+1)! < \left(\frac{n+1}{2}\right)^{n+1}.$$

Using the inductive hypothesis we can reduce the inequality on the left to

$$\left(\frac{n+1}{3}\right)^{n+1} < (n+1)\left(\frac{n}{3}\right)^n,$$

$$\left(1 + \frac{1}{n}\right)^n < 3,$$

while the inequality on the right can be reduced to

$$\left(1 + \frac{1}{n}\right)^n > 2.$$

These are both true for all $n \geq 1$ because the sequence $\left(1 + \frac{1}{n}\right)^n$ is increasing and converges to e, which is less than 3. Hence the conclusion.

17. The left-hand side grows with n, while the right-hand side stays constant, so apparently a proof by induction would fail. It works, however, if we sharpen the inequality to

$$1 + \frac{1}{2^3} + \frac{1}{3^3} + \cdots + \frac{1}{n^3} < \frac{3}{2} - \frac{1}{n}, \quad n \geq 2.$$

As such, the cases $n = 1$ and $n = 2$ need to be treated separately, and they are easy to check.

The base case is for $n = 3$:

$$1 + \frac{1}{2^3} + \frac{1}{3^3} < 1 + \frac{1}{8} + \frac{1}{27} < \frac{3}{2} - \frac{1}{3}.$$

For the inductive step, note that from

$$1 + \frac{1}{2^3} + \frac{1}{3^3} + \cdots + \frac{1}{n^3} < \frac{3}{2} - \frac{1}{n}, \quad \text{for some } n \geq 3,$$

we obtain

$$1 + \frac{1}{2^3} + \frac{1}{3^3} + \cdots + \frac{1}{n^3} + \frac{1}{(n+1)^3} < \frac{3}{2} - \frac{1}{n} + \frac{1}{(n+1)^3}.$$

All we need to check is

$$\frac{3}{2} - \frac{1}{n} + \frac{1}{(n+1)^3} < \frac{3}{2} - \frac{1}{(n+1)},$$

which is equivalent to

$$\frac{1}{(n+1)^3} < \frac{1}{n} - \frac{1}{(n+1)},$$

or

$$\frac{1}{(n+1)^3} < \frac{1}{n(n+1)}.$$

This is true, completing the inductive step. This proves the inequality.

18. We prove both parts by induction on n. For (a), the case $n = 1$ is straightforward. Assume now that we have found an n-digit number m divisible by 2^n made out of the digits 2 and 3 only. Let $m = 2^n k$ for some integer k. If n is even, then

$$2 \times 10^n + m = 2^n(2 \cdot 5^n + k)$$

is an $(n+1)$-digit number written only with 2's and 3's, and divisible by 2^{n+1}. If k is odd, then

$$3 \times 10^n + m = 2^n(3 \cdot 5^n + k)$$

has this property.

The idea of part (b) is the same. The base case is trivial, $m = 5$. Now if we have found an n-digit number $m = 5^n k$ with this property, then looking modulo 5, one of the $(n+1)$-digit numbers

$$5 \times 10^n + m = 5^n(5 \cdot 2^n + k),$$

$$6 \times 10^n + m = 5^n(6 \cdot 2^n + k),$$

$$7 \times 10^n + m = 5^n(7 \cdot 2^n + k),$$

$$8 \times 10^n + m = 5^n(8 \cdot 2^n + k),$$

$$9 \times 10^n + m = 5^n(9 \cdot 2^n + k)$$

has the required property, and the problem is solved.

(USA Mathematical Olympiad, 2003, proposed by T. Andreescu)

19. We proceed by induction on n. The base case is obvious; the decomposition consists of just one piece. For the induction step, let us assume that the tiling is possible for such a $2^n \times 2^n$ board and consider a $2^{n+1} \times 2^{n+1}$ board. Start by placing a piece in the middle of the board as shown in Figure 49. The remaining surface decomposes into four $2^n \times 2^n$ boards with corner squares removed, each of which can be tiled by the induction hypothesis. Hence we are done.

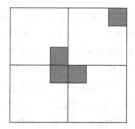

Figure 49

20. The property is clearly true for a single number. Now assume that it is true whenever we have such a sequence of length k and let us prove it for a sequence of length $k + 1$: $x_1, x_2, \ldots, x_{k+1}$. Call a cyclic shift with all partial sums positive "good".

With indices taken modulo $k+1$, there exist two terms x_j and x_{j+1} such that $x_j > 0$, $x_{j+1} \leq 0$. Without loss of generality, we may assume that these terms are x_k and x_{k+1}. Define a new sequence by $y_j = x_j, j \leq k - 1$, $y_k = x_k + x_{k+1}$. By the inductive hypothesis, $y_1, y_2, \ldots, y_k$ has a unique good cyclic shift. Expand y_k into x_k, x_{k+1} to obtain a good cyclic shift of $x_1, x_2, \ldots, x_{k+1}$. This proves the existence. To prove uniqueness, note that a good cyclic shift of $x_1, x_2, \ldots, x_{k+1}$ can start only with one of $x_1, x_2, \ldots, x_k$ (since $x_{k+1} < 0$). It induces a good cyclic shift of $y_1, y_2, \ldots, y_k$ that starts at the same term; hence two good cyclic shifts of the longer sequence would produce two good cyclic shifts of the shorter. This is ruled out by the induction hypothesis, and the uniqueness is proved.

(G. Raney)

21. We induct on $m + n$. The base case $m + n = 4$ can be verified by examining the equalities

$$1 + 1 = 1 + 1 \quad \text{and} \quad 1 + 2 = 1 + 2.$$

Now let us assume that the property is true for $m + n = k$ and prove it for $m + n = k + 1$. Without loss of generality, we may assume that $x_1 = \max_i x_i$ and $y_1 = \max_i y_i$, $x_1 \geq y_1$. If $m = 2$, then

$$y_1 + y_2 = x_1 + x_2 + \cdots + x_n \geq x_1 + n - 1 \geq y_1 + n - 1.$$

It follows that $y_1 = x_1 = n$ or $n - 1$, $y_2 = n - 1$, $x_2 = x_3 = \cdots = x_n = 1$. Consequently, $y_2 = x_2 + x_3 + \cdots + x_n$, and we are done. If $m > 2$, rewrite the original equality as

$$(x_1 - y_1) + x_2 + \cdots + x_n = y_2 + \cdots + y_m.$$

This is an equality of the same type, with the observation that $x_1 - y_1$ could be zero, in which case x_1 and y_1 are the numbers to be suppressed.

We could apply the inductive hypothesis if $y_1 \geq n$, in which case $y_2 + \cdots + y_m$ were less than $mn - y_1 < (m-1)n$. In this situation just suppress the terms provided by the inductive hypothesis; then move y_1 back to the right-hand side.

Let us analyze the case in which this argument does not work, namely when $y_1 < n$. Then $y_2 + y_3 + \cdots + y_m \leq (m-1)y_1 < (m-1)n$, and again the inductive hypothesis can be applied. This completes the solution.

22. Let f be the function. We will construct g and h such that $f = g + h$, with g an odd function and h a function whose graph is symmetric with respect to the point $(1, 0)$.

Let g be any odd function on the interval $[-1, 1]$ for which $g(1) = f(1)$. Define $h(x) = f(x) - g(x)$, $x \in [-1, 1]$. Now we proceed inductively as follows. For $n \geq 1$, let $h(x) = -h(2-x)$ and $g(x) = f(x) - h(x)$ for $x \in (2n-1, 2n+1]$, and then extend these functions such that $g(x) = -g(-x)$ and $h(x) = f(x) - g(x)$ for $x \in [-2n-1, -2n+1)$. It is straightforward to check that the g and h constructed this way satisfy the required condition.

(*Kvant* (*Quantum*))

23. *First solution.* We prove the property by induction on n. For $n = 2$, any number of the form $n = 2t^2$, t an integer, would work.

Let us assume that for $n = k$ there is a number m with the property from the statement, and let us find a number m' that fulfills the requirement for $n = k + 1$.

We need the fact that every integer $p \geq 2$ can be represented as $a^2 + b^2 - c^2$, where a, b, c are positive integers. Indeed, if p is even, say $p = 2q$, then

$$p = 2q = (3q)^2 + (4q - 1)^2 - (5q - 1)^2,$$

while if p is odd, $p = 2q + 1$, then

$$p = 2q + 1 = (3q - 1)^2 + (4q - 4)^2 - (5q - 4)^2,$$

if $q > 1$, while if $q = 1$, then $p = 3 = 4^2 + 6^2 - 7^2$.

Returning to the inductive argument, let

$$m = a_1^2 + a_2^2 = b_1^2 + b_2^2 + b_3^2 = \cdots = l_1^2 + l_2^2 + \cdots + l_k^2,$$

and also $m = a^2 + b^2 - c^2$. Taking $m' = m + c^2$ we have

$$m' = a^2 + b^2 = a_1^2 + a_2^2 + c^2 = b_1^2 + b_2 + c^2 = \cdots = l_1^2 + l_2^2 + \cdots + l_2^2 + c^2.$$

This completes the induction.

Second solution. We prove by induction that $m = 25^{n-1}$ can be written as the sum of $1, 2, \ldots, n$ nonzero perfect squares. Base case: $1 = 1^2$. Inductive step: Suppose 25^{n-1} can be expressed as the sum of $1, 2, \ldots, n$ positive squares. Then 25^n can be written as the sum of p positive squares, for any p in $1, 2, \ldots, n$, by multiplying each addend in the decomposition of 25^{n-1} into p squares by 25. Now let

$$25^{n-1} = (a_1)^2 + \cdots + (a_n)^2.$$

We have

$$25^n = (3a_1)^2 + (4a_1)^2 + (5a_2)^2 + (5a_3)^2 + \ldots + (5a_n)^2,$$

and we're done (for $n = 1$, we simply have $25 = 9 + 16$).

(*Gazeta Matematică* (*Mathematics Gazette, Bucharest*), 1980, proposed by M. Cavachi, second solution by E. Glazer)

24. We will prove a more general inequality namely that for all $m > 1$,

$$\sqrt[m]{a_1} - \sqrt[m]{a_2} + \sqrt[m]{a_3} - \cdots - \sqrt[m]{a_{2n}} + \sqrt[m]{a_{2n+1}} < \sqrt[m]{a_1 - a_2 + a_3 - \cdots - a_{2n} + a_{2n+1}}.$$

The inequality from the statement is the particular case $m = n$.

This more general inequality will be proved by induction on n. For $n = 2$, we have to show that if $a_1 < a_2 < a_3$, then

$$\sqrt[m]{a_1} - \sqrt[m]{a_2} + \sqrt[m]{a_3} < \sqrt[m]{a_1 - a_2 + a_3}.$$

Denote $a = a_1, b = a_3, t = a_2 - a_1 > 0$. The inequality can be written as

$$\sqrt[m]{a + t} - \sqrt[m]{a} > \sqrt[m]{b} - \sqrt[m]{b - t}.$$

Define the function $f : (0, \infty) \to \mathbb{R}, f(x) = \sqrt[m]{x + t} - \sqrt[m]{x}$. Its first derivative is $f'(x) = \frac{1}{m}[(x+t)^{(1-m)/m} - x^{(1-m)/m}]$, which is negative. This shows that f is strictly decreasing, which proves the inequality.

For the induction step, let us assume that the inequality holds for $n \leq k - 1$ and prove it for $n = k$. Using the induction hypothesis we deduce that

$$\sqrt[m]{a_1} - \sqrt[m]{a_2} + \sqrt[m]{a_3} - \cdots - \sqrt[m]{a_{2k}} + \sqrt[m]{a_{2k+1}}$$
$$< \sqrt[m]{a_1 - a_2 + a_3 - \cdots - a_{2k-2} + a_{2k-1}} - \sqrt[m]{a_{2k}} + \sqrt[m]{a_{2k+1}}.$$

Using the base case $n = 2$, we deduce that the latter is less than $\sqrt[m]{a_1 - a_2 + a_3 - \cdots - a_{2k} + a_{2k+1}}$, which completes the induction.

(Balkan Mathematical Olympiad, 1998, proposed by B. Enescu)

25. We will say that the lines of the set X pass through k nodes if there are k points in the plane such that each line in X passes through at least one of them. We denote by $S(n, k)$ the statement which says that from the fact that any n lines of set X pass through k nodes it follows that all the lines of X pass through k nodes. We are supposed to prove $S(k^2 + 1, k)$ for $k \geq 1$. We do this by induction.

First note that $S(3, 1)$ is obvious, if any three lines pass through a point, then all lines pass through a point. Next notice that $S(6, 2)$ is a corollary of $S(3, 1)$ by the following argument:

Consider 6 lines, which, by hypothesis pass through 2 points. Then through one of the points, which we call P, pass at least 3 lines. Denote the set of all lines passing through P by M. We will show that any 3 lines in $A \backslash M$ pass through a point. Consider 3 such lines, and add to them 3 lines in M. Then these six lines pass through 2 points. One of these poins must be P, or else the lines passing through P would generate 3 different nodes. Hence the other 3 lines must themselves pass through a point.

The argument can be adapted to prove that $S(6, 2)$ implies $S(10, 3)$. Basically one starts again with 6 lines outside the similar set M, add the 4 lines in M and argue the same.

This argument can be adapted to prove $S((k - 1)^2 + 1, k - 1)$ implies $S(k^2 + 1, k)$ as follows. Consider $k^2 + 1$ points in A and the k points through which they pass. Through one of these points, which we call P, pass at least $k + 1$ lines. Denote by M the set of lines in A that pass through P. We will show that for the lines in $A \backslash M$ any $(k - 1)^2 + 1$ lines pass through $k - 1$ nodes. Indeed, to each subset of $(k - 1)^2 + 1$ lines in $A \backslash M$ add lines from M, and some other lines if M is exhausted, until we obtain $k^2 + 1$ lines. By hypothesis, these pass through k nodes. One of these nodes is P, for else the more than $k + 1$ lines passing through it would pass through at least that many nodes, contradicting the hypothesis. It follows that the lines in $A \backslash M$ pass throgh the remaining $k - 1$ nodes, proving the claim. By the induction hypothesis, all lines in $A \backslash M$ pass through $k - 1$ nodes. Add P to these nodes to complete the induction step.

(Moscow Mathematical Olympiad, 1995–1996)

26. The property can be checked easily for small integers, which will constitute the base case. Assuming the property true for all integers less than n, let F_k be the largest term of the Fibonacci sequence that does not exceed n. The number $n - F_k$ is strictly less than n, so by the induction hypothesis it can be written as a sum of distinct terms of the Fibonacci sequence, say $n - F_k = \sum_j F_{i_j}$. The assumption on the maximality of F_k implies that $n - F_k < F_k$ (this because $F_{k+1} = F_k + F_{k-1} < 2F_k$ for $k \geq 2$). It follows that $F_k \neq F_{i_j}$, for all j. We obtain $n = \sum_j F_{i_j} + F_k$, which gives a way of writing n as a sum of distinct terms of the Fibonacci sequence.

27. We will prove a more general identity, namely,

$$F_{m+n+1} = F_{m+1}F_{n+1} + F_m F_n, \text{ for } m, n \geq 0.$$

We do so by induction on n. The inductive argument will assume the property to be true for $n = k - 1$ and $n = k$, and prove it for $n = k + 1$. Thus the base case consists of $n = 0$, $F_{m+1} = F_{m+1}$; and $n = 1$, $F_{m+2} = F_{m+1} + F_m$ – both of which are true.

Assuming that $F_{m+k} = F_{m+1}F_k + F_m F_{k-1}$ and $F_{m+k+1} = F_{m+1}F_{k+1} + F_m F_k$, we obtain by addition,

$$F_{m+k} + F_{m+k+1} = F_{m+1}(F_k + F_{k+1}) + F_m(F_{k-1} + F_k),$$

which is, in fact, the same as $F_{m+k+2} = F_{m+1}F_{k+2} + F_m F_{k+1}$. This completes the induction. For $m = n$, we obtain the identity in the statement.

28. Inspired by the previous problem, we generalize the identity to

$$F_{m+n+p} = F_{m+1}F_{n+1}F_{p+1} + F_m F_n F_p - F_{m-1}F_{n-1}F_{p-1},$$

which should hold for $m, n, p \geq 1$. In fact, we can augment the Fibonacci sequence by $F_{-1} = 1$ (so that the recurrence relation still holds), and then the above formula makes sense for $m, n, p \geq 0$. We prove it by induction on p. Again for the base case we consider $p = 0$, with the corresponding identity

$$F_{m+n} = F_{m+1}F_{n+1} - F_{m-1}F_{n-1},$$

and $p = 1$, with the corresponding identity

$$F_{m+n+1} = F_{m+1}F_{n+1} + F_m F_n.$$

Of the two, the second was proved in the solution to the previous problem. And the first identity is just a consequence of the second, obtained by subtracting $F_{m+n-1} = F_m F_n + F_{m-1} F_{n-1}$ from $F_{m+n+1} = F_{m+1} F_{n+1} + F_m F_n$. So the base case is verified. Now we assume that the identity holds for $p = k - 1$ and $p = k$, and prove it for $p = k + 1$. Indeed, adding

$$F_{m+n+k+1} = F_{m+1} F_{n+1} F_k + F_m F_n F_{k-1} - F_{m-1} F_{n-1} F_{k-2}$$

and

$$F_{m+n+k} = F_{m+1} F_{n+1} F_{k+1} + F_m F_n F_k - F_{m-1} F_{n-1} F_{k-1},$$

we obtain

$$
\begin{aligned}
F_{m+n+k+1} &= F_{m+n+k-1} + F_{m+n+k} \\
&= F_{m+1} F_{n+1} (F_k + F_{k+1}) + F_m F_n (F_{k-1} + F_k) - F_{m-1} F_{n-1} (F_{k-2} + F_{k-1}) \\
&= F_{m+1} F_{n+1} F_{k+2} + F_m F_n F_{k+1} - F_{m-1} F_{n-1} F_k.
\end{aligned}
$$

This proves the identity. Setting $m = n = p$, we obtain the identity in the statement.

29. The base case consists of the dissections for $n = 4, 5$, and 6 shown in Figure 50. The induction step jumps from $P(k)$ to $P(k+3)$ by dissecting one of the triangles into four triangles similar to it.

(R. Gelca)

Figure 50

30. First, we explain the inductive step, which is represented schematically in Figure 51. If we assume that such a k-gon exists for all $k < n$, then the n-gon can be obtained by cutting off two vertices of the $(n - 2)$-gon by two parallel lines. The sum of the distances from an interior point to the two parallel sides does not change while the point varies, and of course the sum of distances to the remaining sides is constant by the induction hypothesis. Choosing the parallel sides unequal, we can guarantee that the resulting polygon is not regular.

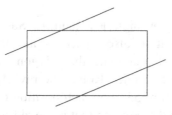

Figure 51

The base case consists of a rectangle ($n = 4$) and an equilateral triangle with two vertices cut off by parallel lines ($n = 5$). Note that to obtain the base case we had to apply the idea behind the inductive step.

31. The property is obviously true for the triangle since there is nothing to dissect. This will be our base case. Let us assume that the property is true for any coloring of a k-gon, for all $k < n$, and let us prove that it is true for an arbitrary coloring of an n-gon. Because at least three colors were used, there is a diagonal whose endpoints have different colors, say red (r) and blue (b). If on both sides of the diagonal a third color appears, then we can apply the induction hypothesis to two polygons and solve the problem.

If this is not the case, then on one side there will be a polygon with an even number of sides and with vertices colored in cyclic order $rbrb \ldots rb$. Pick a blue point among them that is not an endpoint of the initially chosen diagonal and connect it to a vertex colored by a third color (Figure 52). The new diagonal dissects the polygon into two polygons satisfying the property from the statement, and having fewer sides. The induction hypothesis can be applied again, solving the problem.

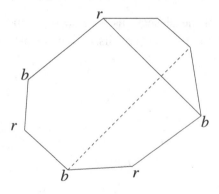

Figure 52

32. We prove the property by induction on the number of vertices. The base case is the triangle, where there is nothing to prove.

Let us assume now that the property holds for polygons with fewer than n vertices and prove it for a polygon with n vertices. The inductive step consists in finding one interior diagonal.

We commence with an interior angle less than π. Such an angle can be found at one of the vertices of the polygon that are also vertices of its convex hull (the convex hull is the smallest convex set in the plane that contains the polygon). Let the polygon be $A_1A_2 \ldots A_n$, with $\angle A_nA_1A_2$ the chosen interior angle. Rotate the ray $|A_1A_n$ toward $|A_1A_2$ continuously inside the angle as shown in Figure 53. For each position of the ray, strictly between A_1A_n and A_1A_2, consider the point on the polygon that is the closest to A_1. If for some position of the ray this point is a vertex, then we have obtained a diagonal that divides the polygon into two polygons with fewer sides. Otherwise, A_2A_n is the diagonal.

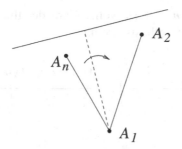

Figure 53

Dividing by the interior diagonal, we obtain two polygons with fewer vertices, which by hypothesis can be divided into triangles. This completes the induction.

33. We induct on the number to be represented. For the base case, we have

$$1 = 1^2$$
$$2 = -1^2 - 2^2 - 3^2 + 4^2,$$
$$3 = -1^2 + 2^2,$$
$$4 = -1^2 - 2^2 + 3^2.$$

The inductive step is "$P(n)$ implies $P(n+4)$"; it is based on the identity

$$m^2 - (m+1)^2 - (m+2)^2 + (m+3)^2 = 4.$$

Remark. This result has been generalized by J. Mitek, who proved that every integer k can be represented in the form $k = \pm 1^s \pm 2^s \pm \cdots \pm m^s$ for a suitable choice of signs, where s is a given integer ≥ 2. The number of such representations is infinite.

(P. Erdős, J. Surányi)

34. First, we show by induction on k that the identity holds for $n = 2^k$. The base case is contained in the statement of the problem. Assume that the property is true for $n = 2^k$ and let us prove it for $n = 2^{k+1}$. We have

$$f\left(\frac{x_1 + \cdots + x_{2^k} + x_{2^k+1} \cdots + x_{2^{k+1}}}{2^{k+1}}\right) = \frac{f\left(\frac{x_1 + \cdots + x_{2^k}}{2^k}\right) + f\left(\frac{x_{2^k+1} + \cdots + x_{2^{k+1}}}{2^k}\right)}{2}$$

$$= \frac{\frac{f(x_1) + \cdots + f(x_{2^k})}{2^k} + \frac{f(x_{2^k+1}) + \cdots + f(x_{2^{k+1}})}{2^k}}{2}$$

$$= \frac{f(x_1) + \cdots + f(x_{2^k}) + f(x_{2^k+1}) + \cdots + f(x_{2^{k+1}})}{2^{k+1}},$$

which completes the induction. Now we work backward, showing that if the identity holds for some n, then it holds for $n - 1$ as well. Consider the numbers $x_1, x_2, \ldots, x_{n-1}$ and $x_n = \frac{x_1 + x_2 + \cdots + x_{n-1}}{n-1}$. Using the hypothesis, we have

$$f\left(\frac{x_1 + \cdots + x_{n-1} + \frac{x_1 + \cdots + x_{n-1}}{n-1}}{n}\right) = \frac{f(x_1) + \cdots + f(x_{n-1}) + f\left(\frac{x_1 + \cdots + x_{n-1}}{n-1}\right)}{n},$$

which is the same as

$$f\left(\frac{x_1 + \cdots + x_{n-1}}{n-1}\right) = \frac{f(x_1) + \cdots + f(x_{n-1})}{n} + \frac{1}{n}f\left(\frac{x_1 + \cdots + x_{n-1}}{n-1}\right).$$

Moving the last term on the right to the other side gives

$$\frac{n-1}{n}f\left(\frac{x_1 + x_2 + \cdots + x_{n-1}}{n-1}\right) = \frac{f(x_1) + f(x_2) + \cdots + f(x_{n-1})}{n}.$$

This is clearly the same as

$$f\left(\frac{x_1 + x_2 + \cdots + x_{n-1}}{n-1}\right) = \frac{f(x_1) + f(x_2) + \cdots + f(x_{n-1})}{n-1},$$

and the argument is complete.

35. This is a stronger form of the inequality discussed in the beginning, which can be obtained from it by applying the AM-GM inequality.

We first prove that the property holds for n a power of 2. The base case

$$(1 + a_1)(1 + a_2) \geq (1 + \sqrt{a_1 a_2})^2$$

reduces to the obvious $a_1 + a_2 \geq 2\sqrt{a_1 a_2}$.

If

$$(1 + a_1)(1 + a_2) + \cdots + (1 + a_{2^k}) \geq \left(1 + \sqrt[2^k]{a_1 a_2 \cdots a_{2^k}}\right)^{2^k}$$

for every choice of nonnegative numbers, then

$$(1 + a_1) \cdots (1 + a_{2^{k+1}}) = (1 + a_1) \cdots (1 + a_{2^k})(1 + a_{2^k+1}) \cdots (1 + a_{2^{k+1}})$$

$$\geq \left(1 + \sqrt[2^k]{a_1 \cdots a_{2^k}}\right)^{2^k} \left(1 + \sqrt[2^k]{a_{2^k+1} \cdots a_{2^{k+1}}}\right)^{2^k}$$

$$\geq \left[\left(1 + \sqrt{\sqrt[2^k]{a_1 \cdots a_{2^k}} \sqrt[2^k]{a_{2^k+1} \cdots a_{2^{k+1}}}}\right)^2\right]^{2^k}$$

$$= \left(1 + \sqrt[2^{k+1}]{a_1 \cdots a_{2^{k+1}}}\right)^{2^{k+1}}.$$

This completes the induction.

Now we work backward. If the inequality holds for $n + 1$ numbers, then choosing $a_{n+1} = \sqrt[n]{a_1 a_2 \cdots a_n}$, we can write

$$(1 + a_1) \cdots (1 + a_n)(1 + \sqrt[n]{a_1 \cdots a_n}) \geq \left(1 + \sqrt[n+1]{a_1 \cdots a_n \sqrt[n]{a_1 \cdots a_n}}\right)^{n+1},$$

which is the same as

$$(1 + a_1) \cdots (1 + a_n)(1 + \sqrt[n]{a_1 \cdots a_n}) \geq (1 + \sqrt[n]{a_1 \cdots a_n})^{n+1}.$$

Canceling the common factor, we obtain the inequality for n numbers. The inequality is proved.

36. The "pigeons" are the numbers. The "holes" are the 49 sets

$$\{1, 98\}, \{2, 97\}, \ldots, \{40, 50\}.$$

Two of the numbers fall in the same set; their sum is equal to 99. We are done.

37. As G. Pólya said, "a trick applied twice becomes a technique". Here we repeat the idea of the Mongolian problem from the 26th International Mathematical Olympiad.

Let $b_1, b_2, \ldots, b_n$ be the sequence, where $b_i \in \{a_1, a_2, \ldots, a_n\}$, $1 \leq i \leq m$. For each $j \leq m$ define the n-tuple $K_j = (k_1, k_2, \ldots, k_n)$, where $k_i = 0$ if a_i appears an even number of times in $b_1, b_2, \ldots, b_j$ and $k_i = 1$ otherwise.

If there exists $j \leq m$ such that $K_j = (0, 0, \ldots, 0)$ then $b_1 b_2 \cdots b_j$ is a perfect square and we are done. Otherwise, there exist $j < l$ such that $K_j = K_l$. Then in the sequence $b_{j+1}, b_{j+2}, \ldots, b_l$ each a_i appears an even number of times. The product $b_{j+1} b_{j+2} \cdots b_l$ is a perfect square.

38. The sequence has the property that for any n the first $n + 1$ terms are less than or equal to $2n$. The problem would be solved if we showed that given a positive integer n, from any $n + 1$ distinct integer numbers between 1 and $2n$ we can choose two whose difference is n. This is true, indeed, since the pigeonhole principle implies that one of the n pairs $(1, n + 1)$, $(2, n + 2), \ldots, (n, 2n)$ contains two terms of the sequence.

(Austrian-Polish Mathematics Competition, 1980)

39. The "holes" will be the residue classes, and the "pigeons", the numbers ax^2, $c - by^2$, $x, y = 0, 1, \ldots, p - 1$. There are $2p$ such numbers. Any residue class, except for 0, can have at most two elements of the form ax^2 and at most two elements of the form $c - by^2$ from the ones listed above. Indeed, $ax_1^2 \equiv ax_2^2$ implies $x_1^2 \equiv x_2^2$, so $(x_1 - x_2)(x_1 + x_2) \equiv 0$. This can happen only if $x_1 = \pm x_2$. Also, $ax^2 \equiv 0$ only when $x = 0$.

We distinguish two cases. If $c - by_0^2 \equiv 0$ for some y_0, then $(0, y_0)$ is a solution. Otherwise, the $2p - 1$ numbers ax^2, $c - by^2$, $x = 1, 2, \ldots, p - 1$, $y = 0, 1, \ldots, p - 1$ are distributed into $p - 1$ "holes", namely the residue classes $1, 2, \ldots, p - 1$. Three of them must lie in the same residue class, so there exist x_0 and y_0 with $ax_0^2 \equiv c - by_0^2 \pmod{p}$. The pair (x_0, y_0) is a solution to the equation from the statement.

Remark. A more advanced solution can be produced based on the theory of quadratic residues.

40. In any 2×2 square, only one of the four numbers can be divisible by 2, and only one can be divisible by 3. Tiling the board by 2×2 squares, we deduce that at most 25 numbers are divisible by 2 and at most 25 numbers are divisible by 3. There are at least 50 remaining numbers that are not divisible by 2 or 3, and thus must equal one of the numbers 1, 5, or 7. By the pigeonhole principle, one of these numbers appears at least 17 times.

(St. Petersburg City Mathematical Olympiad, 2001)

41. A more general property is true, namely that for any positive integer n there exist infinitely many terms of the Fibonacci sequence divisible by n.

We apply now the pigeonhole principle, letting the "objects" be all pairs of consecutive Fibonacci numbers (F_n, F_{n+1}), $n \geq 1$, and the "boxes" the pairs of residue classes modulo n. There are infinitely many objects, and only n^2 boxes, and so there exist indices $i > j > 1$ such that $F_i \equiv F_j \pmod{n}$ and $F_{i+1} \equiv F_{j+1} \pmod{m}$.

In this case

$$F_{i-1} = F_{i+1} - F_i \equiv F_{j+1} - F_j = F_{j-1} \pmod{n},$$

and hence $F_{i-1} \equiv F_{j-1} \pmod{n}$ as well. An inductive argument proves that $F_{i-k} \equiv F_{j-k} \pmod{n}$, $k = 1, 2, \ldots, j$. In particular, $F_{i-j} \equiv F_0 = 0 \pmod{n}$. This means that F_{i-j} is divisible by n. Moreover, the indices i and j range in an infinite family, so the difference $i - j$ can assume infinitely many values. This proves our claim, and as a particular case, we obtain the conclusion of the problem.

(Irish Mathematical Olympiad, 1999)

42. We are allowed by the recurrence relation to set $x_0 = 0$. We will prove that there is an index $k \leq m^3$ such that x_k divides m. Let r_t be the remainder obtained by dividing x_t by m for $t = 0, 1, \ldots, m^3 + 2$. Consider the triples (r_0, r_1, r_2), (r_1, r_2, r_3), $\ldots$, $(r_{m^3}, r_{m^3+1}, r_{m^3+2})$. Since r_t can take m values, the pigeonhole principle implies that at least two triples are equal. Let p be the smallest number such that the triple (r_p, r_{p+1}, r_{p+2}) is equal to another triple (r_q, r_{q+1}, r_{q+2}), $p < q \leq m^3$. We claim that $p = 0$.

Assume by way of contradiction that $p \geq 1$. Using the hypothesis, we have

$$r_{p+2} \equiv r_{p-1} + r_p r_{p+1} \pmod{m} \quad \text{and} \quad r_{q+2} \equiv r_{q-1} + r_q r_{q+1} \pmod{m}.$$

Because $r_p = r_q$, $r_{p+1} = r_{q+1}$, and $r_{p+2} = r_{q+2}$, it follows that $r_{p-1} = r_{q-1}$, so $(r_{p-1}, r_p, r_{p+1}) = (r_{q-1}, r_q, r_{q+1})$, contradicting the minimality of p. Hence $p = 0$, so $r_q = r_0 = 0$, and therefore x_q is divisible by m.

(T. Andreescu, D. Miheţ)

43. We focus on 77 consecutive days, starting on a Monday. Denote by a_n the number of games played during the first n days, $n \geq 1$. We consider the sequence of positive integers

$$a_1, a_2, \ldots, a_{77}, a_1 + 20, a_2 + 20, \ldots, a_{77} + 20.$$

Altogether there are $2 \times 77 = 154$ terms not exceeding $11 \times 12 + 20 = 152$ (here we took into account the fact that during each of the 11 weeks there were at most 12 games). The pigeonhole principle implies right away that two of the above numbers are equal. They cannot both be among the first 77, because by hypothesis, the number of games increases by at least 1 each day. For the same reason the numbers cannot both be among the last 77. Hence there are two indices k and m such that $a_m = a_k + 20$. This implies that in the time interval starting with the $(k + 1)$st day and ending with the nth day, exactly 20 games were played, proving the conclusion.

Remark. In general, if a chess player decides to play d consecutive days, playing at least one game a day and a total of no more than m with $d < m < 2d$, then for each $i \leq 2d - n - 1$ there is a succession of days on which, in total, the chess player played exactly i games.

(D.O. Shklyarskyi, N.N. Chentsov, I.M. Yaglom, *Izbrannye Zadachi i Theoremy Elementarnoy Matematiki (Selected Problems and Theorems in Elementary Mathematics)*, Nauka, Moscow, 1976)

44. The solution combines the induction and pigeonhole principles. We commence with induction. The base case $m = 1$ is an easy check, the numbers can be only $-1, 0, 1$.

Assume now that the property is true for any $2m - 1$ numbers of absolute value not exceeding $2m - 3$. Let A be a set of $2m + 1$ numbers of absolute value at most $2m - 1$. If A contains $2m - 1$ numbers of absolute value at most $2m - 3$, then we are done by the induction hypothesis. Otherwise, A must contain three of the numbers $\pm(2m - 1), \pm(2m - 2)$. By eventually changing signs we distinguish two cases.

Case I. $2m - 1, -2m + 1 \in A$. Pair the numbers from 1 through $2m - 2$ as $(1, 2m - 2)$, $(2, 2m - 3), \ldots, (m - 1, m)$ so that the sum of each pair is equal to $2m - 1$, and the numbers from 0 through $-2m+1$ as $(0, -2m+1), (-1, -2m+2), \ldots, (-m+1, -m)$, so that the sum of each pair is $-2m + 1$. There are $2m - 1$ pairs, and $2m$ elements of A lie in them, so by the pigeonhole principle there exists a pair with both elements in A. Those elements combined with either $2m - 1$ or $-2m + 1$ give a triple whose sum is equal to zero.

Case II. $2m-1, 2m-2, -2m+2 \in A$ and $-2m+1 \notin A$. If $0 \in A$, then $0-2m+2+2m-2 = 0$ and we are done. Otherwise, consider the pairs $(1, 2m - 3), (2, 2m - 4), \ldots$, $(m - 2, m)$, each summing up to $2m - 2$, and the pairs $(1, -2m), \ldots, (-m+1, -m)$, each summing up to $-2m + 1$. Altogether there are $2m - 2$ pairs containing $2m - 1$ elements from A, so both elements of some pair must be in A. Those two elements combined with either $-2m + 2$ or $2m - 1$ give a triple with the sum equal to zero.

This concludes the solution.
(*Kvant (Quantum)*)

45. Denote by Δ the set of ordered triples of people (a, b, c) such that c is either a common acquaintance of both a and b or unknown to both a and b. If c knows exactly k participants, then there exist exactly $2k(n - 1 - k)$ ordered pairs in which c knows exactly one of a and b (the factor 2 shows up because we work with *ordered* pairs). There will be

$$(n - 1)(n - 2) - 2k(n - 1 - k) \geq (n - 1)(n - 2) - 2\left(\frac{n-1}{2}\right)^2 = \frac{(n-1)(n-3)}{2}$$

ordered pairs (a, b) such that c knows either both or neither of a and b. Counting by the c's, we find that the number of elements of Δ satisfies

$$|\Delta| \geq \frac{n(n - 1)(n - 3)}{2}.$$

To apply the pigeonhole principle, we let the "holes" be the ordered pairs of people (a, b), and the "pigeons" be the triples $(a, b, c) \in \Delta$. Put the pigeon (a, b, c) in the hole (a, b) if c

knows either both or neither of a and b. There are $fn(n-1)(n-3)2$ pigeons distributed in $n(n-1)$ holes. So there will be at least

$$\left\lceil \frac{n(n-1)(n-3)}{2} \Big/ n(n-1) \right\rceil = \left\lfloor \frac{n}{2} \right\rfloor - 1$$

pigeons in one hole, where $\lceil x \rceil$ denotes the least integer greater than or equal to x. To the "hole" corresponds a pair of people satisfying the required condition.

(USA Mathematical Olympiad, 1985)

46. The beautiful observation is that if the sequence

$$a_n = \cos(n\pi x_1) + \cos(n\pi x_2) + \cdots + \cos(n\pi x_k), \ n \geq 1,$$

assumes finitely many distinct values, then so does the sequence of k-tuples $u_n = (a_n, a_{2n}, \ldots, a_{kn})$, $n \geq 1$. By the pigeonhole principle there exist $m < n$ such that $a_n = a_m$, $a_{2n} = a_{2m}, \ldots, a_{kn} = a_{km}$. Let us take a closer look at these relations. We know that $\cos(nx)$ is a polynomial of degree n with integer coefficients in $\cos(x)$, namely the Chebyshev polynomial. If $A_i = \cos(n\pi x_i)$ and $B_i = \cos(m\pi x_i)$, then the previous relations combined with this observation show that $A_1^j + A_2^j + \cdots + A_k^j = B_1^j + B_2^j + \cdots + B_k^j$ for all $j = 1, 2, \ldots, k$. Using Newton's formulas, we deduce that the polynomials having the zeros $A_1, A_2, \ldots, A_k$, respectively, $B_1, B_2, \ldots, B_k$ are equal (they have equal coefficients). Hence there is a permutation σ of $1, 2, \ldots, n$ such that $A_i = B_{\sigma(i)}$. Thus $\cos(n\pi x_i) = \cos(m\pi x_{\sigma(i)})$, which means that $nx_i - mx_{\sigma(i)}$ is a rational number r_i for $1 \leq i \leq k$. We want to show that the x_i's are themselves rational. If $\sigma(i) = i$, this is obvious. On the other hand, if we consider a cycle of σ, $(i_1 i_2 i_3, \ldots, i_s)$, we obtain the linear system

$$mx_{i_1} - nx_{i_2} = r_{i_1},$$
$$mx_{i_2} - nx_{i_3} = r_{i_2},$$
$$\cdots$$
$$mx_{i_s} - nx_{i_1} = r_{i_s}.$$

It is not hard to compute the determinant of the coefficient matrix, which is $n^s - m^s$ (for example, by expanding by the first row, then by the first column, and then noting that the new determinants are triangular). The determinant is nonzero; hence the system has a unique solution. By applying Cramer's rule we determine that this solution consists of rational numbers. We conclude that the x_i's are all rational, and the problem is solved.

(V. Pop)

47. Place the circle at the origin of the coordinate plane and consider the rectangular grid determined by points of integer coordinates, as shown in Figure 54. The circle is inscribed in an 8×8 square decomposed into 64 unit squares. Because $3^2 + 3^2 > 4^2$, the four unit squares at the corners lie outside the circle. The interior of the circle is therefore covered by 60 squares, which are our "holes". The 61 points are the "pigeons", and by the pigeonhole principle two lie inside the same square. The distance between them does not exceed the length of the diagonal, which is $\sqrt{2}$. The problem is solved.

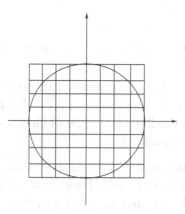

Figure 54

48. If $r = 1$, all lines pass through the center of the square. If $r \neq 1$, a line that divides the square into two quadrilaterals with the ratio of their areas equal to r has to pass through the midpoint of one of the four segments described in Figure 55 (in that figure the endpoints of the segments divide the sides of the square in the ratio r). Since there are four midpoints and nine lines, by the pigeonhole principle three of them have to pass through the same point.

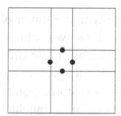

Figure 55

49. Choose a face with maximal number of edges, and let n be this number. The number of edges of each of the n adjacent faces ranges between 3 and n, so by the pigeonhole principle, two of these faces have the same number of edges.

(Moscow Mathematical Olympiad)

50. An n-gon has $\binom{n}{2} - n = \frac{1}{2}n(n - 3)$ diagonals. For $n = 21$ this number is equal to 189. If through a point in the plane we draw parallels to these diagonals, $2 \times 189 = 378$ adjacent angles are formed. The angles sum up to $360°$, and thus one of them must be less than $1°$.

51. The geometric aspect of the problem is only apparent. If we number the vertices of the polygon counterclockwise $1, 2, \ldots, 2n$, then $P_1, P_2, \ldots, P_{2n}$ is just a permutation of these numbers. We regard indices modulo $2n$. Then $P_i P_{i+1}$ is parallel to $P_j P_{j+1}$ if and only if $P_i - P_j \equiv P_{j+1} - P_{i+1} \pmod{2n}$, that is, if and only if $P_i + P_{i+1} \equiv P_j + P_{j+1} \pmod{2n}$. Because

$$\sum_{i=1}^{2n}(P_i + P_{i+1}) \equiv 2\sum_{i=1}^{2n} P_i \equiv 2n(2n - 1) \equiv 0 \pmod{2n}$$

and

$$\sum_{i=1}^{2n} i = n(2n - 1) \equiv n \pmod{2n},$$

it follows that $P_i + P_{i+1}$, $i = 1, 2, \ldots, 2n$, do not exhaust all residues modulo $2n$. By the pigeonhole principle there exist $i \neq j$ such that $P_i + P_{i+1} \equiv P_j + P_{j+1} \pmod{2n}$. Consequently, the sides $P_i P_{i+1}$ and $P_j P_{j+1}$ are parallel, and the problem is solved.

(German Mathematical Olympiad, 1976)

52. Let C be a circle inside the triangle formed by three noncollinear points in S. Then C is contained entirely in S. Set $m = np + 1$ and consider a regular polygon $A_1 A_2 \ldots A_m$ inscribed in C. By the pigeonhole principle, some n of its vertices are colored by the same color. We have thus found a monochromatic n-gon. Now choose α an irrational multiple of π. The rotations of $A_1 A_2 \ldots A_m$ by $k\alpha$, $k = 0, 1, 2, \ldots$, are all disjoint. Each of them contains an n-gon with vertices colored by n colors. Only finitely many incongruent n-gons can be formed with the vertices of $A_1 A_2 \ldots A_m$. So again by the pigeonhole principle, infinitely many of the monochromatic n-gons are congruent. Of course, they might have different colors. But the pigeonhole principle implies that one color occurs infinitely many times. Hence the conclusion.

(Romanian Mathematical Olympiad, 1995)

53. *First solution.* This is an example with the flavor of Ramsey theory (see Section 6.3.3) that applies the pigeonhole principle. Pick two infinite families of lines, $\{A_i, i \geq 1\}$, and $\{B_j, j \geq 1\}$, such that for any i and j, A_i and B_j are orthogonal. Denote by M_{ij} the point of intersection of A_i and B_j. By the pigeonhole principle, infinitely many of the M_{1j}'s, $j \geq 1$, have the same color. Keep only the lines B_j corresponding to these points, and delete all the others. So again we have two families of lines, but such that M_{1j} are all of the same color; call this color c_1.

Next, look at the line A_2. Either there is a rectangle of color c_1, or at most one point M_{2j} is colored by c_1. Again by the pigeonhole principle, there is a color c_2 that occurs infinitely many times among the M_{2j}'s. We repeat the reasoning. Either at some step we encounter a rectangle, or after finitely many steps we exhaust the colors, with infinitely many lines A_i still left to be colored. The impossibility to continue rules out this situation, proving the existence of a rectangle with vertices of the same color.

Second solution. Let there be p colors. Consider a $(p + 1) \times \left(\binom{p+1}{2} + 1\right)$ rectangular grid. By the pigeonhole principle, each of the $\binom{p+1}{2} + 1$ horizontal segments contains two points of the same color. There are $\binom{p+1}{2}$ possible configurations of monochromatic pairs, so two must repeat. The repeating pairs are vertices of a monochromatic rectangle.

54. We place the unit square in standard position. The "boxes" are the vertical lines crossing the square, while the "objects" are the horizontal diameters of the circles (Figure 56). Both the boxes and the objects come in an infinite number, but what we use for counting is length on the horizontal. The sum of the diameters is

$$\frac{10}{\pi} = 3 \times 1 + \varepsilon, \ \varepsilon > 0.$$

Consequently, there is a segment on the lower side of the square covered by at least four diameters. Any vertical line passing through this segment intersects the four corresponding circles.

55. If three points are collinear then we are done. Thus we can assume that no three points are collinear. The convex hull of all points is a polygon with at most n sides, which has therefore an angle not exceeding $\frac{(n-2)\pi}{n}$. All other points lie inside this angle. Ordered counterclockwise around the vertex of the angle they determine $n - 2$ angles that sum up to at most $\frac{(n-2)\pi}{n}$. It follows that one of these angles is less than or equal to $\frac{(n-2)\pi}{n(n-2)} = \frac{\pi}{n}$. The three points that form this angle have the required property.

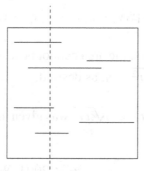

Figure 56

56. Denote by $D(O, r)$ the disk of center O and radius r. Order the disks

$$D(O_1, r_1), D(O_2, r_2), \ldots, D(O_n, r_n),$$

in decreasing order of their radii.

Choose the disk $D(O_1, r_1)$ and then delete all disks that lie entirely inside the disk of center O_1 and radius $3r_1$. The remaining disks are disjoint from $D(O_1, r_1)$. Among them choose the first in line (i.e., the one with maximal radius), and continue the process with the remaining circles.

The process ends after finitely many steps. At each step we deleted less than eight times the area of the chosen circle, so in the end we are left with at least $\frac{1}{9}$ of the initial area. The chosen circles satisfy the desired conditions.

(M. Pimsner, S. Popa, *Probleme de Geometrie Elementară* (*Problems in Elementary Geometry*), Editura Didactică şi Pedagogică, Bucharest, 1979)

57. Given a circle of radius r containing n points of integer coordinates, we must prove that $n < 2\pi \sqrt[3]{r^2}$. Because $r > 1$ and $2\pi > 6$ we may assume $n \geq 7$.

Label the n lattice points counterclockwise $P_1, P_2, \ldots, P_n$. The (counterclockwise) arcs $\overparen{P_1P_3}, \overparen{P_2P_4}, \ldots, \overparen{P_nP_2}$ cover the circle twice, so they sum up to 4π. Therefore, one of them, say $\overparen{P_1P_3}$, measures at most $\frac{4\pi}{n}$.

Consider the triangle $P_1P_2P_3$, which is inscribed in an arc of measure $\frac{4\pi}{n}$. Because $n \geq 7$, the arc is less than a quarter of the circle. The area of $P_1P_2P_3$ will be maximized if P_1 and P_3 are the endpoints and P_2 is the midpoint of the arc. In that case,

$$\text{Area}(P_1 P_2 P_3) = \frac{abc}{4r} = \frac{2r \sin \dfrac{\pi}{n} \cdot 2r \sin \dfrac{\pi}{n} \cdot 2r \sin \dfrac{2\pi}{n}}{4r} \leq \frac{2r \dfrac{\pi}{n} \cdot 2r \dfrac{\pi}{n} \cdot 2r \dfrac{2\pi}{n}}{4r} = \frac{4r^2 \pi^3}{n^3}.$$

And in general, the area of $P_1 P_2 P_3$ cannot exceed $\frac{4r^2\pi^3}{n^3}$. On the other hand, if the coordinates of the points P_1, P_2, P_3 are, respectively, (x_1, y_1), (x_2, y_2), and (x_3, y_3), then

$$\text{Area}(P_1 P_2 P_3) = \pm \frac{1}{2} \begin{vmatrix} 1 & 1 & 1 \\ x_1 & x_2 & x_3 \\ y_1 & y_2 & y_3 \end{vmatrix}$$

$$= \frac{1}{2} |x_1 y_2 - x_2 y_1 + x_2 y_3 - x_3 y_2 + x_3 y_1 - x_1 y_3|$$

Because the coordinates are integers, the area cannot be less than $\frac{1}{2}$. We obtain the inequality $\frac{1}{2} \leq \frac{4r^2\pi^3}{n^3}$, which proves that $2\pi \sqrt[3]{r^2} \geq n$, as desired.

Remark. The weaker inequality $n(r) < 6\sqrt[3]{\pi r^2}$ was given in 1999 at the Iranian Mathematical Olympiad.

58. Order the eight integers $a_1 < a_2 < \cdots < a_8 \leq 2004$. We argue by contradiction. Assume that for any choice of the integers a, b, c, d, either $a + b + c < d + 4$ or $a + b + c > 4d$. Let us look at the situation in which d is a_3 and $a, b,$ and c are a_1, a_2 and a_4. The inequality $a_1 + a_2 + a_4 < 4 + a_3$ is impossible because $a_4 \geq a_3 + 1$ and $a_1 + a_2 \geq 3$. Thus with our assumption, $a_1 + a_2 + a_4 > 4a_3$, or

$$a_4 > 4a_3 - a_2 - a_1.$$

By similar logic,

$$a_5 > 4a_4 - a_2 - a_1 > 16a_3 - 5a_2 - 5a_1,$$
$$a_6 > 4a_5 - a_2 - a_1 > 64a_3 - 21a_2 - 21a_1,$$
$$a_7 > 4a_6 - a_2 - a_1 > 256a_3 - 85a_2 - 85a_1,$$
$$a_8 > 4a_7 - a_2 - a_1 > 1024a_3 - 341a_2 - 341a_1.$$

We want to show that if this is the case, then a_8 should exceed 2004. The expression $1024a_3 - 341a_2 - 341a_1$ can be written as $683a_3 + 341(a_3 - a_2) + 341(a_3 - a_1)$, so to minimize it we have to choose $a_1 = 1, a_2 = 2, a_3 = 3$. But then the value of the expression would be 2049, which, as predicted, exceeds 2004. This contradiction shows that our assumption was false, proving the existence of the desired four numbers.

(Mathematical Olympiad Summer Program, 2004, proposed by T. Andreescu)

59. There is no loss of generality in supposing that $a_1 < a_2 < \cdots < a_n < \cdots$. Now proceed by induction on n. For $n = 1$, $a_1^2 \geq \frac{2 \times 1 + 1}{3} a_1$ follows from $a_1 \geq 1$. The inductive step reduces to

$$a_{n+1}^2 \geq \frac{2}{3}(a_1 + a_2 + \cdots + a_n) + \frac{2n+3}{3} a_{n+1}.$$

An equivalent form of this is

$$3a_{n+1}^2 - (2n+3)a_{n+1} \geq 2(a_1 + a_2 + \cdots + a_n).$$

At this point there is an interplay between the indices and the terms of the sequence, namely the observation that $a_1 + a_2 + \cdots + a_n$ does not exceed the sum of integers from 1 to a_n. Therefore,

$$2(a_1 + a_2 + \cdots + a_n) \leq 2(1 + 2 + \cdots + a_n) = a_n(a_n + 1) \leq (a_{n+1} - 1)a_{n+1}.$$

We are left to prove the sharper, yet easier, inequality

$$3a_{n+1}^2 - (2n+3)a_{n+1} \geq (a_{n+1} - 1)a_{n+1}.$$

This is equivalent to $a_{n+1} \geq n+1$, which follows from the fact that a_{n+1} is the largest of the numbers.

(Romanian Team Selection Test for the International Mathematical Olympiad, proposed by L. Panaitopol)

60. Again, there will be an interplay between the indices and the values of the terms. We start by ordering the a_i's increasingly $a_1 < a_2 < \cdots < a_n$. Because the sum of two elements of X is in X, given a_i in the complement of X, for each $1 \leq m \leq \frac{a_i}{2}$, either m or $a_i - m$ is not in X. There are $\lceil \frac{a_i}{2} \rceil$ such pairs and only $i - 1$ integers less than a_i and not in X, where $\lceil x \rceil$ denotes the least integer greater than or equal to x. Hence $a_i \leq 2i - 1$. Summing over i gives $a_1 + a_2 + \cdots + a_n \leq n^2$ as desired.

(Proposed by R. Stong for the USAMO, 2000)

61. Because $P(P(x)) - x$ is a polynomial of degree n, A is finite. If $a, b \in A$, $a \neq b$, then $a - b$ divides $P(a) - P(b)$ and $P(a) - P(b)$ divides $P(P(a)) - P(P(b)) = a - b$. It follows that $|a - b| = |P(a) - P(b)|$. Let the elements of A be $x_1 < x_2 < \cdots < x_k$. We have

$$x_k - x_1 = \sum_{i=1}^{k-1}(x_{i+1} - x_i) = \sum_{i=1}^{k-1}|P(x_{i+1}) - P(x_i)|$$

$$\geq \left| \sum_{i=1}^{k-1}(P(x_{i+1}) - P(x_i)) \right| = |P(x_k) - P(x_1)| = x_k - x_1.$$

It follows that the inequality in this relation is an equality, and so all the numbers $P(x_{i+1}) - P(x_i)$ have the same sign. So either $P(x_{i+1}) - P(x_i) = x_{i+1} - x_i$, for all i, or $P(x_{i+1}) - P(x_i) = x_i - x_{i+1}$ for all i. It follows that the numbers $x_1, x_2, \ldots, x_k$ are the roots of a polynomial equation of the form $P(x) \pm x = a$. And such an equation has at most n real roots. The conclusion follows.

(Romanian Mathematics Competition, 1986, proposed by Gh. Eckstein)

62. Call the elements of the 4×4 tableau a_{ij}, $i, j = 1, 2, 3, 4$, according to their location. As such, $a_{13} = 2$, $a_{22} = 5$, $a_{34} = 8$ and $a_{41} = 3$. Look first at the row with the *largest* sum, namely, the fourth. The unknown entries sum up to 27; hence all three of them, a_{42}, a_{43}, and a_{44}, must equal 9. Now we consider the column with *smallest* sum. It is the third, with

$$a_{13} + a_{23} + a_{33} + a_{43} = 2 + a_{23} + a_3 + 9 = 13.$$

We see that $a_{23} + a_{33} = 2$; therefore $a_{23} = a_{33} = 1$. We than have

$$a_{31} + a_{32} + a_{33} + a_{34} = a_{31} + a_{32} + 1 + 8 = 26.$$

Therefore, $a_{31} + a_{32} = 17$, which can happen only if one of them is 8 and the other is 9. Checking the two cases separately, we see that only $a_{31} = 8$, $a_{32} = 9$ yields a solution, which is described in Figure 57.

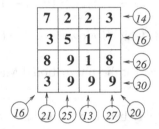

Figure 57

Remark. Such puzzles would appear in the Sunday edition of the *San Francisco Chronicle* at the time of publication of this book.

63. There are only finitely many polygonal lines with these points as vertices. Choose the one of minimal length $P_1 P_2 \ldots P_n$. If two sides, say $P_i P_{i+1}$ and $P_j P_{j+1}$, intersect at some point M, replace them by $P_i P_j$ and $P_{i+1} P_{j+1}$ to obtain the closed polygonal line $P_1 \ldots P_i P_j P_{j-1} \ldots P_{i+1} P_{j+1} \ldots P_n$ (Figure 58). The triangle inequality in triangles $MP_i P_j$ and $MP_{i+1} P_{j+1}$ shows that this polygonal line has shorter length, a contradiction. It follows that $P_1 P_2 \ldots P_n$ has no self-intersections, as desired.

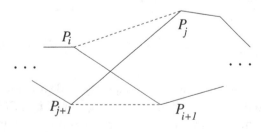

Figure 58

64. Let $A_i A_{i+1}$ be the longest side of the polygon (or one of them if more such sides exist). Perpendicular to it and at the endpoints A_i and A_{i+1} take the lines L and L', respectively. We argue on the configuration from Figure 59.

If all other vertices of the polygon lie to the right of L', then $A_{i-1} A_i > A_i A_{i+1}$, because the distance from A_i to a point in the half-plane determined by L' and opposite to A_i is greater

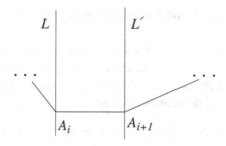

Figure 59

than the distance from A_i to L'. This contradicts the maximality, so it cannot happen. The same argument shows than no vertex lies to the left of L. So there exists a vertex that either lies on one of L and L', or is between them. That vertex projects onto the (closed) side A_iA_{i+1}, and the problem is solved.

Remark. It is possible that no vertex projects in the interior of a side, as is the case with rectangles or with the regular hexagon.

(M. Pimsner, S. Popa, *Probleme de Geometrie Elementară* (*Problems in Elementary Geometry*), Editura Didactică și Pedagogică, Bucharest, 1979)

65. *First solution*: Consider the oriented graph of roads and cities. By hypothesis, the graph has no cycles. Define a partial order of the cities, saying that $A < B$ if one can travel from A to B. A partial order on a finite set has maximal and minimal elements. In a maximal city all roads enter, and from a minimal city all roads exit.

Second solution: Pick an itinerary that travels through a maximal number of cities (more than one such itinerary may exist). No roads enter the starting point of the itinerary, while no roads exit the endpoint.

(*Kvant (Quantum)*)

66. Let b be a boy dancing with the maximal number of girls. There is a girl g' he does not dance with. Choose as b' a boy who dances with g'. Let g be a girl who dances with b but not with b'. Such a girl exists because of the maximality of b, since b' already dances with a girl who does not dance with b. Then the pairs (b, g), (b', g') satisfy the requirement.

(26th W.L. Putnam Mathematical Competition, 1965)

67. Arguing by contradiction, assume that we can have a set of finitely many points with this property. Let $V_1 \subset V$ be the vectors whose x-coordinate is positive or whose x-coordinate is 0 and the y-coordinate is positive. Let $V_2 = V \setminus V_1$.

Order the points marked points lexicographically by their coordinates (x, y). Then examining the largest point we obtain $|V_1| < |V_2|$ and examining the smallest point we obtain $|V_2| < |V_1|$. This impossible. The conclusion follows.

(*Kvant (Quantum)*), proposed by D. Rumynin)

68. Let $(a_{ij})_{ij}$, $1 \le i \le m$, $1 \le j \le n$, be the matrix. Denote the sum of the elements in the ith row by s_i, $i = 1, 2, \ldots, m$. We will show that among all matrices obtained by permuting the elements of each column, the one for which the sum $|s_1| + |s_2| + \cdots + |s_m|$ is minimal has the desired property.

If this is not the case, then $|s_k| \geq 2$ for some k. Without loss of generality, we can assume that $s_k \geq 2$. Since $s_1 + s_2 + \cdots + s_m = 0$, there exists j such that $s_j < 0$. Also, there exists an i such that $a_{ik} > a_{ij}$ for otherwise s_j would be larger than s_k. When exchanging a_{ik} and a_{ij} the sum $|s_1| + |s_2| + \cdots + |s_m|$ decreases. Indeed,

$$|s_k - a_{ik} + a_{ij}| + |s_j + a_{ik} - a_{ij}| = s_k - a_{ik} + a_{ij} + |s_j + a_{ik} - a_{ij}|$$
$$< s_k - a_{ik} + a_{ij} + |s_j| + a_{ik} - a_{ij},$$

where the equality follows from the fact that $s_k \geq 2 \geq a_{ik} - a_{ij}$, while the *strict* inequality follows from the triangle inequality and the fact that s_j and $a_{ik} - a_{ij}$ have opposite signs. This shows that any minimal configuration must satisfy the condition from the statement. Note that a minimal configuration always exists, since the number of possible permutations is finite.

(Austrian-Polish Mathematics Competition, 1984)

69. We call a number *good* if it satisfies the given condition. It is not difficult to see that all powers of primes are good. Suppose n is a good number that has at least two distinct prime factors. Let $n = p^r s$, where p is the smallest prime dividing n and s is not divisible by p. Because n is good, $p+s-1$ must divide n. For any prime q dividing s, $s < p+s-1 < s+q$, so q does not divide $p + s - 1$. Therefore, the only prime factor of $p + s - 1$ is p. Then $s = p^c - p + 1$ for some integer $c > 1$. Because p^c must also divide n, $p^c + s - 1 = 2p^c - p$ divides n. Because $2p^{c-1} - 1$ has no factors of p, it must divide s. But

$$\frac{p-1}{2}(2p^{c-1} - 1) = p^c - p^{c-1} - \frac{p-1}{2} < p^c - p + 1 < \frac{p+1}{2}(2p^{c-1} - 1)$$
$$= p^c + p^{c-1} - \frac{p+1}{2},$$

a contradiction. It follows that the only good integers are the powers of primes.

(Russian Mathematical Olympiad, 2001)

70. Let us assume that no infinite monochromatic sequence exists with the desired property, and consider a maximal white sequence $2k_1 < k_1 + k_2 < \cdots < 2k_n$ and a maximal black sequence $2l_1 < l_1 + l_2 < \cdots < 2l_m$. By maximal we mean that these sequences cannot be extended any further. Without loss of generality, we may assume that $k_n < l_m$.

We look at all white even numbers between $2k_n + 1$ and some arbitrary $2x$; let W be their number. If for one of these white even numbers $2k$ the number $k + k_n$ were white as well, then the sequence of whites could be extended, contradicting maximality. Hence $k + k_n$ must be black. Therefore, the number b of blacks between $2k_n + 1$ and $x + k_n$ is at least W.

Similarly, if B is the number of black evens between $l_m + 1$ and $2x$, the number w of whites between $2l_m + 1$ and $x + l_m$ is at least B. We have $B + W \geq x - l_m$, the latter being the number of even integers between $2l_m + 1$ and $2x$, while $b + w \leq x - k_n$, since $x - k_n$ is the number of integers between $2k_n + 1$ and $x + k_n$. Subtracting, we obtain

$$0 \leq (b - W) + (w - B) \leq l_m - k_n,$$

and this inequality holds for all x. This means that as x varies there is an upper bound for $b - W$ and $w - B$. Hence there can be only a finite number of black squares that cannot be written as $k_n + k$ for some white $2k$ and there can only be a finite number of white squares which cannot be written as $l_m + 1$ for some black $2l$. Consequently, from a point onward all white squares are of the form $l_m + l$ for some black $2l$ and from a point onward all black squares are of the form $k_n + k$ for some white $2k$.

We see that for k sufficiently large, k is black if and only if $2k - 2k_n$ is white, while k is white if and only if $2k - 2l_m$ is black. In particular, for each such k, $2k - 2k_n$ and $2k - 2l_m$ have the same color, opposite to the color of k. So if we let $l_m - k_n = a > 0$, then from some point onward $2x$ and $2x + 2a$ are of the same color. The arithmetic sequence $2x + 2na$, $n \geq 0$, is thus monochromatic. It is not hard to see that it also satisfies the condition from the statement, a contradiction. Hence our assumption was false, and sequences with the desired property do exist.

(Communicated by A. Neguţ)

71. We begin with an observation that will play an essential role in the solution. Given a triangle XYZ, if $\angle XYZ \leq \frac{\pi}{3}$, then either the triangle is equilateral or else $\max\{XY, YZ\} > XZ$, and if $\angle XYZ \leq \frac{\pi}{3}$, then either the triangle is equilateral or else $\min\{YX, YZ\} < XZ$.

Choose vertices A and B that minimize the distance between vertices. If C is a vertex such that $\angle ACB = \frac{\pi}{3}$, then $\max\{CA, CB\} \leq AB$, so by our observation the triangle ABC is equilateral. So there exists an equilateral triangle ABC formed by vertices of the polygon and whose side length is the minimal distance between two vertices of the polygon. By a similar argument there exists a triangle $A_1B_1C_1$ formed by vertices whose side length is the maximal distance between two vertices of the polygon. We will prove that the two triangles are congruent.

The lines AB, BC, CA divide the plane into seven open regions. Denote by R_A the region distinct from the interior of ABC and bounded by side BC, plus the boundaries of this region except for the vertices B and C. Define R_B and R_C analogously. These regions are illustrated in Figure 60. Because the given polygon is convex, each of A_1, B_1, and C_1 lies in one of these regions or coincides with one of A, B, and C.

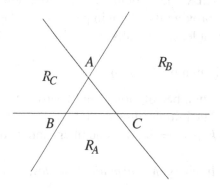

Figure 60

If two of A_1, B_1, C_1, say A_1 and B_1, are in the same region R_X, then $\angle A_1XB_1 < \frac{\pi}{3}$. Hence $\max\{XA_1, XB_1\} > A_1B_1$, contradicting the maximality of the length A_1B_1. Therefore, no two of A_1, B_1, C_1 are in the same region.

Suppose now that one of A_1, B_1, C_1 (say A_1) lies in one of the regions (say R_A). Because $\min\{A_1B, A_1C\} \geq BC$, we have that $\angle BA_1C \leq \frac{\pi}{3}$. We know that B_1 does not lie in R_A. Also, because the polygon is convex, B does not lie in the interior of the triangle AA_1B_1, and C does not lie in the interior of triangle AA_1B_1. It follows that B_1 lies in the closed region bounded by the rays $|A_1B$ and $|A_1C$. So does C_1. Therefore, $\frac{\pi}{3} = \angle B_1A_1C_1 \le \angle BA_1C \le \frac{\pi}{3}$, with equalities if B_1 and C_1 lie on rays $|A_1B$ and $|A_1C$. Because the given polygon is convex, this is possible only if B_1 and C_1 equal B and C in some order, in which case $BC = B_1C_1$. This would imply that triangles ABC and $A_1B_1C_1$ are congruent.

The remaining situation occurs when none of A_1, B_1, C_1 are in $R_A \cup R_B \cup R_C$, in which case they coincide with A, B, C in some order. Again we conclude that the two triangles are congruent.

We have proved that the distance between any two vertices of the given polygon is the same. Therefore, given a vertex, all other vertices are on a circle centered at that vertex. Two such circles have at most two points in common, showing that the polygon has at most four vertices. If it had four vertices, it would be a rhombus, whose longer diagonal would be longer than the side, a contradiction. Hence the polygon can only be the equilateral triangle, the desired conclusion.

(Romanian Mathematical Olympiad, 2000)

72. Because

$$a^2 + b^2 = \left(\frac{a+b}{\sqrt{2}}\right)^2 + \left(\frac{a-b}{\sqrt{2}}\right)^2,$$

the sum of the squares of the numbers in a triple is invariant under the operation. The sum of squares of the first triple is $\frac{13}{2}$ and that of the second is $6 + 2\sqrt{2}$, so the first triple cannot be transformed into the second.

(D. Fomin, S. Genkin, I. Itenberg, *Mathematical Circles*, AMS, 1996)

73. Assign the value i to each white ball, $-i$ to each red ball, and -1 to each green ball. A quick check shows that the given operations preserve the product of the values of the balls in the box. This product is initially $i^{2000} = 1$. If three balls were left in the box, none of them green, then the product of their values would be $\pm i$, a contradiction. Hence, if three balls remain, at least one is green, proving the claim in part (a). Furthermore, because no ball has value 1, the box must contain at least two balls at any time. This shows that the answer to the question in part (b) is *no*.

(Bulgarian Mathematical Olympiad, 2000)

74. Let I be the sum of the number of stones and heaps. An easy check shows that the operation leaves I invariant. The initial value is 1002. But a configuration with k heaps, each containing 3 stones, has $I = k + 3k = 4k$. This number cannot equal 1002, since 1002 is not divisible by 4.

(D. Fomin, S. Genkin, I. Itenberg, *Mathematical Circles*, AMS, 1996)

75. The quantity $I = xv + yu$ does not change under the operation, so it remains equal to $2mn$ throughout the algorithm. When the first two numbers are both equal to $\gcd(m, n)$, the sum of the latter two is $\frac{2mn}{\gcd(m,n)} = 2\mathrm{lcm}(m, n)$.

(St. Petersburg City Mathematical Olympiad, 1996)

76. We can assume that p and q are coprime; otherwise, shrink the size of the chessboard by their greatest common divisor. Place the chessboard on the two-dimensional integer lattice such that the initial square is centered at the origin, and the other squares, assumed to have side length 1, are centered at lattice points. We color the chessboard by the Klein four group

$$K = \{a, b, c, e \mid a^2 = b^2 = c^2 = e, \ ab = c, \ ac = b, \ bc = a\}$$

as follows: if (x, y) are the coordinates of the center of a square, then the square is colored by e if both x and y are even, by c if both are odd, by a if x is even and y is odd, and by b if x is odd and y is even (see Figure 61). If p and q are both odd, then at each jump the color of the location of the knight is multiplied by c. Thus after n jumps the knight is on a square colored by c^n. The initial square was colored by e, and the equality $c^n = e$ is possible only if n is even.

If one of p and q is even and the other is odd, then at each jump the color of the square is multiplied by a or b. After n jumps the color will be $a^k b^{n-k}$. The equality $a^k b^{n-k} = e$ implies $a^k = b^{n-k}$, so both k and $n - k$ have to be even. Therefore, n itself has to be even. This completes the solution.

(German Mathematical Olympiad)

77. The invariant is the 5-colorability of the knot, i.e., the property of a knot to admit a coloring by the residue classes modulo 5 such that

(i) at least two residue classes are used;

(ii) at each crossing, $a + c \equiv 2b \pmod 5$,

where b is the residue class assigned to the overcrossing, and a and c are the residue classes assigned to the other two arcs.

c	b	c	b	c	b
a	e	a	e	a	e
c	b	c	b	c	b
a	e	a	e	a	e
c	b	c	b	c	b
a	e	a	e	a	e

Figure 61

A coloring of the figure eight knot is given in Figure 62, while the trivial knot does not admit 5-colorings since its simplest diagram does not. This proves that the figure eight knot is knotted.

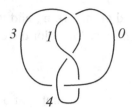

Figure 62

78. The answer is no. The idea of the proof is to associate to the configuration (a) an encoding defined by a pair of vectors $(v, w) \in \mathbb{Z}_2^2$ square contains a $+$ if the ith coordinate of v is equal to the jth coordinate of w, and a $-$ otherwise. A possible encoding for our configuration is $v = w = (1, 1, 0)$. Any other configuration that can be obtained from it admits such an encoding. Thus we choose as the invariant the *possibility* of encoding a configuration in such a manner.

It is not hard to see that the configuration in (b) cannot be encoded this way. A slick proof of this fact is that the configuration in which all signs are negative except for the one in the center can be obtained from this by the specified move, and this latter one cannot be encoded. Hence it is impossible to transform the first configuration into the second.

(Russian Mathematical Olympiad 1983–1984, solution by A. Badev)

79. The answer is no. The essential observation is that

$$99 \ldots 99 \equiv 99 \equiv 3 \pmod 4.$$

When we write this number as a product of two factors, one of the factors is congruent to 1 and the other is congruent to 3 modulo 4. Adding or subtracting a 2 from each factor produces numbers congruent to 3, respectively, 1 modulo 4. We deduce that what stays invariant in this process is the parity of the number of numbers on the blackboard that are congruent to 3 modulo 4. Since initially this number is equal to 1, there will always be at least one number that is congruent to 3 modulo 4 written on the blackboard. And this is not the case with the sequence of nines. This proves our claim.

(St. Petersburg City Mathematical Olympiad, 1997)

80. Without loss of generality, we may assume that the length of the hypotenuse is 1 and those of the legs are p and q. In the process, we obtain homothetic triangles that are in the ratio $p^m q^n$ to the original ones, for some nonnegative integers m and n. Let us focus on the pairs (m, n).

Each time we cut a triangle, we replace the pair (m, n) with the pairs $(m + 1, n)$ and $(m, n + 1)$. This shows that if to the triangle corresponding to the pair (m, n) we associate the weight $\frac{1}{2^{m+n}}$, then the sum I of all the weights is invariant under cuts. The initial value of I is 4. If at some stage the triangles were pairwise incongruent, then the value of I would be strictly less than

$$\sum_{m,n=0}^{\infty} \frac{1}{2^{m+n}} = \sum_{m=0}^{\infty} \frac{1}{2^m} \sum_{n=0}^{\infty} \frac{1}{2^n} = 4,$$

a contradiction. Hence a configuration with all triangles of distinct sizes cannot be achieved.

(Russian Mathematical Olympiad, 1995)

81. *First solution*: Here the invariant is given; we just have to prove its invariance. We first examine the simpler case of a cyclic quadrilateral $ABCD$ inscribed in a circle of radius R. Recall that for a triangle XYZ the radii of the incircle and the circumcircle are related by

$$r = 4R \sin \frac{X}{2} \sin \frac{Y}{2} \sin \frac{Z}{2}.$$

Let $\angle CAD = \alpha_1$, $\angle BAC = \alpha_2$, $\angle ABD = \beta$. Then $\angle DBC = \alpha_1$, and $\angle ACD = \beta$, $\angle BDC = \alpha_2$, and $\angle ACB = \angle ADB = 180° - \alpha_1 - \alpha_2 - \beta$. The independence of the sum of the inradii in the two possible dissections translates, after dividing by $4R$, into the identity

$$\sin \frac{\alpha_1 + \alpha_2}{2} \sin \frac{\beta}{2} \sin \left(90° - \frac{\alpha_1 + \alpha_2 + \beta}{2}\right) + \sin \left(90° - \frac{\alpha_1 + \alpha_2}{2}\right) \sin \frac{\alpha_1}{2} \sin \frac{\alpha_2}{2}$$

$$= \sin \frac{\alpha_1 + \beta_1}{2} \sin \frac{\alpha_2}{2} \sin \left(90° - \frac{\alpha_1 + \alpha_2 + \beta}{2}\right) + \sin \left(90° - \frac{\alpha_1 + \beta_1}{2}\right) \sin \frac{\alpha_1}{2} \sin \frac{\beta}{2}.$$

This is equivalent to

$$\cos \frac{\alpha_1 + \beta_1 + \alpha_2}{2} \left(\sin \frac{\alpha_1 + \alpha_2}{2} \sin \frac{\beta}{2} - \sin \frac{\alpha_1 + \beta}{2} \sin \frac{\alpha_2}{2}\right)$$

$$= \sin \frac{\alpha_1}{2} \left(\sin \frac{\beta}{2} \cos \frac{\alpha_1 + \beta_1}{2} - \sin \frac{\alpha_2}{2} \cos \frac{\alpha_1 + \alpha_2}{2}\right),$$

or

$$\cos \frac{\alpha_1 + \alpha_2 + \beta}{2} \left(\cos \frac{\alpha_1 + \alpha_2 - \beta}{2} - \cos \frac{\alpha_1 - \alpha_2 + \beta}{2}\right)$$

$$= \sin \frac{\alpha_1}{2} \left(\sin \left(\beta_1 + \frac{\alpha_1}{2}\right) - \sin \left(\alpha_2 + \frac{\alpha_1}{2}\right)\right).$$

Using product-to-sum formulas, both sides can be transformed into

$$\cos(\alpha_1 + \alpha_2) + \cos \beta_1 - \cos(\alpha_1 + \beta_1) - \cos \alpha_2.$$

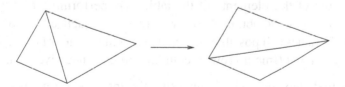

Figure 63

The case of a general polygon follows from the particular case of the quadrilateral. This is a consequence of the fact that any two dissections can be transformed into one another by a sequence of *quadrilateral moves* (Figure 63). Indeed, any dissection can be transformed into a dissection in which all diagonals start at a given vertex, by moving the endpoints of diagonals one by one to that vertex. So one can go from any dissection to any other dissection using this particular type as an intermediate step. Since the sum of the inradii is invariant under quadrilateral moves, it is independent of the dissection.

Second solution: This time we use the trigonometric identity

$$1 + \frac{r}{R} = \cos X + \cos Y + \cos Z.$$

We will check therefore that the sum of $1 + \frac{r_i}{R}$ is invariant, where r_i are the inradii of the triangles of the decomposition. Again we prove the property for a cyclic quadrilateral and then obtain the general case using the quadrilateral move. Using the fact that the sum of cosines of supplementary angles is zero and chasing angles in the cyclic quadrilateral $ABCD$, we obtain

$$\begin{aligned}
\cos\ &\angle DBA + \cos \angle BDA + \cos \angle DAB + \cos \angle BDC + \cos \angle CBD + \cos \angle CDB \\
&= \cos \angle DBA + \cos \angle BDA + \cos \angle CBD + \cos \angle CDB \\
&= \cos \angle DCA + \cos \angle BCA + \cos \angle CAD + \cos \angle CAB \\
&= \cos \angle DCA + \cos \angle CAD + \cos \angle ADC + \cos \angle BCA + \cos \angle CAB + \cos \angle ABC,
\end{aligned}$$

and we are done.

Remark. A more general theorem states that two triangulations of a polygonal surface (not necessarily by diagonals) are related by the move from Figure 63 and the move from Figure 64 or its inverse. These are called Pachner moves.

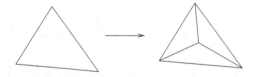

Figure 64

(Indian Team Selection Test for the International Mathematical Olympiad, 2005, second solution by A. Tripathy)

82. Let S be the sum of the elements of the table. By performing moves on the rows or columns with negative sum, we obtain a strictly increasing sequence $S_1 < S_2 < \cdots$. Because S can take at most 2^{n^2} values (all possible sign choices for the entries of the table), the sequence becomes stationary. At that time no row or column will have negative sum.

83. Skipping the first step, we may assume that the integers are nonnegative. The semi-invariant is $S(a, b, c, d) = \max(a, b, c, d)$. Because for nonnegative numbers x, y, we have $|x - y| \le \max(x, y)$, S does not increase under T. If S decreases at every step, then it eventually becomes 0, in which case the quadruple is $(0, 0, 0, 0)$. Let us see in what situation S is preserved by T. If

$$S(a, b, c, d) = S(T(a, b, c, d)) = S(|a - b|, |b - c|, |c - d|, |d - a|),$$

then next to some maximal entry there must be a zero. Without loss of generality, we may assume $a = S(a, b, c, d)$ and $b = 0$. Then

$$(a, 0, c, d) \xrightarrow{T} (a, c, |c - d|, |d - a|)$$
$$\xrightarrow{T} (|a - c|, |c - |c - d||, ||c - d| - |d - a||, |a - |d - a||).$$

Can S stay invariant in both these steps? If $|a - c| = a$, then $c = 0$. If $|c - |c - d|| = a$, then since a is the largest of the four numbers, either $c = d = a$ or else $c = 0$, $d = a$. The equality $||c - d| - |d - a|| = a$ can hold only if $c = 0$, $d = a$, or $d = 0$, $c = a$. Finally, $|a - |d - a|| = a$ if $d = a$. So S remains invariant in two consecutive steps only for quadruples of the form

$$(a, 0, 0, d), (a, 0, 0, a), (a, 0, a, 0), (a, 0, c, a),$$

and their cyclic permutations.

At the third step these quadruples become

$$(a, 0, d, |d - a|), (a, 0, a, 0), (a, a, a, a), (a, c, |c - a|, 0).$$

The second and the third quadruples become $(0, 0, 0, 0)$ in one and two steps, respectively. Now let us look at the first and the last. By our discussion, unless they are of the form $(a, 0, a, 0)$ or $(a, a, 0, 0)$, respectively, the semi-invariant will decrease at the next step. So unless it is equal to zero, S can stay unchanged for at most five consecutive steps. If initially $S = m$, after $5m$ steps it will be equal to zero and the quadruple will then be $(0, 0, 0, 0)$.

84. If a, b are erased and $c < d$ are written instead, we have $c \leq \min(a, b)$ and $d \geq \max(a, b)$. Moreover, $ab = cd$. Using derivatives we can show that the function $f(c) = c + \frac{ab}{c}$ is strictly decreasing on $\left(0, \frac{a+b}{2}\right)$, which implies $a + b \leq c + d$. Thus the sum of the numbers is nondecreasing. It is obviously bounded, for example by n times the product of the numbers, where n is the number of numbers on the board. Hence the sum of the numbers eventually stops changing. At that moment the newly introduced c and d should satisfy $c + d = a + b$ and $cd = ab$, which means that they should equal a and b. Hence the numbers themselves stop changing.

(St. Petersburg City Mathematical Olympiad, 1996)

85. To a configuration of pebbles we associate the number

$$S = \sum \frac{1}{2^{|i|+|j|}},$$

where the sum is taken over the coordinates of all nodes that contain pebbles. At one move of the game, a node (i, j) loses its pebble, while two nodes (i_1, j_1) and (i_2, j_2) gain pebbles. Since either the first coordinate or the second changes by one unit, $|i_k| + |j_k| \leq |i| + |j| + 1$, $k = 1, 2$. Hence

$$\frac{1}{2^{|i|+|j|}} = \frac{1}{2^{|i|+|j|+1}} + \frac{1}{2^{|i|+|j|+1}} \leq \frac{1}{2^{|i_1|+|j_1|}} + \frac{1}{2^{|i_2|+|j_2|}},$$

which shows that S is a nondecreasing semi-invariant. We will now show that at least one pebble is inside or on the boundary of the square R determined by the lines $x \pm y = \pm 5$. Otherwise, the total value of S would be less than

$$\sum_{|i|+|j|>5} \frac{1}{2^{|i|+|j|}} = 1 + 4 \sum_{i=1}^{\infty} \sum_{j=0}^{\infty} \frac{1}{2^{i+j}} - \sum_{|i|+|j|\le5} \frac{1}{2^{|i|+|j|}}$$

$$= 1 + 4 \sum_{i=1}^{\infty} \frac{1}{2^i} \sum_{j=0}^{\infty} \frac{1}{2^j} - 1 - 4 \left(1 \cdot \frac{1}{2} + 2 \cdot \frac{1}{4} + 3 \cdot \frac{1}{8} + 4 \cdot \frac{1}{16} + 5 \cdot \frac{1}{32} \right).$$

This equals $9 - \frac{65}{8} = \frac{7}{8}$ which is impossible, since the original value of S was 1. So there is always a pebble inside R, which is at distance at most 5 from the origin.

Algebra

86. Assume that both numbers are perfect cubes. Then so is their product

$$(n+3)(n^2 + 3n + 3) = n^3 + 6n^2 + 12n + 9.$$

However, this number differs from the perfect cube $(n + 2)^3 = n^3 + 6n^2 + 12n + 8$ by one unit. And this is impossible because no perfect cubes can be consecutive integers (unless one of them is zero). This proves the claim.

87. Let $m = pq$. We use the identity

$$x^m - y^m = (x - y)(x^{m-1} + x^{m-2}y + \cdots + y^{m-1}),$$

which can be applied to the matrices A and $-B$ since they commute. We have

$$(A - (-B))(A^{m-1} + A^{m-2}(-B) + \cdots + (-B)^{m-1})$$

$$= A^m - (-B)^m = (A^p)^q - (-1)^{pq}(B^q)^p = \mathcal{I}_n.$$

Hence the inverse of $A + B = A - (-B)$ is $A^{m-1} + A^{m-2}(-B) + \cdots + (-B)^{m-1}$.

88. *First solution:* Let $F(x)$ be the polynomial in question. If $F(x)$ is the square of a polynomial, then write $F(x) = G(x)^2 + 0^2$. In general, $F(x)$ is nonnegative for all real numbers x if and only if it has even degree and is of the form

$$F(x) = R(x)^2(x^2 + a_1x + b_1)(x^2 + a_2x + b_2) \cdots (x^2 + a_nx + b_n),$$

where the discriminant of each quadratic factor is negative. Completing the square

$$x^2 + a_kx + b_j = \left(x + \frac{a_k}{2}\right)^2 + \Delta_k^2, \quad \text{with} \quad \Delta_k = \sqrt{b_k - \frac{a_k^2}{4}},$$

© Springer International Publishing AG 2017
R. Gelca and T. Andreescu, *Putnam and Beyond*, DOI 10.1007/978-3-319-58988-6

we can write

$$F(x) = (P_1(x)^2 + Q_1(x)^2)(P_2(x)^2 + Q_2(x)^2) \cdots (P_n(x)^2 + Q_n(x)^2),$$

where the factor $R(x)^2$ is incorporated in $P_1(x)^2$ and $Q_1(x)^2$. Using the Lagrange identity

$$(a^2 + b^2)(c^2 + d^2) = (ac + bd)^2 + (ad - bc)^2,$$

we can transform this product in several steps into $P(x)^2 + Q(x)^2$, where $P(x)$ and $Q(x)$ are polynomials.

Second solution: Likewise with the first solution write the polynomial as

$$F(x) = R(x)^2(x^2 + a_1x + b_1)(x^2 + a_2x + b_2) \cdots (x^2 + a_nx + b_n).$$

Factor the quadratics as $(x + \alpha_k + i\beta_k)(x + \alpha_k - i\beta_k)$. Group the factors with $+i\beta_k$ into a polynomial $P(x) + iQ(x)$ and the factors with $-i\beta_k$ into the polynomial $P(x) - iQ(x)$. Then

$$F(x) = (R(x)P(x))^2 + (R(x)Q(x))^2,$$

which proves the conclusion.

Remark. D. Hilbert discovered that not every positive two-variable polynomial can be written as a sum of squares of polynomials. The appropriate generalization to the case of rational functions makes the object of his 16th problem. While Hilbert's proof is nonconstructive, the first examples of such polynomials were discovered surprisingly late, and were quite complicated. Here is a simple example found by T. Motzkin:

$$f(x, y) = 1 + x^2y^2(x^2 + y^2 - 3).$$

89. Simply substitute $x = 5^{5^n}$ in the factorization

$$x^5 + x + 1 = (x^2 + x + 1)(x^3 - x^2 + 1)$$

to obtain a factorization of the number from the statement. It is not hard to prove that both factors are greater than 1.

(T. Andreescu, published in T. Andreescu, D. Andrica, 360 *Problems for Mathematical Contests*, GIL, 2003)

90. Let

$$N = 5^{n-1} - \binom{n}{1}5^{n-2} + \binom{n}{2}5^{n-3} - \cdots + \binom{n}{n-1}.$$

Then $5N - 1 = (5 - 1)^n$. Hence

$$N = \frac{4^n + 1}{5} = \frac{4(2^k)^4 + 1}{5} = \frac{(2^{2k+1} + 2^{k+1} + 1)(2^{2k+1} - 2^{k+1} + 1)}{5},$$

where $k = \frac{n-1}{2}$. Since $n \geq 5$, both factors at the numerator are greater than 5, which shows that after canceling the denominator, the expression on the right can still be written as a product of two numbers. This proves that N is not prime.

(T. Andreescu, published in T. Andreescu, D. Andrica, 360 *Problems for Mathematical Contests*, GIL, 2003)

91. We use the identity

$$a^5 - 1 = (a-1)(a^4 + a^3 + a^2 + a + 1)$$

applied for $a = 5^{397}$. The difficult part is to factor $a^4 + a^3 + a^2 + a + 1$. Note that

$$a^4 + a^3 + a^2 + a + 1 = (a^2 + 3a + 1)^2 - 5a(a+1)^2.$$

Hence

$$\begin{aligned}
a^4 + a^3 + a^2 + a + 1 &= (a^2 + 3a + 1)^2 - 5^{398}(a+1)^2 \\
&= (a^2 + 3a + 1)^2 - (5^{199}(a+1))^2 \\
&= (a^2 + 3a + 1 + 5^{199}(a+1))(a^2 + 3a + 1 - 5^{199}(a+1)).
\end{aligned}$$

It is obvious that $a - 1$ and $a^2 + 3a + 1 + 5^{199}(a+1)$ are both greater than 5100. As for the third factor, we have

$$a^2 + 3a + 1 - 5^{199}(a+1) = a(a - 5^{199}) + 3a - 5^{199} + 1 \geq a + 0 + 1 \geq 5^{100}.$$

Hence the conclusion.

(Proposed by Russia for the 26th International Mathematical Olympiad, 1985)

92. The number from the statement is equal to $a^4 + a^3 + a^2 + a + 1$, where $a - 5^{25}$. As in the case of the previous problem, we rely on the identity

$$a^4 + a^3 + a^2 + a + 1 = (a^2 + 3a + 1)^2 - 5a(a+1)^2,$$

and factor our number as follows:

$$\begin{aligned}
a^4 + a^3 + a^2 + a + 1 &= (a^2 + 3a + 1)^2 - (5^{13}(a+1))^2 \\
&= (a^2 + 3a + 1 + 5^{13}(a+1))(a^2 + a + 1 - 5^{13}(a+1)).
\end{aligned}$$

The first factor is obviously greater than 1. The second factor is also greater than 1, since

$$a^2 + a + 1 - 5^{13}a - 5^{13} = a(a - 5^{13}) + (a - 5^{13}) + 1,$$

and $a > 513$. This proves that the number from the statement of the problem is not prime.

(Proposed by South Korea for the 33rd International Mathematical Olympiad, 1992)

93. The solution is based on the identity

$$a^k + b^k = (a+b)(a^{k-1} + b^{k-1}) - ab(a^{k-2} + b^{k-2}).$$

This identity arises naturally from the fact that both a and b are solutions to the equation $x^2 - (a+b)x + ab = 0$, hence also to $x^k - (a+b)x^{k-1} + abx^{k-2} = 0$.

Assume that the conclusion is false. Then for some n, $a^{2n} + b^{2n}$ is divisible by $a + b$. For $k = 2n$, we obtain that the right-hand side of the identity is divisible by $a + b$, hence so is $ab(a^{2n-2} + b^{2n-2})$. Moreover, a and b are coprime to $a + b$, and therefore $a^{2n-2} + b^{2n-2}$ must be divisible by $a + b$. Through a backward induction, we obtain that $a^0 + b^0 = 2$ is divisible by $a + b$, which is impossible since $a, b > 1$. This contradiction proves the claim.

(R. Gelca)

94. Let n be an integer and let $\frac{n^3-n}{6} = k$. Because $n^3 - n$ is the product of three consecutive integers, $n - 1, n, n + 1$, it is divisible by 6; hence k is an integer. Then

$$n^3 - n = 6k = (k - 1)^3 + (k + 1)^3 - k^3 - k^3.$$

It follows that

$$n = n^3 - (k - 1)^3 - (k + 1)^3 + k^3 + k^3,$$

and thus

$$n = n^3 + \left(1 - \frac{n^3 - n}{6}\right)^3 + \left(-1 - \frac{n^3 + n}{6}\right)^3 + \left(\frac{n^3 - n}{6}\right)^3 + \left(\frac{n^3 - n}{6}\right)^3.$$

Remark. Lagrange showed that every positive integer is a sum of at most four perfect squares. Wieferich showed that every positive integer is a sum of at most nine perfect cubes of positive integers. Waring conjectured that in general, for every n there is a number $w(n)$ such that every positive integer is the sum of at most $w(n)$ nth powers of positive integers. This conjecture was proved by Hilbert.

95. Let $a = (k - 1)^{4/5}$ and $b = (k + 1)^{4/5}$. We have

$$\frac{1}{(k - 1)^{4/5} - k^{4/5} + (k - 1)^{4/5}} < \frac{1}{(k - 1)^{4/5} - (k^2 - 1)^{2/5} + (k - 1)^{4/5}}$$

$$= \frac{1}{a^4 - a^2b^2 + b^4} < \frac{1}{a^4 - a^3b + a^2b^2 - ab^3 + b^4} = \frac{a + b}{a^5 + b^5}$$

$$= \frac{(k - 1)^{1/5} + (k + 1)^{1/5}}{(k - 1) + (k + 1)}.$$

In this computation we used that $a^3b + ab^3 \geq 2a^2b^2$ (a consequence of the AM-GM inequality). It follows that

$$\sum_{k=1}^{31} \frac{k}{(k - 1)^{4/5} - k^{4/5} + (k - 1)^{4/5}} < \frac{1}{2} \sum_{k=1}^{31} \left[(k - 1)^{1/5} + (k + 1)^{1/5}\right]$$

$$= -\frac{1}{2} + \sum_{k=1}^{31} (k - 1)^{(1/5)} + \frac{1}{2} 31^{1/5} + \frac{1}{2} 32^{1/5},$$

and the conclusion follows.

(T. Andreescu)

96. *First solution*: Using the identity

$$a^3 + b^3 + c^3 - 3abc = \frac{1}{2}(a + b + c)((a - b)^2 + (b - c)^2 + (c - a)^2)$$

applied to the (distinct) numbers $a = \sqrt[3]{x-1}$, $b = \sqrt[3]{x}$, and $c = \sqrt[3]{x+1}$, we transform the equation into the equivalent

$$(x-1) + x + (x+1) - 3\sqrt[3]{(x-1)x(x+1)} = 0.$$

We further change this into $x = \sqrt[3]{x^3 - x}$. Raising both sides to the third power, we obtain $x^3 = x^3 - x$. We conclude that the equation has the unique solution $x = 0$.

Second solution: The function $f : \mathbb{R} \to \mathbb{R}$, $f(x) = \sqrt[3]{x-1} + \sqrt[3]{x} + \sqrt[3]{x+1}$ s strictly increasing, so the equation $f(x) = 0$ has at most one solution. Since $x = 0$ satisfies this equation, it is the unique solution.

97. The key observation is that the left-hand side of the equation can be factored as

$$(x+y+z)(x^2 + y^2 + z^2 - xy - yz - zx) = 0.$$

Since $x+y+z > 1$ and p is prime, we must have $x+y+z = p$ and $x^2+y^2+z^2-xy-yz-zx = 1$. The second equality can be written as $(x - y)^2 + (y - z)^2 + (z - x)^2 = 2$. Without loss of generality, we may assume that $x \geq y \geq z$. If $x > y > z$, then $x - y \geq 1$, $y - z \geq 1$, and $x - z \geq 2$, which would imply that $(x - y)^2 + (y - z)^2 + (z - x)^2 \geq 6 > 2$.

Therefore, either $x = y = z+1$ or $x - 1 = y = z$. According to whether the prime p is of the form $3k + 1$ or $3k + 2$, the solutions are $\left(\frac{p-1}{3}, \frac{p-1}{3}, \frac{p+2}{3}\right)$ and the corresponding permutations, or $\left(\frac{p-2}{3}, \frac{p+1}{3}, \frac{p+1}{3}\right)$ and the corresponding permutations.

(T. Andreescu, D. Andrica, *An Introduction to Diophantine Equations*, GIL 2002)

98. The inequality to be proved is equivalent to

$$a^3 + b^3 + c^3 - 3abc \geq 9k.$$

The left-hand side can be factored, and the inequality becomes

$$(a+b+c)(a^2 + b^2 + c^2 - ab - bc - ca) \geq 9k.$$

Without loss of generality, we may assume that $a \geq b \geq c$. It follows that $a-b \geq 1$, $b-c \geq 1$, $a - c \geq 2$; hence $(a - b)^2 + (b - c)^2 + (c - a)^2 \geq 1 + 1 + 4 = 6$. Dividing by 2, we obtain

$$a^2 + b^2 + c^2 - ab - bc - ca \geq 3.$$

The solution will be complete if we show that $a + b + c \geq 3k$. The computation

$$(a+b+c)^2 = a^2 + b^2 + c^2 - ab - bc - ca + 3(ab + bc + ca)$$
$$\geq 3 + 3(3k^2 - 1) = 9k^2$$

completes the proof.

(T. Andreescu)

99. Apply the identity

$$a^3 + b^3 + c^3 - 3abc = \frac{1}{2}(a+b+c)[(a-b)^2 + (b-c)^2 + (c-a)^2]$$
$$= (a+b+c)(a^2 + b^2 + c^2 - ab - bc - ca)$$

to obtain that the expression is equal to

$$\frac{1}{2}(x^2 + y^2 + z^2 - xy - xz - yz)[(x^2 - yz - y^2 + xz)^2 + (y^2 - xz - z^2 + xy)^2$$
$$+ (z^2 - xy - x^2 + yz)^2]$$
$$= \frac{1}{2}(x^2 + y^2 + z^2 - xy - xz - yz)(x + y + z)^2[(x - y)^2 + (x - z)^2 + (y - z)^2]$$
$$= (x + y + z)^2(x^2 + y^2 + z^2 - xy - yz - zx) = (x^3 + y^3 + z^3 - 3xyz)^2.$$

(C. Coşniţă, *Teme şi Probleme Alese de Matematici (Selected Mathematics Themes and Problems)*, Ed. Didactică şi Pedagogică, Bucharest)

100. This is a difficult exercise in completing squares. We have

$$mnp = 1 + \frac{x^2}{z^2} + \frac{z^2}{y^2} + \frac{x^2}{y^2} + \frac{y^2}{x^2} + \frac{y^2}{z^2} + \frac{z^2}{x^2} + 1$$

$$= \left(\frac{x}{y} + \frac{y}{x}\right)^2 + \left(\frac{y}{z} + \frac{z}{y}\right)^2 + \left(\frac{z}{x} + \frac{x}{z}\right)^2 - 4.$$

Hence

$$m^2 + n^2 + p^2 = mnp + 4.$$

Adding $2(mn + np + pm)$ to both sides yields

$$(m + n + p)^2 = mnp + 2(mn + np + pm) + 4.$$

Adding now $4(m + n + p) + 4$ to both sides gives

$$(m + n + p + 2)^2 = (m + 2)(n + 2)(p + 2).$$

It follows that

$$(m + 2)(n + 2)(p + 2) = 2004^2.$$

But $2004 = 2^2 \times 3 \times 167$, and a simple case analysis shows that the only possibilities are $(m+2, n+2, p+2) = (4, 1002, 1002), (1002, 4, 1002), (1002, 1002, 4)$. The desired triples are $(2, 1000, 1000), (1000, 2, 1000), (1000, 1000, 2)$.

(Proposed by T. Andreescu for the 43rd International Mathematical Olympiad, 2002)

101. Let $M(a, b) = \max(a^2 + b, b^2 + a)$. Then $M(a, b) \geq a^2 + b$ and $M(a, b) \geq b^2 + a$, so $2M(a, b) \geq a^2 + b + b^2 + a$. It follows that

$$2M(a, b) + \frac{1}{2} \geq \left(a + \frac{1}{2}\right)^2 + \left(b + \frac{1}{2}\right)^2 \geq 0,$$

hence $M(a, b) \geq -\frac{1}{4}$. We deduce that

$$\min_{a,b\in\mathbb{R}} M(a, b) = -\frac{1}{4},$$

which, in fact, is attained when $a = b = -\frac{1}{2}$.
   (T. Andreescu)

102. Let $a = 2^x$ and $b = 3^x$. We need to show that

$$a + b - a^2 + ab - b^2 \le 1.$$

But this is equivalent to

$$0 \le \frac{1}{2}[(a - b)^2 + (a - 1)^2 + (b - 1)^2].$$

The equality holds if and only if $a = b = 1$, i.e., $x = 0$.
   (T. Andreescu, Z. Feng, 101 *Problems in Algebra*, Birkhäuser, 2001)

103. Clearly, 0 is not a solution. Solving for n yields $\frac{-4x-3}{x^4} \ge 1$, which reduces to $x^4 + 4x + 3 \le 0$. The last inequality can be written in its equivalent form,

$$(x^2 - 1)^2 + 2(x + 1)^2 \le 0,$$

whose only real solution is $x = -1$.
   Hence $n = 1$ is the unique solution, corresponding to $x = -1$.
   (T. Andreescu)

104. If $x = 0$, then $y = 0$ and $z = 0$, yielding the triple $(x, y, z) = (0, 0, 0)$. If $x \ne 0$, then $y \ne 0$ and $z \ne 0$, so we can rewrite the equations of the system in the form

$$1 + \frac{1}{4x^2} = \frac{1}{y},$$

$$1 + \frac{1}{4y^2} = \frac{1}{z},$$

$$1 + \frac{1}{4z^2} = \frac{1}{x}.$$

Summing up the three equations leads to

$$\left(1 - \frac{1}{x} + \frac{1}{4x^2}\right) + \left(1 - \frac{1}{y} + \frac{1}{4y^2}\right) + \left(1 - \frac{1}{z} + \frac{1}{4z^2}\right) = 0.$$

This is equivalent to

$$\left(1 - \frac{1}{2x}\right)^2 + \left(1 - \frac{1}{2y}\right)^2 + \left(1 - \frac{1}{2z}\right)^2 = 0.$$

It follows that $\frac{1}{2x} = \frac{1}{2y} = \frac{1}{2z} = 1$, yielding the triple $(x, y, z) = \left(\frac{1}{2}, \frac{1}{2}, \frac{1}{2}\right)$. Both triples satisfy the equations of the system.
   (Canadian Mathematical Olympiad, 1996)

105. First, note that $\left(x - \frac{1}{2}\right)^2 \geq 0$ implies $x - \frac{1}{4} \leq x^2$, for all real numbers x. Applying this and using the fact that the x_i's are less than 1, we find that

$$\log_{x_k}\left(x_{k+1} - \frac{1}{4}\right) \geq \log_{x_k}(x_{k+1}^2) = 2\log_{x_k} x_{k+1} = 2\frac{\ln x_{k+1}}{\ln x_k}.$$

Therefore,

$$\sum_{k=1}^{n} \log_{x_k}\left(x_{k+1} - \frac{1}{4}\right) \geq 2\sum_{k=1}^{n} \log_{x_k} x_{k+1} \geq 2n\sqrt[n]{\frac{\ln x_2}{\ln x_1} \cdot \frac{\ln x_3}{\ln x_2} \cdots \frac{\ln x_n}{\ln x_1}} = 2n,$$

where for the last step we applied the AM-GM inequality (see Section 2.1.5). So a good candidate for the minimum is $2n$, which is actually attained for $x_1 = x_2 = \cdots = x_n = \frac{1}{2}$.

(Romanian Mathematical Olympiad, 1984, proposed by T. Andreescu)

106. Assume the contrary, namely that $7a + 5b + 12ab > 9$. Then

$$9a^2 + 8ab + 7b^2 - (7a + 5b + 12ab) < 6 - 9.$$

Hence

$$2a^2 - 4ab + 2b^2 + 7\left(a^2 - a + \frac{1}{4}\right) + 5\left(b^2 - b + \frac{1}{4}\right) < 0,$$

or

$$2(a - b)^2 + 7\left(a - \frac{1}{2}\right)^2 + 5\left(b - \frac{1}{2}\right)^2 < 0,$$

a contradiction. The conclusion follows.

(T. Andreescu)

107. We rewrite the inequalities to be proved as $-1 \leq a_k - n \leq 1$. In this respect, we have

$$\sum_{k=1}^{n}(a_k - n)^2 = \sum_{k=1}^{n} a_k^2 - 2n\sum_{k=1}^{n} a_k + n \cdot n^2 \leq n^3 + 1 - 2n \cdot n^2 + n^3 = 1,$$

and the conclusion follows.

(*Math Horizons*, proposed by T. Andreescu)

108. Adding up the two equations yields

$$\left(x^4 + 2x^3 - x + \frac{1}{4}\right) + \left(y^4 + 2y^3 - y + \frac{1}{4}\right) = 0.$$

Here we recognize two perfect squares, and write this as

$$\left(x^2 + x - \frac{1}{2}\right)^2 + \left(y^2 + y - \frac{1}{2}\right)^2 = 0.$$

Equality can hold only if $x^2 + x - \frac{1}{2} = y^2 + y - \frac{1}{2} = 0$, which then gives $\{x, y\} \subset$ $\left\{ -\frac{1}{2} - \frac{\sqrt{3}}{2}, -\frac{1}{2} + \frac{\sqrt{3}}{2} \right\}$. Moreover, since $x \neq y$, $\{x, y\} = \left\{ -\frac{1}{2} - \frac{\sqrt{3}}{2}, -\frac{1}{2} + \frac{\sqrt{3}}{2} \right\}$. A simple verification leads to $(x, y) = \left(-\frac{1}{2} + \frac{\sqrt{3}}{2}, -\frac{1}{2} - \frac{\sqrt{3}}{2} \right)$.

(*Mathematical Reflections*, proposed by T. Andreescu)

109. Let $n = 2k$. It suffices to prove that

$$\frac{1}{2} \pm x + x^2 \pm x^3 + x^4 \pm \ldots \pm x^{2k-1} + x^{2k} > 0,$$

for all 2^k choices of the signs $+$ and $-$. This reduces to

$$\left(\frac{1}{2} \pm x + \frac{1}{2}x^2 \right) + \left(\frac{1}{2}x^2 \pm x^3 + \frac{1}{2}x^4 \right) + \ldots + \left(\frac{1}{2}x^{2k-2} \pm x^{2k-1} \frac{1}{2}x^{2k} \right) + \frac{1}{2}x^{2k} > 0,$$

which is true because $\frac{1}{2}x^{2k-2} \pm x^{2k-1} + \frac{1}{2}x^{2k} = \frac{1}{2}(x^{k-1} \pm x^k)^2 \geq 0$ and $\frac{1}{2}x^{2k} \geq 0$, and the equality cases cannot hold simultaneously.

110. This is the Cauchy-Schwarz inequality applied to the numbers $a_1 = a\sqrt{b}$, $a_2 = b\sqrt{c}$, $a_3 = c\sqrt{a}$ and $b_1 = c\sqrt{b}$, $b_2 = a\sqrt{c}$, $b_3 = b\sqrt{a}$. Indeed,

$$9a^2b^2c^2 = (abc + abc + abc)^2 = (a_1b_1 + a_2b_2 + a_3b_3)^2$$
$$\leq (a_1^2 + a_2^2 + a_3^2)(b_1^2 + b_2^2 + b_3^2) = (a^2b + b^2c + c^2a)(c^2b + a^2c + b^2a).$$

111. By the Cauchy-Schwarz inequality,

$$(a_1 + a_2 + \cdots + a_n)^2 \leq (1 + 1 + \cdots + 1)(a_1^2 + a_2^2 + \cdots + a_n^2).$$

Hence $a_1^2 + a_2^2 + \cdots + a_n^2 \geq n$. Repeating, we obtain

$$(a_1^2 + a_2^2 + \cdots + a_n^2)^2 \geq (1 + 1 + \cdots + 1)(a_1^4 + a_2^4 + \cdots + a_n^4),$$

which shows that $a_1^4 + a_2^4 + \ldots + a_n^4 \geq n$, as desired.

112. Apply Cauchy-Schwarz:

$$(a_1a_{\sigma(a)} + a_2a_{\sigma(2)} + \cdots + a_na_{\sigma(n)})^2 \leq (a_1^2 + a_2^2 + \cdots + a_n^2)(a_{\sigma(1)}^2 + a_{\sigma(2)}^2 + \cdots + a_{\sigma(n)}^2)$$
$$= (a_1^2 + a_2^2 + \cdots + a_n^2)^2.$$

The maximum is $a_1^2 + a_2^2 + \cdots + a_n^2$. The only permutation realizing it is the identity permutation.

113. Applying the Cauchy-Schwarz inequality to the numbers $\sqrt{f_1}x_1, \sqrt{f_2}x_2, \ldots, \sqrt{f_n}x_n$ and $\sqrt{f_1}, \sqrt{f_2}, \ldots, \sqrt{f_n}$, we obtain

$$(f_1x_1^2 + f_2x_2^2 + \cdots + f_nx_n^2)(f_1 + f_2 + \cdots + f_n) \geq (f_1x_1 + f_2x_2 + \cdots + f_nx_n)^2,$$

hence the inequality from the statement.

Remark. In statistics the numbers f_i are integers that record the frequency of occurrence of the sampled random variable x_i, $i = 1, 2, \ldots, n$. If $f_1 + f_2 + \cdots + f_n = N$, then

$$s^2 = \frac{f_1 x_1^2 + f_2 x_2^2 + \cdots + f_n x_n^2 - \dfrac{(f_1 x_1 + f_2 x_2 + \cdots + f_n x_n)^2}{N}}{N - 1}$$

is called the sample variance. We have just proved that the sample variance is nonnegative.

114. By the Cauchy-Schwarz inequality,

$$(k_1 + \cdots + k_n) \left(\frac{1}{k_1} + \cdots + \frac{1}{k_n} \right) \geq n^2.$$

We must thus have $5n - 4 \geq n^2$, so $n \leq 4$. Without loss of generality, we may suppose that $k_1 \leq \cdots \leq k_n$.

If $n = 1$, we must have $k_1 = 1$, which is a solution. Note that hereinafter we cannot have $k_1 = 1$.

If $n = 2$, we have $(k_1, k_2) \in \{(2, 4), (3, 3)\}$, neither of which satisfies the relation from the statement.

If $n = 3$, we have $k_1 + k_2 + k_3 = 11$, so $2 \leq k_1 \leq 3$. Hence $(k_1, k_2, k_3) \in \{(2, 2, 7), (2, 3, 6), (2, 4, 5), (3, 3, 5), (3, 4, 4)\}$, and only $(2, 3, 6)$ works.

If $n = 4$, we must have equality in the Cauchy-Schwarz inequality, and this can happen only if $k_1 = k_2 = k_3 = k_4 = 4$.

Hence the solutions are $n = 1$ and $k_1 = 1$, $n = 3$, and (k_1, k_2, k_3) is a permutation of $(2, 3, 6)$, and $n = 4$ and $(k_1, k_2, k_3, k_4) = (4, 4, 4, 4)$.

(66th W.L. Putnam Mathematical Competition, 2005, proposed by T. Andreescu)

115. One can check that geometric progressions satisfy the identity. A slick proof of the converse is to recognize that we have the equality case in the Cauchy-Schwarz inequality. It holds only if $\frac{a_0}{a_1} = \frac{a_1}{a_2} = \cdots = \frac{a_{n-1}}{a_n}$, i.e., only if $a_0, a_1, \ldots, a_n$ is a geometric progression.

116. Let $P(x) = c_n x^n + c_{n-1} x^{n-1} + \cdots + c_0$. Then

$$P(a)P(b) = (c_n a^n + c_{n-1} a^{n-1} + \cdots + c_0)(c_n b^n + c_{n-1} b^{n-1} + \cdots + c_0)$$
$$\geq (c_n (\sqrt{ab})^n + c_{n-1} (\sqrt{ab})^{n-1} + \cdots + c_0)^2 = (P(\sqrt{ab}))^2,$$

by the Cauchy-Schwarz inequality, and the conclusion follows.

117. *First solution*: If $a_1, a_2, \ldots, a_n$ are positive integers, the Cauchy-Schwarz inequality implies

$$(a_1 + a_2 + \cdots + a_n) \left(\frac{1}{a_1} + \frac{1}{a_2} + \cdots + \frac{1}{a_n} \right) \geq n^2.$$

For $a_1 = x_0 - x_1, a_2 = x_1 - x_2, \ldots, a_n = x_{n-1} - x_n$ this gives

$$\frac{1}{x_0 - x_1} + \frac{1}{x_1 - x_2} + \cdots + \frac{1}{x_{n-1} - x_n} \geq \frac{n^2}{x_0 - x_1 + x_1 - x_2 + \cdots + x_{n-1} - x_n}$$
$$= \frac{n^2}{x_0 - x_n}.$$

The inequality from the statement now follows from

$$x_0 + x_n + \frac{n^2}{x_0 - x_n} \geq 2n,$$

which is rather easy, because it is equivalent to

$$\left(\sqrt{x_0 - x_n} - \frac{n}{\sqrt{x_0 - x_n}} \right)^2 \geq 0.$$

Equality in Cauchy-Schwarz holds if and only if $x_0 - x_1, x_1 - x_2, \ldots, x_{n-1} - x_n$ are proportional to $\frac{1}{x_0 - x_1}, \frac{1}{x_1 - x_2}, \ldots, \frac{1}{x_{n-1} - x_n}$. This happens when $x_0 - x_1 = x_1 - x_2 = \cdots = x_{n-1} - x_n$. Also, $\sqrt{x_0 - x_n} - n/\sqrt{x_0 - x_n} = 0$ only if $x_0 - x_n = n$. This means that the inequality from the statement becomes an equality if and only if $x_0, x_1, \ldots, x_n$ is an arithmetic sequence with common difference 1.

Second solution: As before, let $a_i = x_i - x_{i+1}$. The inequality can be written as

$$\sum_{i=1}^{n-1} \left(a_i + \frac{1}{a_i} \right) \geq 2n.$$

This follows immediately from $x + x^{-1} \geq 2$.

(St. Petersburg City Mathematical Olympiad, 1999, second solution by R. Stong)

118. Because

$$\frac{1}{\sec(a - b)} = \cos(a - b) = \sin a \sin b + \cos a \cos b,$$

it suffices to show that

$$\left(\frac{\sin^3 a}{\sin b} + \frac{\cos^3 a}{\cos b} \right) (\sin a \sin b + \cos a \cos b) \geq 1.$$

This is true because by the Cauchy-Schwarz inequality,

$$\left(\frac{\sin^3 a}{\sin b} + \frac{\cos^3 a}{\cos b} \right) (\sin a \sin b + \cos a \cos b) \geq (\sin^2 a + \cos^2 a)^2 = 1.$$

119. Bring the denominator to the left:

$$(a + b)(b + c)(c + a) \left(\frac{1}{a + b} + \frac{1}{b + c} + \frac{1}{c + a} + \frac{1}{2\sqrt[3]{abc}} \right) \geq (a + b + c + \sqrt[3]{abc})^2.$$

The identity

$$(a + b)(b + c)(c + a) = c^2(a + b) + b^2(c + a) + a^2(b + c) + 2abc$$

enables us to transform this into

$$(c^2(a+b) + b^2(c+a) + a^2(b+c) + 2abc)\left(\frac{1}{a+b} + \frac{1}{b+c} + \frac{1}{c+a} + \frac{1}{2\sqrt[3]{abc}}\right)$$

$$\geq (c+b+a+\sqrt[3]{abc})^2.$$

And now we recognize the Cauchy-Schwarz inequality. Equality holds only if $a = b = c$.
   (Mathematical Olympiad Summer Program, T. Andreescu)

120. Let c be the largest side. By the triangle inequality, $c^n < a^n + b^n$ for all $n \geq 1$. This is equivalent to

$$1 < \left(\frac{a}{c}\right)^n + \left(\frac{b}{c}\right)^n, \ n \geq 1.$$

If $a < c$ and $b < c$, then by letting $n \to \infty$, we obtain $1 < 0$, impossible. Hence one of the other two sides equals c, and the triangle is isosceles.

121. Define $\vec{d} = -\vec{a} - \vec{b} - \vec{c}$. The inequality becomes

$$\|\vec{a}\| + \|\vec{b}\| + \|\vec{c}\| + \|\vec{d}\| \geq \|\vec{a} + \vec{d}\| + \|\vec{b} + \vec{d}\| + \|\vec{c} + \vec{d}\|.$$

If the angles formed by $\vec{a}$ come in increasing order, then the closed polygonal line $\vec{a}$, $\vec{b}$, $\vec{c}$, $\vec{d}$ is a convex quadrilateral. Figure 65 shows how this quadrilateral can be transformed into one that is skew by choosing one angle such that one of the pairs of adjacent angles containing it totals at most 180° and the other at least 180° and then folding that angle in.

   The triangle inequality implies $\|\vec{b}\| + \|\vec{c}\| \geq \|\vec{b} + \vec{d}\| + \|\vec{c} + \vec{d}\|$. To be more convincing, let us explain that the left-hand side is the sum of the lengths of the dotted segments, while the right-hand side can be decomposed into the lengths of some four segments, which together with the dotted segments form two triangles. The triangle inequality also gives $\|\vec{a}\| + \|\vec{d}\| \geq \|\vec{a} + \vec{d}\|$. Adding the two yields the inequality from the statement.
   (*Kvant (Quantum)*)

122. Let $\lambda_1, \lambda_2, \ldots, \lambda_n$ be the roots of the polynomial, $D_1 = \{z, \ |z - c| \leq R\}$ the disk covering them, and $D_2 = \{z, \ |z - c| \leq R + |k|\}$. We will show that the roots of $nP(z) - kP'(z)$ lie inside D_2.

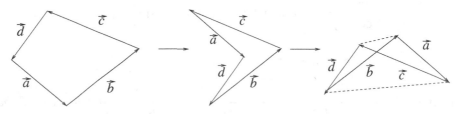

Figure 65

For $u \notin D_2$, the triangle inequality gives

$$|u - \lambda_i| \geq |u - c| - |c - \lambda_i| > R + |k| - R = |k|.$$

Hence $\frac{|k|}{|u-\lambda_i|} < 1$, for $i = 1, 2, \ldots, n$. For such a u we then have

$$|nP(u) - kP'(u)| = \left| nP(u) - kP(u) \sum_{i=1}^{n} \frac{1}{u - \lambda_i} \right| = |P(u)| \left| n - k \sum_{i=1}^{n} \frac{1}{u - \lambda_i} \right|$$

$$\geq |P(u)| \left| n - \sum_{i=1}^{n} \frac{|k|}{|u - \lambda_i|} \right|,$$

where the last inequality follows from the triangle inequality.

But we have seen that

$$n - \sum_{i=1}^{n} \frac{|k|}{|u - \lambda_i|} = \sum_{i=1}^{n} \left(1 - \frac{|k|}{|u - \lambda_i|} \right) > 0,$$

and since $P(u) \neq 0$, it follows that u cannot be a root of $nP(u) - kP'(u)$. Thus all roots of this polynomial lie in D_2.

(17th W.L. Putnam Mathematical Competition, 1956)

123. The inequality in the statement is equivalent to

$$(a^2 + b^2 + c^2)^2 < 4(a^2b^2 + b^2c^2 + c^2a^2).$$

The latter can be written as

$$0 < (2bc)^2 - (a^2 - b^2 - c^2)^2,$$

or

$$(2bc + b^2 + c^2 - a^2)(2bc - b^2 - c^2 + a^2).$$

This is equivalent to

$$0 < (a + b + c)(-a + b + c)(a - b + c)(a - b - c).$$

It follows that $-a + b + c, a - b + c, a - b - c$ are all positive, because $a + b + c > 0$, and no two of the factors could be negative, for in that case the sum of the three numbers would also be negative. Done.

124. The first idea is to simplify the problem and prove separately the inequalities $|AB - |CD| \geq |AC - BD|$ and $|AD - BC| \geq |AC - BD|$. Because of symmetry it suffices to prove the first.

Let M be the intersection of the diagonals AC and BD. For simplicity, let $AM = x$, $BM = y, AB = z$. By the similarity of triangles MAB and MDC there exists a positive number k such that $DM = kx, CM = ky$, and $CD = kz$ (Figure 66). Then

$$|AB - CD| = |k - 1|z$$

and

$$|AC - BD| = |(kx + y) - (ky + x)| = |k - 1| \cdot |x - y|.$$

By the triangle inequality, $|x - y| \leq z$, which implies $|AB - CD| \geq |AC - BD|$, completing the proof.

(USA Mathematical Olympiad, 1999, proposed by T. Andreescu, solution by P.R. Loh)

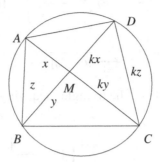

Figure 66

125. We induct on m. When $m = 1$ there is nothing to prove. Now assume that the inequality holds for $m - 1$ isometries and let us prove that it holds for m isometries. Define $V = \prod_{i=1}^{m-1} V_i$ and $W = \prod_{i=1}^{m-1} W_i$. Both V and W are isometries. For a vector x with $\|x\| \leq 1$, we have

$$\left\| \left(\prod_{i=1}^{m} V_i \right) x - \left(\prod_{i=1}^{m} W_i \right) x \right\| = \|V V_m x - W W_m x\|$$
$$= \|V(V_m - W_m)x + (V - W)W_m x\|.$$

Now we use the triangle inequality to increase the value of this expression to

$$\|V(V_m - W_m)x\| + \|(V - W)W_m x\|.$$

From the fact that V is an isometry it follows that

$$\|V(V_m - W_m)x\| = \|(V_m - W_m)x\| \leq 1.$$

From the fact that W_m is an isometry, it follows that $\|W_m x\| \leq 1$, and so $\|(V - W)W_m x\| \leq m - 1$ by the induction hypothesis. Putting together the two inequalities completes the induction, and the inequality is proved.

Remark. In quantum mechanics the vector spaces are complex (not real) and the word *isometry* is replaced by *unitary*. Unitary linear transformations model evolution, and the above property shows that (measurement) errors accumulate linearly.

126. Place triangle ABC in the complex plane such that the coordinates of the vertices A, B, and C are, respectively, the third roots of unity $1, \varepsilon, \varepsilon^2$. Call z the complex coordinate of P. Start with the obvious identity

$$(z - 1) + \varepsilon(z - \varepsilon) + \varepsilon^2(z - \varepsilon^2) = 0.$$

Move one term to the other side:

$$-\varepsilon^2(z - \varepsilon^2) = (z - 1) + \varepsilon(z - \varepsilon).$$

Now take the absolute value and use the triangle inequality:

$$|z - \varepsilon^2| = |(z - 1) + \varepsilon(z - \varepsilon)| \le |z - 1| + |\varepsilon(z - \varepsilon)| = |z - 1| + |z - \varepsilon^2|.$$

Geometrically, this is $PC \le PA + PB$.

Equality corresponds to the equality case in the triangle inequality for complex numbers, which holds if the complex numbers have positive ratio. Specifically, $(z - 1) = a\varepsilon(z - \varepsilon)$ for some positive real number a, which is equivalent to

$$\frac{z - 1}{z - \varepsilon} = a\varepsilon.$$

In geometric terms this means that PA and PB form an angle of $120°$, so that P is on the arc $\overset{\frown}{AB}$. The other two inequalities are obtained by permuting the letters.

(D. Pompeiu)

127. We start with the algebraic identity

$$x^3(y - z) + y^3(z - x) + z^3(x - y) = (x + y + z)(x - y)(y - z)(z - x),$$

where x, y, z are complex numbers. Applying to it the triangle inequality, we obtain

$$|x|^2|y - z| + |y|^3|z - x| + |z|^3|x - y| \ge |x + y + z||x - y||x - z||y - z|.$$

So let us see how this can be applied to our problem. Place the triangle in the complex plane so that M is the origin, and let a, b, and c, respectively, be the complex coordinates of A, B, C. The coordinate of G is $\frac{(a+b+c)}{3}$, and if we set $x = a$, $y = b$, and $z = c$ in the inequality we just derived, we obtain the geometric inequality from the statement.

(M. Dincă, M. Chiriță, *Numere Complexe în Matematica de Liceu* (*Complex Numbers in High School Mathematics*), ALL Educational, Bucharest, 1996)

128. Because $P(x)$ has odd degree, it has a real zero r. If $r > 0$, then by the AM-GM inequality
$$P(r) = r^5 + 1 + 1 + 1 + 2^5 - 5 \cdot 2 \cdot r \ge 0.$$

And the inequality is strict since $1 \ne 2$. Hence $r < 0$, as desired.

129. We must have $x > 0$. From the AM-GM inequality we obtain

$$x \cdot 2^{\frac{1}{x}} + \frac{1}{x} \cdot 2^x \ge 2\sqrt{2^{\frac{1}{x}+x}} = 2 \cdot 2^{\frac{\frac{1}{x}+x}{2}}.$$

Because $x + \frac{1}{x} \ge 2$, it follows that this is greater or equal to 4. Since we should have equality, we deduce that $x = 1$.

(*Gazeta Matematică* (*Mathematics Gazette, Bucharest*), proposed by L. Panaitopol)

130. *First solution:* The inequality is homogeneous in the sense that if we multiply some a_k and b_k simultaneously by a positive number, the inequality does not change. Hence we can assume that $a_k + b_k = 1$, $k = 1, 2, \ldots, n$. In this case, applying the AM-GM inequality, we obtain

$$(a_1 a_2 \cdots a_n)^{1/n} + (b_1 b_2 \cdots b_n)^{1/n} \leq \frac{a_1 + a_2 + \cdots + a_n}{n} + \frac{b_1 + b_2 + \cdots + b_n}{n}$$
$$= \frac{a_1 + b_1 + a_2 + b_2 + \cdots + a_n + b_n}{n} = \frac{n}{n} = 1,$$

and the inequality is proved.

Second solution: There is an approach that uses multivariable differentiable calculus. The case where $a_i = b_i = 0$ for some i is trivial, so let us assume that this does not happen.

We observe that the inequality does not change if for some i we divide both a_i and b_i by the same positive constant λ_i. Let us choose $\lambda_i = (a_i + b_i)$ for each i, and divide the inequality by $\lambda_1 \lambda_2 \cdots \lambda_n$. As such, the inequality becomes

$$(a_1 a_2 \cdots a_n)^{1/n} + (b_1 b_2 \cdots b_n)^{1/n} \leq 1,$$

with the hypothesis that $a_1 + b_1 = a_2 + b_2 = \cdots = a_n + b_n = 1$. We can rewrite this inequality as

$$(a_1 a_2 \cdots a_n)^{1/n} + ((1 - a_1)(1 - a_2) \cdots (1 - a_n))^{1/n} \leq 1,$$

which is supposed to hold for all $a_1, a_2, \ldots, a_n \in [0, 1]$. Let us consider the function $f : [0, 1]^n \to [0, \infty)$,

$$f(x_1, x_2, \ldots, x_n) = (x_1 x_2 \cdots x_n)^{1/n} + ((1 - x_1)(1 - x_2) \cdots (1 - x_n))^{1/n}.$$

We want to find the maximum of f. The critical points inside the domain of definition satisfy the system of equations

$$\frac{\partial f}{\partial x_i} = 0, \quad i = 1, 2, \ldots, n,$$

which translates to

$$\frac{1}{x_i}(x_1 x_2 \cdots x_n)^{1/n} - \frac{1}{1 - x_i}((1 - x_1)(1 - x_2) \cdots (1 - x_n))^{1/n} = 0.$$

This implies that

$$\frac{1 - x_i}{x_i} = \frac{((1 - x_1)(1 - x_2) \cdots (1 - x_n))^{1/n}}{(x_1 x_2 \cdots x_n)^{1/n}}.$$

Since the equation $\frac{1-x}{x} = k$ is linear in x, it has a unique solution. Thus the critical points are disposed along the line: $x_1 = x_2 = \cdots = x_n$.

We have

$$f(x_1, x_1, \ldots, x_1) = x_1 + (1 - x_1) = 1.$$

On the other hand, on the boundary of the domain some x_i is 0 or 1. So on the boundary one of the terms in f is zero, while the other term is a product of numbers not exceeding 1. This proves that on the boundary of the domain f is less than or equal to 1. We conclude that the maximum of f is 1, which proves the inequality.

(64th W.L. Putnam Mathematical Competition, 2003)

131. The inequality from the statement is equivalent to

$$0 < 1 - (a + b + c) + ab + bc + ca - abc < \frac{1}{27},$$

that is,

$$0 < (1 - a)(1 - b)(1 - c) \le \frac{1}{27}.$$

From the triangle inequalities $a + b > c, b + c > a, a + c > b$ and the condition $a + b + c = 2$ it follows that $0 < a, b, c < 1$. The inequality on the left is now evident, and the one on the right follows from the AM-GM inequality

$$\sqrt[3]{xyz} \le \frac{x + y + z}{3}$$

applied to $x = 1 - a, y = 1 - b, z = 1 - c$.

132. It is natural to try to simplify the product, and for this we make use of the AM-GM inequality:

$$\prod_{n=1}^{25} \left(1 - \frac{n}{365}\right) \le \left[\frac{1}{25} \sum_{n=1}^{25} \left(1 - \frac{n}{365}\right)\right]^{25} = \left(\frac{352}{365}\right)^{25} = \left(1 - \frac{13}{365}\right)^{25}.$$

We now use Newton's binomial formula to estimate this power. First, note that

$$\binom{25}{k} \left(\frac{13}{365}\right)^k \ge \binom{25}{k+1} \left(\frac{13}{365}\right)^{k+1},$$

since this reduces to

$$\frac{13}{365} \le \frac{k+1}{25 - k},$$

and the latter is always true for $1 \le k \le 24$. For this reason if we ignore the part of the binomial expansion beginning with the fourth term, we increase the value of the expression. In other words,

$$\left(1 - \frac{13}{365}\right)^{25} \le 1 - \binom{25}{1} \frac{13}{365} + \binom{25}{2} \frac{13^2}{365^2} = 1 - \frac{65}{73} + \frac{169 \cdot 12}{63^2} < \frac{1}{2}.$$

We conclude that the second number is larger.

(Soviet Union University Student Mathematical Olympiad, 1975)

133. The solution is based on the Lagrange identity, which in our case states that if M is a point in space and G is the centroid of the tetrahedron $ABCD$, then

$$AB^2 + AC^2 + CD^2 + AD^2 + BC^2 + BD^2$$
$$= 4(MA^2 + MB^2 + NC^2 + MD^2) - 16MG^2.$$

For $M = O$ the center of the circumscribed sphere, this reads

$$AB^2 + AC^2 + CD^2 + AD^2 + BC^2 + BD^2 = 16 - 16OG^2.$$

Applying the AM-GM inequality, we obtain

$$6\sqrt[3]{AB \cdot AC \cdot CD \cdot AD \cdot BC \cdot BD} \le 16 - 16OG^2.$$

This combined with the hypothesis yields $16 \le 16 - OG^2$. So on the one hand we have equality in the AM-GM inequality, and on the other hand $O = G$. Therefore, $AB = AC = AD = BC = BD = CD$, so the tetrahedron is regular.

134. Adding 1 to all fractions transforms the inequality into

$$\frac{x^2 + y^2 + 1}{2x^2 + 1} + \frac{y^2 + z^2 + 1}{2y^2 + 1} + \frac{z^2 + x^2 + 1}{2z^2 + 1} \ge 3.$$

Applying the AM-GM inequality to the left-hand side gives

$$\frac{x^2 + y^2 + 1}{2x^2 + 1} + \frac{y^2 + z^2 + 1}{2y^2 + 1} + \frac{z^2 + x^2 + 1}{2z^2 + 1}$$

$$\ge \sqrt[3]{\frac{x^2 + y^2 + 1}{2x^2 + 1} \cdot \frac{y^2 + z^2 + 1}{2y^2 + 1} \cdot \frac{z^2 + x^2 + 1}{2z^2 + 1}}.$$

We are left with the simpler but sharper inequality

$$\frac{x^2 + y^2 + 1}{2x^2 + 1} + \frac{y^2 + z^2 + 1}{2y^2 + 1} + \frac{z^2 + x^2 + 1}{2z^2 + 1} \ge 1.$$

This can be proved by multiplying together

$$x^2 + y^2 + 1 = x^2 + \frac{1}{2} + y^2 + \frac{1}{2} \ge 2\sqrt{\left(x^2 + \frac{1}{2}\right)\left(y^2 + \frac{1}{2}\right)},$$

$$y^2 + z^2 + 1 = y^2 + \frac{1}{2} + z^2 + \frac{1}{2} \ge 2\sqrt{\left(y^2 + \frac{1}{2}\right)\left(z^2 + \frac{1}{2}\right)},$$

$$z^2 + x^2 + 1 = z^2 + \frac{1}{2} + x^2 + \frac{1}{2} \ge 2\sqrt{\left(z^2 + \frac{1}{2}\right)\left(y^2 + \frac{1}{2}\right)},$$

and each of these is just the AM-GM inequality.

(Greek Team Selection Test for the Junior Balkan Mathematical Olympiad, 2005)

135. Denote the positive number $1 - (a_1 + a_2 + \cdots + a_n)$ by a_{n+1}. The inequality from the statement becomes the more symmetric

$$\frac{a_1 a_2 \cdots a_n a_{n+1}}{(1 - a_1)(1 - a_2) \cdots (1 - a_n)(1 - a_{n+1})} \leq \frac{1}{n^{n+1}}.$$

But from the AM-GM inequality,

$$1 - a_1 = a_2 + a_3 + \cdots + a_{n+1} \geq n\sqrt[n]{a_2 a_3 \cdots a_{n+1}},$$

$$1 - a_2 = a_1 + a_3 + \cdots + a_{n+1} \geq n\sqrt[n]{a_1 a_3 \cdots a_{n+1}},$$

$$\cdots$$

$$1 - a_{n+1} = a_1 + a_2 + \cdots + a_n \geq n\sqrt[n]{a_1 a_2 \cdots a_n}.$$

Multiplying these $n + 1$ inequalities yields

$$(1 - a_1)(1 - a_2) \cdots (1 - a_{n+1}) \geq n^{n+1} a_1 a_2 \dots a_n,$$

and the conclusion follows.

(Short list of the 43rd International Mathematical Olympiad, 2002)

136. Trick number 1: Use the fact that

$$1 = \frac{n - 1 + x_j}{n - 1 + x_j} = (n - 1)\frac{1}{n - 1 + x_j} + \frac{x_j}{n - 1 + x_j}, \quad j = 1, 2, \ldots, n,$$

to transform the inequality into

$$\frac{x_1}{n - 1 + x_1} + \frac{x_2}{n - 1 + x_2} + \cdots + \frac{x_n}{n - 1 + x_n} \geq 1.$$

Trick number 2: Break this into the n inequalities

$$\frac{x_j}{n - 1 + x_j} \geq \frac{x_j^{1 - \frac{1}{n}}}{x_1^{1 - \frac{1}{n}} + x_2^{1 - \frac{1}{n}} + \cdots + x_n^{1 - \frac{1}{n}}}, \quad j = 1, 2, \ldots, n.$$

We are left with n somewhat simpler inequalities, which can be rewritten as

$$x_1^{1 - \frac{1}{n}} + x_2^{1 - \frac{1}{n}} + x_{j-1}^{1 - \frac{1}{n}} + x_{j+1}^{1 - \frac{1}{n}} + \cdots + x_n^{1 - \frac{1}{n}} \geq (n - 1)x_j^{-\frac{1}{n}}.$$

Trick number 3: Use the AM-GM inequality

$$\frac{x_1^{1 - \frac{1}{n}} + x_2^{1 - \frac{1}{n}} + x_{j-1}^{1 - \frac{1}{n}} + x_{j+1}^{1 - \frac{1}{n}} + \cdots + x_n^{1 - \frac{1}{n}}}{n - 1} \geq \left((x_1 x_2 \cdots x_{j-1} x_{j+1} \cdots x_n)^{\frac{n-1}{n}}\right)^{\frac{1}{n-1}}$$

$$= (x_1 x_2 \cdots x_{j-1} x_{j+1} \cdots x_n)^{\frac{1}{n}} = x_j^{-\frac{1}{n}}.$$

This completes the proof.

(Romanian Team Selection Test for the International Mathematical Olympiad, 1999, proposed by V. Cârtoaje and Gh. Eckstein)

137. *First solution*: Note that the triple (a, b, c) ranges in the closed and bounded set $D = \{(x, y, z) \in \mathbb{R}^3 \mid 0 \leq x, y, z \leq 1, \ x + y + z = 1\}$. The function $f(x, y, z) = 4(xy + yz + zx) - 9xyz - 1$ is continuous; hence it has a maximum on D. Let (a, b, c) be a point in D at which f attains this maximum. By symmetry we may assume that $a \geq b \geq c$. This immediately implies $c \leq \frac{1}{3}$.

Let us apply Sturm's method. Suppose that $b < a$, and let $0 < x < a - b$. We show that $f(a - x, b + x, c) > f(a, b, c)$. The inequality is equivalent to

$$4(a - x)(b + x) - 9(a - x)(b + x)c > 4ab - 9abc,$$

or

$$(4 - 9c)((a - b)x - x^2) > 0,$$

and this is obviously true. But this contradicts the fact that (a, b, c) was a maximum. Hence $a = b$. Then $c = 1 - 2a$, and it suffices to show that $f(a, a, 1 - 2a) \leq 0$. Specifically, this means

$$4a^2 - 8a(1 - 2a) - 9a^2(1 - 2a) - 1 \leq 0.$$

The left-hand side factors as $-(1 - 2a)(3a - 1)^2 = -c(3a - 1)^2$, which is negative or zero. The inequality is now proved. Moreover, we have showed that the only situations in which equality is attained occur when two of the numbers are equal to $\frac{1}{2}$ and the third is 0, or when all three numbers are equal to $\frac{1}{3}$.

Second solution: A solution is possible using the Viète relations. Here it is. Consider the polynomial

$$P(x) = (x - a)(x - b)(x - c) = x^3 - x^2 + (ab + bc + ca)x - abc,$$

the monic polynomial of degree 3 whose roots are a, b, c. Because $a + b + c = 1$, at most one of the numbers a, b, c can be equal to or exceed $\frac{1}{2}$. If any of these numbers is greater than $\frac{1}{2}$, then

$$P\left(\frac{1}{2}\right) = \left(\frac{1}{2} - a\right)\left(\frac{1}{2} - b\right)\left(\frac{1}{2} - c\right) < 0.$$

This implies

$$\frac{1}{8} - \frac{1}{4} + \frac{1}{2}(ab + bc + ca) - abc < 0,$$

and so $4(ab + bc + ca) - 8abc \leq 1$, and the desired inequality holds.

If $\frac{1}{2} - a \geq 0$, $\frac{1}{2} - b \geq 0$, $\frac{1}{2} - c \geq 0$, then

$$2\sqrt{\left(\frac{1}{2} - a\right)\left(\frac{1}{2} - b\right)} \leq \left(\frac{1}{2} - a\right) + \left(\frac{1}{2} - b\right) = 1 - a - b = c.$$

Similarly,

$$2\sqrt{\left(\frac{1}{2} - b\right)\left(\frac{1}{2} - c\right)} \leq a \quad \text{and} \quad 2\sqrt{\left(\frac{1}{2} - c\right)\left(\frac{1}{2} - a\right)} \leq b.$$

It follows that

$$8 \left(\frac{1}{2} - a \right) \left(\frac{1}{2} - b \right) \left(\frac{1}{2} - c \right) \le abc,$$

and the desired inequality follows.

(*Mathematical Reflections*, proposed by T. Andreescu)

138. Define

$$f : \{(x_1, x_2, \ldots, x_n) \in \mathbb{R}^n \mid x_j > 0, \, x_1 + x_2 + \cdots + x_n = 1\} \to (0, \infty),$$

$$f(x_1, x_2, \ldots, x_n) = \left(1 + \frac{1}{x_1} \right) \left(1 + \frac{1}{x_2} \right) \cdots \left(1 + \frac{1}{x_n} \right).$$

The domain is not compact but it is bounded, and f becomes infinite on the boundary. So f has a minimum inside the domain. We will show that the minimum is attained when all x_j are equal.

If $x_i < x_j$ for some i and j, increase x_i and decrease x_j by some number a, $0 < a \le x_j - x_i$. We need to show that

$$\left(1 + \frac{1}{x_i + a} \right) \left(1 + \frac{1}{x_j - a} \right) < \left(1 + \frac{1}{x_i} \right) \left(1 + \frac{1}{x_j} \right),$$

or

$$\frac{(x_i + a + 1)(x_j - a + 1)}{(x_i + a)(x_j - a)} < \frac{(x_i + 1)(x_j + 1)}{x_i x_j}.$$

All denominators are positive, so after multiplying out and canceling terms, we obtain the equivalent inequality

$$-ax_i^2 + ax_j^2 - a^2 x_i - a^2 x_i - ax_i + ax_j - a^2 > 0.$$

This can be rewritten as

$$a(x_j - x_i)(x_j + x_i + 1) > a^2(x_j + x_i + 1),$$

which is true, since $a < x_j - x_i$. So the minimum can only be attained with $x_1 = x_2 = \cdots = x_n = \frac{1}{n}$. The value of the the minimum is $f\left(\frac{1}{n}, \frac{1}{n}, \ldots, \frac{1}{n} \right) = (n+1)^n$. The inequality is proved.

139. Project orthogonally the ellipse onto a plane to make it a circle. Because all areas are multiplied by the same constant, namely the cosine of the angle made by the plane of the ellipse and that of the projection, the problem translates to finding the largest area triangles inscribed in a given circle. We apply Sturm's principle, after we guess that all these triangles have to be equilateral.

Let $\mathcal{C}$ be the circle and let us define $f : \mathcal{C}^3 \to \mathbb{R}$, $f(P_1, P_2, P_3)$ equal to the area of triangle P_1, P_2, P_3. The area depends continuously on the vertices (because you can write it as a determinant with entries the coordinates of the vertices). $\mathcal{C}^3$ is closed and bounded in $\mathbb{R}^3$. So f has a maximum. Let $P_1 P_2 P_3$ be a triangle for which the maximum is achieved.

If $P_1 P_2 P_3$ is not equilateral, two cases can be distinguished. Either the triangle is obtuse, in which case it lies inside a semidisk. Then its area is less than half the area of the disk, and consequently smaller than the area of the inscribed equilateral triangle. Or otherwise the triangle is acute.

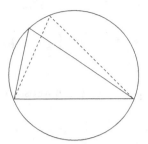

Figure 67

In this case, one of the sides of the triangle is larger than the side of the equilateral triangle and one is smaller (since some side must subtend an arc greater than $\frac{2\pi}{3}$ and another an arc smaller than $\frac{2\pi}{3}$). Moving the vertex on the circle in the direction of the longer side increases the area, as seen in Figure 67. So this is not the maximum. Therefore, the inscribed triangles that maximize the area are the equilateral triangles. These triangles are exactly those whose centroid coincides with the center of the circle. Returning to the ellipse, since the orthogonal projection preserves centroids, we conclude that the maximal-area triangles inscribed an ellipse are those with the centroid at the center of the ellipse.

(12th W.L. Putnam Mathematical Competition, 1952)

140. This is equivalent to asking what is the largest ratio between the area of an inscribed convex n-gon and the area of the circle.

We can assume that the circle has fixed radius R and vary the polygon. Let $\mathcal{C}$ be this circle, viewed as a subspace of $\mathbb{R}^2$. Then $\mathcal{C}^n$ is a closed bounded subset of $\mathbb{R}^{2n}$, and any continuous function defined on $\mathcal{C}^n$ has a maximum.

We define thus

$$f : \mathcal{C}^n \to \mathbb{R},$$

$f(P_1, P_2, \ldots, P_n)$ equal to the area of the convex polygon that has vertices $P_1, P_2, \ldots, P_n$ in some order. Note that some of the vertices might coincide, but that will not affect the final answer, as we will see. The function f is continuous since the area depends continuously on the vertices.

Hence f has a maximum, and let $(P_1, P_2, \ldots, P_n)$ be the point in $\mathcal{C}^n$ where the maximum is attained. Without loss of generality we may assume that the points appear in this order on the circle. Working with indices modulo n, assume that for some $j \in \{1, 2, \ldots, n\}$, $P_j P_{j+1} \neq P_{j+1} P_{j+2}$. Moving P_{j+1} to the midpoint of the arc $\overarc{P_j P_{j+1} P_{j+2}}$ we obtain a polygon with strictly larger area. This is impossible, so $P_1 P_2 \ldots P_n$ must be the regular n-gon.

Since the area of the regular n-gon is related to the area of the circumcircle by $A = \frac{1}{2} n R^2 \sin \frac{2\pi}{n}$, the smallest radius that the circle can have is

$$R = \sqrt{\frac{2}{n \sin \frac{2\pi}{n}}}.$$

141. The first inequality follows easily from $ab \geq abc$ and $bc \geq abc$. For the second, set

$$K = \{(a, b, c) \in \mathbb{R}^3 \mid a, b, c \geq 0, a + b + c = 1\}.$$

Then K is closed and bounded (in other words it is compact). Define the function

$$E : K \to (0, \infty), \quad E(a, b, c) = ab + bc + ac - 2abc.$$

The function E has a maximum on K. We claim that the minimum is attained when $a = b = c = \frac{1}{3}$.

Suppose that the maximum is attained at (a, b, c), without loss of generality, we may assume that $a \leq b \leq c$. Note that $b \leq c$ and $a + b + c = 1$ imply $b \leq \frac{1}{2}$. If $a < c$, choose α a positive number smaller than $c - a$. We have

$$E(a + \alpha, b, c - \alpha) - E(a, b, c) = \alpha(1 - 2b)[(c - a) - \alpha] > 0.$$

So (a, b, c) is not the maximum. This means that the maximum is attained when all three numbers are equal, and the inequality is proved.

Second solution. Let us also give a proof to the inequality on the right that does not rely on real analysis. Define $E(a, b, c) = ab + bc + ac - 2abc$. Assume that $a \leq b \leq c, a < c$, and let $\alpha = \min\left(\frac{1}{3} - a, c - \frac{1}{3}\right)$, which is a positive number. We compute

$$E(a + \alpha, b, c - \alpha) = E(a, b, c) + \alpha(1 - 2b)[(c - a) - \alpha].$$

Since $b \leq c$ and and $a + b + c = 1$, we have $b \leq \frac{1}{2}$. This means that $E(a + \alpha, b, c - \alpha) \geq E(a, b, c)$. So we were able to make one of a and c equal to $\frac{1}{3}$ by increasing the value of the expression. Repeating the argument for the remaining two numbers, we are able to increase $E(a, b, c)$ to $E\left(\frac{1}{3}, \frac{1}{3}, \frac{1}{3}\right) = \frac{7}{27}$. This proves the inequality.
(Communicated by V. Grover)

142. The inequality from the statement can be rewritten as

$$\frac{\prod\limits_{j=1}^{n} x_j}{\prod\limits_{j=1}^{n}(1 - x_j)} \leq \frac{\left(\sum\limits_{j=1}^{n} x_j\right)^n}{\left(\sum\limits_{j=1}^{n}(1 - x_j)\right)^n}.$$

We prove the inequality for a fixed, but arbitrary value S of the sum $x_1 + x_2 + \cdots x_n$. Then the right-hand side is equal to $\left(\frac{S}{n-S}\right)^n$. Define

$$K = \left\{(x_1, x_2, \ldots, x_n) \in \mathbb{R}^n \mid 0 \leq x_j \leq \frac{1}{2}, x_1 + x_2 + \cdots + x_n = 1\right\}.$$

The K is closed and bounded, so the continuous function

$$f : K \to (0, \infty), \quad f(x_1, x_2, \ldots, x_n) = \frac{\prod\limits_{j=1}^{n} x_j}{\prod\limits_{j=1}^{n}(1 - x_j)}$$

has a maximum on K. We are supposed to show that the maximum is attained when all x_j are equal (in which case we have equality in the given inequality).

Let $(x_1, x_2, \ldots, x_n)$ be a point where the maximum is attained. If the x_j's are not all equal, then there exist two of them, x_k and x_i, with $x_k < \frac{S}{n} < x_l$. We would like to show that by adding a small positive number α to x_k and subtracting the same number from x_l the expression grows. This reduces to

$$\frac{(x_k + \alpha)(x_l - \alpha)}{(1 - x_k - \alpha)(1 - x_l + \alpha)} < \frac{x_k x_l}{(1 - x_k)(1 - x_l)}.$$

Some computations transform this into

$$\alpha(1 - x_k - x_l)(x_l - x_k - \alpha) > 0,$$

which is true if $\alpha < x_l - x_k$. So this is not the maximum. Hence the maximum is attained when all numbers are equal. Doing this for all possible values of S proves the inequality.

(Indian Team Selection Test for the International Mathematical Olympiad, 2004)

143. We apply the same kind of reasoning, varying the parameters until we reach the maximum. To find the maximum of $\sqrt{a} + \sqrt{b} + \sqrt{c} + \sqrt{d}$, we increase the sum $a + b + c + d$ until it reaches the upper limit 30. Because $a + b + c \leq 14$ it follows that $d \geq 16$. Now we fix a, b and vary c, d to maximize $\sqrt{c} + \sqrt{d}$. This latter expression is maximal if c and d are closest to $\frac{c+d}{2}$. But since $c + d \leq 30$, $\frac{c+d}{2} \leq 15$. So in order to maximize $\sqrt{c} + \sqrt{d}$, we must choose $d = 16$.

Now we have $a + b + c = 14$, $a + b \leq 5$, and $a \leq 1$. The same argument carries over to show that in order to maximize $\sqrt{a} + \sqrt{b} + \sqrt{c}$ we have to choose $c = 9$. And the reasoning continues to show that a has to be chosen 1 and b has to be 4.

We conclude that under the constraints $a \leq 1$, $a+b \leq 5$, $a+b+c \leq 14$, and $a+b+c+d \leq 30$, the sum $\sqrt{a} + \sqrt{b} + \sqrt{c} + \sqrt{d}$ is maximal when $a = 1$, $b = 4$, $c = 9$, $d = 16$, in which case the sum of the square roots is equal to 10. The inequality is proved.

(V. Cârtoaje)

144. There exist finitely many n-tuples of positive integers with the sum equal to m, so the expression from the statement has indeed a maximal value.

We show that the maximum is not attained if two of the x_i's differ by 2 or more. Without loss of generality, we may assume that $x_1 \leq x_2 - 2$. Increasing x_1 by 1 and decreasing x_2 by 1 yields

$$\sum_{2<i<j} x_i x_j + (x_1 + 1) \sum_{2<i} x_i + (x_2 - 1) \sum_{2<i} x_i + (x_1 + 1)(x_2 - 1)$$

$$= \sum_{2<i<j} x_i x_j + x_1 \sum_{2<i} x_i + x_2 \sum_{2<i} x_i + x_1 x_2 - x_1 + x_2 + 1.$$

The sum increased by $x_2 - x_1 - 1 \geq 1$, and hence the original sum was not maximal.

This shows that the expression attains its maximum for a configuration in which the x_i's differ from each other by at most 1. If $m = rn + s$, with $0 \leq s < n$, then for this to happen

s of the x_i's must be equal to $r + 1$ and the remaining must be equal to r. This gives that the maximal value of the expression must be equal to

$$\frac{1}{2}(n - s)(n - s - 1)r^2 + s(n - s)r(r + 1) + \frac{1}{2}s(s - 1)(r + 1)^2.$$

(Mathematical Olympiad Summer Program 2002, communicated by Z. Sunik)

145. There are finitely many such products, so a smallest product does exist. Examining the $2 \times 2, 3 \times 3$, and 4×4 arrays, we conjecture that the smallest product is attained on the main diagonal and is $1 \cdot 3 \cdot 5 \cdots (2n - 1)$. To prove this, we show that if the permutation σ of $\{1, 2, \ldots, n\}$ has an inversion, then $a_{1\sigma(1)}a_{2\sigma(2)} \cdots a_{n\sigma(n)}$ is not minimal.

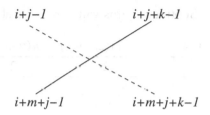

Figure 68

So assume that the inversion gives rise to the factors $i + (j + k) - 1$ and $(i + m) + j - 1$ in the product. Let us replace them with $i + j - 1$ and $(i + m) + (j + k) - 1$, as shown in Figure 68. The product of the first pair is

$$i^2 + ik + i(j - 1) + mi + mk + m(j - 1) + (j - 1)i + (j - 1)k + (j - 1)^2,$$

while the product of the second pair is

$$i^2 + im + ik + i(j - 1) + (j - 1)m + (j - 1)k + (j - 1)^2.$$

We can see that the first of these expressions exceeds the second by mk. This proves that if the permutation has an inversion, then the product is not minimal. The only permutation without inversions is the identity permutation. By Sturm's principle, it is the permutation for which the minimum is attained. This minimum is $1 \cdot 3 \cdot 5 \cdots (2n - 1)$, as claimed.

146. Order the numbers $x_1 < x_2 < \cdots < x_n$ and call the expression from the statement $E(x_1, x_2, \ldots, x_n)$. Note that $E(x_1, x_2, \ldots, x_n) > \frac{x_n^2}{n}$, which shows that as the variables tend to infinity, so does the expression. This means that the minimum exists. Assume that the minimum is attained at the point $(y_1, y_2, \ldots, y_n)$. If $y_n - y_1 > n$ then there exist indices i and j, $i < j$, such that $y_1, \ldots, y_i + 1, \ldots, y_j - 1, \ldots, y_n$ are still distinct integers. When substituting these numbers into E the denominator stays constant while the numerator changes by $3(y_j + y_i)(y_j - y_i - 1)$, a negative number, decreasing the value of the expression. This contradicts the minimality. We now look at the case with no gaps: $y_n - y_1 = n - 1$. Then there exists a such that $y_1 = a + 1$, $y_2 = a + 2$, ..., $y_n = a + n$. We have

$$E(y_1, \ldots, y_n) = \frac{na^3 + 3\dfrac{n(n+1)}{2}a^2 + \dfrac{n(n+1)(2n+1)}{2}a + \dfrac{n^2(n+1)^2}{4}}{na + \dfrac{n(n+1)}{2}}$$

$$= \frac{a^3 + \dfrac{3(n+1)}{2}a^2 + \dfrac{(n+1)(2n+1)}{2}a + \dfrac{n(n+1)^2}{4}}{a + \dfrac{n+1}{2}}.$$

When $a = 0$ this is just $\frac{n(n+1)}{2}$. Subtracting this value from the above, we obtain

$$\frac{a^3 + \dfrac{3(n+1)}{2}a^2 + \left[\dfrac{(n+1)(2n+1)}{2} - \dfrac{n(n+1)}{2}\right]a}{a + \dfrac{n+1}{2}} > 0.$$

We deduce that $\frac{n(n+1)}{2}$ is a good candidate for the minimum.

If $y_n - y_1 = n$, then there exist a and k such that $y_1 = a, \ldots, y_k = a + k - 1$, $y_{k+1} = a + k + 1, \ldots, y_n = a + n$. Then

$$E(y_1, \ldots, y_n) = \frac{a^3 + \cdots + (a+k-1)^3 + (a+k+1)^3 + \cdots + (a+n)^3}{a + \cdots + (a+k-1) + (a+k+1) + \cdots + (a+n)}$$

$$= \frac{\displaystyle\sum_{j=0}^{n}(a+j)^3 - (a+k)^3}{\displaystyle\sum_{j=0}^{n}(a+j) - (a+k)}$$

$$= \frac{na^3 + 3\left[\dfrac{n(n+1)}{2} - k\right]a^2 + 3\left[\dfrac{n(n+1)(2n+1)}{6} - k^2\right]a + \left[\dfrac{n^2(n+1)^2}{4} - k^3\right]}{na + \dfrac{n(n+1)}{2} - k}.$$

Subtracting $\frac{n(n+1)}{2}$ from this expression, we obtain

$$\frac{na^3 + 3\left[\dfrac{n(n+1)}{2} - k\right]a^2 + \left[\dfrac{n(n+1)(2n+1)}{2} - 3k^2 - \dfrac{n^2(n+1)}{2}\right]a - k^3 + \dfrac{n(n+1)}{2}k}{na + \dfrac{n(n+1)}{2} - k}.$$

The numerator is the smallest when $k = n$ and $a = 1$, in which case it is equal to 0. Otherwise, it is strictly positive, proving that the minimum is not attained in that case. Therefore, the desired minimum is $\frac{n(n+1)}{2}$, attained only if $x_k = k$, $k = 1, 2, \ldots, n$.

(*American Mathematical Monthly*, proposed by C. Popescu)

147. First, note that the inequality is obvious if either x or y is at least 1. For the case $x, y \in (0, 1)$, we rely on the inequality

$$a^b \geq \frac{a}{a + b - ab},$$

which holds for $a, b \in (0, 1)$. To prove this new inequality, write it as

$$a^{1-b} \leq a + b - ab,$$

and then use the Bernoulli inequality to write

$$a^{1-b} = (1 + a - 1)^{1-b} \leq 1 + (a - 1)(1 - b) = a + b - ab.$$

Using this, we have

$$x^y + y^x \geq \frac{x}{x + y - xy} + \frac{y}{x + y - xy} > \frac{x}{x + y} + \frac{y}{x + y} = 1,$$

completing the solution to the problem.

(French Mathematical Olympiad, 1996)

148. We have

$$x^5 - x^2 + 3 \geq x^3 + 2,$$

for all $x \geq 0$, because this is equivalent to $(x^3 - 1)(x^2 - 1) \geq 0$. Thus

$$(a^5 - a^2 + 3)(b^5 - b^2 + 3)(c^5 - c^2 + 3) \geq (a^3 + 1 + 1)(1 + b^3 + 1)(1 + 1 + c^3).$$

Let us recall Hölder's inequality, which in its most general form states that for $r_1, r_2, \ldots, r_k > 0$, with $\frac{1}{r_1} + \frac{1}{r_2} + \cdots + \frac{1}{r_k} = 1$ and for positive real numbers $a_{ij}, i = 1, 2, \ldots, k, j = 1, 2, \ldots, n$,

$$\sum_{i=1}^{n} a_{1i} a_{2i} \cdots a_{ki} \leq \left(\sum_{i=1}^{n} a_{1i}^{r_1} \right)^{\frac{1}{r_1}} \left(\sum_{i=1}^{n} a_{2i}^{r_2} \right)^{\frac{1}{r_2}} \cdots \left(\sum_{i=1}^{n} a_{ki}^{r_k} \right)^{\frac{1}{r_k}}.$$

Applying it for $k = n = 3$, $r_1 = r_3 = 3$, and the numbers $a_{11} = a, a_{12} = 1, a_{13} = 1, a_{21} = 1$, $a_{22} = b, a_{23} = 1, a_{31} = 1, a_{32} = 1, a_{33} = c$, we obtain

$$(a + b + c) \leq (a^3 + 1 + 1)^{\frac{1}{3}} (1 + b^3 + 1)^{\frac{1}{3}} (1 + 1 + c)^{\frac{1}{3}}.$$

we thus have

$$(a^3 + 1 + 1)(1 + b^3 + 1)(1 + 1 + c^3) \geq (a + b + c)^3,$$

and the inequality is proved.

(USA Mathematical Olympiad, 2004, proposed by T. Andreescu)

149. Let x_i, $i = 1, 2, \ldots, n$, $x_i > 0$, be the roots of the polynomial. Using the relations between the roots and the coefficients, we obtain

$$\sum x_1 x_2 \cdots x_m = \binom{n}{m} \quad \text{and} \quad \sum x_1 x_2 \cdots x_p = \binom{n}{p}.$$

The generalized Maclaurin inequality

$$\sqrt[m]{\frac{\sum x_1 x_2 \cdots x_m}{\binom{n}{m}}} \geq \sqrt[m]{\frac{\sum x_1 x_2 \cdots x_p}{\binom{n}{p}}}$$

thus becomes equality. This is possible only if $x_1 = x_2 = \cdots = x_n$. Since $\sum x_1 x_2 \cdots x_m = \binom{n}{m}$, it follows that $x_i = 1$, $i = 1, 2, \ldots, n$, and hence $P(x) = (x - 1)^n$.

(*Revista Matematică din Timişoara* (*Timişoara Mathematics Gazette*), proposed by T. Andreescu)

150. The idea of the solution is to reduce the inequality to a particular case of the Huygens inequality,

$$\prod_{i=1}^{n} (a_i + b_i)^{p_i} \geq \prod_{i=1}^{n} a_i^{p_i} + \prod_{i=1}^{n} b_i^{p_i},$$

which holds for positive real numbers $p_1, p_2, \ldots, p_n, a_1, a_2, \ldots, a_n, b_1, b_2, \ldots, b_n$ with $p_1 + p_2 + \cdots + p_n = 1$,

To this end, start with

$$\frac{n - x_i}{1 - x_i} = 1 + \frac{n - 1}{x_1 + \cdots + x_{i-1} + x_{i+1} + \cdots + x_n}$$

and apply the AM-GM inequality to get

$$\frac{n - x_i}{1 - x_i} \leq 1 + \frac{1}{\sqrt[n-1]{x_1 \cdots x_{i-1} x_{i+1} \cdots x_n}}.$$

Multiplying all n inequalities gives

$$\prod_{i=1}^{n} \left(\frac{n - x_i}{1 - x_i} \right) \leq \prod_{i=1}^{n} \left(1 + \frac{1}{\sqrt[n-1]{x_1 \cdots x_{i-1} x_{i+1} \cdots x_n}} \right).$$

Thus we are left to prove

$$\prod_{i=1}^{n} \left(1 + \frac{1}{x_i} \right) \geq \prod_{i=1}^{n} \left(1 + \frac{1}{\sqrt[n-1]{x_1 \cdots x_{i-1} x_{i+1} \cdots x_n}} \right).$$

This inequality is a product of the individual inequalities

$$\prod_{j \neq i} \left(1 + \frac{1}{x_j} \right) \geq \left(1 + \sqrt[n-1]{\prod_{j \neq i} \frac{1}{x_i}} \right)^{n-1}, \quad j = 1, 2, \ldots, n.$$

Each of these is Huygens' inequality applied to the numbers $1, 1, \ldots, 1$ and $\frac{1}{x_1}, \ldots, \frac{1}{x_{i-1}}, \frac{1}{x_{i+1}}, \ldots, \frac{1}{x_n}$, with $p_1 = p_2 = \cdots = p_n = \frac{1}{n-1}$.

(*Crux Mathematicorum*, proposed by W. Janous)

151. We will use

Aczel's inequality. *If $x_1, x_2, \ldots, x_n, y_1, y_2, \ldots, y_m$ are real numbers such that $x_1^2 > x_2^2 + \cdots + x_m^2$, then*

$$(x_1 y_1 - x_2 y_2 - \cdots x_m y_m)^2 \geq (x_1^2 - x_2^2 - \cdots - x_m^2)(y_1^2 - y_2^2 - \cdots y_m^2).$$

Proof. Consider

$$f(t) = (x_1 t + y_1)^2 - \sum_{i=2}^{m} (x_i t + y_i)^2$$

and note that $f\left(-\frac{y_1}{x_1}\right) \leq 0$. It follows that the discriminant of the quadratic function $f(t)$ is nonnegative. This condition that the discriminant is nonnegative is basically Aczel's inequality. $\qquad \square$

Let us return to the problem. It is clear that $a_1^2 + a_2 + \cdots + a_n^2 - 1$ and $b_1^2 + b_2^2 + \cdots + b_n^2 - 1$ have the same sign. If

$$1 > a_1^2 + a_2^2 + \cdots + a_n^2 \quad \text{or} \quad 1 > b_1^2 + b_2^2 + \cdots + b_n^2,$$

then by Aczel's inequality,

$$(1 - a_1 b_1 - \cdots - a_n b_n)^2 \geq (1 - a_1^2 - a_2^2 - \cdots - a_n^2)(1 - b_1^2 - b_2^2 - \cdots - b_n^2),$$

which contradicts the hypothesis. The conclusion now follows.

(USA Team Selection Test for the International Mathematical Olympiad, proposed by T. Andreescu and D. Andrica)

152. The solution is based on the Muirhead inequality.

Theorem. *If $a_1, a_2, a_3, b_1, b_2, b_3$ are real numbers such that*

$$a_1 \geq a_2 \geq a_3 \geq 0, b_1 \geq b_2 \geq b_3 \geq 0, \ a_1 \geq b_1, \ a_1 + a_2 \geq b_1 + b_2,$$

$$a_1 + a_2 + a_3 = b_1 + b_2 + b_3,$$

then for any positive real numbers x, y, z, one has

$$\sum_{\text{sym}} x^{a_1} y^{a_2} z^{a_3} \geq \sum_{\text{sym}} x^{b_1} y^{b_2} z^{b_3},$$

where the index sym *signifies that the summation is over all permutations of x, y, z.*

Using the fact that $abc = 1$, we rewrite the inequality as

$$\frac{1}{a^3(b+c)} + \frac{1}{b^3(c+a)} + \frac{1}{c^3(a+b)} \geq \frac{3}{2(abc)^{4/3}}.$$

Set $a = x^3$, $b = y^3$, $c = z^3$, with $x, y, z > 0$. The inequality becomes

$$\sum_{\text{cyclic}} \frac{1}{x^9(y^3 + z^3)} \geq \frac{3}{2x^4y^4z^4}.$$

Clearing denominators, this becomes

$$\sum_{\text{sym}} x^{12}y^{12} + 2\sum_{\text{sym}} x^{12}y^9z^3 + \sum_{\text{sym}} x^9y^9z^6 \geq 3\sum_{\text{sym}} x^{11}y^8z^5 + 6x^8y^8z^8,$$

or

$$\left(\sum_{\text{sym}} x^{12}y^{12} - \sum_{\text{sym}} x^{11}y^8z^5\right) + 2\left(\sum_{\text{sym}} x^{12}y^9z^3 - \sum_{\text{sym}} x^{11}y^8z^5\right)$$

$$+ \left(\sum_{\text{sym}} x^9y^9z^6 - \sum_{\text{sym}} x^8y^8z^8\right) \geq 0.$$

And every term on the left-hand side is nonnegative by the Muirhead inequality.
(36th International Mathematical Olympiad, 1995)

153. The equation can be transformed into

$$(x^2 + 6x + 5)(x^2 + 6x + 8)(x^2 + 6x + 9) = 360.$$

Substitute $x^2 + 6x = y$. We obtain

$$(y + 5)(y + 8)(y + 9) = 360$$

or

$$y^3 + 22y^2 + 157y = 0.$$

We find $y = 0$, $y = -11 + 6i$, and $y = -11 - 6i$. The equation $x^2 + 6x = 0$ gives $x = 0$, $x = -6$.

Let us solve $x^2 + 6x = -11 + 6i$. This is equivalent to $(x + 3)^2 = -2 + 6i$. Setting $x + 3 = u + iv$, we obtain the system

$$u^2 - v^2 = -2$$
$$2uv = 6.$$

It follows that $(u^2 + v^2)^2 = (u^2 - v^2)^2 + (2uv)^2 = 40$. Hence $u^2 + v^2 = 2\sqrt{10}$. Then $u^2 = \sqrt{10} - 1$, $v^2 = \sqrt{10} + 1$. So $u = \pm\sqrt{\sqrt{10} - 1}$, $v = \pm\sqrt{\sqrt{10} + 1}$, and $x = u + iv - 3$, for all four choices of signs for u and v.

154. Note that $(\sqrt{5}+1)^2 = 6 + 2\sqrt{5}$. The polynomial equation can be written as

$$x^3 - (6 + 2\sqrt{5})x - x + \sqrt{5} + 1 = 0,$$

or

$$x(x^2 - (\sqrt{5}+1)^2) - (x - (\sqrt{5}+1)) = 0.$$

This factors as

$$[x - (\sqrt{5}+1)][x^2 + (\sqrt{5}+1)x - 1] = 0,$$

with solutions

$$x_1 = \sqrt{5}+1, \quad x_{2,3} = \frac{-\sqrt{5}-1 \pm \sqrt{10 + 2\sqrt{5}}}{2}.$$

(Gazeta Matematică (Mathematics Gazette, Bucharest), proposed by A. Eckstein)

155. Moving 3 to the left and distributing it among the three terms we obtain

$$[x - (a + b + c)]\left[\frac{1}{x - a} + \frac{1}{x - b} + \frac{1}{x - c}\right] = 0.$$

One solution is $x = a + b + c$. The other two are the roots of the quadratic equation

$$3x^2 - 2(a + b + c)x + (ab + bc + ac) = 0.$$

The discriminant of this equation is

$$(a + b + c)^2 - 3(ab + bc + ac) = a^2 + b^2 + c^2 - ab - bc - ac.$$

This is further equal to

$$\frac{1}{2}[(a - b)^2 + (b - c)^2 + (c - a)^2],$$

which is nonnegative.

(C. Coşniţă, *Teme şi Probleme Alese de Matematici (Selected Mathematical Themes and Problems)*, Ed. Didactică şi Pedagogică, Bucharest)

156. The relation $(x + 1)P(x) = (x - 10)P(x + 1)$ shows that $P(x)$ is divisible by $(x - 10)$. Shifting the variable, we obtain the equivalent relation $xP(x - 1) = (x - 11)P(x)$, which shows that $P(x)$ is also divisible by x. Hence $P(x) = x(x - 10)P_1(x)$ for some polynomial $P_1(x)$. Substituting in the original equation and canceling common factors, we find that $P_1(x)$ satisfies

$$xP_1(x) = (x - 9)P_1(x + 1).$$

Arguing as before, we find that $P_1(x) = (x - 1)(x - 9)P_2(x)$. Repeating the argument, we eventually find that $P(x) = x(x - 1)(x - 2) \cdots (x - 10)Q(x)$, where $Q(x)$ satisfies $Q(x) = Q(x + 1)$. It follows that $Q(x)$ is constant, and the solution to the problem is

$$P(x) = ax(x - 1)(x - 2) \cdots (x - 10),$$

where a is an arbitrary constant.

157. Subtract the right-hand side from the left-hand side to obtain a polynomial $P(x)$ of degree $n - 1$, with zeros $a_1, a_2, \ldots, a_n$. This polynomial is therefore identically equal to zero.

158. Having odd degree, $P(x)$ is surjective. Hence for every root r_i of $P(x) = 0$ there exists a solution a_i to the equation $P(a_i) = r_i$, and trivially $a_i \neq a_j$ if $r_i \neq r_j$. Then $P(P(a_i)) = 0$, and the conclusion follows.

(Russian Mathematical Olympiad, 2002)

159. Since

$$P(x) = \left(x + \frac{2007}{2}\right)^2 - \frac{2007^2}{4} + 1,$$

the range of P is $A = [-\frac{2007^2}{4} + 1, \infty)$, which contains the interval $D = [-\frac{2007}{2}, \infty)$. Since $D \subset f(D)$, we obtain that the image of $P^{(n)}$ is A, which contains 0.

(Brazilian Mathematical Olympiad, 2007)

160. *First solution*: Let m be the degree of $P(x)$, and write

$$P(x) = a_m x^m + a_{m-1} x^{m-1} + \cdots + a_0.$$

Using the binomial formula for $\left(x \pm \frac{1}{n}\right)^m$ and $\left(x \pm \frac{1}{n}\right)^{m-1}$ we transform the identity from the statement into

$$2a_m x^m + 2a_{m-1} x^{m-1} + 2a_{m-2} x^{m-2} + a_m \frac{m(m - 1)}{n^2} x^{m-2} + Q(x)$$

$$= 2a_m a^m + 2a_{m-1} x^{m-1} + 2a_{m-2} x^{m-2} + R(x),$$

where Q and R are polynomials of degree at most $m - 3$. If we identify the coefficients of the corresponding powers of x, we find that $a_m \frac{m(m-1)}{n^2} = 0$. But $a_m \neq 0$, being the leading coefficient of the polynomial; hence $m(m - 1) = 0$. So either $m = 0$ or $m = 1$. One can check in an instant that all polynomials of degree 0 or 1 satisfy the required condition.

Second solution: Fix a point x_0. The graph of $P(x)$ has infinitely many points in common with the line that has slope

$$m = n\left(P\left(x_0 + \frac{1}{n}\right) - P(x_0)\right)$$

and passes through the point $(x_0, P(x_0))$. Therefore, the graph of $P(x)$ is a line, so the polynomial has degree 0 or 1.

Third solution: If there is such a polynomial of degree $m \geq 2$, differentiating the given relation $m - 2$ times we find that there is a quadratic polynomial that satisfies the given relation. But then any point on its graph would be the vertex of the parabola, which of course is impossible. Hence only linear and constant polynomials satisfy the given relation.

(Romanian Team Selection Test for the International Mathematical Olympiad, 1979, proposed by D. Buşneag)

161. Let $x = \sqrt{2} + \sqrt[3]{3}$. Then $\sqrt[3]{3} = x - \sqrt{2}$, which raised to the third power yields

$$3 = x^3 - 3\sqrt{2}x^2 + 6x - 2\sqrt{2}, \quad \text{or} \quad x^3 + 6x - 3 = (3x^2 + 2)\sqrt{2}.$$

By squaring this equality we deduce that x satisfies the polynomial equation

$$x^6 - 6x^4 - 6x^3 + 12x^2 - 36x + 1 = 0.$$

(Belgian Mathematical Olympiad, 1978, from a note by P. Radovici-Mărculescu)

162. We compute $P(0) = -1$, and then inductively $P(-3k) = -1$ for all positive integers k. We conclude that $P(-1989) = -1$. The polynomial is constant!

(*Gazeta Matematică (Mathematics Gazette, Bucharest)* proposed by A. Szőrös)

163. We know that

$$x^6 + ax^5 + bx^4 + cx^3 + bx^2 + ax + 1 = \prod_{k=1}^{6}(x - x_k).$$

Setting $x = i$ we obtain

$$\prod_{k=1}^{6}(i - x_k) = i^6 + ai^5 + bi^4 + ci^3 + bi^2 + ai + 1 = 2ai - ci.$$

Setting $x = -i$ we obtain

$$\prod_{k=1}^{6}(-i - x_k) = (-i)^6 + a(-i)^5 + b(-i)^4 + c(-i)^3 + b(-i)^2 + a(-i) + 1 = -2ai + ci.$$

Multiplying we obtain

$$\prod_{k=1}^{6}(i - x_k)(-i - x_k) = (2a - c)^2,$$

or

$$\prod_{k=1}^{6}(x_k^2 + 1) = (2a - c)^2,$$

as desired.

(*Gazeta Matematică (Mathematics Gazette, Bucharest)*, proposed by M. Szőrös)

164. First, it suffices to show that for any r, z, w with $0 < r < 1$, $|z| = 1$, and $|w| \geq 1$,

$$\left| \frac{z - w}{rz - w} \right| \leq \frac{2}{1 + r},$$

then replace w by each root of $P(z)$ and multiply the resulting inequalities.

Without loss of generality we may assume $z = 1$. It remains to prove that for $0 < r < 1$ and $|w| \geq 1$,

$$\left| \frac{1 - w}{r - w} \right| \leq \frac{2}{r + 1}.$$

This is equivalent to

$$\frac{2}{|w - 1|} \geq \frac{1 + r}{|w - r|}$$

Let A, B, C, D be the points in the complex plane of coordinates $w, 1, r, -1$ respectively. We have to prove $BD/AB \geq CD/AC$. This is equivalent to $\sin DAB / \sin BDA \geq \sin DAC / \sin BDA$. And this is true because $\angle DAB > \angle DAC$ and both are in the first quadrant. Equality is attained if $A = D$, that is if $w = -1$.

Remark. This simple inequality was used by the author of the problem to prove the 2-dimensional case of a conjecture in functional analysis. The conjecture, due to R. Douglas and V. Paulsen, states that an ideal of polynomials in several complex variables is closed in the topology induced by the Hardy space of the polydisk if and only if every algebraic component of its zero set intersects the closed polydisk.

(R. Gelca)

165. Note that r and s are zeros of both $P(x)$ and $Q(x)$. So on the one hand, $Q(x) = (x - r)(x - s)$. and on the other, r and s are roots of $P(x) - Q(x)$. The assumption that this polynomial is nonnegative implies that the two roots are double; hence

$$P(x) - Q(x) = (x - r)^2(x - s)^2 = Q(x)^2.$$

We find that $P(x) = Q(x)(Q(x) + 1)$. Because the signs of $P(x)$ and $Q(x)$ agree, the quadratic polynomial $Q(x) + 1$ is nonnegative. This cannot happen because its discriminant is $(r - s)^2 - 4 > 0$. The contradiction proves that our assumption was false; hence for some x_0, $P(x_0) < Q(x_0)$.

(Russian Mathematical Olympiad, 2001)

166. Because $P(0) = 0$, there exists a polynomial $Q(x)$ such that $P(x) = xQ(x)$. Then

$$Q(k) = \frac{1}{k + 1}, \quad k = 1, 2, \ldots, n.$$

Let $H(x) = (x + 1)Q(x) - 1$. The degree of $H(x)$ is n and $H(k) = 0$ for $k = 1, 2, \ldots, n$. Hence

$$H(x) = (x + 1)Q(x) - 1 = a_0(x - 1)(x - 2) \cdots (x - n).$$

In this equality $H(-1) = -1$ yields $a_0 = \frac{(-1)^{n+1}}{(n+1)!}$. For $x = m, m > n$, which gives

$$Q(m) = \frac{(-1)^{n+1}(m-1)(m-2)\cdots(m-n)+1}{(n+1)!(m+1)} + \frac{1}{m+1},$$

and so

$$P(m) = \frac{(-1)^{m+1}m(m-1)\cdots(m-n)}{(n+1)!(m+1)} + \frac{m}{m+1}.$$

(D. Andrica, published in T. Andreescu, D. Andrica, 360 *Problems for Mathematical Contests*, GIL, 2003)

167. Adding and subtracting the conditions from the statement, we find that $a_1 + a_2 + \cdots + a_n$ and $a_1 - a_2 + \cdots + (-1)^n a_n$ are both real numbers, meaning that $P(1)$ and $P(-1)$ are real numbers. It follows that $P(1) = \overline{P(1)}$ and $P(-1) = \overline{P(-1)}$. Writing $P(x) = (x - x_1)(x - x_2)\cdots(x - x_n)$, we deduce

$$(1 - x_1)(1 - x_2)\cdots(1 - x_n) = (1 - \overline{x}_1)(1 - \overline{x}_2)\cdots(1 - \overline{x}_n),$$

$$(1 + x_1)(1 + x_2)\cdots(1 + x_n) = (1 + \overline{x}_1)(1 + \overline{x}_2)\cdots(1 + \overline{x}_n).$$

Multiplying, we obtain

$$(1 - x_1^2)(1 - x_2^2)\cdots(1 - x_n^2) = (1 - \overline{x}_1^2)(1 - \overline{x}_2^2)\cdots(1 - \overline{x}_n^2).$$

This means that $Q(1) = \overline{Q(1)}$, and hence $b_1 + b_2 + \cdots + b_n$ is a real number, as desired.

(*Revista Matematică din Timişoara* (*Timişoara Mathematics Gazette*), proposed by T. Andreescu)

168. If such a $Q(x)$ exists, it is clear that $P(x)$ is even. Conversely, assume that $P(x)$ is an even function. Writing $P(x) = P(-x)$ and identifying coefficients, we conclude that no odd powers appear in $P(x)$. Hence

$$P(x) = a_{2n}x^{2n} + a_{2n-2}x^{2n-2} + \cdots + a_2x^2 + a_0 = P_1(x^2).$$

Factoring

$$P_1(y) = a(y - y_1)(y - y_2)\cdots(y - y_n),$$

we have

$$P(x) = a(x^2 - y_1)(x^2 - y_2)\cdots(x^2 - y_n).$$

Now choose complex numbers $b, x_1, x_2, \ldots, x_n$ such that $b^2 = (-1)^n a$ and $x_j^2 = y_j, j = 1, 2, \ldots, n$. We have the factorization

$$\begin{aligned}
P(x) &= b^2(x_1^2 - x^2)(x_2^2 - x^2)\cdots(x_n^2 - x^2) \\
&= b^2(x_1 - x)(x_1 + x)(x_2 - x)(x_2 + x)\cdots(x_n - x)(x_n + x) \\
&= [b(x_1 - x)(x_2 - x)\cdots(x_n - x)][b(x_1 + x)(x_2 + x)\cdots(x_n + x)] \\
&= Q(x)Q(-x),
\end{aligned}$$

where $Q(x) = b(x_1 - x)(x_2 - x)\cdots(x_n - x)$. This completes the proof.

(Romanian Mathematical Olympiad, 1979, proposed by M. Ţena)

169. View $Q(x, y, z)$ as a polynomial in x. It is easy to see that y is a zero of this polynomial; hence $Q(x, y, z)$ is divisible by $x - y$. By symmetry, it is also divisible by $y - z$ and $z - x$.

170. View the expression as a polynomial $P(x, y, z)$. Then

$$P(0, y, z) = (y + z)^3 - (y + z)^3 - (-y + z)^3 - (y - z)^3 = 0.$$

Similarly $P(x, 0, z) = 0$ and $P(x, y, 0) = 0$. So $P(x, y, z) = xyzQ(x, y, z)$. But $P(x, y, z)$ is of third degree, so $Q(x, y, z)$ is constant. In fact by setting $x = y = z$ we obtain $Q(x, y, z) = 24$.
Hence

$$(x + y + z)^3 - (-x + y + z)^3 - (x - y + z)^3 - (x + y - z)^3 = 2 \cdot 2 \cdot 2 \cdot 3 \cdot x \cdot y \cdot z.$$

(C. Coşniţă, *Teme şi Probleme Alese de Matematici (Selected Mathematical Themes and Problems)*, Ed. Didactică şi Pedagogică, Bucharest)

171. Like in the case of the previous problem, we view the expression as a polynomial $P(x, y, z)$. We check $P(0, y, z) = P(x, 0, z) = P(x, y, 0) = 0$, so $P(x, y, z) = xyzQ(x, y, z)$, where $Q(x, y, z)$ is a cuadratic homogeneous symmetric polynomial. Then there exist constants α and β such that

$$Q(x, y, z) = \alpha(x^2 + y^2 + z^2) + \beta(xy + xz + yz).$$

Setting $x = y = z$ in $P(x, y, z) = xyzQ(x, y, z)$ we obtain

$$243x^5 - x^5 - x^5 = 3x^5(\alpha + \beta),$$

so $\alpha + \beta = 80$. Setting $x = y = -z$ in the same equality we obtain

$$x^5 + x^5 + x^5 - 243x^5 = -x^3(3\alpha - \beta)$$

so $3\alpha - \beta = 240$. We obtain $\alpha = 80$ and $\beta = 0$, hence

$$(x + y + z)^5 - (-x + y + z)^5 - (x - y + z)^5 - (x + y - z)^5 = 80xyz(x^2 + y^2 + z^2).$$

(C. Coşniţă, *Teme şi Probleme Alese de Matematici (Selected Mathematical Themes and Problems)*, Ed. Didactică şi Pedagogică, Bucharest)

172. View the expression as a polynomial in a, b, c and write it in decreasing order of the degree of a as

$$E = a^3(b - c) - a(b^3 - c^3) + bc(b^2 - c^2).$$

We can see that $a(b - c)$ can be factored:

$$E = (b - c)[a^3 - a(b^2 + bc + c^2) + bc(b + c)].$$

The expression in the bracket can be viewed as a polynomial in b, which ordered by the degrees of the monomials yields

$$E = (b-c)[b^2(c-a) + bc(c-a) - a(c^2 - a^2)].$$

Now we factor a $c - a$, and finally a $b - a$ to obtain

$$E = -(b-c)(c-a)(a-b)(a+b+c).$$

173. View a, b, c, d as independent variables. The left-hand side is a cubic polynomial $P(x)$. We see that $P(a) = P(b) = P(c) = P(d) = 1$, so $P(x) = 1$, hence we obtain the identity

$$\frac{(x-b)(x-c)(x-d)}{(a-b)(a-c)(a-d)} + \frac{(x-a)(x-c)(x-d)}{(b-a)(b-c)(b-d)} + \frac{(x-a)(x-b)(x-d)}{(c-a)(c-b)(c-d)}$$
$$+ \frac{(x-a)(x-b)(x-c)}{(d-a)(d-b)(d-c)} = 1.$$

Thus the equation has solutions if a, b, c, d are distinct real numbers with $abcd = 1$. Then every real number is a solution.

(C. Coşniţă, *Teme şi Probleme Alese de Matematici (Selected Mathematical Themes and Problems)*, Ed. Didactică şi Pedagogică, Bucharest)

174. The answer is yes. We rely on the identity

$$a^3 + b^3 + c^3 - 3abc = \frac{1}{2}(a+b+c)((a-b)^2 + (b-c)^2 + (c-a)^2),$$

which shows that $a + b + c$ and $a^3 + b^3 + c^3 - 3abc$ have the same sign for all real numbers a, b, c, not all of them equal. With the obvious choice, we want a polynomial such that $f(x, y, z)$ and $x^3 + 2y^3 + 3z^3 - 3\sqrt[3]{6}xyz$ have the same sign. We choose

$$P(x, y, z) = (x^3 + 2y^3 + 3z^3)^3 - 27 \cdot 6(xyz)^3.$$

175. Let $p = x, q = y$ and view

$$x^2 + y^2 + r^2 + xyr = x^2 + y^2 + z^2 + xyz$$

as a quadratic equation in r. Rewrite it as

$$r^2 + (xy)r + (z^2 - xyz) = 0.$$

We already know the solution $r = z$. The other solution is $r = -xy - z$. So the answer to the problem is negative.

176. View the polynomial as a one variable polynomial in x, $P(x)$. The condition from the statement is equivalent to

$$P(-y) = 0, \quad P'(-y) = 0.$$

Explicitly, these conditions are

$$(-y)^p + a(-y)^{p-q}y^q + a(-y)^{p-2q}y^{2q} + y^p = 0$$
$$p(-y)^{p-1} + a(p-q)(-y)^{p-q-1}y^q + a(p-2q)(-y)^{p-2q-1}y^{2q} = 0.$$

Canceling y^p in the first and y^{p-1} in the second yields the equivalent conditions

$$(-1)^p + 1 + a(-1)^p[(-1)^q + 1] = 0$$
$$p + a(p-q)(-1)^q + a(p-2q) = 0.$$

We distinguish the following cases:

- If p and q are odd, then the first condition is satisfied, and the second yields $a = p/q$.

- If p and q are even, then the first condition gives $2 + 2a = 0$, i.e. $a = -1$, and the secod condition implies $p = 3q$.

- If p is even and q is odd, the first condition cannot be satisfied.

- If p is odd and q is even, then the first condition implies $a = 0$. But the second condition gives $p = 0$, so again no solutions.

(C. Coşniţă, *Teme şi Probleme Alese de Matematici (Selected Mathematical Themes and Problems)*, Ed. Didactică şi Pedagogică, Bucharest)

177. With a change of variables, we consider the polynomial

$$Q(x, y) = P\left(\frac{x+y}{2}, \frac{x-y}{2}\right).$$

Then

$$Q(x, y)Q(z, t) = Q(xz, yt).$$

Hence

$$Q(x, y) = Q(x, 1)Q(1, y).$$

The one-variable polynomials $Q(x, 1)$ and $Q(1, y)$ are multiplicative so they are either equal to zero, or of the form x^p. We conclude that either $P(x, y) = 0$ or $P(x, y) = Q(x+y, x-y) = (x+y)^k(x-y)^l$.

(Balkan Mathematical Olympiad, 1988, proposed by B. Enescu and M. Becheanu)

178. The trick is to view this as an equation in a. The discriminant is $\Delta = 4(x-1)^2$, and we get

$$a = \frac{x^2 + x}{x} \text{ or } a = \frac{x^2 - x + 2}{x}.$$

These are quadratic equations that can be solved easily.

(Gh. Andrei, I. Cucurezeanu, C. Caragea, Gh. Bordea, *Exerciţii şi Probleme de Algebră (Exercises and Problems in Algebra)*, 1990)

179. Assume for some n, $x = p/q$ is a rational solution, with $\gcd(p, q) = 1$. Then

$$(n^3 - n + 1)p^2 - (n^5 - n + 1)pq - (n^7 - n + 1)q^2 = 0.$$

Each of the coefficients $(n^3 - n + 1)$, $(n^5 - n + 1)$, $(n^7 - n + 1)$ is odd, and of the numbers p and q, either both are odd, or one is odd and the other is even. It follows that the expression on the left is an odd number, which cannot be equal to 0, as 0 is an even number.

We conclude that the answer to the problem is negative.

(Konhauser Problem Fest, 2010, proposed by R. Gelca)

180. We have $f(x) = (x - r_1)(x - r_2)$, so $f(n) = (n - r_1)(n - r_2)$. Because $n - r_1$ and $n - r_2$ are integers that divide 13, one of them is ± 1 and the other is ± 13 (signs agree). If one is 1 and the other is 13 then $f(n + 1) = (n - r_1 + 1)(n - r_2 + 1) = 2 \cdot 14 = 28$. If one is -1 and the other is -13, then $f(n - 1) = 28$.

(Kvant (Quantum))

181. (a) If $a^2 + b^2 > c^2$ then $(2ab + 1)^2 - 4(a^2 + b^2 - c^2)$ is the square of an odd number less than $2ab + 1$. Hence

$$(2ab + 1)^2 - 4(a^2 + b^2 - c^2) \le (2ab - 1)^2$$

so $c^2 \le (a - b)^2$, a contradiction. The case $c^2 > a^2 + b^2$ is similar. We conclude that $c^2 = a^2 + b^2$; the triangle is right.

(b) If the equation has integer roots, then $(a^2 + b^2 + c^2 + 1)^2 - 4(ab + bc + ac)$ is a perfect square less than $(a^2 + b^2 + c^2 + 1)^2$ and of the same parity with this number. We obtain

$$(a^2 + b^2 + c^2 + 1)^2 - 4(ab + bc + ac) \le (a^2 + b^2 + c^2 - 1)^2.$$

This is equivalent to

$$(a - b)^2 + (b - c)^2 + (c - a)^2 \le 0.$$

This can only happen if $a = b = c$; the triangle is equilateral.

(I. Cucurezeanu)

182. If $a + b + c + d = 0$, then $x = 1$ is a common root. For the converse, let x_0 be the common root. Multiplying the equations respectively by $b - c$, $c - a$ and $a - b$, then adding, we obtain

$$(a^2 + b^2 + c^2 - ab - ac - bc)x_0 = a^2 + b^2 + c^2 - ab - ac - bc.$$

But $a^2 + b^2 + c^2 - ab - ac - bc = (a - b)^2 + (b - c)^2 + (c - a)^2 \ne 0$. Hence $x_0 = 1$ and we are done.

(V. Matrosenco and M. Andronache)

183. Rewrite the inequality as

$$2ax^2 + 2x(2ay - y - 1) + 2ay^2 - y^2 + 1 \ge 0.$$

The left-hand side is a quadratic function in x. A first condition is that $a > 0$. Also the discriminant should be negative, namely

$$(2ay - y - 1)^2 - 2a(2ay^2 - y^2 + 1) \leq 0.$$

This implies $(1 - 2a)(y + 1)^2 \geq 0$, hence a should be at least $1/2$. The answer to the problem is $a \geq \frac{1}{2}$.

(Romanian Mathematical Olympiad, 1986, proposed by I. Mitrache)

184. The discriminant of the second quadratic equation is

$$[a^2 - 2(b - 1)]^2 - 4(a^2 + (b - 1)^2] = a^2(a^2 - 4b) \geq 0.$$

185. It suffices to prove that $\log_2 3 + \log_3 4 + \log_4 5 > 4$, and combine it with $\log_5 6 > 1$. Because $3 \cdot 2^{10} = 2072 < 3125 = 5^5$ it follows that $5 \log_2 5 > 10 + \log_2 3$. Setting $x = \log_2 3$ we are left to prove that

$$x + \frac{2}{x} + \frac{1}{2}\left(2 + \frac{x}{5}\right) > 4.$$

This is equivalent to $11x^2 - 30x + 20 > 0$. The roots of the equation are $\frac{15-\sqrt{5}}{11}$ and $\frac{15+\sqrt{5}}{11}$. It suffices to show that

$$\log_2 3 > \frac{15 + \sqrt{5}}{11}.$$

We have $\frac{15+\sqrt{5}}{11} < \frac{17.6}{11} = 1.6$ and $\log_2 3 > \frac{8}{5}$ is equivalent to $2^8 > 3^5$ i.e. $256 > 243$, true.

186. As a quadratic function in x, the left-hand side has the discriminant equal to

$$\Delta = -my^2 + 2y(2m - 3) - 2m + 1$$

which has to be negative for all y. So $m > 0$ and the discriminant of this new quadratic function in y should be negative. Hence $(2m - 3)^2 - m(2m + 1) \leq 0$. Thus $2m^2 - 11m + 9 \leq 0$, so $m \in [1, 9/2]$.

(Gh. Andrei, I. Cucurezeanu, C. Caragea, Gh. Bordea, *Exerciţii şi Probleme de Algebră (Exercises and Problems in Algebra)*, 1990)

187. If $a = b = 0$ the equation is of first degree, with unique solution $x_0 = 0$. Otherwise we have that $a \neq 0$ or $b \neq 0$. The equation is

$$(a^2 + b^2)x^2 - (4ab + 1)x + a^2 + b^2 = 0;$$

a quadratic function with roots x_1, x_2, where $x_1 \in \mathbb{Z}$. From

$$x_1 = (ax_1 - b)^2 + (bx_1 - a)^2,$$

we deduce that $x_1 > 0$. Now, since the roots are real, the discriminant will be non-negative,

$$(4ab + 1)^2 - 4(a^2 + b^2)^2 \geq 0.$$

This is equivalent to

$$(1 - 2(a - b)^2)(1 + 2(a - b)^2) \geq 0$$

which means that $1 - 2(a - b)^2 \geq 0$. Since $(a - b)^2 \in \mathbb{Z}$ we must have that $(a - b) = 0$, or equivalently, $a = b$. Taking this into account the original expression becomes

$$2a^2 - (4a^2 + 1)x + 2a^2 = 0$$

and using Viète's relations we obtain

$$x_1 + x_2 = 2 + \frac{1}{2a^2}, \quad x_1 x_2 = 1.$$

We observe that x_1 being a non-negative integer, neither $x_1 = 0$ nor $x_1 = 1$ can be roots of the above quadratic equation. Therefore $x_1 \geq 2$. Now, since $x_2 = \frac{1}{x_1} > 0$, it follows that

$$x_1 < x_1 + x_2 = 2 + \frac{1}{2a^2} < 3.$$

Hence $2 \leq x_1 < 3$, but since x_1 is an integer this implies that $x_1 = 2, x_2 = \frac{1}{2}$. By substituting the values of x_1 and x_2 we obtain $a^2 = 1$, thus $a = b = \pm 1$ and the roots are 2 and $\frac{1}{2}$.
(M. Becheanu)

188. Let $f_1(x)$ and $f_2(x)$ be the two quadratic functions, and let

$$f(x) = f_1(x) - f_2(x).$$

Note that the hypothesis implies $p_1 \neq p_2$. The equation $f(x) = 0$ has the root $\gamma = -\frac{q_1 - q_2}{p_1 - p_2}$, so

$$f(x) = (p_1 - p_2)(x - \gamma).$$

Writing

$$(p_1 - p_2)^2 f_2(x) = f(x)g(x) + R,$$

we see that $R = (q_1 - q_2)^2 + (p_1 - p_2)(p_1 q_2 - p_2 q_1)$. Hence $(p_1 - p_2)^2 f_2(\gamma) = R < 0$. So $f_1(\gamma) = f_2(\gamma) < 0$, which shows that the two quadratic functions have distinct real roots. Let α_1, β_1 and α_2, β_2 be these roots. We have

$$f_1(\alpha_2)f_1(\beta_2) = f(\alpha_2)f(\beta_2) = (p_1 - p_2)^2(\alpha_2 - \gamma)(\beta_2 - \gamma) = (p_1 - p_2)^2 f_2(\gamma) < 0.$$

The conclusion follows.
(*Kvant (Quantum)*)

189. Consider the function $f(x) = x^2 + bx + ac$. Then $f(a) < 0$, and since the coefficient of x^2 is positive, it follows that f has real zeros, thus the discriminant is positive. Hence the conclusion.

(*Matematika v Škole (Mathematics in Schools)*, 1988)

190. We compute $a^2 + b^2 = (x_1 + x_2)^2 + (x_1x_2 - 1)^2 = (x_1^2 + 1)(x_2^2 + 1)$. If one root is an integer, so is the other, and they are nonzero. Hence $a^2 + b^2$ is composite, a contradiction.

(All Union Mathematical Olympiad, 1986)

191. Let $AB = c, AC = b, BC = a, BD = x, CD = a - x$. By Stewart's relation in a triangle

$$c^2(a - x) + b^2x = ax(a - x) + aAD^2.$$

With the given hypothesis, $AD^2 = x(a - x)$, so the relation yields the quadratic equation in x:

$$2ax^2 - (2a^2 - b^2 + c^2)x + c^2a = 0.$$

We want to have a real root x, with $0 < x < a$. The discriminant is

$$\Delta = (2a^2 - b^2 + c^2)^2 - 8a^2c^2$$
$$= (2a^2 - b^2 + c^2 - 2\sqrt{2}ac)(2a^2 - b^2 + c^2 + 2\sqrt{2}ac).$$

Factoring further we obtain

$$(\sqrt{2}a - c + b)(\sqrt{2}a - c - b)(\sqrt{2}a + c + b)(\sqrt{2}a - c + b).$$

This has to be positive. The first, third and fourth factors are positive, so a necessary and sufficient condition for the equation to have real roots is that $b + c \leq \sqrt{2}a$. Let us show that in this case the root is in the desired interval. We have $f(0) > 0$ and the vertex of the parabola has the x-coordinate $\frac{2a^2 - b^2 + c^2}{4a}$, which is less than a. We are done.

192. The conditions from the statement imply that for $i, j \in \{1, 2, \ldots, n\}$,

$$(a_i - a_j)(b_i - b_j) \leq 0.$$

Consider the quadratic function $f : \mathbb{R} \to \mathbb{R}$,

$$f(x) = \sum_{1 \leq i < j \leq n} (a_ix - a_j)(b_ix - b_j).$$

If we write $f(x) = Ax^2 - Bx + C$, then

$$A = \sum_{i<j} a_ib_i = \sum_{i=1}^{n} \sum_{j=i+1}^{n} a_ib_i = \sum_{i=1}^{n} (n - i)a_ib_i,$$

$$B = \sum_{i<j} (a_ib_j + a_jb_i),$$

$$C = \sum_{i<j} a_jb_j = \sum_{j=2}^{n} \sum_{i=1}^{j-1} a_jb_j = \sum_{j=1}^{n} (j - 1)a_jb_j.$$

But

$$B = \left(\sum_{i=1}^{n} a_i \right) \left(\sum_{i=1}^{n} b_i \right) - \left(\sum_{i=1}^{n} a_i b_i \right).$$

Because $A > 0$ and $f(1) < 0$, it follows that the discriminant of the quadratic function is positive, namely that $B^2 \geq 4AC$. This is the inequality to be proved.

(*Gazeta Matematică (Mathematics Gazette, Bucharest*, proposed by D. Andrica and M.O. Drimbe)

193. Consider the quadratic function $f : \mathbb{R} \to \mathbb{R}$,

$$f(x) = (ax - b)^2 - (a_1 x - b_1)^2 - (a_2 x - b_2)^2 - \cdots - (a_{2n} x - b_{2n})^2.$$

Then $f(x) = f_1(x) + f_2(x)$, where

$$f_1(x) = \frac{1}{2} A_1 x^2 - Bx + \frac{1}{2} C_1^2 \text{ and } f_2(x) = \frac{1}{2} A_2 x^2 - Bx + \frac{1}{2} C_2^2$$

with

$$A_1 = a^2 - 2a_1^2 - 2a_3^2 - \cdots - 2a_{2n-1}^2, \quad A_2 = a^2 - 2a_2^2 - 2a_4^2 - \cdots - 2a_{2n}^2$$
$$B = ab - a_1 b_1 - a_2 b_2 - \cdots - a_{2n} b_{2n}$$
$$C_1 = b^2 - 2b_1^2 - 2b_3^2 - \cdots - 2b_{2n-1}^2, \quad C_2 = b^2 - 2b_2^2 - 2b_4^2 - \cdots - 2b_{2n}^2.$$

Note that A_1 and A_2 are positive. On the other hand, $f(b/a)$ is the negative of a sum of squares, so $f(b/a) \leq 0$. So either $f_1(b/a) \leq 0$ or $f_2(b/a) \leq 0$. It follows that one of the functions has non-negative discriminant. This implies that either $B^2 \geq A_1 C_1$ or $B^2 \geq A_2 C_2$, and so $B^2 \geq \min(A_1 C_1, A_2 C_2)$. This is the inequality to be proved.

(*Gazeta Matematică (Mathematics Gazette, Bucharest*, proposed by D. Andrica and M.O. Drimbe)

194. Let A be the vertex of the cone, O the center of the sphere, and B the center of the base of the cone. Let C be a point on the base circle, and R the radius of the sphere. If we denote $BAC = \alpha$, then $AB = R(1 + \sin\alpha)/\sin\alpha$, and $BC = R(1 + \sin\alpha)\tan\alpha/\sin\alpha$. Using the formula for the volume of the cone we obtain

$$V_1 = \pi R^3 \frac{(1 + \sin\alpha)^2}{3\sin\alpha(1 - \sin\alpha)}.$$

Also $V_2 = 2\pi R^3$. Consequently

$$k = \frac{(1 + \sin\alpha)^2}{6\sin\alpha(1 - \sin\alpha)}.$$

We can rewrite this as a quadratic equation in $\sin\alpha$

$$(1 + 6k)\sin^2\alpha + 2(1 - 3k)\sin\alpha + 1 = 0.$$

The discriminant of this quadratic must be non-negative. Hence

$$0 \le (1 - 3k)^2 - (1 + 6k) = k(9k - 12).$$

Therefore $k \ge 4/3$. It follows that k cannot be 1. The case $k = 4/3$ yields $\sin \alpha = 1/3$, hence the smallest value that k can take is $4/3$.

195. Denote the zeros of $P(x)$ by x_1, x_2, x_3, x_4, such that $x_1 + x_2 = 4$. The first Viète relation gives $x_1 + x_2 + x_3 + x_4 = 6$; hence $x_3 + x_4 = 2$. The second Viète relation can be written as

$$x_1 x_2 + x_3 x_4 + (x_1 + x_2)(x_3 + x_4) = 18,$$

from which we deduce that $x_1 x_2 + x_3 x_4 = 18 - 2 \cdot 4 = 10$. This, combined with the fourth Viète relation $x_1 x_2 x_3 x_4 = 25$, shows that the products $x_1 x_2$ and $x_3 x_3$ are roots of the quadratic equation $u^2 - 10u + 25 = 0$. Hence $x_1 x_2 = x_3 x_4 = 5$, and therefore x_1 and x_2 satisfy the quadratic equation $x^2 - 4x + 5 = 0$, while x_3 and x_4 satisfy the quadratic equation $x^2 - 2x + 5 = 0$. We conclude that the zeros of $P(x)$ are $2 + i, 2 - i, 1 + 2i, 1 - 2i$.

196. If $a \ge 0, b \ge 0, c \ge 0$, then obviously $a + b + c > 0$, $ab + bc + ca \ge 0$, and $abc \ge 0$. For the converse, let $u = a + b + c$, $v = ab + bc + ca$, and $w = abc$, which are assumed to be positive. Then a, b, c are the three zeros of the polynomial

$$P(x) = x^3 - ux^2 + vx - w.$$

Note that if $t < 0$, that is, if $t = -s$ with $s > 0$, then $P(t) = s^3 + us^2 + vs + w > 0$; hence t is not a zero of $P(x)$. It follows that the three zeros of $P(x)$ are nonnegative, and we are done.

197. Taking the conjugate of the first equation, we obtain

$$\bar{x} + \bar{y} + \bar{z} = 1,$$

and hence

$$\frac{1}{x} + \frac{1}{y} + \frac{1}{z} = 1.$$

Combining this with $xyz = 1$, we obtain

$$xy + yz + xz = 1.$$

Therefore, x, y, z are the roots of the polynomial equation

$$t^3 - t^2 + t - 1 = 0,$$

which are $1, i, -i$. Any permutation of these three complex numbers is a solution to the original system of equations.

198. Let $\alpha = x_1^5 + x_2^3 + x_3^2$. Because x_1 is solution to the original equation, we have

$$x_1^3 = x_1^2 + 2x_1 - 4.$$

Hence

$$x_1^5 = x_1^2 \cdot x_1^3 = x_1^2(x_1^2 + 2x_1 - 4) = x_1 \cdot x_1^3 + 2x_1^3 - 4x_1^2$$
$$= x_1(x_1^2 + 2x_1 - 4) + 2(x_1^2 + 2x_1 - 4) - 4x_1^2 = x_1^3 - 8 = x_1^2 + 2x_1 - 4 - 8$$
$$= x_1^2 + 2x_1 - 12.$$

Similarly

$$x_2^3 = x_2^2 + 2x_2 - 4.$$

Hence

$$\alpha = x_1^5 + x_2^3 + x_3^2 = x_1^2 + 2x_1 - 12 + x_2^2 + 2x_2 - 4 + x_3^2$$
$$= (x_1^2 + x_2^2 + x_3^2) + 2(x_1 + x_2 + x_3) - 16 - 2x_3$$
$$= -2x_3 - 9,$$

where we used the Viète relations:

$$x_1 + x_2 + x_3 = -1,$$
$$x_1^2 + x_2^2 + x_3^2 = (x_1 + x_2 + x_3)^2 - 2(x_1 x_2 + x_1 x_3 + x_2 x_3) = 1 + 4 = 5.$$

It is not hard to check that the original polynomial is irreducible over $\mathbb{Q}$, hence the polynomial obtained by applying $x \mapsto -2x - 9$ is also irreducible. This polynomial has the roots $-2x_i - 9$, $i = 1, 2, 3$, being irreducible, and having integer coefficients, it is a multiple of the desired polynomial. Computing we obtain that the anwer to the problem is the polynomial

$$x^3 + 29x^2 + 271x + 463.$$

(*Gazeta Matematică (Mathematics Gazette, Bucharest)*, proposed by L. Panaitopol)

199. Dividing by the nonzero xyz yields $\frac{x}{z} + \frac{y}{x} + \frac{z}{y} = \frac{y}{z} + \frac{z}{x} + \frac{x}{y} = r$. Let $a = \frac{x}{y}$, $b = \frac{y}{z}$, $c = \frac{z}{x}$. Then $abc = 1$, $\frac{1}{a} + \frac{1}{b} + \frac{1}{c} = r$, $a + b + c = r$. Hence

$$a + b + c = r,$$

$$ab + bc + ca = r,$$

$$abc = 1.$$

We deduce that a, b, c are the solutions of the polynomial equation $t^3 - rt^2 + rt - 1 = 0$. This equation can be written as

$$(t - 1)[t^2 - (r - 1)t + 1] = 0.$$

Since it has three real solutions, the discriminant of the quadratic must be positive. This means that $(r - 1)^2 - 4 \geq 0$, leading to $r \in (-\infty, -1] \cup [3, \infty)$. Conversely, all such r work.

200. Assume first that a, b, c, d are nonzero. Consider the equation

$$x^4 - \left(\sum a\right) x^3 + \left(\sum ab\right) - \left(\sum abc\right) + abcd = 0,$$

with roots a, b, c, d. Substitute x by a, b, c, d and cancel $a, b, c, d \neq 0$, then add the three equations to obtain

$$\sum a^3 - \left(\sum a\right)\left(\sum a^2\right) + \left(\sum ab\right)\left(\sum a\right) - 3 \sum abc = 0.$$

The second and the third term are zero, and the identity follows. The case where one of the numbers is zero is simpler, and we leave it to the reader.

201. Denoting $b = \sum xy$, $c = \sum xyz$, $d = xyzt$, we see that x, y, z, t are the roots of the polynomial equation

$$u^4 + bu^2 - cu + d = 0.$$

If we let $S_k = \sum x^k$, $k \geq 1$, then using Viète's relations we can deduce $S_1 = 0$, $S_2 = -2b$, $S_3 = 3c$, $S_4 = 2b^2 - 4d$. Multiplying the polynomial equation by u^k, substituting u by the x, y, z, t, and then adding, we obtain the recursive relation

$$S_{k+4} = -bS_{k+2} + cS_{k+1} - dS_k.$$

From this recursion we obtain $S_5 = -5bc$ and $S_7 = 7c(b^2 - d)$. So either $c = 0$, or $b^2 = d$.

In the first case, x, y, z, t are the roots of the equation $u^4 + bu^2 + d = 0$, which, being real, are of the form $\pm u_1, \pm u_2$, where u_1, u_2 are the roots of $u^2 + bu + d = 0$. The desired relations clearly holds.

In the second case, $b^2 = d$, so $S_4 = -2b^2$. But S_4 is nonnegative, which means that $b = 0$, and hence $d = 0$. Then x, y, z, t are the roots of the equation $u^4 - cu = 0$. One of the roots is zero, and so the relation holds again.

(Romanian Mathematics Competition, 1989, proposed by M. Becheanu)

202. Consider the polynomial $P(t) = r^5 + qt^4 + rt^3 + st^2 + ut + v$ with roots a, b, c, d, e. The condition from the statement implies that q is divisible by n. Moreover, since

$$\sum ab = \frac{1}{2}\left(\sum a\right)^2 - \frac{1}{2}\left(\sum a^2\right),$$

it follows that r is also divisible by n. Adding the equalities $P(a) = 0$, $P(b) = 0$, $P(c) = 0$, $P(d) = 0$, $P(e) = 0$, we deduce that

$$a^5 + b^5 + c^5 + d^5 + e^5 + s(a^2 + b^2 + c^2 + d^2 + e^2) + u(a + b + c + d + e) + 5v$$

is divisible by n. But since $v = -abcde$, it follows that

$$a^5 + b^5 + c^5 + d^5 + e^5 - 5abcde$$

is divisible by n, and we are done.

(*Kvant (Quantum)*)

203. Let $P(x) = a_n x^n + a_{n-1} x^{n-1} + \cdots + a_0$. Denote its zeros by $x_1, x_2, \ldots, x_n$. The first two of Viète's relations give

$$x_1 + x_2 + \cdots + x_n = -\frac{a_{n-1}}{a_n},$$

$$x_1 x_2 + x_1 x_3 + \cdots + x_{n-1} x_n = \frac{a_{n-2}}{a_n}.$$

Combining them, we obtain

$$x_1^2 + x_2^2 + \cdots + x_n^2 = \left(\frac{a_{n-1}}{a_n}\right)^2 - 2\left(\frac{a_{n-2}}{a_n}\right).$$

The only possibility is $x_1^2 + x_2^2 + \cdots + x_n^2 = 3$. Given that $x_1^2 x_2^2 \cdots x_n^2 = 1$, the AM-GM inequality yields

$$3 - x_1^2 + x_2^2 + \cdots + x_n^2 \geq n\sqrt[n]{x_1^2 x_2^2 \cdots x_n^2} = n.$$

Therefore, $n \geq 3$. Eliminating case by case, we find among linear polynomials $x + 1$ and $x - 1$, and among quadratic polynomials $x^2 + x - 1$ and $x^2 - x - 1$. As for the cubic polynomials, we should have equality in the AM-GM inequality. So all zeros should have the same absolute values. The polynomial should share a zero with its derivative. This is the case only for $x^3 + x^2 - x - 1$ and $x^3 - x^2 - x + 1$, which both satisfy the required property. Together with their negatives, these are all desired polynomials.

(Indian Olympiad Training Program, 2005)

204. The first Viète relation gives

$$r_1 + r_2 + r_3 + r_4 = -\frac{b}{a},$$

so $r_3 + r_4$ is rational. Also,

$$r_1 r_2 + r_1 r_3 + r_1 r_4 + r_2 r_3 + r_2 r_4 + r_3 r_4 = \frac{c}{a}.$$

Therefore,

$$r_1 r_2 + r_3 r_4 = \frac{c}{a} - (r_1 + r_2)(r_3 + r_4).$$

Finally,

$$r_1 r_2 r_3 + r_1 r_2 r_4 + r_1 r_3 r_4 + r_2 r_3 r_4 = -\frac{d}{a},$$

which is equivalent to

$$(r_1 + r_2) r_3 r_4 + (r_3 + r_4) r_1 r_2 = -\frac{d}{a}.$$

We observe that the products $r_1 r_2$ and $r_3 r_4$ satisfy the linear system of equations

$$\alpha x + \beta y = u,$$

$$\gamma x + \delta y = v,$$

where $\alpha = 1$, $\beta = 1$, $\gamma = r_3 + r_4$, $\delta = r_1 + r_2$, $u = \frac{c}{a} - (r_1 + r_2)(r_3 + r_4)$, $v = -\frac{d}{a}$. Because $r_1 + r_2 \neq r_3 + r_4$, this system has a unique solution; this solution is rational. Hence both $r_1 r_2$ and $r_3 r_4$ are rational, and the problem is solved.

(64th W.L. Putnam Mathematical Competition, 2003)

205. Set $\frac{x_1}{x_2} = t$. Let us observe that if either x_1 or x_2 is rational, so is the other, and by Viète's relations x_3 is rational as well. Also, if x_3 is rational, then $x_1 + x_2 = x_2(1 + \frac{x_1}{2})$ is rational, so x_2 is rational, and x_1 is rational as well. Hence it suffices to show that $P(x)$ has a rational root.

Substituting $x_1 = tx_2$ in Viète's relations we obtain

$$(t+1)x_2 + x_3 = -a$$
$$x_2[tx_2 + (t+1)x_3] = b.$$

Substituting x_3 from the first equation we obtain the quadratic equation in x_2,

$$(t^2 + t + 1)x_2^2 + (t+1)ax_2 + b = 0.$$

Thus x_2 is a zero of the quadratic polynomial with rational coefficients $Q(x) = (t^2 + t + 1)x^2 + (t+1)ax + b$. We deduce that the greatest common divisor of $P(x)$ and $Q(x)$ is a non-constant polynomial. Moreover, because both $P(x)$ and $Q(x)$ have rational coefficients their greatest common divisor must have rational coefficients as well. So $P(x)$ can be written as a product of two polynomials with rational coefficients. One of the factors must be a linear polynomial, showing that $P(x)$ has a rational zero. Hence the conclusion.

(Romanian Mathematics Competition, 1995, proposed by B. Enescu)

206. *First solution*: Let $\alpha = \arctan u$, $\beta = \arctan v$, and $\gamma = \arctan w$. We are required to determine the sum $\alpha + \beta + \gamma$. The addition formula for the tangent of three angles,

$$\tan(\alpha + \beta + \gamma) = \frac{\tan\alpha + \tan\beta + \tan\gamma - \tan\alpha \tan\beta \tan\gamma}{1 - (\tan\alpha \tan\beta + \tan\beta \tan\gamma + \tan\alpha \tan\gamma)},$$

implies

$$\tan(\alpha + \beta + \gamma) = \frac{u + v + w - uvw}{1 - (uv + vw + uv)}.$$

Using Viète's relations,

$$u + v + w = 0, \quad uv + vw + uw = -10, \quad uvw = -11,$$

we further transform this into $\tan(\alpha + \beta + \gamma) = \frac{11}{1+10} = 1$. Therefore, $\alpha + \beta + \gamma = \frac{\pi}{4} + k\pi$, where k is an integer that remains to be determined.

From Viète's relations we can see the product of the zeros of the polynomial is negative, so the number of negative zeros is odd. And since the sum of the zeros is 0, two of them are positive and one is negative. Therefore, one of α, β, γ lies in the interval $\left(-\frac{\pi}{2}, 0\right)$ and two of them lie in $\left(0, \frac{\pi}{2}\right)$. Hence k must be equal to 0, and $\arctan u + \arctan v + \arctan w = \frac{\pi}{4}$.

Second solution: Because

$$\text{Im} \ln(1 + ix) = \arctan x,$$

we see that

$$\arctan u + \arctan v + \arctan w = \operatorname{Im} \ln(iP(i)) = \operatorname{Im} \ln(11 + 11i) = \arctan 1 = \frac{\pi}{4}.$$

(*Középiskolai Matematikai Lapok* (*Mathematics Magazine for High Schools, Budapest*), proposed by K. Bérczi).

207. Expanding the binomial $(\cos \alpha + i \sin \alpha)^m$, and using the de Moivre formula,

$$(\cos \alpha + i \sin \alpha)^m = \cos m\alpha + i \sin m\alpha,$$

we obtain

$$\sin m\alpha = \binom{m}{1} \cos^{m-1} \alpha \sin \alpha - \binom{m}{3} \cos^{m-3} \alpha \sin^3 \alpha + \binom{m}{5} \cos^{m-5} \alpha \sin^5 \alpha + \cdots$$

For $m = 2n + 1$, if $\alpha = \frac{\pi}{2n+1}, \frac{2\pi}{2n+1}, \ldots, \frac{n\pi}{2n+1}$ then $\sin(2n + 1)\alpha = 0$, and $\sin \alpha$ and $\cos \alpha$ are both different from zero. Dividing the above relation by $\sin^{2n} \alpha$, we find that

$$\binom{2n + 1}{1} \cot^{2n} \alpha - \binom{2n + 1}{3} \cot^{2n-2} \alpha + \cdots + (-1)^n \binom{2n + 1}{2n + 1} = 0$$

holds true for $\alpha = \frac{\pi}{2n+1}, \frac{2\pi}{2n+1}, \ldots, \frac{n\pi}{2n+1}$. Hence the equation

$$\binom{2n + 1}{1} x^n - \binom{2n + 1}{3} x^{n-1} + \cdots + (-1)^n \binom{2n + 1}{2n + 1} = 0$$

has the roots

$$x_k = \cot^2 \frac{k\pi}{2n + 1}, \quad k = 1, 2, \ldots, n.$$

The product of the roots is

$$x_1 x_2 \cdots x_n = \frac{\binom{2n+1}{2n+1}}{\binom{2n+1}{1}} = \frac{1}{2n + 1}.$$

So

$$\cot^2 \frac{\pi}{2n + 1} \cot^2 \frac{2\pi}{2n + 1} \cdots \cot^2 \frac{n\pi}{2n + 1} = \frac{1}{2n + 1}.$$

Because $0 < \frac{k\pi}{2n+1} < \frac{\pi}{2}$, $k = 1, 2, \ldots, n$, it follows that all these cotangents are positive. Taking the square root and inverting the fractions, we obtain the identity from the statement.

(Romanian Team Selection Test for the International Mathematical Olympiad, 1970)

208. A good guess is that $P(x) = (x - 1)^n$. We want to show that this is the case. To this end, let $x_1, x_2, \ldots, x_n$ be the zeros of $P(x)$. Using Viète's relations, we can write

$$\sum_i (x_i - 1)^2 = \left(\sum_i x_i \right)^2 - 2 \sum_{i<j} x_i x_j - 2 \sum_i x_i + n$$

$$= n^2 - 2\frac{n(n - 1)}{2} - 2n + n = 0.$$

This implies that all squares on the left are zero. So $x_1 = x_2 = \cdots = x_n = 1$, and $P(x) = (x-1)^n$, as expected.

(*Gazeta Matematică* (*Mathematics Gazette, Bucharest*))

209. Let α, β, γ be the zeros of $P(x)$. Without loss of generality, we may assume that $0 \le \alpha \le \beta \le \gamma$. Then

$$x - a = x + \alpha + \beta + \gamma \ge 0 \quad \text{and} \quad P(x) = (x - \alpha)(x - \beta)(x - \gamma).$$

If $0 \le x \le \alpha$, using the AM-GM inequality, we obtain

$$-P(x) = (\alpha - x)(\beta - x)(\gamma - x) \le \frac{1}{27}(\alpha + \beta + \gamma - 3x)^3$$

$$\le \frac{1}{27}(x + \alpha + \beta + \gamma)^3 = \frac{1}{27}(x - a)^3,$$

so that $P(x) \ge -\frac{1}{27}(x-a)^3$. Equality holds exactly when $\alpha - x = \beta - x = \gamma - x$ in the first inequality and $\alpha + \beta + \gamma - 3x = x + \alpha + \beta + \gamma$ in the second, that is, when $x = 0$ and $\alpha = \beta = \gamma$.

If $\beta \le x \le \gamma$, then using again the AM-GM inequality, we obtain

$$-P(x) = (x - \alpha)(x - \beta)(\gamma - x) \le \frac{1}{27}(x + \gamma - \alpha - \beta)^3$$

$$\le \frac{1}{27}(x + \alpha + \beta + \gamma)^3 = \frac{1}{27}(x - a)^3,$$

so that again $P(x) \ge -\frac{1}{27}(x-a)^3$. Equality holds exactly when there is equality in both inequalities, that is, when $\alpha = \beta = 0$ and $\gamma = 2x$.

Finally, when $\alpha < x < \beta$ or $x > \gamma$, then

$$P(x) > 0 \ge -\frac{1}{27}(x-a)^3.$$

Thus the desired constant is $\lambda = -\dfrac{1}{27}$, and the equality occurs when $\alpha = \beta = \gamma$ and $x = 0$, or when $\alpha = \beta = 0$, γ is any nonnegative real, and $x = \frac{\gamma}{2}$.

(Chinese Mathematical Olympiad, 1999)

210. The key idea is to view $a^{n+1} - (a+1)^n - 2001$ as a polynomial in a. Its free term is 2002, so any integer zero divides this number.

From here the argument shifts to number theory and becomes standard. First, note that $2002 = 2 \times 7 \times 11 \times 13$. Since 2001 is divisible by 3, we must have $a \equiv 1 \pmod 3$; otherwise, one of a^{n+1} and $(a+1)^n$ would be a multiple of 3 and the other not, and their difference would not be divisible by 3. We deduce that $a \ge 7$. Moreover, $a^{n+1} \equiv 1 \pmod 3$, so we must have $(a+1)^n \equiv 1 \pmod 3$, which forces n to be even, and in particular at least 2.

If a is even, then $a^{n+1} - (a+1)^n \equiv -(a+1)^n \pmod 4$. Because n is even, $-(a+1)^n \equiv -1 \pmod 4$. But on the right-hand side, $2001 \equiv 1 \pmod 4$, so equality is impossible. Therefore, a must odd, so it divides $1001 = 7 \times 11 \times 13$. Moreover, $a^{n+1} - (a+1)^n \equiv a \pmod 4$, so $a \equiv 1 \pmod 4$.

Of the divisors of $7 \times 11 \times 13$, those congruent to 1 modulo 3 are precisely those not divisible by 11 (since 7 and 13 are both congruent to 1 modulo 3). Thus a divides 7×13. Now $a \equiv 1 \pmod 4$ is possible only if a divides 13.

We cannot have $a = 1$, since $1 - 2^n \neq 2001$ for any n. Hence the only possibility is $a = 13$. One easily checks that $a = 13, n = 2$ is a solution; all that remains to check is that no other n works. In fact, if $n > 2$, then $13^{n+1} \equiv 2001 \equiv 1 \pmod 8$. But $13^{n+1} \equiv 138$ since n is even, a contradiction. We conclude that $a = 13, n = 2$ is the unique solution.

(62nd W.L. Putnam Mathematical Competition, 2001)

211. Let us first consider the case $n \geq 2$. Let $P(x) = a_n x^n + a_{n-1} x^{n-1} + \cdots + a_0, a_n \neq 0$. Then

$$P'(x) = na_n x^{n-1} + (n-1)a_{n-1} x^{n-2} + \cdots + a_1.$$

Identifying the coefficients of $x^{n(n-1)}$ in the equality $P(P'(x)) = P'(P(x))$, we obtain

$$a_n^{n+1} \cdot n^n = a_n^n \cdot n.$$

This implies $a_n n^{n-1} = 1$, and so

$$a_n = \frac{1}{n^{n-1}}.$$

Since a_n is an integer, n must be equal to 1, a contradiction. If $n = 1$, say $P(x) = ax + b$, then we should have $a^2 + b = a$, hence $b = a - a^2$. Thus the answer to the problem is the polynomials of the form $P(x) = ax + a - a^2$.

(*Revista Matematică din Timişoara* (*Timişoara Mathematics Gazette*), proposed by T. Andreescu)

212. Let m be the degree of $P(x)$, so $P(x) = a_m x^m + a_{m-1} x^{m-1} + \cdots + a_0$. If $P(x) = x^k Q(x)$, then

$$x^{kn}(Q(x))^n = x^{kn} Q(x^n),$$

so

$$(Q(x))^n = Q(x^n),$$

which means that $Q(x)$ satisfies the same relation.

Thus we can assume that $P(0) \neq 0$. Substituting $x = 0$, we obtain $a_0^n = a_0$, and since a_0 is a nonzero real number, it must be equal to 1 if n is even, and to ± 1 if n is odd.

Differentiating the relation from the statement, we obtain

$$nP^{n-1}(x)P'(x) = nP'(x^n)x^{n-1}.$$

For $x = 0$ we have $P'(0) = 0$; hence $a_1 = 0$. Differentiating the relation again and reasoning similarly, we obtain $a_2 = 0$, and then successively $a_3 = a_4 = \cdots = a_m = 0$. It follows that $P(x) = 1$ if n is even and $P(x) = \pm 1$ if n is odd.

In general, the only solutions are $P(x) = x^m$ if n is even, and $P(x) = \pm x^m$ if n is odd, m being some nonnegative integer.

(T. Andreescu)

213. Since $a \neq 0$, $P(x)$ and $Q(x)$ are relatively prime. Assuming they are nonconstant (if one is constant then so is the other), differentiate the equations from the statement to obtain

$$3(P(x))^2 P'(x) = 2Q(x)Q'(x).$$

Thus $(P(x))^2$ divides $Q'(x)$, and so $2\deg P(x) < \deg Q(x)$. But from the initial equation $2\deg Q(x) = 3\deg P(x)$. We reached a contradiction which shows that our assumption was false, so $P(x)$ and $Q(x)$ are both constant.

(*Mathematical Reflections*, proposed by M. Athanasios)

214. Assume without loss of generality that $\deg(P(z)) = n \geq \deg(Q(z))$. Consider the polynomial $R(z) = (P(z) - Q(z))P'(z)$. Clearly, $\deg(R(z)) \leq 2n - 1$. If ω is a zero of $P(z)$ of multiplicity k, then ω is a zero of $P'(z)$ of multiplicity $k - 1$. Hence ω is also a zero of $R(z)$, and its multiplicity is at least k. So the n zeros of $P(z)$ produce at least n zeros of $R(z)$, when multiplicities are counted.

Analogously, let ω be a zero of $P(z) - 1$ of multiplicity k. Then ω is a zero of $Q(z) - 1$, and hence of $P(z) - Q(z)$. It is also a zero of $(P(z) - 1)' = P'(z)$ of multiplicity $k - 1$. It follows that ω is a zero of $R(z)$ of multiplicity at least k. This gives rise to at least n more zeros for $R(z)$.

It follows that $R(z)$, which is a polynomial of degree less than or equal to $2n - 1$, has at least $2n$ zeros. This can happen only if $R(z)$ is identically zero, hence if $P(z) \equiv Q(z)$.

(Soviet Union University Student Mathematical Olympiad, 1976)

215. Let $Q(x) = xP(x)$. The conditions from the statement imply that the zeros of $Q(x)$ are all real and distinct. From Rolle's theorem, it follows that the zeros of $Q'(x)$ are real and distinct.

Let $H(x) = xQ'(x)$. Reasoning similarly we deduce that the polynomial $H'(x)$ has all zeros real and distinct. Note that the equation $H'(x) = 0$ is equivalent to the equation

$$x^2 P''(x) + 3xP'(x) + P(x) = 0;$$

the problem is solved.

(D. Andrica, published in T. Andreescu, D. Andrica, 360 *Problems for Mathematical Contests*, GIL, 2003)

216. Let $a \in \mathbb{R}$ such that $P(a) \neq 0$. It suffices to show that the discriminant

$$D = 16[P'(a)]^2 - 20P(a)P''(a)$$

of the quadratic equation $x^2 P(a) + 4xP'(a) + 5P''(a) = 0$ is nonegative, or equivalently that

$$4[P'(a)]^2 \geq nP(a)P''(a).$$

More generally, we will show that if $n \geq 2$ is an integer, $P(x)$ is a polynomial of degree n with real coefficients and real roots, and $a \in \mathbb{R}$ is such that $P(a) \neq 0$, then we have

$$(n - 1)[P'(a)]^2 \geq nP(a)P''(a).$$

Let $P(x) = c(x - r_1)(x - r_2) \cdots (x - r_n)$, where $c, r_1, r_2, \ldots, r_n \in \mathbb{R}$ and $c \neq 0$. If $a \in \mathbb{R}$ is such that $P(a) \neq 0$, then it is easy to see that

$$\frac{P'(a)}{P(a)} = \sum_{i=1}^{n} \frac{1}{a - r_i}$$

and

$$\frac{P''(a)P(a) - [P'(a)]^2}{[P(a)]^2} = -\sum_{i=1}^{n} \frac{1}{(a - r_i)^2}.$$

Hence

$$\frac{P''(a)}{P(a)} = \left[\frac{P'(a)}{P(a)}\right]^2 - \sum_{i=1}^{n} \frac{1}{(a - r_i)^2},$$

or, equivalently,

$$\frac{P''(a)}{P(a)} = \left(\sum_{i=1}^{n} \frac{1}{a - r_i}\right)^2 - \sum_{i=1}^{n} \frac{1}{(a - r_i)^2}.$$

We apply the Cauchy-Schwarz inequality to obtain

$$\left(\frac{1}{a - r_i}\right)^2 \leq \left(\sum_{i=1}^{n} 1^2\right)\left(\sum_{i=1}^{n} \frac{1}{(a - r_i)^2}\right) = n \sum_{i=1}^{n} \frac{1}{(a - r_i)^2}.$$

The above inequality can be written equivalently as

$$(n - 1)\left(\sum_{i=1}^{n} \frac{1}{a - r_i}\right)^2 \geq n\left(\sum_{i=1}^{n} \frac{1}{a - r_i}\right)^2 - n\sum_{i=1}^{n} \frac{1}{(a - r_i)^2},$$

which is the desired inequality.

(*Mathematical Reflections*, proposed by T. Andreescu)

217. Differentiating the product, we obtain

$$P'(x) = \sum_{k=1}^{n} kx^{k-1}(x^n - 1) \cdots (x^{k+1} - 1)(x^{k-1} - 1) \cdots (x - 1).$$

We will prove that each of the terms is divisible by $P_{\lfloor n/2 \rfloor}(x)$. This is clearly true if $k > \lfloor \frac{n}{2} \rfloor$. If $k \leq \lfloor \frac{n}{2} \rfloor$, the corresponding term contains the factor

$$(x^n - 1) \cdots (x^{\lfloor n/2 \rfloor + 2} - 1)(x^{\lfloor n/2 \rfloor + 1} - 1).$$

That this is divisible by $P_{\lfloor n/2 \rfloor}(x)$ follows from a more general fact, namely that for any positive integers k and m, the polynomial

$$(x^{k+m} - 1)(x^{k+m-1} - 1) \cdots (x^{k+1} - 1)$$

is divisible by

$$(x^m - 1)(x^{m-1} - 1) \cdots (x - 1)$$

in the ring of polynomials with integer coefficients. Since the two polynomials are monic and have integer coefficients, it suffices to prove that the zeros of the second are also zeros of the first, with at least the same multiplicity.

Note that if ζ is a primitive rth root of unity, then ζ is a zero of $x^j - 1$ precisely when j is divisible by r. So the multiplicity of ζ as a zero of the polynomial $(x^m - 1)(x^{m-1} - 1) \cdots (x - 1)$ is $\lfloor \frac{m}{r} \rfloor$, while its multiplicity as a zero of $(x^{k+m} - 1)(x^{k+m-1} - 1) \cdots (x^{k+1} - 1)$ is $\lfloor \frac{m+k}{r} \rfloor - \lfloor \frac{k}{r} \rfloor$. The claim now follows from the inequality

$$\left\lfloor \frac{m+k}{r} \right\rfloor - \left\lfloor \frac{k}{r} \right\rfloor \geq \left\lfloor \frac{m}{r} \right\rfloor.$$

This completes the solution.

(Communicated by T.T. Le)

218. The equation $Q(x) = 0$ is equivalent to

$$n \frac{P(x)P''(x) - (P'(x))^2}{P(x)^2} + \left[\frac{P'(x)}{P(x)} \right]^2 = 0.$$

We recognize the first term on the left to be the derivative of $\frac{P'(x)}{P(x)}$. Denoting the roots of $P(x)$ by $x_1, x_2, \ldots, x_n$, the equation can be rewritten as

$$-n \sum_{k=1}^{n} \frac{1}{(x - x_k)^2} + \left(\sum_{k=1}^{n} \frac{1}{x - x_k} \right)^2 = 0,$$

or

$$n \sum_{k=1}^{n} \frac{1}{(x - x_k)^2} = \left(\sum_{k=1}^{n} \frac{1}{x - x_k} \right)^2.$$

If this were true for some real number x, then we would have the equality case in the Cauchy-Schwarz inequality applied to the numbers $a_k = 1$, $b_k = \frac{1}{x - x_k}$, $k = 1, 2, \ldots, n$. This would then further imply that all the x_i's are equal, which contradicts the hypothesis that the zeros of $P(x)$ are distinct. Therefore the equality cannot hold for a real number, meaning that none of the zeros of $Q(x)$ is real.

(D.M. Bătinețu, I.V. Maftei, I.M. Stancu-Minasian, *Exerciții și Probleme de Analiză Matematică* (*Exercises and Problems in Mathematical Analysis*), Editura Didactică și Pedagogică, Bucharest, 1981)

219. We start with the identity

$$\frac{P'(x)}{P(x)} = \frac{1}{x - x_1} + \frac{1}{x - x_2} + \cdots + \frac{1}{x - x_n}, \quad \text{for } x \neq x_j, \ j = 1, 2, \ldots, n.$$

If $P'\left(\frac{x_1 + x_2}{2} \right) = 0$, then this identity gives

$$0 = \frac{1}{\frac{x_1 + x_2}{2} - x_3} + \frac{1}{\frac{x_1 + x_2}{2} - x_4} + \cdots + \frac{1}{\frac{x_1 + x_2}{2} - x_n} < 0 + 0 + \cdots + 0 = 0,$$

a contradiction. Similarly, if $P'\left(\frac{x_{n-1}+x_n}{2}\right) = 0$, then

$$0 = \frac{1}{\frac{x_{n-1}+x_n}{2} - x_1} + \frac{1}{\frac{x_{n-1}+x_n}{2} - x_2} + \cdots + \frac{1}{\frac{x_{n-1}+x_n}{2} - x_{n-2}} > 0 + 0 + \cdots + 0 = 0,$$

another contradiction. The conclusion follows.

(T. Andreescu)

220. The condition $|F(a)| = |F(-a)|$ for all a is equivalent to $F^2(a) = F^2(-a)$ for all a, that is to $F^2(x) = F^2(-x)$. In other words

$$F^2(x) - F^2(-x) = (F(x) - F(-x))(F(x) + F(-x)) = 0.$$

For this to be identically equal to zero, one of the factors must be identically equal to zero. We conclude that $F(x)$ is mirror if and only if it has either just terms of even degree, or just terms of odd degree. So the mirror polynomials are precisely the polynomials of the form $G(x^2)$ or $xG(x^2)$, where $G(x)$ is a polynomial with real coefficients.

The next idea is that $P(x)$ and $Q(x)$ can be computed explicitly in terms of $F(x)$. First note that $P(x)$ is unique, because if $P_1(x) - P_1'(x) = P_2(x) - P_2'(x)$ then $P_1(x) - P_2(x) = P_1'(x) - P_2'(x)$, and a polynomial equals its derivative only if it is identically zero. The sum

$$P(x) = F(x) + F'(x) + F''(x) + \cdots$$

is finite, and

$$P(x) - P'(x) = (F(x) + F'(x) + F''(x) + \cdots) - (F'(x) + F''(x) + \cdots) = F(x).$$

So we have found $P(x)$. A similar argument shows that

$$Q(x) = F(x) - F'(x) + F''(x) - \cdots .$$

So

$$P(x) + Q(x) = F(x) + F''(x) + \cdots .$$

If $F(x)$ is a mirror polynomial, then the monomials appearing in $F(x)$ have only odd degree, and the same is true for its derivatives of even order. The same is true if $F(x)$ has only even degree terms. This implies that $P(x) + Q(x)$ is a mirror polynomial. For the converse, note that

$$F(x) = \frac{1}{2}[(P(x) + Q(x)) - (P(x) + Q(x))''],$$

and by the same reasoning $F(x)$ is a mirror polynomial.

(*Mathematical Reflections*, proposed by I. Boreico)

221. The equation $P(x) = 0$ is equivalent to the equation $f(x) = 1$, where

$$f(x) = \frac{a_1}{x} + \frac{a_2}{x^2} + \cdots + \frac{a_n}{x^n}.$$

Since f is strictly decreasing on $(0, \infty)$, $\lim_{x \to 0^+} f(x) = \infty$ and $\lim_{x \to \infty} f(x) = 0$, the equation has a unique solution.

Remark. A more general principle is true, namely that if the terms of the polynomial are written in decreasing order of their powers, then the number of sign changes of the coefficients is the maximum possible number of positive zeros; the actual number of positive zeros may differ from this by an even number.

222. Assume to the contrary that there is z with $|z| \geq 2$ such that $P(z) = 0$. Then by the triangle inequality,

$$0 = \left| \frac{P(z)}{z^7} \right| = \left| 1 + \frac{7}{z^3} + \frac{4}{z^6} + \frac{1}{z^7} \right| \geq 1 - \frac{7}{|z|^3} - \frac{4}{|z|^6} - \frac{1}{|z|^7}$$

$$\geq 1 - \frac{7}{8} - \frac{4}{64} - \frac{1}{128} = \frac{7}{128} > 0,$$

a contradiction. Hence our initial assumption was false, and therefore all the zeros of $P(z)$ lie inside the disk of radius 2 centered at the origin.

223. Let x_1, x_2 be the two roots of the quadratic equation. Then from $|p| + |q| < 1$ we obtain, using Viète's relations,

$$|x_1 + x_2| + |x_1 x_2| < 1.$$

From the triangle inequality we have

$$|x_1| - |x_2| < |x_1 + x_2|.$$

Combining the two inequalities we obtain

$$|x_1| - |x_2| + |x_1| \cdot |x_2| - 1 < 0.$$

Factoring we obtain

$$(|x_1| - 1)(|x_2| + 1) < 0.$$

Since the second factor is positive, the first factor is negative, so $|x_1| < 1$. By symmetry, $|x_2| < 1$ as well.

(Romanian Mathematics Competition, 1986, proposed by B. Enescu)

224. Let $P(x) = (x - 1)^r (x - 2)^s Q(x)$, where the roots $x_1, x_2, \ldots, x_k$ of $Q(x)$ are different from 1 and 2 (and of course lie in $(0, 3)$). Then $Q(0)Q(1)Q(2)Q(3)$ is a nonzero integer number, so the absolute value of this number is greater than or equal to 1. We thus have

$$1 \leq |Q(0)Q(1)Q(2)Q(3)| = \left| \prod_{j=1}^{k} x_j (1 - x_j)(2 - x_j)(3 - x_j) \right|$$

$$= \left| \prod_{j=1}^{k} (x_j^4 - 6x_j^3 + 11x_j^2 - 6x_j) \right|.$$

We have $t^4 - 6t^3 + 11t^2 - 6t = (t^2 - 3t + 1)^2 - 1$. Notice that the function $f(t) = (t^2 - 3t + 1)^2 - 1$ has the minimum at the zeros of $t^2 - 3t + 1$, which are $\frac{3 \pm \sqrt{5}}{2}$ and the maximum at the vertex of the parabola $t^2 - 3t + 1$, and this maximum is $9/16$. Hence the maximum of $|t^4 - 6t^3 + 11t^2 - 6t|$ on $(0, 3)$ is 1 and is attained precisely when $t = \frac{3 \pm \sqrt{5}}{2}$.

Thus $|Q(0)Q(1)Q(2)Q(3)|$ is strictly less than 1 unless all x_j are equal to $\frac{3 \pm \sqrt{5}}{2}$. The problem is solved.

(Test from the International Mathematical Olympiad training program of Brazil, 2013)

225. Let $z = r(\cos t + i \sin t)$, $\sin t \neq 0$. Using the de Moivre formula, the equality $z^n + az + 1 = 0$ translates to

$$r^n \cos nt + ar \cos t + 1 = 0,$$

$$r^n \sin nt + ar \sin t = 0.$$

View this as a system in the unknowns r^n and ar. Solving the system gives

$$r^n = \frac{\begin{vmatrix} -1 & \cos t \\ 0 & \sin t \end{vmatrix}}{\begin{vmatrix} \cos nt & \cos t \\ \sin nt & \sin t \end{vmatrix}} = \frac{\sin t}{\sin(n-1)t}.$$

An exercise in the section on induction shows that for any positive integer k, $|\sin kt| \leq k|\sin t|$. Then

$$r^n = \frac{\sin t}{\sin(n-1)t} \geq \frac{1}{n-1}.$$

This implies the desired inequality $|z| = r \geq \sqrt[n]{\frac{1}{n-1}}$.

(Romanian Mathematical Olympiad, proposed by I. Chiţescu)

226. By the theorem of Lucas, if the zeros of a polynomial lie in a closed convex domain, then the zeros of the derivative lie in the same domain. In our problem, change the variable to $z = \frac{1}{x}$ to obtain the polynomial $Q(z) = z^n + z^{n-1} + a$. If all the zeros of $ax^n + x + 1$ were outside of the circle of radius 2 centered at the origin, then the zeros of $Q(z)$ would lie in the interior of the circle of radius $\frac{1}{2}$. Applying the theorem of Lucas to the convex hull of these zeros, we deduce that the same would be true for the zeros of the derivative. But $Q'(z) = nz^{n-1} + (n-1)z^{n-2}$ has $z = \frac{n-1}{n} \geq \frac{1}{2}$ as one of its zeros, which is a contradiction. This implies that the initial polynomial has a root of absolute value less than or equal to 2.

227. The problem amounts to showing that the zeros of $Q(z) = zP'(z) - \frac{n}{2}P(z)$ lie on the unit circle. Let the zeros of $P(z)$ be $z_1, z_2, \ldots, z_n$, and let z be a zero of $Q(z)$. The relation $Q(z) = 0$ translates into

$$\frac{z}{z - z_1} + \frac{z}{z - z_2} + \cdots + \frac{z}{z - z_n} = \frac{n}{2},$$

or

$$\left(\frac{2z}{z - z_1} - 1 \right) + \left(\frac{2z}{z - z_2} - 1 \right) + \cdots + \left(\frac{2z}{z - z_n} - 1 \right) = 0,$$

and finally

$$\frac{z + z_1}{z - z_1} + \frac{z + z_2}{z - z_2} + \cdots + \frac{z + z_n}{z - z_n} = 0.$$

The terms of this sum should remind us of a fundamental transformation of the complex plane. This transformation is defined as follows: for a a complex number of absolute value 1, we let $\phi_a(z) = (z + a)/(z - a)$. The map ϕ_a has the important property that it maps the unit circle to the imaginary axis, the interior of the unit disk to the half-plane $\mathrm{Re}\, z < 0$, and the exterior of the unit disk to the half-plane $\mathrm{Re}\, z > 0$. Indeed, since the unit disk is invariant under rotation by the argument of a, it suffices to check this for $a = 1$. Then $\phi(e^{i\theta}) = -i \cot \frac{\theta}{2}$, which proves that the unit circle maps to the entire imaginary axis. The map is one-to-one, so the interior of the unit disk is mapped to that half-plane where the origin goes, namely to $\mathrm{Re}\, z < 0$, and the exterior is mapped to the other half-plane. If z has absolute value less than one, then all terms of the sum

$$\frac{z + z_1}{z - z_1} + \frac{z + z_2}{z - z_2} + \cdots + \frac{z + z_n}{z - z_n}$$

have negative real part, while if z has absolute value greater than 1, all terms in this sum have positive real part. In order for this sum to be equal to zero, z must have absolute value 1. This completes the proof.

An alternative approach to this last step was suggested by R. Stong. Taking the real part of

$$\frac{z + z_1}{z - z_1} + \frac{z + z_2}{z - z_2} + \cdots + \frac{z + z_n}{z - z_n} = 0,$$

we obtain

$$\sum_{j=1}^{n} \mathrm{Re}\left(\frac{z + z_j}{z - z_j}\right) = \sum_{j=1}^{n} \frac{1}{|z - z_j|^2} \mathrm{Re}\left((z + z_j)(\bar{z} - \bar{z}_j)\right) = \sum_{j=1}^{n} \frac{|z|^2 - |z_j|^2}{|z - z_j|^2}.$$

Since $|z_j| = 1$ for all j, we conclude that $|z| = 1$.

Remark. When $a = -i$, ϕ_a is called the Cayley transform.

228. Let the zeros of the polynomial be p, q, r, s. We have $p + q + r + s = 0$, $pq + pr + rs + qr + qs + rs = -2$, and hence $p^2 + q^2 + r^2 + s^2 = 0^2 - 2(-2) = 4$. By the Cauchy-Schwarz inequality, $(1 + 1 + 1)(q^2 + r^2 + s^2) \geq (q + r + s)^2$. Furthermore, because q, r, s must be distinct, the inequality is strict. Thus

$$4 = p^2 + q^2 + r^2 + s^2 > p^2 + \frac{(-p)^2}{3} = \frac{4p^2}{3},$$

or $|p| < \sqrt{3}$. The same argument holds for the other zeros.

(Hungarian Mathematical Olympiad, 1999)

229. We argue by induction on k. For $k = 1$ the property is obviously true.

Assume that the property is true for polynomials of degree $k - 1$ and let us prove it for the polynomials $P_n(z)$, $n \geq 1$, and $P(z)$ of degree k. Subtracting a constant from all polynomials, we may assume that $P(0) = 0$. Order the zeros of $P_n(z)$ such that $|z_1(n)| \leq |z_2(n)| \leq \cdots \leq |z_k(n)|$. The product $z_1(n)z_2(n) \cdots z_k(n)$, being the free term of $P_n(z)$, converges to 0. This can happen only if $z_1(n) \to 0$. So we have proved the property for one of the zeros.

In general, the polynomial obtained by dividing a monic polynomial $Q(z)$ by $z - a$ depends continuously on a and on the coefficients of $Q(z)$. This means that the coefficients of $P_n(z)/(z - z_1(n))$ converge to the coefficients of $P(z)/z$, so we can apply the induction hypothesis to these polynomials. The conclusion follows.

Remark. A stronger result is true, namely that if the coefficients of a monic polynomial are continuous functions of a parameter t, then the zeros are also continuous functions of t.

230. The hypothesis of the problem concerns the coefficients a_m and a_0, and the conclusion is about a zero of the polynomial. It is natural to write the Viète relations for the two coefficients,

$$\frac{a_m}{a_n} = (-1)^m \sum x_1 x_2 \cdots x_m,$$

$$\frac{a_0}{a_n} = (-1)^n x_1 x_2 \cdots x_n.$$

Dividing, we obtain

$$\left| \sum \frac{1}{x_1 x_2 \cdots x_m} \right| = \left| \frac{a_m}{a_0} \right| > \binom{n}{m}.$$

An application of the triangle inequality yields

$$\sum \frac{1}{|x_1||x_2| \cdots |x_m|} > \binom{n}{m}.$$

Of the absolute values of the zeros, let α be the smallest. If we substitute all absolute values in the above inequality by α, we obtain an even bigger left-hand side. Therefore,

$$\binom{n}{m} \frac{1}{\alpha^{n-m}} > \binom{n}{m}.$$

It follows that $\alpha < 1$, and hence the corresponding zero has absolute value less than 1, as desired.

(*Revista Matematică din Timişoara* (*Timişoara Mathematics Gazette*), proposed by T. Andreescu)

231. Let

$$f(x) = \frac{P'(x)}{P(x)} = \frac{1}{x - x_1} + \frac{1}{x - x_2} + \cdots + \frac{1}{x - x_n}.$$

First, note that from Rolle's theorem applied to $\phi(x) = e^{-kx} f(x)$ it follows that all roots of the polynomial $P'(x) - kP(x)$ are real. We need the following lemma.

Lemma. *If for some j, y_0 and y_1 satisfy $y_0 < x_j < y_1 \leq y_0 + \delta(P)$, then y_0 and y_1 are not zeros of f and $f(y_0) < f(y_1)$.*

Proof. Let $d = \delta(P)$. The hypothesis implies that for all i, $y_1 - y_0 \leq d \leq x_{i+1} - x_i$. Hence for $1 \leq i \leq j - 1$ we have $y_0 - x_i \geq y_1 - x_{i+1} > 0$, and so $1/(y_0 - x_i) \leq 1/(y_1 - x_{i+1})$; similarly, for $j \leq i \leq n - 1$ we have $y_1 - x_{i+1} \leq y_0 - x_i < 0$ and again $1/(y_0 - x_i) \leq 1/(y_1 - x_{i+1})$.

Finally, $y_0 - x_n < 0 < y_1 - x_1$, so $1/(y_0 - x_n) < 0 < 1/(y_1 - x_1)$, and the result follows by addition of these inequalities. $\qquad\square$

Returning to the problem, we see that if y_0 and y_1 are zeros of $P'(x) - kP(x)$ with $y_0 < y_1$, then they are separated by a zero of P and satisfy $f(y_0) = f(y_1) = k$. From the lemma it follows that we cannot have $y_1 \leq y_0 + \delta(P(x))$, so $y_1 - y_0 > d$, and we are done.

(*American Mathematical Monthly*, published in a note by P. Walker, solution by R. Gelca)

232. The number 101 is prime, yet we cannot apply Eisenstein's criterion because of the 102. The trick is to observe that the irreducibility of $P(x)$ s equivalent to the irreducibility of $P(x - 1)$. Because the binomial coefficients $\binom{101}{k}$, $1 \leq k \leq 100$, are all divisible by 101, the polynomial $P(x - 1)$ has all coefficients but the first divisible by 101, while the last coefficient is $(-1)^{101} + 101(-1)^{101} + 102 = 202$, which is divisible by 101 but not by 101^2. Eisenstein's criterion proves that $P(x - 1)$ is irreducible; hence $P(x)$ is irreducible as well.

233. Note that $P(x) = (x^p - 1)/(x - 1)$. If $P(x)$ were reducible, then so would be $P(x + 1)$. But

$$P(x + 1) = \frac{(x + 1)^p - 1}{x} = x^{p-1} + \binom{p}{1}x^{p-1} + \cdots + \binom{p}{p-1}.$$

The coefficient $\binom{p}{k}$ is divisible by p for all $1 \leq k \leq p - 1$, and $\binom{p}{p-1} = p$ is not divisible by p^2; thus Eisenstein's criterion applies to show that $P(x + 1)$ is irreducible. It follows that $P(x)$ itself is irreducible, and the problem is solved.

234. Same idea as in the previous problem. We look at the polynomial

$$P(x + 1) = (x + 1)^{2^n} + 1 = x^{2^n} + \binom{2^n}{1}x^{2^n-1} + \binom{2^n}{2}x^{2^{n-1}-2} + \cdots + \binom{2^n}{2^n - 1}x + 2.$$

For $1 \leq k \leq 2^n$, the binomial coefficient $\binom{2^n}{k}$ is divisible by 2. This follows from the equality

$$\binom{2^n}{k} = \frac{2^n}{k}\binom{2^n - 1}{k - 1},$$

since the binomial coefficient on the right is an integer, and 2 appears to a larger power in the numerator than in the denominator. The application of Eisenstein's irreducibility criterion is now straightforward.

235. Arguing by contradiction, assume that $P(x)$ can be factored, and let $P(x) = Q(x)R(x)$. Because $P(a_i) = -1$, $i = 1, 2, \ldots, n$, and $Q(a_i)$ and $R(a_i)$ are integers, either $Q(a_i) = 1$ and $R(a_i) = -1$, or $Q(a_i) = -1$ and $R(a_i) = 1$. In both situations $(Q + R)(a_i) = 0$, $i = 1, 2, \ldots, n$. Since the a_i's are all distinct and the degree of $Q(x) + R(x)$ is at most $n - 1$, it follows that $Q(x) + R(x) \equiv 0$. Hence $R(x) = -Q(x)$, and $P(x) = -Q^2(x)$. But this contradicts the fact that the coefficient of the term of maximal degree in $P(x)$ is 1. The contradiction proves that $P(x)$ is irreducible.

(I. Schur)

236. Assume that the polynomial $P(x)$ is reducible, and write it as a product $Q(x)R(x)$ of monic polynomials with integer coefficients of degree i, respectively, $2n - i$. Both $Q(x)$ and $R(x)$ are positive for any real number x (being monic and with no real zeros), and from $Q(a_k)R(a_k) = 1, k = 1, 2, \ldots, n$, we find $Q(a_k) = R(a_k) = 1, k = 1, 2, \ldots, n$. If, say, $i < n$, then the equation $Q(x) = 1$ has n solutions, which, taking into account the fact that $Q(x)$ has degree less than n, means that $Q(x)$ is identically equal to 1. This contradicts our original assumption. Also, if $i = n$, the polynomial $Q(x) - R(x)$ has n zeros, and has degree less than n, so it is identically equal to 0. Therefore, $Q(x) = R(x)$, which means that

$$(x - a_1)^2 (x - a_2)^2 \cdots (x - a_n)^2 + 1 = Q(x)^2.$$

Substituting integer numbers for x, we obtain infinitely many equalities of the form $p^2 + 1 = q^2$, with p and q integers. But this equality can hold only if $p = 0$ and $q = 1$, and we reach another contradiction. Therefore, the polynomial is irreducible.

(I. Schur)

237. Let $P(x) = a_n x^n + a_{n-1} x^{n-1} + \cdots + a_0$, and assume to the contrary that $P(x) = Q(x)R(x)$, where $Q(x)$ and $R(x)$ are polynomials with integer coefficients of degree at least 1 (the degree zero is ruled out because any factor that divides all coefficients of $P(x)$ divides the original prime).

Because the coefficients of $P(x)$ are nonnegative integers between 0 and 9, and the leading coefficient is positive, it follows that the zeros of $P(x)$ are in the union of the left half-plane $\mathrm{Re}\, z \leq 0$ and the disk $|z| < 4$. Otherwise, if $\mathrm{Re}\, z > 0$ and $|z| \geq 4$, then

$$1 \leq a_n \leq \mathrm{Re}\, (a_n + a_{n-1} z^{-1}) = \mathrm{Re}\, (-a_{n-2} z^{-2} - \cdots - a_0 z^{-n})$$

$$\leq 9(|z|^{-2} + |z|^{-3} + \cdots + |z|^{-n}) < \frac{9|z|^{-2}}{1 - |z|^{-1}} \leq \frac{3}{4},$$

a contradiction.

On the other hand, by hypothesis $P(10)$ is prime; hence either $Q(10)$ or $R(10)$ is 1 (or -1 but then just multiply both polynomials by -1). Assume $Q(10) = 1$, and let $Q(x) = c(x - x_1)(x - x_2) \cdots (x - x_k)$. Then $x_i, i = 1, 2, \ldots, k$, are also zeros of $P(x)$, and we have seen that these lie either in the left half-plane or in the disk of radius 4 centered at the origin. It follows that

$$1 = Q(10) = |Q(10)| = |c| \cdot |10 - x_1| \cdot |10 - x_2| \cdots |10 - x_k| \geq |c| \cdot 6^k,$$

a contradiction. We conclude that P(x) is irreducible.

238. Assume the contrary, and let

$$(x^2 + 1)^n + p = Q(x)R(x),$$

with $Q(x)$ and $R(x)$ of degree at least 1. Denote by $\widehat{Q}(x), \widehat{R}(x)$ the reduction of these polynomials modulo p, viewed as polynomials in $\mathbb{Z}_p[x]$. Then $\widehat{Q}(x)\widehat{R}(x) = (x^2 + 1)^n$. The polynomial $x^2 + 1$ is irreducible in $\mathbb{Z}_p[x]$, since -1 is not a quadratic residue in $\mathbb{Z}_p$. This implies $\widehat{Q}(x) = (x^2 + 1)^k$ and $\widehat{R}(x) = (x^2 + 1)^{n-k}$, with $1 \leq k \leq n - 1$ (the polynomials are monic and their degree is at least 1). It follows that there exist polynomials $Q_1(x)$ and $R_1(x)$ with integer coefficients such that

$$Q(x) = (x^2 + 1)^k + pQ_1(x) \quad \text{and} \quad R(x) = (x^2 + 1)^{n-k} + pR_1(x).$$

Multiplying the two, we obtain

$$(x^2 + 1)^n + p = (x^2 + 1)^n + p((x^2 + 1)^{n-k}Q_1(x) + (x^2 + 1)^k R_1(x)) + p^2 Q_1(x)R_1(x).$$

Therefore,

$$(x^2 + 1)^{n-k}Q_1(x) + (x^2 + 1)^k R_1(x) + pQ_1(x)R_1(x) = 1.$$

Reducing modulo p we see that $x^2 + 1$ divides 1 in $\mathbb{Z}_p[x]$, which is absurd. The contradiction proves that the polynomial from the statement is irreducible.

239. We will show that all the zeros of $P(x)$ have absolute value greater than 1. Let y be a complex zero of $P(x)$. Then

$$0 = (y - 1)P(y) = y^p + y^{p-1} + y^{p-2} + \cdots + y - p.$$

Assuming $|y| \leq 1$, we obtain

$$p = |y^p + y^{p-1} + y^{p-2} + \cdots + y| \leq \sum_{i=1}^{p} |y|^i \leq \sum_{i=1}^{p} 1 = p.$$

This can happen only if the two inequalities are, in fact, equalities, in which case $y = 1$. But $P(1) > 0$, a contradiction that proves our claim.

Next, let us assume that $P(x) = Q(x)R(x)$ with $Q(x)$ and $R(x)$ polynomials with integer coefficients of degree at least 1. Then $p = P(0) = Q(0)R(0)$. Since both $Q(0)$ and $R(0)$ are integers, either $Q(0) = \pm 1$ or $R(0) = \pm 1$. Without loss of generality, we may assume $Q(0) = \pm 1$. This, however, is impossible, since all zeros of $Q(x)$, which are also zeros of $P(x)$, have absolute value greater than 1. We conclude that $P(x)$ is irreducible.

(Proposed by M. Manea for *Mathematics Magazine*)

240. Let n be the degree of $P(x)$. Suppose that we can find polynomials with integer coefficients $R_1(x)$ and $R_2(x)$ of degree at most $2n - 1$ such that $Q(x) = P(x^2) = R_1(x)R_2(x)$. Then we also have $Q(x) = Q(-x) = R_1(-x)R_2(-x)$. Let $F(x)$ be the greatest common divisor of $R_1(x)$ and $R_1(-x)$. Since $F(x) = F(-x)$, we can write $F(x) = G(x^2)$ with the degree of $G(x)$ at most $n - 1$. Since $G(x^2)$ divides $Q(x) = P(x^2)$, we see that $G(x)$ divides $P(x)$ and has lower degree; hence by the irreducibility of $P(x)$, $G(x)$ is constant. Similarly, the greatest common divisor of $R_2(x)$ and $R_2(-x)$ is constant. Hence $R_1(-x)$ divides $R_2(x)$, while $R_2(x)$ divides $R_1(-x)$. It follows that $R_1(x)$ and $R_2(x)$ both have degree n, $R_2(x) = cR_1(-x)$, and $Q(x) = cR_1(x)R_2(x)$. Because $P(x)$ is monic, we compute $c = (-1)^n$ and $P(0) = (-1)^n R_1(0)^2$. Hence $|P(0)|$ is a square, contradicting the hypothesis.

(Romanian Team Selection Test for the International Mathematical Olympiad, 2003, proposed by M. Piticari)

241. These are just direct consequences of the trigonometric identities

$$\cos(n + 1)\theta = \cos\theta \cos n\theta - \sin\theta \sin n\theta$$

and

$$\frac{\sin(n+1)\theta}{\sin\theta} = \cos\theta\frac{\sin n\theta}{\sin\theta} + \cos n\theta.$$

242. Denote the second determinant by D_n. Expanding by the first row, we obtain

$$D_n = 2xD_{n-1} - \begin{vmatrix} 1 & 1 & 0 & \cdots & 0 \\ 0 & 2x & 1 & \cdots & 0 \\ 0 & 1 & 2x & \cdots & 0 \\ \vdots & \vdots & \vdots & \ddots & \vdots \\ 0 & 0 & 0 & \cdots & 2x \end{vmatrix} = 2xD_{n-1} - D_{n-2}.$$

Since $D_1 = 2x$ and $D_2 = 4x^2 - 1$, we obtain inductively $D_n = U_n(x)$, $n \geq 1$. The same idea works for the first determinant, except that we expand it by the last row. With the same recurrence relation and with the values x for $n = 1$ and $2x^2 - 1$ for $n = 2$, the determinant is equal to $T_n(x)$ for all n.

243. Let $P(x) = x^4 + ax^3 + bx^2 + cx + d$ and denote by M the maximum of $|P(x)|$ on $[-1, 1]$. From $-M \leq P(x) \leq M$, we obtain the necessary condition $-M \leq \frac{1}{2}(P(x) + P(-x)) \leq M$ for $x \in [-1, 1]$. With the substitution $y = x^2$, this translates into

$$-M \leq y^2 + by + d \leq M, \text{ for } y \in [0, 1].$$

For a monic quadratic function to have the smallest variation away from 0 on $[0, 1]$, it needs to have the vertex (minimum) at $\frac{1}{2}$. The variation is minimized by $\left(y - \frac{1}{2}\right)^2 - \frac{1}{8}$, and so we obtain $M \geq \frac{1}{8}$. Equality is attained for $\frac{1}{8}T_4(x)$.

Now let us assume that $P(x)$ is a polynomial for which $M = \frac{1}{8}$. Then $b = -1$, $d = \frac{1}{8}$. Writing the double inequality $-\frac{1}{8} \leq P(x) \leq \frac{1}{8}$ for $x = 1$ and -1, we obtain $-\frac{1}{8} \leq \frac{1}{8} + a + c \leq \frac{1}{8}$ and $-\frac{1}{8} \leq \frac{1}{8} - a - c \leq \frac{1}{8}$. So on the one hand, $a + c \geq 0$, and on the other hand, $a + c \leq 0$. It follows that $a = -c$. But then for $x = \frac{1}{\sqrt{2}}$, $0 \leq a\left(\frac{1}{2\sqrt{2}} - \frac{1}{\sqrt{2}}\right) \leq \frac{1}{4}$, and for $x = -\frac{1}{\sqrt{2}}$, $0 \leq -a\left(\frac{1}{2\sqrt{2}} - \frac{1}{\sqrt{2}}\right) \leq \frac{1}{4}$. This can happen only if $a = 0$. Therefore,

$$P(x) = x^4 - x^2 + \frac{1}{8} = \frac{1}{8}T_4(x).$$

244. From the identity

$$x^3 + \frac{1}{x^3} = \left(x + \frac{1}{x}\right)^3 - 3\left(x + \frac{1}{x}\right),$$

it follows that

$$\sqrt{r} + \frac{1}{\sqrt{r}} = 6^3 - 3 \times 6 = 198.$$

Hence

$$\left(\sqrt[4]{r} - \frac{1}{\sqrt[4]{r}}\right)^2 = 198 - 2,$$

and the maximum value of $\sqrt[4]{r} - \frac{1}{\sqrt[4]{r}}$ is 14.

(University of Wisconsin at Whitewater Math Meet, 2003, proposed by T. Andreescu)

245. Let $x_1 = 2\cos\alpha, x_2 = 2\cos 2\alpha, \ldots, x_n = 2\cos n\alpha$. We are to show that the determinant

$$\begin{vmatrix} T_0(x_1) & T_0(x_2) & \cdots & T_0(x_n) \\ T_1(x_1) & T_1(x_2) & \cdots & T_1(x_n) \\ \vdots & \vdots & \ddots & \vdots \\ T_{n-1}(x_1) & T_{n-1}(x_2) & \cdots & T_{n-1}(x_n) \end{vmatrix}$$

is nonzero. Substituting $T_0(x_i) = 1$, $T_1(x_i) = x$, $i = 1, 2, \ldots, n$, and performing row operations to eliminate powers of x_i, we can transform the determinant into

$$2 \cdot 4 \cdots 2^{n-1} \begin{vmatrix} 1 & 1 & \cdots & 1 \\ x_1 & x_2 & \cdots & x_n \\ \vdots & \vdots & \ddots & \vdots \\ x_1^{n-1} & x_2^{n-1} & \cdots & x_n^{n-1} \end{vmatrix}.$$

This is a Vandermonde determinant, and the latter is not zero since $x_i \neq x_j$, for $1 \leq i < j \leq n$, whence the original matrix is invertible. Its determinant is equal to

$$2^{(n-1)(n-2)/2} \prod_{1 \leq i < j \leq n} (\cos j\alpha - \cos i\alpha) \neq 0.$$

246. Because the five numbers lie in the interval $[-2, 2]$ we can find corresponding angles $t_1, t_2, t_3, t_4, t_5 \in [0, \pi]$ such that $x = 2\cos t_1$, $y = 2\cos t_2$, $z = 2\cos t_3$, $v = 2\cos t_4$, and $w = 2\cos t_5$. We would like to translate the third and fifth powers into trigonometric functions of multiples of the angles. For that we use the polynomials $T_n(a)$. For example, $T_5(a) = a^5 - 5a^3 + 5a$. This translates into the trigonometric identity $2\cos 5\theta = (2\cos\theta)^5 - 5(2\cos\theta)^3 + 5(2\cos\theta)$.

Add to the third equation of the system the first multiplied by 5 and the second multiplied by -5, then use the above-mentioned trigonometric identity to obtain

$$2\cos 5t_1 + 2\cos 5t_2 + 2\cos 5t_3 + 2\cos 5t_4 + 2\cos 5t_5 = -10.$$

This can happen only if $\cos 5t_1 = \cos 5t_2 = \cos 5t_3 = \cos 5t_4 = \cos 5t_5 = -1$. Hence

$$t_1, t_2, t_3, t_4, t_5 \in \left\{\frac{\pi}{5}, \frac{3\pi}{5}, \frac{5\pi}{5}\right\}.$$

Using the fact that the roots of $x^5 = 1$, respectively, $x^{10} = 1$, add up to zero, we deduce that

$$\sum_{k=0}^{4} \cos\frac{2k\pi}{5} = 0 \quad \text{and} \quad \sum_{k=0}^{9} \cos\frac{k\pi}{5} = 0.$$

It follows that

$$\cos \frac{\pi}{5} + \cos \frac{3\pi}{5} + \cos \frac{5\pi}{5} + \cos \frac{7\pi}{5} + \cos \frac{9\pi}{5} = 0.$$

Since $\cos \frac{\pi}{5} = \cos \frac{9\pi}{5}$ and $\cos \frac{3\pi}{5} = \cos \frac{7\pi}{5}$, we find that $\cos \frac{\pi}{5} + \cos \frac{3\pi}{5} = \frac{1}{2}$. Also, it is not hard to see that the equation $T_5(a) = -2$ has no rational solutions, which implies that $\cos \frac{\pi}{5}$ is irrational.

The first equation of the system yields $\sum_{i=1}^{5} t_i = 0$, and the above considerations show that this can happen only when two of the t_i are equal to $\frac{\pi}{5}$, two are equal to $\frac{3\pi}{5}$, and one is equal to π. Let us show that in this situation the second equation is also satisfied. Using $T_3(a) = a^3 - 3a$, we see that the first two equations are jointly equivalent to $\sum_{k=1}^{5} \cos t_i = 0$ and $\sum_{k=1}^{5} \cos 3t_i = 0$. Thus we are left to check that this last equality is satisfied. We have

$$2\cos \frac{3\pi}{5} + 2\cos \frac{9\pi}{5} = 2\cos \frac{3\pi}{5} + 2\cos \frac{\pi}{5} + \cos \pi = 0,$$

as desired. We conclude that up to permutations, the solution to the system is

$$\left(2\cos \frac{\pi}{5}, 2\cos \frac{\pi}{5}, 2\cos \frac{3\pi}{5}, 2\cos \frac{3\pi}{5}, 2\cos \pi \right).$$

(Romanian Mathematical Olympiad, 2002, proposed by T. Andreescu)

247. The Lagrange interpolation formula applied to the Chebyshev polynomial $T_{n-1}(x)$ and to the points $x_1, x_2, \ldots, x_n$ gives

$$T_{n-1}(x) = \sum_{k=1}^{n} T_{n-1}(x_k) \frac{(x - x_1) \cdots (x - x_{k-1})(x - x_{k+1}) \cdots (x - x_n)}{(x_k - x_1) \cdots (x_k - x_{k-1})(x_k - x_{k+1}) \cdots (x_k - x_n)}.$$

Equating the leading coefficients on both sides, we obtain

$$2^{n-2} = \sum_{k=1}^{n} \frac{T_{n-1}(x_k)}{(x_k - x_1) \cdots (x_k - x_{k-1})(x_k - x_{k+1}) \cdots (x_k - x_n)}.$$

We know that the maximal variation away from 0 of $T_{n-1}(x)$ is 1; in particular, $|T_{n-1}(x_k)| \leq 1$, $k = 1, 2, \ldots, n$. Applying the triangle inequality, we obtain

$$2^{n-2} \leq \sum_{k=1}^{n} \frac{|T_{n-1}(x_k)|}{|x_k - x_1| \cdots |x_k - x_{k-1}||x_k - x_{k+1}| \cdots |x_k - x_n|} \leq \sum_{k=1}^{n} \frac{1}{t_k}.$$

The inequality is proved.

(T. Andreescu, Z. Feng, 103 *Trigonometry Problems*, Birkhäuser, 2004)

248. We will prove that

$$\sum_{k=1}^{\frac{n-1}{2}} \sec \frac{2k\pi}{n} = \begin{cases} \frac{n-1}{2}, & \text{if } n \equiv 1 \pmod 4, \\ -\frac{n+1}{2}, & \text{if } n \equiv 3 \pmod 4. \end{cases}$$

To prove this, we use the Chebyshev polynomial of the second kind. From $U_{n-1}(\cos\theta) = \frac{\sin n\theta}{\sin\theta}$, it is not hard to guess that the n roots of $U_{n-1}(x)$ are $\cos\frac{k\pi}{n}$, $1 \le k \le n-1$. In fact

$$U_{n-1}(x) = 2^{n-1} \prod_{k=1}^{n-1}\left(x - \cos\frac{k\pi}{n}\right).$$

We have

$$\frac{U'_{n-1}(x)}{U_{n-1}(x)} = \sum_{k=1}^{n-1} \frac{1}{x - \cos\frac{k\pi}{n}}.$$

Since $\cos\frac{k\pi}{n} = \cos\frac{(n-k)\pi}{n}$, we can further write this as

$$\frac{1}{2}\sum_{k=1}^{n-1}\left(\frac{1}{x - \cos\frac{k\pi}{n}} + \frac{1}{x + \cos\frac{k\pi}{n}}\right) = \sum_{k=1}^{n-1}\frac{x}{x^2 - \cos^2\frac{k\pi}{n}} = \sum_{k=1}^{n-1}\frac{2x}{2x^2 - 1 - \cos\frac{2k\pi}{n}}.$$

Substituting $x = \cos\theta$, we obtain

$$\frac{U'_{n-1}(\cos\theta)}{U_{n-1}(\cos\theta)} = \sum_{k=1}^{n-1}\frac{2\cos\theta}{\cos 2\theta - \cos\frac{2k\pi}{n}}.$$

But we also have

$$(-\sin\theta)\frac{U'_{n-1}(\cos\theta)}{U_{n-1}(\cos\theta)} = \frac{(U_{n-1}(\cos\theta))'}{U_{n-1}(\cos\theta)} = \frac{\left(\frac{\sin n\theta}{\sin\theta}\right)'}{\frac{\sin n\theta}{\sin\theta}} = n\cot n\theta - \cot\theta.$$

Therefore

$$\sum_{k=1}^{n-1}\frac{1}{\cos 2\theta - \cos\frac{2k\pi}{n}} = \frac{1}{2\sin^2\theta} - \frac{n\cot n\theta}{\sin 2\theta}.$$

This is equivalent for n odd to

$$\sum_{k=1}^{\frac{n-1}{2}}\frac{1}{\cos\frac{2k\pi}{n} - \cos 2\theta} = \frac{n\cot n\theta}{2\sin 2\theta} - \frac{1}{4\sin^2\theta}.$$

Taking $\theta = \frac{\pi}{4}$, we obtain the desired identity.

   (*Mathematical Reflections*, proposed by T. Andreescu)

249. Let us try to prove the first identity. Viewing both sides of the identity as sequences in n, we will show that they satisfy the same recurrence relation and the same initial condition. For the left-hand side the recurrence relation is, of course,

$$\frac{T_{n+1}(x)}{\sqrt{1-x^2}} = 2x\frac{T_n(x)}{\sqrt{1-x^2}} - \frac{T_{n+1}(x)}{\sqrt{1-x^2}},$$

and the initial condition is $T_1(x)/\sqrt{1-x^2} = x/\sqrt{1-x^2}$. It is an exercise to check that the right-hand side satisfies the same initial condition. As for the recurrence relation, we compute

$$\frac{d^{n+1}}{dx^{n+1}}(1-x^2)^{n+1-\frac{1}{2}} = \frac{d^n}{dx^n}\frac{d}{dx}(1-x^2)^{n+1-\frac{1}{2}}$$

$$= \frac{d^n}{dx^n}\left(n+1-\frac{1}{2}\right)(1-x^2)^{n-\frac{1}{2}}(-2x)$$

Here we apply the Leibniz rule for the differentiation of a product to obtain

$$-(2n+1)x\frac{d^n}{dx^n}(1-x^2)^{n-\frac{1}{2}} - (2n+1)\binom{n}{1}\left(\frac{d}{dx}x\right)\frac{d^{n-1}}{dx^{n-1}}(1-x^2)^{n-\frac{1}{2}}$$

$$= -(2n+1)x\frac{d^n}{dx^n}(1-x^2)^{n-\frac{1}{2}} - n(2n+1)\frac{d^{n-1}}{dx^{n-1}}(1-x^2)^{n-\frac{1}{2}}.$$

So if $t_n(x)$ denotes the right-hand side, then

$$t_{n+1}(x) = xt_n(x) - \frac{(-1)^{n-1}n}{1\cdot3\cdots(2n-1)}\frac{d^{n-1}}{dx^{n-1}}(1-x^2)^{n-1+\frac{1}{2}}.$$

Look at the second identity from the statement! If it were true, then the last term would be equal to $\sqrt{1-x^2}U_{n-1}(x)$. This suggests a simultaneous proof by induction. Call the right-hand side of the second identity $u_n(x)$.

We will prove by induction on n that $t_n(x) = T_n(x)/\sqrt{1-x^2}$ and $u_{n-1}(x) = \sqrt{1-x^2}U_{n-1}(2x)$. Let us assume that this holds true for all $k < n$. Using the induction hypothesis, we have

$$t_n(x) = x\frac{T_{n-1}(x)}{\sqrt{1-x^2}} - \sqrt{1-x^2}U_{n-2}(x).$$

Using the first of the two identities proved in the first problem of this section, we obtain

$$t_n(x) = \frac{T_n(x)}{\sqrt{1-x^2}}.$$

For the second half of the problem we show that $\sqrt{1-x^2}U_{n-1}(x)$ and $u_{n-1}(x)$ are equal by verifying that their derivatives are equal, and that they are equal at $x = 1$. The latter is easy to check: when $x = 1$ both are equal to 0. The derivative of the first is

$$\frac{-x}{\sqrt{1-x^2}}U_{n-1}(x) + 2\sqrt{1-x^2}U'_{n-1}(x).$$

Using the inductive hypothesis, we obtain $u'_{n-1}(x) = -nT_n(x)/\sqrt{1-x^2}$. Thus we are left to prove that

$$-xU_{n-1}(x) + 2(1-x^2)U'_{n-1}(x) = -nT_n(x),$$

which translates to

$$-\cos x \frac{\sin nx}{\sin x} + 2\sin^2 x \frac{n\cos nx \sin x - \cos x \sin nx}{\sin^2 x} \cdot \frac{1}{\sin x} = n\cos nx.$$

This is straightforward, and the induction is complete.

Remark. These are called the formulas of Rodrigues.

250. If $M = A + iB$, then $\overline{M^t} = \overline{A^t} - i\overline{B^t} = A - iB$. So we should take

$$A = \frac{1}{2}(M + \overline{M^t}) \quad \text{and} \quad B = \frac{1}{2i}(M - \overline{M^t}),$$

which are of course both Hermitian.

Remark. This decomposition plays a special role, especially for linear operators on infinite-dimensional spaces. If A and B commute, then M is called normal.

251. The answer is negative. The trace of $AB - BA$ is zero, while the trace of $\mathcal{I}_n$ is n; the matrices cannot be equal.

Remark. The equality cannot hold even for continuous linear transformations on an infinite-dimensional vector space. If P and Q are the linear maps that describe the momentum and the position in Heisenberg's matrix model of quantum mechanics, and if $\hbar$ is reduced Planck's constant, then the equality $PQ - QP = -\hbar\mathcal{I}$ is the canonical commutation relation from which the Heisenberg's uncertainty principle. We now see that the position and the momentum cannot be modeled using finite-dimensional matrices (not even infinite-dimensional continuous linear transformations). Note on the other hand that the matrices whose entries are residue classes in $\mathbb{Z}_4$,

$$A = \begin{pmatrix} 0 & 1 & 0 & 0 \\ 0 & 0 & 1 & 0 \\ 0 & 0 & 0 & 1 \\ 0 & 0 & 0 & 0 \end{pmatrix} \quad \text{and} \quad B = \begin{pmatrix} 0 & 0 & 0 & 0 \\ 1 & 0 & 0 & 0 \\ 0 & 2 & 0 & 0 \\ 0 & 0 & 3 & 0 \end{pmatrix},$$

satisfy $AB - BA = \mathcal{I}_4$, as matrices in $\mathbb{Z}_4$.

252. To simplify our work, we note that in general, for any two square matrices A and B of arbitrary dimension, the trace of $AB - BA$ is zero. We can therefore write

$$AB - BA = \begin{pmatrix} a & b \\ c & -a \end{pmatrix}.$$

But then $(AB - BA)^2 = k\mathcal{I}_2$, where $k = a^2 + bc$. This immediately shows that an odd power of $AB - BA$ is equal to a multiple of this matrix. The odd power cannot equal $\mathcal{I}_2$ since it has trace zero. Therefore, n is even.

The condition from the statement implies that k is a root of unity. But there are only two real roots of unity and these are 1 and -1. The squares of both are equal to 1. It follows that $(AB - BA)^4 = k^2 \mathcal{I}_2 = \mathcal{I}_2$, and the problem is solved.

(*Revista Matematică din Timişoara* (*Timişoara Mathematics Gazette*), proposed by T. Andreescu)

253. Assume that $p \neq q$. The second relation yields $A^2 B^2 = B^2 A^2 = rA^4$ and $rB^2 A = rAB^2 = A^3$. Multiplying the relation $pAB + qBA = \mathcal{I}_n$ on the right and then on the left by B, we obtain

$$pBAB - qB^2 A = B \quad \text{and} \quad pAB^2 = qBAB = B.$$

From these two identities and the fact that $B^2 A = AB^2$ and $p \neq q$ we deduce $BAB = AB^2 = B^2 A$. Therefore, $(p+q)AB^2 = (p+q)B^2 A = B$. This implies right away that $(p+q)A^2 B^2 = AB$ and $(p+q)B^2 A^2 = BA$. We have seen that A^2 and B^2 commute, and so we find that A and B commute as well, which contradicts the hypothesis. Therefore, $p = q$.

(V. Vornicu)

254. For any number t,

$$\begin{pmatrix} 1 & t \\ 0 & 1 \end{pmatrix}\begin{pmatrix} 1 & -t \\ 0 & 1 \end{pmatrix} = \begin{pmatrix} 1 & -t \\ 0 & 1 \end{pmatrix}\begin{pmatrix} 1 & t \\ 0 & 1 \end{pmatrix} = \begin{pmatrix} 1 & 0 \\ 0 & 1 \end{pmatrix}.$$

The equality from the statement can be rewritten

$$\begin{pmatrix} 1 & u \\ 0 & 1 \end{pmatrix}\begin{pmatrix} a & b \\ c & d \end{pmatrix}\begin{pmatrix} 1 & v \\ 0 & 1 \end{pmatrix} = \begin{pmatrix} 1 & 0 \\ c & 1 \end{pmatrix}.$$

This translates to

$$\begin{pmatrix} a + uc & v(a + uc) + b + ud \\ c & cv + d \end{pmatrix} = \begin{pmatrix} 1 & 0 \\ c & 1 \end{pmatrix}.$$

Because $c \neq 0$ we can choose u such that $a + uc = 1$. Then choose $v = -(b + ud)$. The resulting matrix has 1 in the upper left corner and 0 in the upper right corner. In the lower right corner it has

$$cv + d = c(b + ud) + d = -bc - cud + d = 1 - ad - ucd + d$$
$$= 1 - (a + uc)d + d = 1.$$

This also follows from the fact that the determinant of the matrix is 1. The numbers u and v that we have constructed satisfy the required identity.

Remark. This factorization appears in Gaussian optics. The matrices

$$\begin{pmatrix} 1 & \pm u \\ 0 & 1 \end{pmatrix} \quad \text{and} \quad \begin{pmatrix} 1 & \pm v \\ 0 & 1 \end{pmatrix}$$

model a ray of light that travels on a straight line through a homogeneous medium, while the matrix

$$\begin{pmatrix} 1 & 0 \\ c & 1 \end{pmatrix}$$

models refraction between two regions of different refracting indices. The result we have just proved shows that any $SL(2, \mathbb{R})$ matrix with nonzero lower left corner is an optical matrix.

255. *First solution*: Computed by hand, the second, third, and fourth powers of $J_4(\lambda)$ are

$$\begin{pmatrix} \lambda^2 & 2\lambda & 1 & 0 \\ 0 & \lambda^2 & 2\lambda & 1 \\ 0 & 0 & \lambda^2 & 2\lambda \\ 0 & 0 & 0 & \lambda^2 \end{pmatrix}, \begin{pmatrix} \lambda^3 & 3\lambda^2 & 3\lambda & 1 \\ 0 & \lambda^3 & 3\lambda^2 & 3\lambda \\ 0 & 0 & \lambda^3 & 3\lambda^2 \\ 0 & 0 & 0 & \lambda^3 \end{pmatrix}, \begin{pmatrix} \lambda^4 & 4\lambda^3 & 6\lambda^2 & 4\lambda \\ 0 & \lambda^4 & 4\lambda^3 & 6\lambda^2 \\ 0 & 0 & \lambda^4 & 4\lambda^3 \\ 0 & 0 & 0 & \lambda^4 \end{pmatrix}.$$

This suggest that in general, the ijth entry of $J_m(\lambda)^n$ is $(J_m(\lambda)^n)_{ij} = \binom{i}{j-i}\lambda^{n+i-j}$, with the convention $\binom{k}{l} = 0$ if $l < 0$. The proof by induction is based on the recursive formula for binomial coefficients. Indeed, from $J_m(\lambda)^{n+1} = J_m(\lambda)^n J_m(\lambda)$, we obtain

$$(J_m(\lambda)^{n+1})_{ij} = \lambda(J_m(\lambda)^n)_{ij} + (J_m(\lambda)^n)_{i,j-1}$$
$$= \lambda\binom{n}{j-i}\lambda^{n+i-j} + \binom{n}{j-1-i}\lambda^{n+i-j+1} = \binom{n+1}{j-i}\lambda^{n+1+i-j},$$

which proves the claim.

Second solution: Define S to be the $n \times n$ matrix with ones just above the diagonal and zeros elsewhere (usually called a shift matrix), and note that S^k has ones above the diagonal at distance k from it, and in particular $S^n = \mathcal{O}_n$. Hence

$$J_m(\lambda)^n = (\lambda\mathcal{I}_n + S)^n = \sum_{k=0}^{n-1}\binom{n}{k}\lambda^{n-k}S^k.$$

The conclusion follows.

Remark. The matrix $J_m(\lambda)$ is called a Jordan block. It is part of the Jordan canonical form of a matrix. Specifically, given a square matrix A there exists an invertible matrix S such that $S^{-1}AS$ is a block diagonal matrix whose blocks are matrices of the form $J_{m_i}(\lambda_i)$. The numbers λ_i are the eigenvalues of A. As a consequence of this problem, we obtain a standard method for raising a matrix to the nth power. The idea is to write the matrix in the Jordan canonical form and then raise the blocks to the power.

256. There is one property of the trace that we need. For an $n \times n$ matrix X with real entries, $\mathrm{tr}(XX^t)$ is the sum of the squares of the entries of X. This number is nonnegative and is equal to 0 if and only if X is the zero matrix. It is noteworthy to mention that $\|X\| = \sqrt{\mathrm{tr}(CC^t)}$ is a norm known as the Hilbert-Schmidt norm.

We would like to apply the above-mentioned property to the matrix $A - B^t$ in order to show that this matrix is zero. Writing

$$\mathrm{tr}[(A - B^t)(A - B^t)^t] = \mathrm{tr}[(A - B^t)(A^t - B)] = \mathrm{tr}(AA^t + B^tB - AB - B^tA^t)$$
$$= \mathrm{tr}(AA^t + B^tB) - \mathrm{tr}(AB + B^tA^t),$$

we see that we could almost use the equality from the statement, but the factors in two terms come in the wrong order. Another property of the trace comes to the rescue, namely, $\operatorname{tr}(XY) = \operatorname{tr}(YX)$. We thus have

$$\operatorname{tr}(AA^t + B^tB) - \operatorname{tr}(AB + B^tA^t) = \operatorname{tr}(AA^t) + \operatorname{tr}(B^tB) - \operatorname{tr}(AB) - \operatorname{tr}(B^tA^t)$$
$$= \operatorname{tr}(AA^t) + \operatorname{tr}(BB^t) - \operatorname{tr}(AB) - \operatorname{tr}(A^tB^t) = 0.$$

It follows that $\operatorname{tr}[(A - B^t)(A - B^t)^t] = 0$, which implies $A - B^t = \mathcal{O}_n$, as desired.

Remark. The Hilbert-Schmidt norm plays an important role in the study of linear transformations of infinite-dimensional spaces. It was first considered by E. Schmidt in his study of integral equations of the form

$$f(x) - \int_a^b K(x, y)f(y)dy = g(x).$$

Here the linear transformation (which is a kind of infinite-dimensional matrix) is

$$f(x) \to \int_a^b K(x, y)f(y)dy,$$

and its Hilbert-Schmidt norm is

$$\left(\int_a^b \int_a^b |K(x, y)^2 dxdy \right)^{1/2}.$$

For a (finite- or infinite-dimensional) diagonal matrix D, whose diagonal elements are $d_1, d_2, \ldots \in \mathbb{C}$, the Hilbert-Schmidt norm is

$$\sqrt{\operatorname{tr}D\overline{D}^t} = (|d_1|^2 + |d_2|^2 + \cdots)^{1/2}.$$

257. The elegant solution is based on the equality of matrices

$$\begin{pmatrix} (x^2 + 1)^2 & (xy + 1)^2 & (xz + 1)^2 \\ (xy + 1)^2 & (y^2 + 1)^2 & (yz + 1)^2 \\ (xz + 1)^2 & (yz + 1)^2 & (z^2 + 1)^2 \end{pmatrix} = \begin{pmatrix} 1 & x & x^2 \\ 1 & y & y^2 \\ 1 & z & z^2 \end{pmatrix} \begin{pmatrix} 1 & 1 & 1 \\ 2x & 2y & 2z \\ x^2 & y^2 & z^2 \end{pmatrix}.$$

Passing to determinants and factoring a 2, we obtain a product of two Vandermonde determinants, hence the formula from the statement.

(C. Coşniţă, F.Turtoiu, *Probleme de Algebră (Problems in Algebra)*, Editura Tehnică, Bucharest, 1972)

258. Consider the matrix

$$M = \begin{pmatrix} 1 & 1 \\ 1 & 0 \end{pmatrix},$$

which has the property that

$$M^n = \begin{pmatrix} F_{n+1} & F_n \\ F_n & F_{n-1} \end{pmatrix}, \text{ for } n \geq 1.$$

Taking determinants, we have

$$F_{n+1}F_{n-1} - F_n^2 = \det M^n = (\det M)^n = (-1)^n,$$

as desired.

(J.D. Cassini)

259. Subtract the pth row from the $(p+1)$st, then the $(p-1)$st from the pth, and so on. Using the identity $\binom{n}{k} - \binom{n-1}{k} = \binom{n-1}{k-1}$, the determinant becomes

$$\begin{vmatrix} 1 & \binom{m}{1} & \cdots & \binom{m}{p} \\ 0 & \binom{m}{0} & \cdots & \binom{m}{p-1} \\ \vdots & \vdots & \ddots & \vdots \\ 0 & \binom{m-1+p}{0} & \cdots & \binom{m-1+p}{p-1} \end{vmatrix}.$$

Expanding by the first row, we obtain a determinant of the same form but with m replaced by $m - 1$ and p replaced by $p - 1$. For $p = 0$ the determinant is obviously equal to 1, and an induction on p proves that this is also true in the general case.

(C. Năstăsescu, C. Niţă, M. Brandiburu, D. Joiţa, *Exerciţii şi Probleme de Algebră* (*Exercises and Problems in Algebra*), Editura Didactică şi Pedagogică, Bucharest, 1983)

260. The determinant
$$\begin{vmatrix} \binom{x_1}{0} & \binom{x_2}{0} & \cdots & \binom{x_n}{0} \\ \binom{x_1}{1} & \binom{x_2}{1} & \cdots & \binom{x_n}{1} \\ \vdots & \vdots & \ddots & \vdots \\ \binom{x_1}{n-1} & \binom{x_2}{n-1} & \cdots & \binom{x_2}{n-1} \end{vmatrix}$$

is an integer. On the other hand, for some positive integer m and k the binomial coefficient $\binom{m}{k}$ is a linear combination of $m^k, \binom{m}{k-1}, \ldots, \binom{m}{0}$ whose coefficients do not depend on m. In this linear combination the coefficient of m^k is $1/k!$. Hence by performing row operations in the above determinant we can transform it into

$$\begin{vmatrix} 1 & 1 & \cdots & 1 \\ x_1 & x_2 & \cdots & x_n \\ \vdots & \vdots & \ddots & \vdots \\ x_1^{n-1} & x_2^{n-1} & \cdots & x_n^{n-1} \end{vmatrix}.$$

The Vandermonde determinant has the value $\prod_{i>j}(x_i - x_j)$.

It follows that our determinant is equal to $\prod_{i>j}(x_i - x_j)/(1!2! \cdots (n-1)!)$, which therefore must be an integer. Hence the conclusion.

(*Mathematical Mayhem*, 1995)

261. Let us consider a matrix with entries equal to ± 1. Its determinant is clearly an integer. Adding the first row to the second, third, ..., eleventh we transform the elements of these rows in either 0 or ± 2. The entries of these new rows are therefore divisible by 2, and factoring these out we deduce that the determinant of the matrix is a multiple of 2^{10}. There are only three integers that are multiples of 2^{10} in the specified interval, namely $0, \pm 2^{10}$. Let us show that each can be the determinant of such a matrix. To obtain 0, just make two rows equal. To obtain 2^{10} take the matrix that has 1 on and above the main diagonal, and -1 elsewhere. To obtain -2^{10} take the negative of this matrix.

(21st Annual Iowa Collegiate Mathematics Competition, 2015, proposed by R. Gelca)

262. *First solution:* The determinant is an nth-degree polynomial in each of the x_i's. Adding all other columns to the first, we obtain that the determinant is equal to zero when $x_1 + x_2 + \cdots + x_n = 0$, so $x_1 + x_2 + \cdots + x_n$ is a factor of the polynomial. This factor corresponds to $j = 0$ on the right-hand side of the identity from the statement. For some other j, multiply the first column by ζ^j, the second by ζ^{2j}, and so forth; then add all columns to the first. As before, we see that the determinant is zero when $\sum_{k=1}^{n} \zeta^{jk} x_k = 0$, so $\sum_{k=1}^{n} \zeta^{jk} x_k$ is a factor of the determinant. No two of these polynomials are a constant multiple of the other, so the determinant is a multiple of

$$\prod_{j=1}^{n-1} \left(\sum_{k=1}^{n} \zeta^{jk} x_k \right).$$

The quotient of the two is a scalar C, independent of $x_1, x_2, \ldots, x_n$. For $x_1 = 1, x_2 = x_3 = \cdots = x_n = 0$, we obtain

$$x_1^n C \prod_{j=1}^{n-1} (\zeta^j x_1) = C\zeta^{1+2+\cdots+(n-1)} x_1 = C\zeta^{n(n-1)/2} x_1^n = Ce^{(n-1)\pi i} x_1^n = C(-1)^{n-1} x_1.$$

Hence $C = (-1)^{n-1}$, which gives rise to the formula from the statement.

Second solution: We use the discrete Fourier transform:

$$\mathcal{F}_n = \frac{1}{\sqrt{n}} \begin{pmatrix} 1 & 1 & 1 & \cdots & 1 \\ 1 & \varepsilon & \varepsilon^2 & \cdots & \varepsilon^{n-1} \\ 1 & \varepsilon^2 & \epsilon^4 & \cdots & \varepsilon^{2(n-1)} \\ \vdots & \vdots & \vdots & \ddots & \vdots \\ 1 & \varepsilon^{n-1} & \varepsilon^{2(n-1)} & \cdots & \varepsilon^{(n-1)^2} \end{pmatrix}.$$

If we let $f_j(t) = \sum_{j=0}^{n-1} a_j t^j$, then

$$
\begin{pmatrix} a_0 & a_1 & \cdots & a_{n-1} \\ a_{n-1} & a_0 & \cdots & a_{n-2} \\ \vdots & \vdots & \ddots & \vdots \\ a_1 & a_2 & \cdots & a_0 \end{pmatrix} \times \frac{1}{\sqrt{n}} \begin{pmatrix} 1 & 1 & \cdots & 1 \\ 1 & \varepsilon & \cdots & \varepsilon^{n-1} \\ 1 & \varepsilon^2 & \cdots & \varepsilon^{2(n-1)} \\ \vdots & \vdots & \ddots & \vdots \\ 1 & \varepsilon^{n-1} & \cdots & \varepsilon^{(n-1)^2} \end{pmatrix}
$$

$$
= \frac{1}{\sqrt{n}} \begin{pmatrix} f(1) & f(\varepsilon) & \cdots & f(\varepsilon^{n-1}) \\ f(1) & \varepsilon f(\varepsilon) & \cdots & \varepsilon^{n-1} f(\varepsilon^{n-1}) \\ \vdots & \vdots & \ddots & \vdots \\ f(1) & \varepsilon^{n-1} f(\varepsilon) & \cdots & \varepsilon^{(n-1)^2} f(\varepsilon^{n-1}) \end{pmatrix}
$$

$$
= \frac{1}{\sqrt{n}} \begin{pmatrix} 1 & 1 & \cdots & 1 \\ 1 & \varepsilon & \cdots & \varepsilon^{n-1} \\ \vdots & \vdots & \ddots & \vdots \\ 1 & \varepsilon^{n-1} & \cdots & \varepsilon^{(n-1)^2} \end{pmatrix} \begin{pmatrix} f(1) & 0 & \cdots & 0 \\ 0 & f(\varepsilon) & \cdots & 0 \\ 0 & 0 & \cdots & 0 \\ \vdots & \vdots & \ddots & \vdots \\ 0 & 0 & \cdots & f(\varepsilon^{n-1}) \end{pmatrix}.
$$

Taking determinants of both sides, and using the fact that the discrete Fourier transform is invertible and so it has nonzero determinant, we obtain the desired formula.

Remark. This formula for the circulant determinant was first proved by L. Cremona. The second proof of the formula is due to Cremona, and it shows that the discrete Fourier transform diagonalizes the circulant determinant. This is the reason why circulant determinants are important in telecommunications and signal processing.

R. Dedekind has generalized this formula as: Given a finite abelian group $G = g_1, g_2, \ldots, g_n$, consider a sequence $a_{g_1}, a_{g_2}, \ldots, a_{g_n}$. Define the determinant $\det(a_{g_j g_k^{-1}})$, whose jk entry is $a_{g_j g_k^{-1}}$. Dedekind proved that

$$
\det(a_{g_j g_k^{-1}}) = \prod_{\chi \in \widehat{G}} \left(\sum_{g \in G} \chi(g) a_g \right)
$$

where $\widehat{G}$ is the group of homomorphisms of G in $\mathbb{C} \backslash \{0\}$. Some mathematicians consider this formula the birth point of the theory of group representations.

263. Add the second, third, and fourth columns to the first. Now let us examine the first column. Recall that by expanding the circulant determinant

$$
\begin{vmatrix} a & b & c \\ c & a & b \\ b & c & a \end{vmatrix}
$$

in two ways: first with the Sarrus rule, and second with the formula given in the previous problem, we have the following factorization (see also Section 2.1.1)

$$
x^3 + y^3 + z^3 - 3xyz = (x + y + z)(x^2 + y^2 + z^2 - xy - yz - xz).
$$

Using this formula for $x = a, y = b, z = -1$, the first entry on the first column is

$$a^3 + b^3 - 1 + 3ab = a^3 + b^3 + (-1)^3 - 3ab(-1)$$
$$= (a + b - 1)(a^2 + b^2 + 1 - ab + a + b).$$

The second entry is

$$-1 + a^2 + b^2 + 2ab = (a + b)^2 - 1 = (a + b - 1)(a + b + 1)$$

and the third entry is

$$2b - 1 + a^2 - b^2 = a^2 - (b - 1)^2 = (a + b - 1)(a - b + 1).$$

And the fourth entry is $a + b - 1$. It follows that the determinant is divisible by $a + b - 1 = 3 \times 11 \times 61$, and we are done.

(Konhauser Problem Fest, 2014, proposed by R. Gelca)

264. By adding the second row to the first, the third row to the second,..., the nth row to the $(n - 1)$st, the determinant does not change. Hence

$$\det(A) = \begin{vmatrix} 2 & -1 & +1 & \cdots & \pm 1 & \mp 1 \\ -1 & 2 & -1 & \cdots & \mp 1 & \pm 1 \\ +1 & -1 & 2 & \cdots & \pm 1 & \mp 1 \\ \vdots & \vdots & \vdots & \ddots & \vdots & \vdots \\ \mp 1 & \pm 1 & \mp 1 & \cdots & 2 & -1 \\ \pm 1 & \mp 1 & \pm 1 & \cdots & -1 & 2 \end{vmatrix} = \begin{vmatrix} 1 & 1 & 0 & 0 & \cdots & 0 & 0 \\ 0 & 1 & 1 & 0 & \cdots & 0 & 0 \\ 0 & 0 & 1 & 1 & \cdots & 0 & 0 \\ \vdots & \vdots & \vdots & \vdots & \ddots & \vdots & \vdots \\ 0 & 0 & 0 & 0 & \cdots & 1 & 1 \\ \pm 1 & \mp 1 & \pm 1 & \mp 1 & \cdots & -1 & 2 \end{vmatrix}.$$

Now subtract the first column from the second, then subtract the resulting column from the third, and so on. This way we obtain

$$\det(A) = \begin{vmatrix} 1 & 0 & 0 & \cdots & 0 & 0 \\ 0 & 1 & 0 & \cdots & 0 & 0 \\ \vdots & \vdots & \vdots & \ddots & \vdots & \vdots \\ 0 & 0 & 0 & \cdots & 1 & 0 \\ \pm 1 & \mp 2 & \pm 3 & \cdots & -n + 1 & n + 1 \end{vmatrix} = n + 1.$$

(9th International Mathematics Competition for University Students, 2002)

265. View the determinant as a polynomial in the independent variables $x_1, x_2, \ldots, x_n$. Because whenever $x_i = x_j$ the determinant vanishes, it follows that the determinant is divisible by $x_i - x_j$, and therefore by the product $\prod_{1 \le i < j \le n} (x_j - x_i)$. Because the k_i's are positive, the determinant is also divisible by $x_1 x_2 \cdots x_n$. To solve the problem, it suffices to show that for any positive integers $x_1, x_2, \ldots, x_n$, the product

$$x_1 x_2 \cdots x_n \prod_{1 \le i < j \le n} (x_j - x_i)$$

is divisible by $n!$. This can be proved by induction on n. A parity check proves the case $n = 2$. Assume that the property is true for any $n - 1$ integers and let us prove it for n. Either one of the numbers $x_1, x_2, \ldots, x_n$ is divisible by n, or, by the pigeonhole principle, the difference of two of them is divisible by n. In the first case we may assume that x_n is divisible by n, in the latter that $x_n - x_1$ is divisible by n. In either case,

$$x_1 x_2 \cdots x_{n-1} \prod_{1 \le i < j \le n-1} (x_j - x_i)$$

is divisible by $(n - 1)$, by the induction hypothesis. It follows that the whole product is divisible by $n \times (n - 1)! = n!$ as desired. We are done.

(Proposed for the Romanian Mathematical Olympiad by N. Chichirim)

266. Expand the determinant as

$$\det(xA + yB) = a_0(x)y^3 + a_1(x)y^2 + a_2(x)y + a_3(x),$$

where $a_i(x)$ are polynomials of degree at most i, $i = 0, 1, 2, 3$. For $y = 0$ this gives $\det(xA) = x^3 \det A = 0$, and hence $a_3(x) = 0$ for all x. Similarly, setting $y = x$ we obtain $\det(xA + xB) = x^3 \det(A + B) = 0$, and thus

$$a_0(x)x^3 + a_1(x)x^2 + a_2(x)x = 0.$$

Also, for $y = -x$ we obtain $\det(xA - xB) = x^3 \det(A - B) = 0$; thus

$$-a_0(x)x^3 + a_1(x)x^2 - a_x(x)x = 0.$$

Adding these two relations gives $a_1(x) = 0$ for all x. For $x = 0$ we find that $\det(yB) = y^3 \det B = 0$, and hence $a_0(0)y^3 + a_x(0)y = 0$ for all y. Therefore, $a_0(0) = 0$. But $a_0(x)$ is constant, so $a_0(x) = 0$. This implies $a_2(x)x = 0$ for all x, and so $a_2(x) = 0$ for all x. We conclude that $\det(xA + yB)$ is identically equal to zero, and the problem is solved.

(*Romanian Mathematics Competition*, 1979, M. Martin)

267. We reduce the problem to a computation with 4×4 determinants. Expanding according to the rule of Laplace, we see that

$$x^2 = \begin{vmatrix} a & 0 & b & 0 \\ c & 0 & d & 0 \\ 0 & b & 0 & a \\ 0 & d & 0 & c \end{vmatrix} \quad \text{and} \quad x'^2 = \begin{vmatrix} b' & a' & 0 & 0 \\ d' & c' & 0 & 0 \\ 0 & 0 & b' & a' \\ 0 & 0 & d' & c' \end{vmatrix}.$$

Multiplying these determinants, we obtain $(xx')^2$.

(C. Coşniţă, F. Turtoiu, *Probleme de Algebră* (*Problems in Algebra*), Editura Tehnică, Bucharest, 1972)

268. First, suppose that A is invertible. Then we can write

$$\begin{pmatrix} A & B \\ C & D \end{pmatrix} = \begin{pmatrix} A & 0 \\ C & I_n \end{pmatrix} \begin{pmatrix} I_n & A^{-1}B \\ 0 & D - CA^{-1}B \end{pmatrix}.$$

The matrices on the right-hand side are of block-triangular type, so their determinants are the products of the determinants of the blocks on the diagonal, as can be seen on expanding the determinants using the rule of Laplace. Therefore,

$$\det \begin{pmatrix} A & B \\ C & D \end{pmatrix} = (\det A)(\det(D - CA^{-1}B)) = \det(AD - ACA^{-1}B).$$

The equality from the statement now follows form that fact that A and C commute.

If A is not invertible, then since the polynomial $\det(A + \varepsilon \mathcal{I}_n)$ has finitely many zeros, $A + \varepsilon \mathcal{I}_n$ is invertible for any sufficiently small $\varepsilon > 0$. This matrix still commutes with C, so we can apply the above argument to A replaced by $A + \varepsilon \mathcal{I}_n$. The identity from the statement follows by letting $\varepsilon \to 0$.

269. Applying the previous problem, we can write

$$\det(\mathcal{I}_n - XY) = \det \begin{pmatrix} \mathcal{I}_n & X \\ Y & \mathcal{I}_n \end{pmatrix} = (-1)^n \det \begin{pmatrix} Y & \mathcal{I}_n \\ \mathcal{I}_n & X \end{pmatrix}$$

$$= (-1)^{2n} \det \begin{pmatrix} \mathcal{I}_n & Y \\ X & \mathcal{I}_n \end{pmatrix} = \det(\mathcal{I}_n - YX).$$

Note that we performed some row and column permutations in the process, while keeping track of the sign of the determinant.

270. For k even, that is, $k = 2m$, the inequality holds even without the assumption from the statement. Indeed, there exists ε arbitrarily small such that the matrix $B_0 = B + \varepsilon \mathcal{I}_n$ is invertible. Then

$$\det(A^{2m} + B_0^{2m}) = \det B_0^{2m} \det((A^m B_0^{-m})^2 + \mathcal{I}_n),$$

and the latter is nonnegative, as seen in the introduction. Taking the limit with ε approaching zero, we obtain $\det(A^{2m} + B^{2m}) \geq 0$.

For k odd, $k = 2m + 1$, let $x_0 = -1$, $x_1, x_2, \ldots, x_{2m}$ be the zeros of the polynomial $x^{2m+1} + 1$, with $x_{j+m} = \bar{x}_j, j = 1, 2, \ldots, m$. Because A and B commute, we have

$$A^{2m+1} + B^{2m+1} = (A + B) \prod_{j=1}^{m} (A - x_j B)(A - \bar{x}_j B).$$

Since A and B have real entries, by taking determinants we obtain

$$\det(A - x_j B)(A - \bar{x}_j B) = \det(A - x_j B) \det(A - \bar{x}_j B)$$
$$= \det(A - x_j B) \det \overline{(A - x_j B)}$$
$$= \det(A - x_j B) \overline{\det(A - x_j B)} \geq 0,$$

for $j = 1, 2, \ldots, m$. This shows that the sign of $\det(A^{2m+1} + B^{2m+1})$ is the same as the sign of $\det(A + B)$ and we are done.

(Romanian Mathematical Olympiad, 1986)

271. The case $\lambda \geq 0$ was discussed before. If $\lambda < 0$, let $\omega = \sqrt{-\lambda}$. We have

$$\det(\mathcal{I}_n + \lambda A^2) = \det(\mathcal{I}_n - \omega^2 A^2) = \det(\mathcal{I}_n - \omega A)(\mathcal{I}_n + \omega A)$$
$$= \det(\mathcal{I}_n - \omega A) \det(\mathcal{I}_n + \omega A).$$

Because $-A = A^t$, it follows that

$$\mathcal{I}_n - \omega A = \mathcal{I}_n + \omega A^t = {}^t(\mathcal{I}_n + \omega A).$$

Therefore,

$$\det(\mathcal{I}_n + \lambda A^2) = \det(\mathcal{I}_n + \omega A) \det {}^t(\mathcal{I}_n + \omega A) = (\det(\mathcal{I}_n + \omega A))^2 \geq 0,$$

and the inequality is proved.

(Romanian Mathematics Competition, proposed by S. Rădulescu)

272. We can assume that the leading coefficient of $P(t)$ is 1. Let α be a real number such that $P(t) + \alpha$ is strictly positive and let Y be a matrix with negative determinant. Assume that f is onto. Then there exists a matrix X such that $P(X) = Y - \alpha \mathcal{I}_n$.

Because the polynomial $Q(t) = P(t) + \alpha$ has no real zeros, it factors as

$$Q(t) = \prod_{k=1}^{m} [(t + a_k)^2 + b_k^2]$$

with $a_k, b_k \in \mathbb{R}$. It follows that

$$\det Q(X) = \prod_{k=1}^{m} \det[(X + a_k)^2 + b_k^2 \mathcal{I}_n] \geq 0,$$

and the latter is positive, since for all k,

$$\det[(X + a_k)^2 + b_k^2 \mathcal{I}_n] = b_k^{2n} \det \left[\left(\frac{1}{b_k} X + \frac{a_k}{b_k} \right)^2 + \mathcal{I}_n \right] \geq 0.$$

In particular, $Q(X) \neq Y$ and thus the function f is not onto.

(*Gazeta Matematică* (*Mathematics Gazette, Bucharest*), proposed by D. Andrica)

273. If $A^2 = \mathcal{O}_n$, then

$$\det(A + \mathcal{I}_n) = \det \left(\frac{1}{4} A^2 + A + \mathcal{I}_n \right) = \det \left(\frac{1}{2} A + \mathcal{I}_n \right)^2 = \left(\det \left(\frac{1}{2} A + \mathcal{I}_n \right) \right)^2 \geq 0.$$

Similarly,

$$\det(A - \mathcal{I}_n) = \det(-(\mathcal{I}_n - A)) = (-1)^n \det(\mathcal{I}_n - A) = (-1)^n \det\left(\mathcal{I}_n - A + \frac{1}{4}A^2\right)$$

$$= (-1)^n \det\left(\mathcal{I}_n - \frac{1}{2}A\right)^2 = (-1)^n \left(\det\left(\mathcal{I}_n - \frac{1}{2}A\right)\right)^2 \le 0,$$

since n is odd. Hence $\det(A + \mathcal{I}_n) \ge 0 \ge \det(A - \mathcal{I}_n)$.

If $A^2 = \mathcal{I}_n$, then

$$0 \le (\det(A + \mathcal{I}_n))^2 = \det(A + \mathcal{I}_n)^2 = \det(A^2 + 2A + \mathcal{I}_n)$$
$$= \det(2A + 2\mathcal{I}_n) = 2^n \det(A + \mathcal{I}_n).$$

Also,

$$\det(A - \mathcal{I}_n) = (-1)^n \det(\mathcal{I}_n - A) = (-1)^n \det\left(\frac{1}{2}(2\mathcal{I}_n - 2A)\right)$$

$$= \left(-\frac{1}{2}\right)^n \det(\mathcal{I}_n - 2A + \mathcal{I}_n) = \left(-\frac{1}{2}\right)^n \det(A^2 - 2A + \mathcal{I}_n)$$

$$= \left(-\frac{1}{2}\right)^n \det(A - \mathcal{I}_n)^2 = \left(-\frac{1}{2}\right)^n (\det(A - \mathcal{I}_n))^2 \le 0,$$

and the inequality is proved in this case, too.
(Romanian Mathematics Competition, 1987)

274. All the information about the inverse of A is contained in its determinant. If we compute the determinant of A by expanding along the kth column, we obtain a polynomial in x_k, and the coefficient of x_k^{m-1} is exactly the minor used for computing the entry b_{km} of the adjoint matrix multiplied by $(-1)^{k+m}$. Viewing the product $\prod_{i>j}(x_i - x_j)$ as a polynomial in x_k, we have

$$\prod_{i>j}(x_i - x_j) = \Delta(x_1, \ldots, x_{k-1}, x_{k+1}, \ldots, x_n) \times (x_k - x_1)\cdots(x_k - x_{k-1})$$

$$\times (x_{k+1} - x_k)\cdots(x_n - x_k)$$
$$= (-1)^{n-k}\Delta(x_1, \ldots, x_{k-1}, x_{k+1}, \ldots, x_n) \times \prod_{j\neq k}(x_k - x_j).$$

In the product $\prod_{j\neq k}(x_k - x_j)$ the coefficient of x_k^{m-1} is

$$(-1)^{n-m}S_{n-m}(x_1, \ldots, x_{k-1}, x_{k+1}, \ldots, x_n).$$

Combining all these facts, we obtain

$$b_{km} = (-1)^{k+m}\Delta(x_1, x_2, \ldots, x_n)^{-1}(-1)^{k+m}(-1)^{n-k}(-1)^{n-m}$$
$$\times \Delta(x_1, \ldots, x_{k-1}, x_{k+1}, \ldots, x_n)S_m(x_1, \ldots, x_{k-1}, x_{k+1}, \ldots, x_n)$$

$$= (-1)^{k+m} \Delta(x_1, x_2, \ldots, x_n)^{-1} \Delta(x_1, \ldots, x_{k-1}, x_{k+1}, \ldots, x_n)$$
$$\times S_m(x_1, \ldots, x_{k-1}, x_{k+1}, \ldots, x_n),$$

as desired.

275. The inverse of a 2×2 matrix $C = (c_{ij})_{i,j}$ with integer entries is a matrix with integer entries if and only if $\det C = \pm 1$ (one direction of this double implication follows from the formula for the inverse, and the other from $\det C^{-1} = 1/\det C$).

With this in mind, let us consider the polynomial $P(x) \in \mathbb{Z}[x]$, $P(x) = \det(A + xB)$. The hypothesis of the problem implies that $P(0), P(1), P(2), P(3), P(4) \in \{-1, 1\}$. By the pigeonhole principle, three of these numbers are equal, and because $P(x)$ has degree at most 2, it must be constant. Therefore, $\det(A + xB) = \pm 1$, for all x, and in particular for $x = 5$ the matrix $A + 5B$ is invertible and has determinant equal to ± 1. Consequently, the inverse of this matrix has integer entries.

(55th W.L. Putnam Mathematical Competition, 1994)

276. We know that $AA^* = A^*A = (\det A)\mathcal{I}_3$, so if A is invertible then so is A^*, and $A = \det A (A^*)^{-1}$. Also, $\det A \det A^* = (\det A)^3$; hence $\det A^* = (\det A)^2$. Therefore, $A = \pm\sqrt{\det A^*}(A^*)^{-1}$.

Because

$$A^* = (1 - m)\begin{pmatrix} -m-1 & 1 & 1 \\ 1 & -m-1 & 1 \\ 1 & 1 & -m-1 \end{pmatrix},$$

we have

$$\det A^* = (1 - m)^3[-(m+1)^3 + 2 + 3(m+1)] = (1 - m)^4(m + 2)^2.$$

Using the formula with minors, we compute the inverse of the matrix

$$\begin{pmatrix} -m-1 & 1 & 1 \\ 1 & -m-1 & 1 \\ 1 & 1 & -m-1 \end{pmatrix}$$

to be

$$\frac{1}{(1-m)(m+2)^2}\begin{pmatrix} -m^2-m-2 & m+2 & m+2 \\ m+2 & -m^2-m-2 & m+2 \\ m+2 & m+2 & -m^2-m-2 \end{pmatrix}.$$

Then $(A^*)^{-1}$ is equal to this matrix divided by $(1 - m)^3$. Consequently, the matrix we are looking for is

$$A = \pm\sqrt{\det A^*}(A^*)^{-1}$$
$$= \pm\frac{1}{(1-m)^2(m+2)}\begin{pmatrix} -m^2-m-2 & m+2 & m+2 \\ m+2 & -m^2-m-2 & m+2 \\ m+2 & m+2 & -m^2-m-2 \end{pmatrix}.$$

(Romanian Mathematics Competition)

277. The series expansion

$$\frac{1}{1-x} = 1 + x + x^2 + x^3 + \cdots +$$

suggests that

$$(\mathcal{I}_n - A)^{-1} = \mathcal{I}_n + A + A^2 + A^3 + \cdots$$

But does the series on the right converge?

Let

$$\alpha = \max_i \left(\sum_{j=1}^n |a_{ij}| \right) < 1.$$

Then

$$\sum_k \left| \sum_j a_{ij} a_{jk} \right| \leq \sum_{j,k} |a_{ij} a_{jk}| = \sum_j \left(|a_{ij}| \sum_k |a_{jk}| \right) \leq \alpha \sum_j |a_{ij}| \leq \alpha^2.$$

Inductively we obtain that the entries $a_{ij}(n)$ of A^n satisfy $\sum_i |a_{ij}(n)| < \alpha^n$ for all i. Because the geometric series $1 + \alpha + \alpha^2 + \alpha^3 + \cdots$ converges, so does $\mathcal{I}_n + A + A^2 + A^3 + \cdots$ And the sum of this series is the inverse of $\mathcal{I}_n - A$.

(P.N. de Souza, J.N. Silva, *Berkeley Problems in Mathematics*, Springer, 2004)

278. The trick is to compute A^2. The elements on the diagonal are

$$\sum_{k=1}^n \sin^2 km\alpha, \quad m = 1, 2, \ldots, n,$$

which are all nonzero. Off the diagonal, the (m, j)th entry is equal to

$$\sum_{k=1}^n \sin km\alpha \sin kj\alpha = \frac{1}{2} \left[\sum_{k=1}^n \cos k(m-j)\alpha - \sum_{k=1}^n \cos k(m+j)\alpha \right].$$

We are led to the computation of two sums of the form $\sum_{k=1}^n \cos kx$. This is done as follows:

$$\sum_{k=1}^n \cos kx = \frac{1}{2\sin \frac{x}{2}} \sum_{k=1}^n \sin \frac{x}{2} \cos kx = \frac{1}{2\sin \frac{x}{2}} \sum_{k=1}^n \left[\sin \left(k + \frac{1}{2} \right) x - \sin \left(k - \frac{1}{2} x \right) \right].$$

The sum telescopes, and we obtain

$$\sum_{k=1}^n \cos kx = \frac{\sin \left(n + \frac{1}{2} \right) x}{2\sin \frac{x}{2}} - \frac{1}{2}.$$

Note that for $x = (m \pm j)\alpha = \frac{(m \pm j)\pi}{n+1}$,

$$\sin\left(n + \frac{1}{2}\right)x = \sin\left((m \pm j)\pi - \frac{x}{2}\right) = (-1)^{m+j+1}\sin\frac{x}{2}.$$

Hence

$$\sum_{k=1}^{n}\cos(m \pm j)k\alpha = \frac{(-1)^{m+j+1}}{2} - \frac{1}{2}.$$

It follows that for $m \neq j$, the (m, j)entry of the matrix A^2 is zero. Hence A^2 is a diagonal matrix with nonzero diagonal entries. This shows that A^2 is invertible, and so is A.

Remark. This is the discrete sine transform.

279. If $A + iB$ is invertible, then so is $A^\dagger - iB^\dagger$. Let us multiply these two matrices:

$$(A^\dagger - iB^\dagger)(A + iB) = A^\dagger A + B^\dagger B + i(A^\dagger B - B^\dagger A).$$

We have

$$\langle(A^\dagger A + B^\dagger B + i(A^\dagger B - B^\dagger A))v, v\rangle = \langle A^\dagger Av, v\rangle + \langle B^\dagger Bv, v\rangle + \langle i(A^\dagger B - B^\dagger A)v, v\rangle$$
$$= \|Av\|^2 + \|Bv\|^2 + \langle i(A^\dagger B - B^\dagger A)v, v\rangle,$$

which is strictly greater than zero for any vector $v \neq 0$. This shows that the product $(A^\dagger - iB^\dagger)(A + iB)$ is a *positive definite* matrix (i.e., $\langle(A^\dagger - iB^\dagger)(A + iB)v, v\rangle > 0$ for all $v \neq 0$). The linear transformation that it defines is therefore injective, hence an isomorphism. This implies that $(^\dagger - iB^\dagger)(A + iB)$ is invertible, and so $(A + iB)$ itself is invertible.

280. *First solution*: The fact that $A - I_n$ is invertible follows from the Spectral mapping theorem. To find its inverse, we recall the identity

$$1 + x + x^2 + \cdots = x^k = \frac{x^{k+1} - 1}{x - 1},$$

which by differentiation gives

$$1 + 2x + \cdots + kx^{k-1} = \frac{kx^{x+1} - (k + 1)x^k + 1}{(x - 1)^2}.$$

Substituting A for x, we obtain

$$(A - I_n)^2(I_n + 2A + \cdots + kA^{k-1}) = kA^{k+1} - (k + 1)A^k + I_n = I_n.$$

Hence

$$(A - I_n)^{-1} = (A - I_n)(I_n + 2A + \cdots + kA^{k-1}).$$

Second solution: Simply write

$$\mathcal{I}_n = kA^{k+1} - (k+1)A^k + \mathcal{I}_n = (A - \mathcal{I}_n)(kA^k - A^{k-1} - \cdots - A - \mathcal{I}_n),$$

which gives the inverse written in a different form.

(*Mathematical Reflections*, proposed by T. Andreescu)

281. If $\alpha \neq -1$, then

$$\left(A^{-1} - \frac{1}{\alpha+1}A^{-1}BA^{-1}\right)(A+B) = \mathcal{I}_n + A^{-1}B - \frac{1}{\alpha+1}A^{-1}BA^{-1}B - \frac{1}{\alpha+1}A^{-1}B.$$

But $(A^{-1}B)^2 = A^{-1}X(YA^{-1}X)Y = \alpha A^{-1}XY = \alpha A^{-1}B$. Hence in the above equality, the right-hand side is equal to the identity matrix. This proves the claim.

If $\alpha = -1$, then $(A^{-1}B)^2 + A^{-1}B = 0$, that is, $(\mathcal{I}_n + A^{-1}B)A^{-1}B = 0$. This implies that $\mathcal{I}_n + A^{-1}B$ is a zero divisor. Multiplying by A on the right we find that $A + B$ is a zero divisor itself. Hence in this case $A + B$ is not invertible.

(C. Năstăsescu, C. Niţă, M. Brandiburu, D. Joiţa, *Exerciţii şi Probleme de Algebră* (*Exercises and Problems in Algebra*), Editura Didactică şi Pedagogică, Bucharest, 1983)

282. The computation

$$(A - b\mathcal{I}_n)(B - a\mathcal{I}_n) = ab\mathcal{I}_n$$

shows that $A - b\mathcal{I}_n$ is invertible, and its inverse is $\frac{1}{ab}(B - a\mathcal{I}_n)$. Then

$$(B - a\mathcal{I}_n)(A - b\mathcal{I}_n) = ab\mathcal{I}_n,$$

which translates into $BA - aA - bB = \mathcal{O}_n$. Consequently, $BA = aA + bB = AB$, proving that the matrices commute.

283. We have

$$(A + iB^2)(B + iA^2) = AB - B^2A^2 + i(A^3 + B^3) = \mathcal{I}_n.$$

This implies that $A + iB^2$ is invertible, and its inverse is $B + iA^2$. Then

$$\mathcal{I}_n = (B + iA^2)(A + iB^2) = BA - A^2B^2 + i(A^3 + B^3) = BA - A^2B^2,$$

as desired.

(Romanian Mathematical Olympiad, 1982, proposed by I.V. Maftei)

284. Of course, one can prove that the coefficient matrix is nonsingular. But there is a slick solution. Add the equations and group the terms as

$$3(x_1 + x_2 + x_3) + 3(x_4 + x_5 + x_6) + \cdots + 3(x_{97} + x_{98} + x_{99}) + 3x_{100} = 0.$$

The terms in the parentheses are all zero; hence $x_{100} = 0$. Taking cyclic permutations yields $x_1 = x_2 = \cdots = x_{100} = 0$.

285. If y is not an eigenvalue of the matrix

$$\begin{pmatrix} 0 & 1 & 0 & 0 & 1 \\ 1 & 0 & 1 & 0 & 0 \\ 0 & 1 & 0 & 1 & 0 \\ 0 & 0 & 1 & 0 & 1 \\ 1 & 0 & 0 & 1 & 0 \end{pmatrix},$$

then the system has the unique solution $x_1 = x_2 = x_3 = x_4 = x_5 = 0$. Otherwise, the eigenvectors give rise to nontrivial solutions. Thus, we have to compute the determinant

$$\begin{vmatrix} -y & 1 & 0 & 0 & 1 \\ 1 & -y & 1 & 0 & 0 \\ 0 & 1 & -y & 1 & 0 \\ 0 & 0 & 1 & -y & 1 \\ 1 & 0 & 0 & 1 & -y \end{vmatrix}.$$

Adding all rows to the first and factoring $2 - y$, we obtain

$$(2 - y)\begin{vmatrix} 1 & 1 & 1 & 1 & 1 \\ 1 & -y & 1 & 0 & 0 \\ 0 & 1 & -y & 1 & 0 \\ 0 & 0 & 1 & -y & 1 \\ 1 & 0 & 0 & 1 & -y \end{vmatrix}.$$

The determinant from this expression is computed using row-column operations as follows:

$$\begin{vmatrix} 1 & 1 & 1 & 1 & 1 \\ 1 & -y & 1 & 0 & 0 \\ 0 & 1 & -y & 1 & 0 \\ 0 & 0 & 1 & -y & 1 \\ 1 & 0 & 0 & 1 & -y \end{vmatrix} = \begin{vmatrix} 1 & 0 & 0 & 0 & 0 \\ 1 & -y-1 & 0 & -1 & -1 \\ 0 & 1 & -y & 1 & 0 \\ 0 & 0 & 1 & -y & 1 \\ 1 & -1 & -1 & 0 & -y-1 \end{vmatrix}$$

$$= \begin{vmatrix} -y-1 & 0 & -1 & -1 \\ 1 & -y & 1 & 0 \\ 0 & 1 & -y & 1 \\ -1 & -1 & 0 & -y-1 \end{vmatrix} = \begin{vmatrix} -y-1 & 0 & -1 & -1 \\ -y & -y & 0 & -1 \\ 0 & 1 & -y & -1 \\ -1 & 0 & -y & -y \end{vmatrix}$$

$$= \begin{vmatrix} -y-1 & 0 & 0 & -1 \\ 0 & -y & 1 & -1 \\ -1 & 1 & -y-1 & -1 \\ -1 & 0 & 0 & -y \end{vmatrix},$$

which, after expanding with the rule of Laplace, becomes

$$-\begin{vmatrix} -y & -1 \\ 1 & -y-1 \end{vmatrix} \cdot \begin{vmatrix} -y-1 & -1 \\ -1 & -y \end{vmatrix} = -(y^2 + y - 1)^2.$$

Hence the original determinant is equal to $(y-2)(y^2+y-1)^2$. If $y = 2$, the space of solutions is therefore one-dimensional, and it is easy to guess the solution $x_1 = x_2 = x_3 = x_4 = x_5 = \lambda$, $\lambda \in \mathbb{R}$.

If $y = \frac{-1+\sqrt{5}}{2}$ or if $y = \frac{-1-\sqrt{5}}{2}$, the space of solutions is two-dimensional. In both cases, the minor

$$\begin{vmatrix} -y & 1 & 0 \\ 1 & -y & 1 \\ 0 & 1 & -y \end{vmatrix}$$

is nonzero, hence x_3, x_4, and x_5 can be computed in terms of x_1 and x_2. In this case the general solution is

$$(\lambda, \mu, -\lambda + y\mu, -y(\lambda + \mu), y\lambda - \mu), \quad \lambda, \mu \in \mathbb{R}.$$

Remark. The determinant of the system can also be computed using the formula for the determinant of a circulant matrix.

(5th International Mathematical Olympiad, 1963, proposed by the Soviet Union)

286. Taking the logarithms of the four relations from the statement, we obtain the following system of linear equations in the unknowns $\ln a$, $\ln b$, $\ln c$, $\ln d$:

$$-x \ln a + \ln b + \ln c + \ln d = 0,$$

$$\ln a - y \ln b + \ln c + \ln d = 0,$$

$$\ln a + \ln b - z \ln c + \ln d = 0,$$

$$\ln a + \ln b + \ln c - t \ln d = 0.$$

We are given that this system has a nontrivial solution. Hence the determinant of the coefficient matrix is zero, which is what had to be proved.

(Romanian Mathematics Competition, 2004)

287. *First solution:* Suppose there is a nontrivial solution (x_1, x_2, x_3). Without loss of generality, we may assume $x_1 \le x_2 \le x_3$. Let $x_2 = x_1 + m$, $x_3 = x_1 + m + n$, $m, n \ge 0$. The first and the last equations of the system become

$$(a_{11} + a_{12} + a_{13})x_1 + (a_{12} + a_{13})m + a_{13}n = 0,$$

$$(a_{31} + a_{32} + a_{33})x_1 + (a_{32} + a_{33})m + a_{33}n = 0.$$

The hypotheses $a_{31} + a_{32} + a_{33} > 0$ and $a_{31} < 0$ imply $a_{32} + a_{33} \ge 0$, and therefore $(a_{32} + a_{33})m \ge 0$ and $a_{33}n \ge 0$. We deduce that $x_1 \le 0$, which combined with $a_{12} < 0$, $a_{13} < 0$, $a_{11} + a_{12} + a_{13} > 0$ gives

$$(a_{11} + a_{12} + a_{13})x_1 \le 0, \quad (a_{12} + a_{13})m \le 0, \quad a_{13}n \le 0.$$

The sum of these three nonpositive terms can be zero only when they are all zero. Hence $x_1 = 0$, $m = 0$, $n = 0$, which contradicts our assumption. We conclude that the system has the unique solution $x_1 = x_2 = x_3 = 0$.

Second solution: Suppose there is a nontrival solution (x_1, x_2, x_3). Without loss of generality, we may assume that $|x_3| \geq |x_2| \geq |x_1|$. We have $a_{31}, a_{32} < 0$ and $0 < -a_{31} - a_{32} < a_{33}$, so

$$|a_{33}x_3| = |-a_{31}x_1 - a_{32}x_2| \leq (-a_{31} - a_{32})|x_2| \leq (-a_{31} - a_{32})|x_3| < a_{33}|x_3|.$$

This is a contradiction, which proves that the system has no nontrivial solution.

(7th International Mathematical Olympiad, 1965, proposed by Poland)

288. *First solution*: The zeros of $P(x)$ are $\varepsilon, \varepsilon^2, \ldots, \varepsilon^n$, where ε is a primitive $(n+1)$st root of unity. As such, the zeros of $P(x)$ are distinct. Let

$$P(x^{n+1}) = Q(x) \cdot P(x) + R(x),$$

where $R(x) = a_{n-1}x^{n-1} + \cdots + a_1x + a_0$ is the remainder. Replacing x successively by $\varepsilon, \varepsilon^2, \ldots, \varepsilon^n$, we obtain

$$a_n\varepsilon^{n-1} + \cdots + a_1\varepsilon + a_0 = n+1,$$
$$a_n(\varepsilon^2)^{n-1} + \cdots + a_1\varepsilon^2 + a_0 = n+1,$$
$$\ldots$$
$$a_n(\varepsilon^n)^{n-1} + \cdots + a_1\varepsilon^n + a_0 = n+1,$$

or

$$[a_0 - (n+1)] + a_1\varepsilon + \cdots + a_{n-1}\varepsilon^{n-1} = 0,$$
$$[a_0 - (n+1)] + a_1(\varepsilon^2) + \cdots + a_{n-1}(\varepsilon^2)^{n-1} = 0,$$
$$\ldots$$
$$[a_0 - (n+1)] + a_1(\varepsilon^n) + \cdots + a_{n-1}(\varepsilon^n)^{n-1} = 0.$$

This can be interpreted as a homogeneous system in the unknowns $a_0 - (n+1)$, $a_1, a_2, \ldots, a_{n-1}$. The determinant of the coefficient matrix is Vandermonde, thus nonzero, and so the system has the unique solution $a_0 - (n+1) = a_1 = \cdots = a_{n-1} = 0$. We obtain $R(x) = n+1$.

Second solution: Note that
$$x^{n+1} = (x-1)P(x) + 1;$$

hence
$$x^{k(n+1)} = (x-1)(x^{(k-1)(n+1)} + x^{(k-2)(n+1)} + \cdots + 1)P(x) + 1.$$

Thus the remainder of any polynomial $F(x^{n+1})$ modulo $P(x)$ is $F(1)$. In our situation this is $n+1$, as seen above.

(*Gazeta Matematică* (*Mathematics Gazette, Bucharest*), proposed by M. Diaconescu)

289. The function $\phi(t) = \frac{t-3}{t+1}$ has the property that $\phi \circ \phi \circ \phi$ equals the identity function. And $\phi(\phi(t)) = \frac{3+t}{1-t}$. Replace x in the original equation by $\phi(x)$ and $\phi(\phi(x))$ to obtain two more equations. The three equations form a linear system

$$f\left(\frac{x-3}{x+1}\right) + f\left(\frac{3+x}{1-x}\right) = x,$$

$$f\left(\frac{3+x}{1-x}\right)+f(x)=\frac{x-3}{x+1},$$

$$f(x)+f\left(\frac{x-3}{x+1}\right)=\frac{3+x}{1-x},$$

in the unknowns

$$f(x),\quad f\left(\frac{x-3}{x+1}\right),\quad f\left(\frac{3+x}{1-x}\right).$$

Solving, we find that

$$f(t)=\frac{4t}{1-t^2}-\frac{t}{2},$$

which is the unique solution to the functional equation.

(*Kvant* (*Quantum*), also appeared at the S. Korean Mathematical Olympiad, 1999)

290. It is obvious that $\gcd(x,x+y)=\gcd(x,x+z)=1$. So in the equality from the statement, x divides $y+z$. Similarly, y divides $z+x$ and z divides $x+y$. It follows that there exist integers a,b,c with $abc=t$ and

$$x+y=cz,$$

$$y+z=ax,$$

$$z+x=by.$$

View this as a homogeneous system in the variables x,y,z. Because we assume that the system admits nonzero solutions, the determinant of the coefficient matrix is zero. Writing down this fact, we obtain a new Diophantine equation in the unknowns a,b,c:

$$abc-a-b-c-2=0.$$

This can be solved by examining the following cases:

1. $a=b=c$. Then $a=2$ and it follows that $x=y=z$, because these numbers are pairwise coprime. This means that $x=y=z=1$ and $t=8$. We have obtained the solution $(1,1,1,8)$.

2. $a=b,\ a\neq c$. The equation becomes $a^2c-2=2a+c$, which is equivalent to $c(a^2-1)=2(a+1)$, that is, $c(a-1)=2$. We either recover case 1, or find the new solution $c=1,\ a=b=3$. This yields the solution to the original equation $(1,1,2,9)$.

3. $a>b>c$. In this case $abc-2=a+b+c<3a$. Therefore, $a(bc-3)<2$. It follows that $bc-3<2$, that is, $bc<5$. We have the following situations:

   (i) $b=2,\ c=1$, so $a=5$ and we obtain the solution $(1,2,3,10)$.
   (ii) $b=3,\ c=1$, so $a=3$ and we return to case 2.
   (iii) $b=4,\ c=1$, so $3a=7$, which is impossible.

In conclusion, we have obtained the solutions $(1, 1, 1, 8)$, $(1, 1, 2, 9)$, $(1, 2, 3, 10)$, and those obtained by permutations of x, y, z.

(Romanian Mathematical Olympiad, 1995)

291. Note that m comparisons give rise to a homogeneous linear system of m equations with n unknowns, namely the masses, whose coefficients are $-1, 0$, and 1. Determining whether all coins have equal mass is the same as being able to decide whether the solution belongs to the one-dimensional subspace of $\mathbb{R}^n$ spanned by the vector $(1, 1, \ldots, 1)$. Since the space of solutions has dimension at least $n - m$, in order to force the solution to lie in a one-dimensional space one needs at least $n - 1$ equations. This means that we need to perform at least $n - 1$ comparisons.

(Mathematical Olympiad Summer Program, 2006)

292. We are given that $a_0 = a_{n+1} = 0$ and $a_{k-1} - 2a_k + a_{k+1} = b_k$, with $b_k \in [-1, 1]$, $k = 1, 2, \ldots, n$. Consider the linear system of equations

$$a_0 - 2a_1 + a_2 = b_1,$$
$$a_1 - 2a_2 + a_3 = b_2,$$
$$\cdots$$
$$a_{n-1} - 2a_n + a_{n+1} = b_n$$

in the unknowns $a_1, a_2, \ldots, a_n$. To determine a_k for some k, we multiply the first equation by 1, the second by 2, the third by 3, and so on up to the $(k-1)$st, which we multiply by $k-1$, then add them up to obtain

$$-ka_{k-1} + (k-1)a_k = \sum_{j<k} jb_j.$$

Working backward, we multiply the last equation by by 1, the next-to-last by 2, and so on up to the $(k+1)$st, which we multiply by $n - k$, then add these equations to obtain

$$-(n - k + 1)a_{k+1} + (n - k)a_k = \sum_{j>k} (n - j + 1)b_j.$$

We now have a system of three equations,

$$-ka_{k-1} + (k-1)a_k = \sum_{j<k} jb_j,$$
$$a_{k-1} - 2a_k + a_{k+1} = b_k,$$
$$-(n - k + 1)a_{k+1} + (n - k)a_k = \sum_{j>k} (n - j + 1)b_j$$

in the unknowns a_{k-1}, a_k, a_{k+1}. Eliminating a_{k-1} and a_{k+1}, we obtain

$$\left(\frac{k-1}{k} - 2 + \frac{n-k}{n-k+1} \right) a_k = b_k + \frac{1}{k} \sum_{j<k} jb_j + \frac{1}{n-k+1} \sum_{j>k} (n - j + 1)b_j.$$

Taking absolute values and using the triangle inequality and the fact that $|b_j| \leq 1$, for all j, we obtain

$$\left| \frac{-n-1}{k(n-k+1)} \right| |a_k| \leq 1 + \frac{1}{k} \sum_{j<k} j + \frac{1}{n-k+1} \sum_{j>k} (n-j+1)$$

$$= 1 + \frac{k-1}{2} + \frac{n-k}{2} = \frac{n+1}{2}.$$

Therefore, $|a_k| \leq k(n-k+1)/2$, and the problem is solved.

293. The fact that the matrix is invertible is equivalent to the fact that the system of linear equations

$$\frac{x_1}{1} + \frac{x_2}{2} + \cdots + \frac{x_n}{n} = 0,$$

$$\frac{x_1}{2} + \frac{x_2}{3} + \cdots + \frac{x_n}{n+1} = 0,$$

$$\cdots$$

$$\frac{x_1}{n} + \frac{x_2}{n+1} + \cdots + \frac{x_n}{2n-1} = 0$$

has only the trivial solution. For a solution $(x_1, x_2, \ldots, x_n)$ consider the polynomial

$$P(x) = x_1(x+1)(x+2) \cdots (x+n-1) + x_2 x(x+2) \cdots (x+n-1) + \cdots$$
$$+ x_n x(x+1) \cdots (x+n-2).$$

Bringing to the common denominator each equation, we can rewrite the system in short form as $P(1) = P(2) = \cdots = P(n) = 0$. The polynomial $P(x)$ has degree $n-1$; the only way it can have n zeros is if it is identically zero. Taking successively $x = 0, -1, -2, \ldots, -n$, we deduce that $x_i = 0$ for all i. Hence the system has only the trivial solution, and the matrix is invertible.

For the second part, note that the sum of the entries of a matrix A is equal to the sum of the coordinates of the vector $A\mathbf{1}$, where $\mathbf{1}$ is the vector $(1, 1, \ldots, 1)$. Hence the sum of the entries of the inverse matrix is equal to $x_1 + x_2 + \cdots + x_n$, where $(x_1, x_2, \ldots, x_n)$ is the unique solution to the system of linear equations

$$\frac{x_1}{1} + \frac{x_2}{2} + \cdots + \frac{x_n}{n} = 0,$$

$$\frac{x_1}{2} + \frac{x_2}{3} + \cdots + \frac{x_n}{n+1} = 0,$$

$$\cdots$$

$$\frac{x_1}{n} + \frac{x_2}{n+1} + \cdots + \frac{x_n}{2n-1} = 0$$

This time, for a solution to this system, we consider the polynomial

$$Q(x) = x_1(x+1)(x+2) \cdots (x+n-1) + \cdots + x_n x(x+1) \cdots (x+n-2) + \cdots$$
$$+ x(x+1) \cdots (x+n-1).$$

Again we observe that $Q(1) = Q(2) = \cdots = Q(n) = 0$. Because $Q(x)$ has degree n and dominating coefficient -1, it follows that $Q(x) = -(x-1)(x-2)\cdots(x-n)$. So

$$x_1 \frac{(x+1)(x+2)\cdots(x+n-1)}{x^{n-1}} + \cdots + x_n \frac{x(x+1)\cdots(x+n-2)}{x^{n-1}}$$

$$= \frac{x(x+1)\cdots(x+n-1) - (x-1)(x-2)\cdots(x-n)}{x^{n-1}}.$$

The reason for writing this complicated relation is that as $x \to \infty$, the left-hand side becomes $x_1 + x_2 + \cdots + x_n$, while the right-hand side becomes the coefficient of x^{n-1} in the numerator. And this coefficient is

$$1 + 2 + \cdots + (n-1) + 1 + 2 + \cdots + n = \frac{n(n-1)}{2} + \frac{n(n+1)}{2} = n^2.$$

The problem is solved.

Remark. It is interesting to note that the same method allows the computation of the inverse as $(b_{k,m})_{km}$, giving

$$b_{k,m} = \frac{(-1)^{k+m}(n+k-1)!(n+m-1)!}{(k+m-1)[(k-1)!(m-1)!]^2(n-m)!(n-k)!}.$$

294. First, note that the polynomials $\binom{x}{1}$, $\binom{x+1}{3}$, $\binom{x+2}{5}$, ... are odd and have degrees $1, 3, 5, \ldots$, and so they form a basis of the vector space of the odd polynomial functions with real coefficients.

The scalars $c_1, c_2, \ldots, c_m$ are computed successively from

$$P(1) = c_1,$$

$$P(2) = c_1 \binom{2}{1} + c_2,$$

$$P(3) = c_1 \binom{3}{1} + c_2 \binom{4}{3} + c_3.$$

The conclusion follows.

(G. Pólya, G. Szegö, *Aufgaben und Lehrsätze aus der Analysis*, Springer-Verlag, 1964)

295. Inspired by the previous problem we consider the integer-valued polynomials

$$\binom{x}{m} = x(x-1)\cdots(x-m+1)/m!, \quad m = 0, 1, 2, \ldots$$

They form a basis of the vector space of polynomials with real coefficients. The system of equations

$$P(k) = b_0 \binom{x}{n} + b_1 \binom{x}{n-1} + \cdots + b_{n-1} \binom{x}{1} + b_n, \quad k = 0, 1, \ldots, n,$$

can be solved by Gaussian elimination, producing an integer solution $b_0, b_1 \ldots, b_n$. Yes, we do obtain an integer solution because the coefficient matrix is triangular and has ones on the diagonal! Finally, when multiplying $\binom{x}{m}$, $m = 0, 1, \ldots, n$, by $n!$, we obtain polynomials with integer coefficients. We find that $n!P(x)$ has integer coefficients, as desired.

(G. Pólya, G. Szegö, *Aufgaben und Lehrsätze aus der Analysis*, Springer-Verlag, 1964)

296. For $n = 1$ the rank is 1. Let us consider the case $n \geq 2$. Observe that the rank does not change under row/column operations. For $i = n, n-1, \ldots, 2$, subtract the $(i-1)$st row from the ith. Then subtract the second row from all others. Explicitly, we obtain

$$\text{rank} \begin{pmatrix} 2 & 3 & \cdots & n+1 \\ 3 & 4 & \cdots & n+2 \\ \vdots & \vdots & \ddots & \vdots \\ n+1 & n+2 & \cdots & 2n \end{pmatrix} = \text{rank} \begin{pmatrix} 2 & 3 & \cdots & n+1 \\ 1 & 1 & \cdots & 1 \\ \vdots & \vdots & \ddots & \vdots \\ 1 & 1 & \cdots & 1 \end{pmatrix}$$

$$= \text{rank} \begin{pmatrix} 1 & 2 & \cdots & n \\ 1 & 1 & \cdots & 1 \\ 0 & 0 & \cdots & 0 \\ \vdots & \vdots & \ddots & \vdots \\ 0 & 0 & \cdots & 0 \end{pmatrix} = 2.$$

So the rank is 2.

(12th International Competition in Mathematics for University Students, 2005)

297. The polynomials $P_j(x) = (x+j)^k$, $j = 0, 1, \ldots, n-1$, lie in the $(k+1)$-dimensional real vector space of polynomials of degree at most k. Because $k+1 < n$, they are linearly dependent. The columns consist of the evaluations of these polynomials at $1, 2, \ldots, n$, so the columns are linearly dependent. It follows that the determinant is zero.

298. We prove this property by induction on n. For $n = 1$, if f_1 is identically equal to zero, then so is f. Otherwise, pick a vector $e \notin f_1^{-1}(0)$. Note that any other vector $v \in V$ is of the form $\alpha e + w$ with $\alpha \in \mathbb{R}$ and $w \in f_1^{-1}(0)$. It follows that $f = \frac{f(e)}{f_1(e)}f_1$, and the base case is proved.

We now assume that the statement is true for $n = k-1$ and prove it for $n = k$. By passing to a subset, we may assume that $f_1, f_2, \ldots, f_k$ are linearly independent. Because f_k is linearly independent of $f_1, f_2, \ldots, f_{k-1}$, by the induction hypothesis there exists a vector e_k such that $f_1(e_k) = f_2(e_k) = \cdots = f_{k-1}(e_k) = 0$, and $f_k(e_k) \neq 0$. Multiplying e_k by a constant, we may assume that $f_k(e_k) = 1$. The vectors $e_1, e_2, \ldots, e_{k-1}$ are defined similarly, so that $f_j(e_i) = 1$ if $i = j$ and 0 otherwise.

For an arbitrary vector $v \in V$ and for $i = 1, 2, \ldots, k$, we have

$$f_i \left(v - \sum_{j=1}^{k} f_j(v)e_j \right) = f_i(v) - \sum_{j=1}^{k} f_j(v)f_i(e_j) = f_i(v) - f_i(v)f_i(e_i) = 0.$$

By hypothesis

$$f \left(v - \sum_{j=1}^{n} f_j(v)e_j \right) = 0.$$

Since f is linear, this implies

$$f(v) = f(e_1)f_1(v) + f(e_2)f_2(v) + \cdots + f(e_k)f_k(v), \text{ for all } v \in V.$$

This expresses f as a linear combination of $f_1, f_2, \ldots, f_k$, and we are done.

(5th International Competition in Mathematics for University Students, 1998)

299. *First solution*: We will prove this property by induction on n. For $n = 1$ it is obviously true. Assume that it is true for $n - 1$, and let us prove it for n. Using the induction hypothesis, we can find $x_1, x_2, \ldots, x_{n-1} \in S$ such that $a_1 x_1 + a_2 x_2 + \cdots + a_{n-1} x_{n-1}$ is irrational for any nonnegative rational numbers $a_1, a_2, \ldots, a_n$ not all equal to zero. Denote the other elements of S by $x_n, x_{n+1}, \ldots, x_{2n-1}$ and assume that the property does not hold for n. Then for each $k = 0, 1, \ldots, n - 1$ we can find rational numbers r_k such that

$$\left(\sum_{i=1}^{n-1} b_{ik} x_i \right) + c_k x_{n+k} = r_k$$

with b_{ik}, c_k some nonnegative integers, not all equal to zero. Because linear combinations of the x_i's, $i = 1, 2, \ldots, n - 1$, with nonnegative coefficients are irrational, it follows that c_k cannot be equal to zero. Dividing by the appropriate numbers if necessary, we may assume that for all k, $c_k = 1$. We can write

$$x_{n+k} = r_k - \sum_{i=1}^{n-1} b_{ik} x_i.$$

Note that the irrationality of x_{n+k} implies in addition that for a fixed k, not all the b_{ik}'s are zero.

Also, for the n numbers $x_n, x_{n+1}, \ldots, x_{2n-1}$, we can find nonnegative rationals $d_1, d_2, \ldots, d_n$, not all equal to zero, such that

$$\sum_{k=0}^{n-1} d_k x_{n+k} = r,$$

for some rational number r. Replacing each x_{n+k} by the formula found above, we obtain

$$\sum_{k=0}^{n-1} d_k \left(-\sum_{i=1}^{n-1} b_{ik} x_i + r_k \right) = r.$$

It follows that

$$\sum_{i=1}^{n-1} \left(\sum_{k=0}^{n-1} d_k b_{ik} \right) x_i$$

is rational. Note that there exists a nonzero d_k, and for that particular k also a nonzero b_{ik}. We found a linear combination of $x_1, x_2, \ldots, x_{n-1}$ with coefficients that are positive, rational, and not all equal to zero, which is a rational number. This is a contradiction. The conclusion follows.

Second solution: Let V be the span of $1, x_1, x_2, \ldots, x_{2n-1}$ over $\mathbb{Q}$. Then V is a finite-dimensional $\mathbb{Q}$-vector space inside $\mathbb{R}$. Choose a $\mathbb{Q}$-linear function $f : V \to \mathbb{Q}$ such that $f(1) = 0$ and $f(x_i) \neq 0$. Such an f exists since the space of linear functions with $f(1) = 0$ has dimension $\dim V - 1$ and the space of functions that vanish on 1 and x_i has dimension $\dim V - 2$, and because $\mathbb{Q}$ is infinite, you cannot cover an m-dimensional vector space with finitely many $(m - 1)$-dimensional subspaces. By the pigeonhole principle there are n of the x_i for which $f(x_i)$ has the same sign. Since $f(r) = 0$ for all rational r, no linear combination of these n with positive coefficients can be rational.

(Second solution by R. Stong)

300. *First solution*: Assume first that all numbers are integers. Whenever we choose a number, the sum of the remaining ones is even; hence the parity of each number is the same as the parity of the sum of all. And so all numbers have the same parity.

By subtracting one of the numbers from all we can assume that one of them is zero. Hence the numbers have the same parity as zero. After dividing by 2, we obtain $2n + 1$ numbers with the same property. So we can keep dividing by 2 forever, which is possible only if all numbers are zero. It follows that initially all numbers were equal.

The case of rational numbers is resolved by multiplying by the least common multiple of the denominators. Now let us assume that the numbers are real. The reals form an infinite-dimensional vector space over the rationals. Using the Axiom of choice we can find a basis of this vector space (sometimes called a Hammel basis). The coordinates of the $2n + 1$ numbers are rational, and must also satisfy the property from the statement (this follows from the fact that the elements of the basis are linearly independent over the rationals). So for each basis element, the corresponding coordinates of the $2n + 1$ numbers are the same. We conclude that the numbers are all equal, and the problem is solved.

However, this solution works only if we assume the Axiom of choice to be true. The axiom states that given a family of sets, one can choose an element from each. Obvious as this statement looks, it cannot be deduced from the other axioms of set theory and has to be taken as a fundamental truth. A corollary of the axiom is Zorn's lemma, which is the actual result used for constructing the Hammel basis. Zorn's lemma states that if every totally ordered subset of a partially ordered set has an upper bound, then the set has a maximal element. In our situation this lemma is applied to families of linearly independent vectors with the ordering given by the inclusion to yield a basis.

Second solution: The above solution can be improved to avoid the use of the axiom of choice. As before, we prove the result for rational numbers. Arguing by contradiction we assume that there exist $2n + 1$ real numbers, not all equal, such that whenever one is removed the others can be separated into two sets with n elements having the sum of their elements equal. If in each of these equalities we move all numbers to one side, we obtain a homogeneous system of $2n + 1$ equations with $2n + 1$ unknowns. In each row of the coefficient matrix, 1 and -1 each occur n times, and 0 appears once. The solution to the system obviously contains the one-dimensional vector space V spanned by the vector $(1, 1, \ldots, 1)$. By hypothesis, it contains another vector that does not lie in V. Solving the system using Gaussian elimination, we conclude that there must also exist a vector with rational coordinates outside of V. But we already know that this is impossible. The contradiction proves that the numbers must be all equal.

301. Let S be a crucial subset of V. Let V_S be the vector space spanned by $V \backslash S$. By adding any vector of S to V_S, we turn this space into $\mathbb{R}^n$. This implies that V_S is a vector space of dimension $n - 1$ and all the vectors of V but the ones in S are in V_S. Now let $W = \cap V_S$, where the intersection is over all crucial subsets. By the finiteness of dimensions, W can bw written as the intersection of a finite collection of spaces V_S, say $W = V_{S_1} \cap V_{S_2} \cap \cdots \cap V_{S_m}$, and assume m is minimal. Starting with V_{S_1} we add at each step to the intersection $V_{S_1} \cap V_{S_2} \cap \cdots \cap V_{S_j}$ a subspace $V_{S_{j+1}}$ such that

$$\dim(V_{S_1} \cap V_{S_2} \cap \cdots \cap V_{S_j}) > \dim(V_{S_1} \cap V_{S_2} \cap \cdots \cap V_{S_j} \cap V_{S_{j+1}}).$$

Hence $m \leq n$. As all but finitely many vectors of V belong to V_{S_i}, we conclude that all but finitely many vectors of V belong to W. But the vectors from S do not belong to V_S, and hence do not belong to W, for any crucial subset S. The we only have finitely many vectors of V, namely those not in W, to choose from for building a crucial set. Thus there are only finitely many crucial sets.

(*Mathematical Reflections*, proposed by I. Boreico)

302. Let λ_1, λ_2 be the eigenvalues of A. Then $-\lambda_1 \mathcal{I}_2$ and $-\lambda_2 \mathcal{I}_2$ both belong to $C(A)$, so

$$0 = |\det(A_{-}\lambda_i \mathcal{I}_2)| \geq |\lambda_i|^2, \text{ for } i = 1, 2.$$

It follows that $\lambda_1 = \lambda_2 = 0$. Change the basis to v, w with v an eigenvector of A (which does exist because $Av = 0$ has nontrivial solutions). This transforms the matrix into one of the form

$$\begin{pmatrix} 0 & a \\ 0 & 0 \end{pmatrix}.$$

One easily checks that the square of this matrix is zero.

Conversely, assume that $A^2 = \mathcal{O}_2$. By the spectral mapping theorem both eigenvalues of A are zero, so by appropriately choosing the basis we can make A look like

$$\begin{pmatrix} 0 & a \\ 0 & 0 \end{pmatrix}.$$

If $a = 0$, we are done. If not, then

$$C(A) = \left\{ \begin{pmatrix} \alpha & \beta \\ 0 & \alpha \end{pmatrix} \mid \alpha, \beta \in \mathbb{R} \right\}.$$

One verifies immediately that for every $B \in C(A)$, $\det(A + B) = \det B$. So the inequality from the statement is satisfied with equality. This completes the solution.

(Romanian Mathematical Olympiad, 1999, proposed by D. Miheţ)

303. Since $\det B = 1$, B is invertible and B^{-1} as integer entries. From

$$A^3 + B^3 = ((AB^{-1})^3 + \mathcal{I}_2)B^3,$$

it follows that $\det((AB^{-1})^3 + \mathcal{I}_2) = 1$. We will show that $(AB^{-1})^2 = \mathcal{O}_2$. Set $AB^{-1} = C$.

We know that $\det(C^3 + \mathcal{I}_2) = 1$. We have the factorization

$$C^3 + \mathcal{I}_2 = (C + \mathcal{I}_2)(C + \varepsilon\mathcal{I}_2)(C + \varepsilon^2\mathcal{I}_2),$$

where ε is a primitive cubic root. Taking determinants, we obtain

$$P(-1)P(-\varepsilon)P(-\varepsilon^2) = 1,$$

where P is the characteristic polynomial of C.

Let $P(x) = x^2 - mx + n$; clearly m, n are integers. Because $P(-\varepsilon^2) = P(-\bar{\varepsilon}) = \overline{P(\varepsilon)}$, it follows that $P(-\varepsilon)P(-\varepsilon^2)$ is a positive integer. So $P(-1) = P(-\varepsilon)P(-\varepsilon^2) = 1$. We obtain $1 + m + n = 1$ and $(\varepsilon^2 + m\varepsilon + n)(\varepsilon + m\varepsilon^2 + 1) = 1$, which, after some algebra, give $m = n = 0$. So C has just the eigenvalue 0, and being a 2×2 matrix, its square is zero.

Finally, from the fact that $AB = BA$ and $(AB^{-1})^2 = \mathcal{O}_2$, we obtain $A^2 B^{-2} = \mathcal{O}_2$, and multiplying on the right by B^2 we have $A^2 = \mathcal{O}_2$, as desired.

(Romanian Mathematics Competition, 2004, proposed by M. Becheanu)

304. *First solution* The eigenvalues are the zeros of the polynomial $\det(\lambda\mathcal{I}_n - aA - bA')$. The matrix $\lambda\mathcal{I}_n - aA - bA'$ is a circulant matrix, and the determinant of a circulant matrix was the subject of problem 262 in Section 2.3.2. According to that formula,

$$\det(\lambda\mathcal{I}_n - aA - bA') = (-1)^{n-1}\prod_{j=0}^{n-1}(\lambda\zeta^j - a\zeta^{2j} - b),$$

where $\zeta = e^{2\pi i/n}$ is a primitive nth root of unity. We find that the eigenvalues of $aA + bA'$ are $a\zeta^j + b\zeta^{-j}, j = 0, 1, \ldots, n - 1$.

Second solution: Simply note that for $\zeta = e^{2\pi i/n}$ and $j = 0, 1, \ldots, n - 1$, $(1, \zeta^j, \zeta^{2j}, \ldots, \zeta^{(n-1)j})$ is an eigenvector with eigenvalue $a\zeta^j + b\zeta^{-j}$.

305. We have

$$\det(\mathcal{I}_n - A) = \det(A)\det(\mathcal{I}_n - A) = \det(A')\det(\mathcal{I}_n - A)$$
$$= \det(A' - \mathcal{I}_n) = \det(A - \mathcal{I}_n) = -\det(\mathcal{I}_n - A).$$

Hence $\det(\mathcal{I}_n - A) = 0$, showing that 1 is an eigenvalue.

Remark. The matrices with the property from the statement form the special orthogonal group $SO(3)$, which is a Lie group whose Lie algebra will be introduced in Section 4.1.1. This is the group of orientation-preserving isometries of $\mathbb{R}^3$, and as a corollary of what we just proved we obtain the fact that any such isometry is the rotation about an axis (the axis of rotation is specified by the corresponding eigenvector).

306. For $n = 2007$ we may choose the signs so that each row of A sums to zero. This means that $Au = 0$ for u the column vector with all entries equal to 1 and hence $\det(A) = 0$.

For $n = 2008$, let J be the 2008×2008 matrix all whose entries are 1, and let u be the column vector with all entries equal to 1. Note that $J = u^t u$. Then u is an eigenvector of $J - I_{2008}$ with eigenvalue 2007, and a 2007-dimensional eigenspace with eigenvalue -1 (namely the space of all vectors v with $u^t v = 0$). Thus

$$\det(J - I_{2008}) = (-1)^{2007} \times 2007 = -2007.$$

This number is odd. For any matrix A of the given form, $A \equiv J - I_{2008} \pmod 2$. Hence $\det(A) \equiv \det(J - I_{2008}) \pmod 2$, so $\det A$ is an odd number, which therefore cannot be equal to zero.

(*Mathematical Reflections*, proposed by A. Ilič)

307. Since A is skew-symmetric, its eigenvalues are purely imaginary. It follows that the nonzero roots of the equation

$$\det(A + x I_n) = 0$$

come in pairs $(z_j, \bar{z}_j) = (z_j, -z_j), j = 1, 2, \ldots, k$. Then

$$\det(A + x I_n) \cdot \det(A + y I_n) = x^{n-2k} \prod_{j=1}^{k} (x^2 + |z_j|^2) y^{n-2k} \prod j = 1^k (y^2 + |z_j|^2)$$

$$= (xy)^{n-2k} \prod_{j=1}^{k} (x^2 + |z_j|^2)(y^2 + |z_j|^2)$$

$$\geq (xy)^{n-2k} \prod_{j=1}^{k} (xy + |z_j|^2)^2 = \left(\det(A + \sqrt{xy} I_n) \right)^2.$$

Here we used the Cauchy-Schwarz identity:

$$(x^2 + |z_j|^2)(y^2 + |z_j|^2) \geq (xy + |z_j|^2)^2.$$

The problem is solved.

(Romanian Mathematical Olympiad, 2008)

308. Let ϕ be the linear transformation of the space $\mathbb{R}^n$ whose matrix in a certain basis $e_1, e_2, \ldots, e_n$ is A. Consider the orthogonal decompositions of the space $\mathbb{R}^n = \ker \phi \oplus T$ $\mathbb{R}^n = \operatorname{Im} \phi \oplus S$. Set $\phi' = \phi|_T$. Then $\phi' : T \to \operatorname{Im} \phi$ is an isomorphism. Let γ' be its inverse, which we extend to a linear transformation γ of the whole of $\mathbb{R}^n$ by setting $\gamma|_S = 0$. Then $\phi \gamma \phi = \phi' \gamma' \phi' = \phi'$ on T and $\phi \gamma \phi = 0$ on $T^{\perp} = \ker \phi$. Hence $\phi \gamma \phi = \phi$, and we can choose B to be the matrix of γ in the basis $e_1, e_2, \ldots, e_n$.

(Soviet Union University Student Mathematical Olympiad, 1976)

309. The map that associates to the angle the measure of its projection onto a plane is linear in the angle. The process of taking the average is also linear. Therefore, it suffices to check the statement for a particular angle. We do this for the angle of measure π, where it trivially works.

Remark. This lemma allows another proof of Fenchel's theorem, which is the subject of problem 758 in Section 4.1.6. If we defined the total curvature of a polygonal line to be the sum of the "exterior" angles, then the projection of any closed polygonal line in three-dimensional space onto a one-dimensional line has total curvature at least $\pi + \pi = 2\pi$ (two complete turns). Hence the total curvature of the curve itself is at least 2π.

(Communicated by J. Sullivan)

310. The first involution A that comes to mind is the symmetry with respect to a hyperplane. For that particular involution, the operator $B = \frac{1}{2}(A + \mathcal{I})$ is the projection onto the hyperplane, where $\mathcal{I}$ is the identity map. Let us show that in general for any involution A, the operator B defined as such is a projection. We have

$$B^2 = \frac{1}{4}(A + \mathcal{I})^2 = \frac{1}{4}(A^2 + 2A\mathcal{I} + \mathcal{I}^2) = \frac{1}{4}(\mathcal{I} + 2A + \mathcal{I}) = B.$$

There exists a basis of V consisting of eigenvectors of B. Just consider the decomposition of V into the direct sum of the image of B and the kernel of B. The eigenvectors that form the basis are either in the image of B, in which case their eigenvalue is 1, or in the kernel, in which case their eigenvalue is 0. Because $A = 2B - \mathcal{I}$, it has the same eigenvectors as B, with eigenvalues ± 1. This proves (a).

Part (b) is based on the fact that any family of commuting diagonalizable operators on V can be diagonalized simultaneously. Let us prove this property by induction on the dimension of V. If all operators are multiples of the identity, there is nothing to prove. If one of them, say S, is not a multiple of the identity, then consider the eigenspace V_λ of a certain eigenvalue λ. If T is another operator in the family, then since $STv = TSv = \lambda Tv$, it follows that $Tv \in V_\lambda$; hence V_λ is an invariant subspace for all operators in the family. This is true for all eigenspaces of A, and so all operators in the family are diagonal blocks on the direct decomposition of V into eigenvectors of A. By the induction hypothesis, the family can be simultaneously diagonalized on each of these subspaces, and so it can be diagonalized on the entire space V.

Returning to the problem, diagonalize the pairwise commuting involutions. Their diagonal entries may equal $+1$ or -1 only, showing that there are at most 2^n such involutions. The number can be attained by considering all choices of sign on the diagonal.

(3rd International Competition in Mathematics for University Students, 1996)

311. From the orthogonality of Au and u, we obtain

$$\langle Au, u \rangle = \langle u, A^t u \rangle = \langle A^t u, u \rangle = 0.$$

Adding, we obtain that $\langle (A + A^t)u, u \rangle = 0$ for every vector u. But $A + A^t$ is symmetric, hence diagonalizable. For an eigenvector v of eigenvalue λ, we have

$$\langle (A + A^t)v, v \rangle = \langle \lambda v, v \rangle = \lambda \langle v, v \rangle = 0.$$

This shows that all eigenvalues are zero, so $A + A^t = 0$, which proves (a).

As a corollary of this, we obtain that A is of the form

$$A = \begin{pmatrix} 0 & a_{12} & a_{13} \\ -a_{12} & 0 & a_{23} \\ -a_{13} & -a_{23} & 0 \end{pmatrix}.$$

So A depends on only three parameters, which shows that the matrix can be identified with a three-dimensional vector. To choose this vector, we compute

$$Au = \begin{pmatrix} 0 & a_{12} & a_{13} \\ -a_{12} & 0 & a_{23} \\ -a_{13} & -a_{23} & 0 \end{pmatrix} \begin{pmatrix} u_1 \\ u_2 \\ u_3 \end{pmatrix} = \begin{pmatrix} a_{12}u_1 + a_{13}u_2 \\ -a_{12}u_1 + a_{23}u_3 \\ -a_{13}u_1 - a_{23}u_2 \end{pmatrix}.$$

It is easy to see now that if we set $v = (-a_{23}, a_{13}, -a_{12})$, then $Au = v \times u$.

Remark. The set of such matrices is the Lie algebra so(3), and the problem describes two of its well-known properties.

312. There is a more general property, of which the problem is a particular case.

Riesz lemma. *If V is a finite-dimensional vector space with inner product $\langle \cdot, \cdot \rangle$, then any linear functional $f : V \to \mathbb{R}$ is of the form $f(x) = \langle x, z \rangle$ for some unique $z \in V$.*

Proof. This result can be generalized to any (complex) Hilbert space, and it is there where it carries the name of F. Riesz. If f is identically zero, then $f(x) = \langle x, 0 \rangle$. Otherwise, let W be the kernel of f, which has codimension 1 in V. There exists a nonzero vector y orthogonal to W such that $f(y) = 1$. Set $\mu = \langle y, y \rangle$ and define $z = \mu^{-1}y$. Then $\langle z, z \rangle = \mu^{-1}$. Any vector $x \in V$ is of the form $x' + \lambda z$, with $x' \in W$. We compute

$$f(x) = f(x') + \lambda f(z) = \lambda \mu^{-1} = \lambda \langle z, z \rangle = \langle x', z \rangle + \lambda \langle z, z \rangle = \langle x, z \rangle.$$

Note that z is unique, because if $\langle x, z \rangle = \langle x, z' \rangle$ for all x, then $z - z'$ is orthogonal to all vectors, hence is the zero vector. There exists a simpler proof, but the one we gave here can be generalized to infinite-dimensional Hilbert spaces!   □

For our particular case, $V = M_n(\mathbb{R})$ and the inner product is the famous Hilbert-Schmidt inner product $\langle A, B \rangle = \text{tr}(AB^t)$.

For the second part of the problem, the condition from the statement translates to $\text{tr}((AB - BA)C) = 0$ for all matrices A and B. First, let us show that all off-diagonal entries of C are zero. If c_{ij} is an entry of C with $i \neq j$, let A be the matrix whose entry a_{ik} is 1 and all others are 0, and B the matrix whose entry b_{kj} is 1 and all others are 0, for some number k. Then $\text{tr}((AB - BA)C) = c_{ij} = 0$. So C is diagonal. Moreover, choose $a_{ij} = b_{ij} = 1$, with $i \neq j$. Then $AB - BA$ has two nonzero entries, the (i, i) entry, which is 1, and the (j, j) entry, which is -1. Therefore, $\text{tr}((AB - BA)C) = c_{ii} - c_{jj} = 0$. We deduce that all diagonal entries of C are equal to some number λ, and hence

$$f(A) = \text{tr}(AC) = \text{tr}(\lambda A) = \lambda \text{tr}(A),$$

as decided.

Remark. The condition $f(AB) = f(BA)$ gives

$$\text{tr}(AC) = f(A) = f(ABB^{-1}) = f(B^{-1}AB) = \text{tr}(B^{-1}ABC) = \text{tr}(ABCB^{-1});$$

hence by uniqueness of C, we have shown that $C = BCB^{-1}$ for all B, or $BC = CB$. The solution of the problem is essentially a proof that if C commutes with all invertible matrices B, then $C = \lambda \mathcal{I}_n$ for some scalar λ.

313. Fix $x \in \mathbb{R}^n$ with $\|x\| = 1$, and let $y = U^{-1}V^{-1}x$. Because U and V are isometric transformations, $\|y\| = 1$. Then

$$
\begin{aligned}
\|UVU^{-1}V^{-1}x - x\| &= \|UVy - VUy\| \\
&= \|(U - \mathcal{I}_n)(V - \mathcal{I}_n)y - (V - \mathcal{I}_n)(U - \mathcal{I}_n)y\| \\
&\leq \|(U - \mathcal{I}_n)(V - \mathcal{I}_n)y\| + \|(V - \mathcal{I}_n)(U - \mathcal{I}_n)y\|.
\end{aligned}
$$

The claim follows if we prove that $\|(U - \mathcal{I}_n)(V - \mathcal{T}_n)y\|$ and $\|(V - \mathcal{I}_n)(U - \mathcal{I}_n)y\|$ are both less than $\frac{1}{4}$, and because of symmetry, it suffices to check this for just one of them. If $(V - \mathcal{I}_n)y = 0$, then $\|(U - \mathcal{I}_n)(V - \mathcal{I}_n)y\| = 0 < \frac{1}{4}$. Otherwise, using the properties of vector length, we proceed as follows:

$$
\begin{aligned}
\|(U - \mathcal{I}_n)(V - \mathcal{I}_n)y\| &= \left\| (U - \mathcal{I}_n)(V - \mathcal{I}_n)y \frac{(V - \mathcal{I}_n)y}{\|(V - \mathcal{I}_n)y\|} \right\| \\
&= \|(V - \mathcal{I}_n)y\| \times \|(U - \mathcal{I}_n)z\|,
\end{aligned}
$$

where z is the length one vector $\frac{1}{\|(V-\mathcal{I}_n)y\|}(V - \mathcal{I}_n)y$. By the hypothesis, each factor in the product is less than $\frac{1}{2}$. This proves the claim and completes the solution.

314. The equality for general k follows from the case $k = n$, when it is the well-known $\det(AB) = \det(BA)$. Apply this to

$$
\begin{pmatrix} \mathcal{I}_n & A \\ \mathcal{O}_n & \mathcal{I}_n \end{pmatrix} \begin{pmatrix} \lambda \mathcal{I}_n - AB & \mathcal{O}_n \\ B & \mathcal{I}_n \end{pmatrix} = \begin{pmatrix} \lambda \mathcal{I}_n & A \\ B & \mathcal{I}_n \end{pmatrix} = \begin{pmatrix} \mathcal{I}_n & \mathcal{O}_n \\ B & \mathcal{I}_n \end{pmatrix} \begin{pmatrix} \mathcal{I}_n & A \\ \mathcal{O}_n & \lambda \mathcal{I}_n - BA \end{pmatrix}
$$

to obtain

$$\det(\lambda \mathcal{I}_n - AB) = \det(\lambda \mathcal{I}_n - BA).$$

The coefficient of λ^k in the left-hand side is $\phi_k(AB)$, while the coefficient of λ^k in the right-hand side is $\phi_k(BA)$, and they must be equal.

Remark. From the many applications of the functions $\phi_k(A)$, we mention the construction of Chern classes in differential geometry.

315. From

$$\mathcal{I}_2 = (u\mathcal{I}_2 + vA)(u'\mathcal{I}_2 + v'A) = uu'\mathcal{I}_2 + (uv' + vu')A + vv'A^2,$$

by using the Cayley-Hamilton Theorem, we obtain

$$\mathcal{I}_2 = (uu' - vv'\det A)\mathcal{I}_2 + (uv' + vu' + vv'\text{tr}A)A.$$

Thus u' and v' should satisfy the linear system

$$uu' - (v \det A)v' = 1,$$
$$vu' + (u + v\operatorname{tr}A)v' = 0.$$

The determinant of the system is $u^2 + uv\operatorname{tr}A + v^2 \det A$, and an easy algebraic computation shows that this is equal to $\det(u\mathcal{I}_2 + vA)$, which is nonzero by hypothesis. Hence the system can be solved, and its solution determines the desired inverse.

316. Rewriting the matrix equation as

$$X^2(X - 3\mathcal{I}_2) = \begin{pmatrix} -2 & -2 \\ -2 & -2 \end{pmatrix}$$

and taking determinants, we obtain that either $\det X = 0$ or $\det(X - 3\mathcal{I}_2) = 0$. In the first case, the Cayley-Hamilton equation implies that $X^2 = (\operatorname{tr}X)X$, and the equation takes the form

$$[(\operatorname{tr}X)^2 - 3\operatorname{tr}X]X = \begin{pmatrix} -2 & -2 \\ -2 & -2 \end{pmatrix}.$$

Taking the trace of both sides, we find that the trace of X satisfies the cubic equation $t^3 - 3t^2 + 4 = 0$. with real roots $t = 2$ and $t = -1$. In the case $\operatorname{tr}X = 2$, the matrix equation is

$$-2X = \begin{pmatrix} -2 & -2 \\ -2 & -2 \end{pmatrix}$$

with the solution

$$X = \begin{pmatrix} 1 & 1 \\ 1 & 1 \end{pmatrix}.$$

When $\operatorname{tr}X = -1$, the matrix equation is

$$4X = \begin{pmatrix} -2 & -2 \\ -2 & -2 \end{pmatrix}$$

with the solution

$$X = \begin{pmatrix} -\frac{1}{2} & -\frac{1}{2} \\ -\frac{1}{2} & -\frac{1}{2} \end{pmatrix}.$$

Let us now study the case $\det(X - 3\mathcal{I}_2) = 0$. One of the two eigenvalues of X is 3. To determine the other eigenvalue, add $4\mathcal{I}_2$ to the equation from the statement. We obtain

$$X^3 - 3X^2 + 4\mathcal{I}_2 = (X - 2\mathcal{I}_2)(X + \mathcal{I}_2) = \begin{pmatrix} -2 & -2 \\ -2 & -2 \end{pmatrix}.$$

Taking determinants we find that either $\det(X - 2\mathcal{I}_2) = 0$ or $\det(X + \mathcal{I}_2) = 0$. So the second eigenvalue of X is either 2 or -1. In the first case, the Cayley-Hamilton equation for X is

$$X^2 - 5X + 6\mathcal{I}_2 = 0,$$

which can be used to transform the original equation into

$$4X - 12\mathcal{I}_2 = \begin{pmatrix} -2 & -2 \\ -2 & -2 \end{pmatrix}$$

with the solution

$$X = \begin{pmatrix} \frac{5}{2} & -\frac{1}{2} \\ -\frac{1}{2} & \frac{5}{2} \end{pmatrix}.$$

The case in which the second eigenvalue of X is -1 is treated similarly and yields the solution

$$X = \begin{pmatrix} 1 & -2 \\ -2 & 1 \end{pmatrix}.$$

(Romanian competition, 2004, proposed by A. Buju)

317. Because the trace of $[A, B]$ is zero, the Cayley.Hamilton Theorem for this matrix is $[A, B]^2 + (\det[A, B])\mathcal{I}_2 = 0$, which shows that $[A, B]^2$ is a multiple of the identity. The same argument applied to the matrices $[C, D]$ and $[A, B] + [C, D]$ shows that their squares are also multiples of the identity.

We have

$$[A, B] \cdot [C, D] + [C, D] \cdot [A, B] = ([A, B] + [C, D])^2 - [A, B]^2 - [C, D]^2.$$

Hence $[A, B] \cdot [C, D] + [C, D] \cdot [A, B]$ is also a multiple of the identity, and the problem is solved.

(Romanian Mathematical Olympiad, 1981, proposed by C. Năstăsescu)

318. The Cayley-Hamilton Theorem gives

$$(AB - BA)^3 - c_1(AB - BA)^2 + c_2(AB - BA) - c_3\mathcal{I}_3 = \mathcal{O}_3,$$

where $c_1 = \text{tr}(AB - BA) = 0$, and $c_3 = \det(AB - BA)$. Taking the trace and using the fact that the trace of $AB - BA$ is zero, we obtain $\text{tr}((AB - BA)^3) - 3\det(AB - BA) = 0$, and the equality is proved.

(*Revista Matematică din Timişoara (Timişoara Mathematics Gazette)*, proposed by T. Andreescu)

319. Let $F(x) = x^2 - \text{tr}(AB)x + \det(AB)$ be the characteristic polynomial of AB. Using the division algorithm, write $P(x) = F(x)Q(x) + \alpha x + \beta$. By the Cayley-Hamilton theorem, $F(AB) = F(BA) = 0$. So $P(AB) = \alpha(AB) + \beta\mathcal{I}_2$. Similarly $P(BA) = \alpha(BA) + \beta\mathcal{I}_2$. Equating the two we obtain $\alpha(AB - BA) = \mathcal{O}_2$, so $\alpha = 0$. Thus $P(AB) = \beta\mathcal{I}_2$.

(*Mathematical Reflections*, proposed by G. Dospinescu)

320. Let $C = AB - BA$. We have

$$AB^2 + BA^2 = (AB - BA)B + B(AB - BA) = CB + BC = 2BC.$$

Let $P_B(\lambda) = \lambda^2 + r\lambda + s$ be the characteristic polynomial of B. By the Cayley-Hamilton Theorem, $P_B(B) = 0$. We have

$$\mathcal{O}_2 = AP_B(B) - P_B(B)A = AB^2 - B^2A + r(AB - BA) = 2BC + rC.$$

Using this and the fact that C commutes with A and B, we obtain

$$\mathcal{O}_2 = A(2BC + rC) - (2BC + rC)A = 2(AB - BA)C = 2C^2.$$

Therefore, $C^2 = \mathcal{O}_2$. In some basis

$$C = \begin{pmatrix} 0 & \alpha \\ 0 & 0 \end{pmatrix}.$$

Hence C commutes only with polynomials in C. But if A and B are polynomials in C, then $C = \mathcal{O}_2$, a contradiction. So C must be scalar whose square is equal to zero, whence $C = \mathcal{O}_2$ again. This shows that such matrices A and B do not exist.

(*American Mathematical Monthly*, solution by W. Gustafson)

321. Choose $\lambda \in \mathbb{R}$ sufficiently large such that $\lambda \mathcal{I}_n + A$ has positive entries. By the Perron-Frobenius theorem, the largest eigenvalue ρ of $\lambda \mathcal{I}_n + A$ is positive, and all other eigenvalues lie inside the circle of radius ρ centered at the origin. In particular, ρ is real and all other eigenvalues lie strictly to its left. The eigenvalues of A are the horizontal translates by λ of the eigenvalues of $\lambda \mathcal{I}_n + A$, so they enjoy the same property.

Remark. The result is true even for matrices whose off-diagonal entries are nonnegative, the so-called Metzler matrices, where a more general form of the Perron-Frobenius theorem needs to be applied.

322. *First solution*: Define $A = (a_{ij})_{i,j=1}^3$. Then replace A by $B = \alpha \mathcal{I}_3 - A$, where α is chosen large enough so that the entries b_{ij} of the matrix B are all positive. By the Perron-Frobenius theorem, there exist a positive eigenvalue λ and an eigenvector $c = (c_1, c_2, c_3)$ with positive coordinates. The equality $Bc = \lambda c$ yields

$$a_{11}c_1 + a_{12}c_2 + a_{13}c_3 = (\alpha - \lambda)c_1,$$

$$a_{21}c_1 + a_{22}c_2 + a_{23}c_3 = (\alpha - \lambda)c_2,$$

$$a_{31}c_1 + a_{32}c_2 + a_{33}c_3 = (\alpha - \lambda)c_3.$$

The three expressions from the statement have the same sign as $\alpha - \lambda$: they are either all three positive, all three zero, or all three negative.

Second solution: The authors of this problem had a geometric argument in mind. Here it is.

Consider the points $P(a_{11}, a_{21}, a_{31})$, $Q(a_{12}, a_{22}, a_{32})$, $R(a_{13}, a_{23}, a_{33})$ in three-dimensional Euclidean space. It is enough to find a point in the interior of the triangle PQR whose coordinates are all positive, all negative, or all zero.

Let P', Q', R' be the projections of P, Q, R onto the xy-plane. The hypothesis implies that P', Q', and R' lie in the fourth, second, and third quadrant, respectively.

Case 1. The origin O is in the exterior or on the boundary of the triangle $P'Q'R'$ (Figure 69).

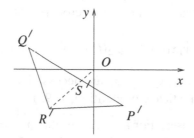

Figure 69

Denote by S' the intersection of the segments $P'Q'$ and OR', and let S be the point on the segment PQ whose projection is S'. Note that the z-coordinate of the point S is negative, since the z-coordinates of P' and Q' are negative. Thus any point in the interior of the segment SR sufficiently close to S has all coordinates negative, and we are done.

Case 2. The origin O is in the interior of the triangle $P'Q'R'$ (Figure 70).

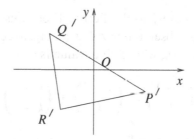

Figure 70

Let T be the point inside the triangle PQR whose projection is O. If $T = O$, we are done. Otherwise, if the z-coordinate of T is negative, choose a point S close to it inside the triangle PQR whose x- and y-coordinates are both negative, and if the z-coordinate of T is positive, choose S to have the x- and y-coordinates positive. Then the coordinates of S are all negative, or all positive, and again we are done.

(Short list of the 44th International Mathematical Olympiad, 2003, proposed by the USA)

323. Let λ be the positive eigenvalue and $v = (v_1, v_2, \ldots, v_n)$ the corresponding eigenvector with positive entries of the transpose of the coefficient matrix. The function $y(t) = v_1 x_1(t) + v_2 x_2(t) + \cdots + v_n x_n(t)$ satisfies

$$\frac{dy}{dt} = \sum_{i,j} v_i a_{ij} x_j = \sum_j \lambda v_j x_j = \lambda y.$$

Therefore, $y(t) = e^{\lambda t} y_0$, for some vector y_0. Because

$$\lim_{t \to \infty} y(t) = \sum_i v_i \lim_{t \to \infty} x_i(t) = 0,$$

and $\lim_{t \to \infty} e^{\lambda t} = \infty$, it follows that y_0 is the zero vector. Hence

$$y(t) = v_1 x_1(t) + v_2 x_2(t) + \cdots + v_n x_n(t) = 0,$$

which shows that the functions $x_1, x_2, \ldots, x_n$ are necessarily linearly dependent.

(56th W.L. Putnam Mathematical Competition, 1995)

324. We try some particular cases. For $n = 2$, we obtain $c = 1$ and the sequence 1, 1, for $n = 3$, $c = 2$ and the sequence 1, 2, 1, and for $n = 4$, $c = 3$ and the sequence 1, 3, 3, 1. We formulate the hypothesis that $c = n - 1$ and $x_k = \binom{n-1}{k-1}$.

The condition $x_{n+1} = 0$ makes the recurrence relation from the statement into a linear system in the unknowns $(x_1, x_2, \ldots, x_n)$. More precisely, the solution is an eigenvector of the matrix $A = (a_{ij})_{ij}$ defined by

$$a_{ij} = \begin{cases} i & \text{if } j = i + 1, \\ n - j & \text{if } j = i - 1, \\ 0 & \text{otherwise.} \end{cases}$$

This matrix has nonnegative entries, so the Perron-Frobenius Theorem as stated here does not really apply. But let us first observe that A has an eigenvector with positive coordinates, namely $x_k = \binom{n-1}{k-1}$, $k = 1, 2, \ldots, n$, whose eigenvalue is $n - 1$. This follows by rewriting the combinatorial identity

$$\binom{n-1}{k} = \binom{n-2}{k} + \binom{n-2}{k-1}$$

as

$$\binom{n-1}{k} = \frac{k+1}{n-1}\binom{n-1}{k+1} + \frac{n-k}{n-1}\binom{n-1}{k-1}.$$

To be more explicit, this identity implies that for $c = n - 1$, the sequence $x_k = \binom{n-1}{k-1}$ satisfies the recurrence relation from the statement, and $x_{n+1} = 0$.

Let us assume that $n - 1$ is not the largest value that c can take. For a larger value, consider an eigenvector v of A. Then $(A + \mathcal{I}_n)v = (c + 1)v$, and $(A + \mathcal{I}_n)^n v = (c + 1)^n v$. The matrix $(A + \mathcal{I}_n)^n$ has positive entries, and so by the Perron-Frobenius Theorem has a unique eigenvector with positive coordinates. We already found one such vector, that for which $x_k = \binom{n-1}{k-1}$. Its eigenvalue has the largest absolute value among all eigenvalues of $(A + \mathcal{I}_n)^n$, which means that $n^n > (c + 1)^n$. This implies $n > c + 1$, contradicting our assumption. So $n - 1$ is the largest value c can take, and the sequence we found is the answer to the problem.

(57th W.L. Putnam Mathematical Competition, 1997, solution by G. Kuperberg published in K. Kedlaya, B. Poonen, R. Vakil, *The William Lowell Putnam Mathematical Competition 1985–2000*, MAA, 2002)

325. Let us first show that if the two numbers are equal, then the product can be found in six steps. For $x \neq -1$, we compute (1) $x \to \frac{1}{x}$, (2) $x \to x + 1$, (3) $x + 1 \to \frac{1}{x+1}$, (4) $\frac{1}{x}, \frac{1}{x+1} \to \frac{1}{x} - \frac{1}{x+1} = \frac{1}{x^2+x}$, (5) $\frac{1}{x^2+x} \to x^2 + x$, (6) $x^2 + x, x \to x^2$.

If $x = -1$, replace step (2) by $x \to x - 1$ and make the subsequent modifications thereon.

If the two numbers are distinct, say x and y, perform the following sequence of operations, where above each arrow we count the steps:

$$x, y \xrightarrow{1} x + y \xrightarrow{7} (x+y)^2,$$

$$x, y \xrightarrow{8} x - y \xrightarrow{14} (x-y)^2,$$

$$(x+y)^2, (x-y)^2 \xrightarrow{15} 4xy \xrightarrow{16} \frac{1}{4xy},$$

$$\frac{1}{4xy}, \frac{1}{4xy} \xrightarrow{17} \frac{1}{4xy} + \frac{1}{4xy} = \frac{2}{xy},$$

$$\frac{2}{4xy}, \frac{2}{4xy} \xrightarrow{18} \frac{2}{4xy} + \frac{2}{4xy} = \frac{4}{4xy} = \frac{1}{xy} \xrightarrow{19} xy.$$

So we are able to compute the product in just 19 steps.

(*Kvant (Quantum)*)

326. Building on the previous problem, we see that it suffices to produce an operation $\circ$, from which the subtraction and reciprocal are derivable. A good choice is $\frac{1}{x-y}$. Indeed, $\frac{1}{x} = \frac{1}{x-0}$, and also $x - y = \frac{1}{(1/(x-y))-0}$. Success!

(D.J. Newman, *A Problem Seminar*, Springer-Verlag)

327. Fix a and c in S and consider the function

$$f_{a,c}(b) = a * (b * c).$$

Because $a * f_{a,c}(b) * c = (a * a) * b * (c * c) = b$, the function is one-to-one. It follows that there are exactly two elements that are not in the image of $f_{a,c}$. These elements are precisely a and c. Indeed, if $a * (b * c) = a$, then $(a * a) * (b * c) = a * a$, so $b * c = a * a$, and then $b * (c * c) = (a * a) * c$, which implies $b = c$. This contradicts the fact that a, b, c are distinct. A similar argument rules out the case $a * (b * c) = c$.

Now choose a', c' different from both a and c. The union of the ranges of $f_{a,c}$ and $f_{a',c'}$, which is contained in the set under discussion, is the entire set S. The conclusion follows.

Remark. An example of such a set is the Klein 4-group.

(R. Gelca)

328. Consider the set

$$U = \{h(x, y) \mid h(-x, -y) = -h(x, y)\}.$$

It is straightforward to check that U is closed under subtraction and taking reciprocals. Because $f(x, y) = x$ and $g(x, y) = y$ are in U, the entire set S is in U. But U does not contain nonzero constant functions, so neither does S.

(*American Mathematical Monthly*, 1987, proposed by I. Gessel, solution by O.P. Lossers)

329. All three parts of the conclusion follow from appropriate substitutions in the identity from the statement. For example,

$$(e * e') \circ (e' * e) = (e \circ e') * (e' \circ e)$$

simplifies to $e' \circ e' = e * e$, which further yields $e' = e$, proving (a). Then, from

$$(x * e) \circ (e * y) = (x \circ e) * (e \circ y),$$

we deduce $x \circ y = x * y$, for every $x, y \in M$, showing that the two binary operations coincide. This further yields

$$(e * x) * (y * e) = (e * x) \circ (y * e) = (e \circ y) * (x \circ e) = (e * y) * (x * e),$$

and so $x * y = y * x$. Thus $*$ is commutative and (c) is proved.
  (Romanian high school textbook)

330. Substituting $x = u * v$ and $y = v$, with $u, v \in S$, in the given condition gives $(u * v) * (v * (u * v)) = v$. But $v * (u * v) = u$, for all $u, v \in S$. So $(u * v) * u = v$, for all $u, v \in S$. Hence the existence and uniqueness of the solution to the equation $a * x = b$ is equivalent to the existence and uniqueness of the solution to the equation $x * a = b$.

The existence of the solution for the equation $a * x = b$ follows from the fact that $x = b * a$ is a solution. To prove the uniqueness, let $c \in S$ be a solution. By hypothesis we have the equalities $a * (b * a) = b$, $b * (c * b) = c$, $c * (a * c) = a$. From $a * c = b$ it follows that $c * (a * c) = c * b = a$. So $a = c * b$, and from $a * c = b$ it follows that $c * (a * c) = c * b = a$. Therefore, $b * a = b * (c * b) = c$, which implies that $b * a = c$. This completes the proof.

331. Substituting $y = e$ in the second relation, and using the first, we obtain $x * z = (x * e) * z = (z * e) * x = z * x$, which proves the commutativity. Using it, the associativity is proved as follows:

$$(x * y) * z = (z * x) * y = (y * z) * x = x * (y * z).$$

  (A. Gheorghe)

332. The answer is yes. Let ϕ be any bijection of F with no fixed points. Define $x * y = \phi(x)$. The first property obviously holds. On the other hand, $x * (y * z) = \phi(x)$ and $(x * y) * z = \phi(x * y) = \phi(\phi(x))$. Again since ϕ has no fixed points, these two are never equal, so the second property also holds.
  (45th W.L. Putnam Mathematical Competition, 1984)

333. From $a * (a * a) = (a * a) * a$ we deduce that $a * a = a$. We claim that

$$a * (b * a) = a \text{ for all } a, b \in S.$$

Indeed, we have $a * (a * (b * a)) = (a * a) * (b * a) = a * (b * a)$ and $(a * (b * a)) * a = (a * b) * (a * a) = (a * b) * a$. Using associativity, we obtain

$$a * (a * (b * a)) = a * (b * a) = (a * b) * a = (a * (b * a)) * a.$$

The "noncommutativity" condition from the statement implies $a * (b * a) = a$, proving the claim.

We apply this property as follows:

$$(a * (b * c)) * (a * c) = (a * b) * (c * (a * c)) = (a * b) * c,$$

$$(a * c) * (a * (b * c)) = (a * (c * a)) * (b * c) = a * (b * c).$$

Since $(a * b) * c = a * (b * c)$ (by associativity), we obtain

$$(a * (b * c)) * (a * c) = (a * c) * (a * (b * c)).$$

This means that $a * (b * c)$ and $a * c$ commute, so they must be equal, as desired.

For an example of such a binary operation consider any set S endowed with the operation $a * b = a$ for any $a, b \in S$.

334. Using the first law we can write

$$y * (x * y) = (x * (x * y)) * (x * y).$$

Now using the second law, we see that this is equal to x. Hence $y * (x * y) = x$. Composing with y on the right and using the first law, we obtain

$$y * x = y * (y * (x * y)) = x * y.$$

This proves commutativity.

For the second part, the set S of all integers endowed with the operation $x * y = -x - y$ provides a counterexample. Indeed,

$$x * (x * y) = -x - (x * y) = -x - (-x - y) = y$$

and

$$(y * x) * x = -(y * x) - x = -(-y - x) - x = y.$$

Also, $(1 * 2) * 3 = 0$ and $1 * (2 * 3) = 4$, showing that the operation is not associative.
(33rd W.L. Putnam Mathematical Competition, 1972)

335. Define $r(x) = 0 * x$, $x \in \mathbb{Q}$. First, note that

$$x * (x + y) = (0 + x) * (y + x) = 0 * y + x = r(y) + x.$$

In particular, for $y = 0$ we obtain $x * x = r(0) + x = 0 * 0 + x = x$.

We will now prove a multiplicative property of $r(x)$, namely that $r\left(\frac{m}{n}x\right) = \frac{m}{n}r(x)$ for any positive integers m and n. To this end, let us show by induction that for all y and all positive integers n, $0 * y * \cdots * ny = nr(y)$. For $n = 0$ we have $0 = 0 \cdot r(y)$, and for $n = 1$ this follows from the definition of $r(y)$. Assume that the property is true for $k \leq n$ and let us show that it is true for $n + 1$. We have

$$0 * y * \cdots * ny * (n+1)y = 0 * y * \cdots * (ny * ny) * (n+1)y$$
$$= (0 * y * \cdots * ny) * (ny * (n+1)y)$$
$$= (n(0 * y)) * ((0 + ny) * (y + ny))$$
$$= (0 * y + (n - 1)(0 * y)) * (0 * y + ny)$$
$$= (n - 1)r(y) * ny + 0 * y.$$

Using the induction hypothesis, $(n-1)r(y) * ny = 0 * y * \cdots * (n-1)y * ny = nr(y)$ (this works even when $n = 1$). Hence $0 * y * \cdots * (n+1)y = nr(y) + r(y) = (n+1)r(y)$, which proves the claim.

Using this and the associativity and commutativity of $*$, we obtain

$$2nr(y) = 0 * y * 2y * \cdots * 2ny$$
$$= (0 * ny) * (y * (n+1)y) * (2y * (n+2)y) * \cdots * (ny * 2ny)$$
$$= r(ny) * (y * (y+ny)) * (2y * (2y+ny)) * \cdots * (ny * (ny+ny)).$$

The first formula we have proved implies that this is equal to

$$(0 + r(ny)) * (y + r(ny)) * \cdots * (ny + r(ny)).$$

The distributive-like property of $*$ allows us to transform this into

$$(0 * y * 2y * \cdots * ny) + r(ny) = nr(y) + r(ny).$$

Hence $2nr(y) = nr(y) + r(ny)$, or $r(ny) = nr(y)$. Replacing y by $\frac{x}{n}$, we obtain $r\left(\frac{x}{n}\right) = \frac{1}{n}r(x)$, and hence $r\left(\frac{m}{n}x\right) = \frac{m}{n}r(x)$, as desired.

Next, note that $r \circ r = r$; hence r is the identity function on its image. Also,

$$r(z) = 0 * z = (-z + z) * (0 + z) = (-z) * 0 + z = r(-z) + z,$$

or $r(z) - r(-z) = z$. Hence for $z \neq 0$, one of the numbers $r(z)$ and $r(-z)$ is nonzero. Let y be this number. Since $r(y) = y$, we have $y = r(y) - r(-y) = y - r(-y)$, so $r(-y) = 0$. Also, if $x = \frac{m}{n}y$, then $r(x) = \frac{m}{n}r(y) = \frac{m}{n}y = x$, and $r(-x) = \frac{m}{n}r(-y) = 0$. If $y > 0$, then $r(y) = \max(y, 0)$ and consequently $r(x) = x = \max(x, 0)$, for all $x > 0$, while $r(x) = 0 = \max(x, 0)$ for all $x < 0$. Similarly, if $y < 0$, then $r(y) = \min(y, 0)$, and then $r(x) = \min(x, 0)$ for all $x \in \mathbb{Q}$. The general case follows $(a-b+b) * (0+b) = (a-b) * 0 + b$.

(*American Mathematical Monthly*, proposed by H. Derksen, solution by J. Dawson)

336. For $x \in G$ and x' its left inverse, let $x'' \in G$ be the left inverse of x', meaning that $x''x' = e$. Then

$$xx' = e(xx') = (x''x')(xx') = x''(x'x)x' = x''(ex') = x''x' = e.$$

So x' is also a right inverse for x. Moreover,

$$xe = x(x'x) = (xx')x = ex = x,$$

which proves that e is both a left and right identity. It follows that G is a group.

337. Let $e \in G$ be the identity element. Set $b = e$ in the relation from the statement. Then

$$a = a * e = (a \perp a) \perp (a \perp e) = (a \perp a) \perp a,$$

and canceling a we obtain $a \perp a = e$, for all $a \in G$. Using this fact, we obtain

$$a * b = (a \perp a) \perp (a \perp b) = e \perp (a \perp b) = a \perp b,$$

which shows that the composition laws coincide. Because $a * a = e$, we see that $a^{-1} = a$, so for $a, b \in G$,

$$ab = (ab)^{-1} = b^{-1}a^{-1} = ba,$$

which proves the commutativity.

(D. Ştefănescu)

338. We can find the integers u and v such that $us + vt = 1$. Since $ab = ba$, we have

$$ab = (ab)^{us+vt} = (ab^{us}((ab)^t)^v = (ab)^{us}e = (ab)^{us} = a^{us}(b^s)^u = a^{us}e = a^{us}.$$

Therefore,

$$b^r = eb^r = a^r b^r = (ab)^r = a^{usr} = (a^r)^{us} = e.$$

Again we can find x, y such that $xr + ys = 1$. Then

$$b = b^{xr+ys} = (b^r)^x (b^s)^y = e.$$

Applying the same argument, mutatis mutandis, we find that $a = e$, so the first part of the problem is solved.

A counterexample for the case of a noncommutative group is provided by the cycles of permutations $a = (123)$ and $b = (34567)$ in the permutation group S_7 of order 7. Then $ab = (1234567)$ and $a^3 = b^5 = (ab)^7 = e$.

(8th International Competition in Mathematics for University Students, 2001)

339. For $g \in G$ the map $T_g : G \to G$, $T_g(x) = gx$ is bijective, because it is easy to check that its inverse is the map $T_{g^{-1}}$. For given g, each of the sets A and

$$T_g(A^{-1}) = \{ga^{-1} \mid a \in A\}$$

contains more than one half of the elements of G, so they overlap. If a_1 and ga_2 are in the overlap, that is if $a_1 = ga_2^{-1}$, with $a_1, a_2 \in A$, then $g = a_1 a_2$. We have proved that every element of G is the product of two elements of A, as desired.

(29th William Lowell Putnam Mathematical Competition, 1968)

340. Note first that if $(M, *)$ is a group, then the product of any two elements of M is again in M. Thus for $x, y \neq 3$, we must have

$$x * y = 3(x - 3)(y - 3) + (m - 27) \neq 3.$$

For $m \neq 30$ this is not always true, for example for $x = 10/3$ and $y = 33 - m$.

To see that $m = 30$ does produce a group, first note that in this case $x * y = 3(x - 3)(y - 3) + 3$. Define the map $f : M \to \mathbb{R}\backslash\{0\}$ by $f(x) = 3(x - 3)$. then f is a bijection whose inverse is $f^{-1}(t) = (t + 9)/3$. Further we compute that

$$f(x * y) = 9(x - 3)(y - 3) = f(x)f(y).$$

Thus for $m = 30$, $(M, *)$ is a group isomorphic to the multiplicative group of nozero real numbers.

(*Mathematical Reflections*, proposed by B. Enescu)

341. Set $c = aba^{-1}$ and observe that $ca = ab$ and $c^n = e$. We have

$$a = ea = c^n a = c^{n-1} ca = c^{n-1} ab = c^{n-2}(ca)b = c^{n-2} ab^2,$$

and, inductively,

$$a = c^{n-k} ab^k, \ 1 \le k \le n.$$

From $a = ab^n$, we obtain the desired conclusion $b^n = e$.

(*Gazeta Matematică* (*Mathematics Gazette, Bucharest*), proposed by D. Bătinețu-Giurgiu)

342. Applying the identity from the statement to the elements x and yx^{-1}, we have

$$xy^2 x^{-1} = x(yx^{-1})x(yx^{-1}) = (yx^{-1})x(yx^{-1})x = y^2.$$

Thus for any x, y, we have $xy^2 = y^2 x$. This means that squares commute with everything. Using this fact, we rewrite the identity from the statement as

$$xyxyx^{-1}y^{-1}x^{-1}y^{-1} = e$$

and proceed as follows:

$$e = xyxyx^{-1}y^{-1}x^{-1}y^{-1} = xyxyx^{-2}xy^{-2}yx^{-2}xy^{-2}y$$
$$= xyxyy^{-2}x^{-2}xyxyy^{-2}x^{-2} = (xyxyy^{-2}x^{-2})^2.$$

Because there are no elements of order 2, it follows that $xyxyy^{-2}x^{-2} = e$ and hence $xyxy = x^2 y^2$. Cancel an x and a y to obtain $yx = xy$. This proves that the group is Abelian, and we are done.

(K.S.Williams, K. Hardy, *The Red Book of Mathematical Problems*, Dover, Mineola, NY, 1996)

343. The first axiom shows that the squares of all elements in M are the same; denote the common value by e. Then $e^2 = e$, and from (ii), $ae = a$ for all $a \in M$. Also, $a * b = a(eb)$ for all $a, b \in M$. Let us verify the associativity of $*$. Using (iii) in its new form $e(bc) = cb$, we obtain

$$a * (b * c) = a[e(b(ec))] = a[(ec)b].$$

Continue using (iv) as follows:

$$a[(ec)b] = [a(eb)][((ec)b)(eb)] = [a(eb)][(ec)e] = [a(eb)](ec) = (a * b) * c.$$

Here we used the fact that $de = d$, for the case $d = ec$. Thus associativity is proved. The element e is a right identity by the following argument:

$$a * e = a(e^2 e) = a(ee) = ae^2 = ae = a.$$

The right inverse of a is ae, since

$$a * (ea) = a[e(ea)] = a(ae) = a^2 = e.$$

So there exists a right identity, and every element has a right inverse, which then implies that $(M, *)$ is a group.

(M. Becheanu, C. Vraciu, *Probleme de Teoria Grupurilor* (*Problems in Group Theory*), University of Bucharest, 1982)

344. The condition from the statement implies that for all integers m and n,

$$f(m\sqrt{2} + n\sqrt{3}) = f(0).$$

Because the ratio $\sqrt{2}/\sqrt{3}$ is irrational, the additive group generated by $\sqrt{2}$ and $\sqrt{3}$ is not cyclic. It means that this group is dense in $\mathbb{R}$. So f is constant on a dense subset of $\mathbb{R}$. Being continuous, it must be constant on the real axis.

345. The conclusion follows from the fact that the additive group

$$S = \{n + 2\pi m; \ m, n \text{ integers}\}$$

is dense in the real numbers. Indeed, by Kronecker's theorem, we only need to check that S is not cyclic. This is so because n and $2m\pi$ cannot both be integer multiples of the same number (they are incommensurable).

346. That 2^k starts with a 7 is equivalent to the existence of an integer m such $\frac{2^k}{10^m} \in [7, 8)$. Let us show that the set $\left\{ \frac{2^k}{10^m} \mid k, m \text{ integers} \right\}$ is dense in the positive real numbers. Canceling the powers of 2, this amounts to showing that $\left\{ \frac{2^n}{5^m} \mid m, n \text{ integers} \right\}$ is dense. We further simplify the problem by applying the function $\log_2$ to the fraction. This function is continuous, so it suffices to prove that $\{n - m \log_2 5 \mid m, n \text{ integers}\}$ is dense on the real axis. This is an additive group, which is not cyclic since $\log_2 5$ is irrational (and so 1 and $\log_2 5$ cannot both be integer multiples of the same number). It follows that this group is dense in the real numbers, and the problem is solved.

(V.I. Arnol'd, *Mathematical Methods of Classical Mechanics*, Springer-Verlag, 1997)

347. If r is the original ratio of the sides, after a number of folds the ratio will be $2^m 3^n r$, where m and n are integer numbers. It suffices to show that the set $\{2^m 3^n r \mid m, n \in \mathbb{Z}\}$ is dense in the positive real axis. This is the same as showing that $\{2^m 3^n \mid m, n \in \mathbb{Z}\}$ is dense. Taking the logarithm, we reduce the problem to the fact that the additive group $\{m + n \log_2 3 \mid m, n \in \mathbb{Z}\}$ is dense in the real axis. And this is true by Kronecker's theorem since the group is not cyclic.

(German Mathematical Olympiad)

348. Call the regular pentagon $ABCDE$ and the set Σ. Composing a reflection across AB with a reflection across BC, we can obtain a 108° rotation around B. The set Σ is invariant under this rotation. There is a similar rotation around C, of the same angle and opposite direction, which also preserves Σ. Their composition is a translation by a vector that makes an angle of 36° with BC and has length $2 \sin 54° BC$. Figure 71 helps us understand why this is so. Indeed, if P rotates to P' around B, and P' to P'' around C, then the triangle $P'BC$ transforms

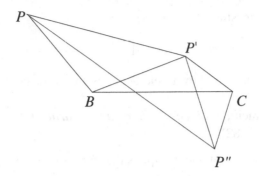

Figure 71

to the triangle $P'PP''$ by a rotation around P' of angle $\angle CP'P'' = 36°$ followed by a dilation of ratio $P'P''/P'C = 2\sin 54°$. Note that the translation preserves the set Σ.

Reasoning similarly with vertices A and D, and taking into account that AD is parallel to BC, we find a translation by a vector of length $2\sin 54° AD$ that makes an angle of $36°$ with BC and preserves Σ. Because $AD/BC = 2\sin 54° = \frac{\sqrt{5}+1}{2}$, the group G_{BC} generated by the two translations is dense in the group of all translations by vectors that make an angle of $36°$ with BC. The same is true if BC is replaced by AB. It follows that Σ is preserved both by the translations in the group G_{BC} and in the analogous group G_{AB}. These generate a group that is dense in the group of all translations of the plane. We conclude that Σ is a dense set in the plane, as desired.

(Communicated by K. Shankar)

349. Assume that A is a 4×4 matrix with $A \neq I_4$ but $A^7 = I_4$. Then the minimal polynomial of A divides $x^7 - 1 = (x - 1)(x^6 + x^5 + x^4 + \cdots + x + 1)$. Also the minimal polynomial divides the characteristic polynomial of A, which has degree at most 4. But in Problem 233 that $x^6 + x^5 + \cdots + 1$ is irreducible, so it has no divisor of degree at most 4. This contradiction shows that no such A exists.

(*Mathematical Reflections*, proposed by J.C. Mathieux)

350. How can we make the sum M interact with the multiplicative structure of Γ? The idea is to square M and use the distributivity of multiplication with respect to the sum of matrices. If $G_1, G_2, \ldots, G_k$ are the elements of Γ, then

$$M^2 = (G_1 + G_2 + \cdots + G_k)^2 = \sum_{i=1}^{k} G_i \left(\sum_{j=1}^{k} G_j \right) = \sum_{i=1}^{k} G_i \left(\sum_{G \in \Gamma} G_i^{-1} G \right)$$

$$= \sum_{G \in \Gamma} \sum_{i=1}^{k} G_i(G_i^{-1} G) = k \sum_{G \in \Gamma} G = kM.$$

Taking determinants, we find that $(\det M)^2 = k^n \det M$. Hence either $\det M = 0$ or $\det M$ is equal to the order of Γ raised to the nth power. The matrix $C = \frac{1}{k}M$, is an idempotent, that is $C^2 = C$. It follows that $\text{rank}(C) = \text{tr}\, C$, and so $\text{tr}\, B = m\text{rank}(C)$ is a multiple of k.

Remark. In fact, much more is true. The determinant of the sum of the elements of a finite multiplicative group of matrices is nonzero only when the group consists of one element, the identity, in which case it is equal to 1. This is the corollary of a basic fact in representation theory.

First notice that the determinant is invariant under change of basis. A representation of a group is a homomorphism of the group into a group of matrices. In our situation the group is already represented as a group of matrices. A representation is called irreducible if there does not exist a basis in which it can be decomposed into blocks. Any representation of a finite group is the block sum of irreducible representations. (Certainly we need to change basis to see the block decomposition, but the determinant is invariant under change of basis). The simplest representation, called the trivial representation, sends all elements of the group to the identity element. A result in representation theory states that for any nontrivial irreducible representation of a finite group, the sum of the matrices of the representation is zero. In an appropriately chosen basis, our group can be written as the block sum of irreducible representations. If the group is nontrivial, then at least one representation is nontrivial. In summing the elements of the group, the diagonal block corresponding to this irreducible representation is the zero matrix. Taking the determinant, we obtain zero.

The trace is also invariant under change of basis. M has nonzero entries only for the blocks that correspond to 1-dimensional irreducible representations, and in those blocks, the element of M is just k (you add a 1 for each matrix in Γ). The trace is either 0 or the sum of several k's, so it is a multiple of k. This shows that the value of the trace can only be a multiple of the order of Γ, and easily constructed examples using permutation matrices show that any such multiple can be obtained.

351. Let G be a finite group with the properties from the hypothesis. Then each element of the group has finite order.

Lemma. *If $A = \mathcal{I}_n + 4B$, where B is a nonzero matrix with integer coefficients, then for no power of A is the identity matrix.*

Proof. Arguing by contradiction, let m be the smallest positive integer such that $A^m = \mathcal{I}_n$. Writing $m = pk$, with p prime, we see that $(A^k)^p = \mathcal{I}_n$, and we can also see that $A^k = \mathcal{I}_n + 4B'$ for some matrix B'. Note that $B' =\neq \mathcal{O}_n$, because $A^k \neq \mathcal{I}_n$. So by substituting A with A^k we may assume that $m = p$, a prime number.

We expand

$$(\mathcal{I}_n + 4B)^p = \mathcal{I}_n + 4pB + 4^2\binom{p}{2}B^2 + \cdots + 4^p B^p = \mathcal{I}_n.$$

Hence

$$4pB + 4^2\binom{p}{2}B^2 + \cdots + 4^p B^p = \mathcal{O}_n,$$

from where

$$4pB = -4^2\binom{p}{2}B^2 - \cdots - 4^p B^p.$$

Let 2^j be the largest power of 2 that divides all entries of B. Then the largest power of 2 dividing the left hand side is either 2^{j+2} or 2^{j+3} (depending on whether $p = 2$ or not), while the largest power of 2 dividing the right-hand side is at least $4^2 \times 2^m = 2^{m+4}$. Thus the equality cannot hold, which is the desired contradiction. The lemma is proved.   □

From the lemma it follows that each element of G that is not the identity matrix has order exactly 2. Indeed, if $A = \mathcal{I}_n + 2B \in G$, then $A^2 = \mathcal{I}_n + 4(B^2 + B)$, and because A has finite order, $B^2 + B$ must be equal to zero. We have seen in the example from our discussion about groups that every such group is Abelian. Because all matrices of G commute, they can be simultaneously diagonalized. In diagonal form they still form a group, so the diagonal entries must be ± 1. There are only 2^n possibilities for choosing these diagonal entries, so G has at most 2^n elements. This is optimal since we can consider the group of all diagonal matrices with diagonal entries ± 1, and this group has 2^n elements and satisfies the condition from the statement.

(*Mathematical Reflections*, proposed by G. Dospinescu)

352. Let $A \in G$, then for all integers k, $A^k \in G$, so $\|A^k - \mathcal{I}_n\| < a$. Let λ be an eigenvalue of A and let x be a corresponding eigenvector. Then

$$\left(A^k - \mathcal{I}_n\right) x = \left(\lambda^k - 1\right) x,$$

and so $|\lambda^k - 1| \le a < 2$ for all integers k. Thus $|\lambda|^k < 3$ for all integers k, positive or negative, which can only happen if $|\lambda| = 1$. Set $\lambda = e^{i\pi r}$, $r \in \mathbb{R}$. Since $|\lambda^k - 1| \le a < 2$, we have

$$\cos(\pi k r) \ge 1 - \frac{a^2}{2} > -1.$$

This means that $\cos \pi k r$ is not dense in $[-1, 1]$. As a consequence of Kronecker's Theorem, this can only happen if r is rational. Moreover, the inequality implies that the denominator of r is bounded in terms of r only. Consequently, there is a positive integer M such that for every $A \in G$ and every eigenvalue λ of A, there is $k < M$ with $\lambda^k = 1$. Setting $N = M!$, we obtain $\lambda^N = 1$ for all $A \in G$, and all eigenvalues λ of A.

Fix $A \in G$, and write $A^N = \mathcal{I}_n + B$. Then B is nilpotent (that is $B^n = \mathcal{O}_n$) because by the Spectral mapping theorem all of its eigenvalues are zero. We know that for all positive integers p, $\|A^{Np} - \mathcal{I}_n\| < 2$. By using the binomial formula, we obtain that

$$\left\| \binom{p}{1} B + \binom{p}{2} B^2 + \cdots + \binom{p}{n-1} B^{n-1} \right\| < 2.$$

Here we used the fact that $B^k = \mathcal{O}_n$ for all $k \ge n$. Assume that B is not zero and let j be the largest positive integer such that $B^j \ne \mathcal{O}_n$. Then, using the triangle inequality, we have

$$2 > \binom{p}{j} \|B^j\| - \binom{p}{j-1} \|B^{j-1}\| - \cdots - \binom{p}{1} \|B\|.$$

This cannot hold for all p, since the right-hand side is a polynomial in p of degree j with dominant coefficient $\frac{1}{j!} \|B^j\|$. We conclude that $B = 0$.

So there is an integer N such that $A^N = \mathcal{I}_n$ for all $A \in G$. The following result will show that G is finite.

Burnside's theorem. If a multiplicative group G of matrices has the property that there exists a positive integer N such that for every $A \in G$, $A^N = \mathcal{I}_N$, then G is finite.

Proof. There are elements $A_1, A_2, \ldots, A_r$ of G that form a basis for the linear space spanned by elements of G in the space of $n \times n$ matrices (they exist because any spanning set of a finite dimensional vector space contains a basis). We claim that the function

$$f : G \to \mathbb{C}^r, \quad f(X) = (\mathrm{tr}(A_1 X), \mathrm{tr}(A_2 X), \ldots, \mathrm{tr}(A_r X)),$$

is injective. Assume to the contrary that there are $A, B \in G$ with $f(A) = f(B)$. Because $A_1, A_2, \ldots, A_r$ are a basis of the linear space spanned by G, and because trace is linear, it follows that $\mathrm{tr}(AX) = \mathrm{tr}(BX)$ for all $X \in G$. Thus if $U = AB^{-1}$, we have $\mathrm{tr}(UX) = \mathrm{tr}(X)$, for all $X \in G$. Set $X = \mathcal{I}_n$ to conclude that $\mathrm{tr}(U) = n$. But we also know that $U^N = \mathcal{I}_n$, so $U = \mathcal{I}_N$, because the trace is the sum of the eigenvalues which are roots of unity. Thus $A = B$.

The image of f consists of sums of roots of unity of order N, so it is finite. Hence G itself is finite, and the theorem is proved. $\square$

Remark. The definition of $\|A\|$ is standard in mathematics, it is called the norm of A, and has the same properties that the norm (length) of a vector has: (i) $\|A\| = 0$ if and only if $A = \mathcal{O}_n$, (ii) $\|\lambda A\| = |\lambda| \|A\|$, (iii) $\|A + B\| \leq \|A\| + \|B\|$. The norm allows us to define a distance between matrices, and so we can define the notion of convergence of a sequence of matrices. The algebra of all $n \times$ matrices endowed with this norm has the properties that addition, scalar multiplication, and multiplication are continuous in the norm, and also $\|AB\| \leq \|A\| \|B\|$. Moreover, every Cauchy sequence is convergent. As such it is a *Banach algebra*. The theory of Banach algebras is an important chapter in functional analysis.

(*Mathematical Reflections*, proposed by G. Dospinescu and A. Thiery)

353. The symmetry groups are, respectively, C_{2v}, D_{2h}, and D_{2d}.

354. (a) In the first example from this section, let $x = ab$, $y = c$.
(b) If

$$A = \begin{pmatrix} 1 & 0 \\ 0 & 0 \end{pmatrix}, \quad B = \begin{pmatrix} 0 & 1 \\ 1 & 0 \end{pmatrix}, \quad C = \begin{pmatrix} 0 & 1 \\ 0 & 0 \end{pmatrix}$$

then $I_n - ABC = I_n$ while

$$I_n - CBA = \begin{pmatrix} 0 & 0 \\ 0 & 1 \end{pmatrix},$$

which is not invertible.

355. If x is an idempotent, then $1 - x$ is an idempotent as well. Indeed,

$$(1 - x)^2 = 1 - 2x + x^2 = 1 - 2x + x = 1 - x.$$

Thus there is an involution on M, $x \mapsto 1 - x$. This involution has no fixed points, since $x = 1 - x$ implies $x^2 = x - x^2$ or $x = x - x = 0$. But then $0 = 1 - 0 = 1$, impossible. Having no fixed points, the involution pairs the elements of M, showing that the cardinality of M is even.

(*Gazeta Matematică* (*Mathematics Gazette, Bucharest*), proposed by V. Zidaru)

356. We have $y = y^6 = (-y)^6 = -y$, hence $2y = 0$ for any $y \in R$. Now let x be an arbitrary element in R. Using the binomial formula, we obtain

$$x + 1 = (x + 1)^6 = x^6 + 6x^5 + 15x^4 + 20x^3 + 15x^2 + 6x + 1 = x^4 + x^2 + x + 1,$$

where we canceled the terms that had even coefficients. Hence $x^4 + x^2 = 0$, or $x^4 = -x^2 = x^2$. We then have
$$x = x^6 = x^2 x^4 = x^2 x^2 = x^4 = x^2,$$

and so $x^2 = x$, as desired. From the equality $(x + y)^2 = x + y$ we deduce $xy + yx = 0$, so $xy = -yx = yx$ for any x, y. This shows that the ring is commutative, as desired.

357. Substituting x by $x + 1$ in the relation from the statement, we find that

$$((x + 1)y)^2 - (x + 1)^2 y^2 = (xy)^2 + xy^2 + yxy + y^2 - x^2 y^2 - 2xy^2 - y^2$$
$$= yxy - xy^2 = 0.$$

Hence $xy^2 = yxy$ for all $x, y \in R$. Substituting in this relation y by $y + 1$, we find that

$$xy^2 + 2xy + x = yxy + yx + xy + x.$$

Using the fact that $xy^2 = yxy$, we obtain $xy = yx$, as desired.

358. (a) First let us notice that for every $a \in R$, $na = 0$ (this is because n is the order of the additive group of R). Now suppose there is a nilpotent element x, and let m be the smallest positive integer such that $x^m = 0$. Then $y = x^{m-1}$ has the property that $y \neq 0$ but $y^2 = 0$. Using the binomial expansion we obtain

$$(1 + y)^n = 1 + ny = 1.$$

So $1 + y = 1$, which forces y to be zero, a contradiction. This proves (a)

(b) Let $x_1, x_2, \ldots, x_n$ be the elements of R. The set of n-tuples

$$\{(x_1^j, x_2^j, \ldots, x_n^j) \mid j > 0\}$$

is finite, since each entry can only take finitely many values. So there are positive integers $p < q$ such that $x^p = x^q$ for all $x \in R$. For each $x \in R$, $x(x^{q-p} - 1)$ is therefore nilpotent, and therefore zero by part (a). If we take $k = q - p + 1$, we have $k \geq 2$ and $x^k = x$ for all $x \in R$, as desired.

Remark. The author of the problem pointed out the following fact. By Jacobson's theorem, mentioned after the second example from the theory, R is commutative. Moreover, R is a product of fields, which can be proved by induction on the number of elements of R as follows.

If R contains no nontrivial idempotent (an idempotent is element $e \neq 0, 1$ such that $e^m = 1$ for some m), then because x^{k-1} is an idempotent, it is either equal to 0 or 1. Consequently, every $x \in R$ is either 0 or is invertible (here we use the fact that there are no nilpotents), making R into a field.

If R has a nontrivial idempotent e, then $1 - e$ is also an idempotent, and R is isomorphic to $Re \times R(1 - e)$. All elements of Re and $R(1 - e)$ satisfy $x^k = x$, so by induction, these rings are products of fields, and so is R.

(Romanian Mathematical Olympiad, 2008)

359. For every $x \in R$, we have $2x = x + x = (1 + 1)x = 0$. Let

$$U = \{x \in R \,|\, x^2 = 0\} \text{ and } V = \{x \in R \,|\, x^2 = 1\}.$$

We define the map $\phi : R \to R$, $\phi(x) = x + 1$. If $x \in U$, then $x^2 = 0$, so $(\phi(x))^2 = (x+1)^2 = x^2 + 2x + 1 = 1$. Hence $\phi(x) \in V$. On the other hand, if $y \in V$, set $x = y - 1$. Then $x^2 = y^2 - 2y + 1 = 1 + 1 = 0$, thus $x \in U$. We conclude that ϕ is a bijection between U and V and we are done.

(*Mathematical Reflections*, proposed by M. Piticari)

360. This problem generalizes the first example from the introduction. The idea of the solution is similar. Let v be the inverse of $1 - (xy)^n$. Then $v(1 - (xy)^n) = (1 - (xy)^n)v = 1$; hence $v(xy)^n = (xy)^n v = v - 1$. We claim that the inverse of $1 - (yx)^n$ is $1 + (yx)^{n-1}yvx$. Indeed, we compute

$$(1 + (yx)^{n-1}yvx)(1 - (yx)^n) = 1 - (yx)^n + (yx)^{n-1}yvx - (yx)^{n-1}yvx(yx)^n$$
$$= 1 - (yx)^n + (yx)^{n-1}yvx - (yx)^{n-1}yv(xy)^n x$$
$$= 1 - (yx)^n + (yx)^{n-1}yvx - (yx)^{n-1}y(v - 1)x = 1.$$

Similarly,

$$(1 - (yx)^n)(1 + (yx)^{n-1}yvx) = 1 - (yx)^n + (yx)^{n-1}yvx - (yx)^n(yx)^{n-1}yvx$$
$$= 1 - (yx)^n + (yx)^{n-1}yvx - (yx)^{n-1}y(xy)^n vx$$
$$= 1 - (yx)^n + (yx)^{n-1}yvx - (yx)^{n-1}y(v - 1)x = 1.$$

It follows that $1 - (yx)^n$ is invertible and its inverse is $1 + (yx)^{n-1}yvx$.

361. (a) Let x and z be as in the statement. We compute

$$(zxz - xz)^2 = (zxz - xz)(zxz - xz)$$
$$= (zxz)(zxz) - (zxz)(xz) - (xz)(zxz) + (xz)(xz)$$
$$= zxz^2xz - zxzxz - xz^2xz + xzxz$$
$$= zxzxz - zxzxz - xzxz - xzxz = 0.$$

Therefore, $(zxz - xz)^2 = 0$, and the property from the statement implies that $zxz - xz = 0$.

(b) We have seen in part (a) that if z is an idempotent, then $zxz - xz = 0$. The same argument works, mutatis mutandis, to prove that $zxz = zx$. Hence $xz = zxz = zx$, which shows that z is in the center of R, and we are done.

362. We will show that the elements

$$ac, a^2c, a^3c, \ldots, a^nc, \ldots$$

are distinct. Let us argue by contradiction assuming that there exist $n > m$ such that $a^nc = a^mc$. Multiplying by c on the left, we obtain $ca(a^{n-1}c) = ca(a^{m-1}c)$, so by (iii), $ba^{n-1}c = ba^{m-1}c$. Cancel b as allowed by hypothesis (ii) to obtain $a^{n-1}c = a^{m-1}c$. Inductively $a^{n-j}c = a^{m-j}c$ for $j \leq m$. Thus $a^kc = c$, where $k = n - m$. Multiplying on the right by a and using $ca = b$, we also obtain $a^kb = b$. The first condition shows that b commutes with a, and so $ba^k = b$; canceling b yields $a^k = 1$. Hence a is invertible and $a^{-1} = a^{k-1}$.

The hypothesis $ca = b$ implies

$$c = ba^{-1} = ba^{k-1} = a^{k-1}b = a^{-1}b,$$

hence $ac = b$, contradicting (iii). The contradiction proves that the elements listed in the beginning of the solution are all distinct, and the problem is solved.

(*Gazeta Matematică* (*Mathematics Gazette, Bucharest*), proposed by C. Guțan)

Real Analysis

363. Examining the sequence, we see that the mth term of the sequence is equal to n exactly for those m that satisfy

$$\frac{n^2 - n}{2} + 1 \leq m \leq \frac{n^2 + n}{2}.$$

So the sequence grows about as fast as the square root of twice the index. Let us rewrite the inequality as

$$n^2 - n + 2 \leq 2m \leq n^2 + n,$$

then try to solve for n. We can almost take the square root. And because m and n are integers, the inequality is equivalent to

$$n^2 - n + \frac{1}{4} < 2m < n^2 + n + \frac{1}{4}.$$

Here it was important that $n^2 - n$ is even. And now we can take the square root. We obtain

$$n - \frac{1}{2} < \sqrt{2m} < n + \frac{1}{2},$$

or

$$n < \sqrt{2m} + \frac{1}{2} < n + 1.$$

Now this happens if and only if $n = \left\lfloor \sqrt{2m} + \frac{1}{2} \right\rfloor$, which then gives the formula for the general term of the sequence

$$a_m = \left\lfloor \sqrt{2m} + \frac{1}{2} \right\rfloor, \quad m \geq 1.$$

(R. Graham, D. Knuth, O. Patashnik, *Concrete Mathematics: A Foundation for Computer Science*, 2nd ed., Addison-Wesley, 1994)

© Springer International Publishing AG 2017
R. Gelca and T. Andreescu, *Putnam and Beyond*, DOI 10.1007/978-3-319-58988-6

364. If we were given the recurrence relation $x_n = x_{n-1} + n$, for all n, the terms of the sequence would be the triangular numbers $T_n = \frac{n(n+1)}{2}$. If we were given the recurrence relation $x_n = x_{n-1} + n - 1$, the terms of the sequence would be $T_{n-1} + 1 = \frac{n^2-n+2}{2}$. In our case,

$$\frac{n^2 - n + 2}{2} \leq x_n \leq \frac{n^2 + n}{2}.$$

We expect $x_n = P(n)/2$ for some polynomial $P(n) = n^2 + an + b$; in fact, we should have $x_n = \lfloor P(n)/2 \rfloor$ because of the jumps. From here one can easily guess that $x_n = \left\lfloor \frac{n^2+1}{2} \right\rfloor$, and indeed

$$\left\lfloor \frac{n^2+1}{2} \right\rfloor = \left\lfloor \frac{(n-1)^2+1}{2} + \frac{2(n-1)+1}{2} \right\rfloor = \left\lfloor \frac{(n-1)^2+1}{2} + \frac{1}{2} \right\rfloor + (n-1),$$

which is equal to $\left\lfloor \frac{(n-1)^2+1}{2} \right\rfloor + (n-1)$ if n is even, and to $\left\lfloor \frac{(n-1)^2+1}{2} \right\rfloor + n$ if n is odd.

Remark. The answer to the problem can also be given in the form

$$x_n = \left\lfloor \frac{n}{2} \right\rfloor^2 + \left\lceil \frac{n}{2} \right\rceil^2.$$

365. From the hypothesis it follows that $a_4 = 12, a_5 = 25, a_6 = 48$. We observe that

$$\frac{a_1}{1} = \frac{a_2}{2} = 1, \ \frac{a_3}{3} = 2, \ \frac{a_4}{4} = 3, \ \frac{a_5}{5} = 5, \ \frac{a_6}{6} = 8$$

are the first terms of the Fibonacci sequence. We conjecture that $a_n = nF_n$, for all $n \geq 1$. This can be proved by induction with the already checked cases as the base case.

The inductive step is

$$\begin{aligned} a_{n+4} &= 2(n+3)Fn + 3 + (n+2)F_{n+2} - 2(n+1)F_{n+1} - nF_n \\ &= 2(n+3)Fn + 3 + (n+2)F_{n+2} - 2(n+1)F_{n+1} - n(F_{n+2} - F_{n+1}) \\ &= 2(n+3)F_{n+3} + 2F_{n+2} - (n+2)(F_{n+3} - F_{n+2}) \\ &= (n+4)(F_{n+3} + F_{n+2}) = (n+4)F_{n+4}. \end{aligned}$$

This proves our claim.

(*Revista Matematică din Timişoara* (*Timişoara Mathematics Gazette*), proposed by D. Andrica)

366. Note that (i) implies $x_i < 2n$ for all i. We will examine the possible values of x_1. It cannot happen that $x_1 = 1$, because then (ii) implies that all numbers less than $2n$ should be terms of the sequence, which is impossible since the sequence has only $n - 1$ terms.

If $x_1 = 2$, then by (ii) the numbers $2, 4, 6, \ldots, 2n-2$ are terms of the sequence, and since the sequence has exactly $n - 1$ terms we get $x_i = 2i, i = 1, 2, \ldots, n - 1$. This sequence satisfies condition (i) as well, so it is a solution to the problem.

Let us examine the case $x_1 \geq 3$. If $n = 2$, the only possibility is $x_1 = 3$, which violates (i). If $n = 3$, then we have the possibilities $x_1 = 3, x_2 = 4; x_1 = 3, x_2 = 5; x_1 = 4, x_2 = 5,$

all three of which violate (a). This suggests that this case yields no solutions to the problem. Assume that for some n there is such a sequence with $x_1 \geq 3$. The numbers

$$x_1, 2x_1, \ldots, \left\lfloor \frac{2n}{x_1} \right\rfloor x_1$$

are terms of the sequence, and no other multiples of x_1 are. Because $x_1 \geq 3$, the above accounts for at most $\frac{2}{3}n$ terms of the sequence, so there must be another term besides these. Let x_j be the smallest term of the sequence that does not appear in the above listing. Then the first j terms of the sequence are

$$x_1, \quad x_2 = 2x_1, \quad \ldots, \quad x_{j-1} = (j-1)x_1, \quad x_j,$$

and we have $x_j < jx_1$. Condition (i) implies that the last j terms of the sequence must be

$$x_{n-1} = 2n - x_1, \quad x_{n-2} = 2n - 2x_1, \quad \ldots, \quad x_{n-j+1} = 2n - (j-1)x_1, \quad x_{n-j} = 2n - x_j.$$

But then $x_1 + x_{n-j} < x_1 + x_{n-1} = 2n$, hence by condition (ii) there exists k such that $x_1 + x_{n-j} = x_k$. We have

$$x_k = x_1 + x_{n-j} = x_1 + 2n - x_j = 2n - (x_j - x_1)$$
$$> 2n - (jx_1 - x_1) = 2n - (j-1)x_1 = x_{n-j+1}$$

on the one hand, and

$$x_k = x_1 + x_{n-j} < x_1 + x_{n-j+1} = x_{n-j+2}.$$

This means that x_k is between x_{n-j+1} and x_{n-j+2} which contradicts the fact that the terms $x_{n-1}, x_{n-2}, \ldots, x_{n-j}$ are the last j terms of the sequence.

We conclude that there is no such sequence with $x_1 \geq 3$, and so the only sequence with the required property is $x_i = 2i$, $i = 1, 2, \ldots, n-1$.

Remark. This problem was inspired by the properties of the Weierstrass gaps in the theory of Riemann surfaces. In short, a Riemann surface is a surface that has local coordinates that look like the complex coordinates of the plane. A meromorphic function on the Riemann surface is locally a quotient of two holomorphic functions (see Section 3.3.3). The points where the meromorphic function has a zero denominator is called a pole. Around a pole p, the meromorphic function can be written as $f(z)/(z-p)^n$ where f is a holomorphic function that is not zero at p; the number n is called the order of the pole. The Weierstrass gaps theorem states that for every compact Riemann surface, which has genus g (meaning that it resembles a sphere with g handles), for every point p there exist g positive integers $1 = n_1 < n_2 < \cdots < n_g < 2g$ such that for no j does there exist a meromorphic function that is holomorphic off p and has a pole of order n_j at p.

(USA Junior Mathematical Olympiad, 2010, proposed by R. Gelca)

367. The relations

$$a_m + a_m = \frac{1}{2}(a_{2m} + a_0) \quad \text{and} \quad a_{2m} + a_0 = \frac{1}{2}(a_{2m} + a_{2m})$$

imply $a_{2m} = 4a_m$, as well as $a_0 = 0$. We compute $a_2 = 4$, $a_4 = 16$. Also, $a_1 + a_3 = (a_2 + a_4)/2 = 10$, so $a_3 = 9$. At this point we guess that $a_k = k^2$ for all $k \geq 1$.

We prove our guess by induction on k. Suppose that $a_j = j^2$ for all $j < k$. The given equation with $m = k - 1$ and $n = 1$ gives

$$a_n = \frac{1}{2}(a_{2n-2} + a_2) - a_{n-2} = 2a_{n-1} + 2a_1 - a_{n-2}$$
$$= 2(n^2 - 2n + 1) + 2 - (n^2 - 4n + 4) = n^2.$$

This completes the proof.

(Russian Mathematical Olympiad, 1995)

368. *First solution*: If we compute some terms, $a_0 = 0, a_1 = 2, a_3 = 8, a_4 = 34, a_5 = 144$, we recognize Fibonacci numbers, namely F_0, F_3, F_6, F_9, and F_{12}. So a good working hypothesis is that $a_n = F_{3n}$ and also that $b_n = (F_n)^3$, for all $n \geq 0$, from which the conclusion would then follow.

We use induction. Everything is fine for $n = 0$ and $n = 1$. Assuming $a_k = F_{3k}$ for all $k \leq n$, we have

$$a_{n+1} = 4F_{3n} + F_{3n-3} = 3F_{3n} + F_{3n} + F_{3n-3}$$
$$= 3F_{3n} + F_{3n-1} + F_{3n-2} + F_{3n-3} = 3F_{3n} + F_{3n-1} + F_{3n-1}$$
$$= F_{3n} + 2F_{3n} + 2F_{3n-1} = F_{3n} + 2F_{3n+1} = F_{3n} + F_{3n+1} + F_{3n+1}$$
$$= F_{3n+2} + F_{3n+1} = F_{3n+3} = F_{3(n+1)},$$

which proves the first part of the claim.

For the second part we deduce from the given recurrence relations that

$$b_{n+1} = 3b_n + 6b_{n-1} - 3b_{n-2} - b_{n-3}, \ n \geq 3.$$

We point out that this is done by substituting $a_n = b_{n+1} + b_n - b_{n-1}$ into the recurrence relation for $(a_n)_n$. On the one hand, $b_n = (F_n)^3$ is true for $n = 0, 1, 2, 3$. The assumption $b_k = (F_k)^3$ for all $k \leq n$ yields

$$b_{n+1} = 3(F_n)^3 + 6(F_{n-1})^3 - 3(F_{n-2})^3 - (F_{n-3})^3$$
$$= 3(F_{n-1} + F_{n-2})^3 + 6(F_{n-1})^3 - 3(F_{n-2})^3 - (F_{n-1} - F_{n-2})^3$$
$$= 8(F_{n-1})^3 + 12(F_{n-1})^2 F_{n-2} + 6F_{n-1}(F_{n-2})^2 + (F_{n-2})^3$$
$$= (2F_{n-1} + F_{n-2})^3 = (F_{n+1})^3.$$

This completes the induction, and with it the solution to the problem.

Second solution: Another way to prove that $b_n = (F_n)^3$ is to observe that both sequences satisfy the same linear recurrence relation. Let

$$M = \begin{pmatrix} 1 & 1 \\ 1 & 0 \end{pmatrix}.$$

We have seen before that

$$M^n = \begin{pmatrix} F_{n+1} & F_n \\ F_n & F_{n-1} \end{pmatrix}.$$

Now the conclusion follows from the equality $M^{3n} = (M^n)^3$.

Remark. A solution based on the Binet formula is possible if we note the factorization

$$\lambda^4 - 3\lambda^3 - 6\lambda^2 + 3\lambda + 1 = (\lambda^2 - 4\lambda - 1)(\lambda^2 + \lambda - 1).$$

Setting the left-hand side equal to 0 gives the characteristic equation for the sequence $(b_n)_n$, while setting the first factor on the right equal to 0 gives the characteristic equation for $(a_n)_n$.

(Proposed by T. Andreescu for a Romanian Team Selection Test for the International Mathematical Olympiad, 2003, remark by R. Gologan)

369. We compute $u_0 = 1 + 1$, $u_1 = 2 + \frac{1}{2}$, $u_2 = 2 + \frac{1}{2}$, $u_3 = 8 + \frac{1}{8}$. A good guess is $u_n = 2^{x_n} + 2^{-x_n}$ for some sequence of positive integers $(x_n)_n$.

The recurrence gives

$$2^{x_{n+1}} + 2^{-x_{n+1}} = 2^{x_n + 2x_{n-1}} + 2^{-x_n - 2x_{n-1}} + 2^{x_n - 2x_{n-1}} + 2^{-x_n + 2x_{n-1}} - 2^{x_1} - 2^{-x_1}.$$

In order to satisfy this we hope that $x_{n+1} = x_n + 2x_{n-1}$ and that $x_n - 2x_{n-1} = \pm x_1 = \pm 1$. The characteristic equation of the first recurrence is $\lambda^2 - \lambda - 2 = 0$, with the roots 2 and -1, and using the fact that $x_0 = 0$ and $x_1 = 1$ we get the general term of the sequence $x_n = (2^n - (-1)^n)/3$. Miraculously this also satisfies $x_n - 2x_{n-1} = (-1)^{n+1}$ so the second condition holds as well. We conclude that $\lfloor u_n \rfloor = 2^{x_n}$, and so $\lfloor u_n \rfloor = 2^{[2^n - (-1)^n]/3}$.

(18th International Mathematical Olympiad, 1976, proposed by the UK)

370. We need to determine m such that $b_m > a_n > b_{m-1}$. It seems that the difficult part is to prove an inequality of the form $a_n > b_m$, which reduces to $3^{a_{n-1}} > 100^{b_{m-1}}$, or $a_{n-1} > (\log_3 100)b_{m-1}$. Iterating, we obtain $3^{a_{n-2}} > (\log_3 100)100^{b_{m-2}}$, that is,

$$a_{n-2} > \log_3(\log_3 100) + (\log_3 100)b_{m-2}.$$

Seeing this we might suspect that an inequality of the form $a_n > u + vb_n$, holding for all n with some fixed u and v, might be useful in the solution. From such an inequality we would derive $a_{n+1} = 3^{a_n} > 3^u(3^v)^{b_m}$. If $3^v > 100$, then $a_{n+1} > 3^u b_{m+1}$, and if $3^u > u + v$, then we would obtain $a_{n+1} > u + vb_{m+1}$, the same inequality as the one we started with, but with $m + 1$ and $n + 1$ instead of m and n.

The inequality $3^v > 100$ holds for $v = 5$, and $3^u > u + 5$ holds for $u = 2$. Thus $a_n > 2 + 5b_m$ implies $a_{n+1} > 2 + 5b_{m+1}$. We have $b_1 = 100$, $a_1 = 3$, $a_2 = 27$, $a_3 = 3^{27}$, and $2 + 5b_1 = 502 > 729 = 3^6$, so $a_3 = 2 + 5b_1$. We find that $a_n > 2 + 5b_{n-2}$ for all $n \geq 3$. In particular, $a_n \geq b_{n-2}$.

On the other hand, $a_n < b_m$ implies $a_{n+1} = 3^{a_n} < 100^{b_m} < b_{m+1}$, which combined with $a_2 < b_1$ yields $a_n < b_{n-1}$ for all $n \geq 2$. Hence $b_{n-2} < a_n < b_{n-1}$, which implies that $m = n - 1$, and for $n = 100$, $m = 99$.

(Short list of the 21st International Mathematical Olympiad, 1979, proposed by Romania, solution by I. Cuculescu)

371. Assume that we have found such numbers for every n. Then $q_{n+1}(x) - xq_n(x)$ must be divisible by $p(x)$. But

$$q_{n+1}(x) - xq_n(x) = x^{n+1} - a_{n+1}x - b_{n+1} - x^{n+1} + a_nx^2 + b_nx$$
$$= -a_{n+1}x - b_{n+1} + a_n(x^2 - 3x + 2) + 3a_nx - 2a_n + b_nx$$
$$= a_n(x^2 - 3x + 2) + (3a_n + b_n - a_{n+1})x - (2a_n + b_{n+1}),$$

and this is divisible by $p(x)$ if and only if $3a_n + b_n - a_{n+1}$ and $2a_n + b_{n+1}$ are both equal to zero. This means that the sequences a_n and b_n are uniquely determined by the recurrences $a_1 = 3$, $b_1 = -2$, $a_{n+1} = 3a_n + b_n$, $b_{n+1} = -2a_n$. The sequences exist and are uniquely defined by the initial condition.

372. Divide through by the product $(n+1)(n+2)(n+3)$. The recurrence relation becomes

$$\frac{x_n}{n+3} = 4\frac{x_{n-1}}{n+2} + 4\frac{x_{n-2}}{n+1}.$$

The sequence $y_n = x_n/(n+3)$ satisfies the recurrence

$$y_n = 4y_{n-1} - 4y_{n-2}.$$

Its characteristic equation has the double root 2. Knowing that $y_0 = 1$ and $y_1 = 1$, we obtain $y_n = 2^n - n2^{n-1}$. follows that the answer to the problem is

$$x_n = (n+3)2^n - n(n+3)2^{n-1}.$$

(D. Buşneag, I. Maftei, *Teme pentru cercurile şi concursurile de matematică* (*Themes for mathematics circles and contests*), Scrisul Românesc, Craiova)

373. Define $c = b/x_1$ and consider the matrix

$$A = \begin{pmatrix} 0 & c \\ x_1 & a \end{pmatrix}.$$

It is not hard to see that

$$A^n = \begin{pmatrix} cx_{n-1} & cx_n \\ x_n & x_{n+1} \end{pmatrix}.$$

Using the equality $\det A^n = (\det A)^n$, we obtain

$$c(x_{n-1}x_{n+1} - x_n^2) = (-x_1c)^n = (-b)^n.$$

Hence $x_n^2 - x_{n+1}x_{n-1} = (-b)^{n-1}x_1$, which does not depend on a.

Remark. In the particular case $a = b = 1$, we obtain the well-known identity for the Fibonacci sequence $F_{n+1}F_{n-1} - F_n^2 = (-1)^{n+1}$.

374. A standard idea is to eliminate the square root. If we set $b_n = \sqrt{1 + 24a_n}$, then $b_n^2 = 1 + 24a_n$, and so

$$
\begin{aligned}
b_{n+1}^2 = 1 + 24a_{n+1} &= 1 + \frac{3}{2}(1 + 4a_n + \sqrt{1 + 24a_n}) \\
&= 1 + \frac{3}{2}\left(1 + \frac{1}{6}(b_n^2 - 1) + b_n\right) \\
&= \frac{1}{4}(b_n^2 + 6b_n + 9) = \left(\frac{b_n + 3}{2}\right)^2.
\end{aligned}
$$

Hence $b_{n+1} = \frac{1}{2}b_n + \frac{3}{2}$, $b_1 = 5$. This is an inhomogeneous first-order linear recursion. We can solve this by analogy with inhomogeneous linear first-order equations. Recall that if a, b are constants, then the equation $f'(x) = af(x) + b$ has the solution

$$
f(x) = e^{ax} \int e^{-ax} b \, dx + c e^{ax}.
$$

In our problem the general term should be

$$
b_n = \frac{1}{2^n} + 3 \sum_{k=1}^{n-1} \frac{1}{2^k}, \quad n \ge 1.
$$

Summing the geometric series, we obtain $b_n = 3 - \frac{1}{2^{n-1}}$, and the answer to our problem is

$$
a_n = \frac{b_n^2 - 1}{24} = \frac{1}{3} - \frac{1}{2^{n+1}} + \frac{1}{3} \cdot \frac{1}{2^{2n+1}}.
$$

(Proposed by Germany for the 22nd International Mathematical Olympiad, 1981)

375. Call the expression from the statement S_n. It is not hard to find a way to write it in closed form. For example, if we let $u = 1 + i\sqrt{a}$, then $S_n = \frac{1}{2}(u^n + \bar{u}^n)$.

Notice that u and $\bar{u}$ are both roots of the quadratic equation $z^2 - 2z + a + 1 = 0$, so they satisfy the recurrence relation $x_{n+1} + 2x_{n+1} - (a+1)x_n$. The same should be true for S_n; hence

$$
S_{n+1} = 2S_{n+1} - (a + 1)S_n, \quad n \ge 1.
$$

One verifies that $S_1 = 1$ and $S_2 = 1 - a$ are divisible by 2. Also, if S_n is divisible by 2^{n-1} and S_{n+1} is divisible by 2^n, then $(a + 1)S_n$ and $2S_{n+1}$ are both divisible by 2^{n+1}, and hence so must be S_{n+2}. The conclusion follows by induction.

(Romanian Mathematical Olympiad, 1984, proposed by D. Miheţ)

376. Denote the vertices of the octagon by $A_1 = A, A_2, A_3, A_4, A_5 = E, A_6, A_7, A_8$ in successive order. Any time the frog jumps back and forth it makes two jumps, so to get from A_1 to any vertex with odd index, in particular to A_5, it makes an even number of jumps. This shows that $a_{2n-1} = 0$.

We compute the number of paths with $2n$ jumps recursively. Consider the case $n > 2$. After two jumps, the frog ends at A_1, A_3, or A_7. It can end at A_1 via A_2 or A_8. Also, the configurations where it ends at A_3 or A_7 are symmetric, so they can be treated simultaneously. If we denote by b_{2n} the number of ways of getting from A_3 to A_5 in $2n$ steps, we obtain the recurrence $a_{2n} = 2a_{2n-2} + 2b_{2n-2}$. On the other hand, if the frog starts at A_3, then it can either return to A_3 in two steps (which can happen in two different ways), or end at A_1 (here it is important that $n > 2$). Thus we can write $b_{2n} = a_{2n-2} + 2b_{2n-2}$. In vector form the recurrence is

$$\begin{pmatrix} a_{2n} \\ b_{2n} \end{pmatrix} = \begin{pmatrix} 2 & 2 \\ 1 & 2 \end{pmatrix} \begin{pmatrix} a_{2n-2} \\ b_{2n-2} \end{pmatrix} = \begin{pmatrix} 2 & 2 \\ 1 & 2 \end{pmatrix}^{n-1} \begin{pmatrix} a_2 \\ b_2 \end{pmatrix}.$$

To find the nth power of the matrix we diagonalize it. The characteristic equation is $\lambda^2 - 4\lambda + 2 = 0$, with roots $x = 2 + \sqrt{2}$ and $y = 2 - \sqrt{2}$. The nth power of the matrix will be of the form

$$X \begin{pmatrix} x^n & 0 \\ 0 & y^n \end{pmatrix} X^{-1},$$

for some matrix X. Consequently, there exist constants α, β determined by the initial condition, such that $a_{2n} = \alpha x^{n-1} + \beta y^{n-1}$. To determine α and β, note that $a_2 = 0$, $b_2 = 1$, and using the recurrence relation, $a_4 = 2$ and $b_4 = 3$. We obtain $\alpha = \frac{1}{\sqrt{2}}$ and $\beta = -\frac{1}{\sqrt{2}}$, whence

$$a_{2n} = \frac{1}{\sqrt{2}} (x^{n-1} - y^{n-1}), \text{ for } n \geq 1.$$

(21st International Mathematical Olympiad, 1979, proposed by Germany)

377. We first try a function of the form $f(n) = n + a$. The relation from the statement yields $a = 667$, and hence $f(n) = n + 667$ is a solution. Let us show that this is the only solution.

Fix some positive integer n and define $a_0 = n$, and $a_k = f(f(\cdots(f(n)\cdots)))$, where the composition is taken k times, $k \geq 1$. The sequence $(a_k)_{k\geq 0}$ satisfies the inhomogeneous linear recurrence relation

$$a_{k+3} - 3a_{k+2} + 6a_{k+1} - 4a_k = 2001.$$

A particular solution is $a_k = 667k$. The characteristic equation of the homogeneous recurrence $a_{k+3} - 3a_{k+2} + 6a_{k+1} - 4a_k = 0$ is

$$\lambda^3 - 3\lambda^2 + 6\lambda - 4 = 0.$$

An easy check shows that $\lambda_1 = 1$ is a solution to this equation. Since $\lambda^3 - 3\lambda^2 + 6\lambda - 4 = (\lambda - 1)(\lambda^2 - 2\lambda + 4)$, the other two solutions are $\lambda_{2,3} = 1 \pm i\sqrt{3}$, that is, $\lambda_{2,3} = 2\left(\cos\frac{\pi}{3} \pm i\sin\frac{\pi}{3}\right)$. It follows that the formula for the general term of a sequence satisfying the recurrence relation is

$$a_k = c_1 + c_2 2^k \cos\frac{k\pi}{3} + c_3 2^k \sin\frac{k\pi}{3} + 667k, \ k \geq 0,$$

with c_1, c_2, and c_3 some real constants.

If $c_2 > 0$, then $a_{3(2m+1)}$ will be negative for large m, and if $c_2 < 0$, then a_{6m} will be negative for large m. Since $f(n)$ can take only positive values, this implies that $c_2 = 0$. A similar argument shows that $c_3 = 0$. It follows that $a_k = c_1 + 667k$. So the first term of the sequence determines all the others. Since $a_0 = n$, we have $c_1 = n$, and hence $a_k = n + 667k$, for all k. In particular, $a_1 = f(n) = n + 667$, and hence this is the only possible solution.

(*Mathematics Magazine*, proposed by R. Gelca)

378. We compute $x_3 = 91$, $x_4 = 436$, $x_5 = 2089$. And we already suggested by placing the problem in this section that the solution should involve some linear recurrence. Let us hope that the terms of the sequence satisfy a recurrence $x_{n+1} = \alpha x_n + \beta x_{n-1}$. Substituting $n = 2$ and $n = 3$ we obtain $\alpha = 5$, $\beta = -1$, and then the relation is also verified for the next term $2089 = 5 \cdot 436 - 91$. Let us prove that this recurrence holds in general.

If y_n is the general term of this recurrence, then $y_n = ar^n + bs^n$, where

$$r = \frac{5 + \sqrt{21}}{2}, \quad s = \frac{5 - \sqrt{21}}{2}, \quad rs = 1, \quad r - s = \sqrt{21};$$

and

$$a = \frac{7 + \sqrt{21}}{14}, \quad b = \frac{7 - \sqrt{21}}{14}, \quad ab = 1.$$

We then compute

$$y_{n+1} - \frac{y_n^2}{y_n - 1} = \frac{y_{n+1}y_{n-1} - y_n^2}{y_{n-1}} = \frac{(ar^{n+1} + bs^{n+1})(ar^{n-1} + bs^{n-1}) - (ar^n + bx^n)^2}{ar^{n-1} + bs^{n-1}}$$

$$= \frac{ab(rs)^{n-1}(r - s)^2}{y_{n-1}} = \frac{3}{y_n - 1}.$$

Of course, $0 < \frac{3}{y_n - 1} < 1$ for $n \geq 2$. Because y_{n+1} is an integer, it follows that

$$y_{n+1} = \left\lceil \frac{y_n^2}{y_n - 1} \right\rceil.$$

Hence x_n and y_n satisfy the same recurrence. This implies that $x_n = y_n$ for all n. The conclusion now follows by induction if we rewrite the recurrence as

$$(x_{n+1} - 1) = 5(x_n - 1) - (x_{n-1} - 1) + 3.$$

(Proposed for the USA Mathematical Olympiad by G. Heuer)

379. From the recurrence relation for $(a_n)_n$, we obtain

$$2a_{n+1} - 3a_n = \sqrt{5a_n^2 - 4},$$

and hence

$$4a_{n+1}^2 - 12a_{n+1}a_n + 9a_n^2 = 5a_n^2 - 4.$$

After canceling similar terms and dividing by 4, we obtain

$$a_{n+1}^2 - 3a_{n+1}a_n + a_n^2 = -1.$$

Subtracting this from the analogous relation for $n - 1$ instead of n yields

$$a_{n+1}^2 - 3a_{n+1}a_n + 3a_na_{n-1} - a_{n-1}^2 = 0.$$

This is the same as

$$(a_{n+1} - a_{n-1})(a_{n+1} - 3a_n + a_{n-1}) = 0,$$

which holds for $n \geq 1$. Looking at the recurrence relation we see immediately that the sequence $(a_n)_n$ is strictly increasing, so in the above product the first factor is different from 0. Hence the second factor must equal to 0, i.e.,

$$a_{n+1} = 3a_n - a_{n-1}, \ n \geq 2.$$

This is a linear recurrence that can, of course, be solved by the usual algorithm. But this is a famous recurrence relation, satisfied by the Fibonacci numbers of odd index. A less experienced reader can simply look at the first few terms, and then prove by induction that $a_n = F_{2n+1}, n \geq 1$.

The sequence $(b_n)_n$ also satisfies a recurrence relation that can be found by substituting $a_n = b_{n+1} - b_n$ in the recurrence relation for $(a_n)_n$. After computations, we obtain

$$b_{n+1} = 2b_n + 2b_{n-1} - b_{n-2}, \ n \geq 3.$$

But now we are told that b_n should be equal to $(F_n)^2, n \geq 1$. Here is a proof by induction on n. It is straightforward to check the equality for $n = 1, 2, 3$. Assuming that $b_k = (F_k)^2$ for all $k \leq n$, it follows that

$$\begin{aligned}
b_{n+1} &= 2(F_n)^2 + 2(F_{n-1})^2 - (F_{n-2})^2 \\
&= (F_n + F_{n-1})^2 + (F_n - F_{n-1})^2 - (F_{n-2})^2 \\
&= (F_{n+1})^2 + (F_{n-2})^2 - (F_{n-2})^2 = (F_{n+1})^2.
\end{aligned}$$

With this the problem is solved.

(*Mathematical Reflections*, proposed by T. Andreescu)

380. Of course, we can find the formula for the general term of the sequence, and then pass to the limit, but here is a clever way to find this particular limit.

Write the numbers in binary form. Then $x_2 = 0.1$, $x_3 = 0.11$, $x_4 = 0.101$, $x_5 = 0.1011$, $x_6 = 0.10101$, and by an easy induction one can prove that $x_{2n} = 0.1010\ldots01$ where there are n ones and $x_{2n+1} = 0.1010\ldots011$ where there are $n + 1$ ones. The limit is therefore the number in binary form $0.10101010\ldots$, which is $\frac{2}{3}$.

381. The function $|\sin x|$ is periodic with period π. Hence

$$\lim_{n \to \infty} |\sin \pi \sqrt{n^2 + n + 1}| = \lim_{n \to \infty} |\sin \pi(\sqrt{n^2 + n + 1} - n)|.$$

But

$$\lim_{n \to \infty} (\sqrt{n^2 + n + 1} - n) = \lim_{n \to \infty} \frac{n^2 + n + 1 - n^2}{\sqrt{n^2 + n + 1} + n} = \frac{1}{2}.$$

It follows that the limit we are computing is equal to $\left|\sin \frac{\pi}{2}\right|$, which is 1.

382. Using the binomial expansion, we see that $(1 + \sqrt{2})^{2n} + (1 - \sqrt{2})^{2n}$ is an integer for all n. Note also that $(1 - \sqrt{2})^{2n} < 1$ for all n. Hence

$$\{(1 + \sqrt{2})^{2n}\} = 1 - (1 - \sqrt{2})^{2n}.$$

Passing to the limit in this equality we obtain

$$\lim_{n \to \infty} \{(1 + \sqrt{2})^{2n}\} = 1 - \lim_{n \to \infty} (1 - \sqrt{2})^{2n} = 1.$$

383. The limit is computed as follows:

$$\lim_{n \to \infty} \binom{n}{k} \left(\frac{\mu}{n}\right)^k \left(1 - \frac{\mu}{n}\right)^{n-k}$$

$$= \lim_{n \to \infty} \frac{n!}{k!(n-k)!} \left(\frac{\frac{\mu}{n}}{1 - \frac{\mu}{n}}\right)^k \left(1 - \frac{\mu}{n}\right)^n$$

$$= \frac{1}{k!} \lim_{n \to \infty} \frac{n(n-1) \cdots (n-k+1)}{\left(\frac{n}{\mu} - 1\right)^k} \cdot \lim_{n \to \infty} \left(1 - \frac{\mu}{n}\right)^{\frac{n}{\mu} \cdot \mu}$$

$$= \frac{e^\mu}{k!} \lim_{n \to \infty} \frac{n^k - (1 + \ldots + (k-1))n^{k-1} + \cdots + (-1)^{k-1}(k-1)!}{\frac{1}{\mu^k} n^k - \binom{k}{1}\frac{1}{\mu^{k-1}} n^{k-1} + \cdots + (-1)^k}$$

$$= \frac{1}{e^\mu \cdot k!} \cdot \frac{1}{\frac{1}{\mu^k}} = \frac{\mu^k}{e^\mu \cdot k!}.$$

Remark. This limit is applied in probability theory in the following context. Consider a large population n in which an event occurs with very low probability p. The probability that the event occurs exactly k times in that population is given by the binomial formula

$$P(k) = \binom{n}{k} p^k (1 - p)^{n-k}.$$

But for n large, the number $(1 - p)^{n-k}$ is impossible to compute. In that situation we set $\mu = np$ (the mean occurrence in that population), and approximate the probability by the Poisson distribution

$$P(k) \approx \frac{\mu^k}{e^k \cdot k!}.$$

The exercise we just solved shows that this approximation is good.

384. Let us assume that the answer is negative. Then the sequence has a bounded subsequence $(x_{n_k})_k$. The set $\{x_{x_{n_k}} \mid k \in \mathbb{Z}\}$ is finite, since the indices x_{n_k} belong to a finite set. But $x_{x_{n_k}} = n_k^4$, and this takes infinitely many values for $k \geq 1$. We reached a contradiction that shows that our assumption was false. So the answer to the question is yes.

(Romanian Mathematical Olympiad, 1978, proposed by S. Rădulescu)

385. Suppose $a \neq 0$. Then $a > 0$ or else the expression would eventually become negative. Let $x_n = \sqrt{a \cdot 2^n + b}$. We have

$$\lim_{n \to \infty} (2x_n - x_{n+2}) = \lim_{n \to \infty} (\sqrt{a \cdot 2^{n+2} + 4b} - \sqrt{a \cdot 2^{n+2} + b})$$

$$= \lim_{n \to \infty} \frac{3b}{\sqrt{a \cdot 2^{n+2} + 4b} + \sqrt{a \cdot 2^{n+2} + b}} = 0.$$

This is a sequence of integers, so there is some N such that $2x_n = x_{n+2}$ for $n \geq N$. But the equality $2x_n = x_{n+2}$ is equivalent to $b = 0$. Then a and $2a$ are both squares, which is impossible, by the prime factor decomposition. So a must be zero.

(Polish Team Selection Test for the International Mathematical Olympiad)

386. Let $x_n = \sqrt{an^2 + bn + c}$. Note that

$$x_n - n\sqrt{a} = \frac{x_n + n\sqrt{a}}{an^2 + bn + c - an^2} = \frac{x_n + n\sqrt{a}}{bn + c}$$

$$= \frac{\sqrt{an^2 + bn + c} + n\sqrt{a}}{bn + c}.$$

And this converges to $\frac{2\sqrt{a}}{b}$. Hence $x_{n+1} - x_n$ converges to $\sqrt{a}$. Because this sequence consists of integers, it eventually becomes constant. So for sufficiently large integers, $x_{n+1} = x_n + \sqrt{a}$. It follows that a is a perfect square, say $a = x^2$. Fix M such that for $n \geq M$, $x_{n+1} = x_n + x$. Then $x_n = x_M + (n - M)x$. So $(x_M - Mx + nx)^2 = x^2n^2 + bn + c$, giving $b = 2(x_M - Mx)x$ and $c = (x_M - Mx)^2$.

387. Define the sequence $(b_n)_n$ by

$$b_n = \max\{|a_k|, \ 2^{n-1} \leq k < 2^n\}.$$

From the hypothesis it follows that $b_n \leq \frac{b_{n-1}}{2}$. Hence $0 \leq b_n \leq \frac{b_1}{2^{n-1}}$, which implies that $(b_n)_n$ converges to 0. We also have that $|a_n| \leq b_n$, for $n \geq 1$, so by applying the squeezing principle, we obtain that $(a_n)_n$ converges to zero, as desired.

(Romanian Mathematical Olympiad, 1975, proposed by R. Gologan)

388. *First solution:* Using the fact that $\lim_{n \to \infty} \sqrt[n]{a} = 1$, we pass to the limit in the relation from the statement to obtain

$$\underbrace{1 + 1 + \cdots + 1}_{k \text{ times}} = \underbrace{1 + 1 + \cdots + 1}_{m \text{ times}}.$$

Hence $k = m$. Using L'Hôpital's theorem, one can prove that $\lim_{x \to 0} x(a^x - 1) = \ln a$, and hence $\lim_{n \to \infty} n(\sqrt[n]{a} - 1) = \ln a$. Transform the relation from the hypothesis into

$$n(\sqrt[n]{a_1} - 1) + \cdots + n(\sqrt[n]{a_k} - 1) = n(\sqrt[n]{b_1} - 1) + \cdots + n(\sqrt[n]{b_k} - 1).$$

Passing to the limit with $n \to \infty$, we obtain

$$\ln a_1 + \ln a_2 + \cdots + \ln a_k = \ln b_1 + \ln b_2 + \cdots + \ln b_k.$$

This implies that $a_1 a_2 \cdots a_k = b_1 b_2 \cdots b_k$, and we are done.

Second solution: Fix $N > k$; then taking $n = \frac{(N!)}{m}$ for $1 \leq m \leq k$, we see that the power-sum symmetric polynomials in $a_i^{1/N!}$ agree with the power-sum symmetric polynomials in $b_i^{1/N!}$. Hence the elementary symmetric polynomials in these variables also agree and hence there is a permutation π such that $b_i = a_{\pi(i)}$.

(*Revista Matematică din Timişoara* (*Timişoara Mathematics Gazette*), proposed by D. Andrica, second solution by R. Stong)

389. It is known that

$$\lim_{x \to 0^+} x^x = 1.$$

Here is a short proof using L'Hôpital's theorem:

$$\lim_{x \to 0^+} x^x = \lim_{x \to 0^+} e^{x \ln x} = e^{\lim_{x \to 0^+} \ln x} = e^{\lim_{x \to 0^+} \frac{\ln x}{\frac{1}{x}}} = e^{\lim_{x \to 0^+} (-x)} = 1.$$

Returning to the problem, fix $\varepsilon > 0$, and choose $\delta > 0$ such that for $0 < x < \delta$,

$$|x^x - 1| < \varepsilon.$$

Then for $n \geq \frac{1}{\delta}$ we have

$$\left| n^2 \int_0^{\frac{1}{n}} (x^{x+1} - x) dx \right| \leq n^2 \int_0^{\frac{1}{n}} |x^{x+1} - x| dx$$

$$= n^2 \int_0^{\frac{1}{n}} x |x^x - 1| dx < \varepsilon n^2 \int_0^{\frac{1}{n}} x \, dx = \frac{\varepsilon}{2}.$$

It follows that

$$\lim_{n \to \infty} \int_0^{\frac{1}{n}} (x^{x+1} - x) dx = 0,$$

and so

$$\lim_{n \to \infty} n^2 \int_0^{\frac{1}{n}} x^{x+1} dx = \lim_{n \to \infty} n^2 \int_0^{\frac{1}{n}} x \, dx = \frac{1}{2}.$$

(*Revista Matematică din Timişoara* (*Timişoara Mathematics Gazette*), proposed by D. Andrica)

390. We will prove by induction on $n \geq 1$ that

$$x_{n+1} > \sum_{k=1}^n k x_k > a \cdot n!,$$

from which it will follow that the limit is ∞.

For $n = 1$, we have $x_2 \geq 3x_1 > x_1 = a$. Now suppose that the claim holds for all values up through n. Then

$$x_{n+2} \geq (n+3)x_{n+1} - \sum_{k=1}^{n} kx_k = (n+1)x_{n+1} + 2x_{n+1} - \sum_{k=1}^{n} kx_k$$

$$> (n+1)x_{n+1} + 2\sum_{k=1}^{n} kx_k - \sum_{k=1}^{n} kx_k = \sum_{k=1}^{n+1} kx_k,$$

as desired. Furthermore, $x_1 > 0$ by definition and $x_2, x_3, \ldots, x_n$ are also positive by the induction hypothesis. Therefore, $x_{n+2} > (n+1)x_{n+1} > (n+1)(a \cdot n!) = a \cdot (n+1)!$. This completes the induction, proving the claim.

(Romanian Team Selection Test for the International Mathematical Olympiad, 1999)

391. Denote $\lambda = \inf_{n \geq 1} \frac{x_n}{n}$ and for simplicity assume that $\lambda > -\infty$. Fix $\varepsilon > 0$. Then there exists n_0 such that $\frac{x_{n_0}}{n_0} \leq \lambda + \varepsilon$. Let $M = \max_{1 \leq i \leq n_0} x_i$.

An integer m can be written as $n_0 q + n_1$, with $0 \leq n_1 < q$ and $q = \left\lfloor \frac{m}{n_0} \right\rfloor$. From the hypothesis it follows that $x_m \leq qx_{n_0} + x_{n_1}$; hence

$$\lambda \leq \frac{x_m}{m} \leq \frac{qx_{n_0}}{m} + \frac{x_{n_1}}{m} \leq \frac{qn_0}{m}(\lambda + \varepsilon) + \frac{M}{m}.$$

Therefore,

$$\lambda \leq \frac{x_m}{m} \leq \frac{\left\lfloor \frac{m}{n_0} \right\rfloor}{\frac{m}{n_0}}(\lambda + \varepsilon) + \frac{M}{m}.$$

Since

$$\lim_{m \to \infty} \frac{\left\lfloor \frac{m}{n_0} \right\rfloor}{\frac{m}{n_0}} = 1 \quad \text{and} \quad \lim_{m \to \infty} \frac{M}{m} = 0,$$

it follows that for large m,

$$\lambda \leq \frac{x_m}{m} \leq \lambda + 2\varepsilon.$$

Since ε was arbitrary, this implies

$$\lim_{n \to \infty} \frac{x_n}{n} = \lambda = \inf_{n \geq 1} \frac{x_n}{n},$$

as desired.

392. We use the fact that

$$\lim_{x \to 0^+} x^x = 1.$$

As a consequence, we have

$$\lim_{x \to 0^+} \frac{x^{x+1}}{x} = 1.$$

For our problem, let $\varepsilon > 0$ be a fixed small positive number. There exists $n(\varepsilon)$ such that for any integer $n \geq n(\varepsilon)$,

$$1 - \varepsilon < \frac{\left(\frac{k}{n^2}\right)^{\frac{k}{n^2}+1}}{\frac{k}{n^2}} < 1 + \varepsilon, \ k = 1, 2, \ldots, n.$$

From this, using properties of ratios, we obtain

$$1 - \varepsilon < \frac{\sum\limits_{k=1}^{n}\left(\frac{k}{n^2}\right)^{\frac{k}{n^2}+1}}{\sum\limits_{k=1}^{n}\frac{k}{n^2}} < 1 + \varepsilon, \ \text{for } n \geq n(\varepsilon).$$

Knowing that $\sum\limits_{k=1}^{n} k = \dfrac{n(n+1)}{2}$, this implies

$$(1 - \varepsilon)\frac{n+1}{2n} < \sum_{k=1}^{n}\left(\frac{k}{n^2}\right)^{\frac{k}{n^2}+1} < (1 + \varepsilon)\frac{n+1}{2n}, \ \text{for } n \geq n(\varepsilon).$$

It follows that

$$\lim_{n\to\infty} \sum_{n\to\infty} \left(\frac{k}{n^2}\right)^{\frac{k}{n^2}+1} = \frac{1}{2}.$$

(D. Andrica)

393. Assume that x_n is a square for all $n > M$. Consider the integers $y_n = \sqrt{x_n}$, for $n \geq M$. Because in base b,

$$\frac{b^{2n}}{b - 1} = \underbrace{11\ldots1}_{2n}.111\ldots,$$

it follows that

$$\lim_{n\to\infty} \frac{\frac{b^{2n}}{b-1}}{x_n} = 1.$$

Therefore,

$$\lim_{n\to\infty} \frac{b^n}{y_n} = \sqrt{b - 1}.$$

On the other hand,

$$(by_n + y_{n+1})(by_n - y_{n+1}) = b^2 x_n - x_{n+1} = b^{n+2} + 3b^2 - 2b - 5.$$

The last two relations imply

$$\lim_{n\to\infty} (by_n - y_{n+1}) = \lim_{n\to\infty} \frac{b^{n+2}}{by_n + y_{n+1}} = \frac{b\sqrt{b - 1}}{2}.$$

Here we used the fact that

$$\lim_{n\to\infty}\frac{b^{n+2}}{by_n}=\lim_{n\to\infty}\frac{b^{n+2}}{y_{n+1}}=b\sqrt{b-1}.$$

Since $by_n - y_{n+1}$ is an integer, if it converges then it eventually becomes constant. Hence there exists $N > M$ such that $by_n - y_{n+1} = \frac{b\sqrt{b-1}}{2}$ for $n > N$. This means that $b - 1$ is a perfect square. If b is odd, then $\frac{\sqrt{b-1}}{2}$ is an integer, and so b divides $\frac{b\sqrt{b-1}}{2}$. Since the latter is equal to $by_n - y_{n+1}$ for $n > N$, and this divides $b^{n+2} + 3b^2 - 2b - 5$, it follows that b divides 5. This is impossible.

If b is even, then by the same argument $\frac{b}{2}$ divides 5. Hence $b = 10$. In this case we have indeed that $x_n = \left(\frac{10^n+5}{3}\right)^2$, and the problem is solved.

(Short list of the 44th International Mathematical Olympiad, 2003)

394. Recall the double inequality

$$\left(1+\frac{1}{n}\right)^n < e < \left(1+\frac{1}{n}\right)^{n+1}, \quad n \ge 1.$$

Taking the natural logarithm, we obtain

$$n\ln\left(1+\frac{1}{n}\right) < 1 < (n+1)\ln\left(1+\frac{1}{n}\right),$$

which yields the double inequality

$$\frac{1}{n+1} < \ln(n+1) - \ln n < \frac{1}{n}.$$

Applying the one on the right, we find that

$$a_n - a_{n-1} = \frac{1}{n} - \ln(n+1) + \ln n > 0, \quad \text{for } n \ge 2,$$

so the sequence is increasing. Adding the inequalities

$$1 \le 1,$$
$$\frac{1}{2} < \ln 2 - \ln 1,$$
$$\frac{1}{3} < \ln 3 - \ln 2,$$
$$\cdots$$
$$\frac{1}{n} < \ln n - \ln(n-1),$$

we obtain

$$1 + \frac{1}{2} + \frac{1}{3} + \cdots + \frac{1}{n} < 1 + \ln n < 1 + \ln(n+1).$$

Therefore, $a_n < 1$, for all n. We found that the sequence is increasing and bounded, hence convergent.

395. The sequence is increasing, so all we need to show is that it is bounded. The main trick is to factor a $\sqrt{2}$. The general term of the sequence becomes

$$a_n = \sqrt{2}\sqrt{\frac{1}{2} + \sqrt{\frac{2}{4} + \sqrt{\frac{3}{8} + \cdots + \sqrt{\frac{n}{2^n}}}}}$$

$$< \sqrt{2}\sqrt{1 + \sqrt{1 + \sqrt{1 + \cdots + \sqrt{1}}}}.$$

Let $b_n = \sqrt{1 + \sqrt{1 + \cdots + \sqrt{1}}}$, where there are n radicals. Then $b_{n+1} = \sqrt{1 + b_n}$. We see that $b_1 = 1 < 2$, and if $b_n < 2$, then $b_{n+1} < \sqrt{1 + 2} < 2$. Inductively we prove that $b_n < 2$ for all n. Therefore, $a_n < 2\sqrt{2}$ for all n. Being monotonic and bounded, the sequence $(a_n)_n$ is convergent.

(*Matematika v Škole*, 1971, solution from R. Honsberger, *More Mathematical Morsels*, Mathematical Association of America, 1991)

396. We examine first the expression under the square root. Its zeros are $\frac{-1 \pm \sqrt{5}}{2}$. In order for the square root to make sense, a_n should be outside the interval $\left(\frac{-1-\sqrt{5}}{2}, \frac{-1+\sqrt{5}}{2}\right)$. Since $a_n \geq 0$ for $n \geq 2$, being the square root of an integer, we must have $a_n \geq \frac{-1+\sqrt{5}}{2}$ for $n \geq 2$. To simplify the notation, let $r = \frac{-1+\sqrt{5}}{2}$.

Now suppose by contradiction that $a_1 \in (-2, 1)$. Then

$$a_2^2 = a_1^2 + a_1 - 1 = \left(a_1 + \frac{1}{2}\right)^2 - \frac{5}{4} < \left(\frac{3}{2}\right)^2 - \frac{5}{4} = 1,$$

so $a_2 \in [r, 1)$. Now if $a_n \in [r, 1)$, then

$$a_{n+1}^2 = a_n^2 + a_n - 1 < a_n^2 < 1.$$

Inductively we prove that $a_n \in [r, 1)$ and $a_{n+1} < a_n$. The sequence $(a_n)_n$ is bounded and strictly decreasing; hence it has a limit L. This limit must lie in the interval $[r, 1)$. Passing to the limit in the recurrence relation, we obtain $L = \sqrt{L^2 + L - 1}$, and therefore $L^2 = L^2 + L - 1$. But this equation has no solution in the interval $[r, 1)$, a contradiction. Hence a_1 cannot lie in the interval $(-2, 1)$.

(Bulgarian Mathematical Olympiad, 2002)

397. This is the Bolzano-Weierstrass theorem. For the proof, let us call a term of the sequence a *giant* if all terms following it are smaller. If the sequence has infinitely many giants, they form a bounded decreasing subsequence, which is therefore convergent. If the sequence has only finitely many giants, then after some rank each term is followed by larger term. These terms give rise to a bounded increasing subsequence, which is again convergent.

Remark. The idea can be refined to show that any sequence of $mn + 1$ real numbers has either a decreasing subsequence with $m + 1$ terms or an increasing subsequence with $n + 1$ terms.

398. Consider the truncations

$$s_n = a_1 - a_2 + a_3 - \cdots \pm a_n, \ n \geq 1.$$

We are to show that the sequence $(s_n)_n$ is convergent. For this we verify that the sequence $(s_n)_n$ is Cauchy. Because $(a_n)_{n \geq 1}$ is decreasing, for all $n > m$,

$$|s_n - s_m| = a_m - a_{m+1} + a_{m+2} - \cdots \pm a_n$$
$$= a_m - (a_{m+1} - a_{m+2}) - (a_{m+3} - a_{m+4}) - \cdots,$$

where the sum ends either in a_n or in $-(a_{n-1} - a_n)$. All terms of this sum, except for the first and maybe the last, are negative. Therefore, $|s_n - s_m| \leq a_m + a_n$, for all $n > m \geq 1$. As $a_n \to 0$, this shows that the sequence $(s_n)_n$ is Cauchy, and hence convergent.
    (The Leibniz criterion)

399. For a triple of real numbers (x, y, z) define $\Delta(x, y, z) = \max(|x - y|, |x - z|, |y - z|)$. Let $\Delta(a_0, b_0, c_0) = \delta$. From the recurrence relation we find that

$$\Delta(a_{n+1}, b_{n+1}, c_{n+1}) = \frac{1}{2}\Delta(a_n, b_n, c_n), \ n \geq 0.$$

By induction $\Delta(a_n, b_n, c_n) = \frac{1}{2^n}\delta$. Also, $\max(|a_{n+1} - a_n|, |b_{n+1} - b_n|, |c_{n+1} - c_n|) = \frac{1}{2}\Delta(a_n, b_n, c_n)$. We therefore obtain that $|a_{n+1} - a_n|, |b_{n+1} - b_n|, |c_{n+1} - c_n|$ are all less than or equal to $\frac{1}{2^n}\delta$. So for $n > m \geq 1$, the absolute values $|a_n - a_m|, |b_n - b_m|$, and $|c_n - c_m|$ are less than

$$\left(\frac{1}{2^m} + \frac{1}{2^{m+1}} + \cdots + \frac{1}{2^n}\right)\delta < \frac{\delta}{2^m}.$$

This proves that the sequences are Cauchy, hence convergent. Because as n tends to infinity $\Delta(a_n, b_n, c_n)$ approaches 0, the three sequences converge to the same limit L. Finally, because for all n, $a_b + b_n + c_n = a_0 + b_0 + c_0$, we should have $3L = a_0 + b_0 + c_0$; hence the common limit is $\frac{(a_0+b_0+c_0)}{3}$.

400. Because $\sum a_n$ converges, Cauchy's criterion implies that

$$\lim_{n \to \infty} (a_{\lfloor n/2 \rfloor+1} + a_{\lfloor n/2 \rfloor+2} + \cdots + a_n) = 0.$$

By monotonicity

$$a_{\lfloor n/2 \rfloor+1} + a_{\lfloor n/2 \rfloor+2} + \cdots + a_n \geq \left\lceil \frac{n}{2} \right\rceil a_n,$$

so $\lim_{n \to \infty} \left\lceil \frac{n}{2} \right\rceil a_n = 0$. Consequently, $\lim_{n \to \infty} \frac{n}{2} a_n = 0$, and hence $\lim_{n \to \infty} na_n = 0$, as desired.
    (Abel's lemma)

401. Think of the larger map as a domain D in the plane. The change of scale from one map to the other is a contraction, and since the smaller map is placed inside the larger, the contraction maps D to D. Translating into mathematical language, a point such as the one described in the statement is a fixed point for this contraction. And by the fixed point theorem the point exists and is unique.

402. Define the function $f(x) = \varepsilon \sin x + t$. Then for any real numbers x_1 and x_2,

$$|f(x_1) - f(x_2)| = |\varepsilon| \cdot |\sin x_1 - \sin x_2| \le 2|\varepsilon| \cdot \left|\sin \frac{x_1 - x_2}{2}\right| \cdot \left|\cos \frac{x_1 + x_2}{2}\right|$$

$$\le 2|\varepsilon| \cdot \left|\sin \frac{x_1 - x_2}{2}\right| \le \varepsilon |x_1 - x_2|.$$

Hence f is a contraction, and there exists a unique x such that $f(x) = \varepsilon \sin x + t = x$. This x is the unique solution to the equation.

(J. Kepler)

403. Define $f : (0, \infty) \to (0, \infty)$, $f(x) = \frac{1}{2}\left(x + \frac{c}{x}\right)$. Then $f'(x) = \frac{1}{2}\left(1 - \frac{c}{x^2}\right)$, which is negative for $x < \sqrt{c}$ and positive for $x > \sqrt{c}$. This shows that $\sqrt{c}$ is a global minimum for f and henceforth $f((0, \infty)) \subset [\sqrt{c}, \infty)$. Shifting indices, we can assume that $x_0 \ge \sqrt{c}$. Note that $|f'(x)| < \frac{1}{2}$ for $x \in [\sqrt{c}, \infty)$, so f is a contraction on this interval. Because $x_n = f(f(\cdots f(x_0)))$, $n \ge 1$, the sequence $(x_n)_n$ converges to the unique fixed point x^* of f. Passing to the limit in the recurrence relation, we obtain $x^* = \frac{1}{2}\left(x^* + \frac{c}{x^*}\right)$, which is equivalent to the quadratic equation $(x^*)^2 - c = 0$. We obtain the desired limit of the sequence $x^* = \sqrt{c}$.

(Hero)

404. Define

$$x_n = \sqrt{1 + \sqrt{1 + \sqrt{1 + \cdots + \sqrt{1}}}}, \quad n \ge 1,$$

where in this expression there are n square roots. Note that x_{n+1} is obtained from x_n by replacing $\sqrt{1}$ by $\sqrt{1 + \sqrt{1}}$ at the far end. The square root function being increasing, the sequence $(x_n)_n$ is increasing. To prove that the sequence is bounded, we use the recurrence relation $x_{n+1} = \sqrt{1 + x_n}$, $n \ge 1$. Then from $x_n < 2$, we obtain that $x_{n+1} = \sqrt{1 + x_n} < \sqrt{1 + 2} < 2$, so inductively $x_n < 2$ for all n. Being bounded and monotonic, the sequence $(x_n)_n$ is convergent. Let L be its limit (which must be greater than 1). Passing to the limit in the recurrence relation, we obtain $L = \sqrt{1 + L}$, or $L^2 - L - 1 = 0$. The only positive solution is the golden ratio $\frac{\sqrt{5}+1}{2}$, which is therefore the limit of the sequence.

405. If the sequence converges to a certain limit L, then $L = \sqrt{a + bL}$, so L is equal to the (unique) positive root α of the equation $x^2 - bx - a = 0$.

The convergence is proved by verifying that the sequence is monotonic and bounded. The condition $x_{n+1} \ge x_n$ translates to $x_n^2 \ge a + bx_n$, which holds if and only if $x_n > \alpha$. On the other hand, if $x_n \ge \alpha$, then $x_{n+1}^2 = a + bx_n \ge a + b\alpha = \alpha^2$; hence $x_{n+1} \ge \alpha$. Similarly, if $x_n \le \alpha$, then $x_{n+1} \le \alpha$. There are two situations. Either $x_1 < \alpha$, and then by induction $x_n < \alpha$ for all n, and hence $x_{n+1} > x_n$ for all n. In this case the sequence is increasing and bounded from above by α; therefore, it is convergent, its limit being of course α. Or $x_1 \ge \alpha$, in which case the sequence is decreasing and bounded from below by the same α, and the limit is again α.

406. By the AM-GM inequality, $a_n < b_n, n \geq 1$. Also,

$$a_{n+1} - a_n = \sqrt{a_n b_n} - a_n = \sqrt{a_n}(\sqrt{b_n} - \sqrt{a_n}) > 0;$$

hence the sequence $(a_n)_n$ is increasing. Similarly,

$$b_{n+1} - b_n = \frac{a_n + b_n}{2} - b_n = \frac{a_n - b_n}{2} < 0,$$

so the sequence b_n is decreasing. Moreover,

$$a_0 < a_1 < a_2 < \cdots < a_n < b_n < \cdots < b_1 < b_0,$$

for all n, which shows that both sequences are bounded. By the Weierstrass theorem, they are convergent. Let $a = \lim_{n\to\infty} a_n$ and $b = \lim_{n\to\infty} b_n$. Passing to the limit in the first recurrence relation, we obtain $a = \sqrt{ab}$, whence $a = b$. Done.

Remark. The common limit, denoted by $M(a, b)$, is called the arithmetic-geometric mean of the numbers a and b. It was Gauss who first discovered, as a result of laborious computations, that the arithmetic-geometric mean is related to elliptic integrals. The relation that he discovered is

$$M(a, b) = \frac{\pi}{4} \cdot \frac{a + b}{K\left(\dfrac{a - b}{a + b}\right)},$$

where

$$K(k) = \int_0^1 \frac{1}{\sqrt{(1 - t^2)(1 - k^2 t^2)}} dt$$

is the elliptic integral of first kind. It is interesting to note that this elliptic integral is used to compute the period of the spherical pendulum. More precisely, for a pendulum described by the differential equation

$$\frac{d^2\theta}{dt^2} + \omega^2 \sin\theta = 0,$$

with maximal angle θ_{max}, the period is given by the formula

$$P = \frac{2\sqrt{2}}{\omega} K\left(\sin\left(\frac{1}{2}\theta_{max}\right)\right).$$

407. The function $f_n(x) = x^n + x - 1$ has positive derivative on $[0, 1]$, so it is increasing on this interval. From $f_n(0) \cdot f_n(1) < 0$ it follows that there exists a unique $x_n \in (0, 1)$ such that $f(x_n) = 0$.

Since $0 < x_n < 1$, we have $x_n^{n+1} + x_n - 1 < x_n^n + x_n - 1 = 0$. Rephrasing, this means that $f_{n+1}(x_n) < 0$, and so $x_{n+1} > x_n$. The sequence $(x_n)_n$ is increasing and bounded, thus it is convergent. Let L be its limit. There are two possibilities, either $L = 1$, or $L < 1$. But L cannot be less than 1, for when passing to the limit in $x_n^n + x_n - 1 = 0$ we obtain $L - 1 = 0$, or $L = 1$, a contradiction. Thus $L = 1$, and we are done.

(*Gazeta Matematică* (*Mathematics Gazette, Bucharest*), proposed by A. Leonte)

408. Let

$$x_n = \sqrt{1 + 2\sqrt{1 + 2\sqrt{1 + \cdots + 2\sqrt{1 + 2\sqrt{1969}}}}}$$

with the expression containing n square root signs. Note that

$$x_1(1 + \sqrt{2}) = \sqrt{1969} - (1 + \sqrt{2}) < 50.$$

Also, since $\sqrt{1 + 2(1 + \sqrt{2})} = 1 + \sqrt{2}$, we have

$$x_{n+1} - (1 + \sqrt{2}) = \sqrt{1 + 2x_n} - \sqrt{1 + 2(1 + \sqrt{2})} = \frac{2(x_n - (1 - \sqrt{2}))}{\sqrt{1 + 2x_n} + \sqrt{1 + 2(1 + \sqrt{2})}}$$

$$< \frac{x_n - (1 + \sqrt{2})}{1 + \sqrt{2}}.$$

From here we deduce that

$$x_{1969} - (1 + \sqrt{2}) < \frac{50}{(1 + \sqrt{2})^{1968}} < 10^{-3},$$

and the approximation of x_{1969} with two decimal places coincides with that of $1 + \sqrt{2} = 2.41$. This argument proves also that the limit of the sequence is $1 + \sqrt{2}$.

(St. Petersburg Mathematical Olympiad, 1969)

409. Write the equation as

$$\sqrt{x + 2\sqrt{x + \cdots + 2\sqrt{x + 2\sqrt{x + 2\sqrt{x + 2x}}}}} = x.$$

We can iterate this equality infinitely many times, always replacing the very last x by its value given by the left-hand side. We conclude that x should satisfy

$$\sqrt{x + 2\sqrt{x + 2\sqrt{x + 2 \cdots}}} = x,$$

provided that the expression on the left makes sense! Let us check that indeed the recursive sequence given by $x_0 = x$, and $x_{n+1} = \sqrt{x + 2x_n}$, $n \geq 0$, converges for any solution x to the original equation. Squaring the equation, we find that $x < x^2$, hence $x > 1$. But then $x_{n+1} < x_n$, because it reduces to $x_n^2 - 2x_n + x > 0$. This is always true, since when viewed as a quadratic function in x_n, the left-hand side has negative discriminant. Our claim is proved, and we can now transform the equation, the one with infinitely many square roots, into the much simpler

$$x = \sqrt{x + 2x}.$$

This has the unique solution $x = 3$, which is also the unique solution to the equation from the statement, and this regardless of the number of radicals.

(D.O. Shklyarski, N.N. Chentsov, I.M. Yaglom, *Selected Problems and Theorems in Elementary Mathematics, Arithmetic and Algebra*, Mir, Moscow)

410. The sequence satisfies the recurrence relation

$$x_{n+2} = \sqrt{7 - \sqrt{7 + x_n}}, \ n \geq 1,$$

with $x_1 = \sqrt{7}$ and $x_2 = \sqrt{7 - \sqrt{7}}$. Let us first determine the possible values of the limit L, assuming that it exists. Passing to the limit in the recurrence relation, we obtain

$$L = \sqrt{7 - \sqrt{7 + L}}.$$

Squaring twice, we obtain the polynomial equation $L^4 - 14L^2 - L + 42 = 0$. Two roots are easy to find by investigating the divisors of 42, and they are $L = 2$ and $L = -3$. The other two are $L = \frac{1}{2} \pm \frac{\sqrt{29}}{2}$. Only the positive roots qualify, and of them $\frac{1}{2} + \frac{\sqrt{29}}{2}$ is not a root of the original equation, since

$$\frac{1}{2} + \frac{\sqrt{29}}{2} > 3 > \sqrt{7 - \sqrt{7 + 3}} > \sqrt{7 - \sqrt{7 + \frac{1}{2} + \frac{\sqrt{29}}{2}}}.$$

So the only possible value of the limit is $L = 2$.

Let $x_n = 2 + \alpha_n$. Then $\alpha_1, \alpha_2 \in (0, 1)$. Also,

$$\alpha_{n+2} = \frac{3 - \sqrt{9 + \alpha_n}}{\sqrt{7 - \sqrt{9 + \alpha_n}} + 2}.$$

If $\alpha_n \in (0, 1)$, then

$$0 > \alpha_{n+2} > \frac{3 - \sqrt{9 + \alpha_n}}{4} \geq -\frac{1}{2}\alpha_n,$$

where the last inequality follows from $3 + 2\alpha_n \geq \sqrt{9 + \alpha_n}$. Similarly, if $\alpha_n \in (-1, 0)$, then

$$0 < \alpha_{n+2} < \frac{3 - \sqrt{9 + \alpha_n}}{4} \leq \frac{1}{2}|\alpha_n|,$$

where the last inequality follows from $3 < \sqrt{9 - |\alpha_n|} + 2\alpha$. Inductively, we obtain that $\alpha_n - (-2^{-\lfloor n/2 \rfloor}, 2^{-\lfloor n/2 \rfloor})$, and hence $\alpha_n \to 0$. Consequently, the sequence $(x_n)_n$ is convergent, and its limit is 2.

(13th W.L. Putnam Mathematics Competition, 1953)

411. (a) The answer is cleary ∞.

(b) We define the sequence

$$x_0 = 1, \quad x_{n+1} = \sqrt{2}^{x_n}, \quad n \geq 0.$$

Then $x_0 < x_1$ and since $x \mapsto \sqrt{2}^x$ is increasing, we obtain inductively that $x_n < x_{n+1}$ for all n. The sequence is increasing, so it has a limit L. We know that $L = \sqrt{2}^L$. One possible solution is $L = \infty$, another is $L = 2$, and another is $L = 4$. Are there other positive solutions to the equation $x = \sqrt{2}^x$?

Squaring we transform the equation into

$$2^x = x^2$$

which after taking the logarithm becomes

$$x \ln 2 - 2 \ln x.$$

Consider the function $f : (0, \infty) \to \mathbb{R}, f(x) = x \ln 2 - 2 \ln x$. We are to find the zeros of f. Differentiating we obtain

$$f'(x) = \ln 2 - \frac{2}{x},$$

which is strictly increasing. The unique zero of the derivative is $2/\ln 2$, and so f' is negative for $x < 2/\ln 2$ and positive for $x > 2/\ln 2$. Note also that $\lim_{x \to 0} f(x) = \lim_{x \to \infty} f(x) = \infty$. There are two possibilities, either $f(2/\ln 2) > 0$ in which case the equation $f(x) = 0$ has no solutions, or $f(2/\ln 2) < 0$ in which case the equation $f(x) = 0$ has exactly two solutions. The latter must be true, as $f(2) = f(4) = 0$. Therefore $x = 2$ and $x = 4$ are the only solutions to $f(x) = 0$, and hence also to the original equation.

Now we have to decide which of $2, 4, \infty$ is the limit of the sequence. Notice that $x_0 < 2$, and if $x_n < 2$, then $x_{n+1} = \sqrt{2}^{x_n} < \sqrt{2}^2 = 2$. So inductively we obtain $x_n < 2$. Hence L, the limit of x_n, must be 2.

(c) Like in the case of (b), the sequence $x_{n+1} = a^{x_n}$, $x_1 = a$ is increasing and it has a limit L. Then $L^{\frac{1}{L}} = a$, that is $\frac{\ln L}{L} = \ln a$. The maximum of the function $f(x) = \frac{\ln x}{x}$ is attained when $f'(x) = \frac{\ln x - 1}{x^2}$ is zero, that is when $x = e$. In that case $a = e^{\frac{1}{e}}$. So we know for sure that $L = \infty$ if $a > e^{\frac{1}{e}}$. On the other hand, if $a < e^{\frac{1}{e}}$, then we can prove by induction that $x_n \le e^{\frac{1}{e}}$. So we have an increasing sequence and bounded sequence, which by the Weierstrass theorem is convergent. Thus the answer to the question is $a \le e^{\frac{1}{e}}$.

Remark. Note that $e^{\frac{1}{e}} \approx 1.44466....$

412. The solution is a direct application of the Cesàro-Stolz theorem. Indeed, if we let $a_n = \ln u_n$ and $b_n = n$, then

$$\ln \frac{u_{n+1}}{u_n} = \ln u_{n_1} - \ln u_n = \frac{a_{n+1} - a_n}{b_{n+1} - b_n}$$

and

$$\ln \sqrt[n]{u_n} = \frac{1}{n} \ln u_n = \frac{a_n}{b_n}.$$

The conclusion follows.

Remark. This gives an easy proof of $\lim_{n \to \infty} \sqrt[n]{n!} = \infty$.

413. In view of the Cesàro-Stolz theorem, it suffices to prove the existence of and to compute the limit

$$\lim_{n \to \infty} \frac{(n+1)^p}{(n+1)^{p+1} - n^{p+1}}.$$

We invert the fraction and compute instead

$$\lim_{n\to\infty} \frac{(n+1)^{p+1} - n^{p+1}}{(n+1)^p}.$$

Dividing both the numerator and denominator by $(n+1)^{p+1}$, we obtain

$$\lim_{n\to\infty} \frac{1 - \left(1 - \dfrac{1}{n+1}\right)^{p+1}}{\dfrac{1}{n+1}},$$

which, with the notation $h = \frac{1}{n+1}$ and $f(x) = (1-x)^{p+1}$, becomes

$$-\lim_{h\to 0} \frac{f(h) - f(0)}{h} = -f'(0) = p+1.$$

We conclude that the required limit is $\frac{1}{p+1}$.

414. An inductive argument shows that $0 < x_n < 1$ for all n. Also, $x_{n+1} = x_n - x_n^2 < x_n$, so $(x_n)_n$ is decreasing. Being bounded and monotonic, the sequence converges; let x be its limit. Passing to the limit in the defining relation, we find that $x = x - x^2$, so $x = 0$.

We now apply the Cesàro-Stolz theorem. We have

$$\lim_{n\to\infty} nx_n = \lim_{n\to\infty} \frac{n}{\dfrac{1}{x_n}} = \lim_{n\to\infty} \frac{n+1-n}{\dfrac{1}{x_{n+1}} - \dfrac{1}{x_n}} = \lim_{n\to\infty} \frac{1}{\dfrac{1}{x_n - x_n^2} - \dfrac{1}{x_n}}$$

$$= \lim_{n\to\infty} \frac{x_n - x_n^2}{1 - (1 - x_n)} = \lim_{n\to\infty} (1 - x_n) = 1,$$

and we are done.

415. It is not difficult to see that $\lim_{n\to\infty} x_n = 0$. Because of this fact,

$$\lim_{n\to\infty} \frac{x_n}{\sin x_n} = 1.$$

If we are able to find the limit of

$$\frac{n}{\dfrac{1}{\sin^2 x_n}},$$

then this will equal the square of the limit under discussion. We use the Cesàro-Stolz theorem.

Suppose $0 < x_0 \le 1$ (the cases $x_0 < 0$ and $x_0 = 0$ being trivial; see above). If $0 < x_n \le 1$, then $0 < \arcsin(\sin^2 x_n) < \arcsin(\sin x_n) = x_n$, so $0 < x_{n+1} < x_n$. It follows by induction on n that $x_n \in (0, 1]$ for all n and x_n decreases to 0. Rewriting the recurrence as $\sin x_{n+1} =$

$\sin x_n \sqrt{1 - \sin^4 x_n} - \sin^2 x_n \cos x_n$ gives

$$\frac{1}{\sin x_{n+1}} - \frac{1}{\sin x_n} = \frac{\sin x_n - \sin x_{n+1}}{\sin x_n \sin x_{n+1}}$$

$$= \frac{\sin x_n - \sin x_n \sqrt{1 - \sin^4 x_n} + \sin^2 x_n \cos x_n}{\sin x_n (\sin x_n \sqrt{1 - \sin^4 x_n} - \sin^2 x_n \cos x_n)}$$

$$= \frac{1 - \sqrt{1 - \sin^4 x_n} + \sin x_n \cos x_n}{\sin x_n \sqrt{1 - \sin^4 x_n} - \sin^2 x_n \cos x_n}$$

$$= \frac{\dfrac{\sin^4 x_n}{1 + \sqrt{1 - \sin^4 x_n}} + \sin x_n \cos x_n}{\sin x_n \sqrt{1 - \sin^4 x_n} - \sin^2 x_n \cos x_n}$$

$$= \frac{\dfrac{\sin^3 x_n}{1 + \sqrt{1 - \sin^4 x_n}} + \cos x_n}{\sqrt{1 - \sin^4 x_n} - \sin x_n \cos x_n}.$$

Hence

$$\lim_{n \to \infty} \left(\frac{1}{\sin x_{n+1}} - \frac{1}{\sin x_n} \right) = 1.$$

From the Cesàro-Stolz theorem it follows that $\lim\limits_{n \to \infty} \frac{1}{n \sin x_n} = 1$, and so we have $\lim\limits_{n \to \infty} n x_n = 1$.

(Gazeta Matematică (Mathematics Gazette, Bucharest), 2002, proposed by T. Andreescu)

416. We compute the square of the reciprocal of the limit, namely $\lim\limits_{n \to \infty} \frac{1}{n x_n^2}$. To this end, we apply the Cesàro-Stolz theorem to the sequences $a_n = \frac{1}{x_n^2}$ and $b_n = n$. First, note that $\lim\limits_{n \to \infty} x_n = 0$. Indeed, in view of the inequality $0 < \sin x < x$ on $(0, \pi)$, the sequence is bounded and decreasing, and the limit L satisfies $L = \sin L$, so $L = 0$. We then have

$$\lim_{n \to \infty} \left(\frac{1}{x_{n+1}^2} - \frac{1}{x_n^2} \right) = \lim_{n \to \infty} \left(\frac{1}{\sin^2 x_n} - \frac{1}{x_n^2} \right) = \lim_{n \to \infty} \frac{x_n^2 - \sin^2 x_n}{x_n^2 \sin^2 x_n}$$

$$= \lim_{x_n \to 0} \frac{x_n^2 - \dfrac{1}{2}(1 - \cos 2x_n)}{\dfrac{1}{2} x_n^2 (1 - \cos 2x_n)} = \lim_{x_n \to 0} \frac{2x_n^2 - \left[\dfrac{(2x_n)^2}{2!} - \dfrac{(2x_n)^4}{4!} + \cdots \right]}{x_n^2 \left[\dfrac{(2x_n)^2}{2!} - \dfrac{(2x_n)^4}{4!} + \cdots \right]}$$

$$= \frac{2^4/4!}{2^2/2!} = \frac{1}{3},$$

where we have used the Taylor series of $\cos 2x$. We conclude that the original limit is $\sqrt{3}$.

(J. Dieudonné, *Infinitesimal Calculus*, Hermann, 1962, solution by Ch. Radoux)

417. Through a change of variable, we obtain

$$b_n = \frac{\displaystyle\int_0^n f(t)\,dt}{n} = \frac{x_n}{y_n},$$

where $x_n = \int_0^n f(t)dt$ and $y_n = n$. We are in the hypothesis of the Cesàro-Stolz theorem, since $(y_n)_n$ is increasing and unbounded and

$$\frac{x_{n+1} - x_n}{y_{n+1} - y_n} = \frac{\int_0^{n+1} f(t)dt - \int_0^n f(t)dt}{(n+1) - n} = \int_n^{n+1} f(t)dt = \int_0^1 f(n+x)dx = a_n,$$

which converges. It follows that the sequence $(b_n)_n$ converges; moreover, its limit is the same as that of $(a_n)_n$.

(Proposed by T. Andreescu for the W.L. Putnam Mathematics Competition)

418. Because $P(x) > 0$, for $x = 1, 2, \ldots, n$, the geometric mean is well defined. We analyze the two sequences separately. First, let

$$S_{n,k} = 1 + 2^k + 3^k + \cdots + n^k.$$

Because

$$\lim_{n\to\infty} \frac{S_{n+1,k} - S_{n,k}}{(n+1)^{k+1} - n^{k+1}} = \lim_{n\to\infty} \frac{(n+1)^k}{\binom{k+1}{1}n^k + \binom{k+1}{2}n^{k-1} + \cdots + 1} = \frac{1}{k+1},$$

by the Cesàro-Stolz theorem we have that

$$\lim_{n\to\infty} \frac{S_{n,k}}{n^{k+1}} = \frac{1}{k+1}.$$

Writing

$$A_n = \frac{P(1) + P(2) + \cdots + P(n)}{n} = a_m \frac{S_{n,m}}{n} + a_{m-1} \frac{S_{n,m-1}}{n} + \cdots + a_m,$$

we obtain

$$\lim_{n\to\infty} \frac{A_n}{n^m} = \frac{a_m}{m+1}.$$

Now we turn to the geometric mean. Applying the Cesàro-Stolz theorem to the sequences

$$u_n = \ln \frac{P(1)}{1^m} + \ln \frac{P(2)}{2^m} + \cdots + \ln \frac{P(n)}{n^m}$$

and $v_n = n$, $n \geq 1$, we obtain

$$\lim_{n\to\infty} \frac{u_n}{v_n} = \lim_{n\to\infty} \ln \frac{G_n}{(n!)^{m/n}} = \lim_{n\to\infty} \ln \frac{P(n)}{n^m} = \ln a_m.$$

We therefore have

$$\lim_{n\to\infty} \frac{A_n}{G_n} \cdot \left(\frac{\sqrt[n]{n!}}{n}\right)^m = \frac{1}{m+1}.$$

Now we can simply invoke Stirling's formula (see Section 3.2.12)

$$n! \approx n^n e^{-n} \sqrt{2\pi n},$$

or we can argue as follows. If we let $u_n = \frac{n!}{n^n}$, then the Cesàro-Stolz theorem applied to $\ln u_n$ and $v_n = n$ shows that if $\frac{u_{n+1}}{u_n}$ converges, then so does $\sqrt[n]{u_n}$, and to the same limit. Because

$$\lim_{n\to\infty} \frac{u_{n+1}}{u_n} = \lim_{n\to\infty} \left(\frac{n}{n+1}\right)^n = \frac{1}{e},$$

we have

$$\lim_{n\to\infty} \frac{\sqrt[n]{n!}}{n} = \frac{1}{e}.$$

Therefore,

$$\lim_{n\to\infty} \frac{A_n}{G_n} = \frac{e^m}{m+1}.$$

(*Gazeta Matematică* (*Mathematics Gazette, Bucharest*), 1937, proposed by T. Popoviciu)

419. Clearly, $(a_n)_{n\geq 0}$ is an increasing sequence. Assume that a_n is bounded. Then it must have a limit L. Taking the limit of both sides of the equation, we have

$$\lim_{n\to\infty} a_{n+1} = \lim_{n\to\infty} a_n + \lim_{n\to\infty} \frac{1}{\sqrt[k]{a_n}},$$

or $L = L + \frac{1}{\sqrt[k]{L}}$, contradiction. Thus $\lim_{n\to\infty} a_n = \infty$ and dividing the equation by a_n, we get $\lim_{n\to\infty} \frac{a_{n+1}}{a_n} = 1$. Let us write

$$\lim_{n\to\infty} \frac{a_n^{k+1}}{n^k} \left(\lim_{n\to\infty} \frac{a_n^{\frac{k+1}{k}}}{n}\right)^k.$$

Using the Cesàro-Stolz theorem, we have

$$\lim_{n\to\infty} \frac{a_n^{\frac{k+1}{k}}}{n} = \lim_{n\to\infty} \frac{a_{n+1}^{\frac{k}{k+1}} - a_n^{\frac{k+1}{k}}}{=} \lim_{n\to\infty} \sqrt[k]{a_{n+1}^{k+1}} - \sqrt[k]{a_n^{k+1}}$$

$$= \lim_{n\to\infty} \frac{a_{n+1}^{k+1} - a_n^{k+1}}{\left(\sqrt[k]{a_{n+1}^{k+1}}\right)^{k-1} + \left(\sqrt[k]{a_{n+1}^{k+1}}\right)^{k-2}\sqrt[k]{a_n^{k+1}} + \cdots + \left(\sqrt[k]{a_n^{k+1}}\right)^{k-1}}$$

$$= \lim_{n\to\infty} \frac{(a_{n+1} - a_n)(a_{n+1}^k + a_{n+1}^{k-1}a_n + \cdots + a_n^k)}{\left(\sqrt[k]{a_{n+1}^{k+1}}\right)^{k-1} + \left(\sqrt[k]{a_{n+1}^{k+1}}\right)^{k-2}\sqrt[k]{a_n^{k+1}} + \cdots + \left(\sqrt[k]{a_n^{k+1}}\right)^{k-1}}$$

$$= \lim_{n\to\infty} \frac{a_{n+1}^k + a_{n+1}^{k-1}a_n + \cdots + a_n^k}{\sqrt[k]{a_n}\left(\left(\sqrt[k]{a_{n+1}^{k+1}}\right)^{k-1} + \left(\sqrt[k]{a_{n+1}^{k+1}}\right)^{k-2}\sqrt[k]{a_n^{k+1}} + \cdots + \left(\sqrt[k]{a_n^{k+1}}\right)^{k-1}\right)}.$$

Dividing both sides by a_n^k, we obtain

$$\lim_{n\to\infty} \frac{a_n^{\frac{k+1}{k}}}{n} = \lim_{n\to\infty} \frac{\left(\frac{a_{n+1}}{a_n}\right)^k + \left(\frac{a_{n+1}}{a_n}\right)^{k-1} + \ldots + 1}{\left(\frac{a_{n+1}}{a_n}\right)^{\frac{(k+1)(k-1)}{k}} + \left(\frac{a_{n+1}}{a_n}\right)^{\frac{(k+1)(k-2)}{k}} + \cdots + 1}.$$

Since $\lim\limits_{n\to\infty} \frac{a_{n+1}}{a_n} = 1$, we obtain

$$\lim_{n\to\infty} \frac{a_n^{\frac{k+1}{k}}}{n} = \frac{k+1}{k}.$$

Hence

$$\lim_{n\to\infty} \frac{a_n^{k+1}}{n^k} = \left(1 + \frac{1}{k}\right)^k.$$

(67th W.L. Putnam Mathematical Competition, proposed by T. Andreescu; the special case $k = 2$ was the object of the second part of a problem given at the regional round of the Romanian Mathematical Olympiad in 2004)

420. Assume no such ξ exists. Then $f(a) > a$ and $f(b) < b$. Construct recursively the sequences $(a_n)_{n\geq 1}$ and $(b_n)_{n\geq 1}$ with $a_1 = a$, $b_1 = b$, and

$$a_{n+1} = a_n \quad \text{and} \quad b_{n+1} = \frac{a_n + b_n}{2} \quad \text{if} \quad f\left(\frac{a_n + b_n}{2}\right) < \frac{a_n + b_n}{2},$$

or

$$a_{n+1} = \frac{a_n + b_n}{2} \quad \text{and} \quad b_{n+1} = b_n \quad \text{if} \quad f\left(\frac{a_n + b_n}{2}\right) > \frac{a_n + b_n}{2}.$$

Because $b_n - a_n = \frac{b-a}{2^n} \to 0$, the intersection of the nested sequence of intervals

$$[a_1, b_1] \supset [a_2, b_2] \supset [a_3, b_3] \supset \cdots \supset [a_n, b_n] \supset \cdots$$

consists of one point; call it ξ. Note that

$$\xi = \lim_{n\to\infty} a_n = \lim_{n\to\infty} b_n.$$

We have constructed the two sequences such that $a_n < f(a_n) < f(b_n) < b_n$ for all n, and the squeezing principle implies that $(f(a_n))_n$ and $(f(b_n))_n$ are convergent, and

$$\lim_{n\to\infty} f(a_n) = \lim_{n\to\infty} f(b_n) = \xi.$$

Now the monotonicity of f comes into play. From $a_n \leq \xi \leq b_n$, we obtain $f(x_n) \leq f(\xi) \leq f(b_n)$. Again, by the squeezing principle,

$$f(\xi) = \lim_{n\to\infty} f(a_n) = \lim_{n\to\infty} f(b_n) = \xi.$$

This contradicts our initial assumption, proving the existence of a point ξ with the desired property.

Remark. This result is known as Knaster's theorem. Its most general form is the Knaster-Tarski theorem: Let L be a complete lattice and let $f : L \to L$ be an order-preserving function. Then the set of fixed points of f in L is also a complete lattice, and in particular this set is nonempty.

421. Let $P_1(x) = x$ and $P_{n+1}(x) = P_n(x)\left(P_n(x) + \frac{1}{n}\right)$, for $n \geq 1$. Then $P_n(x)$ is a polynomial of degree 2^{n-1} with positive coefficients and $x_n = P_n(x_1)$. Because the inequality $x_{n+1} > x_n$ is equivalent to $x_n > 1 - \frac{1}{n}$, it suffices to show that there exists a unique positive real number t such that $1 - \frac{1}{n} < P_n(t) < 1$ for all n. The polynomial function $P_n(x)$ is strictly increasing for $x \geq 0$, and $P_n(0) = 0$, so there exist unique numbers a_n and b_n such that $P_n(a_n) = 1 - \frac{1}{n}$ and $P_n(b_n) = 1$, respectively. We have that $a_n < a_{n+1}$, since $P_{n+1}(a_n) = 1 - \frac{1}{n}$ and $P_{n+1}(a_{n+1}) = 1 - \frac{1}{n+1}$. Similarly, $b_{n+1} < b_n$, since $P_{n+1}(b_{n+1}) = 1$ and $P_{n+1}(b_n) = 1 + \frac{1}{n}$.

It follows by induction on n that the polynomial function $P_n(x)$ is convex for $x \geq 0$, since

$$P''_{n+1}(x) = P''_n(x)\left(2P_n(x) + \frac{1}{n}\right) + (P'_n(x))^2,$$

and $P_n(x) \geq 0$, for $x \geq 0$. Convexity implies

$$P_n(x) \leq \frac{P_n(b_n) - P(0)}{b_n - 0} x = \frac{x}{b_n}, \quad \text{for } 0 \leq x \leq b_n.$$

In particular, $1 - \frac{1}{n} = P_n(a_n) \leq \frac{a_n}{b_n}$. Together with the fact that $b_n \leq 1$, this means that $b_n - a_n \leq \frac{1}{n}$. By Cantor's nested intervals theorem there exists a unique number t such that $a_n < t < b_n$ for every n. This is the unique number satisfying $1 - \frac{1}{n} < P_n(t) < 1$ for all n. We conclude that t is the unique number for which the sequence $x_n = P_n(t)$ satisfies $0 < x_n < x_{n+1} < 1$ for every n.

(26th International Mathematical Olympiad, 1985)

422. The answer to the question is yes. We claim that for any sequence of positive integers n_k, there exists a number $\gamma > 1$ such that $(\lfloor \gamma^k \rfloor)_k$ and $(n_k)_k$ have infinitely many terms in common. We need the following lemma.

Lemma. *For any $\alpha, \beta, 1 < \alpha < \beta$, the set $\bigcup_{k=1}^{\infty} [\alpha^k, \beta^k - 1]$ contains some interval of the form (a, ∞).*

Proof. Observe that $(\beta/\alpha)^k \to \infty$ as $k \to \infty$. Hence for large k, $\alpha^{k+1} < \beta^k - 1$, and the lemma follows. □

Let us return to the problem and prove the claim. Fix the numbers α_1 and β_1, $1 < \alpha_1 < \beta_1$. Using the lemma we can find some k_1 such that the interval $[\alpha_1^{k_1}, \beta_1^{k_1} - 1]$ contains some terms of the sequence $(n_k)_k$. Choose one of these terms and call it t_1. Define

$$\alpha_2 = t_1^{1/k_1}, \quad \beta_2 = \left(t_1 + \frac{1}{2}\right)^{1/k_1}.$$

Then $[\alpha_2, \beta_2] \subset [\alpha_1, \beta_1]$, and for any $x \in [\alpha_2, \beta_2]$, $\lfloor x^{k_1} \rfloor = t_1$. Again by the lemma, there exists k_2 such that $[\alpha_2^{k_2}, \beta_2^{k_2} - 1]$ contains a term of $(n_k)_k$ different from n_1. Call this term t_2.

Let

$$\alpha_3 = t_2^{1/k_2}, \quad \beta_3 = \left(t_2 + \frac{1}{2}\right)^{1/k_2}.$$

As before, $[\alpha_3, \beta_3] \subset [\alpha_2, \beta_2]$ and $\lfloor x^{k_2} \rfloor = t_2$ for any $x \in [\alpha_3, \beta_3]$. Repeat the construction infinitely many times. By Cantor's nested intervals theorem, the intersection of the decreasing sequence of intervals $[\alpha_j, \beta_j]$, $j = 1, 2, \ldots$, is nonempty. Let γ be an element of this intersection. Then $\lfloor \gamma^{k_j} \rfloor = t_j, j = 1, 2, \ldots$, which shows that the sequence $(\lfloor \gamma^j \rfloor)_j$ contains a subsequence of the sequence $(n_k)_k$. This proves the claim.

To conclude the solution to the problem, assume that the sequence $(a_n)_n$ does not converge to 0. Then it has some subsequence $(a_{n_k})_k$ that approaches a nonzero (finite or infinite) limit as $n \to \infty$. But we saw above that this subsequence has infinitely many terms in common with a sequence that converges to zero, namely with some $(a_{\lfloor \gamma^k \rfloor})_k$. This is a contradiction. Hence the sequence $(a_n)_n$ converges to 0.

(Soviet Union University Student Mathematical Olympiad, 1975)

423. The solution follows closely that of the previous problem. Replacing f by $|f|$ we may assume that $f \geq 0$. We argue by contradiction. Suppose that there exists $a > 0$ such that the set

$$A = f^{-1}((a, \infty)) = \{x \in (0, \infty) \mid f(x) > a\}$$

is unbounded. We want to show that there exists $x_0 \in (0, \infty)$ such that the sequence $(nx_0)_{n \geq 1}$ has infinitely many terms in A. The idea is to construct a sequence of closed intervals $I_1 \supset I_2 \supset I_3 \supset \cdots$ with lengths converging to zero and a sequence of positive integers $n_1 < n_2 < n_3 < \cdots$ such that $n_k I_k \subset A$ for all $k \geq 1$.

Let I_1 be any closed interval in A of length less than 1 and let $n_1 = 1$. Exactly as in the case of the previous problem, we can show that there exists a positive number m_1 such that $\bigcup_{m \geq m_1} m I_1$ is a half-line. Thus there exists $n_2 > n_1$ such that $n_2 I_1$ intersects A. Let J_2 be a closed interval of length less than 1 in this intersection. Let $I_2 = \frac{1}{n_2} J_2$. Clearly, $I_2 \subset I_1$, and the length of I_2 is less than $\frac{1}{n_2}$. Also, $n_2 I_2 \subset A$. Inductively, let $n_k > n_{k-1}$ be such that $n_k I_{k-1}$ intersects A, and let J_k be a closed interval of length less than 1 in this intersection. Define $I_k = \frac{1}{n_k} J_k$.

We found the decreasing sequence of intervals $I_1 \supset I_2 \supset I_3 \supset \cdots$ and positive integers $n_1 < n_2 < n_3 < \cdots$ such that $n_k I_k \subset A$. Cantor's nested intervals theorem implies the existence of a number x_0 in the intersection of these intervals. The subsequence $(n_k x_0)_k$ lies in A, which means that $(nx_0)_n$ has infinitely many terms in A. This implies that the sequence $f(nx_0)$ does not converge to 0, since it has a subsequence bounded away from zero. But this contradicts the hypothesis. So our assumption was false, and therefore $\lim_{x \to \infty} f(x) = 0$.

Remark. This result is known as Croft's lemma. It has an elegant proof using the Baire category theorem.

424. Adding a few terms of the series, we can guess the identity

$$\frac{1}{1+x} + \frac{2}{1+x^2} + \cdots + \frac{2^n}{1+x^{2^n}} = \frac{1}{x-1} + \frac{2^{n+1}}{1-x^{2^{n+1}}}, \quad n \geq 1.$$

And indeed, assuming that the formula holds for n, we obtain

$$\frac{1}{1+x} + \frac{2}{1+x^2} + \cdots + \frac{2^n}{1+x^{2^n}} + \frac{2^{n+1}}{1+x^{2^{n+1}}} = \frac{1}{x-1} + \frac{2^{n+1}}{1-x^{2^{n+1}}} + \frac{2^{n+1}}{1+x^{2^{n+1}}}$$

$$= \frac{1}{x-1} + \frac{2^{n+2}}{1-x^{2^{n+2}}}.$$

This completes the inductive proof.

Because

$$\frac{1}{x-1} + \lim_{n\to\infty} \frac{2^{n+1}}{1-x^{2^{n+1}}} = \frac{1}{x-1} + \lim_{m\to\infty} \frac{m}{1-x^m} = \frac{1}{x-1},$$

our series converges to $1/(x-1)$.

(C. Năstăsescu, C. Niță, M. Brandiburu, D. Joița, *Exerciţii şi Probleme de Algebră* (*Exercises and Problems in Algebra*), Editura Didactică şi Pedagogică, Bucharest, 1983)

425. The series clearly converges for $x = 1$. We will show that it does not converge for $x \neq 1$.

The trick is to divide through by $x - 1$ and compare to the harmonic series. By the Mean value theorem applied to $f(t) = t^{1/n}$, for each n there exists c_n between x and 1 such that

$$\frac{\sqrt[n]{x}-1}{x-1} = \frac{1}{n} c_n^{\frac{1}{n}-1}.$$

Note also that $t \mapsto t^{\frac{1}{n}-1}$ is concave, so its minimum on the interval with endpoints 1 and x is attained at one of the endpoints. It follows that

$$\frac{\sqrt[n]{x}-1}{x-1} > \frac{1}{n}(\max(1,x))^{\frac{1}{n}-1} > \frac{1}{n}(\max(1,x))^{-1}.$$

Summing, we obtain

$$\sum_{n=1}^{\infty} \frac{\sqrt[n]{x}-1}{x-1} \geq \max(1,x))^{-1} \sum_{n=1}^{\infty} \frac{1}{n} = \infty,$$

which proves that the series diverges.

(G.T. Gilbert, M.I. Krusemeyer, L.C. Larson, *The Wohascum County Problem Book*, MAA, 1996)

426. Using the AM-GM inequality we have

$$\sum_{n=1}^{\infty} \sqrt{a_n a_{n+1}} \leq \sum_{n=1}^{\infty} \frac{a_n + a_{n+1}}{2} = \frac{1}{2}\sum_{n=1}^{\infty} a_n + \frac{1}{2}\sum_{n=2}^{\infty} a_n < \infty.$$

Therefore, the series converges. Here we can change the order of summation because the terms are positive.

427. There are exactly $8 \cdot 9^{n-1}$ n-digit numbers in S (the first digit can be chosen in 8 ways, and all others in 9 ways). The least of these numbers is 10^n. We can therefore write

$$\sum_{x_j < 10^n} \frac{1}{x_j} = \sum_{i=1}^{n} \sum_{10^{i-1} \le x_j < 10^i} \frac{1}{x_j} < \sum_{i=1}^{n} \sum_{10^{i-1} \le x_j \le 10^i} \frac{1}{10^{i-1}}$$

$$= \sum_{i=1}^{n} \frac{8 \cdot 9^{i-1}}{10^{i-1}} = 80 \left(1 - \left(\frac{9}{10} \right)^n \right).$$

Letting $n \to \infty$, we obtain the desired inequality.

428. Define the sequence

$$y_n = x_n + 1 + \frac{1}{2^2} + \cdots + \frac{1}{(n-1)^2}, \ n \ge 2.$$

By the hypothesis, $(y_n)_n$ is a decreasing sequence; hence it has a limit. But

$$1 + \frac{1}{2^2} + \cdots + \frac{1}{(n-1)^2} + \cdots$$

converges to a finite limit (which is $\frac{\pi^2}{6}$ as shown by Euler, see Problem 595), and therefore

$$x_n = y_n - 1 - \frac{1}{2^2} - \cdots - \frac{1}{(n-1)^2}, \ n \ge 2,$$

has a limit.

(P.N. de Souza, J.N. Silva, *Berkeley Problems in Mathematics*, Springer, 2004)

429. We have

$$\sin \pi \sqrt{n^2 + 1} = (-1)^n \sin \pi (\sqrt{n^2 + 1} - n) = (-1)^n \sin \frac{\pi}{\sqrt{n^2 + 1} + n}.$$

Clearly, the sequence $x_n = \frac{\pi}{\sqrt{n^2+1}+n}$ lies entirely in the interval $\left(0, \frac{\pi}{2} \right)$, is decreasing, and converges to zero. It follows that $\sin x_n$ is positive, decreasing, and converges to zero. By the Leibniz alternating series test, $\sum_{k \ge 1} (-1)^n \sin x_n$, which is the series in question, is convergent.

(Gh. Siretchi, *Calcul Diferenţial şi Integral (Differential and Integral Calculus)*, Editura Ştiinţifică şi Enciclopedică, 1985)

430. (a) We claim that the answer to the first question is yes. We construct the sequences $(a_n)_n$ and $(b_n)_n$ inductively, in a way inspired by the proof that the harmonic series diverges. At step 1, let $a_1 = 1$, $b_1 = \frac{1}{2}$. Then at step 2, let $a_2 = a_3 = \frac{1}{8}$ and $b_2 = b_3 = \frac{1}{2}$. In general, at step k we already know $a_1, a_2, \ldots, a_{n_k}$ and $b_1, b_2, \ldots, b_{n_k}$ for some integer n_k. We want to define the next terms. If k is even, and if

$$b_{n_k} = \frac{1}{2^{r_k}},$$

let

$$b_{n_k+1} = \cdots = b_{n_k+2^{r_k}} = \frac{1}{2^{r_k}}$$

and

$$a_{n_k+1} = \cdots = a_{n_k+2^{r_k}} = \frac{1}{2^k \cdot 2^{r_k}}.$$

If k is odd, we do precisely the same thing, with the roles of the sequences $(a_n)_n$ and $(b_n)_n$ exchanged. As such we have

$$\sum_n b_n \geq \sum_{k \text{ odd}} 2^{r_k} \frac{1}{2^{r_k}} = 1 + 1 + \cdots = \infty,$$

$$\sum_n a_n \geq \sum_{k \text{ even}} 2^{r_k} \frac{1}{2^{r_k}} = 1 + 1 + \cdots = \infty,$$

which shows that both series diverge. On the other hand, if we let $c_n = \min(a_n, b_n)$, then

$$\sum_n c_n = \sum_k 2^{r_k} \frac{1}{2^k 2^{r_k}} = \sum_k \frac{1}{2^k},$$

which converges to 1. The example proves our claim.

(b) The answer to the second question is no, meaning that the situation changes if we work with the harmonic series. Suppose there is a series $\sum_n a_n$ with the given property. If $c_n = \frac{1}{n}$ for only finitely many n's, then for large n, $a_n = c_n$, meaning that both series diverge. Hence $c_n = \frac{1}{n}$ for infinitely many n. Let $(k_m)_m$ be a sequence of integers satisfying $k_{m+1} \geq 2k_m$ and $c_{k_m} = \frac{1}{k_m}$. Then

$$\sum_{k=k_m+1}^{k_{m+1}} c_k \geq (k_{m+1} - k_m)c_{k_{m+1}} = (k_{m+1} - k_m)\frac{1}{k_{m+1}} = \frac{1}{2}.$$

This shows that the series $\sum_n c_n$ diverges, a contradiction.

(Short list of the 44th International Mathematical Olympiad, 2003)

431. For $n \geq 1$, define the function $f_n : (0, 1) \to \mathbb{R}$, $f_n(x) = x - nx^2$. It is easy to see that $0 < f_n(x) \leq \frac{1}{4n}$, for all $x \in (0, 1)$. Moreover, on $\left(0, \frac{1}{2n}\right]$ the function is decreasing. With this in mind, we prove by induction that

$$0 < x_n < \frac{2}{n^2},$$

for $n \geq 2$. We verify the first three cases:

$$0 = f_1(0) < x_2 = f_1(x_1) = x_1 - x_1^2 \leq \frac{1}{4} < \frac{2}{4},$$

$$0 = f_2(0) < x_3 = f_2(x_2) = x_2 - 2x_2^2 \leq \frac{1}{8} < \frac{2}{9},$$

$$0 = f_3(0) < x_4 = f_3(x_3) = x_3 - 3x_3^2 \leq \frac{1}{12} < \frac{2}{16}.$$

Here we used the inequality $x_1 - x_1^2 - \frac{1}{4} = -\left(x_1 - \frac{1}{2}\right)^2 \le 0$ and the like. Now assume that the inequality is true for $n \ge 4$ and prove it for $n + 1$. Since $n \ge 1$, we have $x_n \le \frac{2}{n^2} \le \frac{1}{2n}$. Therefore,

$$0 = f_n(0) < x_{n+1} = f_n(x_n) \le f_n\left(\frac{2}{n^2}\right) = \frac{2}{n^2} - n \cdot \frac{4}{n^4} = \frac{2n - 4}{n^3}.$$

It is an easy exercise to check that

$$\frac{2n - 4}{n^3} < \frac{2}{(n + 1)^2},$$

which then completes the induction.

We conclude that the series $\sum_n x_n$ has positive terms and is bounded from above by the convergent p-series $2 \sum_n \frac{1}{n^2}$, so it is itself convergent.

(*Gazeta Matematică* (*Mathematics Gazette, Bucharest*), 1980, proposed by L. Panaitopol)

432. The series is convergent because it is bounded from above by the geometric series with ratio $\frac{1}{2}$. Assume that its sum is a rational number $\frac{a}{b}$. Choose n such that $b < 2^n$. Then

$$\frac{a}{b} - \sum_{k=1}^{n} \frac{1}{2^{k^2}} = \sum_{k \ge n+1} \frac{1}{2^{k^2}}.$$

But the sum $\sum_{k=1}^{n} \frac{1}{2^{k^2}}$ is equal to $\frac{n}{2^{n^2}}$ for some integer n. Hence

$$\frac{a}{b} - \sum_{k=1}^{n} \frac{1}{2^{k^2}} = \frac{a}{b} - \frac{m}{2^{n^2}} > \frac{1}{2^{n^2} b} > \frac{1}{2^{n^2 + n}} > \frac{1}{2^{(n+1)^2 - 1}} = \sum_{k \ge (n+1)^2} \frac{1}{2^k} > \sum_{k \ge n+1} \frac{1}{2^{k^2}},$$

a contradiction. This shows that the sum of the series is an irrational number.

Remark. In fact, this number is transcendental.

433. The series is bounded from above by the geometric series $|a_0|(1 + |z| + |z|^2 + \ldots)$, so it converges absolutely. Using the discrete version of integration by parts, known as the Abel summation formula, we can write

$$a_0 + a_1 z + a_2 z^2 + \ldots + a_n z^n + \cdots$$
$$= (a_0 - a_1) + (a_1 - a_2)(1 + z) + \cdots + (a_n - a_{n+1})(1 + z + \cdots + z^n) + \cdots$$

Assume that this is equal to zero. Multiplying by $1 - z$, we obtain

$$(a_0 - a_1)(1 - z) + (a_1 - a_2)(1 - z^2) + \cdots + (a_n - a_{n+1})(1 - z^{n+1}) + \cdots = 0.$$

Define the sequence $b_n = a_n - a_{n+1}$, $n \geq 0$. It is positive and $\sum_n b_n = a_0$. Because $|z| < 1$, the series $\sum_n b_n z^n$ converges absolutely. This allows us in the above inequality to split the left-hand side into two series and move one to the right to obtain

$$b_0 + b_1 + \ldots + b_n + \cdots = b_0 z + b_1 z^2 + \cdots + b_n z^{n+1} + \cdots$$

Applying the triangle inequality to the expression on the right gives

$$|b_0 z + b_1 z^2 + \cdots + b_n z^{n+1}| \leq b_0 |z| + b_1 |z^2| + \cdots + b_n |z^n| + \cdots$$
$$< b_0 + b_1 + \cdots + b_n + \cdots ,$$

which implies that equality cannot hold. We conclude that the sum of the series is not equal to zero.

434. If such a sequence existed, then the numbers

$$\frac{1}{p_0 p_1} - \frac{1}{p_0 p_1 p_2} + \frac{1}{p_0 p_1 p_2 p_3} - \cdots \quad \text{and} \quad \frac{1}{p_0 p_1 p_2} - \frac{1}{p_0 p_1 p_2 p_3} + \cdots$$

should both be positive. It follows that

$$0 < \frac{1}{p_0} - w = \frac{1}{p_0 p_1} - \frac{1}{p_0 p_1 p_2} + \frac{1}{p_0 p_1 p_2 p_3} - \cdots < \frac{1}{p_0 p_1} < \frac{1}{p_0(p_0 + 1)}.$$

Hence p_0 has to be the unique integer with the property that

$$\frac{1}{p_0 + 1} < w < \frac{1}{p_0}.$$

This integer satisfies the double inequality

$$p_0 < \frac{1}{w} < p_0 + 1,$$

which is equivalent to $0 < 1 - p_0 w < w$.

Let $w_1 = 1 - p_0 w$. Then

$$w = \frac{1}{p_0} - \frac{w_1}{p_0}.$$

The problem now repeats for w_1, which is irrational and between 0 and 1. Again p_1 has to be the unique integer with the property that

$$\frac{1}{p_1 + 1} < 1 - p_0 w < \frac{1}{p_1}.$$

If we set $w_2 = 1 - p_1 w_1$, then

$$w = \frac{1}{p_0} - \frac{1}{p_0 p_1} + \frac{w_2}{p_0 p_1}.$$

Now the inductive pattern is clear. At each step we set $w_{k+1} = 1 - p_k w_k$, which is an irrational number between 0 and 1. Then choose p_{k+1} such that

$$\frac{1}{p_{k+1} + 1} < w_{k+1} < \frac{1}{p_{k+1}}.$$

Note that

$$w_{k+1} = 1 - p_k w_k < 1 - p_k \frac{1}{p_k + 1} = \frac{1}{p_k + 1},$$

and therefore $p_{k+1} \geq p_k + 1 > p_k$.

Once the numbers $p_0, p_1, p_2, \ldots$ have been constructed, it is important to observe that since $w_k \in (0, 1)$ and $p_0 p_1 \cdots p_k \geq (k + 1)!$, the sequence

$$\frac{1}{p_0} - \frac{1}{p_0 p_1} + \cdots + (-1)^{k+1} \frac{w_{k+1}}{p_1 p_2 \cdots p_k}$$

converges to w. So $p_0, p_1, \ldots, p_k, \ldots$ have the required properties, and as seen above, they are unique.

(13th W.L. Putnam Mathematical Competition, 1953)

435. First, denote by M the set of positive integers greater than 1 that are not perfect powers (i.e., are not of the form a^n, where a is a positive integer and $n \geq 2$). Note that the terms of the series are positive, so we can freely permute them. The series is therefore equal to

$$\sum_{m \in M} \sum_{k=2}^{\infty} \frac{1}{m^k - 1}.$$

Expanding each term as a geometric series, we transform this into

$$\sum_{m \in M} \sum_{k=2}^{\infty} \sum_{j=1}^{\infty} \frac{1}{m^{kj}} = \sum_{m \in M} \sum_{j=1}^{\infty} \sum_{k=2}^{\infty} \frac{1}{m^{kj}}.$$

Again, we can change the order of summation because the terms are positive. The innermost series should be summed as a geometric series to give

$$\sum_{m \in M} \sum_{j=1}^{\infty} \frac{1}{m^j (m^j - 1)}.$$

This is the same as

$$\sum_{n=2}^{\infty} \frac{1}{n(n - 1)} = \sum_{n=2}^{\infty} \left(\frac{1}{n - 1} - \frac{1}{n} \right).$$

The latter is a telescopic series that sums up to 1, and we are done.

(Ch. Goldbach, solution from G.M. Fihtenholts, *Kurs Differentsial'novo i Integral'novo Ischisleniya* (*Course in Differential and Integral Calculus*), Gosudarstvennoe Izdatel'stvo Fiziko-Matematicheskoi Literatury, Moscow 1964)

436. Let us make the convention that the letter p always denotes a prime number. Consider the set $A(n)$ consisting of those positive integers that can be factored into primes that do not exceed n. Then

$$\prod_{p \le n}\left(1 + \frac{1}{p} + \frac{1}{p^2} + \cdots\right) = \sum_{m \in A(n)} \frac{1}{m}.$$

This sum includes $\displaystyle\sum_{m=1}^{n} \frac{1}{m}$, which is known to exceed $\ln n$. Thus, after summing the geometric series, we obtain

$$\prod_{p \le n}\left(1 - \frac{1}{p}\right)^{-1} > \ln n.$$

For the factors of the product we use the estimate

$$e^{t+t^2} > (1 - t)^{-1}, \quad \text{for } 0 \le t \le \frac{1}{2}.$$

To prove this estimate, rewrite it as $f(t) \ge 1$, where $f(t) = (1 - t)e^{t+t^2}$. Because $f'(t) = t(1 - 2t)e^{t+t^2} \ge 0$ on $\left[0, \frac{1}{2}\right], f$ is increasing; thus $f(t) \ge f(0) = 1$.

Returning to the problem, we have

$$\prod_{p \le n} \exp\left(\frac{1}{p} + \frac{1}{p^2}\right) \ge \prod_{p \le n}\left(1 - \frac{1}{p}\right)^{-1} > \ln n.$$

Therefore,

$$\sum_{p \le n} \frac{1}{p} + \sum_{p \le n} \frac{1}{p^2} > \ln \ln n.$$

But

$$\sum_{p \le n} \frac{1}{p^2} < \sum_{n=2}^{\infty} \frac{1}{k^2} = \frac{\pi^2}{6} - 1 < 1.$$

Hence

$$\sum_{p \le n} \frac{1}{p} \ge \ln \ln n - 1,$$

as desired.

Remark. This is another proof that there are infinitely many primes.

(Solution from I. Niven, H.S. Zuckerman, H.L. Montgomery, *An Introduction to the Theory of Numbers*, Wiley, 1991)

437. We have

$$(k^2 + 1)k! = (k^2 + k - k + 1)k! = k(k + 1)k! - (k - 1)k!$$
$$= k(k + 1)! - (k - 1)k! = a_{k+1} - a_k,$$

where $a_k = (k - 1)k!$. The sum collapses to $a_{n+1} - a_1 = n(n + 1)!$.

438. If ζ is an mth root of unity, then all terms of the series starting with the mth are zero. We are left to prove that

$$\zeta^{-1} = \sum_{n=0}^{m-1} \zeta^n (1 - \zeta)(1 - \zeta^2) \cdots (1 - \zeta^n).$$

Multiplying both sides by ζ yields the equivalent identity

$$1 = \sum_{n=0}^{m-1} \zeta^{n+1} (1 - \zeta)(1 - \zeta^2) \cdots (1 - \zeta^n).$$

The sum telescopes as follows:

$$\sum_{n=0}^{m-1} \zeta^{n+1} (1 - \zeta)(1 - \zeta^2) \cdots (1 - \zeta^n) = \sum_{n=0}^{m-1} (1 - (1 - \zeta^{n+1}))(1 - \zeta)(1 - \zeta^2) \cdots (1 - \zeta^n)$$

$$= \sum_{n=0}^{m-1} [(1 - \zeta)(1 - \zeta^2) \cdots (1 - \zeta^n) - (1 - \zeta)(1 - \zeta^2) \cdots (1 - \zeta^{n+1})] = 1 - 0 = 1,$$

and the identity is proved.

439. We have

$$1 + \sum_{k=0}^{r-1} \binom{r}{k} S_k(n) = 1 + \sum_{k=0}^{r-1} \binom{r}{k} \sum_{p=1}^{n} p^k = 1 + \sum_{p=1}^{n} \sum_{k=0}^{r-1} \binom{r}{k} p^k$$

$$= 1 + \sum_{p=1}^{n} [(p+1)^r - p^r] = (n+1)^r.$$

440. Set $b_n = \sqrt{2n-1}$ and observe that $4n = b_{n+1}^2 + b_n^2$. Then

$$a_n = \frac{b_{n+1}^2 + b_n^2 + b_{n+1} b_n}{b_{n+1} + b_n} = \frac{(b_{n+1} - b_n)(b_{n+1}^2 + b_{n+1} b_n + b_{n-1}^2)}{(b_{n+1} - b_n)(b_{n+1} + b_n)}$$

$$= \frac{b_{n+1}^3 - b_n^3}{b_{n+1}^2 - b_n^2} = \frac{1}{2} (b_{n+1}^3 - b_n^3).$$

So the sum under discussion telescopes as

$$a_1 + a_2 + \cdots + a_{40} = \frac{1}{2}(b_2^3 - b_1^3) + \frac{1}{2}(b_3^3 - b_2^3) + \cdots + \frac{1}{2}(b_{41}^3 - b_{40}^3)$$

$$= \frac{1}{2}(b_{41}^3 - b_1^3) = \frac{1}{2}(\sqrt{81}^3 - 1) = 364,$$

and we are done.

(Romanian Team Selection Test for the Junior Balkan Mathematical Olympiad, proposed by T. Andreescu)

441. The important observation is that

$$\frac{(-1)^{k+1}}{1^2 - 2^2 + 3^2 - \cdots + (-1)^{k+1}k^2} = \frac{2}{k(k+1)}.$$

Indeed, this is true for $k = 1$, and inductively, assuming it to be true for $k = l$, we obtain

$$1^2 - 2^2 + 3^2 - \cdots + (-1)^{l+1}l^2 = (-1)^{l+1}\frac{l(l+1)}{2}.$$

Then

$$1^2 - 2^2 + 3^2 - \cdots + (-1)^{l+2}(l+1)^2 = (-1)^{l+1}\frac{l(l+1)}{2} + (-1)^{l+2}(l+1)^2$$

$$= (-1)^{l+2}(l+1)\left(-\frac{l}{2} + l + 1\right),$$

whence

$$\frac{(-1)^{l+2}}{1^2 - 2^2 + 3^3 - \cdots + (-1)^{l+2}(l+1)^2} = \frac{2}{(l+1)(l+2)},$$

as desired. Hence the given sum equals

$$\sum_{k=1}^{n}\frac{2}{k(k+1)} = 2\sum_{k=1}^{n}\left(\frac{1}{k} - \frac{1}{k+1}\right),$$

telescoping to

$$2\left(1 - \frac{1}{n+1}\right) = \frac{2n}{n+1},$$

and we are done.

(T. Andreescu)

442. The sum telescopes once we rewrite the general term as

$$\frac{1}{(\sqrt{n} + \sqrt{n+1})(\sqrt[4]{n} + \sqrt[4]{n+1})} = \frac{\sqrt[4]{n+1} - \sqrt[4]{n}}{(\sqrt{n+1} + \sqrt{n})(\sqrt[4]{n+1} + \sqrt[4]{n})(\sqrt[4]{n+1} - \sqrt[4]{n})}$$

$$= \frac{\sqrt[4]{n+1} - \sqrt[4]{n}}{(\sqrt{n+1} + \sqrt{n})(\sqrt{n+1} - \sqrt{n})}$$

$$= \frac{\sqrt[4]{n+1} - \sqrt[4]{n}}{n+1-n} = \sqrt[4]{n+1} - \sqrt[4]{n}.$$

The sum from the statement is therefore equal to $\sqrt[4]{10000} - 1 = 10 - 1 = 9$.

(*Mathematical Reflections*, proposed by T. Andreescu)

443. As usual, the difficulty lies in finding the "antiderivative" of the general term. We have

$$\frac{1}{\sqrt{1 + \left(1 + \frac{1}{n}\right)^2} + \sqrt{1 + \left(1 - \frac{1}{n}\right)^2}} = \frac{\sqrt{1 + \left(1 + \frac{1}{n}\right)^2} - \sqrt{1 + \left(1 - \frac{1}{n}\right)^2}}{1 + \left(1 + \frac{1}{n}\right)^2 - 1 - \left(1 - \frac{1}{n}\right)^2}$$

$$= \frac{\sqrt{1 + \left(1 + \frac{1}{n}\right)^2} - \sqrt{1 + \left(1 - \frac{1}{n}\right)^2}}{\frac{4}{n}}$$

$$= \frac{1}{4}\left(\sqrt{n^2 + (n+1)^2} - \sqrt{n^2 + (n-1)^2}\right)$$

$$= \frac{1}{4}(b_{n+1} - b_n),$$

where $b_n = \sqrt{n^2 + (n-1)^2}$. Hence the given sum collapses to $\frac{1}{4}(29 - 1) = 7$.
(*Mathematical Reflections*, proposed by T. Andreescu)

444. Let us look at the summation over n first. Multiplying each term by $(m+n+2) - (n+1)$ and dividing by $m + 1$, we obtain

$$\frac{m!}{m+1} \sum_{n=0}^{\infty} \left(\frac{n!}{(m+n+1)!} - \frac{(n+1)!}{(m+n+2)!}\right).$$

This is a telescopic sum that adds up to

$$\frac{m!}{m+1} \cdot \frac{0!}{(m+1)!}.$$

Consequently, the expression we are computing is equal to

$$\sum_{m=0}^{\infty} \frac{1}{(m+1)^2} = \frac{\pi^2}{6},$$

where we have used Euler's formula (see Section 3.2.12 Problem 595).
(*Mathematical Mayhem*, 1995)

445. This problem is similar to the last example from the introduction. We start with

$$a_k - b_k = \frac{1}{2}\left[4k + (k+1) + (k-1) - 4\sqrt{k^2 + k} + 4\sqrt{k^2 - k} + 2\sqrt{k^2 - 1}\right]$$

$$= \frac{1}{2}\left(2\sqrt{k} - \sqrt{k+1} - \sqrt{k-1}\right)^2.$$

From here we obtain

$$\sqrt{a_k - b_k} = \frac{1}{\sqrt{2}}\left(2\sqrt{k} - \sqrt{k+1} - \sqrt{k-1}\right)$$

$$= -\frac{1}{\sqrt{2}}\left(\sqrt{k+1} - \sqrt{k}\right) + \frac{1}{\sqrt{2}}\left(\sqrt{k} - \sqrt{k+1}\right).$$

The sum from the statement telescopes to

$$-\frac{1}{\sqrt{2}}\left(\sqrt{50}-\sqrt{1}\right)+\frac{1}{\sqrt{2}}\left(\sqrt{49}-\sqrt{0}\right)=-5+4\sqrt{2}.$$

(Romanian Mathematical Olympiad, 2004, proposed by T. Andreescu)

446. *First solution:* Let $S_n = \sum_{k=0}^{n}(-1)^k(n-k)!(n+k)!$. Reordering the terms of the sum, we have

$$S_n = (-1)^n \sum_{k=0}^{n}(-1)^k k!(2n-k)!$$

$$= (-1)^n \frac{1}{2}\left((-1)^n n!n! + \sum_{k=0}^{2n}(-1)^k k!(2n-k)!\right)$$

$$= \frac{(n!)^2}{2} + (-1)^n \frac{T_n}{2},$$

where $T_n = \sum_{k=0}^{2n}(-1)^k k!(2n-k)!$. We now focus on the sum T_n. Observe that

$$\frac{T_n}{(2n)!} = \sum_{k=0}^{2n}\frac{(-1)^k}{\binom{2n}{k}}$$

and

$$\frac{1}{\binom{2n}{k}} = \frac{2n+1}{2(n+1)}\left[\frac{1}{\binom{2n+1}{k}}+\frac{1}{\binom{2n+1}{k+1}}\right].$$

Hence

$$\frac{T_n}{(2n)!} = \frac{2n+1}{2(n+1)}\left[\frac{1}{\binom{2n+1}{0}}+\frac{1}{\binom{2n+1}{1}}-\frac{1}{\binom{2n+1}{1}}-\frac{1}{\binom{2n+1}{2}}+\cdots+\frac{1}{\binom{2n+1}{2n}}+\frac{1}{\binom{2n+1}{2n+1}}\right].$$

This sum telescopes to

$$\frac{2n+1}{2(n+1)}\left[\frac{1}{\binom{2n+1}{0}}+\frac{1}{\binom{2n+1}{2n+1}}\right]=\frac{2n+1}{n+1}.$$

Thus $T_n = \frac{(2n+1)!}{n+1}$, and therefore

$$S_n = \frac{(n!)^2}{2}+(-1)^n\frac{(2n+1)!}{2(n+1)}.$$

Second solution: Multiply the kth term in S_n by $(n - k + 1) + (n + k + 1)$ and divide by $2(n + 1)$ to obtain

$$S_n = \frac{1}{2(n+1)} \sum_{k=0}^{n} [(-1)^k (n-k+1)!(n+k)! + (-1)^k (n-k)!(n+k+1)!].$$

This telescopes to

$$\frac{1}{2(n+1)} [n!(n+1)! + (-1)^n (2n+1)!].$$

(T. Andreescu, second solution by R. Stong)

447. The sequence is obviously strictly decreasing. Because $a_k - a_{k+1} = 1 - \frac{1}{a_k+1}$, we have

$$a_n = a_0 + (a_1 - a_0) + \cdots + (a_n - a_{n-1}) = 1994 - n + \frac{1}{a_0 + 1} + \cdots + \frac{1}{a_{n-1} + 1}$$

$$> 1994 - n.$$

Also, because the sequence is strictly decreasing, for $1 \le n \le 998$,

$$\frac{1}{a_0 + 1} + \cdots + \frac{1}{a_{n-1} + 1} < \frac{n}{a_{n-1} + 1} < \frac{998}{a_{997} + 1} < 1,$$

since we have seen above that $a_{997} > 1994 - 997 = 997$. Hence $\lfloor a_n \rfloor = 1994 - n$, as desired.

(Short list of the 35th International Mathematical Olympiad, 1994, proposed by T. Andreescu)

448. Let $x_1 = k + \sqrt{k^2 + 1}$ and $x_2 = k - \sqrt{k^2 + 1}$. We have $|x_2| = \frac{1}{x_1} < \frac{1}{2k} \le \frac{1}{2}$, so $-\left(\frac{1}{2}\right)^2 \le x_2^n \le \left(\frac{1}{2}\right)^n$. Hence

$$x_1^n + x_2^n - 1 < x_1^n + \left(\frac{1}{2}\right)^n - 1 < a_n \le x_1^n - \left(\frac{1}{2}\right)^n + 1 < x_1^n + x_2^n + 1,$$

for all $n \ge 1$. From

$$x_1^{n+1} + x_2^{n+1} = (x_1 + x_2)(x_1^n + x_2^n) - x_1 x_2 (x_1^{n-1} + x_2^{n-1})$$
$$= 2k(x_1^n + x_2^n) + (x_1^{n-1} + x_2^{n-1})$$

for $n \ge 1$, we deduce that $x_1^n + x_2^n$ is an integer for all n. We obtain the more explicit formula $a_n = x_1^n + x_2^n$ for $n \ge 0$, and consequently the recurrence relation $a_{n+1} = 2ka_n + a_{n-1}$, for all $n \ge 1$. Then

$$\frac{1}{a_{n-1}a_{n+1}} = \frac{1}{2ka_n} \cdot \frac{2ka_n}{a_{n-1}a_{n+1}} = \frac{1}{2k} \cdot \frac{a_{n+1} - a_{n-1}}{a_{n-1}a_n a_{n+1}} = \frac{1}{2k} \left(\frac{1}{a_{n-1}a_n} - \frac{1}{a_n a_{n+1}} \right).$$

It follows that

$$\sum_{n=1}^{\infty} \frac{1}{a_{n-1}a_{n+1}} = \frac{1}{2k} \left(\frac{1}{a_0 a_1} - \lim_{N \to \infty} \frac{1}{a_N a_{N+1}} \right) = \frac{1}{2ka_0 a_1} = \frac{1}{8k^2}.$$

449. For $N \geq 2$, define

$$a_N = \left(1 - \frac{4}{1}\right)\left(1 - \frac{4}{9}\right)\left(1 - \frac{4}{25}\right) \cdots \left(1 - \frac{4}{(2N-1)^2}\right).$$

The problem asks us to find $\lim\limits_{N \to \infty} a_N$. The defining product for a_N telescopes as follows:

$$a_N = \left[\left(1 - \frac{2}{1}\right)\left(1 + \frac{2}{1}\right)\right]\left[\left(1 - \frac{2}{3}\right)\left(1 + \frac{2}{3}\right)\right] \cdots \left[\left(1 - \frac{2}{2N-1}\right)\left(1 + \frac{2}{2N-1}\right)\right]$$
$$= (-1 \cdot 3)\left(\frac{1}{3} \cdot \frac{5}{3}\right)\left(\frac{3}{5} \cdot \frac{7}{5}\right) \cdots \left(\frac{2N-3}{2N-1} \cdot \frac{2N+1}{2N-1}\right) = -\frac{2N+1}{2N-1}.$$

Hence the infinite product is equal to

$$\lim_{N \to \infty} a_N = -\lim_{N \to \infty} \frac{2N+1}{2N-1} = -1.$$

450. Define the sequence $(a_N)_N$ by

$$a_N = \prod_{n=1}^{N} (1 + x^{2^n}).$$

Note that $(1 - x)a_N$ telescopes as

$$(1-x)(1+x)(1+x^2)(1+x^4) \cdots (1+x^{2^N}) = (1-x^2)(1+x^2)(1+x^4) \cdots (1+x^{2^N})$$
$$= (1-x^4)(1+x^4) \cdots (1+x^{2^N}) \cdots$$
$$= (1 - x^{2^{N+1}}).$$

Hence $(1-x)a_N \to 1$ as $N \to \infty$, and therefore

$$\prod_{n \geq 0} (1 + x^{2^n}) = \frac{1}{1-x}.$$

451. Let $P_N = \prod_{n=1}^{N} \left(1 - \frac{x^n}{x_{n+1}}\right)$, $N \geq 1$. We want to examine the behavior of P_N as $N \to \infty$. Using the recurrence relation we find that this product telescopes as

$$P_N = \prod_{n=1}^{n} \left(\frac{x_{n+1} - x^n}{x_{n+1}}\right) = \prod_{n=1}^{n} \frac{nx_n}{x_{n+1}} = \frac{N!}{x_{N+1}}.$$

Hence

$$\frac{1}{P_{n+1}} - \frac{1}{P_n} = \frac{x_{n+2}}{(n+1)!} - \frac{x_{n+1}}{n!} = \frac{x_{n+2} - (n+1)x_{n+1}}{(n+1)!} = \frac{x^{n+1}}{(n+1)!}, \quad \text{for } n \geq 1.$$

Adding up these relations for $1 \leq n \leq N + 1$, and using the fact that the sum on the left telescopes, we obtain

$$\frac{1}{P_{N+1}} = \frac{1}{P_1} + \frac{x^2}{2!} + \frac{x^3}{3!} + \cdots + \frac{x^{N+1}}{(N+1)!}$$

$$= 1 + \frac{x}{1!} + \frac{x^2}{2!} + \cdots + \frac{x^{N+1}}{(N+1)!}.$$

Because this last expression converges to e^x, we obtain that $\lim_{N \to \infty} P_N = e^{-x}$, as desired.

(*Revista Matematică din Timişoara* (*Timişoara Mathematics Gazette*), proposed by T. Andreescu and D. Andrica)

452. One can see immediately that $x = 0$ is a solution. The left-hand side is an increasing function, while the right-hand side is a decreasing function. So the solution is unique.

(Romanian Math Olympiad, 1975, proposed by L. Panaitopol)

453. We have $\sqrt[n]{3} + \sqrt[n]{7} > 2\sqrt[2n]{21} > 2\sqrt[n]{4.5}$. On the other hand, for $d > 0$ the function $f : (0, \infty) \to \mathbb{R}$,

$$f(x) = \sqrt[n]{x+d} - \sqrt[n]{x} = \frac{d}{(\sqrt[n]{x+d})^{n-1} + (\sqrt[n]{x+d})^{n-2}\sqrt[n]{x} + \cdots + (\sqrt[n]{x})^{n-1}}$$

is decreasing. Hence $\sqrt[n]{4.5} - \sqrt[n]{4} > \sqrt[n]{5} - \sqrt[n]{4.5}$. The inequality follows.

454. Assume that such a function exists. Remove the 2015 solutions from $\mathbb{R}$ to obtain a set A, and restrict f to A. Then $f(f(x)) = x$ has $5102 - 2015 = 3087$ solutions in A while $f(x) = x$ has none. The 3087 solutions can be grouped in pairs (x, y) such that $f(x) = y$ and $f(y) = x$. But this is impossible since 3087 is an odd number. It follows that such a function does not exist, so the answer to the question is negative.

455. It is not hard to see that $f(0) = 0$, because the graph is invariant under the $180°$ rotation about the origin, which maps $(0, f(0))$ to $(0, -f(0))$.

For $x \neq 0$ we can use the formula

$$f(x) = \frac{x}{|x|} - (-1)^{\lfloor -|x| \rfloor} x.$$

456. If f is such a function, then

$$(f \circ f \circ f \circ f \circ f)(x) = [(f \circ f) \circ (f \circ f \circ f)](x) = (f \circ f)(x^3) = x^5.$$

So

$$(f \circ f)(x) = x^{5/3}.$$

Then

$$(f \circ f \circ f)(x) = [f \circ (f \circ f)](x) = f(x^{5/3}) = x^3.$$

Setting $t = x^{5/3}$ and substituting in the above we obtain

$$f(t) = t^{9/5}.$$

But this function does not satisfy either of the conditions from the statement. Hence the answer is negative.

(Konhauser Problem Fest, 2014, proposed by R. Gelca)

457. Write the system as

$$3^x = 2^y + 3^y$$
$$9^x = 6^y + 19^y.$$

Square the first equation then substitute 9^x from the second to obtain

$$4^y + 2 \cdot 6^y + 9^y = 6^y + 19^y.$$

Rewrite this as

$$4^y + 6^y + 9^y = 19^y.$$

It is easy to see that $y = 1$ is a solution. There are no other solutions to this equation because after dividing by 19^y we obtain

$$\left(\frac{4}{19}\right)^y + \left(\frac{6}{19}\right)^y + \left(\frac{9}{19}\right)^y = 1$$

and the left-hand side is a strictly decreasing function which assumes the value 1 exactly once. We conclude that the only pair of real numbers satisfying the system is $(\log_3 5, 1)$.

(Konhauser Problem Fest, 2014, proposed by R. Gelca)

458. The right-hand side is less than 1, hence so is the left-hand side. This shows that $x \in (0, 1)$. Taking the logarithm in base a, we obtain

$$a^x \log_a x = x^a$$

or $a^x \log_a x - x^a = 0$. The left-hand side is a decreasing functions, so the solution, if it exists, is unique. Of course, $x = a$ is a solution.

(Romanian Mathematical Olympiad, 1983, proposed by T. Andreescu and I.V. Maftei)

459. It is not hard to guess that $x = 2, y = 3, z = 4$ is a solution. Using the logarithms we obtain

$$y \log_2 x = 3, \quad z \log_3 y = 4, \quad x \log_4 z = 2.$$

Furthermore,

$$\log_3 y + \log_3 (\log_2 x) = 1, \quad \log_4 z + \log_4 (\log_3 y) = 1, \quad \log_4 z = \frac{2}{x}.$$

Eliminating y and z we obtain

$$\frac{2}{x} + \log_4(1 - \log_3(\log_2 x)) = 1.$$

The function on the left is decreasing, so the solution $x = 2$ is unique. Hence the system has the unique solution $x = 2, y = 3, z = 4$.

(*Gazeta Matematică (Mathematics Gazette)*, proposed by A. Ene)

460. The functions $x \mapsto x^n + 1$ and $x \mapsto \sqrt[n]{x} - 1$ are strictly increasing on $\mathbb{R}$, hence their composition, $f : \mathbb{R} \to \mathbb{R}$, $f(x) = \sqrt[n]{x^n + 1} - 1$ is also strictly increasing. Note that f is also onto, so it is bijective, and its inverse is $f^{-1} = \sqrt[n]{(x + 1)^n - 1}$. The equation from the statement translates to

$$f(x) = f^{-1}(x).$$

Let α be a solution. Then $f(f(\alpha)) = \alpha$. If $f(\alpha) < \alpha$, then $f(f(\alpha)) < f(\alpha)$ and so $\alpha < f(\alpha)$, a contradiction. If $f(\alpha) > \alpha$, then $f(f(\alpha)) > f(\alpha)$, so $\alpha > f(\alpha)$, again a contradiction. Hence $f(\alpha) = \alpha$, which leads to the simpler equation to solve

$$\sqrt[n]{x^n + 1} = x + 1,$$

which is equivalent to

$$(\alpha + 1)^n = \alpha^n + 1.$$

It is easy to check that $x = 0, -1$ are solutions. Newton's binomial formula shows that there are no positive solutions.

If $\alpha < -1$, set $\beta = -\alpha - 1 > 0$. The equation translates to $(\beta + 1)^n = \beta^n + 1$, which has no solutions. If $\alpha \in (-1, 0)$, then set $\beta = -\alpha$. We have $(1 - \beta)^n = (-\beta)^n + 1$ or $(1 - \beta)^n + \beta^n = 1$. But the left-hand side is less than $1 - \beta + \beta$, so this is again impossible. Thus the equation has the only solutions $x = 0$ and $x = -1$.

Remark. A similar problem: "solve the equation $2\sqrt[3]{2y - 1} = y^3 + 1$" was part of the collection of problems used for the oral admission exam by the Moscow State University during communist times to prevent Jewish people and other undesirables from entering that school (cf. T. Khovanova, A. Radul, *Killer Problems*, *The American Mathematical Monthly*, Vol. 119, No. 10, 2012)

(*Gazeta Matematică (Mathematics Gazette, Bucharest)*, proposed by I. Băetu)

461. Let $g : \mathbb{R} \to \mathbb{R}$, $g(x) = ax^2 + bx + c$. The functional equation from the statement implies that $f \circ g = g \circ f$.

On the other hand the condition $4ac \le (b-1)^2$ is equivalent to the fact that the discriminant of the quadratic equation $g(x) = x$ is nonnegative. Hence g has fixed points. In fact it has either one or two fixed points.

Let u be a fixed point of g. Then

$$f(g(u)) = f(u) = g(f(u)),$$

so $f(u)$ is a fixed point of g as well. From $f(u) = g(f(u))$ we obtain

$$f(f(u)) = f(g(f(u))) = g(f(f(u))),$$

which shows that $f(f(u))$ is a fixed point of g. Among the three fixed points of g, $u, f(u)$, $f(f(u))$, two must coincide.

Case I. $f(u) = u$. Then u is a fixed point for f, and hence for $f \circ f$.

Case II. $f(f(u)) = u$. This is exactly what we desire.

Case III. $f(f(u)) = f(u)$. In this case f has a fixed point, and, as in Case I, so does $f \circ f$. (Romanian Mathematical Olympiad, 1986, proposed by T. Andreescu)

462. We will show that $\cos A < \cos 72°$. Squaring the formula provided by the law of cosines we obtain

$$a^4 = b^4 + c^4 + 4b^2 c^2 \cos^2 A + 2b^2 c^2 - 4b^3 c \cos A - 4bc^3 \cos A.$$

Substituting $a^4 = b^4 + c^4$ and dividing by 2 we obtain

$$2bc \cos^2 A - 2(b^2 + c^2) \cos A + bc = 0.$$

Solving this as a quadratic equation in $\cos A$, we obtain

$$\cos A = \frac{1}{2}\left[\left(\frac{b}{c} + \frac{c}{b} \right) \pm \sqrt{\frac{b^2}{c^2} + \frac{c^2}{b^2}} \right]$$

The larger of the two values exceeds 1, since $\frac{b}{c} + \frac{c}{b} \geq 2$. We thus conclude that

$$\cos A = \frac{1}{2}\left[\left(\frac{b}{c} + \frac{c}{b} \right) - \sqrt{\frac{b^2}{c^2} + \frac{c^2}{b^2}} \right]$$

For simplicity, set $x = \frac{b}{c}$. Consider the function $f : (0, \infty) \to \mathbb{R}$,

$$f(x) = \left(x + \frac{1}{x} - \sqrt{x^2 + \frac{1}{x^2}} \right),$$

and observe that

$$f(x) = \frac{\left(x + \frac{1}{x} \right)^2 - \left(\sqrt{x^2 + \frac{1}{x^2}} \right)^2}{\left(x + \frac{1}{x} + \sqrt{x^2 + \frac{1}{x^2}} \right)} = \frac{2}{\left(x + \frac{1}{x} - \sqrt{x^2 + \frac{1}{x^2}} \right)}.$$

The maximum of f is attained when the denominator is minimal, and this happens when $x = 1$. This shows that $f(x) \leq f(1) = 2 - \sqrt{2}$ for all x, and hence

$$\cos A \leq \frac{2 - \sqrt{2}}{2}.$$

On the other hand,

$$\cos 72° = \frac{\sqrt{5} - 1}{4}.$$

To see why this is true, note that $\cos 72° + i \sin 72°$ is a root of the equation $x^5 - 1 = 0$. Removing the solution $x = 1$, we obtain the reciprocal equation

$$x^4 + x^3 + x^2 + x + 1 = 0.$$

Dividing through by x^2, and substituting $x + \frac{1}{x} = y$, we obtain the quadratic equation $y^2 + y - 1 = 0$, with roots $y_{1,2} = \frac{-1 \pm \sqrt{5}}{2}$. We are interested in the positive root only. We thus have

$$x + \frac{1}{x} = \frac{\sqrt{5} - 1}{2}.$$

Solving we obtain

$$x = \frac{\sqrt{5} - 1}{4} \pm i \frac{\sqrt{10 + 2\sqrt{5}}}{4},$$

which proves the claim. An easy check shows that

$$\frac{2 - \sqrt{2}}{2} < \frac{\sqrt{5} - 1}{4},$$

and we are done.

(Romanian Teams Selection Test, 1983, proposed by I. Tomescu)

463. We are supposed to find m and n such that

$$\lim_{x \to \infty} \sqrt[3]{8x^3 + mx^2} - nx = 1 \quad \text{or} \quad \lim_{x \to -\infty} \sqrt[3]{8x^3 + mx} - nx = 1.$$

We compute

$$\sqrt[3]{8x^3 + mx^2} - nx = \frac{(8 - n^3)x^3 + mx^2}{\sqrt[3]{(8x^3 + mx^2)^2} + nx\sqrt[3]{8x^3 + mx^2} + n^2x^2}.$$

For this to have a finite limit at either $+\infty$ or $-\infty$, $8 - n^3$ must be equal to 0 (otherwise the highest degree of x in the numerator would be greater than the highest degree of x in the denominator). We have thus found that $n = 2$.

Next, factor out and cancel an x^2 to obtain

$$f(x) = \frac{m}{\sqrt[3]{\left(8 + \frac{m}{x}\right)^2} + 2\sqrt[3]{8 + \frac{m}{x}} + 4}.$$

We see that $\lim_{x \to \infty} f(x) = \frac{m}{12}$. For this to be equal to 1, m must be equal to 12. Hence the answer to the problem is $(m, n) = (12, 2)$.

464. This is a limit of the form 1^∞. It can be computed as follows:

$$\lim_{x\to\pi/2} (\sin x)^{\frac{1}{\cos x}} = \lim_{x\to\pi/2} (1 + \sin x - 1)^{\frac{1}{\sin x - 1} \cdot \frac{\sin x - 1}{\cos x}}$$

$$= \left(\lim_{t\to 0} (1+t)^{1/t} \right)^{\lim_{x\to\pi/2} \frac{\sin x - 1}{\cos x}} = \exp\left(\lim_{u\to 0} \frac{\cos u - 1}{\sin u} \right)$$

$$= \exp\left(\frac{\cos u - 1}{u} \cdot \frac{u}{\sin u} \right) = e^{0\cdot 1} = e^0 = 1.$$

The limit therefore exists.

465. Without loss of generality, we may assume that $m > n$. Write the limit as

$$\lim_{x\to 0} \frac{\sqrt[mn]{\cos^n x} - \sqrt[mn]{\cos^m x}}{x^2}.$$

Now we can multiply by the rational conjugate and obtain

$$\lim_{x\to 0} \frac{\cos^n x - \cos^m x}{x^2 \left(\sqrt[mn]{(\cos^n x)^{mn-1}} + \cdots + \sqrt[mn]{(\cos^m x)^{mn-1}} \right)}$$

$$= \lim_{x\to 0} \frac{\cos^n x (1 - \cos^{m-n} x)}{mnx^2} = \lim_{x\to 0} \frac{1 - \cos^{m-n} x}{mnx^2}$$

$$= \lim_{x\to 0} \frac{(1 - \cos x)(1 + \cos x + \cdots + \cos^{m-n-1} x)}{mnx^2}$$

$$= \frac{m-n}{mn} \lim_{x\to 0} \frac{1 - \cos x}{x^2} = \frac{m-n}{2mn}.$$

We are done.

466. For $x > 1$ define the sequence $(x_n)_{n\geq 0}$ by $x_0 = x$ and $x_{n+1} = \frac{x_n^2 + 1}{2}$, $n \geq 0$. The sequence is increasing because of the AM-GM inequality. Hence it has a limit L, finite or infinite. Passing to the limit in the recurrence relation, we obtain $L = \frac{L^2 + 1}{2}$; hence either $L = 1$ or $L = \infty$. Since the sequence is increasing, $L \geq x_0 > 1$, so $L = \infty$. We therefore have

$$f(x) = f(x_0) = f(x_1) = f(x_2) = \cdots = \lim_{n\to\infty} f(x_n) = \lim_{x\to\infty} f(x).$$

This implies that f is constant, which is ruled out by the hypothesis. So the answer to the question is negative.

467. We can assume that $m > 1$; otherwise, we can flip the fraction and change t to $\frac{1}{m}t$. There is an integer n such that $m < 2^n$. Because f is increasing, $f(t) < f(mt) < f(2^n t)$, We obtain

$$1 < \frac{f(mt)}{f(t)} < \frac{f(2^n t)}{f(t)}.$$

The right-hand side is equal to the telescopic product

$$\frac{f(2^n t)}{f(2^{n-1} t)} \cdot \frac{f(2^{n-1} t)}{f(2^{n-2} t)} \cdots \frac{f(2t)}{f(t)},$$

whose limit as t goes to infinity is 1. The squeezing principle implies that

$$\lim_{t \to \infty} \frac{f(mt)}{f(t)} = 1,$$

as desired.

(V. Radu)

468. The sum under discussion is the derivative of f at 0. We have

$$\left| \sum_{k=1}^{n} k a_k \right| = |f'(0)| = \lim_{x \to 0} \left| \frac{f(x) - f(0)}{x - 0} \right|$$

$$= \lim_{x \to 0} \left| \frac{f(x)}{x} \right| = \lim_{x \to 0} \left| \frac{f(x)}{\sin x} \right| \cdot \left| \frac{\sin x}{x} \right| \le 1.$$

The inequality is proved.

(28th W.L. Putnam Mathematics Competition, 1967)

469. The condition from the statement implies that $f(x) = f(-x)$, so it suffices to check that f is constant on $[0, \infty)$. For $x \ge 0$, define the recursive sequence $(x_n)_{n \ge 0}$, by $x_0 = x$, and $x_{n+1} = \sqrt{x_n}$, for $n \ge 0$. Then

$$f(x_0) = f(x_1) = f(x_2) = \cdots = f \left(\lim_{n \to \infty} x_n \right).$$

And $\lim_{n \to \infty} x_n = 1$ if $x > 0$. It follows that f is constant and the problem is solved.

470. The answer is yes, there is a tooth function with this property. We construct f to have local maxima at $\frac{1}{2^{2n+1}}$ and local minima at 0 and $\frac{1}{2^{2n}}$, $n \ge 0$. The values of the function at the extrema are chosen to be $f(0) = f(1) = 0$, $f \left(\frac{1}{2} \right) = \frac{1}{2}$, and $f \left(\frac{1}{2^{2n+1}} \right) = \frac{1}{2^n}$ and $f \left(\frac{1}{2^{2n}} \right) = \frac{1}{2^{n+1}}$ for $n \ge 1$. These are connected through segments. The graph from Figure 72 convinces the reader that f has the desired properties.

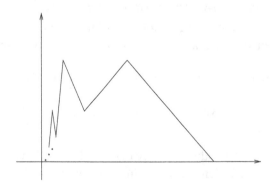

Figure 72

(*Kőzépiskolai Matematikai Lapok* (*Mathematics Gazette for High Schools, Budapest*))

471. We prove by induction on n that $f \left(\frac{m}{3^n} \right) = 0$ for all integers $n \ge 0$ and all integers $0 \le m \le 3^n$. The given conditions show that this is true for $n = 0$. Assuming that it is true for $n - 1 \ge 0$, we prove it for n.

If $m \equiv 0 \pmod 3$, then

$$f\left(\frac{m}{3^n}\right) = f\left(\frac{\frac{m}{3}}{3^{n-1}}\right) = 0$$

by the induction hypothesis.

If $m \equiv 1 \pmod 3$, then $1 \le m \le 3^n - 2$ and

$$3f\left(\frac{m}{3^n}\right) = 2f\left(\frac{\frac{m-1}{3}}{3^{n-1}}\right) + f\left(\frac{\frac{m+2}{3}}{3^{n-1}}\right) = 0 + 0 = 0.$$

Thus $f\left(\frac{m}{3^n}\right) = 0$.

Finally, if $m \equiv 2 \pmod 3$, then $2 \le m \le 3^n - 1$ and

$$3f\left(\frac{m}{3^n}\right) = 2f\left(\frac{\frac{m+1}{3}}{3^{n-1}}\right) + f\left(\frac{\frac{m-2}{3}}{3^{n-1}}\right) = 0 + 0 = 0.$$

Hence $f\left(\frac{m}{3^n}\right) = 0$ in this case, too, finishing our induction.

Because the set $\left\{\frac{m}{3^n}; \; m, n \in \mathbb{N}\right\}$ is dense in $[0, 1]$ and f is equal to zero on this set, f is identically equal to zero.

(Vietnamese Mathematical Olympiad, 1999)

472. We argue by contradiction. Assume that there exist $a < b$ such that $f(a) \ne f(b)$, say, $f(a) > f(b)$.

Let $g : \mathbb{R} \to \mathbb{R}$, $g(x) = f(x) + \lambda x$, where $\lambda > 0$ is chosen very small such that $g(a) > g(b)$. We note that

$$\lim_{h \to 0^+} \frac{g(x + 2h) - g(x + h)}{h} = \lambda > 0, \quad \text{for all } x \in \mathbb{R}.$$

Since g is a continuous function on a closed and bounded interval, g has a maximum. Let $c \in [a, b]$ be the point where g attains its maximum. It is important that this point is not b, since $g(a) > g(b)$. Fix $0 < \varepsilon < \lambda$. Then there exists $\delta = \delta(\varepsilon) > 0$ such that

$$0 < \lambda - \varepsilon < \frac{g(c + 2h) - g(c + h)}{h} < \lambda + \varepsilon, \quad \text{for all } 0 < h < \delta.$$

For $0 < h_0 < \min\left\{\delta, \frac{b-c}{2}\right\}$. The above inequality written for $h = h_0, \frac{h_0}{2}, \frac{h_0}{4}$, etc., yields

$$g(c + 2h_0) > g(c + h_0) > g\left(c + \frac{h_0}{2}\right) > \cdots > g\left(c + \frac{h_0}{2^n}\right) > \cdots$$

Passing to the limit, we obtain that $g(c + 2h) > g(c)$, contradicting the maximality of c. The contradiction proves that our initial assumption was false, and the conclusion follows.

473. From the given condition, it follows that f is one-to-one. Indeed, if $f(x) = f(y)$, then $f(f(x)) = f(f(y))$, so $bx = by$, which implies $x = y$. Because f is continuous and one-to-one, it is strictly monotonic.

We will show that f has a fixed point. Assume by way of contradiction that this is not the case. So either $f(x) > x$ for all x, or $f(x) < x$ for all x. In the first case f must be strictly increasing, and then we have the chain of implications

$$f(x) > x \Rightarrow f(f(x)) > f(x) \Rightarrow af(x) + bx > f(x) \Rightarrow f(x) < \frac{bx}{1-a},$$

for all $x \in \mathbb{R}$. In particular, $f(1) < \frac{b}{1-a} < 1$, contradicting our assumption.

In the second case the simultaneous inequalities $f(x) < x$ and $f(f(x)) < f(x)$ show that f must be strictly increasing again. Again we have a chain of implications

$$f(x) < x \Rightarrow f(f(x)) < f(x) \Rightarrow f(x) > af(x) + bx \Rightarrow f(x) > \frac{bx}{1-a},$$

for all $x \in \mathbb{R}$. In particular, $f(-1) > -\frac{b}{1-a} > -1$, again a contradiction.

In conclusion, there exists a real number c such that $f(c) = c$. The condition $f(f(c)) = af(c) + bc$ implies $c = ac + bc$; thus $c(a + b - 1) = 0$. It follows that $c = 0$ (because $a + b < 1/2 + 1/2 = 1$), and we obtain $f(0) = 0$.

(45th W.L. Putnam Mathematical Competition, 2002, proposed by T. Andreescu)

474. Being continuous on the closed interval $[0, 1]$, the function f is bounded and has a maximum and a minimum. Let M be the maximum and m the minimum. Then $\frac{m}{2^n} \leq \frac{f(x^n)}{2^n} \leq \frac{M}{2^n}$, which implies that the series is absolutely convergent and its limit is a number in the interval $[m, M]$.

Let $a \in (0, 1)$ and m_a and M_a be the minimum and the maximum of f on $[0, a]$. If $\alpha \in [0, a]$ is such that $f(\alpha) = M_a$, then

$$M_a = f(\alpha) = \sum_{n=1}^{\infty} \frac{f(\alpha^n)}{2^n} \leq M_a \sum_{n=1}^{\infty} \frac{1}{2^n} = M_a,$$

whence we must have equality in the above inequality, so $f(\alpha^n) = M_a$. Since $\lim_{n \to \infty} \alpha^n = 0$, it follows that M_a must equal $\lim_{x \to 0} f(x) = f(0)$. Similarly, $m_a = f(0)$, and hence f is constant on $[0, a]$. Passing to the limit with $a \to 1$, we conclude that f is constant on the interval $[0, 1]$. Clearly, constant functions satisfy the property, providing all solutions to the problem.

(*Gazeta Matematică* (*Mathematics Gazette, Bucharest*), proposed by M. Bălună)

475. Let $\phi : [0, 1] \times [0, 1]$ be a continuous surjection. Define ψ to be the composition

$$[0, 1] \xrightarrow{\phi} [0, 1] \times [0, 1] \xrightarrow{\phi \times id} [0, 1] \times [0, 1] \times [0, 1] \xrightarrow{pr_{12}} [0, 1] \times [0, 1],$$

where $pr_{12} : [0, 1] \times [0, 1] \times [0, 1] \to [0, 1] \times [0, 1]$ is the projection of the cube onto the bottom face. Each function in the above chain is continuous and surjective, so the composition is continuous and surjective. Moreover, because the projection takes each value infinitely many times, so does ψ. Therefore, ψ provides the desired example.

476. The first example of such a function was given by Weierstrass. The example we present here, of a function $f : [0, 1] \rightarrow [0, 1]$, was published by S. Marcus in the *Mathematics Gazette, Bucharest*.

If $0 \leq x \leq 1$ and $x = 0.a_1a_2a_3 \dots$ is the *ternary* expansion of x, we let the *binary* representation off (x) be $0.b_1b_2b_3\dots$, where the binary digits $b_1, b_2, b_3, \dots$ are uniquely determined by the conditions

(i) $b_1 = 1$ if and only if $_1 = 1$,

(ii) $b_{n+1} = b_n$ if and only if $a_{n+1} = a_n$, $n \geq 1$.

It is not hard to see that $f(x)$ does not depend on which ternary representation you choose for x. For example

$$f(0.0222\dots) = 0.0111\dots = 0.1000\dots = f(0.1000\dots).$$

Let us prove first that the function is continuous. If x is a number that has a unique ternary expansion and $(x_n)_n$ is a sequence converging to x, then the first m digits of x_n become equal to the first m digits of x for n sufficiently large. It follows from the definition of f that the first m binary digits of $f(x_n)$ become equal to the first m binary digits of $f(x)$ for n sufficiently large. Hence $f(x_n)$ converges to $f(x)$, so f is continuous at x.

If x is a number that has two possible ternary expansions, then in one expansion x has only finitely many nonzero digits $x = 0.a_1a_2\dots a_k00\dots$, with $a_k \neq 0$. The other expansion is $0.a_1a_2\dots a_k'222\dots$, with $a_k' = a_k - 1$ ($=0$ or 1). Given a sequence $(x_n)_n$ that converges to x, for sufficiently large n the first $k - 1$ digits of x_n are equal to $a_1, a_2, \dots, a_{k-1}$, while the next $m - k + 1$ are either $a_k, 0, 0, \dots, 0$, or $a_k', 2, 2, \dots, 2$. If $f(x) = f(0.a_1a_2\dots a_k00\dots) = 0.b_1b_2b_3\dots$, then for n sufficiently large, the first $k - 1$ digits of $f(x_n)$ are $b_1, b_2, \dots, b_{k-1}$, while the next $m - k + 1$ are either $b_k, b_{k+1} = b_{k+2} = \dots = b_m$ (the digits of $f(x)$) or $1 - b_k$, $1 - b_{k+1} = \dots = 1 - b_m$. The two possible binary numbers are $0.b_1b_2\dots b_{k-1}0111\dots$ and $0.b_1b_2\dots b_{k-1}1000\dots$; they differ from $f(x)$ by at most $\frac{1}{2^{m+1}}$. We conclude again that as $n \rightarrow \infty, f(x_n) \rightarrow f(x)$. This proves the continuity of f.

Let us show next that f does not have a finite derivative from the left at any point $x \in (0, 1]$. For such x consider the ternary expansion $x = 0.a_1a_2a_3\dots$ that has infinitely many nozero digits, and, applying the definition of f for this expansion, let $f(x) = 0.b_1b_2b_3\dots$ Now consider an arbitrary positive number n, and let $k_n \geq n$ be such that $a_{k_n} \neq 0$. Construct a number $x' \in (0, 1)$ whose first $k_n - 1$ digits are the same as those of x, whose k_nth digit is zero, and all of whose other digits are equal to 0 if $b_{k_n+1} = 1$ and to 1 if $b_{k_n+1} = 0$. Then

$$0 < x - x' < 2 \cdot 3^{-k_n} + 0.\underbrace{00\dots0}_{k_n}22\dots = 3^{-k_n+1},$$

while in the first case,

$$|f(x) - f(x')| \geq 0.\underbrace{00\dots0}_{k_n}b_{k_n+1} = 0.\underbrace{00\dots0}_{k_n}1,$$

and in the second case,

$$|f(x) - f(x')| \geq 0.\underbrace{00\dots0}_{k_n}11\dots1 - 0.\underbrace{00\dots0}_{k_n}0b_{k_n+2},$$

and these are both greater than or equal to 2^{-k_n-1}. Since $k_n \geq n$, we have $0 < x - x' < 3^{-n+1}$ and

$$\left| \frac{f(x) - f(x')}{x - x'} \right| > \frac{2^{-k_n-1}}{3^{-k_n+1}} = \frac{1}{6} \left(\frac{3}{2} \right)^{k_n} \geq \frac{1}{6} \left(\frac{3}{2} \right)^{n}.$$

Letting $n \to \infty$, we obtain

$$x' \to x, \quad \text{while} \quad \left| \frac{f(x) - f(x')}{x - x'} \right| \to \infty.$$

This proves that f does not have a derivative on the left at x. The argument that f does not have a derivative on the right at x is similar and is left to the reader.

Remark. S. Banach has shown that in some sense, there are far more continuous functions that are not differentiable at any point than continuous functions that are differentiable at least at some point.

477. We apply the intermediate value property to the function $g : [a, b] \to [a, b]$, $g(x) = f(x) - x$. Because $f(a) \geq a$ and $f(b) \leq b$, it follows that $g(a) \leq 0$ and $g(b) \geq 0$. Hence there is $x \in [a, b]$ such that $g(c) = 0$. This c is a fixed point of f.

478. Let L be the length of the trail and T the total duration of the climb, which is the same as the total duration of the descent. Counting the time from the beginning of the voyage, denote by $f(t)$ and $g(t)$ the distances from the monk to the temple at time t on the first and second day, respectively. The functions f and g are continuous; hence so is $\phi : [0, T] \to \mathbb{R}$, $\phi(t) = f(t) - g(t)$. It follows that ϕ has the intermediate value property. Because $\phi(0) = f(0) - g(0) = L - 0 = L > 0$ and $\phi(T) = f(T) - g(T) = 0 - L < 0$, there is a time t_0 with $\phi(t_0) = 0$. At $t = t_0$ the monk reached the same spot on both days.

479. The fact that f is decreasing implies immediately that

$$\lim_{x \to -\infty} (f(x) - x) = \infty \quad \text{and} \quad \lim_{x \to \infty} (f(x) - x) = -\infty.$$

By the intermediate value property, there is x_0 such that $f(x_0) = x_0$. The function cannot have another fixed point because if x and y are fixed points, with $x < y$, then $x = f(x) \geq f(y) = y$, impossible.

The triple (x_0, x_0, x_0) is a solution to the system. And if (x, y, z) is a solution then $f(f(f(x))) = x$. The function $f \circ f \circ f$ is also continuous and decreasing, so it has a unique fixed point. And this fixed point can only be x_0. Therefore, $x = y = z = x_0$, proving that the solution is unique.

480. The inequality from the statement implies right away that f is injective, and also that f transforms unbounded intervals into unbounded intervals. The sets $f((-\infty, 0])$ and $f([0, \infty))$ are unbounded intervals that intersect at one point. They must be two intervals that cover the entire real axis.

(P.N. de Souza, J.N. Silva, *Berkeley Problems in Mathematics*, Springer, 2004)

481. Let x denote the distance along the course, measured in miles from the starting line. For each $x \in [0, 5]$, let $f(x)$ denote the time that elapses for the mile from the point x to the point $x + 1$. Note that f depends continuously on x. We are given that

$$f(0) + f(1) + f(2) + f(3) + f(4) + f(5) = 30.$$

It follows that not all of f $(0), f(1), \ldots, f(5)$ are smaller than 5, and not all of them are larger than 5. Choose $a, b \in \{0, 1, \ldots, 5\}$ such that $f(a) \leq 5 \leq f(b)$. By the intermediate value property, there exists c between a and b such that $f(c) = 5$. The mile between c and $c + 1$ was run in exactly 5 minutes.

(L.C. Larson, *Problem-Solving Through Problems*, Springer-Verlag, 1990)

482. Without loss of generality, we may assume that the cars traveled on one day from A to B keeping a distance of at most one mile between them, and on the next day they traveled in opposite directions in the same time interval, which we assume to be of length one unit of time.

Since the first car travels in both days on the same road and in the same direction, it defines two parametrizations of that road. Composing the motions of both cars during the second day of travel with a homeomorphism (continuous bijection) of the time interval $[0, 1]$, we can ensure that the motion of the first car yields the same parametrization of the road on both days. Let $f(t)$ be the distance from the second car to A when the first is at t on the first day, and $g(t)$ the distance from the second car to A when the first is at t on the second day. These two functions are continuous, so their difference is also continuous. But $f(0) - g(0) = -\text{dist}(A, B)$, and $f(1) - g(1) = \text{dist}(A, B)$, where $\text{dist}(A, B)$ is the distance between the cities.

The intermediate value property implies that there is a moment t for which $f(t) - g(t) = 0$. At that moment the two cars are in the same position as they were the day before, so they are at distance at most one mile. Hence the answer to the problem is no.

483. We compute

$$\sum_{j=0}^{n} P(2^j) = \sum_{j=0}^{n} \sum_{k=1}^{n} a_k 2^{kj} = \sum_{k=1}^{n} \left(\sum_{j=0}^{n} 2^{kj} \right) a_k$$

$$= \sum_{k=1}^{n} \frac{2^{k(n+1)} - 1}{2^k - 1} = Q(2^{n+1}) - Q(1) = 0.$$

It follows that $P(1) + P(2) + \cdots + P(2^n) = 0$. If $P(2^k) = 0$ for some $k < n$, we are done. Otherwise, there exist $1 \leq i, j \leq n$ such that $P(2^i)P(2^j) < 0$, and by the intermediate value property, $P(x)$ must have a zero between 2^i and 2^j.

(USA Team Selection Test for the International Mathematical Olympiad, proposed by R. Gelca)

484. Consider the lines fixed, namely the x- and the y-axes, and vary the position of the surface in the plane. Rotate the surface by an angle ϕ, then translate it in such a way that the x-axis divides it into two regions of equal area. The coordinate axes divide it now into four regions of areas A, B, C, D, counted counterclockwise starting with the first quadrant. Further translate it such that $A = B$. The configuration is now uniquely determined by the angle ϕ.

It is not hard to see that $A = A(\phi)$, $B = B(\phi)$, $C = C(\phi)$, and $D = D(\phi)$ are continuous functions of ϕ.

If $C(0°) = D(0°)$, then the equality of the areas of the regions above and below the x-axis implies $A(0°) = B(0°) = C(0°) = D(0°)$, and we are done.

If $C(0°) > D(0°)$, then the line that divides the region below the x-axis into two polygons of equal area lies to the left of the y-axis (see Figure 73). This means that after a 180°-rotation the line that determines the regions $A(180°)$ and $B(180°)$ will divide the other region into $C(180°)$ and $D(180°)$ in such a way that $C(180°) < D(180°)$. Similarly, if $C(0°) < D(0°)$, then $C(180°) > D(180°)$.

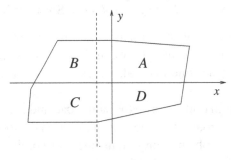

Figure 73

It follows that the continuous function $C(\phi) - D(\phi)$ assumes both positive and negative values on the interval $[0°, 180°]$, so by the intermediate value property there is an angle ϕ_0 for which $C(\phi_0) = D(\phi_0)$. Consequently, $A(\phi_0) = B(\phi_0) = C(\phi_0) = D(\phi_0)$, and the problem is solved.

Remark. This result was called the "Pancake theorem" in W.G. Chinn, N.E. Steenrod, *First Concepts of Topology*, MAA 1966.

485. Assume that f is not continuous at some point a. Then there exists $\varepsilon > 0$ and a sequence $s_n \to a$ such that $|f(x_n) - f(a)| > \varepsilon$ for all $n \geq 1$. Without loss of generality, we may assume that there is a subsequence $(x_{n_k})_k$ such that $f(x_{n_k}) < f(a)$, for all k, in which case $f(x_{n_k}) \leq f(a) - \varepsilon$. Choose γ in the interval $(f(a) - \varepsilon, f(a))$. Since f has the intermediate value property, and $f(x_{n_k}) < \gamma < f(a)$, for each k there exists y_k between x_{n_k} and a such that $f(y_k) = \gamma$. The set $f^{-1}(\gamma)$ contains the sequence $(y_k)_k$, but does not contain its limit a, which contradicts the fact that the set is closed. This contradiction proves that the initial assumption was false; hence f is continuous on the interval I.

(A.M. Gleason)

486. The function is continuous off 0, so it maps any interval that does not contain 0 onto an interval. Any interval containing 0 is mapped onto $[-1, 1]$, which proves that f has the intermediate value property for any $a \in [-1, 1]$.

For the second part of the problem, we introduce the function

$$F(x) = \begin{cases} x^2 \sin \frac{1}{x} & \text{for } x \neq 0, \\ 0 & \text{for } x = 0. \end{cases}$$

One can verify easily that

$$F'(x) = \begin{cases} 2x \sin \frac{1}{x} & \text{for } x \neq 0, \\ 0 & \text{for } x = 0 \end{cases} + \begin{cases} \cos \frac{1}{x} & \text{for } x \neq 0, \\ 0 & \text{for } x = 0. \end{cases}$$

The only place where this computation might pose some difficulty is $x = 0$, which can be done using L'Hôpital's theorem. The first function is continuous; hence it is the derivative of a function. Because the differentiation operator is linear we find that the second function, which is $f_0(x)$ is a derivative. And because when $a \neq 0$,

$$f_a(x) - f_0(x) = \begin{cases} 0 & \text{for } x \neq 0, \\ a & \text{for } x = 0, \end{cases}$$

does not have the intermediate value property, so it is not the derivative of a function, $f_a(x)$ itself cannot be the derivative of a function. This completes the solution.

(Romanian high school textbook)

487. Taking the logarithm, transform the equation into the equivalent $x \ln 2 = 2 \ln x$. Define the function $f : (0, \infty) \to \mathbb{R}, f(x) = x \ln 2 - 2 \ln x$. We are to find the zeros of f. Differentiating, we obtain

$$f'(x) = \ln 2 - \frac{2}{x},$$

which is strictly increasing. The unique zero of the derivative is $\frac{2}{\ln 2}$, and so f' is negative for $x < 2/\ln 2$ and positive for $x > \frac{2}{\ln 2}$. Note also that $\lim_{x \to 0} f(x) = \lim_{x \to \infty} f(x) = \infty$. There are two possibilities: either $f\left(\frac{2}{\ln 2}\right) > 0$, in which case the equation $f(x) = 0$ has no solutions, or $f\left(\frac{2}{\ln 2}\right) < 0$, in which case the equation $f(x) = 0$ has exactly two solutions. The latter must be true, since $f(2) = f(4) = 0$. Therefore, $x = 2$ and $x = 4$ are the only solutions to $f(x) = 0$, and hence also to the original equation.

488. If $f(x) \geq 0$ for all x, then the function $g(x) = (x - a_1)(x - a_2)(x - a_3)$ is increasing, since its derivative is f. It follows that g has only one zero, and we conclude that $a_1 = a_2 = a_3$.

(V. Boskoff)

489. The inequality is homogeneous, so we can transform it into one for a single variable by dividing both sides by b^n and denoting a/b by x. We obtain the equivalent inequality

$$(n-1)x^n + 1 \geq nx^{n-1},$$

or

$$(n-1)x^n - nx^{n-1} + 1 \geq 0.$$

The left-hand side is a differentiable function $f(x)$. Let us find the critical points of f in the interval $(0, \infty)$. The equation $f'(x) = 0$ reads

$$n(n-1)x^{n-1} - n(n-1)x^{n-2} = 0,$$

which yields the unique critical point $x = 1$. We compute $f(1) = 0$, $f(0) = 1$, and $\lim_{x \to \infty} f(x) = \infty$. It follows that 1 is a global minimum, and consequently $f(x)$ is non-negative. This proves the inequality.

Remark. It should be noticed that the inequality is a particular case of the AM-GM inequality applied to the numbers $a^n, a^n, \ldots, a^n, b^n$, where there are $n - 1$ powers of a.

(L. Larson *Problem-Solving through Problems*, Springer, 1983)

490. Let $f : \mathbb{C} \to \mathbb{C}$, $f(x) = z^3 - z + 2$. We have to determine $\max_{|z|=1} |f(z)|^2$. For this, we switch to real coordinates. If $|z| = 1$, then $z = x + iy$ with $y^2 = 1 - x^2$, $-1 \le x \le 1$. View the restriction of $|f(z)|^2$ to the unit circle as a function depending on the real variable x:

$$\begin{aligned}
|f(z)|^2 &= |(x+iy)^3 - (x+iy) + 2|^2 \\
&= |(x^3 - 3xy^2 - x + 2) + iy(3x^2 - y^2 - 1)|^2 \\
&= |(x^3 - 3x(1 - x^2) - x + 2) + iy(3x^2 - (1 - x^2) - 1)|^2 \\
&= (4x^3 - 4x + 2)^2 + (1 - x^2)(4x^2 - 2)^2 \\
&= 16x^3 - 4x^2 - 16x + 8.
\end{aligned}$$

Call this last expression $g(x)$. Its maximum on $[-1, 1]$ is either at a critical point or at an endpoint of the interval. The critical points are the roots of $g'(x) = 48x^2 - 8x - 16 = 0$, namely, $x = \frac{2}{3}$ and $x = -\frac{1}{2}$. We compute $g(-1) = 4$, $g\left(-\frac{1}{2}\right) = 13$, $g\left(\frac{2}{3}\right) = \frac{8}{27}$, $g(1) = 4$. The largest of them is 13, which is therefore the answer to the problem. It is attained when $z = -\frac{1}{2} \pm \frac{\sqrt{3}}{2}i$.

(8th W.L. Putnam Mathematical Competition, 1947)

491. After we bring the function into the form

$$f(x) = \frac{\left(x - 1 + \dfrac{1}{x}\right)^3}{x^3 - 1 + \dfrac{1}{x^3}},$$

the substitution $x + \frac{1}{x} = s$ becomes natural. We are to find the minimum of the function

$$h(s) = \frac{(s-1)^3}{s^3 - 3s - 1} = 1 + \frac{-3s^2 + 6s}{s^3 - 3s - 1}$$

over the domain $(-\infty, -2] \cup [2, \infty)$. Setting the first derivative equal to zero yields the equation

$$3(s-1)(s^3 - 3s^2 + 2) = 0.$$

The roots are $s = 1$ (double root) and $s = 1 \pm \sqrt{3}$. Of these, only $s = 1 + \sqrt{3}$ lies in the domain of the function.

We compute

$$\lim_{x \to \pm\infty} h(s) = 1, \quad h(2) = 1, \quad h(-2) = 9, \quad h(1 + \sqrt{3}) = \frac{\sqrt{3}}{2 + \sqrt{3}}.$$

Of these the last is the least. Hence the minimum of f is $\sqrt{3}/(2 + \sqrt{3})$, which is attained when $x + \frac{1}{x} = 1 + \sqrt{3}$, that is, when $x = (1 + \sqrt{3} \pm \sqrt[4]{12})/2$.

(*Mathematical Reflections*, proposed by T. Andreescu)

492. Let $f(x) = \sin(\sin(\sin(\sin(\sin(x)))))$. The first solution is $x = 0$. We have

$$f'(0) = \cos 0 \cos(\sin 0) \cos(\sin(\sin 0)) \cos(\sin(\sin(\sin 0))) \cos(\sin(\sin(\sin(\sin 0))))$$

$$= 1 > \frac{1}{3}.$$

Therefore, $f(x) > \frac{x}{3}$ in some neighborhood of 0. On the other hand, $f(x) < 1$, whereas $\frac{x}{3}$ is not bounded as $x \to \infty$. Therefore, $f(x_0) = \frac{x_0}{3}$ for some $x_0 > 0$. Because f is odd, $-x_0$ is also a solution. The second derivative of f is

$$- \cos(\sin x) \cos(\sin(\sin x)) \cos(\sin(\sin(\sin x))) \cos(\sin(\sin(\sin(\sin x)))) \sin x$$
$$- \cos^2 x \cos(\sin(\sin x)) \cos(\sin(\sin(\sin x))) \cos(\sin(\sin(\sin(\sin x)))) \sin(\sin x)$$
$$- \cos^2 x \cos^2(\sin x) \cos(\sin(\sin(\sin x))) \cos(\sin(\sin(\sin(\sin x)))) \sin(\sin(\sin x))$$
$$- \cos^2 x \cos^2(\sin x) \cos^2(\sin(\sin x)) \cos(\sin(\sin(\sin(\sin x)))) \sin(\sin(\sin(\sin x)))$$
$$- \cos^2 x \cos^2(\sin x)) \cos^2(\sin(\sin x)) \cos^2(\sin(\sin(\sin x))) \sin(\sin(\sin(\sin(\sin x)))),$$

which is clearly nonpositive for $0 < x < 1$. This means that $f'(x)$ is monotonic. Therefore, $f'(x)$ has at most one root x' in $[0, +\infty)$ Then $f(x)$ is monotonic at $[0, x']$ and $[x', \infty)$ and has at most two nonnegative roots. Because $f(x)$ is an odd function, it also has at most two nonpositive roots. Therefore, $-x_0, 0, x_0$ are the only solutions.

493. Let $f : \mathbb{R} \to \mathbb{R}$,

$$f(x) = e^x + nx - 2.$$

Then $f'(x) = e^x + n > 0$, so f is strictly increasing. Also $f(0) < 0$ and $f(1) > 0$, thus a_n exists and is unique. Moreover, $a_n \in (0, 1)$. Because $2 > na_n$, $(a_n)_n$ converges to 0. Passing to the limit in the definition of a_n we see that $(na_n)_n$ converges to 1. Finally, note that

$$\lim_{n \to \infty} \frac{n(1 - na_n)}{na_n} = \lim_{n \to \infty} \frac{e^{a_n} - 1}{a_n} = 1,$$

the latter being the derivative of e^x at 0. So $(n(1 - na_n))_n$ also converges to 1.

(*Mathematical Reflections*, proposed by T.L. Rădulescu)

494. If both x and y are negative, then $|x|^n > n^n + |y|^{n+1}$ and $|y|^n > n^n + |x|^{n+1}$. Without loss of generality, $|x| \geq |y|$; then $|x|^n > n^n + |x|^{n+1}$. This means that $|x|$, and consequently $|y|$ are are less than 1, which is impossible.

For the inequality $x + y \leq 1$ to hold, one of the numbers has to be negative, or else both are less than or equal to 1 and $n^n < x^n + y^{n+1} < 2$. So let $y < 0$ and $x > 0$, and assume $x + y \leq 1$. Set $y = -a$, $a > 0$. Then $x \in (0, 1 + a)$, and so $x^n < (1 + a)^n$. The relation $y^{n+1} + x^n > n^n$ implies

$$(1 + a)^n - a^{n+1} > n^n.$$

Consider the function $f : (0, \infty) \to \mathbb{R}, f(a) = a^{n+1} - (1 + a)^n + n^n$. We will show that $f > 0$. We have

$$f'(a) = (n + 1)a^n - n(1 + a)^{n-1},$$

and we see that $f'(a) = 0$ at the unique point a_0 satisfying $(1 + a_0)^{n-1} = \frac{n+1}{n}a_0^n$. Because $f(0) = n^n - 1 > 0$ and $\lim_{a \to \infty} f(a) = \infty$, it follows that a_0 is a global minimum, and so it suffices to show that $f(a_0) > 0$. We have $f'(0) < 0$ and $f'(n - 1) = (n + 1)(n - 1)^n - n^n = (n - 1)^n(n + 1 - (1 + \frac{1}{n-1})^n) > 0$, so $a_0 < n - 1$. Hence

$$f(a_0) = a_0^{n+1} - (1 + a_0)\frac{n + 1}{n}a_0^n + n^n = n^n - \frac{a_0^n}{n}(1 + a_0 + n) > n^n - \frac{(n - 1)^n \cdot 2n}{n} > 0,$$

since $n^n - 2(n - 1)^n = (n - 1)^n((1 - \frac{1}{n-1})^n - 2) > 0$. So $f > 0$. Thus our assumption was false, and so $x + y > 1$.

(*Gazeta Matematică (Mathematics Gazette, Bucharest)*, proposed by L. Panaitopol)

495. For $x \neq y$ this is the same as

$$\left|\frac{f(x) - f(y)}{x - y}\right| \leq |x - y|.$$

In the right-hand term we have $\lim_{y \to x} |x - y| = 0$, so by the squeezing principle

$$\lim_{y \to x} \left|\frac{f(x) - f(y)}{x - y}\right| = 0,$$

and so f is differentiable at x and $f'(x) = 0$. This means that f is constant. And every constant function satisfies the given inequality.

(from a note published in *The American Mathematical Monthly* by T. Khovanova and A. Radul)

496. Assume that this is not true. Then there is $\varepsilon > 0$ so that for every $\alpha \in \mathbb{R}$ there is $x > \alpha$ such that $|f(x)| > \varepsilon$. By choosing only those values of f with the same sign, and maybe changing f to $-f$, we may remove the bars in the last equality, so that it reads $f(x) > \varepsilon$. Choose α large enough so that $|f(x) + f'(x)| < \varepsilon/2$ for all $x > \alpha$.

If $f(x) > \varepsilon$ on $[\alpha, \infty)$, then $f'(x) < -\varepsilon/2$, which forces $\lim_{x \to \infty} f(x) = -\infty$, and consequently $\lim_{x \to \infty} f(x) + f'(x) = -\infty$. Hence f takes values smaller than ε on every interval $[\beta, \infty)$. So we can find an interval $[a, b]$ $a > \alpha$ which contains an x such that $f(x) > \varepsilon$ and such that $f(a), f(b) < \varepsilon$. But then f has a point of maximum $c \in (a, b)$. Then $f'(c) = 0$, and of course $f(c) > \varepsilon$ since this inequality already holds for a point in (a, b). But this is impossible since $c > \alpha$ implies $f(c) = |f(c) + f'(c)| < \varepsilon/2$. The conclusion follows.

497. Define the function $G : \mathbb{R} \to \mathbb{R}$, $G(x) = \left(\int_0^x f(t)dt \right)^2$. It satisfies

$$G'(x) = 2f(x) \int_0^x f(t)dt.$$

Because $G'(0) = 0$ and $G'(x) = g(x)$ is nonincreasing it follows that G' is nonnegative on $(-\infty, 0)$ and nonpositive on $(0, \infty)$. This implies that G is nondecreasing on $(-\infty, 0)$ and nonincreasing on $(0, \infty)$. And this, combined with the fact that $G(0) = 0$ and $G(x) \geq 0$ for all x, implies $G(x) = 0$ for all x. Hence $\int_0^x f(t)dt = 0$. Differentiating with respect to x, we conclude that $f(x) = 0$ for all x, and we are done.

(Romanian Mathematical Olympiad, 1978, proposed by S. Rădulescu)

498. Consider the function

$$F(t) = \left[\int_0^t f(x)dx \right]^2 - \int_0^t [f(x)]^3 dx \quad \text{for} \quad t \in [0, 1].$$

We want to show that $F(t) \geq 0$, from which the conclusion would then follow. Because $F(0) = 0$, it suffices to show that F is increasing. To prove this fact we differentiate and obtain

$$F'(t) = f(t) \left[2 \int_0^t f(x)dx - f^2(t) \right].$$

It remains to check that $G(t) = 2 \int_0^t f(x)dx - f^2(t)$ is positive on $[0, 1]$. Because $G(0) = 0$, it suffices to prove that G itself is increasing on $[0, 1]$. We have

$$G'(t) = 2f(t) - 2f(t)f'(t).$$

This function is positive, since on the one hand $f'(0) \leq 1$, and on the other hand f is increasing, having a positive derivative, and so $f(t) \geq f(0) = 0$. This proves the inequality. An example in which equality holds is the function $f : [0, 1] \to \mathbb{R}, f(x) = x$.

(34th W.L. Putnam Mathematical Competition, 1973)

499. Define $h : \mathbb{R} \to \mathbb{R}$,

$$h(t) = f(e^t) - e^t.$$

The given equation becomes

$$h(3t) + h(2t) + h(t) = 0 \quad \text{for all } t.$$

The chain rule shows that this function is differentiable at 0.

$$h(3 \cdot 0) + h(2 \cdot 0) + h(0) = 3h(0) = 0,$$

so $h(0) = 0$. Also

$$3h'(0) + 2h'(0) + h'(0) = 6h'(0) = 0,$$

so $h'(0) = 0$. We can write this as

$$\lim_{t \to 0} \frac{h(t) - h(0)}{t} = \lim_{t \to 0} \frac{h(t)}{t} = 0.$$

So for every $\varepsilon > 0$, there is $\delta > 0$ such that

$$\sup_{0 < |t| < \delta} \left| \frac{h(t)}{t} \right| < \varepsilon.$$

Fix such an ε, and the corresponding δ. We have

$$\left| \frac{h(3t)}{3t} \right| \le \left| \frac{h(2t)}{3t} \right| + \left| \frac{h(t)}{3t} \right| = \frac{2}{3} \left| \frac{h(2t)}{2t} \right| + \frac{1}{3} \left| \frac{h(t)}{t} \right|.$$

Choosing $|t| < \delta/2$, the right-hand side is less than ε. But then $y = 3t$ is between $(-3\delta/2, 3\delta/2)$. And $|h(y)/y|$ is still less than ε. Thus we can increase δ by a factor of $3/2$ and the inequality still holds. Repeating, we can make δ arbitrarily large, so we conclude that $|h(t)/t| < \varepsilon$ for all t. But ε is an arbitrary positive number, so $h(t) = 0$ for all t.

We therefore have $f(x) = x$ as the only solution to the functional equation.

(*Mathematical Reflections*, proposed by M. Piticari)

500. (a) To avoid the complicated exponents, divide the inequality by the right-hand side; then take the natural logarithm. Next, fix positive numbers y and z, and then introduce the function $f : (0, \infty) \to \mathbb{R}$,

$$f(x) = (x + y + z) \ln(x + y + z) + x \ln x + y \ln y + z \ln z$$
$$- (x + y) \ln(x + y) - (y + z) \ln(y + z) - (z + x) \ln(z + x).$$

Differentiating $f(x)$ with respect to x, we obtain

$$f'(x) = \ln \frac{(x + y + z)x}{(x + y)(z + x)} = \ln \frac{x^2 + yx + zx}{x^2 + yx + zx + yz} < \ln 1 = 0,$$

for all positive numbers x. It follows that $f(x)$ is strictly decreasing, so $f(x) < \lim_{t \to 0} f(t) = 0$, for all $x > 0$. Hence $e^{f(x)} < 1$ for all $x > 0$, which is equivalent to the first inequality from the statement.

(b) We apply the same idea, fixing $y, z > 0$ and considering the function $g : (0, \infty) \to \mathbb{R}$,

$$g(x) = (x + y + z)^2 \ln(x + y + z) + x^2 \ln x + y^2 \ln y + z^2 \ln z$$
$$- (x + y)^2 \ln(x + y) - (y + z)^2 \ln(y + z) - (z + x)^2 \ln(z + x).$$

Differentiating with respect to x, we obtain

$$g'(x) = 2 \ln \frac{(x+y+z)^{x+y+z} x^x}{(x+y)^{x+y}(z+x)^{z+x}}.$$

We would like to show this time that g is increasing, for then $g(x) > \lim_{t \to 0} g(t) = 0$, from which the desired inequality is obtained by exponentiation. We are left to prove that $g'(x) > 0$, which is equivalent to

$$(x+y+z)^{x+y+z} x^x > (x+y)^{x+y}(z+x)^{z+x}, \quad \text{for} \quad x, y, z > 0.$$

And we take the same path as in (a). Because we want to make the derivative as simple as possible, we fix $x, y > 0$ and define $h : (0, \infty) \to \mathbb{R}$,

$$h(z) = (x+y+z) \ln(x+y+z) + x \ln x - (x+y) \ln(x+y) - (z+x) \ln(z+x).$$

Then

$$h'(z) = \ln \frac{x+y+z}{z+x} > \ln 1 - 0,$$

for $z > 0$. Hence $h(z) > \lim_{t \to 0} h(t) = 0$, $z > 0$. This implies the desired inequality and completes the solution.

(*American Mathematical Monthly*, proposed by Sz. András, solution by H.-J. Seiffert)

501. Let us examine the function $F(x) = f(x) - g(x)$. Because $F^{(n)}(a) \neq 0$, we have $F^{(n)}(x) \neq 0$ for x in a neighborhood of a. Hence $F^{(n-1)}(x) \neq 0$ for $x \neq a$ and x in a neighborhood of a (otherwise, this would contradict Rolle's theorem). Then $F^{(n-2)}(x)$ is monotonic to the left, and to the right of a, and because $F^{(n-2)}(a) = 0$, $F^{(n-2)}(x) \neq 0$ for $x \neq a$ and x in a neighborhood of a. Inductively, we can decrease the order of the derivative, to botain $F(x) \neq 0$ and so $f(x) \neq g(x)$ in some neighborhood of a.

The limit from the statement can be written as

$$\lim_{x \to a} e^{g(x)} \frac{e^{f(x)-g(x)} - 1}{f(x) - g(x)}.$$

We only have to compute the limit of the fraction, since $g(x)$ is a continuous function. We are in a $\frac{0}{0}$ situation, and can apply L'Hôpital's theorem:

$$\lim_{x \to a} \frac{e^{f(x)-g(x)} - 1}{f(x) - g(x)} = \lim_{x \to a} \frac{(f'(x) - g'(x))e^{f(x)-g(x)}}{f'(x) - g'(x)} = e^0 = 1.$$

Hence the limit from the statement is equal to $e^{g(a)} = e^{\alpha}$.

(N. Georgescu-Roegen)

502. The function $h : [1, \infty) \to [1, \infty)$ given by $h(t) = t(1 + \ln t)$ is strictly increasing, and $h(1) = 1$, $\lim_{t \to \infty} h(t) = \infty$. Hence h is bijective, and its inverse is clearly the function $f : [1, \infty) \to [1, \infty)$, $\lambda \to f(\lambda)$. Since h is differentiable, so is f, and

$$f'(\lambda) = \frac{1}{h'(x(\lambda))} = \frac{1}{2 + \ln f(\lambda)}.$$

Also, since h is strictly increasing and $\lim\limits_{t\to\infty} h(t) = \infty, f(\lambda)$ is strictly increasing, and its limit at infinity is also infinity. Using the defining relation for $f(\lambda)$, we see that

$$\frac{f(\lambda)}{\dfrac{\lambda}{\ln\lambda}} = \ln\lambda \cdot \frac{f(\lambda)}{\lambda} = \frac{\ln\lambda}{1+\ln f(\lambda)}.$$

Now we apply L'Hôpital's theorem and obtain

$$\lim_{\lambda\to\infty} \frac{f(\lambda)}{\dfrac{\lambda}{\ln\lambda}} = \lim_{\lambda\to\infty} \frac{\dfrac{1}{f(\lambda)}}{\dfrac{1}{2+\ln f(\lambda)}} = \lim_{\lambda\to\infty} \frac{f(\lambda)}{\lambda}(2+\ln f(\lambda)) = \lim_{\lambda\to\infty} \frac{2+\ln f(\lambda)}{1+\ln f(\lambda)} = 1,$$

where the next-to-last equality follows again from $f(\lambda)(1+\ln f(\lambda)) = \lambda$. Therefore, the required limit is equal to 1.

(*Gazeta Matematică* (*Mathematics Gazette, Bucharest*), proposed by I. Tomescu)

503. We want to find a formula for $\lim_{x\to 0} f_n(x)/g_n(x)$. Inductively we prove that

$$\lim_{x\to 0} f_n(x) = \lim_{n\to 0} g_n(x) = 0,$$

so we might be able to apply L'Hospital.

Let us check small cases of n. For $n = 1$, L'Hospital can indeed be applied, since

$$\lim_{x\to 0} \frac{f_1'(x)}{g_1'(x)} = \lim_{x\to 0} \frac{\frac{1}{1+3x^2}\cdot 2\cdot 3x}{\frac{1}{1+5x^2}\cdot 2\cdot 5x} = \frac{3}{5}.$$

So $\lim_{x\to 0} f_1(x)/g_1(x) = 3/5$. Also for $n = 2$, we check

$$\lim_{x\to 0} \frac{f_2'(x)}{g_2'(x)} = \lim_{x\to 0} \frac{\frac{1}{1+2(f_1(x))^2}\cdot 2\cdot 3f_1(x)f_1'(x)}{\frac{1}{1+3g_1(x)}\cdot 2\cdot 5\cdot g_1(x)g_1'(x)} = \frac{3^3}{5^3}.$$

So $\lim_{x\to 0} f_1(x)/g_1(x) = 3^3/5^3$. Inductively we prove that

$$\lim_{x\to 0} \frac{f_n(x)}{g_n(x)} = \lim_{x\to 0} \frac{f_n'(x)}{g_n'(x)} = \frac{3^{2^n-1}}{5^{2^n-1}}.$$

Indeed,

$$\lim_{n\to 0} \frac{f_{n+1}'(x)}{g_{n+1}'(x)} = \lim_{n\to 0} \frac{\frac{1}{1+3(f_n(x))^2}\cdot 2\cdot 3\cdot f_n(x)f_n'(x)}{\frac{1}{1+5(g_n(x))^2}\cdot 2\cdot 5g_n(x)g_n'(x)}$$

$$= \frac{3}{5}\lim_{x\to 0} \frac{1+5(g_n(x))^2}{1+3(f_n(x))^2}\cdot \lim_{x\to 0}\frac{f_n(x)}{g_n(x)}\cdot \lim_{x\to 0}\frac{f_n'(x)}{g_n'(x)}$$

$$= \frac{3}{5}\cdot 1\cdot \frac{3^{2^n-1}}{5^{2^n-1}}\cdot \frac{3^{2^n-1}}{5^{2^n-1}} = \frac{3^{2^n+1}}{5^{2^n+1}}.$$

Hence the answer to the problem is

$$\lim_{x \to 0} \frac{f_{2014}(x)}{g_{2014}(x)} = \frac{3^{2^{2014}-1}}{5^{2^{2014}-1}}.$$

(Konhauser Problem Fest, 2014, proposed by R. Gelca)

504. If all four zeros of the polynomial $P(x)$ are real, then by Rolle's theorem all three zeros of $P'(x)$ are real, and consequently both zeros of $P''(x) = 12x^2 - 6\sqrt{7}x + 8$ are real. But this quadratic polynomial has the discriminant equal to -132, which is negative, and so it has complex zeros. The contradiction implies that not all zeros of $P(x)$ are real.

505. Replacing f by $-f$ if necessary, we may assume $f(b) > f(c)$, hence $f(a) > f(c)$ as well. Let ξ be an absolute minimum of f on $[a, b]$, which exists because the function is continuous. Then $\xi \in (a, b)$ and therefore $f'(\xi) = 0$.

506. Consider the function $f : [2, \infty) \to \mathbb{R}, f(x) = x \cos \frac{\pi}{x}$. By the Mean value theorem there exists $u \in [x, x+1]$ such that $f'(u) = f(x+1) - f(x)$. The inequality from the statement will follow from the fact that $f'(u) > 1$. Since $f'(u) = \cos \frac{\pi}{u} + \frac{\pi}{u} \sin \frac{\pi}{u}$, we have to prove that

$$\cos \frac{\pi}{u} + \frac{\pi}{u} \sin \frac{\pi}{u} > 1,$$

for all $u \in [2, \infty)$. Note that $f''(u) = -\frac{\pi^2}{u^3} \cos \frac{\pi}{u} < 0$, for $u \in [2, \infty)$, so f' is strictly decreasing. This implies that $f'(u) > \lim_{v \to \infty} f'(v) = 1$ for all u, as desired. The conclusion follows.

(Romanian college admission exam, 1987)

507. Let α be the slope of the line through the collinear points $(a_i, f(a_i))$, $i = 0, 1, \ldots, n$, on the graph of f. Then

$$\frac{f(a_i) - f(a_{i-1})}{a_i - a_{i-1}} = \alpha, \ i = 1, 2, \ldots, n.$$

From the Mean value theorem it follows that there exist points $c_i \in (a_{i-1}, a_i)$ such that $f'(c_i) = \alpha$, $i = 1, 2, \ldots, n$. Consider the function $F : [a_0, a_n] \to \mathbb{R}$, $F(x) = f'(x) - \alpha$. It is continuous, $(n-1)$-times differentiable, and has n zeros in $[a_0, a_n]$. Applying successively Rolle's theorem, we conclude that $F^{(n-1)} = f^{(n)}$ has a zero in $[a, b]$, and the problem is solved.

(*Gazeta Matematică (Mathematics Gazette, Bucharest)*, proposed by G. Sirețchi)

508. The functions $\phi, \psi : [a, b] \to \mathbb{R}$, $\phi(x) = \frac{f(x)}{x-\alpha}$ and $\psi(x) = \frac{1}{x-\alpha}$ satisfy the conditions of Cauchy's theorem. Hence there exists $c \in (a, b)$ such that

$$\frac{\phi(b) - \phi(a)}{\psi(b) - \psi(a)} = \frac{\phi'(c)}{\psi'(c)}.$$

Replacing ϕ and ψ with their formulas gives

$$\frac{(a - \alpha)f(b) - (b - \alpha)f(a)}{a - b} = f(c) - (c - \alpha)f'(c).$$

On the other hand, since m lies on the line determined by $(a, f(a))$, $(b, f(b))$, the coordinates of M are related by

$$\beta = \frac{(a - \alpha)f(b) - (b - \alpha)f(a)}{a - b}.$$

This implies that $\beta = f'(c)(c - \alpha) + f(c)$, which shows that $M(\alpha, \beta)$ lies on the tangent to the graph of f at $(c, f(c))$, and we are done.

509. Consider the function $F : [a, b] \to \mathbb{R}$,

$$F(x) = f'(x)e^{-\lambda f(x)}, \quad \lambda \in \mathbb{R}.$$

Because f is twice differentiable, F is differentiable. We have $F(a) = F(b)$, which by Rolle's theorem implies that there exists $c \in (a, b)$ with $F'(c) = 0$. But

$$F'(x) = e^{-\lambda f(x)}(f''(x) - \lambda(f'(x))^2),$$

so $f''(c) - \lambda(f'(c))^2 = 0$. We are done.
  (D. Andrica)

510. *First solution*: Let us assume that such numbers do exist. If $x = y$ it follows that $x(2^x + 2^{-x}) = 2x$, which implies $x = y = 0$. This is impossible because x and y are assumed to be positive.

Hence x should be different from y. Let $x_1 > x_2 > x_3 > 0$ be such that $y = x_1 - x_2$ and $x = x_2 - x_3$. The relation from the statement can be written as

$$\frac{2^{x_1 - x_2} - 1}{1 - 2^{x_3 - x_2}} = \frac{x_1 - x_2}{x_2 - x_3},$$

or

$$\frac{2^{x_1} - 2^{x_2}}{x_1 - x_2} = \frac{2^{x_2} - 2^{x_3}}{x_2 - x_3}.$$

Applying the Mean value theorem to the exponential, we deduce the existence of the numbers $\theta_1 \in (x_2, x_1)$ and $\theta_2 \in (x_3, x_2)$ such that

$$\frac{2^{x_1} - 2^{x_2}}{x_1 - x_2} = 2^{\theta_1} \ln 2,$$

$$\frac{2^{x_2} - 2^{x_3}}{x_2 - x_3} = 2^{\theta_2} \ln 2.$$

But this implies $2^{\theta_1} \ln 2 = 2^{\theta_2} \ln 2$, or $\theta_1 = \theta_2$, which is impossible since the two numbers lie in disjoint intervals. This contradiction proves the claim.

Second solution: Define $F(z) = (2^z - 1)/z$. Note that by L'Hôpital's rule, defining $F(0) = \log 2$ extends F continuously to $z = 0$. Rearrange the equality to give

$$F(-x) = \frac{2^{-x} - 1}{-x} = \frac{2^y - 1}{y} = F(y).$$

Thus the lack of solutions will follow if we show that F is strictly increasing. Recall that $e^{-t} > 1 - t$ for $t \neq 0$, hence $2^{-z} > 1 - z \log 2$ for $z \neq 0$. Hence

$$F'(z) = \frac{2^z(z \log 2 - 1 + 2^{-z})}{z^2} > 0$$

for $z \neq 0$ and hence F is strictly increasing.

(T. Andreescu, second solution by R. Stong)

511. Clearly, α is nonnegative. Define $\Delta f(x) = f(x+1) - f(x)$, and $\Delta^{(k)} f(x) = \Delta(\Delta^{(k-1)} f(x))$, $k \geq 2$. By the Mean value theorem, there exists $\theta_1 \in (0, 1)$ such $f(x+1) - f(x) = f'(x+\theta_1)$, and inductively for every k, there exists $\theta_k \in (0, k)$ such that $\Delta^{(k)} f(x) = f^{(k)}(\theta_k)$. Applying this to $f(x) = x^\alpha$ and $x = n$, we conclude that for every k there exists $\theta_k \in (0, k)$ such that $f^{(k)}(n + \theta_k)$ is an integer. Choose $k = \lfloor \alpha \rfloor + 1$. Then

$$f^{(k)}(n + \theta_k) = \frac{\alpha(\alpha - 1) \cdots (\alpha + 1 - k)}{(n + \theta_k)^{k-\alpha}}.$$

This number is an integer by hypothesis. It is not hard to see that it is also positive and less than 1. The only possibility is that it is equal to 0, which means that $\alpha = k - 1$, and the conclusion follows.

(W.L. Putnam Mathematical Competition)

512. The equation is $a^3 + b^3 + c^3 = 3abc$, with $a = 2^x$, $b = -3^{x-1}$, and $c = -1$. Using the factorization

$$a^3 + b^3 + c^3 - 3abc = \frac{1}{2}(a + b + c)[(a - b)^2 + (b - c)^2 + (c - a)^2]$$

we find that $a + b + c = 0$ (the other factor cannot be zero since, for example, 2^x cannot equal -1). This yields the simpler equation

$$2^x = 3^{x-1} + 1.$$

Rewrite this as

$$3^{x-1} - 2^{x-1} = 2^{x-1} - 1.$$

We immediately notice the solutions $x = 1$ and $x = 2$. Assume that another solution exists, and consider the function $f(t) = t^{x-1}$. Because $f(3) - f(2) = f(2) - f(1)$, by the Mean value theorem there exist $t_1 \in (2, 3)$ and $t_2 \in (1, 2)$ such that $f'(t_1) = f'(t_2)$. This gives rise to the impossible equality $(x - 1)t_1^{x-2} = (x - 1)t_2^{x-2}$. We conclude that there are only two solutions: $x = 1$ and $x = 2$.

(*Mathematical Reflections*, proposed by T. Andreescu)

513. We first show that $P(x)$ has rational coefficients. Let k be the degree of $P(x)$, and for each n, let x_n be the rational root of $P(x) = n$. The system of equations in the coefficients

$$P(x_n) = n, \ n = 0, 1, 2, \ldots, k,$$

has a unique solution since its determinant is Vandermonde. Cramer's rule yields rational solutions for this system, hence rational coefficients for $P(x)$. Multiplying by the product of the denominators, we may thus assume that $P(x)$ has integer coefficients, say $P(x) = a_k x^k + \cdots + a_1 x + a_0$, that $a_k > 0$, and that $P(x) = Nn$ has a rational solution x_n for all $n \geq 1$, where N is some positive integer (the least common multiple of the previous coefficients).

Because x_n is a rational number, its representation as a fraction in reduced form has the numerator a divisor of $a_0 - n$ and the denominator a divisor of a_k. If $m \neq n$, then $x_m \neq x_n$, so

$$|x_m - x_n| \geq \frac{1}{a_k}.$$

Let us now show that under this hypothesis the derivative of the polynomial is constant. Assume the contrary. Then $\lim_{|x| \to \infty} |P'(x)| = \infty$. Also, $\lim_{n \to \infty} P(x_n) = \lim_{n \to \infty} n = \infty$. Hence $|x_n| \to \infty$, and so $|P'(x_n)| \to \infty$, for $n \to \infty$.

For some n, among the numbers x_n, x_{n+1}, x_{n+2} two have the same sign, call them x and y. Then, by the Mean value theorem, there exists a c_n between x and y such that

$$P'(c_n) = \frac{P(y) - P(x)}{y - x}.$$

Taking the absolute value, we obtain

$$|P'(c_n)| \leq \frac{(n+2) - n}{|y - x|} \leq 2a_k,$$

where we use the fact that x and y are at least $1/a_k$ apart. But c_n tends to infinity, and so $|P'(c_n)|$ must also tend to infinity, a contradiction. This shows that our assumption was false, so $P'(x)$ is constant. We conclude that $P(x)$ is linear.

(*Gazeta Matematică* (*Mathematics Gazette, Bucharest*), proposed by M. Dădărlat)

514. Divide the inequality by 2, and notice that what you obtain is the inequality

$$\frac{f(x) + f(y)}{2} \leq f\left(\frac{x+y}{2}\right),$$

for the concave function $f : (0, \infty) \to (0, \infty), f(x) = \sqrt[3]{x}$, and for $x = 3 + \sqrt[3]{3}, y = 3 - \sqrt[3]{3}$.

515. Arrange the x_i's in increasing order $x_1 \leq x_2 \leq \ldots \leq x_n$. The function

$$f(a) = |a - x_1| + |a - x_2| + \cdots + |a - x_n|$$

is convex, being the sum of convex functions. It is piecewise linear. The derivative at a point a, in a neighborhood of which f is linear, is equal to the difference between the number of x_i's that are less than a and the number of x_i's that are greater than a. The global minimum is attained where the derivative changes sign. For n odd, this happens precisely at $x_{\lfloor n/2 \rfloor + 1}$. If n is even, the minimum is achieved at any point of the interval $[x_{\lfloor n/2 \rfloor}, x_{\lfloor n/2 \rfloor + 1}]$ at which the first derivative is zero and the function is constant.

So the answer to the problem is $a = x_{\lfloor n/2 \rfloor + 1}$ if n is odd, and a is any number in the interval $[x_{\lfloor n/2 \rfloor}, x_{\lfloor n/2 \rfloor + 1}]$ if n is even.

Remark. The required number x is called the median of $x_1, x_2, \ldots, x_n$. In general, if the numbers $x \in \mathbb{R}$ occur with probability distribution $d\mu(x)$ then their median a minimizes

$$E(|x - a|) = \int_{-\infty}^{\infty} |x - a| d\mu(x).$$

The median is any number such that

$$\int_{-\infty}^{a} d\mu(x) = P(x \le a) \ge \frac{1}{2}$$

and

$$\int_{a}^{\infty} d\mu(x) = P(x \ge a) \ge \frac{1}{2}.$$

In the particular case of our problem, the numbers $x_1, x_2, \ldots, x_n$ occur with equal probability, so the median lies in the middle.

516. The function $f(t) = t^c$ is convex, while $g(t) = x^t$ is convex and increasing. Therefore, $h(t) = g(f(t)) = x^{t^c}$ is convex. We thus have

$$x^{a^c} + x^{b^c} = h(a) + h(b) \ge 2h\left(\frac{a+b}{2}\right) = 2x^{\left(\frac{a+b}{2}\right)^{2c}} \ge 2x^{(ab)^{c/2}}.$$

This completes the solution.
    (P. Alexandrescu)

517. The problem amounts to showing that $\ln(\cos x) - \tan x \ln(\sin x)$ is non-negative for $0 < x < \frac{\pi}{4}$. The concavity of the natural logarithm implies that

$$\ln(\lambda a + (1 - \lambda)b) > \lambda \ln a + (1 - \lambda) \ln b,$$

for all $a, b > 0$ and $\lambda \in (0, 1)$. If we set $a = \sin x$, $b = \sin x + \cos x$ and $\lambda = \tan x$, then

$$\ln(\cos x) > \tan x \ln(\sin x) + (1 - \tan x) \ln(\cos x + \sin x)$$

The last term is positive since $\sin x + \cos x = \sqrt{2} \cos(\pi/4 - x) > 1$. Hence the conclusion.
    (*American Mathematical Monthly*, proposed by W.W. Chao)

518. We can assume that the triangle is inscribed in a circle of diameter 1, so that $a = \sin A$, $b = \sin B$, $c = \sin C$, $A \ge B \ge C$. The sine function is concave on the interval $[0, \pi]$, and since B is between A and C, and all three angles lie in this interval, we have

$$\frac{\sin B - \sin C}{B - C} \ge \frac{\sin A - \sin C}{A - C}.$$

Multiplying out, we obtain

$$(A - C)(\sin B - \sin C) \ge (B - C)(\sin A - \sin C),$$

or

$$A \sin B - A \sin C - C \sin B \geq B \sin A - C \sin A - B \sin C.$$

Moving the negative terms to the other side and substituting the sides of the triangle for the sines, we obtain the inequality from the statement.

519. *First solution:* Consider the function $f(x) = x^2 + \sqrt{x}$. Since $f''(x) = 2 - \frac{1}{4}x^{-3/2}, f'' \geq 0$ for $x \geq \frac{1}{4}$. It follows that on $[\frac{1}{4}, \infty)$ the function f is convex. Hence

$$\frac{f(x_1) + f(x_2)}{2} \geq f\left(\frac{x_1 + x_2}{2}\right),$$

for all $x_1, x_2 \geq \frac{1}{4}$. Substituting $x_1 = a^2, x_2 = b^2$, we obtain the inequality from the statement. *Second solution:* So that the reader can see the advantage of using derivatives, we also give an entirely algebraic solution. The inequality is equivalent to

$$(a^2 - b^2)^2 \geq 2\sqrt{2(a^2 + b^2)} - 2(a + b),$$

which can be transformed into

$$(a^2 - b^2)^2 + 2(a + b) \geq 2\sqrt{2(a^2 + b^2)}.$$

Squaring we obtain the equivalent inequality

$$(a^2 - b^2)^4 + 4(a^2 - b^2)^2(a + b) + 4(a + b)^2 \geq 8(a^2 + b^2).$$

Move everything to the left:

$$(a^2 - b^2)^4 + 4(a^2 - b^2)^2(a + b) - 4(a - b)^2 \geq 0.$$

This can be factored as

$$(a - b)^2((a - b)^2(a + b)^4 + 4(a + b)^3 - 4) \geq 0.$$

The first factor is obviously non-negative. We show that the second factor is non-negative as follows

$$(a - b)^2(a + b)^4 + 4(a + b)^3 - 4 \geq 4(a + b)^3 - 4 \geq 4 \cdot 1^3 - 4 = 0,$$

and the inequality is proved.

(*Kvant (Quantum)*, second solution by A.J.S. Chen)

520. Fix $x_0 \in (a, b)$ and let α and β be two limit points of f: α from the left and β from the right. We want to prove that they are equal. If not, without loss of generality we can assume $\alpha < \beta$. We argue from Figure 74. Choose $x < x_0$ and $y > x_0$ very close to x_0 such that $|f(x) - \alpha|$ and $|f(y) - \beta|$ are both very small. Because β is a limit point of f at x_0, there will exist points on the graph of f close to (x_0, β), hence above the segment joining $(x, f(x))$ and $(y, f(y))$. But this contradicts the convexity of f. Hence $\alpha = \beta$.

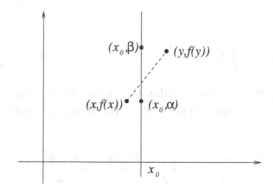

Figure 74

Because all limit points from the left are equal to all limit points from the right, f has a limit at x_0. Now redo the above argument for $x = x_0$ to conclude that the limit is equal to the value of the function at x_0. Hence f is continuous at x_0.

521. The key point of the solution is Cauchy's method of backward induction discussed in the first chapter of the book. We first prove that for any positive integer k and points $x_1, x_2, \ldots, x_{2^k}$, we have

$$f\left(\frac{x_1 + x_2 + \cdots + x_{2^k}}{2^k}\right) \le \frac{f(x_1) + f(x_2) + \cdots + f(x_{2^k})}{2^k}.$$

The base case is contained in the statement of the problem, while the inductive step is

$$f\left(\frac{x_1 + \cdots + x_{2^k} + x_{2^k+1} + \cdots + x_{2^{k+1}}}{2^{k+1}}\right) \le \frac{f\left(\frac{x_1 + \cdots + x_{2^k}}{2^k}\right) + f\left(\frac{x_{2^k+1} + \cdots + x_{2^{k+1}}}{2^k}\right)}{2}$$

$$\le \frac{\dfrac{f(x_1) + \cdots + f(x_{2^k})}{2^k} + \dfrac{f(x_{2^k+1}) + \cdots + f(x_{2^{k+1}})}{2^k}}{2}$$

$$= \frac{f(x_1) + \cdots + f(x_{2^k}) + f(x_{2^k} + 1) + \cdots + f(x_{2^{k+1}})}{2^{k+1}}$$

Next, we show that

$$f\left(\frac{x_1 + x_2 + \cdots + x_n}{n}\right) \le \frac{f(x_1) + f(x_2) + \cdots + f(x_n)}{n}, \quad \text{for all } x_1, x_2, \ldots, x_n.$$

Assuming that the inequality holds for any n points, we prove that it holds for any $n - 1$ points as well. Consider the points $x_1, x_2, \ldots, x_{n-1}$ and define $x_n = \frac{x_1 + x_2 + \cdots + x_{n-1}}{n-1}$. Using the induction hypothesis, we can write

$$f\left(\frac{x_1 + \cdots + x_{n-1} + \dfrac{x_1 + \cdots + x_{n-1}}{n-1}}{n}\right) \le \frac{f(x_1) + \cdots + f(x_{n-1}) + f\left(\dfrac{x_1 + \cdots + x_{n-1}}{n-1}\right)}{n}.$$

This is the same as

$$f\left(\frac{x_1 + \cdots + x_{n-1}}{n-1}\right) \leq \frac{f(x_1) + \cdots + f(x_{n-1})}{n} + \frac{1}{n}f\left(\frac{x_1 + \cdots + x_{n-1}}{n-1}\right).$$

Moving the last term on the right to the other side gives the desired inequality. Starting with a sufficiently large power of 2 we can cover the case of any positive integer n.

In the inequality

$$f\left(\frac{x_1 + x_2 + \cdots + x_n}{n}\right) \leq \frac{f(x_1) + f(x_2) + \cdots + f(x_n)}{n}$$

that we just proved, for some $m < n$ set $x_1 = x_2 = \cdots = x_m = x$ and $x_{m+1} = x_{m+2} = \cdots = x_n = y$. Then

$$f\left(\frac{m}{n}x + \left(1 - \frac{m}{n}\right)y\right) \leq \frac{m}{n}f(x) + \left(1 - \frac{m}{n}\right)f(y).$$

Because f is continuous we can pass to the limit with $\frac{m}{n} \to \lambda$ to obtain the desired

$$f(\lambda x + (1 - \lambda)y) \leq \lambda f(x) + (1 - \lambda)f(y), \quad \text{for every } \lambda \in (0, 1),$$

which characterizes convex functions.

522. *First solution*: Fix $n - 1$. For each integer i, define

$$\Delta_i = f\left(\frac{i+1}{n}\right) - f\left(\frac{i}{n}\right).$$

If in the inequality from the statement we substitute $x = \frac{i+2}{n}$ and $y = \frac{i}{n}$, we obtain

$$\frac{f\left(\frac{i+2}{n}\right) - f\left(\frac{i}{n}\right)}{2} \geq f\left(\frac{i+1}{n}\right) + \frac{2}{n}, \quad i = 1, 2, \ldots, n,$$

or

$$f\left(\frac{i+2}{n}\right) - f\left(\frac{i+1}{n}\right) \geq f\left(\frac{i+1}{n}\right) - f\left(\frac{i}{n}\right) + \frac{4}{n}, \quad i = 1, 2, \ldots, n.$$

In other words, $\Delta_{i+1} \geq \Delta_i + \frac{4}{n}$. Combining this for n consecutive values of i gives

$$\Delta_{i+n} \geq \Delta_i + 4.$$

Summing this inequality for $i = 0$ to $n - 1$ and canceling terms yields

$$f(2) - f(1) \geq f(1) - f(0) + 4n.$$

This cannot hold for all $n \geq 1$. Hence, there are no very convex functions.

Second solution: We show by induction on n that the given inequality implies

$$\frac{f(x) + f(y)}{2} - f\left(\frac{x+y}{2}\right) \geq 2^n|x - y|, \quad \text{for } n \geq 0.$$

This will yield a contradiction, because for fixed x and y the right-hand side gets arbitrarily large, while the left-hand side remains fixed.

The statement of the problem gives us the base case $n = 0$. Now, if the inequality holds for a given n, then for two real numbers a and b,

$$\frac{f(a) + f(a + 2b)}{2} \geq f(a + b) + 2^{n+1}|b|,$$

$$f(a + b) + f(a + 3b) \geq 2(f(a + 2b) + 2^{n+1}|b|),$$

and

$$\frac{f(a + 2b) + f(a + 4b)}{2} \geq f(a + 3b) + 2^{n+1}|b|.$$

Adding these three inequalities and canceling terms yields

$$\frac{f(a) + f(a + 4b)}{2} \geq f(a + 2b) + 2^{n+3}|b|.$$

Setting $x = a$, $y = a + 4b$, we obtain

$$\frac{f(x) + f(y)}{2} \geq f\left(\frac{x + y}{2}\right) + 2^{n+1}|x - y|,$$

completing the induction. Hence the conclusion.

(USA Mathematical Olympiad, 2000, proposed by B. Poonen)

523. The case $x = y = z$ is straightforward, so let us assume that not all three numbers are equal. Without loss of generality, we may assume that $x \leq y \leq z$. Let us first discuss the case $y \leq \frac{x+y+z}{3}$. Then $y \leq \frac{x+z}{2}$, and so

$$\frac{x + y + z}{3} \leq \frac{x + z}{2} \leq z.$$

Obviously $x \leq (x + y + z)/3$, and consequently

$$\frac{x + y + z}{3} \leq \frac{y + z}{2} \leq z.$$

It follows that there exist $s, t \in [0, 1]$ such that

$$\frac{x + z}{2} = s\frac{x + y + z}{3} + (1 - s)z,$$

$$\frac{y + z}{2} = t\frac{x + y + z}{3} + (1 - t)z.$$

Adding up these inequalities and rearranging yields

$$\frac{x + y - 2z}{2} = (s + t)\frac{x + y - 2z}{3}.$$

Since $x + y < 2z$, this equality can hold only if $s + t = \frac{3}{2}$. Writing the fact that f is a convex function, we obtain

$$f\left(\frac{x+z}{2}\right) = f\left(s\frac{x+y+z}{3} + (1-s)z\right) \leq sf\left(\frac{x+y+z}{3}\right) + (1-s)f(z),$$

$$f\left(\frac{y+z}{2}\right) = f\left(t\frac{x+y+z}{3} + (1-t)z\right) \leq tf\left(\frac{x+y+z}{3}\right) + (1-t)f(z),$$

$$f\left(\frac{x+y}{2}\right) \leq \frac{1}{2}f(x) + \frac{1}{2}f(y).$$

Adding the three, we obtain

$$f\left(\frac{x+y}{2}\right) + f\left(\frac{y+z}{2}\right) + f\left(\frac{z+x}{2}\right)$$

$$\leq (s+t)f\left(\frac{x+y+z}{3}\right) + \frac{1}{2}f(x) + \frac{1}{2}f(y) + (2-s-t)f(z)$$

$$= \frac{2}{3}f\left(\frac{x+y+z}{3}\right) + \frac{1}{2}f(x) + \frac{1}{2}f(y) + \frac{1}{2}f(z),$$

and the inequality is proved.

(T. Popoviciu, solution published by Gh. Eckstein in *Timişoara Mathematics Gazette*)

524. The fact that all sequences $(a^n b_n)_n$ are convex implies that for any real number a, $a^{n+1}b_{n+1} - 2a^n b_n + a^{n-1}b_{n-1} \geq 0$. Hence $b_{n+1}a^2 - 2b_n a + b_{n-1} \geq 0$ for all a. Viewing the left-hand side as a quadratic function in a, its discriminant must be less than or equal to zero. This is equivalent to $b_n^2 \leq b_{n+1}b_{n-1}$ for all n. Taking the logarithm, we obtain that $2\ln b_n \leq \ln b_{n+1} + \ln b_{n-1}$, proving that the sequence $(\ln b_n)_n$ is convex.

525. We will show that the largest such constant is $C = \frac{1}{2}$. For example, if we consider the sequence $a_1 = \varepsilon, a_2 = 1, a_3 = \varepsilon$, with ε a small positive number, then the condition from the statement implies

$$C \leq \frac{1}{2} \cdot \frac{(1+2\varepsilon)^2}{1+2\varepsilon^2}.$$

Here if we let $\varepsilon \to 0$, we obtain $C \leq \frac{1}{2}$.

Let us now show that $C = \frac{1}{2}$ satisfies the inequality for all concave sequences. For every i, concavity forces the elements $a_1, a_2, \ldots, a_i$ to be greater than or equal to the corresponding terms in the arithmetic progression whose first term is a_1 and whose ith term is a_i. Consequently,

$$a_1 + a_2 + \cdots + a_i \geq i\left(\frac{a_1 + a_i}{2}\right).$$

The same argument repeated for $a_i, a_{i+1}, \ldots, a_n$ shows that

$$a_i + a_{i+1} + \cdots + a_n \geq (n-i+1)\left(\frac{a_i + a_n}{2}\right).$$

Adding the two inequalities, we obtain

$$a_1 + a_2 + \cdots + a_n \geq i\left(\frac{a_1 + a_i}{2}\right) + (n - i + 1)\left(\frac{a_i + a_n}{2}\right) - a_i$$

$$= i\frac{a_1}{2} + (n - i + 1)\frac{a_n}{2} + \frac{(n-1)a_i}{2}$$

$$\geq \left(\frac{n-1}{2}\right)a_i.$$

Multiplying by a_i and summing the corresponding inequalities for all i gives

$$(a_1 + a_2 + \cdots + a_n)^2 \geq \frac{n-1}{2}(a_1^2 + a_2^2 + \cdots + a_n^2).$$

This shows that indeed $C = \frac{1}{2}$ is the answer to our problem.
(Mathematical Olympiad Summer Program, 1994)

526. We assume that $\alpha \leq \beta \leq \gamma$, the other cases being similar. The expression is a convex function in each of the variables, so it attains its maximum for some $x, y, z = a$ or b.

Now let us fix three numbers $x, y, z \in [a, b]$, with $x \leq y \leq z$. We have

$$E(x, y, z) - E(x, z, y) = (\gamma - \alpha)((z - x)^2 - (y - z)^2) \geq 0,$$

and hence $E(x, y, z) \geq E(x, z, y)$. Similarly, $E(x, y, z) \geq E(y, x, z)$ and $E(z, y, x) \geq E(y, z, x)$. So it suffices to consider the cases $x = a, z = b$ or $x = b$ and $z = a$. For these cases we have

$$E(a, a, b) = E(b, b, a) = (\beta + \gamma)(b - a)^2$$

and

$$E(a, b, b) = E(b, a, a) = (\alpha + \gamma)(b - a)^2.$$

We deduce that the maximum of the expression under discussion is $(\beta + \gamma)(b - a)^2$, which is attained for $x = y = a, z = b$ and for $x = y = b, z = a$.

(*Revista Matematică din Timişoara* (*Timişoara Mathematics Gazette*), proposed by D. Andrica and I. Raşa)

527. The left-hand side of the inequality under discussion is a convex function in each x_i. Hence in order to maximize this expression we must choose some of the x_i's equal to a and the others equal to b. For such a choice, denote by u the sum of the t_i's for which $x_i = a$ and by v the sum of the t_i's for which $x_i = b$. It remains to prove the simpler inequality

$$(ua + bv)\left(\frac{u}{a} + \frac{v}{b}\right) \leq \frac{(a+b)^2}{4ab}(u + b)^2.$$

This is equivalent to

$$4(ua + vb)(ub + va) \leq (ua + vb + ub + va)^2,$$

which is the AM-GM inequality applied to $ua + vb$ and $ub + va$.
(L.V. Kantorovich)

528. Expanding with Newton's binomial formula, we obtain

$$(1+x)^n + (1-x)^n = \sum_{k=0}^{\lfloor \frac{n}{2} \rfloor} \binom{n}{2k} x^{2k}.$$

The coefficients in the expansion are positive, so the expression is a convex function in x (being a sum of power functions that are convex). Its maximum is attained when $|x| = 1$, in which case the value of the expression is 2^n. This proves the inequality.

(C. Năstăsescu, C. Niţă, M. Brandiburu, D. Joiţa, *Exerciţii şi Probleme de Algebră* (*Exercises and Problems in Algebra*), Editura Didactică şi Pedagogică, Bucharest, 1983)

529. Without loss of generality, we may assume that b is the number in the middle. The inequality takes the form

$$a + b + c - 3\sqrt[3]{abc} \le 3(a + c - 2\sqrt{ac}).$$

For fixed a and c, define $f : [a, c] \to \mathbb{R}, f(b) = 3(a + c - 2\sqrt{ac}) - a - b - c + 3\sqrt{abc}$. This function is concave because $f''(b) = -\frac{2}{3}(ac)^{1/3} b^{-5/3} < 0$, so it attains its minimum at one of the endpoints of the interval $[a, c]$. Thus the minimum is attained for $b = a$ or $b = c$. Let us try the case $b = a$. We may rescale the variables so that $a = b = 1$. The inequality becomes

$$\frac{2c + 3c^{1/3} + 1}{6} \ge c^{1/2},$$

and this is just an instance of the generalized AM-GM inequality. The case $a = c$ is similar.

(USA Team Selection Test for the International Mathematical Olympiad, 2002, proposed by T. Andreescu)

530. For (a) we apply Sturm's principle. Given $x \in (a, b)$ choose $h > 0$ such that $a < x - h < x + h < b$. The Mean value theorem implies that $f(x) \le \max_{x-h \le y \le x+y} f(y)$, with equality only when f is constant on $[x - h, x + h]$. Hence $f(x)$ is less than or equal to the maximum of f on $[a, b]$, with equality if and only if f is constant on $[a, b]$. We know that the maximum of f is attained on $[a, b]$. It can be attained at the chosen point x only if f is constant on $[a, b]$. This proves that the maximum is attained at one of the endpoints of the interval.

To prove (b) we define the linear function

$$L(x) = \frac{(x - a)f(b) + (b - x)f(a)}{b - a}.$$

It is straightforward to verify that L itself satisfies the mean value inequality from the statement with equality, and so does $-L$. Therefore, the function $G(x) = f(x) - L(x)$ satisfies the mean value inequality, too. It follows that G takes its maximum value at a or at b. A calculation shows that $G(a) = G(b) = 0$. Therefore, $G(x) \le 0$ for $x \in [a, b]$. This is equivalent to

$$f(x) \le \frac{(x - a)f(b) + (b - x)f(a)}{b - a},$$

which is, in fact, the condition for f to be convex.

(P.N. de Souza, J.N. Silva, *Berkeley Problems in Mathematics*, Springer, 2004)

531. The function $f(t) = \sin t$ is concave on the interval $[0, \pi]$. Jensen's inequality yields

$$\sin A + \sin B + \sin C \geq 3 \sin \frac{A + B + C}{3} = 3 \sin \frac{\pi}{3} = \frac{3\sqrt{3}}{2}.$$

532. If we set $y_i = \ln x_i$, then $x_i \in (0, 1]$ implies $y_i \leq 0$, $i = 1, 2, \ldots, n$. Consider the function $f : (-\infty, 0] \to \mathbb{R}, f(y) = (1 + e^y)^{-1}$. This function is twice differentiable and

$$f''(y) = e^y(e^y - 1)(1 + e^y)^{-3} \leq 0, \text{ for } y \leq 0.$$

It follows that this function is concave, and we can apply Jensen's inequality to the points $y_1, y_2, \ldots, y_n$ and the weights $a_1, a_2, \ldots, a_n$. We have

$$\sum_{i=1}^{n} \frac{a_i}{1 + x_i} = \sum_{i=1}^{n} \frac{a_i}{1 + e^{y_i}} \leq \frac{1}{1 + e^{\sum_{i=1}^{n} a_i y_i}}$$

$$= \frac{1}{1 + \prod_{i=1}^{n} e^{a_i y_i}} = \frac{1}{1 + \prod_{i=1}^{n} x_i^{a_i}},$$

which is the desired inequality.

(D. Buşneag, I. Maftei, *Teme pentru cercurile şi concursurile de matematică* (*Themes for mathematics circles and contests*), Scrisul Românesc, Craiova)

533. *First solution*: Apply Jensen's inequality to the convex function $f(x) = x^2$ and to

$$x_1 = \frac{a_1^2 + a_2^2 + a_3^2}{2a_2 a_3}, \quad x_2 = \frac{a_1^2 + a_2^2 + a_3^2}{2a_3 a_1}, \quad x_3 = \frac{a_1^2 + a_2^2 + a_3^2}{2a_1 a_2},$$

$$\lambda_1 = \frac{a_1^2}{a_1^2 + a_2^2 + a_3^2}, \quad \lambda_2 = \frac{a_2^2}{a_1^2 + a_2^2 + a_3^2}, \quad \lambda_3 = \frac{a_3^2}{a_1^2 + a_2^2 + a_3^2}.$$

The inequality

$$f(\lambda_1 x_2 + \lambda_2 x_2 + \lambda_3 x_3) \leq \lambda_1 f(x_1) + \lambda_2 f(x_2) + \lambda_3 f(x_3)$$

translates to

$$\frac{(a_1^3 + a_2^3 + a_3^3)^2}{4a_1^2 a_2^2 a_3^2} \leq \frac{(a_1^4 + a_2^4 + a_3^4)(a_1^2 + a_2^2 + a_3^2)}{4a_1^2 a_2^2 a_3^2},$$

and the conclusion follows.

Second solution: The inequality from the statement is equivalent to

$$(a_1^2 + a_2^2 + a_3^2)(a_1^4 + a_2^4 + a_3^4) \geq (a_1^3 + a_2^3 + a_3^3)^2.$$

This is just the Cauchy-Schwarz inequality applied to a_1, a_2, a_3 and a_1^2, a_2^2, a_3^2.

(*Gazeta Matematică* (*Mathematics Gazette, Bucharest*))

534. Take the natural logarithm of both sides, which are positive because $x_i \in (0, \pi)$, $i = 1, 2, \ldots, n$, to obtain the equivalent inequality

$$\sum_{i=1}^{n} \ln \frac{\sin x_i}{x_i} \leq n \ln \frac{\sin x}{x}.$$

All we are left to check is that the function $f(t) = \ln \frac{\sin t}{t}$ is concave on $(0, \pi)$.
   Because $f(t) = \ln \sin t - \ln t$, its second derivative is

$$f''(t) = -\frac{1}{\sin^2 t} + \frac{1}{t^2}.$$

The fact that this is negative follows from $\sin t < t$ for $t > 0$, and the inequality is proved.
   (39th W.L. Putnam Mathematical Competition, 1978)

535. The function $f : (0, 1) \to \mathbb{R}, f(x) = \frac{x}{\sqrt{1-x}}$ is convex. By Jensen's inequality,

$$\frac{1}{n} \sum_{i=1}^{n} \frac{x_i}{\sqrt{1 - x_i}} \geq \frac{\frac{1}{n} \sum_{i=1}^{n} x_i}{\sqrt{1 - \frac{1}{n} \sum_{i=1}^{n} x_i}} = \frac{1}{\sqrt{n(n-1)}}.$$

We have thus found that

$$\frac{x_1}{\sqrt{1 - x_1}} + \frac{x_2}{\sqrt{1 - x_2}} + \cdots + \frac{x_n}{\sqrt{1 - x_n}} \geq \sqrt{\frac{n}{n-1}}.$$

On the other hand, by the Cauchy-Schwarz inequality

$$n = n \sum_{i=1}^{n} x_i \geq \left(\sum_{i=1}^{n} \sqrt{x_i} \right)^2,$$

whence $\sum_{i=1}^{n} \sqrt{x_i} \leq \sqrt{n}$. It follows that

$$\frac{\sqrt{x_1} + \sqrt{x_2} + \cdots + \sqrt{x_n}}{\sqrt{n-1}} \leq \sqrt{\frac{n}{n-1}}.$$

Combining the two inequalities, we obtain the one from the statement.

536. We apply Jensen's inequality for the concave function $f(x) = \sqrt{x}$ and $\lambda_1 = \frac{1}{10}, \lambda_2 = \frac{2}{10}, \lambda_3 = \frac{3}{10}$ and $\lambda_4 = \frac{4}{10}$. We have

$$\frac{1}{10} \sqrt{a} + \frac{2}{10} \sqrt{\frac{b}{4}} + \frac{3}{10} \sqrt{\frac{c}{9}} + \frac{4}{10} \sqrt{\frac{d}{16}} \leq \sqrt{\frac{a}{10} + \frac{b}{20} + \frac{c}{30} + \frac{d}{40}}.$$

Hence

$$\sqrt{a} + \sqrt{b} + \sqrt{c} + \sqrt{d} \leq 10\sqrt{\frac{12a + 6b + 4c + 3d}{120}}.$$

But

$$12a + 6b + 4c + 3d = 3(a + b + c + d) + (a + b + c) + 2(a + b) + 6a$$
$$\leq 3 \cdot 30 + 14 + 2 \cdot 5 + 6 \cdot 1 = 120.$$

The inequality follows.

(Romanian Team Selection Test for the International Mathematical Olympiad, proposed by V. Cârtoaje)

537. Split the integral as

$$\int e^{x^2} dx + \int 2x^2 e^{x^2} dx.$$

Denote the first integral by I_1. Then use integration by parts to transform the second integral as

$$\int 2x^2 e^{x^2} dx = xe^{x^2} - \int e^{x^2} dx = xe^{x^2} - I_1.$$

The integral from the statement is therefore equal to

$$I_1 + xe^{x^2} - I_1 = xe^{x^2} + C.$$

538. Adding and subtracting e^x in the numerator, we obtain

$$\int \frac{x + \sin x - \cos x - 1}{x + e^x + \sin x} dx = \int \frac{x + e^x + \sin x - 1 - e^x - \cos x}{x + e^x + \sin x} dx$$
$$= \int \frac{x + e^x + \sin x}{x + e^x + \sin x} dx - \int \frac{1 + e^x + \cos x}{x + e^x + \sin x} dx$$
$$= x + \ln(x + e^x + \sin x) + C.$$

(Romanian college entrance exam)

539. The trick is to bring a factor of x inside the cube root:

$$\int (x^6 + x^3)\sqrt[3]{x^3 + 2} dx = \int (x^5 + x^2)\sqrt[3]{x^6 + 2x^3} dx.$$

The substitution $u = x^6 + 2x^3$ now yields the answer

$$\frac{1}{6}(x^6 + 2x^3)^{4/3} + C.$$

(G.T. Gilbert, M.I. Krusemeyer, L.C. Larson, *The Wohascum County Problem Book*, MAA, 1993)

540. We want to avoid the lengthy method of partial fraction decomposition. To this end, we rewrite the integral as

$$\int \frac{x^2\left(1+\dfrac{1}{x^2}\right)}{x^2\left(x^2-1+\dfrac{1}{x^2}\right)}\,dx = \int \frac{1+\dfrac{1}{x^2}}{x^2-1+\dfrac{1}{x^2}}\,dx.$$

With the substitution $x-\frac{1}{x}=t$ we have $\left(1+\frac{1}{x^2}\right)dx = dt$, and the integral takes the form

$$\int \frac{1}{t^2+1}\,dt = \arctan t + C.$$

We deduce that the integral from the statement is equal to

$$\arctan\left(x-\frac{1}{x}\right)+C.$$

541. Substitute $u = \sqrt{\frac{e^x-1}{e^x+1}}$, $0 < u < 1$. Then $x = \ln(1+u^2)-\ln(1-u^2)$, and $dx = \left(\frac{2u}{1+u^2}+\frac{2u}{1-u^2}\right)du$. The integral becomes

$$\int u\left(\frac{2u}{u^2+1}+\frac{2u}{u^2-1}\right)du = \int\left(4-\frac{2}{u^2+1}+\frac{2}{u^2-1}\right)du$$

$$= 4u - 2\arctan u + \int\left(\frac{1}{u+1}+\frac{1}{1-u}\right)du$$

$$= 4u - 2\arctan u + \ln(u+1)-\ln(u-1)+C.$$

In terms of x, this is equal to

$$4\sqrt{\frac{e^x-1}{e^x+1}} - 2\arctan\sqrt{\frac{e^x-1}{e^x+1}} + \ln\left(\sqrt{\frac{e^x-1}{e^x+1}}+1\right) - \ln\left(\sqrt{\frac{e^x-1}{e^x+1}}-1\right)+C.$$

542. Note that

$$\frac{1+x^2\ln x}{x+x^2\ln x} = \frac{1}{x}+1-\frac{1+\ln x}{1+x\ln x}.$$

We thus have

$$\int \frac{1+x^2\ln x}{x+x^2\ln x}\,dx = \int\frac{1}{x}\,dx + \int dx - \int\frac{1+\ln x}{1+x\ln x}\,dx$$

$$= \ln x + x - \ln(1+x\ln x)+C.$$

(slightly modified version of a *Mathematical Reflections* problem, proposed by Z. Starc)

543. If we naively try the substitution $t = x^3 + 1$, we obtain

$$f(t) = \sqrt{t + 1 - 2\sqrt{t}} + \sqrt{t + 9 - 6\sqrt{t}}.$$

Now we recognize the perfect squares, and we realize that

$$f(x) = \sqrt{(\sqrt{x^3 + 1} - 1)^2} + \sqrt{(\sqrt{x^3 + 1} - 3)^2} = |\sqrt{x^3 + 1} - 1| + |\sqrt{x^3 + 1} - 3|.$$

When $x \in [0, 2]$, $1 \le \sqrt{x^3 + 1} \le 3$. Therefore,

$$f(x) = \sqrt{x^3 + 1} - 1 + 3 - \sqrt{x^3 + 1} = 2.$$

The antiderivatives of f are therefore the linear functions $f(x) = 2x + C$, where C is a constant.
   (Communicated by E. Craina)

544. Let $f_n = 1 + x + \frac{x^2}{2!} + \cdots + \frac{x^n}{n!}$. Then $f'(x) = 1 + x + \cdots + \frac{x^{n-1}}{(n-1)!}$. The integral in the statement becomes

$$I_n = \int \frac{n!(f_n(x) - f_n'(x))}{f_n(x)} dx = n! \int \left(1 - \frac{f_n'(x)}{f_n(x)}\right) dx = n!x - n! \ln f_n(x) + C$$

$$= n!x - n! \ln\left(1 + x + \frac{x^2}{2!} + \cdots + \frac{x^n}{n!}\right) + C.$$

   (C. Mortici, *Probleme Pregătitoare pentru Concursurile de Matematică* (*Training Problems for Mathematics Contests*), GIL, 1999)

545. The substitution is

$$u = \frac{x}{\sqrt[4]{2x^2 - 1}},$$

for which

$$du = \frac{x^2 - 1}{(2x^2 - 1)\sqrt[4]{2x^2 - 1}} dx.$$

We can transform the integral as follows:

$$\int \frac{2x^2 - 1}{-(x^2 - 1)^2} \cdot \frac{x^2 - 1}{(2x^2 - 1)\sqrt[4]{2x^2 - 1}} dx = \int \frac{1}{\dfrac{-x^4 + 2x^2 - 1}{2x^2 - 1}} \cdot \frac{x^2 - 1}{(2x^2 - 1)\sqrt[4]{2x^2 - 1}} dx$$

$$= \int \frac{1}{1 - \dfrac{x^4}{2x^2 - 1}} \cdot \frac{x^2 - 1}{(2x^2 - 1)\sqrt[4]{2x^2 - 1}} dx$$

$$= \int \frac{1}{1 - u^4} du.$$

This is computed using Jacobi's method of partial fraction decomposition, giving the final answer to the problem

$$\frac{1}{4} \ln \frac{\sqrt[4]{2x^2 - 1} + x}{\sqrt[4]{2x^2 - 1} - x} - \frac{1}{2} \arctan \frac{\sqrt[4]{2x^2 - 1}}{x} + C.$$

546. Of course, Jacobi's partial fraction decomposition method can be applied, but it is more laborious. However, in the process of applying it we factor the denominator as $x^6 + 1 = (x^2 + 1)(x^4 - x^2 + 1)$, and this expression can be related somehow to the numerator. Indeed, if we add and subtract an x^2 in the numerator, we obtain

$$\frac{x^4 + 1}{x^6 + 1} = \frac{x^4 - x^2 + 1}{x^6 + 1} + \frac{x^2}{x^6 + 1}.$$

Now integrate as follows:

$$\int \frac{x^4 + 1}{x^6 + 1} dx = \int \frac{x^4 - x^2 + 1}{x^6 + 1} dx + \int \frac{x^2}{x^6 + 1} dx$$

$$= \int \frac{1}{x^2 + 1} dx + \int \frac{1}{3} \frac{(x^3)'}{(x^3)^2 + 1} dx$$

$$= \arctan x + \frac{1}{3} \arctan x^3.$$

To write the answer in the required form we should have

$$3 \arctan x + \arctan x^3 = \arctan \frac{P(x)}{Q(x)}.$$

Applying the tangent function to both sides, we deduce

$$\frac{\frac{3x - x^3}{1 - 3x^2} + x^3}{1 - \frac{3x - x^3}{1 - 3x^2} \cdot x^3} = \tan\left(\arctan \frac{P(x)}{Q(x)}\right).$$

From here

$$\arctan \frac{P(x)}{Q(x)} = \arctan \frac{3x - 3x^5}{1 - 3x^2 - 3x^4 + x^6},$$

and hence $P(x) = 3x - 3x^5$, $Q(x) = 1 - 3x^2 - 3x^4 + x^6$. The final answer is

$$\frac{1}{3} \arctan \frac{3x - 3x^5}{1 - 3x^2 - 3x^4 + x^6} + C.$$

547. The function $f : [-1, 1] \to \mathbb{R}$,

$$f(x) = \frac{\sqrt[3]{x}}{\sqrt[3]{1 - x} + \sqrt[3]{1 + x}},$$

is odd; therefore, the integral is zero.

548. We use the example from the introduction for the particular function $f(x) = \frac{x}{1+x^2}$ to transform the integral into

$$\pi \int_0^{\frac{\pi}{2}} \frac{\sin x}{1 + \sin^2 x} dx.$$

This is the same as

$$\pi \int_0^{\frac{\pi}{2}} -\frac{d(\cos x)}{2 - \cos^2 x},$$

which with the substitution $t = \cos x$ becomes

$$\pi \int_0^1 \frac{1}{2 - t^2} dt = \frac{\pi}{2\sqrt{2}} \ln \frac{\sqrt{2} + t}{\sqrt{2} - t} \Big|_0^1 = \frac{\pi}{2\sqrt{2}} \ln \frac{\sqrt{2} + 1}{\sqrt{2} - 1}.$$

549. We have

$$\int_0^{\sqrt{\frac{\pi}{3}}} \sin x^2 dx + \int_{-\sqrt{\frac{\pi}{3}}}^{\sqrt{\frac{\pi}{3}}} x^2 \cos x^2 dx = \int_0^{\sqrt{\frac{\pi}{3}}} \sin x^2 dx + 2 \int_0^{\sqrt{\frac{\pi}{3}}} x^2 \cos x^2 dx$$

$$= \int_0^{\sqrt{\frac{\pi}{3}}} [\sin x^2 + x(\cos x^2) \cdot 2x] dx = \int_0^{\sqrt{\frac{\pi}{3}}} \frac{d}{dx} (x \sin x^2) dx$$

$$= x \sin x^2 \Big|_0^{\sqrt{\frac{\pi}{3}}} = \frac{\sqrt{\pi}}{2}.$$

(21st Annual Iowa Collegiate Mathematics Competition, proposed by R. Gelca)

550. Denote the value of the integral by I. With the substitution $t = \frac{ab}{x}$ we have

$$I = \int_a^b \frac{e^{\frac{b}{t}} - e^{\frac{t}{a}}}{\frac{ab}{t}} \cdot \frac{-ab}{t^2} dt = -\int_a^b \frac{e^{\frac{t}{a}} - e^{\frac{b}{t}}}{t} dt = -I.$$

Hence $I = 0$.

551. The substitution $t = 1 - x$ yields

$$I = \int_0^1 \sqrt[3]{2(1 - t)^3 - 3(1 - t)^2 - (1 - t) + 1} dt = -\int_0^1 \sqrt[3]{2t^3 - 3t^2 - t + 1} dt = -I.$$

Hence $I = 0$.

(*Mathematical Reflections*, proposed by T. Andreescu)

552. Using the substitutions $x = a \sin t$, respectively, $x = a \cos t$, we find the integral to be equal to both the integral

$$L_1 = \int_0^{\pi/2} \frac{\sin t}{\sin t + \cos t} dt$$

and the integral

$$L_2 = \int_0^{\pi/2} \frac{\cos t}{\sin t + \cos t} dt.$$

Hence the desired integral is equal to

$$\frac{1}{2}(L_1 + L_2) = \frac{1}{2} \int_0^{\pi/2} 1 dt = \frac{\pi}{4}.$$

553. Denote the integral by I. With the substitution $t = \frac{\pi}{4} - x$ the integral becomes

$$I = \int_{\frac{\pi}{4}}^{0} \ln\left(1 + \tan\left(\frac{\pi}{4} - t\right)\right)(-1)dt = \int_{0}^{\frac{\pi}{4}} \ln\left(1 + \frac{1 - \tan t}{1 + \tan t}\right) dt$$

$$= \int_{0}^{\frac{\pi}{4}} \ln \frac{2}{1 + \tan t} dt = \frac{\pi}{4} \ln 2 - I.$$

Solving for I, we obtain $I = \frac{\pi}{8} \ln 2$.

554. With the substitution $\arctan x = t$ the integral takes the form

$$I = \int_{0}^{\frac{\pi}{4}} \ln(1 + \tan t)dt.$$

This we already computed in the previous problem. ("Happiness is longing for repetition", says M. Kundera.) So the answer to the problem is $\frac{\pi}{8} \ln 2$.

(66th W.L. Putnam Mathematical Competition, 2005, proposed by T. Andreescu)

555. The function $\ln x$ is integrable near zero, and the function under the integral sign is dominated by $x^{-3/2}$ near infinity; hence the improper integral converges. We first treat the case $a = 1$. The substitution $x = 1/t$ yields

$$\int_{0}^{\infty} \frac{\ln x}{x^2 + 1} dx = \int_{\infty}^{0} \frac{\ln \frac{1}{t}}{\frac{1}{t^2} + 1}\left(-\frac{1}{t^2}\right) dt = -\int_{0}^{\infty} \frac{\ln t}{t^2 + 1} dt,$$

which is the same integral but with opposite sign. This shows that for $a = 1$ the integral is equal to 0. For general a we compute the integral using the substitution $x = a/t$ as follows

$$\int_{0}^{\infty} \frac{\ln x}{x^2 + a^2} dx = \int_{\infty}^{0} \frac{\ln a - \ln t}{\left(\frac{a}{t}\right)^2 + a^2}\left(-\frac{a}{t^2}\right) dt = \frac{1}{a} \int_{0}^{\infty} \frac{\ln a - \ln t}{1 + t^2} dt$$

$$= \frac{\ln a}{a} \int_{0}^{\infty} \frac{dt}{t^2 + 1} - \frac{1}{a} \int_{0}^{\infty} \frac{\ln t}{t^2 + 1} dt = \frac{\pi \ln a}{2a}.$$

(P.N. de Souza, J.N. Silva, *Berkeley Problems in Mathematics*, Springer, 2004)

556. The statement is misleading. There is nothing special about the limits of integration! The *indefinite* integral can be computed as follows:

$$\int \frac{x \cos x - \sin x}{x^2 + \sin^2 x} dx = \int \frac{\frac{\cos x}{x} - \frac{\sin x}{x^2}}{1 + \left(\frac{\sin x}{x}\right)^2} dx = \int \frac{1}{1 + \left(\frac{\sin x}{x}\right)^2}\left(\frac{\sin x}{x}\right)' dx$$

$$= \arctan\left(\frac{\sin x}{x}\right) + C.$$

Therefore,

$$\int_0^{\frac{\pi}{2}} \frac{x\cos x - \sin x}{x^2 + \sin^2 x}dx = \arctan\frac{2}{\pi} - \frac{\pi}{4}.$$

(Z. Ahmed)

557. If α is a multiple of π, then $I(\alpha) = 0$. Otherwise, use the substitution $x = \cos\alpha + t\sin\alpha$. The indefinite integral becomes

$$\int \frac{\sin\alpha dx}{1 - 2x\cos\alpha + x^2} = \int \frac{dt}{1 + t^2} = \arctan t + C.$$

It follows that the definite integral $I(\alpha)$ has the value

$$\arctan\left(\frac{1 - \cos\alpha}{\sin\alpha}\right) - \arctan\left(\frac{-1 - \cos\alpha}{\sin\alpha}\right),$$

where the angles are to be taken between $-\frac{\pi}{2}$ and $\frac{\pi}{2}$. But

$$\frac{1 - \cos\alpha}{\sin\alpha} \times \frac{-1 - \cos\alpha}{\sin\alpha} = -1.$$

Hence the difference between these angles is $\pm\frac{\pi}{2}$. Notice that the sign of the integral is the same as the sign of $\sin\alpha$.

Hence $I(\alpha) = \frac{\pi}{2}$ if $\alpha \in (2k\pi, (2k+1)\pi)$ and $-\frac{\pi}{2}$ if $\alpha \in ((2k+1)\pi, (2k+2)\pi)$ for some integer k.

Remark. This is an example of an integral with parameter that does not depend continuously on the parameter. (E. Goursat, *A Course in Mathematical Analysis*, Dover, NY, 1904)

558. First, note that $1/\sqrt{x}$ has this property for $p > 2$. We will alter slightly this function to make the integral finite for $p = 2$. Since we know that logarithms grow much slower than power functions, a possible choice might be

$$f(x) = \frac{1}{\sqrt{x}\ln x}.$$

Then

$$\int_2^\infty f^2(x)dx = \int_2^\infty \frac{1}{x\ln^2 x} = -\frac{1}{\ln x}\Big|_2^\infty = \frac{1}{\ln 2} < \infty.$$

Consequently, the integral of f^p is finite for all real numbers $p \geq 2$.

Let us see what happens for $p < 2$. An easy application of L'Hôpital's theorem gives

$$\lim_{x\to\infty} \frac{f(x)^p}{x^{-1}} = \lim_{x\to\infty} \frac{x^{-\frac{p}{2}}\ln^{-p} x}{x^{-1}} = \lim_{x\to\infty} \frac{x^{1-\frac{p}{2}}}{\ln^p x} = \infty,$$

and hence the comparison test implies that for $p < 2$ the integral is infinite. Therefore, $f(x) = \frac{1}{\sqrt{x}\ln x}$ satisfies the required condition.

Remark. Examples like the above are used in measure theory to prove that inclusions between L^p spaces are strict.

559. Suppose $(f(x))^2 + (f'(x))^2 > 1$ for all $x \in \left[-\frac{\pi}{2}, \frac{\pi}{2}\right]$. We can rewrite this inequality as

$$\frac{f'(x)}{\sqrt{1 - f^2(x)}} > 1, \quad x \in \left[-\frac{\pi}{2}, \frac{\pi}{2}\right].$$

Integrating from $-\frac{\pi}{2}$ to $\frac{\pi}{2}$ we obtain

$$\arcsin f\left(\frac{\pi}{2}\right) - \arcsin f\left(-\frac{\pi}{2}\right) > \pi.$$

But the difference of two arcsines is at most π, which is a contradiction. Hence the conclusion.
  (*Mathematical Reflections*, proposed by T. Andreescu)

560. Let n be the degree of $P(x)$. Integrating successively by parts, we obtain

$$\int_0^t e^{-x} P(x) dt = -e^{-x} P(x) \Big|_0^t + \int_0^t e^{-x} P'(x) dx$$

$$= -e^{-x} P(x) \Big|_0^t - e^{-x} P'(x) \Big|_0^t + \int_0^t e^{-x} P'(x) dx = \cdots$$

$$= -e^{-x} P(x) \Big|_0^t - e^{-x} P'(x) \Big|_0^t - \cdots - e^{-x} P^{(n)}(x) \Big|_0^t.$$

Because $\lim_{t \to \infty} e^{-t} P^{(k)}(t) = 0$, $k = 0, 1, \ldots, n$, when passing to the limit we obtain

$$\lim_{t \to \infty} \int_0^t e^{-x} P(x) dx = P(0) + P'(0) + P''(0) + \cdots,$$

hence the conclusion.

561. First, note that by L'Hôpital's theorem,

$$\lim_{x \to 0} \frac{1 - \cos nx}{1 - \cos x} = n^2,$$

which shows that the integrand can be extended continuously to $[0, 1]$. So the integral is well defined.
  Denote the integral by I_n. Then

$$\frac{I_{n+1} + I_{n-1}}{2} = \int_0^\pi \frac{2 - \cos(n+1)x - \cos(n-1)x}{2(1 - \cos x)} dx = \int_0^\pi \frac{1 - \cos nx \cos x}{1 - \cos x} dx$$

$$= \int_0^\pi \frac{(1 - \cos nx) + \cos nx(1 - \cos x)}{1 - \cos x} dx = I_n + \int_0^\pi \cos nx \, dx = I_n.$$

Therefore,

$$I_n = \frac{1}{2}(I_{n+1} + I_{n-1}), \quad n \ge 1.$$

This shows that $I_0, I_1, I_2, \ldots$ is an arithmetic sequence. From $I_0 = 0$ and $I_1 = \pi$ it follows that $I_n = n\pi$, $n \geq 1$.

562. Integration by parts gives

$$I_n = \int_0^{\pi/2} \sin^n x\, dx = \int_0^{\pi/2} \sin^{n-1} x \sin x\, dx$$

$$= -\sin^{n-1} x \cos^2 x \Big|_0^{\pi/2} + (n-1)\int_0^{\pi/2} \sin^{n-2} x \cos^2 x\, dx$$

$$= (n-1)\int_0^{\pi/2} \sin^{n-2} x(1 - \sin^2 x)\, dx = (n-1)I_{n-2} - (n-1)I_n.$$

We obtain the recursive formula

$$I_n = \frac{n-1}{n}I_{n-2}, \ n \geq 2.$$

This combined with $I_0 = \frac{\pi}{2}$ and $I_1 = 1$ yields

$$I_n = \begin{cases} \dfrac{1 \cdot 3 \cdot 5 \cdots (2k-1)}{2 \cdot 4 \cdot 6 \cdots (2k)} \cdot \dfrac{\pi}{2}, & \text{if } n = 2k \\[3mm] \dfrac{2 \cdot 4 \cdot 6 \cdots (2k)}{1 \cdot 3 \cdot 5 \cdots (2k+1)}, & \text{if } n = 2k+1. \end{cases}$$

To prove the Wallis formula, we use the obvious inequality $\sin^{2n+1} x < \sin^{2n} x < \sin^{2n-1} x$, $x \in \left(0, \frac{\pi}{2}\right)$ to deduce that $I_{2n+1} < I_{2n} < I_{2n-1}$, $n \geq 1$. This translates into

$$\frac{2 \cdot 4 \cdot 6 \cdots (2n)}{1 \cdot 3 \cdot 5 \cdots (2n+1)} < \frac{1 \cdot 3 \cdot 5 \cdots (2n-1)}{2 \cdot 4 \cdot 6 \cdots (2n)} \cdot \frac{\pi}{2} < \frac{2 \cdot 4 \cdot 6 \cdots (2n-2)}{1 \cdot 3 \cdot 5 \cdots (2n-1)},$$

which is equivalent to

$$\left[\frac{2 \cdot 4 \cdot 6 \cdots (2n)}{1 \cdot 3 \cdot 5 \cdots (2n-1)}\right]^2 \cdot \frac{2}{2n+1} < \pi < \left[\frac{2 \cdot 4 \cdot 6 \cdots (2n)}{1 \cdot 3 \cdot 5 \cdots (2n-1)}\right]^2 \cdot \frac{2}{2n}.$$

We obtain the double inequality

$$\pi < \left[\frac{2 \cdot 4 \cdot 6 \cdots (2n)}{1 \cdot 3 \cdot 5 \cdots (2n-1)}\right]^2 \cdot \frac{1}{n} < \pi \cdot \frac{2n+1}{2n}.$$

Passing to the limit and using the squeezing principle, we obtain the Wallis formula.

563. Denote the integral from the statement by I_n, $n \geq 0$. We have

$$I_n = \int_{-\pi}^0 \frac{\sin nx}{(1 + 2^x)\sin x}\, dx + \int_0^\pi \frac{\sin nx}{(1 + 2^x)\sin x}\, dx.$$

In the first integral change x to $-x$ to further obtain

$$
\begin{aligned}
I_n &= \int_0^\pi \frac{\sin nx}{(1+2^{-x})\sin x}dx + \int_0^\pi \frac{\sin nx}{(1+2^x)\sin x}dx \\
&= \int_0^\pi \frac{2^x \sin nx}{(1+2^x)\sin x}dx + \int_0^\pi \frac{\sin nx}{(1+2^x)\sin x}dx \\
&= \int_0^\pi \frac{(1+2^x)\sin nx}{(1+2^x)\sin x}dx = \int_0^\pi \frac{\sin nx}{\sin x}dx.
\end{aligned}
$$

And these integrals can be computed recursively. Indeed, for $n \geq 0$ we have

$$
I_{n+2} - I_n = \int_0^\pi \frac{\sin(n+2)x - \sin nx}{\sin x}dx = 2\int_0^\pi \cos(n-1)xdx = 0,
$$

a very simple recurrence. Hence for n even, $I_n = I_0 = 0$, and for n odd, $I_n = I_1 = \pi$.
    (3rd International Mathematics Competition for University Students, 1996)

564. We have

$$
\begin{aligned}
s_n &= \frac{1}{\sqrt{4n^2-1^2}} + \frac{1}{\sqrt{4n^2-2^2}} + \cdots + \frac{1}{\sqrt{4n^2-n^2}} \\
&= \frac{1}{n}\left[\frac{1}{\sqrt{4-\left(\frac{1}{n}\right)^2}} + \frac{1}{\sqrt{4-\left(\frac{2}{n}\right)^2}} + \cdots + \frac{1}{\sqrt{4-\left(\frac{n}{n}\right)^2}}\right].
\end{aligned}
$$

Hence s_n is the Riemann sum of the function $f : [0, 1] \to \mathbb{R}, f(x) = \frac{1}{\sqrt{4-x^2}}$ associated to the subdivision $x_0 = 0 < x_1 = \frac{1}{n} < x_2 = \frac{2}{n} < \cdots < x_n = \frac{n}{n} = 1$, with the intermediate points $\xi_i = \frac{i}{n} \in [x_i, x_{i+1}]$. The answer to the problem is therefore

$$
\lim_{n\to\infty} s_n = \int_0^1 \frac{1}{\sqrt{4-x^2}}dx = \arcsin\frac{x}{2}\Big|_0^1 = \frac{\pi}{6}.
$$

565. Dividing by n and moving the negative term to the right, we can turn the left side inequality into

$$
0.785 < \frac{1}{n} + \frac{1}{n}\sqrt{1-\left(\frac{1}{n}\right)^2} + \frac{1}{n}\sqrt{1-\left(\frac{2}{n}\right)^2} + \cdots + \frac{1}{n}\sqrt{-\left(\frac{n-1}{n}\right)^2},
$$

that is

$$
0.785 < \frac{1}{n}\sum_{k=0}^{n-1}\sqrt{1-\left(\frac{k}{n}\right)^2}.
$$

On the right we have an upper Darboux sum of $\sqrt{1-x^2}$ (i.e. the Riemann sum with the points ξ_k being the maxima of f on the corresponding intervals) on the interval $[0, 1]$. Since

$$\int_0^1 \sqrt{1-x^2} = \frac{\pi}{4} > .785,$$

the inequality on the left is proved.

Write the inequality on the right as

$$\frac{1}{n}\sum_{k=1}^n \sqrt{1-\left(\frac{k}{n}\right)^2} < .79.$$

The term on the left is a lower Darboux sum of $\sqrt{1-x^2}$ on $[0, 1]$, so it is less than the integral of this function which is $\frac{\pi}{4}$. The inequality follows from the fact that $\pi/4 < .79$.

(*Kvant (Quantum)*)

566. Rewrite the formula for the term of the sequence as

$$x_n = \sum_{k=1}^n \frac{\frac{k}{n}}{1+2\left(\frac{k}{n}\right)^2} \cdot \frac{1}{n}.$$

We recognize a Riemann sum for the integral

$$\int_0^1 \frac{x}{1+2x^2}dx.$$

Hence the sequence converges to the value of this integral, which is $\frac{1}{4}\ln 2$.

(Konhauser Problem Fest, 2014, proposed by R. Gelca)

567. Write the inequality as

$$\frac{1}{n}\sum_{i=1}^n \frac{1}{\sqrt{2\frac{i}{n}+5}} < \sqrt{7}-\sqrt{5}.$$

The left-hand side is the Riemann sum of the strictly decreasing function $f(x) = \frac{1}{\sqrt{2x+5}}$. This Riemann sum is computed at the right ends of the intervals of the subdivision of $[0, 1]$ by the points $\frac{i}{n}$, $i = 1, 2, \ldots, n-1$. It follows that

$$\frac{1}{n}\sum_{i=1}^n \frac{1}{\sqrt{2\frac{i}{n}+5}} < \int_0^1 \frac{1}{\sqrt{2x+5}}dx = \sqrt{2x+5}\Big|_0^1 = \sqrt{7}-\sqrt{5},$$

the desired inequality.

(Communicated by E. Craina)

568. We would like to recognize the general term of the sequence as being a Riemann sum. This, however, does not seem to happen, since we can only write

$$\sum_{i=}^{n} \frac{2^{i/n}}{n + \frac{1}{i}} = \frac{1}{n} \sum_{i=1}^{n} \frac{2^{i/n}}{1 + \frac{1}{ni}}.$$

But for $i \geq 2$,

$$2^{i/n} > \frac{2^{i/n}}{1 + \frac{1}{ni}},$$

and, using the inequality $e^x > 1 + x$,

$$\frac{2^{i/n}}{1 + \frac{1}{ni}} = 2^{(i-1)/n} \frac{2^{1/n}}{1 + \frac{1}{ni}} = 2^{(i-1)/n} \frac{e^{\ln 2/n}}{1 + \frac{1}{ni}} > 2^{(i-1)/n} \frac{1 + \frac{\ln 2}{n}}{1 + \frac{1}{ni}} > 2^{(i-1)/n},$$

for $i \geq 2$. By the intermediate value property, for each $i \geq 2$ there exists $\xi_i \in \left[\frac{i-1}{n}, \frac{i}{n}\right]$ such that

$$\frac{2^{i/n}}{1 + \frac{1}{ni}} = 2^{\xi_i}.$$

Of course, the term corresponding to $i = 1$ can be neglected when n is large. Now we see that our limit is indeed a Riemann sum of the function 2^x integrated over the interval $[0, 1]$. We obtain

$$\lim_{n \to \infty} \left(\frac{2^{1/n}}{n+1} + \frac{2^{2/n}}{n + \frac{1}{2}} + \cdots + \frac{2^{n/n}}{n + \frac{1}{n}} \right) = \int_0^1 2^x dx = \frac{1}{\ln 2}.$$

(Soviet Union University Student Mathematical Olympiad, 1976)

569. This is an example of an integral that is computed using Riemann sums. Divide the interval $[0, \pi]$ into n equal parts and consider the Riemann sum

$$\frac{\pi}{n} \left[\ln \left(a^2 - 2a \cos \frac{\pi}{n} + 1 \right) + \ln \left(a^2 - 2a \cos \frac{2\pi}{n} + 1 \right) + \cdots \right.$$
$$\left. + \ln \left(a^2 - 2a \cos \frac{(n-1)\pi}{n} + 1 \right) \right].$$

This expression can be written as

$$\frac{\pi}{n} \ln \left[\left(a^2 - 2a \cos \frac{\pi}{n} + 1 \right) \left(a^2 - 2a \cos \frac{2\pi}{n} + 1 \right) \cdots \left(a^2 - 2a \cos \frac{(n-1)\pi}{n} + 1 \right) \right].$$

The product inside the natural logarithm factors as

$$\prod_{k=1}^{n-1} \left[a - \left(\cos \frac{k\pi}{n} + i \sin \frac{k\pi}{n} \right) \right] \left[a - \left(\cos \frac{k\pi}{n} - i \sin \frac{k\pi}{n} \right) \right].$$

These are exactly the factors in $a^{2n} - 1$, except for $a - 1$ and $a + 1$. The Riemann sum is therefore equal to

$$\frac{\pi}{n} \ln \frac{a^{2n} - 1}{a^2 - 1}.$$

We are left to compute the limit of this expression as n goes to infinity. If $|a| \le 1$, this limit is equal to 0. If $|a| > 1$, the limit is

$$\lim_{n \to \infty} \pi \ln \sqrt[n]{\frac{a^{2n} - 1}{a^2 - 1}} = 2\pi \ln |a|.$$

Try to prove this last limit!
    (S.D. Poisson)

570. The condition $f(x)f(2x) \cdots f(nx) \le an^k$ can be written equivalently as

$$\sum_{j=1}^{n} \ln f(jx) < \ln a + k \ln n, \quad \text{for all } x \in \mathbb{R}, \ n \ge 1.$$

Taking $\alpha > 0$ and $x = \frac{\alpha}{n}$, we obtain

$$\sum_{j=1}^{n} \ln f \left(\frac{j\alpha}{n} \right) \le \ln a + k \ln n,$$

or

$$\sum_{j=1}^{n} \frac{\alpha}{n} \ln f \left(\frac{j\alpha}{n} \right) \le \frac{\alpha \ln a + k\alpha \ln n}{n}.$$

The left-hand side is a Riemann sum for the function $\ln f$ on the interval $[0, \alpha]$. Because f is continuous, so is $\ln f$, and thus $\ln f$ is integrable. Letting n tend to infinity, we obtain

$$\int_0^1 \ln f(x) dx \le \lim_{n \to \infty} \frac{\alpha \ln a + k\alpha \ln n}{n} = 0.$$

The fact that $f(x) \ge 1$ implies that $\ln f(x) \ge 0$ for all x. Hence $\ln f(x) = 0$ for all $x \in [0, \alpha]$. Since α is an arbitrary positive number, $f(x) = 1$ for all $x \ge 0$. A similar argument yields $f(x) = 1$ for $x < 0$. So there is only one such function, the constant function equal to 1.
    (Romanian Mathematical Olympiad, 1999, proposed by R. Gologan)

571. The relation from the statement can be rewritten as

$$\int_0^1 (xf(x) - f(x)^2) dx = \int_0^1 \frac{x^2}{4} dx.$$

Moving everything to one side, we obtain

$$\int_0^1 \left(f(x)^2 - xf(x) + \frac{x^2}{4} \right) dx = 0.$$

We now recognize a perfect square and write this as

$$\int_0^1 \left(f(x) - \frac{x}{2} \right)^2 dx = 0.$$

The integral of the nonnegative continuous function $\left(f(x) - \frac{x}{2} \right)^2$ is strictly positive, unless the function is identically equal to zero. It follows that the only function satisfying the condition from the statement is $f(x) = \frac{x}{2}$, $x \in [0, 1]$.

(*Revista de Matematică din Timişoara* (*Timişoara Mathematics Gazette*), proposed by T. Andreescu)

572. Performing the substitution $x^{\frac{1}{k}} = t$, the given conditions become

$$\int_0^1 (f(t))^{n-k} t^{k-1} dt = \frac{1}{n}, \quad k = 1, 2, \ldots, n-1.$$

Observe that this equality also holds for $k = n$. With this in mind we write

$$\int_0^1 (f(t) - t)^{n-1} dt = \int_0^1 \sum_{k=0}^{n-1} \binom{n-1}{k} (-1)^k (f(t))^{n-1-k} t^k dt$$

$$= \int_0^1 \sum_{k=1}^{n} \binom{n-1}{k-1} (-1)^{k-1} (f(t))^{n-k} t^{k-1} dt$$

$$= \sum_{k=1}^{n} (-1)^{k-1} \binom{n-1}{k-1} \int_0^1 (f(t))^{n-k} t^{k-1} dt$$

$$= \sum_{k=1}^{n} (-1)^{k-1} \binom{n-1}{k-1} \frac{1}{n} = \frac{1}{n}(1 - 1)^{n-1} = 0.$$

Because $n - 1$ is even, $(f(t) - t)^{n-1} \geq 0$. The integral of this function can be zero only if $f(t) - t = 0$ for all $t \in [0, 1]$. Hence the only solution to the problem is $f : [0, 1] \to \mathbb{R}$, $f(x) = x$.

(Romanian Mathematical Olympiad, 2002, proposed by T. Andreescu)

573. Note that the linear function $g(x) = 6x - 2$ satisfies the same conditions as f. Therefore,

$$\int_0^1 (f(x) - g(x))dx = \int_0^1 x(f(x) - g(x))dx = 0.$$

Considering the appropriate linear combination of the two integrals, we obtain

$$\int_0^1 p(x)(f(x) - g(x))dx = 0.$$

We have

$$0 \leq \int_0^1 (f(x) - g(x))^2 dx = \int_0^1 f(x)(f(x) - g(x)) dx - \int_0^1 g(x)(f(x) - g(x)) dx$$

$$= \int_0^1 f^2(x) - f(x)g(x) dx = \int_0^1 f^2(x) dx - 6 \int_0^1 x f(x) dx + 2 \int_0^1 f(x) dx$$

$$= \int_0^1 f^2(x) dx - 4.$$

Here we used the fact that

$$\int_0^1 g(x)(f(x) - g(x)) dx = 6 \int_0^1 x(f(x) - g(x)) dx - 2 \int_0^1 (f(x) - g(x)) dx = 0.$$

The inequality is proved.

(Romanian Mathematical Olympiad, 2004, proposed by I. Raşa)

574. We change this into a minimum problem, and then relate the latter to an inequality of the form $x \geq 0$. Completing the square, we see that

$$xf(x)^2 - x^2 f(x) = \sqrt{x}f(x)^2 - 2\sqrt{x}f(x)\frac{x^{\frac{3}{2}}}{2} = \left(\sqrt{x}f(x) - \frac{x^{\frac{3}{2}}}{2}\right)^2 - \frac{x^3}{4}.$$

Hence, indeed,

$$J(f) - I(f) = \int_0^1 \left(\sqrt{x}f(x) - \frac{x^{\frac{3}{2}}}{2}\right)^2 dx - \int_0^1 \frac{x^3}{4} dx \geq -\frac{1}{16}.$$

It follows that $I(f) - J(f) \leq \frac{1}{16}$ for all f. The equality holds, for example, for $f : [0, 1] \to \mathbb{R}$, $f(x) = \frac{x}{2}$. We conclude that

$$\max_f (I(f) - J(f)) = \frac{1}{16}.$$

(49th W.L. Putnam Mathematical Competition, 2006, proposed by T. Andreescu)

575. We can write the inequality as

$$\sum_{i,j} x_i x_j (a_i + a_j - 2\min(a_i, a_j)) \leq 0.$$

Note that

$$\sum_{i,j} x_i x_j a_i = x_j \sum_{i=1}^n a_i x_i = 0,$$

and the same stays true if we exchange i with j. So it remains to prove that

$$\sum_{i,j} x_i x_j \min(a_i, a_j) \geq 0.$$

If $\chi_{[0,a_i]}$ is the characteristic function of the interval $[0, a_i]$ (equal to 1 on the interval and to 0 outside), then our inequality is, in fact,

$$\int_0^\infty \left(\sum_{i=1}^n x_i \chi_{[0,a_i]}(t) \right)^2 dt \geq 0,$$

which is obvious. Equality holds if and only if $\sum_{i=1}^n x_i \chi_{[0,a_i]} = 0$ everywhere except at finitely many points. It is not hard to see that this is equivalent to the condition from the statement.

(G. Dospinescu)

576. This is just the Cauchy-Schwarz inequality applied to the functions f and g, with $g(t) = 1$ for $t \in [0, 1]$.

577. By Hölder's inequality,

$$\int_0^3 f(x) \cdot 1 \, dx \leq \left(\int_0^3 |f(x)|^3 dx \right)^{\frac{1}{3}} \left(\int_0^3 1^{\frac{3}{2}} dx \right)^{\frac{2}{3}} = 3^{\frac{2}{3}} \left(\int_0^3 |f(x)|^3 dx \right)^{\frac{1}{3}}.$$

Raising everything to the third power and using the fact that f is positive, we obtain

$$\left(\int_0^3 f(x) dx \right)^3 / \int_0^3 f(x)^3 dx \leq 9.$$

To see that the maximum 9 can be achieved, choose f to be constant.

578. The argument relies on Figure 75. The left-hand side is the area of the shaded region (composed of the subgraph of f and the subgraph of f^{-1}). The product ab is the area of the rectangle $[0, a] \times [0, b]$, which is contained inside the shaded region. Equality holds if and only if the two regions coincide, which is the case exactly when $b = f(a)$.

(Young's inequality)

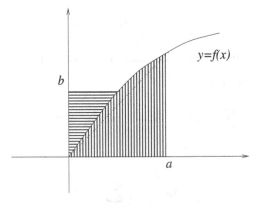

Figure 75

579. Suppose that $x > y$. Transform the inequality successively into

$$mn(x - y)(x^{m+n-1} - y^{m+n-1}) \geq (m + n - 1)(x^m - y^m)(x^n - y^n),$$

and then

$$\frac{x^{m+n-1} - y^{m+n-1}}{(m+n-1)(x-y)} \geq \frac{x^m - y^m}{m(x-y)} \cdot \frac{x^n - y^n}{n(x-y)}.$$

The last one can be written as

$$(x-y) \int_y^x t^{m+n-2} dt \geq \int_y^x t^{m-1} dt \cdot \int_y^x t^{n-1} dt.$$

Here we recognize Chebyshev's inequality applied to the integrals of the functions f, g : $[y, x] \to \mathbb{R}, f(t) = t^{m-1}$ and $g(t) = t^{n-1}$.

(Austrian-Polish Competition, 1995)

580. Observe that f being monotonic, it is automatically Riemann integrable. Taking the mean of f on the intervals $[0, \alpha]$ and $[1 - \alpha, 1]$ and using the monotonicity of the function, we obtain

$$\frac{1}{1-\alpha} \int_\alpha^1 f(x)dx \leq \frac{1}{\alpha} \int_0^\alpha f(x)dx,$$

whence

$$\alpha \int_\alpha^1 f(x)dx \leq (1-\alpha) \int_0^\sigma f(x)dx.$$

Adding $\int_0^\alpha f(x)dx$ to both sides gives

$$\alpha \int_0^1 f(x)dx \leq \int_0^\alpha f(x)dx,$$

as desired.

(Soviet Union University Student Mathematical Olympiad, 1976)

581. For $x \in [0, 1]$, we have $f'(x) \leq f'(1)$, and so

$$\frac{f'(1)}{f(x)^2 + 1} \leq \frac{f'(x)}{f(x)^2 + 1}.$$

Integrating, we obtain

$$f'(1) \int_0^1 \frac{dx}{f(x)^2 + 1} \leq \int_0^1 \frac{f'(x)}{f(x)^2 + 1} = \arctan f(1) - \arctan f(0) = \arctan f(1).$$

Because $f'(1) > 0$ and $\arctan y \leq y$ for $y \geq 0$ (since this is equivalent to $y \leq \tan y$), we further obtain

$$\int_0^1 \frac{dx}{f(x)^2 + 1} \leq \frac{\arctan f(1)}{f'(1)} \leq \frac{f(1)}{f'(1)},$$

proving the inequality. In order for equality to hold we must have $\arctan f(1) = f(1)$, which happens only when $f(1) = 0$. Then $\int_0^1 \frac{dx}{f(x)^2 + 1} = 0$. But this cannot be true since the function that is integrated is strictly positive. It follows that the inequality is strict. This completes the solution.

(Romanian Mathematical Olympiad, 1978, proposed by R. Gologan)

582. The Leibniz-Newton fundamental theorem of calculus gives

$$f(x) = \int_a^x f'(t)dt.$$

Squaring both sides and applying the Cauchy-Schwarz inequality, we obtain

$$f(x)^2 = \left(\int_a^b f'(t)dt \right)^2 \le (b-a) \int_a^b f'(t)^2 dt.$$

The right-hand side is a constant, while the left-hand side depends on x. Integrating the inequality with respect to x yields

$$\int_a^b f(x)^2 dx \le (b-a)^2 \int_a^b f'(t)^2 dt.$$

Substitute t by x to obtain the inequality as written in the statement of the problem.

583. This is an example of a problem in which it is important to know how to organize the data. We start by letting A be the subset of $[0, 1]$ on which f is nonnegative, and B its complement. Let $m(A)$, respectively, $m(B)$ be the lengths (measures) of these sets, and I_A and I_B the integrals of $|f|$ on A, respectively, B. Without loss of generality, we can assume $m(A) \ge \frac{1}{2}$; otherwise, change f to $-f$.

We have

$$\int_0^1 \int_0^1 |f(x) + f(y)| dx dy = \int_A \int_A (f(x) + f(y)) dx dy + \int_B \int_B (|f(x)| + |f(y)|) dx dy$$
$$+ 2 \int_A \int_B |f(x) + f(y)| dx dy.$$

Let us first try a raw estimate by neglecting the last term. In this case we would have to prove

$$2m(A)I_A + 2m(B)I_B \ge I_A + I_B.$$

Since $m(A) + m(B) = 1$, this inequality translates into

$$\left(m(A) - \frac{1}{2} \right) (I_A - I_B) \ge 0,$$

which would be true if $I_A \ge I_B$. However, if this last assumption does not hold, we can return to the term that we neglected, and use the triangle inequality to obtain

$$\int_A \int_B |f(x) + f(y)| dx dy \ge \int_A \int_B |f(x)| - |f(y)| dx dy = m(A)I_B - m(B)I_A.$$

The inequality from the statement would then follow from

$$2m(A)I_A + 2m(B)I_B + 2m(A)I_B - 2m(B)I_A \ge I_A + I_B,$$

which is equivalent to

$$\left(m(A) - \frac{1}{2}\right)(I_A + I_B) + m(B)(I_B - I_A) \geq 0.$$

This is true since both terms are positive.

(64th W.L. Putnam Mathematical Competition, 2003)

584. We have

$$\int_a^b f(x)dx = \int_a^{\frac{a+b}{2}} \left(f(x) + f\left(x + \frac{b-a}{2}\right)\right)dx.$$

Using Jensen's inequality for the function f, we see that this is greater than or equal than

$$2\int_a^{\frac{a+b}{2}} f\left(x + \frac{b-a}{4}\right)dx = 2\int_{\frac{3a+b}{4}}^{\frac{3b+a}{4}} f(x)dx.$$

This proves the inequality on the left. For the inequality on the right, we use the integral form of Jensen's inequality:

Jensen's inequality. If $f : [a, b] \to \mathbb{R}$ and $g : [c, d] \to [a, b]$ are two functions, with f being convex, then

$$f\left(\frac{1}{b-a}\int_a^b g(x)dx\right) \leq \frac{1}{b-a}\int_a^b f(g(x))dx.$$

Applying this inequality we can write

$$\begin{aligned}
2\int_{\frac{3a+b}{4}}^{\frac{3b+a}{4}} f(x)dx &\geq 2\frac{b-a}{2}f\left(\frac{2}{b-a}\int_{\frac{3a+b}{4}}^{\frac{3b+a}{4}} xdx\right) \\
&= (b-a)f\left(\frac{1}{b-a}\left[\left(\frac{3b+a}{4}\right)^2 - \left(\frac{3a+b}{4}\right)^2\right]\right) \\
&= (b-a)f\left(\frac{1}{b-a} \cdot \frac{b-a}{2} \cdot (b+a)\right) \\
&= (b-a)f\left(\frac{a+b}{2}\right).
\end{aligned}$$

The problem is solved.

(*Mathematical Reflections*, proposed by C. Lupu)

585. Combining the Taylor series expansions

$$\cos x = 1 - \frac{x^2}{2!} + \frac{x^4}{4!} - \frac{x^6}{6!} + \frac{x^8}{8!} + \cdots,$$

$$\cosh x = 1 + \frac{x^2}{2!} + \frac{x^4}{4!} + \frac{x^6}{6!} + \frac{x^8}{8!} + \cdots,$$

we see that the given series is the Taylor series of $\frac{1}{2}(\cos x + \cosh x)$.

(The *Mathematics Gazette* Competition, Bucharest, 1935)

586. Denote by p the numerator and by q the denominator of this fraction. Recall the Taylor series expansion of the sine function,

$$\sin x = \frac{x}{1!} - \frac{x^3}{3!} + \frac{x^5}{5!} - \frac{x^7}{7!} + \frac{x^9}{9!} + \cdots$$

We recognize the denominators of these fractions inside the expression that we are computing, and now it is not hard to see that $p\pi - q\pi^3 = \sin \pi = 0$. Hence $p\pi = q\pi^3$, and the value of the expression from the statement is π^2.

(Soviet Union University Student Mathematical Olympiad, 1975)

587. Consider the series expansion

$$\frac{1}{1+x^2} = 1 - x^2 + x^4 - x^6 + x^8 - \cdots,$$

which converges uniformly on any interval of the form $[-a, a]$ with $0 < a < 1$. We can integrate both sides, and obtain the Taylor series expansion for the arctangent:

$$\arctan x = x - \frac{1}{3}x^3 + \frac{1}{5}x^5 - \frac{1}{7}x^7 + \cdots,$$

for $x \in (-1, 1)$. Substituting $x = \frac{1}{\sqrt{3}}$, we obtain that the value of the series from the statement is $\arctan \frac{1}{\sqrt{3}} = \frac{\pi}{6}$.

Remark. The series of the arctangent also converges for $x = 1$ (but not for $x = -1$), giving another proof the Leibniz formula for $\pi/4$, which was proved in Section 3.2.9.

(Communicated by J. Staff)

588. Expand the cosine in a Taylor series,

$$\cos ax = 1 - \frac{(ax)^2}{2!} + \frac{(ax)^4}{4!} - \frac{(ax)^6}{6!} + \cdots$$

Let us forget for a moment the coefficient $\frac{(-1)^n a^{2n}}{(2n)!}$ and understand how to compute

$$\int_{-\infty}^{\infty} e^{-x^2} x^{2n} dx.$$

If we denote this integral by I_n, then integration by parts yields the recursive formula

$$I_n = \frac{2n-1}{2} I_{n-1}.$$

Starting with

$$I_0 = \int_{-\infty}^{\infty} e^{-x^2} dx = \sqrt{\pi},$$

we obtain

$$I_n = \frac{(2n)!\sqrt{\pi}}{4^n n!}.$$

It follows that the integral in question is equal to

$$\sum_{n=0}^{\infty}(-1)^n\frac{a^{2n}}{(2n)!}\cdot\frac{(2n)!\sqrt{\pi}}{4^n n!} = \sqrt{\pi}\sum_{n=0}^{\infty}\frac{\left(-\frac{a^2}{4}\right)^n}{n!},$$

and this is clearly equal to $\sqrt{\pi}e^{-a^2/4}$.

One thing remains to be explained: why are we allowed to perform the expansion and then the summation of the integrals? This is because the series that consists of the integrals of the absolute values of the terms converges itself. Indeed,

$$\sum_{n=1}^{\infty}\frac{a^{2n}}{(2n)!}\int_{-\infty}^{\infty}e^{-x^2}x^{2n}dx = \sqrt{\pi}\sum_{1}^{\infty}\frac{\left(\frac{a^2}{4}\right)^n}{n!} = \sqrt{\pi}e^{a^2/4} < \infty.$$

With this the problem is solved.

(G.B. Folland, *Real Analysis, Modern Techniques and Their Applications*, Wiley, 1999)

589. Consider the Taylor series expansion around 0,

$$\frac{1}{x-4} = -\frac{1}{4} - \frac{1}{16}x - \frac{1}{64}x^2 - \frac{1}{256}x^3 - \cdots$$

A good guess is to truncate this at the third term and let

$$P(x) = \frac{1}{4} + \frac{1}{16}x + \frac{1}{64}x^2.$$

By the residue formula for Taylor series we have

$$\left|P(x)+\frac{1}{x-4}\right| = \frac{x^3}{256} + \frac{1}{(\xi-4)^4}x^5,$$

for some $\xi \in (0, x)$. Since $|x| \le 1$ and also $|\xi| \le 1$, we have $\frac{x^3}{256} \le \frac{1}{256}$ and $x^4/(\xi-4)^5 \le \frac{1}{243}$. An easy numerical computation shows that $\frac{1}{256} + \frac{1}{243} < \frac{1}{100}$, and we are done.

(Romanian Team Selection Test for the International Mathematical Olympiad, 1979, proposed by O. Stănăşilă)

590. By Taylor's formula, one can write

$$\sin t = t - \frac{\sin(\theta t)}{2}t^2,$$

for some $\theta = \theta(t) \in (0, 1)$. In particular, for $t = \frac{1}{x}$, we have

$$\sin\frac{1}{x} = \frac{1}{x} - \frac{\sin\frac{\theta}{x}}{2x^2},$$

so

$$x^2 \sin \frac{1}{x} = x - \frac{1}{2} \sin \frac{\theta}{x}.$$

By substituting this into the original equation, we find that any solution should satisfy

$$x - \frac{1}{2} \sin \frac{\theta}{x} = 2x - 1977$$

or

$$x = 1977 - \frac{1}{2} \sin \frac{\theta}{x}.$$

From here we deduce that $x > 1976$, and so $\frac{\theta}{x} < \frac{1}{1976}$. It follows that $x = 1977 + \varepsilon$, where

$$|\varepsilon| = \frac{1}{2} \sin \frac{\theta}{x} < \frac{1}{2} \sin \frac{1}{1976} < \frac{1}{2 \cdot 1976} < 0.001.$$

It follows that $x = 1977$ with an error less than 0.01.

   (V.A. Sadovnichii, A.S. Podkolzin, *Problems of the University Students Mathematical Olympiad*, Nauka, Moscow, 1978)

591. The Taylor series expansion of $\cos \sqrt{x}$ around 0 is

$$\cos \sqrt{x} = 1 - \frac{x}{2!} + \frac{x^2}{4!} - \frac{x^3}{6!} + \frac{x^4}{8!} - \cdots$$

Integrating term by term, we obtain

$$\int_0^1 \cos \sqrt{x}\, dx = \sum_{n=1}^{\infty} \frac{(-1)^{n-1} x^n}{(n+1)(2n)!} \Big|_0^1 = \sum_{n=0}^{\infty} \frac{(-1)^{n-1}}{(n+1)(2n)!}.$$

Grouping consecutive terms we see that

$$\left(\frac{1}{5 \cdot 8!} - \frac{1}{6 \cdot 10!} \right) + \left(\frac{1}{7 \cdot 12!} - \frac{1}{8 \cdot 14!} \right) + \cdots < \frac{1}{2 \cdot 10^4} + \frac{1}{2 \cdot 10^5} + \frac{1}{2 \cdot 10^6} + \cdots < \frac{1}{10^4}.$$

Also, truncating to the fourth decimal place yields

$$0.7638 < 1 - \frac{1}{4} + \frac{1}{72} - \frac{1}{2880} < 0.7639.$$

We conclude that

$$\int_0^1 \cos \sqrt{x}\, dx \approx 0.763.$$

592. Consider the Newton binomial expansion

$$(x+1)^{-\frac{1}{2}} = \sum_{k=0}^{\infty} \binom{-\frac{1}{2}}{x} x^k$$

$$= \sum_{k=0}^{\infty} \frac{\left(-\frac{1}{2}\right)\left(-\frac{1}{2}-1\right)\left(-\frac{1}{2}-2\right)\cdots\left(-\frac{1}{2}-k+1\right)}{k!} x^k$$

$$= \sum_{k=0}^{\infty} (-1)^k \frac{1\cdot 3\cdots(2k-1)}{2^k\cdot k!} x^k = \sum_{k=0}^{\infty} (-1)^k \frac{(2k)!}{2^{2k}\cdot k!\cdot k!} x^k$$

$$= \sum_{k=0}^{\infty} (-1)^k \frac{1}{2^{2k}} \binom{2k}{x} x^k.$$

Replacing x by $-x^2$ then taking antiderivatives, we obtain

$$\arcsin x = \int_0^x (1-t^2)^{-\frac{1}{2}} dt = \sum_{k=0}^{\infty} \frac{1}{2^{2k}} \binom{2k}{k} \int_0^x t^{2k} dt$$

$$= \sum_{k=0}^{\infty} \frac{1}{2^{2k}(2k+1)} \binom{2k}{k} x^{2k+1},$$

as desired.

593. (a) Differentiating the identity from the second example from the introduction, we obtain

$$\frac{2\arcsin x}{\sqrt{1-x^2}} = \sum_{k\geq 1} \frac{1}{k\binom{2k}{k}} 2^{2k} x^{2k-1},$$

whence

$$\frac{x\arcsin x}{\sqrt{1-x^2}} = \sum_{k\geq 1} \frac{1}{k\binom{2k}{k}} 2^{2k-1} x^{2k}.$$

Differentiating both sides and multiplying by x, we obtain

$$x\frac{\arcsin x + x\sqrt{1-x^2}}{(1-x^2)^{3/2}} = \sum_{k\geq 0} \frac{1}{\binom{2k}{k}} 2^{2k} x^{2k}.$$

Substituting $\frac{x}{2}$ for x, we obtain the desired identity.

Part (b) follows from (a) if we let $x=1$.

(S. Rădulescu, M. Rădulescu, *Teoreme şi Probleme de Analiză Matematică* (*Theorems and Problems in Mathematical Analysis*), Editura Didactică şi Pedagogică, Bucharest, 1982)

594. Consider the function f of period 2π defined by $f(x) = x$ if $0 \leq x < 2\pi$. This function is continuous on $(0, 2\pi)$, so its Fourier series converges (pointwise) on this interval. We compute

$$a_0 = \frac{1}{2\pi} \int_0^{2\pi} x dx = \pi, \ a_m = 0, \ \text{for } m \geq 1,$$

$$b_m = \frac{1}{\pi} \int_0^{2\pi} x \sin mx dx = -\frac{x \cos mx}{m\pi}\Big|_0^{2\pi} + \frac{1}{m\pi}\int_0^{2\pi} \cos mx dx = -\frac{2}{m}, \text{ for } m \geq 1.$$

Therefore,

$$x = \pi - \frac{2}{1}\sin x - \frac{2}{2}\sin 2x - \frac{2}{3}\sin 3x - \cdots$$

Divide this by 2 to obtain the identity from the statement. Substituting $x = \frac{\pi}{2}$, we obtain the Leibniz series (see Section 3.2.9)

$$\frac{\pi}{4} = 1 - \frac{1}{3} + \frac{1}{5} - \frac{1}{7} + \cdots$$

In the series

$$\frac{\pi - x}{2} = \sum_{n=1}^{\infty} \frac{\sin nx}{n},$$

replace x by $2x$, and then divide by 2 to obtain

$$\frac{\pi}{4} - \frac{x}{2} = \sum_{k=1}^{\infty} \frac{\sin 2kx}{2k}, \quad x \in (0, \pi).$$

Subtracting this from the original formula, we obtain

$$\frac{\pi}{4} = \sum_{k=1}^{\infty} \frac{\sin(2k-1)x}{2k-1}, \quad x \in (0, \pi).$$

595. One computes

$$\int_0^1 f(x)dx = 0,$$

$$\int_0^1 f(x)\cos 2\pi nx dx = 0, \quad \text{for all } n \geq 1,$$

$$\int_0^1 f(x)\sin 2\pi nx dx = \frac{1}{2\pi k}, \quad \text{for all } n \geq 1.$$

Recall that for a general Fourier expansion

$$f(x) = a_0 + \sum_{n=1}^{\infty} \left(a_n \cos \frac{2\pi}{T}nx + b_n \sin \frac{2\pi}{T}nx \right),$$

one has

Parseval's identity.

$$\frac{1}{T}\int_0^T |f(x)|^2 dx = a_0^2 + 2\sum_{n=1}^{\infty}(a_n^2 + b_n^2).$$

Geometrically, Parseval's identity is just the Pythagorean theorem in the infinite dimensional Hilbert space of square integrable functions. Our particular function has the Fourier series expansion

$$f(x) = \frac{1}{2\pi} \sum_{n=-\infty}^{\infty} \frac{1}{n} \cos 2\pi nx,$$

and in this case Parseval's identity reads

$$\int_0^1 |f(x)|^2 dx = \frac{1}{2\pi^2} \sum_{n=1}^{\infty} \frac{1}{n^2}.$$

The left-hand side is $\int_0^1 |f(x)|^2 dx = \frac{1}{12}$, and the formula follows.

596. This problem uses the Fourier series expansion of $f(x) = |x|$, $x \in [-\pi, \pi]$. A routine computation yields

$$|x| = \frac{\pi}{2} - \frac{4}{\pi} \sum_{k=0}^{\infty} \frac{\cos(2k+1)x}{(2k+1)^2}, \quad \text{for } x \in [-\pi, \pi].$$

Setting $x = 0$, we obtain the identity from the statement.

597. We will use only trigonometric considerations, and compute no integrals. A first remark is that the function is even, so only terms involving cosines will appear. Using Euler's formula

$$e^{i\alpha} = \cos\alpha + i\sin\alpha$$

we can transform the identity

$$\sum_{k=1}^{n} e^{2ikx} = \frac{e^{2i(n+1)x} - 1}{e^{2ix} - 1}$$

into the corresponding identities for the real and imaginary parts:

$$\cos 2x + \cos 4x + \cdots + \cos 2nx = \frac{\sin nx \cos(n+1)x}{\sin x},$$

$$\sin 2x + \sin 4x + \cdots + \sin 2nx = \frac{\sin nx \sin(n+1)x}{\sin x}.$$

These two relate to our function as

$$\frac{\sin^2 nx}{\sin^2 x} = \left(\frac{\sin nx \cos(n+1)x}{\sin x}\right)^2 + \left(\frac{\sin nx \sin(n+1)x}{\sin x}\right)^2,$$

which allows us to write the function as an expression with no fractions:

$$f(x) = (\cos 2x + \cos 4x + \cdots + \cos 2nx)^2 + (\sin 2x + \sin 4x + \cdots + \sin 2nx)^2.$$

Expanding the squares, we obtain

$$f(x) = n + \sum_{1 \leq l < k \leq n} (2 \sin 2lx \sin 2kx + 2 \cos 2lx \cos 2kx)$$

$$= n + 2 \sum_{1 \leq l < k \leq n} \cos 2(k - l)x = n + \sum_{m=1}^{n-1} 2(n - m) \cos 2mx.$$

In conclusion, the nonzero Fourier coefficients of f are

$$a_{2m} = 2(n - m), \quad m = 1, 2, \ldots, n - 1.$$

(D. Andrica)

598. Expand the function f as a Fourier series

$$f(x) = \sum_{n=1}^{\infty} a_n \sin nx,$$

where

$$a_n = \frac{2}{\pi} \int_0^{\pi} f(t) \sin ntdt.$$

This is possible, for example, since f can be extended to an odd function on $[-\pi, \pi]$.

Fix $n \geq 2$, and consider the function $g : [0, \pi] \to \mathbb{R}$, $g(x) = n \sin x - \sin nx$. The function g is nonnegative because of the inequality $n|\sin x| \geq |\sin nx|$, $x \in \mathbb{R}$, which was proved in the section on induction.

Integrating repeatedly by parts and using the hypothesis, we obtain

$$(-1)^m \int_0^{\pi} f^{(2m)}(t) \sin ntdt = n^{2m} a_n \frac{\pi}{2}, \quad \text{for } m \geq 0.$$

It follows that

$$(-1)^m \int_0^{\pi} f^{(2m)}(x)(n \sin x - \sin nx)dx = (na_1 - n^{2m} a_n)\frac{\pi}{2} \geq 0.$$

Indeed, the first term is the integral of a product of two nonnegative functions. This must hold for any integer m; hence $a_n \leq 0$ for any $n \geq 2$.

Similarly

$$(-1)^m \int_0^{\pi} f^{(2m)}(x)(n \sin x + sinnx)dx = (na_1 - n^{2m} a_n)\frac{\pi}{2} \geq 0.$$

This implies that $a_n \geq 0$, for $n \geq 2$. We deduce that $a_n = 0$ for $n \geq 2$, and so $f(x) = a_1 \sin x$, for $x \in [0, \pi]$.

(S. Rădulescu, M. Rădulescu, *Teoreme şi Probleme de Analiză Matematică* (*Theorems and Problems in Mathematical Analysis*), Editura Didactică şi Pedagogică, Bucharest, 1982)

599. This is an exercise in the product and chain rules. We compute

$$\frac{\partial v}{\partial t}(x, t) = \frac{\partial}{\partial t}\left(t^{-\frac{1}{2}}e^{-\frac{x^2}{4t}}u(xt^{-1}, -t^{-1}) \right)$$

$$= -\frac{1}{2}t^{-\frac{3}{2}}e^{-\frac{x^2}{4t}}v(x, t) + \frac{x^2 t^{-\frac{5}{2}}}{4}e^{-\frac{x^2}{4t}}v(x, t) - xt^{-\frac{5}{2}}e^{-\frac{x^2}{4t}}\frac{\partial u}{\partial x}(xt^{-1}, -t^{-1})$$

$$+ t^{-\frac{5}{2}}e^{-\frac{x^2}{4t}}\frac{\partial u}{\partial t}(xt^{-1}, -t^{-1}),$$

then

$$\frac{\partial v}{\partial x}(x, t) = t^{-\frac{1}{2}}e^{-\frac{x^2}{4t}}\left(-\frac{1}{2}t^{-1}x \right)u(xt^{-1}, -t^{-1}) + t^{-\frac{3}{2}}e^{-\frac{x^2}{4t}}\frac{\partial u}{\partial x}(xt^{-1}, -t^{-1})$$

and

$$\frac{\partial^2 v}{\partial x^2}(x, t) = \frac{1}{4}x^2 t^{-\frac{5}{2}}e^{-\frac{x^2}{4t}}v(x, t) - \frac{1}{2}t^{-\frac{3}{2}}e^{-\frac{x^2}{4t}}v(x, t) - \frac{1}{2}xt^{-\frac{5}{2}}e^{-\frac{x^2}{4t}}\frac{\partial u}{\partial x}(xt^{-1}, -t^{-1})$$

$$- \frac{1}{2}xt^{-\frac{5}{2}}e^{-\frac{x^2}{4t}}\frac{\partial u}{\partial x}(xt^{-1}, -t^{-1}) + t^{-\frac{3}{2}}e^{-\frac{x^2}{4t}}\frac{\partial^2 u}{\partial x^2}(xt^{-1}, -t^{-1}).$$

Comparing the two formulas and using the fact that $\frac{\partial u}{\partial t} = \frac{\partial^2 u}{\partial x^2}$, we obtain the desired equality.

Remark. The equation

$$\frac{\partial u}{\partial t} = \frac{\partial^2 u}{\partial x^2}$$

is called the heat equation. It describes how heat spreads through a long, thin metal bar.

600. We switch to polar coordinates, where the homogeneity condition becomes the simpler

$$u(r, \theta) = r^n g(\theta),$$

where g is a one-variable function of period 2π. Writing the Laplacian

$$\Delta = \frac{\partial^2}{\partial x^2} + \frac{\partial^2}{\partial y^2}$$

in polar coordinates, we obtain

$$\Delta = \frac{\partial^2}{\partial r^2} + \frac{1}{r}\frac{\partial}{\partial r} + \frac{1}{r^2}\frac{\partial^2}{\partial \theta^2}.$$

For our harmonic function,

$$0 = \Delta u = \Delta(r^n g(\theta)) = n(n-1)r^{n-2}g(\theta) + nr^{n-2}g(\theta) + r^{n-2}g''(\theta)$$
$$= r^{n-2}(n^2 g(\theta) + g''(\theta)).$$

Therefore, g must satisfy the differential equation $g'' + n^2 g = 0$. This equation has the general solution $g(\theta) = A \cos n\theta + B \sin n\theta$. In order for such a solution to be periodic of period 2π, n must be an integer.

(P.N. de Souza, J.N. Silva, *Berkeley Problems in Mathematics*, Springer, 2004)

601. Assume the contrary and write $P(x, y) = (x^2 + y^2)^m R(x, y)$, where $R(x, y)$ is not divisible by $x^2 + y^2$. The harmonicity condition can be written explicitly as

$$4m^2 (x^2 + y^2)^{m-1} R + 2m(x^2 + y^2)^{m-1} \left(x \frac{\partial R}{\partial x} + y \frac{\partial R}{\partial y} \right)$$

$$+ (x^2 + y^2)^m \left(\frac{\partial^2 R}{\partial x^2} + \frac{\partial^2 R}{\partial y^2} \right) = 0.$$

If $R(x, y)$ were n-homogeneous for some n, then Euler's formula would allow us to simplify this to

$$(4m^2 + 2mn)(x^2 + y^2)^{m-1} R + (x^2 + y^2)^m \left(\frac{\partial^2 R}{\partial x^2} + \frac{\partial^2 R}{\partial y^2} \right) = 0.$$

If this were true, it would imply that $R(x, y)$ is divisible by $x^2 + y^2$, a contradiction. But the polynomial $x^2 + y^2$ is 2-homogeneous and $R(x, y)$ can be written as a sum of n-homogeneous polynomials, $n = 0, 1, 2, \ldots$. Since the Laplacian $\frac{\partial}{\partial x^2} + \frac{\partial}{\partial y^2}$ maps an n-homogeneous polynomial to an $(n - 2)$-homogeneous polynomial, the nonzero homogeneous parts of $R(x, y)$ can be treated separately to reach the above-mentioned contradiction. Hence $P(x, y)$ is identically equal to zero.

Remark. The solution generalizes in a straightforward manner to the case of n variables, which was the subject of a Putnam problem in 2005. But as I.M. Vinogradov said, "it is the first nontrivial example that counts".

602. Using the Leibniz-Newton fundamental theorem of calculus, we can write

$$f(x, y) - f(0, 0) = \int_0^x \frac{\partial f}{\partial x}(s, 0) ds + \int_0^y \frac{\partial f}{\partial y}(x, t) dt.$$

Using the changes of variables $s = x\sigma$ and $t = y\tau$, and the fact that $f(0, 0) = 0$, we obtain

$$f(x, y) = x \int_0^1 \frac{\partial f}{\partial x}(x\sigma, 0) d\sigma + y \int_0^1 \frac{\partial f}{\partial y}(x, y\tau) d\tau.$$

Hence if we set

$$g_1(x, y) = \int_0^1 \frac{\partial f}{\partial x}(x\sigma, 0), d\sigma \quad \text{and} \quad g_2(x, y) = \int_0^1 \frac{\partial f}{\partial y}(x, y\tau) d\tau,$$

then $f(x, y) = x g_1(x, y) + y g_2(x, y)$. Are g_1 and g_2 continuous? The answer is yes, and we prove it only for g_1, since for g_2 the proof is identical. Our argument uses the uniform continuity of a continuous function on a closed bounded interval.

Lemma. *If* $\phi : [a, b] \to \mathbb{R}$ *is continuous, then for every* $\varepsilon > 0$ *there is* $\delta > 0$ *such that whenever* $|x - y| < \delta$, *we have* $|f(x) - f(y)| < \varepsilon$.

Proof. The property is called uniform continuity; the word "uniform" signifies the fact that the "δ" from the definition of continuity is the same for all points in $[a, b]$.

We argue by contradiction. Assume that the property is not true. Then there exist two sequences $(x_n)_{n \geq 1}$ and $(y_n)_{n \geq 1}$ such that $x_n - y_n \to 0$, but $|f(x_n) - f(x_y)| \geq \varepsilon$ for some $\varepsilon > 0$. Because any sequence in $[a, b]$ has a convergent subsequence, passing to subsequences we may assume that $(x_n)_n$ and $(y_n)_n$ converge to some c in $[a, b]$. Then by the triangle inequality,

$$\varepsilon \leq |f(x_n) - f(y_n)| \leq |f(x_n) - f(c)| + |f(c) - f(y_n)|,$$

which is absurd because the right-hand side can be made arbitrarily close to 0 by taking n sufficiently large. This proves the lemma.   □

Returning to the problem, note that as x' ranges over a small neighborhood of x and δ ranges between 0 and 1, the numbers $x\sigma$ and $x'\sigma$ lie inside a small interval of the real axis. Note also that $|x\sigma - x'\sigma| \leq |x - x'|$ when $0 \leq \sigma \leq 1$. Combining these two facts with the lemma, we see that for every $\varepsilon > 0$, there exists $\delta > 0$ such that for $|x - x'| < \delta$ we have

$$\left| \frac{\partial f}{\partial x}(x\sigma, 0) - \frac{\partial f}{\partial x}(x'\sigma, 0) \right| < \varepsilon.$$

In this case,

$$\int_0^1 \left| \frac{\partial f}{\partial x}(x\sigma, 0) - \frac{\partial f}{\partial x}(x'\sigma, 0) \right| d\sigma < \varepsilon,$$

showing that g_1 is continuous. This concludes the solution.

603. First, observe that if $|x| + |y| \to \infty$, then $f(x, y) \to \infty$, hence the function indeed has a global minimum. The critical points of f are solutions to the system of equations

$$\frac{\partial f}{\partial x}(x, y) = 4x^3 + 12xy^2 - \frac{9}{4} = 0,$$

$$\frac{\partial f}{\partial y}(x, y) = 12x^2 y + 4y^3 - \frac{7}{4} = 0.$$

If we divide the two equations by 4 and then add, respectively, subtract them, we obtain

$$x^3 + 3x^2 y + 3xy^2 + y^3 - 1 = 0 \quad \text{and} \quad x^3 - 3x^2 y + 3xy^3 - y^3 = \frac{1}{8}.$$

Recognizing the perfect cubes, we write these as $(x + y)^3 = 1$ and $(x - y)^3 = \frac{1}{8}$, from which we obtain $x + y = 1$ and $x - y = \frac{1}{2}$. We find a unique critical point $x = \frac{3}{4}$, $y = \frac{1}{4}$. The minimum of f is attained at this point, and it is equal to $f\left(\frac{3}{4}, \frac{1}{4}\right) = -\frac{51}{32}$.

(R. Gelca)

604. Since the function is continuous, it suffices to find its absolute extrema in the domain $[-1, 1]^2$, and the range will be a closed interval with endpoints the values of these extrema.

First, notice that if we extend the function to the entire plane, then as the distance from a point (x, y) to the origin grows to infinity, so does the value of $f(x, y)$. This means that the (extended) function has an absolute minimum on $\mathbb{R}^2$. We will show that this minimum lies inside $[-1, 1]^2$. We compute

$$\frac{\partial f}{\partial x} = 4x^3 + 12xy^2 + 8y$$

$$\frac{\partial f}{\partial y} = 4y^3 + 12x^2y + 8x.$$

Setting these equal to zero we obtain the system of equations

$$4x^3 + 12xy^2 + 8y = 0$$
$$4y^3 + 12x^2y + 8x = 0.$$

Multiply the first equation by x and the second equation by y, then subtract the two equations to obtain $4(x^4 - y^4) = 0$. This can only happen if $x = \pm y$. Returning to the system, one solution is $x = y = 0$, and for any other solution we can only have $x = -y$. Then $4x^3 + 12x^3 - 8x = 0$, so $x = \pm\sqrt{1/2}$. We conclude that the critical points are $(0, 0)$, $(\sqrt{1/2}, -\sqrt{1/2})$ and $(-\sqrt{1/2}, \sqrt{1/2})$. One of these is the point where the function reaches its absolute minimum, and because $f(0, 0) = 0$ and $f(\sqrt{1/2}, -\sqrt{1/2}) = f(-\sqrt{1/2}, \sqrt{1/2}) = -2$, we deduce that the absolute minimum of f on $[-1, 1]^2$ is the same as the absolute minimum of f on $\mathbb{R}^2$, and this is -2.

The maximum of f is attained on the boundary, because $f(1, 1) = 16 > 0 = f(0, 0)$. So let us examine the behaviour of f on the boundary. Because of symmetry we only need to analyze the sides $y = 1$ and $y = -1$ of the square. We have $f(x, 1) = x^4 + 6x^2 + 8x + 1$. Its second derivative with respect to x is $12x^2 + 12$, which is positive, so $f(x, 1)$ is convex. This means that its maximum on $[-1, 1]$ is attained at one of the endpoints of the interval $[-1, 1]$. Repeating the argument we deduce that in order to find the maximum of f we only need to check the four corners: $(1, 1)$, $(-1, 1)$, $(1, -1)$, $(-1, -1)$. We deduce that the maximum of f is 16.

We conclude that the range of f is the interval $[-2, 16]$.

(21st Annual Iowa Collegiate Mathematics Competition, 2015, proposed by R. Gelca)

605. The diameter of the sphere is the segment that realizes the minimal distance between the lines. So if $P(t+1, 2t+4, -3t+5)$ and $Q(4s - 12, -t+8, t+17)$, we have to minimize the function

$$\|PQ\|^2 = (s - 4t + 13)^2 + (2s + t - 4)^2 + (-3s - t - 12)^2$$
$$= 14s^2 + 2st + 18t^2 + 82s - 88t + 329.$$

To minimize this function we set its partial derivatives equal to zero:

$$28s + 2t + 82 = 0,$$

$$2s + 36t - 88 = 0.$$

This system has the solution $t = -782/251$, $s = 657/251$. Substituting into the equation of the line, we deduce that the two endpoints of the diameter are $P\left(-\frac{531}{251}, -\frac{560}{251}, \frac{3601}{251}\right)$ and $Q\left(-\frac{384}{251}, \frac{1351}{251}, \frac{4924}{251}\right)$. The center of the sphere is $\frac{1}{502}(-915, 791, 8252)$, and the radius $\frac{147}{\sqrt{1004}}$.
    The equation of the sphere is

$$(502x + 915)^2 + (502y - 791)^2 + (502z - 8525)^2 = 251(147)^2.$$

(20th W.L. Putnam Competition, 1959)

606. Writing $C = \pi - A - B$, the expression can be viewed as a function in the independent variables A and B, namely,

$$f(A, B) = \cos A + \cos B - \cos(A + B).$$

And because A and B are angles of a triangle, they are constrained to the domain $A, B > 0$, $A + B < \pi$. We extend the function to the boundary of the domain, then study its extrema. The critical points satisfy the system of equations

$$\frac{\partial f}{\partial A}(A, B) = -\sin A + \sin(A + B) = 0,$$

$$\frac{\partial f}{\partial B}(A, B) = -\sin B + \sin(A + B) = 0.$$

From here we obtain $\sin A = \sin B = \sin(A + B)$, which can happen only if $A = B = \frac{\pi}{3}$. This is the unique critical point, for which $f\left(\frac{\pi}{3}, \frac{\pi}{3}\right) = \frac{3}{2}$. On the boundary, if $A = 0$ or $B = 0$, then $f(A, B) = 1$. Same if $A + B - \pi$. We conclude that the maximum of $\cos A + \cos B + \cos C$ is $\frac{3}{2}$, attained for the equilateral triangle, while the minimum is 1, which is attained only for a degenerate triangle in which two vertices coincide.

607. We rewrite the inequality as

$$\sin \alpha \cos \beta \cos \gamma + \cos \alpha \sin \beta \cos \gamma + \cos \alpha \cos \beta \sin \gamma \leq \frac{2}{\sqrt{3}},$$

and prove it for $\alpha, \beta, \gamma \in \left[0, \frac{\pi}{2}\right]$. To this end, we denote the left-hand side by $f(\alpha, \beta, \gamma)$ and find its maximum in the specified region. The critical points in the interior of the domain are solutions to the system of equations

$$\cos \alpha \cos \beta \cos \gamma - \sin \alpha \sin \beta \cos \gamma - \sin \alpha \cos \beta \sin \gamma = 0,$$

$$-\sin \alpha \sin \beta \cos \gamma + \cos \alpha \cos \beta \cos \gamma - \cos \alpha \sin \beta \sin \gamma = 0,$$

$$-\sin \alpha \cos \beta \sin \gamma - \cos \alpha \sin \beta \sin \gamma + \cos \alpha \cos \beta \cos \gamma = 0.$$

Bring this system into the form

$$\cos \alpha \cos \beta \cos \gamma = \sin \alpha \sin(\beta + \gamma),$$
$$\cos \alpha \cos \beta \cos \gamma = \sin \beta \sin(\gamma + \alpha),$$
$$\cos \alpha \cos \beta \cos \gamma = \sin \gamma \sin(\alpha + \beta).$$

From the first two equations, we obtain

$$\frac{\sin \alpha}{\sin(\alpha + \gamma)} = \frac{\sin \beta}{\sin(\beta + \gamma)}.$$

The function $g : \left(0, \frac{\pi}{2}\right)$, $g(t) = \frac{\sin t}{\sin(t+\gamma)}$ is strictly increasing, since

$$g'(t) = \frac{\cos t \sin(t + \gamma) - \sin t \cos(t + \gamma)}{(\sin(t + \gamma))^2} = \frac{\sin \gamma}{(\sin(t + \gamma))^2} > 0.$$

Hence $g(\alpha) = g(\beta)$ implies $\alpha = \beta$. Similarly, $\beta = \gamma$. The condition that (α, α, α) is a critical point is the trigonometric equation $\cos^3 \alpha = \sin \alpha \sin 2\alpha$, which translates into $\cos^3 \alpha = 2(1 - \cos^2 \alpha) \cos \alpha$. We obtain $\cos \alpha = \sqrt{\frac{2}{3}}$, and $f(\alpha, \alpha, \alpha) = \frac{2}{\sqrt{3}}$. This will be the maximum once we check that no value on the boundary of the domain exceeds this number.

But when one of the three numbers, say α, is zero, then $f(0, \beta, \gamma) = \sin(\beta + \gamma) \le 1$. Also, if $\alpha = \frac{\pi}{2}$, then $f\left(\frac{\pi}{2}, \beta, \gamma\right) = \cos \beta \cos \gamma \le 1$. Hence the maximum of f is $\frac{2}{\sqrt{3}}$ and the inequality is proved.

608. Consider a coordinate system in the plane and let the n points be $P_1(x_1, y_1)$, $P_2(x_2, y_2)$, $\dots, P_n(x_n, y_n)$. For an oriented line l, we will denote by $l^\perp$ the oriented line passing through the origin that is the clockwise rotation of l by $90°$. The origin of the coordinate system of the plane will also be the origin of the coordinate system on $l^\perp$.

An oriented line l is determined by two parameters: θ, the angle it makes with the positive side of the x-axis, which should be thought of as a point on the unit circle or an element of $\mathbb{R}/2\pi \mathbb{Z}$; and x, the distance from l to the origin, taken with sign on $l^\perp$. Define $f : \left(\frac{\mathbb{R}}{2\pi\mathbb{Z}}\right) \times \mathbb{R} \to \mathbb{R}$,

$$f(\theta, x) = \sum_{i=1}^{n} \text{dist}(P_i, l),$$

where l is the line determined by the pair (θ, x). The function f is continuous and $\lim_{x \to \pm\infty} f(\theta, x) = \infty$ for all θ; hence f has an absolute minimum $f(\theta_{min}, x_{min})$.

For fixed θ, $f(\theta, x)$ is of the form $\sum_{i=1}^{n} |x - a_i|$, which is a piecewise linear convex function. Here $a_1 \le a_2 \le \cdots \le a_n$ are a permutation of the coordinates of the projections of $P_1, P_2, \dots, P_n$ onto $l^\perp$. It follows from Problem 515 that at the absolute minimum of f, $x_{min} = a_{\lfloor n/2 \rfloor + 1}$ if n is odd and $a_{\lfloor n/2 \rfloor} \le x_{min} \le a_{\lfloor n/2 \rfloor + 1}$ if n is even (i.e., x_{min} is the median of the a_i, $i = 1, 2, \dots, n$).

If two of the points project to $a_{\lfloor n/2 \rfloor + 1}$, we are done. If this is not the case, let us examine the behavior of f in the direction of θ. By applying a translation and a rotation of the original coordinate system, we may assume that $a_i = x_i$, $i = 1, 2, \dots, n$, $x_{min} = x_{\lfloor n/2 \rfloor + 1} = 0$, $y_{\lfloor n/2 \rfloor + 1} = 0$, and $\theta_{min} = 0$. Then $f(0, 0) = \sum_{i} |x_i|$. If we rotate the line by an angle θ

keeping it through the origin, then for small θ,

$$f(\theta, 0) = \sum_{i < \lfloor n/2 \rfloor + 1} (-x_i \cos\theta - y_i \sin\theta) + \sum_{i > \lfloor n/2 \rfloor + 1} (x_i \cos\theta + y_i \sin\theta)$$

$$= \sum_{i=1}^{n} |x_i| \cos\theta + \sum_{i < \lfloor n/2 \rfloor + 1} (-y_i) \sin\theta + \sum_{i > \lfloor n/2 \rfloor + 1} y_i \sin\theta.$$

Of course, the absolute minimum of f must also be an absolute minimum in the first coordinate, so

$$\frac{\partial f}{\partial \theta}(0, 0) = \sum_{i < \lfloor n/2 \rfloor + 1} (-y_i) + \sum_{i > \lfloor n/2 \rfloor + 1} y_i = 0.$$

The second partial derivative of f with respect to θ at $(0, 0)$ should be positive. But this derivative is

$$\frac{\partial^2 f}{\partial \theta^2}(0, 0) = -\sum_{i=1}^{n} |x_i| < 0.$$

Hence the second derivative test fails, a contradiction. We conclude that the line for which the minimum is achieved passes through two of the points. It is important to note that the second derivative is *strictly* negative; the case in which it is zero makes the points collinear, in which case we are done.

Remark. This is the two-dimensional least absolute deviations problem. This method for finding the line that best fits a set of data was used well before Gauss' least squares method, for example by Laplace; its downside is that it can have multiple solutions (for example, if four points form a rectangle, both diagonals give a best approximation). The property proved above also holds in n dimensions, in which case a hyperplane that minimizes the sum of distances from the points passes through n of the given points.

609. We assume that the light ray travels from A to B crossing between media at point P. Let C and D be the projections of A and B onto the separating surface. The configuration is represented schematically in Figure 76.

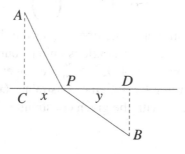

Figure 76

Let $AP = x$, $BP = y$, variables subject to the constraint $g(x, y) = x + y = CD$. The principle that light travels on the fastest path translates to the fact that x and y minimize the function

$$f(x, y) = \frac{\sqrt{x^2 + AC^2}}{v_1} + \frac{\sqrt{y^2 + BD^2}}{v_2}.$$

The method of Lagrange multipliers gives rise to the system

$$\frac{x}{v_1 \sqrt{x^2 + AC^2}} = \lambda,$$

$$\frac{y}{v_2 \sqrt{y^2 + BD^2}} = \lambda,$$

$$x + y = CD.$$

From the first two equations, we obtain

$$\frac{x}{v_1 \sqrt{x^2 + AC^2}} = \frac{y}{v_2 \sqrt{y^2 + BD^2}},$$

which is equivalent to

$$\frac{\cos APC}{\cos BPD} = \frac{v_1}{v_2}.$$

Snell's law follows once we note that the angles of incidence and refraction are, respectively, the complements of $\angle APC$ and $\angle BPD$.

610. Let D, E, F be the projections of the incenter onto the sides BC, AC, and AB, respectively. If we set $x = AF$, $y = BD$, and $z = CE$, then

$$\cot \frac{A}{2} = \frac{x}{r}, \quad \cot \frac{B}{2} = \frac{y}{r}, \quad \cot \frac{C}{2} = \frac{z}{r}.$$

The lengths x, y, z satisfy

$$x + y + z = s,$$

$$x^2 + 4y^2 + 9z^2 = \left(\frac{6s}{7}\right)^2.$$

We first determine the triangle similar to the one in question that has semiperimeter equal to 1. The problem asks us to show that the triangle is unique, but this happens only if the plane $x + y + z = 1$ and the ellipsoid $x^2 + 4y^2 + 9z^2 = \frac{36}{49}$ are tangent. The tangency point must be at an extremum of $f(x, y, z) = x + y + z$ with the constraint $g(x, y, z) = x^2 + 4y^2 + 9z^2 = \frac{36}{49}$.

We determine the extrema of f with the given constraint using Lagrange multipliers. The equation $\nabla f = \lambda \nabla g$ becomes

$$1 = 2\lambda x,$$

$$1 = 8\lambda y,$$

$$1 = 18\lambda z.$$

We deduce that $x = \frac{1}{2\lambda}$, $y = \frac{1}{8\lambda}$, and $z = \frac{1}{18\lambda}$, which combined with the constraint $g(x, y, z) = \frac{36}{49}$ yields $\lambda = \frac{49}{72}$. Hence $x = \frac{36}{49}$, $y = \frac{9}{49}$, and $z = \frac{4}{49}$, and so $f(x, y, z) = 1$. This proves that, indeed, the plane and the ellipsoid are tangent. It follows that the triangle with semiperimeter 1 satisfying the condition from the statement has sides equal to $x + y = \frac{43}{49}$, $x + z = \frac{45}{49}$, and $y + z \frac{13}{49}$.

Consequently, the unique triangle whose sides are integers with common divisor equal to 1 and that satisfies the condition from the statement is 45, 43, 13.

(USA Mathematical Olympiad, 2002, proposed by T. Andreescu)

611. We will show that if a triangle is right or obtuse, then the product of the measures of its angles is at most $\pi^3/32$. Let x, y, z be these measures, and assume that x is obtuse. We consider the domain

$$D = \{(x, y, z) \in \mathbb{R}^3 \mid \pi/2 \le x \le \pi, y \ge 0, z \ge 0, x + y + z = \pi\}$$

and the function

$$f : D \to [0, \infty), \quad f(x, y, z) = xyz.$$

The method of Lagrange multipliers shows that the maxima of f are reached either on the boundary of the planar domain D, or at the points (x, y, z) which arise by solving the system of equations

$$xy = \lambda$$
$$yz = \lambda$$
$$zx = \lambda$$
$$x + y + z = \pi.$$

The only solution to this system is $(\pi/3, \pi/3, \pi/3)$, which is not in the domain. Hence we must examine the boundary. The part of the boundary with $y = 0$ or $z = 0$ yields a minimum for the function, so we only focus on the part where $x = \pi/2$. In this case f becomes the two variable function $f\left(\frac{\pi}{2}, y, z\right) = \frac{\pi}{2}yz$, which, by using the constraint, is turned into a quadratic

$$f(\pi/2, y, \pi/2 - y) = \pi/2 y(\pi/2 - y).$$

The maximum of its quadratic is at its vertex, and is equal to $\pi^3/32$. We conclude that our triangle cannot be right or obtuse, so it is acute.

(Konhauser Problem Fest, 2014, proposed by R. Gelca)

612. Let a, b, c, d be the sides of the quadrilateral in this order, and let x and y be the cosines of the angles formed by the sides a and b, respectively, c and d. The condition that the triangle formed by a and b shares a side with the triangle formed by c and d translates, via the law of cosines, into the relation

$$a^2 + b^2 - 2abx = c^2 + d^2 - 2cdy.$$

We want to maximize the expression $ab\sqrt{1 - x^2} + cd\sqrt{1 - y^2}$, which is twice the area of the rectangle. Let

$$f(x, y) = ab\sqrt{1 - x^2} + cd\sqrt{1 - y^2},$$
$$g(x, y) = a^2 + b^2 - 2abx - c^2 - d^2 + 2cdy.$$

We are supposed to maximize $f(x, y)$ over the square $[-1, 1] \times [-1, 1]$, with the constraint $g(x, y) = 0$. Using Lagrange multipliers we see that any candidate for the maximum that lies in the interior of the domain satisfies the system of equations

$$-ab\frac{2x}{\sqrt{1 - x^2}} = -\lambda 2ab,$$

$$-cd\frac{2y}{\sqrt{1 - y^2}} = \lambda 2cd,$$

for some λ. It follows that $\sqrt{1 - x^2}/x = -\sqrt{1 - y^2}/y$, and so the tangents of the opposite angles are each the negative of the other. It follows that the angles are supplementary. In this case $x = -y$. The constraint becomes a linear equation in x. Solving it and substituting in the formula of the area yields Brahmagupta's formula for the area:

$$A = \sqrt{(s - a)(s - b)(s - c)(s - d)}, \quad \text{where} \quad s = \frac{a + b + c + d}{2}.$$

Is this the maximum? Let us analyze the behavior of f on the boundary. When $x = 1$ or $y = 1$, the quadrilateral degenerates to a segment; the area is therefore 0. Let us see what happens when $y = -1$. Then the quadrilateral degenerates to a triangle, and the area can be computed using Hero's formula

$$A = \sqrt{s(s - a)(s - b)(s - (c + d))}.$$

Since $s(s-(c+d)) < (s-c)(s-d)$ (because this is the same as $s^2 - sc - sd < s^2 - sc - sd + cd$), we conclude that the cyclic quadrilateral maximizes the area.

(E. Goursat, *A Course in Mathematical Analysis*, Dover, New York, 1904)

613. Without loss of generality, we may assume that the circle has radius 1. If a, b, c are the sides, and $A(a, b, c)$ the area, then (because of the formula $S = sr$, where s is the semiperimeter) the constraint reads $A = \frac{a+b+c}{2}$. We will maximize the function $f(a, b, c) = A(a, b, c)^2$ with the constraint $g(a, b, c) = A(a, b, c)^2 - \left(\frac{a+b+c}{2}\right)^2 = 0$. Using Hero's formula, we can write

$$f(a, b, c) = \frac{a + b + c}{2} \cdot \frac{-a + b + c}{2} \cdot \frac{a - b + c}{2} \cdot \frac{a + b - c}{2}$$
$$= \frac{-a^4 - b^4 - c^4 + 2(a^2b^2 + b^2c^2 + a^2c^2)}{16}.$$

The method of Lagrange multipliers gives rise to the system of equations

$$(\lambda - 1)\frac{-a^3 + a(b^2 + c^2)}{4} = \frac{a + b + c}{2},$$

$$(\lambda - 1)\frac{-b^3 + b(a^2 + c^2)}{4} = \frac{a + b + c}{2},$$

$$(\lambda - 1)\frac{-c^3 + c(a^2 + b^2)}{4} = \frac{a+b+c}{2},$$
$$g(a, b, c) = 0.$$

Because $a + b + c \neq 0$, λ cannot be 1, so this further gives

$$-a^3 + a(b^2 + c^2) = -b^3 + b(a^2 + c^2) = -c^3 + c(a^2 + b^2).$$

The first equality can be written as $(b - a)(a^2 + b^2 - c^2) = 0$. This can happen only if either $a = b$ or $a^2 + b^2 = c^2$, so either the triangle is isosceles, or it is right. Repeating this for all three pairs of sides we find that either $b = c$ or $b^2 + c^2 = a^2$, and also that either $a = c$ or $a^2 + c^2 = b^2$. Since at most one equality of the form $a^2 + b^2 = c^2$ can hold, we see that, in fact, all three sides must be equal. So the critical point given by the method of Lagrange multipliers is the equilateral triangle.

Is this the global minimum? We just need to observe that as the triangle degenerates, the area becomes infinite. So the answer is yes, the equilateral triangle minimizes the area.

614. Let $f(x, y, z) = xy + yz + zx - 2xyz$ and $g(x, y, z) = x + y + z$. We want to find the extrema of $f(x, y, z)$ subject to the constraint $g(z, y, z) = 1$. Applying the Lagrange multiplier method we write $\nabla f = \lambda \nabla g$, which yields the system

$$x + y - 2xy - \lambda = 0$$
$$x + z - 2xz - \lambda = 0$$
$$y + z - 2yz - \lambda = 0.$$
$$x + y + z = 1$$

The solutions are $(1/2, 1/2, 0)$, $(0, 1/2, 1/2)$, $(1/2, 0, 1/2)$ and $(1/3, 1/3, 1/3)$. Note that on the boundary $f(x, y, z) = xy$ which lies in the interval $[0, 1/4]$. We compute $f(1/3, 1/3, 1/3) = 7/27$, while the other values are non-negative and smaller than this. Hence the range of f is $[0, 7/27]$ and the inequality follows.

(solution from V. Boju, L. Funar, *The Math Problems Notebook*, Birkhäuser, 2007)

615. Consider the function $f : \{(a, b, c, d) \mid a, b, c, d \geq 1, \ a + b + c + d = 1\} \to \mathbb{R}$,

$$f(a, b, c, d) = \frac{1}{27} + \frac{176}{27}abcd - abc - bcd - cda - dab.$$

Being a continuous function on a closed and bounded set in $\mathbb{R}^4$, f has a minimum. We claim that the minimum of f is nonnegative. The inequality $f(a, b, c, d) \geq 0$ is easy on the boundary, for if one of the four numbers is zero, say $d = 0$, then $f(a, b, c, 0) = \frac{1}{27} - abc$, and this is nonnegative by the AM-GM inequality.

Any minimum in the interior of the domain should arise by applying the method of Lagrange multipliers. This method gives rise to the system

$$\frac{\partial f}{\partial a} = \frac{176}{27}bcd - bc - cd - db = \lambda,$$

$$\frac{\partial f}{\partial b} = \frac{176}{27}acd - ac - cd - ad = \lambda,$$

$$\frac{\partial f}{\partial c} = \frac{176}{27}abd - ab - ad - bd = \lambda,$$

$$\frac{\partial f}{\partial d} = \frac{176}{27}abc - ab - bc - ac = \lambda,$$

$$a + b + c + d = 1.$$

One possible solution to this system is $a = b = c = d = \frac{1}{4}$, in which case $f\left(\frac{1}{4}, \frac{1}{4}, \frac{1}{4}, \frac{1}{4}\right) = 0$. Otherwise, let us assume that the numbers are not all equal. If three of them are distinct, say a, b, and c, then by subtracting the second equation from the first, we obtain

$$\left(\frac{176}{27}cd - c - d\right)(b - a) = 0,$$

and by subtracting the third from the first, we obtain

$$\left(\frac{176}{27}bd - b - d\right)(c - a) = 0.$$

Dividing by the nonzero factors $b - a$, respectively, $c - a$, we obtain

$$\frac{176}{27}cd - c - d = 0,$$

$$\frac{176}{27}bd - b - d = 0;$$

and subtracting the equations we deduce $b = c$, a contradiction. It follows that the numbers a, b, c, d for which a minimum is achieved have at most two distinct values. Modulo permutations, either $a = b = c$ or $a = b$ and $c = d$. In the first case, by subtracting the fourth equation from the third and using the fact that $a = b = c$, we obtain

$$\left(\frac{176}{27}a^2 - 2a\right)(d - a) = 0.$$

Since $a \neq d$, it follows that $a = b = c = \frac{27}{88}$ and $d = 1 - 3a = \frac{7}{88}$. One can verify that

$$f\left(\frac{27}{88}, \frac{27}{88}, \frac{27}{88}, \frac{7}{88}\right) = \frac{1}{27} + \frac{6}{88} \cdot \frac{27}{88} \cdot \frac{27}{88} > 0.$$

The case $a = b$ and $c = d$ yields

$$\frac{176}{27}cd - c - d = 0,$$

$$\frac{176}{27}ab - a - b = 0,$$

which gives $a = b = c = d = \frac{27}{88}$, impossible. We conclude that f is nonnegative, and the inequality is proved.

(Short list of the 34th International Mathematical Olympiad, 1993, proposed by Vietnam)

616. Fix α, β, γ and consider the function

$$f(x, y, z) = \frac{\cos x}{\sin \alpha} + \frac{\cos y}{\sin \beta} + \frac{\cos z}{\sin \gamma}$$

with the constraints $x + y + z = \pi$, $x, y, z \geq 0$. We want to determine the maximum of $f(x, y, z)$. In the interior of the triangle described by the constraints the Lagrange multipliers theorem shows that a maximum satisfies

$$\frac{\sin x}{\sin \alpha} = -\lambda,$$

$$\frac{\sin y}{\sin \beta} = -\lambda,$$

$$\frac{\sin z}{\sin \beta} = -\lambda,$$

$$x + y + z = \pi.$$

By the law of sines, the triangle with angles x, y, z is similar to that with angles α, β, γ, hence $x = \alpha$, $y = \beta$, and $z = \gamma$.

Let us now examine the boundary. If $x = 0$, then $\cos z = -\cos y$. We prove that

$$\frac{1}{\sin \alpha} + \cos y \left(\frac{1}{\sin \beta} - \frac{1}{\sin \gamma} \right) < \cot \alpha + \cot \beta + \cot \gamma.$$

This is a linear function in $\cos y$, so the inequality will follow from the corresponding inequalities at the two endpoints of the interval $[-1, 1]$, namely from

$$\frac{1}{\sin \alpha} + \frac{1}{\sin \beta} - \frac{1}{\sin \gamma} < \cot \alpha + \cot \beta + \cot \gamma$$

and

$$\frac{1}{\sin \alpha} - \frac{1}{\sin \beta} + \frac{1}{\sin \gamma} < \cot \alpha + \cot \beta + \cot \gamma.$$

By symmetry, it suffices to prove just one of these two, the first for example. Eliminating the denominators, we obtain

$$\sin \beta \sin \gamma + \sin \alpha \sin \gamma - \sin \alpha \sin \beta < \sin \beta \sin \gamma \cos \alpha + \sin \alpha \sin \gamma \cos \beta$$
$$+ \sin \alpha \sin \beta \cos \gamma.$$

The laws of sine and cosine allow us to transform this into the equivalent

$$bc + ca - ab < \frac{b^2 + c^2 - a^2}{2} + \frac{a^2 + c^2 - b^2}{2} + \frac{a^2 + b^2 - c^2}{2},$$

and this is equivalent to $(a + b - c)^2 > 0$. Hence the conclusion.

(*Kvant* (*Quantum*), proposed by R.P. Ushakov)

617. The domain is bounded by the hyperbolas $xy = 1$, $xy = 2$ and the lines $y = x$ and $y = 2x$. This domain can mapped into a rectangle by the transformation

$$T : u = xy, \ v = \frac{y}{x}.$$

Thus it is natural to consider the change of coordinates

$$T^{-1} : x = \sqrt{\frac{u}{v}}, \ y = \sqrt{uv}.$$

The domain becomes the rectangle $D^* = \{(u, v) \in \mathbb{R}^2 \mid 1 \leq u \leq 2, \ 1 \leq v \leq 2\}$. The Jacobian of T^{-1} is $\frac{1}{2v} \neq 0$. The integral becomes

$$\int_1^2 \int_1^2 \sqrt{\frac{u}{v}} \frac{1}{2v} du dv = \frac{1}{2} \int_1^2 u^{1/2} du \int_1^2 v^{-3/2} dv = \frac{1}{3}(5\sqrt{2} - 6).$$

(Gh. Bucur, E. Câmpu, S. Găină, *Culegere de Probleme de Calcul Diferenţial şi Integral* (*Collection of Problems in Differential and Integral Calculus*), Editura Tehnică, Bucharest, 1967)

618. Denote the integral by I. The change of variable $(x, y, z) \to (z, y, x)$ transforms the integral into

$$\iiint_B \frac{z^4 + 2y^4}{x^4 + 4y^4 + z^4} dx dy dz.$$

Hence

$$2I = \iiint_B \frac{x^4 + 2y^4}{x^4 + 4y^4 + z^4} dx dy dz + \iiint_B \frac{2y^4 + z^4}{x^4 + 4y^4 + z^4} dx dy dz$$

$$= \iiint_B \frac{x^4 + 4y^4 + z^4}{x^4 + 4y^4 + z^4} dx dy dz = \frac{4\pi}{3}.$$

It follows that $I = \frac{2\pi}{3}$.

619. The domain D is depicted in Figure 77. We transform it into the rectangle $D_1 = \left[\frac{1}{4}, \frac{1}{2}\right] \times \left[\frac{1}{6}, \frac{1}{2}\right]$ by the change of coordinates

$$x = \frac{u}{u^2 + v^2}, \ y = \frac{v}{u^2 + v^2}.$$

The Jacobian is

$$J = -\frac{1}{(u^2 + v^2)^2}.$$

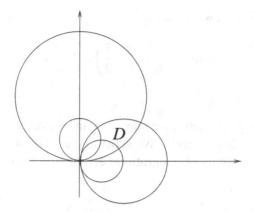

Figure 77

Therefore,

$$\iint_D \frac{dxdy}{(x^2+y^2)^2} = \iint_{D_1} dudv = \frac{1}{12}.$$

(D. Flondor, N. Donciu, *Algebră şi Analiză Matematică* (*Algebra and Mathematical Analysis*), Editura Didactică şi Pedagogică, Bucharest, 1965)

620. In the equation of the curve that bounds the domain

$$\left(\frac{x^2}{a^2} + \frac{y^2}{b^2}\right)^2 = \frac{x^2}{a^2} - \frac{y^2}{b^2},$$

the expression on the left suggests the use of generalized polar coordinates, which are suited for elliptical domains. And indeed, if we set $x = ar\cos\theta$ and $y = br\sin\theta$, the equation of the curve becomes $r^4 = r^2\cos 2\theta$, or $r = \sqrt{\cos 2\theta}$. The condition $x \geq 0$ becomes $-\frac{\pi}{2} \leq \theta \leq \frac{\pi}{2}$, and because $\cos 2\theta$ r4 = r2 cos 2, or r = should be positive we should further have $-\frac{\pi}{4} \leq \theta \leq \frac{\pi}{4}$. Hence the domain of integration is

$$\left\{ (r,\theta); \ 0 \leq r \leq \sqrt{\cos 2\theta}, \ -\frac{\pi}{4} \leq \theta \leq \frac{\pi}{4} \right\}.$$

The Jacobian of the transformation is $J = abr$. Applying the formula for the change of variables, the integral becomes

$$\int_{-\frac{\pi}{4}}^{\frac{\pi}{4}} \int_0^{\sqrt{\cos 2\theta}} a^2 b^2 r^3 \cos\theta |\sin\theta| drd\theta = \frac{a^2 b^2}{4} \int_0^{\frac{\pi}{4}} \cos^2 2\theta \sin 2\theta d\theta = \frac{a^2 b^2}{24}.$$

(Gh. Bucur, E. Câmpu, S. Găină, *Culegere de Probleme de Calcul Diferenţial şi Integral* (*Collection of Problems in Differential and Integral Calculus*), Editura Tehnică, Bucharest, 1967)

621. The method is similar to that for computing the Fresnel integrals, only simpler. If we denote the integral by I, then

$$I^2 = \int_{-\infty}^{\infty} e^{-x^2} dx \int_{-\infty}^{\infty} e^{-y^2} dy = \int_{-\infty}^{\infty} \int_{-\infty}^{\infty} e^{-(x^2+y^2)} dxdy.$$

Switching to polar coordinates, we obtain

$$I^2 = \int_0^{2\pi} \int_0^\infty e^{-e^2} r \, dr \, d\theta = \int_0^{2\pi} \left(-\frac{1}{2}\right) e^{-r^2} \Big|_0^\infty d\theta = \int_0^{2\pi} \frac{1}{2} d\theta = \pi.$$

Hence the desired formula $I = \sqrt{\pi}$.

622. Call the integral I. By symmetry, we may compute it over the domain $\{(u, v, w) \in \mathbb{R}^3 \mid 0 \le v \le u \le 1\}$, then double the result. We substitute $u = r \cos\theta$, $v = r \sin\theta$, $w = \tan\phi$, taking into account that the limits of integration become $0 \le \theta, \phi \le \frac{\pi}{4}$, and $0 \le r \le \sec\theta$. We have

$$I = 2 \int_0^{\frac{\pi}{4}} \int_0^{\frac{\pi}{4}} \int_0^{\sec\theta} \frac{r \sec^2\phi}{(1 + r^2 \cos^2\theta + r^2 \sin^2\theta + \tan^2\phi)^2} dr \, d\theta \, d\phi$$

$$= 2 \int_0^{\frac{\pi}{4}} \int_0^{\frac{\pi}{4}} \int_0^{\sec\theta} \frac{r \sec^2\phi}{(r^2 + \sec^2\phi)^2} dr \, d\theta \, d\phi$$

$$= 2 \int_0^{\frac{\pi}{4}} \int_0^{\frac{\pi}{4}} \sec^2\phi \frac{-1}{2(r^2 + \sec^2\phi)} \Big|_{r=0}^{r=\sec\theta} d\theta \, d\phi$$

$$= - \int_0^{\frac{\pi}{4}} \int_0^{\frac{\pi}{4}} \frac{\sec^2\phi}{\sec^2\theta + \sec^2\phi} d\theta \, d\phi + \left(\frac{\pi}{4}\right)^2.$$

But notice that this is the same as

$$\int_0^{\frac{\pi}{4}} \int_0^{\frac{\pi}{4}} \left(1 - \frac{\sec^2\phi}{\sec^2\theta + \sec^2\phi}\right) d\theta \, d\phi = \int_0^{\frac{\pi}{4}} \int_0^{\frac{\pi}{4}} \frac{\sec^2\theta}{\sec^2\theta + \sec^2\phi} d\theta \, d\phi.$$

If we exchange the roles of θ and ϕ in this last integral we see that

$$- \int_0^{\frac{\pi}{4}} \int_0^{\frac{\pi}{4}} \frac{\sec^2\phi}{\sec^2\theta + \sec^2\phi} d\theta \, d\phi + \left(\frac{\pi}{4}\right)^2 = \int_0^{\frac{\pi}{4}} \int_0^{\frac{\pi}{4}} \frac{\sec^2\phi}{\sec^2\theta + \sec^2\phi} d\theta \, d\phi.$$

Hence

$$\int_0^{\frac{\pi}{4}} \int_0^{\frac{\pi}{4}} \frac{\sec^2\phi}{\sec^2\theta + \sec^2\phi} d\theta \, d\phi = \frac{\pi^2}{32}.$$

Consequently, the integral we are computing is equal to $\frac{\pi^2}{32}$.

(American Mathematical Monthly, proposed by M. Hajja and P. Walker)

623. We have

$$I = \iint_D \ln|\sin(x - y)| dx \, dy = \int_0^\pi \left(\int_0^y \ln|\sin(y - x)| dx\right) dy$$

$$= \int_0^\pi \left(\int_0^y \ln \sin t \, dt\right) dy = y \int_0^y \ln \sin t \, dt \Big|_{y=0}^{y=\pi} - \int_0^\pi y \ln \sin y \, dy$$

$$= \pi A - B,$$

where $A = \int_0^\pi \ln \sin t \, dt$, $B = \int_0^\pi t \ln \sin t \, dt$. Note that here we used integration by parts! We compute further

$$A = \int_0^{\frac{\pi}{2}} \ln \sin t \, dt + \int_{\frac{\pi}{2}}^\pi \ln \sin t \, dt = \int_0^{\frac{\pi}{2}} \ln \sin t \, dt + \int_0^{\frac{\pi}{2}} \ln \cos t \, dt$$

$$= \int_0^{\frac{\pi}{2}} (\ln \sin 2t - \ln 2) dt = -\frac{\pi}{2} \ln 2 + \frac{1}{2} A.$$

Hence $A = -\pi \ln 2$. For B we use the substitution $t = \pi - x$ to obtain

$$B = \int_0^\pi (\pi - x) \ln \sin x \, dx = \pi A - B.$$

Hence $B = \frac{\pi}{2} A$. Therefore, $I = \pi A - B = -\frac{\pi^2}{2} \ln 2$, and we are done.

Remark. The identity

$$\int_0^{\frac{\pi}{2}} \ln \sin t \, dt = -\frac{\pi}{2} \ln 2$$

belongs to Euler.

(S.Rădulescu, M. Rădulescu, *Teoreme și Probleme de Analiză Matematică* (*Theorems and Problems in Mathematical Analysis*), Editura Didactică și Pedagogică, Bucharest, 1982)

624. This problem applies the discrete version of Fubini's theorem. Define

$$f(i, j) = \begin{cases} 1 & \text{for } j \le a_i, \\ 0 & \text{for } j > a_i. \end{cases}$$

The left-hand side is equal to $\sum_{i=1}^n \sum_{j=1}^m f(i, j)$, while the right-hand side is equal to $\sum_{j=1}^m \sum_{i=1}^n f(i, j)$. The equality follows.

625. First, note that for $x > 0$,

$$e^{-sx} x^{-1} |\sin x| < e^{-sx},$$

so the integral that we are computing is finite.

Now consider the two-variable function

$$f(x, y) = e^{-sxy} \sin x.$$

We have

$$\int_0^\infty \int_1^\infty |f(x, y)| dy dx = \int_0^\infty \int_1^\infty e^{-sxy} |\sin x| dy dx = \frac{1}{s} \int_0^\infty e^{-sx} x^{-1} |\sin x| dx,$$

and we just saw that this is finite. Hence we can apply Fubini's theorem, to conclude that on the one hand,

$$\int_0^\infty \int_1^\infty f(x, y) dy dx = \frac{1}{s} \int_0^\infty e^{-sx} x^{-1} \sin x \, dx,$$

and on the other hand,

$$\int_0^\infty \int_1^\infty f(x, y)dydx = \int_1^\infty \frac{1}{s^2 y^2 + 1} dy.$$

Here of course we used the fact that

$$\int_0^\infty e^{-ax} \sin x dx = \frac{1}{a^2 + 1}, \quad a > 0,$$

a formula that can be proved by integrating by parts. Equating the two expressions that we obtained for the double integral, we obtain

$$\int_0^\infty e^{-sx} x^{-1} \sin x dx = \frac{\pi}{2} - \arctan s = \arctan(s^{-1}),$$

as desired.

(G.B. Folland, *Real Analysis*, Modern Techniques and Their Applications, Wiley, 1999)

626. Applying Tonelli's theorem to the function $f(x, y) = e^{-xy}$, we can write

$$\int_0^\infty \frac{e^{-ax} - e^{-bx}}{x} dx = \int_0^\infty \int_a^b e^{-xy} dydx = \int_a^b \int_0^\infty e^{-xy} dxdy$$

$$= \int_a^b \frac{1}{y} dy = \ln \frac{b}{a}.$$

Remark. This is a particular case of integrals of the form $\int_0^\infty \frac{f(ax) - f(bx)}{x} dx$, known as Froullani integrals. In general, if f is continuous and has finite limit at infinity, the value of the integral is $(f(0) - \lim_{x \to \infty} f(x)) \ln \frac{b}{a}$.

627. We do the proof in the case $0 < x < 1$, since for $-1 < x < 0$ the proof is completely analogous, while for $x = 0$ the property is obvious. The function $f : \mathbb{N} \times [0, x] \to \mathbb{R}$, $f(n, t) = t^{n-1}$ satisfies the hypothesis of Fubini's theorem. So integration commutes with summation:

$$\sum_{n=0}^\infty \int_0^x t^{n-1} dt = \int_0^x \frac{dt}{1 - t}.$$

This implies

$$\sum_{n=1}^\infty \frac{x^n}{n} = -\ln(1 - x).$$

Dividing by x, we obtain

$$\sum_{n=1}^\infty \frac{x^{n-1}}{n} = -\frac{1}{x} \ln(1 - x).$$

The right-hand side extends continuously at 0, since $\lim\limits_{x\to 0} \frac{1}{t}\ln(1-t) = -1$. Again we can apply Fubini's theorem to $f(n, t) = \frac{t^{n-1}}{n}$ on $\mathbb{N} \times [0, x]$ to obtain

$$\sum_{n=1}^{\infty} \frac{x^n}{n^2} = \sum_{n=1}^{\infty} \int_0^x \frac{t^{n-1}}{n} dt = \int_0^x \sum_{n=1}^{\infty} \frac{t^{n-1}}{n} dt = -\int_0^x \frac{1}{t}\ln(1-t)dt,$$

as desired.

628. We can apply Tonelli's theorem to the function $f(x, n) = \frac{1}{x^2+n^4}$. Integrating term by term, we obtain

$$\int_0^x F(t)dt = \int_0^x \sum_{n=1}^{\infty} f(t, n)dt = \sum_{n=1}^{\infty} \int_0^x \frac{dt}{t^2 + n^4} = \sum_{n=1}^{\infty} \frac{1}{n^2} \arctan \frac{x}{n^2}.$$

This series is bounded from above by $\sum\limits_{n=1}^{\infty} \frac{1}{n^2} = \frac{\pi^2}{6}$ (see Problem 595). Hence the summation commutes with the limit as x tends to infinity. We have

$$\int_0^{\infty} F(t)dt = \lim_{x\to\infty} \int_0^x F(t)dt = \lim_{x\to\infty} \sum_{n=1}^{\infty} \frac{1}{n^2} \arctan \frac{x}{n^2} = \sum_{n=1}^{\infty} \frac{1}{n^2} \cdot \frac{\pi}{2}.$$

Using the identity $\sum\limits_{n\geq 1} \frac{1}{n^2} = \frac{\pi^2}{6}$, we obtain

$$\int_0^{\infty} F(t)dt = \frac{\pi^3}{12}.$$

(Gh. Sireţchi, *Calcul Diferenţial şi Integral (Differential and Integral Calculus)*, Editura Ştiinţifică şi Enciclopedică, Bucharest, 1985)

629. The integral from the statement can be written as

$$\oint_{\partial D} xdy - ydx.$$

Applying Green's theorem for $P(x, y) = -y$ and $Q(x, y) = x$, we obtain

$$\oint_{\partial D} xdy - ydx = \iint_D (1 + 1)dxdy,$$

which is twice the area of D. The conclusion follows.

630. (a) Because each horizontal side is followed by a vertical side, there are as many horizontal as vertical sides. Thus $n = 2k$, where k is the number of vertical sides. Let the coordinates of the consecutive vertices of the polygon be

$$(a_1, b_1), (a_2, b_1), (a_2, b_2), (a_3, b_2), (a_3, b_3), \ldots,$$
$$(a_{k-1}, b_{k-1}), (a_k, b_{k-1}), (a_k, b_k), (a_1, b_k),$$

and assume that we travel around the polygon counterclockwise.

We know that the numbers $a_{j+1} - a_j$ and $b_{j+1} - b_j$ are odd integers, $j = 1, 2, \ldots, k$ (where $a_{k+1} = a_1$ and $b_{k+1} = b_1$). As all the differences $a_{j+1} - a_j$ are odd, the parities of the numbers

$$a_1, a_2, \ldots, a_k, a_{k+1} = a_1$$

alternate, and this is only possible when k is even, that is $k = 2m$. So $n = 2k = 4m$ is a multiple of 4, which solves the first part of the problem.

(b) For the second part we will use Green's formula for the area of the domain D surrounded by the curve C oriented counterclockwise (see previous problem):

$$\iint_D 1 dx dy = \oint_C x dy.$$

For our particular case,

$$\oint_C x dy = \sum_{j=1}^{2m} a_j (b_{j+1} - b_j).$$

As all the differences $b_{j+1} - b_j$ are odd, the sum has the same parity as $a_1 + a_2 + \cdots + a_{2m}$. On the other hand, as the parity of the numbers $a_1, a_2, \ldots, a_{2m}$ alternate, the sums $a_1 + a_2, a_3 + a_4, \cdots a_{2m-1} + a_{2m}$ are odd. Writing

$$a_1 + a_2 + a_3 + \cdots + a_{2m} = (a_1 + a_2) + (a_3 + a_4) + \cdots + (a_{2m-1} + a_{2m}),$$

we deduce that the sum has the same parity as m, so the area of the polygon has the same parity as m. For $n = 100$, we have $m = 25$, so the area is odd. The problem is solved.

(*Kvant (Quantum)*, proposed by M. Kontsevich)

631. It can be checked that $\operatorname{div} \overrightarrow{F} = 0$ (in fact, $\overrightarrow{F}$ is the curl of the vector field $e^{yx} \overrightarrow{i} + e^{zx} \overrightarrow{j} + e^{xy} \overrightarrow{k}$). If S be the union of the upper hemisphere and the unit disk in the xy-plane, then by the divergence theorem

$$\iint_S \overrightarrow{F} \cdot \overrightarrow{n} \, dS = 0.$$

And on the unit disk $\overrightarrow{F} \cdot \overrightarrow{n} = 0$, which means that the flux across the unit disk is zero. It follows that the flux across the upper hemisphere is zero as well.

632. We simplify the computation using the Kelvin-Stokes theorem:

$$\oint_C y^2 dx + z^2 dy + x^2 dz = -2 \iint_S y dx dy + z dy dz + x dz dx,$$

where S is the portion of the sphere bounded by the Viviani curve. We have

$$-2 \iint_S y dx dy + z dy dz + x dz dx = -2 \iint_S (z, x, y) \cdot \overrightarrow{n} \, d\sigma,$$

where (z, x, y) denotes the three-dimensional vector with coordinates z, x, and y, while $\overrightarrow{n}$ denotes the unit vector normal to the sphere at the point of coordinates (x, y, z). We parametrize the portion of the sphere in question by the coordinates (x, y), which range inside the circle $x^2 + y^2 - ax = 0$. This circle is the projection of the Viviani curve onto the xy-plane.

The unit vector normal to the sphere is

$$\overrightarrow{n} = \left(\frac{x}{a}, \frac{y}{a}, \frac{z}{x} \right) = \left(\frac{x}{a}, \frac{y}{a}, \frac{\sqrt{a^2 - x^2 - y^2}}{a} \right),$$

while the area element is

$$d\sigma = \frac{1}{\cos \alpha} dx dy,$$

α being the angle formed by the normal to the sphere with the xy-plane. It is easy to see that $\cos \alpha = \frac{z}{a} = \frac{\sqrt{a^2 - x^2 - y^2}}{a}$. Hence the integral is equal to

$$-2 \iint_D \left(z\frac{x}{a} + x\frac{y}{a} + y\frac{z}{a} \right) \frac{a}{z} dx dy = -2 \iint_D \left(x + y + \frac{xy}{\sqrt{a^2 - x^2 - y^2}} \right) dx dy,$$

the domain of integration D being the disk $x^2 + y^2 - ax \leq 0$. Split the integral as

$$-2 \iint_D (x + y) dx dy - 2 \iint_D \frac{xy}{\sqrt{a^2 - x^2 - y^2}} dx dy.$$

Because the domain of integration is symmetric with respect to the y-axis, the second double integral is zero. The first double integral can be computed using polar coordinates: $x = \frac{a}{2} + r \cos \theta$, $y = r \sin \theta$, $0 \leq r \leq \frac{a}{2}$, $0 \leq \theta \leq 2\pi$. Its value is $-\frac{\pi a^3}{4}$, which is the answer to the problem.

(D. Flondor, N. Donciu, *Algebră şi Analiză Matematică* (*Algebra and Mathematical Analysis*), Editura Didactică şi Pedagogică, Bucharest, 1965)

633. We will apply the Kelvin-Stokes theorem. We begin with

$$\frac{\partial \phi}{\partial y} \frac{\partial \psi}{\partial z} - \frac{\partial \phi}{\partial z} \frac{\partial \psi}{\partial y} = \frac{\partial \phi}{\partial y} \frac{\partial \psi}{\partial z} + \phi \frac{\partial^2 \psi}{\partial y \partial z} - \frac{\partial \phi}{\partial z} \frac{\partial \psi}{\partial y} - \phi \frac{\partial^2 \psi}{\partial z \partial y}$$

$$= \frac{\partial}{\partial y} \left(\phi \frac{\partial \psi}{\partial z} \right) - \frac{\partial}{\partial z} \left(\phi \frac{\partial \psi}{\partial y} \right),$$

which combined with the two other analogous computations gives

$$\nabla \phi \times \nabla \psi = \text{curl}(\phi \nabla \psi).$$

By the Kelvin-Stokes theorem, the integral of the curl of a vector field on a surface without boundary is zero.

(Soviet University Student Mathematical Competition, 1976)

634. For the solution, recall the following identity.

Green's first identity. *If f and g are twice-differentiable functions on the solid region R bounded by the closed surface S, then*

$$\iiint_R (f\nabla^2 g + \nabla f \cdot \nabla g)dV = \iint_S f\frac{\partial g}{\partial n}dS,$$

where $\frac{\partial g}{\partial n}$ is the derivative of g in the direction of the normal to the surface.

Proof. For the sake of completeness we will prove Green's identity. Consider the vector field $\overrightarrow{F} = f\nabla g$. Then

$$\operatorname{div}\overrightarrow{F} = \frac{\partial}{\partial x}\left(f\frac{\partial g}{\partial x}\right) + \frac{\partial}{\partial y}\left(f\frac{\partial g}{\partial y}\right) + \frac{\partial}{\partial z}\left(f\frac{\partial g}{\partial z}\right)$$

$$= f\left(\frac{\partial^2 g}{\partial x^2} + \frac{\partial^2 g}{\partial y^2} + \frac{\partial^2 g}{\partial z^2}\right) + \left(\frac{\partial f}{\partial x}\frac{\partial g}{\partial x} + \frac{\partial f}{\partial y}\frac{\partial g}{\partial y} + \frac{\partial f}{\partial z}\frac{\partial g}{\partial z}\right).$$

So the left-hand side of the identity is $\iiint_R \operatorname{div}\overrightarrow{F}\,dV$. By the Gauss-Ostrogradsky divergence theorem this is equal to

$$\iint_S (f\nabla g)\cdot\overrightarrow{n}\,dS = \iint_S f(\nabla g\cdot\overrightarrow{n})dS = \iint_S f\frac{\partial g}{\partial n}dS.$$

The conclusion follows. $\square$

Writing Green's first identity for the vector field $g\nabla f$ and then subtracting it from that of the vector field $f\nabla g$, we obtain

Green's second identity.

$$\iiint_R (f\nabla^2 g - g\nabla^2 f)dV = \iint_S \left(f\frac{\partial g}{\partial n} - g\frac{\partial f}{\partial n}\right)dS.$$

Returning to the problem, the fact that f and g are constant along the lines passing through the origin means that on the unit sphere,

$$\frac{\partial f}{\partial n} = \frac{\partial g}{\partial n} = 0.$$

Plug this into the right-hand side of Green's second identity to obtain the equality from the statement.

635. Because $\overrightarrow{F}$ is obtained as an integral of the point-mass contributions of the masses distributed in space, it suffices to prove this equality for a mass M concentrated at one point, say the origin. We will use

Newton's law. The gravitational force between two masses m_1 and m_2 at distance r is equal to

$$\vec{F} = \frac{m_1 m_2 G}{r^2}.$$

By Newton's law, a mass M located at the origin generates the gravitational field

$$\vec{F}(x, y, z) = MG \frac{1}{x^2 + y^2 + z^2} \cdot \frac{x\vec{i} + y\vec{j} + z\vec{k}}{\sqrt{x^2 + y^2 + z^2}} = -MG \frac{x\vec{i} + y\vec{j} + z\vec{k}}{(x^2 + y^2 + z^2)^{3/2}}.$$

One can easily check that the divergence of this field is zero. Consider a small sphere S_0 of radius r centered at the origin, and let V be the solid lying between S_0 and S. By the Gauss-Ostrogradsky divergence theorem,

$$\iint_S \vec{F} \cdot \vec{n}\, dS - \iint_{S_0} \vec{F} \cdot \vec{n}\, dS = \iiint_V \operatorname{div}\vec{F}\, dV = 0.$$

Hence it suffices to prove the Gauss law for the sphere S_0. On this sphere the flow $\vec{F} \cdot \vec{n}$ is constantly equal to $-\frac{GM}{r^2}$. Integrating it over the sphere gives $-4\pi MG$, proving the Gauss' law.

636. The condition $\operatorname{curl}\vec{F} = 0$ suggests the use of the Kelvin-Stokes theorem:

$$\iint_S \operatorname{curl}\vec{F} \cdot \vec{n}\, dS = \oint_{\partial C} \vec{F} \cdot d\vec{R}.$$

We expect the answer to the question to be negative. All we need is to find a surface S whose boundary lies in the xy-plane and such that the integral of $\vec{G}(x, y)$ on ∂S is nonzero.

A simple example that comes to mind is the interior S of the ellipse $x^2 + 4y^2 = 4$. Parametrize the ellipse as $x = 2\cos\theta, y = \sin\theta, \theta \in [0, 2\pi)$. Then

$$\oint_{\partial S} \vec{G} \cdot d\vec{R} = \int_0^{2\pi} \left(\frac{-\sin\theta}{4}, \frac{2\cos\theta}{4}, 0 \right) \cdot (-2\sin\theta, \cos\theta, 0) d\theta = \int_0^{2\pi} \frac{1}{2} d\theta = \pi.$$

By the Kelvin-Stokes theorem this should be equal to the integral of the curl of $\vec{F}$ over the interior of the ellipse. The curl of $\vec{F}$ is zero except at the origin, but we can fix that by adding a smooth tiny upward bump at the origin, which does not alter too much the above computation. The integral should on the one hand be close to 0, and on the other hand close to π, which is impossible. This proves that such a vector field $\vec{F}$ cannot exist.

(48th W.L. Putnam Mathematical Competition, 1987, solution from K. Kedlaya, B. Poonen, R. Vakil, *The William Lowell Putnam Mathematical Competition 1985-2000*, MAA, 2002)

637. Let $D = [a_1, b_1] \times [a_2, b_2]$ be a rectangle in the plane, and $a, b \in \mathbb{R}, a < b$. We consider the three-dimensional parallelepiped $V = D \times [a, b]$. Denote by $\vec{n}$ the outward normal vector field on the boundary ∂V of V (which is defined everywhere except on the edges). By the Leibniz-Newton fundamental theorem of calculus,

$$
\int_a^b \frac{d}{dt} \iint_D G(x, y, t) dx dy dt = \int_a^b \iint_D \frac{\partial}{\partial t} G(x, y, t) dx dy dt
$$

$$
= \iint_D \int_a^b \frac{\partial}{\partial t} G(x, y, t) dt dx dy
$$

$$
= \iint_D G(x, y, b) dx dy - \iint_D G(x, y, a) dx dy
$$

$$
= \int_{D \times \{b\}} G(x, y, t) \vec{k} \cdot d\vec{n} + \int_{D \times \{a\}} G(x, y, t) \vec{k} \cdot d\vec{n},
$$

where $\vec{k}$ denotes the unit vector that points in the z-direction. With this in mind, we compute

$$
0 = \int_a^b \left(\frac{d}{dt} \iint_D G(x, y, t) dx dy + \oint_C \vec{F} \cdot d\vec{R} \right) dt
$$

$$
= \int_{D \times \{b\}} G(x, y, t) \vec{k} \cdot d\vec{n} + \int_{D \times \{a\}} G(x, y, t) \vec{k} \cdot d\vec{n}
$$

$$
+ \int_a^b \int_{a_1}^{b_1} F_1(x, a_2) dx - \int_a^b \int_{b_1}^{a_1} F_1(x, b_2) dx
$$

$$
+ \int_a^b \int_{a_2}^{b_2} F_2(b_1, y) dy - \int_a^b \int_{b_2}^{a_2} F_2(a_1, y) dy.
$$

If we introduce the vector field $\vec{H} = F_2 \vec{i} + F_1 \vec{j} + G \vec{k}$, this equation can be written simply as

$$
\iint_{\partial V} \vec{H} \cdot \vec{n} \, dS = 0.
$$

By the divergence theorem,

$$
\iiint_V \operatorname{div} \vec{H} \, dV = \iint_{\partial V} \vec{H} \cdot \vec{n} \, dS = 0.
$$

Since this happens in every parallelepiped, $\operatorname{div} \vec{H}$ must be identically equal to 0. Therefore,

$$
\operatorname{div} \vec{H} = \frac{\partial F_2}{\partial x} + \frac{\partial F_1}{\partial y} + \frac{\partial G}{\partial t} = 0,
$$

and the relation is proved.

Remark. The interesting case occurs when $\vec{F}$ and G depend on spatial variables (spatial dimensions). Then G becomes a vector field $\vec{B}$, or better a 2-form, called the magnetic flux, while F becomes the electric field strength E. The relation

$$
\frac{d}{dt} \int_S \vec{B} = - \int_{\partial S} E
$$

is Faraday's law of induction. Introducing a fourth dimension (the time), and redoing mutatis mutandis the above computation gives rise to the first group of Maxwell's equations

$$\text{div}\,\vec{B} = 0, \quad \frac{\partial \vec{B}}{\partial t} = \text{curl}E.$$

638. In the solution we ignore the factor $\frac{1}{4\pi}$, which is there only to make the linking number an integer. We will use the more general form of Green's theorem applied to the curve $C = C_1 \cup C_1'$ and surface S,

$$\oint_C Pdx + Wdy + Rdz = \iint_S \left(\frac{\partial Q}{\partial x} - \frac{\partial P}{\partial y}\right)dxdy + \left(\frac{\partial R}{\partial y} - \frac{\partial Q}{\partial z}\right)dydz$$
$$+ \left(\frac{\partial P}{\partial z} - \frac{\partial R}{\partial x}\right)dzdx.$$

Writing the parametrization with coordinate functions $\vec{v}_1(s) = (x(s), y(s), z(s))$, $\vec{v}_2(t) = (x'(t), y'(t), z'(t))$, the linking number of C_1 and C_2 (with the factor $\frac{1}{4\pi}$ ignored) becomes

$$\oint_{C_1} \oint_{C_2} \frac{(x' - x)(dz'dy - dy'dz) + (y' - y)(dx'dz - dz'dx) + (z' - z)(dy'dx - dx'dy)}{((x' - x)^2 + (y' - y)^2 + (z' - z)^2)^{3/2}}.$$

The 1-form $Pdx + Qdy + Rdz$, which we integrate on $C = C_1 \cup C_1'$, is

$$\oint_{C_2} \frac{(x' - x)(dz'dy - dy'dz) + (y' - y)(dx'dz - dz'dx) + (z' - z)(dy'dx - dx'dy)}{((x' - x)^2 + (Y' - y)^2 + (z' - z)^2)^{3/2}}.$$

Note that here we integrate against the variables x', y', z', so this expression depends only on x, y, and z. Explicitly,

$$P(x, y, z) = \oint_{C_2} \frac{-(y' - y)dz + (x' - z)dy'}{((x' - x)^2 + (y' - y)^2 + (z' - z)^2)^{3/2}},$$
$$Q(x, y, z) = \oint_{C_2} \frac{(x' - x)dz' - (z' - z)dx'}{((x' - x)^2 + (y' - y)^2 + (z' - z)^2)^{3/2}},$$
$$R(x, y, z) = \oint_{C_2} \frac{-(x' - x)dy' + (y' - y)dx'}{((x' - x)^2 + (y' - y)^2 + (z' - z)^2)^{3/2}}.$$

By the Kelvin-Stokes theorem, $\text{lk}(C_1, C_2) = \text{lk}(C_1', C_2)$ if

$$\frac{\partial Q}{\partial x} - \frac{\partial P}{\partial y} = \frac{\partial R}{\partial y} - \frac{\partial Q}{\partial z} = \frac{\partial P}{\partial z} - \frac{\partial R}{\partial x} = 0.$$

We will verify only $\frac{\partial Q}{\partial x} - \frac{\partial P}{\partial y} = 0$, the other equalities having similar proofs. The part of it that contains dz' is equal to

$$\oint_{C_2} -2((x'-x)^2 + (y'-y)^2 + (z'-z)^2)^{-3/2}$$
$$+ 3(x'-x)^2((x'-x)^2 + (y'-y)^2 + (z'-z)^2)^{-5/2}$$
$$+ 3(y'-y)^2((x'-x)^2 + (y'-y)^2 + (z'-z)^2)^{-5/2}dz'$$
$$= \oint_{C_2} ((x'-x)^2 + (y'-y)^2 + (z'-z)^2)^{-3/2}$$
$$+ 3(z'-z)^2((x'-z)^2 + (y'-y)^2 + (z'-z)^2)^{-5/2}dz'$$
$$= \oint_{C_2} \frac{\partial}{\partial z'}((x'-x)^2 + (y'-y)^2 + (z'-z)^2)^{-3/2}dz' = 0,$$

where the last equality is a consequence of the Leibniz-Newton fundamental theorem of calculus. Also, of the two terms, only $\frac{\partial Q}{\partial x}$ has a dx' in it, and that part is

$$3\oint_{C_2} ((x-x')^2 + (y-y')^2 + (z-z')^2)^{-5/2}(x-x')(z-z')dx'$$

$$= \oint_{C_2} \frac{\partial}{\partial x'} \frac{z-z'}{((x-x')^2 + (y-y')^2 + (z-z')^2)^{3/2}}dx' = 0.$$

The term involving dy' is treated similarly. This yields

$$\frac{\partial Q}{\partial x} - \frac{\partial P}{\partial y} = 0,$$

and the conclusion follows.

Remark. The linking number is, in fact, an integer, which measures the number of times the curves wind around each other. It was defined by C.F. Gauss, who used it to decide, based on astronomical observations, whether the orbits of certain asteroids were winding around the orbit of the earth. The way Gauss discovered the linking number was by using the Bio-Savart law, which computes the magnetic field $\vec{B}$ at a given point $\vec{r}$ produced by an electric field of a steady current I in a thin closed wire C_1. The Bio-Savart law gives

$$\vec{B}(\vec{r}) = \frac{\mu_0}{4\pi} \int_{C_1} \frac{I}{\|\vec{r} - \vec{r_1}\|^3} \frac{d\vec{r_1}}{ds} \times (\vec{r} - \vec{r_1})ds,$$

where μ_0 is the permeability of the vacuum, and $\vec{r_1}(s)$ is the parametrization of C_1. Gauss made the point $\vec{r}$ vary along C_2.

639. Plugging in $x = y$, we find that $f(0) = 0$, and plugging in $x = -1, y = 0$, we find that $f(1) = -f(-1)$. Also, plugging in $x = a, y = 1$, and then $x = a, y = -1$, we obtain

$$f(a^2 - 1) = (a - 1)(f(a) + f(1)),$$

$$f(a^2 - 1) = (a + 1)(f(a) - f(1)).$$

Equating the right-hand sides and solving for $f(a)$ gives $f(a) = f(1)a$ for all a.

So any such function is linear. Conversely, a function of the form $f(x) = kx$ clearly satisfies the equation.

(S. Korean Mathematical Olympiad, 2000)

640. Replace z by $1 - z$ to obtain

$$f(1 - z) + (1 - z)f(z) = 2 - z.$$

Combine this with $f(z) + zf(1 - z) = 1 + z$, and eliminate $f(1 - z)$ to obtain

$$(1 - z + z^2)f(z) = 1 - z + z^2.$$

Hence $f(z) = 1$ for all z except maybe for $z = e^{\pm \pi i/3}$, when $1 - z + z^2 = 0$. For $\alpha = e^{i\pi/3}$, $\overline{\alpha} = \alpha^2 = 1 - \alpha$; hence $f(\alpha) + \alpha f(\overline{\alpha}) = 1 + \alpha$. We therefore have only one constraint, namely $f(\overline{\alpha}) = [1 + \alpha - f(\alpha)]/\alpha = \overline{\alpha} - 1 - \overline{\alpha}f(\alpha)$. Hence the solution to the functional equation is of the form

$$f(z) = 1 \text{ for } z \ne e^{\pm i\pi/3}, \quad f(e^{i\pi/3}) = \beta, \quad f(e^{-i\pi/3}) = \overline{\alpha} + 1 - \overline{\alpha}\beta,$$

where β is an arbitrary complex parameter.

(20th W.L. Putnam Competition, 1959)

641. Successively, we obtain

$$f(-1) = f\left(-\frac{1}{2}\right) = f\left(-\frac{1}{3}\right) = \cdots = \lim_{n \to \infty} f\left(-\frac{1}{n}\right) = f(0).$$

Hence $f(x) = f(0)$ for $x \in \left\{0, -1, -\frac{1}{2}, \ldots, -\frac{1}{n}, \ldots\right\}$.

If $x \ne 0, -1, \ldots, -\frac{1}{n}, \ldots$ replacing x by $\frac{x}{1+x}$ in the functional equation, we obtain

$$f\left(\frac{x}{1+x}\right) = f\left(\frac{\frac{x}{1+x}}{1 - \frac{x}{1+x}}\right) = f(x).$$

And this can be iterated to yield

$$f\left(\frac{x}{1 + nx}\right) = f(x), \quad n = 1, 2, 3, \ldots$$

Because f is continuous at 0 it follows that

$$f(x) = \lim_{n \to \infty} f\left(\frac{x}{1 + nx}\right) = f(0).$$

This shows that only constant functions satisfy the functional equation.

642. In this problem we denote by f^n the composition of f with itself n times. By induction one can check that

$$f(2^n x_0) = 2^{n+1} x_0, \text{ for all } n \geq 0.$$

Going backwards, we can write

$$f(x_0) - 3x_0 + 2f^{-1}(x_0) = 0,$$

so $f^{-1}(x_0) = x_0/2$. Again by induction

$$f^{-1}(2^{-n} x_0) = 2^{-n+1} x_0, \text{ for all } n \geq 0.$$

As a consequence, for every $x \in (0, \infty)$ there is x_1 such that $x \in [x_1, 2x_1]$ and $f(x_1) = 2x_1$, $f(2x_1) = 4x_1$.

Next, we rewrite the original functional equation in terms of f^{-1} as

$$2f^{-1}(f^{-1}(x)) - 3f^{-1}(x) + x = 0.$$

Fix an arbitrary x, and define recursively the sequence $a_0 = f(x)$, and $a_{n+1} = f^{-1}(a_n)$, $n \geq 0$. Then a_n satisfies the linear recursive relation

$$2a_{n+2} - 3a_{n+1} + a_n = 0, \quad a_0 = f(x), a_1 = x.$$

Solving for the general term we obtain

$$a_n = \frac{1}{2^{n-1}}[(2^n - 1)x - (2^{n-1} - 1)f(x)],$$

that is

$$(f^{-1})^{n-1}(x) = \frac{1}{2^{n-1}}[(2^n - 1)x - (2^{n-1} - 1)f(x)].$$

Because f^{-1} is positive, this gives $f(x)/x \leq 2$. We will actually prove a finer inequality.

Because f is increasing, f^{-1} is increasing, and so is f^{-1} composed with itself $n-1$ times. So for $x < y$, $(f^{-1})^{n-1}(x) < (f^{-1})^{n-1}(y)$, which means that

$$(2^n - 1)x - (2^{n-1} - 1)f(x) < (2^n - 1)y - (2^{n-1} - 1)f(y).$$

We obtain

$$\frac{f(y) - f(x)}{y - x} < \frac{2^n - 1}{2^{n-1} - 1} = 2 + \frac{1}{2^{n-1} - 1} \text{ for all } n = 1, 2, 3, \ldots.$$

Passing to the limit, we conclude that for $x < y$,

$$\frac{f(y) - f(x)}{y - x} \leq 2.$$

Now let $x \in (0, \infty)$ be arbitrary. We want to show that $f(x) = 2x$. If $x = 2^n x_0$, for some $n \in \mathbb{Z}$, then we are done. Otherwise, choose x_1 such that $x_1 < x < 2x_1$ and $f(x_1) = 2x_1$,

$f(2x_1) = 4x_1$. Because f is increasing, $f(x) \in (2x_1, 4x_1)$. Using the above inequality we can write

$$\frac{f(x) - 2x_1}{x - x_1} \le 2 \text{ and } \frac{4x_1 - f(x)}{2x_1 - x} \le 2.$$

The first inequality yields $f(x) - 2x_1 \le 2x - 2x_1$, so $f(x) \le 2x$. The second inequality yields $4x_1 - f(x) \le 4x_1 - 2x$, so $f(x) \ge 2x$. Combining the two inequalities we obtain $f(x) = 2x$. We conclude that

$$f(x) = 2x \text{ for all } x$$

is the only function satisfying the conditions from the statement.

    (R. Gelca)

643. Plugging in $x = t, y = 0, z = 0$ gives

$$f(t) + f(0) + f(t) \ge 3f(t),$$

or $f(0) \ge f(t)$ for all real numbers t. Plugging in $x = \frac{t}{2}, y = \frac{t}{2}, z = -\frac{t}{2}$ gives

$$f(t) + f(0) + f(0) \ge 3f(0),$$

or $f(t) \ge f(0)$ for all real numbers t. Hence $f(t) = f(0)$ for all t, so f must be constant. Conversely, any constant function f clearly satisfies the given condition.

    (Russian Mathematical Olympiad, 2000)

644. No! In fact, we will prove a more general result.

Proposition. *Let S be a set and $g : S \to S$ a function that has exactly two fixed points $\{a, b\}$ and such that $g \circ g$ has exactly four fixed points $\{a, b, c, d\}$. Then there is no function $f : S \to S$ such that $g = f \circ f$.*

Proof. Let $g(c) = y$. Then $c = g(g(c)) = g(y)$; hence $y = g(c) = g(g(y))$. Thus y is a fixed point of $g \circ g$. If $y = a$, then $a = g(a) = g(y) = c$, leading to a contradiction. Similarly, $y = b$ forces $c = b$. If $y = c$, then $c = g(y) = g(c)$, so c is a fixed point of g, again a contradiction. It follows that $y = d$, i.e., $g(c) = d$, and similarly $g(d) = c$.

    Suppose there is $f : S \to S$ such that $f \circ f = g$. Then $f \circ g = f \circ f \circ f = g \circ f$. Then $f(a) = f(g(a)) = g(f(a))$, so $f(a)$ is a fixed point of g. Examining case by case, we conclude that $f(\{a, b\}) \subset \{a, b\}$ and $f(\{a, b, c, d\}) \subset \{a, b, c, d\}$. Because $f \circ f = g$, the inclusions are, in fact, equalities.

    Consider $f(c)$. If $f(c) = a$, then $f(a) = f(f(c)) = g(c) = d$, a contradiction since $f(a)$ is in $\{a, b\}$. Similarly, we rule out $f(c) = b$. Of course, c is not a fixed point of f, since it is not a fixed point of g. We are left with the only possibility $f(c) = d$. But then $f(d) = f(f(c)) = g(c) = d$, and this again cannot happen because d is not a fixed point of g. We conclude that such a function f cannot exist. $\qquad\square$

In the particular case of our problem, $g(x) = x^2 - 2$ has the fixed points -1 and 2, and $g(g(x)) = (x^2 - 2)^2 - 2$ has the fixed points $-1, 2, \frac{-1+\sqrt{5}}{2}$, and $\frac{-1-\sqrt{5}}{2}$. This completes the solution.

(B.J. Venkatachala, *Functional Equations: A Problem Solving Approach*, Prism Books PVT Ltd., 2002)

645. The standard approach is to substitute particular values for x and y. The solution found by the student S.P. Tungare does quite the opposite. It introduces an additional variable z. The solution proceeds as follows:

$$f(x + y + z) = f(x)f(y + z) - c \sin x \sin(y + z)$$
$$= f(x)[f(y)f(z) - c \sin y \sin z] - c \sin x \sin y \cos z - c \sin x \cos y \sin z$$
$$= f(x)f(y)f(z) - cf(x) \sin y \sin z - c \sin x \sin y \cos z - c \sin x \cos y \sin z.$$

Because obviously $f(x + y + z) = f(y + x + z)$, it follows that we must have

$$\sin z[f(x) \sin y - f(y) \sin x] = \sin z[\cos x \sin y - \cos y \sin x].$$

Substitute $z = \frac{\pi}{2}$ to obtain

$$f(x) \sin y - f(y) \sin x = \cos x \sin y - \cos y \sin x.$$

For $x = \pi$ and y not an integer multiple of π, we obtain $\sin y[f(\pi) + 1] = 0$, and hence $f(\pi) = -1$.

Then, substituting in the original equation $x = y = \frac{\pi}{2}$ yields

$$f(\pi) = \left[f\left(\frac{\pi}{2}\right)\right]^2 - c,$$

whence $f\left(\frac{\pi}{2}\right) = \pm\sqrt{c - 1}$. Substituting in the original equation $y = \pi$ we also obtain $f(x + \pi) = -f(x)$. We then have

$$-f(x) = f(x + \pi) = f\left(x + \frac{\pi}{2}\right)f\left(\frac{\pi}{2}\right) - c \cos x$$
$$= f\left(\frac{\pi}{2}\right)\left(f(x)f\left(\frac{\pi}{2}\right) - c \sin x\right) - c \cos x,$$

whence

$$f(x)\left[\left(f\left(\frac{\pi}{2}\right)\right)^2 - 1\right] = cf\left(\frac{\pi}{2}\right) \sin x - x \cos x.$$

It follows that $f(x) = f\left(\frac{\pi}{2}\right) \sin x + \cos x$. We find that the functional equation has two solutions, namely,

$$f(x) = \sqrt{c - 1} \sin x + \cos x \quad \text{and} \quad f(x) = -\sqrt{c - 1} \sin x + \cos x.$$

(Indian Team Selection Test for the International Mathematical Olympiad, 2004)

646. Because $|f|$ is bounded and is not identically equal to zero, its supremum is a positive number M. Using the equation from the statement and the triangle inequality, we obtain that for any x and y,

$$2|f(x)||g(y)| = |f(x+y)+f(x-y)|$$
$$\leq |f(x+y)| + |f(x-y)| \leq 2M.$$

Hence

$$|g(y)| \leq \frac{M}{|f(x)|}.$$

If in the fraction on the right we take the supremum of the denominator, we obtain $|g(y)| \leq \frac{M}{M} = 1$ for all y, as desired.

Remark. The functions $f(x) = \sin x$ and $g(x) = \cos x$ are an example.
(14th International Mathematical Olympiad, 1972)

647. Substituting for f a linear function $ax + b$ and using the method of undetermined coefficients, we obtain $a = 1, b = -\frac{3}{2}$, so $f(x) = x - \frac{3}{2}$ is a solution.

Are there other solutions? Setting $g(x) = f(x) - \left(x - \frac{3}{2}\right)$, we obtain the simpler functional equation

$$3g(2x+1) = g(x), \quad \text{for all } x \in \mathbb{R}.$$

This can be rewritten as

$$g(x) = \frac{1}{3}g\left(\frac{x-1}{2}\right), \quad \text{for all } x \in \mathbb{R}.$$

For $x = -1$ we have $g(-1) = \frac{1}{3}g(-1)$; hence $g(-1) = 0$. In general, for an arbitrary x, define the recursive sequence $x_0 = x, x_{n+1} = \frac{x_n-1}{2}$ for $n \geq 0$. It is not hard to see that this sequence is Cauchy, for example, because $|x_{m+n} - x_m| \leq \frac{1}{2^{m-2}} \max(1, |x|)$. This sequence is therefore convergent, and its limit L satisfies the equation $L = \frac{L-1}{2}$. It follows that $L = -1$. Using the functional equation, we obtain

$$g(x) = \frac{1}{3}g(x_1) = \frac{1}{9}g(x_2) = \cdots = \frac{1}{3^n}g(x_n) = \lim_{n\to\infty} g(x_n) = g(-1) = 0.$$

This shows that $f(x) = x - \frac{3}{2}$ is the unique solution to the functional equation.
(B.J. Venkatachala, *Functional Equations: A Problem Solving Approach*, Prism Books PVT Ltd., 2002)

648. We will first show that $f(x) \geq x$ for all x. From (i) we deduce that $f(3x) \geq 2x$, so $f(x) \geq \frac{2x}{3}$. Also, note that if there exists k such that $f(x) \geq kx$ for all x, then $f(x) \geq \frac{k^3+2}{3}x$ for all x as well. We can iterate and obtain $f(x) \geq k_n x$, where k_n are the terms of the recursive sequence defined by $k_1 = \frac{2}{3}$, and $k_{n+1} = \frac{k_n^3+2}{3}$ for $k \geq 1$. Let us examine this sequence.

By the AM-GM inequality,

$$k_{n+1} = \frac{k_n^3 + 1^3 + 1^3}{3} \geq k_n,$$

so the sequence is increasing. An easy induction shows that $k_n < 1$. Weierstrass' criterion implies that $(k_n)_n$ is convergent. Its limit L should satisfy the equation

$$L = \frac{L^3 + 2}{3},$$

which shows that L is a root of the polynomial equation $L^3 - 3L + 2 = 0$. This equation has only one root in $[0, 1]$, namely $L = 1$. Hence $\lim_{n \to \infty} k_n = 1$, and so $f(x) \geq x$ for all x.

It follows immediately that $f(3x) \geq 2x + f(x)$ for all x. Iterating, we obtain that for all $n \geq 1$,

$$f(3^n x) - f(x) \geq (3^n - 1)x.$$

Therefore, $f(x) - x \leq f(3^n x) - 3^n x$. If we let $n \to \infty$ and use (ii), we obtain $f(x) - x \leq 0$, that is, $f(x) \leq x$. We conclude that $f(x) = x$ for all $x > 0$. Thus the identity function is the unique solution to the functional equation.

(G. Dospinescu)

649. We should keep in mind that $f(x) = \sin x$ and $g(x) = \cos x$ satisfy the condition. As we proceed with the solution to the problem, we try to recover some properties of $\sin x$ and $\cos x$. First, note that the condition $f(t) = 1$ and $g(t) = 0$ for some $t \neq 0$ implies $g(0) = 1$; hence g is nonconstant. Also, $0 = g(t) = g(0)g(t) + f(0)f(t) = f(0)$; hence f is nonconstant. Substituting $x = 0$ in the relation yields $g(-y) = g(y)$, so g is even.

Substituting $y = t$, we obtain $g(x - t) = f(x)$, with its shifted version $f(x + t) = g(x)$. Since g is even, it follows that $f(-x) = g(x + t)$. Now let us combine these facts to obtain

$$f(x - y) = g(x - y - t) = g(x)g(y + t) + f(x)f(y + t)$$
$$= g(x)f(-y) + f(x)g(y).$$

Change y to $-y$ to obtain $f(x + y) = f(x)g(y) + g(x)f(y)$ (the addition formula for sine).

The remaining two identities are consequences of this and the fact that f is odd. Let us prove f odd. From $g(x - (-y)) = g(x + y) = g(-x - y)$, we obtain

$$f(x)f(-y) = f(y)f(-x)$$

for all x and y in $\mathbb{R}$. Setting $y = t$ and $x = -t$ yields $f(-t)^2 = 1$, so $f(-t) = \pm 1$. The choice $f(-t) = 1$ gives $f(x) = f(x)f(-t) = f(-x)f(t) = f(-x)$; hence f is even. But then

$$f(x - y) = f(x)g(-y) + g(x)f(-y) = f(x)g(y) + g(x)f(y) = f(x + y),$$

for all x and y. For $x = \frac{z+w}{2}$, $y = \frac{z-w}{2}$, we have $f(z) = f(w)$, and so f is constant, a contradiction. For $f(-t) = -1$, we obtain $f(-x) = -f(-x)f(-t) = -f(x)f(t) = -f(x)$; hence f is odd. It is now straightforward that

$$f(x - y) = f(x)g(y) + g(x)f(-y) = f(x)g(y) - g(x)f(y)$$

and

$$g(x + y) = g(x - (-y)) = g(x)g(-y) + f(x)f(-y) = g(x)g(y) - f(x)f(y),$$

where in the last equality we also used the fact, proved above, that g is even.

(*American Mathematical Monthly*, proposed by V.L. Klee, solution by P.L. Kannappan)

650. Substituting x and y by $x/2$ we obtain $f(x) = f(x/2)^2 > 0$, the function $g(x) = \ln f(x)$ is well defined. It satisfies Cauchy's equation and is continuous; therefore, $g(x) = \alpha x$ for some constant α. We obtain $f(x) = c^x$, with $c = e^\alpha$.

651. Adding 1 to both sides of the functional equation and factoring, we obtain

$$f(x + y) + 1 = (f(x) + 1)(f(y) + 1).$$

The continuous function $g(x) = f(x) + 1$ satisfies the functional equation $g(x+y) = g(x)g(y)$, and we have seen in the previous problem that $g(x) = c^x$ for some nonnegative constant c. We conclude that $f(x) = c^x - 1$ for all x.

652. If there exists x_0 such that $f(x_0) = 1$, then

$$f(x) = f(x_0 + (x - x_0)) = \frac{1 + f(x - x_0)}{1 + f(x - x_0)} = 1.$$

In this case, f is identically equal to 1. In a similar manner, we obtain the constant solution $f(x) \equiv -1$.

Let us now assume that f is never equal to 1 or -1. Define $g : \mathbb{R} \to \mathbb{R}$, $g(x) = \frac{1+f(x)}{1-f(x)}$. To show that g is continuous, note that for all x,

$$f(x) = \frac{2f\left(\frac{x}{2}\right)}{1 + f\left(\frac{x}{2}\right)} < 1.$$

Now the continuity of g follows from that of f and of the function $h(t) = \frac{1+t}{1-t}$ on $(-\infty, 1)$. Also,

$$
\begin{aligned}
g(x + y) &= \frac{1 + f(x + y)}{1 - f(x + y)} = \frac{f(x)f(y) + 1 + f(x) + f(y)}{f(x)f(y) + 1 - f(x) - f(y)} \\
&= \frac{1 + f(x)}{1 - f(x)} \cdot \frac{1 + f(y)}{1 - f(y)} = g(x)g(y).
\end{aligned}
$$

Hence g satisfies the functional equation $g(x + y) = g(x)g(y)$. As seen in Problem 651, $g(x) = c^x$ for some $c > 0$. We obtain $f(x) = \frac{c^x - 1}{c^x + 1}$. The solutions to the equation are therefore

$$f(x) = \frac{c^x - 1}{c^x + 1}, \quad f(x) = 1, \quad f(x) = -1.$$

Remark. You might have recognized the formula addition formula for the hyperbolic tangent. This explains the choice of g, by expressing the exponential in terms of the hyperbolic tangent.

653. Rewrite the functional equation as

$$\frac{f(xy)}{xy} = \frac{f(x)}{x} + \frac{f(y)}{y}.$$

It now becomes natural to let $g(x) = \frac{f(x)}{x}$. which satisfies the equation

$$g(xy) = g(x) + g(y).$$

The particular case $x = y$ yields $g(x) = \frac{1}{2}g(x^2)$, and hence

$$g(-x) = \frac{1}{2}g((-x)^2) = \frac{1}{2}g(x^2) = g(x).$$

Thus we only need to consider the case $x > 0$.

Note that g is continuous on $(0, \infty)$. If we compose g with the continuous function $h : \mathbb{R} \to (0, \infty)$, $h(x) = e^x$, we obtain a continuous function on $\mathbb{R}$ that satisfies Cauchy's equation. Hence $g \circ h$ is linear, which then implies $g(x) = \log_a x$ for some positive base a. It follows that $f(x) = x \log_a x$ for $x > 0$ and $f(x) = x \log_a |x|$ if $x < 0$.

All that is missing is the value of f at 0. This can be computed directly setting $x = y = 0$, and it is seen to be 0. We conclude that $f(x) = x \cos_a |x|$ if $x \neq 0$, and $f(0) = 0$, where a is some positive number. The fact that any such function is continuous at zero follows from

$$\lim_{x \to 0+} x \log_a x = 0,$$

which can be proved by applying the L'Hôpital's theorem to the functions $\log_a x$ and $\frac{1}{x}$. This concludes the solution.

654. Setting $y = z = 0$ yields $\phi(x) = f(x) + g(0) + h(0)$, and similarly $\phi(y) = g(y) + f(0) + h(0)$. Substituting these three relations in the original equation and letting $z = 0$ gives rise to a functional equation for ϕ, namely

$$\phi(x + y) = \phi(x) + \phi(y) - (f(0) + g(0) + h(0)).$$

This should remind us of the Cauchy equation, which it becomes after changing the function ϕ to $\psi(x) = \phi(x) - (f(0) + g(0) + h(0))$. The relation $\psi(x + y) = \psi(x) + \psi(y)$ together with the continuity of ψ shows that $\psi(x) = cx$ for some constant c. We obtain the solution to the original equation

$$\phi(x) = cx + \alpha + \beta + \gamma, \quad f(x) = cx + \alpha, \quad g(x) = cx + \beta, \quad h(x) = cx + \gamma,$$

where α, β, β are arbitrary real numbers. These functions satisfy the given equation.

(*Gazeta Matematică* (*Mathematics Gazette, Bucharest*), proposed by M. Vlada)

655. This is a generalization of Cauchy's equation. Trying small values of n, one can guess that the answer consists of all polynomial functions of degree at most $n - 1$ with no constant term (i.e., with $f(0) = 0$). We prove by induction on n that this is the case.

The case $n = 2$ is Cauchy's equation. Assume that the claim is true for $n - 1$ and let us prove it for n. Fix x_n and consider the function $g_{x_n} : \mathbb{R} \to \mathbb{R}$, $g_{x_n}(x) = f(x + x_n) - f(x) - f(x_n)$. It is continuous and, more importantly, it satisfies the functional equation for $n - 1$. Hence $g_{x_n}(x)$ is a polynomial of degree $n - 2$. And this is true for all x_n.

It follows that $f(x + x_n) - f(x)$ is a polynomial of degree $n - 2$ for all x_n. In particular, there exist polynomials $P_1(x)$ and $P_2(x)$ such that $f(x + 1) - f(x) = P_1(x)$, and $f(x + \sqrt{2}) -$

$f(x) = P_2(x)$. Note that for any a, the linear map from the vector space of polynomials of degree at most $n - 1$ to the vector space of polynomials of degree at most $n - 2$, $P(x) \mapsto P(x + a) - P(x)$, has kernel the one-dimensional space of constant polynomials (the only periodic polynomials). Because the first vector space has dimension n and the second has dimension $n - 1$, the map is onto. Hence there exist polynomials $Q_1(x)$ and $Q_2(x)$ of degree at most $n - 1$ such that

$$Q_1(x + 1) - Q_1(x) = P_1(x) = f(x + 1) - f(x),$$

$$Q_2(x + \sqrt{2}) - Q_2(x) = P_2(x) = f(x + \sqrt{2}) - f(x).$$

We deduce that the functions $f(x) - Q_1(x)$ and $f(x) - Q_2(x)$ are continuous and periodic, hence bounded. Their difference $Q_1(x) - Q_2(x)$ is a bounded polynomial, hence constant. Consequently, the function $f(x) - Q_1(x)$ is continuous and has the periods 1 and $\sqrt{2}$. Since the additive group generated by 1 and $\sqrt{2}$ is dense in $\mathbb{R}$, $f(x) - Q_1(x)$ is constant. This completes the induction.

That any polynomial of degree at most $n - 1$ with no constant term satisfies the functional equation also follows by induction on n. Indeed, the fact that f satisfies the equation is equivalent to the fact that g_{x_n} satisfies the equation. And g_{x_n} is a polynomial of degree $n - 2$.
   (G. Dospinescu)

656. *First solution*: Assume that such functions do exist. Because $g \circ g$ is a bijection, f is one-to-one and g is onto. Since f is a one-to-one continuous function, it is monotonic, and because g is onto but $f \circ g$ is not, it follows that f maps $\mathbb{R}$ onto an interval I strictly included in $\mathbb{R}$. One of the endpoints of this interval is finite, call this endpoint a. Without loss of generality, we may assume that $I = (a, \infty)$. Then as $g \circ f$ is onto, $g(I) = \mathbb{R}$. This can happen only if $\limsup_{x \to \infty} g(x) = \infty$ and $\liminf_{x \to \infty} g(x) = -\infty$, which means that g oscillates in a neighborhood of infinity. But this is impossible because $f(g(x)) = x^2$ implies that g assumes each value at most twice. Hence the question has a negative answer; such functions do not exist.

Second solution: Since $g \circ f$ is a bijection, f is one-to-one and g is onto. Note that $f(g(0)) = 0$. Since g is onto, we can choose a and b with $g(a) = g(0) - 1$ and $g(b) = g(0) + 1$. Then $f(g(a)) = a^2 > 0$ and $f(g(b)) = b^2 > 0$. Let $c = \min(a^2, b^2)/2 > 0$. The intermediate value property guarantees that there is an $x_0 \in (g(a), g(0))$ with $f(x_0) = c$ and an $x_1 \in (g(0), g(b))$ with $f(x_1) = c$. This contradicts the fact that f is one-to-one. Hence no such functions can exist.
   (R. Gelca, second solution by R. Stong)

657. The relation from the statement implies that f is injective, so it must be monotonic. Let us show that f is increasing. Assuming the existence of a decreasing solution f to the functional equation, we can find x_0 such that $f(x_0) \neq x_0$. Rewrite the functional equation as $f(f(x)) - f(x) = f(x) - x$. If $f(x_0) < x_0$, then $f(f(x_0)) < f(x_0)$, and if $f(x_0) > x_0$, then $f(f(x_0)) > f(x_0)$, which both contradict the fact that f is decreasing. Thus any function f that satisfies the given condition is increasing.

   Pick some $a > b$, and set $\Delta f(a) = f(a) - a$ and $\Delta f(b) = f(b) - b$. By adding a constant to f (which yields again a solution to the functional equation), we may assume that $\Delta f(a)$

and $\Delta f(b)$ are positive. Composing f with itself n times, we obtain $f^n(a) = a + n\Delta f(a)$ and $f^n(b) = b + n\Delta f(b)$. Recall that f is an increasing function, so f^n is increasing, and hence $f^n(a) > f^n(b)$, for all n. This can happen only if $\Delta f(a) \geq \Delta f(b)$.

On the other hand, there exists m such that $b + m\Delta f(b) = f^m(b) > a$, and the same argument shows that $\Delta f(f^{m-1}(b)) > \Delta f(a)$. But $\Delta f(f^{m-1}(b)) = \Delta f(b)$, so $\Delta f(b) \geq \Delta f(a)$. We conclude that $\Delta f(a) = \Delta f(b)$, and hence $\Delta f(a) = f(a) - a$ is independent of a. Therefore, $f(x) = x + c$, with $c \in \mathbb{R}$, and clearly any function of this type satisfies the equation from the statement.

658. The answer is yes! We have to prove that for $f(x) = e^{x^2}$, the equation $f'g + fg' = f'g'$ has nontrivial solutions on some interval (a, b). Explicitly, this is the first-order linear equation in g,

$$(1 - 2x)e^{x^2}g' + 2xe^{x^2}g = 0.$$

Separating the variables, we obtain

$$\frac{g'}{g} = \frac{2x}{2x - 1} = 1 + \frac{1}{2x - 1},$$

which yields by integration $\ln g(x) = x + \frac{1}{2}\ln|2x - 1| + C$. We obtain the one-parameter family of solutions

$$g(x) = ae^x\sqrt{|2x - 1|}, \quad a \in \mathbb{R},$$

on any interval that does not contain $\frac{1}{2}$.

(49th W.L. Putnam Mathematical Competition, 1988)

659. Rewrite the equation $f^2 + g^2 = f'^2 + g'^2$ as

$$(f + g)^2 + (f - g)^2 = (f' + g')^2 + (g' - f')^2.$$

This, combined with $f + g = g' - f'$, implies that $(f - g)^2 = (f' + g')^2$.

Let x_0 be the second root of the equation $f(x) = g(x)$. On the intervals $I_1 = (-\infty, 0)$, $I_2 = (0, x_0)$, and $I_3 = (x_0, \infty)$ the function $f - g$ is nonzero; hence so is $f' + g'$. These two functions maintain constant sign on the three intervals; hence $f - g = \varepsilon_j(f' + g')$ on I_j, for some $\varepsilon_j \in \{-1, 1\}, j = 1, 2, 3$.

If on any of these intervals $f - g = f' + g'$, then since $f + g = g' - f'$ it follows that $f = g'$ on that interval, and so $g' + g = g' - g''$. This implies that g satisfies the equation $g'' + g = 0$, or that $g(x) = A\sin x + B\cos x$ on that interval. Also, $f(x) = g'(x) = A\cos x - B\sin x$.

If $f - g = -f' - g'$ on some interval, then using again $f + g = g' - f'$, we find that $g = g'$ on that interval. Hence $g(x) = C_1e^x$. From the fact that $f = -f'$, we obtain $f(x) = C_2e^{-x}$.

Assuming that f and g are exponentials on the interval $(0, x_0)$, we deduce that $C_1 = g(0) = f(0) = C_2$ and that $C_1e^{x_0} = g(x_0) = f(x_0) = C_2e^{-x}$. These two inequalities cannot hold simultaneously, unless f and g are identically zero, ruled out by the hypothesis of the problem. Therefore, $f(x) = A\cos x - B\sin x$ and $g(x) = A\sin x + B\cos x$ on $(0, x_0)$, and consequently $x_0 = \pi$.

On the intervals $(-\infty, 0]$ and $[x_0, \infty)$ the functions f and g cannot be periodic, since then the equation $f = g$ would have infinitely many solutions. So on these intervals the functions are exponentials. Imposing differentiability at 0 and π, we obtain $B = A$, $C_1 = A$ on I_1 and $C_1 = -Ae^{-\pi}$ on I_3 and similarly $C_2 = A$ on I_1 and $C_2 = -Ae^{\pi}$ on I_3. Hence the answer to the problem is

$$f(x) = \begin{cases} Ae^x & \text{for} \quad x \in (-\infty, 0], \\ A(\sin x + \cos x) & \text{for} \quad x \in (0, \pi], \\ -Ae^{-x+\pi} & \text{for} \quad x \in (\pi, \infty), \end{cases}$$

$$g(x) = \begin{cases} Ae^x & \text{for} \quad x \in (-\infty, 0], \\ A(\sin x - \cos x) & \text{for} \quad x \in (0, \pi], \\ -Ae^{x-\pi} & \text{for} \quad x \in (\pi, \infty), \end{cases}$$

where A is some nonzero constant.

(Romanian Mathematical Olympiad, 1976, proposed by V. Matrosenco)

660. The idea is to integrate the equation using an integrating factor. If instead we had the first-order differential equation $(x^2 + y^2)dx + xydy = 0$, then the standard method finds x as an integrating factor. So if we multiply our equation by f to transform it into

$$(f^3 + fg^2)f' + f^2 gg' = 0,$$

then the new equation is equivalent to

$$\left(\frac{1}{4}f^4 + \frac{1}{2}f^2 g^2\right)' = 0.$$

Therefore, f and g satisfy

$$f^4 + 2f^2 g^2 = C,$$

for some real constant C. In particular, f is bounded.

(R. Gelca)

661. The idea is to write the equation as

$$Bydx + Axdy + x^m y^n (Dydx + Cxdy) = 0,$$

then find an integrating factor that integrates simultaneously $Bydx + Axdy$ and $x^m y^n (Dydx + Cxdy)$. An integrating factor of $Bydx + Axdy$ will be of the form $x^{-1}y^{-1}\phi_1(x^B y^A)$, while an integrating factor of $x^m y^n (Dydx + Cxdy) = Dx^m y^{n+1}dx + Cx^{m+1}y^n dy$ will be of the form $x^{-m-1}y^{-n-1}\phi_2(x^D y^C)$, where ϕ_1 and ϕ_2 are one-variable functions. To have the same integrating factor for both expressions, we should have

$$x^m y^n \phi_1(x^B y^A) = \phi_2(x^D y^C).$$

It is natural to try power functions, say $\phi_1(t) = t^p$ and $\phi_2(t) = t^q$. The equality condition gives rise to the system

$$Ap - Cq = -n,$$
$$Bp - Dq = -m,$$

which according to the hypothesis can be solved for p and q. Using Cramer's rule, we find that

$$p = \frac{Bn - Am}{AD - BC}, \quad q = \frac{Dn - Cm}{AD - BC}.$$

Multiplying the equation by $x^{-1}y^{-1}(x^B y^A)^p = x^{-1-m}y^{-1-n}(x^D y^C)^q$ and integrating, we obtain

$$\frac{1}{p+1}(x^B y^A)^{p+1} + \frac{1}{q+1}(x^D y^C)^{q+1} = \text{constant},$$

which gives the solution in implicit form.

(M. Ghermănescu, *Ecuaţii Diferenţiale* (*Differential Equations*), Editura Didactică şi Pedagogică, Bucharest, 1963)

662. The differential equation can be rewritten as

$$e^{y' \ln y} = e^{\ln x}.$$

Because the exponential function is injective, this is equivalent to $y' \ln y = \ln x$. Integrating, we obtain the algebraic equation $y \ln y - y = x \ln x - x + C$, for some constant C. The initial condition yields $C = 0$. We are left with finding all differentiable functions y such that

$$y \ln y - y = x \ln x - x.$$

Let us focus on the function $f(t) = t \ln t - t$. Its derivative is $f'(t) = \ln t$, which is negative if $t < 1$ and positive if $t > 1$. The minimum of f is at $t = 1$, and is equal to -1. An easy application of L'Hôpital's rule shows that $\lim_{t \to 0} f(t) = 0$. It follows that the equation $f(t) = c$ fails to have a unique solution precisely when $c \in (0, 1) \cup (1, e)$, in which case it has exactly two solutions.

If we solve algebraically the equation $y \ln y - y = x \ln x - x$ on $(1, e)$, we obtain two possible continuous solutions, one that is greater than 1 and one that is less than 1. The continuity of y at e rules out the second, so on the interval $[1, \infty)$, $y(x) = x$. On $(0, 1)$ again we could have two solutions, $y_1(x) = x$, and some other function y_2 that is greater than 1 on this interval. Let us show that y_2 cannot be extended to a solution having continuous derivative at $x = 1$. On $(1, \infty)$, $y_2(x) = x$, hence $\lim_{x \to 1+} y_2'(x) = 1$. On $(0, 1)$, as seen above, $y_2' \ln y_2 = \ln x$, so $y_2' = \ln x / \ln y_2 < 0$, since $x < 1$, and $y_2(x) > 1$. Hence $\lim_{x \to 1-} y_2'(x) \leq 0$, contradicting the continuity of y_2' at $x = 1$. Hence the only solution to the problem is $y(x) = x$ for all $x \in (0, \infty)$.

(R. Gelca)

663. Define

$$g(x) = f(x)f'\left(\frac{a}{x}\right), \quad x \in (0, \infty).$$

We want to show that g is a constant function.

Substituting $x \to \frac{a}{x}$ in the given condition yields

$$f\left(\frac{a}{x}\right) f'(x) = \frac{a}{x},$$

for all $x > 0$. We have

$$g'(x) = f'(x)f\left(\frac{a}{x}\right) + f(x)f'\left(\frac{a}{x}\right)\left(-\frac{a}{x^2}\right) = f'(x)f\left(\frac{a}{x}\right) - \frac{a}{x^2}f\left(\frac{a}{x}\right)f(x)$$
$$= \frac{a}{x} - \frac{a}{x} = 0,$$

so g is identically equal to some positive constant b. Using the original equation we can write

$$b = g(x) = f(x)f\left(\frac{a}{x}\right) = f(x) \cdot \frac{a}{x} \cdot \frac{1}{f'(x)},$$

which gives

$$\frac{f'(x)}{f(x)} = \frac{a}{bx}.$$

Integrating both sides, we obtain $\ln f(x) = \frac{a}{b}\ln x + \ln c$, where $c > 0$. It follows that $f(x) = cx^{\frac{a}{b}}$, for all $x > 0$. Substituting back into the original equation yields

$$c \cdot \frac{a}{b} \cdot \frac{a^{\frac{a}{b}-1}}{x^{\frac{a}{b}}-1} = \frac{x}{cx^{\frac{a}{b}}},$$

which is equivalent to

$$c^2 a^{\frac{a}{b}} = b.$$

By eliminating c, we obtain the family of solutions

$$f_b(x) = \sqrt{b}\left(\frac{x}{\sqrt{a}}\right)^{\frac{a}{b}}, \quad b > 0.$$

All such functions satisfy the given condition.

(66th W.L. Putnam Mathematical Competition, 2005, proposed by T. Andreescu)

664. Let us look at the solution to the differential equation

$$\frac{\partial y}{\partial x} = f(x, y),$$

passing through some point (x_0, y_0). The condition from the statement implies that along this solution, $\frac{df(x,y)}{dx} = 0$, and so along the solution the function f is constant. This means that the solution to the differential equation with the given initial condition is a line $(y - y_0) = f(x_0, y_0)(x - x_0)$. If for some (x_1, y_1), $f(x_1, y_1) \neq f(x_0, y_0)$, then the lines $(y - y_0) = f(x_0, y_0)(x - x_0)$ and $(y - y_1) = f(x_1, y_1)(x - x_1)$ intersect somewhere, providing two solutions passing through the same point, which is impossible. This shows that f is constant, as desired.

(Soviet Union University Student Mathematical Olympiad, 1976)

665. The equation can be rewritten as

$$(xy)'' + (xy) = 0.$$

Solving, we find $xy = C_1 \sin x + C_2 \cos x$, and hence

$$y = C_1 \frac{\sin x}{x} + C_2 \frac{\cos x}{x},$$

on intervals that do not contain 0.

666. The function $f'(x)f''(x)$ is the derivative of $\frac{1}{2}(f'(x))^2$. The equation is therefore equivalent to

$$f'(x)^2 = \text{constant}.$$

And because $f'(x)$ is continuous, $f'(x)$ itself must be constant, which means that $f(x)$ is linear. Clearly, all linear functions are solutions.

667. The relation from the statement implies right away that f is differentiable. Differentiating

$$f(x) + x \int_0^x f(t)dt - \int_0^x tf(t)dt = 1,$$

we obtain

$$f'(x) + \int_0^x f(t)dt + xf(x) - xf(x) = 0,$$

that is

$$f'(x) + \int_0^x f(t)dt = 0.$$

Again we conclude that f is twice differentiable, and so we can transform this equality into the differential equation $f'' + f = 0$. The general solution is $f(x) = A \cos x + B \sin x$. Substituting in the relation from the statement, we obtain $A = 1$, $B = 0$, that is, $f(x) = \cos x$.

(E. Popa, *Analiză Matematică, Culegere de Probleme* (*Mathematical Analysis, Collection of Problems*), Editura GIL, 2005)

668. The equation is of Laplace type, but we can bypass the standard method once we make the following observation. The associated homogeneous equation can be written as

$$x(y'' + 4y' + 4y) - (y'' + 5y' + 6y) = 0,$$

and the equations $y'' + 4y' + 4y = 0$ and $y'' + 5y' + 6y = 0$ have the common solution $y(x) = e^{-2x}$. This will therefore be a solution to the above equation, as well. To find a solution to the inhomogeneous equation, we use the method of variation of the constant. Set $y(x) = C(x)e^{-2x}$. The equation becomes

$$(x - 1)C'' - C' = x,$$

which as a first order equation has the solution

$$C'(x) = \lambda(x - 1) + (x - 1) \ln |x - 1| - 1.$$

Integrating, we obtain

$$C(x) = \frac{1}{2}(x - 1)^2 \ln |x - 1| + \left(\frac{\lambda}{2} - \frac{1}{4} \right)(x - 1)^2 - x + C_1.$$

If we set $C_2 = \frac{\lambda}{2} - \frac{1}{4}$, then the general solution to the equation from the start is

$$y(x) = e^{-2x} \left[C_1 + C_2(x-1)^2 + \frac{1}{2}(x-1)^2 \ln|x-1| - x \right].$$

(D. Flondor, N. Donciu, *Algebră și Analiză Matematică* (*Algebra and Mathematical Analysis*), Editura Didactică și Pedagogică, Bucharest, 1965)

669. Consider the change of variable $x = \cos t$. Then, by the chain rule,

$$\frac{dy}{dx} = \frac{\dfrac{dy}{dt}}{\dfrac{dx}{dt}} = -\frac{\dfrac{dy}{dt}}{\sin t}$$

and

$$\frac{d^2y}{dx^2} = \frac{\dfrac{d^2y}{dt^2} - \dfrac{dy}{dx}\dfrac{d^2x}{dt^2}}{\left(\dfrac{dx}{dt}\right)^2} = \frac{\dfrac{d^2y}{dt^2}}{\sin^2 t} - \frac{\cos t \dfrac{dy}{dt}}{\sin^3 t}.$$

Substituting in the original equation, we obtain the much simpler

$$\frac{d^2y}{dt^2} + n^2 y = 0.$$

This has the function $y(t) = \cos nt$ as a solution. Hence the original equation admits the solution $y(x) = \cos(n \arccos x)$, which is the nth Chebyshev polynomial.

670. We interpret the differential equation as being posed for a function y of x. In this perspective, we need to write $\frac{d^2x}{dy^2}$ in terms of the derivatives of y with respect to x. We have

$$\frac{dx}{dy} = \frac{1}{\dfrac{dy}{dx}},$$

and using this fact and the chain rule yields

$$\frac{d^2x}{dy^2} = \frac{d}{dy}\left(\frac{1}{\dfrac{dy}{dx}}\right) = \frac{d}{dx}\left(\frac{1}{\dfrac{dy}{dx}}\right) \cdot \frac{dx}{dy}$$

$$= -\frac{1}{\left(\dfrac{dy}{dx}\right)^2} \cdot \frac{d^2y}{dx^2} \cdot \frac{dx}{dy} = -\frac{1}{\left(\dfrac{dy}{dx}\right)^3} \cdot \frac{d^2y}{dx^2}.$$

The equation from the statement takes the form

$$\frac{d^2y}{dx^2}\left(1 - \frac{1}{\left(\dfrac{dy}{dx}\right)^3}\right) = 0.$$

This splits into

$$\frac{d^2y}{dx^2} = 0 \quad \text{and} \quad \left(\frac{dy}{dx}\right)^3 = 1.$$

The first of these has the solutions $y = ax + b$, with $a \neq 0$, because y has to be one-to-one, while the second reduces to $y' = 1$, whose family of solutions $y = x + c$ is included in the first. Hence the answer to the problem consists of the nonconstant linear functions.

(M. Ghermănescu, *Ecuaţii Diferenţiale (Differential Equations)*, Editura Didactică şi Pedagogică, Bucharest, 1963)

671. *First solution*: Multiplying the equation by $e^{-x}y'$ and integrating from 0 to x, we obtain

$$y^2(x) - y^2(0) + 2\int_0^x e^{-t}y'y''dt = 0.$$

The integral in this expression is positive. To prove this we need the following lemma.

Lemma. *Let $f : [0, a] \to \mathbb{R}$ be a continuous function and $\phi : [0, a] \to \mathbb{R}$ a positive, continuously differentiable, decreasing function with $\phi(0) = 1$. Then there exists $c \in [0, a]$ such that*

$$\int_0^a \phi(t)f(t)dt = \int_0^c f(t)dt.$$

Proof. Let $F(x) = \int_0^x f(t)dt$, $x \in [0, a]$, and let α be the negative of the derivative of ϕ, which is a positive function. Integrating by parts, we obtain

$$\int_0^a \phi(t)f(t)dt = \phi(a)F(a) + \int_0^a \alpha(t)F(t)dt = F(a) - \int_0^a (F(a) - F(t))\alpha(t)dt.$$

We are to show that there exists a point c such that

$$F(a) - F(c) = \int_0^a (F(a) - F(t))\alpha(t)dt.$$

If $\int_0^a \alpha(t)dt$ were equal to 1, this would be true by the Mean value theorem applied to the function $F(a) - F(t)$ and the probability measure $\alpha(t)dt$. But in general, this integral is equal to some subunitary number θ, so we can find c' such that the integral is equal to $\theta(F(a) - F(c'))$. But this number is between $F(a) - F(a)$ and $F(a) - F(c')$, so by the intermediate value property, there is a c such that $\theta(F(a) - F(c')) = F(a) - F(c)$. This proves the lemma. □

Returning to the problem, we see that there exists $c \in [0, x]$ such that

$$\int_0^x e^{-t}y'y''dt = \int_0^c y'y''dt = \frac{1}{2}[(y'(c))^2 - (y'(0))^2].$$

In conclusion,

$$y(x)^2 + y'(c)^2 = y(0)^2 + y'(0)^2, \text{ for } x > 0,$$

showing that y is bounded as $x \to \infty$.

Second solution: Use an integrating factor as in the previous solution to obtain

$$y(x)^2 - y(0)^2 + 2 \int_0^x e^{-t} y' y'' dt = 0.$$

Then integrate by parts to obtain

$$y(x)^2 + e^{-x} y'(x)^2 + \int_0^x e^{-t} (y'(t))^2 dt = y(0)^2 + y'(0)^2.$$

Because every term on the left is nonnegative, it follows immediately that

$$|y(x)| \le (y(0)^2 + y'(0)^2)^{1/2}$$

is bounded, and we are done.

(27th W.L. Putnam Mathematical Competition, 1966)

672. We have

$$y_1''(t) + y_1(t) = \int_0^\infty \frac{t^2 e^{-tx}}{1 + t^2} dt + \int_0^\infty \frac{e^{-tx}}{1 + t^2} dt = \int_0^\infty e^{-tx} dt = \frac{1}{x}.$$

Also, integrating by parts, we obtain

$$y_2(x) = \left. \frac{-\cos t}{t + x} \right|_0^\infty - \int_0^\infty \frac{\cos t}{(t + x)^2} dt = \frac{1}{x} - \left. \frac{\sin t}{(t + x)^2} \right|_0^\infty - \int_0^\infty \frac{2 \sin t}{(t + x)^3} dt$$

$$= \frac{1}{x} - y_2''(x).$$

Since the functions y_1 and y_2 satisfy the same inhomogeneous equation, their difference $y_1 - y_2$ satisfies the homogeneous equation $y'' + y = 0$, and hence is of the form $A \cos x + B \sin x$. On the other hand,

$$\lim_{x \to \infty} (y_1(x) - y_2(x)) = \lim_{x \to \infty} y_1(x) - \lim_{x \to \infty} y_2(x) = 0,$$

which implies that $A = B = 0$, and therefore $y_1 = y_2$, as desired.

(M. Ghermănescu, *Ecuații Diferențiale (Differential Equations)*, Editura Didactică şi Pedagogică, Bucharest, 1963)

673. Let $F(t) = \int_0^t f(s) ds$ be the antiderivative of f that is 0 at the origin. The inequality from the problem can be written as

$$\frac{F'(t)}{\sqrt{1 + 2F(t)}} \le 1,$$

which now reminds us of the method of separation of variables. The left-hand side is the derivative of $\sqrt{1 + 2F(t)}$, a function whose value at the origin is 1. Its derivative is dominated by the derivative of $g(t) = t + 1$, another function whose value at the origin is also 1. Integrating, we obtain

$$\sqrt{1 + 2F(t)} \leq t + 1.$$

Look at the relation from the statement. It says that $f(t) \leq \sqrt{1 + 2F(t)}$. Hence the conclusion.

(P.N. de Souza, J.N. Silva, *Berkeley Problems in Mathematics*, Springer, 2004)

674. We will use the "integrating factor" e^x. The inequality

$$f''(x)e^x + 2f'(x)e^x + f(x)e^x \geq 0$$

is equivalent to $(f(x)e^x)'' \geq 0$. So the function $f(x)e^x$ is convex, which means that it attains its maximum at one of the endpoints of the interval of definition. We therefore have $f(x)e^x \leq \max(f(0), f(1)e) = 0$, and so $f(x) \leq 0$ for all $x \in [0, 1]$.

(P.N. de Souza, J.N. Silva, *Berkeley Problems in Mathematics*, Springer, 2004)

675. Assume that such a function exists. Because $f'(x) = f(f(x)) > 0$, the function is strictly increasing.

The monotonicity and the positivity of f imply that $f(f(x)) > f(0)$ for all x. Thus $f(0)$ is a lower bound for $f'(x)$. Integrating the inequality $f(0) < f'(x)$ for $x < 0$, we obtain

$$\int_x^0 f(0)dt \leq \int_x^0 f'(t)dt$$

that is $-f(0)x \leq f(0) - f(x)$, so $f(x) < f(0) + f(0)x = (x + 1)f(0)$. But then for $x \leq -1$, we would have $f(x) \leq 0$, contradicting the hypothesis that $f(x) > 0$ for all x. We conclude that such a function does not exist.

(9th International Mathematics Competition for University Students, 2002)

676. We use the separation of variables, writing the relation from the statement as

$$\sum_{i=1}^n \frac{P'(x)}{P(x) - x_i} = \frac{n^2}{x}.$$

Integrating, we obtain

$$\sum_{i=1}^n \ln |P(x) - x_i| = n^2 \ln C|x|,$$

where C is some positive constant. After adding the logarithms on the left we have

$$\ln \prod_{i=1}^n |P(x) - x_i| = \ln C^{n^2} |x|^{n^2},$$

and so

$$\left| \prod_{i=1}^n (P(x) - x_i) \right| = k|x|^{n^2},$$

with $k = C^{n^2}$. Eliminating the absolute values, we obtain

$$P(P(x)) = \lambda x^{n^2}, \ \lambda \in \mathbb{R}.$$

We end up with an algebraic equation. An easy induction can prove that the coefficient of the term of kth degree is 0 for $k < n$. Hence $P(x) = ax^n$, with a some constant, are the only polynomials that satisfy the relation from the statement.

(*Revista Matematică din Timişoara* (*Timişoara Mathematics Gazette*), proposed by T. Andreescu)

677. The idea is to use an "integrating factor" that transforms the quantity under the integral into the derivative of a function. We already encountered this situation in a previous problem, and should recognize that the integrating factor is e^{-x}. We can therefore write

$$\int_0^1 |f'(x) - f(x)| dx = \int_0^1 |f'(x)e^{-x} - f(x)e^{-x}| e^x dx = \int_0^1 |(f(x)e^{-x})'| e^x dx$$

$$\geq \int_0^1 |(f(x)e^{-x})'| dx = f(1)e^{-1} - f(0)e^{-0} = \frac{1}{e}.$$

We have found a lower bound. We will prove that it is the greatest lower bound. Define $f_a : [0, 1] \to \mathbb{R}$,

$$f_a(x) = \begin{cases} \dfrac{e^{a-1}}{a} x & \text{for } x \in [0, a], \\ e^{x-1} & \text{for } x \in [a, 1]. \end{cases}$$

The functions f_a are continuous but not differentiable at a, but we can smoothen this "corner" without altering too much the function or its derivative. Ignoring this problem, we can write

$$\int_0^1 |f_a'(x) - f_a(x)| dx = \int_0^a \left| \frac{e^{a-1}}{a} - \frac{e^{a-1}}{a} x \right| dx = \frac{e^{a-1}}{a} \left(a - \frac{a^2}{2} \right) = e^{a-1} \left(1 - \frac{a}{2} \right).$$

As $a \to 0$, this expression approaches $\frac{1}{e}$. This proves that $\frac{1}{e}$ is the desired greatest lower bound.

(41st W.L. Putnam Mathematical Competition, 1980)

678. Without loss of generality, we can assume that $a < b < c$. Set $\alpha = \frac{a}{c}$ and $\beta = \frac{b}{c}$, and $t = cx$. Choose n such that $g^{(n)} \equiv 0$ and $\alpha^n + \beta^n < 1$. Differentiate the equation from the statement with respect to t to obtain the nth order differential equation

$$f^{(n)}(t) = -\alpha^n f^{(n)}(\alpha t) - \beta^n f^{(n)}(\beta t).$$

Fix $u > 0$ and set M be the maximum of $f^{(n)}$ on the interval $[-u, u]$. If $M > 0$, then

$$|f^n(t)| \leq |\alpha^n f^{(n)}(\alpha t)| + |\beta^n f^{(n)}(\beta t)| \leq (\alpha^n + \beta^n) M < M.$$

Taking the supremum of the left-hand side over $t \in [-u, u]$, we obtain $M < M$. So $M = 0$. Varying u we obtain that $f^{(n)}$ is identically equal to 0 on $\mathbb{R}$. So f is a polynomial function.

(*Mathematical Reflections*, proposed by M. Bǎluna and M. Piticari)

Geometry and Trigonometry

679. *First solution:* This is the famous Jacobi identity. Identifying vectors with matrices in the Lie algebra so(3), we compute

$$\vec{u} \times (\vec{v} \times \vec{w}) + \vec{v} \times (\vec{w} \times \vec{u}) + \vec{w} \times (\vec{u} \times \vec{v})$$
$$= U(VW - WV) - (VW - WV)U + V(WU - UW) - (WU - UW)V$$
$$+ W(UV - VU) - (UV - VU)W$$
$$= UVW - UWV - VWU + WVU + VWU - VUW - WUV + UWV$$
$$+ WUV - WVU - UVW + VUW.$$

All terms of the latter sum cancel, giving the answer zero.

Second solution: We use the BAC-CAB identity

$$\vec{a} \times (\vec{b} \times \vec{c}) = \vec{b}(\vec{a} \cdot \vec{c}) - \vec{c}(\vec{a} \cdot \vec{b}).$$

We write

$$\vec{u} \times (\vec{v} \times \vec{w}) + \vec{v} \times (\vec{w} \times \vec{u}) + \vec{w} \times (\vec{u} \times \vec{v})$$
$$= \vec{v}(\vec{u} \cdot \vec{w}) - \vec{w}(\vec{u} \cdot \vec{v}) + \vec{w}(\vec{v} \cdot \vec{u}) - \vec{u}(\vec{v} \cdot \vec{w}) + \vec{u}(\vec{w} \cdot \vec{v}) - \vec{v}(\vec{w} \cdot \vec{u}).$$

Given that the dot product is commutative, the terms cancel in pairs and so this is equal to zero.

680. One checks easily that $\vec{u} + \vec{v} + \vec{w} = 0$; hence $\vec{u}, \vec{v}, \vec{w}$ form a triangle. We compute

$$\vec{u} \cdot \vec{c} = (\vec{b} \cdot \vec{c})(\vec{a} \cdot \vec{c}) - (\vec{c} \cdot \vec{a})(\vec{b} \cdot \vec{c}) = 0.$$

It follows that $\vec{u}$ and $\vec{c}$ are orthogonal. Similarly, we prove that $\vec{v}$ is orthogonal to $\vec{a}$, and $\vec{w}$ is orthogonal to $\vec{b}$. Hence the sides of the triangle formed with $\vec{u}, \vec{v}, \vec{w}$ are

© Springer International Publishing AG 2017

R. Gelca and T. Andreescu, *Putnam and Beyond*, DOI 10.1007/978-3-319-58988-6

perpendicular to the sides of the triangle formed with $\vec{a}$, $\vec{b}$, $\vec{c}$. This shows that the two triangles have equal angles so they are similar, and we are done.

(Romanian Mathematical Olympiad, 1976, proposed by M. Chiriță)

681. Multiply the second equation on the left by $\vec{a}$ to obtain

$$\vec{a} \times (\vec{x} \times \vec{b}) = \vec{a} \times \vec{c}.$$

Using the BAC-CAB formula, we transform this into

$$(\vec{a} \cdot \vec{b})\vec{x} - (\vec{a} \cdot \vec{x})\vec{b} = \vec{a} \times \vec{c}.$$

Hence the solution to the equation is

$$\vec{x} = \frac{m}{\vec{a} \cdot \vec{b}}\vec{b} + \frac{1}{\vec{a} \cdot \vec{b}}\vec{a} \times \vec{c}.$$

(C. Coşniţă, I. Sager, I. Matei, I. Dragotă, *Culegere de Probleme de Geometrie Analitică* (*Collection of Problems in Analytical Geometry*), Editura Didactică şi Pedagogică, Bucharest, 1963)

682. The vectors $\vec{b} - \vec{a}$ and $\vec{c} - \vec{a}$ belong to the plane under discussion, so the vector $(\vec{b} - \vec{a}) \times (\vec{c} - \vec{a})$ is perpendicular to this plane. Multiplying out, we obtain

$$(\vec{b} - \vec{a}) \times (\vec{c} - \vec{a}) = \vec{b} \times \vec{c} - \vec{a} \times \vec{c} - \vec{b} \times \vec{a}$$
$$= \vec{b} \times \vec{c} + \vec{c} \times \vec{a} + \vec{a} \times \vec{b}.$$

Hence the conclusion.

683. The hypothesis implies that

$$(\vec{a} \times \vec{b}) - (\vec{b} \times \vec{c}) = \vec{0}.$$

It follows that $\vec{b} \times (\vec{a} + \vec{c}) = \vec{0}$, hence $\vec{b} = \lambda(\vec{a} + \vec{c})$, where λ is a scalar. Analogously, we deduce $\vec{c} \times (\vec{a} + \vec{b}) = \vec{0}$, and substituting the formula we found for $\vec{b}$, we obtain

$$\vec{c} \times (\vec{a} + \lambda\vec{a} + \lambda\vec{c}) = \vec{0}.$$

Hence $(1 + \lambda)\vec{c} \times \vec{a} = \vec{0}$. It follows that $\lambda = -1$ and so $\vec{b} = -\vec{a} - \vec{c}$. Therefore, $\vec{a} + \vec{b} + \vec{c} = \vec{0}$.

(C. Coşniţă, I. Sager, I. Matei, I. Dragotă, *Culegere de Probleme de Geometrie Analitică* (*Collection of Problems in Analytical Geometry*), Editura Didactică şi Pedagogică, Bucharest, 1963)

684. Differentiating the equation from the statement, we obtain

$$\vec{u}' \times \vec{u}' + \vec{u} \times \vec{u}'' = \vec{u} \times \vec{u}'' = \vec{v}'.$$

It follows that the vectors $\vec{u}$ and $\vec{v}'$ are perpendicular. But the original equation shows that $\vec{u}$ and $\vec{v}$ are also perpendicular, which means that $\vec{u}$ stays parallel to $\vec{v} \times \vec{v}'$. Then we can write $\vec{u} = f\vec{v} \times \vec{v}'$ for some scalar function $f = f(t)$. The left-hand side of the original equation is therefore equal to

$$f(\vec{v} \times \vec{v}') \times [f'\vec{v} \times \vec{v}' + f\vec{v}' \times \vec{v}' + f\vec{v} \times \vec{v}'']$$

$$= f^2(\vec{v} \times \vec{v}') \times (\vec{v} \times \vec{v}'').$$

By the BAC-CAB formula this is further equal to

$$f^2(\vec{v}'' \cdot (\vec{v} \times \vec{v}')\vec{v} - \vec{v} \cdot (\vec{v} \times \vec{v}')\vec{v}) = f^2((\vec{v} \times \vec{v}') \cdot \vec{v}'')\vec{v}.$$

The equation reduces therefore to

$$f^2((\vec{v} \times \vec{v}') \cdot \vec{v}'')\vec{v} = \vec{v}.$$

By hypothesis $\vec{v}$ is never equal to $\vec{0}$, so the above equality implies

$$f = \frac{1}{\sqrt{(\vec{v} \times \vec{v}) \cdot \vec{v}''}}.$$

So the equation can be solved only if the frame $(\vec{v}, \vec{v}', \vec{v}'')$ consists of linearly independent vectors and is positively oriented and in that case the solution is

$$\vec{u} = \frac{1}{\sqrt{\mathrm{Vol}(\vec{v}, \vec{v}', \vec{v}'')}} \vec{v} \times \vec{v}',$$

where $\mathrm{Vol}(\vec{v}, \vec{v}', \vec{v}'')$ denotes the volume of the parallelepiped determined by the three vectors.

(*Revista Matematică din Timişoara* (*Timişoara Mathematics Gazette*), proposed by M. Ghermănescu)

685. (a) Yes: simply rotate the plane 90° about some axis perpendicular to it. For example, in the xy-plane we could map each point (x, y) to the point $(y, -x)$.

(b) Suppose such a bijection existed. In vector notation, the given condition states that

$$(\vec{a} - \vec{b}) \cdot (f(\vec{a}) - f(\vec{b})) = 0$$

for any three-dimensional vectors $\vec{a}$ and $\vec{b}$.

Assume without loss of generality that f maps the origin to itself; otherwise, $g(\vec{p}) = f(\vec{p}) - f(\vec{0})$ is still a bijection and still satisfies the above equation. Plugging $\vec{b} = (0, 0, 0)$ into the above equation, we obtain that $\vec{a} \cdot f(\vec{a}) = 0$ for all $\vec{a}$. The equation reduces to

$$\vec{a} \cdot f(\vec{b}) + \vec{b} \cdot f(\vec{a}) = 0.$$

Given any vectors $\vec{a}, \vec{b}, \vec{c}$ and any real numbers m, n, we then have

$$m(\vec{a} \cdot f(\vec{b}) + \vec{b} \cdot f(\vec{a})) = 0,$$

$$n(\vec{a} \cdot f(\vec{c}) + \vec{c} \cdot f(\vec{a})) = 0,$$
$$a \cdot f(m\vec{b} + n\vec{c}) + (m\vec{b} + n\vec{c}) \cdot f(\vec{a}) = 0.$$

Adding the first two equations and subtracting the third gives

$$\vec{a} \cdot (mf(\vec{b}) + nf(\vec{c}) - f(m\vec{b} + n\vec{c})) = 0.$$

Because this is true for any vector $\vec{a}$, we must have

$$f(m\vec{b} + n\vec{c}) = mf(\vec{b}) + nf(\vec{c}).$$

Therefore, f is linear, and it is determined by the images of the unit vectors $\vec{i} = (1, 0, 0)$, $\vec{j} = (0, 1, 0)$, and $\vec{k} = (0, 0, 1)$. If

$$f(\vec{i}) = (a_1, a_2, a_3), \quad f(\vec{j}) = (b_1, b_2, b_3), \quad \text{and} \quad f(\vec{k}) = (c_1, c_2, c_3),$$

then for a vector $\vec{x}$ we have

$$f(\vec{x}) = \begin{bmatrix} a_1 & b_1 & c_1 \\ a_2 & b_2 & c_2 \\ a_3 & b_3 & c_3 \end{bmatrix} \vec{x}.$$

Substituting in $f(\vec{a}) \cdot \vec{a} = 0$ successively $\vec{a} = \vec{i}, \vec{j}, \vec{k}$, we obtain $a_1 = b_2 = c_3 = 0$. Then substituting in $\vec{a} \cdot f(\vec{b}) + \vec{b} \cdot f(\vec{a}) = 0$, $(\vec{a}, \vec{b}) = (\vec{i}, \vec{j}), (\vec{j}, \vec{k}), (\vec{k}, \vec{i})$, we obtain $b_1 = -a_2, c_2 = -b_3, c_1 = -a_3$.

Setting $k_1 = c_2, k_2 = -c_1$, and $k_3 = b_1$ yields

$$f(k_1 \vec{i} + k_2 \vec{j} + k_3 \vec{k}) = k_1 f(\vec{i}) + k_2 f(\vec{j}) + k_3 f(\vec{k}) = \vec{0}.$$

Because f is injective and $f(\vec{0}) = \vec{0}$, this implies that $k_1 = k_2 = k_3 = 0$. Then $f(\vec{x}) = 0$ for all $\vec{x}$, contradicting the assumption that f was a surjection. Therefore, our original assumption was false, and no such bijection exists.

(Team Selection Test for the International Mathematical Olympiad, Belarus, 1999)

686. The important observation is that

$$A * B = AB - \frac{1}{2}\text{tr}(AB),$$

which can be checked by hand. The identity is therefore equivalent to

$$CBA - BCA + ABC - ACB = -\frac{1}{2}\text{tr}(AC)B + \frac{1}{2}\text{tr}(AB)C.$$

And this is the BAC-CAB identity once we notice that

$$\vec{a} \cdot \vec{b} = -\frac{1}{2}\text{tr}(AB).$$

687. An easy computation shows that the map $f : \mathbb{R}^3 \to su(2)$,

$$f(x, y, z) = \begin{pmatrix} -iz & y - ix \\ y + ix & iz \end{pmatrix},$$

has the desired property.

688. We will show that for n vectors, if each line has at least $k < n/2$ on each side, then the sum $\vec{s}$ of the vectors does not exceed $n - 2k$.

Choose the positive Ox ray to be in the direction of the sum $\vec{s}$ of the n vectors. Let the vectors in the upper half-plane be numbered clockwise as $\vec{a}_1, \vec{a}_2, \vec{a}_3, \ldots, \vec{a}_k, \ldots$, and the vectors in the lower half-plane be numbered counterclockwise as $\vec{b}_1, \vec{b}_2, \vec{b}_3, \ldots, \vec{b}_k, \ldots$ (there are at least k in each group by hypothesis).

Also there are at least k vectors in the left half-plane, let these be (in counterclockwise order) $\ldots, \vec{a}_\ell, \vec{a}_{\ell-1}, \cdots, \vec{a}_1, \vec{b}_1, \vec{b}_2, \vec{b}_{k-l}, \ldots$. There are k vectors on each side of the line of support of $\vec{a}_{\ell-j}, j = 0, 1, 2, \ldots, \ell - 1$, and for this reason $\vec{b}_{k-\ell+j+1}$ is to the "left" of this line, meaning that $\vec{a}_{\ell-j} + \vec{b}_{k-\ell+j+1}$, has a negative projection on the x-axis. Similarly $\vec{b}_{k-\ell-j} + \vec{a}_{\ell+j+1}$ has a negative projection on the x-axis.

It follows that the sum

$$\vec{s}_0 = \vec{a}_1 + \vec{a}_2 + \cdots + \vec{a}_k + \vec{b}_1 + \vec{b}_2 + \cdots + \vec{b}_k$$

has a negative projection on the x-axis. Let $\vec{s}_1$ be the sum of the remaining $2n - k$ vectors. By the triangle inequality $\|\vec{s}_1\| \le n - 2k$. Then $\vec{s}$, $\vec{s}_0$ and $\vec{s}_1$ form and obtuse triangle, with $\vec{s}_1$ opposite to the obtuse angle. Thus

$$\|\vec{s}\| \le \|\vec{s}_1\| \le n - 2k$$

and we are done.

(*Kvant (Quantum)*, proposed by P.A. Kalugin and V.V. Prasolov)

689. Denoting by $\vec{A}, \vec{B}, \vec{C}, \vec{A}', \vec{B}', \vec{C}'$ the position vectors of the vertices of the two triangles, the condition that the triangles have the same centroid reads

$$\vec{A} + \vec{B} + \vec{C} = \vec{A}' + \vec{B}' + \vec{C}'.$$

Subtracting the left-hand side, we obtain

$$\overrightarrow{AA'} + \overrightarrow{BB'} + \overrightarrow{CC'} = \vec{0}.$$

This shows that $\overrightarrow{AA'}, \overrightarrow{BB'}, \overrightarrow{CC'}$ form a triangle, as desired.

690. Set $\vec{v}_1 = \overrightarrow{AB}, \vec{v}_2 = \overrightarrow{BC}, \vec{v}_3 = \overrightarrow{CD}, \vec{v}_4 = \overrightarrow{DA}, \vec{u}_1 = \overrightarrow{A'B'}, \vec{u}_2 = \overrightarrow{B'C'}, \vec{u}_3 = \overrightarrow{C'D'}, \vec{u}_4 = \overrightarrow{D'A'}$. By examining Figure 78 we can write the system of equations

$$2\vec{v}_2 - \vec{v}_1 = \vec{u}_1,$$

$$2\vec{v}_3 - \vec{v}_2 = \vec{u}_2,$$

$$2\vec{v}_4 - \vec{v}_3 = \vec{u}_3,$$

$$2\vec{v}_1 - \vec{v}_4 = \vec{u}_4,$$

in which the right-hand side is known. Solving, we obtain

$$\vec{v}_1 = \frac{1}{15}\vec{u}_1 + \frac{2}{15}\vec{u}_2 + \frac{4}{15}\vec{u}_3 + \frac{8}{15}\vec{u}_4,$$

and the analogous formulas for $\vec{v}_2$, $\vec{v}_3$, and $\vec{v}_4$.

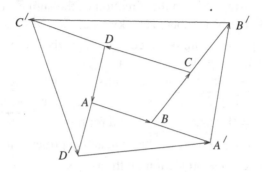

Figure 78

Since the rational multiple of a vector and the sum of two vectors can be constructed with straightedge and compass, we can construct the vectors $\vec{v}_i$, $i = 1, 2, 3, 4$. Then we take the vectors $\overrightarrow{A'B} = -\vec{v}_1$, $\overrightarrow{B'C} = -\vec{v}_2$, $\overrightarrow{C'D} = -\vec{v}_3$, and $\overrightarrow{D'A} = -\vec{v}_4$ from the points A', B', C', and D' to recover the vertices B, C, D, and A.

Remark. Maybe we should elaborate more on how one effectively does these constructions. The sum of two vectors is obtained by constructing the parallelogram they form. Parallelograms can also be used to translate vectors. An integer multiple of a vector can be constructed by drawing its line of support and then measuring several lengths of the vector with the compass. This construction enables us to obtain segments divided into an arbitrary number of equal parts. In order to divide a given segment into equal parts, form a triangle with it and an already divided segment, then draw lines parallel to the third side and use Thales' theorem.

691. Let O be the intersection of the perpendicular bisectors of A_1A_2 and B_1B_2. We want to show that O is on the perpendicular bisector of C_1C_2. This happens if and only if $(\overrightarrow{OC_1} + \overrightarrow{OC_2}) \cdot \overrightarrow{C_1C_2} = 0$.

Set $\overrightarrow{OA} = \vec{l}$, $\overrightarrow{OB} = \vec{m}$, $\overrightarrow{OC} = \vec{n}$, $\overrightarrow{AA_2} = \vec{a}$, $\overrightarrow{BB_2} = \vec{b}$, $\overrightarrow{CC_2} = \vec{c}$. That the perpendicular bisectors of A_1A_2 and B_1B_2 pass through O can be written algebraically as

$$(2\vec{l} + \vec{a} + \vec{c}) \cdot (\vec{c} - \vec{a}) = 0 \quad \text{and} \quad (2\vec{m} + \vec{a} + \vec{b}) \cdot (\vec{a} - \vec{b}) = 0.$$

The orthogonality of the sides of the rectangles translates into formulas as

$$(\vec{m} - \vec{l}) \cdot \vec{a} = 0, \quad (\vec{m} - \vec{n}) \cdot \vec{b} = 0, \quad (\vec{n} - \vec{l}) \cdot \vec{c} = 0.$$

We are required to prove that $(2\vec{n} + \vec{b} + \vec{c}) \cdot (\vec{b} - \vec{c}) = 0$. And indeed,

$$(2\vec{n} + \vec{b} + \vec{c}) \cdot (\vec{c} - \vec{b}) = 2\vec{n} \cdot \vec{c} - 2\vec{n} \cdot \vec{b} + \vec{c}^2 - \vec{b}^2$$
$$= 2(\vec{m} - \vec{l}) \cdot \vec{a} + 2\vec{l} \cdot \vec{c} - 2\vec{m} \cdot \vec{b} + \vec{c}^2 - \vec{b}^2$$
$$= 2\vec{m} \cdot \vec{a} - 2\vec{m} \cdot \vec{b} + \vec{a}^2 - \vec{b}^2 + 2\vec{l} \cdot \vec{c} - 2\vec{l} \cdot \vec{a} - \vec{a}^2 + \vec{c}^2 = 0.$$

Hence the conclusion.

692. Let H' be the orthocenter of triangle ACD. The quadrilaterals $HPBQ$ and $HCH'A$ satisfy $HC \perp BP$, $H'C \perp HP$, $H'A \perp HQ$, $AH \perp BQ$, $AC \perp HB$ (see Figure 79). The conclusion follows from a more general result.

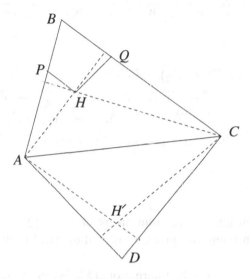

Figure 79

Lemma. *Let $MNPQ$ and $M'N'P'Q'$ be two quadrilaterals such that $MN \perp N'P'$, $NP \perp M'N'$, $PQ \perp Q'M'$, $QM \perp P'Q'$, and $MP \perp N'Q'$. Then $NQ \perp M'P'$.*

Proof. Let $\overrightarrow{MN} = \vec{v}_1$, $\overrightarrow{NP} = \vec{v}_2$, $\overrightarrow{PQ} = \vec{v}_3$, $\overrightarrow{QM} = \vec{v}_4$, and $\overrightarrow{M'N'} = \vec{w}_1$, $\overrightarrow{N'P'} = \vec{w}_2$, $\overrightarrow{P'Q'} = \vec{w}_3$, $\overrightarrow{Q'M'} = \vec{w}_4$. The conditions from the statement can be written in vector form as

$$\vec{v}_1 \cdot \vec{w}_2 = \vec{v}_2 \cdot \vec{w}_1 = \vec{v}_3 \cdot \vec{w}_4 = \vec{v}_4 \cdot \vec{w}_3 = 0,$$

$$\vec{v}_1 + \vec{v}_2 + \vec{v}_3 + \vec{v}_4 = \vec{w}_1 + \vec{w}_2 + \vec{w}_3 + \vec{w}_4 = \vec{0},$$

$$(\vec{v}_1 + \vec{v}_2) \cdot (\vec{w}_2 + \vec{w}_3) = 0.$$

We are to show that

$$(\vec{v}_2 + \vec{v}_3) \cdot (\vec{w}_1 + \vec{w}_2) = 0.$$

First, note that

$$0 = (\vec{v}_1 + \vec{v}_2)(\vec{w}_2 + \vec{w}_3) = \vec{v}_1 \cdot \vec{w}_2 + \vec{v}_1 \cdot \vec{w}_3 + \vec{v}_2 \cdot \vec{w}_2 + \vec{v}_2 \cdot \vec{w}_3$$
$$= \vec{v}_1 \cdot \vec{w}_3 + \vec{v}_2 \cdot \vec{w}_2 + \vec{v}_2 \cdot \vec{w}_3.$$

Also, the dot product that we are supposed to show is zero is equal to

$$(\vec{v}_2 + \vec{v}_3) \cdot (\vec{w}_1 + \vec{w}_2) = \vec{v}_2 \cdot \vec{w}_1 + \vec{v}_2 \cdot \vec{w}_2 + \vec{v}_3 \cdot \vec{w}_1 + \vec{v}_3 \cdot \vec{w}_2$$
$$= \vec{v}_2 \cdot \vec{w}_2 + \vec{v}_3 \cdot \vec{w}_1 + \vec{v}_3 \cdot \vec{w}_2.$$

This would indeed equal zero if we showed that

$$\vec{v}_1 \cdot \vec{w}_3 + \vec{v}_2 \cdot \vec{w}_3 = \vec{v}_3 \cdot \vec{w}_1 + \vec{v}_3 \cdot \vec{w}_2.$$

And indeed,

$$\vec{v}_1 \cdot \vec{w}_3 + \vec{v}_2 \cdot \vec{w}_3 = (\vec{v}_1 + \vec{v}_2) \cdot \vec{w}_3$$
$$= -(\vec{v}_3 + \vec{v}_4) \cdot \vec{w}_3 = -\vec{v}_3 \cdot \vec{w}_3 - \vec{v}_4 \cdot \vec{w}_3 = -\vec{v}_3 \cdot \vec{w}_3$$
$$= -\vec{v}_3 \cdot \vec{w}_3 - \vec{v}_3 \cdot \vec{w}_4 = -\vec{v}_3 \cdot (\vec{w}_3 + \vec{w}_4)$$
$$= \vec{v}_3 \cdot (\vec{w}_1 + \vec{w}_2) = \vec{v}_3 \cdot \vec{w}_1 + \vec{v}_3 \cdot \vec{w}_2.$$

The lemma is proved. $\square$

Remark. A. Dang gave an alternative solution by observing that triangles *AHC* and *QHP* are orthological, and then using the property of orthological triangles proved by us in the introduction.

(Indian Team Selection Test for the International Mathematical Olympiad, 2005, proposed by R. Gelca)

693. Let $\vec{a}, \vec{b}, \vec{c}, \vec{d}$, and $\vec{p}$ denote vectors from a common origin to the vertices A, B, C, D of the tetrahedron and to the point P of concurrency of the four lines. Then the vector equation for the altitude from A is given by

$$\vec{r}_a = \vec{a} + \lambda[(\vec{b} + \vec{c} + \vec{d})/3 - \vec{p}].$$

The position vector of the point corresponding to $\lambda = 3$ is $\vec{a} + \vec{b} + \vec{c} + \vec{d} - 3\vec{p}$, which is the same for all four vertices of the tetrahedron. This shows that the altitudes are concurrent.

For the converse, if the four altitudes are concurrent at a point H with position vector $\vec{h}$, then the line through the centroid of the face *BCD* and perpendicular to that face is described by

$$\vec{r}'_a = [(\vec{b} + \vec{c} + \vec{d})/3] + \lambda'(\vec{a} - \vec{h}).$$

This time the common point of the four lines will correspond, of course, to $\lambda' = \frac{1}{3}$, and the problem is solved.

(Proposed by M. Klamkin for *Mathematics Magazine*)

694. The double of the area of triangle ONQ is equal to

$$\|\overrightarrow{ON} \times \overrightarrow{OQ}\| = \left\|\left(\frac{1}{3}\overrightarrow{OA} + \frac{2}{3}\overrightarrow{OB}\right) \times \left(\frac{2}{3}\overrightarrow{OD} + \frac{1}{3}\overrightarrow{OC}\right)\right\|.$$

Since $\overrightarrow{OA}$ is parallel to $\overrightarrow{OC}$ and $\overrightarrow{OB}$ is parallel to $\overrightarrow{OD}$, this is further equal to

$$\left\|\frac{2}{9}(\overrightarrow{OA} \times \overrightarrow{OD} + \overrightarrow{OB} \times \overrightarrow{OC})\right\|.$$

A similar computation shows that this is also equal to $\|\overrightarrow{OM} \times \overrightarrow{OP}\|$, which is twice the area of triangle OMP. Hence the conclusion.

695. The area of triangle AMN is equal to

$$\frac{1}{2}\|\overrightarrow{AM} \times \overrightarrow{AN}\| = \frac{1}{8}\|(\overrightarrow{AB} + \overrightarrow{AD}) + (\overrightarrow{AE} \times \overrightarrow{AC})\| = \frac{1}{8}\|(\overrightarrow{AB} \times \overrightarrow{AC} - \overrightarrow{AE} \times \overrightarrow{AD})\|.$$

Since $\overrightarrow{AB} \times \overrightarrow{AC}$ and $\overrightarrow{AE} \times \overrightarrow{AD}$ are perpendicular to the plane of the triangle and oriented the same way, this is equal to one-fourth of the area of the quadrilateral $BCDE$. Done.

696. We work in affine coordinates with the diagonals of the quadrilateral as axes. The vertices are $A(a, 0)$, $B(0, b)$, $C(c, 0)$, $D(0, d)$. The midpoints of the sides are $M\left(\frac{a}{2}, \frac{b}{2}\right)$, $N\left(\frac{c}{2}, \frac{b}{2}\right)$, $P\left(\frac{c}{2}, \frac{d}{2}\right)$, and $Q\left(\frac{a}{2}, \frac{d}{2}\right)$. The segments MP and NQ have the same midpoint, namely, the centroid $\left(\frac{a+c}{4}, \frac{b+d}{4}\right)$, of the quadrilateral. Hence $MNPQ$ is a parallelogram.

697. Choose a coordinate system that places M at the origin and let the coordinates of A, B, C, respectively, be (x_A, y_A), (x_B, y_B), (x_C, y_C). Then the coordinates of the centroids of MAB, MAC, and MBC are, respectively,

$$G_A = \left(\frac{x_A + x_B}{3}, \frac{y_A + y_B}{3}\right),$$

$$G_B = \left(\frac{x_A + x_C}{3}, \frac{y_A + y_C}{3}\right),$$

$$G_C = \left(\frac{x_B + x_C}{3}, \frac{y_B + y_C}{3}\right).$$

The coordinates of G_A, G_B, G_C are obtained by subtracting the coordinates of A, B, and C from $(x_A + x_B + x_C, y_A + y_B + y_C)$, then dividing by 3. Hence the triangle $G_A G_B G_C$ is obtained by taking the reflection of triangle ABC with respect to the point $(x_A + x_B + x_C, y_A + y_B + y_C)$, then contracting with ratio $\frac{1}{3}$ with respect to the origin M. Consequently, the two triangles are similar.

698. Denote by $\delta(P, MN)$ the distance from P to the line MN. The problem asks for the locus of points P for which the inequalities

$$\delta(P, AB) < \delta(P, BC) + \delta(P, CA),$$

$$\delta(P, BC) < \delta(P, CA) + \delta(P, AB),$$

$$\delta(P, CA) < \delta(P, AB) + \delta(P, BC)$$

are simultaneously satisfied.

Let us analyze the first inequality, written as $f(P) = \delta(P, BC) + \delta(P, CA) - \delta(P, AB) > 0$. As a function of the coordinates (x, y) of P, the distance from P to a line is of the form $mx + ny + p$. Combining three such functions, we see that $f(P) = f(x, y)$ is of the same form, $f(x, y) = \alpha x + \beta y + \gamma$. To solve the inequality $f(x, y) > 0$ it suffices to find the line $f(x, y) = 0$ and determine on which side of the line the function is positive. The line intersects the side BC where $\delta(P, CA) = \delta(P, AB)$, hence at the point E where the angle bisector from A intersects this side. It intersects side CA at the point F where the bisector from B intersects the side. Also, $f(x, y) > 0$ on side AB, hence on the same side of the line EF as the segment AB.

Arguing similarly for the other two inequalities, we deduce that the locus is the interior of the triangle formed by the points where the angle bisectors meet the opposite sides.

699. Consider an affine system of coordinates such that none of the segments determined by the n points is parallel to the x-axis. If the coordinates of the midpoints are (x_i, y_i), $i = 1, 2, \ldots, m$, then $x_i \neq x_j$ for $i \neq j$. Thus we have reduced the problem to the one-dimensional situation. So let $A_1, A_2, \ldots, A_n$ lie on a line in this order. The midpoints of $A_1 A_2, A_1 A_3, \ldots, A_1 A_n$ are all distinct and different from the (also distinct) midpoints of $A_2 A_n$, $A_3 A_n, \ldots, A_{n-1} A_n$. Hence there are at least $(n-1) + (n-2) = 2n - 3$ midpoints. This bound can be achieved for $A_1, A_2, \ldots, A_n$ the points $1, 2, \ldots, n$ on the real axis.

(*Kőzépiskolai Matematikai Lapok* (*Mathematics Magazine for High Schools, Budapest*), proposed by M. Salát)

700. We consider a Cartesian system of coordinates with BC and AD as the x- and y-axes, respectively (the origin is at D). Let $A(0, a)$, $B(b, 0)$, $C(c, 0)$, $M(0, m)$. Because the triangle is acute, $a, c > 0$ and $b < 0$. Also, $m > 0$. The equation of BM is $mx + by = bm$, and the equation of AC is $ax + cy = ac$. Their intersection is

$$E\left(\frac{bc(a - m)}{ab - cm}, \frac{am(b - c)}{ab - cm}\right).$$

Note that the denominator is strictly negative, hence nonzero. The point E therefore exists.

The slope of the line DE is the ratio of the coordinates of E, namely,

$$\frac{am(b - c)}{bc(a - m)}.$$

Interchanging b and c, we find that the slope of DF is

$$\frac{am(c - b)}{bc(a - m)},$$

which is the negative of the slope of DE. It follows that the lines DE and DF are symmetric with respect to the y-axis, i.e., the angles $\angle ADE$ and $\angle ADF$ are equal.

(18th W.L. Putnam Mathematical Competition, 1958)

701. We refer everything to Figure 80. Let $A(c, 0)$, c being the parameter that determines the variable line. Because B has the coordinates $\left(\frac{a}{2}, \frac{b}{2}\right)$, the line AB is given by the equation

$$y = \frac{b}{a - 2c}x + \frac{bc}{2c - a}.$$

Hence C has coordinates $\left(0, \frac{bc}{2c-a}\right)$.

The slope of the line CM is $\frac{b}{a}$, so the equation of this line is

$$y = \frac{b}{a}x + \frac{bc}{2c - a}.$$

Intersecting it with AP, whose equation is

$$y = \frac{b}{a - c}x + \frac{bc}{c - a},$$

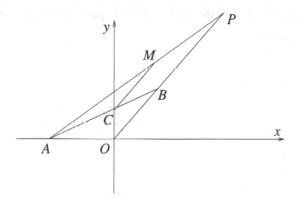

Figure 80

we obtain M of coordinates $\left(\frac{ac}{2c-a}, \frac{2bc}{2c-a}\right)$. This point lies on the line $y = \frac{2b}{a}x$, so this line might be the locus.

One should note, however, that $A = O$ yields an ambiguous construction, so the origin should be removed from the locus. On the other hand, any (x, y) on this line yields a point c, namely, $c = \frac{ax}{2x-a}$, except for $x = \frac{a}{2}$. Hence the locus consists of the line of slope $\frac{2b}{a}$ through the origin with two points removed.

(A. Myller, *Geometrie Analitică* (*Analytical Geometry*), 3rd ed., Editura Didactică şi Pedagogică, Bucharest, 1972)

702. First, assume that $ABCD$ is a rectangle (see Figure 81). Let H be the intersection point of FG and BD. In the right triangles ABC and FBG, the segments BE and BH are altitudes. Then $\angle ABE = \angle ACB$ and $\angle BGF = \angle HBC$. Since $\angle HBC = \angle ACB$, it follows that $\angle GBE = \angle BGF$ and $BE = GE$. This implies that in the right triangle BGF, $GE = EF$.

For the converse, we employ coordinates. We reformulate the problem as follows:

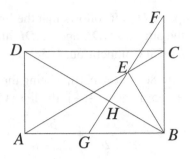

Figure 81

Alternative problem. *Given a triangle ABC with AB ≠ BC, let BE be the altitude from B and O the midpoint of side AC. The perpendicular from E to BO intersects AB at G and BC at F. Show that if the segments GE and EF are equal, then the angle ∠B is right.*

Let E be the origin of the rectangular system of coordinates, with line EB as the y-axis. Let also $A(a, 0)$, $B(0, b)$, $C(-c, 0)$, where $a, b, c > 0$. We have to prove that $b^2 = ac$, and then use the reciprocal of the Right triangle altitude theorem.

By standard computations, we obtain the following equations and coordinates:

line GF: $y = \dfrac{a - c}{2b}x$;

line BC: $-\dfrac{x}{c} + \dfrac{y}{b} = 1$;

point F: $x_F = \dfrac{2b^2c}{-2b^2 - c^2 + ac}$, $y_F = \dfrac{bc(a - c)}{-2b^2 - c^2 + ac}$;

line AB: $\dfrac{x}{a} + \dfrac{y}{b} = 1$;

point G: $x_G = \dfrac{2ab^2}{2b^2 + a^2 - ac}$, $y_G = \dfrac{ab(a - c)}{2b^2 + a^2 - ac}$.

The condition $EG = EF$ is equivalent to $x_G = -x_F$, that is,

$$\frac{2b^2c}{2b^2 + c^2 - ac} = \frac{2ab^2}{2b^2 + a^2 - ac}.$$

This yields $2(b^2 - ac)(c - a) = 0$, hence $b^2 = ac$ or $a = c$, and since the latter is ruled out by hypothesis, $b^2 = ac$ and this completes the solution.
(Romanian Mathematics Competition, 2004, proposed by M. Becheanu)

703. If we let $D = (0, -1)$, $F = (x_1, y_1)$, $E = (x_2, y_2)$ with $x_k^2 + y_k^2 = 1$ then

$$K = (x_1 + x_2, y_1 + y_2 - 1), \quad B = \left(\frac{1 + y_1}{x_1}, -1\right), \quad C = \left(\frac{1 + y_2}{x_2}, -1\right).$$

We then compute

$$BE^2 - CF^2 = 2x_1\frac{1+y_2}{x_2} - 2x_2\frac{1+y_1}{x_1} + 2y_2 - 2y_1 = BK^2 - CK^2,$$

and we are done.

(proposed for 2015 USAMO by T. Andreescu and C. Pohoață)

704. The inequality from the statement can be rewritten as

$$-\frac{\sqrt{2}-1}{2} \le \sqrt{1-x^2} - (px+q) \le \frac{\sqrt{2}-1}{2},$$

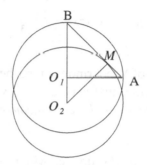

Figure 82

or

$$\sqrt{1-x^2} - \frac{\sqrt{2}-1}{2} \le px+q \le \sqrt{1-x^2} + \frac{\sqrt{2}-1}{2}.$$

Let us rephrase this in geometric terms. We are required to include a segment $y = px + q$, $0 \le x \le 1$, between two circular arcs.

The arcs are parts of two circles of radius 1 and of centers $O_1\left(0, \frac{\sqrt{2}-1}{2}\right)$ and $O_2\left(0, -\frac{\sqrt{2}-1}{2}\right)$. By examining Figure 82 we will conclude that there is just one such segment. On the first circle, consider the points $A\left(1, \frac{\sqrt{2}-1}{2}\right)$ and $B\left(0, \frac{\sqrt{2}+1}{2}\right)$. The distance from B to O_2 is $\sqrt{2}$, which is equal to the length of the segment AB. In the isosceles triangle BO_2A, the altitudes from O_2 and A must be equal. The altitude from A is equal to the distance from A to the y-axis, hence is 1. Thus the distance from O_2 to AB is 1 as well. This shows that the segment AB is tangent to the circle centered at O_2. This segment lies between the two arcs, and above the entire interval [0, 1]. Being inscribed in one arc and tangent to the other, it is the only segment with this property.

This answers the problem, by showing that the only possibility is $p = -1$, $q = \frac{\sqrt{2}+1}{2}$.

(Romanian Team Selection Test for the International Mathematical Olympiad, 1983)

705. The fact that the points $\left(x_i, \frac{1}{x_i}\right)$ lie on a circle means that there exist numbers A, B, and C such that

$$x_i^2 + \frac{1}{x_i^2} + 2x_iA + 2\frac{1}{x_i}B + C = 0, \text{ for } i = 1, 2, 3, 4.$$

View this as a system in the unknowns $2A$, $2B$, C. The system admits a solution only if the determinant of the augmented matrix of the system is zero. This determinant is equal to

$$\begin{vmatrix} x_1^2 + \frac{1}{x_1^2} & x_1 & \frac{1}{x_1} & 1 \\ x_2^2 + \frac{1}{x_2^2} & x_2 & \frac{1}{x_2} & 1 \\ x_3^2 + \frac{1}{x_3^2} & x_3 & \frac{1}{x_3} & 1 \\ x_4^2 + \frac{1}{x_4^2} & x_4 & \frac{1}{x_4} & 1 \end{vmatrix} = \begin{vmatrix} x_1^2 & x_1 & \frac{1}{x_1} & 1 \\ x_2^2 & x_2 & \frac{1}{x_2} & 1 \\ x_2^2 & x_3 & \frac{1}{x_3} & 1 \\ x_4^2 & x_4 & \frac{1}{x_4} & 1 \end{vmatrix} + \begin{vmatrix} \frac{1}{x_1^2} & x_1 & \frac{1}{x_1} & 1 \\ \frac{1}{x_2^2} & x_2 & \frac{1}{x_2} & 1 \\ \frac{1}{x_3^2} & x_3 & \frac{1}{x_3} & 1 \\ \frac{1}{x_4^2} & x_4 & \frac{1}{x_4} & 1 \end{vmatrix}$$

$$= \left(-\frac{1}{x_1 x_2 x_3 x_4} + \frac{1}{x_1^2 x_2^2 x_3^2 x_4^2} \right) \begin{vmatrix} x_1^3 & x_1^2 & x_1 & 1 \\ x_2^3 & x_2^2 & x_2 & 1 \\ x_3^3 & x_3^2 & x_3 & 1 \\ x_3^3 & x_4^2 & x_4 & 1 \end{vmatrix}.$$

One of the factors is a determinant of Vandermonde type, hence it cannot be 0. Thus the other factor is equal to 0. From this we infer that $x_1 x_2 x_3 x_4 = 1$, which is what had to be proved.

(A. Myller, *Geometrie Analitică* (*Analytical Geometry*), 3rd ed., Editura Didactică şi Pedagogică, Bucharest, 1972)

706. Choosing a Cartesian system of coordinates with origin at A and axes AB and AC, we have $A(0, 0)$, $B(1, 0)$, $C(0, 1)$, $D(1/2, -1/2)$. Let $M(0, t)$, $t \in [0, 1]$. Then

$$BC: \quad y = -x + 1$$
$$DM: \quad y = -(2t + 1)x + t.$$

Hence $N = \left(\frac{t-1}{2t}, \frac{t+1}{2t} \right)$. We have

$$BM: \quad y = -tx + t$$
$$AN: \quad y = \frac{t+1}{t-1}x,$$

and so

$$P = \left(\frac{t^2 - t}{t^2 + 1}, \frac{t^2 + t}{t^2 + 1} \right).$$

Let us find the equation in Cartesian coordinates of the arc that P describes. We want to eliminate t from the equations $x = (t^2 - t)/(t^2 + 1)$ and $y = (t^2 + t)/(t^2 + 1)$. We have

$$t^2 - t = (t^2 + 1)x$$
$$t^2 + t = (t^2 + 1)y.$$

Adding and subtracting we get

$$2t^2 = (t^2 + 1)(x + y)$$
$$2t = (t^2 + 1)(x - y).$$

Dividing we obtain

$$t = \frac{x+y}{x-y}.$$

After substituting in $2t = (t^2 + 1)(x - y)$ and performing the algebraic computations we obtain

$$x^2 + y^2 - x - y = 0,$$

which is the equation of the circle of radius $\sqrt{2}/2$ centered at $(1/2, 1/2)$. The arc of curve in question is the arc of this circle with endpoints A and C; its length is $1/4$ of the total circle, hence $\pi\sqrt{2}/4$.

(Kohnauser Problem Fest, 2014, proposed by R. Gelca)

707. Consider complex coordinates with the origin O at the center of the circle. The coordinates of the vertices, which we denote correspondingly by $\alpha, \beta, \gamma, \delta, \eta, \phi$, have absolute value $|r|$. Moreover, because the chords AB, CD, and EF are equal to the radius, $\angle AOB = \angle COD = \angle EOF = \frac{\pi}{3}$. It follows that $\beta = \alpha e^{i\pi/3}$, $\delta = \gamma e^{i\pi/3}$, and $\phi = \eta e^{i\pi/3}$. The midpoints P, Q, R of BC, DE, FA, respectively, have the coordinates

$$p = \frac{1}{2}(\alpha e^{i\pi/3} + \gamma), \quad q = \frac{1}{2}\gamma e^{i\pi/3} + \eta), \quad r = \frac{1}{2}(\eta e^{i\pi/3} + \alpha).$$

We compute

$$\frac{r-q}{p-q} = \frac{\alpha e^{i\pi/3} + \gamma(1 - e^{i\pi/3}) - \eta}{\alpha - \gamma e^{i\pi/3} + \eta(e^{i\pi/3} - 1)}$$
$$= \frac{\alpha e^{i\pi/3} - \gamma e^{2i\pi/3} + \eta e^{3i\pi/3}}{\alpha - \gamma e^{i\pi/3} + \eta e^{2i\pi/3}} = e^{i\pi/3}.$$

It follows that RQ is obtained by rotating PQ around Q by $60°$. Hence the triangle PQR is equilateral, as desired.

(28th W.L. Putnam Mathematical Competition, 1967)

708. We work in complex coordinates such that the circumcenter is at the origin. Let the vertices be a, b, c on the unit circle. Since the complex coordinate of the centroid is $\frac{a+b+c}{3}$, we have to show that the complex coordinate of the orthocenter is $a + b + c$. By symmetry, it suffices to check that the line passing through a and $a + b + c$ is perpendicular to the line passing through b and c. This is equivalent to the fact that the argument of $\frac{b-c}{b+c}$ is $\pm\frac{\pi}{2}$. This is true because the vector $b + c$ is constructed as one of the diagonals of the rhombus determined by the vectors (of the same length) b and c, while $b - c$ is the other diagonal of the rhombus. And the diagonals of a rhombus are perpendicular. This completes the solution.

(L. Euler)

709. With the convention that the lowercase letter denotes the complex coordinate of the point denoted by the same letter in uppercase, we translate the geometric conditions from the statement into the algebraic equations

$$\frac{m-a}{b-a} = \frac{n-c}{b-c} = \frac{p-c}{d-c} = \frac{q-a}{d-a} = \varepsilon,$$

where $\varepsilon = \cos\frac{\pi}{3} + i\sin\frac{\pi}{3}$. Therefore,

$$m = a + (b - a)\varepsilon, \quad n = c + (b - c)\varepsilon,$$

$$p = c + (d - c)\varepsilon, \quad q = a + (d - a)\varepsilon.$$

It is now easy to see that $\frac{1}{2}(m + p) = \frac{1}{2}(n + q)$, meaning that MP and NQ have the same midpoint. So either the four points are collinear, or they form a parallelogram.

(Short list of the 23rd International Mathematical Olympiad, 1982)

710. We refer everything to Figure 83. The triangle BAQ is obtained by rotating the triangle PAC around A by the angle α. Hence the angle between the lines PC and BQ is equal to α. It follows that in the circumcircle of BRC, the measure of the arc $\overset{\frown}{BRC}$ is equal to 2α, and this is also the measure of $\angle BOC$. We deduce that O is obtained from B through the counterclockwise rotation about C by the complement of α followed by contraction by a factor of $2\sin\alpha$.

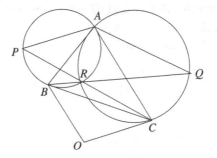

Figure 83

Now we introduce complex coordinates with the origin at A, with the coordinates of B and C being b and c. Set $\omega = e^{i\alpha}$, so that the counterclockwise rotation by α is multiplication by ω, and hence rotation by the complement of α is multiplication by $i/\omega = i\bar{\omega}$. Then the coordinate z of O satisfies

$$\frac{z - c}{b - c} = \frac{1}{2\sin\alpha} \cdot \frac{i}{\omega},$$

from which we compute

$$z = \frac{b - c}{2\sin\alpha} \cdot \frac{i}{\omega} + c = \frac{b - c}{-i(\omega - \bar{\omega})} \cdot \frac{i}{\omega} + c = \frac{b - c}{1 - \omega^2}.$$

On the other hand, P is obtained by rotating B around A by $-\alpha$, so its coordinate is $p = b\bar{\omega}$. Similarly, the coordinate of Q is $q = c\omega$. It is now straightforward to check that

$$\frac{q - p}{z - 0} = \omega - \frac{1}{\omega},$$

a purely imaginary number. Hence the lines PQ and AO form a 90° angle, which is the desired result.

(USA Team Selection Test for the International Mathematical Olympiad, 2006, solution by T. Leung)

711. In the language of complex numbers we are required to find the maximum of

$$\prod_{k=1}^{n} |z - \varepsilon^k|$$

as z ranges over the unit disk, where $\varepsilon = \cos \frac{2\pi}{n} + i \sin \frac{2\pi}{n}$. We have

$$\prod_{k=1}^{n} |z - \varepsilon^k| = \left| \prod_{k=1}^{n} (z - \varepsilon^k) \right| = |z^n - 1| \leq |z^n| + 1 = 2.$$

The maximum is 2, attained when z is an nth root of -1.

(Romanian Mathematics Competition "Grigore Moisil", 1992, proposed by D. Andrica)

712. *First solution*: In a system of complex coordinates, place each vertex A_k, $k = 0, 1, \ldots, n-1$, at ε^k, where $\varepsilon = e^{2i\pi/n}$. Then

$$A_0 A_1 \cdot A_0 A_2 \cdots A_0 A_{n-1} = |(1 - \varepsilon)(1 - \varepsilon^2) \cdots (1 - \varepsilon^{n-1})|.$$

Observe that, in general,

$$(z - \varepsilon)(z - \varepsilon^2) \cdots (z - \varepsilon^{n-1}) = \frac{1}{z - 1}(z - 1)(z - \varepsilon) \cdots (z - \varepsilon^{n-1})$$

$$= \frac{1}{z - 1}(z^n - 1) = z^{n-1} + z^{n-2} + \cdots + 1.$$

By continuity, this equality also holds for $z = 1$. Hence

$$A_0 A_1 \cdot A_0 A_2 \cdots A_0 A_{n-1} = 1^{n-1} + 1^{n-2} + \cdots + 1 = n,$$

and the identity is proved.

Second solution: Choose a point P on the ray $|OA_0$, where O is center of the circumcircle of the polygon, such that A_0 is between O and P. If $OP = x$, then the last problem in the introduction showed that $A_0 A_1 \cdot A_0 A_2 \cdots A_0 A_{n-1} = x^n - 1$. Hence

$$A_0 A_1 \cdot A_0 A_2 \cdots A_0 A_{n-1} = \lim_{x \to 1} \frac{x^n - 1}{x - 1} = n.$$

Remark. Let us show how this geometric identity can be used to derive a trigonometric identity. For $n = 2m + 1$, m an integer,

$$A_0 A_1 \cdot A_0 A_2 \cdots A_0 A_m = A_0 A_{2m} \cdot A_0 A_{2m-1} \cdots A_0 A_{m+1};$$

hence

$$A_0 A_1 \cdot A_0 A_2 \cdots A_0 A_m = \sqrt{2m + 1}.$$

On the other hand, for $i = 1, 2, \ldots, m$, in triangle $A_0 O A_i$,

$$AA_i = 2 \sin \frac{2\pi}{2m + 1}.$$

We conclude that

$$\sin \frac{2\pi}{2m+1} \sin \frac{4\pi}{2m+1} \cdots \sin \frac{2m\pi}{2m+1} = \frac{1}{2^m} \sqrt{2m+1}.$$

(J. Dürschák, *Matemaikai Versenytételek*, Harmadik Kiadás Tankönyviadó, Budapest, 1965)

713. Note that the positive integers $a_1, a_2, \ldots, a_n$, are the side-lengths of an equiangular polygon, in this order, if and if for

$$\varepsilon = \cos \frac{2\pi}{n} + i \sin \frac{2\pi}{n},$$

one has

$$a_n \varepsilon^{n-1} + a_{n-1} \varepsilon^{n-2} + \cdots + a_2 \varepsilon + a_1 = 0.$$

Let us assume that p is prime, and that we are given an equiangular polygon with rational side-lengths, and let these side-lengths be $a_1, a_2, \ldots, a_n$. Then the polynomial

$$P(x) = a_p x^{p-1} + a_{p-1} x^{p-2} + \cdots + a_1$$

has ε as a root. But ε is also a root of

$$Q(x) = x^{p-1} + x^{p-2} + \cdots + 1.$$

It follows that $P(x)$ and $Q(x)$ have a non-constant common divisor. But $Q(x)$ is irreducible, which can be shown by applying the Eisenstein irreducibility criterion to $Q(x+1) = x^{p-1} + \binom{p}{1} x^{p-2} + \cdots + \binom{p}{p-1}$. So $P(x)$ must be a multiple of $Q(x)$, in which case all coefficients of $P(x)$ are equal. So the polygon is regular.

Conversely, assume that p is not prime. Let $p = mn$, with $m, n > 1$. It follows that ε^n is an mth root of unity, that is

$$1 + \varepsilon^n + \varepsilon^{2n} + \cdots \varepsilon^{(m-1)n} = 0.$$

But ε is an mnth root of unity, so

$$1 + \varepsilon + \cdots + \varepsilon^{mn} = 0.$$

Adding these equalities we obtain a polynomial with some coefficients equal to 1 and the others equal to 2 that has ε as a root. It follows that there is an equiangular polygon with some of the sides equal to 1 and the other equal to 2, which is not equiangular.

(*Revista Matematică din Timişoara (Timişoara Mathematics Gazette)*, proposed by M. Piticari)

714. *First solution*: We assume that the radius of the circle is equal to 1. Set the origin at B with BA the positive x-semiaxis and t the y-axis (see Figure 84). If $\angle BOM = \theta$, then $BP = PM = \tan\frac{\theta}{2}$. In triangle PQM, $PQ = \tan\frac{\theta}{2}/\sin\theta$. So the coordinates of Q are

$$\left(\frac{\tan\dfrac{\theta}{2}}{\sin\theta}, \tan\frac{\theta}{2}\right) = \left(\frac{1}{1+\cos\theta}, \frac{\sin\theta}{1+\cos\theta}\right).$$

The x and y coordinates are related as follows:

$$\left(\frac{\sin\theta}{1+\cos\theta}\right)^2 = \frac{1-\cos^2\theta}{(1+\cos\theta)^2} = \frac{1-\cos\theta}{1+\cos\theta} = 2\frac{1}{1+\cos\theta} - 1.$$

Hence the locus of Q is the parabola $y^2 = 2x - 1$.

Second solution: With $\angle BOM = \theta$ we have $\angle POM = \angle POB = \frac{\theta}{2}$. Since PQ is parallel to OB, it follows that $\angle OPQ = \frac{\theta}{2}$. So the triangle OPQ is isosceles, and therefore $QP = OQ$. We conclude that Q lies on the parabola of focus O and directrix t. A continuity argument shows that the locus is the entire parabola.

(A. Myller, *Geometrie Analitică (Analytical Geometry)*, 3rd ed., Editura Didactică și Pedagogică, Bucharest, 1972, solutions found by the students from the Mathematical Olympiad Summer Program, 2004)

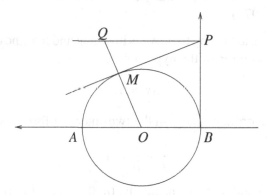

Figure 84

715. We will use the equation of the tangent with prescribed slope. Write the parabola in standard form

$$y^2 = 4px.$$

The tangent of slope m to this parabola is given by

$$y = mx + \frac{p}{m}.$$

If $A(p+a, 0)$ and $B(p-a, 0)$ are the two fixed points, $(p, 0)$ being the focus, then the distances to the tangent are

$$\left|\frac{m(p \pm a) + \dfrac{p}{m}}{\sqrt{1+m^2}}\right|.$$

The difference of their squares is

$$\frac{\left(m^2(p+a)^2 + 2p(p+a) + \frac{p^2}{m^2}\right) - \left(m^2(p-a)^2 + 2p(p-a) + \frac{p^2}{m^2}\right)}{1+m^2}.$$

An easy computation shows that this is equal to $4pa$, which does not depend on m, meaning that it does not depend on the tangent.

(A. Myller, *Geometrie Analitică* (*Analytical Geometry*), 3rd ed., Editura Didactică şi Pedagogică, Bucharest, 1972)

716. The statement of the problem is invariant under affine transformations, so we can assume the hyperbola to have the equation $xy = 1$, such that the asymptotes are the coordinate axes. If $P(x_1, y_1)$ and $Q(x_2, y_2)$ are two of the vertices, then the other two vertices of the parallelogram are (x_1, y_2) and (x_2, y_1). The line they determine has the equation

$$y - y_1 = \frac{y_2 - y_1}{x_1 - x_2}(x - x_2).$$

Substituting the coordinates of the origin in this equation yields $-y_1 = \frac{y_2-y_1}{x_1-x_2}(-x_2)$, or $x_1 y_1 - x_2 y_1 = x_2 y_2 - x_2 y_1$. This clearly holds, since $x_1 y_1 = x_2 y_2 = 1$, and the property is proved.

(A. Myller, *Geometrie Analitică* (*Analytical Geometry*), 3rd ed., Editura Didactică şi Pedagogică, Bucharest, 1972)

717. Since the property we are trying to prove is invariant under affine changes of coordinates, we can assume that the equation of the hyperbola is

$$xy = 1.$$

The asymptotes are the coordinate axes. In the two-intercept form, the equation of the line is

$$\frac{x}{a} + \frac{y}{b} = 1.$$

Then the coordinates of A and B are, respectively, $(a, 0)$ and $(0, b)$. To find the coordinates of P and Q, substitute $y = \frac{1}{x}$ in the equation of the line. This gives rise to the quadratic equation

$$x^2 - ax + \frac{a}{b} = 0.$$

The roots x_1 and x_2 of this equation satisfy $x_1 + x_2 = a$. Similarly, substituting $x = \frac{1}{y}$ in the same equation yields

$$y^2 - by + \frac{b}{a} = 0,$$

and the two roots y_1 and y_2 satisfy $y_1 + y_2 = b$. The coordinates of P and Q are, respectively, (x_1, y_1) and (x_2, y_2). We have

$$AP^2 = (x_1 - a)^2 + y_1^2 = (a - x_2 - a)^2 + (b - y_2)^2 = x_2^2 + (b - y_2)^2 = BQ^2.$$

The property is proved.

(L.C. Larson, *Problem Solving through Problems*, Springer-Verlag, 1983)

718. The condition that a line through (x_0, y_0) be tangent to the parabola is that the system

$$y^2 = 4px,$$

$$y - y_0 = m(x - x_0)$$

have a unique solution. This means that the discriminant of the quadratic equation in x obtained by eliminating y,

$$(mx - mx_0 + y_0)^2 - 4px = 0,$$

is equal to zero. This translates into the condition

$$m^2 x_0 - my_0 + p = 0.$$

The slopes m of the two tangents are therefore the solutions to this quadratic equation. They satisfy

$$m_1 + m_2 = \frac{y_0}{x_0},$$

$$m_1 m_2 = \frac{p}{x_0}.$$

We also know that the angle between the tangents is ϕ. We distinguish two situations.

First, if $\phi = 90°$, then $m_1 m_2 = -1$. This implies $\frac{p}{x_0} = -1$, so the locus is the line $x = -p$, which is the directrix of the parabola.

If $\phi \neq 90°$, then

$$\tan \phi = \frac{m_1 - m_2}{1 + m_1 m_2} = \frac{m_1 - m_2}{1 + \dfrac{p}{x_0}}.$$

We thus have

$$m_1 + m_2 = \frac{y_0}{x_0},$$

$$m_1 - m_2 = \tan \phi + \frac{p}{x_0} \tan \phi.$$

We can compute $m_1 m_2$ by squaring the equations and then subtracting them, and we obtain

$$m_1 m_2 = \frac{y_0^2}{4x_0^2} - \left(1 + \frac{p}{x_0}\right)^2 \tan^2 \phi.$$

This must equal $\frac{p}{x_0}$. We obtain the equation of the locus to be

$$-y^2 + (x + p)^2 \tan^2 \phi + 4px = 0,$$

which is a hyperbola. One branch of the hyperbola contains the points from which the parabola is seen under the angle ϕ, and one branch contains the points from which the parabola is seen under an angle equal to the suplement of ϕ.

(A. Myller, *Geometrie Analitică (Analytical Geometry)*, 3rd ed., Editura Didactică și Pedagogică, Bucharest, 1972)

719. Choose a Cartesian system of coordinates such that the equation of the parabola is $y^2 = 4px$. The coordinates of the three points are $T_i(4p\alpha_i^2, 4p\alpha_i)$, for appropriately chosen α_i, $i = 1, 2, 3$. Recall that the equation of the tangent to the parabola at a point (x_0, y_0) is $yy_0 = 2p(x + x_0)$. In our situation the three tangents are given by

$$2\alpha_i y = x + 4p\alpha_i^2, \quad i = 1, 2, 3.$$

If P_{ij} is the intersection of t_i and t_j, then its coordinates are $(4p\alpha_i\alpha_j, 2p(\alpha_i + \alpha_j))$. The area of triangle $T_1 T_2 T_3$ is given by a Vandermonde determinant:

$$\pm \frac{1}{2} \begin{vmatrix} 4p\alpha_1^2 & 4p\alpha_1 & 1 \\ 4p\alpha_2^2 & 4p\alpha_2 & 1 \\ 4p\alpha_3^2 & 4p\alpha_3 & 1 \end{vmatrix} = \pm 8p^2 \begin{vmatrix} \alpha_1^2 & \alpha_1 & 1 \\ \alpha_2^2 & \alpha_2 & 1 \\ \alpha_3^2 & \alpha_3 & 1 \end{vmatrix} = 8p^2 |(\alpha_1 - \alpha_2)(\alpha_1 - \alpha_3)(\alpha_2 - \alpha_3)|.$$

The area of the triangle $P_{12}P_{23}P_{31}$ is given by

$$\pm \begin{vmatrix} 4p\alpha_1\alpha_2 & 2p(\alpha_1 + \alpha_2) & 1 \\ 4p\alpha_3\alpha_3 & 2p(\alpha_2 + \alpha_3) & 1 \\ 4p\alpha_3\alpha_1 & 2p(\alpha_3 + \alpha_1) & 1 \end{vmatrix} = \pm 4p^2 \begin{vmatrix} \alpha_1\alpha_2 & (\alpha_1 + \alpha_2) & 1 \\ \alpha_2\alpha_3 & (\alpha_2 + \alpha_3) & 1 \\ \alpha_3\alpha_1 & (\alpha_3 + \alpha_1) & 1 \end{vmatrix}$$

$$= \pm 4p^2 \begin{vmatrix} (\alpha_1 - \alpha_3)\alpha_2 & (\alpha_1 - \alpha_3) & 0 \\ (\alpha_2 - \alpha_1)\alpha_3 & (\alpha_2 - \alpha_1) & 0 \\ \alpha_3\alpha_1 & (\alpha_3 + \alpha_1) & 1 \end{vmatrix} = 4p^2 |(\alpha_1 - \alpha_3)(\alpha_1 - \alpha_2)(\alpha_2 - \alpha_3)|.$$

We conclude that the ratio of the two areas is 2, regardless of the location of the three points or the shape of the parabola.

(Gh. Călugărița, V. Mangu, *Probleme de Matematică pentru Treapta I și a II-a de Liceu* (*Mathematics Problems for High School*), Editura Albatros, Bucharest, 1977)

720. Choose a Cartesian system of coordinates such that the focus is $F(p, 0)$ and the directrix is $x = -p$, in which case the equation of the parabola is $y^2 = 4px$. Let the three points be $A\left(\frac{a^2}{4p}, a\right)$, $B\left(\frac{b^2}{4p}, b\right)$, $C\left(\frac{c^2}{4p}, c\right)$.

(a) The tangents NP, PM, and MN to the parabola are given, respectively, by

$$ay = 2px + \frac{a^2}{2}, \quad by = 2px + \frac{b^2}{2}, \quad cy = 2px + \frac{c^2}{2},$$

from which we deduce the coordinates of the vertices

$$M\left(\frac{bc}{4p}, \frac{b+c}{2}\right), \quad N\left(\frac{ca}{4p}, \frac{c+a}{2}\right), \quad P\left(\frac{ab}{4p}, \frac{a+b}{2}\right).$$

The intersection of the line AC of equation $4px - (c + a)y + ca = 0$ with the parallel to the symmetry axis through B, which has equation $y = b$, is $L\left(\frac{ab+bc-ca}{4p}, b\right)$. It is straightforward to verify that the segments MP and LN have the same midpoint, the point with coordinates $\left(\frac{b(c+a)}{8p}, \frac{a+2b+c}{4}\right)$. Consequently, $LMNP$ is a parallelogram.

(b) Writing that the equation of the circle $x^2 + y^2 + 2\alpha x + 2\beta x + \gamma = 0$ is satisfied by the points M, N, P helps us determine the parameters α, β, γ. We obtain the equation of the circumcircle of MNP,

$$x^2 + y^2 - \frac{ab + bc + ca + 4p^2}{4p}x + \frac{abc - 4p^2(a + b + c)}{8p^2}y + \frac{ab + bc + ca}{4} = 0.$$

This equation is satisfied by $(p, 0)$, showing that the focus F is on the circle.

(c) Substituting the coordinates of L in the equation of the circle yields

$$(ac + 4p^2)(a - b)(c - b) = 0.$$

Since $a \neq b \neq c$, we must have $ac = -4p^2$. Thus the x-coordinate of N is $-p$, showing that this point is on the directrix.

(d) The condition for F to be on AC is $4p^2 + ac = 0$, in which case N is on the directrix. The slope of BF is $m = \frac{4pb}{b^2 - 4p^2}$. The orthogonality condition is

$$\frac{4pb}{b^2 - 4p^2} \cdot \frac{4p}{c + a} = -1,$$

which is equivalent to

$$(b^2 - 4p^2)(c + a) + 16p^2 b = 0.$$

The locus is obtained by eliminating a, b, c from the equations

$$4px - (c + a)y + ca = 0,$$
$$y = b,$$
$$4p^2 + ac = 0,$$
$$(b^2 - 4p^2)(c + a) + 16p^2 b = 0.$$

The answer is the cubic curve

$$(y^2 - 4p^2)x + 3py^2 + 4p^3 = 0.$$

(The *Mathematics Gazette* Competition, Bucharest, 1938)

721. An equilateral triangle can be inscribed in any closed, non-self-intersecting curve, therefore also in an ellipse. The argument runs as follows. Choose a point A on the ellipse. Rotate the ellipse around A by $60°$. The image of the ellipse through the rotation intersects the original ellipse once in A, so it should intersect it at least one more time. Let B an be intersection point different from A. Note that B is on both ellipses, and its preimage C through rotation is on the original ellipse. The triangle ABC is equilateral.

A square can also be inscribed in the ellipse. It suffices to vary an inscribed rectangle with sides parallel to the axes of the ellipse and use the intermediate value property.

Let us show that these are the only possibilities. Up to a translation, a rotation, and a dilation, the equation of the ellipse has the form

$$x^2 + ay^2 = b, \text{ with } a, b > 0, a \neq 1.$$

Assume that a regular n-gon, $n \geq 5$, can be inscribed in the ellipse. Its vertices (x_i, y_i) satisfy the equation of the circumcircle:

$$x^2 + y^2 + cx + dy + e = 0, \quad i = 1, 2, \ldots, n.$$

Writing the fact that the vertices also satisfy the equation of the ellipse and subtracting, we obtain

$$(1 - a)y_i^2 + cx_i + dy_i + (e + b) = 0.$$

Hence

$$y_i^2 = -\frac{c}{1-a}x_i - \frac{d}{1-a}y_i - \frac{e+b}{1-a}.$$

The number c cannot be 0, for otherwise the quadratic equation would have two solutions y_i and each of these would yield two solutions x_i, so the polygon would have four or fewer sides, a contradiction. This means that the regular polygon is inscribed in a parabola. Change the coordinates so that the parabola has the standard equation $y^2 = 4px$. Let the new coordinates of the vertices be (ξ_i, η_i) and the new equation of the circumcircle be $x^2 + y^2 + c'x + d'y + e' = 0$. That the vertices belong to both the parabola and the circle translates to

$$\eta_i^2 = 4p\xi_i \quad \text{and} \quad \xi_i^2 + \eta_i^2 + c'\xi + d'\eta + e' = 0, \quad \text{for } i = 1, 2, \ldots, n.$$

So the η_i's satisfy the fourth-degree equation

$$\frac{1}{16p^2}\eta_i^4 + \eta_i^2 + \frac{c'}{4}\eta_i^2 + d'\eta_i + e' = 0.$$

This equation has at most four solutions, and each solution yields a unique x_i. So the regular polygon can have at most four vertices, a contradiction. We conclude that no regular polygon with five or more vertices can be inscribed in an ellipse that is not also a circle.

722. Set $FB_k = t_k$, $k = 1, 2, \ldots, n$. Also, let α be the angle made by the ray $|FB_1$ with the x-axis and $\alpha_k = \alpha + \frac{2(k-1)\pi}{n}$, $k = 2, \ldots, n$. The coordinates of the focus F are $\left(\frac{p}{2}, 0\right)$.

In general, the coordinates of the points on a ray that originates in F and makes an angle β with the x-axis are $\left(\frac{p}{2} + t\cos\beta, t\sin\beta\right)$, $t > 0$ (just draw a ray from the origin of the coordinate system that makes an angle β with the x-axis; then translate it to F). It follows that the coordinates of B_k are $\left(\frac{p}{2} + t_k\cos\alpha_k, t_k\sin\alpha_k\right)$, $k = 1, 2, \ldots, n$.

The condition that B_k belongs to the parabola is written as $t_k^2\sin^2\alpha_k = p^2 + 2pt_k\cos\alpha_k$. The positive root of this equation is $t_k = p/(1 - \cos\alpha_k)$. We are supposed to prove that $t_1 + t_2 + \cdots + t_k > np$, which translates to

$$\frac{1}{1 - \cos\alpha_1} + \frac{1}{1 - \cos\alpha_2} + \cdots + \frac{1}{1 - \cos\alpha_n} > n.$$

To prove this inequality, note that

$$(1 - \cos\alpha_1) + (1 - \cos\alpha_2) + \cdots + (1 - \cos\alpha_n) = n - \sum_{k=1}^{n}\cos\left(\alpha + \frac{2(k-1)\pi}{n}\right)$$

$$= n - \cos \alpha \sum_{k=1}^{n} \cos \left(\frac{2(k-1)\pi}{n} \right) + \sin \alpha \sum_{k=1}^{n} \sin \left(\frac{2(k-1)\pi}{n} \right) = n.$$

By the Cauchy-Schwarz inequality,

$$\left(\frac{1}{1 - \cos \alpha_1} + \frac{1}{1 - \cos \alpha_2} + \cdots + \frac{1}{1 - \cos \alpha_n} \right)$$

$$\geq \frac{n^2}{(1 - \cos \alpha_1) + (1 - \cos \alpha_2) + \cdots + (1 - \cos \alpha_n)} = \frac{n^2}{n} = n.$$

The equality case would imply that all α_k's are equal, which is impossible. Hence the inequality is strict, as desired.

(Romanian Mathematical Olympiad, 2004, proposed by C. Popescu)

723. We solve part (e). Choose a coordinate system such that $B = (-1, 0)$, $C = (1, 0)$, $S = (0, \sqrt{3})$, $S' = (0, -\sqrt{3})$. Assume that the ellipse has vertices $(0, \pm k)$ with $k > \sqrt{3}$, so its equation is

$$\frac{x^2}{k^2 - 3} + \frac{y^2}{k^2} = 1.$$

If we set $r = \sqrt{k^2 - 3}$, then the ellipse is parametrized by $A = (r \cos \theta, k \sin \theta)$. Parts (a) through (d) are covered by the degenerate situation $k = \sqrt{3}$, when the ellipse becomes the line segment SS'.

Let $A = (r \cos \theta, k \sin \theta)$ with θ not a multiple of π. Consider the points D, E, F, respectively, on BC, AC, AB, given by

$$D = ((r + k) \cos \theta, 0),$$

$$E = \left(\frac{(2k^2 + rk - 3) \cos \theta + k - r}{r + 2k + 3 \cos \theta}, \frac{k(2r + k) \sin \theta}{r + 2k + 3 \cos \theta} \right),$$

$$F = \left(\frac{(2k^2 + rk - 3) \cos \theta - k + r}{r + 2k - 3 \cos \theta}, \frac{k(2r + k) \sin \theta}{r + 2k - 3 \cos \theta} \right).$$

The denominators are never zero since $r \geq 0$ and $k \geq \sqrt{3}$. The lines AD, BE, and CF intersect at the point

$$P = \left(\frac{r + 2k}{3} \cos \theta, \frac{2r + k}{3} \sin \theta \right),$$

as one can verify, using $r^2 = k^2 - 3$, that, coordinate-wise,

$$P = \frac{k + 2r}{3k} A + \frac{2k - 2r}{3k} D$$

$$= \frac{k - r - 3 \cos \theta}{3k} B + \frac{2k + r + 3 \cos \theta}{3k} E$$

$$= \frac{k - r + 3 \cos \theta}{3k} C + \frac{2k + r - 3 \cos \theta}{3k} F.$$

An algebraic computation shows that $AD = BE = CF = k$, so P is an equicevian point, and $\frac{AP}{PD} = \frac{(2k - 2r)}{(k + 2r)}$ is independent of A.

To find the other equicevian point note that if we replace k by $-k$ and θ by $-\theta$, then A remains the same. In this new parametrization, we have the points

$$D' = ((r-k)\cos\theta, 0),$$

$$E' = \left(\frac{(2k^2 - rk - 3)\cos\theta - k - r}{r - 2k + 3\cos\theta}, \frac{k(2r-k)\sin\theta}{r - 2k + 3\cos\theta} \right),$$

$$F' = \left(\frac{(2k^2 - rk - 3)\cos\theta + k + r}{r - 2k - 3\cos\theta}, \frac{k(2r-k)\sin\theta}{r - 2k - 3\cos\theta} \right),$$

$$P' = \left(\frac{r - 2k}{3}\cos\theta, \frac{k - 2r}{3}\sin\theta \right).$$

Of course, P' is again an equicevian point, and $\frac{AP'}{P'D'} = \frac{(2k+2r)}{(k-2r)}$, which is also independent of A. When $r \neq 0$ the points P and P' are distinct, since $\sin\theta \neq 0$. When $r = 0$, the two points P and P' coincide when $A = S$, a case ruled out by the hypothesis. As θ varies, P and P' trace an ellipse. Moreover, since

$$\left(\frac{r \pm 2k}{3} \right)^2 - \left(\frac{k \pm 2r}{3} \right)^2 = 1,$$

this ellipse has foci at B and C.

(*American Mathematical Monthly*, proposed by C.R. Pranesachar)

724. The interesting case occurs of course when b and c are not both equal to zero. Set $d = \sqrt{b^2 + c^2}$ and define the angle α by the conditions

$$\cos\alpha = \frac{b}{\sqrt{b^2 + c^2}} \quad \text{and} \quad \sin\alpha = \frac{c}{\sqrt{b^2 + c^2}}.$$

The integral takes the form

$$\int \frac{dx}{a + d\cos(x - \alpha)},$$

which, with the substitution $u = x - \alpha$, becomes the simpler

$$\int \frac{du}{a + d\cos u}.$$

The substitution $t = \tan\frac{u}{2}$ changes this into

$$\frac{2}{a+d} \int \frac{dt}{1 + \dfrac{a - d}{a + d}t^2}.$$

If $a = d$ the answer to the problem is $\frac{1}{a}\tan\frac{x-\alpha}{2} + C$. If $\frac{a-d}{a+d} > 0$, the answer is

$$\frac{2}{\sqrt{a^2 - d^2}} \arctan\left(\sqrt{\frac{a - d}{a + d}} \tan\frac{x - \alpha}{2} + C \right),$$

while if $\frac{a-d}{a+d} < 0$, the answer is

$$\frac{1}{\sqrt{d^2 - a^2}} \ln \left| \frac{1 + \sqrt{\dfrac{d-a}{d+a}} \tan \dfrac{x-\alpha}{2}}{1 - \sqrt{\dfrac{d-a}{d+a}} \tan \dfrac{x-\alpha}{2}} \right| + C.$$

725. The first equation is linear, so it is natural to solve for one of the variables, say u, and substitute in the second equation. We obtain

$$2xy = z(x + y - z),$$

or

$$z^2 - xz - yz + 2xy = 0.$$

This is a homogeneous equation. Instead of looking for its integer solutions, we can divide through by one of the variables, and then search for the rational solutions of the newly obtained equation. In fancy language, we switch from a projective curve to an affine curve. Dividing by y^2 gives

$$\left(\frac{z}{y}\right)^2 - \left(\frac{z}{y}\right)\left(\frac{x}{y}\right) - \left(\frac{z}{y}\right) + 2\left(\frac{x}{y}\right) = 0.$$

The new equation is

$$Z^2 - ZX - Z + 2X = 0,$$

which defines a hyperbola in the XZ-plane. Let us translate the original problem into a problem about this hyperbola. The conditions $x \geq y$ and $m \leq \frac{x}{y}$ become $X \geq 1$ and $X \geq m$. We are asked to find the largest m such that any point (X, Z) with rational coordinates lying on the hyperbola and in the half-plane $X \geq 1$ has $X \geq m$.

There is a standard way to see that the points of rational coordinates are dense in the hyperbola, which comes from the fact that the hyperbola is *rational*. Substituting $Z - tX$, we obtain

$$X(t^2 X - tX - t + 2) = 0.$$

The root $X = 0$ corresponds to the origin. The other root $X = \frac{t-2}{t^2-t}$ gives the desired parametrization of the hyperbola by rational functions $\left(\frac{t-2}{t^2-t}, \frac{t^2-2t}{t^2-t}\right)$, t real. So the problem has little to do with number theory, and we only need to find the leftmost point on the hyperbola that lies in the half-plane $X \geq 1$. Write the equation of the hyperbola as

$$\left(Z - \frac{X}{2}\right)^2 - \left(\frac{X}{2} - 2\right)^2 = 6.$$

The center is at $(4, 2)$, and the asymptotes are $Z = 2$ and $Z = X - 2$. Let us first minimize X for the points on the hyperbola and in the half-plane $X \geq 4$. We thus minimize the function $f(X, Z) = X$ on the curve $g(X, Z) = Z^2 - ZX - Z + 2X = 0$. The Lagrange multipliers method gives

$$1 = \lambda(-Z + 2),$$
$$0 = \lambda(2Z - X - 1).$$

From the second equation we obtain $Z = \frac{X+1}{2}$. Substitute in $g(X, Z) = 0$ to obtain $X = 3 \pm 2\sqrt{2}$. The further constraint $X \geq 1$ shows that $X = 3 + 2\sqrt{2}$ gives the actual minimum. The same argument shows that the other branch of the hyperbola lies in the half-plane $X < 1$, and so the answer to the problem is $m = 3 + 2\sqrt{2}$.

(Short list of the 42nd International Mathematical Olympiad, 2001)

726. We start with (b) and show that the only possible value of $a_{2015} \cdot a_{2016}$ is 0. By translating the sequence we may assume instead that a_0 and a_1 are the only terms that are squares of integers. Say $a_0 = p^2$ and $a_1 = q^2$.

Now consider the cubic curve

$$y^2 = x^3 + bx^2 + cx + d.$$

Let us first assume that the curve is elliptic, and consider the Abelian group structure defined in the example from the introduction. $P + Q + R = 0$ if R is the third intersection point of PQ with the cubic.

Now consider the subgroup generated by $P = (0, p)$ and $Q = (1, q)$. The identity of this subgroup is the point at infinity of vertical lines, and the subgroup also contains $-P = (0, -p)$ and $-Q = (1, -q)$. We observe that points $P+Q$ and $P-Q$ have integer coordinates, because the line through P and Q has equation $y = (q - p)x + p$ and the line through P and $-Q$ has equation $y = -(q + p)x + p$. Consequently the x-coordinate of $P + Q$ is a root of

$$x^3 + (b - (q - p)^2)x^2 + (c - 2(q - p)p)x + (d - p^2) = 0$$

with the other two roots being 0 and 1. So the x-coordinate of $P+Q$ is $(q-p)^2 - b - 1$ (from the first Viète's relation). The y coordinate is obtained by plugging this in the equation of a line. But this would provide a third perfect square term of the sequence, unless $(q - p)^2 = b + 1$ or $(q - p)^2 = b + 2$, in which case the line through P and Q is tangent to the cubic (either at P or at Q). Using the other pair of points, we find that $(q + p)^2 = b + 1$ or $(q + p)^2 = b + 2$. Since $(q + p)^2$ and $(q - p)^2$ have the same parity, they must simmultaneously be equal to either $b + 1$ or two $b + 2$, which can only happen if either p or q is zero. Thus $a_0 \cdot a_1 = 0$, and (b) is proved for elliptic curves. But all the geometric constructs above apply even when the curve is not elliptic, as we always "add" points that are distinct, so no tangent line is taken, and then we have a well defined line that we intersect with the curve. So removing the sophisticated language of group theory, we can repeat the argument for singular curves.

Inspired by the above discussion, we choose a curve such that $(0, 0)$ and $(1, 1)$ are on the curve, with $(1, 1)$ a double point. Then $a_n = n^3 - n^2 + n$. If $n(n^2 - n + 1)$ is a perfect square for $n \geq 2$, then both factors must be perfect squares (since they are coprime as the second is a multiple of the first plus 1). But $(n - 1)^2 < n^2 - n + 1 < n^2$, so the second factor is not a perfect square. This answers (a).

(Romanian Master of Mathematics, 2016)

727. The curve from the statement is a cubic. In fact it is a singular cubic. Indeed, we can write it in the form

$$f(x, y) = (x + y + 3)^3 - 27(x^2 + xy + y^2)$$

and set

$$\frac{\partial f}{\partial x} = 3(x + y + 3)^2 - 27(2x + y) = 0$$

$$\frac{\partial f}{\partial y} = 3(x + y + 3)^2 - 27(x + 2y) = 0.$$

The difference between the two equations is $27(x - y)$, which should be zero, and this gives $x = y$. But then we easily solve to obtain $x = y = 3$. The graph of the cubic is shown in Figure 85.

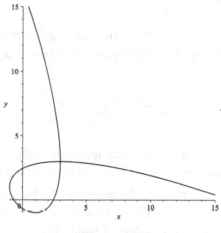

Figure 85

We can move the node to the origin by the substitution $x = a + 3, y = b + 3$ to obtain the equation

$$(a + b)^3 = -27ab.$$

We want to turn this into the standard equation

$$v^2 = \alpha u^3 + \beta u^2.$$

It is natural to require $(a + b)^3$ to be equal to u^3, for example by using $u = a + b$. Then the product ab can be obtain by subtracting $(a - b)^2$ from $(a + b)^2$ and then dividing by 4. Set $u = a + b$, $v = a - b$, and note that they are both even since a and b must have the same parity (or else the equation that they satisfy has an even term on the right and an odd term on the left). We obtain the equation

$$u^3 = -\frac{27}{4}(u^2 - v^2).$$

In other words the cubic in standard form is

$$v^2 = \frac{4}{27}u^3 + u^2.$$

We are to find the points of integer coordinates on this cubic. We repeat the trick that leads to Euler's substitutions. We draw a line passing through the origin and the point of coordinates (u, v), and let t be the slope of the line. We want to parametrize the cubic by t. Since $t = u/v$, we obtain

$$1 = \frac{4}{27} ut^2 + t^2,$$

and this gives

$$u = \frac{27}{4} \cdot \frac{1 - t^2}{t^2}, \quad v = \frac{u}{t} = \frac{27}{4} \cdot \frac{1 - t^2}{t^3}.$$

So this cubic is a rational curve! The geometric part of the problem is over. We now want to find the values of t for which u and v are integers. We know that t is rational, so let $t = p/q$, with p coprime with q. Then

$$v = \frac{27}{4} \cdot \frac{q(q^2 - p^2)}{p^3}.$$

So p^3 divides 27. Thus $p = 1$ or $p = 3$ (we can incorporate the sign in q).

We can do both cases simultaneously by allowing q to be a multiple of 3. Thus $t = \frac{3}{q}$. Then $u = \frac{3}{4}(q^2 - 9)$ and $v = \frac{1}{4}q(q^2 - 9)$. We deduce that q is an odd number, say $q = 2n + 1$. We compute $u = 3(n^2 + n - 2)$, $v = (2n + 1)(n^2 + n - 2)$. So $a = (n + 2)^2(n - 1)$ and $b = -(n - 1)^2(n + 2)$, and we obtain family of solutions:

$$x = (n + 2)^2(n - 1) + 3 = n^3 + 3n^2 - 1,$$
$$y = -(n - 1)^2(n + 2) + 3 = -n^3 + 3n + 1, \quad n \in \mathbb{Z}.$$

Remark. Note that the substitution $n \to -n - 1$ turns the formula for x into that for y and viceversa, as expected from the symmetry of the original equation.

(USA Mathematical Olympiad, 2015, proposed by T. Andreescu)

728. Let $x + y = s$. Then $x^3 + y^3 + 3xys = s^3$, so $3xys - 3xy = s^3 - 1$. It follows that the locus is described by

$$(s - 1)(s^2 + s + 1 - 3xy) = 0.$$

Recalling that $s = x + y$, we notice that the cubic is degenerate: it consists of the line $x + y = 1$ and the conic $(x + y)^2 + x + y + 1 - 3xy = 0$.

The equation of the conic is

$$\frac{1}{2}[(x - y)^2 + (x + 1)^2 + (y + 1)^2] = 0,$$

i.e., $x = y = -1$. So the conic itself degenerates to one point! Thus the cubic in the problem consists of the line $x + y = 1$ and the point $(-1, -1)$, which we will call A. Points B and C are on the line $x + y = 1$ such that they are symmetric to one another with respect to the point $D \left(\frac{1}{2}, \frac{1}{2}\right)$ and such that $BC\frac{\sqrt{3}}{2} = AD$. It is clear that there is only one set $\{B, C\}$ with this property, so we have justified the uniqueness of the triangle ABC (up to the permutation of vertices). Because

$$AD = \sqrt{\left(\frac{1}{2}+1\right)^2 + \left(\frac{1}{2}+1\right)^2} = \frac{3}{2}\sqrt{2},$$

it follows that $BC = \sqrt{6}$; hence Area$(ABC) = \frac{6\sqrt{3}}{4} = \frac{3\sqrt{3}}{2}$.

(49th W.L. Putnam Mathematical Competition, 2006, proposed by T. Andreescu)

729. If $P(x) = x^3 + ax^2 + bx + c$ has a double rational root, then this root is necessarily an integer, so $P(x) = (x-p)^2(x-q)$ with $p, q \in \mathbb{Z}$. We can choose an integer n such that $n - q$ is not a perfect square which then makes $\sqrt{P(n)}$ irrational.

If $P(x)$ is squarefree over $\mathbb{Q}$, then $y^2 = P(x)$ is an elliptic curve E over $\mathbb{Q}$. Now we use a deep result about height functions in algebraic geometry. The height of a rational number x written in lowest terms as u/v is $H(r) = \max(|u|, |v|)$. For $B > 0, N(E, B)$ be the number of points of rational coordinates (x, y) on E with the property that $H(x) < B$. A theorem of A. Néron implies that

$$N(E, B) \sim (\ln B)^{r/2},$$

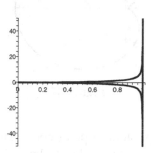

Figure 86

where r is the rank of the elliptic curve over $\mathbb{Q}$. For defining the rank consider the Abelian group $E(\mathbb{Q})$ consisting only of points of rational coordinates. The Mordell-Weil theorem implies that this group is finitely generated. The rank is the number of copies of $\mathbb{Z}$ in $E(\mathbb{Q})$.

Returning to the problem, for sufficiently large B, $N(E, B) < B$, so not all the integers between 1 and B yield a point of integer coordinates on the curve. That is not all of the numbers $\sqrt{P(n)}$, $n = 1, 2 \ldots, B$ are integers.

We challenge the reader to find an elementary proof!

(59th W.L. Putnam Mathematical Competition, 1998, solution from K. Kedlaya, B. Poonen, R. Vakil, *The William Lowell Putnam Mathematical Competition, 1985–2000*, MAA, 2002)

730. We convert to Cartesian coordinates, obtaining the equation of the cardioid

$$\sqrt{x^2 + y^2} = 1 + \frac{x}{\sqrt{x^2 + y^2}},$$

or

$$x^2 + y^2 = \sqrt{x^2 + y^2} + x.$$

By implicit differentiation, we obtain

$$2x + 2y\frac{dy}{dx} = (x^2 + y^2)^{-1/2}\left(x + y\frac{dy}{dx}\right) + 1,$$

which yields

$$\frac{dy}{dx} = \frac{-2x + x(x^2 + y^2)^{-1/2} + 1}{2y - y(x^2 + y^2)^{-1/2}}.$$

The points where the tangent is vertical are among those where the denominator cancels. Solving $2y - y(x^2 + y^2)^{-1/2} = 0$, we obtain $y = 0$ or $x^2 + y^2 = \frac{1}{4}$. Combining this with the equation of the cardioid, we find the possible answers to the problem as $(0, 0)$, $(2, 0)$, $\left(-\frac{1}{4}, \frac{\sqrt{3}}{4}\right)$, and $\left(-\frac{1}{4}, -\frac{\sqrt{3}}{4}\right)$. Of these the origin has to be ruled out, since there the cardioid has a corner, while the other three are indeed points where the tangent to the cardioid is vertical.

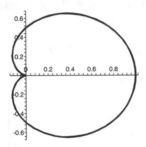

Figure 87

731. Let $AB = a$ and consider a system of polar coordinates with pole A and axis AB. The equation of the curve traced by M is obtained as follows. We have $AM = r$, $AD = \frac{a}{\cos\theta}$, and $AC = a\cos\theta$. The equality $AM = AD - AC$ yields the equation

$$r = \frac{a}{\cos\theta} - a\cos\theta.$$

The equation of the locus is therefore $r = \frac{a\sin^2\theta}{\cos\theta}$. This curve is called the cisoid of Diocles (Figure 86).

732. Let O be the center and a the radius of the circle, and let M be the point on the circle. Choose a system of polar coordinates with M the pole and MO the axis. For an arbitrary tangent, let I be its intersection with MO, T the tangency point, and P the projection of M onto the tangent. Then

$$OI = \frac{OT}{\cos\theta} = \frac{a}{\cos\theta}.$$

Hence

$$MP = r = (MO + OI)\cos\theta = \left(a + \frac{a}{\cos\theta}\right)\cos\theta.$$

We obtain $r = a(1 + \cos\theta)$, which is the equation of a cardioid (Figure 87).

733. Working in polar coordinates we place the pole at O and axis OA. Denote by a the radius of the circle. We want to find the relation between the polar coordinates (r, θ) of the point L. We have $AM = AL = 2a \sin \frac{\theta}{2}$. In the isosceles triangle LAM, $\angle LMA = \frac{\pi}{2} - \frac{\theta}{2}$; hence

$$LM = 2AM \cos\left(\frac{\pi}{2} - \frac{\theta}{2}\right) = 2 \cdot 2a \sin \frac{\theta}{2} \cdot \sin \frac{\theta}{2} = 4a \sin^2 \frac{\theta}{2}.$$

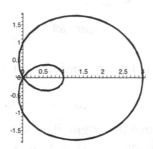

Figure 88

Substituting this in the relation $OL = OM - LM$, we obtain

$$r = a - 4a \sin^2 \frac{\theta}{2} = a[1 - 2(1 - \cos\theta)].$$

The equation of the locus is therefore

$$r = a(2\cos\theta - 1),$$

a curve known as Pascal's snail, or limaçon, whose shape is described in Figure 88.

734. As before, we work with polar coordinates, choosing O as the pole and OA as the axis. Denote by a the length of the segment AB and by $P(r, \theta)$ the projection of O onto this segment. Then $OA = \frac{r}{\cos\theta}$ and $OA = AB \sin\theta$, which yield the equation of the locus

$$r = a \sin\theta \cos\theta = \frac{a}{2}\sin 2\theta.$$

This is a four-leaf rose.

735. Choosing a Cartesian system of coordinates whose axes are the asymptotes, we can bring the equation of the hyperbola into the form $xy = a^2$. The equation of the tangent to the hyperbola at a point (x_0, y_0) is $x_0 y + y_0 x - 2a^2 = 0$. Since $a^2 = x_0 y_0$, the x and y intercepts of this line are $2x_0$ and $2y_0$, respectively.

Let (r, θ) be the polar coordinates of the foot of the perpendicular from the origin to the tangent. In the right triangle determined by the center of the hyperbola and the two intercepts we have $2x_0 \cos\theta = r$ and $2y_0 \sin\theta = r$. Multiplying, we obtain the polar equation of the locus

$$r^2 = 2a^2 \sin 2\theta.$$

This is the lemniscate of Bernoulli, shown in Figure 89.

(1st W.L. Putnam Mathematical Competition, 1938)

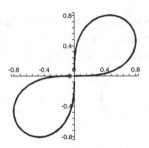

Figure 89

736. The solution uses complex and polar coordinates. Our goal is to map the circle onto a cardioid of the form

$$r = a(1 + \cos \theta), \ a > 0.$$

Because this cardioid passes through the origin, it is natural to work with a circle that itself passes through the origin, for example $|z - 1| = 1$. If $\phi : \mathbb{C} \to \mathbb{C}$ maps this circle into the cardioid, then the equation of the cardioid will have the form

$$|\phi^{-1}(z) - 1| = 1.$$

So we want to bring the original equation of the cardioid into this form. First, we change it to

$$r = a \cdot 2 \cos^2 \frac{\theta}{2};$$

then we take the square root,

$$\sqrt{r} = \sqrt{2a} \cos \frac{\theta}{2}.$$

Multiplying by $\sqrt{r}$, we obtain

$$r = \sqrt{2a}\sqrt{r} \cos \frac{\theta}{2},$$

or

$$r - \sqrt{2a}\sqrt{r} \cos \frac{\theta}{2} = 0.$$

This should look like the equation of a circle. We modify the expression as follows:

$$r - \sqrt{2a}\sqrt{r} \cos \frac{\theta}{2} = r \left(\cos^2 \frac{\theta}{2} + \sin^2 \frac{\theta}{2} \right) - \sqrt{2a}\sqrt{r} \cos \frac{\theta}{2} + 1 - 1$$

$$= \left(\sqrt{r} \cos \frac{\theta}{2} \right)^2 - \sqrt{2a}\sqrt{r} \cos \frac{\theta}{2} + 1 + \left(\sqrt{r} \sin \frac{\theta}{2} \right)^2 - 1.$$

If we set $a = 2$, we have a perfect square, and the equation becomes

$$\left(\sqrt{r} \cos \frac{\theta}{2} - 1 \right)^2 + \left(\sqrt{r} \sin \frac{\theta}{2} \right)^2 = 1,$$

which in complex coordinates reads $|\sqrt{z} - 1| = 1$. Of course, there is an ambiguity in taking the square root, but we are really interested in the transformation ϕ, not in ϕ^{-1}. Therefore, we can choose $\phi(z) = z^2$, which maps the circle $|z - 1| = 1$ into the cardioid $r = 2(1 + \cos \theta)$.

Remark. Of greater practical importance is the Zhukovski transformation $z \rightarrow \frac{1}{2}\left(z + \frac{1}{z}\right)$, which maps the unit circle onto the profile of the airplane wing (the so-called aerofoil). Because the Zhukovski map preserves angles, it helps reduce the study of the air flow around an airplane wing to the much simpler study of the air flow around a circle.

737. View the parametric equations of the curve as a linear system in the unknowns t^n and t^p:

$$a_1 t^n + b_1 t^p = x - c_1,$$

$$a_2 t^n + b_2 t^p = x - c_2,$$

$$a_3 t^n + b_3 t^p = x - c_3.$$

This system admits solutions; hence the augmented matrix of the system is singular. We thus have

$$\begin{vmatrix} a_1 & b_1 & x - c_1 \\ a_2 & b_2 & x - c_2 \\ a_3 & b_3 & x - c_3 \end{vmatrix} = 0.$$

This is the equation of a plane that contains the given curve.

(C. Ionescu-Bujor, O. Sacter, *Exerciţii şi Probleme de Geometrie Analitica şi Diferenţială (Exercises and Problems in Analytic and Differential Geometry)*, Editura Didactică şi Pedagogică, Bucharest, 1963)

738. Let the equation of the curve be $y(x)$. Let $T(x)$ be the tension in the chain at the point $(x, y(x))$. The tension acts in the direction of the derivative $y'(x)$. Let $H(x)$ and $V(x)$ be, respectively, the horizontal and vertical components of the tension. Because the chain is in equilibrium, the horizontal component of the tension is constant at all points of the chain (just cut the chain mentally at two different points). Thus $H(x) = H$. The vertical component of the tension is then $V(x) = Hy'(x)$.

On the other hand, for two infinitesimally close points, the difference in the vertical tension is given by $dV = \rho ds$, where ρ is the density of the chain and ds is the length of the arc between the two poins. Since $ds = \sqrt{1 + (y'(x))^2}\, dx$, it follows that y satisfies the differential equation

$$Hy'' = \rho\sqrt{1 + (y')^2}.$$

If we set $z(x) = y'(x)$, we obtain the separable first-order equation

$$Hz' = \rho\sqrt{1 + z^2}.$$

By integration, we obtain $z = \sinh\left(\frac{\rho}{H}x + C_1\right)$. The answer to the problem is therefore

$$y(x) = \frac{H}{\rho}\cosh\left(\frac{\rho}{H} + C_1\right) + C_2.$$

Remark. Galileo claimed that the curve was a parabola, but this was later proved to be false. The correct equation was derived by G.W. Leibniz, Ch. Huygens, and Johann Bernoulli. The curve is called a "catenary" and plays an important role in the theory of minimal surfaces.

739. An edge adjacent to the main diagonal describes a cone. For an edge not adjacent to the main diagonal, consider an orthogonal system of coordinates such that the rotation axis is the z-axis and, in its original position, the edge is parallel to the y-plane (Figure 90). In the appropriate scale, the line of support of the edge is $y = 1$, $z = \sqrt{3}x$.

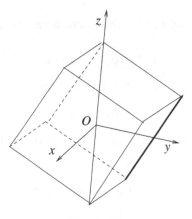

Figure 90

The locus of points on the surface of revolution is given in parametric form by

$$(x, y, z) = (t \cos \theta + \sin \theta, \cos \theta - t \sin \theta, \sqrt{3}t), \ t \in \mathbb{R}, \ \theta \in [0, 2\pi).$$

A glimpse at these formulas suggests the following computation:

$$x^2 + y^2 - \frac{1}{3}z^2 = t^2 \cos^2 \theta + \sin^2 \theta + 2t \sin \theta \cos \theta + \cos^2 \theta + t^2 \sin^2 \theta - 2t \cos \theta \sin \theta - t^2$$
$$= t^2(\cos^2 \theta + \sin^2 \theta) + \cos^2 \theta + \sin^2 \theta - t^2 = 1.$$

The locus is therefore a hyperboloid of one sheet, $x^2 + y^2 - \frac{1}{3}z^3 = 1$.

Remark. The fact that the hyperboloid of one sheet is a ruled surface makes it easy to build. It is a more resilient structure than the cylinder. This is why the cooling towers of power plants are built as hyperboloids of one sheet.

740. The equation of the plane tangent to the hyperboloid at a point $M(x_0, y_0, z_0)$ is

$$\frac{x_0 x}{a^2} + \frac{y_0 y}{b^2} - \frac{z_0 z}{c^2} = 1.$$

This plane coincides with the one from the statement if and only if

$$\frac{\frac{x_0}{a^2}}{\frac{1}{a}} = \frac{\frac{y_0}{b^2}}{\frac{1}{b}} = \frac{\frac{z_0}{c^2}}{\frac{1}{c}}.$$

We deduce that the point of contact has coordinates (a, b, c), and therefore the given plane is indeed tangent to the hyperboloid.

741. The area of the ellipse given by the equation

$$\frac{x^2}{A^2} + \frac{y^2}{B^2} = R^2$$

is $\pi A B R^2$. The section perpendicular to the x-axis is the ellipse

$$\frac{y^2}{b^2} + \frac{z^2}{c^2} = 1 - \frac{x_0^2}{a^2}$$

in the plane $x = x_0$. Hence $S_x = \pi bc \left(1 - \frac{x_0^2}{a^2}\right)$. Similarly, $S_y = \pi ac \left(1 - \frac{y_0^2}{b^2}\right)$ and $S_x = \pi ab \left(1 - \frac{z_0^2}{c^2}\right)$. We thus have

$$aS_x + bS_y + cS_z = \pi abc \left(3 - \frac{x_0^2}{a^2} + \frac{y_0^2}{b^2} + \frac{z_0^2}{c^2}\right) = 2\pi abc,$$

which, of course, is independent of M.

742. Figure 91 describes a generic ellipsoid. Since parallel cross-sections of the ellipsoid are always similar ellipses, any circular cross-section can be increased in size by taking a parallel cutting plane passing through the origin. Because of the condition $a > b > c$, a circular cross-section cannot lie in the xy-, xz-, or yz-plane. Looking at the intersection of the ellipsoid with the yz-plane, we see that some diameter of the circular cross-section is a diameter (segment passing through the center) of the ellipse $x = 0$, $\frac{y^2}{b^2} + \frac{z^2}{c^2} = 1$. Hence the radius of the circle is at most b. The same argument for the xy-plane shows that the radius is at least b, whence b is a good candidate for the maximum radius.

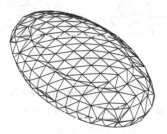

Figure 91

To show that circular cross-sections of radius b actually exist, consider the intersection of the plane $(c\sqrt{a^2 - b^2})x = (a\sqrt{b^2 - c^2})z$ with the ellipsoid. We want to compute the distance from a point (x_0, y_0, z_0) on this intersection to the origin. From the equation of the plane, we obtain by squaring

$$x_0^2 + z_0^2 = b^2 \left(\frac{x_0^2}{a^2} + \frac{z_0^2}{c^2}\right).$$

The equation of the ellipsoid gives

$$y_0^2 = b^2 \left(1 - \frac{x_0^2}{a^2} - \frac{z_0^2}{c^2}\right).$$

Adding these two, we obtain $x_0^2 + y_0^2 + z_0^2 = 1$; hence (x_0, y_0, z_0) lies on the circle of radius 1 centered at the origin and contained in the plane $(c\sqrt{a^2 - b^2})x + (a\sqrt{b^2 - c^2})z = 0$. This completes the proof.

(31st W.L. Putnam Mathematical Competition, 1970)

743. Without loss of generality, we may assume $a < b < c$. Fix a point (x_0, y_0, z_0) and let us examine the equation in λ,

$$f(\lambda) = \frac{x_0^2}{a^2 - \lambda} + \frac{y_0^2}{b^2 - \lambda} + \frac{z_0^2}{c^2 - \lambda} - 1 = 0.$$

For the function $f(\lambda)$ we have the following table of signs:

$f(-\infty)$	$f(a^2 - \varepsilon)$	$f(a^2 + \varepsilon)$	$f(b^2 - \varepsilon)$	$f(b^2 + \varepsilon)$	$f(c^2 - \varepsilon)$	$f(c^2 + \varepsilon)$	$f(+\infty)$
$+$	$+$	$-$	$+$	$-$	$+$	$-$	$-$

where ε is a very small positive number. Therefore, the equation $f(\lambda) = 0$ has three roots, $\lambda_1, \lambda_2, \lambda_3$, with $\lambda_1 < a^2 < \lambda_2 < b^2 < \lambda_3 < c^2$. These provide the three surfaces, which are an ellipsoid for $\lambda = \lambda_1$ (Figure 91), a hyperboloid of one sheet for $\lambda = \lambda_2$, and a hyperboloid of two sheets for $\lambda = \lambda_3$ (Figure 92).

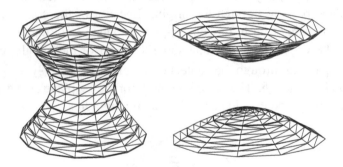

Figure 92

To show that the surfaces are pairwise orthogonal we have to compute the angle between the normals at an intersection point. We do this for the roots λ_1 and λ_2 the other cases being similar. The normal to the ellipsoid at a point (x, y, z) is parallel to the vector

$$\vec{v}_1 = \left(\frac{x}{a^2 - \lambda_1}, \frac{y}{b^2 - \lambda_1}, \frac{z}{c^2 - \lambda_1} \right),$$

while the normal to the hyperboloid of one sheet is parallel to the vector

$$\vec{v}_2 = \left(\frac{x}{a^2 - \lambda_2}, \frac{y}{b^2 - \lambda_2}, \frac{z}{c^2 - \lambda_2} \right).$$

The dot product of these vectors is

$$\vec{v}_1 \cdot \vec{v}_2 = \frac{x}{a^2 - \lambda_1} \cdot \frac{x}{a^2 - \lambda_2} + \frac{y}{b^2 - \lambda_1} \cdot \frac{y}{b^2 - \lambda_2} + \frac{z}{c^2 - \lambda_1} \cdot \frac{z}{c^2 - \lambda_2}.$$

To prove that this is equal to 0, we use the fact that the point (x, y, z) belongs to both quadrics, which translates into the relation

$$\frac{x^2}{a^2 - \lambda_1} + \frac{y^2}{b^2 - \lambda_1} + \frac{z^2}{c^2 - \lambda_1} = \frac{x^2}{a^2 - \lambda_2} + \frac{y^2}{b^2 - \lambda_2} + \frac{z^2}{c^2 - \lambda_2}.$$

If we write this as

$$\left(\frac{x^2}{a^2 - \lambda_1} - \frac{x^2}{a^2 - \lambda_2}\right) + \left(\frac{y^2}{b^2 - \lambda_1} - \frac{y^2}{b^2 - \lambda_2}\right) + \left(\frac{z^2}{c^2 - \lambda_1} - \frac{x^z}{c^2 - \lambda_2}\right) = 0,$$

we recognize immediately the left-hand side to be $(\lambda_1 - \lambda_2)\overrightarrow{v}_1 \cdot \overrightarrow{v}_2$. We obtain the desired $\overrightarrow{v}_1 \cdot \overrightarrow{v}_2 = 0$, which proves the orthogonality of the two surfaces. This completes the solution.

(C. Ionescu-Bujor, O. Sacter, *Exerciţii şi Probleme de Geometrie Analitică şi Diferenţială* (*Exercises and Problems in Analytic and Differential Geometry*), Editura Didactică şi Pedagogică, Bucharest, 1963)

744. Using the algebraic identity (see Section 2.1.1):

$$(u^3 + v^3 + w^3 - 3uvw) = \frac{1}{2}(u + v + w)[3(u^2 + v^2 + w^2) - (u + v + w)^2],$$

we obtain

$$z - 3 = \frac{3}{2}xy - \frac{1}{2}x^3,$$

or

$$x^3 - 3xy + 2z - 6 = 0.$$

This is the cubic surface from Figure 93.

Figure 93

(C. Coşniţă, I. Sager, I. Matei, I. Dragotă, *Culegere de Probleme de Geometrie Analitică* (*Collection of Problems in Analytical Geometry*), Editura Didactică şi Pedagogică, Bucharest, 1963)

745. By the $(2n + 1)$-dimensional version of the Pythagorean theorem, the edge L of the cube is the square root of an integer. The volume of the cube is computed as a determinant in coordinates of vertices; hence it is also an integer. We conclude that L^2 and L^{2n+1} are both integers. It follows that $L^{2n+1}/(L^2)^n = L$ is a rational number. Because its square is an integer, L is actually an integer, as desired.

746. The equation of the locus can be expressed in a simple form using determinants as

$$\begin{vmatrix} x_1 & x_2 & \cdots & x_n \\ x_n & x_1 & \cdots & x_{n-1} \\ \vdots & \vdots & \ddots & \vdots \\ x_2 & x_3 & \cdots & x_1 \end{vmatrix} = 0.$$

Adding all rows to the first, we see that the determinant has a factor of $x_1 + x_2 + \cdots + x_n$. Hence the hyperplane $x_1 + x_2 + \cdots + x_n = 0$ belongs to the locus.

747. Without loss of generality, we may assume that the edges of the cube have length equal to 2, in which case the cube consists of the points $(x_1, x_2, \ldots, x_n)$ with $\max |x_i| \leq 1$. The intersection of the cube with the plane determined by $\vec{a}$ and $\vec{b}$ is

$$P = \left\{ s\vec{a} + t\vec{b} \mid \max_k \left| s\cos\frac{2k\pi}{n} + t\sin\frac{2k\pi}{n} \right| \leq 1 \right\}.$$

This set is a convex polygon with at most $2n$ sides, being the intersection of n strips determined by parallel lines, namely the strips

$$P_k = \left\{ s\vec{a} + t\vec{b} \mid \left| s\cos\frac{2k\pi}{n} + t\sin\frac{2k\pi}{n} \right| \leq 1 \right\}.$$

Adding $\frac{2\pi}{n}$ to all arguments in the coordinates of $\vec{a}$ and $\vec{b}$ permutes the P_k's, leaving P invariant. This corresponds to the transformation

$$\vec{a} \mapsto \cos\frac{2\pi}{n}\vec{a} - \sin\frac{2\pi}{n}\vec{b},$$

$$\vec{b} \mapsto \sin\frac{2\pi}{n}\vec{a} + \cos\frac{2\pi}{n}\vec{b},$$

which is a rotation by $\frac{2\pi}{n}$ in the plane of the two vectors. Hence P is invariant under a rotation by $\frac{2\pi}{n}$, and being a polygon with at most $2n$ sides, it must be a regular $2n$-gon.

(V.V. Prasolov, V.M. Tikhomirov, *Geometry*, AMS, 2001)

748. We consider the case of the unit cube, whose vertices are of the form $(x_1, x_2, \ldots, x_n)$, $x_n \in \{0, 1\}$.

First, consider the hyperplanes

$$H_k: \quad x_1 + x_2 + \cdots + x_n = k + \frac{1}{2}, \quad 0 \leq k \leq n - 1.$$

Then H_k crosses every edge joining one of the $\binom{n}{k}$ points with k nonzero coordinates to its $n - k$ neighbors with $k + 1$ nonzero coordinates and only these edges. thus such a plane crosses

$$(n - k)\binom{n}{k} = \frac{n!}{k!(n - k - 1)!}.$$

By examining the monotonicity and the minimum of the function $f(x) = x(n - x - 1)$, we deduce that the above fraction is maximized when n is as close as possible to $\frac{n-1}{2}$. So the maximum number of crossings of edges with one of the hyperplanes H_k is

$$\frac{n!}{\lfloor (n-1)/2 \rfloor! \lceil (n-1)/2 \rceil!},$$

where $\lfloor \cdot \rfloor$ and $\lceil \cdot \rceil$ are the greatest integer and the least integer functions. We will prove that this is the desired maximum.

Let us consider a hyperplane

$$H : \sum_{j=1}^{n} a_j x_j = b.$$

Using the symmetries of the cube, we may assume $a_j \geq 0$ for all j, and hence $b \geq 0$. Now cosider any of the $n!$ paths from $(0, 0, \ldots, 0)$ to $(1, 1, \ldots, 1)$, in which we increase one coordinate at a time, from 0 to 1. The function $\sum_j a_j x_j$ is nondecreasing on such a path, hence there is at most one edge on the path that is cut by H. This happens when $\sum_j a_j x_j$ transitions from less than b to more than b (we do not count the edges contained in H, where this function is constantly equal to b).

If an edge $e = (x, y)$ joins the vertex x that has k coordinates equal to 1 to a vertex y that has $k + 1$ coordinates equal to 1, that edge lies on $k!(n - k - 1)!$ such paths. Indeed, along such a path the k locations that correspond to 1's in x are turned first one-by-one from 0 into 1's, and this can be done in $k!$ ways, and then we move to y, after which the remaining 0's are turned one-by-one into 1's, and again this can be done in $(n - k - 1)!$ ways. As explained above,

$$k!(n - k - 1)! \geq \lfloor (n-1)/2 \rfloor! \lceil (n-1)/2 \rceil!.$$

There are exactly $n!$ paths, and one edge appears in at least $\lfloor (n-1)/2 \rfloor! \lceil (n-1)/2 \rceil!$ paths, and since each path is cut at most once by the hyperplane P, this hyperplane intersects at most

$$\frac{n!}{\lfloor (n-1)/2 \rfloor! \lceil (n-1)/2 \rceil!}.$$

The conclusion follows.

(*Mathematical Reflections*, proposed by G. Dospinescu)

749. Consider the unit sphere in $\mathbb{R}^n$,

$$S^{n-1} = \left\{ (x_1, x_2, \ldots, x_n) \in \mathbb{R}^n \mid \sum_{k=1}^{n} x_k^2 = 1 \right\}.$$

The distance between two points $X = (x_1, x_2, \ldots, x_n)$ and $Y = (y_1, y_2, \ldots, y_n)$ is given by

$$d(X, Y) = \left(\sum_{k=1}^{n} (x_k - y_k)^2 \right)^{1/2}.$$

Note that $d(X, Y) > \sqrt{2}$ if and only if

$$d^2(X, Y) = \sum_{k=1}^{n} x_k^2 + \sum_{k=1}^{n} y_k^2 - 2 \sum_{k=1}^{n} x_k y_k > 2.$$

Therefore, $d(X, Y) > \sqrt{2}$ implies $\sum_{k=1}^{n} x_k y_k < 0$.

Now let $A_1, A_2, \ldots, A_{m_n}$ be points satisfying the condition from the hypothesis, with m_n maximal. Using the symmetry of the sphere we may assume that $A_1 = (-1, 0, \ldots, 0)$. Let $A_i = (x_1, x_2, \ldots, x_n)$ and $A_j = (y_1, y_2, \ldots, y_n)$, $i, j \geq 2$. Because $d(A_1, A_i)$ and $d(A_1, A_j)$ are both greater than $\sqrt{2}$, the above observation shows that x_1 and y_1 are positive.

The condition $d(A_i, A_j) > \sqrt{2}$ implies $\sum_{k=1}^{n} x_k y_k < 0$, and since $x_1 y_1$ is positive, it follows that

$$\sum_{k=2}^{n} x_k y_k < 0.$$

This shows that if we normalize the last $n - 1$ coordinates of the points A_i by

$$x_k' = \frac{x_k}{\sqrt{\sum_{k=1}^{n-1} x_k^2}}, \quad k = 1, 2, \ldots, n - 1,$$

we obtain the coordinates of point B_i in S^{n-2}, and the points $B_2, B_3, \ldots, B_n$ satisfy the condition from the statement of the problem for the unit sphere in $\mathbb{R}^{n-1}$.

It follows that $m_n \leq 1 + m_{n-1}$, and $m_1 = 2$ implies $m_n \leq n + 1$. The example of the n-dimensional regular simplex inscribed in the unit sphere shows that $m_n = n + 1$. To determine explicitly the coordinates of the vertices, we use the additional information that the distance from the center of the sphere to a hyperface of the n-dimensional simplex is $\frac{1}{n}$ and then find inductively

$$A_1 = (-1, 0, 0, 0, \ldots, 0, 0),$$

$$A_2 = \left(\frac{1}{n}, -c_1, 0, 0, \ldots, 0, 0 \right),$$

$$A_3 = \left(\frac{1}{n}, \frac{1}{n-1} \cdot c_1, -c_2, 0, \ldots, 0, 0 \right),$$

$$A_4 = \left(\frac{1}{n}, \frac{1}{n-1} \cdot c_1, \frac{1}{n-2} \cdot c_2, c_3, \ldots, 0, 0 \right),$$

$$\cdots$$

$$A_{n-1} = \left(\frac{1}{n}, \frac{1}{n-1} \cdot c_1, \ldots, \frac{1}{3} \cdot c_{n-3}, -c_{n-2}, 0 \right),$$

$$A_n = \left(\frac{1}{n}, \frac{1}{n-1} \cdot c_1, \ldots, \frac{1}{3} \cdot c_{n-3}, \frac{1}{2} \cdot c_{n-2}, -c_{n-1}\right),$$

$$A_{n=1} = \left(\frac{1}{n}, \frac{1}{n-1} \cdot c_1, \ldots, \frac{1}{3} \cdot c_{n-3}, \frac{1}{2} \cdot c_{n-2}, c_{n-1}\right),$$

where

$$c_k = \sqrt{\left(1 + \frac{1}{n}\right)\left(1 - \frac{1}{n-k+1}\right)}, \quad k = 1, 2, \ldots, n-1.$$

One computes that the distance between any two points is

$$\sqrt{2}\sqrt{1 + \frac{1}{n}} > \sqrt{2},$$

and the problem is solved.

(8th International Mathematics Competition for University Students, 2001)

750. View the ring as the body obtained by revolving about the x-axis the surface that lies between the graphs of $f, g : [-h/2, h/2] \to \mathbb{R}, f(x) = \sqrt{R^2 - x^2}, g(x) = \sqrt{R^2 - h^2/4}$. Here R denotes the radius of the sphere. Using the washer method we find that the volume of the ring is

$$\pi \int_{-h/2}^{h/2} (\sqrt{R^2 - x^2})^2 - (\sqrt{R^2 - h^2/4})^2 dx = \pi \int_{-h/2}^{h/2} (h^2/4 - x^2) dx = \frac{h^3\pi}{12},$$

which does not depend on R.

751. Let the inscribed sphere have radius R and center O. For each big face of the polyhedron, project the sphere onto the face to obtain a disk D. Then connect D with O to form a cone. Because the interiors of the cones are pairwise disjoint, the cones intersect the sphere in several nonoverlapping regions. Each circular region is a slice of the sphere, of width $R\left(1 - \frac{1}{2}\sqrt{2}\right)$. Recall the lemma used in the solution to the first problem from the introduction. We apply it to the particular case in which one of the planes is tangent to the sphere to find that the area of a slice is $2\pi R^2\left(1 - \frac{1}{2}\sqrt{2}\right)$, and this is greater than $\frac{1}{7}$ of the sphere's surface. Thus each circular region takes up more than $\frac{1}{7}$ of the total surface area of the sphere. So there can be at most six big faces.

(Russian Mathematical Olympiad, 1999)

752. Keep the line of projection fixed, for example the x-axis, and rotate the segments in A and B simultaneously.

Now, given a segment with one endpoint at the origin, the length of its projection onto the z-axis is $r|\cos\phi|$, where (r, θ, ϕ) are the spherical coordinates of the second endpoint, i.e., r is the length of the segment, ϕ is the angle it makes with the semiaxis Oz, and θ is the oriented angle that its projection onto the xy-plane makes with Ox. If we average the lengths of the projections onto the x-axis of the segment over all possible rotations, we obtain

$$\frac{1}{4\pi} \int_0^\pi \int_0^{2\pi} r|\cos\phi|\sin\phi \, d\theta \, d\phi = \frac{r}{2}.$$

Denote by a and b the sums of the lengths of the segments in A and B, respectively. Then the average of the sum of the lengths of the projections of segments in A is $\frac{r}{2}a$, and the average of the same sum for B is $\frac{r}{2}b$. The second is smaller, proving that there exists a direction such that the sum of the lengths of the projections of the segments from A onto that direction is larger that the corresponding sum for B.

753. This is just a two-dimensional version of the previous problem. If we integrate the length of the projection of a segment onto a line over all directions of the line, we obtain twice the length of the segment. Doing this for the sides of a convex polygon, we obtain the perimeter (since the projection is double covered by the polygon). Because the projection of the inner polygon is always smaller than the projection of the outer, the same inequality will hold after integration. Hence the conclusion.

754. For $i = 1, 2, \ldots, n$, let a_i be the lengths of the segments and let ϕ_i be the angles they make with the positive x-axis ($0 \le \phi_i \le \pi$). The length of the projection of a_i onto some line that makes an angle ϕ with the x-axis is $f_i(\phi) = a_i |\cos(\phi - \phi_i)|$; denote by $f(\phi)$ the sum of these lengths. The integral mean of f over the interval $[0, \pi]$ is

$$\frac{1}{\pi} \int_0^\pi f(\phi)d\phi = \frac{1}{\pi} \sum_{i=1}^n \int_0^\pi f_i(\phi)d\phi = \frac{1}{\pi} \sum_{i=1}^n a_i \int_0^\pi |\cos(\phi - \phi_i)|d\phi$$

$$= \frac{2}{\pi} \sum_{i=1}^n a_i = \frac{2}{\pi}.$$

Here we used the fact that $|\cos x|$ is periodic with period π. Since the integral mean of f is $\frac{2}{\pi}$ and since f is continuous, by the intermediate value property there exists an angle ϕ for which $f(\phi) = \frac{2}{\pi}$. This completes the proof.

755. The law of cosines in triangle APB gives

$$AP^2 = x^2 + c^2 - 2xc \cos B$$

and

$$x^2 = c^2 + AP^2 = x^2 + c^2 - 2xc \cos B - 2c\sqrt{x^2 + c^2 - 2xc \cos B} \cos t,$$

whence

$$\cos t = \frac{c - x \cos B}{\sqrt{x^2 + c^2 - 2xc \cos B}}.$$

The integral from the statement is

$$\int_0^a \cos t(x)dx = \int_0^a \frac{c - x \cos B}{\sqrt{x^2 + c^2 - 2xc \cos B}}dx.$$

Using the standard integration formulas

$$\int \frac{dx}{\sqrt{x^2 + \alpha x + \beta}} = \ln\left(2x + \alpha + 2\sqrt{x^2 + \alpha x + \beta}\right),$$

$$\int \frac{xdx}{\sqrt{x^2 + \alpha x + \beta}} = \sqrt{x^2 + \alpha x + \beta} - \frac{\alpha}{2}\ln\left(2x + \alpha + 2\sqrt{x^2 + \alpha x + \beta}\right),$$

we obtain

$$\int_0^a \cos t(x)dx = c\sin^2 B \ln\left(2x + 2c\cos B + 2\sqrt{x^2 - 2cx\cos B + c^2}\right)\Big|_0^a$$

$$- \cos B\sqrt{x^2 - 2cx\cos B + c^2}\Big|_0^a$$

$$= c\sin^2 B \ln\frac{a - c\cos B + b}{c(1 - \cos B)} + \cos B(c - b).$$

756. It is equivalent to ask that the volume of the dish be half of that of the solid of revolution obtained by rotating the rectangle $0 \le x \le a$ and $0 \le y \le f(a)$. Specifically, this condition is

$$\int_0^a 2\pi x f(x)dx = \frac{1}{2}\pi a^2 f(a).$$

Because the left-hand side is differentiable with respect to a for all $a > 0$, the right-hand side is differentiable, too. Differentiating, we obtain

$$2\pi a f(a) = \pi a f(a) + \frac{1}{2}\pi a^2 f'(a).$$

This is a differential equation in f, which can be written as $f'(a)/f(a) = \frac{2}{a}$. Integrating, we obtain $\ln f(a) = 2\ln a$, or $f(a) = ca^2$ for some constant $c > 0$. This solves the problem.
  (*Math Horizons*)

757. Parametrize the curve by its length as $(x(s), y(s), z(s))$, $0 \le s \le L$. Then the coordinates (ξ, η, ζ) of its spherical image are given by

$$\xi = \frac{dx}{ds}, \quad \eta = \frac{dy}{ds}, \quad \zeta = \frac{dz}{ds}.$$

The fact that the curve is closed implies that

$$\int_0^L \xi ds = \int_0^L \eta ds = \int_0^L \zeta ds = 0.$$

Pick an arbitrary great circle of the unit sphere, lying in some plane $\alpha x + \beta y + \gamma z = 0$. To show that the spherical image of the curve intersects the circle, it suffices to show that it intersects the plane. We compute

$$\int_0^L (\alpha\xi + \beta\eta + \gamma\zeta) = 0,$$

which implies that the continuous function $\alpha\xi + \beta\eta + \gamma\zeta$ vanishes at least once (in fact, at least twice since it takes the same value at the endpoints of the interval). The equality

$$\alpha\xi(s) + \beta\eta(x) + \gamma\zeta(s) = 0$$

is precisely the condition that $(\xi(s), \eta(x), \zeta(s))$ is in the plane. The problem is solved.

Remark. The spherical image of a curve was introduced by Gauss.
  (K. Löwner)

758. We use Löwner's theorem, which was the subject of the previous problem. The total curvature is the length of the spherical image of the curve. In view of Löwner's theorem, it suffices to show that a curve $\gamma(t)$ that intersects every great circle of the unit sphere has length at least 2π.

For each t, let H_t be the hemisphere centered at $\gamma(t)$. The fact that the curve intersects every great circle implies that the union of all the H_t's is the entire sphere. We prove the conclusion under this hypothesis. Let us analyze how the covered area adds up as we travel along the curve. Looking at Figure 94, we see that as we add to a hemisphere H_{t_0} the hemisphere H_{t_1}, the covered surface increases by the portion of the sphere contained within the dihedral angle formed by two planes. The area of such a "wedge" is directly proportional to the length of the arc of the great circle passing through $\gamma(t_0)$ and $\gamma(t_1)$. When the arc is the whole great circle the area is 4π, so in general, the area is numerically equal to twice the length of the arc. This means that as we move along the curve from t to $t + \Delta t$, the covered area increases by at most $2\|\gamma'(t)\|$. So after we have traveled along the entire curve, the covered area has increased by at most $2\displaystyle\int_C \|\gamma'(t)\| dt$ (C denotes the curve). For the whole sphere, we should have $2\displaystyle\int_C \|\gamma'(t)\| dt \geq 4\pi$. This implies that the length of the spherical image, which is equal to $\displaystyle\int_C \|\gamma'(t)\| dt$, is at least 2π, as desired.

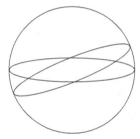

Figure 94

Remark. More is true, namely that the total curvature is equal to 2π if and only if the curve is planar and convex. A result of Milnor and Fáry shows that the total curvature of a *knotted* curve in space exceeds 4π.

(W. Fenchel)

759. Consider a coordinate system with axes parallel to the sides of R (and hence to the sides of all rectangles of the tiling). It is not hard to see that if $D = [a, b] \times [c, d]$ rectangle whose sides are parallel to the axes, then the four integrals

$$\iint_D \sin 2\pi x \sin 2\pi y\, dxdy, \qquad \iint_D \sin 2\pi x \cos 2\pi y\, dxdy,$$

$$\iint_D \cos 2\pi x \sin 2\pi y\, dxdy, \qquad \iint_D \cos 2\pi x \cos 2\pi y\, dxdy$$

are simultaneously equal to zero if and only if either $b - a$ or $d - c$ is an integer. Indeed, this is equivalent to the fact that

$$(\cos 2\pi b - \cos 2\pi a)(\cos 2\pi d - \cos 2\pi c) = 0,$$

$$(\cos 2\pi b - \cos 2\pi a)(\sin 2\pi d - \sin 2\pi c) = 0,$$

$$(\sin 2\pi b - \sin 2\pi a)(\cos 2\pi d - \cos 2\pi c) = 0,$$

$$(\sin 2\pi b - \sin 2\pi a)(\sin 2\pi d - \sin 2\pi c) = 0,$$

and a case check shows that either $\cos 2\pi b = \cos 2\pi a$ and $\sin 2\pi b = \sin 2\pi a$, or $\cos 2\pi d = \cos 2\pi c$ and $\sin 2\pi d = \sin 2\pi c$, which then implies that either a and b or c and d differ by an integer. Because the four integrals are zero on each rectangle of the tiling, by adding they are zero on R. Hence at least one of the sides of R has integer length.

(Short list of the 30th International Mathematical Olympiad, 1989, proposed by France)

760. For the solution we will use the following result

Isoperimetric Inequality. Let $\gamma : [0, 2\pi] \to \mathbb{R}^2$ be a smooth simple closed curve of length L that encloses a domain Ω of area A. Then

$$L^2 \geq 4\pi A.$$

Equality holds if and only if γ is a circumference.

Proof. We present the proof given by Hurwitz. Consider a parametrization $\gamma(t) = (x(t), y(t))$ of the curve with constant velocity $|\gamma'(t)| = \frac{L}{2\pi}$, $t \in [0, 2\pi]$. The functions x and y can be extended periodically over $\mathbb{R}$ by smooth functions. Expand in Fourier series:

$$x(t) = \sum_{n=-\infty}^{\infty} a_n e^{int} \text{ and } y(t) = \sum_{n=-\infty}^{\infty} b_n e^{int}.$$

We have

$$x'(t) = \sum_{n=-\infty}^{\infty} ina_n e^{int} \text{ and } y'(t) = \sum_{n=-\infty}^{\infty} inb_n e^{int}.$$

An application of the Parseval identity yields

$$\frac{L^2}{2\pi} = \int_0^{2\pi} (x'(t))^2 + (y'(t))^2 dt = 2\pi \sum_{n=-\infty}^{\infty} n^2(|a_n|^2 + |b_n|^2).$$

On the other hand, from Green's formula we obtain

$$A = \frac{1}{2} \left| \int_0^{2\pi} (x(t)y'(t) - x'(t)y(t)) dt \right| = \pi \left| \sum_{n=-\infty}^{\infty} n(a_n \overline{b_n} - \overline{a_n} b_n) \right|.$$

Hence

$$4\pi A = 4\pi^2 \left| \sum_{n=-\infty}^{\infty} n(a_n \overline{b_n} - \overline{a_n} b_n) \right| \leq 4\pi^2 \sum_{n=-\infty}^{\infty} |n| |a_n \overline{b_n} - \overline{a_n} b_n|$$

$$\leq 4\pi^2 \sum_{n=-\infty}^{\infty} |n|(|a_n|^2 + |b_n|^2) \leq 4\pi^2 \sum_{n=-\infty}^{\infty} n^2(|a_n|^2 + |b_n|^2) = L^2,$$

where we used the fact that

$$|a_n \overline{b_n} - \overline{a_n} b_n| \leq 2|a_n||b_n| \leq |a_n|^2 + |b_n|^2.$$

Equality holds if and only if $a_n = b_n = 0$, for $n \geq 2$, which then implies $|a_1| = |b_1| = 1$. But this only happens when γ is a circle. $\square$

Returning to the problem, for an arbitrary line ℓ with olar angle θ, denote by $r(\theta)$ the length of the projection of the polygon K on the line ℓ. Then the perimeter of K can be expressed as

$$P = \int_0^{\pi} r(\theta) d\theta.$$

But since $r(\theta) \leq 1$, we have $P \leq \pi$. Using the isoperimetric inequality we deduce that the area of K is less than $\frac{\pi}{4}$.

Remark. The original problem was given at a Romanian Team Selection Test for the International Mathematical Olympiad in 2005, with $\pi/4$ replaced by the larger number $\sqrt{3}/2$. The solution and the better upper bound were found by the contestant G. Kreindler during the competition.

(communicated by C. Lupu)

761. We denote by $A(XYZ)$ the area of triangle XYZ. Look first at the degenerate situation described in Figure 95, when P is on one side of the triangle. With the notation from that figure, we have

$$\frac{A(BMP)}{A(ABC)} = \left(\frac{BP}{BC} \right)^2 \quad \text{and} \quad \frac{A(CNP)}{A(ABC)} = \left(\frac{PC}{BC} \right)^2.$$

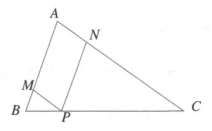

Figure 95

Adding up, we obtain

$$\frac{A(BMP) + A(CNP)}{A(ABC)} = \frac{BP^2 + PC^2}{(BP + PC)^2} \geq \frac{1}{2}.$$

The last inequality follows from the AM-GM inequality:

$$BP^2 + PC^2 \geq 2BP \cdot PC.$$

Note that in the degenerate case the inequality is even stronger, with $\frac{1}{3}$ replaced by $\frac{1}{2}$.

Let us now consider the general case, with the notation from Figure 96. By what we just proved, we know that the following three inequalities hold:

$$S_1 + S_2 \geq \frac{1}{2} A(A_1 B_2 C),$$

$$S_1 + S_3 \geq \frac{1}{2} A(A_2 B C_1),$$

$$S_2 + S_3 \geq \frac{1}{2} A(A B_1 C_2).$$

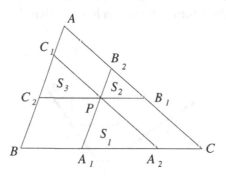

Figure 96

Adding them up, we obtain

$$2S_1 + 2S_2 + 2S_3 \geq \frac{1}{2}(A(ABC) + S_1 + S_2 + S_3).$$

The inequality follows.

(M. Pimsner, S. Popa, *Probleme de Geometrie Elementară* (*Problems in Elementary Geometry*), Editura Didactică și Pedagogică, Bucharest, 1979)

762. Assume that the two squares do not overlap. Then at most one of them contains the center of the circle. Take the other square. The line of support of one of its sides separates it from the center of the circle. Looking at the diameter parallel to this line, we see that the square is entirely contained in a half-circle, in such a way that one of its sides is parallel to the diameter. Translate the square to bring that side onto the diameter, then translate it further so that the center of the circle is the middle of the side (see Figure 97).

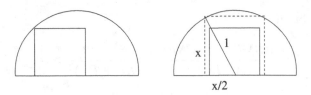

Figure 97

The square now lies inside another square with two vertices on the diameter and two vertices on the circle. From the Pythagorean theorem compute the side of the larger square to be $\sqrt{\frac{4}{5}}$. This is smaller than 0.9, a contradiction. Therefore, the original squares overlap.
(R. Gelca)

763. The Möbius band crosses itself if the generating segments at two antipodal points of the unit circle intersect. Let us analyze when this can happen. We refer everything to Figure 98. By construction, the generating segments at the antipodal points M and N are perpendicular. Let P be the intersection of their lines of support. Then the triangle MNP is right, and its acute angles are $\frac{\alpha}{2}$ and $\frac{\pi}{2} - \frac{\alpha}{2}$. The generating segments intersect if they are longer than twice the longest leg of this triangle. The longest leg of this triangle attains its shortest length when the triangle is isosceles, in which case its length is $\sqrt{2}$. We conclude that the maximal length that the generating segment of the Möbius band can have so that the band does not cross itself is $2\sqrt{2}$.

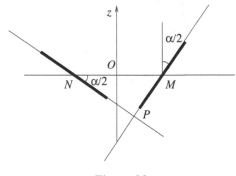

Figure 98

Remark. Even if we allow the Möbius band to be flexible, there is a maximal width it can have before crossing itself.

764. Comparing the perimeters of AOB and BOC, we find that $\|AB\| + \|AO\| = \|CB\| + \|CO\|$, and hence A and C belong to an ellipse with foci B and O. The same argument applied to triangles AOD and COD shows that A and C belong to an ellipse with foci D and O. The foci of the two ellipses are on the line BC; hence the ellipses are symmetric with respect to this line. It follows that A and C are symmetric with respect to BC, hence $AB = BC$ and $AD = DC$. Exchanging the roles of A and C with B and D, we find that $AB = AD$ and $BC = CD$. Therefore, $AB = BC = CD = DE$ and the quadrilateral is a rhombus.

The property is no longer true if O is not the intersection of the diagonals. A counterexample consists of a quadrilateral with $AB = BC = 3$, $BC = CD = 4$, $BD = 5$, and O on BD such that $OB = 3$ and $OD = 2$.

(Romanian Team Selection Test for the International Mathematical Olympiad, 1978, proposed by L. Panaitopol)

765. Assume by way of contradiction that the interiors of finitely many parabolas cover the plane. The intersection of a line with the interior of a parabola is a half-line if that line is parallel to the axis of the parabola, and it is void or a segment otherwise. There is a line that is not parallel to the axis of any parabola. The interiors of the parabolas cover the union of finitely many segments on this line, so they do not cover the line entirely. Hence the conclusion.

766. Without loss of generality, we may assume that $AC = 1$, and let as usual $AB = c$. We have

$$BC^2 = AB^2 + AC^2 - 2AB \cdot AC \cos \angle BAC \geq AB^2 + AC^2 - AB = c^2 + 1 - c,$$

because $\angle BAC \geq 60°$. On the other hand,

$$CD^2 = AC^2 + AD^2 - 2AC \cdot AD \cos \angle CAD \geq 1 + c^6 + c^3,$$

because $\angle CAD \leq 120°$ (so $2AC \cdot \cos \angle BAC < 1$). We are left to prove the inequality

$$c^6 + c^3 + 1 \leq 3(c^2 - c + 1)^3,$$

which, after dividing both sides by $c^3 > 0$, takes the form

$$c^3 + 1 + \frac{1}{c^3} \leq 3 \left(c - 1 + \frac{1}{c} \right)^3.$$

With the substitution $c + \frac{1}{c} = x$, the inequality becomes

$$x^3 - 3x + 1 \leq 3(x - 1)^3, \text{ for } x \geq 2.$$

But this reduces to

$$(x - 2)^2 (2x - 1) \geq 0,$$

which is clearly true. Equality holds if and only if $\angle A = 60°$ and $c = 1$ ($AB = AC$), i.e., when the triangle ABC is equilateral.

(Proposed by T. Andreescu for the USA Mathematical Olympiad, 2006)

767. *First solution:* Denote by a, b, c, d, e, f, g, h the lengths of the sides of the octagon. Its angles are all equal to $135°$ (see Figure 99). If we project the octagon onto a line perpendicular to side d, we obtain two overlapping segments. Writing the equality of their lengths, we obtain

$$a\frac{\sqrt{2}}{2} + b + c\frac{\sqrt{2}}{2} = e\frac{\sqrt{2}}{2} + f + g\frac{\sqrt{2}}{2}.$$

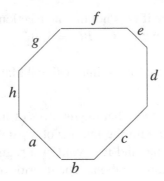

Figure 99

Because a, b, c, e, f, g are rational, equality can hold only if $b = f$. Repeating the argument for all sides, we see that the opposite sides of the octagon have equal length. The opposite sides are also parallel. This means that any two consecutive main diagonals intersect at their midpoints, so all main diagonals intersect at their midpoints. The common intersection is the center of symmetry.

Second solution: Note that the positive integers $a_1, a_2, \ldots, a_n$, are the side-lengths of an equiangular polygon, in this order, if and if for

$$\varepsilon = \cos \frac{2\pi}{n} + i \sin \frac{2\pi}{n},$$

one has

$$a_n \varepsilon^{n-1} + a_{n-1} \varepsilon^{n-2} + \cdots + a_2 \varepsilon + a_1 = 0.$$

In our case the side-lengths would therefore satisfy

$$a_8 \varepsilon^7 + a_7 \epsilon^6 + \cdots + a_1 = 0.$$

Using the fact that $\varepsilon^4 = 1$, we obtain

$$(a_4 - a_8)\varepsilon^3 + (a_3 - a_7)\varepsilon^2 + (a_2 - a_6)\varepsilon + (a_1 - a_5) = 0.$$

Thus ε is the root of a cubic equation with integer coefficients. But also ε is a zero of the polynomial $x^4 + 1$, which is irreducible, by Eisenstein's criterion (see Section 2.2.7) applied to $(x+1)^4 + 1 = x^4 + 4x^3 + 6x^2 + 4x + 2$. It follows that the cubic polynomial is identically equal to zero, so $a_1 = a_5, a_2 = a_6, a_3 = a_7, a_4 = a_8$. The octagon being equiangular, opposite sides are parallel, so they pairwise form parallelograms. Hence all three diagonals meet at their midpoints. The common intersection point is the center of symmetry.

(Russian Mathematical Olympiad)

768. Let us assume that the three diagonals do not intersect. Denote by M the intersection of AD with CF, by N the intersection of BE with CF, and by P the intersection of AD with BE. There are two possibilities: either M is between A and P, or P is between A and M. We discuss only the first situation, shown in Figure 100, and leave the second, which is analogous, to the reader.

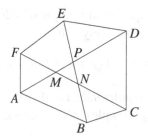

Figure 100

Let $A(x)$ denote the area of the polygon x. From $A(BCDE) = A(ABEF)$ it follows that

$$A(EPD) + A(NPDC) + A(BNC) = A(ENF) + A(AMF) + A(MNBA).$$

Adding $A(MNP)$ to both sides, we obtain

$$A(EPD) + A(DMC) + A(BNC) = A(ENF) + A(AMF) + A(APB).$$

Writing the other two similar relations and then subtracting these relations two by two, we obtain

$$A(AMF) = A(DMC), \quad A(APB) = A(EPD), \quad A(BNC) = A(ENF).$$

The equality $A(AMF) = A(DMC)$ implies that $MF \cdot MA \cdot \angle AMF = MC \cdot MD \cdot \sin \angle CMD$, hence $MF \cdot MA = MC \cdot MD$. Similarly, $BN \cdot CN = EN \cdot FN$ and $AP \cdot BP = DP \cdot EP$. If we write $AM = a, AP = \alpha, BN = b, BP = \beta, CN = c, CM = \gamma, DP = d, DM = \delta, EP = e, EN = \eta, FM = f, FN = \phi$, then

$$\frac{a}{\delta} = \frac{\gamma}{f}, \quad \frac{b}{\eta} = \frac{\phi}{c}, \quad \frac{e}{\beta} = \frac{\alpha}{d}.$$

Also, any Latin letter is smaller than the corresponding Greek letter. Hence

$$\frac{a}{\delta} = \frac{\gamma}{f} > \frac{c}{\phi} = \frac{\eta}{b} > \frac{e}{\beta} = \frac{\alpha}{d} > \frac{a}{\delta}.$$

This is a contradiction. The study of the case in which P is between A and M yields a similar contradiction, since M is now between D and P, and D can take the role of A above, showing that the three main diagonals must intersect.

(*Revista Matematică din Timişoara* (*Timişoara Mathematics Gazette*))

769. (a) Define $f : \mathbb{Z} \to [0, 1), f(x) = \sqrt{3} - \lfloor x\sqrt{3} \rfloor$. By the pigeonhole principle, there exist distinct integers x_1 and x_2 such that $|f(x_1) - f(x_2)| < 0.001$. Set $a = |x_1 - x_2|$. Then the distance either between $(a, a\sqrt{3})$ and $(a, \lfloor a\sqrt{3} \rfloor)$ or between $(a, a\sqrt{3})$ and $(a, \lfloor a\sqrt{3} \rfloor + 1)$ is less than 0.001. Therefore, the points $(0, 0), (2a, 0), (a, a\sqrt{3})$ lie in different disks and form an equilateral triangle.

(b) Suppose that $P'Q'R'$ is an equilateral triangle of side $l \leq 96$, whose vertices P', Q', R' lie in disks with centers P, Q, R, respectively. Then

$$l - 0.002 \leq PQ, PR, RP \leq l + 0.002.$$

On the other hand, since there is no equilateral triangle whose vertices have integer coordinates (which can be proved easily using complex coordinates), we may assume that $PQ \neq QR$. Therefore,

$$\begin{aligned}
|PQ^2 - QR^2| &= (PQ + QR)|PQ - QR| \\
&\leq ((l + 0.002) + (l + 0.002))((l + 0.002) - (l - 0.002)) \\
&\leq 2 \times 96.002 \times 0.004 < 1.
\end{aligned}$$

However, $PQ^2 - QR^2$ is an integer. This contradiction proves the claim.

(Short list of the 44th International Mathematical Olympiad, 2003)

770. Imagine instead that the figure is fixed and the points move on the cylinder, all rigidly linked to each other. Let P be one of the n points; when another point traces S, P itself will trace a figure congruent to S. So after all the points have traced S, P alone has traced a surface F of area strictly less than n.

On the other hand, if we rotate P around the cylinder or translate it back and forth by $\frac{n}{4\pi r}$, we trace a surface of area exactly equal to n. Choose on this surface a point P' that does not lie in F, and consider the transformation that maps P to P'. The fact that P' is not in F means that at this moment none of the points lies in S. This transformation, therefore, satisfies the required condition.

(M. Pimsner, S. Popa, *Probleme de Geometrie Elementară* (*Problems in Elementary Geometry*), Editura Didactică şi Pedagogică, Bucharest, 1979)

771. The left-hand side is equal to

$$\cos 20° \sin 40° - \sin 10° \cos 10° = 2 \sin 20° \cos^2 20° - \frac{\sin 20°}{2}$$

$$= \frac{1}{2}(3 \sin 20° - 4 \sin^3 20°) = \frac{1}{2} \sin 60° = \frac{\sqrt{3}}{4}.$$

(Romanian Mathematical Olympiad, 1967, proposed by C. Ionescu- Ţiu)

772. Because $-\frac{\pi}{2} < 1 \leq \sin x \leq 1 < \frac{\pi}{2}$, $\cos(\sin x) > 0$. Hence $\sin(\cos x) > 0$, and so $\cos x > 0$. We deduce that the only possible solutions can lie in the interval $\left(-\frac{\pi}{2}, \frac{\pi}{2}\right)$. Note that if x is a solution, then $-x$ is also a solution; thus we can restrict our attention to the first quadrant. Rewrite the equation as

$$\sin(\cos x) = \sin\left(\frac{\pi}{2} - \sin x\right).$$

Then $\cos x = \frac{\pi}{2} - \sin x$, and so $\sin x + \cos x = \frac{\pi}{2}$. This equality cannot hold, since the range of the function $f(x) = \sin x + \cos x = \sqrt{2}\cos\left(\frac{\pi}{4} - x\right)$ is $[-\sqrt{2}, \sqrt{2}]$, and $\frac{\pi}{2} > \sqrt{2}$.

773. The relation from the statement can be transformed into

$$\tan^2 b = \frac{\tan^2 a + 1}{\tan^2 a - 1} = -\frac{1}{\cos 2a}.$$

This is further equivalent to

$$\frac{\sin^2 b}{1 - \sin^2 b} = \frac{1}{2\sin^2 a - 1}.$$

Eliminating the denominators, we obtain

$$2\sin^2 a \sin^2 b = 1,$$

which gives the desired $\sin a \sin b = \pm\frac{\sqrt{2}}{2} = \pm\sin 45°$.
   (Romanian Mathematical Olympiad, 1959)

774. We have

$$f(x) = \sin x \cos x + \sin x + \cos x + 1 = \frac{1}{2}(\sin x + \cos x)^2 - \frac{1}{2} + \sin x + \cos x + 1$$
$$= \frac{1}{2}[(\sin x + \cos x)^2 + 2(\sin x + \cos x) + 1] = \frac{1}{2}[(\sin x + \cos x) + 1]^2.$$

This is a function of $y = \sin x + \cos x$, namely $f(y) = \frac{1}{2}(y + 1)^2$. Note that

$$y - \cos\left(\frac{\pi}{2} - x\right) + \cos x = 2\cos\frac{\pi}{4}\cos\left(x - \frac{\pi}{4}\right) = \sqrt{2}\cos\left(x - \frac{\pi}{4}\right).$$

So y ranges between $-\sqrt{2}$ and $\sqrt{2}$. Hence $f(y)$ ranges between 0 and $\frac{1}{2}(\sqrt{2} + 1)^2$.

775. Relate the secant and the cosecant to the tangent and cotangent:

$$\sec^2 x = \tan^2 x + 1 \geq 2\tan x \quad \text{and} \quad \csc^2 x = \cot^2 x + 1 \geq 2\cot x,$$

where the inequalities come from the most particular case of AM-GM. It follows that

$$\sec^{2n} x + \csc^{2n} x \geq 2^n(\tan^n x + \cot^n x).$$

Now observe that

$$\tan^n x + \cot^n x = \tan^n x + \frac{1}{\tan^n x} \geq 2,$$

again by the AM-GM inequality. We obtain

$$\sec^{2x} x + \csc^{2n} x \geq 2^{n+1},$$

as desired.

(*Gazeta Matematică* (*Mathematics Gazette, Bucharest*), proposed by D. Andrica)

776. We would like to eliminate the square root, and for that reason we recall the trigonometric identity

$$\frac{1 - \sin t}{1 + \sin t} = \frac{\cos^2 t}{(1 + \sin t)^2}.$$

The proof of this identity is straightforward if we express the cosine in terms of the sine and then factor the numerator. Thus if we substitute $x = \sin t$, then $dx = \cos t \, dt$ and the integral becomes

$$\int \frac{\cos^2 t}{1 + \sin t} dt = \int 1 - \sin t \, dt = t + \cos t + C.$$

Since $t = \arcsin x$, this is equal to $\arcsin x + \sqrt{1 - x^2} + C$.

(Romanian high school textbook)

777. We will prove that a function of the form $f(x, y) = \cos(ax + by)$, a, b integers, can be written as a polynomial in $\cos x$, $\cos y$, and $\cos(x + ky)$ if and only if b is divisible by k.

For example, if $b = k$, then from

$$\cos(ax + ky) = 2 \cos x \cos((a \pm 1)x + ky) - \cos((a \pm 2)x + ky),$$

we obtain by induction on the absolute value of a that $\cos(ax + by)$ is a polynomial in $\cos x$, $\cos y$, $\cos(x + ky)$. In general, if $b = ck$, the identity

$$\cos(ax + cky) = 2 \cos y \cos(ax + (c \pm 1)ky) - \cos(ax + (c \pm 2)ky)$$

together with the fact that $\cos ax$ is a polynomial in $\cos x$ allows an inductive proof of the fact that $\cos(ax + by)$ can be written as a polynomial in $\cos x$, $\cos y$, and $\cos(x + ky)$ as well.

For the converse, note that by using the product-to-sum formula we can write any polynomial in cosines as a linear combination of cosines. We will prove a more general statement, namely that if a linear combination of cosines is a polynomial in $\cos x$, $\cos y$, and $\cos(x + ky)$, then it is of the form

$$\sum_m \left[b_m \cos mx + \sum_{0 \leq q < |p|} c_{m,p,q} (\cos(mx + (pk + q)y) + \cos(mx + (pk - q)y)) \right].$$

This property is obviously true for polynomials of degree one, since any such polynomial is just a linear combination of the three functions. Also, any polynomial in $\cos x$, $\cos y$, $\cos(x + ky)$ can be obtained by adding polynomials of lower degrees, and eventually multiplying them by one of the three functions.

Hence it suffices to show that the property is invariant under multiplication by $\cos x$, $\cos y$, and $\cos(x + ky)$. It can be verified that this follows from

$$2 \cos(ax + by) \cos x = \cos((a + 1)x + by) + \cos((a - 1)x + by),$$

$$2\cos(ax + by)\cos y = \cos(ax + (b+1)y) + \cos(ax + (b-1)y),$$

$$2\cos(ax + by)\cos(x + ky) = \cos((a+1)x + (b+k)y) + \cos((a-1)x + (b-k)y).$$

So for $\cos(ax + by)$ to be a polynomial in $\cos x$, $\cos y$, and $\cos(x + ky)$, it must be such a sum with a single term. This can happen only if b is divisible by k.

The answer to the problem is therefore $k = \pm 1, \pm 3, \pm 9, \pm 11, \pm 33, \pm 99$.

(Proposed by R. Gelca for the USA Mathematical Olympiad, 1999)

778. Clearly, this problem is about the addition formula for the cosine. For it to show up we need products of sines and cosines, and to obtain them it is natural to square the relations. Of course, we first separate a and d from b and c. We have

$$(2\cos a + 9\cos d)^2 = (6\cos b + 7\cos c)^2,$$

$$(2\sin a - 9\sin d)^2 = (6\sin b - 7\sin c)^2.$$

This further gives

$$4\cos^2 a + 36\cos a \cos d + 81\cos^2 d = 36\cos^2 b + 84\cos b \cos c + 49\cos^2 c,$$

$$4\sin^2 a - 36\sin a \sin d + 81\sin^2 d = 36\sin^2 b - 84\sin b \sin c + 49\sin^2 c.$$

After adding up and using $\sin^2 x + \cos^2 x = 1$, we obtain

$$85 + 36(\cos a \cos d - \sin a \sin d) = 85 + 84(\cos b \cos c - \sin b \sin c).$$

Hence $3\cos(a + d) = 7\cos(b + c)$, as desired.

(S. Korean Mathematics Competition, 2002, proposed by T. Andreescu)

779. The first equality can be written as

$$\sin^3 a + \cos^3 a + \left(-\frac{1}{5}\right)^3 - 3(\sin a)(\cos a)\left(-\frac{1}{5}\right) = 0.$$

We have seen in Section 2.1.1 that the expression $x^3 + y^3 + z^3 - 3xyz$ factors as

$$\frac{1}{2}(x + y + z)[(x - y)^2 + (y - z)^2 + (z - x)^2].$$

Here $x = \sin a$, $y = \cos a$, $z = -\frac{1}{5}$. It follows that either $x + y + z = 0$ or $x = y = z$. The latter would imply $\sin a = \cos a = -\frac{1}{5}$, which violates the identity $\sin^2 a + \cos^2 a = 1$. Hence $x + y + z = 0$, implying $\sin a + \cos a = \frac{1}{5}$. Then $5(\sin a + \cos a) = 1$, and so

$$\sin^2 a + 2\sin a \cos a + \cos^2 a = \frac{1}{25}.$$

It follows that $1 + 2\sin a \cos a = 0.04$; hence

$$5(\sin a + \cos a) + 2\sin a \cos a = 0.04,$$

as desired.

Conversely,

$$5(\sin a + \cos a) + 2\sin a \cos a = 0.04$$

implies

$$125(\sin a + \cos a) = 1 - 50 \sin a \cos a.$$

Squaring both sides and setting $2 \sin a \cos a = b$ yields

$$125^2 + 125^2 b = 1 - 50b + 25^2 b^2,$$

which simplifies to

$$(25b + 24)(25b - 651) = 0.$$

We obtain $2 \sin a \cos a = -\frac{24}{25}$, or $2 \sin a \cos a = \frac{651}{25}$. The latter is impossible because $\sin 2a \le 1$. Hence $2 \sin a \cos a = -0.96$, and we obtain $\sin a + \cos a = 0.2$. Then

$$5(\sin^3 a + \cos^3 a) + 3 \sin a \cos a$$
$$= 5(\sin a + \cos a)(\sin^2 a - \sin a \cos a + \cos^2 a) + 3 \sin a \cos a$$
$$= \sin^2 a - \sin a \cos a + \cos^2 a + 3 \sin a \cos a = (\sin a + \cos a)^2 = (0.2)^2 = 0.04,$$

as desired.

(*Mathematical Reflections*, proposed by T. Andreescu)

780. If we set $b_k = \tan\left(a_k - \frac{\pi}{4}\right)$, $k = 0, 1, \ldots, n$, then

$$\tan\left(a_k - \frac{\pi}{4} + \frac{\pi}{4}\right) = \frac{1 + \tan\left(a_k - \dfrac{\pi}{4}\right)}{1 - \tan\left(a_k - \dfrac{\pi}{4}\right)} = \frac{1 + b_k}{1 - b_k}.$$

So we have to prove that

$$\prod_{k=0}^{n} \frac{1 + b_k}{1 - b_k} \ge n^{n+1}.$$

The inequality from the statement implies

$$1 + b_k \ge \sum_{l \ne k}(1 - b_l), \quad k = 0, 1, \ldots, n.$$

Also, the condition $a_k \in \left(0, \frac{\pi}{2}\right)$ implies $-1 < b_k < 1$, $k = 0, 1, \ldots, n$, so the numbers $1 - b_k$ are all positive. To obtain their product, it is natural to apply the AM-GM inequality to the right-hand side of the above inequality, and obtain

$$1 + b_k \ge n \sqrt[n]{\prod_{l \ne k}(1 - b_l)}, \quad k = 0, 1, \ldots, n.$$

Multiplying all these inequalities yields

$$\prod_{k=0}^{n}(1 + b_k) \ge n^{n+1} \sqrt[n]{\prod_{l=0}^{n}(1 - b_l)^n}.$$

Hence

$$\prod_{k=0}^{n} \frac{1 + b_k}{1 - b_k} \geq n^{n+1},$$

as desired.

(USA Mathematical Olympiad, 1998, proposed by T. Andreescu)

781. If we multiply the denominator and the numerator of the left-hand side by $\cos t$, and of the right-hand side by $\cos nt$, we obtain the obvious equality

$$\left(\frac{e^{it}}{e^{-it}} \right)^n = \frac{e^{int}}{e^{-int}}.$$

782. Using the de Moivre formula, we obtain

$$(1 + i)^n - \left[\sqrt{2} \left(\cos \frac{\pi}{4} + i \sin \frac{\pi}{4} \right) \right]^n = 2^{n/2} \left(\cos \frac{n\pi}{4} + i \sin \frac{n\pi}{4} \right).$$

Expanding $(1 + i)^n$ and equating the real parts on both sides, we deduce the identity from the statement.

783. Denote the sum in question by S_1 and let

$$S_2 = \binom{n}{1} \sin x + \binom{n}{2} \sin 2x + \cdots + \binom{n}{n} \sin nx.$$

Using Euler's formula, we can write

$$1 + S_1 + iS_2 = \binom{n}{0} + \binom{n}{1} e^{ix} + \binom{n}{2} e^{2ix} + \cdots + \binom{n}{n} e^{inx}.$$

By the multiplicative property of the exponential we see that this is equal to

$$\sum_{k=0}^{n} \binom{n}{k} (e^{ix})^k = (1 + {}^{ix})^n = \left(2 \cos \frac{x}{2} \right)^n (e^{i\frac{x}{2}})^n.$$

The sum in question is the real part of this expression less 1, which is equal to

$$2^n \cos^n \frac{x}{2} \cos \frac{nx}{2} - 1.$$

784. Combine $f(x)$ with the function $g(x) = e^{x \cos \theta} \sin(\sin x \sin \theta)$ to write

$$f(x) + ig(x) = e^{x \cos \theta} (\cos(x \sin \theta) + i \sin(x \sin \theta))$$
$$= e^{x \cos \theta} \cdot e^{ix \sin \theta} = e^{x(\cos \theta + i \sin \theta)}.$$

Using the de Moivre formula we expand this in a Taylor series as

$$1 + \frac{x}{1!} (\cos \theta + i \sin \theta) + \frac{x^2}{2!} (\cos 2\theta + i \sin 2\theta) + \cdots + \frac{x^n}{n!} (\cos n\theta + i \sin n\theta) + \cdots$$

Consequently, the Taylor expansion of $f(x)$ around 0 is the real part of this series, i.e.,

$$f(x) = 1 + \frac{\cos\theta}{1!}x + \frac{\cos 2\theta}{2!}x^2 + \cdots + \frac{\cos n\theta}{n!}x^n + \cdots$$

785. Let $z_j = r(\cos t_j + i\sin t_j)$, with $r \neq 0$ and $t_j \in (0, \pi) \cup (\pi, 2\pi)$, $j = 1, 2, 3$. By hypothesis,

$$\sin t_1 + r\sin(t_2 + t_3) = 0,$$
$$\sin t_2 + r\sin(t_3 + t_1) = 0,$$
$$\sin t_3 + r\sin(t_1 + t_2) = 0.$$

Let $t = t_1 + t_2 + t_3$. Then

$$\sin t_j = -r\sin(t - t_i) = -r\sin t\cos t_j - r\cos t\sin t_j, \text{ for } j = 1, 2, 3,$$

which means that

$$\cot t_j \sin t = \frac{1}{r} - \cos t, \text{ for } j = 1, 2, 3.$$

If $\sin t \neq 0$, then $\cot t_1 = \cot t_2 = \cot t_3$. There are only two possible values that t_1, t_2, t_3 can take between 0 and 2π, and so two of the t_j are equal, which is ruled out by the hypothesis. It follows that $\sin t = 0$. Then on the one hand, $r\cos t - 1 = 0$, and on the other, $\cos t = \pm 1$. This can happen only if $\cos t = 1$ and $r = 1$. Therefore, $z_1z_2z_3 = r^3\cos t = 1$, as desired.

786. Consider the complex number $\omega = \cos\theta + i\sin\theta$. The roots of the equation

$$\left(\frac{1+ix}{1-ix}\right)^n = \omega^{2n}$$

are precisely $a_k = \tan\left(\theta + \frac{k\pi}{n}\right)$, $k = 1, 2, \ldots, n$. Rewriting this as a polynomial equation of degree n, we obtain

$$0 = (1 + ix)^2 - \omega^{2n}(1 - ix)^n$$
$$= (1 - \omega^{2n}) + ni(1 + \omega^{2n})x + \cdots + ni^{n-1}(1 - \omega^{2n})x^{n-1} + i^n(1 + \omega^{2n})x^n.$$

The sum of the zeros of the latter polynomial is

$$\frac{-ni^{n-1}(1 - \omega^{2n})}{i^n(1 + \omega^{2n})},$$

and their product

$$\frac{-(1 - \omega^{2n})}{i^n(1 + \omega^{2n})}.$$

Therefore,

$$\frac{a_1 + a_2 + \cdots + a_n}{a_1a_2\cdots a_n} = ni^{n-1} = n(-1)^{(n-1)/2}.$$

(67th W.L. Putnam Competition, 2006, proposed by T. Andreescu)

787. More generally, for an odd integer n, let us compute

$$S = (\cos \alpha)(\cos 2\alpha) \cdots (\cos n\alpha)$$

with $\alpha = \frac{2\pi}{2n+1}$. We can let $\zeta = e^{i\alpha}$ and then

$$S = 2^{-n} \prod_{k=1}^{n} (\zeta^k + \zeta^{-k}).$$

Since $\zeta^k + \zeta^{-k} = \zeta^{2n+1-k} + \zeta^{-(2n+1-k)}$, $k = 1, 2, \ldots, n$, we obtain

$$S^2 = 2^{-2n} \prod_{k=1}^{2n} (\zeta^k + \zeta^{-k}) = 2^{-2n} \times \prod_{k=1}^{2n} \zeta^{-k} \times \prod_{k=1}^{2n} (1 + \zeta^{2k}).$$

The first of the two products is just $\zeta^{-(1+2+\cdots+2n)}$. Because $1 + 2 + \cdots + 2n = n(2n + 1)$, which is a multiple of $2n + 1$, this product equals 1.

As for the product $\prod_{k=1}^{2n} (1 + \zeta^{2k})$, note that it can be written as $\prod_{k=1}^{2n} (1 + \zeta^k)$, since the numbers ζ^{2k} range over the $(2n+1)$st roots of unity other than 1 itself, taking each value exactly once. We compute this using the factorization

$$z^{n+1} - 1 = (z - 1) \prod_{k=1}^{2n} (z - \zeta^k).$$

Substituting $z = -1$ and dividing both sides by -2 gives

$$\prod_{k=1}^{2n} (-1 - \zeta^k) = 1,$$

so

$$\prod_{k=1}^{2n} (1 + \zeta^k) = 1.$$

Hence $S^2 = 2^{-2n}$, and so $S = \pm 2^{-n}$. We need to determine the sign.

For $1 \le k \le n$, $\cos k\alpha < 0$ when $\frac{\pi}{2} < k\alpha < \pi$. The values of k for which this happens are $\lceil \frac{n+1}{2} \rceil$ through n. The number of such k is odd if $n \equiv 1$ or 2 (mod 4), and even if $n \equiv 0$ or 3 (mod 4). Hence

$$S = \begin{cases} +2^{-n} & \text{if } n \equiv 1 \text{ or } 2 \pmod 4, \\ -2^{-n} & \text{if } n \equiv 0 \text{ or } 3 \pmod 4. \end{cases}$$

Taking $n = 999 \equiv 3 \pmod 4$, we obtain the answer to the problem, -2^{-999}.

(Proposed by J. Propp for the USA Mathematical Olympiad, 1999)

788. Define the complex numbers $p = xe^{iA}$, $q = ye^{iB}$, and $r = ze^{iC}$ and consider $f(n) = p^n + q^n + r^n$. Then $F(n) = \text{Im}(f(n))$. We claim by induction that $f(n)$ is real for all n, which would imply that $F(n) = 0$. We are given that $f(1)$ and $f(2)$ are real, and $f(0) = 3$ is real as well.

Now let us assume that $f(k)$ is real for all $k \leq n$ for some $n \geq 3$, and let us prove that $f(n+1)$ is also real. Note that $a = p+q+r = f(1)$, $b = pq+qr+rp = \frac{1}{2}(f(1)^2 - f(2))$, and $c = pqr = xyze^{i(A+B+C)}$ are all real. The numbers p, q, r are the zeros of the cubic polynomial $P(t) = t^3 - at^2 + bt - c$, which has real coefficients. Using this fact, we obtain

$$
\begin{aligned}
f(n+1) &= p^{n+1} + q^{n+1} + r^{n+1} \\
&= a(p^n + q^n + r^n) - b(p^{n-1} + q^{n-1} + r^{n-1}) + c(p^{n-2} + q^{n-2} + r^{n-2}) \\
&= af(n) - bf(n-1) + cf(n-2).
\end{aligned}
$$

Since $f(n)$, $f(n-1)$ and $f(n-2)$ are real by the induction hypothesis, it follows that $f(n+1)$ is real, and we are done.

789. By eventually changing $\phi(t)$ to $\phi(t) + \frac{\theta}{2}$, where θ is the argument of $4P^2 - 2Q$, we may assume that $4P^2 - 2Q$ is real and positive. We can then ignore the imaginary parts and write

$$
4P^2 - 2Q = 4\left(\int_0^\infty e^{-t} \cos \phi(t)dt\right)^2 - 4\left(\int_0^\infty e^{-t} \sin \phi(t)dt\right)^2
$$
$$
- 2\int_0^\infty e^{-2t} \cos \phi(t)dt.
$$

Ignore the second term. Increase the first term using the Cauchy-Schwarz inequality:

$$
\left(\int_0^\infty e^{-t} \cos \phi(t)dt\right)^2 = \left(\int_0^\infty e^{-\frac{1}{2}t}e^{-\frac{1}{2}t} \cos \phi(t)dt\right)^2
$$
$$
\leq \left(\int_0^\infty e^{-t}dt\right)\left(\int_0^\infty e^{-t} \cos^2 \phi(t)dt\right)
$$
$$
= \int_0^\infty e^{-t} \cos^2 \phi(t)dt.
$$

We then have

$$
4P^2 - 2Q \leq 4\int_0^\infty e^{-t} \cos^2 \phi(t)dt - 2\int_0^\infty e^{-2t} \cos 2\phi(t)dt
$$
$$
= 4\int_0^\infty (e^{-t} - e^{-2t}) \cos^2 \phi(t)dt + 1
$$
$$
\leq 4\int_0^\infty (e^{-t} - e^{-2t})dt + 1 = 3.
$$

Equality holds only when $\cos^2 \phi(t) = 1$ for all t, and in general if $\phi(t)$ is constant.

(K. Löwner, from G. Pólya, G. Szegö, *Aufgaben und Lehrsätze aus der Analysis*, Springer-Verlag, 1964)

790. The given inequality follows from the easier

$$
\sqrt{ab} + \sqrt{(1-a)(1-b)} \leq 1.
$$

To prove this one, let $a = \sin^2 \alpha$ and $b = \sin^2 \beta$, $\alpha, \beta \in \left[0, \frac{\pi}{2}\right]$. The inequality becomes $\sin \alpha \sin \beta + \cos \alpha \cos \beta \leq 1$, or $\cos(\alpha - \beta) \leq 1$, and this is clearly true.

791. First, note that if $x > 2$, then $x^3 - 3x > 4x - 3x = x > \sqrt{x+2}$, so all solutions x should satisfy $-2 \leq x \leq 2$. Therefore, we can substitute $x = 2 \cos a$ for some $a \in [0, \pi]$. Then the given equation becomes

$$2 \cos 3a = \sqrt{2(1 + \cos a)} = 2 \cos \frac{a}{2},$$

so

$$2 \sin \frac{7a}{4} \sin \frac{5a}{4} = 0,$$

meaning that $a = 0, \frac{4\pi}{7}, \frac{4\pi}{5}$. It follows that the solutions to the original equation are $x = 2$, $2 \cos \frac{4\pi}{7}, -\frac{1}{2}(1 + \sqrt{5})$.

792. The points (x_1, x_2) and (y_1, y_2) lie on the circle of radius c centered at the origin. Parametrizing the circle, we can write $(x_1, x_2) = (c \cos \phi, c \sin \phi)$ and $(y_1, y_2) = (c \cos \psi, c \sin \psi)$. Then

$$S = 2 - c(\cos \phi + \sin \phi + \cos \psi + \sin \psi) + c^2(\cos \phi \cos \psi + \sin \phi \sin \psi)$$
$$= 2 + c\sqrt{2}\left(-\sin\left(\phi + \frac{\pi}{4}\right) - \sin\left(\psi + \frac{\pi}{4}\right)\right) + c^2 \cos(\phi - \psi).$$

We can simultaneously increase each of $-\sin\left(\phi + \frac{\pi}{4}\right)$, $-\sin\left(\psi + \frac{\pi}{4}\right)$, and $\cos(\phi - \psi)$ to 1 by choosing $\phi = \psi = \frac{5\pi}{4}$. Hence the maximum of S is $2 + 2c\sqrt{2} + c^2 = (c + \sqrt{2})^2$.

(Proposed by C. Rousseau for the USA Mathematical Olympiad, 2002)

793. Let $a = \tan \alpha$, $b = \tan \beta$, $c = \tan \gamma$, $\alpha, \beta, \gamma \in \left(-\frac{\pi}{2}, \frac{\pi}{2}\right)$. Then $a^2 + 1 = \sec^2 \alpha$, $b^2 + 1 = \sec^2 \beta$, $c^2 + 1 = \sec^2 \gamma$, and the inequality takes the simpler form

$$|\sin(\alpha - \beta)| \leq |\sin(\alpha - \gamma)| + |\sin(\beta - \gamma)|.$$

This is proved as follows:

$$|\sin(\alpha - \beta)| = |\sin(\alpha - \gamma + \gamma - \beta)| = |\sin(\alpha - \gamma)\cos(\gamma - \beta) + \sin(\gamma - \beta)\cos(\alpha - \gamma)|$$
$$\leq |\sin(\alpha - \gamma)||\cos(\gamma - \beta)| + |\sin(\gamma - \beta)||\cos(\alpha - \gamma)| \leq |\sin(\alpha - \gamma)| + |\sin(\gamma - \beta)|.$$

(N.M. Sedrakyan, A.M. Avoyan, *Neravenstva, Metody Dokazatel'stva* (*Inequalities, Methods of Proof*), FIZMATLIT, Moscow, 2002)

794. Expressions of the form $x^2 + 1$ suggest a substitution by the tangent. We let $a = \tan u$, $b = \tan v$, $c = \tan w$, $u, v, w \in \left(-\frac{\pi}{2}, \frac{\pi}{2}\right)$. The product on the right-hand side becomes $\sec^2 u \sec^2 v \sec^2 w$, and the inequality can be rewritten as

$$-1 \leq (\tan u \tan v + \tan u \tan w + \tan v \tan w - 1) \cos u \cos v \cos w \leq 1.$$

The expression in the middle is simplified as follows:

$$(\tan u \tan v + \tan u \tan w + \tan v \tan w - 1) \cos u \cos v \cos w$$
$$= \sin u \sin v \cos w + \sin u \cos v \sin w + \cos u \sin v \sin w - \cos u \cos v \cos w$$
$$= \sin u \sin(v+w) - \cos u \cos(v+w) = -\cos(u+v+w).$$

And of course this takes values in the interval $[-1, 1]$. The inequality is proved.
    (T. Andreescu, Z. Feng, 103 *Trigonometry Problems*, Birkhäuser 2004)

795. The denominators suggest the substitution based on tangents. This idea is further enforced by the identity $x + y + z = xyz$, which characterizes the tangents of the angles of a triangle. Set $x = \tan A$, $y = \tan B$, $z = \tan C$, with A, B, C the angles of an acute triangle. Note that

$$\frac{\tan A}{\sqrt{1 + \tan^2 A}} = \frac{\tan A}{\sec A} = \sin A,$$

so the inequality is equivalent to

$$\sin A + \sin B + \sin C \leq \frac{3\sqrt{3}}{2}.$$

This is Jensen's inequality (see Section 3.2.7) applied to the function $f(x) = \sin x$, which is concave on $\left(0, \frac{\pi}{2}\right)$.

796. If we multiply the inequality through by 2, thus obtaining

$$\frac{2x}{1 - x^2} + \frac{2y}{1 - y^2} + \frac{2z}{1 - z^2} \geq 3\sqrt{3},$$

the substitution by tangents becomes transparent. This is because we should recognize the double-angle formulas on the left-hand side.

The conditions $0 < x, y, z < 1$ and $xy + xz + yz = 1$ characterize the tangents of the half-angles of an acute triangle. Indeed, if $x = \tan \frac{A}{2}$, $y = \tan \frac{B}{2}$, and $z = \tan \frac{C}{2}$, then $0 < x, y, z < 1$ implies $A, B, C \in \left(0, \frac{\pi}{2}\right)$. Also, the equality $xy + xz + yz = 1$, which is the same as

$$\frac{1}{z} = \frac{x+y}{1-xy},$$

implies

$$\cot \frac{C}{2} = \tan \frac{A+B}{2}.$$

And this is equivalent to $\frac{\pi}{2} - \frac{C}{2} = \frac{A+B}{2}$, or $A + B + C = \pi$.

Returning to the problem, with the chosen trigonometric substitution the inequality assumes the much simpler form

$$\tan A + \tan B + \tan C \geq 3\sqrt{3}.$$

And this is Jensen's inequality applied to the tangent function, which is convex on $\left(0, \frac{\pi}{2}\right)$, and to the points $A, B, C \in \left(0, \frac{\pi}{2}\right)$, with $A + B + C = \pi$.

797. The numbers x, y, z have all the same sign, and if (x, y, z) is a solution, then so is $(-x, -y, -z)$. Thus it suffices to find the positive solutions.

If we denote $x = \tan\alpha$, $y = \tan\beta$, and $z = \tan\gamma$, with $\alpha, \beta, \gamma \in (0, \frac{\pi}{2})$, then the last equation implies that α, β, γ add up to $\frac{\pi}{2}$. Indeed, this equation can be rewritten as

$$\frac{1}{z} = \frac{x+y}{1-xy},$$

which is the same as $\cot\alpha = \tan(\alpha + \beta)$.

On the other hand, the first group of equations can be rewritten as

$$\frac{x}{3(1+x^2)} = \frac{y}{4(1+y^2)} = \frac{z}{5(1+z^2)}.$$

Using the trigonometric identity

$$\sin 2t = \frac{2\tan t}{1 + \tan^2 t},$$

can rewrite this as

$$\frac{\sin 2\alpha}{3} = \frac{\sin 2\beta}{4} = \frac{\sin 2\gamma}{5}.$$

Given that 2α, 2β and 2γ are the angles of a triangle, we deduce using the law of sines that the side-lengths of this triangle are proportional to 3, 4, and 5. This implies $\gamma = 45°$, so $z = 1$. Also $\sin 2\alpha = \frac{3}{5}$ and $\sin 2\beta = \frac{4}{5}$ and hence $x = \tan\alpha = \frac{1}{3}$ and $y = \tan\beta = \frac{1}{2}$. We obtain the solutions

$$\left(\frac{1}{3}, \frac{1}{2}, 1\right) \quad \text{and} \quad \left(-\frac{1}{3}, -\frac{1}{2}, -1\right).$$

798. From the first equation, it follows that if x is 0, then so is y, making x^2 indeterminate; hence x, and similarly y and z, cannot be 0. Solving the equations, respectively, for y, z, and x, we obtain the equivalent system

$$y = \frac{3x - x^3}{1 - 3x^2},$$

$$z = \frac{3y - y^3}{1 - 3y^2},$$

$$x = \frac{3z - z^3}{1 - 3z^2},$$

where x, y, z are real numbers different from 0.

There exists a unique number u in the interval $\left(-\frac{\pi}{2}, \frac{\pi}{2}\right)$ such that $x = \tan u$. Then

$$y = \frac{3\tan u - \tan^3 u}{1 - 3\tan^2 u} = \tan 3u,$$

$$z = \frac{3\tan 3u - \tan^3 3u}{1 - 3\tan^2 3u} = \tan 9u,$$

$$x = \frac{3\tan 9u - \tan^3 9u}{1 - 3\tan^2 9u} = \tan 27u.$$

The last equality yields $\tan u = \tan 27u$, so u and $27u$ differ by an integer multiple of π. Therefore, $u = \frac{k\pi}{26}$ for some k satisfying $-\frac{\pi}{2} < \frac{k\pi}{26} < \frac{\pi}{2}$. Besides, k must not be 0, since $x \neq 0$. Hence the possible values of k are $\pm 1, \pm 2, \ldots, \pm 12$, each of them generating the corresponding triple

$$x = \tan \frac{k\pi}{26}, \quad y = \tan \frac{3k\pi}{26}, \quad z = \tan \frac{9k\pi}{26}.$$

It is immediately checked that all of these triples are solutions of the initial system.

799. In the case of the sequence $(a_n)_n$, the innermost square root suggests one of the substitutions $a_n = 2\sin t_n$ or $a_n = 2\cos t_n$, with $t_n \in \left[0, \frac{\pi}{2}\right]$, $n \geq 0$. It is the first choice that allows a further application of a half-angle formula:

$$2\sin t_{n+1} = a_{n+1} = \sqrt{2 - \sqrt{4 - 4\sin^2 t_n}} = \sqrt{2 - 2\cos t_n} = 2\sin \frac{t_n}{2}.$$

It follows that $t_{n+1} = \frac{t_n}{2}$, which combined with $t_0 = \frac{\pi}{4}$ gives $t_n = \frac{\pi}{2^{n+2}}$ for $n \geq 0$. Therefore, $a_n = 2\sin \frac{\pi}{2^{n+2}}$ for $n \geq 0$.

For $(b_n)_n$, the innermost square root suggests a trigonometric substitution as well, namely $b_n = 2\tan u_n$, $n \geq 0$. An easy induction shows that the sequence $(b_n)_n$ is positive, so we can choose $u_n \in \left[0, \frac{\pi}{2}\right)$. Substituting in the recursive formula, we obtain

$$2\tan u_{n+1} = b_{n+1} = \frac{2\tan u_n}{2 + \sqrt{4 + 4\tan u_n}} = \frac{4\tan u_n}{2 + \dfrac{2}{\cos u_n}}$$

$$= 2 \cdot \frac{\sin u_n}{1 + \cos u_n} = 2\tan \frac{u_n}{2}.$$

Therefore, $u_{n+1} = \frac{u_n}{2}$, which together with $u_0 = \frac{\pi}{4}$ implies $u_n = \frac{\pi}{2^{n+2}}$, $n \geq 0$. Hence $b_n = 2\tan \frac{\pi}{2^{n+2}}$ for $n \geq 0$.

Returning to the problem, we recall that sine and tangent are decreasing on $\left(0, \frac{\pi}{2}\right)$ and their limit at 0 is 0. This takes care of (a).

For (b), note that the functions $\sin x / x$ and $\tan x / x$ are increasing, respectively, decreasing, on $\left(0, \frac{\pi}{2}\right)$ (this can be checked using derivatives). Hence $2^n a_n = \frac{\pi}{2}\sin\frac{\pi}{2^{n+2}} / \frac{\pi}{2^{n+2}}$ is increasing, and $2^n b_n = \frac{\pi}{2}\tan\frac{\pi}{2^{n+2}} / \frac{\pi}{2^{n+2}}$ is decreasing. Also, since

$$\lim_{x \to 0} \frac{\sin x}{x} = \lim_{x \to 0} \frac{\tan x}{x} = 1,$$

it follows that

$$\lim_{n \to \infty} 2^n a_n = \frac{\pi}{2} \lim_{n \to \infty} \frac{\sin \dfrac{\pi}{2^{n+2}}}{\dfrac{\pi}{2^{n+2}}} = \frac{\pi}{2},$$

and similarly $\lim_{n \to \infty} 2^n b_n = \frac{\pi}{2}$. This answers (b).

The first inequality in (c) follows from the fact that $\tan x > \sin x$ for $x \in (0, \frac{\pi}{2})$. For the second inequality we use Taylor series expansions. We have

$$\tan x - \sin x = x - \frac{x^3}{12} + o(x^4) - x + \frac{x^3}{6} + o(x^4) = \frac{x^3}{12} + o(x^4).$$

Hence

$$b_n - a_n = 2\left(\tan \frac{\pi}{2^{n+2}} - \sin \frac{\pi}{2^{n+2}}\right) = \frac{\pi^3}{6 \cdot 2^6} \cdot \frac{1}{8^n} + o\left(\frac{1}{2^{4n}}\right).$$

It follows that for $C > \frac{\pi^3}{6 \cdot 2^6}$ we can find n_0 such that $b_n - a_n < \frac{C}{8^n}$ for $n \geq n_0$. Choose C such that the inequality also holds for (the finitely many) $n < n_0$. This concludes (c).

(8th International Competition in Mathematics for University Students, 2001)

800. With the substitution $y = 1/x$ we obtain the equation $y^3 - 3y + 1 = 0$. The expression on the left should remind us of the trigonometric identity

$$2 \sin 3t = 3(2 \sin t) - (2 \sin t)^3.$$

With the trigonometric substitution $2 \sin t = y$ we obtain the equation $2 \sin 3t = 1$. Hence $3t = k \cdot 360° + 30°$ and $3t = k \cdot 360° + 150°$, $k \in \mathbb{Z}$. The solutions in $[0°, 360°]$ are $t = 10°, 50°, 130°, 170°, 250°, 290°$. Checking cases, we obtain that the solutions to the equation $y^3 - 3y + 1 = 0$ are $2 \cos 10°$, $2 \cos 50°$, $2 \cos 250°$. Hence the solutions to the original equation are $\frac{1}{2 \sin 10°}, \frac{1}{2 \sin 50°}, \frac{1}{2 \sin 250°} = -\frac{1}{2 \sin 70°}$. Of these the last is negative and the first two are positive, with the largest being $\frac{1}{2 \sin 10°}$. Thus $\alpha = \frac{1}{2 \sin 10°}$. Set $\beta = \frac{1}{2 \sin 50°}$ and $\gamma = -\frac{1}{2 \sin 70°}$. Clearly $\beta > |\gamma|$. Also, since $2 \sin 50° > 2 \sin 45° = \sqrt{2}$, $\beta < \frac{\sqrt{2}}{2} < 0.75$. Thus for $n \geq 3$,

$$0 < \beta^n + \gamma^n < 1.$$

Let us consider the sequence $u_n = \alpha^n + \beta^n + \gamma^n$, $n \geq 0$. Then $u_{n+3} = 3u_{n+2} - u_n$, $u_0 = 3$, $u_1 = 3$, $u_2 = 9$ (the recursive relation for Newton's polynomials). In particular it follows that u_n is an integer, for all n, and consequently, if $n \geq 3$, $\lfloor \alpha^n \rfloor = u_n - 1$.

Modulo 17 the sequence must be periodic, by the Pigeonhole Principle. We compute $u_0 \equiv 3$, $u_1 \equiv 3$, $u_2 \equiv 9$, $u_3 \equiv 7$, $u_4 \equiv 1$, $u_5 \equiv 11$, $u_6 \equiv 9$, $u_7 \equiv 9$, $u_8 \equiv 16$, $u_9 \equiv 5$, $u_{10} \equiv 6$, $u_{11} \equiv 2$, $u_{12} \equiv 1$, $u_{13} \equiv 14$, $u_{14} \equiv 6$, $u_{15} \equiv 0$, $u_{16} \equiv 3$, $u_{17} \equiv 3$, $u_{18} \equiv 9$. This shows that u_n is periodic modulo 17 with period 16. Since $1788 \equiv 12 (\bmod 17)$ and $1988 \equiv 4 (\bmod 17)$, we obtain $u_{1788} \equiv u_{12} \equiv 1 (\bmod 17)$ and $u_{1988} \equiv u_4 \equiv 1 (\bmod 17)$. The problem is solved.

(Short list, 29th International Mathematical Olympiad, 1988)

801. Writing $x_n = \tan a_n$ for $0° < a_n < 90°$, we have

$$x_{n+1} = \tan a_n + \sqrt{1 + \tan^2 a_n} = \tan a_n + \sec a_n = \frac{1 + \sin a_n}{\cos a_n} = \tan\left(\frac{90° + a_n}{2}\right).$$

Because $a_1 = 60°$, we have $a_2 = 75°$, $a_3 = 82.5°$, and in general $a_n = 90° - \frac{30°}{2^{n-1}}$, whence

$$x_n = \tan\left(90° - \frac{30°}{2^{n-1}}\right) = \cot\left(\frac{30°}{2^{n-1}}\right) = \cot \theta_n, \quad \text{where} \quad \theta_n = \frac{30°}{2^{n-1}}.$$

A similar calculation shows that

$$y_n = \tan 2\theta_n = \frac{2\tan\theta_n}{1 - \tan^2\theta_n},$$

which implies that

$$x_n y_n = \frac{2}{1 - \tan^2\theta_n}.$$

Because $0° < \theta_n < 45°$, we have $0 < \tan^2\theta_n < 1$ and $x_n y_n > 2$. For $n > 1$, we have $\theta_n < 30°$, and hence $\tan^2\theta_n < \frac{1}{3}$. It follows that $x_n y_n < 3$, and the problem is solved.

(Team Selection Test for the International Mathematical Olympiad, Belarus, 1999)

802. Let $a = \tan x$, $b = \tan y$, $c = \tan z$, where $x, y, z \in \left(0, \frac{\pi}{2}\right)$. From the identity

$$\tan(x + y + z) = \frac{\tan x + \tan y + \tan z - \tan x \tan y \tan z}{1 - \tan x \tan y - \tan y \tan z - \tan x \tan z}$$

it follows that $abc = a + b + c$ only if $x + y + z = k\pi$, for some integer k. In this case $\tan(3x + 3y + 3z) = \tan 3k\pi = 0$, and from the same identity it follows that

$$\tan 3x \tan 3y \tan 3z = \tan 3x + \tan 3y + \tan 3z.$$

This is the same as

$$\frac{3a - a^3}{3a^2 - 1} \cdot \frac{3b - b^3}{3b^2 - 1} \cdot \frac{3c - c^3}{3c^2 - 1} = \frac{3a - a^3}{3a^2 - 1} + \frac{3b - b^3}{3b^2 - 1} + \frac{3c - c^3}{3c^2 - 1},$$

and we are done.

(Mathematical Olympiad Summer Program, 2000, proposed by T. Andreescu)

803. Rewrite the second equation as

$$\frac{a^2}{yz} + \frac{b^2}{zx} + \frac{c^2}{xy} + \frac{abc}{xyz} = 4.$$

We recognize

$$u^2 + v^2 + w^2 + uvw = 4.$$

Then there is an acute triangle ABC with $u = 2\cos A$, $v = 2\cos B$, $w = 2\cos C$. The first equation reads

$$x + y + z = 2\sqrt{xy}\cos C + 2\sqrt{yz}\cos A + 2\sqrt{zx}\cos B.$$

We solve this as a quadratic in $\sqrt{x}$. The discriminant is $-4(\sqrt{y}\sin C - \sqrt{z}\sin B)^2$. Hence $\sqrt{y}\sin C = \sqrt{z}\sin B$. It follows that

$$\sqrt{x} = \sqrt{y}\cos C + \sqrt{z}\cos B = \sqrt{y}\cos C + \frac{\sqrt{y}}{\sin B}\sin C\cos B$$

$$= \frac{\sqrt{y}}{\sin B}(\sin B\cos C + \sin C\cos B) = \frac{\sqrt{y}}{\sin B}\sin(B + C) = \frac{\sqrt{y}}{\sin B}\sin A.$$

Combining the last two relations we find that

$$\frac{\sqrt{x}}{\sin A} = \frac{\sqrt{y}}{\sin B} = \frac{\sqrt{z}}{\sin C}.$$

Using the fact that $\cos A = \frac{a}{2\sqrt{yz}}$, $\cos B = \frac{b}{2\sqrt{zx}}$, $\cos C = \frac{c}{2\sqrt{xy}}$ we get

$$x = \frac{b+c}{2}, \quad y = \frac{c+a}{2}, \quad z = \frac{a+b}{2}.$$

804. With the substitution $x = \cosh t$, the integral becomes

$$\int \frac{1}{\sinh t + \cosh t} \sinh t\, dt = \int \frac{e^t - e^{-t}}{2e^t} dt = \frac{1}{2}\int (1 - e^{-2t})dt = \frac{1}{2}t + \frac{e^{-2t}}{4} + C$$
$$= \frac{1}{2}\ln(x + \sqrt{x^2 - 1}) + \frac{1}{4}\cdot\frac{1}{2x^2 - 1 + 2x\sqrt{x^2 - 1}} + C.$$

805. Suppose by contradiction that there exists an irrational a and a positive integer n such that the expression from the statement is rational. Substitute $a = \cosh t$, where t is an appropriately chosen real number. Then

$$\sqrt[n]{a + \sqrt{a^2 - 1}} + \sqrt[n]{a - \sqrt{a^2 - 1}} = \sqrt[n]{\cosh t + \sinh t} + \sqrt[n]{\cosh t - \sinh t}$$
$$= \sqrt[n]{e^t} + \sqrt[n]{e^{-t}} = e^{t/n} + e^{-t/n} = 2\cosh\frac{t}{n}.$$

It follows that $\cosh\frac{t}{n}$ is rational. From the recurrence relation

$$\cosh(k+1)\alpha = 2\cosh\alpha\cosh k\alpha - \cosh(k-1)\alpha, \ k \geq 1,$$

applied to $\alpha = \frac{t}{n}$, we can prove inductively that $\cosh k\frac{t}{n}$ is rational for all positive integers k. In particular, $\cosh n\frac{t}{n} = \cosh t = a$ is rational. This contradicts the hypothesis. Hence our assumption was false and the conclusion follows.

(Romanian Team Selection Test for the International Mathematical Olympiad, 1979, proposed by T. Andreescu)

806. We use the triple-angle formula

$$\sin 3x = 3\sin x - 4\sin^3 x,$$

which we rewrite as

$$\sin^3 x = \frac{1}{4}(3\sin x - \sin 3x).$$

The expression on the left-hand side of the identity from the statement becomes

$$3\cdot\frac{3\sin 9° - \sin 27°}{4} + 9\cdot\frac{3\sin 27° - \sin 81°}{4} + 3\cdot\frac{3\sin 81° - \sin 243°}{4}$$
$$+ \frac{3\sin 243° - \sin 729°}{4}.$$

This collapses to

$$\frac{81 \sin 9° - \sin 729°}{4} = \frac{81 \sin 9° - \sin 9°}{4} = 20 \sin 9°.$$

(T. Andreescu)

807. The triple-angle formula for the tangent gives

$$3 \tan 3x = \frac{3(3 \tan x - \tan^2 x)}{1 - 3 \tan^2 x} = \frac{3 \tan^3 x - 9 \tan x}{3 \tan^2 x - 1} = \tan x - \frac{8 \tan x}{3 \tan^2 x - 1}.$$

Hence

$$\frac{1}{\cot x - 3 \tan x} = \frac{\tan x}{1 - 3 \tan^2 x} = \frac{1}{8}(3 \tan 3x - \tan x) \text{ for all } x \neq k\frac{\pi}{2},\ k \in \mathbb{Z}.$$

It follows that the left-hand side telescopes as

$$\frac{1}{8}(3 \tan 27° - \tan 9° + 9 \tan 81° - 3 \tan 27° + 27 \tan 243° - 9 \tan 81° + 81 \tan 729°$$

$$- 27 \tan 243°) = \frac{1}{8}(81 \tan 9° - \tan 9°) = 10 \tan 9°,$$

and we are done.

(T. Andreescu)

808. Multiply the left-hand side by $\sin 1°$ and transform it using the identity

$$\frac{\sin((k+1)° - k°)}{\sin k° \sin(k+1)°} = \cot k° - \cot(k+1)°.$$

We obtain

$$\cot 45° - \cot 46° + \cot 47° - \cot 48° + \cdots + \cot 131° - \cot 132° + \cot 133° - \cot 134°.$$

At first glance this sum does not seem to telescope. It does, however, after changing the order of terms. Indeed, if we rewrite the sum as

$$\cot 45° - (\cot 46° + \cot 134°) + (\cot 47° + \cot 133°) - (\cot 48° + \cot 132°)$$

$$+ \cdots + (\cot 89° + \cot 91°) - \cot 90°,$$

then the terms in the parentheses cancel, since they come from supplementary angles. The conclusion follows from $\cot 45° = 1$ and $\cot 90° = 0$.

(T. Andreescu)

809. The formula

$$\tan(a - b) = \frac{\tan a - \tan b}{1 + \tan a \tan b}$$

translates into

$$\arctan \frac{x - y}{1 + xy} = \arctan x - \arctan y.$$

Applied to $x = n + 1$ and $y = n - 1$, it gives

$$\arctan \frac{2}{n^2} \arctan \frac{(n+1) - (n-1)}{1 + (n+1)(n-1)} = \arctan(n + 1) - \arctan(n - 1).$$

The sum in part (a) telescopes as follows:

$$\sum_{n=1}^{\infty} \arctan \frac{2}{n^2} = \lim_{N \to \infty} \sum_{n=1}^{N} \arctan \frac{2}{n^2} = \lim_{N \to \infty} \sum_{n=1}^{N} (\arctan(n + 1) - \arctan(n - 1))$$

$$= \lim_{N \to \infty} (\arctan(N + 1) + \arctan N - \arctan 1 - \arctan 0)$$

$$= \frac{\pi}{2} + \frac{\pi}{2} - \frac{\pi}{4} = \frac{3\pi}{4}.$$

The sum in part (b) is only slightly more complicated. In the above-mentioned formula for the difference of arctangents we have to substitute $x = \left(\frac{n+1}{\sqrt{2}}\right)^2$ and $y = \left(\frac{n-1}{\sqrt{2}}\right)^2$. This is because

$$\frac{8n}{n^4 - 2n^2 + 5} = \frac{8n}{4 + (n^2 - 1)^2} = \frac{2[(n+1)^2 - (n-1)^2]}{4 - (n+1)^2(n-1)^2} = \frac{\left(\frac{n+1}{\sqrt{2}}\right)^2 - \left(\frac{n-1}{\sqrt{2}}\right)^2}{1 - \left(\frac{n+1}{\sqrt{2}}\right)^2 \left(\frac{n-1}{\sqrt{2}}\right)^2}.$$

The sum telescopes as

$$\sum_{n=1}^{\infty} \arctan \frac{8n}{n^4 - 2n^2 + 5} = \lim_{N \to \infty} \sum_{n=1}^{N} \arctan \frac{8n}{n^4 - 2n^2 + 5}$$

$$= \lim_{N \to \infty} \sum_{n=1}^{N} \left[\arctan \left(\frac{n+1}{\sqrt{2}}\right)^2 - \arctan \left(\frac{n-1}{\sqrt{2}}\right)^2 \right]$$

$$= \lim_{N \to \infty} \left[\arctan \left(\frac{N+1}{\sqrt{2}}\right)^2 + \arctan \left(\frac{N}{\sqrt{2}}\right)^2 - \arctan 0 - \arctan \frac{1}{2} \right] = \pi - \arctan \frac{1}{2}.$$

(*American Mathematical Monthly*, proposed by J. Anglesio)

810. In order for the series to telescope, we wish to write the general term in the form $\arcsin b_n - \arcsin b_{n+1}$. To determine b_n let us apply the sine function and write

$$\frac{\sqrt{n+1} - \sqrt{n}}{\sqrt{n+2}\sqrt{n+1}} = \sin u_n = b_n \sqrt{1 - b_{n+1}^2} - b_{n+1} \sqrt{1 - b_n^2}.$$

If we choose $b_n = \frac{1}{\sqrt{n+1}}$, then this equality is satisfied. Therefore,

$$S = \lim_{N \to \infty} \sum_{n=0}^{N} \left(\arcsin \frac{1}{\sqrt{n+1}} - \arcsin \frac{1}{\sqrt{n+2}} \right) = \arcsin 1 - \lim_{N \to \infty} \arcsin \frac{1}{\sqrt{N+2}} = \frac{\pi}{2}.$$

(The *Mathematics Gazette Competition*, Bucharest, 1927)

811. The radii of the circles satisfy the recurrence relation

$$R_1 = 1, \quad R_{n+1} = R_n \cos \frac{\pi}{2^{n+1}}.$$

Hence

$$\lim_{n \to \infty} R_n = \prod_{n=1}^{\infty} \cos \frac{\pi}{2^n}.$$

The product can be made to telescope if we use the double-angle formula for sine written $\cos x = \frac{\sin 2x}{2 \sin x}$. We then have

$$\prod_{n=2}^{\infty} \cos \frac{\pi}{2^n} = \lim_{N \to \infty} \prod_{n=2}^{N} \cos \frac{\pi}{2^n} = \lim_{N \to \infty} \prod_{n=2}^{N} \frac{1}{2} \cdot \frac{\sin \frac{\pi}{2^{n-1}}}{\sin \frac{\pi}{2^n}}$$

$$= \lim_{N \to \infty} \frac{1}{2^N} \frac{1}{2^N} \frac{\sin \frac{\pi}{2}}{\sin \frac{\pi}{2^N}} = \frac{2}{\pi} \lim_{N \to \infty} \frac{\frac{\pi}{2^N}}{\sin \frac{\pi}{2^N}} = \frac{2}{\pi}.$$

Thus the answer to the problem is $\frac{2}{\pi}$.

Remark. As a corollary, we obtain the formula

$$\frac{2}{\pi} = \frac{\sqrt{2}}{2} \cdot \frac{\sqrt{2 + \sqrt{2}}}{2} \cdot \frac{\sqrt{2 + \sqrt{2 + \sqrt{2}}}}{2} \cdots .$$

This formula is credited to F. Viète, although Archimedes already used this approximation of the circle by regular polygons to compute π.

812. For $k = 1, 2, \ldots, 59$,

$$1 - \frac{\cos(60° + k°)}{\cos k°} = \frac{\cos k° - \cos(60° + k°)}{\cos k°} = \frac{2 \sin 30° \sin(30° + k°)}{\cos k°}$$

$$= \frac{\cos(90° - 30° - k°)}{\cos k°} = \frac{\cos(60° - k°)}{\cos k°}.$$

So

$$\prod_{k=1}^{59} \left(1 - \frac{\cos(60° + k°)}{\cos k°} \right) = \frac{\cos 59° \cos 58° \cdots \cos 1°}{\cos 1° \cos 2° \cdots \cos 59°} = 1.$$

813. We have

$$(1 - \cot 1°)(1 - \cot 2°) \cdots (1 - \cot 44°)$$

$$= \left(1 - \frac{\cos 1°}{\sin 1°}\right)\left(1 - \frac{\cos 2°}{\sin 2°}\right) \cdots \left(1 - \frac{\cos 44°}{\sin 44°}\right)$$

$$= \frac{(\sin 1° - \cos 1°)(\sin 2° - \cos 2°) \cdots (\sin 44° - \cos 44°)}{\sin 1° \sin 2° \cdots \sin 44°}.$$

Using the identity $\sin a - \cos a = \sqrt{2}\sin(a - 45°)$ in the numerators, we transform this further into

$$\frac{\sqrt{2}\sin(1° - 45°) \cdot \sqrt{2}\sin(2° - 45°) \cdots \sqrt{2}\sin(44° - 45°)}{\sin 1° \sin 2° \cdots \sin 44°}$$

$$= \frac{(\sqrt{2})^{44}(-1)^{44}\sin 44° \sin 43° \cdots \sin 1°}{\sin 44° \sin 43° \ldots \sin 1°}.$$

After cancellations, we obtain 2^{22}.

814. We can write

$$\sqrt{3} + \tan n° = \tan 60° + \tan n° = \frac{\sin 60°}{\cos 60°} + \frac{\sin n°}{\cos n°}$$

$$= \frac{\sin(60° + n°)}{\cos 60° \cos n°} = 2 \cdot \frac{\sin(60° + n°)}{\cos n°} = 2 \cdot \frac{\cos(30° - n°)}{\cos n°}.$$

And the product telescopes as follows:

$$\prod_{n=1}^{29}(\sqrt{3} + \tan n°) = 2^{29}\prod_{n=1}^{29}\frac{\cos(30° - n°)}{\cos n°} = 2^{29} \cdot \frac{\cos 29° \cos 28° \cdots \cos 1°}{\cos 1° \cos 2° \cdots \cos 29°} = 2^{29}.$$

(T. Andreescu)

815. (a) Note that

$$1 - 2\cos 2x = 1 - 2(2\cos^2 x - 1) = 3 - 4\cos^2 x = -\frac{\cos 3x}{\cos x}.$$

The product becomes

$$\left(-\frac{1}{2}\right)^3 \frac{\cos\frac{3\pi}{7}}{\cos\frac{\pi}{7}} \cdot \frac{\cos\frac{9\pi}{7}}{\cos\frac{3\pi}{7}} \cdot \frac{\cos\frac{27\pi}{7}}{\cos\frac{9\pi}{7}} = -\frac{1}{8} \cdot \frac{\cos\frac{27\pi}{7}}{\cos\frac{\pi}{7}}.$$

Taking into account that $\cos\frac{27\pi}{7} = \cos\left(2\pi - \frac{\pi}{7}\right) = \cos\frac{\pi}{7}$, we obtain the desired identity.
    (b) Analogously,

$$1 + 2\cos 2x = 1 + 2(1 - 2\sin^2 x) = 3 - 4\sin^2 x = \frac{\sin 3x}{\sin x},$$

and the product becomes

$$\frac{1}{2^4} \cdot \frac{\sin\frac{3\pi}{20}}{\sin\frac{\pi}{20}} \cdot \frac{\sin\frac{9\pi}{20}}{\sin\frac{3\pi}{20}} \cdot \frac{\sin\frac{27\pi}{20}}{\sin\frac{9\pi}{20}} \cdot \frac{\sin\frac{81\pi}{20}}{\sin\frac{27\pi}{20}} = \frac{1}{16} \cdot \frac{\sin\frac{81\pi}{20}}{\sin\frac{\pi}{20}}.$$

Because $\sin\frac{81\pi}{20} = \sin\left(4\pi + \frac{\pi}{20}\right) = \sin\frac{\pi}{20}$, this is equal to $\frac{1}{16}$.
(T. Andreescu)

816. (a) We observe that

$$\sec x = \frac{1}{\cos x} = \frac{2 \sin x}{2 \sin x \cos x} = 2 \frac{\sin x}{\sin 2x}.$$

Applying this to the product in question yields

$$\prod_{n=1}^{24} \sec(2^n)^\circ = 2^{24} \prod_{n=1}^{24} \frac{\sin(2^n)^\circ}{\sin(2^{n+1})^\circ} = 2^{24} \frac{\sin 2^\circ}{\sin(2^{25})^\circ}.$$

We want to show that $\sin(2^{25})^\circ = \cos 2^\circ$. To this end, we prove that $2^{25} - 2 - 90$ is an odd multiple of 180. This comes down to proving that $2^{23} - 23$ is an odd multiple of $45 = 5 \times 9$. Modulo 5, this is $2 \cdot (2^2)^{11} - 3 = 2 \cdot (-1)^{11} - 3 = 0$, and modulo 9, $4 \cdot (2^3)^7 - 5 = 4 \cdot (-1)^7 - 5 = 0$. This completes the proof of the first identity.

(b) As usual, we start with a trigonometric computation

$$2 \cos x - \sec x = \frac{2 \cos^2 x - 1}{\cos x} = \frac{\cos 2x}{\cos x}.$$

Using this, the product becomes

$$\prod_{n=2}^{25} \frac{\cos(2^{n+1})^\circ}{\cos(2^n)^\circ} = \frac{\cos(2^{26})^\circ}{\cos 4^\circ}.$$

The statement of the problem suggests that $\cos(2^{26})^\circ = -\cos 4^\circ$, which is true only if $2^{26} - 4$ is a multiple of 180, but not of 360. And indeed, $2^{26} - 2^2 = 4(2^{24} - 1)$, which is divisible on the one hand by $2^4 - 1$ and on the other by $2^6 - 1$. This number is therefore an odd multiple of $4 \times 5 \times 9 = 180$, and we are done.

(T. Andreescu)

Number Theory

817. Because $a_{n-1} \equiv n - 1 \pmod{k}$, the first positive integer greater than a_{n-1} that is congruent to n modulo k must be $a_{n-1} + 1$. The nth positive integer greater than a_{n-1} that is congruent to n modulo k is simply $(n - 1)k$ more than the first positive integer greater than a_{n-1} that satisfies this condition. Therefore, $a_n = a_{n-1} + 1 + (n - 1)k$. Solving this recurrence gives

$$a_n = n + \frac{(n - 1)nk}{2}.$$

(Austrian Mathematical Olympiad, 1997)

818. First, let us assume that none of the progressions contains consecutive numbers, for otherwise the property is obvious. Distributing the eight numbers among the three arithmetic progressions shows that either one of the progressions contains at least four of the numbers, or two of them contain exactly three of the numbers. In the first situation, if one progression contains $2, 4, 6, 8$, then it consists of all positive even numbers, and we are done. If it contains $1, 3, 5, 7$, then the other two contain $2, 4, 6, 8$ and again we have two possibilities: either a progression contains two consecutive even numbers, whence it contains all even numbers thereafter, or one progression contains $2, 6$, the other $4, 8$, and hence the latter contains 1980.

Let us assume that two progressions each contain exactly three of the numbers $1, 2, 3, 4, 5, 6, 7, 8$. The numbers 3 and 6 must belong to different progressions, for otherwise all multiples of 3 occur in one of the progressions and we are done. If 3 belongs to one of the progressions containing exactly three of the numbers, then these numbers must be $3, 5, 7$. But then the other two progressions contain $2, 4, 6, 8$, and we saw before that 1980 occurs in one of them. If 6 belongs to one of the progressions containing exactly three of the numbers, then these numbers must be $4, 6, 8$, and 1980 will then belong to this progression. This completes the proof.

(Austrian-Polish Mathematics Competition, 1980)

819. From $f(1) + 2f(f(1)) = 8$ we deduce that $f(1)$ is an even number between 1 and 6, that is, $f(1) = 2, 4$, or 6. If $f(1) = 2$ then $2 + 2f(2) = 8$, so $f(2) = 3$. Continuing with $3 + 2f(3) = 11$,

© Springer International Publishing AG 2017
R. Gelca and T. Andreescu, *Putnam and Beyond*, DOI 10.1007/978-3-319-58988-6

we obtain $f(3) = 4$, and formulate the conjecture that $f(n) = n+1$ for all $n \geq 1$. And indeed, in an inductive manner we see that $f(n) = n + 1$ implies $n + 1 + 2f(n + 1) = 3n + 5$; hence $f(n + 1) = n + 2$.

The case $f(1) = 4$ gives $4 + 2f(4) = 8$, so $f(4) = 2$. But then $2 + 2f(f(4)) = 17$, which cannot hold for reasons of parity. Also, if $f(1) = 6$, then $6 + 2f(6) = 8$, so $f(6) = 1$. This cannot happen, because $f(6) + 2f(f(6)) = 1 + 2 \cdot 6$, which does not equal $3 \cdot 6 + 5$.

We conclude that $f(n) = n + 1$, $n \geq 1$, is the unique solution to the functional equation.

820. Let $g(x) = f(x) - x$. The given equation becomes $g(x) = 2g(f(x))$. Iterating, we obtain that $g(x) = 2^n f^{(n)}(x)$ for all $x \in \mathbb{Z}$, where $f^{(n)}(x)$ means f composed n times with itself. It follows that for every $x \in \mathbb{Z}$, $g(x)$ is divisible by all powers of 2, so $g(x) = 0$. Therefore, the only function satisfying the condition from the statement is $f(x) = x$ for all x.

(*Revista Matematică din Timişoara* (*Timişoara Mathematics Gazette*), proposed by L. Funar)

821. Assume such a function exists, and define $g : \mathbb{N} \to 3\mathbb{N} + 1$, $g(x) = 3f(x) + 1$. Then g is bijective and satisfies $g(xy) = g(x)g(y)$. This implies in particular that $g(1) = 1$.

We will need the following fact. If x is such that $g(x) = n$, where $n = pq$, and p, q are prime numbers congruent to 2 modulo 3, then x is prime. Indeed, if $x = yz$, $y, z \geq 2$, then $g(x) = g(y)g(z)$. This implies that n can be factored as the product of two numbers in $3\mathbb{N} + 1$ which is not true.

Now choose two distinct numbers p and q that are congruent to 2 modulo 3 (for example, 2 and 5). Then pq, p^2, and q^2 are all in the image of g. Let $g(a) = p^2$, $g(b) = q^2$, and $g(c) = pq$. We have

$$g(c^2) = g(c)^2 = p^2 q^2 = g(a)g(b) = g(ab).$$

It follows that $c^2 = ab$, with a, b, and c distinct prime numbers, and this is impossible. Therefore, such a function f does not exist.

(Balkan Mathematical Olympiad, 1991)

822. We will prove that a sequence of positive integers satisfying the double inequality from the statement terminates immediately. Precisely, we show that if $a_1, a_2, \ldots, a_N$ satisfy the relation from the statement for $n = 1, 2, \ldots, N$, then $N \leq 5$.

Arguing by contradiction, let us assume that the sequence has a sixth term a_6. Set $b_n = a_{n+1} - a_n$, $n = 1, \ldots, 5$. The relation from the statement implies $a_n \geq a_{n-1}$ for $n \geq 2$, and so b_n is a nonnegative integer for $n = 1, \ldots, 5$. For $n = 2, 3, 4$ we have

$$-a_n \leq -b_n^2 \leq -a_{n-1},$$

$$a_n \leq b_{n+1}^2 \leq a_{n+1}.$$

Adding these two inequalities, we obtain

$$0 \leq b_{n+1}^2 - b_n^2 \leq b_n + b_{n-1},$$

or

$$0 \leq (b_{n+1} - b_n)(b_{n+1} + b_n) \leq b_n + b_{n-1}.$$

Therefore, $b_{n+1} \leq b_n$ for $n = 2, 3, 4$. If for $n = 3$ or $n = 4$ this inequality were strict, then for that specific n we would have

$$0 < b_{n+1}^2 - b_n^2 \leq b_n + b_{n-1} < b_{n+1} + b_n,$$

with the impossible consequence $0 < b_{n+1} - b_n < 1$. It follows that $b_3 = b_4 = b_5$. Combining this with the inequality from the statement, namely with

$$b_3^2 \leq a_3 \leq b_4^2 \leq a_4 \leq b_5^2,$$

we find that $a_3 = a_4$. But then $b_3 = a_4 - a_3 = 0$, which would imply $a_2 \leq b_3^2 = 0$, a contradiction. We conclude that the sequence can have at most *five* terms. This limit is sharp, since $a_1 = 1, a_2 = 3, a_3 = 4, a_4 = 6, a_5 = 8$ satisfies the condition from the statement.

(Romanian Team Selection Test for the International Mathematical Olympiad, 1985, proposed by L. Panaitopol)

823. Setting $x = y = z = 0$ we find that $f(0) = 3(f(0))^3$. This cubic equation has the unique integer solution $f(0) = 0$. Next, with $y = -x$ and $z = 0$ we have $f(0) = (f(x))^3 + (f(-x))^3 + (f(0))^3$, which yields $f(-x) = -f(x)$ for all integers x; hence f is an odd function. Now set $x = 1$, $y = z = 0$ to obtain $f(1) = (f(1))^3 + 2(f(0))^3$; hence $f(1) = f(1)^3$. Therefore, $f(1) \in \{-1, 0, 1\}$. Continuing with $x = y = 1$ and $z = 0$ and $x = y = z = 1$ we find that $f(2) = 2(f(1))^3 = 2f(1)$ and $f(3) = 3(f(1))^3 = 3f(1)$. We conjecture that $f(x) = xf(1)$ for all integers x. We will do this by strong induction on the absolute value of x, and for that we need the following lemma.

Lemma. *If x is an integer whose absolute value is greater than 3, then x^3 can be written as the sum of five cubes whose absolute values are less than x.*

Proof. We have

$$4^3 = 3^3 + 3^3 + 2^3 + 1^3 + 1^3, \ 5^3 = 4^3 + 4^3 + (-1)^3 + (-1)^3 + (-1)^3,$$
$$6^3 = 5^3 + 4^3 + 3^3 + 0^3 + 0^3, \ 7^3 = 6^3 + 5^3 + 1^3 + 1^3 + 0^3,$$

and if $x = 2k + 1$ with $k > 3$, then

$$x^3 = (2k + 1)^3 = (2k - 1)^3 + (k + 4)^3 + (4 - k)^3 + (-5)^3 + (-1)^3.$$

In this last case it is not hard to see that $2k - 1, k + 4, |4 - k|, 5$, and 1 are all less than $2k + 1$. If $x > 3$ is an arbitrary integer, then we write $x = my$, where y is 4, 6, or an odd number greater than 3, and m is an integer. If we express $y^3 = y_1^3 + y_2^3 + y_3^3 + y_4^3 + y_5^3$, then $x^3 = (my_1)^3 + (my_2)^3 + (my_3)^3 + (my_4)^3 + (my_5)^3$, and the lemma is proved. $\square$

Returning to the problem, using the fact that f is odd and what we proved before, we see that $f(x) = xf(1)$ for $|x| \leq 3$. For $x > 4$, suppose that $f(y) = yf(1)$ for all y with $|y| < |x|$. Using the lemma write $x^3 = x_1^3 + x_2^3 + x_3^3 + x_4^3 + x_5^3$, where $|x_i| < |x|$, $i = 1, 2, 3, 4, 5$. After writing

$$x^3 + (-x_4)^3 + (-x_5)^3 = x_1^3 + x_2^3 + x_3^3,$$

we apply f to both sides and use the fact that f is odd and the condition from the statement to obtain

$$(f(x))^3 - (f(x_4))^3 - (f(x_5))^3 = f(x_1)^3 + f(x_2)^3 + f(x_3)^3.$$

The inductive hypothesis yields

$$(f(x))^3 - (x_4 f(1))^3 - (x_5 f(1))^3 = (x_1 f(1))^3 + (x_2 f(1))^3 + (x_3 f(1))^3;$$

hence

$$(f(x))^3 = (x_1^3 + x_2^3 + x_3^3 + x_4^3 + x_5^3)(f(1))^3 = x^3 (f(1))^3.$$

Hence $f(x) = x f(1)$, and the induction is complete. Therefore, the only answers to the problem are $f(x) = -x$ for all x, $f(x) = 0$ for all x, and $f(x) = x$ for all x. That these satisfy the given equation is a straightforward exercise.

(*American Mathematical Monthly*, proposed by T. Andreescu)

824. The number on the left ends in a 0, 1, 4, 5, 6, or 9, while the one on the right ends in a 0, 2, 3, 5, 7, or 8. For equality to hold, both x and z should be multiples of 5, say $x = 5x_0$ and $z = 5z_0$. But then $25x_0^2 + 10y^2 = 3 \cdot 25z^2$. It follows that y is divisible by 5 as well, $y = 5y_0$. The positive integers x_0, y_0, z_0 satisfy the same equation, and continuing we obtain an infinite descent. Since this is not possible, the original equation does not have positive integer solutions.

825. It suffices to show that there are no positive solutions. Adding the two equations, we obtain

$$6(x^2 + y^2) = z^2 + t^2.$$

So 3 divides $z^2 + t^2$. Since the residue of a square modulo 3 is either 0 or 1, this can happen only if both z and t are divisible by 3, meaning that $z = 3z_1$, $t = 3t_1$. But then

$$6(x^2 + y^2) = 9(z_1^2 + t_1^2),$$

and hence $x^2 + y^2$ is divisible by 3. Again, this can happen only if $x = 3x_1$, and $y = 3y_1$, with x_1, y_1 positive integers. So (x_1, y_1, z_1, t_1) is another solution. We construct inductively a decreasing infinite sequence of positive solutions, which, of course, cannot exist. Hence the system does not admit nontrivial solutions.

(W. Sierpiński, 250 *Problems in Elementary Number Theory*, Państwowe Wydawnictwo Naukowe, Warszawa, 1970)

826. Assume that the positive integers x, y, z satisfy the given equation, and let $d = xy$. If $d = 1$, then $x = y = 1$ and $z = 0$, which cannot happen. Hence $d > 1$. Let p be a prime divisor of d. Because

$$(x + y)(x - y) = x^2 - y^2 = 2xyz \equiv 0 \pmod{p},$$

either $x \equiv y \pmod{p}$ or $x \equiv -y \pmod{p}$. But p divides one of x and y, so p must divide the other, too. Hence $x_1 = x/p$ and $y_1 = y/p$ are positive integers, and x_1, y_1, z satisfy the given equation as well. Repeating the argument, we construct an infinite sequence of solutions (x_n, y_n, z_n), $n \geq 1$, to the original equation, with $x_1 > x_2 > x_3 > \cdots$. This is, of course, impossible; hence the equation has no solutions.

(T. Andreescu, D. Andrica, *An Introduction to Diophantine Equations*, GIL, 2002)

827. If $(a_n^2)_n$ is an infinite arithmetic progression, then

$$a_{k+1}^2 - a_k^2 = a_k^2 - a_{k-1}^2, \text{ for } k \geq 2.$$

Such an arithmetic progression must be increasing, so $a_{k+1} + a_k > a_k + a_{k-1}$. Combining the two relations, we obtain $a_{k+1} - a_k < a_k - a_{k-1}$, for all $k \geq 2$. We have thus obtained an infinite descending sequence of positive integers

$$a_2 - a_1 > a_3 - a_2 > a_4 - a_3 > \cdots$$

Clearly, such a sequence cannot exist. Hence there is no infinite arithmetic progression whose terms are perfect squares.

Remark. In fact, much more is true. No four perfect squares can form an arithmetic progression.

(T.B. Soulami, *Les olympiades de mathématiques: Réflexes et stratégies*, Ellipses, 1999)

828. Assume that the property does not hold, and fix a. Only finitely many numbers of the form $f(a + k)$ can be less than a, so we can choose r such that $f(a + nr) > f(a)$ for all n. By our assumption $f(a + 2^{m+1}r) < f(a + 2^m r)$ for all m, for otherwise a and $d = 2^m r$ would satisfy the desired property. We have constructed an infinite descending sequence of positive integers, a contradiction. Hence the conclusion.

(British Mathematical Olympiad, 2003)

829. We will apply Fermat's infinite descent method to the prime factors of n.

Let p_1 be a prime divisor of n, and q the smallest positive integer for which p_1 divides $2^q - 1$. From Fermat's little theorem (Section 5.2.4 below) it follows that p_1 also divides 2^{p_1-1}. Hence $q \leq p_1 - 1 < p_1$.

Let us prove that q divides n. If not, let $n = kq + r$, where $0 < r < q$. Then

$$2^n - 1 = 2^{kq} \cdot 2^r - 1 = (2^q)^k \cdot 2^r - 1 = (2^q - 1 + 1)^k \cdot 2^r - 1$$

$$= \sum_{j=1}^{k} \binom{k}{j} (2^q - 1)^j \cdot 2^r - 1 \equiv 2^r - 1 \pmod{p_1}.$$

It follows that p_1 divides $2^r - 1$, contradicting the minimality of q. Hence q divides n, and $1 < q < p_1$. Let p_2 be a prime divisor of q. Then p_2 is also a divisor of n, and $p_2 < p_1$. Repeating the argument, we construct an infinite sequence of prime divisors of n, $p_1 > p_2 > \cdots$, which is impossible. Hence the conclusion.

(1st W.L. Putnam Mathematical Competition, 1939)

830. The divisibility condition can be written as

$$k(ab + a + b) = a^2 + b^2 + 1,$$

where k is a positive integer. The small values of k are easy to solve. For example, $k = 1$ yields $ab + a + b = a^2 + b^2 + 1$, which is equivalent to $(a - b)^2 + (a - 1)^2 + (b - 1)^2 = 0$, whose only solution is $a = b = 1$. Also, for $k = 2$ we have $2ab + 2a + 2b = a^2 + b^2 + 1$. This can be rewritten either as $4a = (b - a - 1)^2$ or as $4b = (b - a + 1)^2$, showing that both a and b are perfect squares. Assuming that $a \leq b$, we see that $(b - a + 1) - (b - a - 1) = 2$, and hence a and b are consecutive squares. We obtain as an infinite family of solutions the pairs of consecutive perfect squares.

Now let us examine the case $k \geq 3$. This is where we apply Fermat's infinite descent method. Again we assume that $a \leq b$. A standard approach is to interpret the divisibility condition as a quadratic equation in b:

$$b^2 - k(a + 1)b + (a^2 - ka + 1) = 0.$$

Because one of the roots, namely b, is an integer, the other root must be an integer, too (the sum of the roots is $k(a+1)$). Thus we may substitute the pair (a, b) by the smaller pair (r, a), provided that $0 < r < a$.

Let us verify first that $0 < r$. Assume the contrary. Since $br = a^2 - ka + 1$, we must have $a^2 - ka + 1 \leq 0$. The equality case is impossible, since $a(k - a) = 1$ cannot hold if $k \geq 3$. If $a^2 - ka + 1 < 0$, the original divisibility condition implies $b(b - ak - k) = ak - a^2 - 1 > 0$, hence $b - ak - k > 0$. But then $b(b - ak - k) > (ak + k) \cdot 1 > ak - a^2 - 1$, a contradiction. This proves that r is positive. From the fact that $br = a^2 - ka + 1 < a^2$ and $b \geq a$, it follows that $r < a$. Successively, we obtain the sequence of pairs of solutions to the original problem $(a_1, b_1) = (a, b)$, $(a_2, b_2) = (r, a)$, (a_3, b_3), ..., with $a_i \leq b_i$ and $a_1 > a_2 > a_3 > \cdots$, $b_1 > b_2 > b_3 > \cdots$, which of course is impossible. This shows that the ratio of $a^2 + b^2 + 1$ to $ab + a + b$ cannot be greater than or equal to 3, and so the answer to the problem consists of the pair $(1, 1)$ together with all pairs of consecutive perfect squares.

(*Mathematics Magazine*)

831. We argue by contradiction: assuming the existence of one triple that does not satisfy the property from the statement, we construct an infinite decreasing sequence of such triples. Let (x_0, y_0, z_0) be a triple such that $x_0 y_0 - z_0^2 = 1$, but for which there do not exist nonnegative integers a, b, c, d such that $x_0 = a^2 + b^2$, $y_0 = c^2 + d^2$, and $z_0 = ac + bd$. We can assume that $x_0 \leq y_0$, and also $x_0 \geq 2$, for if $x_0 = 1$, then $x_0 = 1^2 + 0^2$, $y_0 = z_0^2 + 1^2$, and $z_0 = z_0 \cdot 1 + 0 \cdot 1$. We now want to construct a new triple (x_1, y_1, z_1) satisfying $x_1^2 y_1^2 - z_1^2 = 1$ such that $x_1 + y_1 + z_1 < x_0 + y_0 + z_0$. To this end, set $z_0 = x_0 + u$. Then

$$1 = x_0 y_0 - (x_0 + u)^2 = x_0 y_0 - x_0^2 - 2x_0 u + u^2$$
$$= x_0(y - x_0 - 2u) - u^2 = x_0(x_0 + y_0 - 2z_0) - (z_0 - x_0)^2.$$

A good candidate for the new triple is (x_1, y_1, z_1) with $x_1 = \min(x_0, x_0 + y_0 - 2z_0)$, $y_1 = \max(x_0, x_0 + y_0 - 2z_0)$, $z_1 = z_0 - x_0$. Note that $x_1 + y_1 + z_1 = x_0 + y_0 - z_0 < x_0 + y_0 + z_0$.

First, let us verify that x_1, y_1, z_1 are positive. From

$$z_0^2 = x_0 y_0 - 1 < x_0 y_0 \leq \left(\frac{x_0 + y_0}{2} \right)^2$$

we deduce that $x_0 + y_0 \geq 2z_0$, which means that $x_0 + y_0 - 2z_0 > 0$. It follows that both x_1 and y_1 are positive. Also,

$$z_0^2 = x_0 y_0 - 1 \geq x_0^2 - 1,$$

which implies $(z_0 - x_0)(z_0 + x_0) \geq -1$. Since $z_0 + x_0 \geq 3$, this can happen only if $z_0 \geq x_0$. Equality would yield $x_0(y_0 - x_0) = 1$, which cannot hold in view of the assumption $x_0 \geq 2$. Hence $z_1 = z_0 - x_0 > 0$. If for the new triple we could find nonnegative integers m, n, p, q such that

$$x_0 = m^2 + n^2, \quad x_0 + y_0 - 2z_0 = p^2 + q^2, \quad z_0 - x_0 = mp + nq.$$

In that case,

$$y_0 = p^2 + q^2 + 2z_0 - x_0 = p^2 + q^2 + 2mp + 2nq + m^2 + n^2 = (m+p)^2 + (n+q)^2$$

and

$$z_0 = m(m+p) + n(n+q),$$

contradicting our assumption.

We therefore can construct inductively an infinite sequence of triples of positive numbers (x_n, y_n, z_n), $n \geq 0$, none of which admits the representation from the statement, and such that $x_n + y_n + z_n > x_{n+1} + y_{n+1} + z_{n+1}$ for all n. This is of course impossible, and the claim is proved.

(Short list of the 20th International Mathematical Olympiad, 1978)

832. *First solution*: Choose k such that

$$\lfloor x \rfloor + \frac{k}{n} \leq x < \lfloor x \rfloor + \frac{k+1}{n}.$$

Then $\left\lfloor x + \frac{j}{n} \right\rfloor$ is equal to $\lfloor x \rfloor$ for $j = 0, 1, \ldots, n-k-1$, and to $\lfloor x \rfloor + 1$ for $x = n-k, \ldots, n-1$. It follows that the expression on the left is equal to $n\lfloor x \rfloor + k$. Also, $\lfloor nx \rfloor = n\lfloor x \rfloor + k$, which shows that the two sides of the identity are indeed equal.

Second solution: Define $f : \mathbb{R} \to \mathbb{N}$,

$$f(x) = \lfloor x \rfloor + \left\lfloor x + \frac{1}{n} \right\rfloor + \cdots + \left\lfloor x + \frac{n-1}{n} \right\rfloor - \lfloor nx \rfloor.$$

We have

$$f\left(x + \frac{1}{n}\right) = \left\lfloor x + \frac{1}{n} \right\rfloor + \cdots + \left\lfloor x + \frac{n-1}{n} \right\rfloor + \left\lfloor x + \frac{n}{n} \right\rfloor - \lfloor nx + 1 \rfloor = f(x).$$

Therefore, f is periodic, with period $\frac{1}{n}$. Also, since $f(x) = 0$ for $x \in \left[0, \frac{1}{n}\right)$, it follows that f is identically 0, and the identity is proved.

(Ch. Hermite)

833. Denote the sum in question by S_n. Observe that

$$S_n - S_{n-1} = \left\lfloor \frac{x}{n} \right\rfloor + \left\lfloor \frac{x+1}{n} \right\rfloor + \cdots + \left\lfloor \frac{x+n-1}{n} \right\rfloor$$

$$= \left\lfloor \frac{x}{n} \right\rfloor + \left\lfloor \frac{x}{n} + \frac{1}{n} \right\rfloor + \cdots + \left\lfloor \frac{x}{n} + \frac{n-1}{n} \right\rfloor,$$

and, according to Hermite's identity,

$$S_n - S_{n-1} = \left\lfloor x\frac{x}{n} \right\rfloor = \lfloor x \rfloor.$$

Because $S_1 = \lfloor x \rfloor$, it follows that $S_n = n\lfloor x \rfloor$ for all $n \geq 1$.

(S. Savchev, T. Andreescu, *Mathematical Miniatures*, MAA, 2002)

834. Let $t = \lfloor x \rfloor^3 + 4\lfloor x \rfloor$. The equation becomes

$$t^2 + 3t = \lfloor y \rfloor^2.$$

The number t is an integer. If $t > 1$, then $(t + 1)^2 < t^2 + 3t = \lfloor y \rfloor^2 < (t + 2)^2$ which is impossible. It $t \leq 1$, then $\lfloor x \rfloor^2 + 4\lfloor x \rfloor \leq 1$, which only leaves the posibilities $\lfloor x \rfloor \in \{-4, -3, -2, -1, 0\}$. Checking cases we find that the set of solutions is

$$[-4, -2) \times [0, 1) \cup [-2, -1) \cup ([-2, -1) \cup [2, 3)) \cup [-1, 1) \times [0, 1).$$

835. Set $k = \lfloor \sqrt{n} \rfloor$. We want to prove that

$$k = \left\lfloor \sqrt{n} + \frac{1}{\sqrt{n} + \sqrt{n+2}} \right\rfloor,$$

which amounts to proving the double inequality

$$k \leq \sqrt{n} + \frac{1}{\sqrt{n} + \sqrt{n+2}} < k + 1.$$

The inequality on the left is obvious. For the other, note that $k \leq \sqrt{n} < k + 1$, which implies $k^2 \leq n \leq (k + 1)^2 - 1$. Using this we can write

$$\sqrt{n} + \frac{1}{\sqrt{n} + \sqrt{n+2}} = \sqrt{n} + \frac{\sqrt{n+2} - \sqrt{n}}{2} = \frac{\sqrt{n+2} + \sqrt{n}}{2}$$

$$\leq \frac{\sqrt{(k+1)^2 + 1} + \sqrt{(k+1)^2 - 1}}{2} < k + 1.$$

The last inequality in this sequence needs to be explained. Rewriting it as

$$\frac{1}{2}\sqrt{(k+1)^2 + 1} + \frac{1}{2}\sqrt{(k+1)^2 - 1} < \sqrt{(k+1)^2},$$

we recognize Jensen's inequality for the (strictly) concave function $f(x) = \sqrt{x}$. This completes the solution.

(Gh. Eckstein)

836. We will show that this equality holds for all $x \geq 1$. Let n be the positive integer for which

$$n^4 \leq x < (n+1)^4.$$

Then $n^2 \leq \sqrt{x} < (n+1)^2$, and so on the one hand

$$n \leq \sqrt{\sqrt{x}} < n+1$$

and therefore

$$\lfloor \sqrt{\sqrt{x}} \rfloor = n,$$

and on the other hand

$$n^2 \leq \lfloor \sqrt{x} \rfloor \leq (n+1)^2,$$

and therefore

$$n \leq \sqrt{\lfloor \sqrt{x} \rfloor} \leq n+1.$$

This proves that

$$\left\lfloor \sqrt{\lfloor \sqrt{x} \rfloor} \right\rfloor = n,$$

and we are done.

(*Kvant (Quantum)*, proposed by V. Prasolov)

837. We apply the identity proved in the introduction to the function $f : [1, n] \rightarrow [1, \sqrt{n}]$, $f(x) = \sqrt{x}$. Because $n(G_f) = \lfloor \sqrt{n} \rfloor$, the identity reads

$$\sum_{k=1}^{n} \lfloor \sqrt{k} \rfloor + \sum_{k=1}^{\lfloor \sqrt{n} \rfloor} \lfloor k^2 \rfloor - \lfloor \sqrt{n} \rfloor = n \lfloor n \rfloor.$$

Hence the desired formula is

$$\sum_{k=1}^{n} \lfloor \sqrt{k} \rfloor = (n+1)a - \frac{a(a+1)(2a+1)}{6}.$$

(S. Korean Mathematical Olympiad, 1997)

838. The function $f : \left[1, \frac{n(n+1)}{2}\right] \rightarrow \mathbb{R}$,

$$f(x) = \frac{-1 + \sqrt{1 + 8x}}{2},$$

is, in fact, the inverse of the increasing bijective function $g : [1, n] \rightarrow \left[1, \frac{n(n+1)}{2}\right]$,

$$g(x) = \frac{x(x+1)}{2}.$$

We apply the identity proved in the introduction to g in order to obtain

$$\sum_{k=1}^{n}\left\lfloor\frac{k(k+1)}{2}\right\rfloor + \sum_{k=1}^{\frac{n(n+1)}{2}}\left\lfloor\frac{-1+\sqrt{1+8k}}{2}\right\rfloor - n = \frac{n^2(n+1)}{2}.$$

Note that $\frac{k(k+1)}{2}$ is an integer for all k, and so

$$\sum_{k=1}^{n}\left\lfloor\frac{k(k+1)}{2}\right\rfloor = \sum_{k=1}^{n}\frac{k(k+1)}{2} = \frac{1}{2}\sum_{k=1}^{n}(k^2+k) = \frac{n(n+1)}{4} + \frac{n(n+1)(2n+1)}{12}$$

$$= \frac{n(n+1)(n+2)}{6}.$$

The identity follows.

839. The property is clearly satisfied if $a = b$ or if $ab = 0$. Let us show that if neither of these is true, and a and b satisfy the property from the statement, then a and b are integers.

First, note that for an integer x, $\lfloor 2x\rfloor = 2\lfloor x\rfloor$ if $x - \lfloor x\rfloor \in \left[0, \frac{1}{2}\right)$ and $\lfloor 2x\rfloor = 2\lfloor x\rfloor + 1$ if $x - \lfloor x\rfloor \in \left[\frac{1}{2}, 1\right)$. Let us see which of the two holds for a and b. If $\lfloor 2a\rfloor = 2\lfloor a\rfloor + 1$, then

$$a\lfloor 2b\rfloor = b\lfloor 2a\rfloor = 2\lfloor a\rfloor b + b = 2a\lfloor b\rfloor + b.$$

This implies $\lfloor 2b\rfloor = 2\lfloor b\rfloor + \frac{b}{a}$, and so $\frac{b}{a}$ is either 0 or 1, which contradicts our working hypothesis. Therefore, $\lfloor 2a\rfloor = 2\lfloor a\rfloor$ and also $\lfloor 2b\rfloor = 2\lfloor b\rfloor$. This means that the fractional parts of both a and b are less than $\frac{1}{2}$. With this as the base case, we will prove by induction that $\lfloor 2^m a\rfloor = 2^m\lfloor a\rfloor$ and $\lfloor 2^m b\rfloor = 2^m\lfloor b\rfloor$ for all $m \geq 1$.

The inductive step works as follows. Assume that the property is true for m and let us prove it for $m + 1$. If $\lfloor 2^{m+1}a\rfloor = 2\lfloor 2^m a\rfloor$, we are done. If $\lfloor 2^{m+1}a\rfloor = 2\lfloor 2^m a\rfloor + 1$, then

$$a\lfloor 2^{m+1}b\rfloor = b\lfloor 2^{m+1}a\rfloor = 2\lfloor 2^m a\rfloor b + b = 2^{m+1}\lfloor a\rfloor b + b = 2^{m+1}a\lfloor b\rfloor + 1.$$

As before, we deduce that $\lfloor 2^{m+1}b\rfloor = 2^{m+1}\lfloor b\rfloor + \frac{b}{a}$. Again this is an impossibility. Hence the only possibility is that $\lfloor 2^{m+1}a\rfloor = 2^{m+1}\lfloor a\rfloor$ and by a similar argument $\lfloor 2^{m+1}b\rfloor = 2^{m+1}\lfloor b\rfloor$. This completes the induction.

From $\lfloor 2^m a\rfloor = 2^m\lfloor a\rfloor$ and $\lfloor 2^m b\rfloor = 2^m\lfloor b\rfloor$ we deduce that the fractional parts of a and b are less than $\frac{1}{2^m}$. Taking $m \to \infty$, we conclude that a and b are integers.

(Short list of the 39th International Mathematical Olympiad, 1998)

840. We compute

$$\frac{x}{\lfloor x\rfloor} + \frac{x}{\{x\}} = \frac{\lfloor x\rfloor + \{x\}}{\lfloor x\rfloor} + \lfloor x\rfloor + \frac{\{x\}}{\{x\}} = 2 + \frac{\{x\}}{\lfloor x\rfloor} + \frac{\lfloor x\rfloor}{\{x\}} > 4.$$

Hence

$$\frac{1}{\lfloor x \rfloor} + \frac{1}{\{x\}} > \frac{4}{x} > \frac{7}{2x}.$$

(*Gazeta Matematică (Mathematics Gazette, Bucharest)*, proposed by R. Ghiţă and I. Ghiţă)

841. Ignoring the "brackets" we have

$$\frac{p}{q} + \frac{2p}{q} + \cdots + \frac{(q-1)p}{q} = \frac{(q-1)p}{2}.$$

The difference between kp/q and $\lfloor kp/q \rfloor$ is r/q, where r is the remainder obtained on dividing kp by q. Since p and q are coprime, $p, 2p, \ldots, (q-1)p$ form a complete set of residues modulo q. So for $k = 1, 2, \ldots, q-1$, the numbers $k/p - \lfloor kp/q \rfloor$ are a permutation of $1, 2, \ldots, q-1$. Therefore,

$$\sum_{k=1}^{q-1} \left\lfloor \frac{kp}{q} \right\rfloor = \sum_{k=1}^{q-1} \frac{kp}{q} - \sum_{k=1}^{q-1} \frac{k}{q} = \frac{(q-1)p}{2} - \frac{q-1}{2} = \frac{(p-1)(q-1)}{2},$$

and the reciprocity formula follows.

Remark. This identity can be used to prove Gauss' quadratic reciprocity law.

842. The function

$$f(x) = \lfloor nx \rfloor - \frac{\lfloor x \rfloor}{1} - \frac{\lfloor 2x \rfloor}{2} - \frac{\lfloor 3x \rfloor}{3} - \cdots - \frac{\lfloor nx \rfloor}{n}$$

satisfies $f(x) = f(x+1)$ for all x and $f(0) = 0$. Moreover, the function is constant on subintervals of $[0, 1)$ that do not contain numbers of the form p/q, $2 \le q \le n$ and $1 \le p \le q-1$. Thus it suffices to verify the inequality for $x = p/q$, where p and q are coprime positive integers, $2 \le q \le n$, $1 \le p \le q-1$. Subtracting the inequality from

$$nx = \frac{x}{1} + \frac{2x}{2} + \cdots + \frac{nx}{n},$$

we obtain the equivalent inequality for the fractional part $\{\cdot\}$ ($\{x\} = x - \lfloor x \rfloor$),

$$\{nx\} \le \frac{\{x\}}{1} + \frac{\{2x\}}{2} + \frac{\{3x\}}{3} + \cdots + \frac{\{nx\}}{n},$$

which we prove for the particular values of x mentioned above. If r_k is the remainder obtained on dividing kp by q, then $\{kx\} = \frac{r_k}{q}$, and so the inequality can be written as

$$\frac{r_n}{q} \le \frac{r_1/q}{1} + \frac{r_2/q}{2} + \frac{r_3/q}{3} + \cdots + \frac{r_n/q}{n},$$

or

$$r_n \le \frac{r_1}{1} + \frac{r_2}{2} + \frac{r_3}{3} + \cdots + \frac{r_n}{n}.$$

Truncate the sum on the right to the $(q-1)$st term. Since p and q are coprime, the numbers $r_1, r_2, \ldots, r_{q-1}$ are a permutation of $1, 2, \ldots, q-1$. Applying this fact and the AM-GM inequality, we obtain

$$\frac{r_1}{1} + \frac{r_2}{2} + \frac{r_3}{3} + \cdots + \frac{r_{q-1}}{q-1} \geq (q-1)\left(\frac{r_1}{1} \cdot \frac{r_2}{2} \cdot \frac{r_3}{3} \cdots \frac{r_{q-1}}{q-1}\right)^{1/(q-1)} = (q-1) \geq r_n.$$

This proves the (weaker) inequality

$$\frac{r_1}{1} + \frac{r_2}{2} + \frac{r_3}{3} + \cdots + \frac{r_n}{n} \geq r_n,$$

and consequently the inequality from the statement of the problem.

    (O.P. Lossers)

843. Let x_1 be the golden ratio, i.e., the (unique) positive root of the equation $x^2 - x - 1 = 0$. We claim that the following identity holds:

$$\left\lfloor x_1 \left\lfloor x_1 n + \frac{1}{2} \right\rfloor + \frac{1}{2} \right\rfloor = \left\lfloor n x_1 + \frac{1}{2} \right\rfloor + n.$$

If this were so, then the function $f(n) = \lfloor x_1 n + \frac{1}{2} \rfloor$ would satisfy the functional equation. Also, since $\alpha = \frac{1+\sqrt{5}}{2} > 1$, f would be strictly increasing, and so it would provide an example of a function that satisfies the conditions from the statement.

    To prove the claim, we only need to show that

$$\left\lfloor (x_1 - 1)\left\lfloor x_1 n + \frac{1}{2}\right\rfloor + \frac{1}{2}\right\rfloor = n.$$

We have

$$\left\lfloor (x_1 - 1)\left\lfloor x_1 n + \frac{1}{2}\right\rfloor + \frac{1}{2}\right\rfloor \leq \left\lfloor (x_1 - 1)\left(x_1 n + \frac{1}{2}\right) + \frac{1}{2}\right\rfloor$$

$$= \left\lfloor x_1 n + n - x_1 n + \frac{x_1}{2}\right\rfloor = n.$$

Also,

$$n = \left\lfloor n + \frac{2 - x_1}{2}\right\rfloor \leq \left\lfloor (x_1 - 1)\left(x_1 n - \frac{1}{2}\right) + \frac{1}{2}\right\rfloor \leq \left\lfloor (x_1 - 1)\left\lfloor x_1 n + \frac{1}{2}\right\rfloor + \frac{1}{2}\right\rfloor.$$

This proves the claim and completes the solution.

    (34th International Mathematical Olympiad, 1993)

844. Suppose first that the pair (f, g) is not unique and that there is a second pair of functions (f', g') subject to the same conditions. Write the sets $\{f(n), n \geq 1\} \cup \{g(n), n \geq 1\}$, respectively, $\{f'(n), n \geq 1\} \cup \{g'(n), n \geq 1\}$, as increasing sequences, and let n_0 be the smallest number where a difference occurs in the values of $f(n)$ and $g(n)$ versus $f'(n)$ and $g'(n)$. Because the pairs of functions exhaust the positive integers, either $f(n_1) = g'(n_0)$ or $f'(n_0) = g(n_1)$. The situations are symmetric, so let us assume that the first occurs. Then

$$f(n_1) = g'(n_0) = f'(f'(kn_0)) + 1 = f(f(kn_0)) + 1 = g(n_0).$$

We stress that the third equality occurs because $f'(kn_0)$ occurs earlier in the sequence (since it is smaller than $f(n_1)$), so it is equal to $f(kn_0)$, and the same is true for $f'(f'(kn_0))$. But the equality $f(n_1) = g(n_0)$ is ruled out by the hypothesis, which shows that our assumption was false. Hence the pair (f, g) is unique.

Inspired by the previous problems we take α to be the positive root of the quadratic equation $kx^2 - kx - 1 = 0$, and set $\beta = k\alpha^2$. Then $\frac{1}{\alpha} + \frac{1}{\beta} = 1$, and because k is an integer, both α and β are irrational. By Beatty's theorem the sequences $f(n) = \lfloor \alpha n \rfloor$ and $g(n) = \lfloor \beta n \rfloor$ are strictly increasing and define a partition of the positive integers into two disjoint sets. Let us show that f and g satisfy the functional equation from the statement.

Because $k\alpha^2 = k\alpha + 1$,

$$g(n) = \lfloor k\alpha^2 n \rfloor = \lfloor (k\alpha + 1)n \rfloor = \lfloor k\alpha n \rfloor + n,$$

and we are left to prove that $\lfloor \alpha kn \rfloor + n = \lfloor \alpha \lfloor \alpha kn \rfloor \rfloor + 1$, the latter being $f(f(kn)) + 1$. Reduce this further to

$$\lfloor (\alpha - 1)\lfloor \alpha kn \rfloor \rfloor = n - 1.$$

Since $(\alpha - 1)\alpha k = 1$ and α is irrational, $\lfloor (\alpha - 1)\lfloor \alpha kn \rfloor \rfloor < n$. Also,

$$(\alpha - 1)\lfloor \alpha kn \rfloor > (\alpha - 1)(\alpha kn - 1) = (\alpha^2 k - \alpha k)n + 1 - \alpha = n + 1 - \alpha > n - 1,$$

since $\alpha < 2$ (which can be checked by solving the quadratic equation that defines α). Hence

$$g(n) = \lfloor \alpha kn \rfloor + n = \lfloor \alpha \lfloor \alpha n \rfloor \rfloor + 1 = f(f(kn)) + 1,$$

and the problem is solved.

Remark. The case $k = 1$ was given at the 20th International Mathematical Olympiad, 1978; the idea of the solution was taken from I.J. Schoenberg, *Mathematical Time Exposures* (MAA, 1982).

845. If we multiplied the fraction by 8, we would still get an integer. Note that

$$8\frac{n^3 - 3n^2 + 4}{2n - 1} = 4n^2 - 10n - 5 + \frac{27}{2n - 1}.$$

Hence $2n - 1$ must divide 27. This happens only when $2n - 1 = \pm 1, \pm 3, \pm 9, \pm 27$, that is, when $n = -13, -4, -1, 1, 2, 5, 14$. An easy check shows that for each of these numbers the original fraction is an integer.

846. The factor to be erased is $50!$. Indeed, using the equality $(2k)! = (2k - 1)! \cdot 2k$, we see that

$$P = (1!)^2 \cdot 2 \cdot (3!)^2 \cdot 4 \cdot (5!)^2 \cdot 6 \cdots (99!)^2 \cdot 100 = (1! \cdot 3! \cdot 5! \cdots 99!)^2 \cdot 2 \cdot 4 \cdot 6 \cdots 100$$

$$= (1! \cdot 3! \cdot 5! \cdots 99!)^2 \cdot 2^{50} \cdot 50! = (1! \cdot 3! \cdot 5! \cdots 99! \cdot 2^{25})^2 \cdot 50!.$$

It is noteworthy that P itself is not a perfect square, since $50!$ is not, the latter because 47 appears to the first power in $50!$.

(First stage of the Moscow Mathematical Olympiad, 1995-1996)

847. For any integer m, we have $\gcd(a_m, a_{2m}) = \gcd(2m, m) = m$, and so m divides a_m. It follows that for any other integer n, m divides a_n if and only if it divides $\gcd(a_m, a_n) = \gcd(m, n)$. Hence a_n has exactly the same divisors as n, so it must equal n, for all n.

(Russian Mathematical Olympiad, 1995)

848. Because $\gcd(a, b)$ divides both a and b, we can factor $n^{\gcd(a,b)} - 1$ from both $n^a - 1$ and $n^b - 1$. Therefore, $n^{\gcd(a,b)} - 1$ divides $\gcd(n^a - 1, n^b - 1)$.

On the other hand, using Euclid's algorithm we can find positive integers x and y such that $ax - by = \gcd(a, b)$. Then $n^a - 1$ divides $n^{ax} - 1$ and $n^b - 1$ divides $n^{by} - 1$. In order to combine these two, we use the equality

$$n^{by}(n^{\gcd(a,b)} - 1) = (n^{ax} - 1) - (n^{by} - 1).$$

Note that $\gcd(n^a - 1, n^b - 1)$ divides the right-hand side, and has no common factor with n^{by}. It therefore must divide $n^{\gcd(a,b)} - 1$. We conclude that $n^{\gcd(a,b)-1} = \gcd(n^a - 1, n^b - 1)$, as desired.

849. We use the particular case $n = 2$ of the previous problem as a lemma. To obtain the negative signs we incorporate $2^a + 1$ and $2^b + 1$ into $2^{2a} - 1$ and $2^{2b} - 1$, then apply the lemma to these two numbers. We have

$$2^{\gcd(2a,2b)} - 1 = \gcd(2^{2a} - 1, 2^{2b} - 1) = \gcd((2^a - 1)(2^a + 1), (2^b - 1)(2^b + 1)).$$

Because $2^a - 1$ and $2^a + 1$ are coprime, and so are $2^b - 1$ and $2^b + 1$, this is further equal to

$$\gcd(2^a - 1, 2^b - 1) \cdot \gcd(2^a - 1, 2^b + 1) \cdot \gcd(2^a + 1, 2^b - 1) \cdot \gcd(2^a + 1, 2^b + 1).$$

It follows that $\gcd(2^a + 1, 2^b + 1)$ divides $2^{\gcd(2a,2b)} - 1$. Of course,

$$2^{\gcd(2a,2b)} - 1 = 2^{2\gcd(a,b)} - 1 = (2^{\gcd(a,b)} - 1)(2^{\gcd(a,b)} + 1),$$

so $\gcd(2^a + 1, 2^b + 1)$ divides the product $(2^{\gcd(a,b)} - 1)(2^{\gcd(a,b)} + 1)$. Again because $\gcd(2^a + 1, 2^a - 1) = \gcd(2^b + 1, 2^b - 1) = 1$, it follows that $\gcd(2^a + 1, 2^b + 1)$ and $2^{\gcd(a,b)} - 1$ do not have common factors. We conclude that $\gcd(2^a + 1, 2^b + 1)$ divides $2^{\gcd(a,b)} + 1$.

850. We compute

$$a_2 = (k + 1)^2 - k(k + 1) + k = (k + 1) + k = a_1 + k,$$

$$a_3 = a_2(a_2 - k) + k = a_2 a_1 + k,$$

$$a_4 = a_3(a_3 - k) + k = a_3 a_2 a_1 + k,$$

and in general if $a_n = a_{n-1} a_{n-2} \cdots a_1 + k$, then

$$a_{n+1} = a_n(a_n - k) + k = a_n a_{n-1} a_{n-2} \cdots a_1 + k.$$

Therefore, $a_n - k$ is divisible by a_m for $1 \le m < n$. On the other hand, inductively we obtain that a_m and k are relatively prime. It follows that a_m and $a_n = (a_n - k) + k$ are also relatively prime. This completes the solution.

(Polish Mathematical Olympiad, 2002)

851. By hypothesis, all coefficients of the quadratic polynomial

$$P(x) = (x+a)(x+b)(x+c) - (x-d)(x-e)(x-f)$$
$$= (a+b+c+d+e+f)x^2 + (ab+bc+ca-de-ef-fd)x + (abc+def)$$

are divisible by $S = a+b+c+d+e+f$. Evaluating $P(x)$ at d, we see that $P(d) = (a+d)(b+d)(c+d)$ is a multiple of S. This readily implies that S is composite because each of $a+d$, $b+d$, and $c+d$ is less than S.

(Short list of 46th International Mathematical Olympiad, 2005)

852. The polynomial

$$P(n) = n(n-1)^4 + 1 = n^5 - 4n^4 + 6n^3 - 4n^2 + n + 1$$

does not have integer zeros, so we should be able to factor it as a product a quadratic and a cubic polynomial. This means that

$$P(n) = (n^2 + an + 1)(n^3 + bn^2 + cn + 1),$$

for some integers a, b, c. Identifying coefficients, we must have

$$a + b = -4,$$

$$c + ab + 1 = 6,$$

$$b + ac + 1 = -4,$$

$$a + c = 1.$$

From the first and last equations, we obtain $b - c = -5$, and from the second and the third, $(b-c)(a-1) = 10$. It follows that $a - 1 = -2$; hence $a = -1$, $b = -4 + 1 = -3$, $c = 1 + 1 = 2$. Therefore,

$$n(n-1)^4 + 1 = (n^2 - n + 1)(n^3 - 3n^2 + 2n + 1),$$

a product of integers greater than 1.

(T. Andreescu)

853. Setting $n = 0$ in (i) gives

$$f(1)^2 = f(0)^2 + 6f(0) + 1 = (f(0) + 3)^2 - 8.$$

Hence
$$(f(0) + 3)^2 - f(1)^2 = (f(0) + 3 + f(1))(f(0) + 3 - f(1)) = 4 \times 2.$$

The only possibility is $f(0) + 3 + f(1) = 4$ and $f(0) + 3 - f(1) = 2$. It follows that $f(0) = 0$ and $f(1) = 1$.

In general,
$$(f(2n + 1) - f(2n))(f(2n + 1) + f(2n)) = 6f(n) + 1.$$

We claim that $f(2n+1) - f(2n) = 1$ and $f(2n+1) + f(2n) = 6f(n) + 1$. To prove our claim, let $f(2n + 1) - f(2n) = d$. Then $f(2n + 1) + f(2n) = d + 2f(2n)$. Multiplying, we obtain

$$6f(n) + 1 = d(d + 2f(2n)) \geq d(d + 2f(n)),$$

where the inequality follows from condition (ii). Moving everything to one side, we obtain the inequality
$$d^2 + (2d - 6)f(n) - 1 \leq 0,$$

which can hold only if $d \leq 3$. The cases $d = 2$ and $d = 3$ cannot hold, because d divides $6f(n) + 1$. Hence $d = 1$, and the claim is proved. From it we deduce that f is computed recursively by the rule
$$f(2n + 1) = 3f(n) + 1,$$

$$f(2n) = 3f(n).$$

At this moment it is not hard to guess the explicit formula for f; it associates to a number in binary representation the number with the same digits but read in ternary representation. For example, $f(5) = f(101_2) = 101_3 = 10$. The formula is easily proved by induction.

854. It is better to rephrase the problem and prove that there are infinitely many prime numbers of the form $4m - 1$. Euclid's proof of the existence of infinitely many primes, presented in the first section of the book, works in this situation, too. Assume that there exist only finitely many prime numbers of the form $4m - 1$, and let these numbers be $p_1, p_2, \ldots, p_n$. Consider $M = 4p_1p_2p_3 \cdots p_n - 1$. This number is of the form $4m - 1$, so it has a prime divisor of the same form, for otherwise M would be a product of numbers of the form $4m + 1$ and itself would be of the form $4m + 1$. But M is not divisible by any of the primes $p_1, p_2, \ldots, p_n$ so it must be divisible by some other prime of the form $4m - 1$. This contradicts our assumption that $p_1, p_2, \ldots, p_n$ are all primes of the form $4m - 1$, showing that it was false. We conclude that there exist infinitely many prime numbers of the form $4m + 3$, m an integer.

Remark. A theorem of Dirichlet shows that for any two coprime numbers a and b, the arithmetic progression $an + b$, $n \geq 0$ contains infinitely many prime terms.

855. We have

$$
\frac{m}{n} = \frac{1}{1} - \frac{1}{2} + \frac{1}{3} - \frac{1}{4} + \cdots + \frac{1}{2k-1} - \frac{1}{2k}
$$

$$
= 1 + \frac{1}{2} + \frac{1}{3} + \cdots + \frac{1}{2k} - 2\left(\frac{1}{2} + \frac{1}{4} + \cdots + \frac{1}{2k}\right)
$$

$$
= 1 + \frac{1}{2} + \frac{1}{3} + \cdots + \frac{1}{2k} - \left(1 + \frac{1}{2} + \cdots + \frac{1}{k}\right)
$$

$$
= \frac{1}{k+1} + \frac{1}{k+2} + \cdots + \frac{1}{2k-1} + \frac{1}{2k}
$$

$$
= \left(\frac{1}{k+1} + \frac{1}{2k}\right) + \left(\frac{1}{k+2} + \frac{1}{2k-1}\right) + \cdots
$$

$$
= \frac{3k+1}{(k+1)2k} + \frac{3k+1}{(k+2)(2k-1)} + \cdots
$$

For a proof by induction of this equality, see Problem 12. It follows that $m(2k)! = n(3k+1)q$ for some positive integer q; hence $p = 3k+1$ divides $m(2k)!$. But p is a prime greater than $2k$, so it is coprime to $(2k)!$. Thus p divides m, and we are done.

(*Mathematical Reflections*, proposed by T. Andreescu)

856. The numbers x and y have the same prime factors,

$$
x = \prod_{i=1}^{k} p_i^{\alpha_i}, \quad y = \prod_{i=1}^{k} p_i^{\beta_i}.
$$

The equality from the statement can be written as

$$
\prod_{i=1}^{k} p_i^{\alpha_i(x+y)} = \prod_{i=1}^{k} p_i^{\beta_i(y-x)};
$$

hence $\alpha_i(y+x) = \beta_i(y-x)$ for $i = 1, 2, \ldots, k$. From here we deduce that $\alpha_i < \beta_i$, $i = 1, 2, \ldots, k$, and therefore x divides y. Writing $y = zx$, the equation becomes $x^{x(z+1)} = (xz)^{x(z-1)}$, which implies $x^2 = z^{z-1}$ and then $y^2 = (xz)^2 = z^{z+1}$. A power is a perfect square if either the base is itself a perfect square or if the exponent is even. For $z = t^2$, $t \geq 1$, we have $x = t^{t^2-1}$, $y = t^{t^2+1}$, which is one family of solutions. For $z - 1 = 2s$, $s \geq 0$, we obtain the second family of solutions $x = (2s+1)^s$, $y = (2s+1)^{s+1}$.

(Austrian-Polish Mathematics Competition, 1999, communicated by I. Cucurezeanu)

857. If n is even, then we can write it as $(2n) - n$. If n is odd, let p be the smallest odd prime that does not divide n. Then write $n = (pn) - ((p-1)n)$. The number pn contains exactly one more prime factor than n. As for $(p-1)n$, it is divisible by 2 because $p-1$ is even, while its odd factors are less than p, so they all divide n. Therefore, $(p-1)n$ also contains exactly one more prime factor than n, and therefore pn and $(p-1)n$ have the same number of prime factors.

(Russian Mathematical Olympiad, 1999)

858. The only numbers that do not have this property are the products of two distinct primes.

Let n be the number in question. If $n = pq$ with p, q primes and $p \neq q$, then any cycle formed by p, q, pq will have p and q next to each other. This rules out numbers of this form.

For any other number $n = p_1^{\alpha_1} p_2^{\alpha_2} \cdots p_k^{\alpha_k}$, with $k \geq 1$, $\alpha_i \geq 1$ for $i = 1, 2, \ldots, k$ and $\alpha_1 + \alpha_2 \geq 3$ if $k = 2$, arrange the divisors of n around the circle according to the following algorithm. First, we place $p_1, p_2, \ldots, p_k$ arranged clockwise around the circle in increasing order of their indices. Second, we place $p_i p_{i+1}$ between p_i and p_{i+1} for $i = 1, \ldots, k-1$. Third, we place n between p_k and p_1. Note that at this point every pair of consecutive numbers has a common factor and each prime p_i occurs as a common factor for some pair of adjacent numbers. Now for any remaining divisor of n we choose a prime p_i that divides it and place it between p_i and one of its neighbors.

(USA Mathematical Olympiad, 2005, proposed by Z. Feng)

859. The answer is negative. To motivate our claim, assume the contrary, and let $a_0, a_1, \ldots, a_{1995} = a_0$ be the integers. Then for $i = 1, 2, \ldots, 1995$, the ratio a_{k-1}/a_k is either a prime, or the reciprocal of a prime. Suppose the former happens m times and the latter $1995 - m$ times. The product of all these ratios is $a_0/a_{1995} = 1$, which means that the product of some m primes equals the product of some $1995 - m$ primes. This can happen only when the primes are the same (by unique factorization), and in particular they must be in the same number on both sides. But the equality $m = 1995 - m$ is impossible, since 1995 is odd, a contradiction. This proves our claim.

(Russian Mathematical Olympiad, 1995)

860. *First solution*: The cases $p = 2, 3, 5$ are done as before. Let $p \geq 7$. The numbers p, $2p, \ldots, 9999999999p$ have distinct terminating ten-digit sequences. Indeed, the difference $mp - np = (m-n)p$ is not divisible by 10^{10}, since p is relatively prime to 10 and $m - n < 10^{10}$. There are $10^{10} - 1$ ten-digit terminating sequences, so all possible combinations of digits should occur. Many of these sequences consist of distinct digits, providing solutions to the problem.

Second solution: The statement is true for $p = 2$ and $p = 5$. Suppose that $p \neq 2, 5$. Then p is relatively prime to 10. From Fermat's little theorem, $10^{p-1} \equiv 1 \pmod{p}$ and hence $10^{k(p-1)} \equiv 1 \pmod{p}$ for all positive integers k. Let a be a 10-digit number with distinct digits, and let $a \equiv n \pmod{p}$, with $0 \leq n \leq p - 1$. Since $p \geq 3$, $10^{6(p-1)} > 10^{10}$. Therefore,

$$N_a = 10^{(p-n+5)(p-1)} + \cdots + 10^{6(p-1)} + a \equiv 1 + \cdots + 1 + n \equiv 0 \pmod{p}.$$

For all positive integers k, the numbers of the form

$$10^{10}kp + N_a,$$

end in a and are divisible by p.

(Proposed by T. Andreescu for the 41st International Mathematical Olympiad, 2000, first solution by G. Galperin, second solution by Z. Feng)

861. The case $p = 2$ is easy, so assume that p is an odd prime. Note that if $p^2 = a^2 + 2b^2$, then $2b^2 = (p-a)(p+a)$. In particular, a is odd. Also, a is too small to be divisible by p. Hence $\gcd(p-a, p+a) = \gcd(p-a, 2p) = 2$. By changing the sign of a we may assume that $p - a$ is not divisible by 4, and so we must have $|p+a| = m^2$ and $|p-a| = 2n^2$ for some integers m and n.

Because $|a| < p$, both $p+a$ and $p-a$ are actually positive, so $p+a = m^2$ and $p-a = 2n^2$. We obtain $2p = m^2 + 2n^2$. This can happen only if m is even, in which case $p = n^2 + 2\left(\frac{m}{2}\right)^2$, as desired.

(Romanian Mathematical Olympiad, 1997)

862. Note that if d is a divisor of n, then so is $\frac{n}{d}$. So the sum s is given by

$$s = \sum_{i=1}^{k-1} d_i d_{i+1} = n^2 \sum_{i=1}^{k-1} \frac{1}{d_i d_{i+1}} \leq n^2 \sum_{i=1}^{k-1} \left(\frac{1}{d_i} - \frac{1}{d_{i+1}}\right) < \frac{n^2}{d_1} = n^2.$$

For the second part, note also that $d_2 = p$, $d_{k-1} = \frac{n}{p}$, where p is the least prime divisor of n. If $n = p$, then $k = 2$, and $s = p$, which divides n^2. If n is composite, then $k > 2$, and $s > d_{k-1}d_k = \frac{n^2}{p}$. If such an s were a divisor of n^2, then $\frac{n^2}{s}$ would also be a divisor of n^2. But $1 < \frac{n^2}{s} < p$, which is impossible, because p is the least prime divisor of n^2. Hence the given sum is a divisor of n^2 if and only if n is a prime.

(43rd International Mathematical Olympiad, 2002, proposed by M. Manea (Romania))

863. We look instead at composite odd positive numbers. Each such number can be written as $(2a + 3)(2b + 3)$, for a and b nonnegative integers. In fact, n is composite if and only if it can be written this way. We only need to write this product as a difference of two squares. And indeed,

$$(2a + 3)(2b + 3) = (a + b + 3)^2 - (a - b)^2.$$

Thus we can choose $f(a, b) = (a + b + 3)^2$ and $g(a, b) = (a - b)^2$.

(Nea Mărin)

864. Arguing by contradiction, assume that there is some k, $0 \leq k \leq n-2$, such that $k^2 + k + n$ is not prime. Choose s to be the smallest number with this property, and let p be the smallest prime divisor of $s^2 + s + n$. First, let us notice that p is rather small, in the sense that $p \leq 2s$. For if $p \geq 2s + 1$, one has

$$s^2 + s + n \geq p^2 \geq (2s + 1)^2 = s^2 + s + 3s^2 + 3s + 1 \geq s^2 + s + n + 3s + 1$$
$$> s^2 + s + n,$$

which is so because $s > \sqrt{\frac{n}{3}}$. This is clearly impossible, which proves $p \leq 2s$.

It follows that either $p = s - k$ or $p = s + k + 1$ for some $0 \leq k \leq s - 1$. But then for this k,

$$s^2 + s + n - k^2 - k - n = (s - k)(s + k + 1).$$

Because p divides $s^2 + s + n$ and the product $(s - k)(s + k + 1)$, it must also divide $k^2 + k + n$. Now, this number cannot be equal to p, because $s - k < n - k < k^2 + k + n$ and $s + k + 1 < n - 1 + k + 1 < k^2 + k + n$. It follows that the number $k^2 + k + n$ is composite, contradicting the minimality of s. Hence the conclusion.

Remark. Euler noticed that 41 has the property that $k^2 + k + 41$ is a prime number for all $0 \leq k \leq 39$. Yet $40^2 + 40 + 41 = 41^2$ is not prime!

865. There are clearly more 2's than 5's in the prime factorization of $n!$, so it suffices to solve the equation

$$\left\lfloor \frac{n}{5} \right\rfloor + \left\lfloor \frac{n}{5^2} \right\rfloor + \left\lfloor \frac{n}{5^3} \right\rfloor + \cdots = 1000.$$

On the one hand,

$$\left\lfloor \frac{n}{5} \right\rfloor + \left\lfloor \frac{n}{5^2} \right\rfloor + \left\lfloor \frac{n}{5^3} \right\rfloor + \cdots < \frac{n}{5} + \frac{n}{5^2} + \frac{n}{5^3} + \cdots = \frac{n}{5} \cdot \frac{1}{1 - \frac{1}{5}} = \frac{n}{4},$$

and hence $n > 4000$. On the other hand, using the inequality $\lfloor a \rfloor > a - 1$, we have

$$1000 > \left(\frac{n}{5} - 1 \right) + \left(\frac{n}{5^2} - 1 \right) + \left(\frac{n}{5^3} - 1 \right) + \left(\frac{n}{5^4} - 1 \right) + \left(\frac{n}{5^5} - 1 \right)$$

$$= \frac{n}{5} \left(1 + \frac{1}{5} + \frac{1}{5^2} + \frac{1}{5^3} + \frac{1}{5^4} \right) - 5 = \frac{n}{5} \cdot \frac{1 - \left(\frac{1}{5} \right)^5}{1 - \frac{1}{5}} - 5,$$

so

$$n < \frac{1005 \cdot 4 \cdot 3125}{3124} < 4022.$$

We have narrowed down our search to $\{4001, 4002, \ldots, 4021\}$. Checking each case with Polignac's formula, we find that the only solutions are $n = 4005, 4006, 4007, 4008,$ and 4009.

866. Polignac's formula implies that the exponent of the number 2 in $n!$ is

$$\left\lfloor \frac{n}{2} \right\rfloor + \left\lfloor \frac{n}{2^2} \right\rfloor + \left\lfloor \frac{n}{2^3} \right\rfloor + \cdots$$

Because

$$\frac{n}{2} + \frac{n}{2^2} + \frac{n}{2^3} + \cdots = n$$

and not all terms in this infinite sum are integers, it follows that n is strictly greater than the exponent of 2 in $n!$, so 2^n does not divide $n!$.

(Mathematics Competition, Soviet Union, 1971)

867. Let p be a prime number. The power of p in $\operatorname{lcm}\left(1, 2, \ldots, \left\lfloor \frac{n}{i} \right\rfloor\right)$ is equal to k if and only if

$$\left\lfloor \frac{n}{p^{k+1}} \right\rfloor < i \le \left\lfloor \frac{n}{p^k} \right\rfloor.$$

Hence the power of p in the expression on the right-hand side is

$$\sum_{k \ge 1} k \left(\left\lfloor \frac{n}{p^k} \right\rfloor - \left\lfloor \frac{n}{p^{k+1}} \right\rfloor \right) = \sum_{k \ge 1} (k - (k-1)) \left\lfloor \frac{n}{p^k} \right\rfloor = \sum_{k \ge 1} \left\lfloor \frac{n}{p^k} \right\rfloor.$$

By Polignac's formula this is the exponent of p in $n!$ and we are done.

(64th W.L. Putnam Mathematical Competition, 2003)

868. *First solution*: We will show that for any prime number p the power to which it appears in the numerator is greater than or equal to the power to which it appears in the denominator, which solves the problem.

Assume that p appears to the power α in n and to the power β in m, $\alpha \geq \beta \geq 0$. Then among the inequalities

$$\left\lfloor \frac{n}{p^k} \right\rfloor \geq \left\lfloor \frac{m}{p^k} \right\rfloor + \left\lfloor \frac{n-m}{p^k} \right\rfloor, \quad k = 1, 2, \ldots$$

those with $\beta < k \leq \alpha$ are strict. Using this fact when applying Polignac's formula to $n!$, $m!$, and $(n-m)!$, we deduce that the power of p in $\binom{n}{m}$ is at least $\alpha - \beta$. Of course, the power of p in $\gcd(m, n)$ is β. Hence p appears to a nonnegative power in $\dfrac{\gcd(m, n)}{n}\binom{n}{m}$, and we are done.

Second solution: A solution that does not involve prime numbers is also possible. Since $\gcd(m, n)$ is an integer linear combination of m and n, it follows that $\dfrac{\gcd(m, n)}{n}\binom{n}{m}$ is an integer linear combination of the integers

$$\frac{m}{n}\binom{n}{m} = \binom{n-1}{m-1} \quad \text{and} \quad \frac{n}{n}\binom{n}{m} = \binom{n}{m},$$

and hence is itself an integer.

(61st W.L. Putnam Mathematical Competition, 2000)

869. Let p be a prime divisor of k. Then $p \leq n$, so p is also a divisor of $n!$. Denote the powers of p in k by α and in $n!$ by β. The problem amounts to showing that $\alpha \leq \beta$ for all prime divisors p of k.

By Polignac's formula, the power of p in $n!$ is

$$\beta = \sum_{i=1}^{\infty} \left\lfloor \frac{n}{p^i} \right\rfloor.$$

Of course, the sum terminates at the mth term, where m is defined by $p^m \leq n < p^{m+1}$.

Write $\gamma = \left\lfloor \frac{\alpha}{2} \right\rfloor$, so that α equals either 2γ or $2\gamma + 1$. From the hypothesis,

$$n^2 \geq 4k \geq 4p^\alpha,$$

and hence $n \geq 2p^{\alpha/2} \geq 2p^\gamma$. Since $n < p^{m+1}$, this leads to $p^{m+1-\gamma} > 2$. If means that if $p = 2$, then $\gamma < m$, and if $p \geq 3$, then $\gamma \leq m$.

If $p = 2$, we will show that $\beta \geq m + \gamma$, from which it will follow that $\beta \geq 2\gamma + 1 \geq \alpha$. The coefficient of 2 in $n!$ is

$$\left\lfloor \frac{n}{2} \right\rfloor + \left\lfloor \frac{n}{2^2} \right\rfloor + \cdots + \left\lfloor \frac{n}{2^m} \right\rfloor.$$

All terms in this sum are greater than or equal to 1. Moreover, we have seen that $n \geq 2 \cdot 2^\gamma$, so the first term is greater than or equal to 2^γ, and so this sum is greater than or equal to $2^\gamma + m - 1$. It is immediate that this is greater than or equal to $\gamma + m$ for any $\gamma \geq 1$.

If $p \geq 3$, we need to show that

$$\left\lfloor \frac{n}{p} \right\rfloor + \left\lfloor \frac{n}{p^2} \right\rfloor + \cdots + \left\lfloor \frac{n}{p^m} \right\rfloor \geq m + \gamma + 1.$$

This time $m \geq \gamma$, and so $m + \gamma + 1 \geq \gamma + \gamma + 1 \geq \alpha$. Again, since $n \geq 2p^\gamma$, the first term of the left-hand side is greater than or equal to $2p^{\gamma-1}$. So the inequality can be reduced to $2p^{\gamma-1} + m - 1 \geq m + \gamma + 1$, or $2p^{\gamma-1} \geq \gamma + 2$. This again holds true for any $p \geq 3$ and $\gamma \geq 2$. For $\gamma = 1$, if $\alpha = 2$, then we have $2p^{\gamma-1} + m - 1 \geq m + \gamma > \alpha$. If $\alpha = 3$, then $n^2 \geq 2p^3$ implies $n \geq 2\lfloor\sqrt{p}\rfloor p \geq 3p$, and hence the first term in the sum is greater than or equal to 3, so again it is greater than or equal to α.

We have thus showed that any prime appears to a larger power in $n!$ than in k, which means that k divides $n!$.

(Austrian-Polish Mathematics Competition, 1986)

870. Define

$$E(a, b) = a^3 b - ab^3 = ab(a - b)(a + b).$$

Since if a and b are both odd, then $a + b$ is even, it follows that $E(a, b)$ is always even. Hence we only have to prove that among any three integers we can find two, a and b, with $E(a, b)$ divisible by 5. If one of the numbers is a multiple of 5, the property is true. If not, consider the pairs $\{1, 4\}$ and $\{2, 3\}$ of residue classes modulo 5. By the pigeonhole principle, the residues of two of the given numbers belong to the same pair. These will be a and b. If $a \equiv b \pmod 5$, then $a - b$ is divisible by 5, and so is $E(a, b)$. If not, then by the way we defined our pairs, $a + b$ is divisible by 5, and so again $E(a, b)$ is divisible by 5. The problem is solved.

(Romanian Team Selection Test for the International Mathematical Olympiad, 1980, proposed by I. Tomescu)

871. Observe that $2002 = 10^3 + 10^3 + 1^3 + 1^3$. so that

$$2002^{2002} = 2002^{2001} \cdot 2002 = ((2002)^{667})^3 (10^3 + 10^3 + 1^3 + 1^3)$$
$$= (10 \cdot 2002^{667})^3 + (10 \cdot 2002^{667})^3 + (2002^{667})^3 + (2002^{667})^3.$$

This proves the first claim. For the second, note that modulo 9, a perfect cube can be only ± 1 or 0. Therefore, the sum of the residues modulo 9 of three perfect cubes can be only 0, ± 1, ± 2, or ± 3. We verify that

$$2002^{2002} \equiv 4^{2002} \equiv (4^3)^{667} \cdot 4 \equiv 1 \cdot 4 \equiv 4 \pmod 9.$$

It is easy now to see that 2002^{2002} cannot be written as the sum of three cubes.

(Communicated by V.V. Acharya)

872. Denote the perfect square by k^2 and the digit that appears in the last four positions by a. Then $k^2 \equiv a \cdot 1111 \pmod{10000}$. Perfect squares end in 0, 1, 4, 5, 6, or 9, so a can only be one of these digits.

Now let us examine case by case. If $a = 0$, we are done. The cases $a \in \{1, 5, 9\}$ can be ruled out by working modulo 8. Indeed, the quadratic residues modulo 8 are 0, 1, and 4, while as a ranges over the given set, $a \cdot 1111$ has the residues 7 or 3.

The cases $a = 2$ or 4 are ruled out by working modulo 16, since neither $4 \cdot 1111 \equiv 12 \pmod{16}$ nor $6 \cdot 1111 \equiv 10 \pmod{16}$ is a quadratic residue modulo 16.

873. Reducing modulo 4, the right-hand side of the equation becomes equal to 2. So the left-hand side is not divisible by 4, which means that $x = 1$. If $y > 1$, then reducing modulo 9 we find that z has to be divisible by 6. A reduction modulo 6 makes the left-hand side 0, while the right-hand side would be $1 + (-1)^z = 2$. This cannot happen. Therefore, $y = 1$, and we obtain the unique solution $x = y = z = 1$.

(*Matematika v Škole* (*Mathematics in Schools*), 1979, proposed by I. Mihailov)

874. Note that a perfect square is congruent to 0 or to 1 modulo 3. Using this fact we can easily prove by induction that $a_n \equiv 2 \pmod{3}$ for $n \geq 1$. Since $2 \cdot 2 \equiv 1 \pmod{3}$, the question has a negative answer.

(Indian International Mathematical Olympiad Training Camp, 2005)

875. By hypothesis, there exist integers t and N such that $aN + b = t^k$. Choose m arbitrary positive integers $s_1, s_2, \ldots, s_m$, and consider the number

$$s = (as_1 + t)^k + \sum_{j=2}^{m}(as_j)^k.$$

Then

$$s \equiv t^k \equiv aN + b \equiv b \pmod{a}.$$

Since $s \equiv b \pmod{a}$, there exists n such that $s = an + b$, and so s is a term of the arithmetic progression that can be written as a sum of m kth powers of integers. Varying the parameters $s_1, s_2, \ldots, s_n$, we obtain infinitely many terms with this property.

(Proposed by E. Just for Mathematics Magazine)

876. Denote the sum from the statement by S_n. We will prove a stronger inequality, namely,

$$S_n > \frac{n}{2}(\log_2 n - 4).$$

The solution is based on the following obvious fact: no odd number but 1 divides 2^n evenly. Hence the residue of 2^n modulo such an odd number is nonzero. From here we deduce that the residue of 2^n modulo a number of the form $2^m(2k + 1)$, $k > 1$. is at least 2^m. Indeed, if $2^{n-m} = (2k + 1)q + r$, with $1 \leq r < 2k + 1$, then $2^n = 2^m(2k + 1)q + 2^m r$, with $2^m < 2^m r < 2^m(2k + 1)$. And so $2^m r$ is the remainder obtained by dividing 2^n by $2^m(2k + 1)$.

Therefore, $S_n \geq 1 \times$(the number of integers of the form $2k + 1$, $k > 1$, not exceeding n)$+2\times$(the number of integers of the form $2(2k + 1)$, $k > 1$, not exceeding n)$+2^2\times$(the number of integers of the form $2^2(2k + 1)$, $k > 1$, not exceeding n)$+\cdots$.

Let us look at the $(j + 1)$st term in this estimate. This term is equal to 2^j multiplied by the number of odd numbers between 3 and $\frac{n}{2^j}$, and the latter is at least $\frac{1}{2}\left(\frac{n}{2^j} - 3\right)$. We deduce that

$$S_n \geq \sum_j 2^j \frac{n - 3 \cdot 2^j}{2^{j+1}} = \sum_j \frac{1}{2}(n - 3 \cdot 2^j),$$

where the sums stop when $2^j \cdot 3 > n$, that is, when $j = \lfloor \log_2 \frac{n}{3} \rfloor$. Setting $l = \lfloor \log_2 \frac{n}{3} \rfloor$, we have

$$S_n \geq (l+1)\frac{n}{2} - \frac{3}{2} \sum_{j=0}^{l} 2^i > (l+1)\frac{n}{2} - \frac{3 \cdot 2^{l+1}}{2}.$$

Recalling the definition of l, we conclude that

$$S_n > \frac{n}{2} \log_2 \frac{n}{3} - n = \frac{n}{2} \left(\log_2 \frac{n}{3} - 2 \right) > \frac{n}{2}(\log_2 n - 4),$$

and the claim is proved. The inequality from the statement follows from the fact that for $n > 1000$, $\frac{1}{2}(\log_2 n - 4) > \frac{1}{2}(\log_2 1000 - 4) > 2$.

(*Kvant* (*Quantum*), proposed by A. Kushnirenko, solution by D. Grigoryev)

877. First, observe that all terms of the progression must be odd. Let $p_1 < p_2 < \cdots < p_k$ be the prime numbers less than n. We prove the property true for p_i by induction on i. For $i = 1$ the property is obviously true, since $p_1 = 2$ and the consecutive terms of the progression are odd numbers. Assume the property is true for $p_1, p_2, \ldots, p_{i-1}$ and let us prove it for p_i.

Let $a, a + d, a + 2d, \ldots, a + (n-1)d$ be the arithmetic progression consisting of prime numbers. Using the inequality $d \geq p_1 p_2 \cdots p_{i-1} > p_i$, we see that if a term of the progression is equal to p_i, then this is exactly the first term (in the special case of $p_2 = 3$, for which the inequality does not hold, the claim is also true because 3 is the first odd prime). But if $a = p_i$, then $a + p_i d$, which is a term of the progression, is divisible by p_i, and the problem states that this number is prime. This means that $a \neq p_i$, and consequently the residues of the numbers a, $a+d, \ldots, a + (p_i - 1)d$ modulo p_i range over $\{1, 2, \ldots, p_i - 1\}$. By the pigeonhole principle, two of these residues must be equal, i.e.,

$$a + sd \equiv a + td \pmod{p_i},$$

for some $0 \leq i < j \leq p_i - 1$. Consequently, $a + sd - a - td = (s - t)d$ is divisible by p_i, and since $|s - t| < p_i$. it follows that d is divisible by p_i. This completes the induction, and with it the solution to the problem.

(G. Cantor)

878. We reduce everything modulo 3; thus we work in the ring of polynomials with $\mathbb{Z}_3$ coefficients. The coefficients of both $P(x)$ and $Q(x)$ are congruent to 1, so the reduced polynomials are $\widehat{P}(x) = \frac{x^{m+1}-1}{x-1}$ and $\widehat{Q}(x) = \frac{x^{n+1}-1}{x-1}$. The polynomial $\widehat{P}(x)$ still divides $\widehat{Q}(x)$; therefore $x^{m+1} - 1$ divides $x^{n+1} - 1$.

Let g be the greatest common divisor of $m+1$ and $n+1$. Then there exist positive integers a and b such that $a(m+1) - b(n+1) = g$. The polynomial $x^{m+1} - 1$ divides $x^{a(m+1)} - 1$, while the polynomial $x^{n+1} - 1$ divides $x^{b(n+1)} - 1$ and so does $x^{m+1} - 1$. It follows that $x^{m+1} - 1$ divides

$$x^{a(m+1)} - 1 - (x^{b(n+1)} - 1) = x^{b(n+1)}(x^{a(m+1)-b(n+1)} - 1) = x^{b(n+1)}(x^g - 1).$$

Hence $x^{m+1} - 1$ divides $x^g - 1$. Because g divides $m + 1$, this can happen only if $g = m + 1$. Therefore, $m + 1$ is a divisor of $n + 1$, and we are done.

(Romanian Mathematical Olympiad, 2002)

879. We use complex coordinates, and for this, let

$$\varepsilon = \cos\frac{2\pi}{n} + i\sin\frac{2\pi}{n}.$$

The vertices of the equiangular polygon should have coordinates

$$\sum_{i=0}^{k}\sigma(i)\varepsilon^i, \ k = 1, 2, \ldots, n-1,$$

where σ is a certain permutation of $1, 2, \ldots, n$. The sides are parallel to the rays $\sigma(k)\varepsilon^k$, so the angle between two consecutive sides is indeed $\frac{2\pi}{n}$, except for maybe the first and the last! For these two sides to form the appropriate angle, the equality

$$\sum_{i=0}^{n-1}\sigma(i)\varepsilon^i = 0$$

must hold. We are supposed to find a permutation σ for which this relation is satisfied. It is here that residues come into play.

Let $n = ab$ with a and b coprime. Represent the nth roots of unity as

$$\varepsilon^{aj+bk}, \ j = 0, 1, \ldots, b-1, \ k = 0, 1, \ldots, a-1.$$

Note that there are $ab = n$ such numbers altogether, and no two coincide, for if $aj + bk \equiv aj' + bk' \pmod{n}$, then $a(j - j') \equiv b(k' - k) \pmod{n}$, which means that $j - j'$ is divisible by b and $k - k'$ is divisible by a, and so $j = j'$ and $k = k'$. Thus we have indeed listed all nth roots of unity.

Order the roots of unity in the lexicographic order of the pairs (j, k). This defines the permutation σ. We are left with proving that

$$\sum_{j=0}^{b-1}\sum_{k=0}^{a-1}(aj + k)\varepsilon^{aj+bk} = 0.$$

And indeed,

$$\sum_{j=0}^{b-1}\sum_{k=0}^{a-1}(aj + k)\varepsilon^{aj+bk} = \sum_{j=0}^{b-1}aj\varepsilon^{aj}\sum_{k=0}^{a-1}(\varepsilon^b)^k + \sum_{k=0}^{a-1}b\varepsilon^{bk}\sum_{j=0}^{b-1}(\varepsilon^a)^j = 0.$$

880. Let S be the set of all primes with the desired property. We claim that $S = \{2, 3, 5, 7, 13\}$.

It is easy to verify that these primes are indeed in S. So let us consider a prime p in S, $p > 7$. Then $p - 4$ can have no factor q larger than 4, for otherwise $p - \lfloor\frac{p}{q}\rfloor q = 4$. Since $p - 4$ is odd, $p - 4 = 3^a$ for some $a \geq 2$. For a similar reason, $p - 8$ cannot have prime factors larger than 8, and so $p - 8 = 3^a - 4 = 5^b7^c$. Reducing the last equality modulo 24, we find that a is even and b is odd.

If $c \neq 0$, then $p - 9 = 5^b 7^c - 1 = 2^d$. Here we used the fact that $p - 9$ has no prime factor exceeding 8 and is not divisible by 3, 5, or 7. Reduction modulo 7 shows that the last equality is impossible, for the powers of 2 are 1, 2, and 4 modulo 7. Hence $c = 0$ and $3^a - 4 = 5^b$, which, since $3^{a/2} - 2$ and $3^{a/2} + 2$ are relatively prime, gives $3^{a/2} - 2 = 1$ and $3^{a/2} + 2 = 5^b$. Thus $a = 2$, $b = 1$, and $p = 13$. This proves the claim.

(*American Mathematical Monthly*, 1987, proposed by M. Cipu and M. Deaconescu, solution by L. Jones)

881. Note that

$$n = 1 + 10 + \cdots + 10^{p-2} = \frac{10^{p-1} - 1}{10 - 1}.$$

By Fermat's little theorem the numerator is divisible by p, while the denominator is not. Hence the conclusion.

882. We have the factorization

$$16320 = 2^6 \cdot 3 \cdot 5 \cdot 17.$$

First, note that $p^{ab} - 1 = (p^a)^b - 1$ is divisible by $p^a - 1$. Hence $p^{32} - 1$ is divisible by $p^2 - 1$, $p^4 - 1$, and $p^{16} - 1$. By Fermat's little theorem, $p^2 - 1 = p^{3-1} - 1$ is divisible by 3, $p^4 - 1 = p^{5-1} - 1$ is divisible by 5, and $p^{16} - 1 = p^{17-1} - 1$ is divisible by 17. Here we used the fact that p, being prime and greater than 17, is coprime to 3, 5, and 17.

We are left to show that $p^{32} - 1$ is divisible by 2^6. Of course, p is odd, say $p = 2m + 1$, m an integer. Then $p^{32} - 1 = (2m + 1)^{32} - 1$. Expanding with Newton's binomial formula, we get

$$(2m)^{32} + \binom{32}{1}(2m)^{31} + \cdots + \binom{32}{2}(2m)^2 + \binom{32}{1}(2m).$$

In this sum all but the last five terms contain a power of two greater than or equal to 6. On the other hand, it is easy to check that in

$$\binom{32}{5}(2m)^5 + \binom{132}{4}(2m)^4 + \binom{32}{3}(2m)^3 + \binom{32}{2}(2m)^2 + \binom{32}{1}(2m)$$

the first binomial coefficient is divisible by 2, the second by 2^2, the third by 2^3, the fourth by 2^4, and the fifth by 2^5. So this sum is divisible by 2^6, and hence $(2m + 1)^{32} - 1 = p^{32} - 1$ is itself divisible by 2^6. This completes the solution.

(*Gazeta Matematică* (*Mathematics Gazette, Bucharest*), proposed by I. Tomescu)

883. If x is a solution to the equation from the statement, then using Fermat's little theorem, we obtain

$$1 \equiv x^{p-1} \equiv a^{\frac{p-1}{2}} \pmod{p}.$$

If m is an integer, then every odd prime factor p of $m^2 + 1$ must be of the form $4m + 1$, with m an integer. Indeed, in this case $m^2 \equiv -1 \pmod{p}$, and by what we just proved,

$$(-1)^{\frac{p-1}{2}} = 1,$$

which means that $p - 1$ is divisible by 4.

Now assume that there are only finitely many primes of the form $4m+1$, m an integer, say $p_1, p_2, \ldots, p_n$. The number $(2p_1p_2 \cdots p_n)^2 + 1$ has only odd prime factors, and these must be of the form $4m+1$, m an integer. Yet these are none of $p_1, p_2, \ldots, p_n$, a contradiction. Hence the conclusion.

884. Assume a solution (x, y) exists. If y were even, then $y^3 + 7$ would be congruent to 3 modulo 4. But a square cannot be congruent to 3 modulo 4. Hence y must be odd, say $y = 2k + 1$. We have

$$x^2 + 1 = y^3 + 2^3 = (y+2)[(y-1)^2 + 3] = (y+2)(4k^2 + 3).$$

We deduce that $x^2 + 1$ is divisible by a number of the form $4m+3$, namely, $4k^2 + 3$. It must therefore be divisible by a prime number of this form. But we have seen in the solution to the previous problem that this is impossible. Hence the equation has no solutions.

   (V.A. Lebesgue)

885. Assume that the equation admits a solution (x, y). Let p be the smallest prime number that divides n. Because $(x+1)^n - x^n$ is divisible by p, and x and $x+1$ cannot both be divisible by p, it follows that x and $x+1$ are relatively prime to p. By Fermat's little theorem, $(x+1)^{p-1} \equiv 1 \equiv x^{p-1} \pmod{p}$. Also, $(x+1)^n \equiv x^n \pmod{p}$ by hypothesis.

   Additionally, because p is the smallest prime dividing n, the numbers $p-1$ and n are coprime. Then there exist integers a and b such that $a(p-1) + bn = 1$. It follows that

$$x + 1 = (x+1)^{a(p-1)+bn} \equiv x^{a(p-1)+bn} \equiv x \pmod{p},$$

which is impossible. Hence the equation has no solutions.

   (I. Cucurezeanu)

886. We construct the desired subsequence $(x_n)_n$ inductively. Suppose that the prime numbers that appear in the prime factor decompositions of $x_1, x_2, \ldots, x_{k-1}$ are $p_1, p_2, \ldots, p_m$. Because the terms of the sequence are odd, none of these primes is equal to 2. Define

$$x_k = 2^{(p_1-1)(p_2-1)\cdots(p_m-1)} - 3.$$

By Fermat's little theorem, $2^{(p_1-1)(p_2-1)\cdots(p_m-1)} - 1$ is divisible by each of the numbers $p_1, p_2, \ldots, p_n$. It follows that x_k is not divisible by any of these primes. Hence x_k is relatively prime to $x_1, x_2, \ldots, x_{k-1}$, and thus it can be added to the sequence. This completes the solution.

887. The recurrence relation is linear. Using the characteristic equation we find that $x_n = A \cdot 2^n + B \cdot 3^n$, where $A = 3x_0 - x_1$ and $B = x_1 - 2x_0$. We see that A and B are integers.

   Now let us assume that all but finitely many terms of the sequence are prime. Then $A, B \neq 0$, and

$$\lim_{n \to \infty} x_n = \lim_{n \to \infty} 3^n \left(A \left(\frac{2}{3} \right)^n + B \right) = \infty.$$

Let n be sufficiently large so that x_n is a prime number different from 2 and 3. Then for $k \geq 1$,

$$x_{n+k(p-1)} = A \cdot 2^{n+k(p-1)} + B \cdot 3^{n+k(p-1)} = A \cdot 2^n \cdot (2^{p-1})^k + B \cdot 3^n \cdot (3^{p-1})^k.$$

By Fermat's little theorem, this is congruent to $A \cdot 2^n + B \cdot 3^n$ modulo p, hence to x_n which is divisible by p. So the terms of the subsequence $x_{n+k(p-1)}$, $k \geq 1$, are divisible by p, and increase to infinity. This can happen only if the terms become composite at some point, which contradicts our assumption. The problem is solved.

888. All congruences in this problem are modulo 13. First, let us show that for $0 \leq k < 12$,

$$\sum_{x=0}^{12} x^k \equiv 0 \pmod{13}.$$

The case $k = 0$ is obvious, so let us assume $k > 0$. First, observe that 2 is a primitive root modulo 13, meaning that 2^m, $m \geq 1$, exhaust all nonzero residues modulo 13. So on the one hand, $2^k \not\equiv 1$ for $1 \leq k < 12$, and on the other hand, the residue classes $2, 4, 6, \ldots, 24$ are a permutation of the residue classes $1, 2, \ldots, 12$. We deduce that

$$\sum_{x=0}^{12} x^k \equiv \sum_{x=0}^{12} (2x)^k = 2^k \sum_{x=0}^{12} x^k,$$

and because $2^k \not\equiv 1$, we must have $\sum_{x=0}^{12} x^k \equiv 0$.

Now let $S = \{(x_1, x_2, \ldots, x_n) \mid 0 \leq x_i \leq 12\}$. Because $|S| = 13^n$ is divisible by 13, it suffices to show that the number of n-tuples $(x_1, x_2, \ldots, x_n) \in S$ such that $f(x_1, x_2, \ldots, x_n) \not\equiv 0$ is divisible by 13. Consider the sum

$$\sum_{(x_1, x_2, \ldots, x_n) \in S} (f(x_1, x_2, \ldots, x_n))^{12}.$$

This sum is congruent modulo 13 to the number of n-tuples $(x_1, x_2, \ldots, x_n) \in S$ such that $f(x_1, x_2, \ldots, x_n) \not\equiv 0$, since by Fermat's little theorem,

$$(f(x_1, x_2, \ldots, x_n))^{12} = \begin{cases} 1 & \text{if } f(x_1, x_2, \ldots, x_n) \not\equiv 0, \\ 0 & \text{if } f(x_1, x_2, \ldots, x_n) \equiv 0. \end{cases}$$

On the other hand, $(f(x_1, x_2, \ldots, x_n))^{12}$ can be expanded as

$$(f(x_1, x_2, \ldots, x_n))^{12} = \sum_{j=1}^{m} c_j \prod_{j=1}^{n} x_i^{\alpha_{ji}},$$

for some integers m, c_j, α_{ji}. Because f is a polynomial of total degree less than n, we have $\alpha_{j1} + \alpha_{j2} + \cdots + \alpha_{jn} < 12n$ for every j, so for each j there exists i such that $\alpha_{ji} \leq 12$. Using what we proved above, we obtain for $1 \leq j \leq m$,

$$\sum_{(x_1, x_2, \ldots, x_n) \in S} c_j \prod_{i=1}^{n} x_i^{\alpha_{ji}} = c_j \prod_{i=1}^{n} \sum_{x_i=0}^{12} x_i^{\alpha_{ji}} \equiv 0,$$

since one of the sums in the product is congruent to 0. Therefore,

$$\sum_{(x_1,x_2,\ldots,x_n)\in S} (f(x_1, x_2, \ldots, x_n))^{12} = \sum_{(x_1,x_2,\ldots,x_n)\in S} \sum_{j=1}^{m} c_j \prod_{i=1}^{n} x_i^{\alpha_{ji}} \equiv 0.$$

This implies that the number of n-tuples $(x_1, x_2, \ldots, x_n)$ in S with the property that $f(x_1, x_2, \ldots, x_n) \not\equiv 0 \pmod{13}$ is divisible by 13, and we are done.

(Turkish Mathematical Olympiad, 1998)

889. We have $12321 = (111)^2 = 3^2 \times 37^2$. It becomes natural to work modulo 3 and modulo 37. By Fermat's little theorem,

$$a^2 \equiv 1 \pmod{3},$$

and since we must have $a^k \equiv -1 \pmod{3}$, it follows that k is odd. Fermat's little theorem also gives

$$a^{36} \equiv 1 \pmod{37}.$$

By hypothesis $a^k \equiv -1 \pmod{37}$. Using Euclid's algorithm we find integers x and y such that $kx + 36y = \gcd(k, 36)$. Since the $\gcd(k, 36)$ is odd, x is odd. We obtain that

$$a^{\gcd(k,36)} \equiv a^{kx+36y} \equiv (-1) \cdot 1 = -1 \pmod{37}.$$

Since $\gcd(k, 36)$ can be 1, 3, or 9, we see that a must satisfy $a \equiv -1$, $a^3 \equiv -1$, or $a^9 \equiv -1$ modulo 37. Thus a is congruent to -1 modulo 3 and to 3, 4, 11, 21, 25, 27, 28, 30, or 36 modulo 37. These residue classes modulo 37 are precisely those for which a is a perfect square but not a perfect fourth power. Note that if these conditions are satisfied, then $a^k \equiv -1 \pmod{3 \times 37}$, for some odd integer k.

How do the 3^2 and 37^2 come into the picture? The algebraic identity

$$x^n - y^n = (x - y)(x^{n-1} + x^{n-2}y + \cdots + xy^{n-2} + y^{n-1})$$

shows that if $x \equiv y \pmod{n}$, then $x^n \equiv y^n \pmod{n^2}$. Indeed, modulo n, the factors on the right are 0, respectively, nx^{n-1}, which is again 0.

We conclude that if a is a perfect square but not a fourth power modulo 37, and is -1 modulo 3, then $a^k \equiv -1 \pmod{3 \times 37}$ and $a^{k \times 3 \times 37} \equiv -1 \pmod{3^2 \times 37^2}$. The answer to the problem is the residue classes

$$11, 41, 62, 65, 77, 95, 101, 104, 110$$

modulo 111.

(Indian Team Selection Test for the International Mathematical Olympiad, 2004, proposed by S.A. Katre)

890. More generally, for an integer $n \geq 3$, we are supposed to find the greatest common divisor of the numbers

$$2(2^{n-1} - 1), 3(3^{n-1} - 1), \ldots, n(n^{n-1} - 1).$$

Let p be a prime dividing all these numbers. If $p > n$, then p divides $k^{n-1} - 1$ for $k = 1, 2, \ldots, n$. So the residue classes of $0, 1, 2, \ldots, p - 1$ are zeros of the polynomial $x^{n-1} - 1$ in $\mathbb{Z}_p[x]$. This is impossible, because it would imply that this is the null polynomial in $\mathbb{Z}_p[x]$ and it is not.

If $p \leq n$, then p does not divide the numbers $2, 3, \ldots, p - 1$, so p divides the numbers

$$2^{n-1} - 1, 3^{n-1} - 1, \ldots, (p - 1)^{n-1} - 1.$$

It also divides $1^{n-1} - 1$, so the residue classes of $1, 2, \ldots, p - 1$ are zeros of the polynomial $x^{n-1} - 1$ in $\mathbb{Z}_p[x]$. But they are also zeros of $x^{p-1} - 1$, by Fermat's little theorem. Hence $x^{p-1} - 1$ divides $x^{n-1} - 1$. From here we deduce that $p - 1$ divides $n - 1$. Indeed, if $n - 1 = q(p-1) + r$, with $0 \leq r < p - 1$, then $x^{n-1} - 1 = x^r(x^{q(p-1)} - 1) + (x^r - 1)$, so $x^{p-1} - 1$ divides $x^r - 1$, whence $r = 0$.

Conversely, if $p - 1$ divides $n - 1$, then by Fermat's little theorem p divides $k^n - k$ for all k, so p divides the numbers from the statement.

Finally, since p^2 does not divide $p(p^{n-1} - 1)$, p^2 does not divide all numbers. So the greatest common divisor is the product of all primes p such that $p - 1$ divides $n - 1$.

In our case $n - 1 = 560$ whose divisors are

$$1, 2, 4, 5, 7, 8, 10, 14, 16, 20, 28, 35, 40, 56, 70, 80, 112, 140, 280.$$

Add 1 to each and notice that the only primes are $2, 3, 5, 11, 17, 29, 41, 71, 113, 281$, so the answer to the problem is

$$2 \cdot 3 \cdot 5 \cdot 11 \cdot 17 \cdot 29 \cdot 41 \cdot 71 \cdot 113 \cdot 281.$$

Remark. The number 561 was chosen by the author of the problem because it has a special property: it is the smallest Carmichael number (a Carmichael number is an odd nonprime number n such that $a^{n-1} \equiv 1 \pmod{n}$ for all integers a with $\gcd(a, n) = 1$). The solution to the problem has as consequence a result by A. Korselt from 1899 that n is Carmichael if and only if it is square-free and has the property that $p - 1$ divides $n - 1$ for each prime that divides n.

(Romanian Team Selection Test for the International Mathematical Olympiad, 2008)

891. If $n + 1$ is composite, then each prime divisor of $(n + 1)!$ is less than n, which also divides $n!$. Then it does not divide $n! + 1$. In this case the greatest common divisor is 1.

If $n + 1$ is prime, then by the same argument the greatest common divisor can only be a power of $n + 1$. Wilson's theorem implies that $n + 1$ divides $n! + 1$. However, $(n + 1)^2$ does not divide $(n + 1)!$, and thus the greatest common divisor is $(n + 1)$.

(Irish Mathematical Olympiad, 1996)

892. We work modulo 7. None of the six numbers is divisible by 7, since otherwise the product of the elements in one set would be divisible by 7, while the product of the elements in the other set would not.

By Wilson's theorem, the product of the six consecutive numbers is congruent to -1 modulo 7. If the partition existed, denote by x the product of the elements in one set. Then

$$x^2 = n(n+1) \cdots (n+5) \equiv -1 \pmod 7.$$

But this is impossible since -1 is not a quadratic residue modulo 7.

(12th International Mathematical Olympiad, 1970)

893. Consider all pairs of numbers i and j with $ij \equiv a \pmod p$. Because the equation $x^2 \equiv a \pmod p$ has no solutions, i is always different from j. Since every nonzero element is invertible in $\mathbb{Z}_p$, the pairs exhaust all residue classes modulo p. Taking the product of all such pairs, we obtain

$$a^{\frac{p-1}{2}} \equiv (p-1)! \pmod p,$$

which by Wilson's theorem is congruent to -1, as desired.

894. We claim that if $p \equiv 1 \pmod 4$, then $x = \left(\frac{p-1}{2}\right)!$ is a solution to the equation $x^2 \equiv -1 \pmod p$. Indeed, by Wilson's theorem,

$$-1 \equiv (p-1)! = 1 \cdot 2 \cdots \left(\frac{p-1}{2}\right)\left(\frac{p+1}{2}\right) \cdots (p-1)$$

$$= 1 \cdot 2 \cdots \left(\frac{p-1}{2}\right)\left(p - \frac{p-1}{2}\right)\left(p - \frac{p-3}{2}\right) \cdots (p-1)$$

$$\equiv (-1)^{\frac{p-1}{2}} \left[\left(\frac{p-1}{2}\right)!\right]^2 \pmod p.$$

Hence

$$\left[\left(\frac{p-1}{2}\right)!\right]^2 \equiv -1 \pmod p,$$

as desired.

To show that the equation has no solution if $p \equiv 3 \pmod 4$, assume that such a solution exists. Call it a. Using Fermat's little theorem, we obtain

$$1 \equiv a^{p-1} \equiv a^{2 \cdot \frac{p-1}{2}} \equiv (-1)^{\frac{p-1}{2}} = -1 \pmod p.$$

This is impossible. Hence the equation has no solution.

895. Multiplying the obvious congruences

$$1 \equiv -(p-1) \pmod p,$$
$$2 \equiv -(p-2) \pmod p,$$
$$\cdots$$
$$n-1 \equiv -(p-n+1) \pmod p,$$

we obtain

$$(n-1)! \equiv (-1)^{n-1}(p-1)(p-2) \cdots (p-n+1) \pmod p.$$

Multiplying both sides by $(p - n)!$ further gives

$$(p - n)!(n - 1)! \equiv (-1)^{n-1}(p - 1)! \pmod{p}.$$

Because by Wilson's theorem $(p - 1)! \equiv -1 \pmod{p}$, this becomes

$$(p - n)!(n - 1)! \equiv (-1)^n \pmod{p},$$

as desired.

(A. Simionov)

896. Because the common difference of the progression is not divisible by p, the numbers $a_1, a_2, \ldots, a_p$ represent different residue classes modulo p. One of them, say a_i, is divisible by p, and the others give the residues $1, 2, \ldots, p - 1$ in some order. Applying Wilson's theorem, we have

$$\frac{a_1 a_2 \cdots a_p}{a_i} \equiv (p - 1)! \equiv -1 \pmod{p};$$

hence $\frac{a_1 a_2 \cdots a_p}{a_i} + 1$ is divisible by p. Since a_i is divisible by p, we find that $a_1 a_2 \cdots a_p + a_i$ is divisible by p^2, as desired.

(I. Cucurezeanu)

897. We use strong induction. The property is true for $n = 1$. Let $n = p^q$, where p is a prime number and q is relatively prime to p (q is allowed to be 1). Assume that the formula holds for q. Any number k that divides n is of the form $p^j m$, where $0 \leq j \leq \alpha$, and m divides q. Hence we can write

$$\sum_{j=0}^{\alpha} \sum_{m|q} \phi(p^j m) = \sum_{j=0}^{\alpha} \sum_{m|q} \phi(p^j)\phi(m) = \sum_{j=0}^{\alpha} \phi(p^j) \sum_{m|q} \phi(m)$$

$$= \left(1 + \sum_{j=1}^{\alpha} p^{j-1}(p - 1)\right) q = p^{\alpha} q = n.$$

This completes the induction.

(C.F. Gauss)

898. If $n = 2^m$, $m \geq 2$, then

$$\phi(n) = 2^m - 2^{m-1} = 2^{m-1} \geq \sqrt{2^m} = \sqrt{n}.$$

If $n = p^m$, where $m \geq 2$ and p is an odd prime, then

$$\phi(n) = p^{m-1}(p - 1) \geq \sqrt{p^m} = \sqrt{n}.$$

Observe, moreover, that if $n = p^m$, $m \geq 2$, where p is a prime greater than or equal to 5, then $\phi(n) \geq \sqrt{2n}$.

Now in general, if n is either odd or a multiple of 4, then

$$\phi(n) = \phi(p_1^{\alpha_1}) \cdots \phi(p_k^{\alpha_k}) \geq \sqrt{p_1^{\alpha_1}} \cdots \sqrt{p_k^{\alpha_k}} = \sqrt{n}.$$

We are left with the case $n = 2t$, with t odd and different from 1 or 3. If any prime factor of t is greater than or equal to 5, then $\phi(n) = \phi(t) \geq \sqrt{2t}$. It remains to settle the case $n = 2 \cdot 3^i$, $i \geq 2$. For $i = 2$, $\phi(18) = 6 > \sqrt{18}$. For $i \geq 3$, $\phi(n) = 2 \cdot 3^{i-1}$, and the inequality reduces to $\sqrt{2} \cdot 3^{\frac{i}{2}-1} > 1$, which is obvious.

899. An example is $n = 15$. In that case $\phi(15) = \phi(3 \cdot 5) = 2 \cdot 4 = 8$, and $8^2 + 15^2 = 17^2$. Observe that for $\alpha, \beta \geq 1$,

$$\phi(3^\alpha \cdot 5^\beta) = 3^{\alpha-1} \cdot 5^{\beta-1}(3-1)(5-1) = 3^{\alpha-1} \cdot 5^{\beta-1} \cdot 8$$

and

$$(3^{\alpha-1} \cdot 5^{\beta-1} \cdot 8)^2 + (3^\alpha \cdot 5^\beta)^2 = (3^{\alpha-1} \cdot 5^{\beta-1} \cdot 17)^2,$$

so any number of the form $n = 3^\alpha \cdot 5^\beta$ has the desired property.

900. We will prove that if $m = 2 \cdot 7^r$, $r \geq 1$, then the equation $\phi(n) = m$ has no solutions. If $n = p_1^{\alpha_1} \cdots p_k^{\alpha_k}$, then

$$\phi(n) = p_1^{\alpha_1-1} \cdots p_k^{\alpha_k-1}(p_1 - 1) \cdots (p_k - 1).$$

If at least two of the primes $p_1, \ldots, p_k$ are odd, then $\phi(n)$ is divisible by 4, so is not equal to m.

If $m = 2^\alpha$, or $n = 2^\alpha p^\beta$, with $\alpha > 2$, then $\phi(n)$ is again divisible by 4, so again $\phi(n) \neq m$. The only cases left are $n = 2^\alpha p^\beta$, with $\alpha = 0$, $\alpha = 1$, or $\alpha = 2$. In the first case,

$$\phi(n) = p^{\beta-1}(p-1).$$

This implies $p = 7$, but even then equality cannot hold. For the other two cases,

$$\phi(n) = 2^{\alpha-1} p^{\beta-1}(p-1).$$

The equality $\phi(n) = m$ implies right away that $\alpha = 1$, $p = 7$, but $7^{\beta-1} \cdot 6$ cannot equal $2 \cdot 7^r$. Hence the conclusion.

901. Let $s = 2^\alpha 5^\beta t$, where t is coprime to 10. Define

$$n = 10^{\alpha+\beta}(10^{\phi(t)} + 10^{2\phi(t)} + \cdots + 10^{s\phi(t)}).$$

The sum of the digits of n is $1 + 1 + \cdots + 1 = s$. By Euler's theorem, $10^{\phi(t)} \equiv 1 \pmod{t}$, and so $10^{k\phi(t)} \equiv 1 \pmod{t}$, $k = 1, 2, \ldots, s$. It follows that

$$n \equiv 10^{\alpha+\beta}(1 + 1 + \cdots + 1) = s \cdot 10^{\alpha+\beta} \pmod{t},$$

so n is divisible by t. This number is also divisible by $2^\alpha 5^\beta$ and therefore has the desired property.

(W. Sierpiński)

902. To have few residues that are cubes, 3 should divide the Euler totient function of the number. This is the case with 7, 9, and 13, since $\phi(7) = 6$, $\phi(9) = 6$, and $\phi(13) = 6$. The cubes modulo 7 and 9 are 0, 1, and -1; those modulo 13 are 0, 1, -1, 8, and -8.

So let us assume that the equation admits a solution x, z. Reducing modulo 7, we find that $x = 3k + 2$, with k a positive integer. The equation becomes $4 \cdot 8^k + 3 = z^3$. A reduction modulo 9 implies that k is odd, $k = 2n + 1$, and the equation further changes into $32 \cdot 64^n + 3 = z^3$. This is impossible modulo 13. Hence, no solutions.

(I. Cucurezeanu)

903. *First solution*: Here is a proof by induction on n. The case $n = 1$ is an easy check. Let us verify the inductive step from n to $n + 1$. We transform the left-hand side as

$$\sum_{k=1}^{n+1} \phi(k) \left\lfloor \frac{n+1}{k} \right\rfloor = \sum_{k=1}^{n+1} \phi(k) \left\lfloor \frac{n}{k} \right\rfloor + \sum_{k=1}^{n+1} \phi(k) \left(\left\lfloor \frac{n+1}{k} \right\rfloor - \left\lfloor \frac{n}{k} \right\rfloor \right).$$

The last term in the first sum can be ignored since it is equal to zero. To evaluate the second sum, we observe that

$$\left\lfloor \frac{n+1}{k} \right\rfloor - \left\lfloor \frac{n}{k} \right\rfloor = \begin{cases} 1 & \text{if } k \text{ divides } n, \\ 0 & \text{otherwise.} \end{cases}$$

Therefore,

$$\sum_{k=1}^{n+1} \phi(k) \left\lfloor \frac{n+1}{k} \right\rfloor = \sum_{k=1}^{n} \phi(k) \left\lfloor \frac{n}{k} \right\rfloor + \sum_{k \mid n+1} \phi(k).$$

Using the induction hypothesis and Gauss' identity $\sum_{k \mid n} \phi(k) = n$ (Problem 897), we find that this is equal to $\frac{n(n+1)}{2} + (n + 1)$, which is further equal to the desired answer $\frac{(n+1)(n+2)}{2}$. This completes the induction, and the solution to the problem.

Second solution: Using the Gauss identity for Euler's totient function (Problem 897), we can write

$$\frac{n(n+1)}{2} = \sum_{m=1}^{n} m = \sum_{m=1}^{n} \sum_{k \mid m} \phi(k) = \sum_{k=1}^{n} \phi(k) \sum_{m=1}^{\lfloor n/k \rfloor} 1.$$

This is clearly equal to the left-hand side of the identity from the statement, and we are done.

(M.O. Drimbe, 200 *de Identități și Inegalități cu "Partea Întreagă"* (200 *Identities and Inequalities about the "Greatest Integer Function"*), GIL, 2004, second solution by R. Stong)

904. We may assume $\gcd(a, d) = 1$, $d \geq 1$, $a > d$. Since $a^{\phi(d)} \equiv 1 \pmod{d}$, it follows that $a^{k\phi(d)} \equiv 1 \pmod{d}$ for all integers k. Hence for all $k \geq 1$,

$$a^{k\phi(d)} = 1 + m_k d,$$

for some positive integers m_k. If we let $n_k = am_k$, $k \geq 1$, then

$$a + n_k d = a^{k\phi(d)+1},$$

so the prime factors of $a + n_k d$, $k \geq 1$, are exactly those of a.

(G. Pólya, G. Szegö, *Aufgaben und Lehrsätze aus der Analysis*, Springer-Verlag, 1964)

905. The customer picks a number k and transmits it securely to the bank using the algorithm described in the essay. Using the two large prime numbers p and q, the bank finds m such that

$km \equiv 1 \pmod{(p-1)(q-1)}$. If α is the numerical information that the customer wants to receive, the bank computes $\alpha^m \pmod n$, then transmits the answer β to the customer. The customer computes $\beta^k \pmod n$. By Euler's theorem, this is α. Success!

906. As before, let p and q be two large prime numbers known by the United Nations experts alone. Let also k be an arbitrary secret number picked by these experts with the property that $\gcd(k, (p-1)(q-1)) = 1$. The number $n = pq$ and the inverse m of k modulo $\phi(n) = (p-1)(q-1)$ are provided to both the country under investigation and to the United Nations.

The numerical data α that comprises the findings of the team of experts is raised to the power k, then reduced modulo n. The answer β is handed over to the country. Computing β^m modulo n, the country can read the data. But it cannot encrypt fake data, since it does not know the number k.

907. We are to find the smallest positive solution to the system of congruences

$$x \equiv 1 \pmod{60},$$
$$x \equiv 0 \pmod 7.$$

The general solution is $7b_1 + 420t$, where b_1 is the inverse of 7 modulo 60 and t is an integer. Since b_1 is a solution to the Diophantine equation $7b_1 + 60y = 1$, we find it using Euclid's algorithm. Here is how to do it: $60 = 8 \cdot 7 + 4, 7 = 1 \cdot 3 + 3, 4 = 1 \cdot 3 + 1$. Then

$$1 = 4 - 1 \cdot 3 = 4 - 1 \cdot (7 - 1 \cdot 4) = 2 \cdot 4 - 7 = 2 \cdot (60 - 8 \cdot 7) - 7$$
$$= 2 \cdot 60 - 17 \cdot 7.$$

Hence $b_1 = -17$, and the smallest positive number of the form $7b_1 + 420t$ is $-7 \cdot 17 + 420 \cdot 1 = 301$.

(Brahmagupta)

908. Let $p_1, p_2, \ldots, p_{2n}$ be different primes. By the Chinese remainder theorem there exists x such that

$$x \equiv 0 \pmod{p_1 p_2},$$
$$x \equiv -1 \pmod{p_3 p_4},$$
$$\ldots$$
$$x \equiv -n + 1 \pmod{p_{2n-1}, p_{2n}}.$$

Then the numbers $x + k, 0 \le k \le n - 1$, are each divisible by $p_{2k+1} p_{2k+2}$, and we are done.

Remark. This problem shows that there exist arbitrarily long arithmetic progressions containing no prime numbers.

909. Let $m = m_1 m_2$. If $x \in \{0, 1, \ldots, m - 1\}$ is such that $P(x) \equiv 0 \pmod n$, then $P(x) \equiv 0 \pmod{m_1}$. Let a_1 be the residue of x modulo m_1. Then $P(a_1) \equiv 0 \pmod{m_1}$. Similarly, if a_2 is the residue of x modulo m_2, then $P(a_2) \equiv 0 \pmod{m_2}$. Thus for each solution x to $P(x) \equiv 0 \pmod{m}$, we have constructed a pair (a_1, a_2) with a_i a solution to $P(x) \equiv 0 \pmod{m_i}$, $i = 1, 2$.

Conversely, given the residues a_i such that $P(a_i) \equiv 0 \pmod{m_i}$, $i = 1, 2$, by the Chinese remainder theorem there exists a unique $x \in \{0, 1, \ldots, m - 1\}$ such that $x \equiv a_i \pmod{m_i}$, $i = 1, 2$. Then $P(x) \equiv 0 \pmod{m_i}$, $i = 1, 2$, and consequently $P(x) \equiv 0 \pmod{m}$. We have established a bijection from the set of solutions to the equation $P(x) \equiv 0 \pmod{m}$ to the Cartesian product of the sets of solutions to $P(x) \equiv 0 \pmod{m_i}$, $i = 1, 2$. The conclusion follows.

(I. Niven, H.S. Zuckerman, H.L. Montgomery, *An Introduction to the Theory of Numbers*, Wiley, 1991)

910. Since this is a game with finite number of possibilities, there is always a winning strategy, either for the first player, or for the second. Arguing by contradiction, let us assume that there are only finitely many n's, say $n_1, n_2, \ldots, n_m$ for which Bob has a winning strategy. Then for every other nonnegative integer n, Alice must have some move on a heap of n stones leading to a position in which the second player wins. This means that any other integer n is of the form $p - 1 + n_k$ for some prime p and some $1 \le k \le m$.

We will prove that this is not the case. Choose an integer N greater than all the n_k's and let $p_1, p_2, \ldots, p_N$ be the first N prime numbers. By the Chinese remainder theorem, there exists a positive integer x such that

$$x \equiv -1 \pmod{p_1^2},$$
$$x \equiv -2 \pmod{p_2^2},$$
$$\cdots$$
$$x \equiv -N \pmod{p_r^2}.$$

Then the number $x + N + 1$ is not of the form $p - 1 + n_k$, because each of the numbers $x + N + 1 - n_k - 1$ is composite, being a multiple of a square of a prime number. We have reached a contradiction, which proves the desired conclusion.

(67th W.L. Putnam Mathematical Competition, 2006)

911. Let $p_1 < p_2 < p_3 < \cdots$ be the sequence of all prime numbers. Set $a_1 = 2$. Inductively, for $n \ge 1$, let a_{n+1} be the least integer greater than a_n that is congruent to $-k$ modulo p_{k+1}, for all $k \le n$. The existence of such an integer is guaranteed by the Chinese remainder theorem. Observe that for all $k \ge 0$, $k + a_n \equiv 0 \pmod{p_{k+1}}$ for $n \ge k + 1$. Then at most $k + 1$ values in the sequence $k + a_n$, $n \ge 1$, can be prime, since from the $(k + 2)$nd term onward, the terms of the sequence are nontrivial multiples of p_{k+1}, and therefore must be composite. This completes the proof.

(Czech and Slovak Mathematical Olympiad, 1997)

912. We construct such a sequence recursively. Suppose that $a_1, a_2, \ldots, a_m$ have been chosen. Set $s = a_1 + a_2 + \cdots + a_m$, and let n be the smallest positive integer that is not yet a term of the sequence. By the Chinese remainder theorem, there exists t such that $t \equiv -s \pmod{m + 1}$, and $t \equiv -s - n \pmod{m+2}$. We can increase t by a suitably large multiple of $(m+1)(m+2)$ to ensure that it does not equal any of $a_1, a_2, \ldots, a_m$. Then $a_1, a_2, \ldots, a_m, t, n$ is also a sequence with the desired property. Indeed, $a_1 + a_2 + \cdots + a_m + t = s + t$ is divisible by $m + 1$ and

$a_1 + \cdots + a_m + t + n = s + t + n$ is divisible by $m + 2$. Continue the construction inductively. Observe that the algorithm ensures that $1, \ldots, m$ all occur among the first $2m$ terms.

(Russian Mathematical Olympiad, 1995)

913. First, let us fulfill a simpler task, namely to find a k such that $k \cdot 2^n + 1$ is composite for every n in an infinite arithmetic sequence. Let p be a prime, and b some positive integer. Choose k such that $k \cdot 2^b \equiv -1 \pmod{p}$ (which is possible since 2^b has an inverse modulo p), and such that $k \cdot 2^b + 1 > p$. Also, let a be such that $2^a \equiv 1 \pmod{p}$. Then $k \cdot 2^{am+b} + 1$ is divisible by p for all $m \geq 0$, hence is composite.

Now assume that we were able to find a finite set of triples (a_i, b_j, p_j), $1 \leq j \leq s$, with $2^{a_j} \equiv 1 \pmod{p_j}$ and such that for any positive integer n there exist m and j with $n = a_j m + b_j$. We would like to determine a k such that $k \cdot 2^{a_j m + b_j} + 1$ is divisible by p_j, $1 \leq j \leq s$, $m \geq 0$. Using the Chinese remainder theorem we can use k as a sufficiently large solution to the system of equations

$$k \equiv -2^{-b_j} \pmod{p_j}, \ 0 \leq j \leq s.$$

Then for every n, $k \cdot 2^n + 1$ is divisible by one of the p_j's, $j = 0, 1, \ldots, s$, hence is composite.

An example of such a family of triples is $(2, 0, 3)$, $(3, 0, 7)$, $(4, 1, 5)$, $(8, 3, 17)$, $(12, 7, 13)$, $(24, 23, 241)$.

(W. Sierpiński, 250 *Problems in Elementary Number Theory*, Państwowe Wydawnictwo Naukowe, Warszawa, 1970)

914. Assume the contrary and consider a prime p that does not divide $b - a$. By the Chinese remainder theorem we can find a positive integer n such that

$$n \equiv 1 \pmod{p - 1},$$
$$n \equiv -a \pmod{p}.$$

Then by Fermat's little theorem,

$$a^n + n \equiv a + n \equiv a - a \equiv 0 \pmod{p}$$

and

$$b^n + n \equiv b_n \equiv b - a \pmod{p}.$$

It follows that p divides $a^n + n$ but does not divide $b^n + n$, a contradiction. Hence $a = b$, as desired.

(Short list of the 46th International Mathematical Olympiad, 2005)

915. The idea is to place (a, b) at the center of a square of size $(2n + 1) \times (2n + 1)$ having the property that all lattice points in its interior and on its sides are not visible from the origin. To this end, choose $(2n + 1)^2$ distinct primes p_{ij}, $-n \leq i, j \leq n$. Apply the Chinese remainder theorem to find an a with $a + i \equiv 0 \pmod{p_{ij}}$ for all i, j and a b with $b + j \equiv 0 \pmod{p_{ij}}$ for all i, j. For any i and j, $a + i$ and $b + j$ are both divisible by p_{ij}. Hence none of the points $(a + i, b + j)$ are visible from the origin. We conclude that any point visible from the origin lies outside the square of size $(2n + 1) \times (2n + 1)$ centered at (a, b), hence at distance greater than n from (a, b).

(*The American Mathematical Monthly*, 1977, proposed by A.A. Mullin)

916. We claim the answer is $b = 6$.

Lemma. *If n, d are integers and p is prime, and if p divides $P(n)$ and $P(n-d)$, then p is odd and p divides $d^2 + 3$ or p divides d.*

Proof. Note that if p is a prime and $p|n^2 + n + 1$, then p is odd, since $n^2 + n$ is even.

If $p|n^2+n+1$ and $p|n^2-2dn+d^2+n-d+1$, then p divides their difference $-2dn+d^2-d$. If p does not divide d, then p divides $-2n + d - 1$, so $n \equiv \frac{d-1}{2} \pmod{p}$ (note that p is odd). So p divides

$$\left(\frac{d-1}{2}\right)^2 + \left(\frac{d-1}{2}\right) + 1 = \frac{1}{4}(d^2 + 3),$$

and hence p divides $d^2 + 3$.   □

In view of the lemma, given an integer d, we will call a prime p for which there exists n such that p divides $P(n)$ and $P(n-d)$ a d-prime. All d-primes are odd. One can see that there are no 1-primes, because there is no odd prime that divides either 1 or $1^2 + 3 = 4$, impossible. So there are no 1-primes. If p is a 2-prime, then $p|2$ or $p|2^2 + 3 = 7$. Thus $p = 7$ is the only 2-prime. If p is a 3-prime, $p|3$ or $p|3^2 + 3 = 12$, so $p = 3$ is the only 3-prime.

Now let us return to the problem. We will show first that $b = 1, 2, 3, 4, 5$ cannot yield a fragrant set.

If $b = 2$, then some p divides $P(a+1)$ and $P(a+2)$, meaning that p is a 1-prime, impossible.

If $b = 3$ the some p divides either $P(a+1)$ and $P(a+2)$ or $P(a+2)$ and $P(a+3)$, impossible for the same reason.

If $b = 4$, then $P(a+2)$ musth share a prime divisor with $P(a+4)$, so the prime p must be 7. Thus 7 divides $P(a+2)$ and $P(a+4)$. For the same reason 7 divides $P(a+1)$ and $P(a+3)$. Thus 7 is a 1-prime (with $P(a+1)$ and $P(a+2)$), impossible.

If $b = 5$, then $P(a+3)$ shares a prime divisor with either $P(a+1)$ or $P(a+5)$. Thus this prime is 7. But then $P(a+2)$ and $P(a+4)$ cannot share a prime factor because then this factor would be 7 as well, ans then 7 would be a 1-prime (with $P(a+2)$ and $P(a+3)$). So $P(a+2)$ must share a factor with $P(a+5)$. The only 3-prime is 3, so 3 divides both $P(a+2)$ and $P(a+5)$. A similar argument shows that 3 divides $P(a+1)$ and $P(a+4)$, so 3 is a 1-prime, impossible.

We conclude that $b \geq 6$. Now take $b = 6$. By the Chinese remainder theorem, there exists a positive integer a such that

$$a \equiv 6 \pmod{19}$$
$$a \equiv 0 \pmod{7}$$
$$a \equiv 1 \pmod{3}.$$

Then, because $P(x) \equiv P(y) \pmod{p}$ if $x \equiv y \pmod{p}$,

$$P(a+1) \equiv P(7) = 49 + 7 + 1 \equiv 0 \pmod{19}$$
$$P(a+2) \equiv P(2) = 4 + 2 + 1 \equiv 0 \pmod{7}$$
$$P(a+3) \equiv P(1) = 1 + 1 + 1 \equiv 0 \pmod{3}$$
$$P(a+4) \equiv P(4) = 16 + 4 + 1 \equiv 0 \pmod{7}$$
$$P(a+5) \equiv P(11) = 121 + 11 + 1 \equiv 0 \pmod{19}$$
$$P(a+6) \equiv P(1) = 1 + 1 + 1 \equiv 0 \pmod{3}.$$

So $\{P(a+1), P(a+2), \dots, P(a+6)\}$ is fragrant.

(57th International Mathematical Olympiad, 2016, solution by M. Kural)

917. As seen in the solution to Problem 883 every prime factor of $n^2 + 1$ is of the form $4k + 1$. Thus every such prime factor is the sum of two squares. Thus m is a product of sums of two squares. Using the Lagrange identity

$$(x^2 + y^2)(z^2 + w^2) = (xz + yw)^2 + (xw - yz)^2,$$

we can inductively reduce the number of factors until m itself will be a sum two squares.

918. First let us discuss the case x odd. Then the equation can be written as

$$(2 + xi)(2 - xi) = y^3.$$

Let us show that $2 + xi$ and $2 - xi$ are coprime in $\mathbb{Z}[i]$. Indeed, if $c + di = \gcd(2 + xi, 2 - xi)$, then $c + id$ divides $2 + xi + 2 - xi = 4$ in $\mathbb{Z}[i]$. It follows that $c - di$ divides 4, and so their product $(c+di)(c-di) = c^2 + d^2$ divides $4 \times 4 = 16$. On the other hand, $c+di$ dividing $2+xi$ implies $c-di$ divides $2-xi$, so $(c+di)(c-di) = c^2 + d^2$ divides $(2+xi)(2-xi) = 4+x^2$. But x is odd, so the greatest common divisor of 16 and $4+x^2$ in $\mathbb{Z}$ is 1. Consequently $c^2 + d^2 = 1$, and hence $c + di$ is a unit in $\mathbb{Z}[i]$.

So $2 + xi$ and $2 - xi$ are coprime, and so $(2 + xi)(2 - xi) = y^3$ implies $2 + xi = (a + bi)^3$ for some integers a and b. Identifying the real and imaginary parts, we obtain

$$a(a^2 - 3b^2) = 2 \text{ and } 3a^2 b - b^3 = x.$$

The first equation yields $a = \pm 1$ or $a = \pm 2$, and this gives the solutions $x = \pm 11$, $y = 5$.

If x is even, then y is even. Substitute in the equation $x = 2u$, $y = 2v$, then divide the equation by 4 to obtain $u^2 + 1 = 2v^3$, that is

$$(u + i)(u - i) = 2v^3.$$

The numbers $u + i$ and $u - i$ differ by $2i$. If they had a common divisor, then this should be a divisor of $2 = (1 + i)(1 - i)$. But $u + i$ and $1 \pm i$ are coprime (since $1 \pm i$ are prime), unless $u = 1$. Then $1 + i$ and $1 - i$ are coprime. So $u + i$ and $u - i$ are coprime in $\mathbb{Z}[i]$. Using again the uniqueness of prime factorization, we deduce that there are integers a and b such that

$$u + i = (1 + i)(a + bi)^3, \quad u - i = (1 - i)(a - bi)^3,$$

(here we can enforce the signs to match because $(1 - i)(a + bi)^3 = (1 + i)(a - bi)$). We identify the real and imaginary parts

$$a^3 - 3a^2b - 3ab^2 + b^3 = u \quad \text{and} \quad a^3 + 3a^2b - 3ab^2 - b^3 = 1.$$

The last relation can be written as $(a - b)(a^2 + 4ab + b^2) = 1$, and yields the systems of equations

$$\begin{cases} a - b = 1 \\ a^2 + 4ab + b^2 = 1 \end{cases} \quad \text{and} \quad \begin{cases} a - b = -1 \\ a^2 + 4ab + b^2 = -1 \end{cases}$$

The first sistem yields $a = b + 1$, $(b + 1)^2 + 4(b + 1)b + b^2 = 6b^2 + 6b + 1 = 1$. This gives $(a, b) = (1, 0)$ or $(a, b) = (0, -1)$. The second system yields $a = b - 1$, $(b - 1)^2 + 4(b - 1)b + b^2 = 6b^2 - 6b + 1 = -1$, with no solutions. We thus obtain the solutions to the equation $(x, y) = (2, 2)$ and $(x, y) = (-2, 2)$.

Remark. The more general equation $x^2 + k = y^3$ with k a nonzero integer is called the Mordell equation. L. Mordell proved that for every nonzero k this equation has only finitely many integral solutions.

   (P. Fermat)

919. Because x, y are coprime, they have different parities. Indeed, if they are both odd, $x^2 + y^2$ is 2 modulo 4, which cannot be a perfect square. Hence z is odd. The equation is equivalent to

$$(x + yi)(x - yi) = z^{2m}.$$

Let $d = \gcd(x + yi, x - yi)$. Then $d | (x + yi) + (x - yi) = 2x$, and $d | (x + yi) - (x - yi) = 2iy$; hence $d | 2x$ and $d | 2y$. Since x and y are coprime, d must divide 2. But we also know that d divides the product of $x + yi$ and $x - yi$, so it divides z^m. And z is odd, so d must be a unit in $\mathbb{Z}[i]$, that is $x + yi$ and $x - yi$ are coprime. From the uniqueness of prime factorization, we must have

$$x + iy = i^k(a + bi)^{2m}$$

that is $x + yi$ is a unit times a $2m$th power. Here a, b, k are integers. We compute

$$\begin{aligned} (a + ib)^{4m} = (a + bi)^{p+1} &= (a + bi)^p(a + bi) \\ &\equiv (a^p + (bi)^p)(a + bi) \pmod{p} \\ &= (a^p - b^pi)(a + bi) \pmod{p}. \end{aligned}$$

By Fermat's little theorem, the last expression is further congruent, modulo p, to $(a - bi)(a + bi) = a^2 + b^2$. Therefore

$$(a + bi)^{4m} \equiv a^2 + b^2 \pmod{p}.$$

On the other hand, from $x + yi = i^k(a + bi)^{2m}$, it follows, by squaring, that

$$x^2 - y^2 + 2xyi = (-1)^k(a + bi)^{4m},$$

and hence

$$x^2 - y^2 + 2xyi \equiv (-1)^k(a^2 + b^2) \pmod{p}.$$

We deduce that there is $u + vi \in \mathbb{Z}[i]$ such that

$$p(u + vi) = pu + pvi = x^2 - y^2 - (-1)^k(a^2 + b^2) + 2xyi.$$

We conclude that p divides $2xy$, and since p is odd, p divides xy.
    (*The American Mathematical Monthly*)

920. First note that y must be odd, because no cube is 2 modulo 4. Write $x^3 = y^2 + 2 = (y + \sqrt{-2})(y - \sqrt{-2})$. Let $\delta = \gcd(y + \sqrt{-2}), y - \sqrt{-2})$. Then

$$\delta | (y + \sqrt{-2}) - (y - \sqrt{-2}) = 2\sqrt{-2}.$$

Notice that $N(\sqrt{-2}) = 2$, which is prime in the integer ring, hence $\sqrt{-2}$ is prime. Thus δ is a power of $\sqrt{-2}$. On the other hand,

$$\delta | (y + \sqrt{-2})(y - \sqrt{-2}) = y^2 + 2,$$

which is odd. This is impossible unless δ is a unit (because $N(\delta)$, which is even unless δ is a unit, divides $N(y^2 + 2)$ which is odd).

So $y + \sqrt{-2}$ and $y - \sqrt{-2}$ are coprime. Because $\mathbb{Z}[\sqrt{-2}]$ is a unique factorization domain, this implies that each of $y + \sqrt{-2}$ and $y - \sqrt{-2}$ is a cube up to multiplication by a unit. But the units are ± 1, which are cubes as well, so we can write

$$y + \sqrt{-2} = (a + b\sqrt{-2})^3.$$

Equating the real and imaginary parts, we obtain $y = a^3 - 6ab^2$ and $1 = 3a^2b - 2b^3$. From the second equation we see that b, which divides the right-hand side, must be ± 1. Then $a = \pm 1$ as well. Consequently $y = \pm 1$, and $x = 3$ are the only solutions.
    (P. Fermat)

921. First we look modulo 13, and notice that $3^y \equiv 1, 3,$ or $9 \pmod{13}$ while $x^2 + 11 \equiv 11, 12, 2, 7, 1, 10,$ or $8 \pmod{13}$. For the two to be equal modulo 13, 3^y must be congruent to 1, so y must be a multiple of 3, that is $y = 3k$ for some positive integer k. Let $z = 3^k$. The equation becomes $x^2 + 11 = z^3$. The right-hand side is odd (z is a power of 3), so x is even. Using the uniqueness of the prime factorization in the ring of integers of $\mathbb{Q}[\sqrt{-11}]$ we can write

$$x \pm \sqrt{-11} = \left(\frac{a + b\sqrt{-11}}{2}\right)^3,$$

where a and b are either both even or both odd (here we can check using the norm that the only units of $\mathbb{Q}[\sqrt{-11}]$ are ± 1, so if a unit appears in front of the left-hand side, it can be incorporated into the cube).

Identifying the imaginary parts, we obtain $\pm 2^3 = 3a^2b - 11b^3$, hence $b|2^3$. Analyzing cases we obtain that the only solutions are $a = \pm 1, b = \pm 1$. Hence $(4, 3)$ is the only solution to the original equation.

(T. Andreescu, D. Andrica, I. Cucurezeanu, *An Introduction to Diophantine Equations (A Problem-Based Approach)*, Birkhäuser, 2010)

922. This problem tests whether you really understood our discussion of the procedure of writing the elements of $SL(2, \mathbb{Z})$ in terms of the generators.

Call the first matrix from the statement $\overline{S}$. This matrix is no longer in $SL(2, \mathbb{Z})$! Let us see again where the linear equation is. The determinant of the matrix

$$\begin{bmatrix} 12 & 5 \\ 7 & 3 \end{bmatrix}$$

is equal to $12 \cdot 3 - 7 \cdot 5 = 1$, so $(3, 5)$ is a solution to the linear equation $12x - 7y = 1$. Note that

$$\overline{S}\begin{pmatrix} p \\ q \end{pmatrix} = \begin{pmatrix} q \\ p \end{pmatrix}, \quad T^n\begin{pmatrix} p \\ q \end{pmatrix} = \begin{pmatrix} p + nq \\ q \end{pmatrix}.$$

So $\overline{S}$ flips a fraction, and T^k adds k to it. This time it is the continued fraction expansion

$$\frac{12}{7} = 1 = \cfrac{1}{1 + \cfrac{1}{2 + \cfrac{1}{2}}}$$

(no negatives !). All we need to do is start with $\overline{S}$ and apply to it T^2, then $\overline{S}$, then again T^2, and so on, following the continued fraction expansion from bottom to top. We thus obtain

$$\begin{bmatrix} 12 & 5 \\ 7 & 3 \end{bmatrix} = T\overline{S}T\overline{S}T^2\overline{S}T^2\overline{S},$$

and the problem is solved.

923. Consider first the case $a = 0$. Since $by = m$ always has solutions, it follows that $b = \pm 1$. From this we deduce that $y = \pm m$. The second equation becomes a linear equation in x, $cx = n \mp dm$, which is supposed always to have an integer solution. This implies $c = \pm 1$, and hence $ad - bc = bc = \pm 1$. The same argument applies if any of b, c, or d is 0.

If none of them is zero, set $\Delta = ab - cd$. Again we distinguish two cases. If $\Delta = 0$, then $\frac{a}{c} = \frac{b}{d} = \lambda$. Then $m = ax + by = \lambda(cx + dy) = \lambda n$, which restricts the range of m and n. Hence $\Delta \neq 0$.

Solving the system using Cramer's rule, we obtain

$$x = \frac{dm - bn}{\Delta}, \quad y = \frac{an - cm}{\Delta}.$$

These numbers are integers for any m and n. In particular, for $(m, n) = (1, 0)$, $x_1 = \frac{d}{\Delta}$, $y_1 = -\frac{c}{\Delta}$, and for $(m, n) = (0, 1)$, $x_2 = -\frac{b}{\Delta}$, $y_2 = \frac{a}{\Delta}$. The number

$$x_1 y_2 - x_2 y_1 = \frac{ad - bc}{\Delta^2} = \frac{1}{\Delta}$$

is therefore an integer. Since Δ is an integer, this can happen only if $\Delta = \pm 1$, and the problem is solved.

Remark. A linear map $T : \mathbb{R}^2 \rightarrow \mathbb{R}^2$ is called orientation preserving if its determinant is positive, and orientation reversing otherwise. As a consequence of what we just proved, we obtain that $SL(2, \mathbb{Z})$ consists of precisely those orientation-preserving linear transformations of the plane that map $\mathbb{Z}^2$ onto itself.

924. Because $\gcd(a, b) = 1$, the equation $au - bv = 1$ has infinitely many positive solutions (u, v). Let (t, z) be a solution. Consider now the system in (x, y),

$$\begin{cases} ax - yz - c = 0, \\ bx - yt + d = 0. \end{cases}$$

The determinant of its coefficient matrix is -1, so the system admits integer solutions. Solving, we obtain

$$\begin{pmatrix} x \\ y \end{pmatrix} = \begin{pmatrix} t & -z \\ b & -a \end{pmatrix} \begin{pmatrix} c \\ -d \end{pmatrix} = \begin{pmatrix} tc + zd \\ bc + ad \end{pmatrix}.$$

So each positive solution (t, z) to the equation $au - bv = 1$ yields a positive solution $(tc + zd, bc + ad, z, t)$ to the original system of equations. This solves the problem.

925. At each cut we add 7 or 11 new pieces. Thus after cutting x times in 8 and y times in 12 we have $7x + 11y + 1$ pieces. The problem amounts to showing that the equation $7x + 11y = n$ has nonnegative solutions for every $n \geq 60$, but no nonnegative solution for $n = 59$. This is of course a corollary to Sylvester's theorem, but let us see how the proof works for this particular situation.

The numbers $11 \cdot 0, 11 \cdot 1, \ldots, 11 \cdot 6$ form a complete set of residues modulo 7. This means that for n equal to one of the numbers $60 = 11 \cdot 6 - 6, 61 = 11 \cdot 6 - 5, \ldots, 66 = 11 \cdot 6$, one can find nonnegative x and y such that $7x + 11y = n$. Indeed,

$$60 = 7 \cdot 7 + 11 \cdot 1,$$

$$61 = 7 \cdot 4 + 11 \cdot 3,$$

$$62 = 7 \cdot 1 + 11 \cdot 5,$$

$$63 = 7 \cdot 9 + 11 \cdot 0,$$

$$64 = 7 \cdot 6 + 11 \cdot 2,$$

$$65 = 7 \cdot 3 + 11 \cdot 4,$$

$$66 = 7 \cdot 0 + 11 \cdot 6.$$

Since if we are able to cut the sheet of paper into n pieces we are also able to cut it into $n + 7$, we can prove by induction that the cut is possible for any $n \geq 61$.

Let us now show that the equation $7x + 11y = 59$ has no solution. Rewrite it as $7x + 11(y - 5) = 4$. This implies $7x \equiv 4 \pmod{11}$. But this means $x \equiv 10 \pmod{11}$, hence

$x \geq 10$. This is impossible since $7x + 11y = 59$ implies $x \leq 8$. Hence we cannot obtain 60 pieces, and the problem is solved.

(German Mathematical Olympiad, 1970/71)

926. Multiply the geometric series

$$\frac{1}{1 - x^a} = 1 + x^a + x^{2a} + \cdots \quad \text{and} \quad \frac{1}{1 - x^b} = 1 + x^b + x^{2b} + \cdots$$

The coefficient of x^n in the product counts the number of ways exponents of the form ka and mb add up to n. And this is $s(n)$.

927. The number n can be represented as $4m$, $4m + 1$, $4m + 2$, or $4m + 3$. The required solution is provided by one of the following identities:

$$4m = (2m - 1) + (2m + 1),$$
$$4m + 1 = 2m + (2m + 1),$$
$$4m + 2 = (2m - 1) + (2m + 3),$$
$$4m + 3 = (2m + 1) + (2m + 2).$$

The two terms on the right are coprime because either they differ by 1, or they are odd and differ by 2 or 4.

928. Note that for any integer k, we can dissect the d-dimensional cube into k^d pieces. If we do this for two integers a and b, then performing the appropriate dissections we can obtain $(a^d - 1)x + (b^d - 1)y + 1$ cubes.

By Sylvester's theorem for coprime positive numbers α and β, the equation $\alpha x + \beta y = n$ has nonnegative solutions provided that n is sufficiently large.

To complete the solution, we just have to find a and b such that $a^d - 1$ and $b^d - 1$ are coprime. We can choose any a and then let $b = a^d - 1$. Indeed, $(a^d - 1)^d - 1$ differs from a power of $a^d - 1$ by 1, so the two numbers cannot have a common divisor.

929. There exist integers u and v such that the two sides in question are $a = u^2 - v^2$ and $b = 2uv$. We are also told that $a + b = k^2$, for some integer k. Then

$$a^3 + b^3 = (a + b)(a^2 - ab + b^2) = k^2((u^2 - v^2)^2 - 2uv(u^2 - v^2) + 4u^2v^2)$$
$$= k^2(u^4 + v^4 - 2u^3v + 2uv^3 + 2u^2v^2) = [k(u^2 - uv)]^2 + [k(v^2 + uv)]^2,$$

and the problem is solved.

930. We use the characterization of Pythagorean triples as

$$a = k(u^2 - v^2), \quad b = 2kuv, \quad c = k(u^2 + v^2)$$

for some k, u, v positive integers, $gcd(u, v) = 1$. The condition from the statement translates into

$$k^2uv(u^2 - v^2) = 2k(u^2 + uv).$$

Dividing by ku and moving $2v$ to the left we obtain

$$v(ku^2 - kv^2 - 2) = 2u.$$

Since $\gcd(u, v) = 1$, $v = 1$ or 2. If $v = 1$ we obtain $ku^2 - k - 2 = 2u$, that is $ku(u-2) = k-2$. This can only happen for small values of k and u, and an easy check yields $k = u = 2$. We then obtain the Pythagorean triple $(6, 8, 10)$.

If $v = 2$, then $ku(u - 1) = 4k + 2$, and again this can only happen for small values of k and u. An easy check yields $k = 1$, $u = 3$, and we obtain the Pythagorean triple $(5, 12, 13)$.

931. We guess immediately that $x = 2$, $y = 4$, and $z = 2$ is a solution because of the Pythagorean triple $3, 4, 5$. This gives us a hint as to how to approach the problem. Checking parity, we see that y has to be even. A reduction modulo 4 shows that x must be even, while a reduction modulo 3 shows that z must be even. Letting $x = 2m$ and $z = 2n$, we obtain a Pythagorean equation

$$(3^m)^2 + y^2 = (5^n)^2.$$

Because y is even, in the usual parametrization of the solution we should have $3^m = u^2 - v^2$ and $5^n = u^2 + v^2$. From $(u - v)(u + v) = 3^m$ we find that $u - v$ and $u + v$ are powers of 3. Unless $u - v$ is 1, $u = (u - v + u + v)/2$ and $v = (u + v - u + v)/2$ are both divisible by 3, which cannot happen because $u^2 + v^2$ is a power of 5. So $u - v = 1$, $u + v = 3^m$, and $u^2 + v^2 = 5^n$. Eliminating the parameters u and v, we obtain the simpler equation

$$2 \cdot 5^n = 9^m + 1.$$

First, note that $n = 1$ yields the solution $(3, 4, 5)$. If $n > 1$, then looking at the equation modulo 25, we see that m has to be an odd multiple of 5, say $m = 5(2k + 1)$. But then

$$2 \cdot 5^n = (9^5)^{2k+1} + 1 = (9^5 + 1)((9^5)^{2k} - (9^5)^{2k-1} + \cdots + 1),$$

which implies that $2 \cdot 5^n$ is a multiple of $9^5 + 1 = 2 \cdot 5^2 \cdot 1181$. This is of course impossible; hence the equation does not have other solutions.

(I. Cucurezeanu)

932. The last digit of a perfect square cannot be 3 or 7. This implies that x must be even, say $x = 2x'$. The condition from the statement can be written as

$$(2^{x'})^2 + (5^y)^2 = z^2,$$

for integers x', y, and z. It follows that there exist integers u and v such that $5^y = u^2 - v^2$ and $2^{x'} = 2uv$ (looking at parity, we rule out the case $5^y = 2uv$ and $2^{x'} = u^2 - v^2$). From the first equality we see that any common factor of u and v is a power of 5. From the second we find that u and v are powers of 2. Thus $u = 2^{x'-1}$ and $v = 1$. It follows that x' and y satisfy the simpler Diophantine equation

$$5^y = 2^{2x'-2} - 1.$$

But then $5^y = (2^{x'-1} - 1)(2^{x'-1} + 1)$, and the factors on the right differ by 2, which cannot happen since no powers of 5 differ by 2. Hence no such numbers can exist.

933. Here is how to transform the equation from the statement into a Pythagorean equation:

$$x^2 + y^2 = 1997(x - y),$$

$$2(x^2 + y^2) = 2 \cdot 1997(x - y),$$

$$(x + y)^2 + (x - y)^2 - 2 \cdot 1997(x - y) = 0,$$

$$(x + y)^2 + (1997 - x + y)^2 = 1997^2.$$

Because x and y are positive integers, $0 < x + y < 1997$, and for the same reason $0 < 1997 - x + y < 1997$. The problem reduces to solving the Pythagorean equation $a^2 + b^2 = 1997^2$ in positive integers. Since 1997 is prime, the greatest common divisor of a and b is 1. Hence there exist coprime positive integers $u > v$ with the greatest common divisor equal to 1 such that

$$1997 = u^2 + v^2, \quad a = 2uv, \quad b = u^2 - v^2.$$

Because u is the larger of the two numbers, $\frac{1997}{2} < u^2 < 1997$; hence $33 \le u \le 44$. There are 12 cases to check. Our task is simplified if we look at the equality $1997 = u^2 + v^2$ and realize that neither u nor v can be divisible by 3. Moreover, looking at the same equality modulo 5, we find that u and v can only be 1 or -1 modulo 5. We are left with the cases $m = 34, 41,$ or 44. The only solution is $(m, n) = (34, 29)$. Solving $x + y = 2 \cdot 34 \cdot 29$ and $1997 - x + y = 34^2 - 29^2$, we obtain $x = 1827$, $y = 145$. Solving $x + y = 34^2 - 29^2$, $1997 - x + y = 2 \cdot 34 \cdot 29$, we obtain $(x, y) = (170, 145)$. These are the two solutions to the equation.

(Bulgarian Mathematical Olympiad, 1997)

934. One can verify that $x = 2m^2 + 1$ and $y = 2m$ is a solution.

(Diophantus)

935. We will search for numbers x and y for which $2^{x^2} = a^2$ and $2^{y^2} = 2a$, so that $1 + 2^{x^2} + 2^{y^2} = (a + 1)^2$. Then $x = 2z$ for some positive integer z, and

$$a = 2^{2z^2} = 2^{y^2 - 1}.$$

This leads to the Pell equation

$$y^2 - 2z^2 = 1.$$

This equation has infinitely many solutions, given by

$$y_n + z_n\sqrt{2} = (3 + 2\sqrt{2})^n,$$

and we are done.

(*Revista Matematică din Timişoara* (*Timişoara Mathematics Gazette*), proposed by M. Burtea)

936. The Pell equation $x^2 - 2y^2 = 1$ has infinitely many solutions. Choose $n = x^2 - 1$. Then $n = y^2 + y^2$, $n + 1 = x^2 + 0^2$, and $n + 2 = x^2 + 1^2$, and we are done.

(61st W.L. Putnam Mathematical Competition, 2000)

937. In other words, the problem asks us to show that the Diophantine equation $x^2 - 2 = 7^y$ has no positive solutions. A reduction modulo 8 makes the right-hand side equal to $(-1)^y$, while the left-hand side could only be equal to $-2, -1, 2$. This means that y must be odd, $y = 2z + 1$, with z an integer. Multiplying by $7^y = 7^{2z+1}$ and completing the square, we obtain the equivalent equation

$$(7^{2z+1} + 1)^2 - 7(7^z x)^2 = 1.$$

Let us analyze the associated Pell equation

$$X^2 - 7Y^2 = 1.$$

Its fundamental solution is $X_1 = 8$, $Y_1 = 3$, and its general solution is given by

$$X_k + Y_k \sqrt{7} = (8 + 3\sqrt{7})^k, \quad k = 1, 2, \ldots$$

Substituting $X = 7^{2z+1} + 1$ and $Y = 7^z x$, we obtain

$$7^{2z+1} + 1 = 8^k + \binom{k}{2} 8^{k-2} \cdot 3^2 \cdot 7 + \binom{k}{4} 8^{k-4} \cdot 3^4 \cdot 7^2 + \cdots,$$

$$7^z x = \binom{k}{1} 8^{k-1} \cdot 3 + \binom{k}{3} 8^{k-3} \cdot 3^3 \cdot 7 + \binom{k}{5} 8^{k-5} \cdot 3^5 \cdot 7^2 + \cdots$$

Let us compare the power of 7 in $k = \binom{k}{1}$ with the power of 7 in $\binom{k}{2m+1} 7^m$, $m > 1$. Writing

$$\binom{k}{2m+1} 7^m = \frac{7^m k(k-1) \cdots (k-2m-1)}{1 \cdot 2 \cdots k},$$

we see that the power of 7 in the numerator grows faster than it can be canceled by the denominator. Consequently, in the second equality from above, the power of 7 in the first term is less than in the others. We thus obtain that 7^z divides k. But then $8^k > 8^{7^z} > 7^{2z+1}$, and the first inequality could not hold. This shows that the equation has no solutions.
   (I. Cucurezeanu)

938. Expanding the cube, we obtain the equivalent equation $3x^2 + 3x + 1 = y^2$. After multiplying by 4 and completing the square, we obtain $(2y)^2 - 3(2x+1)^2 = 1$, a Pell equation, namely, $u^2 - 3v^2 = 1$ with u even and v odd. The solutions to this equation are generated by $u_n \pm v_n \sqrt{3} = (2 \pm \sqrt{3})^n$, and the parity restriction shows that we must select every other solution. So the original equation has infinitely many solutions generated by

$$2y_n \pm (2x_n + 1)\sqrt{3} = (2 \pm \sqrt{3})(5 \pm 4\sqrt{3})^n,$$

or, explicitly,

$$x_n = \frac{(2 + \sqrt{3})(5 + 4\sqrt{3})^n - (2 - \sqrt{3})(5 - 4\sqrt{3})^n - 1}{2},$$

$$y_n = \frac{(2 + \sqrt{3})(5 + 4\sqrt{3})^n + (2 - \sqrt{3})(5 - 4\sqrt{3})^n}{2}.$$

939. One family of solutions is of course (n, n), $n \in \mathbb{N}$. Let us see what other solutions the equation might have. Denote by t the greatest common divisor of x and y, and let $u = \frac{x}{t}$, $v = \frac{y}{t}$. The equation becomes $t^5(u - v)^5 = t^3(u^3 - v^3)$. Hence

$$t^2(u - v)^4 = \frac{u^3 - v^3}{u - v} = u^2 + uv + v^2 = (u - v)^2 + 3uv,$$

or

$$(u - v)^2[t^2(u - v)^2 - 1] = 3uv.$$

It follows that $(u - v)^2$ divides $3uv$, and since u and v are relatively prime and $u > v$, this can happen only if $u - v = 1$. We obtain the equation $3v(v + 1) = t^2 - 1$, which is the same as

$$(v + 1)^3 - v^3 = t^2.$$

This was solved in the previous problem. The solutions to the original equation are then given by $x = (v + 1)t$, $y = vt$, for any solution (v, t) to this last equation.
    (A. Rotkiewicz)

940. It is easy to guess that $(x, y, z, t) = (10, 10, -1, 0)$ is a solution. Because quadratic Diophantine equations are usually simpler than cubic equations, we try to reduce the given equation to a quadratic. We do this by *perturbing* the particular solution that we already know.
    We try to find numbers u and v such that $\left(10 + u, 10 - u, -\frac{1}{2} + v, -\frac{1}{2} - v\right)$ is a solution. Of course, v has to be a half-integer, so it is better to replace it by $\frac{w}{2}$, where w is an odd integer. The equation becomes

$$(2000 + u^2) - \frac{1 + 3w^2}{4} = 1999,$$

which is the same as

$$w^2 - 80u^2 = 1.$$

This is a Pell equation. The smallest solution is $(w_1, u_1) = (9, 1)$, and the other positive solutions are generated by

$$w_n + u_n\sqrt{80} = \left(w_1 + u_1\sqrt{80}\right)^n.$$

This gives rise to the recurrence

$$(w_{n+1}, u_{n+1}) = (9w_n + 80u_n, w_n + 9u_n), \quad n \geq 1.$$

It is now easy to prove by induction that all the w_n's are odd, and hence any solution (w_n, u_n) to Pell's equation yields the solution

$$(x_n, y_n, z_n, t_n) = \left(10 + u_n, 10 - u_n, -\frac{1}{2} + \frac{w_n}{2}, -\frac{1}{2} - \frac{w_n}{2}\right)$$

to the original equation.
    (Bulgarian Mathematical Olympiad, 1999)

941. Consider first the case that n is even, $n = 2k$, k an integer. We have

$$\left(\sqrt{m} + \sqrt{m-1}\right)^{2k} = \left(2m - 1 + 2\sqrt{m(m-1)}\right)^k.$$

The term on the right-hand side generates the solution to Pell's equation

$$X^2 - m(m-1)Y^2 = 1.$$

If for a certain n, (X_n, Y_n) is the corresponding solution, then choose $p = X_n^2$. Since $p - 1 = X_n^2 - 1 = m(m-1)Y_n^2$, it follows that

$$\left(\sqrt{m} + \sqrt{m-1}\right)^{2k} = \left(2m - 1 + 2\sqrt{m(m-1)}\right)^k = X_n + Y_n\sqrt{m(m-1)}$$
$$= \sqrt{p} + \sqrt{p-1},$$

as desired.

This now suggests the path we should follow in the case that n is odd. Write

$$\left(\sqrt{m} + \sqrt{m-1}\right)^n = U_n\sqrt{m} + V_n\sqrt{m-1}.$$

This time, (U_n, V_n) is a solution to the generalized Pell equation

$$mU^2 - (m-1)V^2 = 1.$$

In a similar manner we choose $p = mU_n^2$ and obtain the desired identity.
    (I. Tomescu, *Problems in Combinatorics*, Wiley, 1985)

942. *First solution*: This solution is based on an idea that we have already encountered in the section on factorizations and divisibility. Solving for y, we obtain

$$y = -\frac{x^2 + 4006x + 2003^2}{3x + 4006}.$$

To make the expression on the right easier to handle we multiply both sides by 9 and write

$$9y = -3x - 8012 - \frac{2003^2}{3x + 4006}.$$

If (x, y) is an integer solution to the given equation, then $3x + 4006$ divides 2003^2. Because 2003 is a prime number, we have $3x + 4006 \in \{\pm 1, \pm 2003, \pm 2003^2\}$. Working modulo 3 we see that of these six possibilities, only 1, -2003, and 2003^2 yield integer solutions for x. We deduce that the equation from the statement has three solutions: $(-1334, -446224)$, $(-2003, 0)$, and $(1336001, -446224)$.

Second solution: Rewrite the equation as

$$(3x + 4006)(3x + 9y + 8012) = -2003^2.$$

This yields a linear system

$$3x + 4006 = d,$$

$$3x + 9y + 8012 = -\frac{2003^2}{d},$$

where d is a divisor of -2003^2. Since 2003 is prime, one has to check the cases $d = \pm 1$, ± 2003, $\pm 2003^2$, which yield the above solutions.

(*American Mathematical Monthly*, proposed by Wu Wei Chao)

943. Divide through by $x^2 y^2$ to obtain the equivalent equation

$$\frac{1}{y^2} + \frac{1}{xy} + \frac{1}{x^2} = 1.$$

One of the denominators must be less than or equal to 3. The situations $x = 1$ and $y = 1$ are ruled out. Thus we can have only $xy = 2$ or 3. But then again either x or y is 1, which is impossible. Hence the equation has no solutions.

944. Note that $2002 = 3^4 + 5^4 + 6^4$. It suffices to consider

$$x_k = 3 \cdot 2002^k, \quad y_k = 5 \cdot 2002^k, \quad z_k = 6 \cdot 2002^k, \quad w_k = 4k + 1,$$

with k a positive integer. Indeed,

$$x_k^4 + y_k^4 + z_k^4 = (3^4 + 5^4 + 6^4)2002^{4k} = 2002^{4k+1},$$

for all $k \geq 1$.

945. If $x \leq y \leq z$, then since $4^x + 4^y + 4^z$ is a perfect square, it follows that the number $1 + 4^{y-x} + 4^{z-x}$ is also a perfect square. Then there exist an odd integer t and a positive integer m such that

$$1 + 4^{y-x} + 4^{z-x} = (1 + 2^m t)^2.$$

It follows that

$$4^{y-x}(1 + 4^{z-x}) = 2^{m+1} t (1 + 2^{m-1} t);$$

hence $m = 2y - 2x - 1$. From $1 + 4^{z-x} = t + 2^{m-1} t^2$, we obtain

$$t - 1 = 4^{y-x-1}(4^{z-2y+x+1} - t^2) = 4^{y-x-1}(2^{z-2y+x+1} + t)(2^{z-2y+x+1} - t).$$

Since $2^{z-2y+x+1} + t > t$, this equality can hold only if $t = 1$ and $z = 2y - x - 1$. The solutions are of the form $(x, y, 2y - x - 1)$ with x, y nonnegative integers.

946. With the substitution $u = 2x + 3$, $v = 2y + 3$, $w = 2z + 3$, the equation reads

$$u^2 + v^2 + w^2 = 7.$$

By eliminating the denominators, it is equivalent to show that the equation

$$U^2 + V^2 + W^2 = 7T^2$$

has no integer solution $(U, V, W, T) \neq (0, 0, 0, 0)$. Assuming the contrary, pick a solution for which $|U| + |V| + |W| + |T|$ is minimal. Reducing the equality modulo 4, we find that

$|U|$, $|V|$, $|W|$, $|T|$ are even, hence $\left(\frac{U}{2}, \frac{V}{2}, \frac{W}{2}, \frac{T}{2}\right)$ is also an integer solution, contradicting minimality. Hence the equation does not have solutions.

(Bulgarian Mathematical Olympiad, 1997)

947. *First solution*: One can see immediately that $x = 1$ is a solution. Assume that there exists a solution $x > 1$. Then $x!$ is even, so $3^{x!}$ has residue 1 modulo 4. This implies that the last digit of the number $2^{3^{x!}}$ is 2, so the last digit of $3^{2^{x!}} = 2^{3^{x!}} + 1$ is 3. But this is impossible because the last digit of an even power of 3 is either 1 or 9. Hence $x = 1$ is the only solution.

Second solution: We will prove by induction the inequality

$$3^{2^{x!}} < 2^{3^{x!}},$$

for $x \geq 2$. The base case $x = 2$ runs as follows: $3^{2^2} = 3^4 = 81 < 512 < 2^9 = 2^{3^2}$. Assume now that $3^{2^{x!}} < 2^{3^{x!}}$ and let us show that $3^{2^{(x+1)!}} < 2^{3^{(x+1)!}}$.

Raising the inequality $3^{2^{x!}} < 2^{3^{x!}}$ to the power $2^{x! \cdot x}$, we obtain

$$\left(3^{2^{x!}}\right)^{2^{x! \cdot x}} < \left(2^{3^{x!}}\right)^{2^{x! \cdot x}} < \left(2^{3^{x!}}\right)^{3^{x! \cdot x}}.$$

Therefore, $3^{2^{(x+1)!}} < 2^{3^{(x+1)!}}$, and the inequality is proved. The inequality we just proved shows that there are no solutions with $x \geq 2$. We are done.

Remark. The proof by induction can be avoided if we perform some computations. Indeed, the inequality can be reduced to

$$3^{2^{x!}} < 2^{3^{x!}}$$

and then to

$$x! < \frac{\log \log 3 - \log \log 2}{\log 3 - \log 2} = 1.13588\ldots$$

(Romanian Mathematical Olympiad, 1985)

948. *First solution*: The solutions are

$$(v + 1, v, 1, 1), \text{ for all } v; \quad (2, 1, 1, y), \text{ for all } y; \quad (2, 3, 2, 1), \ (3, 2, 2, 3).$$

To show that these are the only solutions, we consider first the simpler case $v = u + 1$. Then $u^x - (u + 1)^y = 1$. Considering this equation modulo u, we obtain $-1 \equiv u^x - (u + 1)^y \equiv 1 \pmod{u}$. So $u = 1$ or 2. The case $u = 1$ is clearly impossible, since then $v^y = 0$, so we have $u = 2$, $v = 3$. We are left with the simpler equation $2^x - 3^y = 1$. Modulo 3 it follows that x is even, $x = 2x'$. The equality $2^{2x'} - 1 = (2^{x'} - 1)(2^{x'} + 1) = 3^y$ can hold only if $x' = 1$ (the only consecutive powers of 3 that differ by 2 are 1 and 3). So $x = 2$, $y = 1$, and we obtain the solution $(2, 3, 2, 1)$.

Now suppose that $u = v + 1$. If $v = 1$, then $u = 2$, $x = 1$, and y is arbitrary. We have found the solution $(2, 1, 2, y)$. If $v = 2$, the equation reduces to $3^x - 2^y = 1$. If $y \geq 2$, then modulo 4 we obtain that x is even, $x = 2x'$, and so $3^{2x'} - 1 = (3^{x'} - 1)(3^{x'} + 1) = 2^y$. Two consecutive powers of 2 differ by 2 if they are 2 and 4. We find that either $x = y = 1$ or $x = 2$, $y = 3$. This gives the solutions $(2, 1, 1, 1)$ and $(3, 2, 2, 3)$.

Next let us assume $v \geq 3$. The case $y = 1$ gives the solutions $(v + 1, v, 1, 1)$. If $y > 1$, then v^2 divides u^y, so $1 \equiv (v + 1)^x \equiv 0 + \binom{x}{1}v + 1 \pmod{v^2}$, and therefore v divides x. Considering the equation modulo $v+1$, we obtain $1 \equiv (v+1)^x - v^y \equiv -(-1)^y \pmod{(v+1)}$. Since $v + 1 > 2$, $1 \not\equiv -1 \pmod{(v + 1)}$, so y must be odd. Now if $x = 1$, then $v^y = v$, so $v = 1$, giving again the family of solutions $(v + 1, v, 1, 1)$. So let us assume $x > 1$. Then $(v + 1)^2$ divides $(v + 1)^x$, so

$$1 \equiv (v + 1)^x - v^y \equiv -(v + 1 - 1)^y$$

$$\equiv 0 - \binom{y}{1}(v + 1)(-1)^{y-1} - (-1)^y$$

$$\equiv -y(v + 1) + 1 \pmod{(v + 1)^2}.$$

Hence $v + 1$ divides y. Since y is odd, $v + 1$ is odd and v is even. Since v divides x, x is also even. Because v is even and $v \geq 3$, it follows that $v \geq 4$. We will need the following result.

Lemma. *If a and q are odd, if $1 \leq m < t$, and if $a^{2^m}q \equiv 1 \pmod{2^t}$, then $a \equiv \pm 1 \pmod{2^{t-m}}$.*

Proof. First, let us prove the property for $q = 1$. We will do it by induction on m. For $m = 1$ we have $a^2 = (a - 1)(a + 1)$, so one of the factors is divisible by 2^{t-1}. Assume that the property is true for $m - 1$ and let us prove it for m. Factoring, we obtain $(a^{2^{m-1}} + 1)(a^{2^{m-1}} - 1)$. For $m \geq 2$, the first factor is 2 modulo 4, hence $a^{2^{m-1}}$ is 1 modulo 2^{t-1}. From the induction hypothesis it follows that $a \equiv \pm 1 \pmod{2^{t-m}}$ (note that $t - m = (t - 1) - (m - 1)$).

For arbitrary q, from what we have proved so far it follows that $a^q \equiv \pm 1 \pmod{2^{t-m}}$. Because $\phi(2^{t-m}) = 2^{t-m-1}$, by Euler's theorem $a^{2^{t-m-1}} \equiv 1 \pmod{2^{t-m}}$. Since q is odd, we can find a positive integer c such that $cq \equiv 1 \pmod{2^{t-m-1}}$. Then $a \equiv a^{cq} \equiv (\pm 1)^c \equiv \pm 1 \pmod{2^{t-m}}$, and the lemma is proved. $\square$

Let us return to the problem. Let $x = 2^m q$, where $m \geq 1$ and q is odd. Because $(v + 1)^x - v^y = 1$, clearly $y \geq x$. We have shown that $v + 1$ divides y, so $y \geq v + 1$. Let us prove that $y \geq 2m + 1$. Indeed, if $m \leq 2$ this holds since $y \geq v + 1 \geq 5 \geq 2m + 1$; otherwise, $y \geq x = 2^m q \geq 2^m \geq 2m + 1$.

Looking at the equation modulo 2^y, we have $(v + 1)^{2^m q} \equiv 1 \pmod{2^y}$, because 2^y divides v^y. Using this and the lemma we obtain that $v + 1 \equiv \pm 1 \pmod{2^{y-m}}$. But $v + 1 \equiv 1 \pmod{2^{y-m}}$ would imply that 2^{m+1} divides v, which is impossible since v divides x. Therefore, $v + 1 \equiv -1 \pmod{2^{y-m}}$ and $v \equiv -2 \pmod{2^{y-m}}$. In particular, $v \geq 2^{y-m} - 2$, so $y \geq 2^{y-m} - 1$. But since $y \geq 2m + 1$ and $y \geq 5$, it follows that $2^{y-m} - 1 > y$, a contradiction. This shows that there are no other solutions.

Second solution: Begin as before until we reduce to the case $u = v + 1$ and $v \geq 3$. Then we use the following lemma.

Lemma. *Suppose $p^s \geq 3$ is a prime power, $r \geq 1$, and $a \equiv 1 \pmod{p^s}$, but not $\mod p^{s+1}$. If $a^k \equiv 1 \pmod{p^{r+s}}$, then p^r divides k.*

Proof. Write $a = 1 + cp^s + dp^{s+1}$, where $1 \leq c \leq p - 1$. Then we compute $a^k \equiv 1 + kcp^s \pmod{p^{s+1}}$, and

$$a^p = 1 + cp^{s+1} + dp^{s+2} + \binom{p}{2}p^{2s}(c + dp) + \binom{p}{3}p^{3s}(c + dp)^3 + \cdots$$

Since either $s \geq 2$ or p is odd, p^{s+2} divides $\binom{p}{2}p^{2s}$; hence the fourth term is zero modulo p^{s+2}. Since $s + 2 \leq 3s$, the latter terms are also zero mod p^{s+2}; hence $a^p \equiv 1 \pmod{p^{s+1}}$, but not modulo p^{s+2}.

We now proceed by induction on r. Since $r \geq 1$, the first equation above shows that p divides k, which is the base case. For the inductive step, we note that the second calculation above lets us apply the previous case to $(a^p)^{k/p}$. $\qquad\square$

To use this lemma, let $p^s \geq 3$ be the highest power of the prime p that divides v. Then $u = v + 1 \equiv 1 \pmod{p^s}$, but not modulo p^{s+1}, and $u^x = v^y + 1 \equiv 1 \pmod{p^{sy}}$. Hence by the lemma, $p^{s(y-1)}$ divides x, and in particular, $x \geq p^{s(y-1)} \geq 3^{y-1}$. Thus either $x > y$ or $y = 1$.

Similarly, let $q^t \geq 3$ be the highest power of the prime q that divides u. Then $(-v) = 1 - u \equiv 1 \pmod{q^t}$, but not modulo q^{t+1}. Since $(-v)^y \equiv 1 \pmod{q^t}$ and $(-v)^y = (-1)^y - (-1)^y u^x \equiv (-1)^y \pmod{q^t}$, we see that y is even. Hence $(-v)^y = 1 - u^x \equiv 1 \pmod{q^{tx}}$. Thus by the lemma, $q^{t(x-1)}$ divides y, and in particular, $y \geq q^{t(x-1)} \geq 3^{x-1}$, so either $y > x$ or $x = 1$.

Combining these, we see that we must have either $x = 1$ or $y = 1$. Either of these implies the other and gives the solution $(v + 1, v, 1, 1)$.

Remark. Catalan conjectured in 1844 a more general fact, namely that the Diophantine equation $u^x - v^y = 1$ subject to the condition $x, y \geq 2$ has the unique solution $3^2 - 2^3 = 1$. This would mean that 8 and 9 are the only consecutive powers. Catalan's conjecture was proved by P. Mihăilescu in 2002.

(*Kvant* (*Quantum*), first solution by R. Barton, second solution by R. Stong)

Combinatorics and Probability

949. The relation from the statement implies

$$(A \cap X) \cup (B \cap X) = A \cap B.$$

Applying de Morgan's law, we obtain

$$(A \cup B) \cap X = A \cap B.$$

But the left-hand side is equal to $(A \cup B \cup X) \cap X$. and this is obviously equal to X. Hence $X = A \cap B$.

(Russian Mathematics Competition, 1977)

950. We prove the property by induction on the number of elements of the set. For a set with one element the property clearly holds. Let us assume that we could find the required list $A_1, A_2, \ldots, A_{2^n}$ of the subsets of the set with n elements, $n \geq 1$. Add the element x to obtain a set with $n+1$ elements. The list for this new set is

$$A_1, A_2, \ldots, A_{2^n}, A_{2^n} \cup \{x\}, \ldots, A_2 \cup \{x\}, A_1 \cup \{x\},$$

and the induction is complete.

951. Fix $A \in \mathcal{F}$ and consider the function $f : \mathcal{P}(S) \to \mathcal{P}(S)$ on the subsets of S, $f(X) = X \triangle A$. Because

$$f(f(X)) = (X \triangle A) \triangle A = ((X \triangle A) \backslash A) \cup (A \backslash (X \triangle A))$$
$$= (X \backslash A) \cup (X \cap A) = X,$$

f is one-to-one. Therefore, $f(\mathcal{F})$ has at least m elements. The conclusion follows.

(I. Tomescu, *Problems in Combinatorics*, Wiley, 1985)

© Springer International Publishing AG 2017

R. Gelca and T. Andreescu, *Putnam and Beyond*, DOI 10.1007/978-3-319-58988-6

952. If all functions f_n, $n = 1, 2, 3, \ldots$, are onto, then the property is obvious. We will reduce the general situation to this particular one. For some k and n, define

$$B_{n,k} = (f_n \circ f_{n+1} \circ \cdots \circ f_{n+k-1})(A_{n+k}).$$

We have the descending sequence of sets

$$A_n \supset B_{n,1} \supset B_{n,2} \supset \cdots$$

Because all these sets are finite, the sequence is stationary, so there exists k_0 such that $B_{n,k} = B_{n,k+1}$, for $k \geq k_0$. Let $B_n = B_{n,k_0}$. It is not hard to see that $f_n(B_{n+1}) = B_n$, and in this way we obtain a sequence of sets and surjective maps. For these the property holds; hence it holds for the original sets as well.

(C. Năstăsescu, C. Niţă, M. Brandiburu, D. Joiţa, *Exerciţii şi Probleme de Algebră* (*Exercises and Problems in Algebra*), Editura Didactică şi Pedagogică, Bucharest, 1983)

953. For a person X we will denote by m_X the number of people he knows. Let A and B be two people who know each other. We denote by M_A and M_B the set of acquaintances of A, respectively, B. By hypothesis M_A and M_B are disjoint. If $X \in M_A$, then X has exactly one acquaintance in M_B. Indeed, either $X = A$, in which case he only knows B in M_B, or $X \neq A$, in which case he does not know B, so he has exactly one common acquaintance with B. This latter person is the only one he knows in M_B. Similarly, any person in M_B has exactly one acquaintance in M_A. This allows us to establish a bijection between M_A and M_B, and conclude that $m_A = m_B$.

Finally, if A and B do not know each other, then they have a common acquaintance C. The above argument shows that $m_A = m_B = m_C$ and we are done.

(*Kvant* (*Quantum*))

954. The answer is positive. Because 1% of 20 million is $200,000 > 3^{11} = 177147$, we can actually assume that the number of supporters of the president is 3^{11}.

First, call a group of type O if all of its members are opponents of Miraflores and of type S if the group contains some supporters. Divide the voters in 5 groups of 4 million each, such that the first two groups are of type O. Then divide the two groups of type O arbitrarily in 5 subgroups, then again in 5 until we obtain groups of size 2^8, then divide each of these in 16 groups of 16 people.

We now have to design the groups of type S. Divide the supporters evenly among these groups, 3^{10} in each. Then divide each of the three groups of type S into five groups, two of which are of type O and three of which are of type S. Now repeat until we reach the groups of size 2^8. Divide each into 2^4 groups, make 7 of them be of type O and 9 of type S. In each of the type S groups of size 16 put 9 supporters. At this moment the entire population has been evenly divided and in each group of type S that shows up in the successive division the supporters of Miraflores win.

(23rd Moscow Mathematical Olympiad)

955. (a) The first player starts by writing 6 on the blackboard. In what follows only the numbers 4, 5, 7, 8, 9, 10 can be written. Split them in pairs $(4, 5)$, $(7, 9)$, $(8, 10)$. At each move, whenever the second player writes one of the numbers from the pair, the first player writes the second number of that pair.

(b) The first player has a winning strategy. Assume that this is not the case, and so that for every number the player chooses, the second can continue to a win. Then the first player can just write 1, because then whatever number the second player writes, the first can continue to a win.

(*Kvant (Quantum)*, proposed by D. Ivanov)

956. Note that the product of the three elements in each of the sets $\{1, 4, 9\}$, $\{2, 6, 12\}$, $\{3, 5, 15\}$, and $\{7, 8, 14\}$ is a square. Hence none of these sets is a subset of M. Because they are disjoint, it follows that M has at most $15 - 4 = 11$ elements.

Since 10 is not an element of the aforementioned sets, if $10 \notin M$, then M has at most 10 elements. Suppose $10 \in M$. Then none of $\{2, 5\}$, $\{6, 15\}$, $\{1, 4, 9\}$, and $\{7, 8, 14\}$ is a subset of M. If $\{3, 12\} \not\subset M$, it follows again that M has at most 10 elements. If $\{3, 12\} \subset M$, then none of $\{1\}$, $\{4\}$, $\{9\}$, $\{2, 6\}$, $\{5, 15\}$, and $\{7, 8, 14\}$ is a subset of M, and then M has at most 9 elements. We conclude that M has at most 10 elements in any case.

Finally, it is easy to verify that the subset

$$M = \{1, 4, 5, 6, 7, 10, 11, 12, 13, 14\}$$

has the desired property. Hence the maximum number of elements in M is 10.

(Short list of the 35th International Mathematical Olympiad, 1994, proposed by Bulgaria)

957. Let us try to find a counterexample. We represent the numbers as the six vertices of a graph. If two numbers are not coprime, we connect them by an edge labeled by their greatest common divisor. The labels must be pairwise coprime, or else we could find three (actually four) numbers with the greatest common divisor greater than 1. A second condition is that from any three vertices, two must be connected by an edge, for otherwise the three vertices would correspond to numbers that are pairwise coprime. We can cosider the complete graph with six vertices color its edges by the first 15 prime numbers. Place at each vertex the product of the labels of the edges adjacent to the vertex. The set of six numbers obtained this way is a counterexample.

Remark. One should contrast this with the property that given a complete graph with six vertices whose edges are colored by two colors, there is a monochromatic triangle.

(*Kvant (Quantum)*)

958. Let us try to construct such a set S. Chosing all numbers greater than $n/2$ solves the first requirement. To fullfil the second, it suffices to choose all even numbers. It is not difficult to see that the number of elements in this S is $\lfloor \frac{n+2}{4} \rfloor$.

We claim that this is the maximum number. Indeed, let S be a set with the required properties and let a be its least element. If $a \leq \frac{n}{2}$, replace a by $2a$. Clearly the new set still has the required properties. Repeat this until all elements are greater than $n/2$. And there should be no consecutive elements.

(Balkan Mathematical Olympiad, 2005)

959. We will prove this by induction on the number of digits of the given number. For a 3-digit number, two digits have the same parity. Delete the third and you are done.

Now let us assume that the property is true for any $2k - 1$-digit number and let us prove it for a $2k + 1$ digit number N. N either has two consecutive digits of the same parity, or the first and the last have the same parity. In either case we can ignore those digits for a moment, to obtain a $2k - 1$ digit number. Delete a digit of this number to obtain an acceptable number. Adding back the two digits we obtain a number that is still acceptable. This completes the induction.

(*Kvant (Quantum)*, proposed by A. Sidorenko)

960. Label initially the people in order, by $0, 1, \ldots, 2n - 1$. Since the table is round, we consider the labels modulo $2n$. Assume that after the rearrangement, the person on position j moved to the position $j + f(j)$, for some $f(j) \in \{0, 1, \ldots, 2n - 1\}$ (recall that everything is considered modulo $2n$). We have to show that there are two people j and k, such that $f(j) = f(k)$. If this is not true, then $f(1), f(2), \ldots, f(2n)$ is a permutation of $1, 2, \ldots, 2n$. Then on the one hand

$$\sum_{j=1}^{2n} (j + f(j)) \equiv 2 \sum_{j=1}^{2n} j = 2n(2n + 1) \equiv 0(\bmod 2n),$$

and on the other hand

$$\sum_{j=1}^{2n} (j + f(j)) \equiv \sum_{j=1}^{2n} j = n(2n + 1) \equiv n(\bmod 2n),$$

the latter because the residule classes of the numbers $j + f(j)$ are a permutation of $0, 1, 2, \ldots, 2n - 1$. Since $n \not\equiv 0(\bmod 2n)$, this cannot happen, so two of the $f(j)$'s are equal.

If the number of people is odd, say $n = 2n + 1$, let $\sigma(j) = 2j - 1$ if $1 \leq j \leq n$ and $\sigma(j) = 2(j - n)$ if $j \geq n + 1$, where $\sigma(j)$ is the position of the jth person after the break. It is not hard to see that if $j < k$, then $k - j \neq |\sigma(k) - \sigma(j)|$, hence for any two people, the number of people sitting between them changes after the break.

(Romanian Mathematics Competition, 1989, see also German Math Olympiad, 1976)

961. Because the sum of the elements in the given set is $nk(nk + 1)/2$, for the partition to exist it is necessary that either k is even or nk is odd. Let us show that this is also a sufficient condition.

For easy reference and intuition, we will write the sets one underneath the other as the rows of an $n \times k$ table.

Case I. k even. We write the numbers $1, 2, \ldots, nk$ in the table in a snake-like fashion, as shown in Figure 101 for $n = 5$ and $k = 6$.

5	6	15	16	25	26
4	7	14	17	24	27
3	8	13	18	23	28
2	9	12	19	22	29
1	10	11	20	21	30

Figure 101

Case II. Let us first see how this can be done for $k = 3$, namely for a table of the form $(2m + 1) \times 3$. In this table we first set $a_{i1} = i$, $i = 1, 2, \ldots, 2m + 1$. Then we set $a_{i2} = 3m + 1 + i$, $1 \le i \le m + 1$, and $a_{i2} = m + i$ if $m + 2 \le i \le 2m + 1$. Finally, $a_{i3} = 6m + 4 - 2i$, $1 \le i \le m + 1$ and $a_{i3} = 8m + 6 - 2i$ if $m + 2 \le i \le 2m + 1$. It is not hard to check that the sum of the elements on each row is $9m + 6$.

For an arbitrary k, write the first three rows in this fashion, and then arrange the remaining $n(k - 3)$ numbers in the snake-like fashion from Case I to the right of the table (here $k - 3$ is even, so this is possible). The rows of the resulting table have the same sum of elements.

(Kvant (Quantum), proposed by S. Berkolaiko)

962. (a) Let us assume that 1 and n are placed at A_i respectively A_j. The sum of the absolute values of the differences of neighboring numbers one either of the two arcs joining A_i to A_j is at least $n - 1$, so the sum of all these absolute values is at least $2n - 2$.

(b) In order for the value $2n - 2$ to be reached, the numbers must be written in increasing order on each of the arcs. Each such configuration is uniquely determined if we specify the location of the number 1, and the numbers on the forward arc from 1 to n. These numbers form a subset of $\{2, 3, \cdots, n - 1\}$, so they can be chosen in 2^{n-2} ways. The location of 1 can be chosen in n ways. So the answer to the question is $n2^{n-2}$.

(Kvant (Quantum), proposed by A. Razborov)

963. For $m = 2^n$ and $k = 2^{n-1} + n$, we will construct a sequence of positive integers with the required property. Let A be the set consisting of all odd numbers and all powers of 2 less or equal to 2^n. Then A has $2^{n-1} + n$ elements, and we will show that it has the desired property.

If $p < q$, then $2^p + 2^q = 2^p(2^{q-p} + 1)$, which does not divide $2^p 2^q$ because it has an odd divisor greater than 1. A number of the form $2^p + 2k + 1$ is odd and greater than $2k + 1$, so it cannot divide $2^p(2k + 1)$. Finally, the sum of two odd numbers is even, so it cannot divide their product. It follows that A has the desired property.

(Romanian Team Selection Test for the International Mathematical Olympiad, 1983, proposed by I. Tomescu)

964. Assume we can place the numbers so that the difference between the numbers in adjacent triangles is at most 3. Divide the big triangle into 4 equal equilateral triangles T_1, T_2, T_3, T_4, with T_4 at the center. Consider the sets $S_1 = \{1, 2, 3, 4, 5\}$ and $S_2 = \{12, 13, 14, 15, 16\}$.

None of these 4 triangles can contain numbers from both sets. So there will be 2 of these triangles that contain only numbers from S_1 (say T_1 and T_2) and 2 of these triangles that contain only numbers from S_2 (say T_3 and T_4). Now consider the division of T_4 into 4 small triangles t_1, t_2, t_3, t_4 (again with the convention that t_4 lies at the center). Notice that between a number from S_1 and a number from S_2 must lie at least 2 squares.

If one number from S_2 lies in the center of t_4, it follows that the numbers from S_1 must lie in the 3 rhombi formed at the vertices of the big triangle. But only the rhombi from T_1 and T_2 are allowed, and there are 4 equilateral triangles to be filled with 5 numbers. Impossible. So the number must be in t_1, t_2 or t_3. A second number from S_2 lying in t_4 would allow only at most 4 triangles in T_1 and T_2 to be filled with numbers from S_1, again a contradiction. Hence t_4 contains just one number from S_2, and this number is not in the center. In fact the triangle containing it must touch T_3, for if it touches T_1 or T_2, there won't be enough fields in these triangles to fill with numbers from S_1.

So the five elements of S_2 fill completely T_3 and one triangle of T_4 that is adjacent to T_3 (say t_3). For the numbers from S_1 the only available fields are the 4 triangles that touch the side opposite to T_3 and one of the triangles that fills a parallelogram in the corner (the one that is furthest away from t_3). The number 1 is in one of these fields, and it is adjacent to a triangle that contains a number that is not in S_1. That number is greater than 5, a contradiction. We are done.

(*Mathematical Reflections*, proposed by I. Borsenco)

965. If the lines are parallel then the integers are all zero and we violate (i). So we have at least two non-parallel lines.

To each half-plane we associate either a $+1$ or a -1. We color each point of the plane with the product of the numbers associated to the half-planes that cover it. Finally associate to each region its color multiplied by the number of vertices that the region has.

Note that neighboring regions are colored by numbers of opposite signs, and two such numbers a and b satisfy $ab < a + b$. To see that the sum of the numbers in one half-plane is zero, count by intersection points of lines, and note that at each such intersection points one has the colors $(+1, -1, +1, -1)$ around the point if the point lies on the boundary, and $(+1, -1)$ if the point lies in the interior.

(Balkan Mathematical Olympiad, 2004)

966. We solve the more general case of the permutations of the first $2n$ positive integers, $n \geq 1$. The average of the sum

$$\sum_{k=1}^{n} |a_{2k-1} - a_{2k}|$$

is just n times the average value of $|a_1 - a_2|$, because the average value of $|a_{2i-1} - a_{2i}|$ is the same for all $i = 1, 2, \ldots, n$. When $a_1 = k$, the average value of $|a_1 - a_2|$ is

$$\frac{(k-1) + (k-2) + \cdots + 1 + 1 + 2 + \cdots + (2n-k)}{2n-1}$$

$$= \frac{1}{2n-1} \left[\frac{k(k-1)}{2} + \frac{(2n-k)(2n-k+1)}{2} \right] = \frac{k^2 - (2n+1)k + n(2n+1)}{2n-1}.$$

It follows that the average value of the sum is

$$n \cdot \frac{1}{2n} \sum_{k=1}^{2n} \frac{k^2 - (2n+1)k + n(2n+1)}{2n-1}$$

$$= \frac{1}{4n-2} \left[\frac{2n(2n+1)(4n+1)}{6} - (2n+1)\frac{2n(2n+1)}{2} + 2n^2(2n+1) \right]$$

$$= \frac{n(2n+1)}{3}.$$

For our problem $n = 5$ and the average of the sums is $\frac{55}{3}$.

(American Invitational Mathematics Examination, 1996)

967. The condition from the statement implies that any such permutation has exactly two disjoint cycles, say $(a_{i_1}, \ldots, a_{i_r})$ and $(a_{i_{r+1}}, \ldots, a_{i_6})$. This follows from the fact that in order to transform a cycle of length r into the identity, $r - 1$ transpositions are needed. Moreover, we can only have $r = 5, 4,$ or 3.

When $r = 5$, there are $\binom{6}{1}$ choices for the number that stays unpermuted. There are $(5-1)!$ possible cycles, so in this case we have $6 \times 4! = 144$ possibilities.

When $r = 4$, there are $\binom{6}{4}$ ways to split the numbers into the two cycles (two cycles are needed and not just one). One cycle is a transposition. There are $(4-1)! = 6$ choices for the other. Hence in this case the number is 90. Note that here exactly four transpositions are needed.

Finally, when $r = 3$, then there are $\binom{6}{3} \times (3-1)! \times (3-1)! = 40$ cases. Therefore, the answer to the problem is $144 + 90 + 40 = 274$.

(S. Korean Mathematical Olympiad, 1999)

968. We would like to find a recursive scheme for $f(n)$. Let us attempt the less ambitious goal of finding a recurrence relation for the number $g(n)$ of permutations of the desired form satisfying $a_n = n$. In that situation either $a_{n-1} = n - 1$ or $a_{n-1} = n - 2$, and in the latter case we necessarily have $a_{n-2} = n - 1$ and $a_{n-3} = n - 3$. We obtain the recurrence relation

$$g(n) = g(n-1) + g(n-3), \text{ for } n \geq 4.$$

In particular, the values of $g(n)$ modulo 3 are $1, 1, 1, 2, 0, 1, 0, 0, \ldots$ repeating with period 8.

Now let $h(n) = f(n) - g(n)$. We see that $h(n)$ counts permutations of the desired form with n occurring in the middle, sandwiched between $n - 1$ and $n - 2$. Removing n leaves an acceptable permutation, and any acceptable permutation on $n-1$ symbols can be so produced, except those ending in $n - 4, n - 2, n - 3, n - 1$. So for $h(n)$, we have the recurrence

$$h(n) = h(n-1) + g(n-1) - g(n-4) = h(n-1) + g(n-2), \text{ for } n \geq 5.$$

A routine check shows that modulo 3 $h(n)$ repeats with period 24.

We find that $f(n)$ repeats with period equal to the least common multiple of 8 and 24, which is 24. Because $1996 \equiv 4 \pmod{24}$, we have $f(1996) \equiv f(4) = 4 \pmod{3}$. So $f(1996)$ is not divisible by 3.

(Canadian Mathematical Olympiad, 1996)

969. To solve this problem we will apply Sturm's principle, a method discussed in Section 2.1.6. The fact is that as σ ranges over all permutations, there are $n!$ sums of the form

$$\sum_{i=1}^{n}(x_i - y_{\sigma(i)})^2,$$

and one of them must be the smallest. If σ is not the identity permutation, then it must contain an inversion, i.e., a pair (i, j) with $i < j$ and $\sigma(i) > \sigma(j)$. We have

$$(x_i - y_{\sigma(i)})^2 + (x_j - y_{\sigma(j)})^2 - (x_i - y_{\sigma(j)})^2 - (x_j - y_{\sigma(i)})^2 = (x_j - x_i)(y_{\sigma(i)} - y_{\sigma(j)}).$$

This product is positive, so by exchanging $y_{\sigma(i)}$ and $y_{\sigma(j)}$ we decrease the sum. This means that this permutation does not minimize the sum. Therefore, the sum is minimal for the identity permutation. The inequality follows.

970. Let $N(\sigma)$ be the number we are computing. Denote by $N_i(\sigma)$ the average number of large integers a_i. Taking into account the fact that after choosing the first $i - 1$ numbers, the ith is completely determined by the condition of being large, for any choice of the first $i - 1$ numbers there are $(n - i + 1)!$ choices for the last $n - i + 1$, from which $(n - i)!$ contain a large integer in the ith position. We deduce that $N_i(\sigma) = \frac{1}{n-i+1}$.

The answer to the problem is therefore

$$N(\sigma) = \sum_{i=1}^{n} N_i(\sigma) = 1 + \frac{1}{2} + \cdots + \frac{1}{n}.$$

(19th W.L. Putnam Mathematical Competition, 1958)

971. We will show that σ is the identity permutation. Assume the contrary and let $(i_1, i_2, \ldots, i_k)$ be a cycle, i.e., $\sigma(i_1) = i_2$, $\sigma(i_2) = i_3, \ldots, \sigma(i_k) = i_1$. We can assume that i_1 is the smallest of the i_j's, $j = 1, 2, \ldots, k$. From the hypothesis,

$$a_{i_1} a_{i_2} = a_{i_1} a_{\sigma(i_1)} < a_{i_k} a_{\sigma(i_k)} = a_{i_k} a_{i_1},$$

so $a_{i_2} < a_{i_k}$ and therefore $i_2 < i_k$. Similarly,

$$a_{i_2} a_{i_3} = a_{i_2} a_{\sigma(i_2)} < a_{i_k} a_{\sigma(i_k)} = a_{i_k} a_{i_1},$$

and since $a_{i_2} > a_{i_1}$ it follows that $a_{i_3} < a_{i_k}$, so $i_3 < i_k$. Inductively, we obtain that $i_j < i_k$, $j = 1, 2, \ldots, k - 1$. But then

$$a_{i_{k-1}} a_{i_k} = a_{i_{k-1}} a_{\sigma(i_{k-1})} < a_{i_k} a_{\sigma(i_k)} = a_{i_k} a_{i_1},$$

hence $i_{k-1} < i_1$, a contradiction. This proves that σ is the identity permutation, and we are done.

(C. Năstăsescu, C. Niţă, M. Brandiburu, D. Joiţa, *Exerciţii şi Probleme de Algebră* (*Exercises and Problems in Algebra*), Editura Didactică şi Pedagogică, Bucharest, 1983)

972. Let $S = \{1, 2, \ldots, 2004\}$. Write the permutation as a function $f : S \to S, f(n) = a_n$, $n = 1, 2, \ldots, 2004$. We start by noting three properties of f:

(i) $f(i) \neq i$ for any i,

(ii) $f(i) \neq f(j)$ if $i \neq j$,

(iii) $f(i) = j$ implies $f(j) = i$.

The first two properties are obvious, while the third requires a proof. Arguing by contradiction, let us assume that $f(i) = j$ but $f(j) \neq i$. We discuss first the case $j > i$. If we let $k = j - i$, then $f(i) = i + k$. Since $k = |f(i) - i| = |f(j) - j|$ and $f(j) \neq i$, it follows that $f(j) = i + 2k$, i.e., $f(i + k) = i + 2k$. The same reasoning yields $f(i + 2k) = i + k$ or $i + 3k$. Since we already have $f(i) = i + k$, the only possibility is $f(i + 2k) = i + 3k$. And the argument can be repeated to show that $f(i + nk) = i + (n + 1)k$ for all n. However, this then forces f to attain ever increasing values, which is impossible since its range is finite. A similar argument takes care of the case $j < i$. This proves (iii).

The three properties show that f is an involution on S with no fixed points. Thus f partitions S into 1002 distinct pairs (i, j) with $i = f(j)$ and $j = f(i)$. Moreover, the absolute value of the difference of the elements in any pair is the same. If $f(1) - 1 = k$ then $f(2) = k + 1, \ldots,$ $f(k) = 2k$, and since f is an involution, the values of f on $k + 1, k + 2, \ldots, 2k$ are already determined, namely $f(k + 1) = 1, f(k + 2) = 2, \ldots, f(2k) = k$. So the first block of $2k$ integers is invariant under f. Using similar reasoning, we obtain $f(2k + 1) = 3k + 1$, $f(2k + 2) = 3k + 2, \ldots, f(3k) = 4k, f(3k + 1) = 2k + 1, \ldots, f(4k) = 3k$. So the next block of $2k$ integers is invariant under f. Continuing this process, we see that f partitions S into blocks of $2k$ consecutive integers that are invariant under f. This can happen only if $2k$ divides 2004, hence if k divides 1002. Furthermore, for each such k we can construct f following the recipe given above. Hence the number of such permutations equals the number of divisors of 1002, which is 8.

(Australian Mathematical Olympiad, 2004, solution by L. Field)

973. Expanding $|\sigma(k) - k|$ as $\pm\sigma(k) \pm k$ and reordering, we see that

$$|\sigma(1) - 1| + |\sigma(2) - 2| + \cdots + |\sigma(n) - n| = \pm 1 \pm 1 \pm 2 \pm 2 \pm \cdots \pm n \pm n,$$

for some choices of signs. The maximum of $|\sigma(1) - 1| + |\sigma(2) - 2| + \ldots + |\sigma(n) - n|$ is reached by choosing the smaller of the numbers to be negative and the larger to be positive, and is therefore equal to

$$2\left(-1 - 2 - \cdots - \frac{n-1}{2}\right) - \frac{n+1}{2} + \frac{n+1}{2} + 2\left(\frac{n+3}{2} + \cdots + n\right)$$

$$= -\left(1 + \frac{n-1}{2}\right)\frac{n-1}{2} + \left(\frac{n+3}{2} + n\right)\frac{n-1}{2} = \frac{n^2 - 1}{2}.$$

Therefore, in order to have $|\sigma(1) - 1| + \cdots + |\sigma(n) - n| = \frac{n^2-1}{2}$, we must have

$$\left\{\sigma(1), \cdots, \sigma\left(\frac{n-1}{2}\right)\right\} \subset \left\{\frac{n+1}{2}, \frac{n+3}{2}, \ldots, n\right\}$$

and

$$\left\{\sigma\left(\frac{n+3}{2}\right), \sigma\left(\frac{n+5}{2}\right), \ldots, \sigma(n)\right\} \subset \left\{1, 2, \ldots, \frac{n+1}{2}\right\}.$$

Let $\sigma\left(\frac{n+1}{2}\right) = k$. If $k \leq \frac{n+1}{2}$, then

$$\left\{\sigma(1), \ldots, \sigma\left(\frac{n-1}{2}\right)\right\} = \left\{\frac{n+3}{2}, \frac{n+5}{2}, \ldots, n\right\}$$

and

$$\left\{\sigma\left(\frac{n+3}{2}\right), \sigma\left(\frac{n+5}{2}\right), \ldots, \sigma(n)\right\} = \left\{1, 2, \ldots, \frac{n+1}{2}\right\} - \{k\}.$$

If $k \geq \frac{n+1}{2}$, then

$$\left\{\sigma(1), \ldots, \sigma\left(\frac{n-1}{2}\right)\right\} = \left\{\frac{n+1}{2}, \frac{n+3}{2}, \ldots, n\right\} - \{k\}$$

and

$$\left\{\sigma\left(\frac{n+3}{2}\right), \sigma\left(\frac{n+5}{2}\right), \ldots, \sigma(n)\right\} = \left\{1, 2, \ldots, \frac{n-1}{2}\right\}.$$

For any value of k, there are $\left[\left(\frac{n-1}{2}\right)!\right]^2$ choices for the remaining values of σ, so there are

$$n\left[\left(\frac{n-1}{2}\right)!\right]^2$$

such permutations.

(T. Andreescu)

974. Color the triangles black and white in a chessboard pattern (or equivalently look at the triangles that are oriented upwards and those that are oriented downwards). One color exceeds the other by n, and two consecutive triangles have opposite colors. So a chain cannot have more than $n^2 - n + 1$ triangles. A chain with $n^2 - n + 1$ triangles is shown in Figure 102.

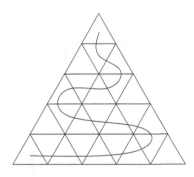

Figure 102

975. The grid is made up of $\frac{n(n+1)}{2}$ small equilateral triangles of side length 1. In each of these triangles, at most 2 segments can be marked, so we can mark at most $\frac{2}{3} \cdot \frac{3n(n+1)}{2} = n(n+1)$ segments in all. Every segment points in one of three directions, so we can achieve the maximum $n(n+1)$ by marking all the segments pointing in two of the three directions.

(Russian Mathematical Olympiad, 1999)

976. To each point we associate a triple of coordinates $(x_1, x_2, x_3) \in \{0, 1, \ldots, n\}^3$, where x_i is the number of units one has to travel in the direction of the ith side in order to meet the $i + 1st$ side (with i taken modulo 3). Then $x_1 + x_2 + x_3 = 2n$. Reformulating, the problem asks us to find the largest number of non-negative integers solutions to the equation

$$x_1 + x_2 + x_3 = 2n,$$

that can be chosen so that in no two solutions the variable x_i is the same ($i = 1, 2, 3$). In other words, what is the largest number m such that one can produce a $3 \times m$ array consisting of non-negative integers, so that the sum of the numbers on each row is m and the numbers on each column are different.

We will show that $m = \lfloor \frac{4n}{3} \rfloor + 1$. To prove that m cannot exceed $\lfloor \frac{4n}{3} \rfloor + 1$, we argue as follows.

Because the numbers in each column do not repeat, the sum of the numbers in that column is at least

$$0 + 1 + \cdots + (m - 1) = \frac{m(m - 1)}{2}.$$

In the array there are 3 columns, so the sum of all numbers is at least $\frac{3m(m-1)}{2}$. On the other hand, because the sum on a row is $2n$, the total sum is $2mn$. So

$$2mn \geq \frac{3m(m - 1)}{2}.$$

It follows that $\frac{4n}{3} \geq m - 1$, and hence the inequality. Let us show that the bound can be attained.

If $2n = 3k$, then $m = 2k + 1$, and we can choose the following solutions

$$(2j, k - j, 2k - j), \quad j = 0, 1, \ldots, k$$
$$(2j + 1, 2k - j, k - j - 1), \quad j = 0, 1, \ldots, k - 1.$$

If $2n = 3k + 1$, then $m = 2k + 1$, and we can choose the following solutions

$$(2j, k - j, 2k + 1 - j), \quad j = 0, 1, \ldots, k$$
$$(2j + 1, 2k - j, k - j), \quad j = 0, 1, \ldots, k - 1.$$

If $2n = 3k - 1$, then $m = 2k$ and we choose the following solutions

$$(2j, k - j - 1, 2k - j), \quad j = 0, 1, \ldots, k - 1$$
$$(2j + 1, 2k - 1 - j, k - 1 - j), \quad j = 0, 1, \ldots, k - 1.$$

(*Kvant (Quantum)*, proposed by M.L. Gerver)

977. Let us consider a red square R_0 and the 3×3 square S centered at R_0. None of the squares next to R_0 on the vertical and horizontal can be red, or else the 2×3 rectangle that lies in S and does not contain this square has only one red square, namely R_0. Then two opposite corners of S must be red, and the other two are blue. Analyzing all 3×3 squares, we conclude that the lattice consist of "diagonals" of red squares separated by two diagonals of "blue" squares. As such, every 3×3 square contains exactly three red squares.

As a 9×11 rectangle can be dissected into nine 3×3 squares and three 2×3 rectangles, it contains precisely $9 \cdot 3 + 3 \cdot 2 = 33$ red squares.

(*Kvant (Quantum)*, proposed by N. Kartashov)

978. Assume that such a tiling exists. In this case each of the 10 interior segments that join opposite sides in the 6×6 lattice intersects one tile. In fact each such segment must intersect an even number of tiles, because on each side of the segment there is a rectangle of even area. And no tile is cut by two such segments. But this would imply that there are at least 20 tiles, contrary to the fact that we only have 18 tiles at hand. Hence such a tiling does not exist.

(*Kvant (Quantum)*, proposed by A.A. Kirilov, solution by N.B. Vassiliev)

979. (a) Divide the rectangle into unit squares and color red the unit squares at odd locations on odd rows. Then each 1×4 rectangle covers an even number of red squares (two or zero), while a 2×2 square covers one red square. So the answer to the problem is negative.

(b) Sometimes it is possible, as shown in Figure 103.

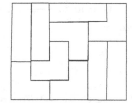 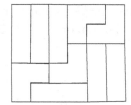

Figure 103

980. *First solution*: We will prove that the maximum value of n is 11. Figure 104 describes an arrangement of 12 dominoes such that no additional domino can be placed on the board. Therefore, $n \leq 11$.

Figure 104

Let us show that for any arrangement of 11 dominoes on the board one can add one more domino. Arguing by contradiction, let us assume that there is a way of placing 11 dominoes on the board so that no more dominoes can be added. In this case there are $36 - 22 = 14$ squares not covered by dominoes.

Denote by S_1 the upper 5×6 subboard, by S_2 the lower 1×6 subboard, and by S_3 the lower 5×6 subboard of the given chessboard as shown in Figure 105.

Because we cannot place another domino on the board, at least one of any two neighboring squares is covered by a domino. Hence there are at least three squares in S_2 that are covered by dominoes, and so in S_2 there are at most three uncovered squares. If A denotes the set of uncovered squares in S_1, then $|A| \geq 14 - 3 = 11$.

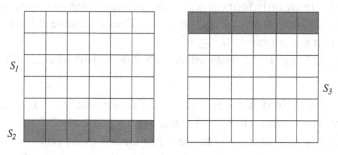

Figure 105

Let us also denote by B the set of dominoes that lie completely in S_3. We will construct a one-to-one map $f : A \rightarrow B$. First, note that directly below each square s in S_1 there is a square t of the chessboard (see Figure 106). If s is in A, then t must be covered by a domino d in B, since otherwise we could place a domino over s and t. We define $f(s) = d$. If f were not one-to-one, that is, if $f(s_1) = f(s_2) = d$, for some $s_1, s_2 \in A$, then d would cover squares directly below s_1 and s_2 as described in Figure 106. Then s_1 and s_2 must be neighbors, so a new domino can be placed to cover them. We conclude that f is one-to-one, and hence $|A| \leq |B|$. It follows that $|B| \geq 11$. But there are only 11 dominoes, so $|B| = 11$. This means that all 11 dominoes lie completely in S_3 and the top row is not covered by any dominoes! We could then put three more dominoes there, contradicting our assumption on the maximality of the arrangement. Hence the assumption was wrong; one can always add a domino to an arrangement of 11 dominoes. The answer to the problem is therefore $n = 11$.

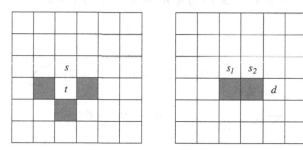

Figure 106

Second solution: Suppose we have an example with k dominoes to which no more can be added. Let X be the number of pairs of an uncovered square and a domino that covers an adjacent square. Let $m = 36 - 2k$ be the number of uncovered squares, let m_∂ be the number of uncovered squares that touch the boundary (including corner squares), and m_c the number of uncovered corner squares. Since any neighbor of an uncovered square must be covered by some domino, we have $X = 4m - m_\partial - m_c$. Similarly, let k_∂ be the number of dominoes that touch the boundary and k_c the number of dominoes that contain a corner square. A domino in the center of the board can have at most four unoccupied neighbors, for otherwise, we could place a new domino adjacent to it. Similarly, a domino that touches the boundary can have at most three unoccupied neighbors, and a domino that contains a corner square can have at most two unoccupied neighbors. Hence $X \leq 4k - k_\partial - k_c$. Also, note that $k_\partial \geq m_\partial$, since as we go around the boundary we can never encounter two unoccupied squares in a row, and $m_c + k_c \leq 4$, since there are only four corners. The inequality $4m - m_\partial - m_c = X \leq 4k - k_\partial - k_c$ gives $4m - 4 \leq 4k$; hence $35 - 3k \leq k$ and $4k \geq 35$. Thus k must be at least 12. This argument also shows that on an $n \times n$ board, $3k^2 \geq n^2 - 1$.

(T. Andreescu, Z. Feng, 102 *Combinatorial Problems*, Birkhäuser, 2000, second solution by R. Stong)

981. Let $f(n)$ be the desired number. We count immediately $f(1) = 2, f(2) = 4$. For the general case we argue inductively. Assume that we already have constructed n circles. When adding the $(n + 1)$st, it intersects the other circles in $2n$ points. Each of the $2n$ arcs determined by those points splits some region in two. This produces the recurrence relation $f(n + 1) = f(n) + 2n$. Iterating, we obtain

$$f(n) = 2 + 2 + 4 + 6 + \cdots + 2(n - 1) = n^2 - n + 2.$$

(25th W.L. Putnam Mathematical Competition, 1965)

982. Again we try to derive a recursive formula for the number $F(n)$ of regions. But this time counting the number of regions added by a new sphere is not easy at all. The previous problem comes in handy. The first n spheres determine on the $(n+1)$st exactly $n^2 - n + 2$ regions. This is because the conditions from the statement give rise on the last sphere to a configuration of circles in which any two, but no three, intersect. And this is the only condition that we used in the solution to the previous problem. Each of the $n^2 - n + 2$ spherical regions divides some spatial region into two parts. This allows us to write the recursive formula

$$F(n + 1) = F(n) + n^2 - n + 2, \quad F(1) = 2.$$

Iterating, we obtain

$$F(n) = 2 + 4 + 8 + \cdots + [(n - 1)^2 - (n - 1) + 2] = \sum_{k=1}^{n-1} (k^2 - k + 2)$$

$$\sum_{k=1}^{n} k^2 - \sum_{k=1}^{n} k + 2(n - 1) = \frac{n^3 - 3n^2 + 8n}{3},$$

where we have used the formulas for the sum of the first $n - 1$ integers and the sum of the first $n - 1$ perfect squares.

983. Choose three points A, B, C of the given set that lie on the boundary of its convex hull. There are $\binom{n-3}{2}$ ways to select two more points from the set. The line DE cuts two of the sides of the triangle ABC, say, AB and AC. Then B, C, D, E form a convex quadrilateral. Making all possible choices of the points D and E, we obtain $\binom{n-3}{2}$ convex quadrilaterals.

(11th International Mathematical Olympiad, 1969)

984. Assume by way of contradiction that the distance between any two points is greater than or equal to 1. Then the spheres of radius $\frac{1}{2}$ with centers at these 1981 points have disjoint interiors, and are included in the cube of side length 10 determined by the six parallel planes to the given cube's faces and situated in the exterior at distance $\frac{1}{2}$. It follows that the sum of the volumes of the 1981 spheres is less than the volume of the cube of side 10, meaning that

$$1981 \cdot \frac{4\pi \cdot \left(\frac{1}{2}\right)^3}{3} = 1981 \cdot \frac{\pi}{6} > 1000,$$

a contradiction. This completes the proof.

Remark. If we naively divide each side of the cube into $\left\lfloor \sqrt[3]{1981} \right\rfloor = 12$ congruent segments, we obtain $12^3 = 1728$ small cubes of side $\frac{9}{12} = \frac{3}{4}$. The pigeonhole principle guarantees that some small cube contains two of the points, but unfortunately the upper bound that we get for the distance between the two points is $\frac{3}{4}\sqrt[3]{3}$, which is greater than 1.

(*Revista Matematică din Timişoara* (*Timişoara Mathematics Gazette*), proposed by T. Andreescu)

985. We examine separately the cases $n = 3, 4, 5$. A triangle can have at most one right angle, a quadrilateral four, and a pentagon three (if four angles of the pentagon were right, the fifth would have to be equal to $180°$).

Let us consider an n-gon with $n \geq 6$ having k internal right angles. Because the other $n-k$ angles are less than $360°$ and because the sum of all angles is $(n-2) \cdot 180°$, the following inequality holds:

$$(n - k) \cdot 360° + k \cdot 90° > (n - 2) \cdot 180°.$$

This readily implies that $k < \frac{2n+4}{3}$, and since k and n are integers, $k \leq \left\lfloor \frac{2n}{3} \right\rfloor + 1$.

We will prove by induction on n that this upper bound can be reached. The basic cases $n = 6, 7, 8$ are shown in Figure 107.

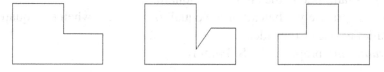

Figure 107

We assume that the construction is done for n and prove that it can be done for $n + 3$. For our method to work, we assume in addition that at least one internal angle is greater than $180°$. This is the case with the polygons from Figure 107. For the inductive step we replace the internal angle greater than $180°$ as shown in Figure 108. This increases the angles by 3 and the right angles by 2. The new figure still has an internal angle greater than $180°$, so the induction works. This construction proves that the bound can be reached.

(Short list of the 44th International Mathematical Olympiad, 2003)

986. It seems that the situation is complicated by successive colorings. But it is not! Observe that each time the moving circle passes through the original position, a new point will be

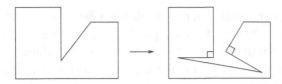

Figure 108

colored. But this point will color the same points on the fixed circle. In short, only the first colored point on one circle contributes to newly colored points on the other; all other colored points follow in its footsteps. So there will be as many colored points on the small circle as there are points of coordinate $2\pi k$, k an integer, on the segment $[0, 200\pi\sqrt{2}]$. Their number is

$$\left\lfloor \frac{200\pi\sqrt{2}}{2\pi} \right\rfloor = \left\lfloor 100\sqrt{2} \right\rfloor = 141.$$

(Ukrainian Mathematical Olympiad)

987. The solution is based on the pigeonhole principle. Let us assume that the sum of lengths of the chords is greater than or equal to $k\pi$. Then the sum of the lengths of the arcs subtended by these chords is greater than $k\pi$. Add to these arcs their reflections about the center of the circle. The sum of the lengths of all arcs is greater than $2k\pi$, so there exists a point covered by at least $k+1$ arcs. The diameter through that point intersects at least $k+1$ chords, contradicting our assumption. Hence the conclusion.

(*Kvant (Quantum)*, proposed by A.T. Kolotov)

988. Assume that the side-length of the square is 1. Let $a_k \le b_k$, $1 \le k \le n$ be the side-lengths of the rectangles of the decomposition. Using the fact that $b_k \le 1$, we can write

$$\frac{a_1}{b_1} + \frac{a_2}{b_2} + \cdots + \frac{a_k}{b_k} \ge a_1 b_1 + a_2 b_2 + \cdots + a_k b_k = 1,$$

because the right-hand side is the area of the square.

Equality holds precisely when all b_k are equal to 1, namely when the square is divided by segments parallel to one of the sides.

(*Kvant (Quantum)*, proposed by S. Fomin)

989. The center of the desired circle must lie at distance at least 1 from the boundary of the square. We will be able to find it somewhere inside the square whose sides are parallel to those of the initial square and at distance 1 from them. The side length of this smaller square is 36.

The locus of all points that lie at distance less than 1 from a convex polygonal surface P is a polygonal surface Q with sides parallel to those of P and whose corners are rounded. The areas of P and Q are related by

$$S[Q] = S[P] + (\text{perimeter of } P) \times 1 + \pi.$$

This is because the circular sectors from the corners of Q complete themselves to a disk of radius 1.

So the locus of the points at distance less than 1 from a polygon of area at most π and perimeter at most 2π is less than or equal to $\pi + 2\pi + \pi = 4\pi$. It follows that the area of the region of all points that are at distance less than 1 from any of the given 100 polygons is at most 400π. But

$$400\pi \leq 400 \cdot 3.2 = 40 \cdot 32 = 36^2 - 4^2 < 36^2.$$

So the set of these points does not cover entirely the interior of the square of side length 36. Pick a point that is not covered; the unit disk centered at that point is disjoint from any of the polygons, as desired.

(M. Pimsner, S. Popa, *Probleme de Geometrie Elementară* (*Problems in Elementary Geometry*), Editura Didactică şi Pedagogică, Bucharest, 1979)

990. The sum of the perimeters in which the square is divided is $40 = 2 \times 18 + 4$. Assume that the square is divided in n regions, let x_i, y_i be the sum of the vertical respectively horizontal sides of the region, and let σ_i be the area of the ith region. Then $x_i y_i \geq \sigma_i$, so

$$x_i + y_i \geq 2\sqrt{x_i y_i} \geq 2\sqrt{\sigma_i}, \quad 1 \leq i \leq n.$$

Then

$$40 = \sum_{i=1}^{n}(2x_i + 2y_i) \geq 4\sum_{i=1}^{n}\sqrt{\sigma_i},$$

whence $\sum_{i=1}^{n}\sqrt{\sigma_i} \leq 10$.

On the other hand, if for all i, $\sigma_i < 0.01$, then

$$1 = \sum_{i=1}^{n}\sigma_i = \sum_{i=1}^{n}\sqrt{\sigma_i}\sqrt{\sigma_i} < \sum_{i=1}^{n}0.1\sqrt{\sigma_i}.$$

This implies $\sum_{i=1}^{n}\sqrt{\sigma_i} > 10$, a contradiction. Hence the conclusion.

(*Kvant* (*Quantum*), proposed by A. Andjan)

991. Place n disks of radius 1 with the centers at the given n points. The problem can be reformulated in terms of these disks as follows.

Alternative problem. *Given $n \geq 3$ disks in the plane such that any 3 intersect, show that the intersection of all disks is nontrivial.*

This is a well-known property, true in d-dimensional space, where "disks" becomes "d-dimensional balls" and the number 3 is replaced by $d + 1$. The case $d = 1$ is rather simple. Translating the problem for the real axis, we have a finite family of intervals $[a_i, b_i]$, $1 \leq i \leq n$, such that any two intersect. Then $a_i < b_j$ for any i, j, and hence

$$[\max a_i, \min b_i] \subset \cap_i[a_i, b_i],$$

proving the claim. In general, we proceed by induction on d. Assume that the property is not true, and select the d-dimensional balls (disks in the two-dimensional case) $B_1, B_2, \ldots, B_{k-1}, B_k$ such that

$$B_1 \cap B_2 \cap \cdots \cap B_{k-1} = G \neq \emptyset \quad \text{and} \quad B_1 \cap B_2 \cap \cdots \cap B_{k-1} \cap B_k = \emptyset.$$

Let H be a hyperplane (line in the two-dimensional case) that separates G from B_k. Since B_k intersects each of the balls $B_1, B_2, \ldots, B_{k-1}$, the sets $X_i = B_i \cap H$, $i = 1, 2, \ldots, k-1$, are nonempty. Moreover, since by hypothesis B_k and any d of the other $k-1$ balls have nontrivial intersection, any collection of d sets X_i has nontrivial intersection. But then, by the induction hypothesis, all X_i have nontrivial intersection. Therefore,

$$H \cap B_1 \cap B_2 \cap \cdots \cap B_{k-1} \neq \emptyset,$$

i.e., $H \cap G \neq \emptyset$, a contradiction. Our assumption was false, which proves the inductive step. So the property is true in general, in particular in the two-dimensional case.

992. We will prove the property by induction on the number n of circles. The cases $n \leq 4$ are trivial since we can color all circles by different colors.

Now let us assume that the property holds for any choice of $n = k$ circles with the required property, and let us prove it for $n = k + 1$. Choose a point P in the plane and consider a circle from our collection whose center is at maximal distance from P. This circle is tangent to at most three other circles, because the centers of these circles must be inside or on the circle of center P and radius OP. Leave this circle aside, and color the remaining k circles by four colors such that any two tangent circles have different colors. Then add this circle, and color it differently than the (at most three circles) that are tangent to it. The new configuration has the desired property, and so the induction is complete.

Three colors do not suffice, for example for the configuration in Figure 109. If this configuration could be colored by only three colors, then since any two circles that are tangent to two others must be of the same color, it follows that M, D, F, H, L must be of the same color. But then the circles M and L have the same color and are tangent, which is not allowed.

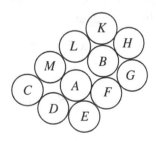

Figure 109

(Kvant (Quantum), proposed by G. Ringel)

993. This is an easy application of the pigeonhole principle. Let n be the number of vertices. Associate to each vertex the set of vertices connected to it by edges. There are n such sets, and each of them has at most $n - 1$ elements. Hence there are two sets with the same number of elements. Their corresponding vertices are endpoints of the same number of edges.

994. We set $f^0 = 1_A, f^{n+1} = f^n \circ f, n \geq 0$. Define on A the relation $x \sim y$ if there exist m and n such that $f^n(x) = f^m(y)$. One verifies immediately that $\sim$ is an equivalence relation, and that equivalence classes are invariant under f. An equivalence class resembles a spiral galaxy, with a cycle into which several branches enter. Such an equivalence class is illustrated in Figure 110, where the dots are elements of E and the arrows describe the action of f.

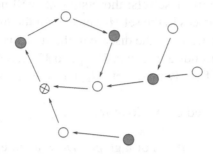

Figure 110

Thus f defines a directed graph whose connected components are the equivalence classes. We color the vertices of this graph by 0, 1, 2, 3 according to the following rule. All fixed points are colored by 0. Each cycle is colored alternately $1, 2, 1, 2, \ldots$ with its last vertex colored by 3. Finally, each branch is colored alternately so that no consecutive vertices have the same color. The coloring has the property that adjacent vertices have different colors. If we let A_i consist of those elements of A colored by $i, i = 0, 1, 2, 3$, then these sets have the required property. The construction works also in the case that the cycle has length one, that is, when it is a fixed points of f. Note that in general the partition is not unique.

This argument can be easily adapted to the case in which A is infinite. All cycles are finite and they are taken care of as in the case of a finite set. The coloring can be done provided that we can choose one element from each cycle to start with, thus we have to assume the Axiom of choice. This axiom states that given a family of sets one can choose one element from each of them. Now let us consider an equivalence class as defined above, and look at the dynamic process of repeated applications of f. It either ends in A_0 or in a cycle, or it continues forever. In the equivalence class we pick a reference point x_0, which is either the point where the equivalence class enters A_0 or a cycle, or otherwise is an arbitrary point. Either x_0 has been colored, by 0 or as part of a cycle, or if not, we color it by the color of our choice. Say the color of x_0 is i, and let j and k be two other colors chosen from 1, 2, and 3. If $x \sim x_0$ then $f^n(x) = f^m(x_0)$ for some integers m and n. For that particular x, choose m and n to be minimal with this property. Color x by j if $m - n$ is even, and by k if $m - n$ is odd.

Note that x and $f(x)$ cannot have the same color, for otherwise in the equalities $f^n(x) = f^m(x_0)$ and $f^{n+1}(x) = f^{m'}(x_0)$ the minimality of m and m' implies that $m = m'$, and then $n - m$ and $n + 1 - m$ would have the same parity, which is impossible. Again, the coloring partitions A into four sets with the desired properties.

995. (a) Let us consider the model in which only three students A, B and C live in the dorm, and they are friends with each other. Assume that on the first day A is imune, B is sick, and C is healthy but not immune. On the second day B is immune, C is sick, and A is healthy but not immune. The epidemics continues in this cyclic fashion, so it never dies out.

(b) Let us represent the student population by a graph, vertices being the students and edges connecting pairs of friends. We define the distance between two students to be the minimal number of edges that path connecting them can have. We partition the vertices in the sets $M_0, M_1, M_2, \ldots$ with M_0 being the vertices corresponding to students that were sick on the first day, and for $k \geq 1$, M_k being the vertices at distance k from M_0. We ignore the vertices who are in none of these sets; those students will never get sick. Note that the students corresponding to vertices in the set M_k get sick for the first time on the kth day of the epidemics, and they can only transmit the disease to the students corresponding to vertices in the set M_{k+1} (because their friends are only in M_{k-1} and M_{k+1}, and those in M_{k-1} are immune at that time). Since there are finitely many sets M_k, the epidemics ends once the largest index is reached.

(*Kvant (Quantum)*, proposed by A. Kolotov)

996. Let the $2n$ teams be the vertices of a graph. Draw a red edge for pairs that competed on the first day, and a blue edge for the pairs that competed on the second day. Then each vertex belongs to exactly a red edge and exactly a blue edge. We now have a graph consisting of cycles, each of which having an even number of vertices. Choose every other vertex from each cycle to obtain the desired set of n teams that have not played with each other.

(*Kvant (Quantum)*, proposed by M. Bona)

997. Recall that the degree of a vertex is the number of edges containing it. If G has some vertices of odd degree, the number of such vertices is even because the sum of the degrees of all vertices equals twice the number of edges. In this situation we add a vertex to G and connect it by edges to all vertices of odd degree. The new graph, G' has all vertices of even degree, therefore G' has an Eulerian cycle. We label the edges of G by $1, 2, \ldots, k$ in the order in which we encounter these when traveling on the Eulerian cycle.

When passing through a vertex of G of even degree, two edges are labeled by consecutive numbers, hence this vertex will have the desired property. On the other hand, we are only interested in vertices of G of odd degree that have a degree greater than or equal to 3. Through one such vertex we pass at least twice, and only once do we pass through it on edges that don't belong to G (since there is only one such edge). Hence again there are two edges labeled by consecutive integers. The problem is solved.

(32nd International Mathematical Olympiad, 1992, solution by R.A. Todor)

998. Label the vertices 1,2,3,4 such that 1 and 4 are not connected, and denote by $N_{i,j}(n)$ the number of paths of length n that join vertices i, j. By symmetry:

$$N_{1,2}(n) = N_{1,3}(n) = N_{4,2}(n) = N_{4,3}(n) = N_{2,1}(n) = N_{2,4}(n) = N_{3,1}(n) = N_{3,4}(n),$$
$$N_{2,3}(n) = N_{3,2}(n), N_{1,4}(n) = N_{4,1}(n), N_{1,1}(n) = N_{4,4}(n), N_{2,2}(n) = N_{3,3}(n).$$

Denote this numbers, respectively, by a_n, b_n, c_n, x_n, and y_n. We should also observe that, since 1 and 4 can be reached only from 2 or 3, then $c_n = x_n$ for $n \geq 1$, but $c_0 = 0$, and $x_0 = 1$. The problem asks for $2x_n + 2y_n$.

Vertex 1 can be reached in one step only from vertices 2 and 3, so a circuit starting at 1 is a path from 1 to either 2 or 3 followed by a step back to 1. Thus $x_n = 2a_{n-1}$. Vertex 2 can be reached from any of the other 3 vertices, and a path of length n from 1 to 2 is the sum of the numbers of the paths from 1 to 1,3, or 4 followed by a step to 2. So $a_n = a_{n-1} + 2x_{n-1}$. The number of paths from 2 to 1 of length n is the sum of the numbers of paths from 2 to itself and from 2 to 3, followed by a step to 1, so $a_n = y_{n-1} + b_{n-1}$. Thus $a_n + 2x_n = b_n + y_n$. Similarly $y_n = b_{n-1} + 2a_{n-1} = 3a_{n-1} + 2x_{n-1} - y_{n-1}$. We obtain a recurrence relation of the form

$$\begin{pmatrix} a_n \\ x_n \\ y_n \end{pmatrix} = \begin{pmatrix} 1 & 2 & 0 \\ 2 & 0 & 0 \\ 3 & 2 & -1 \end{pmatrix} \begin{pmatrix} a_{n-1} \\ x_{n-1} \\ y_{n-1} \end{pmatrix}.$$

Now we apply the techniques from Section 3.1.2. The characteristic polynomial of the matrix that defines the recursion is $(\lambda + 1)(\lambda^2 - \lambda - 4)$, with roots -1, $\frac{1 \pm \sqrt{17}}{2}$. The initial condition $a_1, x_1 = y_1 = 0$ give $a_2 = 1$, $x_2 = 2$, $y_2 = 3$, and $x_3 = 2$, $y_3 = 4$. Thus the desired number of circuits $z_n = 2(x_n + y_n)$ satisfies the recursive relation with the same characteristic polynomial, so

$$z_n = A(-1)^n + B\left(\frac{1+\sqrt{17}}{2}\right)^n + C\left(\frac{1-\sqrt{17}}{2}\right)^n,$$

where A, B, C are computed from and $z_1 = 0$, $z_2 = 10$, $z_3 = 12$ (and the latter are computed from the corresponding values of x_n, y_n). We obtain a system of 3 equations with 3 unknowns in A, B, C, with the unique solution $A = B = C = 1$. We conclude that the answer to the problem is

$$(-1)^n + \left(\frac{1+\sqrt{17}}{2}\right)^n + \left(\frac{1-\sqrt{17}}{2}\right)^n.$$

(Mathematical Reflections, proposed by I. Borsenco)

999. We will prove that if G is not a complete graph, then it is a cycle. If G is not complete, then there are distinct vertices v_1, v_2 that are not connected by an edge. Then from the hypothesis we obtain that v_1 and v_2 are connected by at least 2 disjoint paths. So $G \setminus \{v_1, v_2\}$ is disconnected. Let $H_1, H_2, \ldots, H_k$ be its connected components, $k \geq 2$. Because the trivial graph with one vertex is trivially connected, by removing either v_1 or v_2 we deduce that both v_1 and v_2 are connected to at least one vertex of each $H_j, j = 1, 2, \ldots, k$.

Now assume $k > 2$. Then $\{v_1, v_2\} \cup H_1$ is connected and by removing it there remain at least $k - 1 \geq 2$ connected components (namely $H_2, H_3, \ldots, H_k$). This contradicts the hypothesis, thus $k = 2$.

Now since $\{v_1, v_2\} \cup H_1$ and $\{v_1, v_2\} \cup H_2$ are connected, we can find $P_k, k = 1, 2$, paths of minimal lengths connecting v_1 to v_2 in these two graphs. Since these paths have minimal lengths, they do not have repeated vertices. If P_1 does not use every vertex in H_1, then in $G - P_1$ the remaining vertices of H_1 will be disconnected from H_2, a contradiction. So P_1 uses all vertices of $\{v_1, v_2\} \cup H_1$. By minimality, there cannot be any edges in $\{v_1, v_2\} \cup H_1$ that are not in P_1, or else we can use such an edge to shorten P_1. We draw the same conclusion about P_2. So $H_1 = P_1$ and $H_2 = P_2$, and hence G is a cycle.

(*Mathematical Reflections*, proposed by C. Pohoață)

1000. We turn the problem into a question about graphs by allowing the triangles to have curved edges. In this case we can use an inductive argument on the number of triangles in the decomposition. The base case is where there is only one triangle in the decomposition, the original triangle.

Now let us assume that the property is true for all decompositions and colorings with less than n triangle, and let us consider some configuration with n triangles. If no edge has the endpoints colored by the same color, then we are done. If there is an edge with endpoints colored by the same color, contract that edge in such a way that the triangles that contain that edge degenerate into edges. There are two triangles if this edge is in the interior of the original triangle, or just one if the edge is on the side. In either situation we arrived at a configuration with fewer triangles, which by the induction hypothesis has one triangle with vertices colored by different colors. This triangle was present in the original configuration, and the problem is solved.

Note that we had to allow curved edges, because some edges might bend in the process of contraction.

(A.M. Yaglom, I.M. Yaglom, *Neelementarnye Zadachi v Elementarnom izlozhenii (Non-Elementary Problems with Elementary Solutions)*, Government Publication House for Technical-Theoretical Literature, Moscow, 1954)

1001. To prove the claim, we will slightly generalize it; namely, we show that if in a planar graph every vertex belongs to an even number of edges, then the faces of the graph and its exterior can be colored black and white such that neighboring regions are of different colors. Once we allow edges to bend, and faces to be bigons, we can induct on the number of faces.

The base case consists of a face bounded by two edges, for which the property obviously holds. Assume that the property holds true for all graphs with at most k faces and let us prove it for an arbitrary graph with $k + 1$ faces. Choose a face of the graph, which may look as in Figure 111. Shrink it to a point. Color the new graph as permitted by the inductive hypothesis. Blow up the face back into the picture. Because an even number of edges meet at each vertex, all the faces that share an edge with the chosen one are colored by the same color (when moving clockwise around the chosen face we get from one neighboring face to the next in an even number of steps). Hence the face can be given the opposite color. This completes the argument.

(*Kvant (Quantum)*)

1002. Let the vertices of the graph be $v_1, v_2, \ldots, v_n$. Double each vertex and consider the graph with $2n$ vertices $v_1', v_1'', v_2', v_2'', \ldots, v_n', v_n''$, and if v_i and v_j were connected, join v_i' and v_j''

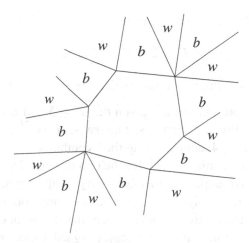

Figure 111

by an edge, as well as v'_j and v''_i. Let $A = \{v'_1, v'_2, \ldots, v'_n\}$ and $B = \{v''_1, v''_2, \ldots, v''_n\}$.

In the new graph impose that at the first step only vertices in A change their color, at the second step only vertices in B, then again in A, and so on, with the same conditions. We will show that at some moment the doubled graph stops changing colors. If this is true, then in the original graph, the vertices for which v'_i and v''_i have the same color stay unchanged, those for which v'_i and v''_i have different colors change at every step.

Given a vertex that changes colors, because it assumes the color of the majority of its neighbors, the number of monochromatic edges grows. Hence, at a step where some vertices change colors, the number of monochromatic edges grows. This cannot happen forever, because there are only finitely many edges. So from some moment on, the colorings won't change anymore. The problem is solved.

(*Kvant (Quantum)*, proposed by O. Kozlov)

1003. For finding the upper bound we employ Euler's formula. View the configuration as a planar graph, and complete as many curved edges as possible, until a triangulation of the plane is obtained. If $V = n$ is the number of vertices, E the number of edges and F the number of faces (with the exterior infinite face counted among them), then $V - E + F = 2$, so $E - F = n + 2$. On the other hand, since every edge belongs to two faces and every face has three edges, $2E = 3F$. Solving, we obtain $E = 3n - 6$. Deleting the "alien" curved edges, we obtain the inequality $E \leq 3n - 6$. That the bound can be reached is demonstrated in Figure 112.

(German Mathematical Olympiad, 1976)

1004. If this were possible, then the configuration would determine a planar graph with $V = 6$ vertices (the 3 neighbors and the 3 wells) and $E = 9$ edges (the paths). Each of its F faces would have 4 or more edges because there is no path between wells or between neighbors. So

$$F \leq \frac{2}{4}E = \frac{9}{2}.$$

On the other hand, by Euler's relation we have

$$F = 2 + E - V = 5.$$

We have reached a contradiction, which shows that the answer to the problem is negative.

1005. With the standard notation, we are given that $F \geq 5$ and $E = \frac{3V}{2}$. We will show that not all faces of the polyhedron are triangles. Otherwise, $E = \frac{3F}{2}$ and Euler's formula yields $F - \frac{3F}{2} + F = 2$, that is, $F = 4$, contradicting the hypothesis.

We will indicate now the game strategy for the two players. The first player writes his/her name on a face that is not a triangle; call this face $A_1 A_2 \ldots A_n$, $n \geq 4$. The second player, in an attempt to obstruct the first, will sign a face that has as many common vertices with the face signed by the first as possible, thus claiming a face that shares an edge with the one chosen by the first player. Assume that the second player signed a face containing the edge $A_1 A_2$. The first player will now sign a face containing the edge $A_3 A_4$. Regardless of the play of the second player, the first can sign a face containing either A_3 or A_4, and wins!

(64th W.L. Putnam Mathematical Competition, 2003, proposed by T. Andreescu)

1006. Start with Euler's relation $V - E + F = 2$. and multiply it by 2π to obtain $2\pi V - 2\pi E + 2\pi F = 4\pi$. If n_k, $k \geq 3$, denotes the number of faces that are k-gons, then $F = n_3 + n_4 + n_5 + \cdots$. Also, counting edges by the faces, and using the fact that each edge belongs to two faces, we have $2E = 3n_3 + 4n_4 + 5n_5 + \cdots$. Euler's relation becomes

$$2\pi V - \pi(n_3 + 2n_4 + 3n_5 + \cdots) = 4\pi.$$

Because the sum of the angles of a k-gon is $(k - 2)\pi$, the sum in the above relation is equal to Σ. Hence the conclusion.

Remark. In general, if a polyhedron P resembles a sphere with g handles, then $2\pi V - \Sigma = 2\pi(2 - 2g)$. As mentioned before, the number $2 - 2g$, denoted by $\chi(P)$, is called the Euler characteristic of the polyhedron. The difference between 2π and the sum of the angles around a vertex is the curvature K_v at that vertex. Our formula then reads

$$\sum_v K_v = 2\pi \chi(P).$$

This is the piecewise linear version of the Gauss-Bonnet theorem.

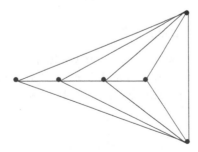

Figure 112

In the differential setting, the Gauss-Bonnet theorem is expressed as

$$\int_S KdA = 2\pi \chi(S),$$

or in words, the integral of the Gaussian curvature K over a closed surface S is equal to the Euler characteristic of the surface multiplied by 2π. This means that no matter how we deform a surface, although locally its Gaussian curvature will change, the total curvature remains unchanged.

1007. (a) We use an argument by contradiction. The idea is to start with Euler's formula

$$V - E + F = 2$$

and obtain a relation that is manifestly absurd. By our assumption each vertex belongs to at least 6 edges. Counting the vertices by the edges, we obtain $2E$ (each edge has two vertices). But we overcounted the vertices at least 6 times. Hence $2E \geq 6V$. Similarly, counting faces by the edges and using the fact that each face has at least three edges, we obtain $2E \geq 3F$. We thus have

$$2 = V - E + F \leq \frac{1}{3}E - E + \frac{2}{3}E = 0,$$

an absurdity. It follows that our assumption was false, and hence there is a vertex belonging to at most five edges.

(b) We use the first part. To the map we associate a connected planar graph G. The vertices of G are the regions. The edges cross the boundary arcs (see Figure 113). For a border consisting of consecutive segments that separates two neighboring regions we add just one edge! The constructed graph satisfies the conditions from part (a). We claim that it can be colored by 5 colors so that whenever two vertices are joined by an edge, they have different colors.

We prove the claim by induction on the number of vertices. The result is obvious if G has at most 5 vertices. Now assume that the coloring exists for any graph with $V - 1$ vertices and let us prove that it exists for graphs with V vertices.

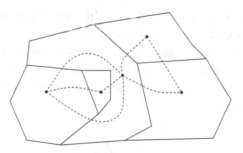

Figure 113

By (a), there is a vertex v that has at most 5 adjacent vertices. Remove v and the incident edges, and color the remaining graph by 5 colors. The only situation that poses difficulties for extending the coloring to v is if v has exactly 5 adjacent vertices and they are colored by different colors. Call these vertices w_1, w_2, w_3, w_4, w_5 in clockwise order, and assume they are colored A, B, C, D, E, respectively. Look at the connected component containing w_1 of the subgraph of G consisting of only those vertices colored by A and C. If w_3 does not belong to this component, switch the colors A and C on this component, and then color v by A. Now let us examine the case in which w_3 belongs to this component. There is a path of vertices colored by A and C that connects w_1 and w_3.

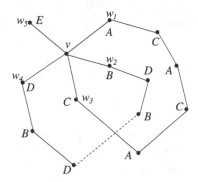

Figure 114

Next, let us focus on w_2 and w_4 (Figure 114). The only case in which we would not know how to perform the coloring is again the one in which there is a path of vertices colored by B and D that joins w_2 to w_4. Add v to the two paths (from w_1 to w_3 and from w_2 to w_4) to obtain two cycles. Because of how we ordered the w_i's and because the graph is planar, the two cycles will intersect at a vertex that must be simultaneously colored by one of A or C and by one of B or D. This is impossible, so this situation cannot occur. This completes the solution.

Remark. The famous Four color theorem states that four colors suffice. This was first conjectured by F. Guthrie in 1853, and proved by K. Appel and W. Haken in 1977 with the aid of a computer. The above Five-color theorem was proved in 1890 by P.J. Heawood using ideas of A. Kempe.

1008. We will prove a more precise result. To this end, we need to define one more type of singularity. A vertex is called a (multi)saddle of index $-k$, $k \geq 1$, if it belongs to some incoming and to some outgoing edge, and if there are $k+1$ changes from incoming to outgoing edges in making a complete turn around the vertex. The name is motivated by the fact that if the index is -1, then the arrows describe the way liquid flows on a horse saddle. Figure 115 depicts a saddle of index -2.

Figure 115

Call a vertex that belongs only to outgoing edges a source, a vertex that belongs only to incoming edges a sink, and a face whose edges form a cycle a circulation. Denote by n_1 the number of sources, by n_2 the number of sinks, by n_3 the number of circulations, and by n_4 the sum of the indices of all (multi)saddles.

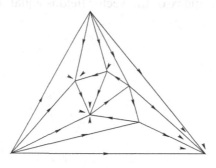

Figure 116

We refer everything to Figure 116. We start with the count of vertices by incoming edges; thus for each incoming edge we count one vertex. Sources are not counted. With the standard notation, if we write

$$E = V - n_1,$$

we have overcounted on the left-hand side. To compensate this, let us count vertices by faces. Each face that is not a circulation has two edges pointing toward the same vertex. In that case, for that face we count that vertex. All faces but the circulations count, and for vertices that are not singularities this takes care of the overcount. So we can improve our "equality" to

$$E = V - n_1 + F - n_3.$$

Each sink is overcounted by 1 on the right. We improve again to

$$E = V - n_1 + F - n_3 - n_2.$$

Still, the right-hand side undercounts saddles, and each saddle is undercounted by the absolute value of its index. We finally reach equality with

$$E = V - n_1 + F - n_3 - n_2 + |n_4| = V + F - n_1 - n_2 - n_3 - n_4.$$

Using Euler's formula, we obtain

$$n_1 + n_2 + n_3 + n_4 = V - E + F = 2.$$

Because $n_4 \leq 0$, we have $n_1 + n_2 + n_3 \geq 2$, which is what we had to prove.

Remark. The polyhedron can be thought of as a discrete approximation of a surface. The orientation of edges is the discrete analogue of a smooth vector field on the surface. The number $n_1 + n_2 + n_3 + n_4$ is called the total index of the vector field. The result we just proved shows that if the polyhedron resembles a (triangulated) sphere, the total index of any vector field is 2. This is a particular case of the Poincaré-Hopf index theorem, which in its general setting states that given a smooth vector field with finitely many zeros on a compact, orientable manifold, the total index of the vector field is equal to the Euler characteristic of the manifold.

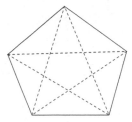

Figure 117

1009. Figure 117 shows that this number is greater than or equal to 5.

Let us show that any coloring by two colors of the edges of a complete graph with 6 vertices has a monochromatic triangle. Assume the contrary. By the pigeonhole principle, 3 of the 5 edges starting at some point have the same color (see Figure 118). Each pair of such edges forms a triangle with another edge. By hypothesis, this third edge must be of the other color. The three pairs produce three other edges that are of the same color and form a triangle. This contradicts our assumption. Hence any coloring of a complete graph with six vertices contains a monochromatic triangle. We conclude that $n = 5$.

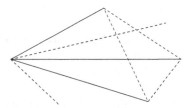

Figure 118

Remark. This shows that the number $R(3, 3)$ is equal to 6.

1010. Let $n = R(p - 1, q) + R(p, q - 1)$. We will prove that for any coloring of the edges of a complete graph with n vertices by red or blue, there is a red complete subgraph with p vertices or a blue complete subgraph with q vertices. Fix a vertex x and consider the $n - 1$ edges starting at x. Among them there are either $R(p - 1, q)$ red edges, or $R(p, q - 1)$ blue edges. Without loss of generality, we may assume that the first case is true, and let X be the set of vertices connected to x by red edges. The complete graph on X has $R(p - 1, q)$ vertices. It either has a blue complete subgraph with q edges, in which case we are done, or it has a red complete subgraph with $p - 1$ edges, to which we add the red edges joining

x to X to obtain a red complete subgraph with p edges of the original graph. This proves $R(p, q) \leq R(p-1, q) + R(p, q-1)$.

To prove the upper bound for the Ramsey numbers we argue by induction on $p + q$. The base case consists of all configurations with $p = 2$ or $q = 2$, in which case $R(p, 2) = R(2, p) = p = \binom{p}{p-1}$, since any graph with p vertices either has an edge colored red, or is entirely colored blue. Let us assume that the inequality is true for all $p, q \geq 2, p + q = n$. Either $p = 2$, or $q = 2$, or otherwise

$$R(p, q) \leq R(p-1, q) + R(p, q-1) \leq \binom{p+q-3}{p-2} + \binom{p+q-3}{p-1} = \binom{p+q-2}{p-1}$$

(P. Erdős, G. Szekeres)

1011. The number from the statement is $R(3, n) - 1$, where $R(3, n)$ is the Ramsey number. By the Erdős-Szekeres inequality from the previous problem, $R(3, n) \leq \binom{n+1}{2}$.

So for $n = 3$, the estimate is 5. An example where equality is attained is a cycle of length 5. This is the same as the graph with vertices the elements of $\mathbb{Z}_5$ with two vertices j, k connected if and only if $j - k$ is ± 1 modulo 5.

For $n = 4$, the number cannot exceed $\binom{5}{2} - 1 = 9$. If we had equality, then (see solution to the previous problem) from each vertex should start exactly 3 red edges and 5 blue edges. This is impossible since, in the graph whose edges are the blue edges, the number of vertices of odd degree must be even. So for $n = 4$ we can have at most 8 vertices. A model is the graph with vertices the elements of $\mathbb{Z}_8$ such that j and k are connected by an edge if and only if $j - k$ is either ± 1 or ± 2 modulo 8.

Finally, $p_5 \leq 13$, and an example is the graph with vertices the elements of $\mathbb{Z}_{13}$ and j, k connected by an edge if and only if $j - k$ is $\pm 1, \pm 2, \pm 3 \pm 5$ modulo 13.

Remark. The only other Ramsey numbers $R(3, n)$ that are known are $R(3, 6) = 18, R(3, 7) = 23, R(3, 8) = 28, R(3, 9) = 36$.

1012. We prove the property by induction on k. First, observe that

$$\lfloor k!e \rfloor = \frac{k!}{1} + \frac{k!}{1!} + \frac{k!}{2!} + \cdots + \frac{k!}{k!}.$$

For $k = 2$, $\lfloor k!e \rfloor + 1 = 6$, and the property was proved in the previous problem. Assume that the property is true for a complete graph replaced with $\lfloor (k-1)!e \rfloor + 1$ vertices colored by $k - 1$ colors, and let us prove it for a complete graph with $\lfloor k!e \rfloor + 1$ vertices colored by k colors. Choose a vertex v of the graph. By the pigeonhole principle, v is connected to $\lfloor (\lfloor k!e \rfloor + 1)/k \rfloor + 1$ vertices by edges of the same color c. Note that

$$\left\lfloor \frac{\lfloor k!e \rfloor + 1}{k} \right\rfloor = \left\lfloor \frac{1}{k}\left(\frac{k!}{1} + \frac{k!}{1!} + \frac{k!}{2!} + \cdots + \frac{k!}{k!}\right) \right\rfloor + 1$$

$$= \frac{(k-1)!}{1} + \frac{(k-1)!}{1!} + \frac{(k-1)!}{2!} + \cdots + \frac{(k-1)!}{(k-1)!} + 1$$

$$= \lfloor (k-1)!e \rfloor + 1.$$

If two of these vertices are connected by an edge of color c, then a c-colored triangle is formed. If not, the complete graph on these $\lfloor (k-1)!e \rfloor + 1$ vertices is colored by the remaining $k-1$ colors, and by the induction hypothesis a monochromatic triangle is formed. This completes the proof.

Remark. This proves that the k-color Ramsey number $R(3, 3, \ldots, 3)$ is bounded from above by $\lfloor k!e \rfloor + 1$.

(F.P. Ramsey)

1013. Yet another Olympiad problem related to Schur numbers. We can reformulate the problem as follows:

Alternative problem. *Show that if the set $\{1, 2, \ldots, 1978\}$ is partitioned into six sets, then in one of these sets there are a, b, c (not necessarily distinct) such that $a + b = c$.*

The germs of the solution have already been glimpsed in the Bielorussian problem from the introduction. Observe that by the pigeonhole principle, one of the six sets, say A, has at least $\lfloor \frac{1978}{6} \rfloor + 1 = 330$ elements; call them $a_1 < a_2 < \cdots < a_{330}$. If any of the 329 differences

$$b_1 = a_{330} - a_{329}, \quad b_2 = a_{330} - a_{328}, \ldots, b_{329} = a_{330} - a_1$$

is in A, then we are done, because $a_{330} - a_m = a_n$ means $a_m + a_n = a_{330}$. So let us assume that none of these differences is in A. Then one of the remaining sets, say B, contains at least $\lfloor \frac{329}{5} \rfloor + 1 = 66$ of these differences. By eventually renumbering them, we may assume that they are $b_1 < b_2 < \cdots < b_{66}$. We repeat the argument for the common differences

$$c_1 = b_{66} - b_{65}, \quad 2 = b_{66} - b_{64}, \ldots, c_{65} = b_{66} - b_1.$$

Note that

$$c_j = b_{66} - b_{66-j} = (a_{330} - a_m) - (a_{330} - a_n) = a_n - a_m.$$

So if one of the c_j's is in A or B, then we are done. Otherwise, there is a fourth set D, which contains $\lfloor \frac{65}{4} \rfloor + 1 = 17$ of the c_j's. We repeat the argument and conclude that either one of the sets A, B, C, D contains a Schur triple, or there is a fifth set E containing $\lfloor \frac{17}{3} \rfloor + 1 = 6$ of the common differences $d_k = c_{17} - c_{17-k}$. Again either we find a Schur triple in A, B, C, or D, or there is a set E containing $\lfloor \frac{5}{2} \rfloor + 1 = 3$ of the five differences $e_i = d_5 - d_{5-k}$. If any of the three differences $e_2 - e_1, e_3 - e_2, e_3 - e_1$ belongs to A, B, C, D, E, then we have found a Schur triple in one of these sets. Otherwise, they are all in the sixth set F, and we have found a Schur triple in F.

Remark. Look at the striking similarity with the proof of Ramsey's theorem, which makes the object of the previous problem. And indeed, Ramsey's theorem can be used to prove Schur's theorem in the general case: $S(n)$ is finite and is bounded above by the k-color Ramsey number $R(3, 3, \ldots, 3)$.

Here is how the proof runs. Think of the partition of the set of the first N positive integers into n subsets as a coloring $c : \{1, 2, \ldots, N\} \to \{1, 2, \ldots, n\}$. Consider the complete graph with vertices $1, 2, \ldots, N$ and color its edges so that for $i > j$, (i, j) is colored by $c(i - j)$. If $N \geq R(3, 3, \ldots, 3)$ (the k-color Ramsey number), then there is a monochromatic triangle. If $i < j < k$ are the vertices of this triangle, then the numbers $x = j - i, y = k - j$, and $z - k - i$ form a Schur triple. The fact that they have the same color means that they belong to the same set of the partition. The theorem is proved.

(20th International Mathematical Olympiad, 1978)

1014. Let

$$I_k = \int_0^{\frac{\pi}{2}} (\sin \theta)^{2k} d\theta, \ k \geq 0.$$

Integrating by parts, we obtain

$$I_k = \int_0^{\frac{\pi}{2}} (2 \sin \theta)^{2k-1} (2 \sin \theta) d\theta$$

$$= (2 \sin \theta)^{2k} (-2 \cos \theta) \Big|_0^{\frac{\pi}{2}} + \int_0^{\frac{\pi}{2}} (2k - 1)(2 \sin \theta)^{2k-2} 4 \cos^2 \theta d\theta$$

$$= (2k - 1) \int_0^{\frac{\pi}{2}} (2 \sin \theta)^{2k-2} (4 - 4 \sin^2 \theta) d\theta$$

$$= 4(2k - 1)I_{k-1} - (2k - 1)I_k.$$

Hence $I_k = \frac{4k-2}{k} I_{k-1}, k \geq 1$. Comparing this with

$$\binom{2k}{k} = \frac{(2k)(2k-1)(2k-2)!}{k^2((k-1)!)^2} = \frac{4k - 2}{k} \binom{2k - 2}{k},$$

we see that all that remains to check is the equality $\frac{2}{\pi} I_0 = 1$, and that is obvious.

1015. We compute

$$A^2 = \begin{pmatrix} 1 & 2 & 3 & \cdots & n \\ 0 & 1 & 2 & \cdots & n-1 \\ 0 & 0 & 1 & \cdots & n-2 \\ \vdots & \vdots & \vdots & \ddots & \vdots \\ 0 & 0 & 0 & \cdots & 1 \end{pmatrix} = \begin{pmatrix} \binom{1}{1} & \binom{2}{1} & \binom{3}{1} & \cdots & \binom{n}{1} \\ 0 & \binom{1}{1} & \binom{2}{1} & \cdots & \binom{n-1}{1} \\ 0 & 0 & \binom{1}{1} & \cdots & \binom{n-2}{1} \\ \vdots & \vdots & \vdots & \ddots & \vdots \\ 0 & 0 & 0 & \cdots & \binom{1}{1} \end{pmatrix}.$$

Also,

$$A^3 = \begin{pmatrix} \binom{2}{2} & \binom{3}{2} & \binom{4}{2} & \cdots & \binom{n+1}{2} \\ 0 & \binom{2}{2} & \binom{3}{2} & \cdots & \binom{n}{2} \\ 0 & 0 & \binom{2}{2} & \cdots & \binom{n-1}{2} \\ \vdots & \vdots & \vdots & \ddots & \vdots \\ 0 & 0 & 0 & \cdots & \binom{2}{2} \end{pmatrix}.$$

In general,

$$A^k = \begin{pmatrix} \binom{k-1}{k-1} & \binom{k}{k-1} & \binom{k+1}{k-1} & \cdots & \binom{k+n-2}{k-1} \\ 0 & \binom{k-1}{k-1} & \binom{k}{k-1} & \cdots & \binom{k+n-3}{k-1} \\ 0 & 0 & \binom{k-1}{k-1} & \cdots & \binom{k+n-4}{k-1} \\ \vdots & \vdots & \vdots & \ddots & \vdots \\ 0 & 0 & 0 & \cdots & \binom{k-1}{k-1} \end{pmatrix}.$$

This formula follows inductively from the combinatorial identity

$$\binom{m}{m} + \binom{m+1}{m} + \cdots + \binom{m+p}{m} = \binom{m+p+1}{m+1},$$

which holds for $m, p \geq 0$. This identity is quite straightforward and can be proved using Pascal's triangle as follows:

$$\begin{aligned} \binom{m}{m} + \binom{m+1}{m} + \cdots + \binom{m+p}{m} &= \binom{m+1}{m+1} + \binom{m+1}{m} + \cdots + \binom{m+p}{m} \\ &= \binom{m+2}{m+1} + \binom{m+2}{m} + \cdots + \binom{m+p}{m} \\ &= \binom{m+3}{m+1} + \binom{m+3}{m} + \cdots + \binom{m+p}{m} \\ &= \cdots = \binom{m+p}{m+1} + \binom{m+p}{m} = \binom{m+p+1}{m+1}. \end{aligned}$$

1016. The general term of the Fibonacci sequence is given by the Binet formula

$$F_n = \frac{1}{\sqrt{5}} \left[\left(\frac{1+\sqrt{5}}{2} \right)^n - \left(\frac{1-\sqrt{5}}{2} \right)^n \right], \quad n \geq 0.$$

Note that because $F_0 = 0$, we can start the summation at the 0th term. We therefore have

$$\sum_{i=0}^{n} F_i \binom{n}{i} = \frac{1}{\sqrt{5}} \left[\sum_{i=0}^{n} \binom{n}{i} \left(\frac{1+\sqrt{5}}{2} \right)^i - \sum_{i=0}^{n} \binom{n}{i} \left(\frac{1-\sqrt{5}}{2} \right)^i \right]$$

$$= \frac{1}{\sqrt{5}} \left[\left(\frac{1+\sqrt{5}}{2} + 1 \right)^n - \left(\frac{1-\sqrt{5}}{2} + 1 \right)^n \right]$$

$$= \frac{1}{\sqrt{5}} \left[\left(\frac{3+\sqrt{5}}{2} \right)^n - \left(\frac{3-\sqrt{5}}{2} \right)^n \right].$$

But

$$\frac{3 \pm \sqrt{5}}{2} = \left(\frac{1 \pm \sqrt{5}}{2} \right)^2.$$

So the sum is equal to

$$\frac{1}{\sqrt{5}} \left[\left(\frac{1+\sqrt{5}}{2} \right)^{2n} - \left(\frac{1-\sqrt{5}}{2} \right)^{2n} \right],$$

and this is F_{2n}. The identity is proved.

(E. Cesàro)

1017. Note that for $k = 0, 1, \ldots, n$,

$$(a_{k+1} + a_{n-k+1})(n+1) = 2S_{n+1}.$$

If we add the two equal sums $\sum_k \binom{n}{k} a_{k+1}$ and $\sum_k \binom{n}{n-k} a_{n-k+1}$, we obtain

$$\sum_{k=0}^{n} \binom{n}{k} (a_{k+1} + a_{n-k+1}) = \frac{2S_{n+1}}{n+1} \sum_{k=0}^{n} \binom{n}{k} = \frac{2^{n+1}}{n+1} S_{n+1}.$$

The identity follows.

1018. Newton's binomial expansion can be used to express our sum in closed form as

$$S_n = \frac{1}{4} \left[\left(2 + \sqrt{3} \right)^{2n+1} + \left(2 - \sqrt{3} \right)^{2n+1} \right].$$

The fact that $S_n = (k-1)^2 + k^2$ for some positive integer k is equivalent to

$$2k^2 - 2k + 1 - S_n = 0.$$

View this as a quadratic equation in k. Its discriminant is

$$\Delta = 4(2S_n - 1) = 2 \left[\left(2 + \sqrt{3} \right)^{2n+1} + \left(2 - \sqrt{3} \right)^{2n+1} - 2 \right].$$

Is this a perfect square? The numbers $\left(2 + \sqrt{3}\right)$ and $\left(2 - \sqrt{3}\right)$ are one the reciprocal of the other, and if they were squares, we would have a perfect square. In fact, $\left(4 \pm 2\sqrt{3}\right)$ are the squares of $\left(1 \pm \sqrt{3}\right)$. We find that

$$\Delta = \left(\frac{\left(1 + \sqrt{3}\right)^{2n+1} + \left(1 - \sqrt{3}\right)^{2n+1}}{2^n}\right)^2.$$

Solving the quadratic equation, we find that

$$k = \frac{1}{2} + \frac{\left(1 + \sqrt{3}\right)^{2n+1} + \left(1 - \sqrt{3}\right)^{2n+1}}{2^{2n+2}}$$

$$= \frac{1}{2} + \frac{1}{4}\left[\left(1 + \sqrt{3}\right)\left(2 + \sqrt{3}\right)^n + \left(1 - \sqrt{3}\right)\left(2 - \sqrt{3}\right)^n\right].$$

This is clearly a rational number, but is it an integer? The numbers $2 + \sqrt{3}$ and $2 - \sqrt{3}$ are the roots of the equation

$$\lambda^2 - 4\lambda + 1 = 0,$$

which can be interpreted as the characteristic equation of a recursive sequence $x_{n+1} - 4x_n + x_{n-1} = 0$. Given that the general formula of the terms of the sequence is $\left(1 + \sqrt{3}\right)$ $\left(2 + \sqrt{3}\right)^n + \left(1 - \sqrt{3}\right)\left(2 - \sqrt{3}\right)^n$, we also see that $x_0 = 2$ and $x_1 = 10$. An induction based on the recurrence relation shows that x_n is divisible by 2 but not by 4. It follows that k is an integer and the problem is solved.

(Romanian Team Selection Test for the International Mathematical Olympiad, 1999, proposed by D. Andrica)

1019. We have

$$a_n + b_n\sqrt[3]{2} + c_n\sqrt[3]{4} = \frac{\sqrt[3]{2}\left(1 + \sqrt[3]{2} + \sqrt[3]{4}\right)^n}{\left(\sqrt[3]{2}\right)^n} = 2^{-\frac{n}{3}}\left(\sqrt[3]{2} + \sqrt[3]{4} + 2\right)^n$$

$$= 2^{-\frac{n}{3}}\left(1 + \left(1 + \sqrt[3]{2} + \sqrt[3]{4}\right)\right)^n$$

$$= 2^{-\frac{n}{3}}\sum_{k=0}^{n}\binom{n}{k}\left(a_k + b_k\sqrt[3]{2} + c_k\sqrt[3]{4}\right).$$

Hence

$$a_n + b_n\sqrt[3]{2} + c_n\sqrt[3]{4} = 2^{-\frac{n}{3}}\sum_{k=0}^{n}\binom{n}{k}a_k + 2^{-\frac{n}{3}}\sum_{k=0}^{n}\binom{n}{k}b_k\sqrt[3]{2} + 2^{-\frac{n}{3}}\sum_{k=0}^{n}\binom{n}{k}c_k\sqrt[3]{4}.$$

The conclusion follows from the fact that $2^{-n/3}$ is an integer if n is divisible by 3, is an integer times $\sqrt[3]{4}$ if n is congruent to 1 modulo 3, and is an integer times $\sqrt[3]{2}$ if n is congruent to 2 modulo 3.

(*Revista Matematică din Timişoara* (*Timişoara Mathematics Gazette*), proposed by T. Andreescu and D. Andrica)

1020. *First solution*: We prove the formula by induction on n. The case $n = 1$ is straightforward. Now let us assume that the formula holds for n and let us prove it for $n + 1$. Using the induction hypothesis, we can write

$$[x + y]_{n+1} = (x + y - n)[x + y]_n = (x + y - n) \sum_{k=0}^{n} \binom{n}{k} [x]_{n-k}[y]_k$$

$$= \sum_{k=0}^{n} \binom{n}{k} ((x - k) + (y - n + k))[x]_k[y]_{n-k}$$

$$= \sum_{k=0}^{n} \binom{n}{k} (x - k)[x]_k[y]_{n-k} + \sum_{k=0}^{n} \binom{n}{k} (y - (n - k))[x]_k[y]_{n-k}$$

$$= \sum_{k=0}^{n} \binom{n}{k} [x]_{k+1}[y]_{n-k} + \sum_{k=0}^{n} \binom{n}{k} [x]_k[y]_{n-k+1}$$

$$= \sum_{k=1}^{n+1} \binom{n}{k-1} [x]_k[y]_{n-k+1} + \sum_{k=0}^{n} \binom{n}{k} [x]_k[y]_{n-k+1}$$

$$= \sum_{k=0}^{n+1} \left(\binom{n}{k} + \binom{n}{k-1} \right) [x]_k[y]_{n-k+1}$$

$$= \sum_{k=0}^{n+1} \binom{n+1}{k} [x]_k[y]_{n+1-k}.$$

The induction is complete.

Second solution: The identity can also be proved by computing $\left(\frac{d}{dt} \right)^n t^{x+y}$ in two different ways. First,

$$\left(\frac{d}{dt} \right)^n t^{x+y} = (x + y)(x + y - 1) \cdots (x + y - n + 1)t^{x+y-n} = [x + y]_n t^{x+y-n}.$$

Second, by the Leibniz rule for differentiating the product,

$$\left(\frac{d}{dt} \right)^n (t^x \cdot t^y) = \sum_{k=0}^{n} \binom{n}{k} \left(\left(\frac{d}{dt} \right)^k t^x \right) \left(\left(\frac{d}{dt} \right)^{n-k} t^y \right) = \sum_{k=0}^{n} \binom{n}{k} [x]_k[y]_{n-k} t^{x+y-n}.$$

The conclusion follows.

1021. The binomial formula $(Q + P)^n = \sum_{k=0}^{n} \binom{n}{k}_q Q^k P^{n-k}$ is of no use because the variables Q and P do not commute, so we cannot set $P = -Q$. The solution relies on the q-Pascal triangle. The q-Pascal triangle is defined by

$$\binom{n}{k}_q = q^k \binom{n-1}{k}_q + \binom{n-1}{k-1}_q.$$

With the standard convention that $\binom{n}{k}_q = 0$ if $k < 0$ or $k > n$, we have

$$\sum_k (-1)^k q^{\frac{k(k-1)}{2}} \binom{n}{k}_q = \sum_k (-1)^k q^{\frac{k(k-1)}{2}} \left(q^k \binom{n-1}{k}_q + \binom{n-1}{k-1}_q \right)$$

$$= \sum_k (-1)^k q^{\frac{k(k+1)}{2}} \binom{n-1}{k}_q - \sum_k (-1)^{k-1} q^{\frac{k(k-1)}{2}} \binom{n-1}{k-1}_q.$$

Now just shift the index in the second sum $k \mapsto k+1$ to obtain the difference of two equal sums. The identity follows.

1022. Let $G(x) = \sum_n y_n x^n$ be the generating function of the sequence. It satisfies the functional equation

$$(1 - ax)G(x) = 1 + bx + bx^2 + \cdots = \frac{1}{1 - bx}.$$

We find that

$$G(x) = \frac{1}{(1 - ax)(1 - bx)} = \frac{A}{1 - ax} + \frac{B}{1 - bx} = \sum_n (Aa^n + Bb^n)x^n,$$

for some A and B. It follows that $y_n = Aa^n + Bb^n$, and because $y_0 = 1$ and $y_1 = a + b$, $A = \frac{a}{a-b}$ and $B = -\frac{b}{a-b}$. The general term of the sequence is therefore

$$\frac{1}{a - b}(a^{n+1} - b^{n+1}).$$

1023. The first identity is obtained by differentiating $(x + 1)^n = \sum_{k=1}^n \binom{n}{k} x^k$, then setting $x = 1$. The answer is $n2^{n-1}$. The second identity is obtained by integrating the same equality and then setting $x = 1$, in which case the answer is $\frac{2^{n+1}}{n+1}$.

1024. The identity in part (a) is the Vandermonde formula. It is proved using the generating function of the binomial coefficients, by equating the coefficients of x^k on the two sides of the equality $(x + 1)^{m+n} = (x + 1)^m (x + 1)^n$.

The identity in part (b) is called the Chu-Vandermonde formula. This time the generating function in question is $(Q+P)^n$, where Q and P are the noncommuting variables that describe the time evolution of the position and the momentum of a quantum particle. They are noncommuting variables satisfying $PQ = qQP$, the exponential form of the canonical commutation relations which lead to the Heisenberg uncertainty principle. The Chu-Vandermonde formula is obtained by identifying the coefficients of $Q^k P^{m+n-k}$ on the two sides of the equality

$$(Q + P)^{m+n} = (Q + P)^m (Q + P)^n.$$

Observe that the powers of q arise when we switch P's and Q's as follows:

$$\binom{m}{j}_q Q^j P^{m-j} \binom{n}{k-j}_q Q^{k-j} P^{n-k+j} = \binom{m}{j}_q \binom{n}{k-j}_q Q^j P^{m-j} Q^{k-j} P^{n-k+j}$$

$$= q^{(m-j)(k-j)} \binom{m}{j}_q \binom{n}{k-j}_q Q^k P^{m+n-k}.$$

1025. The sum is equal to the coefficient of x^n in the expansion of

$$x^n(1-x)^n + x^{n-1}(1-x)^n + \cdots + x^{n-m}(1-x)^n.$$

This expression is equal to

$$x^{n-m} \cdot \frac{1 - x^{m+1}}{1 - x}(1-x)^n,$$

which can be written as $(x^{n-m} - x^{n+1})(1-x)^{n-1}$. Hence the sum is equal to $(-1)^m \binom{n-1}{m}$ if $m < n$, and to 0 if $m = n$.

1026. The sum from the statement is equal to the coefficient of x^k in the expansion of

$$(1+x)^n + (1+x)^{n+1} + \cdots + (1+x)^{n+m}.$$

This expression can be written in compact form as

$$\frac{1}{x}\left((1+x)^{n+m+1} - (1+x)^n\right).$$

We deduce that the sum is equal to $\binom{n+m+1}{k+1} - \binom{n}{k+1}$ for $k < n$ and to $\binom{n+m+1}{n+1}$ for $k = n$.

1027. The generating function of the Fibonacci sequence is

$$\phi(x) = \frac{1}{1 - x - x^2}.$$

Expanding like a geometric series, we obtain

$$\frac{1}{1 - x - x^2} = \frac{1}{1 - x(1+x)} = 1 + x(1+x) + x^2(1+x)^2 + \cdots + x^n(1+x)^n + \cdots$$

The coefficient of x^n is on the one hand F_{n+1} and on the other hand

$$\binom{n}{0} + \binom{n-1}{1} + \binom{n-2}{2} + \cdots.$$

The identity follows.

1028. We introduce some additional parameters and consider the expansion

$$\frac{1}{(1 - a_1 x)(1 - a_2 x^2)(1 - a_3 x^3) \cdots}$$
$$= (1 + a_1 x + a_1^2 x^2 + \cdots)(1 + a_2 x^2 + a_2^2 x^4 + \cdots)(1 + a_3 x^3 + a_3^2 x^6 + \cdots) \cdots$$
$$= 1 + a_1 x + (a_1^2 + a_2)x^2 + \cdots + (a_1^{\lambda_1} a_2^{\lambda_2} \cdots a_k^{\lambda_k} + \cdots)x^n + \cdots.$$

The term $a_1^{\lambda_1} a_2^{\lambda_2} \cdots a_k^{\lambda_k}$ that is part of the coefficient of x^n has the property that $\lambda_1 + 2\lambda_2 + \cdots + k\lambda_k = n$; hence it defines a partition of n, namely,

$$n = \underbrace{1 + 1 + \cdots + 1}_{\lambda_1} + \underbrace{2 + 2 + \cdots + 2}_{\lambda_2} + \cdots + \underbrace{k + k + \cdots + k}_{\lambda_k}.$$

So the terms that appear in the coefficient of x^n generate all partitions of n. Setting $a_1 = a_2 = a_3 = \cdots = 1$, we obtain for the coefficient of x^n the number $P(n)$ of the partitions of n. And we are done.

1029. The argument of the previous problem can be applied mutatis mutandis to show that the number of ways of writing n as a sum of odd positive integers is the coefficient of x^n in the expansion of

$$\frac{1}{(1 - x)(1 - x^3)(1 - x^5)(1 - x^7)\cdots},$$

while the number of ways of writing n as a sum of distinct positive integers is the coefficient of x^n in

$$(1 + x)(1 + x^2)(1 + x^3)(1 + x^4)\cdots$$

We have

$$\frac{1}{(1 - x)(1 - x^3)(1 - x^5)(1 - x^7)\cdots} = \frac{1 - x^2}{1 - x} \cdot \frac{1 - x^4}{1 - x^2} \cdot \frac{1 - x^6}{1 - x^3} \cdot \frac{1 - x^8}{1 - x^4} \cdot \frac{1 - x^{10}}{1 - x^5}\cdots$$
$$= (1 + x)(1 + x^2)(1 + x^3)(1 + x^4)\cdots$$

This proves the desired equality.

Remark. This property is usually phrased as follows: Prove that the number of partitions of n into distinct parts is equal to the number of partitions of n into odd parts.
   (L. Euler)

1030. The number of subsets with the sum of the elements equal to n is the coefficient of x^n in the product

$$G(x) = (1 + x)(1 + x^2) \cdots (1 + x^p).$$

We are asked to compute the sum of the coefficients of x^n for n divisible by p. Call this number $s(p)$. There is no nice way of expanding the generating function; instead we compute $s(p)$ using particular values of G. It is natural to try pth roots of unity.

   The first observation is that if ξ is a pth root of unity, then $\sum_{k=1}^{p} \xi^p$ is zero except when $\xi = 1$. Thus if we sum the values of G at the pth roots of unity, only those terms with exponent divisible by p will survive. To be precise, if ξ is a pth root of unity different from 1, then

$$\sum_{k=1}^{p} G(\xi^k) = ps(p).$$

We are left with the problem of computing $G(\xi^k)$, $k = 1, 2, \ldots, p$. For $k = p$, this is just 2^p. For $k = 1, 2, \ldots, p-1$,

$$G(\xi^k) = \prod_{j=1}^{p}(1 + \xi^{kj}) = \prod_{j=1}^{p}(1 + \xi^{j}) = (-1)^p \prod_{j=1}^{p}((-1) - \xi^{j}) = (-1)^p((-1)^p - 1) = 2.$$

We therefore have $ps(p) = 2^p + 2(p-1) = 2^p + 2p - 2$. The answer to the problem is $s(p) = \frac{2^p - 2}{p} + 2$. The expression is an integer because of Fermat's little theorem.

(T. Andreescu, Z. Feng, *A Path to Combinatorics for Undergraduates*, Birkhäuser 2004)

1031. We introduce the generating function

$$G_n(x) = \left(x + \frac{1}{x}\right)\left(x^2 + \frac{1}{x^2}\right) \cdots \left(x^n + \frac{1}{x^n}\right).$$

Then $S(n)$ is the term not depending on x in $G_n(x)$. If in the expression

$$\left(x + \frac{1}{x}\right)\left(x^2 + \frac{1}{x^2}\right) \cdots \left(x^n + \frac{1}{x^n}\right) = S(n) + \sum_{k \neq 0} c_k x^k$$

we set $x = e^{it}$ and then integrate between 0 and 2π, we obtain

$$\int_0^{2\pi} (2\cos t)(2\cos 2t) \cdots (2\cos nt)dt = 2\pi S(n) + 0,$$

whence the desired formula

$$S(n) = \frac{2^{n-1}}{\pi} \int_0^{2\pi} \cos t \cos 2t \cdots \cos nt\, dt.$$

(Communicated by D. Andrica)

1032. Let us assume that n is not a power of 2. We consider a more exotic kind of generating function where the sequence is encoded in the exponents, not in the coefficients:

$$f(x) = x^{a_1} + x^{a_2} + \cdots + x^{a_n} \quad \text{and} \quad g(x) = x^{b_1} + x^{b_2} + \cdots + x^{b_n}.$$

In fact, these are the generating functions of the characteristic functions of the sets A and B. By assumption,

$$f(x)^2 - f(x^2) = 2\sum_{i<j} x^{a_i + a_j} = 2\sum_{i<j} x^{b_i + b_j} = g(x)^2 - g(x^2).$$

Therefore,

$$(f(x) - g(x))(f(x) + g(x)) = f(x^2) - g(x^2).$$

Let $h(x) = f(x) - g(x)$ and $p(x) = f(x) + g(x)$. We want to prove that if n is not a power of 2, then h is identically 0. Note that $h(1) = 0$. We will prove by strong induction that all derivatives of h at 1 are zero, which will make the Taylor series of h identically zero. Note that

$$h'(x)p(x) + h(x)p'(x) = 2xh'(x^2),$$

and so $h'(1)p(1) = 2h'(1)$. Since $p(1) = f(1) + g(1) = 2n$, which is not a power of 2, it follows that $h'(1) = 0$. Assuming that all derivatives of h of order less than k at 1 are zero, by differentiating the functional equation k times and substituting $x = 1$, we obtain

$$h^{(k)}(1)p(1) = 2^k h^{(k)}(1).$$

Hence $h^{(k)}(1) = 0$. This completes the induction, leading to a contradiction. It follows that n is a power of 2, as desired.

(Communicated by A. Neguţ)

1033. We use the same generating functions as in the previous problem. So to the set A_n we associate the function

$$a_n(x) = \sum_{a=1}^{\infty} c_a x^a,$$

with $c_a = 1$ if $a \in A_n$ and $c_a = 0$ if $a \notin A_n$. To B_n we associate the function $b_n(x)$ in a similar manner. These functions satisfy the recurrence $a_1(x) = 0$, $b_1(x) = 1$,

$$a_{n+1}(x) = xb_n(x),$$
$$b_{n+1} \equiv a_n(x) + b_n(x) \pmod 2.$$

From now on we understand all equalities modulo 2. Let us restrict our attention to the sequence of functions $b_n(x)$, $n = 1, 2, \ldots$. It satisfies $b_1(x) = b_2(x) = 1$,

$$b_{n+1}(x) = b_n(x) + xb_{n-1}(x).$$

We solve this recurrence the way one usually solves second-order recurrences, namely by finding two linearly independent solutions $p_1(x)$ and $p_2(x)$ satisfying

$$p_i(x)^{n+1} = p_i(x)^n + xp_i(x)^{n-1}, \quad i = 1, 2.$$

Again the equality is to be understood modulo 2. The solutions $p_1(x)$ and $p_2(x)$ are formal power series whose coefficients are residue classes modulo 2. They satisfy the "characteristic" equation

$$p(x)^2 = p(x) + x,$$

which can be rewritten as

$$p(x)(p(x) + 1) = x.$$

So $p_1(x)$ and $p_2(x)$ can be chosen as the factors of this product, and thus we may assume that $p_1(x) = xh(x)$ and $p_2(x) = 1 + p_1(x)$, where $h(x)$ is again a formal power series. Writing $p_1(x) = \sum \alpha_a x^a$ and substituting in the characteristic equation, we find that $\alpha_1 = 1$, $\alpha_{2k} = \alpha_k^2$, and $\alpha_{2k+1} = 0$ for $k > 1$. Therefore,

$$p_1(x) = \sum_{k=0}^{\infty} x^{2^k}.$$

Since $p_1(x) + p_2(x) = p_1(x)^2 + p_2(x)^2 = 1$, it follows that in general,

$$b_n(x) = p_1(x)^n + p_2(x)^n = \left(\sum_{k=0}^{\infty} x^{2^k}\right)^n + \left(1 + \sum_{k=0}^{\infty} x^{2^k}\right)^n, \text{ for } n \geq 1.$$

We emphasize again that this is to be considered modulo 2. In order for $b_n(x)$ to be identically equal to 1 modulo 2, we should have

$$\left(\left(\sum_{k=0}^{\infty} x^{2^k}\right) + 1\right)^n = \left(\sum_{k=0}^{\infty} x^{2^k}\right)^n + 1 \pmod{2}.$$

This obviously happens if n is a power of 2, since all binomial coefficients in the expansion are even.

If n is not a power of 2, say $n = 2^i(2j + 1)$, $j \geq 1$, then the smallest m for which $\binom{n}{m}$ is odd is 2^j. The left-hand side will contain an x^{2^j} with coefficient equal to 1, while the smallest nonzero power of x on the right is n. Hence in this case equality cannot hold.

We conclude that $B_n = \{0\}$ if and only if n is a power of 2.

(Chinese Mathematical Olympiad)

1034. We will count the number of committees that can be chosen from n people, each committee having a president and a vice-president.

Choosing first a committee of k people, the president and the vice-president can then be elected in $k(k-1)$ ways. It is necessary that $k \geq 2$. The committees with president and vice-president can therefore be chosen in

$$1 \cdot 2\binom{n}{2} + 2 \cdot 3\binom{n}{3} + \cdots + (n-1) \cdot n\binom{n}{n}$$

ways.

But we can start by selecting first the president and the vice-president, and then adding the other members to the committee. From the n people, the president and the vice-president can be selected in $n(n-1)$ ways. The remaining members of the committee can be selected in 2^{n-2} ways, since they are some subset of the remaining $n-2$ people. We obtain

$$1 \cdot 2\binom{n}{2} + 2 \cdot 3\binom{n}{3} + \cdots + (n-1) \cdot n\binom{n}{n} = n(n-1)2^{n-2}.$$

1035. Rewrite the identity as

$$\sum_{k=1}^{n} k\binom{n}{k}\binom{n}{n-k} = n\binom{2n-1}{n-1}.$$

We claim that both sides count the number of n-member committees with a physicist president that can be elected from a group of n mathematicians and n physicists. Indeed, on the left-hand side we first elect k physicists and $n - k$ mathematicians, then elect the president among the k physicists, and do this for all k. On the right-hand side we first elect the president and then elect the other members of the committee from the remaining $2n - 1$ people.

1036. We will prove that both terms of the equality count the same thing. To this end, we introduce two disjoint sets M and N containing m, respectively, n elements.

For the left-hand side, choose first k elements in M. This can be done in $\binom{m}{k}$ ways. Now add these k elements to N and choose m elements from the newly obtained set. The number of ordered pairs of sets (X, Y) with $X \subset M$, $Y \subset N \cup X$, $|X| = k$, and $|Y| = m$ is equal to $\binom{m}{k}\binom{n+k}{m}$. Varying k, we obtain, for the total number of pairs (X, Y),

$$\sum_{k=0}^{m} \binom{m}{k}\binom{n+k}{m}.$$

The same problem can be solved differently, namely choosing Y first. If we fix the cardinality of $Y \cap N$, say $|Y \cap N| = j$, $0 \le j \le m$, then $|Y \cap M| = m - j$, and so there are $\binom{n}{j}\binom{m}{m-j} = \binom{n}{j}\binom{m}{j}$ ways to choose Y. Now X contains the set $Y \cap M$, the union with some (arbitrary) subset of $M \backslash Y$. There are j elements in $M \backslash Y$, so there are 2^j possible choices for X. Consequently, the number of pairs with the desired property is

$$\sum_{j=0}^{n} \binom{n}{j}\binom{m}{j}2^j.$$

Setting the two numbers equal yields the identity from the statement.

(I. Tomescu, *Problems in Combinatorics*, Wiley, 1985)

1037. *First solution:* We prove the identity by counting, in two different ways, the cardinality of the set of words of length n using the alphabet $\{A, B, C\}$ and satisfying the condition that precisely k of the letters are A, and all of the letters B must be among the first m letters as read from the left.

The first count is according to the number of B's. Place m symbols X in a row and following them $n - m$ symbols Y:

$$\underbrace{XX \ldots XX}_{m} \underbrace{YY \ldots YY}_{n-m} .$$

Choose i of the X's (in $\binom{m}{i}$ ways), and replace them by B's. Choose k of the $n - i$ remaining symbols (in $\binom{n-i}{k}$ ways), and replace them by A's. Any remaining X's or Y's are now replaced by C's. We have constructed $\binom{m}{i}\binom{n-i}{k}$ words satisfying the conditions. Summing over i, we have the sum on the left.

The second count is according to the number of A's among the first m letters of the word. We start with the same sequence of X's and Y's as before. Choose i of the m X's (in $\binom{m}{i}$ ways), replace each of them by A and replace each of the other $m - i$ X's by B or C (this can be done in 2^{m-i} ways). Then choose $k - i$ of the $n - m$ Y's (in $\binom{n-m}{k-i}$ ways) and replace each of them by A, and replace the remaining Y's by C. We have constructed $\binom{m}{i}\binom{n-m}{k-i}2^{m-i}$ words satisfying the conditions. Summing over i, we obtain the right-hand side of the identity.

Second solution: There is also a solution by generating functions. Fix m and consider the two expressions that are to be shown equal as functions of n, and let the expression on the left be A_n and the one on the right B_n. Then, consider the generating function for the numbers A_n:

$$F(x) = \sum_{n} A_n x^n = \sum_{n} \sum_{i=0}^{m} \binom{m}{i}\binom{n-i}{k}x^n.$$

We compute

$$F(x) = \sum_i \binom{m}{i} \sum_n \binom{n-i}{k} x^n = \sum_i \binom{m}{i} x^i \sum_n \binom{n-i}{k} x^{n-i}$$

$$= \sum_i \binom{m}{i} x^i \frac{x^k}{(1-x)^{k+1}} = \frac{x^k(1+x)^m}{(1-x)^{k+1}}.$$

Considering the generating function for the numbers B_n:

$$G(x) = \sum_n B_n x^n = \sum_n \sum_{i=0}^{m} \binom{m}{i} \binom{n-m}{k-i} 2^{m-i} x^n,$$

we compute

$$G(x) = \sum_i \binom{m}{i} x^m 2^{m-i} \sum_n \binom{n-m}{k-i} x^{n-m} = \sum_i \binom{m}{i} x^m 2^{m-i} \frac{x^{k-i}}{(1-x)^{k-i+1}}$$

$$= \sum_i \binom{m}{i} \left(\frac{1-x}{2x}\right)^i \cdot \frac{(2x)^m x^k}{(1-x)^{k+1}} = \frac{(2x)^m x^k}{(1-x)^{k+1}} \left(1 - \frac{1-x}{2x}\right)^m = \frac{x^k(1+x)^m}{(1-x)^{k+1}}.$$

We see that $F(x) = G(x)$, and hence $A_n = B_n$, proving the combinatorial identity.

(*Mathematics Magazine*, the case $m = k - 1$ proposed by D. Callan, first solution and generalization by W. Moser, second solution by M.C. Zanarella)

1038. *First solution:* For a counting argument to work, the identity should involve only integers. Thus it is sensible to write it as

$$\sum_{k=0}^{q} 2^{q-k} \binom{p+k}{k} + \sum_{k=0}^{p} 2^{p-k} \binom{q+k}{k} = 2^{p+q+1}.$$

This looks like the count of the elements of a set partitioned into two subsets. The right-hand side counts the number of subsets of a set with $p + q + 1$ elements. It is better to think of it as the number of elements of $\{0, 1\}^{p+q+1}$. We partition this set into two disjoint sets A and B such that A is the set of $(p + q + 1)$-tuples with at least $p + 1$ entries equal to 1, and B, its complement, is the set of $(p + q + 1)$-tuples with at least $q + 1$ entries equal to 0. If the position of the $(p + 1)$st 1 is $p + k + 1$, $0 \le k \le q$, then there are $\binom{p+k}{p} = \binom{p+k}{k}$ ways of choosing the positions of the first p ones. Several subsequent coordinates can also be set to 1, and this can be done in 2^{q-k} ways. It follows that $2^{q-k} \binom{p+k}{k}$ elements in A have the $(p+1)$st 1 in position $p + k + 1$. Therefore, the first sum counts the elements of A. Similarly, the second sum counts the elements of B, and the conclusion follows.

Second solution: Like with most combinatorial identities, there is a solution with generating functions. Denote the expression on the left-hand side of the identity by $A_{p,q}$, and consider the generating function of the numbers A_{ij}:

$$F(x, y) = \sum_{i,j \ge 0} A_{i,j} x^i y^j$$

We compute

$$F(x, y) = \sum_i \sum_j \left[\sum_{k=0}^{j} \frac{1}{2^{i+k}} \binom{i+k}{k} + \sum_{k=0}^{i} \frac{1}{2^{j+k}} \binom{j+k}{k} \right] x^i y^j$$

$$= \sum_j y^j \sum_{k=0}^{j} \sum_i \frac{1}{2^{i+k}} \binom{i+k}{k} x^i + \sum_i x^i \sum_{k=0}^{i} \sum_j \frac{1}{2^{j+k}} \binom{j+k}{k} y^j.$$

Since

$$\sum_i \binom{i+k}{k} x^i = \frac{1}{(1-x)^{k+1}}$$

we have

$$\sum_i \frac{1}{2^{i+k}} \binom{i+k}{k} x^i = \frac{1}{2^k \left(1 - \dfrac{x}{2}\right)^{k+1}}.$$

From there we obtain

$$\sum_k y^k \sum_i \binom{i+k}{k} \frac{x^i}{2^{i+k}} = \sum_k \frac{y^k}{2^k \left(1 - \dfrac{x}{2}\right)^{k+1}} = \frac{1}{1 - \dfrac{x}{2}} \cdot \frac{1}{1 - \dfrac{y}{2-x}}.$$

Notice that in general, if $L(y) = \sum_k y^k B_k$, then

$$\frac{L(y)}{1-y} = \sum_k y^k \sum_{j=0}^{k} B_j.$$

Using this we deduce that

$$\sum_j y^j \sum_{k=0}^{j} \sum_i \frac{1}{2^{i+k}} \binom{i+k}{k} x^i = \frac{1}{1-y} \sum_k y^k \sum_i \binom{i+k}{k} \frac{x^i}{2^{i+k}}$$

$$= \frac{1}{1-y} \cdot \frac{1}{1 - \dfrac{x}{2}} \cdot \frac{1}{1 - \dfrac{y}{2-x}} = \frac{2}{(1-y)(2-x-y)}.$$

By symmetry,

$$\sum_i x^i \sum_{k=0}^{i} \sum_j \frac{1}{2^{j+k}} \binom{j+k}{k} y^j = \frac{2}{(1-x)(2-x-y)}.$$

Therefore

$$F(x, y) = \frac{2}{(1-y)(2-x-y)} + \frac{2}{(1-x)(2-x-y)} = \frac{2}{(1-x)(1-y)} = \sum_i \sum_j 2 x^i y^j.$$

We conclude that $A_{p,q} = 2$ for all p, q.

(French Contest, 1985, solution from T.B. Soulami, *Les olympiades de mathématiques: Réflexes et stratégies*, Ellipses, 1999, second solution by M.C. Zanarella)

1039. A group of $2n + 1$ people, consisting of n male/female couples and one extra male, wish to split into two teams. Team 1 should have n people, consisting of $\lfloor \frac{n}{2} \rfloor$ women and $\lfloor \frac{n+1}{2} \rfloor$ men, while Team 2 should have $n + 1$ people, consisting of $\lceil \frac{n}{2} \rceil$ women and $\lceil \frac{n+1}{2} \rceil$ men, where $\lceil x \rceil$ denotes the least integer greater than or equal to x. The number of ways to do this is counted by the first team, and is $c_n c_{n+1}$.

There is a different way to count this, namely by the number k of couples that are split between the two teams. The single man joins Team 1 if and only if k and n have opposite parity. The split couples can be chosen in $\binom{n}{k}$ ways. From the remaining $n - k$ couples, the number to join Team 1 is $\lfloor \frac{n-k}{2} \rfloor$, which can be chosen in c_{n-k} ways. Since these couples contribute $\lfloor \frac{n-k}{2} \rfloor$ women to Team 1, the number of women from the k split couples that join Team 1 must be $\lfloor \frac{n}{2} \rfloor - \lfloor \frac{n-k}{2} \rfloor$, which equals either $\lfloor \frac{k}{2} \rfloor$ for n odd or $\lceil \frac{k}{2} \rceil$ for n even. Since $\binom{k}{\lfloor k/2 \rfloor} = \binom{k}{\lceil k/2 \rceil}$, these women can be chosen in c_k ways. Thus the left side also counts the choices.

(*American Mathematical Monthly*, proposed by D.M. Bloom, solution by Ch.N. Swanson)

1040. We count the points of integer coordinates in the rectangle

$$1 \le x \le p', \quad 1 \le y \le q'.$$

Their total number is $p'q'$. Now let us look at the expression in the first set of parentheses. The terms count the number of points with integer coordinates that lie below the line $y = \frac{q}{p}x$ and on the lines $x = 1, x = 2, \ldots, x = p'$. Here it is important to remark that since p and q are coprime, none of these points lie on the line $y = \frac{q}{p}x$. Similarly, the expression in the second parentheses counts the number of points with integer coordinates that lie above the line $y = \frac{q}{p}x$ and on the lines $y = 1, y = 2, \ldots, y = q'$. Together, these are all the points of the rectangle. That there are no others follows from the inequalities

$$\left\lfloor \frac{p'q}{p} \right\rfloor \le q' \quad \text{and} \quad \left\lfloor \frac{q'p}{q} \right\rfloor \le p'.$$

Indeed,

$$\left\lfloor \frac{p'q}{p} \right\rfloor = \left\lfloor \frac{p'(2q'+1)}{2p'+1} \right\rfloor = \left\lfloor \frac{q' + \frac{1}{2}}{1 + \frac{1}{2p'}} \right\rfloor \le \left\lfloor q' + \frac{1}{2} \right\rfloor = q',$$

and the other inequality is similar.

Thus both sides of the identity in question count the same points, so they are equal.

(G. Eisenstein)

1041. *First solution*: For each pair of students, consider the set of those problems not solved by them. There are $\binom{200}{2}$ such sets, and we have to prove that at least one of them is empty.

For each problem there are at most 80 students who did not solve it. From these students at most $\binom{80}{2} = 3160$ pairs can be selected, so the problem can belong to at most 3160 sets. The 6 problems together can belong to at most $6 \cdot 3160$ sets.

Hence at least $19900 - 18960 = 940$ sets must be empty, and the conclusion follows.

Second solution: Since each of the six problems was solved by at least 120 students, there were at least 720 correct solutions in total. Since there are only 200 students, there is some student who solved at least four problems. If a student solved five or six problems, we are clearly done. Otherwise, there is a student who solved exactly four. Since the two problems he missed were solved by at least 120 students, there must be a student (in fact, at least 40) who solved both of them.

(9th International Mathematical Competition for University Students, 2002)

1042. *First solution*: We prove the formula by induction on m. For $m = 1$ it clearly is true, since there is only one solution, $x_1 = n$. Assume that the formula is valid when the number of unknowns is $k \leq m$, and let us prove it for $m + 1$ unknowns. Write the equation as

$$x_1 + x_2 + \cdots + x_m = n - x_{m+1}.$$

As x_{m+1} ranges between 0 and n, the right-hand sides assumes all values between 0 and n. Using the induction hypothesis for all these cases and summing up, we find that the total number of solutions is

$$\sum_{r=0}^{n} \binom{m+r-1}{m-1}.$$

As before, this sums up to $\binom{m+n}{m}$, proving the formula for $m + 1$ unknowns. This completes the solution.

Second solution: Let $y_i = x_i + 1$. Then $y_1, \ldots, y_m$ is a solution in positive integers to the equation $y_1 + y_2 + \cdots + y_m = n + m$. These solutions were counted in one of the examples discussed at the beginning of this section.

1043. We associate to such a subset S a word $a_1 a_2 \ldots a_n$ with $a_i = 1$ if $i \in S$ and $a_i = 0$ if $i \notin S$. It suffices to count the number of words that do not contain two consecutive ones.

Let us first count the number of such words that contain precisely k ones. Such a word is obtained by starting with a sequence of $n - k$ zeros. This sequence has $n - k + 1$ slots where ones can be inserted: one between every two consecutive zeros, one at the beginning, and one at the end. The word is the obtained by choosing k slots, and inserting a one in each of them. It follows that the number of such words is $\binom{n-k+1}{k}$.

So the total number of words is

$$f(n) = \sum_{k} \binom{n-k+1}{k}.$$

To write this in short form note that

$$\binom{n-k+1}{k} = \binom{n-k}{k} + \binom{n-k}{k-1} = \binom{(n-1)-k+1}{k} + \binom{(n-2)-(k-1)+1}{(k-1)}$$

hence $f(n) = f(n-1) + f(n-2)$ for all $n \geq 2$. It is easy to count $f(1) = 2, f(2) = 3$, and therefore $f(n) = F_{n+1}$, where $(F_n)_{n \geq 0}$ is the Fibonacci sequence.

Remark. Note the similarity with the second solution given to the previous problem.

1044. Let M_1 be the set of edges whose endpoints have different colors, M_2 the set of faces with vertices colored by three different colors, and M_3 the set of faces with exactly two vertices colored by the same color. As usually, we denote by $|A|$ the number of elements of the set A.

Each triangle in M_2 contains exactly 3 edges in M_1, and each triangle in M_3 contains exaclty two edges in M_1. Every edge in M_1 belongs to either two faces in M_2, two faces in M_3, or a face in M_2 and one in M_3. Counting the edges by faces, we obtain

$$2|M_1| = 3|M_2| + 2|M_3|.$$

Hence $|M_2|$ is even, as desired.

(Romanian Team Selection Test for the International Mathematical Olympiad, 1983, propsosed by I. Tomescu, solution by O. Bucikovski)

1045. Since each tennis player played $n-1$ games, $x_i + y_i = n-1$ for all i. Altogether there are as many victories as losses; hence $x_1 + x_2 + \cdots + x_n = y_1 + y_2 + \cdots + y_n$. We have

$$\begin{aligned}
x_1^2 + x_2^2 + \cdots + x_n^2 - y_1^2 - y_2^2 - \cdots - y_n^2 &= (x_1^2 - y_1^2) + (x_2^2 - y_2^2) + \cdots + (x_n^2 - y_n^2) \\
&= (x_1 + y_1)(x_1 - y_1) + (x_2 + y_2)(x_2 - y_2) + \cdots + (x_n + y_n)(x_n - y_n) \\
&= (n-1)(x_1 - y_1 + x_2 - y_2 + \cdots + x_n - y_n) \\
&= (n-1)(x_1 + x_2 + \cdots + x_n - y_1 - y_2 - \cdots - y_n) = 0,
\end{aligned}$$

and we are done.

(L. Panaitopol, D. Şerbănescu, *Probleme de Teoria Numerelor şi Combinatorică pentru Juniori* (*Problems in Number Theory and Combinatorics for Juniors*), GIL, 2003)

1046. Let $B = \{b_1, b_2, \ldots, b_p\}$ be the union of the ranges of the two functions. For $b_i \in B$, denote by n_{b_i} the number of elements $x \in A$ such that $f(x) = b_i$, and by k_{b_i} the number of elements $x \in A$ such that $g(x) = b_i$. Then the number of pairs $(x, y) \in A \times A$ for which $f(x) = g(x) = b_i$ is $n_{b_i} k_{b_i}$, the number of pairs for which $f(x) = f(y) = b_i$ is $n_{b_i}^2$, and the number of pairs for which $g(x) = g(y) = b_i$ is $k_{b_i}^2$. Summing over i, we obtain

$$\begin{aligned}
m &= n_{b_1} k_{b_1} + n_{b_2} k_{b_2} + \cdots + n_{b_p} k_{b_p}, \\
n &= n_{b_1}^2 + n_{b_2}^2 + \cdots + n_{b_p}^2, \\
k &= k_{b_1}^2 + k_{b_2}^2 + \cdots + k_{b_p}^2.
\end{aligned}$$

The inequality from the statement is a consequence of the AM-GM inequality $2ab \leq a^2 + b^2$.

(T.B. Soulami, *Les Olympiades de Mathématiques: Réflexes et stratégies*, Ellipses, 1999)

1047. Let $a < b < c < d$ be the members of a connected set S. Because $a - 1$ does not belong to the set, it follows that $a + 1 \in S$, hence $b = a + 1$. Similarly, since $d + 1 \notin S$, we deduce that $d - 1 \in S$; hence $c = d - 1$. Therefore, a connected set has the form $\{a, a+1, d-1, d\}$, with $d - a > 2$.

(a) There are 10 connected subsets of the set $\{1, 2, 3, 4, 5, 6, 7\}$, namely,

$$\{1, 2, 3, 4\}; \ \{1, 2, 4, 5\}; \ \{1, 2, 5, 6\}; \ \{1, 2, 6, 7\}; \ \{2, 3, 4, 5\};$$
$$\{2, 3, 5, 6\}; \ \{2, 3, 6, 7\}; \ \{3, 4, 5, 6\}; \ \{2, 4, 6, 7\}; \ \text{and} \ \{4, 5, 6, 7\}.$$

(b) Call $D = d - a + 1$ the diameter of the set $\{a, a + 1, d - 1, d\}$. Clearly, $D > 3$ and $D \leq n - 1 + 1 = n$. For $D = 4$ there are $n - 3$ connected sets, for $D = 5$ there are $n - 4$ connected sets, and so on. Adding up yields

$$C_n = 1 + 2 + 3 + \cdots + n - 3 = \frac{(n-3)(n-2)}{2},$$

which is the desired formula.

(Romanian Mathematical Olympiad, 2006)

1048. The solution involves a counting argument that shows that the total number of colorings exceeds those that make some 18-term arithmetic sequence monochromatic.

There are 2^{2005} colorings of a set with 2005 elements by two colors. The number of colorings that make a fixed 18-term sequence monochromatic is $2^{2005-17}$, since the terms not belonging to the sequence can be colored without restriction, while those in the sequence can be colored either all black or all white.

How many 18-term arithmetic sequences can be found in the set $\{1, 2, \ldots, 2005\}$? Such a sequence $a, a + r, a + 2r, \ldots, a + 17r$ is completely determined by a and r subject to the condition $a + 17r \leq 2005$. For every a there are $\left\lfloor \frac{2005-a}{17} \right\rfloor$ arithmetic sequences that start with a. Altogether, the number of arithmetic sequences does not exceed

$$\sum_{a=1}^{2005} \frac{2005 - a}{17} = \frac{2004 \cdot 2005}{2 \cdot 17}.$$

So the total number of colorings that makes an arithmetic sequence monochromatic does not exceed

$$2^{2005-17} \cdot \frac{2004 \cdot 2005}{34},$$

which is considerably smaller than 2^{2005}. The conclusion follows.

(Communicated by A. Neguţ)

1049. Let us consider the collection of all subsets with 2 elements of $A_1, A_2, \ldots, A_m$. We thus have a collection of $6m$ subsets with two elements of A. But the number of distinct subsets of cardinal 2 in A is 4950. By the pigeonhole principle, there exist distinct elements $x, y \in A$ that belong to at least 49 subsets. Let these subsets be $A_1, A_2, \ldots, A_{49}$. Then the conditions of the problem imply that the union of these subsets has $2 + 49 \times 2 = 100$ elements, so the union is A. However, the union of any 48 subsets among the 49 has at most $2 + 2 \times 48 = 98$ elements, and therefore it is different from A.

(G. Dospinescu)

1050. First, it is not hard to see that a configuration that maximizes the number of partitions should have no three collinear points. After examining several cases we guess that the maximal number of partitions is $\binom{n}{2}$. This is exactly the number of lines determined by two points, and we will use these lines to count the number of partitions. By pushing such a line slightly so that the two points lie on one side or the other, we obtain a partition. Moreover, each partition can be obtained this way. There are $2\binom{n}{2}$ such lines, obtained by pushing the lines through the n points to one side or the other. However, each partition is counted at least twice this way, except for the partitions that come from the sides of the polygon that is the convex hull of the n points, but those can be paired with the partitions that cut out one vertex of the convex hull from the others. Hence we have at most $2\binom{n}{2}/2 = \binom{n}{2}$ partitions.

Equality is achieved when the points form a convex n-gon, in which case $\binom{n}{2}$ counts the pairs of sides that are intersected by the separating line.

(67th W.L. Putnam Mathematical Competition, 2006)

1051. *First solution*: Consider the set of differences $D = \{x - y \mid x, y \in A\}$. It contains at most $101 \times 100 + 1 = 10101$ elements. Two sets $A + t_i$ and $A + t_j$ have nonempty intersection if and only if $t_i - t_j$ is in D. We are supposed to select the 100 elements in such a way that no two have the difference in D. We do this inductively.

First, choose one arbitrary element. Then assume that k elements have been chosen, $k \leq 99$. An element x that is already chosen prevents us from selecting any element from the set $x + D$. Thus after k elements are chosen, at most $10101k \leq 10101 \times 99 = 999999$ elements are forbidden. This allows us to choose the $(k + 1)$st element, and induction works. With this the problem is solved.

Second solution: The first solution can be improved if we look instead at the set of positive differences $P = \{x - y, \mid x, y \in A, \ x \geq y\}$. The set P has $\binom{101}{2} + 1 = 5051$ elements. The inductive construction has to be slightly modified, by choosing at each step the *smallest* element that is not forbidden. In this way we can obtain far more elements than the required 100. In fact, in the general situation, the argument proves that if A is a k-element subset of $S = \{1, 2, \ldots, n\}$ and m is a positive integer such that $n > (m - 1)\left(\binom{k}{2} + 1\right)$, then there exist $t_1, t_2, \ldots, t_m \in S$ such that the sets $A_j = \{x + t_j \mid x \in A\}, j = 1, 2, \ldots, m$, are pairwise disjoint.

(44th International Mathematical Olympiad, 2003, proposed by Brazil)

1052. (a) For fixed $x \in A$, denote by $k(x)$ the number of sets $B \in \mathcal{F}$ that contain x. List these sets as $B_1, B_2, \ldots, B_{k(x)}$. Then $B_1\backslash\{x\}, B_2\backslash\{x\}, \ldots, B_{k(x)}\backslash\{x\}$ are disjoint subsets of $A\backslash\{x\}$. Since each $B_i\backslash\{x\}$ has $n - 1$ elements, and $A\backslash\{x\}$ has $n^2 - 1$ elements, $k(x) \leq \frac{n^2-1}{n-1} = n + 1$. Repeating the argument for all $x \in A$ and adding, we obtain

$$\sum_{x \in A} k(x) \leq n^2(n + 1).$$

But

$$\sum_{x \in A} k(x) = \sum_{B \in \mathcal{F}} |B| = n|\mathcal{F}|.$$

Therefore, $n|\mathcal{F}| \leq n^2(n + 1)$, which implies $|\mathcal{F}| \leq n^2 + n$, proving (a).

For (b) arrange the elements $1, 2, \ldots, 9$ in a matrix

$$
\begin{matrix}
1 & 2 & 3 \\
4 & 5 & 6 \\
7 & 8 & 9
\end{matrix}
$$

and choose the sets of $\mathcal{F}$ as the rows, columns, and the "diagonals" that appear in the expansion of the 3×3 determinant by the Sarrus rule:

$$\{1, 2, 3\}, \ \{4, 5, 6\}, \ \{7, 8, 9\}, \ \{1, 4, 7\}, \ \{2, 5, 8\}, \ \{3, 6, 9\},$$

$$\{1, 5, 9\}, \ \{2, 6, 7\}, \ \{3, 4, 8\}, \ \{3, 5, 7\}, \ \{2, 4, 9\}, \ \{1, 6, 8\}.$$

It is straightforward to check that they provide the required counterexample.

(Romanian Team Selection Test for the International Mathematical Olympiad, 1985)

1053. At every cut the number of pieces grows by 1, so after n cuts we will have $n + 1$ pieces.

Let us evaluate the total number of vertices of the polygons after n cuts. After each cut the number of vertices grows by 2 if the cut went through two vertices, by 3 if the cut went through a vertex and a side, or by 4 if the cut went through two sides. So after n cuts there are at most $4n + 4$ vertices.

Assume now that after N cuts we have obtained the one hundred polygons with 20 sides. Since altogether there are $N + 1$ pieces, besides the one hundred polygons there are $N + 1 - 100$ other pieces. Each of these other pieces has at least 3 vertices, so the total number of vertices is $100 \cdot 20 + (N - 99) \cdot 3$. This number does not exceed $4N + 4$. Therefore,

$$4N + 4 \geq 100 \cdot 20 + (N - 99) \cdot 3 = 3N + 1703.$$

We deduce that $N \geq 1699$.

We can obtain one hundred polygons with twenty sides by making 1699 cuts in the following way. First, cut the square into 100 rectangles (99 cuts needed). Each rectangle is then cut through 16 cuts into a polygon with twenty sides and some triangles. We have performed a total of $99 + 100 \cdot 16 = 1699$ cuts.

(*Kvant (Quantum)*, proposed by I. Bershtein)

1054. We give a proof by contradiction. Let us assume that the conclusion is false. We can also assume that no problem was solved by at most one sex. Denote by b_i and g_i the number of boys, respectively, girls, that solved problem i, and by p the total number of problems. Then since $b_i, g_i \geq 1$, it follows that $(b_i - 2)(g_i - 2) \leq 1$, which is equivalent to

$$b_i g_i \leq 2(b_i + g_i) - 3.$$

Let us sum this over all problems. Note that condition (ii) implies that $441 \leq \sum b_i g_i$. We thus have

$$441 \leq \sum b_i g_i \leq 2(b_i + g_i) - 3 \leq 2(6 \cdot 21 + 6 \cdot 21) - 3p = 504 - 3p.$$

This implies that $p \leq 21$, so 21 is an upper bound for p.

We now do a different count of the problems that will produce a lower bound for p. Pairing a girl with each of the 21 boys, and using the fact that she solved at most six problems, by the pigeonhole principle we conclude that some problem was solved by that girl and 4 of the boys. By our assumption, there are at most two girls who solved that problem. This argument works for any girl, which means that there are at least 11 problems that were solved by at least 4 boys and at most 2 girls. Symmetrically, 11 other problems were solved by at least 4 girls and at most 2 boys. This shows that $p \leq 22$, a contradiction. The problem is solved.

(42nd International Mathematical Olympiad, 2001)

1055. By examining a few cases, we are led to believe that this is impossible. Nevertheless, let us assume that there is an $m \times n$ rectangle for which the coloring is possible. We associate to a row/column the color the dominating color. Let m_1, m_2 be the number of black respectively white rows, and n_1, n_2 be the number of black respectively white columns. Then $m_1 + m_2 = m$ and $n_1 + n_2 = n$. Without loss of generality we may assume that $m_1 \leq m_2$.

All squares at the intersection of a row and a column of different colors differ from the color of either their row or of their column. There are $m_1 n_2 + m_2 n_1$ such squares. Again without loss of generality we may assume that the color of more than half of these squares differs from that of the rows they are in.

By hypothesis, in the white rows there are more than $\frac{3}{4} m_2 n$ white squares. Hence in the black rows there are less than $\frac{mn}{2} - \frac{3}{4} m_2 n$ white squares, so that the total number of white squares is $\frac{mn}{2}$. On the other hand, in the white rows there are less than $\frac{1}{4} m_2 n$ black squares. So the number of squares whose color differs from that of the rows they are in is less than

$$\frac{mn}{2} - \frac{3}{4} m_2 n + \frac{1}{4} m_2 n = \frac{m_1 n}{2}.$$

We obtain the inequality

$$\frac{1}{2}(m_1 n_2 + m_2 n_1) < \frac{m_1 n}{2} = \frac{1}{2}(m_1 n_1 + m_1 n_2).$$

It follows that $m_1 > m_2$, a contradiction. Hence the answer to the question is negative, as claimed.

(*Kvant (Quantum)*, proposed by S. Konyagyn)

1056. First, let us forget about the constraint and count the number of paths from $(0, 0)$ and (m, n) such that at each step one of the coordinates increases by 1. There are a total of $m + n$ steps, out of which n go up. These n can be chosen in $\binom{m+n}{n}$ ways from the total of $m + n$. Therefore, the number of paths is $\binom{m+n}{n}$.

How many of these go through (p, q)? There are $\binom{p+q}{q}$ paths from $(0, 0)$ to (p, q) and $\binom{m+n-p-q}{n-q}$ paths from (p, q) to (m, n). Hence

$$\binom{p+q}{q} \cdot \binom{m+n-p-q}{n-q}$$

of all the paths pass through (p, q). And, of course,

$$\binom{r+s}{s} \cdot \binom{m+n-r-s}{n-s}$$

paths pass through (r, s). To apply the inclusion-exclusion principle, we also need to count the number of paths that go simultaneously through (p, q) and (r, s). This number is

$$\binom{p+q}{q} \cdot \binom{r+s-p-q}{s-q} \cdot \binom{m+n-r-s}{n-s}.$$

Hence, by the inclusion-exclusion principle, the number of paths avoiding (p, q) and (r, s) is

$$\binom{m+n}{n} - \binom{p+q}{q} \cdot \binom{m+n-p-q}{n-q} - \binom{r+s}{s} \cdot \binom{m+n-r-s}{n-s}$$
$$+ \binom{p+q}{q} \cdot \binom{r+s-p-q}{s-q} \cdot \binom{m+n-r-s}{n-s}.$$

1057. Let $E = \{1, 2, \ldots, n\}$ and $F = \{1, 2, \ldots, p\}$. There are p^n functions from E to F. The number of surjective functions is $p^n - N$, where N is the number of functions that are not surjective. We compute N using the inclusion-exclusion principle.

Define the sets

$$A_i = \{f : E \to F \mid i \notin f(E)\}.$$

Then

$$N = \left| \bigcup_{i=1}^{p} A_i \right| = \sum_i |A_i| - \sum_{i \neq j} |A_i \cap A_j| + \cdots + (-1)^{p-1} \left| \bigcap_{i=1}^{p} A_i \right|.$$

But A_i consists of the functions E to $F \setminus \{i\}$; hence $|A_i| = (p-1)^n$. Similarly, for all k, $2 \leq k \leq p - 1$, $A_{i_1} \cap A_{i_2} \cap \cdots \cap A_{i_k}$ is the set of functions from E to $F \setminus \{i_1, i_2, \ldots, i_k\}$; hence $|A_{i_1} \cap A_{i_2} \cap \cdots \cap A_{i_k}| = (p-k)^n$. Also, note that for a certain k, there are $\binom{p}{k}$ terms of the form $|A_{i_1} \cap A_{i_2} \cap \cdots \cap A_{i_k}|$. It follows that

$$N = \binom{p}{1}(p-1)^n - \binom{p}{2}(p-2)^n + \cdots + (-1)^{p-1} \binom{p}{p-1}.$$

We conclude that the total number of surjections from E to F is

$$p^n - \binom{p}{1}(p-1)^n + \binom{p}{2}(p-2)^n + \cdots + (-1)^{p-1} \binom{p}{p-1}.$$

1058. We count instead the permutations that are not derangements. Denote by A_i the set of permutations σ with $\sigma(i) = i$. Because the elements in A_i have the value at i already prescribed, it follows that $|A_i| = (n-1)!$. And for the same reason, $|A_{i_1} \cup A_{i_2} \cup \cdots \cup A_{i_k}| = (n-k)!$ for any distinct $i_1, i_2, \ldots, i_k, 1 \leq k \leq n$. Applying the inclusion-exclusion principle, we find that

$$|A_1 \cup A_2 \cup \cdots \cup A_n| = \binom{n}{1}(n-1)! - \binom{n}{2}(n-2)! + \cdots + (-1)^n \binom{n}{n}1!.$$

The number of derangements is therefore $n! - |A_1 \cup A_2 \cup \cdots \cup A_n|$, which is

$$n! - \binom{n}{1}(n-1)! + \binom{n}{2}(n-2)! + \cdots + (-1)^n \binom{n}{n}0!.$$

This number can also be written as

$$n! \left[1 - \frac{1}{1!} + \frac{1}{2!} - \cdots + \frac{(-1)^n}{n!} \right].$$

This number is approximately equal to $\frac{n!}{e}$.

1059. For a vertex x, denote by A_x the set of vertices connected to x by an edge. Assume that $|A_x| \geq \lfloor \frac{n}{2} \rfloor + 1$ for all vertices x.

Now choose two vertices x and y such that $y \in A_x$. Counting with the inclusion-exclusion principle, we get

$$|A_x \cup A_y| = |A_x| + |A_y| - |A_x \cap A_y|.$$

Rewrite this as

$$|A_x \cap A_y| = |A_x| + |A_y| - |A_x \cup A_y|.$$

From the fact that $|A_x \cup A_y| \leq n$ we find that $|A_x \cap A_y|$ is greater than or equal to

$$2 \left\lfloor \frac{n}{2} \right\rfloor + 2 - n \geq 1.$$

If follows that the set $A_x \cap A_y$ contains some vertex z, and so x, y, z are the vertices of a triangle.

(D. Buşneag, I. Maftei, *Teme pentru Cercurile şi Concursurile de Matematică* (*Themes for Mathematics Circles and Contests*), Scrisul Românesc, Craiova)

1060. Let the polygonal lines be P_1 and P_2. The case where a side of P_1 is parallel to a side of P_2 is obvious, since then they form a parallelogram. So let us consider the case where no side of P_1 is parallel to a side of P_2.

The number N of intersections of the lines of support of the sides of P_1 with the lines of support of the sides of P_2 is odd. The pairs of sides that do not form a convex quadrilateral are those for which the line of support of one side crosses the other side. We denote by

- n_1 the number of pairs of sides, one from each polygonal line, such that the line of support of the side of P_1 crosses the side of P_2,

- n_2 be the number of pairs of sides, one from each polygonal line, such that the line of support of the side of P_2 crosses the side of P_1,

- n_{12} be the number of pairs of sides, one from each polygonal line, that cross each other.

Then, by the inclusion-exclusion principle, the number of pairs of sides, one from each polygonal line, such that the line of support of one crosses the other side is

$$n_1 + n_2 - n_{12}.$$

We will prove that n_1, n_2, n_{12} are all even. To prove that n_1 is even, we can choose a point P in the plane that is not on any line of support and consider a homothety of center P with very small ratio. This homothety maps the second polygon to one in which no line of support of P_1 cuts a side of P_2. Now view the homothety as a continuous process. The only events where the number of intersections changes is where one vertex of P_2 goes from one half-plane of the line of support of a side of P_1 to the other half-plane. But since the vertex belongs to two sides, the number of intersections grows by 2 or decreases by 2, so the parity does not change. Thus the parity of n_1 is the same as the parity of 0, showing that n_1 is even. Similarly n_2 and n_{12} are even.

The number of pairs of sides that determine a convex quadrilateral is

$$N - (n_1 + n_2 - n_{12});$$

this is an odd number, so it is nonzero.

(*Kvant (Quantum)*, proposed by Yu. Khokhlov)

1061. If the m-gon has three acute angles, say at vertices A, B, C, then with a fourth vertex D they form a cyclic quadrilateral $ABCD$ that has three acute angles, which is impossible. Similarly, if the m-gon has two acute angles that do not share a side, say at vertices A and C, then they form with two other vertices B and D of the m-gon a cyclic quadrilateral $ABCD$ that has two opposite acute angles, which again is impossible. Therefore, the m-gon has either exactly one acute angle, or has two acute angles and they share a side.

To count the number of such m-gons we employ the principle of inclusion and exclusion. Thus we first find the number of m-gons with at least one acute angle, then subtract the number of m-gons with two acute angles (which were counted twice the first time).

If the acute angle of the m-gon is $A_k A_1 A_{k+r}$, the condition that this angle is acute translates into $r \leq n$. The other vertices of the m-gon lie between A_k and A_{k+r}; hence $m - 2 \leq r$, and these vertices can be chosen in $\binom{r-1}{m-3}$ ways. Note also that $1 \leq k \leq 2n - r$. Thus the number of m-gons with an acute angle at A_1 is

$$\sum_{r=m-2}^{n} \sum_{k=1}^{2n-r} \binom{r-1}{m-3} = 2n \sum_{m-2}^{n} \binom{r-1}{m-3} - \sum_{r=m-2}^{n} r \binom{r-1}{m-3}$$

$$= 2n \binom{n}{m-2} - (m-2) \binom{n+1}{m-1}.$$

There are as many polygons with an acute angle at $A_2, A_3, \dots, A_{2n+1}$.

To count the number of m-gons with two acute angles, let us first assume that these acute angles are $A_s A_1 A_k$ and $A_1 A_k A_r$. The other vertices lie between A_r and A_s. We have the restrictions $2 \leq k \leq 2n - m + 3, n + 2 \leq r < s \leq k + n$ if $k \leq n$ and no restriction on r and s otherwise. The number of such m-gons is

$$\sum_{k=1}^{n} \binom{k-1}{m-2} + \sum_{k=n+1}^{2n+1-(m-2)} \binom{2n+1-k}{m-2} = \sum_{k=m-1}^{n} \binom{k-1}{m-2} + \sum_{s=m-2}^{n} \binom{s}{m-2}$$

$$= \binom{n+1}{m-1} + \binom{n}{m-1}.$$

This number has to be multiplied by $2n + 1$ to take into account that the first acute vertex can be at any other vertex of the regular n-gon.

We conclude that the number of m-gons with at least one acute angle is

$$(2n+1) \left(2n \binom{n}{m-2} - (m-1) \binom{n+1}{m-1} - \binom{n}{m-1} \right).$$

1062. Denote by U_n the set of $z \in S^1$ such that $f^n(z) = z$. Because $f^n(z) = z^{m^n}$, U_n is the set of the roots of unity of order $m^n - 1$. In our situation $n = 1989$, and we are looking for those elements of U_{1989} that do not have period less than 1989. The periods of the elements of U_{1989} are divisors of 1989. Note that $1989 = 3^2 \times 13 \times 17$. The elements we are looking for lie in the complement of $U_{1989/3} \cup U_{1989/13} \cup U_{1989/17}$. Using the inclusion-exclusion principle, we find that the answer to the problem is

$$|U_{1989}| - |U_{1989/3}| - |U_{1989/13}| - |U_{1989/17}| + |U_{1989/3} \cap U_{1989/13}| + |U_{1989/3} \cap U_{1989/17}|$$
$$+ |U_{1989/13} \cap U_{1989/17}| + |U_{1989/3} \cap U_{1989/13} \cap U_{1989/17}|,$$

i.e.,

$$|U_{1989}| - |U_{663}| - |U_{153}| - |U_{117}| + |U_{51}| + |U_{39}| + |U_9| - |U_3|.$$

This number is equal to

$$m^{1989} - m^{663} - m^{153} - m^{117} + m^{51} + m^{39} + m^9 - m^3,$$

since the -1's in the formula for the cardinalities of the U_n's cancel out.
   (Chinese Mathematical Olympiad, 1989)

1063. Here we apply a "multiplicative" inclusion-exclusion formula for computing the least common multiple of several integers, which states that the least common multiple $[x_1, x_2, \ldots, x_n]$ of the numbers $x_1, x_2, \ldots, x_n$ is equal to

$$x_1 x_2 \cdots x_n \frac{1}{(x_1, x_2)(x_1, x_3) \cdots (x_{n-1}, x_n)} (x_1, x_2, x_3) \cdots (x_{n-2}, x_{n-1}, x_n) \cdots$$

For three numbers, this formula reads

$$[a, b, c] = abc \frac{1}{(a, b)(b, c)(c, a)} (a, b, c),$$

while for two numbers, it reads

$$[a, b] = ab \frac{1}{(a, b)}.$$

Let us combine the two. Square the first formula; then substitute the products ab, bc, and ca using the second. In detail,

$$[a, b, c]^2 = ab \cdot bc \cdot ca \frac{1}{(a, b)^2(b, c)^2(c, a)^2} (a, b, c)^2$$
$$= [a, b][b, c][c, a](a, b)(b, c)(c, a) \frac{1}{(a, b)^2(b, c)^2(c, a)^2} (a, b, c)^2$$
$$= [a, b][b, c][c, a] \frac{(a, b, c)^2}{(a, b)(b, c)(c, a)}.$$

The identity follows.

1064. We solve the problem for the general case of a rectangular solid of width w, length l, and height h, where w, l, and h are positive integers. Orient the solid in space so that one vertex is at $O = (0, 0, 0)$ and the opposite vertex is at $A = (w, l, h)$. Then OA is the diagonal of the solid.

The diagonal is transversal to the planes determined by the faces of the small cubes, so each time it meets a face, edge, or vertex, it leaves the interior of one cube and enters the interior of another. Counting by the number of interiors of small cubes that the diagonal leaves, we find that the number of interiors that OA intersects is equal to the number of points on OA having at least one integer coordinate.

We count these points using the inclusion-exclusion principle. The first coordinate of the current point $P = (tw, tl, th)$, $0 < t \leq 1$, on the diagonal is a positive integer for exactly w values of t, namely, $t = \frac{1}{w}, \frac{2}{w}, \ldots, \frac{w}{w}$. The second coordinate is an integer for l values of t, and the third coordinate is an integer for h values of t. However, the sum $w + l + h$ doubly counts the points with two integer coordinates, and triply counts the points with three integer coordinates. The first two coordinates are integers precisely when $t = \frac{k}{\gcd(w,l)}$, for some integer k, $1 \leq k \leq \gcd(w, l)$. Similarly, the second and third coordinates are positive integers for $\gcd(l, h)$, respectively, $\gcd(h, w)$ values of t, and all three coordinates are positive integers for $\gcd(w, l, h)$ values of t.

The inclusion-exclusion principle shows that the diagonal passes through the interiors of

$$w + l + h - \gcd(w, l) - \gcd(l, h) - \gcd(h, w) + \gcd(w, l, h)$$

small cubes. For $w = 150$, $l = 324$, $h = 375$ this number is equal to 768.

(American Invitational Mathematics Examination, 1996)

1065. Because the 1997 roots of the equation are symmetrically distributed in the complex plane, there is no loss of generality to assume that $v = 1$. We are required to find the probability that

$$|1 + w|^2 = |(1 + \cos\theta) + i\sin\theta|^2 = 2 + 2\cos\theta \geq 2 + \sqrt{3}.$$

This is equivalent to $\cos\theta \geq \frac{1}{2}\sqrt{3}$, or $|\theta| \leq \frac{\pi}{6}$. Because $w \neq 1$, θ is of the form $\pm\frac{2k\pi}{1997}k$, $1 \leq k \leq \left\lfloor \frac{1997}{12} \right\rfloor$. There are $2 \cdot 166 = 332$ such angles, and hence the probability is

$$\frac{332}{1996} = \frac{83}{499} \approx 0.166.$$

(American Invitational Mathematics Examination, 1997)

1066. It is easier to compute the probability that no two people have the same birthday. Arrange the people in some order. The first is free to be born on any of the 365 days. But only 364 dates are available for the second, 363 for the third, and so on. The probability that no two people have the same birthday is therefore

$$\frac{364}{365} \cdot \frac{363}{365} \cdots \frac{365 - n + 1}{365} = \frac{365!}{(365 - n)!365^n}.$$

And the probability that two people have the same birthday is

$$1 - \frac{365!}{(365 - n)!365^n}.$$

Remark. Starting with $n = 23$ the probability becomes greater than $\frac{1}{2}$, while when $n > 365$ the probability is clearly 1 by the pigeonhole principle.

1067. Denote by $P(n)$ the probability that a bag containing n distinct pairs of tiles will be emptied, $n \geq 2$. Then $P(n) = Q(n)P(n-1)$ where $Q(n)$ is the probability that two of the first three tiles selected make a pair. The latter one is

$$Q(n) = \frac{\text{number of ways to select three tiles, two of which match}}{\text{number of ways to select three tiles}}$$

$$= \frac{n(2n-2)}{\binom{2n}{3}} = \frac{3}{2n-1}.$$

The recurrence relation

$$P(n) = \frac{3}{2n-1}P(n-1)$$

yields

$$P(n) = \frac{3^{n-2}}{(2n-1)(2n-3)\cdots 5}P(2).$$

Clearly, $P(2) = 1$, and hence the answer to the problem is

$$P(6) = \frac{3^4}{11 \cdot 9 \cdot 7 \cdot 5} = \frac{9}{385} \approx 0.023.$$

(American Invitational Mathematics Examination, 1994)

1068. Because there are two extractions each of with must contain a certain ball, the total number of cases is $\binom{n-1}{m-1}^2$. The favorable cases are those for which the balls extracted the second time differ from those extracted first (except of course the chosen ball). For the first extraction there are $\binom{n-1}{m-1}$ cases, while for the second there are $\binom{n-m}{m-1}$, giving a total number of cases $\binom{n-1}{m-1}\binom{n-m}{m-1}$. Taking the ratio, we obtain the desired probability as

$$P = \frac{\binom{n-1}{m-1}\binom{n-m}{m-1}}{\binom{n-1}{m-1}^2} = \frac{\binom{n-m}{m-1}}{\binom{n-1}{m-1}}.$$

(*Gazeta Matematică (Mathematics Gazette, Bucharest)*, proposed by C. Marinescu)

1069. First, observe that since at least one ball is removed during each stage, the process will eventually terminate, leaving no ball or one ball in the bag. Because red balls are removed 2 at a time and since we start with an odd number of red balls, the number of red balls in the bag at any time is odd. Hence the process will always leave red balls in the bag, and so it must terminate with exactly one red ball. The probability we are computing is therefore 1.

(*Mathematics and Informatics Quarterly*, proposed by D. Macks)

1070. Consider the dual cube to the octahedron. The vertices A, B, C, D, E, F, G, H of this cube are the centers of the faces of the octahedron (here $ABCD$ is a face of the cube and $(A, G), (B, H), (C, E), (D, F)$ are pairs of diagonally opposite vertices). Each assignment of the numbers $1, 2, 3, 4, 5, 6, 7$, and 8 to the faces of the octahedron corresponds to a permutation of $ABCDEFGH$, and thus to an octagonal circuit of these vertices. The cube has 16 diagonal segments that join nonadjacent vertices. The problem requires us to count octagonal circuits that can be formed by eight of these diagonals.

Six of these diagonals are edges of the tetrahedron $ACFH$, six are edges of the tetrahedron $DBEG$, and four are long diagonals, joining opposite vertices of the cube. Notice that each vertex belongs to exactly one long diagonal. It follows that an octagonal circuit must contain either 2 long diagonals separated by 3 tetrahedron edges (Figure 119a), or 4 long diagonals (Figure 119b) alternating with tetrahedron edges.

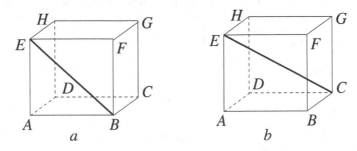

Figure 119

When forming a (skew) octagon with 4 long diagonals, the four tetrahedron edges need to be disjoint; hence two are opposite edges of $ACFH$ and two are opposite edges of $DBEG$. For each of the three ways to choose a pair of opposite edges from the tetrahedron $ACFH$, there are two possible ways to choose a pair of opposite edges from tetrahedron $DBEG$. There are $3 \cdot 22 = 6$ octagons of this type, and for each of them, a circuit can start at 8 possible vertices and can be traced in two different ways, making a total of $6 \cdot 8 \cdot 2 = 96$ permutations.

An octagon that contains exactly two long diagonals must also contain a three-edge path along the tetrahedron $ACFH$ and a three-edge path along tetrahedron the $DBEG$. A three-edge path along the tetrahedron the $ACFH$ can be chosen in $4! = 24$ ways. The corresponding three-edge path along the tetrahedron $DBEG$ has predetermined initial and terminal vertices; it thus can be chosen in only 2 ways. Since this counting method treats each path as different from its reverse, there are $8 \cdot 24 \cdot 2 = 384$ permutations of this type.

In all, there are $96 + 384 = 480$ permutations that correspond to octagonal circuits formed exclusively from cube diagonals. The probability of randomly choosing such a permutation is $\frac{480}{8!} = \frac{1}{84}$.

(American Invitational Mathematics Examination, 2001)

1071. The total number of permutations is of course $n!$. We will count instead the number of permutations for which 1 and 2 lie in different cycles.

If the cycle that contains 1 has length k, we can choose the other $k-1$ elements in $\binom{n-2}{k-1}$ ways from the set $\{3, 4, \ldots, n\}$. There exist $(k-1)!$ circular permutations of these elements, and $(n-k)!$ permutations of the remaining $n-k$ elements. Hence the total number of permutations for which 1 and 2 belong to different cycles is equal to

$$\sum_{k=1}^{n-1} \binom{n-2}{k-1}(k-1)!(n-k)! = (n-2)! \sum_{k=1}^{n-1}(n-k) = (n-2)!\frac{n(n-1)}{2} = \frac{n!}{2}.$$

It follows that exactly half of all permutations contain 1 and 2 in different cycles, and so half contain 1 and 2 in the same cycle. The probability is $\frac{1}{2}$.

(I. Tomescu, *Problems in Combinatorics*, Wiley, 1985)

1072. There are $\binom{n}{k}$ ways in which exactly k tails appear, and in this case the difference is $n - 2k$. Hence the expected value of $|H - T|$ is

$$\frac{1}{2^n} \sum_{k=0}^{n} |n - 2k| \binom{n}{k}.$$

Evaluate the sum as follows:

$$\frac{1}{2^n} \sum_{m=0}^{n} |n - 2m| \binom{n}{m} = \frac{1}{2^{n-1}} \sum_{m=0}^{\lfloor n/2 \rfloor} (n - 2m) \binom{n}{m}$$

$$= \frac{1}{2^{n-1}} \left(\sum_{m=0}^{\lfloor n/2 \rfloor} (n - m) \binom{n}{m} - \sum_{m=0}^{\lfloor n/2 \rfloor} m \binom{n}{m} \right)$$

$$= \frac{1}{2^{n-1}} \left(\sum_{m=0}^{\lfloor n/2 \rfloor} n \binom{n-1}{m} - \sum_{m=0}^{\lfloor n/2 \rfloor} n \binom{n-1}{m-1} \right)$$

$$= \frac{n}{2^{n-1}} \binom{n-1}{\lfloor \frac{n}{2} \rfloor}.$$

(35th W.L. Putnam Mathematical Competition, 1974)

1073. Use n cards with the numbers $1, 2, \ldots, n$ on them. Shuffle the cards and stack them with the numbered faces down. Then pick cards from the top of this pack, one at a time. We say that a matching occurs at the ith draw if the number on the card drawn is i. The probability that no matching occurs is

$$\sum_{i=0}^{n} \frac{(-1)^i}{i!} = p(n),$$

which follows from the derangements formula (see Section 6.4.4.). The probability that exactly k matches occur is

$$\binom{n}{k} \frac{p(n-k)(n-k)!}{n!} = \frac{1}{k!} p(n-k) = \frac{1}{k!} \sum_{i=0}^{n-k} \frac{(-1)^i}{i!}.$$

Denote by X the number of matchings in this n-card game. The expected value of X is

$$E(X) = \sum_{k=0}^{n} k P(X = k) = \sum_{k=0}^{n} k \frac{1}{k!} \sum_{i=0}^{n-k} \frac{(-1)^i}{i!} = \sum_{k=1}^{n} \frac{1}{(k-1)!} \sum_{i=0}^{n-k} \frac{(-1)^i}{i!},$$

because

$$P(X = k) = \frac{1}{k!} \sum_{i=0}^{n-k} \frac{(-1)^i}{i!}.$$

Let us compute $E(X)$ differently. Set

$$X_i = \begin{cases} 1 & \text{if there is a match at the } i\text{th draw,} \\ 0 & \text{if there is no match at the } i\text{th draw.} \end{cases}$$

Then

$$E(X) = E(X_1 + \cdots + X_n) = \sum_{i=1}^{n} E(X_i) = n\frac{1}{n} = 1,$$

because

$$E(X_i) = 1 \cdot P(X_i = 1) = \frac{(n-1)!}{n!} = \frac{1}{n}.$$

Combining the two, we obtain

$$\sum_{k=1}^{n} \frac{1}{(k-1)!} \sum_{i=0}^{n-k} \frac{(-1)^i}{i!} = 1,$$

which proves the first identity. The proof of the second identity is similar. We have

$$E(X^2) = E\left(\left(\sum_{i=1}^{n} X_i\right)^2\right) = \sum_{i=1}^{n} E(X_i^2) + 2\sum_{i<j} E(X_i X_j).$$

But

$$E(X_i^2) = E(X_i) = \frac{1}{n} \quad \text{and} \quad E(X_i X_j) = 1 \cdot 1 \cdot P(X_i = 1, X_j = 1) = \frac{1}{n(n-1)}.$$

Hence $E(X^2) = 1 + 1 = 2$. On the other hand,

$$E(X^2) = \sum_{k=1}^{n} k^2 \frac{1}{k!} \sum_{i=0}^{n-k} \frac{(-1)^i}{i!},$$

which proves the second identity.

(Proposed for the USA Mathematical Olympiad by T. Andreescu)

1074. Denote by A_i the event "the student solves correctly exactly i of the three proposed problems", $i = 0, 1, 2, 3$. The event A whose probability we are computing is

$$A = A_2 \cup A_3,$$

and its probability is

$$P(A) = P(A_2) + P(A_3),$$

since A_2 and A_3 exclude each other.

Because the student knows how to solve half of all the problems,

$$P(A_0) = P(A_3) \quad \text{and} \quad P(A_1) = P(A_2).$$

The equality

$$P(A_0) + P(A_1) + P(A_2) + P(A_3) = 1$$

becomes

$$2[P(A_2) + P(A_3)] = 1.$$

It follows that the probability we are computing is

$$P(A) = P(A_2) + P(A_3) = \frac{1}{2}.$$

(N. Negoescu, *Probleme cu...Probleme (Problems with...Problems)*, Editura Facla, 1975)

1075. For the solution we will use Bayes' theorem for conditional probabilities. We denote by $P(A)$ the probability that the event A holds, and by $P\left(\frac{B}{A}\right)$ the probability that the event B holds given that A in known to hold. Bayes' theorem states that

$$P(B/A) = \frac{P(B)}{P(A)} \cdot P(A/B).$$

For our problem A is the event that the mammogram is positive and B the event that the woman has breast cancer. Then $P(B) = 0.01$, while $P(A/B) = 0.60$. We compute $P(A)$ from the formula

$$P(A) = P(A/B)P(B) + P(A/\text{not } B)P(\text{not } B) = 0.6 \cdot 0.01 + 0.07 \cdot 0.99 = 0.0753.$$

The answer to the question is therefore

$$P(B/A) = \frac{0.01}{0.0753} \cdot 0.6 = 0.795 \approx 0.08$$

The chance that the woman has breast cancer is only 8%!

1076. We call a *successful string* a sequence of H's and T's in which $HHHHH$ appears before TT does. Each successful string must belong to one of the following three types:

  (i) those that begin with T, followed by a successful string that begins with H;

 (ii) those that begin with H, HH, HHH, or $HHHH$, followed by a successful string that begins with T;

(iii) the string $HHHHH$.

Let P_H denote the probability of obtaining a successful string that begins with H, and let P_T denote the probability of obtaining a successful string that begins with T. Then

$$P_T = \frac{1}{2}P_H,$$

$$P_H = \left(\frac{1}{2} + \frac{1}{4} + \frac{1}{8} + \frac{1}{16}\right)P_T + \frac{1}{32}.$$

Solving these equations simultaneously, we find that

$$P_H = \frac{1}{17} \quad \text{and} \quad P_T = \frac{1}{34}.$$

Hence the probability of obtaining five heads before obtaining two tails is $\frac{3}{34}$.
    (American Invitational Mathematics Examination, 1995)

1077. Let us denote the events $x = 70°$, $y = 70°$, $\max(x°, y°) = 70°$, $\min(x°, y°) = 70°$ by A, B, C, D, respectively. We see that $A \cup B = C \cup D$ and $A \cap B = C \cap D$. Hence

$$P(A) + P(B) = P(A \cup B) + P(A \cap B) = P(C \cup D) + P(C \cap D) = P(C) + P(D).$$

Therefore, $P(D) = P(A) + P(B) - P(C)$, that is,

$$P(\min(x°, y°) = 70°) = P(x° = 70°) + P(y° = 70°) - P(\max(x°, y°) = 70°)$$
$$= a + b - c.$$

(29th W.L. Putnam Mathematical Competition, 1968)

1078. In order for n black marbles to show up in $n + x$ draws, two independent events should occur. First, in the initial $n + x - 1$ draws exactly $n - 1$ black marbles should be drawn. Second, in the $(n + x)$th draw a black marble should be drawn. The probability of the second event is simply q. The probability of the first event is computed using the Bernoulli scheme; it is equal to

$$\binom{n + x - 1}{x} p^x q^{n-1}.$$

The answer to the problem is therefore

$$\binom{n + m - 1}{m} p^m q^{n-1} q = \binom{n + m - 1}{m} p^m q^n.$$

(Romanian Mathematical Olympiad, 1971)

1079. *First solution*: Denote by p_1, p_2, p_3 the three probabilities. By hypothesis,

$$P(X = 0) = \prod_i (1 - p_i) = 1 - \sum_i p_i + \sum_{i \neq j} p_i p_j - p_1 p_2 p_3 = \frac{2}{5},$$

$$P(X = 1) = \sum_{\{i,j,k\}=\{1,2,3\}} p_i(1 - p_j)(1 - p_k) = \sum_i p_i - 2\sum_{i \neq j} p_i p_j + 3 p_1 p_2 p_3 = \frac{13}{30},$$

$$P(X = 2) = \sum_{\{i,j,k\}=\{1,2,3\}} p_i p_j(1 - p_k) = \sum_{i \neq j} p_i p_j - 3 p_1 p_2 p_3 = \frac{3}{20},$$

$$P(X = 3) = p_1 p_2 p_3 = \frac{1}{60}.$$

This is a linear system in the unknowns $\sum_i p_i$, $\sum_{i \neq j} p_i p_j$, and $p_1 p_2 p_3$ with the solution

$$\sum_i p_i = \frac{47}{60}, \quad \sum_{i \neq j} p_i p_j = \frac{1}{5}, \quad p_1 p_2 p_3 = \frac{1}{60}.$$

It follows that p_1, p_2, p_3 are the three solutions to the equation

$$x^3 - \frac{47}{60} x^2 + \frac{1}{5} x - \frac{1}{60} = 0.$$

Searching for solutions of the form $\frac{1}{q}$ with q dividing 60, we find the three probabilities to be equal to $\frac{1}{3}, \frac{1}{4},$ and $\frac{1}{5}$.

Second solution: Using the Poisson scheme

$$(p_1x + 1 - p_1)(p_2x + 1 - p_2)(p_3x + 1 - p_3) = \frac{2}{5} + \frac{13}{30}x + \frac{3}{20}x^2 + \frac{1}{60}x^3,$$

we deduce that $1 - \frac{1}{p_i}, i = 1, 2, 3$, are the roots of $x^3 + 9x^2 + 26x + 24 = 0$ and $p_1p_2p_3 = \frac{1}{60}$. The three roots are $-2, -3, -4$, which again gives p_i's equal to $\frac{1}{3}, \frac{1}{4},$ and $\frac{1}{5}$.

(N. Negoescu, *Probleme cu...Probleme* (*Problems with...Problems*), Editura Facla, 1975)

1080. Set $q_i = 1 - p_i, i = 1, 2, \ldots, n$, and consider the generating function

$$Q(x) = \prod_{i=1}^{n}(p_ix + q_i) = Q_0 + Q_1x + \cdots + Q_nx^n.$$

The probability for exactly k of the independent events $A_1, A_2, \ldots, A_n$ to occur is equal to the coefficient of x^k in $Q(x)$, hence to Q_k. The probability P for an odd number of events to occur is thus equal to $Q_1 + Q_3 + \cdots$. Let us compute this number in terms of $p_1, p_2, \ldots, p_n$.
    We have

$$Q(1) = Q_0 + Q_1 + \cdots + Q_n \quad \text{and} \quad Q(-1) = Q_0 - Q_1 + \cdots + (-1)^nQ_n.$$

Therefore,

$$P = \frac{Q(1) - Q(-1)}{2} = \frac{1}{2}\left(1 - \prod_{i=1}^{n}(1 - 2p_i)\right).$$

(Romanian Mathematical Olympiad, 1975)

1081. It is easier to compute the probability of the contrary event, namely that the batch passes the quality check. Denote by A_i the probability that the ith checked product has the desired quality standard. We then have to compute $P\left(\cap_{i=1}^{5}A_i\right)$. The events are not independent, so we use the formula

$$P\left(\cap_{i=1}^{5}A_i\right) = P(A_1)P(A_2/A_1)(A_3/A_1 \cap A_2)P(A_4/A_1 \cap A_2 \cap A_3)$$
$$\times P(A_5/A_1 \cap A_2 \cap A_3 \cap A_4).$$

We find successively $P(A_1) = \frac{95}{100}$, $P(A_2/A_1) = \frac{94}{99}$ (because if A_1 occurs then we are left with 99 products out of which 94 are good), $P(A_3/A_1 \cap A_2) = \frac{93}{98}$, $P(A_4/A_1 \cap A_2 \cap A_3) = \frac{92}{97}$, $P(A_5/A_1 \cap A_2 \cap A_3 \cap A_4) = \frac{91}{96}$. The answer to the problem is

$$1 - \frac{95}{100} \cdot \frac{94}{99} \cdot \frac{93}{98} \cdot \frac{92}{97} \cdot \frac{91}{96} \approx 0.230.$$

1082. We apply Bayes' formula. Let B be the event that the plane flying out of Eulerville is a jet plane and A_1, respectively, A_2, the events that the plane flying between the two cities is a jet, respectively, a propeller plane. Then

$$P(A_1) = \frac{2}{3}, \quad P(A_2) = \frac{1}{3}, \quad P(B/A_1) = \frac{2}{7}, \quad P(B/A_2) = \frac{1}{7}.$$

Bayes formula gives

$$P(A_2/B) = \frac{P(A_2)P(B/A_2)}{P(A_1)P(B/A_1) + P(A_2)P(B/A_2)} = \frac{\frac{1}{3} \cdot \frac{1}{7}}{\frac{2}{3} \cdot \frac{2}{7} + \frac{1}{3} \cdot \frac{1}{7}} = \frac{1}{5}.$$

Thus the answer to the problem is $\frac{1}{5}$.

Remark. Without the farmer seeing the jet plane flying out of Eulerville, the probability would have been $\frac{1}{3}$. What you know affects your calculation of probabilities.

1083. We find instead the probability $P(n)$ for no consecutive heads to appear in n throws. We do this recursively. If the first throw is tails, which happens with probability $\frac{1}{2}$, then the probability for no consecutive heads to appear afterward is $P(n-1)$. If the first throw is heads, the second must be tails, and this configuration has probability $\frac{1}{4}$. The probability that no consecutive heads appear later is $P(n-2)$. We obtain the recurrence

$$P(n) = \frac{1}{2}P(n-1) + \frac{1}{4}P(n-2),$$

with $P(1) = 1$, and $P(2) = \frac{3}{4}$. Make this relation more homogeneous by substituting $x_n = 2^n P(n)$. We recognize the recurrence for the Fibonacci sequence $x_{n+1} = x_n + x_{n-1}$, with the remark that $x_1 = F_3$ and $x_2 = F_4$. It follows that $x_n = F_{n+2}$, $P(n) = \frac{F_{n+2}}{2^n}$, and the probability required by the problem is $P(n) = 1 - \frac{F_{n+2}}{2^n}$.

(L.C. Larson, *Problem-Solving Through Problems*, Springer-Verlag, 1990)

1084. Fix $N = m+n$, the total amount of money, and vary m. Denote by $P(m)$ the probability that A wins all the money when starting with m dollars. Clearly, $P(0) = 0$ and $P(N) = 1$. We want a recurrence relation for $P(m)$.

Assume that A starts with k dollars. During the first game, A can win, lose, or the game can be a draw. If A wins this game, then the probability of winning all the money afterward is $P(k + 1)$. If A loses, the probability of winning in the end is $P(k - 1)$. Finally, if the first game is a draw, nothing changes, so the probability of A winning in the end remains equal to $P(k)$. These three situations occur with probabilities p, q, r, respectively; hence

$$P(k) = pP(k + 1) + qP(k - 1) + rP(k).$$

Taking into account that $p + q + r = 1$, we obtain the recurrence relation

$$pP(k + 1) - (p + q)P(k) + qP(k - 1) = 0.$$

The characteristic equation of this recurrence is $p\lambda^2 - (p+q)\lambda + q = 0$. There are two cases. The simpler is $p = q$. Then the equation has the double root $\lambda = 1$, in which case the general term is a linear function in k. Since $P(0) = 0$ and $P(N) = 1$, it follows that $P(m) = \frac{m}{N} = \frac{m}{n+m}$. If $p \neq q$, then the distinct roots of the equation are $\lambda_1 = 1$ and $\lambda_2 = \frac{p}{q}$, and the general term must be of the form $P(k) = c_1 + c_2 \left(\frac{q}{p}\right)^k$. Using the known values for $k = 0$ and N, we compute

$$c_1 = -c_2 = \frac{1}{1 - \left(\dfrac{q}{p}\right)^N}.$$

Hence the required probability is

$$\frac{m}{m+n} \text{ if } p = q \quad \text{and} \quad \frac{1 - \left(\dfrac{q}{p}\right)^m}{1 - \left(\dfrac{q}{p}\right)^{m+n}} \text{ if } p \neq q.$$

(K.S. Williams, K. Hardy, *The Red Book of Mathematical Problems*, Dover, Mineola, NY, 1996)

1085. Seeking a recurrence relation, we denote by $E(m, n)$ this expected length. What happens, then, after one toss? Half the time you win, and then the parameters become $m + 1$, $n - 1$; the other half of the time you lose, and the parameters become $m - 1$, $n + 1$. Hence the recurrence relation is

$$E(m, n) = 1 + \frac{1}{2}E(m - 1, n + 1) + \frac{1}{2}E(m + 1, n - 1),$$

the 1 indicating the first toss. Of course, this assumes $m, n > 0$. The boundary conditions are that $E(0, n) = 0$ and $E(m, 0) = 0$, and these, together with the recurrence formula, do determine uniquely the function $E(m, n)$.

View $E(m, n)$ as a function of one variable, say n, along the line $m + n = $ constant. Solving the inhomogeneous second-order recurrence relation, we obtain $E(m, n) = mn$. Alternately, the recursive formula says that the second difference is the constant (-2), and so $E(m, n)$ is a quadratic function. Vanishing at the endpoints forces it to be cmn, and direct evaluation shows that $c = 1$.

(D.J. Newman, *A Problem Seminar*, Springer-Verlag, 1982)

1086. Let x and y be the two numbers. The set of all possible outcomes is the unit square

$$D = \{(x, y) \mid 0 \leq x \leq 1, \ 0 \leq y \leq 1\}.$$

The favorable cases consist of the region

$$D_f = \left\{(x, y) \in D \mid x + y \leq 1, \ xy \leq \frac{2}{9}\right\}.$$

This is the set of points that lie below both the line $f(x) = 1 - x$ and the hyperbola $g(x) = \frac{2}{9x}$. The required probability is $P = \frac{\text{Area}(D_f)}{\text{Area}(D)}$. The area of D is 1. The area of D_f is equal to

$$\int_0^1 \min(f(x), g(x))dx.$$

The line and the hyperbola intersect at the points $\left(\frac{1}{3}, \frac{2}{3}\right)$ and $\left(\frac{2}{3}, \frac{1}{3}\right)$. Therefore,

$$\text{Area}(D_f) \int_0^{1/3} (1-x)dx + \int_{1/3}^{2/3} \frac{2}{9x}dx + \int_{2/3}^1 (1-x)dx = \frac{1}{3} + \frac{2}{9}\ln 2.$$

We conclude that $P = \frac{1}{3} + \frac{2}{9}\ln 2 \approx 0.487$.

(C. Reischer, A. Sâmboan, *Culegere de Probleme de Teoria Probabilităţilor şi Statistică Matematică* (*Collection of Problems of Probability Theory and Mathematical Statistics*), Editura Didactică şi Pedagogică, Bucharest, 1972)

1087. The total region is a square of side β. The favorable region is the union of the two triangular regions shown in Figure 120, and hence the probability of a favorable outcome is

$$\frac{(\beta - \alpha)^2}{\beta^2} = \left(1 - \frac{\alpha}{\beta}\right)^2.$$

Figure 120

(22nd W.L. Putnam Mathematical Competition, 1961)

1088. Denote by x, respectively, y, the fraction of the hour when the husband, respectively, wife, arrive. The configuration space is the square

$$D = \{(x, y) \mid 0 \le x \le 1, \ 0 \le y \le 1\}.$$

In order for the two people to meet, their arrival time must lie inside the region

$$D_f = \left\{(x, y) \mid |x - y| \le \frac{1}{4}\right\}.$$

The desired probability is the ratio of the area of this region to the area of the square.

The complement of the region consists of two isosceles right triangles with legs equal to $\frac{3}{4}$, and hence of areas $\frac{1}{2}\left(\frac{3}{4}\right)^2$. We obtain for the desired probability

$$1 - 2 \cdot \frac{1}{2} \cdot \left(\frac{3}{4}\right)^2 = \frac{7}{16} \approx 0.44.$$

(B.V. Gnedenko)

1089. The set of possible events is modeled by the square $[0, 24] \times [0, 24]$. It is, however, better to identify the 0th and the 24th hours, thus obtaining a square with opposite sides identified, an object that in mathematics is called a torus (which is, in fact, the Cartesian product of two circles). The favorable region is outside a band of fixed thickness along the curve $x = y$ on the torus as depicted in Figure 121. On the square model this region is obtained by removing the points (x, y) with $|x - y| \leq 1$ together with those for which $|x - y - 1| \leq 1$ and $|x - y + 1| \leq 1$. The required probability is the ratio of the area of the favorable region to the area of the square, and is

$$P = \frac{24^2 - 2 \cdot 24}{24^2} = \frac{11}{12} \approx 0.917.$$

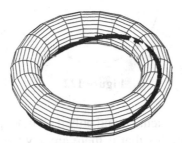

Figure 121

1090. Let $-\vec{y}$ denote the antipode of $\vec{y}$. The Pythagorean theorem gives

$$\|\vec{x} - \vec{y}\|^2 + \|\vec{x} - (-\vec{y})\|^2 = \|\vec{y} - (-\vec{y})\|^2 = 4\|\vec{y}\|^2 = 4.$$

Also note that if $\vec{x}$ and $\vec{y}$ are randomly chosen, then so are $\vec{x}$ and $-\vec{y}$. So we have an equality of expected values:

$$E[\|\vec{x} - \vec{y}\|^2] = E[\|\vec{x} + \vec{y}\|^2]$$

So

$$2E[\|\vec{x} - \vec{y}\|^2] = E[\|\vec{x} - \vec{y}\|^2] + E[\|\vec{x} + \vec{y}\|^2] = E[4] = 4.$$

The answer to the problem is

$$E[\|\vec{x} - \vec{y}\|^2] = 2.$$

(*Mathematical Reflections*, proposed by I. Borsenco)

1091. We assume that the circle of the problem is the unit circle centered at the origin O. The space of all possible choices of three points P_1, P_2, P_3 is the product of three circles; the volume of this space is $2\pi \times 2\pi \times 2\pi = 8\pi^3$.

Let us first measure the volume of the configurations (P_1, P_2, P_3) such that the arc $\overset{\frown}{P_1 P_2 P_3}$ is included in a semicircle and is oriented counterclockwise from P_1 to P_3. The condition that the arc is contained in a semicircle translates to $0 \leq \angle P_1 O P_2 \leq \pi$ and $0 \leq \angle P_2 O P_3 \leq$

$\pi - \angle P_1 O P_2$ (see Figure 122). The point P_1 is chosen randomly on the circle, and for each P_1 the region of the angles θ_1 and θ_2 such that $0 \le \theta_1 \le \pi$ and $0 \le \theta_1 \le \pi - \theta_1$ is an isosceles right triangle with leg equal to π. Hence the region of points (P_1, P_2, P_3) subject to the above constraints has volume $2\pi \cdot \frac{1}{2}\pi^2 = \pi^3$. There are $3! = 6$ such regions and they are disjoint. Therefore, the volume of the favorable region is $6\pi^3$. The desired probability is therefore equal to $\frac{6\pi^3}{8\pi^3} = \frac{3}{4}$.

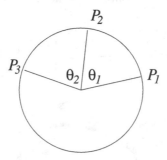

Figure 122

1092. The angle at the vertex P_i is acute if and only if all other points lie on an open semicircle facing P_i. We first deduce from this that if there are any two acute angles at all, they must occur consecutively. Otherwise, the two arcs that these angles subtend would overlap and cover the whole circle, and the sum of the measures of the two angles would exceed $180°$.

So the polygon has either just one acute angle or two consecutive acute angles. In particular, taken in counterclockwise order, there exists *exactly* one pair of consecutive angles the second of which is acute and the first of which is not.

We are left with the computation of the probability that for one of the points P_j, the angle at P_j is not acute, but the following angle is. This can be done using integrals. But there is a clever argument that reduces the geometric probability to a probability with a finite number of outcomes. The idea is to choose randomly $n - 1$ pairs of antipodal points, and then among these to choose the vertices of the polygon. A polygon with one vertex at P_j and the other among these points has the desired property exactly when $n - 2$ vertices lie on the semicircle to the clockwise side of P_j and one vertex on the opposite semicircle. Moreover, the points on the semicircle should include the counterclockwise-most to guarantee that the angle at P_j is not acute. Hence there are $n - 2$ favorable choices of the total 2^{n-1} choices of points from the antipodal pairs. The probability for obtaining a polygon with the desired property is therefore $(n - 2)2^{-n+1}$.

Integrating over all choices of pairs of antipodal points preserves the ratio. The events $j = 1, 2, \ldots, n$ are independent, so the probability has to be multiplied by n. The answer to the problem is therefore $n(n - 2)2^{-n+1}$.

(66th W.L. Putnam Mathematical Competition, 2005, solution by C. Lin)

1093. The pair (p, q) is chosen randomly from the three-dimensional domain $C \times \text{int } C$, which has a total volume of $2\pi^2$ (here int C denotes the interior of C). For a fixed p, the locus of points q for which R does not have points outside of C is the rectangle whose diagonal is the diameter through p and whose sides are parallel to the coordinate axes (Figure 123). If the coordinates of p are $(\cos\theta, \sin\theta)$, then the area of the rectangle is $2|\sin 2\theta|$.

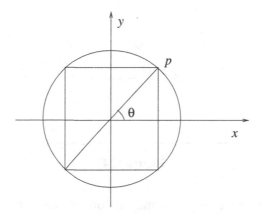

Figure 123

The volume of the favorable region is therefore

$$V = \int_0^{2\pi} 2|\sin 2\theta|\,d\theta = 4\int_0^{\pi/2} 2\sin 2\theta\,d\theta = 8.$$

Hence the probability is

$$P = \frac{8}{2\pi^2} = \frac{4}{\pi^2} \approx 0.405.$$

(46th W.L. Putnam Mathematical Competition, 1985)

1094. Mark an endpoint of the needle. Translations parallel to the given (horizontal) lines can be ignored; thus we can assume that the marked endpoint of the needle always falls on the same vertical. Its position is determined by the variables (x, θ), where x is the distance to the line right above and θ the angle made with the horizontal (Figure 124).

The pair (x, θ) is randomly chosen from the region $[0, 2) \times [0, 2\pi)$. The area of this region is 4π. The probability that the needle will cross the upper horizontal line is

$$\frac{1}{4\pi}\int_0^{\pi}\int_0^{\sin\theta} dx\,d\theta = \int_0^{\pi}\frac{\sin\theta}{4\pi}\,d\theta = \frac{1}{2\pi},$$

which is also equal to the probability that the needle will cross the lower horizontal line. The probability for the needle to cross either the upper or the lower horizontal line is therefore $\frac{1}{\pi}$.

Remark. This gives an experimental way of approximating π.

(G.-L. Leclerc, Comte de Buffon)

1095. *First solution*: We will prove that the probability is $1 - \frac{35}{12\pi^2}$. To this end, we start with some notation and simplifications. The area of a triangle XYZ will be denoted by $A(XYZ)$. For simplicity, the circle is assumed to have radius 1. Also, let E denote the expected value of a random variable over all choices of P, Q, R.

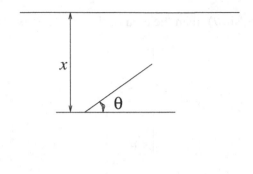

Figure 124

If P, Q, R, S are the four points, we may ignore the case in which three of them are collinear, since this occurs with probability zero. Then the only way they can fail to form the vertices of a convex quadrilateral is if one of them lies inside the triangle formed by the other three. There are four such configurations, depending on which point lies inside the triangle, and they are mutually exclusive. Hence the desired probability is 1 minus four times the probability that S lies inside triangle PQR. That latter probability is simply $E(A(PQR))$ divided by the area of the disk.

Let O denote the center of the circle, and let P', Q', R' be the projections of P, Q, R onto the circle from O. We can write

$$A(PQR) = \pm A(OPQ) \pm A(OQR) \pm A(ORP)$$

for a suitable choice of signs, determined as follows. If the points P', Q', R' lie on no semicircle, then all of the signs are positive. If P', Q', R' lie on a semicircle in that order and Q lies inside the triangle OPR, then the sign on $A(OPR)$ is positive and the others are negative. If P', Q', R' lie on a semicircle in that order and Q lies outside the triangle OPR, then the sign on $A(OPR)$ is negative and the others are positive.

We first calculate

$$E(A(OPQ) + A(OQR) + A(ORP)) = 3E(A(OPQ)).$$

Write $r_1 = OP, r_2 = OQ, \theta = \angle POQ$, so that

$$A(OPQ) = \frac{1}{2}r_1 r_2 \sin \theta.$$

The distribution of r_1 is given by $2r_1$ on $[0, 1]$ (e.g., by the change of variable formula to polar coordinates, or by computing the areas of annuli centered at the origin), and similarly for r_2. The distribution of θ is uniform on $[0, \pi]$. These three distributions are independent; hence

$$E(A(OPQ)) = \frac{1}{2}\left(\int_0^1 2r^2 dr\right)^2 \left(\frac{1}{\pi}\int_0^\pi \sin\theta d\theta\right) = \frac{4}{9\pi},$$

and

$$E(A(OPQ) + A(OQR) + A(ORP)) = \frac{4}{3\pi}.$$

We now treat the case in which P', Q', R' lie on a semicircle in that order. Set $\theta_1 = \angle POQ$ and $\theta_2 = \angle QOR$; then the distribution of θ_1, θ_2 is uniform on the region

$$0 \leq \theta_1, \quad 0 \leq \theta_2, \quad \theta_1 + \theta_2 \leq \pi.$$

In particular, the distribution on $\theta = \theta_1 + \theta_2$ is $\frac{2\theta}{\pi^2}$ on $[0, \pi]$. Set $r_P = OP$, $r_Q = OQ$, $r_R = OR$. Again, the distribution on r_P is given by $2r_P$ on $[0, 1]$, and similarly for r_Q, r_R; these are independent of each other and the joint distribution of θ_1, θ_2. Write $E'(X)$ for the expectation of a random variable X restricted to this part of the domain.

Let χ be the random variable with value 1 if Q is inside triangle OPR and 0 otherwise. We now compute

$$E'(A(OPR)) = \frac{1}{2} \left(\int_0^1 2r^2 dr \right)^2 \left(\int_0^\pi \frac{2\theta}{\pi^2} \sin \theta d\theta \right) = \frac{4}{9\pi}$$

and

$$E'(\chi A(OPR)) = E' \left(\frac{2A(OPR)^2}{\theta} \right)$$

$$= \frac{1}{2} \left(\int_0^1 2r^3 dr \right)^2 \left(\int_0^\pi \frac{2\theta}{\pi^2} \theta^{-1} \sin^2 \theta d\theta \right) = \frac{1}{8\pi}.$$

Also, recall that given any triangle XYZ, if T is chosen uniformly at random inside XYZ, the expectation of $A(TXY)$ is the area of triangle bounded by XY and the centroid of XYZ, namely, $\frac{1}{3}A(XYZ)$.

Let χ be the random variable with value 1 if Q is inside triangle OPR and 0 otherwise. Then

$$E'(A(OPQ) + A(OQR) + A(ORP) - A(PQR))$$
$$= 2E'(\chi(A(OPQ) + A(OQR))) + 2E'((1 - \chi)A(OPR))$$
$$= 2E' \left(\frac{2}{3}\chi A(OPR) \right) + 2E'(A(OPR)) - 2E'(\chi A(OPR))$$
$$= 2E'(A(OPR)) - \frac{2}{3}E'(\chi A(OPR)) = \frac{29}{36\pi}.$$

Finally, note that the case in which P', Q', R' lie on a semicircle in some order occurs with probability $\frac{3}{4}$. (The case in which they lie on a semicircle proceeding clockwise from P' to its antipode has probability $\frac{1}{4}$; this case and its two analogues are exclusive and exhaustive.) Hence

$$E(A(PQR)) = E(A(OPQ) + A(OQR) + A(ORP))$$
$$- \frac{3}{4}E'(A(OPQ) + A(OQR) + A(ORP) - A(PQR))$$
$$= \frac{4}{3\pi} - \frac{29}{48\pi} = \frac{35}{48\pi}.$$

We conclude that the original probability is

$$1 - \frac{4E(A(PQR))}{\pi} = 1 - \frac{35}{12\pi^2}.$$

Second solution: As in the first solution, it suffices to check that for P, Q, R chosen uniformly at random in the disk, $E(A(PQR)) = \frac{35}{48\pi}$. Draw the lines PQ, QR, RP, which with probability 1 divide the interior of the circle into seven regions. Set $a = A(PQR)$, let b_1, b_2, b_3 denote the areas of the other three regions sharing a side with the triangle, and let c_1, c_2, c_3 denote the areas of the other three regions. Set $A = E(a)$, $B = E(b_1)$, $C = E(c_1)$, so that $A + 3B + 3C = \pi$.

Note that $c_1 + c_2 + c_3 + a$ is the area of the region in which we can choose a fourth point S such that the quadrilateral $PQRS$ fails to be convex. By comparing expectations we find that $3C + A = 4A$, so $A = C$ and $4A + 3B = \pi$.

We will compute $B + 2A = B + 2C$, which is the expected area of the part of the circle cut off by a chord through two random points D, E, on the side of the chord not containing a third random point F. Let h be the distance from the center O of the circle to the line DE. We now determine the distribution of h.

Set $r = OD$. As seen before, the distribution of r is $2r$ on $[0, 1]$. Without loss of generality, we may assume that O is the origin and D lies on the positive x-axis. For fixed r, the distribution of h runs over $[0, r]$, and can be computed as the area of the infinitesimal region in which E can be chosen so the chord through DE has distance to O between h and $h + dh$, divided by π. This region splits into two symmetric pieces, one of which lies between chords making angles of $\arcsin\left(\frac{h}{r}\right)$ and $\arcsin\left(\frac{h+dh}{r}\right)$ with the x-axis. The angle between these is $d\theta = \frac{dh}{r^2 - h^2}$. Draw the chord through D at distance h to O, and let L_1, L_2 be the lengths of the parts on opposite sides of D; then the area we are looking for is $\frac{1}{2}(L_1^2 + L_2^2)d\theta$. Because

$$\{L_1, L_2\} = \left\{\sqrt{1 - h^2} + \sqrt{r^2 - h^2}, \sqrt{1 - h^2} - \sqrt{r^2 - h^2}\right\},$$

the area we are seeking (after doubling) is

$$2\frac{1 + r^2 - 2h^2}{\sqrt{r^2 - h^2}}.$$

Dividing by π, then integrating over r, we compute the distribution of h to be

$$\frac{1}{\pi}\int_h^1 2\frac{1 + r^2 - 2h^2}{\sqrt{r^2 - h^2}}2r dr = \frac{16}{3\pi}(1 - h^2)^{3/2}.$$

Let us now return to the computation of $B + 2A$. Denote by $A(h)$ the smaller of the two areas of the disk cut off by a chord at distance h. The chance that the third point is in the smaller (respectively, larger) portion is $\frac{A(h)}{\pi}$ (respectively, $1 - \frac{A(h)}{\pi}$), and then the area we are trying to compute is $\pi - A(h)$ (respectively, $A(h)$). Using the distribution on h, and the fact that

$$A(h) = 2\int_h^1 \sqrt{1 - h^2}dh = \frac{\pi}{2} - \arcsin(h) - h\sqrt{1 - h^2},$$

we obtain

$$B + 2A = \frac{2}{\pi} \int_0^1 A(h)(\pi - A(h))\frac{16}{3\pi}(1 - h^2)^{3/2}dh = \frac{35 + 24\pi^2}{72\pi}.$$

Using the fact that $4A + 3B = \pi$, we obtain $A = \frac{35}{48\pi}$ as in the first solution.

Remark. This is a particular case of the Sylvester four-point problem, which asks for the probability that four points taken at random inside a convex domain D form a non-convex quadrilateral. Nowadays the standard method for computing this probability uses Crofton's theorem on mean values. We have seen above that when D is a disk the probability is $\frac{35}{12\pi^2}$. When D is a triangle, square, regular hexagon, or regular octagon, the probability is, respectively, $\frac{1}{3}$, $\frac{11}{36}$, $\frac{289}{972}$, and $\frac{1181+867\sqrt{2}}{4032+2880\sqrt{2}}$ (cf. H. Solomon, *Geometric Probability*, SIAM, 1978).

(First solution by D. Kane, second solution by D. Savitt)

Index of Notation

$\mathbb{N}$	the set of positive integers $1, 2, 3, \ldots$
$\mathbb{Z}$	the set of integers
$\mathbb{Q}$	the set of rational numbers
$\mathbb{R}$	the set of real numbers
$\mathbb{C}$	the set of complex numbers
$[a, b]$	closed interval, i.e., all x such that $a \le x \le b$
(a, b)	open interval, i.e., all x such that $a < x < b$
$[a, b)$	half-open interval, i.e., all x such that $a \le x < b$
$|x|$	absolute value of x
$\bar{z}$	complex conjugate of z
Re z	real part of z
Im z	imaginary part of z
$\vec{v}$	the vector v
$\|\vec{x}\|$	norm of the vector $\vec{x}$
$\langle \vec{v}, \vec{w} \rangle$	inner (dot) product of vectors $\vec{v}$ and $\vec{w}$
$\vec{v} \cdot \vec{w}$	dot product of vectors $\vec{v}$ and $\vec{w}$
$\vec{v} \times \vec{w}$	cross-product of vector $\vec{v}$ and $\vec{w}$
$\lfloor x \rfloor$	greatest integer not exceeding x
$\{x\}$	fractional part of x, equal to $x - \lfloor x \rfloor$
$\displaystyle\sum_{i=1}^{n} a_i$	the sum $a_1 + a_2 + \ldots + a_n$
$\displaystyle\prod_{i=1}^{n} a_i$	the product $a_1 \cdot a_2 \ldots a_n$
$n!$	n factorial, equal to $n(n-1) \ldots 1$
$x \in A$	element x is in set A
$A \subset B$	A is a subset of B
$A \cup B$	the union of the sets A and B
$A \cap B$	the intersection of the sets A and B

© Springer International Publishing AG 2017
R. Gelca and T. Andreescu, *Putnam and Beyond*, DOI 10.1007/978-3-319-58988-6

$A \backslash B$	the set of the elements of A that are not in B
$A \times B$	the Cartesian product of the sets A and B
$\mathcal{P}(A)$	the family of all subsets of the set A
$\emptyset$	the empty set
$a \equiv b \pmod{c}$	a is congruent to b modulo c, i.e., $a - b$ is divisible by c
$a \mid b$	a divides b
$\gcd(a, b)$	greatest common divisor of a and b
$\binom{n}{k}$	binomial coefficient n choose k
$\mathcal{O}_n$	the $n \times n$ zero matrix
$\mathcal{I}_n$	the $n \times n$ identity matrix
$\det A$	determinant of the matrix A
$\operatorname{tr} A$	trace of the matrix A
A^{-1}	inverse of A
A^t	transpose of the matrix A
$A^\dagger$	transpose conjugate of the matrix A
$f \circ g$	f composed with g
$\lim\limits_{x \to a}$	limit as x approaches a
$f'(x)$	derivative of $f(x)$
$\frac{df}{dx}$	derivative of $f(x)$
$\frac{\partial f}{\partial x}$	partial derivative of f with respect to x
$f^{(n)}(x)$	nth derivative of $f(x)$ with respect to x
$\int f(x)dx$	indefinite integral of $f(x)$
$\int_a^b f(x)dx$	definite integral of $f(x)$ from a to b
$\int_D f(x)dx$	integral of $f(x)$ over the domain D
$\phi(x)$	Euler's totient function of x
$\angle ABC$	angle ABC
$\overset{\frown}{AB}$	arc of a circle with extremities A and B
$\operatorname{sign}(\sigma)$	signature of the permutation σ
$\operatorname{div} \overrightarrow{F}$	divergence of the vector field $\overrightarrow{F}$
$\operatorname{curl} \overrightarrow{F}$	curl of the vector field $\overrightarrow{F}$
∇f	gradient of f
$\oint_C f(x)dx$	integral of f along the closed path C

Index

© Springer International Publishing AG 2017
R. Gelca and T. Andreescu, *Putnam and Beyond*, DOI 10.1007/978-3-319-58988-6